JACARANDA

MATHS QUEST 10

STAGE 5 NSW SYLLABUS | THIRD EDITION

T0360069

JACARANDA
MATHS QUEST 10

STAGE 5 NSW SYLLABUS | THIRD EDITION

BEVERLY LANGSFORD WILLING

CATHERINE SMITH

CONTRIBUTING AUTHORS

Vanessa Christman | Simon Baird

jacaranda
A Wiley Brand

Third edition published 2023 by
John Wiley & Sons Australia, Ltd
Level 4, 600 Bourke Street, Melbourne, Vic 3000

First edition published 2011
Second edition published 2014

Typeset in 10.5/13 pt TimesLTStd

© John Wiley & Sons Australia, Ltd 2023

The moral rights of the authors have been asserted.

ISBN: 978-0-7303-8641-4

Reproduction and communication for educational purposes

The Australian *Copyright Act 1968* (the Act) allows a maximum
of one chapter or 10% of the pages of this work, whichever is
the greater, to be reproduced and/or communicated by any
educational institution for its educational purposes provided that
the educational institution (or the body that administers it) has
given a remuneration notice to Copyright Agency Limited
(CAL).

Reproduction and communication for other purposes

Except as permitted under the Act (for example, a fair dealing
for the purposes of study, research, criticism or review), no part
of this book may be reproduced, stored in a retrieval system,
communicated or transmitted in any form or by any means
without prior written permission. All inquiries should be made
to the publisher.

Trademarks

Jacaranda, the JacPLUS logo, the learnON, assessON and
studyON logos, Wiley and the Wiley logo, and any related trade
dress are trademarks or registered trademarks of John Wiley &
Sons Inc. and/or its affiliates in the United States, Australia and
in other countries, and may not be used without written
permission. All other trademarks are the property of their
respective owners.

Front cover images: © vectorstudi/Shutterstock
© Marish/Shutterstock; © Irina Strelnikova/Shutterstock
© Visual Generation/Shutterstock

Illustrated by various artists, diacriTech and Wiley Composition
Services

Typeset in India by diacriTech

A catalogue record for this
book is available from the
National Library of Australia

Printed in Singapore
M WEP222654 310823

The Publishers of this series acknowledge and pay their respects
to Aboriginal Peoples and Torres Strait Islander Peoples as the
traditional custodians of the land on which this resource was
produced.

This suite of resources may include references to (including
names, images, footage or voices of) people of Aboriginal
and/or Torres Strait Islander heritage who are deceased. These
images and references have been included to help Australian
students from all cultural backgrounds develop a better
understanding of Aboriginal and Torres Strait Islander Peoples'
history, culture and lived experience.

It is strongly recommended that teachers examine resources on
topics related to Aboriginal and/or Torres Strait Islander
Cultures and Peoples to assess their suitability for their own
specific class and school context. It is also recommended that
teachers know and follow the guidelines laid down by the
relevant educational authorities and local Elders or community
advisors regarding content about all First Nations Peoples.

All activities in this resource have been written with the safety
of both teacher and student in mind. Some, however, involve
physical activity or the use of equipment or tools. **All due care
should be taken when performing such activities.** To the
maximum extent permitted by law, the author and publisher
disclaim all responsibility and liability for any injury or loss that
may be sustained when completing activities described in this
resource.

The Publisher acknowledges ongoing discussions related to
gender-based population data. At the time of publishing, there
was insufficient data available to allow for the meaningful
analysis of trends and patterns to broaden our discussion of
demographics beyond male and female gender identification.

Contents

About this resource

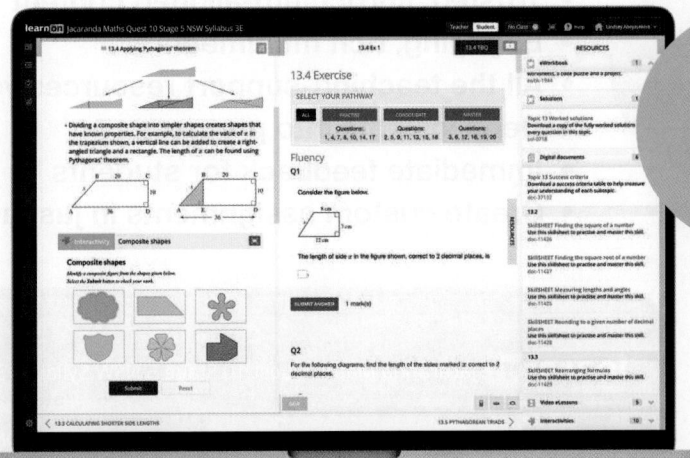

NEW FOR
2024 NSW SYLLABUS

JACARANDA
MATHS QUEST 10
NSW SYLLABUS
THIRD EDITION

Developed by teachers for students

Tried, tested and trusted. The third edition of the *Jacaranda Maths Quest series*, continues to focus on helping teachers achieve learning success for every student — ensuring no student is left behind, and no student is held back.

Because both what and how students learn matter

Learning is personal

Whether students need a challenge or a helping hand, you'll find what you need to create engaging lessons.

Whether in class or at home, students can get unstuck and progress! Scaffolded lessons, with detailed worked examples, are all supported by teacher-led video eLessons. Automatically marked, differentiated question sets are all supported by detailed worked solutions. And Brand-new Quick Quizzes support in-depth skill acquisition.

Learning is effortful

Learning happens when students push themselves. With learnON, Australia's most powerful online learning platform, students can challenge themselves, build confidence and ultimately achieve success.

Learning is rewarding

Through real-time results data, students can track and monitor their own progress and easily identify areas of strength and weakness.

And for teachers, Learning Analytics provide valuable insights to support student growth and drive informed intervention strategies.

Learn online with Australia's most

Everything you need for each of your lessons in one simple view

- Trusted, curriculum-aligned content
- Engaging, rich multimedia
- All the teaching-support resources you need
- Deep insights into progress
- Immediate feedback for students
- Create custom assignments in just a few clicks.

Practical teaching advice and ideas for each lesson provided in teachON

Teaching videos for all lessons

Reading content and rich media including embedded videos and interactivities

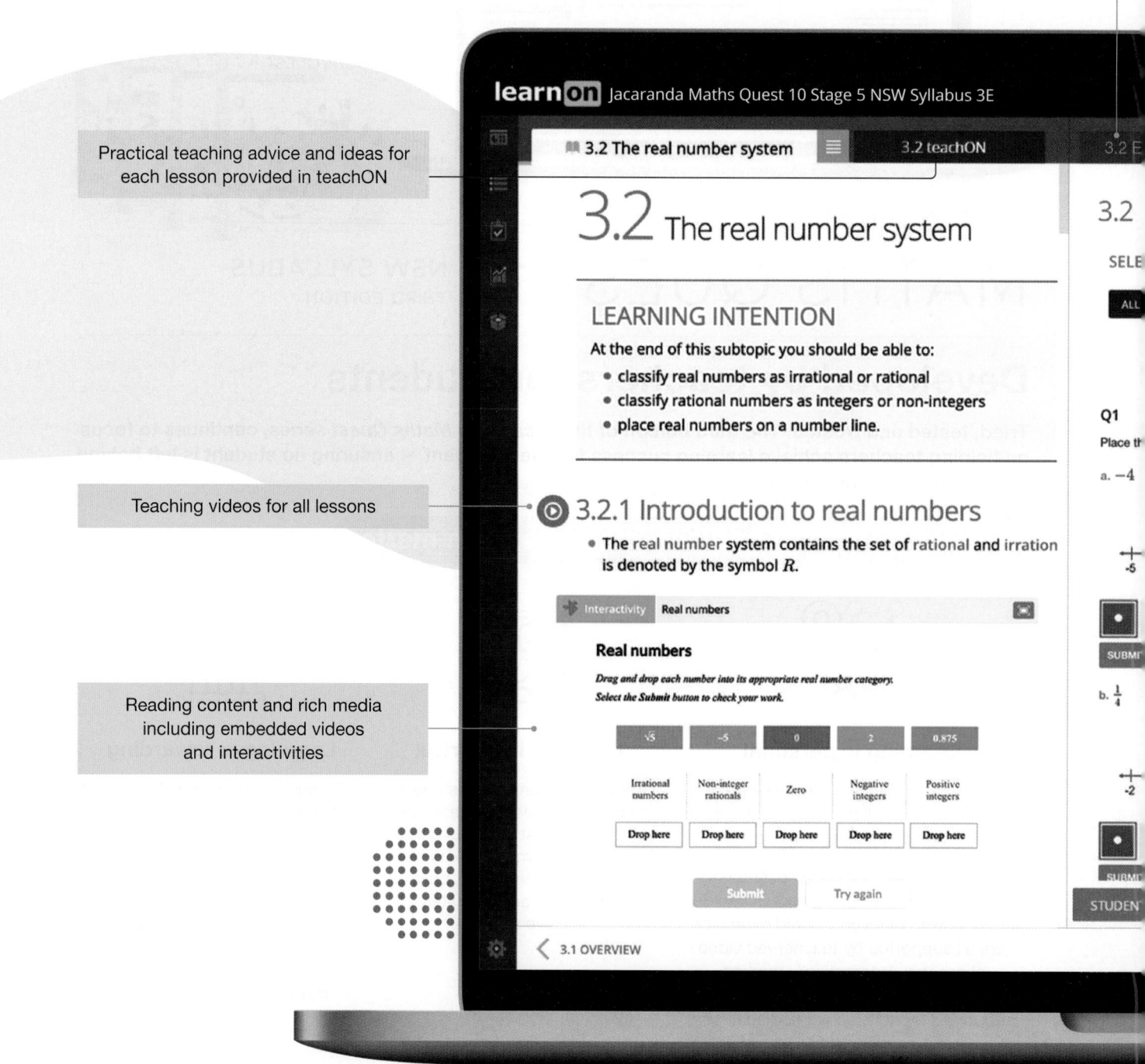

learn**on** Jacaranda Maths Quest 10 Stage 5 NSW Syllabus 3E

📖 3.2 The real number system   3.2 teachON   3.2 E

3.2 The real number system

LEARNING INTENTION
At the end of this subtopic you should be able to:
- classify real numbers as irrational or rational
- classify rational numbers as integers or non-integers
- place real numbers on a number line.

⊙ 3.2.1 Introduction to real numbers
- The real number system contains the set of rational and irration is denoted by the symbol R.

✳ Interactivity   Real numbers

Real numbers

Drag and drop each number into its appropriate real number category.
Select the Submit button to check your work.

| $\sqrt{5}$ | −5 | 0 | 2 | 0.875 |

| Irrational numbers | Non-integer rationals | Zero | Negative integers | Positive integers |

| Drop here | Drop here | Drop here | Drop here | Drop here |

Submit   Try again

❮ 3.1 OVERVIEW

3.2

SELE

ALL

Q1
Place th
a. −4

+
-5

SUBMI

b. $\frac{1}{4}$

+
-2

SUBM

STUDEN

powerful learning tool, learnON

New! **Quick Quiz questions** for skill acquisition

Differentiated question sets

Teacher and student views

Textbook questions

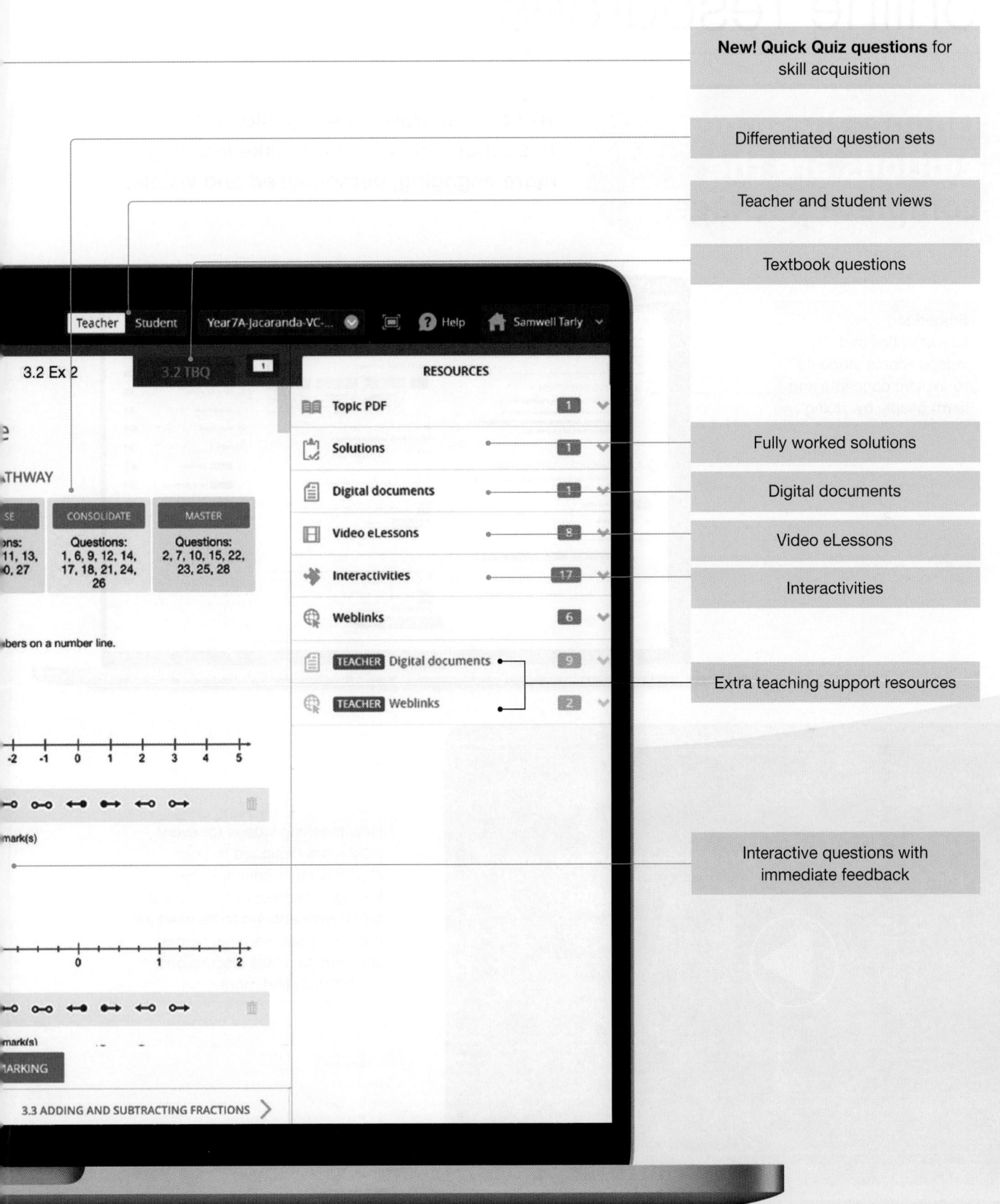

Fully worked solutions

Digital documents

Video eLessons

Interactivities

Extra teaching support resources

Interactive questions with immediate feedback

Get the most from your online resources

Online, these new editions are the complete package

Trusted Jacaranda theory, plus tools to support teaching and make learning more engaging, personalised and visible.

Embedded interactivities and videos enable students to explore concepts and learn deeply by 'doing'.

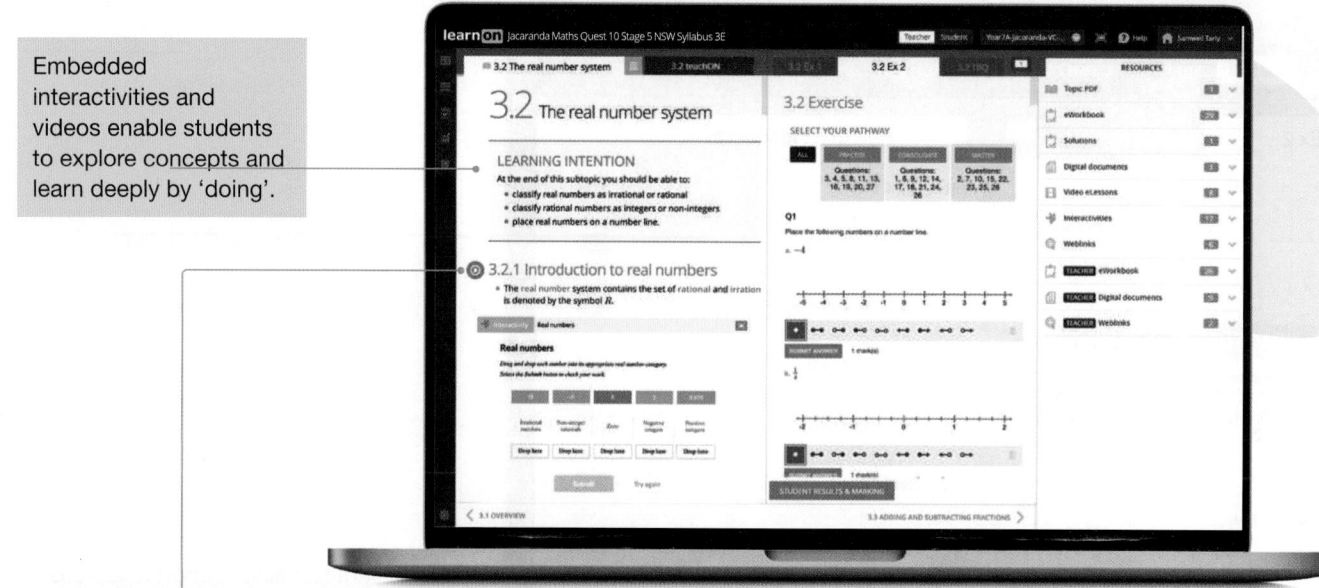

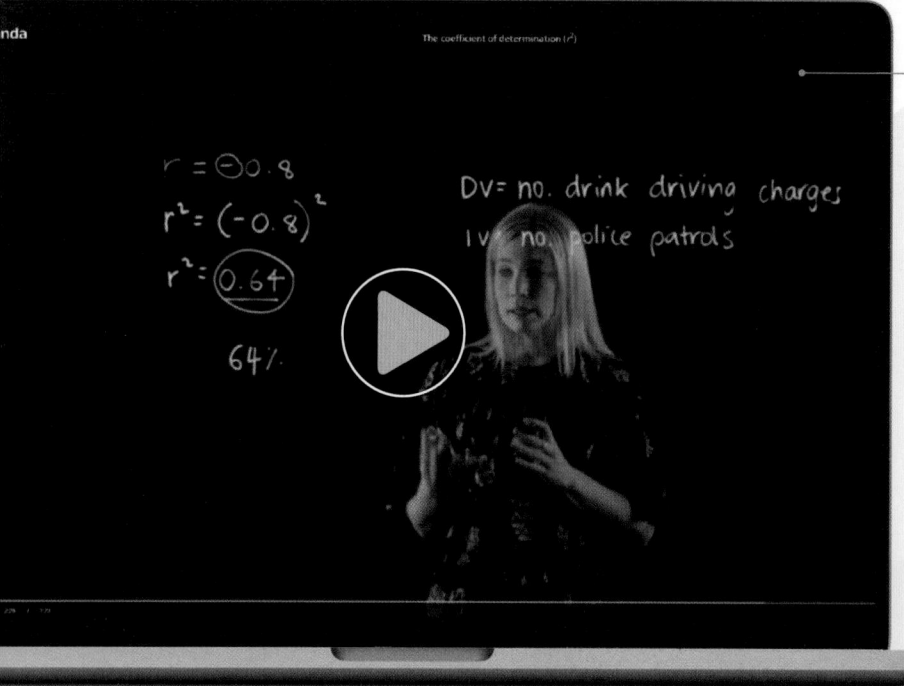

New teaching videos for every lesson are designed to help students learn concepts by having a 'teacher at home', and are flexible enough to be used for pre- and post-learning, flipped classrooms, class discussions, remediation and more.

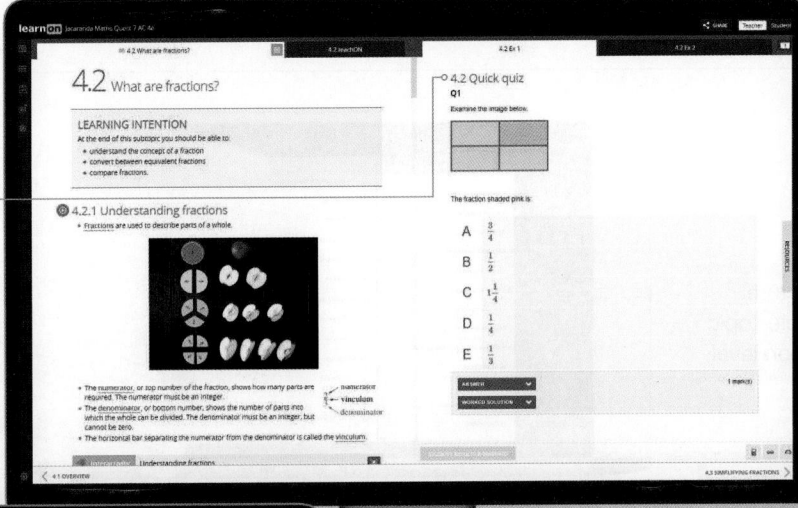

Brand new! Quick Quiz questions for skill acquisition in every lesson.

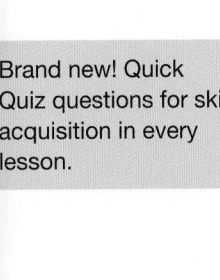

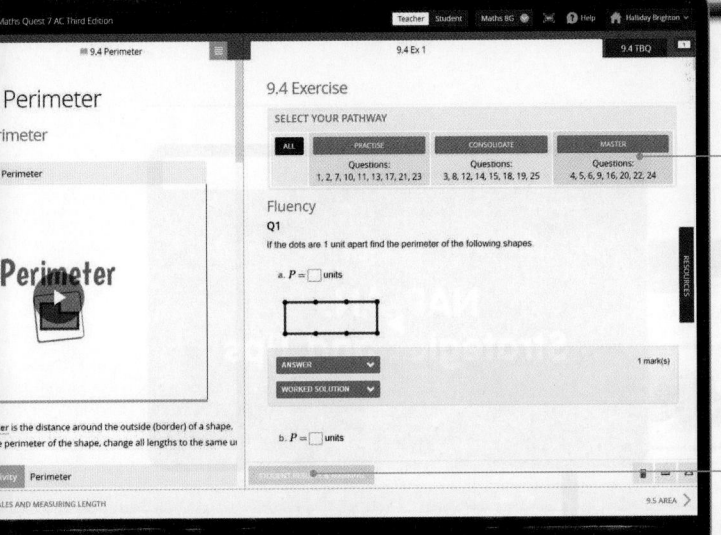

Three differentiated question sets, with immediate feedback in every lesson, enable students to challenge themselves at their own level.

Instant reports give students visibility into progress and performance.

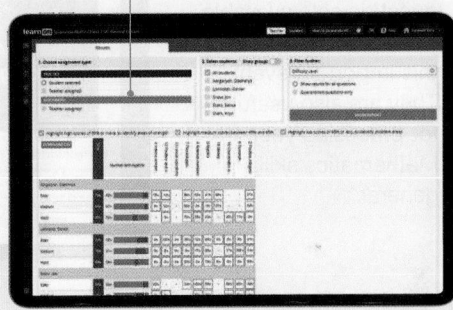

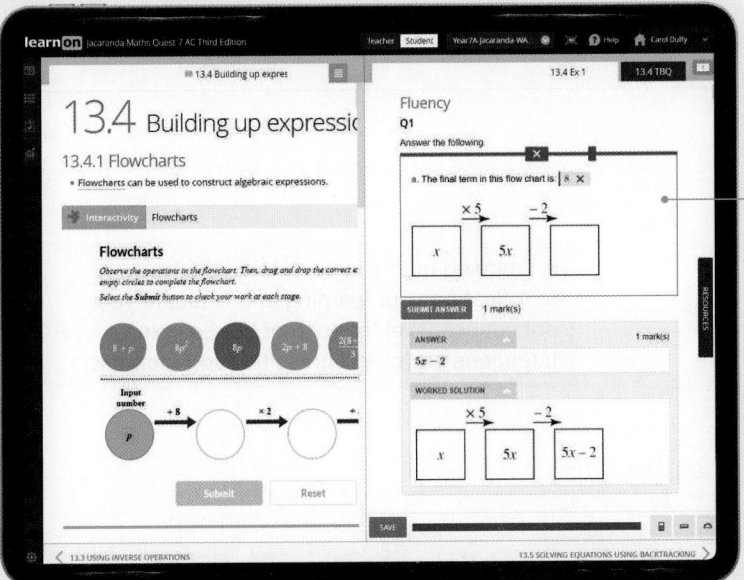

Every question has immediate, corrective feedback to help students overcome misconceptions as they occur and get unstuck as they study independently — in class and at home.

Core–Paths structure made visible

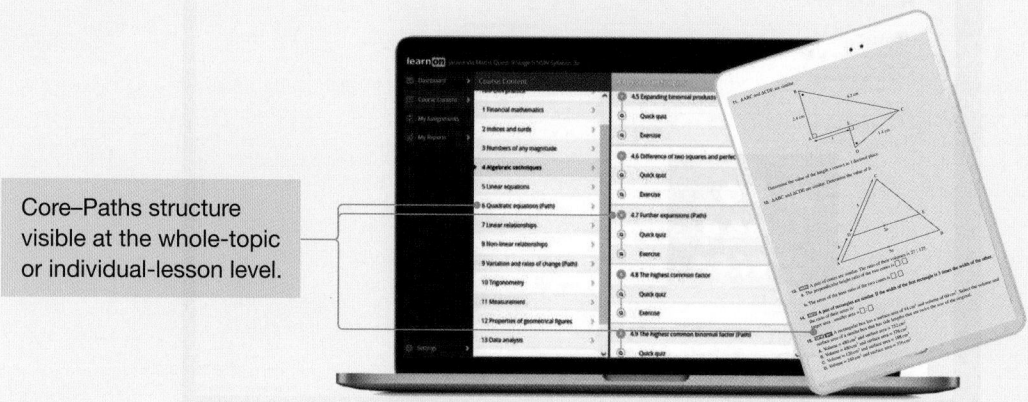

Core–Paths structure visible at the whole-topic or individual-lesson level.

NAPLAN Online Practice

Go online to complete practice NAPLAN tests. There are 6 NAPLAN-style question sets available to help you prepare for this important event. They are also useful for practising your Mathematics skills in general.

Also available online is a video that provides strategies and tips to help with your preparation.

Learning matrix

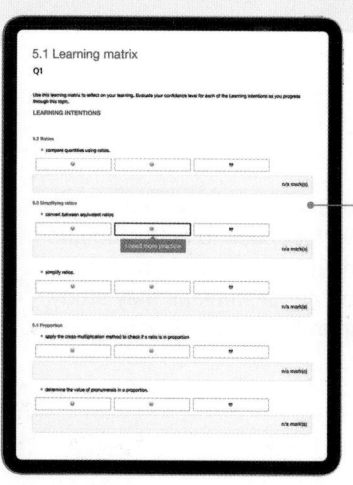

A Learning matrix in each topic enables you to reflect on your learning and evaluate your confidence level for each of the Learning Intentions as you progress through the topic.

A wealth of teacher resources

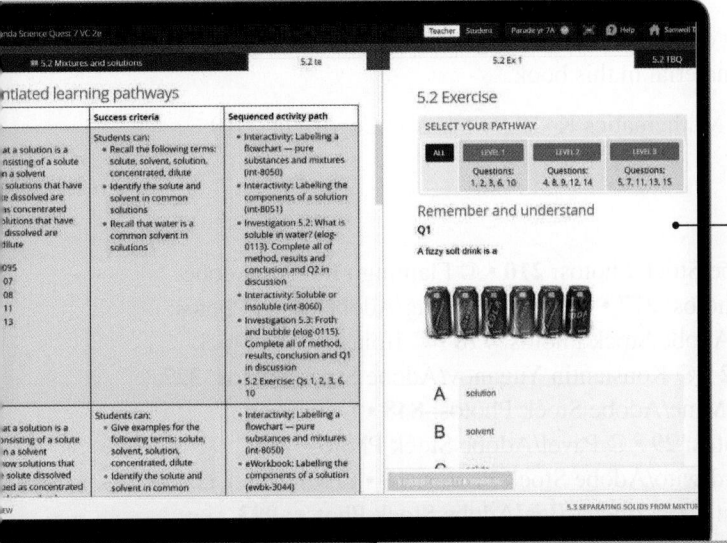

Enhanced teaching-support resources for every lesson, including:

- work programs and curriculum grids
- practical teaching advice
- three levels of differentiated teaching programs
- quarantined topic tests (with solutions)

Customise and assign

An inbuilt testmaker enables you to create custom assignments and tests from the complete bank of thousands of questions for immediate, spaced and mixed practice.

Reports and results

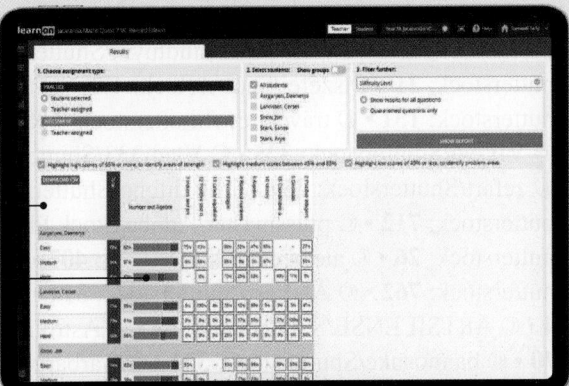

Data analytics and instant reports provide data-driven insights into progress and performance within each lesson and across the entire course.

Show students (and their parents or carers) their own assessment data in fine detail. You can filter their results to identify areas of strength and weakness.

Acknowledgements

The authors and publisher would like to thank the following copyright holders, organisations and individuals for their assistance and for permission to reproduce copyright material in this book.

Subject Outcomes, Objectives and Contents from the NSW Mathematics K–10 Syllabus © Copyright 2019 NSW Education Standards Authority.

Images

• © simon/Adobe Stock Photos: **355** • © Drobot Dean/Adobe Stock Photos: **210** • © Flamingo Images/Adobe Stock Photos: **405** • © R. Gino Santa Maria/Adobe Stock Photos: **377** • © saran_poroong/Adobe Stock Photos: **623** • © Dusan Kostic/Adobe Stock Photos: **405** • © jabiru/Adobe Stock Photos: **778** • © ffolas/Adobe Stock Photos: **602** • © Farknot Architect/Adobe Stock Photos: **392** • © Konstantin Yuganov/Adobe Stock Photos: **327** • © Koto Amatsukami/Adobe Stock Photos: **173** • © VectorMine/Adobe Stock Photos: **848** • © onlyyouqj/Adobe Stock Photos: **1018** • © Damianbn/Adobe Stock Photos: **29** • © Pavel/Adobe Stock Photos: **184** • © Krakenimages.com/Adobe Stock Photos: **12** • © Ekahardiwito/Adobe Stock Photos: **10** • © Michele Ursi/Adobe Stock Photos: **784** • © See Less/Adobe Stock Photos: **2** • © Buch&Bee/Adobe Stock Photos: **993** • © Lustre/Adobe Stock Photos: **711** • © Azee Jacobs/peopleimages.com/Adobe Stock Photos: **790** • © Michael Brown/Adobe Stock Photos: **790** • © AMR Image/Vetta/Getty Images: **211** • © fsachs78/Getty Images: **491, 654** • © vgajic/Getty Images: **217** • © Piter Lenk/Alamy Stock Photo: **89** • © Szepy/Getty Images: **213** • © Martin Ruegner/Getty Images: **253** • © A.D.S.Portrait/Shutterstock: **280** • © Africa Studio/Shutterstock: **694** • © aiyoshi597/Shutterstock: **884** • © Aleks Kend/Shutterstock: **834** • © Alex Mit/Adobe Stock Photo: **166** • © Alexander Raths/Adobe Stock Photo: **143** • © alexandre zveiger/Shutterstock: **258** • © ALPA PROD/Shutterstock: **440** • © Barone Firenze/Shutterstock: **268** • © benjaminec/Shutterstock: **858** • © Bobex-73/Shutterstock: **761** • © BSG_1974/Shutterstock: **940** • © Cheryl Casey/Shutterstock: **738** • © David Malik/Shutterstock: **24** • © DenisProduction.com/Shutterstock: **167** • © doomu/Shutterstock: **704** • © Elena Noeva/Adobe Stock Photos: **911** • © en Rozhnovsky/Shutterstock: **378** • © enterphoto/Shutterstock: **753** • © ESB Professional/Shutterstock: **152, 680** • © FamVeld/Shutterstock: **782** • © Gordon Bell/Shutterstock: **564** • © Inked Pixels/Shutterstock: **18** • © iravgustin/Shutterstock: **344** • © karamysh/Shutterstock: **19** • © kavione/Shutterstock: **6** • © l i g h t p o e t/Shutterstock: **218** • © Lukas Gojda/Shutterstock: **484** • © Maksym Dykha/Shutterstock: **286** • © Maridav/Shutterstock: **484, 514** • © Mijolanta/Shutterstock: **514** • © Monkey Business Images/Shutterstock: **783** • © mw2st/Shutterstock: **310** • © OlegDoroshin/Shutterstock: **748** • © Olga Danylenko/Shutterstock: **449** • © Olga Listopad/Shutterstock: **749** • © Omegafoto/Shutterstock: **11** • © P Meybruck/Shutterstock: **475** • © PabloBenii/Shutterstock: **239** • © Palatinate Stock/Shutterstock: **614** • © Patryk Kosmider/Shutterstock: **744** • © Pavel Nesvadba/Shutterstock: **749** • © Pavlo Baliukh/Shutterstock: **804** • © PHONT/Shutterstock: **547** • © Photo Melon/Shutterstock: **649** • © pressmaster/Adobe Stock Photo: **163** • © Racheal Grazias/Shutterstock: **906** • © Rashevskyi Viacheslav/Shutterstock: **9** • © Scanrail1/Shutterstock: **11** • © Serg64/Shutterstock: **574** • © SIAATH/Shutterstock: **750** • © sirtravelalot/Shutterstock: **224** • © Skylines/Shutterstock: **134** • © Soloviova Liudmyla/Shutterstock: **691** • © SpeedKingz/Shutterstock: **682** • © stefan11/Shutterstock: **21** • © szefei/Shutterstock: **473** • © theskaman306/Shutterstock: **211** • © ThomasLENNE/Shutterstock: **131** • © travelview/Shutterstock: **608** • © Triff/Shutterstock: **863** • © vikky/Shutterstock: **762** • © VILevi/Shutterstock: **325** • © Vladi333/Shutterstock: **98, 504** • © WAYHOME studio/Shutterstock: **857** • © zefart/Shutterstock: **23** • © zhu difeng/Shutterstock: **378** • © Zurijeta/Shutterstock: **694** • © fuyu liu/Shutterstock: **712** • © pressmaster/Adobe Stock Photos: **674** • © 9lives/Shutterstock: **175** • © a1b2c3/Shutterstock: **26** • © alean che/Shutterstock: **407** • © Aleix Ventayol Farrés/Shutterstock: **472** • © Alexey Repka/Shutterstock: **762** • © Andresr/Shutterstock: **814** • © Andrew Burgess/Shutterstock: **217** • © arbit/Shutterstock: **74** • © ARTSILENSE/Shutterstock: **62** • © Astronoman/Shutterstock: **762** • © Balaguta Evgeniya/Shutterstock: **670** • © bannosuke/Shutterstock: **321** • © Barbara Dudzinska/Shutterstock: **708** • © Bildagentur Zoonar GmbH/Shutterstock: **713** • © BlueSkyImage/Shutterstock: **289** • © BrunoRosa/Shutterstock: **805** • © CandyBox images/Shutterstock: **696** • © Cavan for Adobe/Adobe Stock Photos: **495** • © CHEN WS/Shutterstock: **243**

• © CLS Design/Shutterstock: **210** • © Crok Photography/Shutterstock: **194** • © Dario Sabljak/Shutterstock: **926** • © Darios/Shutterstock: **655** • © Darren Brode/Shutterstock: **173** • © davidundderriese/Shutterstock; **175** • © Denis Kuvaev/Shutterstock: **695** • © Diego Barbieri/Shutterstock: **247** • © dizain/Adobe Stock Photos: **845** • © Dmitry Morgan/Shutterstock: **601** • © DrHitch/Shutterstock: **925** • © Edward Westmacott/Shutterstock: **209** • © EsHanPhot/Shutterstock: **842** • © Gino Santa Maria/Shutterstock: **36** • © GlobalTravelPro/Shutterstock: **864** • © goodluz/Adobe Stock Photos: **719** • © gpointstudio/Shutterstock: **856** • © Hallgerd/Shutterstock: **659** • © Hydromet/Shutterstock: **209** • © Ildar Akhmerov/Shutterstock: **784** • © Irina Strelnikova/Shutterstock; Marish/Shutterstock; vectorstudi/Shutterstock; Visual Generation/Shutterstock: **1, 33, 97, 151, 183, 223, 267, 343, 439, 503, 563, 613, 669, 737, 799, 841, 883, 939, 981** • © J.D.S/Shutterstock: **134** • © jabiru/Shutterstock: **748** • © Jayme Burrows/Shutterstock: **319** • © Jirsak/Shutterstock: **55** • © Johan Larson/Shutterstock: **36** • © JUN KAWAGUCHI/Shutterstock: **139** • © Keo/Shutterstock: **831** • © koya979/Shutterstock: **599** • © kurhan/Shutterstock: **166** • © Larry-Rains/Shutterstock: **137** • © Lee Torrens/Shutterstock: **198** • © Liv friis-larsen/Shutterstock: **163** • © Lucky Business/Shutterstock: **756** • © Maksim Kabakou/Shutterstock: **136** • © Maria Maarbes/Shutterstock: **705** • © max blain/Shutterstock: **683** • © Maxim Blinkov/Shutterstock: **115** • © Meder Lorant/Shutterstock: **786** • © Megapixel/Shutterstock: **123** • © michaklootwijk/Adobe Stock Photo: **256** • © Natali Glado/Shutterstock: **705** • © Natykach Nataliia/Shutterstock: **20** • © Neale Cousland/Shutterstock: **231** • © nito/Adobe Stock Photo: **256** • © nito/Shutterstock: **372** • © Nomad_Soul/Shutterstock: **549** • © Oleksandr Khoma/Shutterstock: **587** • © Oleksiy Mark/Shutterstock: **207** • © Orla/Shutterstock: **364** • © Paulo M. F. Pires/Shutterstock: **864** • © petrroudny/Adobe Stock Photos: **714** • © PhotoDisc: **234, 235, 355, 397, 448, 449, 483, 582, 656, 761** • © photoiconix/Shutterstock: **598** • © Popartic/Shutterstock: **598** • © Poznyakov/Shutterstock: **601** • © Robyn Mackenzie/Shutterstock: **163** • © royaltystockphoto.com/Shutterstock: **386** • © Ruth Peterkin/Shutterstock: **235** • © Source: Wiley Art.: **763** • © Stephen Frink/Digital Vision: **595** • © Sunset Blue/Shutterstock: **140** • © Suzanne Tucker/Shutterstock: **166, 782** • © Syda Productions/Shutterstock: **826** • © Sylverarts Vectors/Shutterstock: **600** • © Tepikina Nastya/Shutterstock: **173** • © theromb/Shutterstock: **834** • © TnT Designs/Shutterstock: **873** • © trekandshoot/Shutterstock: **205** • © VectorMine/Shutterstock: **973** • © Viaceslav/Shutterstock: **777** • © WilleeCole Photography/Shutterstock: **659** • © wonlopcolors/Shutterstock: **450** • © worldswildlifewonders/Shutterstock: **80** • © XiXinXing/Shutterstock: **364** • © zamanbeku/Shutterstock: **24** • © Amy Johansson/Shutterstock: **348** • © John Carnemolla/Shutterstock: **348** • © Paul D Smith/Shutterstock: **348** • © Petar Milevski/Shutterstock: **5** • © del-Mar/Shutterstock: **34** • © Nic Vilceanu/Shutterstock: **800** • © Strelciuc/Adobe Stock Photos: **712** • © Source: Aussies add four to contract list, Carey turns down deal | cricket.com.au, Source: Australia Cricket Players Salary, Rules, Central Contract 2022-23, Highest Paid Cricketers - Sporty Report, Source: How much money do social media companies make from advertising? - Zippia: **710** • © Source: Australian Communications and Media Authority, The digital lives of Younger Australians, 2021: **716** • © Source: How Much Time Do Children Spend on TikTok? New Report Reveals Staggering Stats (movieguide.org): **711** • © Source: Sydney Metro surveys likened to push polling, The Sydney Morning Herald, Sean Nicholls, February 6 2017: **702** • © Source: Woolworths Limited: **706** • © nmedia/Shutterstock: **82** • © Tada Images/Adobe Stock Photo: **83** • © agsandrew/Shutterstock: **83**

Every effort has been made to trace the ownership of copyright material. Information that will enable the publisher to rectify any error or omission in subsequent reprints will be welcome. In such cases, please contact the Permissions Section of John Wiley & Sons Australia, Ltd.

1 Financial mathematics

LESSON SEQUENCE

LESSON
1.1 Overview

Why learn this?

Everyone requires food, housing, clothing and transport, and a fulfilling social life. Money allows us to purchase the things we need and desire. The ability to manage money is key to a financially secure future and a reasonable retirement with some fun along the way. Each individual is responsible for managing his or her own finances; therefore, it is imperative that everyone is financially literate.

In this topic, you will investigate different investment options for saving your money. It is important to understand how investments work to be able to decide whether an investment is a good idea or not. You will also investigate the options available to purchase items such as computers or other small items on terms.

Understanding the principal concepts of depreciation is necessary in many business situations. A sound knowledge of financial mathematics is essential in a range of careers, including financial consultancy, accountancy and business management.

Hey students! Bring these pages to life online

▶ Watch videos

Engage with interactivities

A+ Answer questions and check solutions

Find all this and MORE in jacPLUS ▶

Reading content and rich media, including interactivities and videos for every concept

Extra learning resources

Differentiated question sets

Questions with immediate feedback, and fully worked solutions to help students get unstuck

Exercise 1.1 Pre-test

1. **MC** Select the formula used to calculate the simple interest on $3000 at 18% for 18 months.

   A. $I = \dfrac{3000 \times 18 \times 18}{100}$ **B.** $I = 3000 \times 18 \times 18$ **C.** $I = \dfrac{3000 \times 18 \times 1.8}{100}$ **D.** $I = \dfrac{3000 \times 18 \times 1.5}{100}$

2. Jett uses his credit card to purchase a PS4 console and games for $500. At the end of 1 month, the credit company charges 22% p.a. Calculate the amount of interest Jett must pay on his credit card after 1 month, to the nearest cent.

3. Calculate the total cost of a $2500 purchase on the following terms: 15% deposit and weekly payments of $15 over 5 years.

4. **MC** Lisa purchases a car for $6500 on the following terms: 20% deposit with the balance plus simple interest paid monthly at 12% p.a. over 5 years. Select the correct amount of each monthly repayment.
   A. $86.67   **B.** $73.67   **C.** $108.33   **D.** $138.67

5. Calculate the simple interest on $3000 invested at 3.75% p.a. for 3 years.

6. **MC** A clothing store offers a discount of 20% during a sale. A further 5% discount is offered to members. If the original price of the jacket was $150, what was the member's sale price?
   A. $112.50   **B.** $114   **C.** $120   **D.** $111

7. Calculate the amount of interest for $3000 compounded annually at 3% p.a. for 3 years, correct to the nearest cent.

8. **MC** Select the value of an investment of $8000 compounded quarterly at 8% p.a. for 5 years.
   A. $3887.58   **B.** $11 754.60   **C.** $11 887.60   **D.** $37 287.70

9. A computer costs $5500. The value of the computer depreciates by 15% p.a. Calculate the value of the computer after 3 years, to the nearest cent.

10. **MC** An industrial machine purchased for $128 000 will have a value of $5000 in 7 years. Select the approximate rate at which the machine is depreciating per annum.
    A. 17%   **B.** 25%   **C.** 37%   **D.** 40%

11. **MC** An airplane depreciates at a rate of 12% p.a. Select how many years will it take for the airplane to reduce to half its initial value?
    A. 5.4   **B.** 6   **C.** 6.5   **D.** 12

12. **MC** The value of a tractor is worth $160 000. The value of the tractor depreciates by 20% p.a. What percentage of its initial value is the tractor worth after 5 years? Select the correct answer.
    A. 20%   **B.** 23.8%   **C.** 32.8%   **D.** 42.8%

13. Calculate the interest payable on a loan of $500 000 to be repaid at 8% p.a. flat rate interest over 5 years.

14. Calculate the balance, at the start of the third year, on a loan of $25 000 that is charged at 10% p.a. reducible over 3 years. The loan is repaid in two annual instalments of $8500.

15. **MC** A monthly credit card statement shows that if $3125 is paid by the due date, no interest will be charged. If only $500 is paid by the due date and the interest rate is 23%, what is the balance owing at the end of the month?
    A. $1906.25   **B.** $2406.25   **C.** $2625.00   **D.** $3228.75

LESSON
1.2 Simple interest

LEARNING INTENTION

At the end of this lesson you should be able to:
- calculate the simple interest on a loan or an investment
- apply the simple interest formula to determine the time, the rate or the principal.

1.2.1 The simple interest formula

eles-6248

- The **simple interest formula** can be used to calculate the interest charged on borrowed money. The formula is:

> ### Formula for simple interest
>
> $$I = Prn$$
>
> where:
> - I = amount of interest earned or paid
> - P = principal
> - r = interest rate as a percentage per annum (yearly), written as a decimal (e.g. 2% p.a. is equal to 0.02 p.a.)
> - n = the duration of the investment in years

WORKED EXAMPLE 1 Calculating simple interest

Calculate the simple interest on $4000 invested at 4.75% p.a. for 4 years.

THINK	WRITE
1. Write the formula and the known values of the variables.	$I = Prn$, where $P = \$4000, r = 0.0475, n = 4$
2. Substitute known values to calculate I.	$I = 4000 \times 0.0475 \times 4$ $\quad = 760$
3. Write the answer.	The simple interest is $760.

1.2.2 Purchasing goods and simple interest

eles-6249

- There are many different payment options when purchasing major goods, such as flat screen televisions and computers. Payment options include:
  - cash
  - credit cards
  - **lay-by**
  - deferred payment
  - buying on terms
  - loans.
- The cost of purchasing an item can vary depending on the method of payment used.
- Some methods of payment involve borrowing money and, as such, mean that interest is charged on the money borrowed.

• What are the ways of purchasing the item shown in the advertisement below?

120 cm HD TV	
 $800	5-year warranty • High definition • HDMI ports • 16 : 9 aspect ratio • 1080i

Cash

• With cash, the marked price is paid on the day of purchase with nothing more to pay.
• A cash-paying customer can often negotiate with the retailer to obtain a lower price for the item.

Lay-by

• With lay-by, the item is held by the retailer while the customer makes regular payments towards paying off the marked price.
• In some cases, a small administration fee may be charged.

Credit cards

• With a credit card, the retailer is paid by the credit card provider, generally a financial lender.
• The customer takes immediate possession of the goods.
• The financial lender collates all purchases over a monthly period and bills the customer accordingly. The entire balance shown on the bill can often be paid with no extra charge, but if the balance is not paid in full, interest is charged on the outstanding amount, generally at a very high rate.

WORKED EXAMPLE 2 Calculating interest charged

The ticketed price of a mobile phone is $600. Andrew decides to purchase the phone using his credit card. At the end of 1 month the credit card company charges interest at a rate of 15% p.a. Calculate the amount of interest that Andrew must pay on his credit card after 1 month.

THINK	WRITE
1. Write the formula and the known values of the variables. Remember that 1 month = $\frac{1}{12}$ year.	$I = Prn$ $P = \$600, r = 0.15, n = \frac{1}{12}$
2. Substitute known values to calculate I.	$I = 600 \times 0.15 \times \frac{1}{12}$ $= 7.50$
3. Write the answer.	The interest Andrew pays is $7.50.

Exercise 1.2 Simple interest

learnon

| **1.2 Quick quiz** on | **1.2 Exercise** |

Individual pathways

■ PRACTISE	■ CONSOLIDATE	■ MASTER
1, 2, 4, 5, 9	3, 6, 8, 10	7, 11, 12, 13

Fluency

1. **WE1** Calculate the simple interest payable on a loan of $8000 at 6% p.a. for 5 years.

2. Calculate the simple interest on each of the following loans.

   a. $5000 at 9% p.a. for 4 years.
   b. $4000 at 7.5% p.a. for 3 years.
   c. $12 000 at 6.4% p.a. for $2\frac{1}{2}$ years.
   d. $6000 at 8% p.a. for $1\frac{1}{2}$ years.

3. Calculate the simple interest on each of the following investments.

   a. $50 000 at 6% p.a. for 6 months.
   b. $12 500 at 12% p.a. for 1 month.
   c. $7500 at 15% p.a. for 3 months.
   d. $4000 at 18% p.a. for 18 months.

4. Calculate the monthly interest charged on each of the following outstanding credit card balances.

   a. $1500 at 15% p.a.
   b. $4000 at 16.5% p.a.
   c. $2750 at 18% p.a.
   d. $8594 at 17.5% p.a.
   e. $5690 at 21% p.a.

Understanding

5. **WE2** The ticketed price of a mobile phone is $800. Elena decides to purchase the phone using her credit card. After 1 month the credit card company charges interest at a rate of 15% p.a. Calculate the amount of interest that Elena must pay on her credit card after 1 month.

6. Arup decides to purchase a new sound system using her credit card. The ticketed price of the sound system is $900. When Arup's credit card statement arrives, it shows that she will pay no interest if she pays the full amount by the due date.

   a. If Arup pays $200 by the due date, what is the balance owing?
   b. If the interest rate on the credit card is 18% p.a., how much interest will Arup be charged in the month?
   c. What will be the balance that Arup owes at the end of the month?
   d. At this time Arup pays another $500 off her credit card. How much interest is Arup then charged for the next month?
   e. Arup then pays off the entire remaining balance of her card. What was the true cost of the sound system, including all the interest payments?

7. Carly has an outstanding balance of $3000 on her credit card for June and is charged interest at a rate of 21% p.a.

   a. Calculate the amount of interest that Carly is charged for June.
   b. Carly makes the minimum repayment of $150 and makes no other purchases using the credit card in the next month. Calculate the amount of interest that Carly will be charged for July.
   c. If Carly had made a repayment of $1000 at the end of June instead of $150, calculate the amount of interest that Carly would then have been charged for July.
   d. How much would Carly save in July had she made the higher repayment at the end of June?

8. Shane buys a new home theatre system using his credit card. The ticketed price of the bundle is $7500. The interest rate that Shane is charged on his credit card is 18% p.a.
   Shane pays off the credit card at a rate of $1000 each month.

   a. Complete the table below.

Month	Balance owing	Interest	Payment	Closing balance
January	$7500.00	$112.50	$1000.00	$6612.50
February	$6612.50	$99.19	$1000.00	
March			$1000.00	
April			$1000.00	
May			$1000.00	
June			$1000.00	
July			$1000.00	
August			$1015.86	$0

   b. What is the total amount of interest that Shane pays?
   c. What is the total cost of purchasing the home theatre system using his credit card?

Communicating, reasoning and problem solving

9. Design a table that compares the features of each method of payment: cash, lay-by and credit card.

10. Choose the most appropriate method of payment for each of the described scenarios below. Explain your choice.

   Scenario 1: Andy has no savings and will not be paid for another 2 weeks. Andy would like to purchase an HD television and watch tomorrow's football final.

   Scenario 2: In September, Lena spots on special a home theatre system which she would like to purchase for her family for Christmas.

11. Merchant banks offer simple interest on all investments. Merchant bank A had an investor invest $10 000 for 5 years. Merchant bank B had a different investor invest $15 000 for 3 years.
    Investor B obtained $2500 more in interest than investor A because the rate of interest per annum she received was 6% greater than the interest obtained by investor A.
    Calculate the simple interest and rate of interest for each investor.

12. Compare the following two investments where simple interest is paid.

	Rate	Principal	Time	Interest
Investment A	r_A	$8000	4 years	SI_A
Investment B	r_B	$7000	5 years	SI_B

It is known that $r_A : r_B = 2 : 3$ and that investment B earned $2000 more interest than investment A. Determine the values of r_A, r_B, SI_A and SI_B. Give your answers correct to 2 decimal places. (Use unrounded calculations to determine subsequent values.)

13. What can you do to remember the simple interest formula?

LESSON
1.3 Buying on terms

LEARNING INTENTION

At the end of this lesson you should be able to:
- understand the concept of buying on terms
- calculate the total cost of buying an item on terms
- apply concepts of buying on terms to small loans to purchase items.

1.3.1 Buying on terms

eles-6250

- When a customer buys an item on terms:
  - the customer pays a deposit
  - the customer pays off the balance over an agreed period of time with set payments
  - the set payments may be calculated as a stated arbitrary amount or interest rate
  - the total monies paid will exceed the initial cash price.

WORKED EXAMPLE 3 Calculating total costs

The cash price of a computer is $2400. It can also be purchased on the following terms: 25% deposit and payments of $16.73 per week for 3 years.
Calculate the total cost of the computer purchased on terms as described.

THINK	WRITE
1. Calculate the deposit.	Deposit = 25% of $2400 $= 0.25 \times \$2400$ $= \$600$

2. Calculate the total of the weekly repayments.	Total repayment $= \$16.73 \times 52 \times 3$ $= \$2609.88$
3. Add these two amounts together to calculate the total cost.	Total cost $= \$600 + \$2609.88$ $= \$3209.88$

WORKED EXAMPLE 4 Calculating repayments

A diamond engagement ring has a purchase price of $2500. Michael buys the ring on the following terms: 10% deposit with the balance plus simple interest paid monthly at 12% p.a. over 3 years.
a. Calculate the amount of the deposit.
b. Determine the balance owing after the initial deposit.
c. Calculate the interest payable.
d. Determine the total amount to be repaid.
e. Calculate the amount of each monthly repayment.

THINK	WRITE
a. Calculate the deposit by finding 10% of $2500.	**a.** Deposit $= 10\%$ of $\$2500$ $= 0.1 \times \$2500$ $= \$250$
b. Determine the balance owing by subtracting the deposit from the purchase price.	**b.** Balance $= \$2500 - \$250$ $= \$2250$
c. Calculate the simple interest on $2250 at 12% p.a. for 3 years.	**c.** $I = Prn$, where $P = \$2250$, $r = 0.12$, $n = 3$ $I = 2250 \times 0.12 \times 3$ $= \$810$
d. Determine the total repayment by adding the balance owing with the interest payable.	**d.** Total repayment $= \$2250 + \$810$ $= \$3060$
e. Calculate the monthly repayment by dividing the total repayment by the number of months over which the ring is to be repaid.	**e.** Monthly repayment $= \$3060 \div 36$ $= \$85$

Loans

- Money can be borrowed from a bank or other financial institutions.
- Interest is charged on the amount of money borrowed.
- Both the money borrowed and the interest charged must be paid back.
- The interest rate on a loan is generally lower than the interest rate offered on a credit card or when buying on terms.
- The calculation of loan payments is done in the same way as for buying on terms; that is, calculate the interest and add it to the principal before dividing into equal monthly repayments.

DISCUSSION: BUY NOW, PAY LATER

Many large department stores offer white goods and furniture on plans described in terms such as, "Take the product home today and don't pay anything for two years" For many people this is a very tempting offer, as it means they can have the goods they need and defer payment until they have the money.

Working in small groups, use the internet to investigate one of these plans and find out what happens if the customer is unable to pay at the end of the interest-free period. Prepare a report to present to the class.

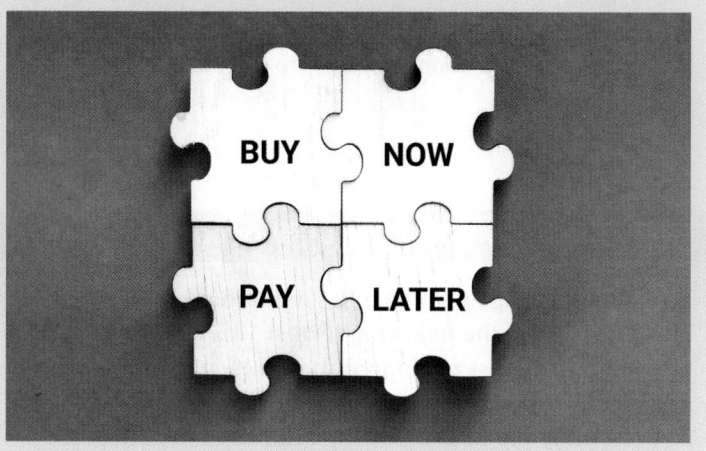

Exercise 1.3 Buying on terms

learn on

1.3 Quick quiz on	1.3 Exercise

Individual pathways

■ PRACTISE	■ CONSOLIDATE	■ MASTER
1, 2, 4, 8, 10, 12	3, 5, 6, 7, 11, 13	9, 14, 15, 16, 17, 18

Fluency

1. Calculate the total cost of a $3000 purchase given the terms described below.
   a. i. 12% deposit and monthly payments of $60 over 5 years.
      ii. 20% deposit and weekly payments of $20 over 3 years.
      iii. 15% deposit and annual payments of $700 over 5 years.
   b. Which of these options is the best deal for a purchaser?

2. Calculate the amount of each repayment for a $5000 purchase given the terms described below.
   a. 10% deposit with the balance plus simple interest paid monthly at 15% p.a. over 5 years.
   b. 10% deposit with the balance plus simple interest paid fortnightly at 12% p.a. over 5 years.
   c. 20% deposit with the balance plus simple interest paid monthly at 10% p.a. over 3 years.

3. Calculate the total repayment and the amount of each monthly repayment for each of the following loans.
   a. $10 000 at 9% p.a. repaid over 4 years.
   b. $25 000 at 12% p.a. repaid over 5 years.
   c. $4500 at 7.5% p.a. repaid over 18 months.
   d. $50 000 at 6% p.a. repaid over 10 years.
   e. $200 000 at 7.2% p.a. repaid over 20 years.

4. **WE3** The cash price of a bedroom suite is $4200. The bedroom suite can be purchased on the following terms: 20% deposit and weekly repayments of $43.94 for 2 years. Calculate the total cost of the bedroom suite if you buy it on terms.

5. Guy purchases a computer that has a cash price of $3750 on the following terms: $500 deposit with the balance plus interest paid over 2 years at $167.92 per month.
What is the total amount that Guy pays for the computer?

6. Dmitry wants to buy a used car with a cash price of $12 600. The dealer offers terms of 10% deposit and monthly repayments of $812.70 for 2 years.

   a. Calculate the amount of the deposit.
   b. Calculate the total amount to be paid in monthly repayments.
   c. What is the total amount Dmitry pays for the car if he buys it on terms?
   d. How much more than the cash price of the car does Dmitry pay? (This is the interest charged by the dealer.)

7. Alja wants to purchase an entertainment system that has a cash price of $5800. She purchases the entertainment system on terms of no deposit and monthly repayments of $233.61 for 3 years.

   a. Calculate the total amount that Alja pays for the entertainment system.
   b. Calculate the amount that Alja pays in interest.
   c. Calculate the amount of interest that Alja pays each year.
   d. Calculate this amount as a percentage of the cash price of the entertainment system.

Understanding

8. **WE4** A used car has a purchase price of $9500. Dayna buys the car on the following terms: 25% deposit with balance plus interest paid at 12% p.a. over 3 years.

   a. Calculate the amount of the deposit.
   b. What is the balance owing?
   c. Calculate the interest payable.
   d. What is the total amount to be repaid?
   e. Determine the amount of each monthly repayment.

9. A department store offers the following terms: one-third deposit with the balance plus interest paid in equal, monthly instalments over 18 months. The interest rate charged is 9% p.a. Ming buys a lounge suite with a ticketed price of $6000.

   a. Calculate the amount of the deposit.
   b. What is the balance owing?
   c. Calculate the interest payable.
   d. What is the total amount to be repaid?
   e. Determine the amount of each monthly repayment.

10. Calculate the monthly payment on each of the following items bought on terms. (*Hint:* Use the steps shown in question 8.)

    a. Dining suite: cash price $2700, deposit 10%, interest rate 12% p.a., term 1 year.

    b. Smartphone: cash price $990, deposit 20%, interest rate 15% p.a., term 6 months.

    c. Car: cash price $16 500, deposit 25%, interest rate 15% p.a., term 5 years.

    d. Mountain bike: cash price $3200, one-third deposit, interest rate 9% p.a., term $2\frac{1}{2}$ years.

    e. Watch: cash price $675, no deposit, interest rate 18% p.a., term 9 months.

11. Samir wants to purchase his first car. He has saved $1000 as a deposit but the cost of the car is $5000. Samir takes out a loan from the bank to cover the balance of the car plus $600 worth of on-road costs.

    a. How much will Samir need to borrow from the bank?

    b. Samir takes the loan out over 4 years at 9% p.a. interest. How much interest will Samir need to pay?

    c. What will be the amount of each monthly payment that Samir makes?

    d. What is the total cost of the car after paying off the loan, including the on-road costs? Give your answer to the nearest dollar.

Communicating, reasoning and problem solving

12. **MC** Kelly wants to borrow $12 000 for some home improvements. Which of the following loans will lead to Kelly making the lowest total repayment?

    A. Interest rate 6% p.a. over 4 years   **B.** Interest rate 7% p.a. over 3 years

    C. Interest rate 5.5% p.a. over 3 years   **D.** Interest rate 6.5% p.a. over 5 years

13. **MC** Without completing any calculations, explain which of the following loans will be the best value for the borrower.

    A. Interest rate 8.2% p.a. over 5 years   **B.** Interest rate 8.2% over 4 years

    C. Interest rate 8% over 5 years   **D.** Interest rate 8% over 4 years

14. Explain how, when purchasing an item, making a deposit using existing savings and taking out a loan for the balance can be an advantage.

15. Gavin borrows $18 000 over 5 years from the bank. The loan is charged at 8.4% p.a. flat rate interest. The loan is to be repaid in equal monthly instalments. Calculate the amount of each monthly repayment.

16. Andrew purchased a new car valued at $32 000. He paid a 10% deposit and was told that he could have 4 years to pay off the balance of the car price plus interest.
An alternative scheme was also offered to him. It involved paying off the balance of the car price plus interest in 8 years. If he chose the latter scheme, he would end up paying $19 584 more. The interest rate for the 8-year scheme was 1% more than for the 4-year scheme.

    a. How much deposit did he pay?

    b. What was the balance to be paid on the car?

    c. Determine the interest rate for each of the two schemes.

    d. Determine the total amount paid for the car for each of the schemes.

    e. What were the monthly repayments for each of the schemes?

17. When buying on terms, what arrangements are the most beneficial to the buyer?

18. Ingrid offered to pay her brother $2 for doing her share of the housework each day, but fined him $5 if he forgot to do it. After 4 weeks, Ingrid discovered that she did not owe her brother any money.
For how many days did Ingrid's brother do her share of the housework?

LESSON
1.4 Compound interest

LEARNING INTENTION

At the end of this lesson you should be able to:
- calculate the final amount received on an investment with compound interest
- calculate the interest earned on an investment
- apply the compound interest formula to determine the time, the rate or the present value of an investment.

1.4.1 Compound interest

eles-6251

- Interest on the principal in a savings account or a short-term or long-term deposit is generally calculated using **compound interest** rather than simple interest.
- Compound interest is when interest is added to the principal at regular intervals, increasing the balance of the account, and each successive interest payment is calculated on the new balance.
- Compound interest can be calculated by calculating simple interest one period at a time.
- The amount to which the initial investment grows is called the **compounded value** or **future value (FV)**.

WORKED EXAMPLE 5 Calculating the future value using simple interest

Kyna invests $8000 at 8% p.a. for 3 years with interest paid at the end of each year. Calculate the compounded value of the investment by calculating the simple interest on each year separately.

THINK	WRITE
1. Write the initial (first year) principal.	Initial principal = $8000
2. Calculate the interest for the first year.	Interest for year 1 = 8% of $8000 = $640
3. Calculate the principal for the second year by adding the first year's interest to the initial principal.	Principal for year 2 = $8000 + $640 = $8640
4. Calculate the interest for the second year.	Interest for year 2 = 8% of $8640 = $691.20
5. Calculate the principal for the third year by adding the second year's interest to the second year's principal.	Principal for year 3 = $8640 + $691.20 = $9331.20
6. Calculate the interest for the third year.	Interest for year 3 = 8% of $9331.20 = $746.50
7. Calculate the future value of the investment by adding the third year's interest to the third year's principal.	Compounded value after 3 years = $9331.20 + $746.50 = $10 077.70

- To calculate the actual amount of interest received, we subtract the initial principal or the present value (*PV*) from the future value.
- In the example above:

$$\text{compound interest} = \$10\,077.70 - \$8000$$
$$= \$2077.70$$

- We can compare this with the simple interest earned at the same rate.

$$I = Prn$$
$$= 800 \times 0.08 \times 3$$
$$= \$1920$$

- The table below shows a comparison between the total interest earned on an investment of $8000 earning 8% p.a. at both simple interest (*I*) and compound interest (*CI*) over an 8-year period.

YEAR	1	2	3	4	5	6	7	8
Total (*I*)	$640.00	$1280.00	$1920.00	$2560.00	$3200.00	$3840.00	$4480.00	$5120.00
Total (*CI*)	$640.00	$1331.20	$2077.70	$2883.91	$3754.62	$4694.99	$5710.59	$6807.44

▶ 1.4.2 Using the compound interest formula
eles-6252

- We can develop a formula for the future value of an investment rather than repeated use of simple interest. Consider Worked example 5. Let the compounded value after year n be A_n.

  After 1 year, $A_1 = 8000 \times 1.08$ (increasing $8000 by 8%)

  After 2 years, $A_2 = A_1 \times 1.08$

  $= 8000 \times 1.08 \times 1.08$ (substituting the value of A_1)

  $= 8000 \times 1.08^2$

  After 3 years, $A_3 = A_2 \times 1.08$

  $= 8000 \times 1.08^2 \times 1.08$ (substituting the value of A_2)

  $= 8000 \times 1.08^3$

- The pattern then continues such that the value of the investment after n years equals
$$\$8000 \times 1.08^n.$$
- This can be generalised for any investment:

Formula for compound interest

$$FV = PV(1 + r)^n$$

where:
- FV = future value of the investment
- PV = present value of the investment
- r = interest rate per time period (or compounding period), expressed as a decimal
- n = number of time periods

- To calculate the amount of compound interest (*CI*) we then use the formula

> ### Amount of compound interest
>
> $$\text{Compound interest} = FV - PV$$
>
> **where:**
> - *FV* = future value of the investment
> - *PV* = present value of the investment

⊙ 1.4.3 Using technology

eles-6253

- Digital technologies, such as spreadsheets, can be used to draw graphs in order to compare interest accrued through simple interest and compound interest.

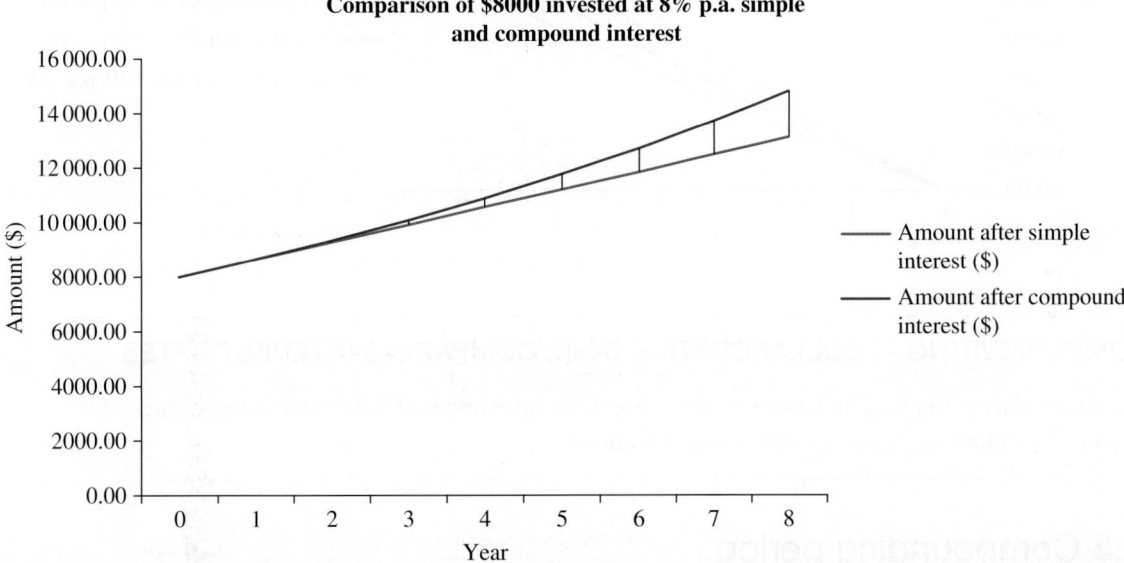

Comparison of $8000 invested at 8% p.a. simple and compound interest

—— Amount after simple interest ($)

—— Amount after compound interest ($)

WORKED EXAMPLE 6 Calculating future value of an investment

William has $14 000 to invest. He invests the money at 9% p.a. for 5 years with interest compounded annually.

a. Use the formula $FV = PV(1 + r)^n$ to calculate the amount to which this investment will grow.

b. Calculate the compound interest earned on the investment.

THINK	WRITE
a. 1. Write the compound interest formula.	**a.** $FV = PV(1 + r)^n$
2. Write the values of *PV* (Present Value), *r* and *n*.	$PV = \$14\,000, \ r = 0.09, \ n = 5$
3. Substitute the values into the formula.	$FV = 14\,000 \times 1.09^5$
4. Calculate the amount.	$= 21\,540.74$
	The investment will grow to $21 540.74.
b. Calculate the compound interest earned.	**b.** Compound interest $= FV - PV$
	$= 21\,540.74 - 14\,000$
	$= 7540.74$
	The compound interest earned is $7540.74.

⏵ 1.4.4 Comparison of fixed principal at various interest rates over a period of time

eles-6254

- It is often helpful to compare the future value (*FV*) of the principal or present value (*PV*) at different compounding interest rates over a fixed period of time.
- Spreadsheets are very useful tools for making comparisons. The graph shown, generated from a spreadsheet, shows the comparisons for $14 000 invested for 5 years at 7%, 8%, 9% and 10% compounding annually.
- There is a significant difference in the future value depending on which interest rate is applied.

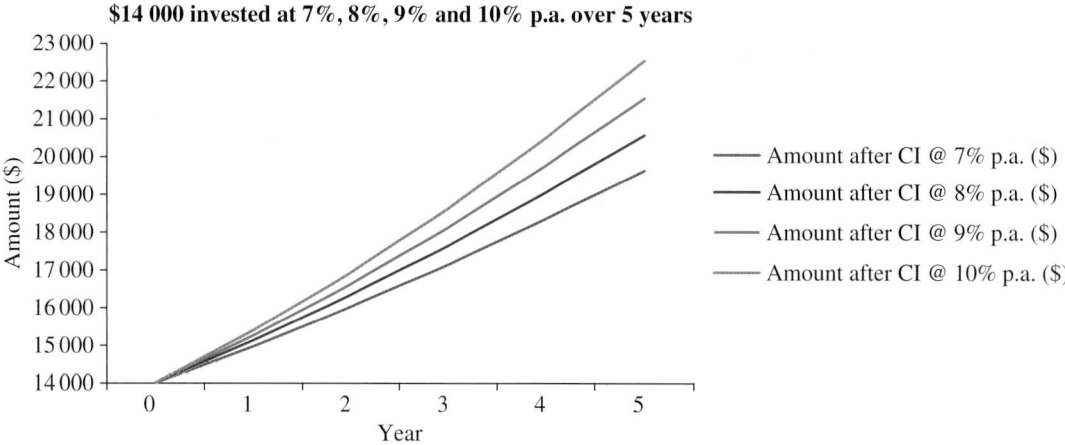

COMMUNICATING — COLLABORATIVE TASK: COMPARING INTEREST RATES

Use a spreadsheet to graph and compare the value of an investment of $20 000 at interest rates of 5%, 6%, 7% and 8% for 8 years, compounding monthly.

⏵ 1.4.5 Compounding period

eles-6255

- In Worked example 6, interest is paid annually.
- Interest can be paid more regularly — it may be paid 6-monthly (twice a year), quarterly (4 times a year), monthly or even daily. This is called the **compounding period**.
- The time and interest rate on an investment must reflect the compounding period. For example, an investment over 5 years at 6% p.a. compounding *quarterly* will have:

$$n = 20 \text{ (4 times each year for 5 years, } 4 \times 5)$$
$$r = 1.5\% \text{ (6\% divided by 4)}$$

Compounding period

To calculate *n*:

  n = **number of years × compounding periods per year**

To calculate *r*:

  r = **interest rate per annum ÷ compounding periods per year**

WORKED EXAMPLE 7 Calculating *FV* with interest compounded quarterly

Calculate the future value of an investment of $4000 at 6% p.a. for 2 years with interest compounded quarterly.

THINK	WRITE
1. Write the compound interest formula.	$FV = PV(1 + r)^n$
2. Write the values of *PV*, *r* and *n*.	$PV = \$4000$, $r = 0.015$, $n = 8$
3. Substitute the values into the formula.	$FV = 4000 \times 1.015^8$
4. Calculate the future value.	$= \$4505.97$
	The future value of the investment is $4505.97.

1.4.6 Guess and refine

eles-6256

- Sometimes, it is useful to know approximately how long it will take to reach a particular future value once an investment has been made.
- Mathematical formulas can be applied to determine when a particular future value will be reached. In this section, a 'guess and refine' method will be shown.
  For example, to determine the number of years required for an investment of $1800 at 9% compounded quarterly to reach a future value of $2500, the following method can be used.
- Let *n* = the number of compounding periods (quarters) and *FV* = the future value in $.
  Therefore, it will take approximately 15 quarters, or 3 years and 9 months, to reach the desired amount.

n	$FV = PV(1 + r)^n$	Comment
1	$1840.50	It is useful to know how the principal is growing after 1 quarter, but the amount is quite far from $2500.
3	$1924.25	The amount is closer to $2500 but still a long way off, so jump to a higher value for *n*.
10	$2248.57	The amount is much closer to $2500.
12	$2350.89	The amount is much closer to $2500.
14	$2457.87	The amount is just below $2500.
15	$2513.17	The amount is just over $2500.

on Resources

Interactivities Compound interest (int-2791)
Compounding periods (int-6186)

Exercise 1.4 Compound interest

1.4 Quick quiz on	1.4 Exercise

Individual pathways

■ PRACTISE	■ CONSOLIDATE	■ MASTER
1, 2, 3, 4, 5, 6, 10, 11, 16	7, 8, 9, 12, 13, 15, 17, 20	14, 18, 19, 21, 22

Fluency

1. Use the formula $FV = PV(1 + r)^n$ to calculate the amount to which each of the following investments will grow with interest compounded annually.
   a. $3000 at 4% p.a. for 2 years.
   b. $9000 at 5% p.a. for 4 years.
   c. $16 000 at 9% p.a. for 5 years.
   d. $12 500 at 5.5% p.a. for 3 years.
   e. $9750 at 7.25% p.a. for 6 years.
   f. $100 000 at 3.75% p.a. for 7 years.

2. Calculate the compounded value of each of the following investments.
   a. $870 for 2 years at 3.50% p.a. with interest compounded 6-monthly.

   b. $9500 for $2\frac{1}{2}$ years at 4.6% p.a. with interest compounded quarterly.

   c. $148 000 for $3\frac{1}{2}$ years at 9.2% p.a. with interest compounded 6-monthly.

   d. $16 000 for 6 years at 8% p.a. with interest compounded monthly.

   e. $130 000 for 25 years at 12.95% p.a. with interest compounded quarterly.

3. **WE5** Danielle invests $6000 at 10% p.a. for 4 years with interest paid at the end of each year. Calculate the compounded value of the investment by calculating the simple interest on each year separately.

4. Ben is to invest $13 000 for 3 years at 8% p.a. with interest paid annually. Determine the amount of interest earned by calculating the simple interest for each year separately.

Year	1	2	3	Total
Simple Interest				

5. **WE6** Simon has $2000 to invest. He invests the money at 6% p.a. for 6 years with interest compounded annually.
   a. Use the formula $FV = PV(1 + r)^n$ to calculate the amount to which this investment will grow.
   b. Calculate the compound interest earned on the investment.

6. **WE7** Calculate the future value of an investment of $14 000 at 7% p.a. for 3 years with interest compounded quarterly.

7. A passbook savings account pays interest of 0.3% p.a. Jill has $600 in such an account. Calculate the amount in Jill's account after 3 years, if interest is compounded quarterly.

Understanding

8. Damien is to invest $35 000 at 7.2% p.a. for 6 years with interest compounded 6-monthly. Calculate the compound interest earned on the investment. If the account compounded annually, how much less interest would Damien's investment have earned?

9. Sam invests $40 000 in a 1-year fixed deposit at an interest rate of 7% p.a. with interest compounding monthly.
   a. Convert the interest rate of 7% p.a. to a rate per month.
   b. Calculate the value of the investment upon maturity.

10. **MC** A sum of $7000 is invested for 3 years at the rate of 5.75% p.a., compounded quarterly. The interest paid on this investment, to the nearest dollar, is:
    A. $1208   **B.** $1308   **C.** $8208   **D.** $8308

11. **MC** After selling their house and paying off their mortgage, Mr and Mrs Fong have $73 600. They plan to invest it at 7% p.a. with interest compounded annually. The value of their investment will first exceed $110 000 after:
    A. 5 years
    B. 6 years
    C. 8 years
    D. 10 years

12. **MC** Maureen wishes to invest $15 000 for a period of 7 years. The following investment alternatives are suggested to her. The best investment would be:
    A. simple interest at 8% p.a.
    B. compound interest at 6.7% p.a. with interest compounded annually
    C. compound interest at 6.6% p.a. with interest compounded 6-monthly
    D. compound interest at 6.5% p.a. with interest compounded quarterly

13. **MC** An amount is to be invested for 5 years and compounded semi-annually at 7% p.a. Which of the following investments will have a future value closest to $10 000?
    A. $700   **B.** $6500   **C.** $7400   **D.** $9000

14. Jake invests $120 000 at 9% p.a. for a 1-year term. For such large investments, interest is compounded daily.
    a. Calculate the daily percentage interest rate, correct to 4 decimal places. Use 1 year = 365 days.
    b. Hence, calculate the compounded value of Jake's investment on maturity.
    c. Calculate the amount of interest paid on this investment.
    d. Calculate the extra amount of interest earned compared with the case where the interest is calculated only at the end of the year.

Communicating, reasoning and problem solving

15. Daniel has $15 500 to invest. An investment over a 2-year term will pay interest of 7% p.a.
    a. Calculate the compounded value of Daniel's investment if the compounding period is:
       i. 1 year   ii. 6 months   iii. 3 months   iv. 1 month.
    b. Explain why it is advantageous to have interest compounded on a more frequent basis.

16. Jasmine invests $6000 for 4 years at 8% p.a. simple interest. David also invests $6000 for 4 years, but his interest rate is 7.6% p.a. with interest compounded quarterly.

   a. Calculate the value of Jasmine's investment on maturity.
   b. Show that the compounded value of David's investment is greater than Jasmine's investment.
   c. Explain why David's investment is worth more than Jasmine's investment despite receiving a lower rate of interest.

17. Quan has $20 000 to invest over the next 3 years. He has the choice of investing his money at 6.25% p.a. simple interest or 6% p.a. compound interest.

   a. Calculate the amount of interest that Quan will earn if he selects the simple interest option.
   b. Calculate the amount of interest that Quan will earn if the interest is compounded:

      i. annually   ii. 6-monthly   iii. quarterly.

   c. Clearly, Quan's decision will depend on the compounding period. Under what conditions should Quan accept the lower interest rate on the compound interest investment?
   d. Consider an investment of $10 000 at 8% p.a. simple interest over 5 years. Use a trial-and-error method to determine an equivalent rate of compound interest over the same period.
   e. Will this equivalent rate be the same if we change:

      i. the amount of the investment
      ii. the period of the investment?

18. A building society advertises investment accounts at the following rates.
   Account 1: 3.875% p.a. compounding daily
   Account 2: 3.895% p.a. compounding monthly
   Account 3: 3.9% p.a. compounding quarterly
   Peter thinks the first account is the best one because the interest is calculated more frequently. Paul thinks the last account is the best one because it has the highest interest rate.
   Explain whether either is correct.

19. Two banks offer the following investment packages.
   Bankwest: 7.5% p.a. compounded annually, fixed for 7 years.
   Bankeast: 5.8% p.a. compounded annually, fixed for 9 years.

   a. Which bank's package will yield the greater interest?
   b. If a customer invests $20 000 with Bankwest, how much would she have to invest with Bankeast to produce the same amount as Bankwest at the end of the investment period?

20. How is compound interest calculated differently to simple interest?

21. How long will it take for a sum of money to double if it is invested at a rate of 15% p.a. compounded monthly?

22. The kangaroo population in a region is 520 and each year increases by 4.5% of the previous year's population. Determine the expected population in 15 years time.

LESSON
1.5 Depreciation

LEARNING INTENTION

At the end of this lesson you should be able to:
- calculate the final value of an item when depreciated at a given rate
- calculate the amount of depreciation over a period of time
- apply the depreciation formula to determine the time, the rate or the present value of an asset.

1.5.1 Depreciation

eles-6257

- **Depreciation** is the reduction in the value of an item as it ages over a period of time. For example, a car that is purchased new for $45 000 will be worth less than that amount 1 year later and less again each year.
- Depreciation is usually calculated as a percentage of the yearly value of the item.
- To calculate the depreciated value, or the **salvage** value, of an item, use the depreciation formula

> ### Formula for depreciation
>
> $$S = V_0(1-r)^n$$
>
> **where:**
> - S = salvage value of the asset
> - V_0 = initial value of the asset
> - r = depreciation rate per time period, expressed as a decimal
> - n = number of time periods

- This formula is almost the same as the compound interest formula except that it subtracts a percentage of the value each year instead of adding.
- In many cases, depreciation can be a tax deduction.
- When the value of an item falls below a certain value, it is said to be *written off*. That is to say, for tax purposes, the item is considered to be worthless.
- Trial-and-error methods can be used to calculate the length of time that the item will take to reduce to this value.

WORKED EXAMPLE 8 Calculating the salvage value

A farmer purchases a tractor for $115 000. The value of the tractor depreciates by 12% p.a. Calculate the value of the tractor after 5 years.

THINK	WRITE
1. Write the depreciation formula.	$S = V_0(1-r)^n$
2. Write the values of V_0, r and n.	$V_0 = \$115\,000, r = 0.12, n = 5$
3. Substitute the values into the formula.	$S = 115\,000 \times (1-0.12)^5$ $= 115\,000 \times (0.88)^5$
4. Calculate the value of the tractor.	$= \$60\,689.17$ The value of the tractor after 5 years is \$60\,689.17.

WORKED EXAMPLE 9 Calculating time using trial and error

A truck driver buys a new prime mover for \$500 000. The prime mover depreciates at the rate of 15% p.a. and is written off when its value falls below \$100 000. How long will it take for the prime mover to be written off?

THINK	WRITE
1 Make an estimate of, say, $n = 5$. Use the depreciation formula to calculate the value of the prime mover after 5 years.	Consider $n = 5$. $S = V_0(1-r)^n$ $= 500\,000 \times (0.85)^5$ $= \$221\,852.66$
2 Since the value will still be greater than \$100 000, try a larger estimate, say, $n = 10$.	Consider $n = 10$. $S = V_0(1-r)^n$ $= 500\,000 \times (0.85)^{10}$ $= \$98\,437.20$
3 As the value is below \$100 000, check $n = 9$.	Consider $n = 9$. $S = V_0(1-r)^n$ $= 500\,000 \times (0.85)^9$ $= \$115\,808.47$
4 Since $n = 10$ is the first time that the value falls below \$100 000, conclude that it takes 10 years to be written off.	The prime mover will be written off in 10 years.

on Resources

Interactivities Different rates of depreciation (int-1155)

Depreciation (int-1155)

Video eLesson What is depreciation? (eles-0182)

Exercise 1.5 Depreciation

Individual pathways

■ PRACTISE	■ CONSOLIDATE	■ MASTER
1, 2, 7, 9	3, 5, 6, 8, 11, 12, 14	4, 10, 13, 15, 16, 17, 18

Fluency

1. Calculate the depreciated value of an item for the initial values, depreciation rates and times given below.
   a. Initial value of $30 000 depreciating at 16% p.a. over 4 years.
   b. Initial value of $5 000 depreciating at 10.5% p.a. over 3 years.
   c. Initial value of $12 500 depreciating at 12% p.a. over 5 years.

2. **WE8** A laundromat installs washing machines and clothes dryers to the value of $54 000. If the value of the equipment depreciates at a rate of 20% p.a., calculate the value of the equipment after 5 years.

3. A drycleaner purchases a new machine for $38 400. The machine depreciates at 16% p.a.
   a. Calculate the value of the machine after 4 years.
   b. Calculate the amount by which the machine has depreciated over this period of time.

4. A tradesman values his new tools at $10 200. For tax purposes, their value depreciates at a rate of 15% p.a.
   a. Calculate the value of the tools after 6 years.
   b. Calculate the amount by which the value of the tools has depreciated over these 6 years.
   c. Calculate the percentage of the initial value that the tools are worth after 6 years.

5. A taxi is purchased for $52 500 with its value depreciating at 18% p.a.
   a. Determine the value of the taxi after 10 years.
   b. Calculate the accumulated depreciation over this period.

6. A printer depreciates the value of its printing presses by 25% p.a. Printing presses are purchased new for $2.4 million.
   What is the value of the printing presses after:
   a. 1 year   b. 5 years   c. 10 years?

Understanding

7. **MC** A new computer workstation costs $5490.
   With its value depreciating at 26% p.a., the workstation's value at the end of the third year will be close to:
   A. $1684   B. $2225   C. $2811   D. $3082

8. **MC** The value of a new photocopier is $8894. Its value depreciates by 26% in the first year, 21% in the second year and 16% p.a. in the remaining 7 years. The value of the photocopier after this time, to the nearest dollar, is:
   A. $1534   B. $1851   C. $2624   D. $3000

9. **MC** A company was purchased 8 years ago for $2.6 million. With a depreciation rate of 12% p.a., the total amount by which the company has depreciated is closest to:
   A. $0.6 million   B. $1.0 million   C. $1.7 million   D. $2.0 million

10. **MC** Equipment is purchased by a company and is depreciated at the rate of 14% p.a. The number of years that it will take for the equipment to reduce to half of its initial value is:

   A. 4 years   **B.** 5 years   **C.** 6 years   **D.** 7 years

11. **MC** An asset that was bought for $12 300 has a value of $6920 after 5 years. The depreciation rate is close to:

   A. 10.87%   **B.** 16.76%   **C.** 18.67%   **D.** 21.33%

12. **WE9** A farmer buys a light aeroplane for crop dusting. The aeroplane costs $900 000. The aeroplane depreciates at the rate of 18% p.a. and is written off when its value falls below $150 000.
   How long will it take for the aeroplane to be written off? Give your answer in whole years.

13. A commercial airline buys a jumbo jet for $750 million. The value of this aircraft depreciates at a rate of 12.5% p.a.

   a. Calculate the value of the plane after 5 years, correct to the nearest million dollars.
   b. How many years will it take for the value of the jumbo jet to fall below $100 million?

Communicating, reasoning and problem solving

14. A machine purchased for $48 000 will have a value of $3000 in 9 years.

   a. Use a trial-and-error method to determine the rate at which the machine is depreciating per annum.
   b. Consider the equation $x = a^n$, $a = \sqrt[n]{x}$. Verify your answer to part **a** using this relationship.

15. Camera equipment purchased for $150 000 will have a value of $9000 in 5 years.

   a. Determine the rate of annual depreciation using trial and error first and then algebraically with the relationship 'if $x = a^n$, then $a = \sqrt[n]{x}$'.
   b. Compare and contrast each method.

16. The value of a new tractor is $175 000. The value of the tractor depreciates by 22.5% p.a.

   a. Determine the value of the tractor after 8 years.
   b. What percentage of its initial value is the tractor worth after 8 years?

17. Anthony has a home theatre valued at $P. The value of the home theatre depreciates by r% annually over a period of 5 years.
   At the end of the 5 years, the value of the home theatre has been reduced by $\dfrac{P}{12}$. Determine the value of r correct to 3 decimal places.

18. How and why is the formula for depreciation different to compound interest?

LESSON
1.6 Review

1.6.1 Topic summary

Simple interest

- The simple interest formula is $I = Prn$, where P = principal, r = interest rate per annum, as a decimal, and n = duration of the investment in years.
- There are alternatives to consider when deciding on how to pay for a major purchase.
- Credit card companies calculate interest on a monthly basis.

Buying on terms

- When we buy an item on terms, we usually pay a deposit with the balance plus interest paid in weekly or monthly instalments over an agreed period of time.
- To calculate the total cost of a purchase, add the deposit to the total of the regular repayments.
- The amount of each repayment is found by following these steps:
    1. Calculate the deposit.
    2. Calculate the balance owing by subtracting the deposit from the cash price.
    3. Determine the total repayments by adding the interest to the balance owing.
    4. Divide the total amount to be repaid by the number of regular repayments that must be made.
- Loan repayments may be calculated in the same way, except that no deposit is made.

FINANCIAL MATHEMATICS

Compound interest

- The future value of an investment under compound interest can be found by calculating the simple interest for each year separately.
- The compound interest formula is
$$FV = PV(1 + r)^n$$
where FV is the future value of the investment, PV is the present value of the investment.
    - In the formula, n is the number of compounding periods over the term of the investment: n = number of years × compounding periods per year.
    - In the formula, r is the interest rate (as a decimal) per compounding period: r = interest rate per annum ÷ compounding periods per year.
- The amount of compound interest earned is then calculated using the formula:
Compound Interest = $FV - PV$

Depreciation

- Depreciation is the reducing value of a major asset over time.
- Depreciation is usually calculated as a percentage of the yearly value of the item.
- The depreciation formula is
$$S = V_0 (1 - r)^n$$
where S is the depreciated value of the item, or salvage value, V_0 is the initial value of the asset, r is the depreciation rate per time period, expressed as a decimal and n is the number of time periods.

Consumer price index

The Consumer Price Index (CPI) measures price movements in Australia. Let's investigate this further to gain an understanding of how this index is calculated.

A collection of goods and services is selected as representative of a high proportion of household expenditure. The prices of these goods are recorded each quarter. The collection on which the CPI is based is divided into eight groups, which are further divided into subgroups. The groups are food, clothing, tobacco/alcohol, housing, health/personal care, household equipment, transportation, and recreation/education.

Weights are attached to each of these subgroups to reflect the importance of each in relation to the total household expenditure. The table shows the weights of the eight groups.

The weights indicate that a typical Australian household spends 19% of its income on food purchases, 7% on clothing and so on. The CPI is regarded as an indication of the cost of living as it records changes in the level of retail prices from one period to another.

CPI group	Weight (% of total)
Food	19
Clothing	7
Tobacco/alcohol	8.2
Housing	14.1
Health/personal care	5.6
Household equipment	18.3
Transportation	17
Recreation/education	10.8

Consider a simplified example showing how this CPI is calculated and how we are able to compare prices between one period and another. Take three items with prices as follows: a pair of jeans costing $75, a hamburger costing $3.90 and a CD costing $25.

Let us say that during the next period of time, the jeans sell for $76, the hamburger for $4.20 and the CD for $29. This can be summarised in the following table.

Item	Weight (W)	Period 1 Price (P)	Period 1 W × P	Period 2 Price (P)	Period 2 W × P
Jeans	7	$75.00	525		
Hamburger	19	$3.90	74.1		
CD	10.8	$25.00	270		
Total			869.1		

In order to calculate the CPI for Period 2, we regard the first period as the base and allocate it an index number of 100 (it is classed as 100%). We compare the second period with the first by expressing it as a percentage of the first period.

$$\text{CPI} = \frac{\text{weighted expenditure for Period 2}}{\text{weighted expenditure for Period 1}} \times 100\%$$

1. Complete the table to determine the total weighted price for Period 2.
2. a. Calculate the CPI for the above example, correct to 1 decimal place.
   b. This figure is over 100%. The amount over 100% is known as the inflation factor. What is the inflation factor in this case?
3. Now apply this procedure to a more varied basket of goods. Complete the following table, then calculate the CPI and inflation factor for the second period.

Item	Weight (W)	Period 1		Period 2	
		Price (P)	W × P	Price (P)	W × P
Bus fare		$4.80		$4.95	
Rent		$220.00		$240.00	
Movie ticket		$10.50		$10.80	
Air conditioner		$1200.00		$1240.00	
Haircut		$18.50		$21.40	
Bread		$2.95		$3.20	
Shirt		$32.40		$35.00	
Bottle of scotch		$19.95		$21.00	
Total					

on Resources

📄 **Digital Document** Investigation — Consumer price index (doc-15944)

🧩 **Interactivities** Crossword (int-2869)
    Sudoku (int-3602)

Fluency

1. Calculate the simple interest that is earned on $5000 at 5% p.a. for 4 years.

2. **MC** Jim invests a sum of money at 9% p.a. Which one of the following statements is true?
   A. Simple interest will earn Jim more money than if compound interest is paid annually.
   B. Jim will earn more money if interest is compounded annually rather than monthly.
   C. Jim will earn more money if interest is compounded quarterly rather than 6-monthly.
   D. Jim will earn more money if interest is compounded annually rather than 6-monthly.

3. Benito has a credit card with an outstanding balance of $3600. The interest rate charged on the loan is 18% p.a. Calculate the amount of interest that Benito will be charged on the credit card for the next month.

4. An LCD television has a cash price of $5750. It can be purchased on terms of 20% deposit plus weekly repayments of $42.75 for 3 years. Calculate the total cost of the television if it is purchased on terms.

5. Erin purchases a new entertainment unit that has a cash price of $6400. Erin buys the unit on the following terms: 10% deposit with the balance plus interest to be repaid in equal monthly repayments over 4 years. The simple interest rate charged is 12% p.a.
   a. Calculate the amount of the deposit.
   b. Calculate the balance owing after the deposit has been paid.
   c. Calculate the interest that will be charged.
   d. What is the total amount that Erin has to repay?
   e. Calculate the amount of each monthly repayment.

6. A new car has a marked price of $40 000. The car can be purchased on terms of 10% deposit and monthly repayments of $1050 for 5 years.
   a. Determine the total cost of the car if it is purchased on terms.
   b. Calculate the amount of interest paid.
   c. Calculate the amount of interest paid per year.
   d. Calculate the interest rate charged.

Understanding

7. Ryan invests $12 500 for 3 years at 8% p.a. with interest paid annually. By calculating the amount of simple interest earned each year separately, determine the amount to which the investment will grow.

8. Calculate the compound interest earned on $45 000 at 12% p.a. over 4 years if interest is compounded:
   a. annually
   b. 6-monthly
   c. quarterly
   d. monthly.

9. **MC** A new computer server costs $7290. With 22% p.a. reducing-value depreciation, the server's value at the end of the third year will be closest to:
   A. $1486
   B. $2257
   C. $2721
   D. $3460

10. **MC** An asset that was bought for $34 100 has a value of $13 430 after 5 years. The depreciation rate is closest to:
    A. 11%
    B. 17%
    C. 18%
    D. 21%

11. The value of a new car depreciates by 15% p.a. Calculate the value of the car after 5 years if it was purchased for $55 000.

Communicating, reasoning and problem solving

12. Virgin Australia buys a new plane so that extra flights can be arranged between Sydney, Australia and Wellington, New Zealand.

The plane costs $1 200 000. It depreciates at a rate of 16.5% p.a. and is written off when its value falls below $150 000. How long can Virgin Australia use this plane before it is written off?

13. Thomas went to an electronics store to buy a flat screen HD TV together with some accessories. The store offered him two different loans to buy the television and equipment.
Loan 1: $7000 for 3 years at 10.5% p.a. compounding yearly
Loan 2: $7000 for 5 years at 8% p.a. compounding yearly
The following agreement was struck with the store.
- Thomas will not be penalised for paying off the loans early.
- Thomas does not have to pay the principal and interest until the end of the loan period.
a. Explain which loan Thomas should choose if he decides to pay off the loan at the end of the first, second or third year.
b. Explain which loan Thomas should choose for these two options.
Paying off loan 1 at term.
Paying off loan 2 at the end of 4 years.
c. Thomas considers the option to pay off the loans at the end of their terms. Explain how you can determine the better option without further calculations.
d. Why would Thomas decide to choose loan 2 instead of loan 1 (paying over its full term), even if it cost him more money?

14. Jan bought a computer for her business at a cost of $2500. Her accountant told her that she was entitled to depreciate the cost of the computer over 5 years at 40% per year.
a. How much was the computer worth at the end of the first year?
b. By how much could Jan reduce her taxable income at the end of the first year? (The amount by which Jan can reduce her taxable income is equal to how much value the asset lost from one year to the next.)
c. Explain whether the amount she can deduct from her taxable income will increase or decrease at the end of the second year.

on To test your understanding and knowledge of this topic, go to your learnON title at www.jacplus.com.au and complete the **post-test**.

Answers

Topic 1 Financial mathematics

1.1 Pre-test

1. D
2. $9.17
3. $4275
4. D
5. $337.50
6. B
7. $278.18
8. C
9. $3377.69
10. C
11. A
12. C
13. $200 000
14. $12 400
15. D

1.2 Simple interest

1. $2400
2. a. $1800   b. $900   c. $1920   d. $720
3. a. $1500   b. $125   c. $281.25   d. $1080
4. a. $18.75   b. $55.00   c. $41.25
   d. $125.33   e. $99.58
5. $10

6. a. $700   b. $10.50   c. $710.50
   d. $3.16   e. $913.66
7. a. $52.50   b. $50.79   c. $35.92   d. $14.87
8. a. See the table at the bottom of the page*
   b. $515.86
   c. $8015.86
9. See the table at the bottom of the page*
10. S1: Credit card — payment is delayed, but possession is immediate.
    S2: Lay-by, or cash if she has savings, would like to negotiate a lower price and has somewhere to store it.
11. $r_A = 4\%$, $r_B = 10\%$, $SI_A = \$2000$ and $SI_B = \$4500$
12. $r_A = 9.76\%$, $r_B = 14.63\%$, $SI_A = \$3121.95$ and $SI_B = \$5121.95$
13. Sample responses can be found in the worked solutions in the online resources.

1.3 Buying on terms

1. a. i. $3960   ii. $3720   iii. $3950
   b. The best deal is the one with the lowest cost — 20% deposit and weekly payments of $20 over 3 years.
2. a. $131.25   b. $55.38   c. $144.44
3. a. $13 600, $283.33   b. $40 000, $666.67
   c. $5006.25, $278.13   d. $80 000, $666.67
   e. $488 000, $2033.33
4. $5409.76
5. $4530.08
6. a. $1260   b. $19 504.80
   c. $20 764.80   d. $8164.80

---

*8.a.

Month	Balance owing	Interest	Payment	Closing balance
January	$7500.00	$112.50	$1000.00	$6612.00
February	$6612.50	$99.19	$1000.00	$5711.69
March	$5711.69	$85.68	$1000.00	$4797.37
April	$4797.37	$71.96	$1000.00	$3869.33
May	$3869.33	$58.04	$1000.00	$2927.37
June	$2927.37	$43.91	$1000.00	$1971.28
July	$1971.28	$29.57	$1000.00	$1000.85
August	$1000.85	$15.01	$1015.86	$0

*9.

Payment option	Immediate payment	Immediate possession	Possible extra cost	Possible price negotiation
Cash	✓	✓		✓
Lay-by	Possible deposit		✓	
Credit card		✓	✓	

Payment option	Payment	Possession	Extra cost	Price
Cash	Immediate	Immediate	Nil	Negotiable
Lay-by	Intervals	Delayed	Limited	–
Credit card	Delayed	Immediate	Possible	–

7. a. $8409.96   b. $2609.96
   c. $869.99   d. 15%

8. a. $2375   b. $7125   c. $2565
   d. $9690   e. $269.17

9. a. $2000   b. $4000   c. $540
   d. $4540   e. $252.22

10. a. $226.80   b. $141.90   c. $360.94
    d. $87.11   e. $85.13

11. a. $4600   b. $1656
    c. $130.33   d. $7256

12. C

13. D

14. The larger the deposit, the smaller the loan and hence the interest charged. Loans generally offer a lower rate than buying on terms.

15. $426

16. a. $3200

    b. $28 800

    c. $r_4 = 15\%, r_8 = 16\%$

    d. 4 years: $49 280, 8 years: $68 864

    e. 4 years: $960, 8 years: $684

17. The most advantageous terms are those which minimise the total monies paid. This can include accepting a larger deposit to reduce the interest paid.

18. 20 days

1.4 Compound interest

1. a. $3244.80   b. $10 939.56   c. $24 617.98
   d. $14 678.02   e. $14 838.45   f. $129 394.77

2. a. $932.52   b. $10 650.81   c. $202 760.57
   d. $25 816.04   e. $3 145 511.41

3. $8784.60

4. $3376.26

Year	1	2	3	Total
Simple Interest	$1040	$1123.20	$1213.06	$3376.26

5. a. $2837.04   b. $837.04

6. $17 240.15

7. $605.42

8. $18 503.86; $386.47

9. a. 0.5833%   b. $42 891.60

10. B

11. B

12. C

13. C

14. a. 0.0247%   b. $\approx$ $131 319.80
    c. $\approx$ $11 319.80   d. $\approx$ $519.80

15. a. i. $17 745.95   ii. $17 786.61
    iii. $17 807.67   iv. $17 821.99

    b. The interest added to the principal also earns interest.

16. a. $7920

    b. David's investment = $8108.46

    c. Because David's interest is compounded, the interest is added to the principal each quarter and earns interest itself.

17. a. $3750

    b. i. $3820.32

       ii. $3881.05

       iii. $3912.36

    c. Compounding quarterly gives the best return.

    d. If we assume that interest is compounded annually, an equivalent return of $R = 7\%$ would be achieved.

    e. i. Yes

       ii. No

18. Neither is correct. The best option is to choose 3.895% p.a. compounding monthly.

19. a. Bankeast   b. $19 976.45

20. Compound interest is added to the total at the end of each compounding period. Simple interest is a fixed amount.

21. 4 years, 8 months

22. 1006 kangaroos

1.5 Depreciation

1. a. $14 936.14   b. $3584.59   c. $6596.65

2. $17 694.72

3. a. $19 118.26   b. $19 281.74

4. a. $3846.93   b. $6353.07   c. 38%

5. a. $7216.02   b. $45 283.98

6. a. $1.8 million   b. $569 531.25   c. $135 152.44

7. B

8. A

9. C

10. B

11. A

12. 10 years

13. a. $385 million   b. 16 years

14. a. 27%

    b.
    $$S = V_0(1 - r)^n$$
    $$3000 = 48\,000(1 - r)^9$$
    $$0.0625 = (1 - r)^9$$
    $$\sqrt[9]{0.0625} = 1 - r$$
    $$r = 1 - \sqrt[9]{0.0625}$$
    $$r = 0.265132$$
    $$r = 26.5\ldots\%$$
    $$r = 27\%$$

15. a. Approximately 43%

    b. Trial and error: can be time-consuming, answer is often an estimate; algebraic solution: correct answer calculated immediately from equation.

16. a. $22 774.65   b. 13%

17. 1.725%

18. The depreciation formula is different from the compound interest formula in that it has a subtraction sign instead of an addition sign. This is because the value is decreasing, not increasing.

Project

1. See the table at the bottom of the page*
2. a. 106.4%   b. 6.4%
3. 104%; 4%
   See the table at the bottom of the page*

1.6 Review questions

1. $1000
2. C
3. $54
4. $7819
5. a. $640   b. $5760   c. $2764.80
   d. $8524.80   e. $177.60
6. a. $67 000   b. $27 000   c. $5400
   d. 13.5% p.a.
7. $15 746.40
8. a. $25 808.37   b. $26 723.16
   c. $27 211.79   d. $27 550.17

9. D
10. B
11. $24 403.79
12. 12 years
13. a. Since the interest rate is lower for loan 2 than for loan 1, Thomas should choose loan 2 if he decides to pay the loan off at the end of the first, second or third year.
    b. Loan 1 at term (3 years) amounts to $9444.63. Loan 2 at the end of 4 years amounts to $9523.42. Thomas should choose loan 1.
    c. Thomas should choose loan 1. At the end of its term (3 years), it amounts to less than loan 2 at 4 years, 1 year before its term is finished.
    d. Thomas may not have the money to pay off loan 1 in 3 years. He may need the extra 2 years to accumulate his funds.
14. a. $1500
    b. $1000
    c. Since the depreciation of 40% is on a lower value each year, the amount Jan can deduct from her taxable income decreases every year.

*1.

Item	Weight (W)	Period 1 Price (P)	Period 1 W × P	Period 2 Price (P)	Period 2 W × P
Jeans	7	$75.00	525	$76.00	532
Hamburger	19	$3.90	74.1	$4.20	79.8
CD	10.8	$25.00	270	$29.00	313.2
Total	36.8	$103.90	869.1	$109.20	925

*3.

Item	Weight (W)	Period 1 Price (P)	Period 1 W × P	Period 2 Price (P)	Period 2 W × P
Bus fare	17	$4.80	81.6	$4.95	84.15
Rent	14.1	$220.00	3102	$240.00	3384
Movie ticket	10.8	$10.50	113.4	$10.80	116.64
Air conditioner	18.3	$1200.00	21 960	$1240.00	22 692
Haircut	5.6	$18.50	103.6	$21.40	119.84
Bread	19	$2.95	56.05	$3.20	60.8
Shirt	7	$32.40	226.8	$35.00	245
Bottle of scotch	8.2	$19.95	163.59	$21.00	172.2
Total	100	$1509.10	25 807.04	$1576.35	25 874.63

2 Indices and surds

LESSON SEQUENCE

LESSON
2.1 Overview

Why learn this?

We often take for granted the amount of time and effort that has gone into developing the number system we use on a daily basis. In ancient times, numbers were used for bartering and trading goods between people. Thus, numbers were always attached to an object; for example, 5 cows, 13 sheep or 20 gold coins. Consequently, it took a long time before more abstract concepts such as the number 0 were introduced and widely used. It took even longer for negative numbers or irrational numbers such as surds to be accepted as their own group of numbers. Historically, there has always been resistance to these changes and updates. In folk law, Hippasus — the man first credited with the discovery of irrational numbers — was drowned at sea for angering the gods with his discovery.

A good example of how far we have come is to look at an ancient number system most people are familiar with: Roman numerals. Not only is there no symbol for 0 in Roman numerals, but they are extremely clumsy to use when adding or subtracting. Consider trying to add 54 (LIV) to 12 (XII). We know that to determine the answer we add the ones together and then the tens to get 66. Adding the Roman numeral is more complex; do we write LXVIII or LIVXII or LVXI or LXVI?

Having a better understanding of our number system makes it easier to understand how to work with concepts such as indices and surds. By building our understanding of these concepts, it is possible to more accurately model real-world scenarios and extend our understanding of number systems to more complex sets, such as complex numbers and quaternions.

Hey students! Bring these pages to life online

▶ Watch videos

🧩 Engage with interactivities

A+ Answer questions and check solutions

Find all this and MORE in jacPLUS ▶

Reading content and rich media, including interactivities and videos for every concept

Extra learning resources

Differentiated question sets

Questions with immediate feedback, and fully worked solutions to help students get unstuck

Exercise 2.1 Pre-test

learn on

1. Simplify: $\left(2ab^3\right)^4$

2. State whether $\sqrt{36}$ is a rational or irrational number.

3. PATH Simplify the following: $3n^{\frac{1}{5}} \times 5n^{\frac{1}{3}}$.

4. PATH Simplify the following: $\sqrt[5]{32p^{10}q^{15}}$.

5. PATH Determine the exact value of $81^{-\frac{3}{4}}$.

6. MC Select which of the numbers of the set $\left\{\sqrt{0.25},\ \pi,\ 0.\overline{261},\ -5,\ \dfrac{2}{3}\right\}$ are rational.

   A. $\left\{\sqrt{0.25},\ \pi,\ 0.\overline{261}\right\}$ B. $\left\{0.\overline{261},\ -5,\ \dfrac{2}{3}\right\}$

   C. $\left\{\pi, 0.\overline{261}\right\}$ D. $\left\{\sqrt{0.25},\ 0.\overline{261},\ -5,\ \dfrac{2}{3}\right\}$

7. MC $\dfrac{12x^8 \times 3x^7}{9x^{10} \times x^3}$ simplifies to:

   A. $\dfrac{5x^2}{3}$ B. $4x^2$ C. $4x^{26}$ D. $\dfrac{5x^{26}}{3}$

8. PATH Simplify the following expression: $3\sqrt{2} \times \sqrt{10}$.

9. PATH Simplify the following expression: $5\sqrt{2} + 12\sqrt{2} - 3\sqrt{2}$.

10. MC PATH Choose the most simplified form of the following expression: $\sqrt{8a^3} + \sqrt{18a} + \sqrt{a^5}$

    A. $5\sqrt{2a} + a\sqrt{a}$ B. $2a\sqrt{2a^2} + 3\sqrt{2a} + a^4\sqrt{a}$

    C. $2a^2\sqrt{2a} + 2\sqrt{3a} + a^4\sqrt{a}$ D. $2a\sqrt{2a} + 3\sqrt{2a} + a^2\sqrt{a}$

11. True or false? $8^{-2} = \dfrac{-1}{64}$

12. True or false? $5^3 \times 5^{-3} = 0$

13. PATH True or false? $\sqrt[3]{8x^9} = 2x^6$

14. Simplify: $\dfrac{4a^2b^3 \times 5a^4b^5}{6a^6b^7}$

15. PATH Simplify: $2^{\frac{3}{2}} \times 4^{-\frac{1}{4}} \div 16^{\frac{3}{4}}$

LESSON
2.2 Rational and irrational numbers (Path)

LEARNING INTENTION

At the end of this lesson you should be able to:
- define and describe real, rational, irrational, integer and natural numbers
- determine whether a number is rational or irrational.

▶ 2.2.1 The real number system

eles-4661

- The number systems used today evolved from a basic and practical need of primitive people to count and measure magnitudes and quantities such as livestock, people, possessions, time and so on.
- As societies grew and architecture and engineering developed, number systems became more sophisticated. Number use developed from solely whole numbers to fractions, decimals and irrational numbers.

- The real number system contains the set of rational and irrational numbers. It is denoted by the symbol R. The set of real numbers contains a number of subsets which can be classified as shown in the chart below.

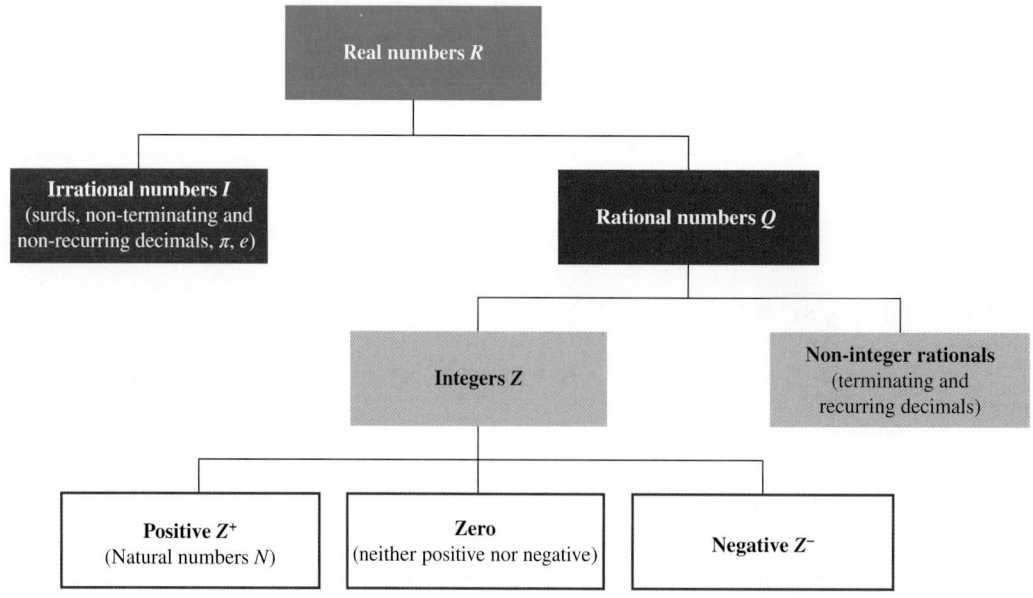

Integers (Z)

- The set of **integers** consists of whole positive and negative numbers and 0 (which is neither positive nor negative).
- The set of integers is denoted by the symbol Z and can be visualised as:

$$Z = \{..., -3, -2, -1, 0, 1, 2, 3, ...\}$$

- The set of positive integers are known as the **natural numbers** (or counting numbers) and is denoted Z^+ or N. That is:

$$Z^+ = N = \{1, 2, 3, 4, 5, 6, ...\}$$

- The set of negative integers is denoted Z^-.

$$Z^- = \{... -6, -5, -4, -3, -2, -1\}$$

- Integers may be represented on the number line as illustrated below.

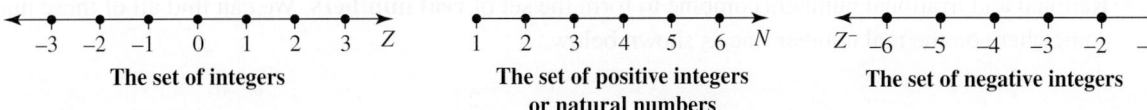

The set of integers   **The set of positive integers or natural numbers**   **The set of negative integers**

Rational numbers (Q)

- A **rational number** is a number that can be expressed as a ratio of two integers in the form $\dfrac{a}{b}$, where $b \neq 0$.
- The set of rational numbers are denoted by the symbol Q.
- Rational numbers include all whole numbers, fractions and all terminating and recurring decimals.
- **Terminating decimals** are decimal numbers which terminate after a specific number of digits. Examples are:

$$\frac{1}{4} = 0.25, \frac{5}{8} = 0.625, \frac{9}{5} = 1.8.$$

- **Recurring decimals** do not terminate but have a specific digit (or number of digits) repeated in a pattern. Examples are:

$$\frac{1}{3} = 0.333\,333\,... = 0.\dot{3} \text{ or } 0.\overline{3}$$

$$\frac{133}{666} = 0.199\,699\,699\,6... = 0.1\dot{9}9\dot{6} \text{ or } 0.1\overline{996}$$

- Recurring decimals are represented by placing a dot or line above the repeating digit/s.
- Using set notations, we can represent the set of rational numbers as:

$$Q = \left\{ \frac{a}{b} : a, b \in Z, b \neq 0 \right\}$$

- This can be read as 'Q is all numbers of the form $\dfrac{a}{b}$ given a and b are integers and b is not equal to 0'.

Irrational numbers (I)

- An **irrational number** is a number that cannot be expressed as a ratio of two integers in the form $\dfrac{a}{b}$, where $b \neq 0$.

- All irrational numbers have a decimal representation that is non-terminating and non-recurring. This means the decimals do not terminate and do not repeat in any particular pattern or order.
  For example:

$$\sqrt{5} = 2.236\,067\,997\,5\ldots$$
$$\pi = 3.141\,592\,653\,5\ldots$$
$$e = 2.718\,281\,828\,4\ldots$$

- The set of irrational numbers is denoted by the symbol I. Some common irrational numbers that you may be familiar with are $\sqrt{2}, \pi, e, \sqrt{5}$.
- The symbol π **(pi)** is used for a particular number that is the circumference of a circle whose diameter is 1 unit.
- In decimal form, π has been calculated to more than 29 million decimal places with the aid of a computer.

Rational or irrational

- Rational and irrational numbers combine to form the set of **real numbers**. We can find all of these number somewhere on the real number line as shown below.

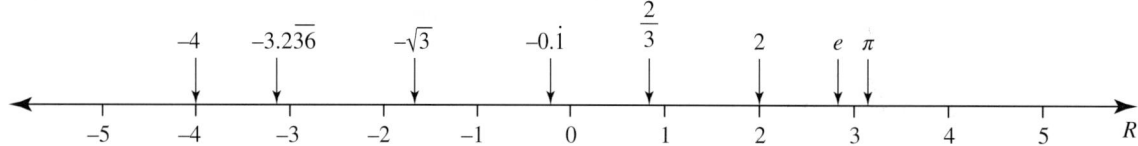

- To classify a number as either rational or irrational:
  1. Determine whether it can be expressed as a whole number, a fraction, or a terminating or recurring decimal.
  2. If the answer is yes, the number is rational. If no, the number is irrational.

WORKED EXAMPLE 1 Classifying numbers as rational or irrational

Classify whether the following numbers are rational or irrational.

a. $\dfrac{1}{5}$ b. $\sqrt{25}$ c. $\sqrt{13}$ d. 3π

e. 0.54 f. $\sqrt[3]{64}$ g. $\sqrt[3]{32}$ h. $\sqrt[3]{\dfrac{1}{27}}$

THINK

a. $\dfrac{1}{5}$ is already a rational number.

b. 1. Evaluate $\sqrt{25}$.

   2. The answer is an integer, so classify $\sqrt{25}$.

c. 1. Evaluate $\sqrt{13}$.

   2. The answer is a non-terminating and non-recurring decimal; classify $\sqrt{13}$.

WRITE

a. $\dfrac{1}{5}$ is rational.

b. $\sqrt{25} = 5$

   $\sqrt{25}$ is rational.

c. $\sqrt{13} = 3.605\,551\,275\,46\ldots$

   $\sqrt{13}$ is irrational.

d. 1. Use your calculator to calculate the value of 3π.	**d.** $3\pi = 9.424\,777\,960\,77\,...$
2. The answer is a non-terminating and non-recurring decimal; classify 3π.	3π is irrational.
e. 0.54 is a terminating decimal; classify it accordingly.	**e.** 0.54 is rational.
f. 1. Evaluate $\sqrt[3]{64}$.	**f.** $\sqrt[3]{64} = 4$
2. The answer is a whole number, so classify $\sqrt[3]{64}$.	$\sqrt[3]{64}$ is rational.
g. 1. Evaluate $\sqrt[3]{32}$.	**g.** $\sqrt[3]{32} = 3.17480210394\,...$
2. The result is a non-terminating and non-recurring decimal; classify $\sqrt[3]{32}$.	$\sqrt[3]{32}$ is irrational.
h. 1. Evaluate $\sqrt[3]{\dfrac{1}{27}}$.	**h.** $\sqrt[3]{\dfrac{1}{27}} = \dfrac{1}{3}$.
2. The result is a number in a rational form.	$\sqrt[3]{\dfrac{1}{27}}$ is rational.

 Resources

Interactivities The number system (int-6027)
Recurring decimals (int-6189)

Exercise 2.2 Rational and irrational numbers (Path)

learn

2.2 Quick quiz on

2.2 Exercise

Individual pathways

■ PRACTISE	■ CONSOLIDATE	■ MASTER
1, 4, 7, 10, 13, 14, 17, 20, 23	2, 5, 8, 11, 15, 18, 21, 24	3, 6, 9, 12, 16, 19, 22, 25

Fluency

For questions **1** to **6**, classify whether the following numbers are rational (Q) or irrational (I).

1.
   a. $\sqrt{4}$ b. $\dfrac{4}{5}$ c. $\dfrac{7}{9}$ d. $\sqrt{2}$

2. a. $\sqrt{7}$ b. $\sqrt{0.04}$ c. $2\dfrac{1}{2}$ d. $\sqrt{5}$

3. a. $\dfrac{9}{4}$ b. 0.15   c. -2.4 d. $\sqrt{100}$

4. a. $\sqrt{14.4}$ b. $\sqrt{1.44}$ c. π d. $\sqrt{\dfrac{25}{9}}$

5. a. 7.32 b. $-\sqrt{21}$ c. $\sqrt{1000}$ d. $7.216\,349\,157\ldots$

6. a. $-\sqrt{81}$ b. 3π c. $\sqrt[3]{62}$ d. $\sqrt{\dfrac{1}{16}}$

For questions **7** to **12**, classify the following numbers as rational (Q), irrational (I) or neither.

7. a. $\dfrac{1}{8}$ b. $\sqrt{625}$ c. $\dfrac{11}{4}$ d. $\dfrac{0}{8}$

8. a. $-6\dfrac{1}{7}$ b. $\sqrt[3]{81}$ c. $-\sqrt{11}$ d. $\sqrt{\dfrac{1.44}{4}}$

9. a. $\sqrt{\pi}$ b. $\dfrac{8}{0}$ c. $\sqrt[3]{21}$ d. $\dfrac{\pi}{7}$

10. a. $\sqrt[3]{(-5)^2}$ b. $-\dfrac{3}{11}$ c. $\sqrt{\dfrac{1}{100}}$ d. $\dfrac{64}{16}$

11. a. $\sqrt{\dfrac{2}{25}}$ b. $\dfrac{\sqrt{6}}{2}$ c. $\sqrt[3]{27}$ d. $\dfrac{1}{\sqrt{4}}$

12. a. $\dfrac{22\pi}{7}$ b. $\sqrt[3]{-1.728}$ c. $6\sqrt{4}$ d. $4\sqrt{6}$

13. **MC** Identify a rational number from the following.

     A. π B. $\sqrt{\dfrac{4}{9}}$ C. $\sqrt{\dfrac{9}{12}}$ D. $\sqrt[3]{3}$

14. **MC** Identify which of the following best represents an irrational number from the following numbers.

     A. $-\sqrt{81}$ B. $\dfrac{6}{5}$ C. $\sqrt[3]{343}$ D. $\sqrt{22}$

15. **MC** Select which one of the following statements regarding the numbers $-0.69, \sqrt{7}, \dfrac{\pi}{3}, \sqrt{49}$ is correct.

     A. $\dfrac{\pi}{3}$ is the only rational number.   B. $\sqrt{7}$ and $\sqrt{49}$ are both irrational numbers.

     C. -0.69 and $\sqrt{49}$ are the only rational numbers.   D. -0.69 is the only rational number.

16. **MC** Select which one of the following statements regarding the numbers $2\dfrac{1}{2}, -\dfrac{11}{3}, \sqrt{624}, \sqrt[3]{99}$ is correct.

     A. $-\dfrac{11}{3}$ and $\sqrt{624}$ are both irrational numbers.

     B. $\sqrt{624}$ is an irrational number and $\sqrt[3]{99}$ is a rational number.

     C. $\sqrt{624}$ and $\sqrt[3]{99}$ are both irrational numbers.

     D. $2\dfrac{1}{2}$ is a rational number and $-\dfrac{11}{3}$ is an irrational number.

Understanding

17. Simplify $\sqrt{\dfrac{a^2}{b^2}}$.

18. **MC** If $p < 0$, then $\sqrt{p}$ is:

    A. positive   **B.** negative   **C.** rational   **D.** none of these

19. **MC** If $p < 0$, then $\sqrt{p^2}$ must be:

    A. positive   **B.** negative   **C.** rational   **D.** irrational

Communicating, reasoning and problem solving

20. Simplify $\left(\sqrt{p} - \sqrt{q}\right) \times \left(\sqrt{p} + \sqrt{q}\right)$. Show full working.

21. Prove that if $c^2 = a^2 + b^2$, it does not follow that $a = b + c$.

22. Assuming that x is a rational number, for what values of k will the expression $\sqrt{x^2 + kx + 16}$ always be rational? Justify your response.

23. Determine the value of m and n if $\dfrac{36}{11}$ is written as:

    a. $3 + \dfrac{1}{\frac{m}{n}}$ b. $3 + \dfrac{1}{3 + \frac{m}{n}}$ c. $3 + \dfrac{1}{3 + \frac{1}{\frac{m}{n}}}$ d. $3 + \dfrac{1}{3 + \frac{1}{1 + \frac{m}{n}}}$

24. If x^{-1} means $\dfrac{1}{x}$, determine the value of $\dfrac{3^{-1} - 4^{-1}}{3^{-1} + 4^{-1}}$.

25. If $x^{-n} = \dfrac{1}{x^n}$, evaluate $\dfrac{3^{-n} - 4^{-n}}{3^{-n} + 4^{-n}}$ when $n = 3$.

LESSON
2.3 Surds (Path)

LEARNING INTENTION

At the end of this lesson you should be able to:
- identify and apply surd notation
- determine whether a number under a root or radical sign is a surd

---

▶ 2.3.1 Identifying surds

eles-4662

- A **surd** is an irrational number that is represented by a root sign or a radical sign, for example:
  $\sqrt{}, \sqrt[3]{}, \sqrt[4]{}$.

  Examples of surds include: $\sqrt{7}, \sqrt{5}, \sqrt[3]{11}, \sqrt[4]{15}$.

- In general, a surd can be written in the form $\sqrt[n]{x}$ where x is a rational number and n is an integer such that $n \geq$, and $x > 0$, when n is even.

- The numbers $\sqrt{9}, \sqrt{16}, \sqrt[3]{125}$, and $\sqrt[4]{81}$ are not surds as they can be simplified to rational numbers, that is: $\sqrt{9} = 3, \sqrt{16} = 4, \sqrt[3]{125} = 5, \sqrt[4]{81} = 3, \sqrt[3]{-8} = -2$.

Digital technology

Use a scientific calculator to verify that attempting to determine the square root and fourth root of a negative number results in an error or undefined response.

Use a scientific calculator to verify that $\sqrt{0} = 0$.

WORKED EXAMPLE 2 Identifying surds

Determine which of the following numbers are surds.

a. $\sqrt{16}$ b. $\sqrt{13}$ c. $\sqrt{\dfrac{1}{16}}$ d. $\sqrt[3]{17}$ e. $\sqrt[4]{63}$ f. $\sqrt[3]{-1728}$

THINK	WRITE
a. 1. Evaluate $\sqrt{16}$.	a. $\sqrt{16} = 4$
2. The answer is rational (since it is a whole number), so state your conclusion.	$\sqrt{16}$ is not a surd.
b. 1. Evaluate $\sqrt{13}$.	b. $\sqrt{13} = 3.60555127546\ldots$
2. The answer is irrational (since it is a non-recurring and non-terminating decimal), so state your conclusion.	$\sqrt{13}$ is a surd.
c. 1. Evaluate $\sqrt{\dfrac{1}{16}}$.	c. $\sqrt{\dfrac{1}{16}} = \dfrac{1}{4}$
2. The answer is rational (a fraction); state your conclusion.	$\sqrt{\dfrac{1}{16}}$ is not a surd.
d. 1. Evaluate $\sqrt[3]{17}$.	d. $\sqrt[3]{17} = 2.57128159066\ldots$
2. The answer is irrational (a non-terminating and non-recurring decimal), so state your conclusion.	$\sqrt[3]{17}$ is a surd.
e. 1. Evaluate $\sqrt[4]{63}$.	e. $\sqrt[4]{63} = 2.81731324726\ldots$
2. The answer is irrational, so classify $\sqrt[4]{63}$ accordingly.	$\sqrt[4]{63}$ is a surd.
f. 1. Evaluate $\sqrt[3]{-1728}$.	f. $\sqrt[3]{-1728} = -12$
2. The answer is rational; state your conclusion.	$\sqrt[3]{-1728}$ is not a surd. So **b**, **d** and **e** are surds.

- *Note:* An irrational number written in surd form gives an exact value of the number; whereas the same number written in decimal form (for example, to 4 decimal places) gives an approximate value.

DISCUSSION

Explaining why an error occurs when attempting to calculate a value of $\sqrt{-4}$ as opposed to $\sqrt{4}$.

 Resources

Interactivity Surds on the number line (int-6029)

Exercise 2.3 Surds (Path)

learn

2.3 Quick quiz on   2.3 Exercise

Individual pathways

PRACTISE	CONSOLIDATE	MASTER
1, 4, 7, 8, 11, 14	2, 5, 9, 12, 15, 16	3, 6, 10, 13, 17, 18

Fluency

WE2 For questions **1** to **6**, determine which of the following numbers are surds.

1. **a.** $\sqrt{81}$ **b.** $\sqrt{48}$ **c.** $\sqrt{16}$ **d.** $\sqrt{1.6}$

2. **a.** $\sqrt{0.16}$ **b.** $\sqrt{11}$ **c.** $\sqrt{\dfrac{3}{4}}$ **d.** $\sqrt[3]{\dfrac{3}{27}}$

3. **a.** $\sqrt{1000}$ **b.** $\sqrt{1.44}$ **c.** $4\sqrt{100}$ **d.** $2+\sqrt{10}$

4. **a.** $\sqrt[3]{32}$ **b.** $\sqrt{361}$ **c.** $\sqrt[3]{100}$ **d.** $\sqrt[3]{125}$

5. **a.** $\sqrt{6}+\sqrt{6}$ **b.** 2π **c.** $\sqrt[3]{169}$ **d.** $\sqrt{\dfrac{7}{8}}$

6. **a.** $\sqrt[4]{16}$ **b.** $\left(\sqrt{7}\right)^{2}$ **c.** $\sqrt[3]{33}$ **d.** $\sqrt{0.0001}$
   e. $\sqrt[5]{32}$ **f.** $\sqrt{80}$

7. **MC** The correct statement regarding the set of numbers $\left\{\sqrt{\dfrac{6}{9}}, \sqrt{20}, \sqrt{54}, \sqrt[3]{27}, \sqrt{9}\right\}$ is:

   A. $\sqrt[3]{27}$ and $\sqrt{9}$ are the only rational numbers of the set.

   B. $\sqrt{\dfrac{6}{9}}$ is the only surd of the set.

   C. $\sqrt{\dfrac{6}{9}}$ and $\sqrt{20}$ are the only surds of the set.

   D. $\sqrt{20}$ and $\sqrt{54}$ are the only surds of the set.

8. **MC** Identify the numbers from the set $\left\{\sqrt{\dfrac{1}{4}}, \sqrt[3]{\dfrac{1}{27}}, \sqrt{\dfrac{1}{8}}, \sqrt{21}, \sqrt[3]{-8}\right\}$ that are surds.

   A. $\sqrt{21}$ only

   B. $\sqrt{\dfrac{1}{8}}$ only

   C. $\sqrt{\dfrac{1}{8}}$ and $\sqrt[3]{-8}$

   D. $\sqrt{\dfrac{1}{8}}$ and $\sqrt{21}$ only

9. **MC** Select a statement regarding the set of numbers $\left\{\pi, \sqrt{\dfrac{1}{49}}, \sqrt{12}, \sqrt{16}, \sqrt{3}, +1\right\}$ that is *not* true.

   A. $\sqrt{12}$ is a surd.

   B. $\sqrt{12}$ and $\sqrt{16}$ are surds.

   C. π is irrational but not a surd.

   D. $\sqrt{12}$ and $\sqrt{3}+1$ are not rational.

10. **MC** Select a statement regarding the set of numbers $\left\{6\sqrt{7}, \sqrt{\dfrac{144}{16}}, 7\sqrt{6}, 9\sqrt{2}, \sqrt{18}, \sqrt{25}\right\}$ that is *not* true.

    A. $\sqrt{\dfrac{144}{16}}$ when simplified is an integer.

    B. $\sqrt{\dfrac{144}{16}}$ and $\sqrt{25}$ are not surds.

    C. $7\sqrt{6}$ is smaller than $9\sqrt{2}$.

    D. $9\sqrt{2}$ is smaller than $6\sqrt{7}$.

Understanding

11. Complete the following statement by selecting appropriate words, suggested in brackets:
    $\sqrt{a}$ is definitely not a surd, if a is… (any multiple of 4; a perfect square; cube).

12. Determine the smallest value of m, where m is a positive integer, so that $\sqrt[3]{16m}$ is not a surd.

13. **a.** Determine any combination of m and n, where m and n are positive integers with $m < n$, so that
    $\sqrt[4]{(m+4)(16-n)}$ is not a surd.
    b. If the condition that $m < n$ is removed, how many possible combinations are there?

Communicating, reasoning and problem solving

14. Determine whether the following are rational or irrational.
    a. $\sqrt{5}+\sqrt{2}$
    b. $\sqrt{5}-\sqrt{2}$
    c. $\left(\sqrt{5}+\sqrt{2}\right)\left(\sqrt{5}-\sqrt{2}\right)$

15. Many composite numbers have a variety of factor pairs. For example, factor pairs of 24 are 1 and 24, 2 and 12, 3 and 8, 4 and 6.
    a. Use each pair of possible factors to simplify the following surds.
       i. $\sqrt{48}$
       ii. $\sqrt{72}$
    b. Explain if the factor pair chosen when simplifying a surd affect the way the surd is written in simplified form.
    c. Explain if the factor pair chosen when simplifying a surd affect the value of the surd when it is written in simplified form.

16. Consider the expression $(\sqrt{p} + \sqrt{q})(\sqrt{m} - \sqrt{n})$. Determine under what conditions will the expression produce a rational number.

17. π is an irrational number and so is $\sqrt{3}$. Therefore, determine whether $\left(\pi - \sqrt{3}\right)\left(\pi + \sqrt{3}\right)$ is an irrational number.

18. Solve $\sqrt{3}x - \sqrt{12} = \sqrt{3}$ and indicate whether the result is rational or irrational.

LESSON
2.4 Operations with surds (Path)

LEARNING INTENTION

At the end of this lesson you should be able to:
- expand and simplify expressions with surds using algebraic operations
- rationalise the denominators of surds of the form $\dfrac{a\sqrt{b}}{c\sqrt{d}}$.

⊳ 2.4.1 Multiplying and simplifying surds

eles-4664

Multiplication of surds

- To multiply surds, multiply the expressions under the radical sign.
  For example: $\sqrt{8} \times \sqrt{3} = \sqrt{8 \times 3} = \sqrt{24}$
- If there are coefficients in front of the surds that are being multiplied, multiply the coefficients and then multiply the expressions under the radical signs.
  For example: $2\sqrt{3} \times 5\sqrt{7} = (2 \times 5)\sqrt{3 \times 7} = 10\sqrt{21}$

Multiplication of surds

In order to multiply two or more surds, use the following:

- $\sqrt{a} \times \sqrt{b} = \sqrt{a \times b}$
- $m\sqrt{a} \times n\sqrt{b} = mn\sqrt{a \times b}$

where a and b are positive real numbers.

Simplification of surds

- To simplify a surd means to make the number under the radical sign as small as possible.
- Surds can only be simplified if the number under the radical sign has a factor which is a perfect square $(4, 9, 16, 25, 36, ...)$.
- Simplification of a surd uses the method of multiplying surds in reverse.

- The process is summarised in the following steps:
  1. Split the number under the radical into the product of two factors, one of which is a perfect square.
  2. Write the surd as the product of two surds multiplied together. The two surds must correspond to the factors identified in step 1.
  3. Simplify the surd of the perfect square and write the surd in the form $a\sqrt{b}$.
- The example below shows the how the surd $\sqrt{45}$ can be simplified by following the steps 1 to 3.

$$
\begin{aligned}
\sqrt{45} &= \sqrt{9 \times 5} &\text{(Step 1)}\\
&= \sqrt{9} \times \sqrt{5} &\text{(Step 2)}\\
&= 3 \times \sqrt{5} = 3\sqrt{5} &\text{(Step 3)}
\end{aligned}
$$

- If possible, try to factorise the number under the radical sign so that the largest possible perfect square is used. This will ensure the surd is simplified in 1 step.

Simplification of surds

$$
\begin{aligned}
\sqrt{a^2 \times b} &= \sqrt{a^2} \times \sqrt{b}\\
&= a \times \sqrt{b}\\
&= a\sqrt{b}
\end{aligned}
$$

WORKED EXAMPLE 3 Simplifying surds

Simplify the following surds. Assume that x and y are positive real numbers.

a. $\sqrt{384}$ b. $3\sqrt{405}$ c. $-\dfrac{1}{8}\sqrt{175}$ d. $5\sqrt{180x^3y^5}$

THINK	WRITE
a. 1. Express 384 as a product of two factors where one factor is the largest possible perfect square.	a. $\sqrt{384} = \sqrt{64 \times 6}$
2. Express $\sqrt{64 \times 6}$ as the product of two surds.	$= \sqrt{64} \times \sqrt{6}$
3. Simplify the square root from the perfect square (that is, $\sqrt{64} = 8$).	$= 8\sqrt{6}$
b. 1. Express 405 as a product of two factors, one of which is the largest possible perfect square.	b. $3\sqrt{405} = 3\sqrt{81 \times 5}$
2. Express $\sqrt{81 \times 5}$ as a product of two surds.	$= 3\sqrt{81} \times \sqrt{5}$
3. Simplify $\sqrt{81}$.	$= 3 \times 9\sqrt{5}$
4. Multiply together the whole numbers outside the square root sign (3 and 9).	$= 27\sqrt{5}$
c. 1. Express 175 as a product of two factors in which one factor is the largest possible perfect square.	c. $-\dfrac{1}{8}\sqrt{175} = -\dfrac{1}{8}\sqrt{25 \times 7}$

2. Express $\sqrt{25 \times 7}$ as a product of 2 surds. $= -\dfrac{1}{8} \times \sqrt{25} \times \sqrt{7}$

3. Simplify $\sqrt{25}$. $= -\dfrac{1}{8} \times 5\sqrt{7}$

4. Multiply together the numbers outside the square root sign. $= -\dfrac{5}{8}\sqrt{7}$

d. 1. Express each of 180, x^3 and y^5 as a product of two factors where one factor is the largest possible perfect square.   d. $5\sqrt{180x^3y^5} = 5\sqrt{36 \times 5 \times x^2 \times x \times y^4 \times y}$

2. Separate all perfect squares into one surd and all other factors into the other surd. $= 5 \times \sqrt{36x^2y^4} \times \sqrt{5xy}$

3. Simplify $\sqrt{36x^2y^4}$. $= 5 \times 6 \times x \times y^2 \times \sqrt{5xy}$

4. Multiply together the numbers and the pronumerals outside the square root sign. $= 30xy^2\sqrt{5xy}$

WORKED EXAMPLE 4 Multiplying surds

Multiply the following surds, expressing answers in the simplest form. Assume that x and y are positive real numbers.

a. $\sqrt{11} \times \sqrt{7}$ b. $5\sqrt{3} \times 8\sqrt{5}$ c. $6\sqrt{12} \times 2\sqrt{6}$ d. $\sqrt{15x^5y^2} \times \sqrt{12x^2y}$

THINK

a. Multiply the surds together, using $\sqrt{a} \times \sqrt{b} = \sqrt{ab}$ (that is, multiply expressions under the square root sign). *Note:* This expression cannot be simplified any further.

b. Multiply the coefficients together and then multiply the surds together.

c. 1. Simplify $\sqrt{12}$.

2. Multiply the coefficients together and multiply the surds together.

3. Simplify the surd.

WRITE

a. $\sqrt{11} \times \sqrt{7} = \sqrt{11 \times 7}$
$= \sqrt{77}$

b. $5\sqrt{3} \times 8\sqrt{5} = 5 \times 8 \times \sqrt{3} \times \sqrt{5}$
$= 40 \times \sqrt{3 \times 5}$
$= 40\sqrt{15}$

c. $6\sqrt{12} \times 2\sqrt{6} = 6\sqrt{4 \times 3} \times 2\sqrt{6}$
$= 6 \times 2\sqrt{3} \times 2\sqrt{6}$
$= 12\sqrt{3} \times 2\sqrt{6}$

$= 24\sqrt{18}$

$= 24\sqrt{9 \times 2}$
$= 24 \times 3\sqrt{2}$
$= 72\sqrt{2}$

d. **1.** Simplify each of the surds.

$$\text{d. } \sqrt{15x^5y^2} \times \sqrt{12x^2y}$$
$$= \sqrt{15 \times x^4 \times x \times y^2} \times \sqrt{4 \times 3 \times x^2 \times y}$$
$$= x^2 \times y \times \sqrt{15 \times x} \times 2 \times x \times \sqrt{3 \times y}$$
$$= x^2y\sqrt{15x} \times 2x\sqrt{3y}$$

2. Multiply the coefficients together and the surds together.

$$= x^2y \times 2x\sqrt{15x \times 3y}$$
$$= 2x^3y\sqrt{45xy}$$
$$= 2x^3y\sqrt{9 \times 5xy}$$

3. Simplify the surd.

$$= 2x^3y \times 3\sqrt{5xy}$$
$$= 6x^3y\sqrt{5xy}$$

- When working with surds, it is sometimes necessary to multiply surds by themselves; that is, square them. Consider the following examples:

$$\left(\sqrt{2}\right)^2 = \sqrt{2} \times \sqrt{2} = \sqrt{4} = 2$$
$$\left(\sqrt{5}\right)^2 = \sqrt{5} \times \sqrt{5} = \sqrt{25} = 5$$

- Observe that squaring a surd produces the number under the radical sign. This is not surprising, because squaring and taking the square root are *inverse operations* and, when applied together, leave the original unchanged.

> ### Squaring surds
>
> **When a surd is squared, the result is the expression under the radical sign; that is:**
>
> $$\left(\sqrt{a}\right)^2 = a$$
>
> **where a is a positive real number.**

WORKED EXAMPLE 5 Squaring surds

Simplify each of the following.

a. $\left(\sqrt{6}\right)^2$

b. $\left(3\sqrt{5}\right)^2$

THINK

a. Use $\left(\sqrt{a}\right)^2 = a$, where $a = 6$.

b. **1.** Square 3 and apply $\left(\sqrt{a}\right)^2 = a$ to square $\sqrt{5}$.

2. Simplify.

WRITE

a. $\left(\sqrt{6}\right)^2 = 6$

b. $\left(3\sqrt{5}\right)^2 = 3^2 \times \left(\sqrt{5}\right)^2$
$$= 9 \times 5$$

$$= 45$$

▶ 2.4.2 Addition and subtraction of surds

eles-4665

- Surds may be added or subtracted only if they are *alike*.
  Examples of *like* surds include $\sqrt{7}, 3\sqrt{7}$ and $-5\sqrt{7}$.

  Examples of *unlike* surds include $\sqrt{11}, \sqrt{5}, 2\sqrt{13}$ and $-2\sqrt{3}$.
- In some cases surds will need to be simplified before you decide whether they are like or unlike, and then addition and subtraction can take place. The concept of adding and subtracting surds is similar to adding and subtracting like terms in algebra.

WORKED EXAMPLE 6 Adding and subtracting surds

Simplify each of the following expressions containing surds. Assume that a and b are positive real numbers.

a. $3\sqrt{6} + 17\sqrt{6} - 2\sqrt{6}$

b. $5\sqrt{3} + 2\sqrt{12} - 5\sqrt{2} + 3\sqrt{8}$

c. $\dfrac{1}{2}\sqrt{100a^3b^2} + ab\sqrt{36a} - 5\sqrt{4a^2b}$

THINK	WRITE
a. All 3 terms are alike because they contain the same surd ($\sqrt{6}$). Simplify.	a. $3\sqrt{6} + 17\sqrt{6} - 2\sqrt{6} = (3 + 17 - 2)\sqrt{6}$ $\qquad\qquad\qquad\qquad\quad = 18\sqrt{6}$
b. 1. Simplify surds where possible.	b. $5\sqrt{3} + 2\sqrt{12} - 5\sqrt{2} + 3\sqrt{8}$ $= 5\sqrt{3} + 2\sqrt{4 \times 3} - 5\sqrt{2} + 3\sqrt{4 \times 2}$ $= 5\sqrt{3} + 2 \times 2\sqrt{3} - 5\sqrt{2} + 3 \times 2\sqrt{2}$
2. Add like terms to obtain the simplified answer.	$= 5\sqrt{3} + 4\sqrt{3} - 5\sqrt{2} + 6\sqrt{2}$ $= 9\sqrt{3} + \sqrt{2}$
c. 1. Simplify surds where possible.	c. $\dfrac{1}{2}\sqrt{100a^3b^2} + ab\sqrt{36a} - 5\sqrt{4a^2b}$ $= \dfrac{1}{2} \times 10\sqrt{a^2 \times a \times b^2} + ab \times 6\sqrt{a} - 5 \times 2 \times a\sqrt{b}$ $= \dfrac{1}{2} \times 10 \times a \times b\sqrt{a} + ab \times 6\sqrt{a} - 5 \times 2 \times a\sqrt{b}$
2. Add like terms to obtain the simplified answer.	$= 5ab\sqrt{a} + 6ab\sqrt{a} - 10a\sqrt{b}$ $= 11ab\sqrt{a} - 10a\sqrt{b}$

▶ 2.4.3 Dividing surds

eles-4666

- To divide surds, divide the expressions under the radical signs.

> **Dividing surds**
>
> $$\frac{\sqrt{a}}{\sqrt{b}} = \sqrt{\frac{a}{b}}$$
>
> where a and b are positive real numbers.

- When dividing surds it is best to simplify them (if possible) first. Once this has been done, the coefficients are divided next and then the surds are divided.

$$\frac{m\sqrt{a}}{n\sqrt{b}} = \frac{m}{n}\sqrt{\frac{a}{b}}$$

WORKED EXAMPLE 7 Dividing surds

Divide the following surds, expressing answers in the simplest form. Assume that x and y are positive real numbers.

a. $\dfrac{\sqrt{55}}{\sqrt{5}}$ b. $\dfrac{\sqrt{48}}{\sqrt{3}}$ c. $\dfrac{9\sqrt{88}}{6\sqrt{99}}$ d. $\dfrac{\sqrt{36xy}}{\sqrt{25x^9y^{11}}}$

THINK

WRITE

a. 1. Rewrite the fraction, using $\dfrac{\sqrt{a}}{\sqrt{b}} = \sqrt{\dfrac{a}{b}}$.

a. $\dfrac{\sqrt{55}}{\sqrt{5}} = \sqrt{\dfrac{55}{5}}$

2. Divide the numerator by the denominator (that is, 55 by 5). Check if the surd can be simplified any further.

$= \sqrt{11}$

b. 1. Rewrite the fraction, using $\dfrac{\sqrt{a}}{\sqrt{b}} = \sqrt{\dfrac{a}{b}}$.

b. $\dfrac{\sqrt{48}}{\sqrt{3}} = \sqrt{\dfrac{48}{3}}$

2. Divide 48 by 3.

$= \sqrt{16}$

3. Evaluate $\sqrt{16}$.

$= 4$

c. 1. Rewrite surds, using $\dfrac{\sqrt{a}}{\sqrt{b}} = \sqrt{\dfrac{a}{b}}$.

c. $\dfrac{9\sqrt{88}}{6\sqrt{99}} = \dfrac{9}{6}\sqrt{\dfrac{88}{99}}$

2. Simplify the fraction under the radical by dividing both numerator and denominator by 11.

$= \dfrac{9}{6}\sqrt{\dfrac{8}{9}}$

3. Simplify surds.

$= \dfrac{9 \times 2\sqrt{2}}{6 \times 3}$

4. Multiply the whole numbers in the numerator together and those in the denominator together.

$= \dfrac{18\sqrt{2}}{18}$

5. Cancel the common factor of 18.

$= \sqrt{2}$

d. 1. Simplify each surd.

d. $\dfrac{\sqrt{36xy}}{\sqrt{25x^9y^{11}}} = \dfrac{6\sqrt{xy}}{5\sqrt{x^8 \times x \times y^{10} \times y}}$

$= \dfrac{6\sqrt{xy}}{5x^4y^5\sqrt{xy}}$

2. Cancel any common factors — in this case $\sqrt{xy}$.

$= \dfrac{6}{5x^4y^5}$

▶ 2.4.4 Expanding binomial products involving surds

eles-6273

- In previous topics, you have studied the expansion of two binomial products, for example, the expansion of $(x + 3)(x - 5)$ is $x^2 - 2x - 15$.
- A common method for expanding two binomial factors is called FOIL.

FOIL – First Outer Inner Last

$$(a + b)(c + d)$$

$$(a + b)(c + d) = ac + ad + bc + bd$$

Expanding binomial factors

Using FOIL, the expanded product of two binomial factors is given by:

$$(a + b)(c + d) = a \times c + a \times d + b \times c + b \times d = ac + ad + bc + bd$$

It is expected that the result will have 4 terms.

Note: **It may be possible to simplify like terms after expanding.**

- This method can be used to expand two binomial products involving surds.

WORKED EXAMPLE 8 Expanding binomial products involving surds

Expand and simplify the following expressions.

a. $\left(2 + \sqrt{3}\right)\left(1 - \sqrt{2}\right)$

b. $\left(\sqrt{5} + 5\sqrt{2}\right)\left(3\sqrt{5} - \sqrt{10}\right)$

THINK	WRITE
a. 1. Expand the brackets using FOIL.	a. $\left(2 + \sqrt{3}\right)\left(1 - \sqrt{2}\right)$
2. There are no like terms to collect.	$= 2 - 2\sqrt{2} + \sqrt{3} - \sqrt{6}$
b. 1. Expand the brackets using FOIL.	b. $\left(\sqrt{5} + 5\sqrt{2}\right)\left(3\sqrt{5} - \sqrt{10}\right)$
2. Simplify the expressions.	$= 3\sqrt{25} - \sqrt{50} + 15\sqrt{10} - 5\sqrt{20}$
	$= 3 \times 5 - \sqrt{25 \times 2} + 15\sqrt{10} - 5\sqrt{4 \times 5}$
	$= 15 - 5\sqrt{2} + 15\sqrt{10} - 10\sqrt{5}$

▶ 2.4.5 Rationalising denominators

- If the **denominator** of a fraction is a surd, it can be changed into a rational number through multiplication. In other words, it can be rationalised.
- As discussed earlier in this chapter, squaring a simple surd (that is, multiplying it by itself) results in a rational number. This fact can be used to rationalise denominators as follows.

> **Rationalising the denominator**
>
> $$\frac{\sqrt{a}}{\sqrt{b}} = \frac{\sqrt{a}}{\sqrt{b}} \times \frac{\sqrt{b}}{\sqrt{b}} = \frac{\sqrt{ab}}{b}$$

- If both numerator and denominator of a fraction are multiplied by the surd contained in the denominator, the denominator becomes a rational number. The fraction takes on a different appearance, but its numerical value is unchanged, because multiplying the numerator and denominator by the same number is equivalent to multiplying by 1.

WORKED EXAMPLE 9 Rationalising the denominator

Express the following in their simplest form with a rational denominator.

a. $\dfrac{\sqrt{6}}{\sqrt{13}}$ b. $\dfrac{2\sqrt{12}}{3\sqrt{54}}$ c. $\dfrac{\sqrt{17} - 3\sqrt{14}}{\sqrt{7}}$

THINK

WRITE

a. 1. Write the fraction.

a. $\dfrac{\sqrt{6}}{\sqrt{13}}$

2. Multiply both the numerator and denominator by the surd contained in the denominator (in this case $\sqrt{13}$). This has the same effect as multiplying the fraction by 1, because $\dfrac{\sqrt{13}}{\sqrt{13}} = 1$.

$= \dfrac{\sqrt{6}}{\sqrt{13}} \times \dfrac{\sqrt{13}}{\sqrt{13}}$

$= \dfrac{\sqrt{78}}{13}$

b. 1. Write the fraction.

b. $\dfrac{2\sqrt{12}}{3\sqrt{54}}$

2. Simplify the surds. (This avoids dealing with large numbers.)

$\dfrac{2\sqrt{12}}{3\sqrt{54}} = \dfrac{2\sqrt{4 \times 3}}{3\sqrt{9 \times 6}}$

$= \dfrac{2 \times 2\sqrt{3}}{3 \times 3\sqrt{6}}$

$= \dfrac{4\sqrt{3}}{9\sqrt{6}}$

3. Multiply both the numerator and denominator by $\sqrt{6}$. This has the same effect as multiplying the fraction by 1, because $\dfrac{\sqrt{6}}{\sqrt{6}} = 1$.

Note: We need to multiply only by the surd part of the denominator (that is, by $\sqrt{6}$ rather than by $9\sqrt{6}$.)

$$= \frac{4\sqrt{3}}{9\sqrt{6}} \times \frac{\sqrt{6}}{\sqrt{6}}$$

$$= \frac{4\sqrt{18}}{9 \times 6}$$

4. Simplify $\sqrt{18}$.

$$= \frac{4\sqrt{9 \times 2}}{9 \times 6}$$

$$= \frac{4 \times 3\sqrt{2}}{54}$$

$$= \frac{12\sqrt{2}}{54}$$

5. Divide both the numerator and denominator by 6 (cancel down).

$$= \frac{2\sqrt{2}}{9}$$

c. 1. Write the fraction.

c. $\dfrac{\sqrt{17} - 3\sqrt{14}}{\sqrt{7}}$

2. Multiply both the numerator and denominator by $\sqrt{7}$. Use grouping symbols (brackets) to make it clear that the whole numerator must be multiplied by $\sqrt{7}$.

$$= \frac{(\sqrt{17} - 3\sqrt{14})}{\sqrt{7}} \times \frac{\sqrt{7}}{\sqrt{7}}$$

3. Apply the Distributive Law in the numerator.
$a(b + c) = ab + ac$

$$= \frac{\sqrt{17} \times \sqrt{7} - 3\sqrt{14} \times \sqrt{7}}{\sqrt{7} \times \sqrt{7}}$$

$$= \frac{\sqrt{119} - 3\sqrt{98}}{7}$$

4. Simplify $\sqrt{98}$.

$$= \frac{\sqrt{119} - 3\sqrt{49 \times 2}}{7}$$

$$= \frac{\sqrt{119} - 3 \times 7\sqrt{2}}{7}$$

$$= \frac{\sqrt{119} - 21\sqrt{2}}{7}$$

on Resources

▶ **Video eLessons** Surds (eles-1906)
Rationalisation of surds (eles-1948)

Interactivities Addition and subtraction of surds (int-6190)
Multiplying surds (int-6191)
Dividing surds (int-6192)
Simplifying surds (int-6028)

Exercise 2.4 Operations with surds (Path)

2.4 Quick quiz on	2.4 Exercise

Individual pathways

■ PRACTISE	■ CONSOLIDATE	■ MASTER
1, 4, 7, 10, 12, 15, 18, 21, 24, 28, 31, 34	2, 5, 8, 11, 13, 16, 19, 22, 25, 27, 29, 32, 35	3, 6, 9, 14, 17, 20, 23, 26, 30, 33, 36

Fluency

WE3a For questions **1** to **3**, simplify the following surds.

1. a. $\sqrt{12}$ b. $\sqrt{24}$ c. $\sqrt{27}$ d. $\sqrt{125}$

2. a. $\sqrt{54}$ b. $\sqrt{112}$ c. $\sqrt{68}$ d. $\sqrt{180}$

3. a. $\sqrt{88}$ b. $\sqrt{162}$ c. $\sqrt{245}$ d. $\sqrt{448}$

WE3b,c For questions **4** to **6**, simplify the following surds.

4. a. $2\sqrt{8}$ b. $8\sqrt{90}$ c. $9\sqrt{80}$ d. $7\sqrt{54}$

5. a. $-6\sqrt{75}$ b. $-7\sqrt{80}$ c. $16\sqrt{48}$ d. $\frac{1}{7}\sqrt{392}$

6. a. $\frac{1}{9}\sqrt{162}$ b. $\frac{1}{4}\sqrt{192}$ c. $\frac{1}{9}\sqrt{135}$ d. $\frac{3}{10}\sqrt{175}$

WE3d For questions **7** to **9**, simplify the following surds. Assume that a, b, c, d, e, f, x and y are positive real numbers.

7. a. $\sqrt{16a^2}$ b. $\sqrt{72a^2}$ c. $\sqrt{90a^2b}$ d. $\sqrt{338a^4}$

8. a. $\sqrt{338a^3b^3}$ b. $\sqrt{68a^3b^5}$ c. $\sqrt{125x^6y^4}$ d. $5\sqrt{80x^3y^2}$

9. a. $6\sqrt{162c^7d^5}$ b. $2\sqrt{405c^7d^9}$ c. $\frac{1}{2}\sqrt{88ef}$ d. $\frac{1}{2}\sqrt{392e^{11}f^{11}}$

10. **WE4a** Simplify the following expressions containing surds. Assume that x and y are positive real numbers.
    a. $3\sqrt{5}+4\sqrt{5}$
    b. $2\sqrt{3}+5\sqrt{3}+\sqrt{3}$
    c. $8\sqrt{5}+3\sqrt{3}+7\sqrt{5}+2\sqrt{3}$
    d. $6\sqrt{11}-2\sqrt{11}$

11. Simplify the following expressions containing surds. Assume that x and y are positive real numbers.
    a. $7\sqrt{2}+9\sqrt{2}-3\sqrt{2}$
    b. $9\sqrt{6}+12\sqrt{6}-17\sqrt{6}-7\sqrt{6}$
    c. $12\sqrt{3}-8\sqrt{7}+5\sqrt{3}-10\sqrt{7}$
    d. $2\sqrt{x}+5\sqrt{y}+6\sqrt{x}-2\sqrt{y}$

Understanding

WE4b For questions **12** to **14**, simplify the following expressions containing surds. Assume that a and b are positive real numbers.

12. a. $\sqrt{200}-\sqrt{300}$
    b. $\sqrt{125}-\sqrt{150}+\sqrt{600}$
    c. $\sqrt{27}-\sqrt{3}+\sqrt{75}$
    d. $2\sqrt{20}-3\sqrt{5}+\sqrt{45}$

13. a. $6\sqrt{12} + 3\sqrt{27} - 7\sqrt{3} + \sqrt{18}$
    c. $3\sqrt{90} - 5\sqrt{60} + 3\sqrt{40} + \sqrt{100}$
    b. $\sqrt{150} + \sqrt{24} - \sqrt{96} + \sqrt{108}$
    d. $5\sqrt{11} + 7\sqrt{44} - 9\sqrt{99} + 2\sqrt{121}$

14. a. $2\sqrt{30} + 5\sqrt{120} + \sqrt{60} - 6\sqrt{135}$
    c. $\dfrac{1}{2}\sqrt{98} + \dfrac{1}{3}\sqrt{48} + \dfrac{1}{3}\sqrt{12}$
    b. $6\sqrt{ab} - \sqrt{12ab} + 2\sqrt{9ab} + 3\sqrt{27ab}$
    d. $\dfrac{1}{8}\sqrt{32} - \dfrac{7}{6}\sqrt{18} + 3\sqrt{72}$

WE4c For questions **15** to **17**, simplify the following expressions containing surds. Assume that a and b are positive real numbers.

15. a. $7\sqrt{a} - \sqrt{8a} + 8\sqrt{9a} - \sqrt{32a}$
    c. $\sqrt{150ab} + \sqrt{96ab} - \sqrt{54ab}$
    b. $10\sqrt{a} - 15\sqrt{27a} + 8\sqrt{12a} + 14\sqrt{9a}$
    d. $16\sqrt{4a^2} - \sqrt{24a} + 4\sqrt{8a^2} + \sqrt{96a}$

16. a. $\sqrt{8a^3} + \sqrt{72a^3} - \sqrt{98a^3}$
    c. $\sqrt{9a^3} + \sqrt{3a^5}$
    b. $\dfrac{1}{2}\sqrt{36a} + \dfrac{1}{4}\sqrt{128a} - \dfrac{1}{6}\sqrt{144a}$
    d. $6\sqrt{a^5b} + \sqrt{a^3b} - 5\sqrt{a^5b}$

17. a. $ab\sqrt{ab} + 3ab\sqrt{a^2b} + \sqrt{9a^3b^3}$
    c. $\sqrt{32a^3b^2} - 5ab\sqrt{8a} + \sqrt{48a^5b^6}$
    b. $\sqrt{a^3b} + 5\sqrt{ab} - 2\sqrt{ab} + 5\sqrt{a^3b}$
    d. $\sqrt{4a^2b} + 5\sqrt{a^2b} - 3\sqrt{9a^2b}$

WE5 For questions **18** to **20**, multiply the following surds, expressing answers in the simplest form. Assume that a, b, x and y are positive real numbers.

18. a. $\sqrt{2} \times \sqrt{7}$
    d. $\sqrt{10} \times \sqrt{10}$
    b. $\sqrt{6} \times \sqrt{7}$
    e. $\sqrt{21} \times \sqrt{3}$
    c. $\sqrt{8} \times \sqrt{6}$
    f. $\sqrt{27} \times 3\sqrt{3}$

19. a. $5\sqrt{3} \times 2\sqrt{11}$
    d. $10\sqrt{6} \times 3\sqrt{8}$
    b. $10\sqrt{15} \times 6\sqrt{3}$
    e. $\dfrac{1}{4}\sqrt{48} \times 2\sqrt{2}$
    c. $4\sqrt{20} \times 3\sqrt{5}$
    f. $\dfrac{1}{9}\sqrt{48} \times 2\sqrt{3}$

20. a. $\dfrac{1}{10}\sqrt{60} \times \dfrac{1}{5}\sqrt{40}$
    d. $\sqrt{12a^7b} \times \sqrt{6a^3b^4}$
    b. $\sqrt{xy} \times \sqrt{x^3y^2}$
    e. $\sqrt{15x^3y^2} \times \sqrt{6x^2y^3}$
    c. $\sqrt{3a^4b^2} \times \sqrt{6a^5b^3}$
    f. $\dfrac{1}{2}\sqrt{15a^3b^3} \times 3\sqrt{3a^2b^6}$

WE6 For questions **21** to **23**, simplify each of the following.

21. a. $\left(\sqrt{2}\right)^2$
    b. $\left(\sqrt{5}\right)^2$
    c. $\left(\sqrt{12}\right)^2$

22. a. $\left(\sqrt{15}\right)^2$
    b. $\left(3\sqrt{2}\right)^2$
    c. $\left(4\sqrt{5}\right)^2$

23. a. $\left(2\sqrt{7}\right)^2$
    b. $\left(5\sqrt{8}\right)^2$

WE7 For questions **24** to **26**, divide the following surds, expressing answers in the simplest form. Assume that a, b, x and y are positive real numbers.

24. a. $\dfrac{\sqrt{15}}{\sqrt{3}}$
    b. $\dfrac{\sqrt{8}}{\sqrt{2}}$
    c. $\dfrac{\sqrt{60}}{\sqrt{10}}$
    d. $\dfrac{\sqrt{128}}{\sqrt{8}}$

25. a. $\dfrac{\sqrt{18}}{4\sqrt{6}}$
    b. $\dfrac{\sqrt{65}}{2\sqrt{13}}$
    c. $\dfrac{\sqrt{96}}{\sqrt{8}}$
    d. $\dfrac{7\sqrt{44}}{14\sqrt{11}}$

26. a. $\dfrac{9\sqrt{63}}{15\sqrt{7}}$ **b.** $\dfrac{\sqrt{2040}}{\sqrt{30}}$ **c.** $\dfrac{\sqrt{x^4y^3}}{\sqrt{x^2y^5}}$

d. $\dfrac{\sqrt{16xy}}{\sqrt{8x^7y^9}}$ **e.** $\dfrac{\sqrt{xy}}{\sqrt{x^5y^7}}\times\dfrac{\sqrt{12x^8y^{12}}}{\sqrt{x^2y^3}}$ **f.** $\dfrac{2\sqrt{2a^2b^4}}{\sqrt{5a^3b^6}}\times\dfrac{\sqrt{10a^9b^3}}{3\sqrt{a^7b}}$

27. `WE8` Expand and simplify the following expressions.

a. $\left(\sqrt{2}+\sqrt{3}\right)\left(\sqrt{5}-\sqrt{2}\right)$ **b.** $\left(\sqrt{10}-2\sqrt{5}\right)\left(3\sqrt{5}-1\right)$

c. $\left(4\sqrt{5}+\sqrt{2}\right)\left(3\sqrt{5}+2\sqrt{2}\right)$ **d.** $\left(\sqrt{15}+2\sqrt{3}\right)\left(5\sqrt{8}-\sqrt{10}\right)$

e. $\left(\sqrt{6}+2\sqrt{2}\right)^2$ **f.** $\left(2\sqrt{5}+5\sqrt{3}\right)\left(2\sqrt{5}-5\sqrt{3}\right)$

`WE9a,b` For questions **28** to **30**, express the following in their simplest form with a rational denominator.

28. a. $\dfrac{5}{\sqrt{2}}$ **b.** $\dfrac{7}{\sqrt{3}}$ **c.** $\dfrac{4}{\sqrt{11}}$ **d.** $\dfrac{8}{\sqrt{6}}$ **e.** $\dfrac{\sqrt{12}}{\sqrt{7}}$

29. a. $\dfrac{\sqrt{15}}{\sqrt{6}}$ **b.** $\dfrac{2\sqrt{3}}{\sqrt{5}}$ **c.** $\dfrac{3\sqrt{7}}{\sqrt{5}}$ **d.** $\dfrac{5\sqrt{2}}{2\sqrt{3}}$ **e.** $\dfrac{4\sqrt{3}}{3\sqrt{5}}$

30. a. $\dfrac{5\sqrt{14}}{7\sqrt{8}}$ **b.** $\dfrac{16\sqrt{3}}{6\sqrt{5}}$ **c.** $\dfrac{8\sqrt{3}}{7\sqrt{7}}$ **d.** $\dfrac{8\sqrt{60}}{\sqrt{28}}$ **e.** $\dfrac{2\sqrt{35}}{3\sqrt{14}}$

`WE9c` For questions **31** to **33**, express the following in their simplest form with a rational denominator.

31. a. $\dfrac{\sqrt{6}+\sqrt{12}}{\sqrt{3}}$ **b.** $\dfrac{\sqrt{15}-\sqrt{22}}{\sqrt{6}}$ **c.** $\dfrac{6\sqrt{2}-\sqrt{15}}{\sqrt{10}}$ **d.** $\dfrac{2\sqrt{18}+3\sqrt{2}}{\sqrt{5}}$

32. a. $\dfrac{3\sqrt{5}+6\sqrt{7}}{\sqrt{8}}$ **b.** $\dfrac{4\sqrt{2}+3\sqrt{8}}{2\sqrt{3}}$ **c.** $\dfrac{3\sqrt{11}-4\sqrt{5}}{\sqrt{18}}$ **d.** $\dfrac{2\sqrt{7}-2\sqrt{5}}{\sqrt{12}}$

33. a. $\dfrac{7\sqrt{12}-5\sqrt{6}}{6\sqrt{3}}$ **b.** $\dfrac{6\sqrt{2}-\sqrt{5}}{4\sqrt{8}}$ **c.** $\dfrac{6\sqrt{3}-5\sqrt{5}}{7\sqrt{20}}$ **d.** $\dfrac{3\sqrt{5}+7\sqrt{3}}{5\sqrt{24}}$

Communicating, reasoning and problem solving

34. Calculate the area of a triangle with base length $\dfrac{3}{\sqrt{2}}$ and perpendicular height $\dfrac{5}{\sqrt{10}}$. Express your answer with a rational denominator. Show full working.

35. Determine the average of $\dfrac{1}{2\sqrt{x}}$ and $\dfrac{1}{3\sqrt{x}}$, writing your answer with a rational denominator. Show full working.

36. a. Show that $\left(\sqrt{a}+\sqrt{b}\right)^2 = a+b+2\sqrt{ab}$.
   b. Use this result to evaluate:

     i. $\sqrt{8+2\sqrt{15}}$ **ii.** $\sqrt{8-2\sqrt{15}}$ **iii.** $\sqrt{7+4\sqrt{3}}$.

LESSON
2.5 Review of index laws

LEARNING INTENTION

At the end of this lesson you should be able to:
- recall and apply the index laws to simplify algebraic expressions involving products and quotients
- simplify algebraic expressions involving a zero index
- simplify expressions involving raising a power to another power.

▶ 2.5.1 Review of index laws

eles-4669

Index notation

- When a number or pronumeral is repeatedly multiplied by itself, it can be written in a shorter form called **index form**.
- A number written in index form has two parts, the **base** and the **index**, and is written as:
- In the example shown, a is the base and x is the index.
- Another name for an index is *exponent* or *power*.

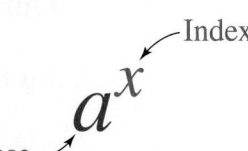

Base $\nearrow$ a^x $\nwarrow$ Index

Index laws

- Performing operations on numbers or pronumerals written in index form requires application of the index laws. There are six index laws.

First Index Law

When terms with the same base are multiplied, the indices are added.

$$a^m \times a^n = a^{m+n}$$

Second Index Law

When terms with the same base are divided, the indices are subtracted.

$$a^m \div a^n = a^{m-n}$$

WORKED EXAMPLE 10 Simplifying using the first two index laws

Simplify each of the following.

a. $m^4 n^3 p \times m^2 n^5 p^3$ b. $2a^2 b^3 \times 3ab^4$ c. $\dfrac{2x^5 y^4}{10x^2 y^3}$

THINK	WRITE
a. 1. Write the expression.	a. $m^4 n^3 p \times m^2 n^5 p^3$
2. Multiply the terms with the same base by adding the indices. *Note:* $p = p^1$.	$= m^{4+2} n^{3+5} p^{1+3}$ $= m^6 n^8 p^4$

▶

b. **1.** Write the expression.

2. Simplify by multiplying the coefficients, then multiply the terms with the same base by adding the indices.

c. **1.** Write the expression.

2. Simplify by dividing both of the coefficients by the same factor, then divide terms with the same base by subtracting the indices.

b. $2a^2b^3 \times 3ab^4$

$= 2 \times 3 \times a^{2+1} \times b^{3+4}$

$= 6a^3b^7$

c. $\dfrac{2x^5y^4}{10x^2y^3}$

$= \dfrac{1x^{5-2}y^{4-3}}{5}$

$= \dfrac{x^3y}{5}$

Third Index Law

Any term (excluding 0) with an index of 0 is equal to 1.

$$a^0 = 1, \; a \neq 0$$

WORKED EXAMPLE 11 Simplifying terms with indices of zero

Simplify each of the following.

a. $\left(2b^3\right)^0$

b. $-4\left(a^2b^5\right)^0$

THINK

a. **1.** Write the expression.

2. Apply the Third Index Law, which states that any term (excluding 0) with an index of 0 is equal to 1.

b. **1.** Write the expression.

2. The entire term inside the brackets has an index of 0, so the bracket is equal to 1.

3. Simplify.

WRITE

a. $\left(2b^3\right)^0$

$= 1$

b. $-4\left(a^2b^5\right)^0$

$= -4 \times 1$

$= -4$

Fourth Index Law

When a power (a^m) is raised to a power, the indices are multiplied.

$$(a^m)^n = a^{mn}$$

Fifth Index Law

When the base is a product, raise every part of the product to the index outside the brackets.

$$(ab)^m = a^m b^m$$

Sixth Index Law

When the base is a fraction, raise both the numerator and denominator to the index outside the brackets.

$$\left(\frac{a}{b}\right)^m = \frac{a^m}{b^m}$$

WORKED EXAMPLE 12 Simplifying terms in index form raised to a power

Simplify each of the following.

a. $\left(2n^4\right)^3$ b. $\left(3a^2b^7\right)^3$ c. $\left(\frac{2x^3}{y^4}\right)^4$ d. $(-4)^3$

THINK

a. 1. Write the term.

   2. Apply the Fourth Index Law and simplify.

b. 1. Write the expression.

   2. Apply the Fifth Index Law and simplify.

c. 1. Write the expression.

   2. Apply the Sixth Index Law and simplify.

d. 1. Write the expression.

   2. Write in expanded form.

   3. Simplify, taking careful note of the negative sign.

WRITE

a. $\left(2n^4\right)^3$

$= 2^{1\times3} \times n^{4\times3}$

$= 2^3 n^{12}$

$= 8n^{12}$

b. $\left(3a^2b^7\right)^3$

$= 3^{1\times3} \times a^{2\times3} \times b^{7\times3}$

$= 3^3 a^6 b^{21}$

$= 27a^6 b^{21}$

c. $\left(\frac{2x^3}{y^4}\right)^4$

$= \frac{2^{1\times4} \times x^{3\times4}}{y^{4\times4}}$

$= \frac{16x^{12}}{y^{16}}$

d. $(-4)^3$

$= -4 \times -4 \times -4$

$= -64$

Resources

▶ **Video eLesson** Index laws (eles-1903)

🧩 **Interactivities** First Index Law (int-3709)

          Second Index Law (int-3711)

          Third Index Law (int-3713)

          Fourth Index Law — Multiplication (int-3716)

          Fifth and sixth index laws (int-6063)

TOPIC 2 Indices and surds **59**

Exercise 2.5 Review of index laws

2.5 Quick quiz on	2.5 Exercise

Individual pathways

■ PRACTISE	■ CONSOLIDATE	■ MASTER
1, 4, 7, 10, 13, 15, 18, 21, 22, 26	2, 5, 8, 11, 14, 16, 19, 23, 24, 27	3, 6, 9, 12, 17, 20, 25, 28

Fluency

WE10a,b For questions **1** to **3**, simplify each of the following.

1. a. $a^3 \times a^4$
   b. $a^2 \times a^3 \times a$
   c. $b \times b^5 \times b^2$
   d. $ab^2 \times a^3 b^5$

2. a. $m^2 n^6 \times m^3 n^7$
   b. $a^2 b^5 c \times a^3 b^2 c^2$
   c. $mnp \times m^5 n^3 p^4$
   d. $2a \times 3ab$

3. a. $4a^2 b^3 \times 5a^2 b \times \dfrac{1}{2} b^5$
   b. $3m^3 \times 2mn^2 \times 6m^4 n^5$
   c. $4x^2 \times \dfrac{1}{2} xy^3 \times 6x^3 y^3$
   d. $2x^3 y^2 \times 4x \times \dfrac{1}{2} x^4 y^4$

WE10c For questions **4** to **6**, simplify each of the following.

4. a. $a^4 \div a^3$
   b. $a^7 \div a^2$
   c. $b^6 \div b^3$
   d. $\dfrac{4a^7}{3a^3}$

5. a. $\dfrac{21b^6}{7b^2}$
   b. $\dfrac{48m^8}{12m^3}$
   c. $\dfrac{m^7 n^3}{m^4 n^2}$
   d. $\dfrac{2x^4 y^3}{4x^4 y}$

6. a. $7ab^5 c^4 \div ab^2 c^4$
   b. $\dfrac{20m^5 n^3 p^4}{16m^3 n^3 p^2}$
   c. $\dfrac{14x^3 y^4 z^2}{28x^2 y^2 z^2}$

WE11 For questions **7** to **9**, simplify each of the following.

7. a. a^0
   b. $(2b)^0$
   c. $\left(3m^2\right)^0$

8. a. $3x^0$
   b. $4b^0$
   c. $-3 \times (2n)^0$

9. a. $4a^0 - \left(\dfrac{a}{4}\right)^0$
   b. $5y^0 - 12$
   c. $5x^0 - \left(5xy^2\right)^0$

WE12 For questions **10** to **12**, simplify each of the following.

10. a. $\left(a^2\right)^3$
    b. $\left(2a^5\right)^4$
    c. $\left(\dfrac{m^2}{3}\right)^4$
    d. $\left(\dfrac{2n^4}{3}\right)^2$
    e. $(-7)^2$

11. a. $\left(a^2 b\right)^3$
    b. $\left(3a^3 b^2\right)^2$
    c. $\left(2m^3 n^5\right)^4$
    d. $\left(\dfrac{3m^2 n}{4}\right)^3$
    e. $\left(\dfrac{a^2}{b^3}\right)^2$

12. a. $\left(\dfrac{5m^3}{n^2}\right)^4$
    b. $\left(\dfrac{7x}{2y^5}\right)^3$
    c. $\left(\dfrac{3a}{5b^3}\right)^4$
    d. $(-3)^5$
    e. $(-2)^5$

13. **MC** a. $2m^{10}n^5$ is the simplified form of:

A. $m^5n^3 \times 2m^4n^2$

B. $\dfrac{6m^{10}n^4}{3n}$

C. $\left(2m^5n^2\right)^2$

D. $2n\left(m^5\right)^2 \times n^4$

b. The value of $4 - (5a)^0$ is:

A. -1

B. 9

C. 1

D. 3

14. **MC** a. $4a^3b \times b^4 \times 5a^2b^3$ simplifies to:

A. $9a^5b^8$

B. $20a^5b^7$

C. $20a^5b^8$

D. $9a^5b^7$

b. $\dfrac{15x^9 \times 3x^6}{9x^{10} \times x^4}$ simplifies to:

A. $5x^9$

B. $9x$

C. $5x^{29}$

D. $5x$

c. $\dfrac{3p^7 \times 8q^9}{12p^3 \times 4q^5}$ simplifies to:

A. $2q^4$

B. $\dfrac{p^4q^4}{2}$

C. $\dfrac{q^4}{2}$

D. $\dfrac{p^4q^4}{24}$

d. $\dfrac{7a^5b^3}{5a^6b^2} \div \dfrac{7b^3a^2}{5b^5a^4}$ simplifies to:

A. $\dfrac{49a^3b}{25}$

B. $\dfrac{25a^3b}{49}$

C. a^3b

D. ab^3

Understanding

For questions **15** to **17**, evaluate each of the following.

15. a. $2^3 \times 2^2 \times 2$

b. $2 \times 3^2 \times 2^2$

c. $\left(5^2\right)^2$

16. a. $\dfrac{3^5 \times 4^6}{3^4 \times 4^4}$

b. $\left(2^3 \times 5\right)^2$

c. $\left(\dfrac{3}{5}\right)^3$

17. a. $\dfrac{4^4 \times 5^6}{4^3 \times 5^5}$

b. $\left(3^3 \times 2^4\right)^0$

c. $4\left(5^2 \times 3^5\right)^0$

For questions **18** to **20**, simplify each of the following.

18. a. $(x^y)^{3z}$

b. $a^b \times (p^q)^0$

19. a. $m^a \times n^b \times (mn)^0$

b. $\left(\dfrac{a^2}{b^3}\right)^x$

20. a. $\dfrac{n^3m^2}{n^pm^q}$

b. $(a^{m+n})^p$

Communicating, reasoning and problem solving

21. Explain why $a^3 \times a^2 = a^5$ and not a^6.

22. Is $2x$ ever the same as x^2? Explain your reasoning using examples.

23. Explain the difference between $3x^0$ and $(3x)^0$.

24. **a.** Complete the table for $a = 0$, 1, 2 and 3.

a	0	1	2	3
$3a^2$				
$5a$				
$3a^2 + 5a$				
$3a^2 \times 5a$				

    b. Analyse what would happen as a becomes very large.

25. Evaluate algebraically the exact value of x if $4^{x+4} = 2^{x^2}$. Justify your answer.

26. Binary numbers (base 2 numbers) are used in computer operations. As the name implies, binary uses only two types of numbers, 0 and 1, to express all numbers.

A binary number such as 101 (read one, zero, one) means $(1 \times 2^2) + (0 \times 2^1) + (1 \times 2^0) = 4 + 0 + 1 = 5$ (in base 10, the base we are most familiar with).

The number 1010 (read one, zero, one, zero) means

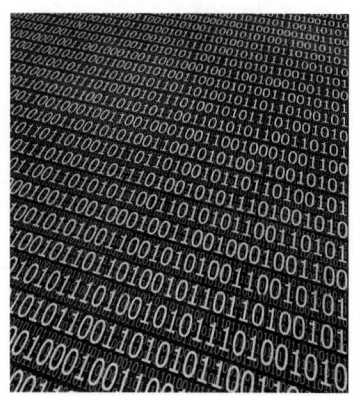

$$(1 \times 2^3) + (0 \times 2^2) + (1 \times 2^1) + (0 \times 2^0) = 8 + 0 + 2 + 0 = 10.$$

If we read the binary number from right to left, the index of 2 increases by one each time, beginning with a power of zero.

Using this information, write out the numbers 1 to 10 in binary (base 2) form.

27. Solve for x:

    a. $\dfrac{7^x \times 7^{1+2x}}{(7^x)^2} = 16\,807$ **b.** $2^{2x} - 5(2^x) = -4$

28. For the following:

    a. determine the correct answer

    b. identify the error in the solution.

$$\left(\frac{a^2 b^3 c}{a^2 b^2}\right)^3 \times \left(\frac{a^3 b^2 c^2}{a^2 b^3}\right)^2 = \left(\frac{b^3 c}{b^2}\right)^3 \times \left(\frac{ab^2 c^2}{b^2}\right)^2$$

$$= \left(\frac{bc}{1}\right)^3 \times \left(\frac{ac^2}{b}\right)^2$$

$$= \left(\frac{abc^3}{b}\right)^6$$

$$= \left(\frac{ac^3}{1}\right)^6$$

$$= a^6 c^{18}$$

LESSON
2.6 Negative indices (Path)

LEARNING INTENTION

At the end of this lesson you should be able to:
- apply index laws to manipulate algebraic expressions involving negative-integer indices
- apply index laws to simplify algebraic products and quotients involving negative-integer indices.

▶ 2.6.1 Negative indices and the Seventh Index Law

eles-4670

- Consider the expression. $\dfrac{a^3}{a^5}$. This expression can be simplified in two different ways.

  1. Written in expanded form: $\dfrac{a^3}{a^5} = \dfrac{a \times a \times a}{a \times a \times a \times a \times a}$

  $$= \dfrac{1}{a \times a}$$

  $$= \dfrac{1}{a^2}$$

  2. Using the Second Index Law: $\dfrac{a^3}{a^5} = a^{3-5}$

  $$= a^{-2}$$

- Equating the results of both of these simplifications we get $a^{-2} = \dfrac{1}{a^2}$.

- In general, $\dfrac{1}{a^n} = \dfrac{a^0}{a^n} \ \left(1 = a^0\right)$

  $$= a^{0-n} \text{ (using the Second Index Law)}$$
  $$= a^{-n}$$

  This statement is the Seventh Index Law.

> ### Seventh Index Law
>
> **A term raised to a negative index is equivalent to 1 over the original term with a positive index.**
>
> $$a^{-n} = \frac{1}{a^n}$$

- The converse of this law can be used to rewrite terms with positive indices only.

$$\frac{1}{a^{-n}} = a^n$$

- It is also worth noting that applying a negative index to a fraction has the effect of swapping the numerator and denominator.

$$\left(\frac{a}{b}\right)^{-n} = \frac{b^n}{a^n}$$

Note: It is proper mathematical convention for an algebraic term to be written with each variable in alphabetical order with positive indices only.

For example: $\dfrac{b^3 a^2 c^{-4}}{y^6 x^{-5}}$ should be written as $\dfrac{a^2 b^3 x^5}{c^4 y^6}$.

WORKED EXAMPLE 13 Writing terms with positive indices only

Express each of the following with positive indices.

a. x^{-3}

b. $2m^{-4}n^2$

c. $\dfrac{4}{a^{-3}}$

THINK	WRITE
a. 1. Write the expression.	a. x^{-3}
2. Apply the Seventh Index Law.	$= \dfrac{1}{x^3}$
b. 1. Write the expression.	b. $2m^{-4}n^2$
2. Apply the Seventh Index Law to write the expression with positive indices.	$= \dfrac{2n^2}{m^4}$
c. 1. Write the expression and rewrite the fraction, using a division sign.	c. $\dfrac{4}{a^{-3}} = 4 \div a^{-3}$
2. Apply the Seventh Index Law to write the expression with positive indices.	$= 4 \div \dfrac{1}{a^3}$
3. To divide the fraction, change fraction division into multiplication.	$= 4 \times \dfrac{a^3}{1}$
	$= 4a^3$

WORKED EXAMPLE 14 Simplifying expressions with negative indices

Simplify each of the following, expressing the answers with positive indices.

a. $a^2 b^{-3} \times a^{-5} b$

b. $\dfrac{2x^4 y^2}{3xy^5}$

c. $\left(\dfrac{2m^3}{n^{-2}}\right)^{-2}$

THINK	WRITE
a. 1. Write the expression.	a. $a^2 b^{-3} \times a^{-5} b$
2. Apply the First Index Law. Multiply terms with the same base by adding the indices.	$= a^{2+-5} b^{-3+1}$
	$= a^{-3} b^{-2}$
3. Apply the Seventh Index Law to write the answer with positive indices.	$= \dfrac{1}{a^3 b^2}$

b. **1.** Write the expression.

b. $\dfrac{2x^4y^2}{3xy^5}$

2. Apply the Second Index Law. Divide terms with the same base by subtracting the indices.

$= \dfrac{2x^{4-1}y^{2-5}}{3}$

$= \dfrac{2x^3y^{-3}}{3}$

3. Apply the Seventh Index Law to write the answer with positive indices.

$= \dfrac{2x^3}{3y^3}$

c. **1.** Write the expression.

c. $\left(\dfrac{2m^3}{n^{-2}}\right)^{-2}$

2. Apply the Sixth Index Law. Multiply the indices of both the numerator and denominator by the index outside the brackets.

$= \dfrac{2^{-2}m^{-6}}{n^4}$

3. Apply the Seventh Index Law to express all terms with positive indices.

$= \dfrac{1}{2^2m^6n^4}$

4. Simplify.

$= \dfrac{1}{4m^6n^4}$

WORKED EXAMPLE 15 Evaluating expressions containing negative indices

Evaluate 6×3^{-3} without using a calculator.

THINK	WRITE
1. Write the multiplication.	6×3^{-3}
2. Apply the Seventh Index Law to write 3^{-3} with a positive index.	$= 6 \times \dfrac{1}{3^3}$
3. Multiply the numerator of the fraction by the whole number.	$= \dfrac{6}{3^3}$
4. Evaluate the denominator.	$= \dfrac{6}{27}$
5. Cancel by dividing both the numerator and denominator by the highest common factor (3).	$= \dfrac{2}{9}$

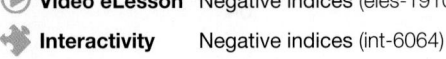 **Resources**

▶ **Video eLesson** Negative indices (eles-1910)

✦ **Interactivity** Negative indices (int-6064)

Exercise 2.6 Negative indices (Path)

2.6 Quick quiz on   **2.6 Exercise**

Individual pathways

■ PRACTISE	■ CONSOLIDATE	■ MASTER
1, 4, 7, 10, 13, 15, 17, 18, 28, 31	2, 5, 8, 11, 14, 16, 19, 20, 23, 26, 29, 32	3, 6, 9, 12, 21, 22, 24, 25, 27, 30, 33

Fluency

WE13 For questions **1** to **3**, express each of the following with positive indices.

1. a. x^{-5} b. y^{-4} c. $2a^{-9}$ d. $\frac{4}{5}a^{-3}$

2. a. $3x^2y^{-3}$ b. $2^{-2}m^{-3}n^{-4}$ c. $6a^3b^{-1}c^{-5}$ d. $\frac{1}{a^{-6}}$

3. a. $\frac{2}{3a^{-4}}$ b. $\frac{6a}{3b^{-2}}$ c. $\frac{7a^{-4}}{2b^{-3}}$ d. $\frac{2m^3n^{-5}}{3a^{-2}b^4}$

WE14 For questions **4** to **6**, simplify each of the following, expressing the answers with positive indices.

4. a. $a^3b^{-2} \times a^{-5}b^{-1}$ b. $2x^{-2}y \times 3x^{-4}y^{-2}$ c. $3m^2n^{-5} \times m^{-2}n^{-3}$
   d. $4a^3b^2 \div a^5b^7$ e. $2xy^6 \div 3x^2y^5$

5. a. $5x^{-2}y^3 \div 6xy^2$ b. $\frac{6m^4n}{2n^3m^6}$ c. $\frac{4x^2y^9}{x^7y^{-3}}$
   d. $\frac{2m^2n^{-4}}{6m^5n^{-1}}$ e. $\left(2a^3m^4\right)^{-5}$

6. a. $4\left(p^7q^{-4}\right)^{-2}$ b. $3\left(a^{-2}b^{-3}\right)^4$ c. $\left(\frac{2p^2}{3q^3}\right)^{-3}$
   d. $\left(\frac{a^{-4}}{2b^{-3}}\right)^2$ e. $\left(\frac{6a^2}{3b^{-2}}\right)^{-3}$

WE15 For questions **7** to **9**, evaluate each of the following without using a calculator.

7. a. 2^{-3} b. 6^{-2} c. 3^{-4} d. $3^{-2} \times 2^3$

8. a. $4^{-3} \times 2^2$ b. 5×6^{-2} c. $\frac{6}{2^{-3}}$ d. $\frac{4 \times 3^{-3}}{2^{-3}}$

9. a. $\frac{1}{3} \times 5^{-2} \times 3^4$ b. $\frac{16^0 \times 2^4}{8^2 \times 2^{-4}}$ c. $\frac{5^3 \times 25^0}{25^2 \times 5^{-4}}$ d. $\frac{3^4 \times 4^2}{12^3 \times 15^0}$

10. Write each of these numbers as a power of 2.
    a. 8   b. $\frac{1}{8}$ c. 32   d. $\frac{1}{64}$

11. Solve each of the following for x.

   a. $125 = 5^x$
   b. $\dfrac{1}{16} = 4^x$
   c. $\dfrac{1}{7} = 7^x$
   d. $216 = 6^x$
   e. $0.01 = 10^x$

12. Solve each of the following for x.

   a. $1 = 8^x$
   b. $64 = 4^x$
   c. $\dfrac{1}{64} = 4^x$
   d. $\dfrac{1}{64} = 2^x$
   e. $\dfrac{1}{64} = 8^x$

13. Evaluate the following expressions.

   a. $\left(\dfrac{2}{3}\right)^{-1}$
   b. $\left(\dfrac{5}{4}\right)^{-1}$
   c. $\left(3\dfrac{1}{2}\right)^{-1}$
   d. $\left(\dfrac{1}{5}\right)^{-1}$

14. Write the following expressions with positive indices.

   a. $\left(\dfrac{a}{b}\right)^{-1}$
   b. $\left(\dfrac{a^2}{b^3}\right)^{-1}$
   c. $\left(\dfrac{a^{-2}}{b^{-3}}\right)^{-1}$
   d. $\left(\dfrac{m^3}{n^{-2}}\right)^{-1}$

15. Evaluate each of the following, using a calculator.

   a. 3^{-6}
   b. 12^{-4}
   c. 7^{-5}

16. Evaluate each of the following, using a calculator.

   a. $\left(\dfrac{1}{2}\right)^{-8}$
   b. $\left(\dfrac{3}{4}\right)^{-7}$
   c. $(0.04)^{-5}$

Understanding

17. **MC** $\dfrac{1}{a^{-4}}$ is the same as:

   A. $4a$
   B. $-4a$
   C. a^4
   D. $\dfrac{1}{a^4}$

18. **MC** $\dfrac{1}{8}$ is the same as:

   A. 2^3
   B. 2^{-3}
   C. 3^2
   D. 3^{-2}

19. **MC** Select which of the following, when simplified, gives $\dfrac{3m^4}{4n^2}$.

   A. $\dfrac{3m^{-4}n^{-2}}{4}$
   B. $3 \times 2^{-2} \times m^4 \times n^{-2}$
   C. $\dfrac{3n^{-2}}{2^{-2}m^{-4}}$
   D. $\dfrac{2^2 n^{-2}}{3^{-1}m^{-4}}$

20. **MC** When simplified, $3a^{-2}b^{-7} \div \left(\dfrac{3}{4}a^{-4}b^6\right)$ is equal to:

   A. $\dfrac{4}{a^6 b^{13}}$
   B. $\dfrac{9b}{4a^6}$
   C. $\dfrac{9a^2}{4b}$
   D. $\dfrac{4a^2}{b^{13}}$

21. **MC** When $(2x^6 y^{-4})^{-3}$ is simplified, it is equal to:

   A. $\dfrac{2x^{18}}{y^{12}}$
   B. $\dfrac{x^{18}}{8y^{12}}$
   C. $\dfrac{y^{12}}{8x^{18}}$
   D. $\dfrac{8y^{12}}{x^{18}}$

22. **MC** If $\left(\dfrac{2a^x}{b^y}\right)^3$ is equal to $\dfrac{8b^9}{a^6}$, then x and y (in that order) are:

   A. -3 and -6 **B.** -6 and -3 **C.** -3 and 2 **D.** -2 and -3

23. Simplify, expressing your answer with positive indices.

   a. $\dfrac{m^{-3}n^{-2}}{m^{-5}n^6}$

   b. $\dfrac{\left(m^3n^{-2}\right)^{-7}}{\left(m^{-5}n^3\right)^4}$

   c. $\dfrac{5\left(a^3b^{-3}\right)^2}{\left(ab^{-4}\right)^{-1}} \div \dfrac{\left(5a^{-2}b\right)^{-1}}{\left(a^{-4}b\right)^3}$

24. Simplify, expanding any expressions in brackets.

   a. $\left(r^3 + s^3\right)\left(r^3 - s^3\right)$

   b. $\left(m^5 + n^5\right)^2$

   c. $\dfrac{\left(x^{a+1}\right)^b \times x^{a+b}}{x^{a(b+1)} \times x^{2b}}$

   d. $\left(\dfrac{p^{x+1}}{p^{x-1}}\right)^{-4} \times \dfrac{p^{8(x+1)}}{\left(p^{2x}\right)^4} \times \dfrac{p^2}{\left(p^{12x}\right)^0}$

25. Write $\left(\dfrac{2^r \times 8^r}{2^{2r} \times 16}\right)$ in the form 2^{ar+b}.

26. Write $2^{-m} \times 3^{-m} \times 6^{2m} \times 3^{2m} \times 2^{2m}$ as a power of 6.

27. Solve for x if $4^x - 4^{x-1} = 48$.

Communicating, reasoning and problem solving

28. Consider the equation $y = \dfrac{6}{x}$. Clearly $x \neq 0$ as $\dfrac{6}{x}$ would be undefined.

   Explain what happens to the value of y as x gets closer to zero coming from:

   a. the positive direction

   b. the negative direction.

29. Consider the expression 2^{-n}. Explain what happens to the value of this expression as n increases.

30. Explain why each of these statements is false. Illustrate each answer by substituting a value for the pronumeral.

   a. $5x^0 = 1$

   b. $9x^5 \div (3x^5) = 3x$

   c. $a^5 \div a^7 = a^2$

   d. $2c^{-4} = \dfrac{1}{2c^4}$

31. Solve the following pair of simultaneous equations.

   $3^{y+1} = \dfrac{1}{9}$ and $\dfrac{5^y}{125^x} = 125$

32. Simplify $\dfrac{x^{n+2} + x^{n-2}}{x^{n-4} + x^n}$.

33. Solve for x and y if $5^{x-y} = 625$ and $3^{2x} \times 3^y = 243$.

   Hence, evaluate $\dfrac{35^x}{7^{-2y} \times 5^{-3y}}$.

LESSON
2.7 Fractional indices (Path)

LEARNING INTENTION

At the end of this lesson you should be able to:
- evaluate numerical expressions involving fractional indices and surds using digital tools
- simplify algebraic expressions involving fractional indices and surds.

▶ 2.7.1 Fractional indices and the Eighth Index Law

eles-4671

- Consider the expression $a^{\frac{1}{2}}$. Now consider what happens if we square that expression.

$$\left(a^{\frac{1}{2}} \right)^2 = a \quad \text{(Using the Fourth Index Law, } (a^m)^n = a^{m\times n}\text{)}$$

- From our work on surds, we know that $\left(\sqrt{a} \right)^2 = a$.

- Equating the two facts above, $\left(a^{\frac{1}{2}} \right)^2 = \left(\sqrt{a} \right)^2$. Therefore, $a^{\frac{1}{2}} = \sqrt{a}$.

- Similarly, $b^{\frac{1}{3}} \times b^{\frac{1}{3}} \times b^{\frac{1}{3}} = \left(b^{\frac{1}{3}} \right)^3 = b$ implying that $b^{\frac{1}{3}} = \sqrt[3]{b}$.

- This pattern can be continued and generalised to produce $a^{\frac{1}{n}} = \sqrt[n]{a}$.

- Now consider: $a^{\frac{m}{n}} = a^{m\times\frac{1}{n}}$ or $a^{\frac{m}{n}} = a^{\frac{1}{n}\times m}$

$$\begin{aligned} &= (a^m)^{\frac{1}{n}} &&= \left(a^{\frac{1}{n}} \right)^m \\ &= \sqrt[n]{a^m} &&= \left(\sqrt[n]{a} \right)^m \end{aligned}$$

Eighth Index Law

A term raised to a fractional index $\dfrac{m}{n}$ is equivalent to the nth root of the term raised to the power m.

$$a^{\frac{m}{n}} = \sqrt[n]{a^m} = \left(\sqrt[n]{a} \right)^m$$

WORKED EXAMPLE 16 Converting fractional indices to surd form

Write each of the following expressions in simplest surd form.

a. $10^{\frac{1}{2}}$

b. $5^{\frac{3}{2}}$

THINK

WRITE

a. Since an index of $\dfrac{1}{2}$ is equivalent to taking the square root, this term can be written as the square root of 10.

a. $10^{\frac{1}{2}} = \sqrt{10}$

▶

b. **1.** A power of $\frac{3}{2}$ means the square root of the number cubed.

b. $5^{\frac{3}{2}} = \sqrt{5^3}$

2. Evaluate 5^3.

$= \sqrt{125}$

3. Simplify $\sqrt{125}$.

$= 5\sqrt{5}$

WORKED EXAMPLE 17 Evaluating fractional indices without a calculator

Evaluate each of the following without using a calculator.

a. $9^{\frac{1}{2}}$

b. $16^{\frac{3}{2}}$

THINK	WRITE
a. **1.** Rewrite the number using Eighth Index Law.	**a.** $9^{\frac{1}{2}} = \sqrt{9}$
2. Evaluate.	$= 3$
b. **1.** Rewrite the number using $a^{\frac{m}{n}} = \left(\sqrt[n]{a}\right)^m$.	**b.** $16^{\frac{3}{2}} = \left(\sqrt{16}\right)^3$
	$= 4^3$
2. Simplify and evaluate the result.	$= 64$

WORKED EXAMPLE 18 Evaluating fractional indices with a calculator

Use a calculator to determine the value of the following, correct to 1 decimal place.

a. $10^{\frac{1}{4}}$

b. $200^{\frac{1}{5}}$

THINK	WRITE
a. Use a calculator to produce the answer.	**a.** $10^{\frac{1}{4}} = 1.77827941$
	≈ 1.8
b. Use a calculator to produce the answer.	**b.** $200^{\frac{1}{5}} = 2.885399812$
	≈ 2.9

WORKED EXAMPLE 19 Simplifying expressions with fractional indices

Simplify each of the following.

a. $m^{\frac{1}{5}} \times m^{\frac{2}{5}}$

b. $(a^2 b^3)^{\frac{1}{6}}$

c. $\left(\dfrac{x^{\frac{2}{3}}}{y^{\frac{3}{4}}}\right)^{\frac{1}{2}}$

THINK	WRITE
a. 1. Write the expression.	**a.** $m^{\frac{1}{5}} \times m^{\frac{2}{5}}$
2. Multiply numbers with the same base by adding the indices.	$= m^{\frac{3}{5}}$
b. 1. Write the expression.	**b.** $(a^2 b^3)^{\frac{1}{6}}$
2. Multiply each index inside the grouping symbols (brackets) by the index on the outside.	$= a^{\frac{2}{6}} b^{\frac{3}{6}}$
3. Simplify the fractions.	$= a^{\frac{1}{3}} b^{\frac{1}{2}}$
c. 1. Write the expression.	**c.** $\left(\dfrac{x^{\frac{2}{3}}}{y^{\frac{3}{4}}} \right)^{\frac{1}{2}}$
2. Multiply the index in both the numerator and denominator by the index outside the grouping symbols.	$= \dfrac{x^{\frac{1}{3}}}{y^{\frac{3}{8}}}$

on Resources

▶ **Video eLesson** Fractional indices (eles-1950)

🧩 **Interactivity** Fractional indices (int-6107)

Exercise 2.7 Fractional indices (Path)   learn on

2.7 Quick quiz on	**2.7 Exercise**

Individual pathways

■ PRACTISE	■ CONSOLIDATE	■ MASTER
1, 3, 7, 10, 13, 15, 17, 20, 23, 25, 28, 30	2, 4, 5, 8, 11, 14, 18, 21, 26, 29, 31	6, 9, 12, 16, 19, 22, 24, 27, 32

Fluency

WE16 For questions **1** to **4**, write the following in surd form.

1. a. $15^{\frac{1}{2}}$ **b.** $m^{\frac{1}{4}}$ **c.** $7^{\frac{2}{5}}$ **d.** $7^{\frac{5}{2}}$

2. a. $w^{\frac{3}{8}}$ **b.** $w^{1.25}$ **c.** $5^{3\frac{1}{3}}$ **d.** $a^{0.3}$

3. a. $\sqrt{t}$ **b.** $\sqrt[4]{5^7}$ **c.** $\sqrt[6]{6^{11}}$ **d.** $\sqrt[7]{x^6}$

4. a. $\sqrt[6]{x^7}$ b. $\sqrt[5]{w^{10}}$ c. $\sqrt[10]{w^5}$ d. $\sqrt[x]{11^n}$

5. **WE17** Evaluate each of the following without using a calculator.

     a. $16^{\frac{1}{2}}$ b. $25^{\frac{1}{2}}$ c. $81^{\frac{1}{2}}$

6. Evaluate each of the following without using a calculator.

     a. $8^{\frac{1}{3}}$ b. $64^{\frac{1}{3}}$ c. $81^{\frac{1}{4}}$

WE18 For questions **7** to **9**, use a calculator to evaluate each of the following, correct to 1 decimal place.

7. a. $5^{\frac{1}{2}}$ b. $7^{\frac{1}{5}}$ c. $8^{\frac{1}{9}}$

8. a. $12^{\frac{3}{8}}$ b. $100^{\frac{5}{9}}$ c. $50^{\frac{2}{3}}$

9. a. $(0.6)^{\frac{4}{5}}$ b. $\left(\dfrac{3}{4}\right)^{\frac{3}{4}}$ c. $\left(\dfrac{4}{5}\right)^{\frac{2}{3}}$

For questions **10** to **19**, simplify each of the expressions.

10. a. $4^{\frac{3}{5}} \times 4^{\frac{1}{5}}$ b. $2^{\frac{1}{8}} \times 2^{\frac{3}{8}}$ c. $a^{\frac{1}{2}} \times a^{\frac{1}{3}}$

11. a. $x^{\frac{3}{4}} \times x^{\frac{2}{5}}$ b. $5m^{\frac{1}{3}} \times 2m^{\frac{1}{5}}$ c. $\dfrac{1}{2}b^{\frac{3}{7}} \times 4b^{\frac{2}{7}}$

12. a. $-4y^2 \times y^{\frac{2}{9}}$ b. $\dfrac{2}{5}a^{\frac{3}{8}} \times 0.05a^{\frac{3}{4}}$ c. $5x^3 \times x^{\frac{1}{2}}$

13. a. $a^{\frac{2}{3}}b^{\frac{3}{4}} \times a^{\frac{1}{3}}b^{\frac{3}{4}}$ b. $x^{\frac{3}{5}}y^{\frac{2}{9}} \times x^{\frac{1}{5}}y^{\frac{1}{3}}$ c. $2ab^{\frac{1}{3}} \times 3a^{\frac{3}{5}}b^{\frac{4}{5}}$

14. a. $6m^{\frac{3}{7}} \times \dfrac{1}{3}m^{\frac{1}{4}}n^{\frac{2}{5}}$ b. $x^3y^{\frac{1}{2}}z^{\frac{1}{3}} \times x^{\frac{1}{6}}y^{\frac{1}{3}}z^{\frac{1}{2}}$ c. $2a^{\frac{2}{5}}b^{\frac{3}{8}}c^{\frac{1}{4}} \times 4b^{\frac{3}{4}}c^{\frac{3}{4}}$

15. a. $x^3y^2 \div x^{\frac{4}{3}}y^{\frac{3}{5}}$ b. $a^{\frac{5}{9}}b^{\frac{2}{3}} \div a^{\frac{2}{5}}b^{\frac{2}{5}}$ c. $m^{\frac{3}{8}}n^{\frac{4}{7}} \div 3n^{\frac{3}{8}}$

16. a. $10x^{\frac{4}{5}}y \div 5x^{\frac{2}{3}}y^{\frac{1}{4}}$ b. $\dfrac{5a^{\frac{3}{4}}b^{\frac{3}{5}}}{20a^{\frac{1}{5}}b^{\frac{1}{4}}}$ c. $\dfrac{p^{\frac{7}{8}}q^{\frac{1}{4}}}{7p^{\frac{2}{3}}q^{\frac{1}{6}}}$

17. a. $\left(2^{\frac{3}{4}}\right)^{\frac{3}{5}}$ b. $\left(5^{\frac{2}{3}}\right)^{\frac{1}{4}}$ c. $\left(7^{\frac{1}{5}}\right)^6$

18. a. $\left(a^3\right)^{\frac{1}{10}}$ b. $\left(m^{\frac{4}{9}}\right)^{\frac{3}{8}}$ c. $\left(2b^{\frac{1}{2}}\right)^{\frac{1}{3}}$

19. a. $4\left(p^{\frac{3}{7}}\right)^{\frac{14}{15}}$ b. $\left(x^{\frac{m}{n}}\right)^{\frac{n}{p}}$ c. $\left(3m^{\frac{a}{b}}\right)^{\frac{b}{c}}$

Understanding

WE19 For questions **20** to **22**, simplify each of the following.

20. a. $\left(a^{\frac{1}{2}}b^{\frac{1}{3}}\right)^{\frac{1}{2}}$
    b. $\left(a^4 b\right)^{\frac{3}{4}}$
    c. $\left(x^{\frac{3}{5}}y^{\frac{7}{8}}\right)^{2}$

21. a. $\left(3a^{\frac{1}{3}}b^{\frac{3}{5}}c^{\frac{3}{4}}\right)^{\frac{1}{3}}$
    b. $5\left(x^{\frac{1}{2}}y^{\frac{2}{3}}z^{\frac{2}{5}}\right)^{\frac{1}{2}}$
    c. $\left(\dfrac{a^{\frac{3}{4}}}{b}\right)^{\frac{2}{3}}$

22. a. $\left(\dfrac{m^{\frac{4}{5}}}{n^{\frac{7}{8}}}\right)^{2}$
    b. $\left(\dfrac{b^{\frac{3}{5}}}{c^{\frac{4}{9}}}\right)^{\frac{2}{3}}$
    c. $\left(\dfrac{4x^7}{2y^{\frac{3}{4}}}\right)^{\frac{1}{2}}$

23. **MC** a. $y^{\frac{2}{5}}$ is equal to:

    A. $\left(y^{\frac{1}{2}}\right)^{5}$
    B. $y \times \dfrac{2}{5}$
    C. $\left(y^5\right)^{\frac{1}{2}}$
    D. $\left(y^{\frac{1}{5}}\right)^{2}$

    b. $k^{\frac{2}{3}}$ is not equal to:

    A. $\left(k^{\frac{1}{3}}\right)^{2}$
    B. $\sqrt[3]{k^2}$
    C. $\left(k^{\frac{1}{2}}\right)^{3}$
    D. $\left(\sqrt[3]{k}\right)^{2}$

    c. $\dfrac{1}{\sqrt[5]{g^2}}$ is equal to:

    A. $g^{\frac{2}{5}}$
    B. $g^{-\frac{2}{5}}$
    C. $g^{\frac{5}{2}}$
    D. $g^{-\frac{5}{2}}$

24. **MC** a. If $\left(a^{\frac{3}{4}}\right)^{\frac{m}{n}}$ is equal to $a^{\frac{1}{4}}$, then m and n could not be:

    A. 1 and 3
    B. 3 and 8
    C. 4 and 9
    D. both **B** and **C**

    b. When simplified, $\left(\dfrac{a^{\frac{m}{n}}}{b^{\frac{n}{p}}}\right)^{\frac{p}{m}}$ is equal to:

    A. $\dfrac{a^{\frac{m}{p}}}{b^{\frac{n}{m}}}$
    B. $\dfrac{a^{\frac{p}{n}}}{b^{\frac{n}{m}}}$
    C. $\dfrac{a^{\frac{mp}{n}}}{b^{\frac{n}{m}}}$
    D. $\dfrac{a^{p}}{b^{m}}$

25. Simplify each of the following.
    a. $\sqrt{a^8}$
    b. $\sqrt[3]{b^9}$
    c. $\sqrt[4]{m^{16}}$

26. Simplify each of the following.
    a. $\sqrt{16x^4}$
    b. $\sqrt[3]{8y^9}$
    c. $\sqrt[4]{16x^8y^{12}}$

27. Simplify each of the following.
    a. $\sqrt[3]{27m^9n^{15}}$
    b. $\sqrt[5]{32p^5q^{10}}$
    c. $\sqrt[3]{216a^6b^{18}}$

Communicating, reasoning and problem solving

28. The relationship between the length of a pendulum (L) in a grandfather clock and the time it takes to complete one swing (T) in seconds is given by the following rule. Note that g is the acceleration due to gravity and will be taken as 9.8.

$$T = 2\pi \left(\frac{L}{g} \right)^{\frac{1}{2}}$$

   a. Calculate the time it takes a 1 m long pendulum to complete one swing.
   b. Determine the time it takes the pendulum to complete 10 swings.
   c. Determine how many swings will be completed after 10 seconds.

29. Using the index laws, show that $\sqrt[5]{32a^5b^{10}} = 2ab^2$.

30. Simplify:

$$\sqrt[5]{\frac{t^2}{\sqrt{t^3}}}$$

31. Expand $\left(m^{\frac{3}{4}} + m^{\frac{1}{2}}n^{\frac{1}{2}} + m^{\frac{1}{4}}n + n^{\frac{3}{2}} \right) \left(m^{\frac{1}{4}} - n^{\frac{1}{2}} \right)$.

32. A scientist has discovered a piece of paper with a complex formula written on it. She thinks that someone has tried to disguise a simpler formula. The formula is:

$$\frac{\sqrt[4]{a^{13}}a^2\sqrt{b^3}}{\sqrt{a^1 b}} \times b^3 \times \left(\frac{\sqrt{a^3 b}}{ab^2} \right)^2 \times \left(\frac{b^2}{a^2\sqrt{b}} \right)^3$$

   a. Simplify the formula using index laws so that it can be worked with.
   b. From your simplified formula, can a take a negative value? Explain.
   c. Evaluate the smallest value for a for which the expression will give a rational answer. Consider only integers.

LESSON
2.8 Combining index laws (Path)

LEARNING INTENTION

At the end of this lesson you should be able to:
• simplify algebraic expressions involving brackets, fractions, multiplication and division using appropriate index laws.

▶ 2.8.1 Combining index laws

eles-4672

• When it is clear that multiple steps are required to simplify an expression, expand brackets first.
• When fractions are involved, it is usually easier to carry out all multiplications first, leaving division as the final process.

Summary of index laws

$$a^m \times a^n = a^{m+n}$$

$$a^m \div a^n = a^{m-n}$$

$$a^0 = 1, a \neq 0$$

$$\left(a^m\right)^n = a^{mn}$$

$$(ab)^m = a^m b^m$$

$$\left(\frac{a}{b}\right)^m = \frac{a^m}{b^m}$$

$$a^{-n} = \frac{1}{a^n}$$

$$a^{\frac{m}{n}} = \sqrt[n]{a^m} = \left(\sqrt[n]{a}\right)^m$$

- Make sure to simplify terms to a common base, before attempting to apply the index laws.
  For example: $5^{2x} \times 25^3 = 5^{2x} \times \left(5^2\right)^3 = 5^{2x} \times 5^6 = 5^{2x+6}$.
- Finally, write the answer with positive indices and variables in alphabetical order, as is convention.

WORKED EXAMPLE 20 Simplifying expressions in multiple steps

Simplify each of the following.

a. $\dfrac{(2a)^4 b^4}{6a^3 b^2}$

b. $\dfrac{3^{n-2} \times 9^{n+1}}{81^{n-1}}$

THINK	WRITE
a. 1. Write the expression.	a. $\dfrac{(2a)^4 b^4}{6a^3 b^2}$
2. Apply the Fourth Index Law to remove the bracket.	$= \dfrac{16a^4 b^4}{6a^3 b^2}$
3. Apply the Second Index Law for each number and pronumeral to simplify.	$= \dfrac{8a^{4-3} b^{4-2}}{3}$
4. Write the answer.	$= \dfrac{8ab^2}{3}$
b. 1. Write the expression.	b. $\dfrac{3^{n-2} \times 9^{n+1}}{81^{n-1}}$
2. Rewrite each term in the expression so that it has a base of 3.	$= \dfrac{3^{n-2} \times \left(3^2\right)^{n+1}}{\left(3^4\right)^{n-1}}$

3. Apply the Fourth Index Law to expand the brackets.

$$= \frac{3^{n-2} \times 3^{2n+2}}{3^{4n-4}}$$

4. Apply the First and Second Index Laws to simplify and write your answer.

$$= \frac{3^{3n}}{3^{4n-4}}$$

$$= \frac{1}{3^{n-4}}$$

WORKED EXAMPLE 21 Simplifying complex expressions involving multiple steps

Simplify each of the following.

a. $(2a^3b)^4 \times 4a^2b^3$

b. $\dfrac{7xy^3}{(3x^3y^2)^2}$

c. $\dfrac{2m^5n \times 3m^7n^4}{7m^3n^3 \times mn^2}$

THINK

a. 1. Write the expression.

2. Apply the Fourth Index Law. Multiply each index inside the brackets by the index outside the brackets.

3. Evaluate the number.

4. Multiply coefficients and multiply pronumerals. Apply the First Index Law to multiply terms with the same base by adding the indices.

b. 1. Write the expression.

2. Apply the Fourth Index Law in the denominator. Multiply each index inside the brackets by the index outside the brackets.

3. Apply the Second Index Law. Divide terms with the same base by subtracting the indices.

4. Use $a^{-m} = \dfrac{1}{a^m}$ to express the answer with positive indices.

c. 1. Write the expression.

2. Simplify each numerator and denominator by multiplying coefficients and then terms with the same base.

WRITE

a. $(2a^3b)^4 \times 4a^2b^3$

$= 2^4a^{12}b^4 \times 4a^2b^3$

$= 16a^{12}b^4 \times 4a^2b^3$

$= 16 \times 4 \times a^{12+2}b^{4+3}$
$= 64a^{14}b^7$

b. $\dfrac{7xy^3}{(3x^3y^2)^2}$

$= \dfrac{7xy^3}{9x^6y^4}$

$= \dfrac{7x^{-5}y^{-1}}{9}$

$= \dfrac{7}{9x^5y}$

c. $\dfrac{2m^5n \times 3m^7n^4}{7m^3n^3 \times mn^2}$

$= \dfrac{6m^{12}n^5}{7m^4n^5}$

3. Apply the Second Index Law. Divide terms with the same base by subtracting the indices.

$$= \frac{6m^8 n^0}{7}$$

4. Simplify the numerator using $a^0 = 1$.

$$= \frac{6m^8 \times 1}{7}$$

$$= \frac{6m^8}{7}$$

WORKED EXAMPLE 22 Simplifying expressions with multiple fractions

Simplify each of the following.

a. $\dfrac{\left(5a^2 b^3\right)^2}{a^{10}} \times \dfrac{a^2 b^5}{\left(a^3 b\right)^7}$

b. $\dfrac{8m^3 n^{-4}}{\left(6mn^2\right)^3} \div \dfrac{4m^{-2} n^{-4}}{6m^{-5} n}$

THINK

WRITE

a. 1. Write the expression.

a. $\dfrac{\left(5a^2 b^3\right)^2}{a^{10}} \times \dfrac{a^2 b^5}{\left(a^3 b\right)^7}$

2. Remove the brackets in the numerator of the first fraction and in the denominator of the second fraction.

$$= \frac{25a^4 b^6}{a^{10}} \times \frac{a^2 b^5}{a^{21} b^7}$$

3. Multiply the numerators and then multiply the denominators of the fractions. (Simplify across.)

$$= \frac{25a^6 b^{11}}{a^{31} b^7}$$

4. Divide terms with the same base by subtracting the indices. (Simplify down.)

$$= 25a^{-25} b^4$$

5. Express the answer with positive indices.

$$= \frac{25b^4}{a^{25}}$$

b. 1. Write the expression.

b. $\dfrac{8m^3 n^{-4}}{\left(6mn^2\right)^3} \div \dfrac{4m^{-2} n^{-4}}{6m^{-5} n}$

2. Remove the brackets.

$$= \frac{8m^3 n^{-4}}{216m^3 n^6} \div \frac{4m^{-2} n^{-4}}{6m^{-5} n}$$

3. Multiply by the reciprocal.

$$= \frac{8m^3 n^{-4}}{216m^3 n^6} \times \frac{6m^{-5} n}{4m^{-2} n^{-4}}$$

4. Multiply the numerators and then multiply the denominators. (Simplify across.)

$$= \frac{48m^{-2} n^{-3}}{864mn^2}$$

5. Cancel common factors and divide pronumerals with the same base. (Simplify down.)

$$= \frac{m^{-3} n^{-5}}{18}$$

6. Simplify and express the answer with positive indices.

$$= \frac{1}{18m^3 n^5}$$

Exercise 2.8 Combining index laws (Path)

learn on

2.8 Quick quiz	2.8 Exercise

Individual pathways

■ PRACTISE	■ CONSOLIDATE	■ MASTER
1, 4, 7, 10, 13, 16, 17, 23	2, 5, 8, 11, 14, 18, 19, 21, 24	3, 6, 9, 12, 15, 20, 22, 25, 26, 27

Fluency

WE21a,b For questions **1** to **3**, simplify each of the following.

1. a. $\left(3a^2b^2\right)^3 \times 2a^4b^3$
   b. $\left(4ab^5\right)^2 \times 3a^3b^6$
   c. $2m^3n^{-5} \times \left(m^2n^{-3}\right)^{-6}$

2. a. $\left(2pq^3\right)^2 \times \left(5p^2q^4\right)^3$
   b. $\left(2a^7b^2\right)^2 \times \left(3a^3b^3\right)^2$
   c. $5\left(b^2c^{-2}\right)^3 \times 3(bc^5)^{-4}$

3. a. $6x^{\frac{1}{2}}y^{\frac{1}{3}} \times \left(4x^{\frac{3}{4}}y^{\frac{4}{5}}\right)^{\frac{1}{2}}$
   b. $\left(16m^3n^4\right)^{\frac{3}{4}} \times \left(m^{\frac{1}{2}}n^{\frac{1}{4}}\right)^3$

   c. $2\left(p^{\frac{2}{3}}q^{\frac{1}{3}}\right)^{-\frac{3}{4}} \times 3\left(p^{\frac{1}{4}}q^{-\frac{3}{4}}\right)^{-\frac{1}{3}}$
   d. $\left(8p^{\frac{1}{5}}q^{\frac{2}{3}}\right)^{-\frac{1}{3}} \times \left(64p^{\frac{1}{3}}q^{\frac{3}{4}}\right)^{\frac{2}{3}}$

WE20 For questions **4** to **6**, simplify each of the following.

4. a. $\dfrac{5a^2b^3}{\left(2a^3b\right)^3}$
   b. $\dfrac{4x^5y^6}{\left(2xy^3\right)^4}$
   c. $\dfrac{\left(3m^2n^3\right)^3}{\left(2m^5n^5\right)^7}$

5. a. $\left(\dfrac{4x^3y^{10}}{2x^7y^4}\right)^6$
   b. $\dfrac{3a^3b^{-5}}{\left(2a^7b^4\right)^{-3}}$
   c. $\left(\dfrac{3g^2h^5}{2g^4h}\right)^3$

6. a. $\dfrac{\left(5p^6q^{\frac{1}{3}}\right)^2}{25\left(p^{\frac{1}{2}}q^{\frac{1}{4}}\right)^{\frac{2}{3}}}$
   b. $\left(\dfrac{3b^2c^3}{5b^{-3}c^{-4}}\right)^{-4}$
   c. $\dfrac{\left(x^{\frac{1}{2}}y^{\frac{1}{4}}z^{\frac{1}{2}}\right)^2}{\left(x^{\frac{2}{3}}y^{-\frac{1}{4}}z^{\frac{1}{3}}\right)^{-\frac{3}{2}}}$

WE21c For questions **7** to **9**, simplify each of the following.

7. a. $\dfrac{2a^2b \times 3a^3b^4}{4a^3b^5}$
   b. $\dfrac{4m^6n^3 \times 12mn^5}{6m^7n^6}$
   c. $\dfrac{10m^6n^5 \times 2m^2n^3}{12m^4n \times 5m^2n^3}$

8. a. $\dfrac{6x^3y^2 \times 4x^6y}{9xy^5 \times 2x^3y^6}$
   b. $\dfrac{\left(6x^3y^2\right)^4}{9x^5y^2 \times 4xy^7}$
   c. $\dfrac{5x^2y^3 \times 2xy^5}{10x^3y^4 \times x^4y^2}$

9. a. $\dfrac{a^3b^2 \times 2(ab^5)^3}{6(a^2b^3)^3 \times a^4b}$

b. $\dfrac{(p^6q^2)^{-3} \times 3pq}{2p^{-4}q^{-2} \times (5pq^4)^{-2}}$

c. $\dfrac{6x^{\frac{3}{2}}y^{\frac{1}{2}} \times x^{\frac{4}{5}}y^{\frac{3}{5}}}{2\left(x^{\frac{1}{2}}y\right)^{\frac{1}{5}} \times 3x^{\frac{1}{2}}y^{\frac{1}{5}}}$

WE22a For questions **10** to **12**, simplify each of the following.

10. a. $\dfrac{a^3b^2}{5a^4b^7} \times \dfrac{2a^6b}{a^9b^3}$

b. $\dfrac{(2a^6)^2}{10a^7b^3} \times \dfrac{4ab^6}{6a^3}$

c. $\dfrac{(m^4n^3)^2}{(m^6n)^4} \times \dfrac{(m^3n^3)^3}{(2mn)^2}$

11. a. $\left(\dfrac{2m^3n^2}{3mn^5}\right)^3 \times \dfrac{6m^2n^4}{4m^3n^{10}}$

b. $\left(\dfrac{2xy^2}{3x^3y^5}\right)^4 \times \left(\dfrac{x^3y^9}{2y^{10}}\right)^2$

c. $\dfrac{4x^{-5}y^{-3}}{(x^2y^2)^{-2}} \times \dfrac{3x^5y^6}{2^{-2}x^{-7}y}$

12. a. $\dfrac{5p^6q^{-5}}{3q^{-4}} \times \left(\dfrac{5p^6q^4}{3p^5}\right)^{-2}$

b. $\dfrac{2a^{\frac{1}{2}}b^{\frac{1}{3}}}{6a^{\frac{1}{3}}b^{\frac{1}{2}}} \times \dfrac{\left(4a^{\frac{1}{4}}b\right)^{\frac{1}{2}}}{b^{\frac{1}{4}}a}$

c. $\dfrac{3x^{\frac{2}{3}}y^{\frac{1}{5}}}{9x^{\frac{1}{3}}y^{\frac{1}{4}}} \times \dfrac{4x^{\frac{1}{2}}}{x^{\frac{3}{4}}y}$

WE22b For questions **13** to **15**, simplify each of the following.

13. a. $\dfrac{5a^2b^3}{6a^7b^5} \div \dfrac{a^9b^4}{3ab^6}$

b. $\dfrac{7a^2b^4}{3a^6b^7} \div \left(\dfrac{3ab}{2a^6b^4}\right)^3$

14. a. $\left(\dfrac{4a^9}{b^6}\right)^3 \div \left(\dfrac{3a^7}{2b^5}\right)^4$

b. $\dfrac{5x^2y^6}{(2x^4y^5)^2} \div \dfrac{(4x^6y)^3}{10xy^3}$

c. $\left(\dfrac{x^5y^{-3}}{2xy^5}\right)^{-4} \div \dfrac{4x^6y^{-10}}{(3x^{-2}y^2)^{-3}}$

15. a. $\dfrac{3m^3n^4}{2m^{-6}n^{-5}} \div \left(\dfrac{2m^4n^6}{m^{-1}n}\right)^{-2}$

b. $4m^{\frac{1}{2}}n^{\frac{3}{4}} \div \dfrac{6m^{\frac{1}{3}}n^{\frac{1}{4}}}{8m^{\frac{3}{4}}n^{\frac{1}{2}}}$

c. $\left(\dfrac{4b^3c^{\frac{1}{3}}}{6c^{\frac{1}{5}}b}\right)^{\frac{1}{2}} \div (2b^3c^{-\frac{1}{5}})^{-\frac{3}{2}}$

Understanding

16. Evaluate each of the following.

a. $(5^2 \times 2)^0 \times (5^{-3} \times 2^0)^5 \div (5^6 \times 2^{-1})^{-3}$

b. $(2^3 \times 3^3)^{-2} \div \dfrac{(2^6 \times 3^9)^0}{2^6 \times (3^{-2})^{-3}}$

17. Evaluate the following for $x = 8$. (*Hint:* Simplify first.)

$(2x)^{-3} \times \left(\dfrac{x}{2}\right)^2 \div \dfrac{2x}{(2^3)^4}$

18. a. Simplify the following fraction: $\dfrac{a^{2y} \times 9b^y \times (5ab)^y}{(a^y)^3 \times 5(3b^y)^2}$

b. Determine the value of y if the fraction is equal to 125.

19. **MC** Select which of the following is not the same as $(4xy)^{\frac{3}{2}}$.

A. $8x^{\frac{3}{2}}y^{\frac{3}{2}}$

B. $(\sqrt{4xy})^3$

C. $\sqrt{64x^3y^3}$

D. $4xy^{\frac{1}{2}} \times (2xy^2)^{\frac{1}{2}}$

20. **MC** The expression $\dfrac{x^2y}{(2xy^2)^3} \div \dfrac{xy}{16x^0}$ is equal to:

A. $\dfrac{2}{x^2y^6}$ 　　　　　 B. $\dfrac{2x^2}{b^6}$ 　　　　　 C. $2x^2y^6$ 　　　　　 D. $\dfrac{2}{xy^6}$

21. Simplify the following.

　a. $\sqrt[3]{m^2n} \div \sqrt{mn^3}$ 　　　　　 b. $(g^{-2}h)^3 \times \left(\dfrac{1}{n^{-3}}\right)^{\frac{1}{2}}$ 　　　　　 c. $\dfrac{45^{\frac{1}{3}}}{9^{\frac{3}{4}} \times 15^{\frac{3}{2}}}$

22. Simplify the following.

　a. $2^{\frac{3}{2}} \times 4^{-\frac{1}{4}} \times 16^{-\frac{3}{4}}$ 　　　　　 b. $\left(\dfrac{a^3b^{-2}}{3^{-3}b^{-3}}\right)^{-2} \div \left(\dfrac{3^{-3}a^{-2}b}{a^4b^{-2}}\right)^2$ 　　c. $\left(\sqrt[5]{d^2}\right)^{\frac{3}{2}} \times \left(\sqrt[3]{d^5}\right)^{\frac{1}{5}}$

Communicating, reasoning and problem solving

23. The population of the number of bacteria on a petri dish is modelled by $N = 6 \times 2^{t+1}$, where N is the number of bacteria after t days.

　a. Determine the initial number of bacteria.
　b. Determine the number of bacteria after one week.
　c. Calculate when the number of bacteria will first exceed 100 000.

24. In a controlled breeding program at the Melbourne Zoo, the population (P) of koalas at t years is modelled by $P = P_0 \times 10^{kt}$. Given $P_0 = 20$ and $k = 0.3$:

　a. Evaluate the number of koalas after 2 years.
　b. Determine when the population will be equal to 1000. Show full working.

25. The decay of uranium is modelled by $D = D_0 \times 2^{-kt}$. It takes 6 years for the mass of uranium to halve. Giving your answers to the nearest whole number, determine the percentage remaining after:

　a. 2 years 　　　　　 b. 5 years 　　　　　 c. 10 years.

26. Solve the following for x: $2^{2x+2} - 2^{2x-1} - 28 = 0$.

27. Simplify $\dfrac{7^{2x+1} - 7^{2x-1} - 48}{36 \times 7^{2x} - 252}$.

LESSON
2.9 Review

2.9.1 Topic summary

Number sets (Path)

- Natural numbers: $N = \{1, 2, 3, 4, 5 \ldots\}$
- Integers: $Z = \{\ldots, -2, -1, 0, 1, 2 \ldots\}$
- Rational numbers: $Q = \left\{\dfrac{1}{4}, -0.36, 2, \dfrac{9}{7}, 0.\dot{3}\dot{6}\right\}$
- Irrational numbers: $I = \{\sqrt{2}, \pi, e\}$
- Real numbers: $R = Q + I$

Index notation

- Index notation is a short way of writing a repeated multiplication.

  e.g. $2 \times 2 \times 2 \times 2 \times 2 \times 2$ can be written as 2^6, which is read '2 to the power of 6'.
- The **base** is the number that is being repeatedly multiplied and the **index** is the number of times it is multiplied.

  e.g. $2^6 = 2 \times 2 \times 2 \times 2 \times 2 \times 2 = 64$

Surds (Path)

- A surd is any number that requires a $\sqrt{x}$ or $\sqrt[n]{x}$ symbol and does not simplify to a whole number.

  e.g. $\sqrt{2}$ is a surd, but $\sqrt{9} = 3$ is not.
- To simplify a surd look for the highest square factor.

  e.g. $\sqrt{48} = \sqrt{16 \times 3} = 4\sqrt{3}$
- Only like surds can be added and subtracted.

  e.g. $\sqrt{5}$, $3\sqrt{5}$ and $-6\sqrt{5}$ are like surds whereas $\sqrt{7}$ and $2\sqrt{11}$ are not like surds.
- Surds are added and subtracted the same way like terms are combined in algebra. You may need to simplify first.

  e.g. $3\sqrt{2} + 7\sqrt{2} - 2\sqrt{2} = 10\sqrt{2} - 2\sqrt{2} = 8\sqrt{2}$

  e.g. $\sqrt{12} + \sqrt{75} = 2\sqrt{3} + 5\sqrt{3} = 7\sqrt{3}$

INDICES AND SURDS

Index laws

- 1st law: $a^m \times a^n = a^{m+n}$
- 2nd law: $a^m \div a^n = a^{m-n}$
- 3rd law: $a^0 = 1, a \neq 0$
- 4th law: $(a^m)^n = a^{m \times n} = a^{mn}$
- 5th law: $(ab)^n = a^n b^n$
- 6th law: $\left(\dfrac{a}{b}\right)^n = \dfrac{a^n}{b^n}$
- 7th law: $a^{-n} = \dfrac{1}{a^n}$
- 8th law: $a^{\frac{1}{n}} = \sqrt[n]{a}$

Multiplying and dividing surds (Path)

- $\sqrt{a} \times \sqrt{b} = \sqrt{ab}$
- $m\sqrt{a} \times n\sqrt{b} = mn\sqrt{ab}$
- $\sqrt{a} \div \sqrt{b} = \dfrac{\sqrt{a}}{\sqrt{b}} = \sqrt{\dfrac{a}{b}}$
- $m\sqrt{a} \div (n\sqrt{b}) = \dfrac{m\sqrt{a}}{n\sqrt{b}} = \dfrac{m}{n}\sqrt{\dfrac{a}{b}}$

Rationalising the denominator (Path)

- Involves re-writing a fraction with a rational denominator.

  e.g. $\dfrac{2}{\sqrt{5}} = \dfrac{2}{\sqrt{5}} \times \dfrac{\sqrt{5}}{\sqrt{5}} = \dfrac{2\sqrt{5}}{5}$

2.9.2 Project

Digital world: 'A bit of this and a byte of that'

'The digital world of today is run by ones and zeros.' What does this mean? Data is represented on a modern digital computer using a base two (binary) system, that is, using the two digits 1 and 0, thought of as 'on' and 'off'. The smallest unit of data that is transferred on a computer is a **bit** (an abbreviation of binary digit). Computer and storage mechanisms need to hold much larger values than a bit. Units such as bytes, kilobytes (KB), megabytes (MB), gigabytes (GB), and terabytes (TB) are based on the conversion of 8 bits to 1 byte. Your text messages, graphics, music and photos are files stored in sequences of bytes, each byte being 8 bits ($8b = 1B$).

You may have heard the terms 'meg' and 'gig'. In computer terminology, these refer to gigabytes and megabytes. In the digital world, the prefixes kilo-, mega- and giga- express powers of two, where kilo- means 2^{10}, mega- means $(2^{10})^2$ and so on. Thus the number of bytes in a computer's memory builds in powers of 2, for example 1 kilobyte = 1024 bytes(210 bytes). (This differs from the decimal system, in which the prefixes kilo-, mega- and giga- express powers of ten, with kilo- meaning 10^3, mega- meaning $(10^3)^2$ and so on.)

A byte (8 bits) is used to represent a single character. For example the letter 'A' is represented in binary as 01000001. A book of a thousand pages in print can be stored in millions of bits, but more commonly it would be described as being stored in megabytes with one byte per character.

1. Complete the table below to show the difference in value between the binary and decimal

Unit	Symbol	Power of 2 and value in bytes	Power of 10 and value in bytes
Byte	B	$2^0 = 1$	$10^0 = 1$
Kilobyte	KB	$2^{10} = 1024$	$10^3 = 1000$
Megabyte		$2^{20} =$	
Gigabyte			
Terabyte			

2. The two numbering systems have led to some confusion, with some manufacturers of digital products thinking of a kilobyte as 1000 bytes rather than 1024 bytes. Similar confusion arises with megabytes, gigabytes, terabytes and so on. This means you might not be getting exactly the amount of storage that you think.

   If you bought a device quoted as having 16 GB memory, what would be the difference in memory storage if the device had been manufactured using the decimal value of GB as opposed to the binary system?

Many devices allow you to check the availability of storage. On one such device, the iPhone, available storage is found by going to 'General' under the heading 'Settings'.

3. How much storage is left in MB on the following iPhone?
4. If each photo uses 3.2 MB of memory, how many photos can be added?

Have you ever wondered about the capacity of our brain to store information and the speed at which information is transmitted inside it?

5. Discuss how the storage and speed of our brains compares to our current ability to send and store information in the digital world. The capacity of the human brain is 10–100 terabytes. On average, 20 million billion bits of information are transmitted within the brain per second.
6. Investigate which country has the fastest internet speed and compare this to Australia.

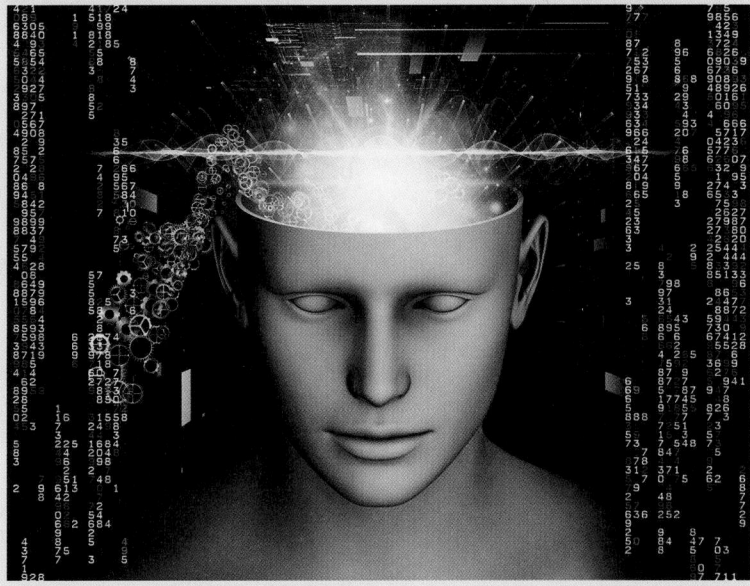

 Resources

Interactivities Crossword (int-2872)
              Sudoku puzzle (int-3891)

Fluency

1. **MC** Identify which of the given numbers are rational.

$$\sqrt{\frac{6}{12}}, \ \sqrt{0.81}, \ 5, \ -3.26, \ 0.5, \ \frac{\pi}{5}, \ \sqrt{\frac{3}{12}}$$

A. $\sqrt{0.81}, 5, -3.26, 0.5$ and $\sqrt{\frac{3}{12}}$

B. $\sqrt{\frac{6}{12}}$ and $\frac{\pi}{5}$

C. $\sqrt{\frac{6}{12}}, \sqrt{0.81}$ and $\sqrt{\frac{3}{12}}$

D. $5, -3.26$ and $\sqrt{\frac{6}{12}}$

2. For each of the following, state whether the number is rational or irrational and give the reason for your answer:

a. $\sqrt{12}$ b. $\sqrt{121}$ c. $\frac{2}{9}$ d. $0.\dot{6}$ e. $\sqrt[3]{0.08}$

3. **MC** **PATH** Identify which of the numbers of the given set are surds.

$$\left\{ 3\sqrt{2}, 5\sqrt{7}, 9\sqrt{4}, 6\sqrt{10}, 7\sqrt{12}, 12\sqrt{64} \right\}$$

A. $9\sqrt{4}, 12\sqrt{64}$

B. $3\sqrt{2}$ and $7\sqrt{12}$ only

C. $3\sqrt{2}, 5\sqrt{7}$ and $6\sqrt{10}$ only

D. $3\sqrt{2}, 5\sqrt{7}, 6\sqrt{10}$ and $7\sqrt{12}$

4. **PATH** Identify which of $\sqrt{2m}, \sqrt{25m}, \sqrt{\frac{m}{16}}, \sqrt{\frac{m}{20}}, \sqrt[3]{m}, \sqrt[3]{8m}$ are surds:

a. if $m = 4$

b. if $m = 8$

5. **PATH** Simplify each of the following.

a. $\sqrt{50}$ b. $\sqrt{180}$ c. $2\sqrt{32}$ d. $5\sqrt{80}$

6. **MC** **PATH** The expression $\sqrt{392x^8y^7}$ may be simplified to:

A. $196x^4y^3\sqrt{2y}$

B. $2x^4y^3\sqrt{14y}$

C. $14x^4y^3\sqrt{2y}$

D. $14x^4y^3\sqrt{2}$

7. **PATH** Simplify the following surds. Give the answers in the simplest form.

a. $4\sqrt{648x^7y^9}$

b. $-\frac{2}{5}\sqrt{\frac{25}{64}x^5y^{11}}$

8. **PATH** Simplify the following, giving answers in the simplest form.

a. $7\sqrt{12} + 8\sqrt{147} - 15\sqrt{27}$

b. $\frac{1}{2}\sqrt{64a^3b^3} - \frac{3}{4}ab\sqrt{16ab} + \frac{1}{5ab}\sqrt{100a^5b^5}$

9. **PATH** Simplify each of the following.

   a. $\sqrt{3} \times \sqrt{5}$
   b. $2\sqrt{6} \times 3\sqrt{7}$
   c. $3\sqrt{10} \times 5\sqrt{6}$
   d. $\left(\sqrt{5}\right)^2$

10. **PATH** Simplify the following, giving answers in the simplest form.

   a. $\dfrac{1}{5}\sqrt{675} \times \sqrt{27}$
   b. $10\sqrt{24} \times 6\sqrt{12}$

11. **PATH** Simplify the following.

   a. $\dfrac{\sqrt{30}}{\sqrt{10}}$
   b. $\dfrac{6\sqrt{45}}{3\sqrt{5}}$
   c. $\dfrac{3\sqrt{20}}{12\sqrt{6}}$
   d. $\dfrac{\left(\sqrt{7}\right)^2}{14}$

12. **PATH** Rationalise the denominator of each of the following.

   a. $\dfrac{2}{\sqrt{6}}$
   b. $\dfrac{\sqrt{3}}{2\sqrt{6}}$
   c. $\dfrac{2\sqrt{3}}{3\sqrt{6}}$
   d. $\dfrac{\sqrt{8} + \sqrt{12}}{\sqrt{2}}$

13. **PATH** Evaluate each of the following, correct to 1 decimal place if necessary.

   a. $64^{\frac{1}{3}}$
   b. $20^{\frac{1}{2}}$
   c. $10^{\frac{1}{3}}$
   d. $50^{\frac{1}{4}}$

14. **PATH** Evaluate each of the following, correct to 1 decimal place.

   a. $20^{\frac{2}{3}}$
   b. $2^{\frac{3}{4}}$
   c. $(0.7)^{\frac{3}{5}}$
   d. $\left(\dfrac{2}{3}\right)^{\frac{2}{3}}$

15. **PATH** Write each of the following in simplest surd form.

   a. $2^{\frac{1}{2}}$
   b. $18^{\frac{1}{2}}$
   c. $5^{\frac{3}{2}}$
   d. $8^{\frac{4}{3}}$

16. **PATH** Evaluate each of the following, without using a calculator. Show all working.

   a. $\dfrac{16^{\frac{3}{4}} \times 81^{\frac{1}{4}}}{6 \times 16^{\frac{1}{2}}}$
   b. $\left(125^{\frac{2}{3}} - 27^{\frac{2}{3}}\right)^{\frac{1}{2}}$

17. Evaluate each of the following, giving your answer as a fraction.

   a. 4^{-1}
   b. 9^{-1}
   c. 4^{-2}
   d. 10^{-3}

18. Determine the value of each of the following, correct to 3 significant figures.

   a. 12^{-1}
   b. 7^{-2}
   c. $(1.25)^{-1}$
   d. $(0.2)^{-4}$

19. Write down the value of each of the following.

   a. $\left(\dfrac{2}{3}\right)^{-1}$
   b. $\left(\dfrac{7}{10}\right)^{-1}$
   c. $\left(\dfrac{1}{5}\right)^{-1}$
   d. $\left(3\dfrac{1}{4}\right)^{-1}$

20. **MC** **PATH** a. The expression $\sqrt{250}$ may be simplified to:

   A. $25\sqrt{10}$ B. $5\sqrt{10}$ C. $10\sqrt{5}$ D. $5\sqrt{50}$

   b. When expressed in its simplest form, $2\sqrt{98} - 3\sqrt{72}$ is equal to:

   A. $-4\sqrt{2}$ B. -4 C. $-2\sqrt{4}$ D. $4\sqrt{2}$

   c. When expressed in its simplest form, $\sqrt{\dfrac{8x^3}{32}}$ is equal to:

   A. $\dfrac{x\sqrt{x}}{2}$ B. $\dfrac{\sqrt{x^3}}{4}$ C. $\dfrac{\sqrt{x^3}}{2}$ D. $\dfrac{x\sqrt{x}}{4}$

21. Determine the value of the following, giving your answer in fraction form.

   a. $\left(\dfrac{2}{5}\right)^{-1}$ b. $\left(\dfrac{2}{3}\right)^{-2}$

22. Determine the value of each of the following, leaving your answer in fraction form.

   a. 2^{-1} b. 3^{-2}

   c. 4^{-3} d. $\left(\dfrac{1}{2}\right)^{-1}$

Understanding

23. **MC** $3d^{10}e^4$ is the simplified form of:

   A. $d^6e^2 \times 3d^4e^3$ B. $\dfrac{6d^{10}e^5}{2e^2}$ C. $\left(3d^5e^2\right)^2$ D. $3e\left(d^5\right)^2 \times e^3$

24. **MC** $8m^3n \times n^4 \times 2m^2n^3$ simplifies to:

   A. $10m^5n^8$ B. $16m^5n^7$ C. $16m^5n^8$ D. $10m^5n^7$

25. **MC** **PATH** $8x^3 \div (4x^{-3})$ is equal to:

   A. 2 B. $2x^0$ C. $2x^6$ D. $2x^{-1}$

26. **MC** $\dfrac{12x^8 \times 2x^7}{6x^9 \times x^5}$ simplifies to:

   A. $4x^5$ B. $8x$ C. $4x$ D. $8x^5$

27. **MC** The expression $\dfrac{\left(a^2b^3\right)^5}{\left(2a^2b\right)^2}$ is equal to:

   A. $\dfrac{a^6b^{13}}{4}$ B. $2a^6b^{13}$ C. $\dfrac{a^3b^6}{2}$ D. $\dfrac{a^6b^{13}}{2}$

28. **MC** $\dfrac{\left(p^2q\right)^4}{\left(2p^5q^2\right)^3} \div \dfrac{\left(p^5q^2\right)^2}{2pq^5}$ can be simplified to:

   A. $\dfrac{1}{4p^{16}q}$ B. $\dfrac{2^2}{p^{16}q}$ C. $\dfrac{1}{4p^8}$ D. $\dfrac{1}{2p^{16}q}$

29. `MC` `PATH` $16^{-\frac{3}{4}} \div 9^{\frac{3}{2}}$ can be simplified to:

A. 2 **B.** $\dfrac{1}{216}$ **C.** $\dfrac{8}{27}$ **D.** $3\dfrac{3}{8}$

30. `MC` `PATH` $\dfrac{\left(2l^{\frac{2}{9}}m^{-1}\right)^{-3}}{8\left(\frac{1}{16}lm^{-2}\right)^{2}}$ can be simplified to:

A. $\dfrac{8m^7}{l^{\frac{11}{3}}}$ **B.** $\dfrac{2m^7}{l^{\frac{7}{3}}}$ **C.** $\dfrac{4m^7}{l^{\frac{8}{3}}}$ **D.** $\dfrac{16m^7}{l^{\frac{5}{3}}}$

31. `MC` `PATH` $\sqrt[5]{32i^{\frac{10}{7}}j^{\frac{5}{11}}k^2}$ can be simplified to:

A. $\dfrac{32i^{\frac{2}{7}}j^{\frac{1}{11}}k^{\frac{2}{5}}}{5}$ **B.** $2i^{\frac{2}{7}}j^{\frac{1}{11}}k^{\frac{2}{5}}$ **C.** $\dfrac{32i^{\frac{10}{7}}j^{\frac{5}{11}}k^2}{5}$ **D.** $2i^{\frac{50}{7}}j^{\frac{25}{11}}k^{10}$

32. Simplify each of the following.

a. $5x^3 \times 3x^5y^4 \times \dfrac{3}{5}x^2y^6$

b. $\dfrac{26a^4b^6c^5}{12a^3b^3c^3}$

c. $\left(\dfrac{20m^5n^2}{6}\right)^3$

d. $\left(\dfrac{14p^7}{21q^3}\right)^4$

33. Evaluate each of the following.

a. $5a^0 - \left(\dfrac{2a}{3}\right)^0 + 12$

b. $-(3b)^0 - \dfrac{(4b)^0}{2}$

34. `PATH` Simplify each of the following and express your answer with positive indices.

a. $2a^{-5}b^2 \times 4a^{-6}b^{-4}$

b. $4x^{-5}y^{-3} \div 20x^{12}y^{-5}$

c. $\left(2m^{-3}n^2\right)^{-4}$

35. Evaluate each of the following without using a calculator.

a. $\left(\dfrac{1}{2}\right)^{-3}$

b. $2 \times (3)^{-3} \times \left(\dfrac{9}{2}\right)^2$

c. $4^{-3} \times \dfrac{5}{8^{-2}} - 5$

36. `PATH` Simplify each of the following.

a. $2a^{\frac{4}{5}}b^{\frac{1}{2}} \times 3a^{\frac{1}{2}}b^{\frac{3}{4}} \times 5a^{\frac{3}{4}}b^{\frac{2}{5}}$

b. $\dfrac{4^3x^{\frac{3}{4}}y^{\frac{1}{9}}}{16x^{\frac{4}{5}}y^{\frac{1}{3}}}$

c. $\left(\dfrac{4a^{\frac{1}{3}}}{b^3}\right)^{\frac{1}{2}}$

37. `PATH` Evaluate each of the following without using a calculator. Show all working.

a. $\dfrac{16^{\frac{3}{4}} \times 81^{\frac{1}{4}}}{6 \times 16^{\frac{1}{2}}}$

b. $\left(125^{\frac{2}{3}} - 27^{\frac{2}{3}}\right)^{\frac{1}{2}}$

38. `PATH` Simplify:

a. $\sqrt[3]{a^9} + \sqrt[4]{16a^8b^2} - 3\left(\sqrt[5]{a}\right)^{15}$

b. $\sqrt[5]{32x^5y^{10}} + \sqrt[3]{64x^3y^6}$

39. `PATH` Simplify each of the following.

a. $\dfrac{\left(5a^{-2}b\right)^{-3} \times 4a^6b^{-2}}{2a^2b^3 \times 5^{-2}a^{-3}b^{-6}}$

b. $\dfrac{2x^4y^{-5}}{3y^6x^{-2}} \times \left(\dfrac{4xy^{-2}}{3x^{-6}y^3}\right)^{-3}$

c. $\left(\dfrac{2m^3n^4}{5m^{\frac{1}{2}}n}\right)^{\frac{1}{3}} \div \left(\dfrac{4m^{\frac{1}{3}}n^{-2}}{5^{-\frac{2}{3}}}\right)^{-\frac{1}{2}}$

40. `PATH` Simplify each of the following and then evaluate.

a. $\left(3 \times 5^6\right)^{\frac{1}{2}} \times 3^{\frac{3}{2}} \times 5^{-2} + \left(3^6 \times 5^{-\frac{1}{2}}\right)^0$

b. $\left(6 \times 3^{-2}\right)^{-1} \div \dfrac{\left(3^{\frac{1}{2}} \times 6^{\frac{1}{3}}\right)^6}{-6^2 \times \left(3^{-3}\right)^0}$

Communicating, reasoning and problem solving

41. Answer the following. Explain how you reached your answer.

a. What is the hundred's digit in 3^{3^3}?
b. What is the one's digit in 6^{704}?
c. What is the thousand's digit in 9^{1000}?

42. `PATH` Simplify $\left(\left(\dfrac{\left(a^2\right)^{-1}}{b^{\frac{1}{2}}}\right)^{-1}\right)^{-1}$

43. `PATH` If $m = 2$, determine the value of:

$$\dfrac{6a^{3m} \times 2b^{2m} \times (3ab)^{-m}}{(4b)^m \times \left(9a^{4m}\right)^{\frac{1}{2}}}$$

44. Answer the following and explain your reasoning.

a. Identify the digit in the tens of 3^{3^3}.
b. Identify the digit in the ones of 6^{309}.
c. Identify the digit in the ones of 8^{1007}.

45. For the work shown below:

  a. calculate the correct answer

  b. identify where the student has made mistakes.

$$\left(\frac{3a^3b^5c^3}{5a^2b}\right)^2 \div \left(\frac{2ab}{c}\right) = \frac{3a^6b^{10}c^6}{10a^4b^2} \div \frac{2ab}{c}$$

$$= \frac{3a^6b^{10}c^6}{10a^4b^2} \times \frac{c}{2ab}$$

$$= \frac{3a^6b^{10}c^7}{20a^5b^3}$$

$$= \frac{3ab^7c^7}{20}$$

46. PATH A friend is trying to calculate the volume of water in a reservoir amid fears there may be a severe water shortage.

She comes up with the following expression:

$$W = \frac{r^4u^2}{r^{\frac{3}{2}}d^2\sqrt{u}} \times \frac{ru \times d^2}{dr^3u^4},$$

where r is the amount of rain, d is how dry the area is, u is the usage of water by the townsfolk, and W is the volume of water in kL.

  a. Help your friend simplify the expression by simplifying each pronumeral one at a time.

  b. Explain whether the final expression contain any potential surds.

  c. Express the fraction with a rational denominator.

  d. List the requirements for the possible values of r, u and d to give a rational answer.

  e. Calculate the volume of water in the reservoir when $r = 4$, $d = 60$ and $u = 9$. Write your answer in:

  i. kL

  ii. L

  iii. mL.

  f. Does a high value for d mean the area is dry? Explain using working.

on To test your understanding and knowledge of this topic, go to your learnON title at www.jacplus.com.au and complete the **post-test**.

Answers

Topic 2 Indices and surds

2.1 Pre-test

1. $16a^4b^{12}$
2. Rational
3. $15\,n^{\frac{8}{15}}$
4. $2p^2q^3$
5. $\dfrac{1}{27}$
6. D
7. B
8. $6\sqrt{5}$
9. $14\sqrt{2}$
10. D
11. False
12. False
13. False
14. $\dfrac{10b}{3}$
15. $\dfrac{1}{4}$

2.2 Rational and irrational numbers (Path)

1. a. Q b. Q c. Q d. I
2. a. I b. Q c. Q d. I
3. a. Q b. Q c. Q d. Q
4. a. I b. Q c. I d. Q
5. a. Q b. I c. I d. I
6. a. Q b. I c. I d. Q
7. a. Q b. Q c. Q d. Q
8. a. Q b. I c. I d. Q
9. a. I b. Undefined
    c. I d. I
10. a. I b. Q c. Q d. Q
11. a. I b. I c. Q d. Q
12. a. I b. Q c. Q d. I
13. B
14. D
15. C
16. C
17. $\pm\dfrac{a}{b}$
18. D
19. A
20. $p - q$
21. Sample responses can be found in the worked solutions in the online resources.
22. $8,\ -8$

23.
a. $m = 11, n = 3$ b. $m = 2, n = 3$
c. $m = 3, n = 2$ d. $m = 1, n = 2$
24. $\dfrac{1}{7}$ or 7^{-1}
25. $\dfrac{37}{91}$

2.3 Surds (Path)

1. b and d
2. b, c and d
3. a and d
4. a and c
5. a, c and d
6. c and f
7. A
8. D
9. B
10. C
11. a perfect square
12. $m = 4$
13. a. $m = 5$, $n = 7$ and $m = 4$, $n = 14$
    b. 15
14. a. Irrational   b. Irrational   c. Rational
15. a. i. $4\sqrt{3}$ ii. $6\sqrt{2}$
    b. Yes. If you don't choose the largest perfect square, then you will need to simplify again.
    c. No
16. $p = m$ and $q = n$
17. Irrational
18. $x = 3$, rational

2.4 Operations with surds (Path)

1. a. $2\sqrt{3}$ b. $2\sqrt{6}$
   c. $3\sqrt{3}$ d. $5\sqrt{5}$
2. a. $3\sqrt{6}$ b. $4\sqrt{7}$
   c. $2\sqrt{17}$ d. $6\sqrt{5}$
3. a. $2\sqrt{22}$ b. $9\sqrt{2}$
   c. $7\sqrt{5}$ d. $8\sqrt{7}$
4. a. $4\sqrt{2}$ b. $24\sqrt{10}$
   c. $36\sqrt{5}$ d. $21\sqrt{6}$
5. a. $-30\sqrt{3}$ b. $-28\sqrt{5}$
   c. $64\sqrt{3}$ d. $2\sqrt{2}$
6. a. $\sqrt{2}$ b. $2\sqrt{3}$
   c. $\dfrac{1}{3}\sqrt{15}$ d. $\dfrac{3}{2}\sqrt{7}$
7. a. $4a$ b. $6a\sqrt{2}$
   c. $3a\sqrt{10b}$ d. $13a^2\sqrt{2}$
8. a. $13ab\sqrt{2ab}$ b. $2ab^2\sqrt{17ab}$
   c. $5x^3y^2\sqrt{5}$ d. $20xy\sqrt{5x}$

9. a. $54c^3d^2\sqrt{2cd}$ b. $18c^3d^4\sqrt{5cd}$
   c. $\sqrt{22ef}$ d. $7e^5f^5\sqrt{2ef}$

10. a. $7\sqrt{5}$ b. $8\sqrt{3}$
   c. $15\sqrt{5}+5\sqrt{3}$ d. $4\sqrt{11}$

11. a. $13\sqrt{2}$ b. $-3\sqrt{6}$
   c. $17\sqrt{3}-18\sqrt{7}$ d. $8\sqrt{x}+3\sqrt{y}$

12. a. $10\left(\sqrt{2}-\sqrt{3}\right)$ b. $5\left(\sqrt{5}+\sqrt{6}\right)$
   c. $7\sqrt{3}$ d. $4\sqrt{5}$

13. a. $14\sqrt{3}+3\sqrt{2}$ b. $3\sqrt{6}+6\sqrt{3}$
   c. $15\sqrt{10}-10\sqrt{15}+10$ d. $-8\sqrt{11}+22$

14. a. $12\sqrt{30}-16\sqrt{15}$ b. $12\sqrt{ab}+7\sqrt{3ab}$

   c. $\dfrac{7}{2}\sqrt{2}+2\sqrt{3}$ d. $15\sqrt{2}$

15. a. $31\sqrt{a}-6\sqrt{2a}$ b. $52\sqrt{a}-29\sqrt{3a}$
   c. $6\sqrt{6ab}$ d. $32a+2\sqrt{6a}+8a\sqrt{2}$

16. a. $a\sqrt{2a}$ b. $\sqrt{a}+2\sqrt{2a}$
   c. $3a\sqrt{a}+a^2\sqrt{3a}$ d. $\left(a^2+a\right)\sqrt{ab}$

17. a. $4ab\sqrt{ab}+3a^2b\sqrt{b}$
   b. $3\sqrt{ab}\left(2a+1\right)$
   c. $-6ab\sqrt{2a}+4a^2b^3\sqrt{3a}$
   d. $-2a\sqrt{b}$

18. a. $\sqrt{14}$ b. $\sqrt{42}$ c. $4\sqrt{3}$
   d. 10 e. $3\sqrt{7}$ f. 27

19. a. $10\sqrt{33}$ b. $180\sqrt{5}$ c. 120
   d. $120\sqrt{3}$ e. $2\sqrt{6}$ f. $2\dfrac{2}{3}$

20. a. $\dfrac{2}{5}\sqrt{6}$ b. $x^2y\sqrt{y}$ c. $3a^4b^2\sqrt{2ab}$

   d. $6a^5b^2\sqrt{2b}$ e. $3x^2y^2\sqrt{10xy}$ f. $\dfrac{9}{2}a^2b^4\sqrt{5ab}$

21. a. 2 b. 5 c. 12

22. a. 15 b. 18 c. 80

23. a. 28 b. 200

24. a. $\sqrt{5}$ b. 2
   c. $\sqrt{6}$ d. 4

25. a. $\dfrac{\sqrt{3}}{4}$ b. $\dfrac{\sqrt{5}}{2}$

   c. $2\sqrt{3}$ d. 1

26. a. $1\dfrac{4}{5}$ b. $2\sqrt{17}$ c. $\dfrac{x}{y}$

   d. $\dfrac{\sqrt{2}}{x^3y^4}$ e. $2xy\sqrt{3y}$ f. $\dfrac{4\sqrt{a}}{3}$

27. a. $\sqrt{10}-2+\sqrt{15}-\sqrt{6}$ b. $15\sqrt{2}-\sqrt{10}-30+2\sqrt{5}$
   c. $64+11\sqrt{10}$ d. $8\sqrt{30}+15\sqrt{6}$
   e. $14+8\sqrt{3}$ f. -55

28. a. $\dfrac{5\sqrt{2}}{2}$ b. $\dfrac{7\sqrt{3}}{3}$ c. $\dfrac{4\sqrt{11}}{11}$

   d. $\dfrac{4\sqrt{6}}{3}$ e. $\dfrac{2\sqrt{21}}{7}$

29. a. $\dfrac{\sqrt{10}}{2}$ b. $\dfrac{2\sqrt{15}}{5}$ c. $\dfrac{3\sqrt{35}}{5}$

   d. $\dfrac{5\sqrt{6}}{6}$ e. $\dfrac{4\sqrt{15}}{15}$

30. a. $\dfrac{5\sqrt{7}}{14}$ b. $\dfrac{8\sqrt{15}}{15}$ c. $\dfrac{8\sqrt{21}}{49}$

   d. $\dfrac{8\sqrt{105}}{7}$ e. $\dfrac{\sqrt{10}}{3}$

31. a. $\sqrt{2}+2$ b. $\dfrac{3\sqrt{10}-2\sqrt{33}}{6}$

   c. $\dfrac{12\sqrt{5}-5\sqrt{6}}{10}$ d. $\dfrac{9\sqrt{10}}{5}$

32. a. $\dfrac{3\sqrt{10}+6\sqrt{14}}{4}$ b. $\dfrac{5\sqrt{6}}{3}$

   c. $\dfrac{3\sqrt{22}-4\sqrt{10}}{6}$ d. $\dfrac{\sqrt{21}-\sqrt{15}}{3}$

33. a. $\dfrac{14-5\sqrt{2}}{6}$ b. $\dfrac{12-\sqrt{10}}{16}$

   c. $\dfrac{6\sqrt{15}-25}{70}$ d. $\dfrac{\sqrt{30}+7\sqrt{2}}{20}$

34. $\dfrac{15\left(3\sqrt{2}-2\right)}{28}$

35. $\dfrac{9\sqrt{x}+6x}{36x-16x^2}$

36. a. Sample responses can be found in the worked solutions in the online resources.
   b. i. $\sqrt{5}+\sqrt{3}$ ii. $\sqrt{5}+\sqrt{3}$ iii. $\sqrt{3}+2$

2.5 Review of index laws

1. a. a^7 b. a^6 c. b^8 d. a^4b^7

2. a. m^5n^{13} b. $a^5b^7c^3$ c. $m^6n^4p^5$ d. $6a^2b$

3. a. $10a^4b^9$ b. $36m^8n^7$ c. $12x^6y^6$ d. $4x^8y^6$

4. a. a b. a^5 c. b^3 d. $\dfrac{4}{3}a^4$

5. a. $3b^4$ b. $4m^5$ c. m^3n d. $\dfrac{1}{2}y^2$

6. a. $7b^3$ b. $\dfrac{5}{4}m^2p^2$ c. $\dfrac{1}{2}xy^2$

7. a. 1 b. 1 c. 1

8. a. 3 b. 4 c. -3

9. a. 3 b. -7 c. 4

10. a. a^6 b. $16a^{20}$ c. $\dfrac{1}{81}m^8$

d. $\dfrac{4}{9}n^8$ e. 49

11. a. a^6b^3 b. $9a^6b^4$ c. $16m^{12}n^{20}$

d. $\dfrac{27}{64}m^6n^3$ e. $\dfrac{a^4}{b^6}$

12. a. $\dfrac{625m^{12}}{n^8}$ b. $\dfrac{343x^3}{8y^{15}}$ c. $\dfrac{81a^4}{625b^{12}}$

d. -243 e. -32

13. a. D   b. D

14. a. C   b. D
c. B   d. D

15. a. 64   b. 72   c. 625

16. a. 48   b. 1600   c. $\dfrac{27}{125}$

17. a. 20   b. 1   c. 4

18. a. x^{3yz} b. a^b

19. a. $m^a n^b$ b. $\dfrac{a^{2x}}{b^{3x}}$

20. a. $n^{3-p}m^{2-q}$ b. a^{mp+np}

21. $\quad a^3 = a \times a \times a$

$\quad a^2 = a \times a$

$a^3 \times a^2 = a \times a \times a \times a \times a$

$\qquad = a^5$, not a^6

Explanations will vary.

22. They are equal when $x = 2$. Explanations will vary.

23. $3x^0 = 3$ and $(3x)^0 = 1$. Explanations will vary.

24. a.

a	0	1	2	3
$3a^2$	0	3	12	27
$5a$	0	5	10	15
$3a^2 + 5a$	0	8	22	42
$3a^2 \times 5a$	0	15	120	405

b. $3a^2 \times 5a$ will become much larger than $3a^2 + 5a$.

25. $x = -2$ or 4

26. $\quad 1 \equiv 1$
$\quad 2 \equiv 10$
$\quad 3 \equiv 11$
$\quad 4 \equiv 100$
$\quad 5 \equiv 101$
$\quad 6 \equiv 110$
$\quad 7 \equiv 111$
$\quad 8 \equiv 1000$
$\quad 9 \equiv 1001$
$\quad 10 \equiv 1010$

27. a. $x = 4$ b. $x = 0, 2$

28. a. a^2bc^7

b. The student made a mistake when multiplying the two brackets in line 3. Individual brackets should be expanded first.

2.6 Negative indices (Path)

1. a. $\dfrac{1}{x^5}$ b. $\dfrac{1}{y^4}$ c. $\dfrac{2}{a^9}$ d. $\dfrac{4}{5a^3}$

2. a. $\dfrac{3x^2}{y^3}$ b. $\dfrac{1}{4m^3n^4}$ c. $\dfrac{6a^3}{bc^5}$ d. a^6

3. a. $\dfrac{2a^4}{3}$ b. $2ab^2$ c. $\dfrac{7b^3}{2a^4}$ d. $\dfrac{2m^3a^2}{3b^4n^5}$

4. a. $\dfrac{1}{a^2b^3}$ b. $\dfrac{6}{x^6y}$ c. $\dfrac{3}{n^8}$

d. $\dfrac{4}{a^2b^5}$ e. $\dfrac{2y}{3x}$

5. a. $\dfrac{5y}{6x^3}$ b. $\dfrac{3}{m^2n^2}$ c. $\dfrac{4y^{12}}{x^5}$

d. $\dfrac{1}{3m^3n^3}$ e. $\dfrac{1}{32a^{15}m^{20}}$

6. a. $\dfrac{4q^8}{p^{14}}$ b. $\dfrac{3}{a^8b^{12}}$ c. $\dfrac{27q^9}{8p^6}$

d. $\dfrac{b^6}{4a^8}$ e. $\dfrac{1}{8a^6b^6}$

7. a. $\dfrac{1}{8}$ b. $\dfrac{1}{36}$ c. $\dfrac{1}{81}$ d. $\dfrac{8}{9}$

8. a. $\dfrac{1}{16}$ b. $\dfrac{5}{36}$ c. 48   d. $\dfrac{32}{27}$

9. a. $\dfrac{27}{25} = 1\dfrac{2}{25}$ b. 4

c. 125   d. $\dfrac{3}{4}$

10. a. 2^3 b. 2^{-3} c. 2^5 d. 2^{-6}

11. a. $x = 3$ b. $x = -2$ c. $x = -1$
d. $x = 3$ e. $x = -2$

12. a. $x = 0$ b. $x = 3$ c. $x = -3$
d. $x = -6$ e. $x = -2$

13. a. $\dfrac{3}{2}$ b. $\dfrac{4}{5}$ c. $\dfrac{2}{7}$ d. 5

14. a. $\dfrac{b}{a}$ b. $\dfrac{b^3}{a^2}$ c. $\dfrac{a^2}{b^3}$ d. $\dfrac{1}{m^3n^2}$

15. a. $\dfrac{1}{729}$

b. $\dfrac{1}{20\,736}$

c. 0.000059499 or $\dfrac{1}{16807}$

16. a. 256   b. $\dfrac{16\,384}{2187}$ c. $9\,765\,625$

17. C

18. B

19. B

20. D

21. C

22. D

23. a. $\dfrac{m^2}{n^8}$ b. $\dfrac{n^2}{m}$ c. $\dfrac{25}{a^7 b^6}$

24. a. $r^6 - s^6$ b. $m^{10} + 2m^5 n^5 + n^{10}$
    c. 1 d. p^2

25. 2^{2r-4}

26. 6^{3m}

27. $x = 3$

28. a. As x gets closer to 0 coming from the positive direction, y gets more and more positive, approaching ∞.

    b. As x gets closer to 0 coming from the negative direction, y gets more and more negative, approaching $-\infty$.

29. $2^{-n} = \dfrac{1}{2^n}$

    A n increases, the value of 2^n increases, so the value of 2^{-n} gets closer to 0.

30. Sample responses can be found in the worked solutions of your online resources.

31. $x = -2$, $y = -3$

32. x^2

33. $x = 3$, $y = -1$; 7

2.7 Fractional indices (Path)

1. a. $\sqrt{15}$ b. $\sqrt[4]{m}$
   c. $\sqrt[5]{7^2}$ d. $\sqrt{7^5}$

2. a. $\sqrt[8]{w^3}$ b. $\sqrt[4]{w^5}$
   c. $\sqrt[3]{5^{10}}$ d. $\sqrt[10]{a^3}$

3. a. $t^{\frac{1}{2}}$ b. $5^{\frac{7}{4}}$
   c. $6^{\frac{11}{6}}$ d. $x^{\frac{6}{7}}$

4. a. $x^{\frac{7}{6}}$ b. w^2
   c. $w^{\frac{1}{2}}$ d. $11^{\frac{n}{x}}$

5. a. 4 b. 5 c. 9

6. a. 2 b. 4 c. 3

7. a. 2.2 b. 1.5 c. 1.3

8. a. 2.5 b. 12.9 c. 13.6

9. a. 0.7 b. 0.8 c. 0.9

10. a. $4^{\frac{4}{5}}$ b. $2^{\frac{1}{2}}$ c. $a^{\frac{5}{6}}$

11. a. $x^{\frac{23}{20}}$ b. $10m^{\frac{8}{15}}$ c. $2b^{\frac{5}{7}}$

12. a. $-4y^{\frac{20}{9}}$ b. $0.02a^{\frac{9}{8}}$ c. $5x^{\frac{7}{2}}$

13. a. $ab^{\frac{3}{2}}$ b. $x^{\frac{4}{5}}y^{\frac{5}{9}}$ c. $6a^{\frac{8}{5}}b^{\frac{17}{15}}$

14. a. $2m^{\frac{19}{28}}n^{\frac{2}{5}}$ b. $x^{\frac{19}{6}}y^{\frac{5}{6}}z^{\frac{5}{6}}$ c. $8a^{\frac{2}{5}}b^{\frac{9}{8}}c$

15. a. $x^{\frac{5}{3}}y^{\frac{7}{5}}$ b. $a^{\frac{7}{45}}b^{\frac{4}{15}}$ c. $\dfrac{1}{3}m^{\frac{3}{8}}n^{\frac{11}{56}}$

16. a. $2x^{\frac{2}{15}}y^{\frac{3}{4}}$ b. $\dfrac{1}{4}a^{\frac{11}{20}}b^{\frac{7}{20}}$ c. $\dfrac{1}{7}p^{\frac{5}{24}}q^{\frac{1}{12}}$

17. a. $2^{\frac{9}{20}}$ b. $5^{\frac{1}{6}}$ c. $7^{\frac{6}{5}}$

18. a. $a^{\frac{3}{10}}$ b. $m^{\frac{1}{6}}$ c. $2^{\frac{1}{3}}b^{\frac{1}{6}}$

19. a. $4p^{\frac{2}{5}}$ b. $x^{\frac{m}{p}}$ c. $3^{\frac{b}{c}}m^{\frac{a}{c}}$

20. a. $a^{\frac{1}{4}}b^{\frac{1}{6}}$ b. $a^3 b^{\frac{3}{4}}$ c. $x^{\frac{6}{5}}y^{\frac{7}{4}}$

21. a. $3^{\frac{1}{3}}a^{\frac{1}{9}}b^{\frac{1}{5}}c^{\frac{1}{4}}$ b. $5x^{\frac{1}{4}}y^{\frac{1}{3}}z^{\frac{1}{5}}$ c. $\dfrac{a^{\frac{1}{2}}}{b^{\frac{2}{3}}}$

22. a. $\dfrac{m^{\frac{8}{5}}}{n^{\frac{7}{4}}}$ b. $\dfrac{b^{\frac{2}{5}}}{c^{\frac{8}{27}}}$ c. $\dfrac{2^{\frac{1}{2}}x^{\frac{7}{2}}}{y^{\frac{3}{8}}}$

23. a. D b. C c. B

24. a. D b. B

25. a. a^4 b. b^3 c. m^4

26. a. $4x^2$ b. $2y^3$ c. $2x^2 y^3$

27. a. $3m^3 n^5$ b. $2pq^2$ c. $6a^2 b^6$

28. a. 2.007 s b. 20.07 s c. 4.98 swings

29. $\left(2^5 a^5 b^{10}\right)^{\frac{1}{5}} = 2ab^2$

30. $t^{\frac{1}{10}}$

31. $m - n^2$

32. a. $a^{-\frac{1}{4}} \times b^{\frac{13}{2}}$

    b. No, because you can't take the fourth root of a negative number.

    c. $a = 1$

2.8 Combining index laws (Path)

1. a. $54a^{10}b^9$ b. $48a^5 b^{16}$ c. $\dfrac{2n^{13}}{m^9}$

2. a. $500p^8 q^{18}$ b. $36a^{20}b^{10}$ c. $\dfrac{15b^2}{c^{26}}$

3. a. $12x^{\frac{7}{8}}y^{\frac{11}{15}}$ b. $8m^{\frac{15}{4}}n^{\frac{15}{4}}$ c. $\dfrac{6}{p^{\frac{7}{12}}}$

   d. $8p^{\frac{7}{45}}q^{\frac{5}{18}}$

4. a. $\dfrac{5}{8a^7}$ b. $\dfrac{x}{4y^6}$ c. $\dfrac{27}{128m^{29}n^{26}}$

5. a. $\dfrac{64y^{36}}{x^{24}}$ b. $24a^{24}b^7$ c. $\dfrac{27h^{12}}{8g^6}$

6. a. $p^{\frac{35}{3}}q^{\frac{1}{2}}$ b. $\dfrac{625}{81b^{20}c^{28}}$ c. $x^{\frac{5}{3}}y^{\frac{1}{8}}z^{\frac{3}{2}}$

7. a. $\dfrac{3a^2}{2}$ b. $8n^2$ c. $\dfrac{m^2 n^4}{3}$

8. a. $\dfrac{4x^5}{3y^8}$ b. $\dfrac{36x^6}{y}$ c. $\dfrac{y^2}{x^4}$

9. a. $\dfrac{b^7}{3a^4}$ b. $\dfrac{75q^5}{2p^{11}}$ c. $x^{\frac{17}{10}}y^{\frac{7}{10}}$

10. a. $\dfrac{2}{5a^4b^7}$ b. $\dfrac{4a^3b^3}{15}$ c. $\dfrac{n^9}{4m^9}$

11. a. $\dfrac{4m^5}{9n^{15}}$ b. $\dfrac{4}{81x^2y^{14}}$ c. $48x^{11}y^6$

12. a. $\dfrac{3p^4}{5q^9}$ b. $\dfrac{2b^{\frac{1}{12}}}{3a^{\frac{17}{24}}}$ c. $\dfrac{4x^{\frac{1}{12}}}{3y^{\frac{21}{20}}}$

13. a. $\dfrac{5}{2a^{13}}$ b. $\dfrac{56a^{11}b^6}{81}$

14. a. $\dfrac{1024b^2}{81a}$ b. $\dfrac{25}{128x^{23}y^4}$ c. $\dfrac{4y^{36}}{27x^{16}}$

15. a. $6m^{19}n^{19}$ b. $\dfrac{16m^{\frac{11}{12}}n}{3}$ c. $\dfrac{4b^{\frac{11}{2}}}{3^{\frac{1}{2}}c^{\frac{7}{30}}}$

16. a. $\dfrac{125}{8}$ b. 1

17. 1

18. a. 5^{y-1} b. $y = 4$

19. D

20. A

21. a. $m^{\frac{1}{6}}n^{-\frac{7}{6}}$ or $\sqrt[6]{\dfrac{m}{n^7}}$

   b. $g^{-6}h^3n^{\frac{3}{2}}$

   c. $3^{-\frac{7}{3}} \times 5^{-\frac{7}{6}}$

22. a. 2^{-2} or $\dfrac{1}{4}$ b. a^6b^{-8} or $\dfrac{a^6}{b^8}$

   c. $d^{\frac{14}{15}}$ or $\sqrt[15]{d^{14}}$

23. a. 12   b. 1536   c. 14 days

24. a. 80 koalas

   b. During the 6th year.

25. a. 79%   b. 56%   c. 31%

26. $\dfrac{3}{2}$

27. $\dfrac{4}{21}$

Project

1. See the table at the bottom of the page*
2. Approximately 1.1 GB
3. 3993.6 MB
4. 1248 photos
5. Discuss with your teacher.
6. Discuss with your teacher. The discussion will depend on the latest information from the internet.

2.9 Review questions

1. A
2. a. Irrational, since equal to non-recurring and non-terminating decimal
   b. Rational, since can be expressed as a whole number
   c. Rational, since given in a rational form
   d. Rational, since it is a recurring decimal
   e. Irrational, since equal to non-recurring and non-terminating decimal
3. D
4. a. $\sqrt{2m}, \sqrt{\dfrac{20}{m}}, \sqrt[3]{m}, \sqrt[3]{8m}$
   b. $\sqrt{25m}, \sqrt{\dfrac{m}{16}}, \sqrt{\dfrac{20}{m}}$
5. a. $5\sqrt{2}$ b. $6\sqrt{5}$ c. $8\sqrt{2}$ d. $20\sqrt{5}$
6. C
7. a. $72x^3y^4\sqrt{2xy}$ b. $-\dfrac{1}{4}x^2y^5\sqrt{xy}$
8. a. $25\sqrt{3}$ b. $3ab\sqrt{ab}$
9. a. $\sqrt{15}$ b. $6\sqrt{42}$
   c. $30\sqrt{15}$ d. 5
10. a. 27   b. $720\sqrt{2}$
11. a. $\sqrt{3}$ b. 6
   c. $\dfrac{\sqrt{10}}{4\sqrt{3}}$ or $\dfrac{\sqrt{30}}{12}$ d. $\dfrac{1}{2}$
12. a. $\dfrac{\sqrt{6}}{3}$ b. $\dfrac{\sqrt{2}}{4}$
   c. $\dfrac{\sqrt{2}}{3}$ d. $2 + \sqrt{6}$

*1.

Unit	Symbol	Power of 2 and value in bytes	Power of 10 and value in bytes
Byte	B	$2^0 = 1$	$100 = 1$
Kilobyte	KB	$2^{10} = 1024$	$10^3 = 1000$
Megabyte		$2^{20} = 1\,048\,576$	$10^6 = 1\,000\,000$
Gigabyte		$2^{30} = 1\,073\,741\,824$	$10^9 = 1\,000\,000\,000$
Terabyte		$2^{40} = 1\,099\,511\,627\,776$	$10^{12} = 1\,000\,000\,000\,000$

13. a. 4 **b.** 4.5
c. 2.2 **d.** 2.7

14. a. 7.4 **b.** 1.7
c. 0.8 **d.** 0.8

15. a. $\sqrt{2}$ **b.** $3\sqrt{2}$
c. $5\sqrt{5}$ **d.** 16

16. a. 1 **b.** 4

17. a. $\dfrac{1}{4}$ **b.** $\dfrac{1}{9}$

c. $\dfrac{1}{16}$ **d.** $\dfrac{1}{1000}$

18. a. 0.0833 **b.** 0.0204
c. 0.800 **d.** 625

19. a. $1\dfrac{1}{2}$ **b.** $1\dfrac{3}{7}$

c. 5 **d.** $\dfrac{4}{13}$

20. a. B **b.** A **c.** A

21. a. $2\dfrac{1}{2}$ **b.** $2\dfrac{1}{4}$

22. a. $\dfrac{1}{2}$ **b.** $\dfrac{1}{9}$

c. $\dfrac{1}{64}$ **d.** $\dfrac{2}{1}$

23. D
24. C
25. C
26. C
27. A
28. A
29. B
30. C
31. B

32. a. $9x^{10}y^{10}$ **b.** $\dfrac{13ab^3c^2}{6}$

c. $\dfrac{1000m^{15}n^6}{27}$ **d.** $\dfrac{16p^{28}}{81q^{12}}$

33. a. 16 **b.** $-\dfrac{3}{2}$

34. a. $\dfrac{8}{a^{11}b^2}$ **b.** $\dfrac{y^2}{5x^{17}}$ **c.** $\dfrac{m^{12}}{16n^8}$

35. a. 8 **b.** $\dfrac{3}{2}$ **c.** 0

36. a. $30a^{\frac{41}{20}}b^{\frac{33}{20}}$ **b.** $\dfrac{4}{x^{\frac{1}{20}}y^{\frac{2}{9}}}$ **c.** $\dfrac{2a^{\frac{1}{6}}}{b^{\frac{3}{2}}}$

37. a. 1 **b.** 4

38. a. $-2a^3 + 2a^2b^{\frac{1}{2}}$

b. $6xy^2$

39. a. $\dfrac{2a^{13}}{5b^2}$ **b.** $\dfrac{9y^4}{32x^{15}}$ **c.** $2^{\frac{4}{3}}m$

40. a. 46 **b.** $-\dfrac{1}{18}$

41. a. 9 **b.** 6 **c.** 0

42. $\dfrac{1}{a^2b^{\frac{1}{2}}}$

43. $\dfrac{1}{36}$

44. a. 8 **b.** 6 **c.** 2

45. a. $\dfrac{9ab^7c^7}{50}$

b. The student has made two mistakes when squaring the left-hand bracket in line 1 : $3^2 = 9, 5^2 = 25$.

46. a. $\dfrac{\sqrt{r}}{d\sqrt{u^3}}$

b. Yes, $\sqrt{r}$, $\sqrt{u^3}$

c. $\dfrac{\sqrt{ru^3}}{du^3}$

d. r should be a perfect square, u should be a perfect cube and d should be a rational number.

e. i. $0.0012346\,\text{kL}$
   ii. $1.2346\,\text{L}$
   iii. $1234.6\,\text{mL}$

f. A high value for d causes the expression to be smaller, as d only appears on the denominator of the fraction. This means that when d is high there is less water in the reservoir and the area is dry.

3 Algebraic techniques (Path)

LESSON SEQUENCE

LESSON
3.1 Overview

Why learn this?

How is your algebraic tool kit? Is there some room to expand your skills? As expressions become more complex, more power will be needed to manipulate them and to carry out basic algebraic skills such as adding, multiplying, expanding and factorising.

We have sent humans into space to live for months at a time on the International Space Station and even landed people on the moon. We have satellites circling our globe which enable us to communicate with friends and family around the world. Satellites also send out weather information, allowing meteorologists to study the patterns and changes and predict the upcoming weather conditions.

Technology connects us with others via our mobile phones or computers, often using a digital social media platform. All of these things that have become such important parts of our modern lives are only possible due to the application of mathematical techniques that we will begin to explore in this topic.

Many careers require a strong understanding of algebraic techniques. An example is a structural engineer. When designing structures, bridges, high rise buildings or domestic homes, they need to make decisions based on mathematics to ensure the structure is strong, durable and passes all regulations. This topic will help you develop your knowledge of algebraic expressions, so that you can apply them to real-life situations.

Hey students! Bring these pages to life online

Watch videos

Engage with interactivities

Answer questions and check solutions

Find all this and MORE in jacPLUS

Reading content and rich media, including interactivities and videos for every concept

Extra learning resources

Differentiated question sets

Questions with immediate feedback, and fully worked solutions to help students get unstuck

1. Expand the following.
   a. $(3x - 1)(3x + 1)$
   b. $(2a - 5b)(2a + 5b)$

2. Expand and simplify the following.
   a. $(2a + 3)^2$
   b. $(2 - 5x)^2$

3. Expand and simplify the following expressions.
   a. $4(m + 3n) - (2m - n)$
   b. $-x(y - 3) + y(5 - x)$

4. Is this statement true or false?
   $(a + 3)^2$ expanded is equal to $a^2 + 9$.

5. Factorise the following expressions.
   a. $7a(b + 3) - (b + 3)$
   b. $10xy + 5x - 4y - 2$

6. **MC** The expression $81x^2 - 36y^2$ factorises to:
   A. $(81x - 36y)(81x - 36y)$
   B. $(81x - 36y)^2$
   C. $(9x - 6y)^2$
   D. $(9x - 6y)(9x + 6y)$

7. **MC** The expression $(a - 4)^2 - 25$ factorises to:
   A. $(a - 9)(a + 1)$
   B. $(a + 9)(a - 1)$
   C. $(a - 7)(a + 3)$
   D. $(a - 4)(a + 4) - 25$

8. **MC** Which of the following expressions are perfect squares? There may be more than one selection.
   A. $p^2 - 6p + 9$
   B. $p^2 + 9p + 9$
   C. $p^2 + 9$
   D. $p^2 - 9$

9. Factorise $m^2 + 2m - 8$.

10. Factorise the equation $5x^2 + 9x - 2$

11. **MC** $\dfrac{9x + 18}{3x + 6}$ simplifies to:

   A. $3x + 3$
   B. $\dfrac{3x + 3}{x + 1}$
   C. $\dfrac{9x + 3}{3x}$
   D. 3

12. **MC** The expression $\dfrac{2}{x + 1} - \dfrac{1}{(x + 1)^2}$ can be simplified to:

   A. $\dfrac{1}{(x + 1)^2}$
   B. $\dfrac{x - 1}{(x + 1)^2}$
   C. $\dfrac{2x}{(x + 1)^2}$
   D. $\dfrac{2x + 1}{(x + 1)^2}$

13. **MC** The expression $\dfrac{5}{2x} + \dfrac{1}{3x}$ simplified is:

   A. $\dfrac{6}{5x}$
   B. $\dfrac{17}{6x}$
   C. $\dfrac{5}{6x^2}$
   D. $\dfrac{17}{6x^2}$

14. Simplify the following expressions.
   a. $\dfrac{m^2 + 4m - 5}{m^2 - 1}$
   b. $\dfrac{8}{s - 3} \div \dfrac{s + 2}{s - 3}$

15. Simplify the following.
   $\dfrac{a^2 + 14a + 24}{a^2 - 16a - 36}$

LESSON
3.2 Special binomial products

LEARNING INTENTION

At the end of this lesson you should be able to:
- recognise and use the difference of two squares rule to expand and simplify binomial products in the form $(a+b)(a-b)$
- recognise and use the perfect square rule to expand and simplify binomial products in the form $(a \pm b)^2$.

▶ 3.2.1 Difference of two squares

eles-4598

- Consider the expansion of $(x+4)(x-4)$:

$$(x+4)(x-4) = x^2 - 4x + 4x - 16 = x^2 - 16$$

- Now consider the expansion of $(x-6)(x+6)$:

$$(x-6)(x+6) = x^2 + 6x - 6x - 36 = x^2 - 36$$

- In both cases the middle two terms cancel each other out, leaving two terms, both of which are perfect squares.
- The two terms that are left are the first value squared minus the second value squared. This is where the phrase **difference of two squares** originates.
- In both cases the binomial terms can be written in the form $(x+a)$ and $(x-a)$. If we can recognise expressions that have this form, we can use the pattern above to quickly expand those expressions. For example: $(x+12)(x-12) = x^2 - 12^2 = x^2 - 144$

Difference of two squares

The difference of two squares rule is used to expand certain binomial products, as long as they are in the forms shown below:

$$(a+b)(a-b) = a^2 - b^2$$

$$(a-b)(a+b) = a^2 - b^2$$

Because the two binomial brackets are being multiplied, the order of the brackets does not affect the final result.

WORKED EXAMPLE 1 Expanding using difference of two squares

Use the difference of two squares rule to expand and simplify each of the following expressions.
a. $(x+8)(x-8)$ b. $(6-3)(6+x)$ c. $(2x-3)(2x+3)$ d. $(3x+5)(5-3x)$

THINK	WRITE
a. 1. Write the expression.	a. $(x+8)(x-8)$
2. This expression is in the form $(a+b)(a-b)$, so the difference of two squares rule can be used. Expand using the formula.	$= x^2 - 8^2$ $= x^2 - 64$

b. 1. Write the expression.

2. This expression is in the form $(a + b)(a - b)$, so the difference of two squares rule can be used. Expand using the formula. *Note:* $36 - x^2$ is not the same as $x^2 - 36$.

c. 1. Write the expression.

2. This expression is in the form $(a - b)(a + b)$, so the difference of two squares rule can be used. Expand using the formula. *Note:* $(2x)^2$ and $2x^2$ are not the same. In this case $a = 2x$, so $a^2 = (2x)^2$.

d. 1. Write the expression.

2. The difference of two squares rule can be used if we rearrange the terms, since $3x + 5 = 5 + 3x$. Expand using the formula.

b. $(6 - x)(6 + x)$
$= 6^2 - x^2$
$= 36 - x^2$

c. $(2x - 3)(2x + 3)$
$= (2x)^2 - 3^2$
$= 4x^2 - 9$

d. $(3x + 5)(5 - 3x)$
$(5 + 3x)(5 - 3x)$
$= 5^2 - (3x)^2$
$= 25 - 9x^2$

▶ 3.2.2 Perfect squares

eles-4599

- A **perfect square** is the result of the square of a whole number. $1 \times 1 = 1, 2 \times 2 = 4$ and $3 \times 3 = 9$, showing that $1, 4$ and 9 are all perfect squares.
- Similarly, $(x + 3)(x + 3) = (x + 3)^2$ is a perfect square because it is the result of a binomial factor multiplied by itself.
- Consider the diagram illustrating $(x + 3)^2$. What shape is it?

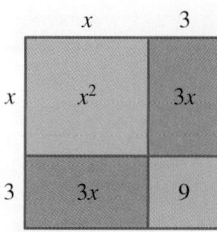

- The area is given by $x^2 + 3x + 3x + 9 = x^2 + 6x + 9$.
- We can see from the diagram that there are two squares produced $\left(x^2 \text{ and } 3^2 = 9\right)$ and two rectangles that are identical to each other $(2 \times 3x)$.
- Compare this with the expansion of $(x + 6)^2$:

$$(x + 6)\,(x + 6) = x^2 + 6x + 6x + 36 = x^2 + 12x + 36$$

- A pattern begins to emerge after comparing these two expansions. The square of a binomial equals the square of the first term, plus double the product of the two terms plus the square of the second term. For example: $(x + 10)(x + 10) = x^2 + 2 \times 10 \times x + 10^2 = x^2 + 20x + 100$

> **Perfect squares**
>
> **The rule for the expansion of the square of a binomial is given by:**
>
> $$(a + b)^2 = (a + b)(a + b) = a^2 + 2ab + b^2$$
>
> $$(a - b)^2 = (a - b)(a - b) = a^2 - 2ab + b^2$$

WORKED EXAMPLE 2 Expanding perfect squares

Use the rules for expanding perfect binomial squares to expand and simplify the following.

a. $(x+1)(x+1)$

b. $(x-2)^2$

c. $(2x+5)^2$

d. $(4x-5y)^2$

THINK	WRITE
a. 1. This expression is the square of a binomial.	**a.** $(x+1)(x+1)$
2. Apply the formula for perfect squares: $(a+b)^2 = a^2 + 2ab + b^2$	$= x^2 + 2 \times x \times 1 + 1^2$ $= x^2 + 2x + 1$
b. 1. This expression is the square of a binomial.	**b.** $(x-2)^2$
2. Apply the formula for perfect squares: $(a-b)^2 = a^2 - 2ab + b^2$.	$= (x-2)(x-2)$ $= x^2 - 2 \times x \times 2 + 2^2$ $= x^2 - 4x + 4$
c. 1. This expression is the square of a binomial.	**c.** $(2x+5)^2$
2. Apply the formula for perfect squares: $(a+b)^2 = a^2 + 2ab + b^2$.	$= (2x)^2 + 2 \times 2x \times 5 + 5^2$ $= 4x^2 + 20x + 25$
d. 1. This expression is the square of a binomial.	**d.** $(4x-5y)^2$
2. Apply the formula for perfect squares: $(a-b)^2 = a^2 - 2ab + b^2$.	$= (4x)^2 - 2 \times 4x \times 5y + (5y)^2$ $= 16x^2 - 40xy + 25y^2$

 Resources

➤ **Interactivity** Difference of two squares (int-6036)

Exercise 3.2 Special binomial products

learn

3.2 Quick quiz **on**	3.2 Exercise

Individual pathways

■ PRACTISE	■ CONSOLIDATE	■ MASTER
1, 3, 5, 8, 13, 16, 20, 21	2, 6, 9, 11, 14, 17, 22, 23, 25	4, 7, 10, 12, 15, 18, 19, 24

Fluency

1. **WE1** Use the difference of two squares rule to expand and simplify each of the following.

   a. $(x+2)(x-2)$
   b. $(y+3)(y-3)$
   c. $(m+5)(m-5)$
   d. $(a+7)(a-7)$

2. Use the difference of two squares rule to expand and simplify each of the following.

   a. $(x+6)(x-6)$
   b. $(p-12)(p+12)$
   c. $(a+10)(a-10)$
   d. $(m-11)(m+11)$

3. Use the difference of two squares rule to expand and simplify each of the following.
   a. $(2x+3)(2x-3)$
   b. $(3y-1)(3y+1)$
   c. $(5d-2)(5d+2)$
   d. $(7c+3)(7c-3)$
   e. $(2+3p)(2-3p)$

4. Use the difference of two squares rule to expand and simplify each of the following.
   a. $(d-9x)(d+9x)$
   b. $(5-12a)(5+12a)$
   c. $(3x+10y)(3x-10y)$
   d. $(2b-5c)(2b+5c)$
   e. $(10-2x)(2x+10)$

5. **WE2** Use the rule for the expansion of the square of a binomial to expand and simplify each of the following.
   a. $(x+2)(x+2)$
   b. $(a+3)(a+3)$
   c. $(b+7)(b+7)$
   d. $(c+9)(c+9)$

6. Use the rule for the expansion of the square of a binomial to expand and simplify each of the following.
   a. $(m+12)^2$
   b. $(n+10)^2$
   c. $(x-6)^2$
   d. $(y-5)^2$

7. Use the rule for the expansion of the square of a binomial to expand and simplify each of the following.
   a. $(9-c)^2$
   b. $(8+e)^2$
   c. $2(x+y)^2$
   d. $(u-v)^2$

8. Use the rule for expanding perfect binomial squares rule to expand and simplify each of the following.
   a. $(2a+3)^2$
   b. $(3x+1)^2$
   c. $(2m-5)^2$
   d. $(4x-3)^2$

9. Use the rule for expanding perfect binomial squares to expand and simplify each of the following.
   a. $(5a-1)^2$
   b. $(7p+4)^2$
   c. $(9x+2)^2$
   d. $(4c-6)^2$

10. Use the rule for expanding perfect binomial squares to expand and simplify each of the following.
    a. $(5+3p)^2$
    b. $(2-5x)^2$
    c. $(9x-4y)^2$
    d. $(8x-3y)^2$

Understanding

11. Use the difference of two squares rule to expand and simplify each of the following.
    a. $(x+3)(x-3)$
    b. $(2x+3)(2x-3)$
    c. $(7x-4)(7x+4)$
    d. $(2x+7y)(2x-7y)$
    e. $(x^2+y^2)(x^2-y^2)$

12. Expand and simplify the following perfect squares.
    a. $(4x+5)^2$
    b. $(7x-3y)^2$
    c. $(5x^2-2y)^2$
    d. $2(x-y)^2$
    e. $\left(\dfrac{2}{x}+4x\right)^2$

13. A square has a perimeter of $4x+12$. Calculate its area.

14. Francis has fenced off a square in her paddock for spring lambs. The area of the paddock is $(9x^2+6x+1)\,\text{m}^2$.
    Using pattern recognition, determine the side length of the paddock in terms of x.

15. A square has an area of $x^2+18x+81$. Determine an expression for the perimeter of this square.

Communicating, reasoning and problem solving

16. Show that $a^2-b^2=(a+b)(a-b)$ is true for each of the following.
    a. $a=5, b=4$
    b. $a=9, b=1$
    c. $a=2,\ b=7$
    d. $a=-10, b=-3$

17. Lin has a square bedroom. Her sister Tasneem has a room that is 1 m shorter in length than Lin's room, but 1 m wider.
    a. Show that Lin has the larger bedroom.
    b. Determine how much bigger Lin's bedroom is than Tasneem's bedroom.

18. Expand each of the following pairs of expressions.
    a. i. $(x-4)(x+4)$ and $(4-x)(4+x)$
       ii. $(x-11)(x+11)$ and $(11-x)(11+x)$
       iii. $(2x-9)(2x+9)$ and $(9-2x)(9+2x)$
    b. State what you notice about the answers to the pairs of expansions above.
    c. Explain how this is possible.

19. Answer the following questions.
    a. Expand $(10k+5)^2$.
    b. Show that $(10k+5)^2 = 100k(k+1)+25$.
    c. Using part b, evaluate 25^2 and 85^2.

20. A large square has been subdivided into two squares and two rectangles.

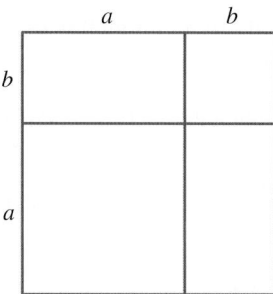

    a. Write formulas for the areas of these four pieces, using the dimensions a and b marked on the diagram.
    b. Write an equation that states that the area of the large square is equal to the combined area of its four pieces. Do you recognise this equation?

21. Expand each of the following pairs of expressions.
    a. i. $(x-3)^2$ and $(3-x)^2$
       ii. $(x-15)^2$ and $(15-x)^2$
       iii. $(3x-7)^2$ and $(7-3x)^2$
    b. State what you notice about the answers to the pairs of expansions above.
    c. Explain how this is possible.

22. The expansion of perfect squares $(a+b)^2 = a^2 + 2ab + b^2$ and $(a-b)^2 = a^2 - 2ab + b^2$ can be used to simplify some arithmetic calculations. For example:

$$97^2 = (100-3)^2$$
$$= 100^2 - 2 \times 100 \times 3 + 3^2$$
$$= 9409$$

Use this method to evaluate the following.
    a. 103^2 b. 62^2 c. 997^2 d. 1012^2 e. 53^2 f. 98^2

23. Use the perfect squares rule to quickly evaluate the following.
    a. 27^2 b. 33^2 c. 39^2 d. 47^2

24. Allen is creating a square deck with a square pool installed in the middle of it. The side length of the deck is $(2x+3)$ m and the side length of the pool is $(x-2)$ m. Evaluate the area of the decking around the pool.

25. Ram wants to create a rectangular garden that is 10 m longer than it is wide. Write an expression for the area of the garden in terms of x, where x is the average length of the two sides.

LESSON
3.3 Further expansions

LEARNING INTENTION

At the end of this lesson you should be able to:
- expand multiple sets of brackets and simplify the result.

⊙ 3.3.1 Expanding multiple sets of brackets

eles-4600

- When expanding expressions with more than two sets of brackets, expand all brackets first before simplifying the like terms.
- It is important to be careful with signs, particularly when subtracting one expression from another. For example:

$$3x(x+6) - (x+5)(x-3)$$
$$= 3x^2 + 18x - (x^2 + 5x - 3x - 15)$$
$$= 3x^2 + 18x - (x^2 + 2x - 15)$$
$$= 3x^2 + 18x - x^2 - 2x + 15$$
$$= 2x^2 + 16x + 15$$

- When expanding this, since the entire expression $(x+5)(x-3)$ is being subtracted from $3x(x+6)$, its expanded form needs to be kept inside its own set of brackets. This helps us to remember that, as each term is being subtracted, the sign of each term will switch when opening up the brackets.
- When expanding, it can help to treat the expression as though it has a -1 at the front, as shown.

$$-(x^2 + 2x - 15) = -1(x^2 + 2x - 15) = -x^2 - 2x + 15$$

WORKED EXAMPLE 3 Expanding and simplifying multiple sets of brackets

Expand and simplify each of the following expressions.
a. $(x+3)(x+4) + 4(x-2)$
b. $(x-2)(x+3) - (x-1)(x+2)$

THINK

WRITE

a. 1. Expand each set of brackets.

a.
$$(x + 3)(x + 4) + 4(x - 2)$$

2. Simplify by collecting like terms.
$$= x^2 + 4x + 3x + 12 + 4x - 8$$
$$= x^2 + 11x + 4$$

b. 1. Expand and simplify each pair of brackets. Because the second expression is being subtracted, keep it in a separate set of brackets.

b.
$$(x - 2)(x + 3) - (x - 1)(x + 2)$$
$$= x^2 + 3x - 2x - 6 - (x^2 + 2x - x - 2)$$

2. Subtract all of the second result from the first result. Remember that $-(x^2 + x - 2) = -1(x^2 + x - 2)$. Simplify by collecting like terms.
$$= x^2 + x - 6 - (x^2 + x - 2)$$
$$= x^2 + x - 6 - x^2 - x + 2$$
$$= -4$$

WORKED EXAMPLE 4 Further expanding with multiple sets of brackets

Consider the expression $(4x+3)(4x+3)-(2x-5)(2x-5)$.
a. **Expand and simplify this expression.**
b. **Apply the difference of perfect squares rule to verify your answer.**

THINK	WRITE
a. 1. Expand each set of brackets.	a. $(4x+3)(4x+3)-(2x-5)(2x-5)$
2. Simplify by collecting like terms.	$=16x^2+12x+12x+9-\left(4x^2-10x-10x+25\right)$
	$=16x^2+24x+9-\left(4x^2-20x+25\right)$
	$=16x^2+24x+9-4x^2+20x-25$
	$=12x^2+44x-16$
b. 1. Rewrite as the difference of two squares.	b. $(4x+3)(4x+3)-(2x-5)(2x-5)$
2. Apply the difference of two squares rule, where $a=(4x+3)$ and $b=(2x-5)$.	$=(4x+3)^2-(2x-5)^2$
	$=((4x+3)+(2x-5))((4x+3)-(2x-5))$
3 Simplify, then expand the new brackets.	$=(6x-2)(2x+8)$
	$=12x^2+48x-4x-16$
4. Simplify by collecting like terms.	$=12x^2+44x-16$

Exercise 3.3 Further expansions

learnon

3.3 Quick quiz on	3.3 Exercise

Individual pathways

■ PRACTISE	■ CONSOLIDATE	■ MASTER
1, 4, 8, 11, 14	2, 5, 7, 9, 12, 15	3, 6, 10, 13, 16

Fluency

1. WE3 Expand and simplify each of the following expressions.
   a. $(x+3)(x+5)+(x+2)(x+3)$
   b. $(x+4)(x+2)+(x+3)(x+4)$
   c. $(x+5)(x+4)+(x+3)(x+2)$
   d. $(x+1)(x+3)+(x+2)(x+4)$

2. Expand and simplify each of the following expressions.
   a. $(p-3)(p+5)+(p+1)(p-6)$
   b. $(a+4)(a-2)+(a-3)(a-4)$
   c. $(p-2)(p+2)+(p+4)(p-5)$
   d. $(x-4)(x+4)+(x-1)(x+20)$

3. Expand and simplify each of the following expressions.
   a. $(y-1)(y+3)+(y-2)(y+2)$
   b. $(d+7)(d+1)+(d+3)(d-3)$
   c. $(x+2)(x+3)+(x-4)(x-1)$
   d. $(y+6)(y-1)+(y-2)(y-3)$

4. Expand and simplify each of the following expressions.
   a. $(x+2)^2 + (x-5)(x-3)$
   b. $(y-1)^2 + (y+2)(y-4)$
   c. $(p+2)(p+7) + (p-3)^2$
   d. $(m-6)(m-1) + (m+5)^2$

5. Expand and simplify each of the following expressions.
   a. $(x+3)(x+5) - (x+2)(x+5)$
   b. $(x+5)(x+2) - (x+1)(x+2)$
   c. $(x+3)(x+2) - (x+4)(x+3)$
   d. $(m-2)(m+3) - (m+2)(m-4)$

6. Expand and simplify each of the following expressions.
   a. $(b+4)(b-6) - (b-1)(b+2)$
   b. $(y-2)(y-5) - (y+2)(y+6)$
   c. $(p-1)(p+4) - (p-2)(p-3)$
   d. $(x+7)(x+2) - (x-3)(x-4)$

7. Expand and simplify each of the following expressions.
   a. $(m+3)^2 - (m+4)(m-2)$
   b. $(a-6)^2 - (a-2)(a-3)$
   c. $(p-3)(p+1) - (p+2)^2$
   d. $(x+5)(x-4) - (x-1)^2$

Understanding

8. **WE4** Consider the expression $(x+3)(x+3) - (x-5)(x-5)$.
   a. Expand and simplify this expression.
   b. Apply the difference of perfect squares rule to verify your answer.

9. Consider the expression $(4x-5)(4x-5) - (x+2)(x+2)$.
   a. Expand and simplify this expression.
   b. Apply the difference of perfect squares rule to verify your answer.

10. Consider the expression $(3x-2y)(3x-2y) - (y-x)(y-x)$.
   a. Expand and simplify this expression.
   b. Apply the difference of perfect squares rule to verify your answer.

Communicating, reasoning and problem solving

11. Determine the value of x for which $(x+3) + (x+4)^2 = (x+5)^2$ is true.

12. Show that $(p-1)(p+2) + (p-3)(p+1) = 2p^2 - p - 5$.

13. Show that $(x+2)(x-3) - (x+1)2 = -3x - 7$.

14. Answer the following questions.
   a. Show that $(a^2 + b^2)(c^2 + d^2) = (ac - bd)^2 + (ad + bc)^2$.
   b. Using part a, write $(2^2 + 1^2)(3^2 + 4^2)$ as the sum of two squares and evaluate.

15. Answer the following questions.
   a. Expand $(x^2 + x - 1)^2$.
   b. Show that $(x^2 + x - 1)^2 = (x-1)x(x+1)(x+2) + 1$.
   c. i. Evaluate $4 \times 3 \times 2 \times 1 + 1$.
      ii. Determine the value of x if $4 \times 3 \times 2 \times 1 + 1 = (x-1)x(x+1)(x+2) + 1$.

16. Answer the following questions.
   a. Expand $(a+b)(d+e)$.
   b. Expand $(a+b+c)(d+e+f)$. Draw a diagram to illustrate your answer.

LESSON
3.4 Factorising by grouping in pairs

LEARNING INTENTION

At the end of this lesson you should be able to:
- factorise algebraic expressions by finding common binomial factors
- factorise algebraic expressions by grouping terms.

▶ 3.4.1 The common binomial factor
eles-4603
- When factorising an expression, we look for the highest common factor(s) first.
- It is possible for the HCF to be a binomial expression.
- Consider the expression $7(a - b) + 8x(a - b)$. The binomial expression $(a - b)$ is a common factor to both terms. Factorising this expression looks like this:

$$7(a - b) + 8x(a - b) = 7 \times (a - b) + 8x \times (a - b)$$
$$= (a - b)(7 + 8x)$$

WORKED EXAMPLE 5 Factorising by finding a binomial common factor

Factorise each of the following expressions.

a. $5(x + y) + 6b(x + y)$

b. $2b(a - 3b) - (a - 3b)$

Note: In both of these expressions the HCF is a binomial factor.

THINK	WRITE
a. 1. The HCF is $(x + y)$.	**a.** $\dfrac{5(x + y)}{x + y} = 5, \dfrac{6b(x + y)}{x + y} = 6b$
2. Divide each term by $(x + y)$ to determine the binomial.	Therefore, $5(x + y) + 6b(x + y)$ $= (x + y)(5 + 6b)$
b. 1. The HCF is $(a - 3b)$.	**b.** $\dfrac{2b(a - 3b)}{a - 3b} = 2b, \dfrac{-1(a - 3b)}{a - 3b} = -1$
2. Divide each term by $(a - 3b)$ to determine the binomial.	Therefore, $2b(a - 3b) - (a - 3b)$ $= 2b(a - 3b) - 1(a - 3b)$ $= (a - 3b)(2b - 1)$

▶ 3.4.2 Factorising by grouping in pairs
eles-4604
- If an algebraic expression has four terms and no common factors in any of its terms, it may be possible to group the terms in pairs and find a common factor in each pair.
- Consider the expression $10x + 15 - 6ax - 9a$.
- We can attempt to factorise by grouping the first two terms and the last two terms:

$$10x + 15 - 6ax - 9a = 5 \times 2x + 5 \times 3 - 3a \times 2x - 3a \times 3$$
$$= 5(2x + 3) - 3a(2x + 3)$$

- Once a common factor has been taken out from each pair of terms, a common binomial factor will appear. This common binomial factor can also be factorised out.

$$5(2x+3) - 3a(2x+3) = (2x+3)(5-3a)$$

- Thus the expression $10x + 15 - 6ax - 9a$ can be factorised to become $(2x+3)(5-3a)$.
- It is worth noting that it doesn't matter which terms are paired up first — the final result will still be the same.

$$
\begin{aligned}
10x + 15 - 6ax - 9a &= 10x - 6ax + 15 - 9a \\
&= 2x \times 5 + 2x \times -3a + 3 \times 5 + 3 \times -3a \\
&= 2x(5 - 3a) + 3(5 - 3a) \\
&= (5 - 3a)(2x + 3) \\
&= (2x + 3)(5 - 3a)
\end{aligned}
$$

WORKED EXAMPLE 6 Factorising by grouping in pairs

Factorise each of the following expressions by grouping the terms in pairs.
a. $5a + 10b + ac + 2bc$ 　　　　b. $x - 3y + ax - 3ay$ 　　　　c. $5p + 6q + 15pq + 2$

THINK	WRITE
a. 1. Write the expression.	a. $5a + 10b + ac + 2bc$ $5a + 10b = 5(a + 2b)$ $ac + 2bc = c(a + 2b)$
2. Take out the common factor $a + 2b$.	$= 5(a + 2b) + c(a + 2b)$ $= (a + 2b)(5 + c)$
b. 1. Write the expression.	b. $x - 3y + ax - 3ay$ $x - 3y = 1(x - 3y)$ $ax - 3ay = a(x - 3y)$
2. Take out the common factor $x - 3y$.	$= 1(x - 3y) + a(x - 3y)$ $= (x - 3y)(1 + a)$
c. 1. Write the expression.	c. $5p + 6q + 15pq + 2$
2. There are no simple common factors. Write the terms in a different order.	$= 5p + 15pq + 6q + 2$ $5p + 15pq = 5p(1 + 3q)$ $6q + 2 = 2(3q + 1)$
3. Take out the common factor $1 + 3q$. Note: $1 + 3q = 3q + 1$.	$= 5p(1 + 3q) + 2(3q + 1)$ $= 5p(1 + 3q) + 2(1 + 3q)$ $= (1 + 3q)(5p + 2)$

on Resources

Interactivity Common binomial factor (int-6038)

3.4 Quick quiz on	3.4 Exercise

Individual pathways

■ PRACTISE	■ CONSOLIDATE	■ MASTER
1, 3, 6, 9	2, 4, 7, 10, 12	5, 8, 11, 13

Fluency

1. **WE5** Factorise each of the following expressions.

   a. $2(a+b)+3c(a+b)$
   d. $4a(3b+2)-b(3b+2)$
   b. $4(m+n)+p(m+n)$
   e. $z(x+2y)-3(x+2y)$
   c. $7x(2m+1)-y(2m+1)$

2. Factorise each of the following expressions.

   a. $12p(6-q)-5(6-q)$
   d. $p^2(q+2p)-5(q+2p)$
   b. $3p^2(x-y)+2q(x-y)$
   e. $6(5m+1)+n^2(5m+1)$
   c. $4a^2(b-3)+3b\,(b-3)$

3. **WE6** Factorise each of the following expressions by grouping the terms in pairs.

   a. $xy+2x+2y+4$
   d. $2xy+x+6y+3$
   b. $ab+3a+3b+9$
   e. $3ab+a+12b+4$
   c. $xy-4y+3x-12$
   f. $ab-2a+5b-10$

4. Factorise each of the following expressions by grouping the terms in pairs.

   a. $m-2n+am-2an$
   d. $10pq-q-20p+2$
   b. $5+3p+15a+9ap$
   e. $6x-2-3xy+y$
   c. $15mn-5n-6m+2$
   f. $16p-4-12pq+3q$

5. Factorise each of the following expressions by grouping the terms in pairs.

   a. $10xy+5x-4y-2$
   d. $4x+12y-xz-3yz$
   b. $6ab+9b-4a-6$
   e. $5pr+10qr-3p-6q$
   c. $5ab-10ac-3b+6c$
   f. $ac-5bc-2a+10b$

Understanding

6. Simplify the following expressions using factorising.

   a. $\dfrac{ax+2ay+3az}{bx+2by+3bz}$

   b. $\dfrac{10\,(3x-4)+2y\,(3x-4)}{7a\,(10+2y)-5\,(10+2y)}$

7. Use factorising by grouping in pairs to simplify the following expressions.

   a. $\dfrac{3x+6+xy+2y}{6+2y+18x+6xy}$

   b. $\dfrac{5xy+10x+3ay+6a}{15bx-10x+9ab-6a}$

8. Use factorising by grouping in pairs to simplify the following expressions.

   a. $\dfrac{6x^2+15xy-4x-10y}{6xy+4x+15y^2+10y}$

   b. $\dfrac{mp+4mq-4np-16nq}{mp+4mq+4np+16nq}$

Communicating, reasoning and problem solving

9. Using the method of rectangles to expand, show how $a(m+n)+3(m+n)$ equals $(a+3)(m+n)$.

10. Fully factorise $6x+4x^2+6x+9$ by grouping in pairs. Discuss what you noticed about this factorisation.

11. a. Write out the product $5(x+2)(x+3)$ and show that it also corresponds to the diagram shown.

   b. Explain why $5(x+2)(x+3)$ is equivalent to $(5x+10)(x+3)$. Use bracket expansion and a labelled diagram to support your answer.

   c. Explain why $5(x+2)(x+3)$ is equivalent to $(x+2)(5x+15)$. Use bracket expansion and a labelled diagram to support your answer.

12. Fully factorise both sets and brackets in the expression $(-12xy+27x+8y-18)-(-8xy+18x+12y-27)$ using grouping in pairs, then factorise the result.

13. The area formulas shown relate to either squares or rectangles.

   i. $9s^2+48s+64$
   ii. $25s^2-4$
   iii. s^2+4s+3
   iv. $4s^2-28s-32$

     a. Without completing any algebraic operations, examine these formulas and work out which ones belong to squares and which ones belong to rectangles. Explain your answer.

     b. Factorise each formula and classify it as a square or rectangle. Check your classifications against your answer to part **a.**

LESSON
3.5 Factorising special products

LEARNING INTENTION

At the end of this lesson you should be able to:
- recognise and factorise the difference of two squares expressions of the form a^2-b^2
- recognise and factorise perfect squares of the form $a^2 \pm 2ab + b^2$.

▶ 3.5.1 Difference of two squares

eles-6260

- Recall the rule for the difference of two squares from section 3.2.1:

$$(a+b)(a-b) = a^2 - b^2$$

This rule can be used in reverse to factorise the difference of two squares.

> **Factorising the difference of two squares**
>
> $$a^2 - b^2 = (a-b)(a+b)$$

- Factorise by taking out any common factor first.

  For example: $5x^2 - 80$

  $$= 5(x^2 - 16)$$
  $$= 5(x-4)(x+4)$$

WORKED EXAMPLE 7 Factorising expressions that are the difference of two squares

Factorise each of the following expressions.

a. $x^2 - 9$ b. $64m^2 - 25n^2$ c. $(c+7)^2 - 16$ d. $3x^2 - 48$

THINK	WRITE
a. 1. The expression is a difference of two squares.	a. $x^2 - 9$
2. Rewrite, showing the two squares.	$= x^2 - 3^2$
3. Factorise using the formula $a^2 - b^2 = (a+b)(a-b)$ where $a = x$ and $b = 3$.	$= (x+3)(x-3)$
b. 1. The expression is a difference of two squares.	b. $64m^2 - 25n^2$
2. Rewrite, showing the two squares.	$= (8m)^2 - (5n)^2$
3. Factorise.	$= (8m+5n)(8m-5n)$
c. 1. The expression is a difference of perfect squares.	c. $(c+7)^2 - 16$
2. Rewrite, showing the two squares.	$= (c+7)^2 - 4^2$
3. Put $c+7$ in each bracket and complete the factors.	$= (c+7+4)(c+7-4)$
4. Simplify.	$= (c+11)(c+3)$
d. 1. There is a common factor.	d. $3x^2 - 48$
2. Take out the common factor.	$= 3(x^2 - 16)$
3. $x^2 - 16$ is a difference of two squares.	$= 3(x^2 - 4^2)$
4. Factorise.	$= 3(x+4)(x-4)$

▶ 3.5.2 Perfect squares

eles-6261

- Recall from section 3.2.2 that a perfect square is an expression such as $(x+2)^2$ or $(t-2)^2$. The expansion of perfect squares produces the following patterns or identities.

$$(a+b)^2 = a^2 + 2ab + b^2$$
$$(a-b)^2 = a^2 - 2ab + b^2$$

- The identities can also be used in reverse to factorise the quadratic expressions.

Factorising perfect squares

$$a^2 + 2ab + b^2 = (a+b)^2$$
$$a^2 - 2ab + b^2 = (a-b)^2$$

- Is $x^2 + 8x + 16$ a perfect square?

  The constant term must be half the coefficient of x, squared if the expression is a perfect square. Because half of 8 is 4 and $4^2 = 16$, $x^2 + 8x + 16$ is a perfect square.

$$x^2 + 8x + 16 = (x+4)^2$$

WORKED EXAMPLE 8 Factorising expressions that are perfect squares

Factorise each of the following expressions.

a. $x^2 - 6x + 9$ b. $64m^2 + 80mn + 25n^2$ c. $3x^2 - 24x + 48$

THINK	WRITE
a. 1. x^2 and 9 are perfect squares, so $x^2 - 6x + 9$ might be $(x-3)^2$. Check as follows.	**a.** $x^2 - 6x + 9$
2. Halve the coefficient of the x-term and then square the result.	$\dfrac{-6}{2} = -3,\ (-3)^2 = 9$
3. The middle term fits the pattern, so factorise using the formula $a^2 - 2ab + b^2 = (a-b)^2$, where $a = x$ and $b = 3$.	$x^2 - 6x + 9 = (x-3)^2$
b. 1. $64m^2$ and $25n^2$ are perfect squares, so $64m^2 + 80mn + 25n^2$ might be $(8m+5n)^2$.	**b.** $64m^2 + 80mn + 25n^2$
2. Check the middle term.	$8m \times 5n \times 2 = 80mn$
3. The middle term fits the pattern, so write down the factors.	$64m^2 + 80mn + 25n^2 = (8m+5n)^2$
c. 1. There is a common factor.	**c.** $3x^2 - 24x + 48$
2. Factorise.	$= 3(x^2 - 8x + 16)$
3. Factorise $x^2 - 8x + 16$.	$= 3(x-4)^2$

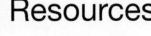 Resources

 Interactivity Difference of two squares (int-6091)

Exercise 3.5 Factorising special products

learn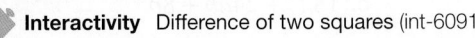

3.5 Quick quiz on	**3.5 Exercise**

Individual pathways

■ PRACTISE	■ CONSOLIDATE	■ MASTER
1, 4, 7, 10, 12, 15, 17, 20, 22, 26, 28	2, 5, 8, 11, 13, 16, 19, 21, 23, 24, 27, 29	3, 6, 9, 14, 18, 25, 30

Fluency

1. **WE7** Factorise each of the following expressions using the difference of two squares rule.

    a. $x^2 - 25$ b. $x^2 - 81$ c. $a^2 - 16$

2. Factorise each of the following expressions using the difference of two squares rule.

    a. $25 - p^2$ b. $121 - a^2$ c. $36 - y^2$

3. Factorise each of the following expressions using the difference of two squares rule.

    a. $4b^2 - 25$ b. $9a^2 - 16$ c. $25d^2 - 1$

4. Factorise the following expressions.
   a. $x^2 - y^2$
   b. $a^2 - b^2$
   c. $p^2 - q^2$

5. Factorise the following expressions.
   a. $25m^2 - n^2$
   b. $81x^2 - y^2$
   c. $p^2 - 36q^2$

6. Factorise the following expressions.
   a. $36m^2 - 25n^2$
   b. $16q^2 - 9p^2$
   c. $4m^2 - 49n^2$

7. Factorise the following expressions.
   a. $(x + 9)^2 - 16$
   b. $(p + 8)^2 - 25$
   c. $(p - 2)^2 - q^2$

8. Factorise the following expressions.
   a. $(c - 6)^2 - d^2$
   b. $(x + 7)^2 - y^2$
   c. $(p + 5)^2 - q^2$

9. Factorise the following expressions.
   a. $(a - 3)^2 - 1$
   b. $(b - 1)^2 - 36$
   c. $(p + 5)^2 - 25$

10. Factorise the following expressions completely after removing any common factors.
    a. $2m^2 - 32$
    b. $5y^2 - 45$
    c. $6p^2 - 24$

11. Factorise the following expressions completely after removing any common factors.
    a. $ax^2 - 9a$
    b. $8b - 2bx^2$
    c. $3(b + 5)^2 - 48$

12. **WE8** Factorise the following expressions by recognising the perfect square rule.
    a. $x^2 + 10x + 25$
    b. $p^2 - 24p + 144$
    c. $n^2 + 20n + 100$

13. Fully factorise the following expressions by recognising the perfect square rule.
    a. $u^2 - 2uv + v^2$
    b. $64 + 16e + e^2$
    c. $4m^2 - 20m + 25$

14. Fully factorise the following expressions by recognising the perfect square rule.
    a. $2x^2 + 24x + 72$
    b. $3x^2 - 24xy + 48y^2$
    c. $18a^2 + 24ab + 8b^2$

Understanding

15. **MC** The expression that can be factorised using the difference of two squares is:
    A. $x^2 - 36$
    B. $x^2 + 36$
    C. $x^2 - 12x + 36$
    D. $x^2 + 12x - 36$

16. **MC** Which of the following expressions can be factorised by the perfect squares rule? Select all possible answers from the options shown.
    A. $x^2 + 6x + 36$
    B. $x^2 - 12x + 36$
    C. $x^2 + 12x - 36$
    D. $x^2 + 12x + 36$

17. **MC** The expression $x^2 - 121$ factorises to:
    A. $(x - 11)^2$
    B. $(x - 11)(x + 12)$
    C. $(x + 11)(x - 11)$
    D. $(x - 11)(x - 11)$

18. **MC** What does the expression $36m^2 - 12m + 1$ factorise to?
    A. $(6m - 1)^2$
    B. $36(m + 1)(m - 1)$
    C. $(6m + 1)^2$
    D. $6(6m - 1)^2$

19. **MC** What does the expression $16a^2 - 25b^2$ factorise to?
    A. $16(a + 5b)(a - 5b)$
    B. $4(4a - 5b)^2$
    C. $(16a + 25b)(16a - 25b)$
    D. $(4a + 5b)(4a - 5b)$

20. **MC** The expression $5c^2 - 20c + 20$ factorises to:
    A. $5(c - 2)^2$
    B. $5(c + 4)(c - 4)$
    C. $5(c - 4)^2$
    D. $(5c + 20)(5c - 20)$

21. **MC** The expression $(x + 4)^2 - 9$ factorises to:
    A. $9(x + 4)(x - 4)$
    B. $(x + 13)(x - 5)$
    C. $(x + 7)(x + 1)$
    D. $(x - 1)(x + 5)$

Communicating, reasoning and problem solving

22. A circular pool with a radius of r metres is surrounded by a circular path 1 m wide.

   a. Calculate the area of the surface of the pool.
   b. Determine, in terms of r, the distance from the centre of the pool to the outer edge of the path.
   c. Evaluate the area of the circle that includes the path and the pool. Give your answer in fully factorised form.
   d. Write an expression for the area of the path. Do not fully simplify.
   e. Simplify the expression from part d.
   f. If the pool had a radius of 5 m, what would be the area of the path correct to the nearest square metre?
   g. If the pool had a radius of 7 m, what would be the area of the path correct to the nearest square metre?

23. An L-shaped piece of farm land is to be planted with wheat.

   a. Write an expression for the area of the wheat. Give your answer in the form $aw^2 + bw + c$.
   b. Calculate how much wheat is to be planted if w is 5.
   c. Calculate how much wheat is to be planted if w is 10.
   d. Calculate how much wheat is to be planted if w is 20.

24. Factorise the following using the difference of two squares rule.

   a. $x^2 - 10$
   b. $4x^2 - 32$
   c. $x^4 - y^2$

25. Factorise the following using the difference of two squares rule.

   a. $(x + 2)^2 - (x - 3)^2$
   b. $y^2 - 4x^4$
   c. $10^4 - x^4$

26. Is $(x + 4)^2$ equal to $x^2 + 16$? Explain using a numerical example.

27. Show that $(a + b)(a - b) = (a - b)(a + b)$.

28. Five squares of increasing size and five area formulas are given below.

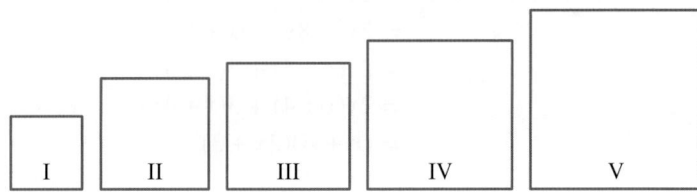

$A_a = x^2 + 6x + 9$

$A_b = x^2 + 10x + 25$

$A_c = x^2 + 16x + 64$

$A_d = x^2 - 6x + 9$

$A_e = x^2 + 12x + 36$

   a. Use factorisation to determine the side length that correlates to each area formula.
   b. Using the areas given and the side lengths found in part a, state the factorised algebraic expressions for the area of the squares.
   c. If $x = 5$ cm, use the formulas given to calculate the area of each square.

29. The area of a rectangle is represented by $(x^2 - 49)$ mm^2.

   a. Factorise the expression.
   b. Using the factorised expression from part **a**, determine a possible length and width for the rectangle. (Length is the longer measurement).
   c. If $x = 25$ mm, determine the dimensions for the rectangle.
   d. If $x = 50$ mm, how much bigger is the area of the rectangle than in part **c**?

30. Use the difference of two squares to completely factorise $x^8 - 1$. Show your working.

LESSON
3.6 Factorising non-monic quadratics

LEARNING INTENTION

At the end of this lesson you should be able to:
   • express $ax^2 + bx + c$ in the form $ax^2 + mx + nc + c$ where $m + n = b$
   • factorise non-monic quadratic trinomial expressions.

⊙ 3.6.1 Factorising quadratic trinomials where $a \neq 1$

eles-6262

• When a quadratic trinomial in the form $ax^2 + bx + c$ is written as $ax^2 + mx + nx + c$, where $m + n = b$, the four terms can be factorised by grouping.
• For example, consider $2x^2 + 11x + 12$

$$2 \times 12 = 24$$

Numbers that multiply to give 24
and add to 11
therefore $\boxed{8}$ and $\boxed{3}$.

$$2x^2 + 11x + 12 = 2x^2 + \boxed{8x} + \boxed{3x} + 12$$
$$= 2x^2 + 8x + 3x + 12$$
$$= 2x(x + 4) + 3(x + 4)$$
$$= 2x(x + 4) + 3(x + 4)$$
$$= (x + 4)(2x + 3)$$

• There are many combinations of numbers that satisfy $m + n = b$, however only one particular combination can be grouped and factorised.
For example,

$$2x^2 + 11x + 12 = 2x^2 + 7x + 4x - 12 \qquad\qquad \text{or} \qquad 2x^2 + 11x + 12 = 2x^2 + 8x + 3x + 12$$
$$= 2x^2 + 7x + 4x + 12 \qquad\qquad\qquad\qquad\qquad\qquad = 2x^2 + 8x + 3x + 12$$
$$= x(2x + 7) + 4(x + 3) \qquad\qquad\qquad\qquad\qquad\qquad = 2x(x + 4) + 3(x + 4)$$
$$\text{cannot be factorised further} \qquad\qquad\qquad\qquad\qquad = (x + 4)(2x + 3)$$

• In examining the general binomial expansion, a pattern emerges that can be used to help identify which combination to use for $m + n = b$.

$$(dx + e)(fx + g) = dfx^2 + dgx + efx + eg$$
$$= dfx^2 + (dg + ef)x + eg$$

$$m + n = dg + ef \quad \text{and} \quad m \times n = dg \times ef$$
$$= b \qquad\qquad\qquad\qquad = dgef$$
$$\qquad\qquad\qquad\qquad\qquad = dfeg$$
$$\qquad\qquad\qquad\qquad\qquad = ac$$

- Therefore, m and n are factors of ac that sum to b.
- To factorise a general quadratic where $a \neq 1$, look for factors of ac that sum to b. Then rewrite the quadratic trinomial with four terms that can then be grouped and factorised.
- Remember to look for any common factors of the quadratic trinomial.

Factorising general quadratic expressions

$$ax^2 + bx + c = ax^2 + mx + nx + c$$

Factors of ac that sum to b

WORKED EXAMPLE 9 Factorising non-monic quadratic expressions

Factorise $6x^2 - 11x - 10$.

THINK

1. Write the expression and look for common factors and special patterns. The expression is a general quadratic with $a = 6$, $b = -11$ and $c = -10$.

2. Since $a \neq 1$, rewrite $ax^2 + bx + c$ as $ax^2 + mx + nx + c$, where m and n are factors of $ac(6 \times -10)$ that sum to $b(-11)$. Calculate the sums of factor pairs of -60.

   As shown in pink, 4 and -15 are factors of -60 that add to -11.

3. Rewrite the quadratic expression: $ax^2 + bx + c = ax^2 + mx + nx + c$ with $m = 4$ and $n = -15$.

4. Factorise using the grouping method.

5. Write the answer.

WRITE

$6x^2 - 11x - 10 = 6x^2 + -11x + -10$

Factors of -60	Sum of factors
-60, 1	-59
-20, 3	-17
-30, 2	-28
15, -4	11
-15, 4	-11

$6x^2 - 11x - 10 = 6x^2 + 4x + -15x - 10$

$6x^2 - 11x - 10 = 2x(3x + 2) + -5(3x + 2)$
$\qquad\qquad\qquad = (3x + 2)(2x - 5)$

---

on Resources

▶ **Video eLesson** Factorisation of trinomials (eles-1921)

🧩 **Interactivitiy** Factorising quadratic trinomials where $y = ax^2 + bx + c$, $a > 1$ (int-6093)

Exercise 3.6 Factorising non-monic quadratics

learnon

| 3.6 Quick quiz   on | 3.6 Exercise |

Individual pathways

■ PRACTISE	■ CONSOLIDATE	■ MASTER
1, 5, 8, 11, 14, 16	2, 3, 6, 9, 12, 15, 17	4, 7, 10, 13, 18

Fluency

1. Fully factorise each of the following quadratic trinomials, by first taking out a common factor.
   a. $2x^2 + 10x + 12$
   b. $3x^2 + 15x + 12$
   c. $4x^2 - 4x - 24$

2. Fully factorise each of the following quadratic trinomials, by first taking out a common factor.
   a. $3x^2 + 9x - 30$
   b. $2x^2 + 8x - 42$
   c. $5x^2 + 20x - 60$

3. WE9 Factorise each of the following quadratic trinomials.
   a. $5x^2 + 3x - 2$
   b. $7x^2 - 17x + 6$
   c. $10x^2 - 11x - 6$

4. Factorise each of the following quadratic trinomials.
   a. $2x^2 + 7x + 3$
   b. $2x^2 - 7x + 3$
   c. $3x^2 - x - 2$

5. MC The expression $3x^2 + 21x + 36$ is equal to:
   A. $3(x + 6)(x + 2)$
   B. $3(x + 4)(x + 3)$
   C. $(3x + 9)(x + 4)$
   D. $(3x + 2)(x + 18)$

6. MC The expression $2x^2 - 16x + 14$ is equal to:
   A. $2(x - 1)(x - 7)$
   B. $2(x + 1)(x + 7)$
   C. $(2x - 7)(x - 2)$
   D. $(2x - 2)(x - 7)$

7. MC The expression $12x^2 + 58x - 10$ is equal to:
   A. $(12x - 2)(x - 5)$
   B. $2(6x + 1)(x - 5)$
   C. $2(6x + 5)(x - 1)$
   D. $2(6x - 1)(x + 5)$

Understanding

8. Factorise the following trinomials
   a. $2x^2 + 3x - 2$
   b. $8x^2 + 2x - 3$
   c. $6x^2 + x - 2$

9. Factorise
   a. $3x^2 - x - 4$
   b. $2x^2 + 9x + 10$

10. Factorise
    a. $6x^2 + 5x - 21$
    b. $2x^2 + 5x - 12$

11. Factorise the following expressions.
    a. $6x^2 - 5x - 6$
    b. $4x^2 + 13x + 10$

12. Factorise the following expressions.
    a. $4x^2 - 19x + 15$
    b. $6x^2 + 5x - 4$

13. Factorise $3x^2 + 15x + 12$:
    a. by first taking out a common factor.
    b. by first separating the middle term, then grouping terms.

14. Factorise the following expressions. Check your answer by expanding.
    a. $4x^2 - 17x - 15$
    b. $3x^2 + 10x + 3$
    c. $10x^2 - 9x - 9$

Communicating, reasoning and problem solving

15. Answer the following questions about the quadratic $4x^2 - 9x - 9$.
    a. Explain why the quadratic can be written as $4x^2 + 3x - 12x - 9$.
    b. Factorise the quadratic.

16. Answer the following questions.
    a. Factorise $12x^2 - 4x - 5$.
    b. Assume that the factors from part a represent the length and width of a rectangle. What is the area of the rectangle?
    c. For what value of x is the rectangle a square?
    d. For what values of x is one side twice as long as the other? Give both possible answers, putting the lowest value first.

17. The area expressions below relate to either squares or rectangles.
    a. Without completing any algebraic operations, examine the expressions and determine which belong to squares and which to rectangles. Explain your answer.
       i. $9s^2 + 48s + 64$ ii. $25s^2 - 4$ iii. $s^2 + 4s + 3$ iv. $4s^2 - 24s - 36$
    b. Fully factorise each expression.
       i. $9s^2 + 48s + 64$ ii. $25s^2 - 4$ iii. $s^2 + 4s + 3$ iv. $4s^2 - 24s - 36$

18. Factorise the following expressions.
    a. $9b^2 + 48b + 64$ b. $(ac)^2 - d^2$ c. $a^2 + ab - 182b^2$ d. $b^2 - 2abc + a^2c^2$

LESSON
3.7 Adding and subtracting algebraic fractions

LEARNING INTENTION

At the end of this lesson you should be able to:
- determine the lowest common denominator of two or more fractions with pronumerals in the denominator
- add and subtract fractions involving algebraic expressions.

⊙ 3.7.1 Algebraic fractions

eles-4698

- In an algebraic fraction, the denominator, the numerator or both contain pronumerals.

  For example, $\dfrac{x}{2}$, $\dfrac{3x+1}{2x-5}$ and $\dfrac{1}{x^2+5}$ are all **algebraic fractions**.

- As with all fractions, algebraic fractions must have a common denominator if they are to be added or subtracted, so an important step is to determine the lowest common denominator (LCD).

Simplify the following expressions.

a. $\dfrac{2x}{3} - \dfrac{x}{2}$

b. $\dfrac{x+1}{6} + \dfrac{x+4}{4}$

THINK	WRITE
a. 1. Write the expression.	a. $\dfrac{2x}{3} - \dfrac{x}{2}$
2. Rewrite each fraction as an equivalent fraction using the LCD of 3 and 2, which is 6.	$= \dfrac{2x}{3} \times \dfrac{2}{2} - \dfrac{x}{2} \times \dfrac{3}{3}$ $= \dfrac{4x}{6} - \dfrac{3x}{6}$
3. Express as a single fraction.	$= \dfrac{4x - 3x}{6}$
4. Simplify the numerator and write the answer.	$= \dfrac{x}{6}$
b. 1. Write the expression.	b. $\dfrac{x+1}{6} + \dfrac{x+4}{4}$
2. Rewrite each fraction as an equivalent fraction using the LCD of 6 and 4, which is 12.	$= \dfrac{x+1}{6} \times \dfrac{2}{2} + \dfrac{x+4}{4} \times \dfrac{3}{3}$ $= \dfrac{2(x+1)}{12} + \dfrac{3(x+4)}{12}$
3. Express as a single fraction.	$= \dfrac{2(x+1) + 3(x+4)}{12}$
4. Simplify the numerator by expanding brackets and collecting like terms.	$= \dfrac{2x + 2 + 3x + 12}{12}$
5. Write the answer.	$= \dfrac{5x + 14}{12}$

3.7.2 Pronumerals in the denominator

eles-4699

- If pronumerals appear in the denominator, the process involved in adding and subtracting the fractions is to determine a lowest common denominator as usual.
- When there is an algebraic expression in the denominator of each fraction, a common denominator can be obtained by writing the product of the denominators. For example, if $x + 3$ and $2x - 5$ are in the denominator of each fraction, then a common denominator of the two fractions will be $(x + 3)(2x - 5)$.

Simplify $\dfrac{2}{3x} - \dfrac{1}{4x}$.

THINK	WRITE
1. Write the expression.	$\dfrac{2}{3x} - \dfrac{1}{4x}$

2. Rewrite each fraction as an equivalent fraction using the LCD of $3x$ and $4x$, which is $12x$.
Note: $12x^2$ is not the lowest LCD.

$$= \frac{2}{3x} \times \frac{4}{4} - \frac{1}{4x} \times \frac{3}{3}$$

$$= \frac{8}{12x} - \frac{3}{12x}$$

3. Express as a single fraction.

$$= \frac{8-3}{12x}$$

4. Simplify the numerator and write the answer.

$$= \frac{5}{12x}$$

WORKED EXAMPLE 12 Simplifying by finding the LCD of two algebraic expressions

Simplify $\dfrac{x+1}{x+3} + \dfrac{2x-1}{x+2}$ by writing it first as a single fraction.

THINK

1. Write the expression.

2. Rewrite each fraction as an equivalent fraction using the LCD of $x+3$ and $x+2$, which is the product $(x+3)(x+2)$.

3. Express as a single fraction.

4. Simplify the numerator by expanding brackets and collecting like terms.
Note: The denominator is generally kept in factorised form. That is, it is not expanded.

5. Write the answer.

WRITE

$$\frac{x+1}{x+3} + \frac{2x-1}{x+2}$$

$$= \frac{(x+1)}{(x+3)} \times \frac{(x+2)}{(x+2)} + \frac{(2x-1)}{(x+2)} \times \frac{(x+3)}{(x+3)}$$

$$= \frac{(x+1)(x+2)}{(x+3)(x+2)} + \frac{(2x-1)(x+3)}{(x+3)(x+2)}$$

$$= \frac{(x+1)(x+2) + (2x-1)(x+3)}{(x+3)(x+2)}$$

$$= \frac{(x^2+2x+x+2) + (2x^2+6x-x-3)}{(x+3)(x+2)}$$

$$= \frac{x^2+3x+2+2x^2+5x-3}{(x+3)(x+2)}$$

$$= \frac{3x^2+8x-1}{(x+3)(x+2)}$$

WORKED EXAMPLE 13 Simplification involving repeated linear factors

Simplify $\dfrac{x+2}{x-3} + \dfrac{x-1}{(x-3)^2}$ by writing it first as a single fraction.

THINK

1. Write the expression.

2. Rewrite each fraction as an equivalent fraction using the LCD of $x-3$ and $(x-3)^2$, which is $(x-3)^2$.

WRITE

$$\frac{x+2}{x-3} + \frac{x-1}{(x-3)^2}$$

$$= \frac{x+2}{x-3} \times \frac{x-3}{x-3} + \frac{x-1}{(x-3)^2}$$

$$= \frac{(x+2)(x-3)}{(x-3)^2} + \frac{x-1}{(x-3)^2}$$

$$= \frac{x^2-x-6}{(x-3)^2} + \frac{x-1}{(x-3)^2}$$

3. Express as a single fraction.

$$= \frac{x^2 - x - 6 + x - 1}{(x-3)^2}$$

4. Simplify the numerator and write the answer.

$$= \frac{x^2 - 7}{(x-3)^2}$$

 Resources

Interactivities Adding and subtracting algebraic fractions (int-6113)
Lowest common denominators with pronumerals (int-6114)

Exercise 3.7 Adding and subtracting algebraic fractions   learn on

| 3.7 Quick quiz | 3.7 Exercise |

Individual pathways

| ■ PRACTISE | ■ CONSOLIDATE | ■ MASTER |
| 1, 4, 7, 10, 13, 16, 19 | 2, 5, 8, 11, 14, 17, 20 | 3, 6, 9, 12, 15, 18 |

Fluency

For questions 1 to 3, simplify each of the following.

1. a. $\dfrac{4}{7} + \dfrac{2}{3}$
   b. $\dfrac{1}{8} + \dfrac{5}{9}$
   c. $\dfrac{3}{5} + \dfrac{6}{15}$

2. a. $\dfrac{4}{9} - \dfrac{3}{11}$
   b. $\dfrac{3}{7} - \dfrac{2}{5}$
   c. $\dfrac{1}{5} - \dfrac{x}{6}$

3. a. $\dfrac{5x}{9} - \dfrac{4}{27}$
   b. $\dfrac{3}{8} - \dfrac{2x}{5}$
   c. $\dfrac{5}{x} - \dfrac{2}{3}$

WE10 For questions 4 to 6, simplify the following expressions.

4. a. $\dfrac{2y}{3} - \dfrac{y}{4}$
   b. $\dfrac{y}{8} - \dfrac{y}{5}$
   c. $\dfrac{4x}{3} - \dfrac{x}{4}$
   d. $\dfrac{8x}{9} + \dfrac{2x}{3}$

5. a. $\dfrac{2w}{14} - \dfrac{w}{28}$
   b. $\dfrac{y}{20} - \dfrac{y}{4}$
   c. $\dfrac{12y}{5} + \dfrac{y}{7}$
   d. $\dfrac{10x}{5} + \dfrac{2x}{15}$

6. a. $\dfrac{x+1}{5} + \dfrac{x+3}{2}$
   b. $\dfrac{x+2}{4} + \dfrac{x+6}{3}$
   c. $\dfrac{2x-1}{5} - \dfrac{2x+1}{6}$
   d. $\dfrac{3x+1}{2} + \dfrac{5x+2}{3}$

WE11 For questions 7 to 9, simplify the following.

7. a. $\dfrac{2}{4x} + \dfrac{1}{8x}$
   b. $\dfrac{3}{4x} - \dfrac{1}{3x}$
   c. $\dfrac{5}{3x} + \dfrac{1}{7x}$

8. a. $\dfrac{12}{5x} + \dfrac{4}{15x}$
   b. $\dfrac{1}{6x} + \dfrac{1}{8x}$
   c. $\dfrac{9}{4x} - \dfrac{9}{5x}$

9. a. $\dfrac{2}{100x} + \dfrac{7}{20x}$ **b.** $\dfrac{1}{10x} + \dfrac{5}{x}$ **c.** $\dfrac{4}{3x} - \dfrac{3}{2x}$

WE12&13 For questions **10** to **12**, simplify the following by writing as single fractions.

10. a. $\dfrac{2}{x+4} + \dfrac{3x}{x-2}$ **b.** $\dfrac{2x}{x+5} + \dfrac{5}{x-1}$ **c.** $\dfrac{5}{2x+1} + \dfrac{x}{x-2}$ **d.** $\dfrac{2x}{x+1} - \dfrac{3}{2x-7}$

11. a. $\dfrac{4x}{x+7} + \dfrac{3x}{x-5}$ **b.** $\dfrac{x+2}{x+1} + \dfrac{x-1}{x+4}$ **c.** $\dfrac{x+8}{x+1} - \dfrac{2x+1}{x+2}$ **d.** $\dfrac{x+5}{x+3} - \dfrac{x-1}{x-2}$

12. a. $\dfrac{x+1}{x+2} - \dfrac{2x-5}{3x-1}$ **b.** $\dfrac{2}{x-1} - \dfrac{3}{1-x}$ **c.** $\dfrac{4}{(x+1)^2} + \dfrac{3}{x+1}$ **d.** $\dfrac{3}{x-1} - \dfrac{1}{(x-1)^2}$

Understanding

13. A classmate attempted to complete an algebraic fraction subtraction problem.

$$\dfrac{x}{x-1} - \dfrac{3}{x-2} = \dfrac{x}{x-1} \times \dfrac{(x-2)}{(x-2)} - \dfrac{3}{x-2} \times \dfrac{(x-1)}{(x-1)}$$

$$= \dfrac{x(x-2) - 3(x-1)}{(x-1)(x+2)}$$

$$= \dfrac{x^2 - 2x - 3x - 1}{(x-1)(x+2)}$$

$$= \dfrac{x^2 - 5x - 1}{(x-1)(x+2)}$$

a. Identify the mistake(s) she made.

b. Determine the correct answer.

14. Simplify the following.

a. $\dfrac{y-x}{x-y}$ **b.** $\dfrac{3}{x-2} + \dfrac{4}{2-x}$

15. Simplify the following.

a. $\dfrac{3}{3-x} + \dfrac{3x}{(x-3)^2}$ **b.** $\dfrac{1}{x-2} - \dfrac{2x}{(2-x)^2} + \dfrac{x^2}{(x-2)^3}$

Communicating, reasoning and problem solving

16. Simplify the following.

a. $\dfrac{1}{x+2} + \dfrac{2}{x+1} + \dfrac{1}{x+3}$

b. $\dfrac{1}{x-1} + \dfrac{4}{x+2} + \dfrac{2}{x-4}$

17. Simplify the following.

a. $\dfrac{3}{x+1} + \dfrac{2}{x+3} - \dfrac{1}{x+2}$

b. $\dfrac{2}{x-4} - \dfrac{3}{x-1} + \dfrac{5}{x+3}$

c. Explain why the process that involves determining the lowest common denominator is important in parts **a** and **b**.

18. The reverse process of adding or subtracting algebraic fractions is quite complex. Use trial and error, or technology, to determine the value of a if $\dfrac{7x-4}{(x-8)(x+5)} = \dfrac{a}{x-8} + \dfrac{3}{x+5}$.

19. Simplify $\dfrac{3}{x^2+7x+12} - \dfrac{1}{x^2+x-6} + \dfrac{2}{x^2+2x-8}$.

20. Simplify $\dfrac{x^2+3x-18}{x^2-x-42} - \dfrac{x^2-3x+2}{x^2-5x+4}$.

LESSON
3.8 Multiplying and dividing algebraic fractions

LEARNING INTENTION

At the end of this lesson you should be able to:
- multiply and divide fractions involving algebraic expressions and simplify the result
- simplify complex algebraic expressions involving algebraic fractions using factorisation.

▶ 3.8.1 Simplifying algebraic fractions

eles-6263

- Algebraic fractions can be simplified using the index laws and by cancelling factors common to the numerator and denominator.
- A fraction can only be simplified if:
  - there is a common factor in the numerator and the denominator
  - the numerator and denominator are both written in factorised form, that is, as the *product* of two or more factors.

$$\dfrac{3ab}{12a} = \dfrac{^1\cancel{3} \times {}^1\cancel{a} \times b}{^4\cancel{12} \times {}^1\cancel{a}} \begin{array}{l} \leftarrow \text{product of factors} \\ \leftarrow \text{product of factors} \end{array} \qquad \dfrac{3a+b}{12a} = \dfrac{3 \times a + b}{12 \times a} \begin{array}{l} \leftarrow \text{not a product of factors} \\ \leftarrow \text{product of factors} \end{array}$$

$$= \dfrac{b}{4} \qquad\qquad\qquad\qquad\qquad\qquad \textbf{Cannot be simplified}$$

WORKED EXAMPLE 14 Factorising the numerator and denominator to simplify

Simplify each of the following fractions by factorising the numerator and the denominator, and then cancelling as appropriate.

a. $\dfrac{3x+9}{15}$ b. $\dfrac{5}{10x+20}$ c. $\dfrac{10x+15}{6x+9}$ d. $\dfrac{x^2+4}{x^2-5x}$

THINK   **WRITE**

a. 1. Write the fraction.   a. $\dfrac{3x+9}{15}$

2. Factorise both the numerator and the denominator. $$= \frac{3(x+3)}{15}$$

3. Cancel the common factors. $$= \frac{{}^1\cancel{3}(x+3)}{{}_1\cancel{3} \times 5}$$

4. Simplify the resulting fraction. $$= \frac{(x+3)}{5}$$

b. 1. Write the fraction. b. $\dfrac{5}{10x+20}$

2. Factorise both the numerator and the denominator. $$= \frac{5 \times 1}{10(x+2)}$$

3. Cancel the common factors. $$= \frac{{}^1\cancel{5} \times 1}{{}_2\cancel{10}(x+2)}$$

4. Simplify the resulting fraction. $$= \frac{1}{2(x+2)}$$

c. 1. Write the fraction. c. $\dfrac{10x+15}{6x+9}$

2. Factorise both the numerator and the denominator. $$= \frac{5\,\cancel{(2x+3)}^{\,1}}{3\,\cancel{(2x+3)}^{\,1}}$$

3. Cancel the common factors. $$= \frac{5}{3}$$

d. 1. Write the fraction. d. $\dfrac{x^2+4x}{x^2-5x}$

2. Factorise both the numerator and the denominator. $$= \frac{{}^1\cancel{x}(x+4)}{{}_1\cancel{x}(x-5)}$$

3. Cancel the common factors. $$= \frac{x+4}{x-5}$$

WORKED EXAMPLE 15 Factorising quadratic expressions to find common factors

Simplify each of the following algebraic fractions.

a. $\dfrac{x^2+3x-4}{x-1}$ b. $\dfrac{x^2-7x-8}{x^2+3x+2}$ c. $\dfrac{x^2-6x-8}{2x^2-16x+30}$

THINK

WRITE

a. 1. Write the expression.

a. $\dfrac{x^2+3x-4}{x-1}$

2. Factorise the numerator and denominator and cancel the common factors. $$= \frac{{}^1\cancel{(x-1)}(x+4)}{\cancel{x-1}^{\,1}}$$

3. Simplify. $$= x+4$$

b. 1. Write the original fraction.

b. $\dfrac{x^2 - 7x - 8}{x^2 + 3x + 2}$

2. Factorise the numerator and denominator.

$= \dfrac{^1(x+1)(x-8)}{^1(x+1)(x+2)}$

3. Cancel any common factors and simplify.

$= \dfrac{x-8}{x+2}$

c. 1. Write the original fraction.

c. $\dfrac{x^2 - 6x + 5}{2x^2 - 16 + 30}$

2. Factorise the numerator and denominator.

$= \dfrac{(x-1)\,(x-5)^1}{2\,(x-3)\,(x-5)^1}$

3. Cancel any common factors and simplify.

$= \dfrac{(x-1)}{2(x-3)}$

⏵ 3.8.2 Multiplying algebraic fractions

eles-6264

- Multiplication of algebraic fractions follows the same rules as multiplication of numerical fractions: multiply the numerators, then multiply the denominators.

WORKED EXAMPLE 16 Multiplying algebraic fractions and simplifying the result

Simplify each of the following.

a. $\dfrac{5y}{3x} \times \dfrac{6z}{7y}$

b. $\dfrac{2x}{(x+1)\,(2x-3)} \times \dfrac{x+1}{x}$

THINK

WRITE

a. 1. Write the expression.

a. $\dfrac{5y}{3x} \times \dfrac{6z}{7y}$

2. Cancel common factors in the numerator and denominator. The y can be cancelled in the denominator and the numerator. Also, the 3 in the denominator can divide into the 6 in the numerator.

$= \dfrac{5y^1}{_1 3x} \times \dfrac{6^2 z}{7y^1}$

$= \dfrac{5}{x} \times \dfrac{2z}{7}$

3. Multiply the numerators, then multiply the denominators and write the answer.

$= \dfrac{10z}{7x}$

b. 1. Write the expression.

b. $\dfrac{2x}{(x+1)\,(2x-3)} \times \dfrac{x+1}{x}$

2. Cancel common factors in the numerator and the denominator. $(x+1)$ and the x are both common in the numerator and the denominator and can therefore be cancelled.

$= \dfrac{2x^1}{^1(x+1)(2x-3)} \times \dfrac{x+1^1}{x^1}$

$= \dfrac{2}{2x-3} \times \dfrac{1}{1}$

3. Multiply the numerators, then multiply the denominators and write the answer.

$= \dfrac{2}{2x-3}$

⏵ 3.8.3 Dividing algebraic fractions

eles-4701

- When dividing algebraic fractions, follow the same rules as for division of numerical fractions: write the division as a multiplication and invert the second fraction.
- This process is sometimes known as multiplying by the **reciprocal**.

WORKED EXAMPLE 17 Dividing algebraic fractions

Simplify the following expressions.

a. $\dfrac{3xy}{2} \div \dfrac{4x}{9y}$

b. $\dfrac{4}{(x+1)(3x-5)} \div \dfrac{x-7}{x+1}$

THINK	WRITE
a. 1. Write the expression.	a. $\dfrac{3xy}{2} \div \dfrac{4x}{9y}$
2. Change the division sign to a multiplication sign and write the second fraction as its reciprocal.	$= \dfrac{3xy}{2} \times \dfrac{9y}{4x}$
3. Cancel common factors in the numerator and denominator. The pronumeral x is common to both the numerator and denominator and can therefore be cancelled.	$= \dfrac{3y}{2} \times \dfrac{9y}{4}$
4. Multiply the numerators, then multiply the denominators and write the answer.	$= \dfrac{27y^2}{8}$
b. 1. Write the expression.	b. $\dfrac{4}{(x+1)(3x-5)} \div \dfrac{x-7}{x+1}$
2. Change the division sign to a multiplication sign and write the second fraction as its reciprocal.	$= \dfrac{4}{(x+1)(3x-5)} \times \dfrac{x+1}{x-7}$
3. Cancel common factors in the numerator and denominator. $(x+1)$ is common to both the numerator and denominator and can therefore be cancelled.	$= \dfrac{4}{3x-5} \times \dfrac{1}{x-7}$
4. Multiply the numerators, then multiply the denominators and write the answer.	$= \dfrac{4}{(3x-5)(x-7)}$

---

on Resources

Interactivities Simplifying algebraic fractions (int-6115)
Multiplying algebraic fractions (int-6116)
Dividing algebraic fractions (int-6117)
Cancelling common factors (int-6094)

Exercise 3.8 Multiplying and dividing algebraic fractions

3.8 Quick quiz on	3.8 Exercise

Individual pathways

■ PRACTISE	■ CONSOLIDATE	■ MASTER
1, 5, 8, 11, 14, 17, 19, 22	2, 3, 6, 9, 12, 15, 20, 23	4, 7, 10, 13, 16, 18, 21, 24, 25

Fluency

1. **WE14** Simplify each of the following by first factorising the numerator and then cancelling as appropriate.

   a. $\dfrac{2a+2}{2}$ b. $\dfrac{3a+6}{9}$ c. $\dfrac{4a-4}{4}$

2. Simplify each of the following by first factorising the numerator and then cancelling as appropriate.

   a. $\dfrac{4x+8}{8}$ b. $\dfrac{3x-12}{6}$ c. $\dfrac{6x+36}{2}$

3. **WE15** Simplify each of the following by first factorising the numerator and then cancelling as appropriate.

   a. $\dfrac{x^2+5x+6}{x+3}$ b. $\dfrac{x^2+7x+12}{x+4}$ c. $\dfrac{x^2-9x+20}{x-5}$

4. Simplify each of the following by first factorising the numerator and the denominator.

   a. $\dfrac{a^2-7a+12}{a^2-16}$ b. $\dfrac{p^2-4p-5}{p^2-25}$ c. $\dfrac{x^2+6x+9}{x^2+2x-3}$

WE16a For questions 5 to 7, simplify each of the following.

5. a. $\dfrac{x}{5}\times\dfrac{20}{y}$ b. $\dfrac{x}{4}\times\dfrac{12}{y}$

   c. $\dfrac{y}{4}\times\dfrac{16}{x}$ d. $\dfrac{x}{2}\times\dfrac{9}{2y}$

6. a. $\dfrac{x}{10}\times\dfrac{-25}{2y}$ b. $\dfrac{3w}{-14}\times\dfrac{-7}{x}$

   c. $\dfrac{3y}{4x}\times\dfrac{8z}{7y}$ d. $\dfrac{-y}{3x}\times\dfrac{6z}{-7y}$

7. a. $\dfrac{x}{3z}\times\dfrac{-9z}{2y}$ b. $\dfrac{5y}{3x}\times\dfrac{x}{8y}$

   c. $\dfrac{-20y}{7x}\times\dfrac{-21z}{5y}$ d. $\dfrac{y}{-3w}\times\dfrac{x}{2y}$

WE16b For questions 8 to 10, simplify the following expressions.

8. a. $\dfrac{2x}{(x-1)(3x-2)}\times\dfrac{x-1}{x}$ b. $\dfrac{5x}{(x-3)(4x+7)}\times\dfrac{4x+7}{x}$

   c. $\dfrac{9x}{(5x+1)(x-6)}\times\dfrac{5x+1}{2x}$ d. $\dfrac{(x+4)}{(x+1)(x+3)}\times\dfrac{x+1}{x+4}$

9. **a.** $\dfrac{2x}{x+1} \times \dfrac{x-1}{(x+1)(x-1)}$

b. $\dfrac{2}{x(2x-3)} \times \dfrac{x(x+1)}{4}$

c. $\dfrac{2x}{4(a+3)} \times \dfrac{3a}{15x}$

d. $\dfrac{15c}{12(d-3)} \times \dfrac{21d}{6c}$

10. **a.** $\dfrac{6x^2}{20(x-2)^2} \times \dfrac{15(x-2)}{16x^4}$

b. $\dfrac{7x^2(x-3)}{5x(x+1)} \times \dfrac{3(x-3)(x+1)}{14(x-3)^2(x-1)}$

WE17a For questions **11** to **13**, simplify the following expressions.

11. **a.** $\dfrac{3}{x} \div \dfrac{5}{x}$ **b.** $\dfrac{2}{x} \div \dfrac{9}{x}$ **c.** $\dfrac{4}{x} \div \dfrac{12}{x}$ **d.** $\dfrac{20}{y} \div \dfrac{20}{3y}$

12. **a.** $\dfrac{1}{5w} \div \dfrac{5}{w}$ **b.** $\dfrac{7}{2x} \div \dfrac{3}{5x}$ **c.** $\dfrac{3xy}{7} \div \dfrac{3x}{4y}$ **d.** $\dfrac{2xy}{5} \div \dfrac{5x}{y}$

13. **a.** $\dfrac{6y}{9} \div \dfrac{3x}{4xy}$ **b.** $\dfrac{8wx}{5} \div \dfrac{3w}{4y}$ **c.** $\dfrac{2xy}{5} \div \dfrac{3xy}{5}$ **d.** $\dfrac{10xy}{7} \div \dfrac{20x}{14y}$

WE17b For questions **14** to **16**, simplify the following expressions.

14. **a.** $\dfrac{9}{(x-1)(3x-7)} \div \dfrac{x+3}{x-1}$

b. $\dfrac{1}{(x+2)(2x-5)} \div \dfrac{x-9}{2x-5}$

15. **a.** $\dfrac{12(x-3)^2}{(x+5)(x-9)} \div \dfrac{4(x-3)}{7(x-9)}$

b. $\dfrac{13}{6(x-4)^2(x-1)} \div \dfrac{3(x+1)}{2(x-4)(x-1)}$

16. **a.** $\dfrac{16(x+5)(x-4)^2}{(x+3)(x+4)} \div \dfrac{8(x+3)(x+5)}{(x-4)(x+4)}$

b. $\dfrac{(x+2)(x+3)(x+4)}{(x-2)(x+3)^2} \div \dfrac{x^2+x-12}{x^2-4}$

Understanding

17. Match the following expressions with their simplified forms.

   i. $\dfrac{3x^2+x}{9x-x^2}$ **ii.** $\dfrac{7x-2x^2}{x^2+5x}$ **iii.** $\dfrac{9x^2+18x}{12x^2-3x}$ **iv.** $\dfrac{32x-16x^2}{24x+16x^2}$

   a. $\dfrac{3(x+2)}{4x-1}$ **b.** $\dfrac{2(2-x)}{3+2x}$ **c.** $\dfrac{7-2x}{x+5}$ **d.** $\dfrac{3x+1}{9-x}$

18. Match the following expressions with their simplified forms.

   i. $\dfrac{3x+6}{2x-6} \div \dfrac{x+3}{8x-24}$ **ii.** $\dfrac{10x-5}{4x+28} \div \dfrac{20x-10}{6x+1}$

   iii. $\dfrac{x^2+8x+15}{x^2+7x+10} \times \dfrac{x^2+6x+8}{x^2+10x+21}$ **iv.** $\dfrac{x^2+6x+8}{x^2+5x+6} \times \dfrac{x^2+8x+15}{x^2+7x+12}$

   a. $\dfrac{x+4}{x+7}$ **b.** $\dfrac{x+5}{x+3}$ **c.** $\dfrac{12(x+2)}{x+3}$ **d.** $\dfrac{6x+1}{8(x+7)}$

For questions **19** to **21**, determine the missing fraction.

19. a. $\dfrac{x+2}{3} \times \boxed{} = 5$

b. $\dfrac{3}{x^2} \div \boxed{} = \dfrac{1}{4}$

20. a. $\dfrac{(x+3)(x+2)}{(x-4)} \times \boxed{} = \dfrac{x-5}{x+2}$

b. $\dfrac{x^2(x-3)}{(x+4)(x-5)} \div \boxed{} = \dfrac{3x}{2(x+4)}$

21. a. $\dfrac{x^2+8x+15}{x^2-4x-21} \times \boxed{} = \dfrac{x^2-25}{x^2-11x+28}$

b. $\boxed{} \div \dfrac{x^2-2x-24}{x^2-36} = \dfrac{x^2+12x+36}{x^2}$

Communicating, reasoning and problem solving

22. Explain whether $\dfrac{3}{x+2}$ is the same as $\dfrac{1}{x+2} + \dfrac{1}{x+2} + \dfrac{1}{x+2}$.

23. Does $\dfrac{12xy+16yz^2}{20xyz}$ simplify to $\dfrac{3+4z}{5}$? Explain your reasoning.

24. a. Simplify $\dfrac{(x-4)(x+3)}{4x-x^2} \times \dfrac{x^2-x}{(x+3)(x-1)}$.

b. Identify and explain the error in the following reasoning.

$$\dfrac{(x-4)(x+3)}{4x-x^2} \times \dfrac{x^2-x}{(x+3)(x-1)}$$
$$= \dfrac{(x-4)(x+3)}{x(4-x)} \times \dfrac{x(x-1)}{(x+3)(x-1)} = 1$$

25. Simplify $\left(\dfrac{\dfrac{x^2+1}{x-1}-x}{\dfrac{x^2-1}{x+1}+1} \right) \times \left(1 - \dfrac{2}{1+\dfrac{1}{x}} \right)$.

LESSON
3.9 Applications of binomial products

LEARNING INTENTION

At the end of this lesson you should be able to:
- write an algebraic expression for worded problems
- apply the skills covered in this topic to solve worded problems.

▶ 3.9.1 Converting worded problems into algebraic expressions

eles-4605

- It is important to be able to convert worded problems (or 'real-world problems') into algebraic expressions.

> ### Key language used in worded problems
>
> **In order to be able to turn a worded problem into an algebraic expression, it helps to look out for the following kinds of words.**
>
> - **Words for addition: sum, altogether, add, more than, and, in total**
> - **Words for subtraction: difference, less than, take away, take off, fewer than**
> - **Words for multiplication: product, groups of, times, of, for each, double, triple**
> - **Words for division: quotient, split into, halve, thirds**

- Drawing diagrams is a useful strategy for understanding and solving worded problems.
- If a worded problem uses units in its questions, make sure that any answers also use those units.

WORKED EXAMPLE 18 Converting worded problems

A rectangular swimming pool measures 30 m by 20 m. A path around the
edge of the pool is x m wide on each side.
a. Determine the area of the pool.
b. Write an expression for the area of the pool plus the area of the path.
c. Write an expression for the area of the path.
d. If the path is 1.5 m wide, calculate the area of the path.

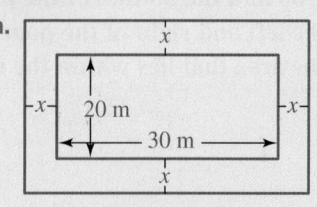

THINK	WRITE
a. 1. Construct a drawing of the pool.	a.

a. 2. Calculate the area of the pool.

$$\text{Area} = \text{length} \times \text{width}$$
$$= 20 \times 30$$
$$= 600 \, \text{m}^2$$

b. 1. Write an expression for the total length from one edge of the path to the other. Write another expression for the total width from one edge of the path to the other.

b. Length $= 30 + x + x$
$= 30 + 2x$
Width $= 20 + x + x$
$= 20 + 2x$

2. Area $=$ length $\times$ width

Area $=$ length $\times$ width
$= (30 + 2x)(20 + 2x)$
$= 600 + 60x + 40x + 4x^2$
$= (600 + 100x + 4x^2)\,\text{m}^2$

c. Determine an expression for the area of the path by subtracting the area of the pool from the total area of the pool and the path combined.

c. Area of path $=$ total area $-$ area of pool
$= 600 + 100x + 4x^2 - 600$
$= (100x + 4x^2)\text{m}^2$

d. Substitute 1.5 for x in the expression you have found for the area of the path.

d. When $x = 1.5$,
Area of path $= 100\,(1.5) + 4(1.5)^2$
$= 159\,\text{m}^2$

- The algebraic expression found in part **c** of Worked example 18 allows us to calculate the area of the path for any given width.

WORKED EXAMPLE 19 Writing expressions for worded problems

Suppose that the page of a typical textbook is 24 cm high by 16 cm wide. Each page has margins of x cm at the top and bottom, and margins of y cm on the left and right.

a. Write an expression for the height of the section of the page that lies inside the margins.
b. Write an expression for the width of the section of the page that lies inside the margins.
c. Write an expression for the area of the section of the page that lies inside the margins.
d. Show that, if the margins on the left and right are doubled, the area within the margins is reduced by $(48y - 4xy)\,\text{cm}$.
e. If the margins at the top and the bottom of the page are 1.5 cm and the margins on the left and right of the page are 1 cm, calculate the size of the area that lies within the margins.

THINK	**WRITE**
a. 1. Construct a drawing that shows the key dimensions of the page and its margins.	**a.** 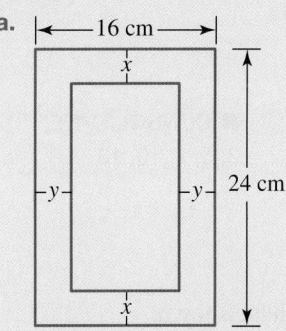
2. The total height of the page (24 cm) is effectively reduced by x cm at the top and x cm at the bottom.	$\text{Height} = 24 - x - x$ $\quad\quad\quad = (24 - 2x)\,\text{cm}$
b. The total width of the page (16 cm) is reduced by y cm on the left and y cm on the right.	**b.** $\text{Width} = 16 - y - y$ $\quad\quad\quad = (16 - 2y)\,\text{cm}$
c. The area of the page that lies within the margins is the product of the width and height.	**c.** $\text{Area}_1 = (24 - 2x)(16 - 2y)$ $\quad\quad\quad = (384 - 48y - 32x + 4xy)\,\text{cm}^2$
d. 1. If the left and right margins are doubled they both become $2y$ cm. Determine the new expression for the new width of the page.	**d.** $\text{Width} = 16 - 2y - 2y$ $\quad\quad\quad = (16 - 4y)\,\text{cm}$
2. Determine the new expression for the reduced area.	$\text{Area}_2 = (24 - 2x)(16 - 4y)$ $\quad\quad\quad = 24 \times 16 + 24 \times -4y - 2x \times 16 - 2x \times -2y$ $\quad\quad\quad = (384 - 96y - 32x + 8xy)\,\text{cm}^2$
3. Determine the difference in area by subtracting the reduced area obtained in part **d** from the original area obtained in part **c**.	$\begin{aligned}\text{Difference in area} &= \text{Area}_1 - \text{Area}_2\\ &= 384 - 48y - 32x + 4xy\\ &\quad -(384 - 96y - 32x + 8xy)\\ &= 384 - 48y - 32x + 4xy\\ &\quad -384 + 96y + 32x - 8xy\\ &= 48y - 4xy\end{aligned}$ So the amount by which the area is reduced is $(48y - 4xy)\,\text{cm}^2$.
e. Using the area found in **c**, substitute 1.5 for x and 1 for y. Solve the expression.	**e.** $\text{Area} = (384 - 48y - 32x + 4xy)\,\text{cm}^2$ $\quad\quad\quad = 384 - 48(1) - 32(1.5) + 4(1.5)(1)$ $\quad\quad\quad 294\,\text{cm}^2$

Exercise 3.9 Applications of binomial products

3.9 Quick quiz on	3.9 Exercise

Individual pathways

■ PRACTISE	■ CONSOLIDATE	■ MASTER
1, 2, 6, 9, 12	3, 5, 7, 10, 13	4, 8, 11, 14, 15

Fluency

1. Answer the following for each shape shown.
    i. Determine an expression for the perimeter.
    ii. Determine the perimeter when $x = 5$.
    iii. Determine an expression for the area. If necessary, simplify the expression by expanding.
    iv. Determine the area when $x = 5$.

a. 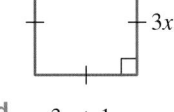 $3x$

b. 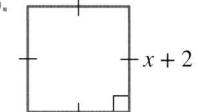 $x + 2$

c. 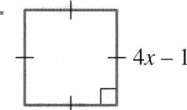 $4x - 1$

d. $3x + 1$ 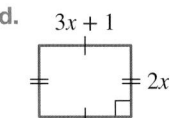 $2x$

e. $5x + 2$ 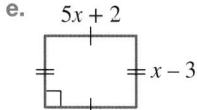 $x - 3$

f. $3x + 5$ 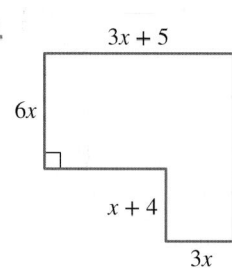 $6x$, $x + 4$, $3x$

Understanding

2. **WE18** A rectangular swimming pool measures 50 m by 25 m. A path around the edge of the pool is x m wide on each side.

    a. Determine the area of the pool.
    b. Write an expression for the area of the pool plus the area of the path.
    c. Write an expression for the area of the path.
    d. If the path is 2.3 m wide, calculate the area of the path.
    e. If the area of the path is $200\,\text{m}^2$, write an equation that can be solved to calculate the width of the path.

3. **WE19** The pages of a book are 20 cm high by 15 cm wide. The pages have margins of x cm at the top and bottom and margins of y cm on the left and right.

    a. Write an expression for the height of the section of the pages that lie inside the margins.
    b. Write an expression for the width of the section of the pages that lie inside the margins.
    c. Write an expression for the area of the section of the pages that lie inside the margins.
    d. Show that, if the margins on the left and right are doubled, the area between the margins is reduced by $(40y - 4xy)$ cm.

4. A rectangular book cover is 8 cm long and 5 cm wide.

   a. Calculate the area of the book cover.
   b. **i.** If the length of the book cover is increased by v cm, write an expression for its new length.
       ii. If the width of the book cover is increased by v cm, write an expression for its new width.
       iii. Write an expression for the new area of the book cover. Expand this expression.
       iv. Calculate the increased area of the book cover if $v = 2$ cm.
   c. **i.** If the length of the book cover is decreased by d cm, write an expression for its new length.
       ii. If the width of the book cover is decreased by d cm, write an expression for its new width.
       iii. Write an expression for the new area of the book cover. Expand this expression.
       iv. Calculate the new area of the book cover if $d = 2$ cm.
   d. **i.** If the length of the book cover is made x times longer, write an expression for its new length.
       ii. If the width of the book cover is increased by x cm, write an expression for its new width.
       iii. Write an expression for the new area of the book cover. Expand this expression.
       iv. Calculate the new area of the book cover if $x = 5$ cm.

5. A square has sides of length $5x$ m.

   a. Write an expression for its perimeter.
   b. Write an expression for its area.
   c. **i.** If the square's length is decreased by 2 m, write an expression for its new length.
       ii. If the square's width is decreased by 3 m, write an expression for its new width.
       iii. Write an expression for the square's new area. Expand this expression.
       iv. Calculate the square's area when $x = 6$ m.

6. A rectangular sign has a length of $2x$ cm and a width of x cm.

   a. Write an expression for the sign's perimeter.
   b. Write an expression for the sign's area.
   c. **i.** If the sign's length is increased by y cm, write an expression for its new length.
       ii. If the sign's width is decreased by y cm, write an expression for its new width.
       iii. Write an expression for the sign's new area and expand.
       iv. Calculate the sign's area when $x = 4$ cm and $y = 3$ cm using your expression.

7. A square has a side length of x cm.

   a. Write an expression for its perimeter.
   b. Write an expression for its area.
   c. **i.** If the square's side length is increased by y cm, write an expression for its new side length.
       ii. Write an expression for the square's new perimeter. Expand this expression.
       iii. Calculate the square's perimeter when $x = 5$ cm and $y = 9$ cm.
       iv. Write an expression for the square's new area and expand.
       v. Calculate the square's area when $x = 3.2$ cm and $y = 4.6$ cm.

8. A swimming pool with length $(4p + 2)$ m and width $3p$ m is surrounded by a path of width p m.
   Write the following in expanded form.

   a. An expression for the perimeter of the pool.
   b. An expression for the area of the pool.
   c. An expression for the length of the pool and path.
   d. An expression for the width of the pool and path.
   e. An expression for the perimeter of the pool and path.
   f. An expression for the area of the pool and path.
   g. An expression for the area of the path.
   h. The area of the path when $p = 2$ m.

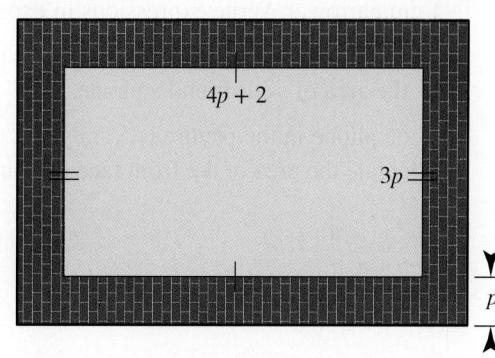

Communicating, reasoning and problem solving

9. The area of a rectangular playground is given by the general expression $(6x^2 + 11x + 3)$ m^2 where x is a positive whole number.

   a. Determine the length and width of the playground in terms of x.
   b. Write an expression for the perimeter of the playground.
   c. If the perimeter of a particular playground is 88 metres, evaluate x.

10. Students decide to make Science Day invitation cards.
    The total area of each card is equal to $(x^2 - 4x - 5)$ cm^2.

    a. Factorise the expression to determine the dimensions of the cards in terms of x.
    b. Write down the length of the shorter side in terms of x.
    c. If the shorter sides of a card is 10 cm in length and the longer sides is 16 cm in length, determine the value of x.
    d. Evaluate the area of the card proposed in part c.
    e. If the students want to make 3000 Science Day invitation cards, determine how much cardboard will be required. Give your answer in terms of x.

11. Can you evaluate 997^2 without a calculator in less than 90 seconds? It is possible to work out the answer using long multiplication, but it would take a fair amount of time and effort. Mathematicians are always looking for quick and simple ways of solving problems.
    What if we consider the expanding formula that produces the difference of two squares?
    $(a + b)(a - b) = a^2 - b^2$

    Adding b^2 to both sides gives $(a + b)(a - b) + b^2 = a^2 - b^2 + b^2$.
    Simplifying and swapping sides gives $a^2 = (a + b)(a - b) + b^2$.

    We can use this formula, combined with the fact that multiplying by 1000 is an easy operation, to evaluate 997^2.

    a. If $a = 997^2$, determine what the value of b should be so that $(a + b)$ equals 1000.
    b. Substitute these a and b values into the formula to evaluate 997^2.
    c. Try the above method to evaluate the following.

       i. 995^2
       ii. 990^2

12. In the picture of the phone shown, the dimensions are in cm.

    a. Use the information in the picture to write an expression for the area of:

       i. the viewing screen
       ii. the entire front face of the phone.

    b. Your friend has a phone that is 4 cm longer than the one shown, but also 1 cm narrower. Write expressions in expanded form for:

       i. the length and width of your friend's phone
       ii. the area of your friend's phone.

    c. If the phone in the picture is 5 cm wide, use your answer to part b to calculate the area of the front face of your friend's phone.

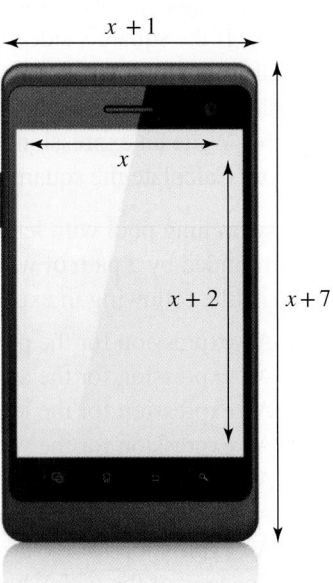

13. A proposed new flag for Australian schools will have the Australian flag in the top left-hand corner. The dimensions of this new flag are given in metres.

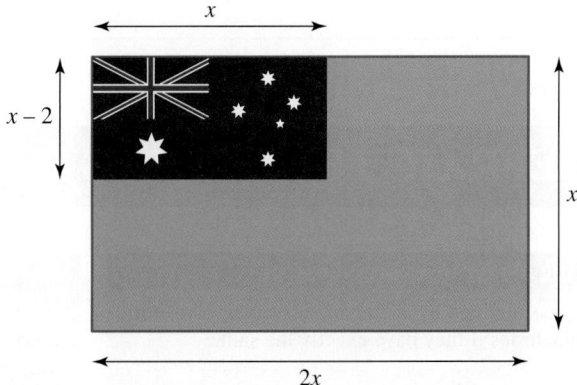

a. Write an expression in factorised form for the area of:
   i. the Australian flag section of the proposed flag   ii. the whole area of the proposed flag.
b. Use the answers from part **a** to write the area of the Australian flag as a fraction of the school flag. Simplify this fraction.
c. Use the fraction from part **b** to express the area of the Australian flag as a percentage of the proposed school flag.
d. Use the formula for the percentage of the area taken up by the Australian flag to calculate the percentages for the following suggested widths for the proposed school flag.
   i. 4 m   ii. 4.5 m   iii. 4.8 m
e. If the percentage of the school flag taken up by the Australian flag measures the importance a school places on Australia, determine what can be said about the three suggested flag widths.

14. Cameron wants to build an in-ground 'endless' pool. Basic models have a depth of 2 metres and a length triple the width. A spa will also be attached to the end of the pool.

a. The pool needs to be tiled. Write an expression for the surface area of the empty pool (that is, the floor and walls only).
b. The spa needs an additional $16\,\text{m}^2$ of tiles. Write an expression for the total area of tiles needed for both the pool and the spa.
c. Factorise this expression.
d. Cameron decides to use tiles that are selling at a discount price, but there is only $280\,\text{m}^2$ of the tile available. Determine the maximum dimensions of the pool he can build if the width is in whole metres. Assume the spa is to be included in the tiling.
e. Evaluate the area of tiles that is actually needed to construct the spa and pool.
f. Determine the volume of water the pool can hold.

15. Cubic expressions are the expanded form of expressions with three linear factors. The expansion process for three linear factors is called trinomial expansion.

a. Explain how the area model can be altered to show that binomial expansion can also be altered to become a model for trinomial expansion.
b. Investigate the powers in cubic expressions by expanding the following expressions.
   i. $(3x+2)(-x+1)(x+1)$
   ii. $(x+5)(2x-2)(4x-8)$
   iii. $(3-x)(x+8)(5-x)$
c. Describe the patterns in the powers in a cubic expression.

LESSON
3.10 Review

3.10.1 Topic summary

ALGEBRAIC TECHNIQUES (PATH)

Simplifying algebraic expressions

- An algebraic expression can be simplified by adding or subtracting like terms.
- Two terms are considered like terms if they have exactly the same pronumeral component.
  - $3x$ and $-10x$ are **like** terms.
  - $5x^2$ and $11x$ are **not like** terms.
  - $4abc$ and $-3cab$ are **like** terms.

e.g. The expression $3xy + 10xy - 5xy$ has 3 like terms and can be simplified to $8xy$.

The expression $4a \times 2b \times a$ can be simplified to $8a^2b$.

The expression $12xy \div 4xz$ can be simplified to $\dfrac{3y}{z}$.

Expanding brackets

- We can expand a single set of brackets using the **Distributive Law**:

$$a(b + c) = ab + ac$$

- The product of binomial factors can be expanded:

$$(a + b)(c + d) = ac + ad + bc + bd$$

e.g. $(x + 3)(y - 4) = xy - 4x + 3y - 12$

Factorising

- Factorising is the opposite process to expanding.
- An expression is factorised by determining the highest common factor of each term.

e.g. $9xy + 15xz - 21x^2$
$= 3x(3y + 5z - 7x)$

The highest common factor of each term is $3x$.

Special cases

- We can also expand if we recognise the following special cases.
  - Difference of two squares:

  $$(a + b)(a - b) = a^2 - b^2$$

  - Perfect squares:

  $$(a + b)^2 = a^2 + 2ab + b^2$$
  $$(a - b)^2 = a^2 - 2ab + b^2$$

e.g. $(x - 6)(x + 6) = x^2 - 36$
$(2x + 5)^2 = 4x^2 + 20x + 25$
$(x - 2)^2 = x^2 - 4x + 4$

Grouping in pairs

- When presented with an expression that has 4 terms with no common factor, we can factorise by grouping the terms in pairs.

e.g. $6xy + 8y - 12xz - 16z$
$= 2y(3x + 4) - 4z(3x + 4)$
$= (3x + 4)(2y - 4z)$

Setting up worded problems

- In order to turn a worded problem into an algebraic expression it is important to look out for the following words.
  - **Words that mean addition:** sum, altogether, add, more than, and, in total
  - **Words that mean subtraction:** difference, less than, take away, take off, fewer than,
  - **Words that mean multiplication:** product, groups of, times, of, for each, double, triple
  - **Words that mean division:** quotient, split into, halve, thirds.

Solving application questions

- Drawing diagrams will help set up expressions to solve worded problems.
- Make sure you include units in your answers.

Non-monic quadratics

- Factorising general quadratic expressions

$$ax^2 + bx + c = ax^2 + mx + nx + c$$

Factors of ac that sum to b

Algebraic fractions: + and −

- Fractions can be added and subtracted if they have the same common denominator.

e.g.

$$\frac{5}{2x} + \frac{4}{3y} = \frac{15y}{6xy} + \frac{8x}{6xy} = \frac{15y + 8x}{6xy}$$

Or

$$\frac{3}{(x + 2)} - \frac{2}{(x - 2)} = \frac{3(x - 2)}{x^2 - 4} - \frac{2(x + 2)}{x^2 - 4}$$
$$= \frac{3x - 6 - 2x - 4}{x^2 - 4}$$
$$= \frac{x - 10}{x^2 - 4}$$

Algebraic fractions: ×

- When multiplying fractions, multiply the numerators together and the denominators together.
- Cancel any common factors in the numerator and denominator.

e.g.

$$\frac{5y}{12x} \times \frac{7x^2}{15z} = \frac{{}^1 5y}{12x} \times \frac{7x^{2^1}}{{}^3 15z} = \frac{7xy}{36z}$$

Cancel common factors from top and bottom.
Write variables in alphabetical order.

Algebraic fractions: ÷

- When dividing two fractions, multiply the first fraction by the reciprocal of the second.
- The reciprocal of $\dfrac{a}{b}$ is $\dfrac{b}{a}$.

e.g.

$$\frac{10x^2}{33z} \div \frac{6x^2}{11y} = \frac{{}^5 10x^2}{{}^3 33 z} \times \frac{{}^1 11y}{{}^3 6x^2} = \frac{5y}{9z}$$

3.10.2 Project

Suspension bridges

When you hold up a chain at both ends, it forms a curve. Its shape is actually a *catenary* — a trigonometric function with a shape similar to that of a quadratic equation whose shape is a parabola. When constructing a suspension bridge, the main cables are attached to a tower at both ends and the curve is a catenary. However, when vertical cables are attached from the main cable to the deck, the curve takes the shape of a parabola. A catenary curves under its own weight; the bridge's main cables are curving not just under their own weight, but also curving from holding up the weight of the deck.

There are many well-known suspension bridges around the world, for example the Golden Gate Bridge in San Francisco.

A sketch of one side of the bridge can be represented as follows (we will consider just one side of the bridge in our calculations). Note that there are no vertical cables at the ends where the main cable is attached to the towers.

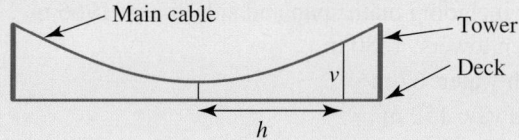

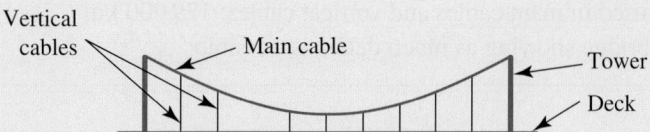

The shape of the cable can be represented by the equation $v = 0.000\,832h^2 + 3$ where v represents the vertical distance between the main cable and the deck (in metres) and h represents the horizontal distance from the centre point of the main cable (in metres). The curve is symmetrical around this central point.

Let us consider a suspension bridge with 55 m high towers at the ends and vertical cables evenly spaced every 50 m.

1. Substitute into the equation to calculate the horizontal distance of the towers from the centre of the main cable.
2. Determine the length of the deck.
3. Draw a diagram displaying the spacing of all the vertical cables on the bridge.
4. How many vertical cables are there in the structure?
5. Use the equation to calculate the length (to the nearest metre) of each of these vertical cables.
6. What is the total length of cable required for these vertical cables?

The Golden Gate Bridge consists of a main suspension span in the centre with side suspension spans at both ends.

Here are some facts about the structure of the bridge.
- Length of suspension span including main span and side spans: 1966 m
- Horizontal distance between towers: 1280 m
- Clearance above mean high water: 67 m
- Height of tower above roadway: 152 m
- The bridge has two main cables that pass over the tops of the two main towers and are secured at either end. Length of each main cable: 2332 m
- Total length of wire used in main cables and vertical cables: 129 000 km
7. Draw a sketch of the bridge showing as much detail as possible.

 Resources

 Interactivities Crossword (int-0707)
Sudoku puzzle (int-3214)

Exercise 3.10 Review questions

Fluency

1. **MC** The equivalent of $(3-a)(3+a)$ is:

   A. $9+a^2$ **B.** $9-a^2$ **C.** $3+a^2$ **D.** $3-a^2$

2. **MC** The equivalent of $(2y+5)^2$ is:

   A. $4y^2+20y+25$ **B.** $4y^2+10y+25$ **C.** $2y^2+20y+25$ **D.** $2y^2+20y+5$

3. **MC** When factorised, $6(a+2b)-x(a+2b)$ equals:

   A. $6(a+2b)-x(a+2b)$ **B.** $6-x(a+2b)$

   C. $(6-x)(a+2b)$ **D.** $6(a+2b-x)$

4. Expand and simplify these expressions

   a. $(x+4)(x-4)$ **b.** $(9-m)(9+m)$ **c.** $(x+y)(x-y)$ **d.** $(1-2a)(1+2a)$

5. Expand and simplify these expressions

   a. $(x+5)^2$ **b.** $(m-3)^2$ **c.** $(4x+1)^2$ **d.** $(2-3y)^2$

6. Expand and simplify the following.

   a. $(x+2)^2+(x+3)^2$ **b.** $(x-2)^2-(x-3)^2$ **c.** $(x+4)^2-(x-4)^2$

7. **MC** The expression $4a^2-12a+9$ factorises to:

   A. $(4a+9)(4a+9)$ **B.** $(4a-9)(4a-9)$ **C.** $(2a-3)(2a-3)$ **D.** $(2a-3)(2a+3)$

8. **MC** What does the expression $9m^2-16n^2$ factorise to?

   A. $9(m-4n)(m+4n)$ **B.** $3(3m-4n)^2$

   C. $(9m-16n)(9m+16n)$ **D.** $(3m-4n)(3m+4n)$

9. **MC** What does the expression $5x^2+20x-60$ factorise to?

   A. $(5x-12)(x+5)$ **B.** $(5x+15)(x-4)$ **C.** $5(x+6)(x-2)$ **D.** $5(x-6)(x+2)$

10. **MC** What does the expression $3x^2-13x+4$ factorise to?

    A. $(3x-1)(x-4)$ **B.** $(3x-2)(x+2)$ **C.** $(3x-2)^2$ **D.** $(3x-4)(x-1)$

11. **MC** $\dfrac{x+3}{x+2} \div \dfrac{(x+3)^2}{x+2}$ is equivalent to:

    A. $\dfrac{(x+3)^2}{(x+2)^2}$ **B.** $\dfrac{1}{x+3}$ **C.** $\dfrac{x+2}{x+3}$ **D.** $\dfrac{-1}{2(x+2)}$

Understanding

12. Factorise each of the following using the difference of two squares rule.

    a. x^2-64 **b.** a^2-144 **c.** $49b^2-1$

    d. $4f^2-9g^2$ **e.** $(n+1)^2-m^2$ **f.** $(r-1)^2-4s^2$

13. Factorise each of the following quadratic trinomials.

   a. $c^2 + 5c + 4$
   b. $p^2 + 10p - 24$
   c. $y^2 - 10y + 24$

   d. $x^2 + 3x + 2$
   e. $m^2 - 7m + 10$
   f. $m^2 + 24m + 44$

14. Factorise each of the following.

   a. $2a^2 + 16a + 24$
   b. $3b^2 - 24b + 36$
   c. $4c^2 - 16c - 48$

   d. $2x^2 + 3x + 1$.
   e. $3x^2 - x - 2$
   f. $6x^2 + 13x + 6$

15. Simplify the following.

   a. $\dfrac{5y}{3} - \dfrac{y}{2}$
   b. $\dfrac{x+4}{5} + \dfrac{x+2}{2}$
   c. $\dfrac{5}{3x} - \dfrac{1}{5x}$
   d. $\dfrac{x-1}{x+3} + \dfrac{2x-5}{x+2}$

16. Simplify the following.

   a. $\dfrac{y}{4} \times \dfrac{32}{x}$
   b. $\dfrac{20y}{7x} \times \dfrac{35z}{16y}$
   c. $\dfrac{x+6}{(x+1)(x+3)} \times \dfrac{5(x+1)}{x+6}$

   d. $\dfrac{25}{x} \div \dfrac{30}{x}$
   e. $\dfrac{xy}{5} \div \dfrac{10x}{y}$
   f. $\dfrac{2x}{(x+8)(x-1)} \div \dfrac{9x+1}{x+8}$

17. Simplify each of the following.

   a. $\dfrac{(x-3)(x+4)}{x+1} \times \dfrac{x+1}{x-3}$
   b. $\dfrac{x(x+5)}{x-5} \div \dfrac{x+5}{x(x-5)}$
   c. $\dfrac{4x-12}{2x+2} \times \dfrac{x+1}{x-3}$

   d. $\dfrac{10x+20}{3x-3} \div \dfrac{5x+10}{6x+6}$
   e. $\dfrac{x^2+7x+12}{x^2+4x+3} \times \dfrac{x^2-6x-7}{x^2+2x-8}$
   f. $\dfrac{x^2-25}{x^2+4x-5} \div \dfrac{x^2-6x+5}{x^2+5x-6}$

Communicating, reasoning and problem solving

18. David owned land in the shape of a square with side length p. He decided to sell part of this land by reducing it by 50 metres in one direction and 90 metres in the other.

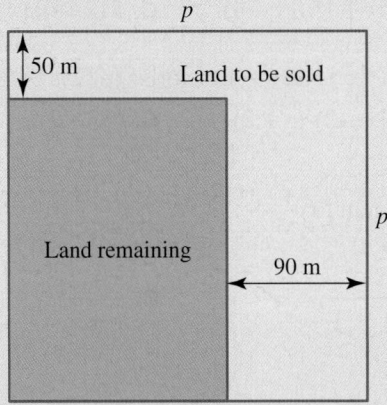

   a. Determine an expression in terms of p for the area of the original land.

   b. Determine an expression in terms of p, in its simplest form, for the area of land David has remaining after he sells the section of land shown.

   c. Determine an expression in terms of p for the area of land he sold.

19. A new computer monitor is made up of a rectangular screen surrounded by a hard plastic frame in which speakers can be inserted.

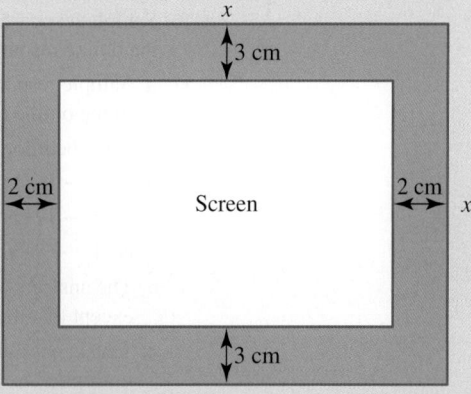

a. Write expressions for the length and width of the screen in terms of x.
b. Write an expression for the area of the screen, using expanded form.
c. Evaluate the area of the screen if $x = 30$ cm.

20. A scientist tried to use a mathematical formula to predict people's moods based on the number of hours of sleep they had the previous night. One formula that he used was what he called the 'grumpy formula', $g = 0.16(h - 8)^2$, which was valid on a 'grumpy scale' from 0 to 10 (least grumpy to most grumpy).

a. Calculate the number of hours needed to not be grumpy.
b. Evaluate the grumpy factor for somebody who has had:
  i. 4 hours of sleep
  ii. 6 hours of sleep
  iii. 10 hours of sleep.
c. Determine the number of hours of sleep required to be most grumpy.

Another scientist already had his own grumpy formula and claims that the scientist above stole his idea and has just simplified it. The second scientist's grumpy formula was

$$g = \frac{0.16\,(h - 8)}{8 - h} \times \frac{2\,(8 - h)}{3\,(h - 8)} \div \frac{2h}{3(h - 8)^2}$$

d. Write the second scientist's formula in simplified form.
e. Are the second scientist's claims justified? Explain.

 To test your understanding and knowledge of this topic, go to your learnON title at www.jacplus.com.au and complete the **post-test**.

Answers

Topic 3 Algebraic techniques (Path)

3.1 Pre-test

1. a. $9x^2 - 1$ b. $4a^2 - 25b^2$

2. a. $4a^2 + 12a + 9$
   b. $4 - 20x + 25x^2$

3. a. $2m + 13n$ b. $-2xy + 3x + 5y$

4. False

5. a. $(7a - 1)(b + 3)$
   b. $(5x - 2)(2y + 1)$

6. D

7. A

8. A

9. $(m - 2)(m + 4)$

10. $(5x - 1)(x + 2)$

11. D

12. D

13. B

14. a. $\dfrac{m + 5}{m - 1}$ b. $\dfrac{8}{s + 2}$

15. $\dfrac{a + 12}{a - 18}$

3.2 Special binomial products

1. a. $x^2 - 4$ b. $y^2 - 9$
   c. $m^2 - 25$ d. $a^2 - 49$

2. a. $x^2 - 36$ b. $p^2 - 144$
   c. $a^2 - 100$ d. $m^2 - 121$

3. a. $4x^2 - 9$ b. $9y^2 - 1$ c. $25d^2 - 4$
   d. $49c^2 - 9$ e. $4 - 9p^2$

4. a. $d^2 - 81x^2$ b. $25 - 144a^2$ c. $9x^2 - 100y^2$
   d. $4b^2 - 25c^2$ e. $100 - 4x^2$

5. a. $x^2 + 4x + 4$ b. $a^2 + 6a + 9$
   c. $b^2 + 14b + 49$ d. $c^2 + 18c + 81$

6. a. $m^2 + 24m + 144$ b. $n^2 + 20n + 100$
   c. $x^2 - 12x + 36$ d. $y^2 - 10y + 25$

7. a. $81 - 18c + c^2$ b. $64 + 16e + e^2$
   c. $2x^2 + 4xy + 2y^2$ d. $u^2 - 2uv + v^2$

8. a. $4a^2 + 12a + 9$ b. $9x^2 + 6x + 1$
   c. $4m^2 - 20m + 25$ d. $16x^2 - 24x + 9$

9. a. $25a^2 - 10a + 1$ b. $49p^2 + 56p + 16$
   c. $81x^2 + 36x + 4$ d. $16c^2 - 48c + 36$

10. a. $25 + 30p + 9p^2$ b. $4 - 20x + 25x^2$
    c. $81x^2 - 72xy + 16y^2$ d. $64x^2 - 48xy + 9y^2$

11. a. $x^2 - 9$ b. $4x^2 - 9$ c. $49x^2 - 16$
    d. $4x^2 - 49y^2$ e. $x^4 - y^4$

12. a. $16x^2 + 40x + 25$ b. $49x^2 - 42xy + 9y^2$
    c. $25x^4 - 20x^2y + 4y^2$ d. $2x^2 - 4xy + 2y^2$
    e. $\dfrac{4}{x^2} + 16 + 16x^2$

13. $(x^2 + 6x + 9)$ units^2

14. $(3x + 1)$ m

15. Perimeter $= 4x + 36$

16. Sample responses can be found in the worked solutions in the online resources.

17. a. Sample responses can be found in the worked solutions in the online resources.
    b. Lin's bedroom is larger by 1 m^2.

18. a.   i. $x^2 - 16$ and $16 - x^2$
        ii. $x^2 - 121$ and $121 - x^2$
        iii. $4x^2 - 81$ and $81 - 4x^2$
    b. The answers to the pairs of expansions are the same, except that the negative and positive signs are reversed.
    c. This is possible because when a negative number is multiplied by a positive number, it becomes negative. When expanding a DOTS in which the expressions have different signs, the signs will be reversed.

19. a. $100k^2 + 100k + 25$
    b. $(10k + 5)^2 = 100 \times k \times k + 100 \times k + 25$
    $$= 100k(k + 1) + 25$$
    c. $25^2 = (10 \times 2 + 5)^2$
    Let $k = 2$.
    $25^2 = 100k(k + 1) + 25$
    $= 100 \times 2 \times (2 + 1) + 25$
    $= 625$
    $85^2 = (10 \times 8 + 5)^2$
    Let $k = 8$.
    $85^2 = 100k(k + 1) + 25$
    $= 100 \times 8 \times (8 + 1) + 25$
    $= 7225$

20. a. $A_1 = a^2$ units^2
    $A_2 = ab$ units^2
    $A_3 = ab$ units^2
    $A_4 = b^2$ units^2
    b. $A = a^2 + ab + ab + b^2$
    $$= a^2 + 2ab + b^2$$
    This is the equation for perfect squares.

21. a.   i. $x^2 - 6x + 9$ and $9 - 6x + x^2$
        ii. $x^2 - 30x + 225$ and $225 - 30x + x^2$
        iii. $9x^2 - 42x + 49$ and $49 - 42x + 9x^2$
    b. The answers to the pairs of expansions are the same.
    c. This is possible because when a negative number is multiplied by itself, it becomes positive. When expanding a perfect square in which the two expressions are the same, the negative signs cancel out and result in the same answer.

22. a. $10\,609$ b. 3844 c. $994\,009$
    d. $1\,024\,144$ e. 2809 f. 9604

23. a. 729 b. 1089
    c. 1521 d. 2209

24. $(3x^2 + 16x + 5)$ m^2

25. $(x^2 - 25)$ m^2

3.3 Further expansions

1. a. $2x^2 + 13x + 21$ b. $2x^2 + 13x + 20$
   c. $2x^2 + 14x + 26$ d. $2x^2 + 10x + 11$

2. a. $2p^2 - 3p - 21$ b. $2a^2 - 5a + 4$
   c. $2p^2 - p - 24$ d. $2x^2 + 19x - 36$

3. a. $2y^2 + 2y - 7$ b. $2d^2 + 8d - 2$
   c. $2x^2 + 10$ d. $2y^2$

4. a. $2x^2 - 4x + 19$ b. $2y^2 - 4y - 7$
   c. $2p^2 + 3p + 23$ d. $2m^2 + 3m + 31$

5. a. $x + 5$ b. $4x + 8$
   c. $-2x - 6$ d. $3m + 2$

6. a. $-3b - 22$ b. $-15y - 2$
   c. $8p - 10$ d. $16x + 2$

7. a. $4m + 17$ b. $-7a + 30$
   c. $-6p - 7$ d. $3x - 21$

8. a. $16x - 16$
   b. Sample responses can be found in the worked solutions in the online resources.

9. a. $15x^2 - 44x + 21$
   b. Sample responses can be found in the worked solutions in the online resources.

10. a. $8x^2 - 10xy + 3y^2$
    b. Sample responses can be found in the worked solutions in the online resources.

11. $x = -4$

12. $(p^2 + p - 2) + (p^2 - 2p - 3)$
    $$= p^2 + p^2 + p - 2p - 2 - 3$$
    $$= 2p^2 - p - 5$$

13. $(x + 2)(x - 3) - (x + 1)^2 = (x^2 - x - 6) - (x^2 + 2x + 1)$
    $$= -3x - 7$$

14. a. Sample responses can be found in the worked solutions in the online resources.
    b. $a = 2, b = 1, c = 3, d = 4$
    $(2^2 + 1^2)(3^2 + 4^2) = (2 \times 3 - 1 \times 4)^2 + (2 \times 4 - 1 \times 3)^2$
    $$= 4 + 121$$
    $$= 125$$

15. a. $x^4 + 2x^3 - x^2 - 2x + 1$
    b. Sample responses can be found in the worked solutions in the online resources.
    c. i. 25
       ii. $x = 2$

16. a. $ad + ae + bd + be$
    b. $(a + b + c)(d + e + f) = ad + ae + bd + be + cd + ce + af + bf + cf$

	a	b	c
d	ad	bd	cd
e	ae	be	ce
f	af	bf	cf

3.4 Factorising by grouping in pairs

1. a. $(a + b)(2 + 3c)$ b. $(m + n)(4 + p)$
   c. $(2m + 1)(7x - y)$ d. $(3b + 2)(4a - b)$
   e. $(x + 2y)(z - 3)$

2. a. $(6 - q)(12p - 5)$ b. $(x - y)(3p^2 + 2q)$
   c. $(b - 3)(4a^2 + 3b)$ d. $(q + 2p)(p^2 - 5)$
   e. $(5m + 1)(6 + n^2)$

3. a. $(y + 2)(x + 2)$ b. $(b + 3)(a + 3)$
   c. $(x - 4)(y + 3)$ d. $(2y + 1)(x + 3)$
   e. $(3b + 1)(a + 4)$ f. $(b - 2)(a + 5)$

4. a. $(m - 2n)(1 + a)$ b. $(5 + 3p)(1 + 3a)$
   c. $(3m - 1)(5n - 2)$ d. $(10p - 1)(q - 2)$
   e. $(3x - 1)(2 - y)$ f. $(4p - 1)(4 - 3q)$

5. a. $(2y + 1)(5x - 2)$ b. $(2a + 3)(3b - 2)$
   c. $(b - 2c)(5a - 3)$ d. $(x + 3y)(4 - z)$
   e. $(p + 2q)(5r - 3)$ f. $(a + 5b)(c - 2)$

6. a. $\dfrac{a}{b}$ b. $\dfrac{3x - 4}{7a - 5}$

7. a. $\dfrac{x + 2}{2 + 6x}$ b. $\dfrac{y + 2}{3b - 2}$

8. a. $\dfrac{3x - 2}{3y + 2}$ b. $\dfrac{m - 4n}{m + 4n}$

9. Sample responses can be found in the worked solutions in the online resources.

10. $(2x + 3)^2$ This is a perfect square.

11. a. $5(x + 2)(x + 3) = 5(x^2 + 5x + 6) = 5x^2 + 25x + 30$

	x	3	x	3	x	3	x	3	x	3
x	x^2	$3x$	x^2	$3x$	x^2	$3x$	x^2	$3x$	x^2	$3x$
2	$2x$	6	$2x$	6	$2x$	6	$2x$	6	$2x$	6

b. $5(x + 2)(x + 3) = (5 \times x + 2 \times 5)(x + 3)$
   $$= (5x + 10)(x + 3)$$

	x	2	x	2	x	2	x	2	x	2
x	x^2	$2x$	x^2	$2x$	x^2	$2x$	x^2	$2x$	x^2	$2x$
3	$3x$	6	$3x$	6	$3x$	6	$3x$	6	$3x$	6

c. $5(x + 2)(x + 3) = (5 \times x + 3 \times 5)(x + 2)$
   $$= (5x + 15)(x + 2)$$

	$5x$	15
x	$5x^2$	$15x$
2	$10x$	30

12. $-(x + 1)(4y - 9)$

13. a. i. Square, because it is a perfect square.
       ii. Rectangle, because it is a DOTS.
       iii. Rectangle, because it is a trinomial.
       iv. Rectangle, because it is a trinomial.
    b. i. $(3s + 8)^2$ ii. $(5s + 2)(5s - 2)$
       iii. $(s + 1)(s + 3)$ iv. $4(s - 8)(s + 1)$

3.5 Factorising special products

1. a. $(x+5)(x-5)$
   b. $(x+9)(x-9)$
   c. $(a+4)(a-4)$
2. a. $(5+p)(5-p)$
   b. $(11+a)(11-a)$
   c. $(6+y)(6-y)$
3. a. $(2b+5)(2b-5)$
   b. $(3a+4)(3a-4)$
   c. $(5d+1)(5d-1)$
4. a. $(x+y)(x-y)$
   b. $(a+b)(a-b)$
   c. $(p+q)(p-q)$
5. a. $(5m+n)(5m-n)$
   b. $(9x+y)(9x-y)$
   c. $(p+6q)(p-6q)$
6. a. $(6m+5n)(6m-5n)$
   b. $(4q+3p)(4q-3p)$
   c. $(2m+7n)(2m-7n)$
7. a. $(x+13)(x+5)$
   b. $(p+13)(p+3)$
   c. $(p-2-q)(p-2+q)$
8. a. $(c-6+d)(c-6-d)$
   b. $(x+7+y)(x+7-y)$
   c. $(p+5-q)(p+5+q)$
9. a. $(a-2)(a-4)$
   b. $(b+5)(b-7)$
   c. $p(p+10)$
10. a. $2(m+4)(m-4)$
    b. $5(y+3)(y-3)$
    c. $6(p+2)(p-2)$
11. a. $a(x+3)(x-3)$
    b. $2b(2+x)(2-x)$
    c. $3(b+9)(b+1)$
12. a. $(x+5)^2$ b. $(p-12)^2$ c. $(n+10)^2$
13. a. $(u-v)^2$ b. $(8+e)^2$ c. $(2m-5)^2$
14. a. $2(x+6)^2$ b. $3(x-4y)^2$ c. $2(3a+2b)^2$
15. A
16. B and D
17. C
18. A
19. D
20. A
21. C
22. a. $\pi r^2 \, \text{m}^2$ b. $(r+1)\,\text{m}$
    c. $\pi(r+1)^2 \, \text{m}^2$ d. $[\pi(r+1)^2 - \pi r^2]\,\text{m}^2$
    e. $\pi(2r+1)\,\text{m}^2$ f. $35\,\text{m}^2$
    g. $47\,\text{m}^2$
23. a. $w^2 + 10w + 25$ b. $100\,\text{units}^2$
    c. $225\,\text{units}^2$ d. $625\,\text{units}^2$

24. a. $(x+\sqrt{10})(x-\sqrt{10})$
    b. $4(x+\sqrt{8})(x-\sqrt{8})$
    c. $(x^2+y)(x^2-y)$
25. a. $5(2x-1)$
    b. $(y+2x^2)(y-2x^2)$
    c. $(10+x)(10-x)(100+x^2)$
26. The two expressions are not equal.
27. $a^2 - b^2$
28. a. Side length$_a$ $= x+3$
    Side length$_b$ $= x+5$
    Side length$_c$ $= x+8$
    Side length$_d$ $= x-3$
    Side length$_e$ $= x+6$
    b. I $= (x-3)^2$
       II $= (x+3)^2$
       III $= (x+5)^2$
       IV $= (x+6)^2$
       V $= (x+8)^2$
    c. $A_a = 64\,\text{cm}^2$
       $A_b = 100\,\text{cm}^2$
       $A_c = 169\,\text{cm}^2$
       $A_d = 4\,\text{cm}^2$
       $A_e = 121\,\text{cm}^2$
29. a. $(x+7)(x-7)$
    b. Length $= x+7$
       Width $= x-7$
    c. Length $= 32\,\text{mm}$
       Width $= 18\,\text{mm}$
    d. $1875\,\text{mm}^2$
30. $(x^4+1)(x^2+1)(x+1)(x-1)$

3.6 Factorising non-monic quadratics

1. a. $2(x+2)(x+3)$
   b. $3(x+4)(x+1)$
   c. $4(x+2)(x-3)$
2. a. $3(x+5)(x-2)$
   b. $2(x+7)(x-3)$
   c. $5(x+6)(x-2)$
3. a. $(x+1)(5x-2)$
   b. $(x-2)(7x-3)$
   c. $(2x-3)(5x+2)$
4. a. $(x+3)(2x+1)$
   b. $(x-3)(2x-1)$
   c. $(x-1)(3x+2)$
5. B
6. A
7. D
8. a. $(2x-1)(x+2)$
   b. $(4x+3)(2x-1)$
   c. $(3x+2)(2x-1)$

9. a. $(3x-4)(x+1)$ b. $(x+2)(2x+5)$

10. a. $(3x+7)(2x-3)$ b. $(2x-3)(x+4)$

11. a. $(2x-3)(3x+2)$ b. $(4x+5)(x+2)$

12. a. $(x-1)(4x-15)$ b. $(3x+4)(2x-1)$

13. a. $3(x+4)(x+1)$ b. $3(x+1)(x+4)$

14. a. $(4x+3)(x-5)$
    b. $(x+3)(3x+1)$
    c. $(5x+3)(2x-3)$

15. a. The middle term, $-9x$, can be re-written as $3x-12x$.
    b. $(4x+3)(x-3)$

16. a. $(6x-5)(2x+1)$ b. $12x^2-4x-5$
    c. 1.5 d. 1.1 or 3.5

17. a. Formula i belongs to a square, as the formula is a perfect square.
    Formula ii belongs to a rectangle, as the formula is a difference of squares.
    Formula iii belongs to a rectangle, as the formula is a trinomial but not a perfect square.
    Formula iv belongs to a rectangle, as the formula is a trinomial but not a perfect square.
    b. i. $(3s+8)^2$ ii. $(5s+2)(5s-2)$
       iii. $(s+1)(s+3)$ iv. $4(s^2-6s-9)$

18. a. $(3b+8)^2$ b. $(ac+d)(ac-d)$
    c. $(a-13b)(a+14b)$ d. $(b-ac)^2$

3.7 Adding and subtracting algebraic fractions

1. a. $\dfrac{26}{21}$ or $1\dfrac{5}{21}$ b. $\dfrac{49}{72}$ c. 1

2. a. $\dfrac{17}{99}$ b. $\dfrac{1}{35}$ c. $\dfrac{6-5x}{30}$

3. a. $\dfrac{15x-4}{27}$ b. $\dfrac{15-16x}{40}$ c. $\dfrac{15-2x}{3x}$

4. a. $\dfrac{5y}{12}$ b. $-\dfrac{3y}{40}$ c. $\dfrac{13x}{12}$ d. $\dfrac{14x}{9}$

5. a. $\dfrac{3w}{28}$ b. $-\dfrac{y}{5}$ c. $\dfrac{89y}{35}$ d. $\dfrac{32x}{15}$

6. a. $\dfrac{7x+17}{10}$ b. $\dfrac{7x+30}{12}$
   c. $\dfrac{2x-11}{30}$ d. $\dfrac{19x+7}{6}$

7. a. $\dfrac{5}{8x}$ b. $\dfrac{5}{12x}$ c. $\dfrac{38}{21x}$

8. a. $\dfrac{8}{3x}$ b. $\dfrac{7}{24x}$ c. $\dfrac{9}{20x}$

9. a. $\dfrac{37}{100x}$ b. $\dfrac{51}{10x}$ c. $-\dfrac{1}{6x}$

10. a. $\dfrac{3x^2+14x-4}{(x+4)(x-2)}$ b. $\dfrac{2x^2+3x+25}{(x+5)(x-1)}$
    c. $\dfrac{2x^2+6x-10}{(2x+1)(x-2)}$ d. $\dfrac{4x^2-17x-3}{(x+1)(2x-7)}$

11. a. $\dfrac{7x^2+x}{(x+7)(x-5)}$ b. $\dfrac{2x^2+6x+7}{(x+1)(x+4)}$
    c. $\dfrac{-x^2+7x+15}{(x+1)(x+2)}$ d. $\dfrac{x-7}{(x+3)(x-2)}$

12. a. $\dfrac{x^2+3x+9}{(x+2)(3x-1)}$ b. $\dfrac{5-5x}{(x-1)(1-x)}=\dfrac{5}{x-1}$
    c. $\dfrac{3x+7}{(x+1)^2}$ d. $\dfrac{3x-4}{(x-1)^2}$

13. a. The student transcribed the denominator incorrectly and wrote $(x+2)$ instead of $(x-2)$ in line 2.
    Also, the student forgot that multiplying a negative number by a negative number gives a positive number. Line 3 should have $+3$ in the numerator, not -1. They didn't multiply.
    b. $\dfrac{x^2-5x+3}{(x-1)(x-2)}$

14. a. -1 b. $\dfrac{-1}{(x-2)}$

15. a. $\dfrac{9}{(x-3)^2}$ b. $\dfrac{4}{(x-2)^3}$

16. a. $\dfrac{4x^2+17x+17}{(x+2)(x+1)(x+3)}$ b. $\dfrac{7x^2-20x+4}{(x-1)(x+2)(x-4)}$

17. a. $\dfrac{4x^2+17x+19}{(x+1)(x+3)(x+2)}$ b. $\dfrac{2\left(2x^2-9x+25\right)}{(x-4)(x-1)(x+3)}$

    c. The lowest common denominator may not always be the product of the denominators. Each fraction must be multiplied by the correct multiple.

18. $a=4$

19. $\dfrac{4(x-1)}{(x+3)(x+4)(x-2)}$

20. $\dfrac{2(x-1)}{(x-7)(x-4)}$

3.8 Multiplying and dividing algebraic fractions

1. a. $a+1$ b. $\dfrac{a+2}{3}$ c. $a-1$

2. a. $\dfrac{x+2}{2}$ b. $\dfrac{x-4}{2}$ c. $3(x+6)$

3. a. $x+2$ b. $x+3$ c. $x-4$

4. a. $\dfrac{a-3}{a+4}$ b. $\dfrac{p+1}{p+5}$ c. $\dfrac{x+3}{x-1}$

5. a. $\dfrac{4x}{y}$ b. $\dfrac{3x}{y}$ c. $\dfrac{4y}{x}$ d. $\dfrac{9x}{4y}$

6. a. $\dfrac{-5x}{4y}$ b. $\dfrac{3w}{2x}$ c. $\dfrac{6z}{7x}$ d. $\dfrac{2z}{7x}$

7. a. $\dfrac{-3x}{2y}$ b. $\dfrac{5}{24}$ c. $\dfrac{12z}{x}$ d. $\dfrac{-x}{6w}$

8. a. $\dfrac{2}{3x-2}$ b. $\dfrac{5}{x-3}$ c. $\dfrac{9}{2(x-6)}$ d. $\dfrac{1}{x+3}$

9. a. $\dfrac{2x}{(x+1)^2}$ b. $\dfrac{x+1}{2(2x-3)}$

c. $\dfrac{a}{10(a+3)}$ d. $\dfrac{35d}{8(d-3)}$

10. a. $\dfrac{9}{32x^2(x-2)}$ b. $\dfrac{3x}{10(x-1)}$

11. a. $\dfrac{3}{5}$ b. $\dfrac{2}{9}$ c. $\dfrac{1}{3}$ d. 3

12. a. $\dfrac{1}{25}$ b. $\dfrac{35}{6}$ or $5\dfrac{5}{6}$

c. $\dfrac{4y^2}{7}$ d. $\dfrac{2y^2}{25}$

13. a. $\dfrac{8y^2}{9}$ b. $\dfrac{32xy}{15}$ c. $\dfrac{2}{3}$ d. y^2

14. a. $\dfrac{9}{(3x-7)(x+3)}$ b. $\dfrac{1}{(x+2)(x-9)}$

15. a. $\dfrac{21(x-3)}{x+5}$ b. $\dfrac{13}{9(x-4)(x+1)}$

16. a. $\dfrac{2(x-4)^3}{(x+3)^2}$ b. $\dfrac{(x+2)^2}{(x+3)(x-3)}$

17. a. iii. $\dfrac{9x^2+18x}{12x^2-3x}=\dfrac{3(x+2)}{4x-1}$

b. iv. $\dfrac{32x-16x^2}{24x+16x^2}=\dfrac{2(2-x)}{3+2x}$

c. ii. $\dfrac{7x-2x^2}{x^2+5x}=\dfrac{7-2x}{x+5}$

d. i. $\dfrac{3x^2+x}{9x-x^2}=\dfrac{3x+1}{9-x}$

18. a. iii. $\dfrac{x^2+8x+15}{x^2+7x+10}\times\dfrac{x^2+6x+8}{x^2+10x+21}=\dfrac{x+4}{x+7}$

b. iv. $\dfrac{x^2+6x+8}{x^2+5x+6}\times\dfrac{x^2+8x+15}{x^2+7x+12}=\dfrac{x+5}{x+3}$

c. i. $\dfrac{3x+6}{2x-6}\div\dfrac{x+3}{8x-24}=\dfrac{12(x+2)}{x+3}$

d. ii. $\dfrac{10x-5}{4x+28}\div\dfrac{20x-10}{6x+1}=\dfrac{6x+1}{8(x+7)}$

19. a. $\dfrac{15}{(x+2)}$ b. $\dfrac{12}{x^2}$

20. a. $\dfrac{(x-4)(x-5)}{(x+3)(x+2)^2}$ b. $\dfrac{2x(x-3)}{3(x-5)}$

21. a. $\dfrac{(x-5)}{(x-4)}$ b. $\dfrac{(x+4)(x+6)}{x^2}$

22. Yes, because all of the fractions have the same denominator and therefore can be added together.

23. No, x and z are not common to all terms so cannot be cancelled down.

24. a. -1

b. $4-x$ considered to be the same as $x-4$.

25. $\dfrac{-1}{x}$

3.9 Applications of binomial products

1. a. i. $12x$ ii. 60
   iii. $9x^2$ iv. 225

   b. i. $4x+8$ ii. 28
   iii. x^2+4x+4 iv. 49

   c. i. $16x-4$ ii. 76
   iii. $16x^2-8x+1$ iv. 361

   d. i. $10x+2$ ii. 52
   iii. $6x^2+2x$ iv. 160

   e. i. $12x-2$ ii. 58
   iii. $5x^2-13x-6$ iv. 54

   f. i. $20x+18$ ii. 118
   iii. $21x^2+42x$ iv. 735

2. a. $1250\,\text{m}^2$
   b. $(1250+150x+4x^2)\,\text{m}^2$
   c. $(150x+4x^2)\,\text{m}^2$
   d. $366.16\,\text{m}^2$
   e. $150x+4x^2=200$

3. a. $(20-2x)$ cm
   b. $(15-2y)$ cm
   c. $(300-40y-30x+4xy)\,\text{cm}^2$
   d. Sample responses can be found in the worked solutions in the online resources.

4. a. $40\,\text{cm}^2$
   b. i. $(8+v)$ cm
      ii. $(5+v)$ cm
      iii. $(v^2+13v+40)\,\text{cm}^2$
      iv. $70\,\text{cm}^2$
   c. i. $(8-d)$ cm
      ii. $(5-d)$ cm
      iii. $(40-13d+d^2)\,\text{cm}^2$
      iv. $18\,\text{cm}^2$
   d. i. $8x\,\text{cm}^2$
      ii. $(5+x)$ cm
      iii. $(8x^2+40x)\,\text{cm}^2$
      iv. $400\,\text{cm}^2$

5. a. $20x$ m
   b. $25x^2\text{m}^2$
   c. i. $(5x-2)$ m
      ii. $(5x-3)$ m
      iii. $(25x^2-25x+6)\,\text{m}^2$
      iv. 756m^2

6. a. $6x$ cm
   b. $2x^2\text{cm}^2$
   c. i. $(2x+y)$ cm
      ii. $(x-y)$ cm
      iii. $(2x^2-xy-y^2)\,\text{cm}^2$
      iv. $11\,\text{cm}^2$

7. a. $4x$ cm
   b. x^2 cm^2
   c. i. $(x + y)$ cm
      ii. $(4x + 4y)$ cm
      iii. 56 cm
      iv. $(x^2 + 2xy + y^2)$ cm^2
      v. 60.84 cm^2
8. a. $(14p + 4)$ m   b. $(12p^2 + 6p)$ m^2
   c. $(6p + 2)$ m   d. $5p$ m
   e. $(22p + 4)$ m   f. $(30p^2 + 10p)$ m^2
   g. $(18p^2 + 4p)$ m^2 h. 80 m^2
9. a. $(2x + 3)(3x + 1)$
   b. $P = 10x + 8$ m
   c. $x = 8$ metres
10. a. $(x - 5)(x + 1)$ cm^2
    b. $(x - 5)$ cm^2
    c. 15 cm
    d. 160 cm^2
    e. $3000 (x - 5)(x + 1)$ cm^2
11. a. 3
    b. 994 009
    c. i. 990 025   ii. 980 100
12. a. i. $A_{\text{screen}} = x(x + 2) = x^2 + 2x$
       ii. $A_{\text{phone 1}} = (x + 1)(x + 7) = x^2 + 8x + 7$
    b. i. $l = x + 11, w = x$
       ii. $A_{\text{phone 2}} = x(x + 11) = x^2 + 11x$
    c. $x = 4$ cm, $A_{\text{phone 2}} = 60$ cm^2
13. a. i. $x(x - 2)$ m^2 ii. $x(2x)$ m^2
    b. $\dfrac{x - 2}{2x}$
    c. $\dfrac{50(x - 2)}{x}\%$
    d. i. 25%   ii. 27.8%   iii. 29.17%
    e. The third flag places the most importance on Australia.
14. a. $(3w2 + 15w)$ m^2 b. $(3w2 + 16w + 16)$ m^2
    c. $(3w + 4)(w + 4)$ d. 21 m, 7 m and 2 m
    e. 275 m^2 f. 294 m^3
15. a. Sample responses can be found in the worked solutions in the online resources.
    b. i. $-3x^3 - 2x^2 + 3x + 2$
       ii. $8x^3 + 16x^2 - 104x + 80$
       iii. $x^3 - 49x + 120$
    c. In a cubic expression there are four terms with descending powers of x and ascending values of the pronumeral.
       For example, $(x + a)^3 = x^3 + 3ax^2 + 3a^2x + a^3$, where the powers of x are descending and the values of a are ascending.

Project

1. $h = 250$ m
2. 500m

3.
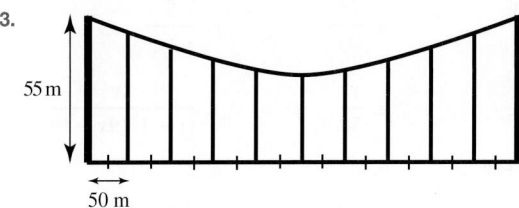

4. 18
5. 3 m; 5 m; 11 m; 22 m; 36 m
6. 305 m
7.
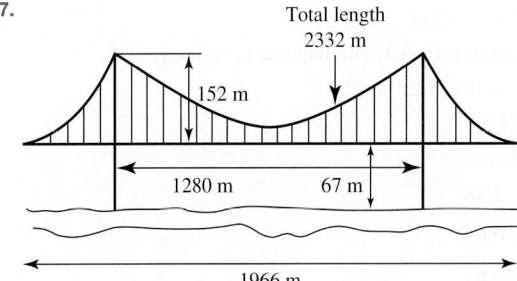

3.10 Review questions

1. B
2. A
3. B
4. a. $x^2 - 16$ b. $81 - m^2$
   c. $x^2 - y^2$ d. $1 - 4a^2$
5. a. $x^2 + 10x + 25$ b. $m^2 - 6m + 9$
   c. $16x^2 + 8x + 1$ d. $4 - 12y + 9y^2$
6. a. $2x^2 + 10x + 13$
   b. $2x - 5$
   c. $16x$
7. C
8. D
9. C
10. A
11. B
12. a. $(x + 8)(x - 8)$
    b. $(a + 12)(a - 12)$
    c. $(7b + 1)(7b - 1)$
    d. $(2f + 3g)(2f - 3g)$
    e. $(n + 1 + m)(n + 1 - m)$
    f. $(r - 1 + 2s)(r - 1 - 2s)$
13. a. $(c + 1)(c + 4)$ b. $(p - 2)(p + 12)$
    c. $(y - 6)(y - 4)$ d. $(x + 1)(x + 2)$
    e. $(m - 2)(m - 5)$ f. $(m + 22)(m + 2)$
14. a. $2(a + 6)(a + 2)$ b. $3(b - 6)(b - 2)$
    c. $4(c - 6)(c + 2)$ d. $(x + 1)(2x + 1)$
    e. $(x - 1)(3x + 2)$ f. $(2x + 3)(3x + 2)$
15. a. $\dfrac{7y}{6}$ b. $\dfrac{7x + 18}{10}$
    c. $\dfrac{22}{15x}$ d. $\dfrac{3x^2 + 2x - 17}{(x + 3)(x + 2)}$

16. a. $\dfrac{8y}{x}$ b. $\dfrac{25z}{4x}$ c. $\dfrac{5}{x+3}$

d. $\dfrac{5}{6}$ e. $\dfrac{y^2}{50}$ f. $\dfrac{2x}{(x-1)(9x+1)}$

17. a. $x+4$ b. x^2 c. 2

d. $\dfrac{4(x+1)}{x-1}$ e. $\dfrac{x+1}{x-2}$ f. $\dfrac{x+6}{x-1}$

18. a. p^2

b. $p^2 - 140p + 4500$

c. $140p - 4500$

19. a. Length $= (x-4)$ cm, width $= (x-6)$ cm

b. $(x^2 - 10x + 24)$ cm^2

c. 624 cm^2

20. a. 8 hours

b. i. 2.56

ii. 0.64

iii. 0.64

c. 0.094 hours or 15.9 hours

d. $g = \dfrac{0.16(h-8)^2}{h}$

e. No, the formula is not the same.

4 Linear equations and inequalities

LESSON SEQUENCE

LESSON
4.1 Overview

Why learn this?

Algebra is like the language of maths; it holds the key to understanding the rules, formulae and relationships that summarise much of our understanding of the universe. Every maths student needs this set of skills in order to process mathematical information and move on to more challenging concepts.

To some extent, this explains why those who want to pursue a career in maths need algebra. Every maths teacher is faced with the question 'Why do *I* need to study algebra, I'm never going to use it?' and yet no one asks why a professional footballer would lift weights when they don't lift any weight in their sport. The obvious answer for the footballer is that they are training their muscles to be fitter and stronger for upcoming matches. Learning algebra is no different, in that you are training your mind to better handle abstract concepts. Abstraction is the ability to consider concepts beyond what we observe. Spatial reasoning, complex reasoning, understanding verbal and non-verbal ideas, recognising patterns, analysing ideas and solving problems all involve abstract thinking to some degree. If some food were to fall on the ground, an adult would think about how long the food has been there, whether the ground is clean, whether the food surface can be washed; whereas a young child would just pick up the food and eat it off the ground, because they lack the ability to think abstractly. Being able to think about all these considerations is just a simple example of abstract thinking. We use abstract thinking every day, and develop this skill over our life. Those who have strong abstract reasoning skills tend to perform highly on intelligence tests and are more likely to be successful in later life. Algebra helps us develop our abstract reasoning skills and thus is of use to all students!

Exercise 4.1 Pre-test

1. **MC** The value of x in the linear equation $3(2x + 5) = 12$ is:
   A. -1.5 B. -1 C. -0.5 D. 2

2. Solve the equation: $\dfrac{3x}{4} + 6 = 10$

3. Solve the equation: $\dfrac{5 - 2x}{3} + 3 = 0$

4. **MC** When the linear equation $y = -6 + 3x$ is rearranged to make x the subject, the equation becomes:
   A. $x = \dfrac{y - 6}{3}$ B. $x = \dfrac{y + 6}{3}$ C. $x = 3y - 6$ D. $x = 3y + 6$

5. **MC** **PATH** Solve for a: $8 - 4a \geq 32$
   A. $a \geq -10$ B. $a \leq -10$ C. $a \geq -6$ D. $a \leq -6$

6. **PATH** Solve the inequality: $3x + 5 < x + 3$

7. **PATH** Solve the inequality: $\dfrac{x + 1}{2} - x > 4$

8. **PATH** Solve the inequality: $4(2 + 3x) > 8 - 3(2x + 1)$

9. **PATH** Solve the equation $\dfrac{2(4r + 3)}{5} = \dfrac{3(2r + 5)}{4}$.

10. **PATH** Solve the equation $\dfrac{8x + 3}{5} - \dfrac{3(x - 1)}{2} = \dfrac{1}{2}$.

11. **PATH** Solve the equation $4(2x + 3) - 5x + 1 = 2(x - 3) - 3(7 - 2x)$.

12. **PATH** At a charity fundraising event, three-eighths of the profit came from sales of tickets, one-fifth came from donations. A third of the profit came from the major raffle and a pop up stall raised $2200. Determine the amount of money raised at the event.

13. **MC** **PATH** If $\dfrac{x + 4}{(x + 1)(x - 2)} = \dfrac{a}{x + 1} + \dfrac{b}{x - 2}$, the values of a and b respectively are:

   A. $a = x$ and $b = 4$ B. $a = 1$ and $b = 2$ C. $a = -1$ and $b = 2$ D. $a = 1$ and $b = -2$

14. **MC** **PATH** Solve the literal equation $\dfrac{1}{a} + \dfrac{1}{b} = \dfrac{1}{c}$ for a.

   A. $a = \dfrac{bc}{b - c}$ B. $a = \dfrac{1}{b - c}$ C. $a = \dfrac{bc}{b + c}$ D. $a = c - b$

15. **MC** **PATH** Rearrange the literal equation $m = \dfrac{pa + qb}{p - q}$ to make p the subject.

   A. $p = \dfrac{qb}{-a}$ B. $p = \dfrac{q(m - b)}{m + a}$ C. $p = \dfrac{q(m + b)}{m + a}$ D. $p = \dfrac{q(m + b)}{m - a}$

LESSON
4.2 Solving linear equations review

LEARNING INTENTION

At the end of this lesson you should be able to:
- solve one- and two-step linear equations using inverse operations
- solve linear equations with pronumerals on both sides of the equals sign
- solve linear equations involving one algebraic fraction.

▶ 4.2.1 Solving equations using inverse operations

eles-4702

- **Equations** show the equivalence of two expressions.
- Equations can be solved using inverse operations.
- Determining the solution of an equation involves calculating the value or values of a variable that, when substituted into that equation, produces a true statement.
- When solving equations, the last operation performed on the pronumeral when building the equation is the first operation undone by applying inverse operations to both sides of the equation.

For example, the equation $2x + 3 = 5$ is built from x by:

First operation: multiplying by 2 to give $2x$

Second operation: adding 3 to give $2x + 3$.

Inverse operations

+ and − are inverse operations

× and ÷ are inverse operations

^{2} and $\sqrt{}$ are inverse operations

- In order to solve the equation, undo the second operation of adding 3 by subtracting 3, then undo the first operation of multiplying by 2 by dividing by 2.

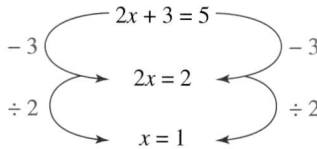

$$2x + 3 = 5$$
$$-3 \qquad -3$$
$$2x = 2$$
$$\div 2 \qquad \div 2$$
$$x = 1$$

- Equations that require one step to solve are called one-step equations.

WORKED EXAMPLE 1 Solving equations using inverse operations

Solve the following equations.

a. $a + 27 = 71$

b. $\dfrac{d}{16} = 3\dfrac{1}{4}$

THINK

a. 1. Write the equation.

   2. 27 has been added to a resulting in 71. The addition of 27 has to be reversed by subtracting 27 from both sides of the equation to obtain the solution.

b. 1. Write the equation.

   2. Express $3\dfrac{1}{4}$ as an improper fraction.

   3. The pronumeral d has been divided by 16 resulting in $\dfrac{13}{4}$. Therefore the division has to be reversed by multiplying both sides of the equation by 16 to obtain d.

WRITE

a. $\qquad a + 27 = 71$

$\qquad a + 27 - 27 = 71 - 27$

$\qquad\qquad\qquad a = 44$

b. $\qquad \dfrac{d}{16} = 3\dfrac{1}{4}$

$\qquad \dfrac{d}{16} = \dfrac{13}{4}$

$\qquad \dfrac{d}{\cancel{16}} \times \cancel{16} = \dfrac{13}{\cancel{4}} \times \cancel{16}^{4}$

$\qquad\qquad d = 52$

▶ 4.2.2 Two-step equations

eles-4704

- Two-step equations involve the inverse of two operations in their solutions.

WORKED EXAMPLE 2 Solving two-step equations

Solve the following equations.

a. $5y - 6 = 79$

b. $\dfrac{4x}{9} = 5$

THINK

a. 1. Write the equation.

   2. Step 1: Add 6 to both sides of the equation.

   3. Step 2: Divide both sides of the equation by 5 to obtain y.

   4. Write the answer.

b. 1. Write the equation.

   2. Step 1: Multiply both sides of the equation by 9.

   3. Step 2: Divide both sides of the equation by 4 to obtain x.

   4. Express the answer as a mixed number.

WRITE

a. $\qquad 5y - 6 = 79$

$\qquad 5y - 6 + 6 = 79 + 6$

$\qquad\qquad\quad 5y = 85$

$\qquad\qquad \dfrac{5y}{5} = \dfrac{85}{5}$

$\qquad\qquad\quad y = 17$

b. $\qquad \dfrac{4x}{9} = 5$

$\qquad \dfrac{4x}{9} \times 9 = 5 \times 9$

$\qquad\qquad 4x = 45$

$\qquad\qquad \dfrac{4x}{4} = \dfrac{45}{4}$

$\qquad\qquad\quad x = \dfrac{45}{4}$

$\qquad\qquad\quad x = 11\dfrac{1}{4}$

⏵ 4.2.3 Equations where the pronumeral appears on both sides

eles-4705

- In solving equations where the pronumeral appears on both sides, add or subtract the smaller pronumeral term so that it is eliminated from both sides of the equation.

WORKED EXAMPLE 3 Solving equations with multiple pronumeral terms

Solve the following equations.

a. $5h + 13 = 2h - 2$ b. $14 - 4d = 27 - d$ c. $2(x - 3) = 5(2x + 4)$

THINK	WRITE
a. 1. Write the equation.	a. $5h + 13 = 2h - 2$
2. Eliminate the pronumeral from the right-hand side by subtracting $2h$ from both sides of the equation.	$3h + 13 = -2$
3. Subtract 13 from both sides of the equation.	$3h = -15$
4. Divide both sides of the equation by 3 and write the answer.	$h = -5$
b. 1. Write the equation.	b. $14 - 4d = 27 - d$
2. Add $4d$ to both sides of the equation.	$14 = 27 + 3d$
3. Subtract 27 from both sides of the equation.	$-13 = 3d$
4. Divide both sides of the equation by 3.	$-\dfrac{13}{3} = d$
5. Express the answer as a mixed number.	$-4\dfrac{1}{3} = d$
6. Write the answer so that d is on the left-hand side.	$d = -4\dfrac{1}{3}$
c. 1. Write the equation.	c. $2(x - 3) = 5(2x + 4)$
2. Expand the brackets on both sides of the equation.	$2x - 6 = 10x + 20$
3. Subtract $2x$ from both sides of the equation.	$-6 = 8x + 20$
4. Subtract 20 from both sides of the equation.	$-26 = 8x$
5. Divide both sides of the equation by 8.	$-\dfrac{26}{8} = x$
6. Simplify and write the answer with the pronumeral on the left-hand side.	$x = -\dfrac{13}{4}$ or $-3\dfrac{1}{4}$

 Resources

⏵ **Video eLessons** Solving linear equations (eles-1895)
Solving linear equations with pronumerals on both sides (eles-1901)

Interactivities Using algebra to solve problems (int-3805)
One-step equations (int-6118)
Two-step equations (int-6119)

Exercise 4.2 Solving linear equations review

4.2 Quick quiz on	4.2 Exercise

Individual pathways

■ PRACTISE	■ CONSOLIDATE	■ MASTER
1, 4, 7, 10, 12, 15, 19, 20, 24, 27, 30	2, 5, 8, 11, 13, 16, 17, 21, 22, 25, 28, 31	3, 6, 9, 14, 18, 23, 26, 29, 32, 33

Fluency

WE1a For questions **1** to **3**, solve the following equations.

1. a. $a + 61 = 85$ b. $k - 75 = 46$ c. $g + 9.3 = 12.2$

2. a. $r - 2.3 = 0.7$ b. $h + 0.84 = 1.1$ c. $i + 5 = 3$

3. a. $t - 12 = -7$ b. $q + \dfrac{1}{3} = \dfrac{1}{2}$ c. $x - 2 = -2$

WE1b For questions **4** to **6**, solve the following equations.

4. a. $\dfrac{f}{4} = 3$ b. $\dfrac{i}{10} = -6$ c. $6z = -42$

5. a. $9v = 63$ b. $6w = -32$ c. $\dfrac{k}{12} = \dfrac{5}{6}$

6. a. $4a = 1.7$ b. $\dfrac{m}{19} = \dfrac{7}{8}$ c. $\dfrac{y}{4} = 5\dfrac{3}{8}$

WE2a For questions **7** to **14**, solve the following.

7. a. $5a + 6 = 26$ b. $6b + 8 = 44$ c. $8i - 9 = 15$

8. a. $7f - 18 = 45$ b. $8q + 17 = 26$ c. $10r - 21 = 33$

9. a. $6s + 46 = 75$ b. $5t - 28 = 21$ c. $8a + 88 = 28$

10. a. $\dfrac{f}{4} + 6 = 16$ b. $\dfrac{g}{6} + 4 = 9$ c. $\dfrac{r}{10} + 6 = 5$

11. a. $\dfrac{m}{9} - 12 = -10$ b. $\dfrac{n}{8} + 5 = 8.5$ c. $\dfrac{p}{12} - 1.8 = 3.4$

12. a. $6(x + 8) = 56$ b. $7(y - 4) = 35$ c. $5(m - 3) = 7$

13. a. $3(2k + 5) = 24$ b. $5(3n - 1) = 80$ c. $6(2c + 7) = 58$

14. a. $2(x - 5) + 3(x - 7) = 19$ b. $3(x + 5) - 5(x - 1) = 12$ c. $3(2x - 7) - (x + 3) = -60$

WE2b For questions **15** to **18**, solve the following.

15. a. $\dfrac{3k}{5} = 15$ b. $\dfrac{9m}{8} = 18$ c. $\dfrac{7p}{10} = -8$

16. a. $\dfrac{8u}{11} = -3$ b. $\dfrac{11x}{4} = 2$ c. $\dfrac{4v}{15} = 0.8$

17. a. $\dfrac{x - 5}{3} = 7$ b. $\dfrac{2m + 1}{3} = -3$ c. $\dfrac{3w - 1}{4} = 6$

18. a. $\dfrac{t-5}{2}=0$ **b.** $\dfrac{6-x}{3}=-1$ **c.** $\dfrac{3n-5}{4}=-6$

19. **MC** **a.** The solution to the equation $\dfrac{p}{5}+2=7$ is:

   A. $p=25$ **B.** $p=45$ **C.** $p=10$ **D.** $p=1$

   b. If $5h+8=53$, then h is equal to:

   A. 12.2   **B.** 225   **C.** 10   **D.** 9

   c. The exact solution to the equation $14x=75$ is:

   A. $x=5.357\,142\,857$ **B.** $x=5.357$ (to 3 decimal places)

   C. $x=5\dfrac{5}{14}$ **D.** $x=5.4$

For questions **20** to **23**, solve the following equations.

20. a. $-5h=10$ **b.** $2-d=3$ **c.** $5-p=-2$ **d.** $-7-x=4$

21. a. $-6t=-30$ **b.** $-\dfrac{v}{5}=4$ **c.** $-\dfrac{r}{12}=\dfrac{1}{4}$ **d.** $-4g=3.2$

22. a. $6-2x=8$ **b.** $10-3v=7$ **c.** $9-6l=-3$ **d.** $-3-2g=1$

23. a. $-5-4t=-17$ **b.** $-\dfrac{3e}{5}=14$ **c.** $-\dfrac{k}{4}-3=6$ **d.** $-\dfrac{4f}{7}+1=8$

WE3a For questions **24** to **26**, solve the following equations.

24. a. $6x+5=5x+7$ **b.** $7b+9=6b+14$ **c.** $11w+17=6w+27$

25. a. $8f-2=7f+5$ **b.** $10t-11=5t+4$ **c.** $12r-16=3r+5$

26. a. $12g-19=3g-31$ **b.** $7h+5=2h-6$ **c.** $5a-2=3a-2$

WE3b For questions **27** to **29**, solve the following equations.

27. a. $5-2x=6-x$ **b.** $10-3c=8-2c$ **c.** $3r+13=9r-3$

28. a. $k-5=2k-6$ **b.** $5y+8=13y+17$ **c.** $17-3g=3-g$

29. a. $14-5w=w+8$ **b.** $4m+7=8-m$ **c.** $14-5p=9-2p$

WE3c For questions **30** to **32**, solve the following equations.

30. a. $3(x+5)=2x$ **b.** $8(y+3)=3y$ **c.** $6(t-5)=4(t+3)$

31. a. $10(u+1)=3(u-3)$ **b.** $12(f-10)=4(f-5)$ **c.** $2(4r+3)=3(2r+7)$

32. a. $5(2d+9)=3(3d+13)$ **b.** $5(h-3)=3(2h-1)$ **c.** $2(4x+1)=5(3-x)$

33. **MC** **a.** The solution to $8-4k=-2$ is:

   A. $k=2\dfrac{1}{2}$ **B.** $k=-2\dfrac{1}{2}$ **C.** $k=1\dfrac{1}{2}$ **D.** $k=-1\dfrac{1}{2}$

   b. The solution to $-\dfrac{6n}{5}+3=-7$ is:

   A. $n=3\dfrac{1}{3}$ **B.** $n=-3\dfrac{1}{3}$ **C.** $n=\dfrac{1}{3}$ **D.** $n=8\dfrac{1}{3}$

   c. The solution to $p-6=8-4p$ is:

   A. $p=\dfrac{2}{5}$ **B.** $p=2\dfrac{4}{5}$ **C.** $p=4\dfrac{2}{3}$ **D.** $p=\dfrac{2}{3}$

LESSON
4.3 Solving multi-step equations, including fractions (Path)

LEARNING INTENTION

At the end of this lesson you should be able to:
- expand brackets and collect like terms in order to solve a multi-step equation
- solve complex linear equations involving algebraic fractions by determining the LCM of the denominators.

4.3.1 Equations with multiple brackets

eles-4706

- Equations can be simplified by expanding brackets and collecting like terms before they are solved.

WORKED EXAMPLE 4 Solving equations with brackets

Solve each of the following linear equations.

a. $6(x+1) - 4(x-2) = 0$

b. $7(5-x) = 3(x+1) - 10$

THINK	WRITE
a. 1. Write the equation.	a. $6(x+1) - 4(x-2) = 0$
2. Expand all the brackets. (Be careful with the -4.)	$6x + 6 - 4x + 8 = 0$
3. Collect like terms.	$2x + 14 = 0$
4. Subtract 14 from both sides of the equation.	$2x = -14$
5. Divide both sides of the equation by 2 to obtain the value of x.	$x = -7$
b. 1. Write the equation.	b. $7(5-x) = 3(x+1) - 10$
2. Expand all the brackets.	$35 - 7x = 3x + 3 - 10$
3. Collect like terms.	$35 - 7x = 3x - 7$
4. Create a single pronumeral term by adding $7x$ to both sides of the equation.	$35 = 10x - 7$
5. Add 7 to both sides of the equation.	$42 = 10x$
6. Divide both sides of the equation by 10 to solve for x and simplify.	$\dfrac{42}{10} = x$ $\dfrac{21}{5} = x$
7. Express the improper fraction as a mixed number fraction.	$4\dfrac{1}{5} = x$
8. Rewrite the equation so that x is on the left-hand side.	$x = 4\dfrac{1}{5}$

▶ 4.3.2 Equations involving algebraic fractions

eles-4707

- To solve an equation containing algebraic fractions, multiply both sides of the equation by the lowest common multiple (LCM) of the denominators. This gives an equivalent form of the equation without fractions.

WORKED EXAMPLE 5 Solving equations with algebraic fractions

Solve the equation $\dfrac{x-5}{3} = \dfrac{x+7}{4}$ and verify the solution.

THINK	WRITE
1. Write the equation.	$\dfrac{x-5}{3} = \dfrac{x+7}{4}$
2. The LCM is $3 \times 4 = 12$. Multiply both sides of the equation by 12.	$\dfrac{^4\cancel{12}\,(x-5)}{^1\cancel{3}} = \dfrac{^3\cancel{12}\,(x+7)}{^1\cancel{4}}$
3. Simplify the fractions.	$4\,(x-5) = 3\,(x+7)$
4. Expand the brackets.	$4x - 20 = 3x + 21$
5. Subtract $3x$ from both sides of the equation.	$x - 20 = 21$
6. Add 20 to both sides of the equation and write the answer.	$x = 41$
7. To verify, check that the answer $x = 41$ is true for both the left-hand side (LHS) and the right-hand side (RHS) of the equation by substitution.	
Substitute $x = 41$ into the LHS.	$\text{LHS} = \dfrac{41-5}{3}$
	$= \dfrac{36}{3}$
	$= 12$
Substitute $x = 41$ into the RHS.	$\text{RHS} = \dfrac{41+7}{4}$
	$= \dfrac{48}{4}$
	$= 12$
8. Write the answer.	Because the LHS = RHS, the solution $x = 41$ is correct.

- In equations with algebraic fractions and multiple terms, determine the lowest common multiple (LCM) of the denominators. Then express each term as an equivalent fraction with the LCM as its denominator.

WORKED EXAMPLE 6 Solving involving algebraic fractions

Solve each of the following equations.

a. $\dfrac{5(x+3)}{6} = 4 + \dfrac{3(x-1)}{5}$

b. $\dfrac{4}{3(x-1)} = \dfrac{1}{x+1}$

THINK

WRITE

a. 1. Write the equation.

a. $\dfrac{5(x+3)}{6} = 4 + \dfrac{3(x-1)}{5}$

2. The lowest common denominator of 5 and 6 is 30. Write each term as an equivalent fraction with a denominator of 30.

$\dfrac{25(x+3)}{30} = \dfrac{120}{30} + \dfrac{18(x-1)}{30}$

3. Multiply each term by 30. This effectively removes the denominator.

$25(x+3) = 120 + 18(x-1)$

4. Expand the brackets and collect like terms.

$25x + 75 = 120 + 18x - 18$

$25x + 75 = 102 + 18x$

5. Subtract $18x$ from both sides of the equation.

$7x + 75 = 102$

6. Subtract 75 from both sides of the equation.

$7x = 27$

7. Divide both sides of the equation by 7 to solve for x.

$x = \dfrac{27}{7}$

8. Express the answer as a mixed number.

$x = 3\dfrac{6}{7}$

b. 1. Write the equation.

b. $\dfrac{4}{3(x-1)} = \dfrac{1}{x+1}$

2. The lowest common denominator of 3, $x+1$ and $x-1$ is $3(x-1)(x+1)$. Write each term as an equivalent fraction with a common denominator of $3(x-1)(x+1)$.

$\dfrac{4(x+1)}{3(x-1)(x+1)} = \dfrac{3(x-1)}{3(x-1)(x+1)}$

3. Multiply each term by the common denominator.

$4(x+1) = 3(x-1)$

4. Expand the brackets.

$4x + 4 = 3x - 3$

5. Subtract $3x$ from both sides of the equation.

$x + 4 = -3$

6. Subtract 4 from both sides of the equation to solve for x.

$x + 4 - 4 = -3 - 4$

7. Write the answer.

$x = -7$

---

on Resources

▶ **Video eLesson** Solving linear equations with algebraic fractions (eles-1857)

✦ **Interactivity** Expanding brackets: Distributive Law (int-3774)

Exercise 4.3 Solving multi-step equations, including fractions (Path)

4.3 Quick quiz on	4.3 Exercise

Individual pathways

■ PRACTISE	■ CONSOLIDATE	■ MASTER
1, 4, 7, 10, 13, 17, 20	2, 5, 8, 11, 14, 18, 21	3, 6, 9, 12, 15, 16, 19, 22

Fluency

WE4 1 to **3**, solve each of the following linear equations.

1. a. $6(4x-3)+7(x+1)=9$ b. $9(3-2x)+2(5x+1)=0$

2. a. $8(5-3x)-4(2+3x)=3$ b. $9(1+x)-8(x+2)=2x$

3. a. $6(4+3x)=7(x-1)+1$ b. $10(4x+2)=3(8-x)+6$

WE5 For questions **4** to **6**, solve each of the following equations and verify the solutions.

4. a. $\dfrac{x+1}{2}=\dfrac{x+3}{3}$ b. $\dfrac{x-7}{5}=\dfrac{x-8}{4}$ c. $\dfrac{x-6}{4}=\dfrac{x-2}{2}$

5. a. $\dfrac{8x+3}{5}=2x$ b. $\dfrac{2x-1}{5}=\dfrac{x-3}{4}$ c. $\dfrac{4x+1}{3}=\dfrac{x+2}{4}$

6. a. $\dfrac{6-x}{3}=\dfrac{2x-1}{5}$ b. $\dfrac{8-x}{9}=\dfrac{2x+1}{3}$ c. $\dfrac{2(x+1)}{5}=\dfrac{3-2x}{4}$

For questions **7** to **9**, solve each of the following linear equations.

7. a. $\dfrac{x}{3}+\dfrac{4x}{5}=\dfrac{1}{3}$ b. $\dfrac{x}{4}-\dfrac{x}{5}=\dfrac{3}{4}$ c. $\dfrac{x}{4}-\dfrac{4x}{7}=2$ d. $\dfrac{-3x}{5}+\dfrac{x}{8}=\dfrac{1}{4}$

8. a. $\dfrac{2x}{3}-\dfrac{x}{6}=-\dfrac{3}{4}$ b. $\dfrac{5x}{8}-8=\dfrac{2x}{3}$ c. $\dfrac{2}{7}-\dfrac{x}{8}=\dfrac{3x}{8}$ d. $\dfrac{4}{x}-\dfrac{1}{6}=\dfrac{2}{x}$

9. a. $\dfrac{15}{x}-4=\dfrac{2}{x}$ b. $\dfrac{1}{3}+\dfrac{4}{x}=\dfrac{5}{x}$ c. $\dfrac{2x-4}{5}+6=\dfrac{x}{2}$ d. $\dfrac{4x-1}{2}-\dfrac{2x+5}{3}=0$

WE6 For questions **10** to **12**, solve each of the following linear equations.

10. a. $\dfrac{3(x+1)}{2}+\dfrac{5(x+1)}{3}=4$ b. $\dfrac{2(x+1)}{7}+\dfrac{3(2x-5)}{8}=0$

   c. $\dfrac{2(4x+3)}{5}-\dfrac{6(x-2)}{2}=\dfrac{1}{2}$ d. $\dfrac{8(x+3)}{5}=\dfrac{3(x+2)}{4}$

11. a. $\dfrac{5(7-x)}{2}=\dfrac{2(2x-1)}{7}+1$ b. $\dfrac{2(6-x)}{3}=\dfrac{9(x+5)}{6}+\dfrac{1}{3}$

   c. $\dfrac{-5(x-2)}{3}-\dfrac{6(2x-1)}{5}=\dfrac{1}{3}$ d. $\dfrac{9(2x-1)}{7}=\dfrac{4(x-5)}{3}$

12. a. $\dfrac{1}{x-1}+\dfrac{3}{x+1}=\dfrac{8}{x+1}$ b. $\dfrac{3}{x+1}+\dfrac{5}{x-4}=\dfrac{5}{x+1}$

   c. $\dfrac{1}{x-1}-\dfrac{3}{x}=\dfrac{-1}{x-1}$ d. $\dfrac{4}{2x-1}-\dfrac{5}{x}=\dfrac{-1}{x}$

Understanding

13. Last week Maya broke into her money box. She spent one-quarter of the money on a birthday present for her brother and one-third of the money on an evening out with her friends, leaving her with $75.
Determine the amount of money in her money box.

14. At work Keith spends one-fifth of his time in planning and buying merchandise. He spends seven-twelfths of his time in customer service and one-twentieth of his time training the staff. This leaves him ten hours to deal with the accounts.
Determine the number of hours he works each week.

15. Last week's school fete was a great success, raising a good deal of money. Three-eighths of the profit came from sales of food and drink, and the market stalls recorded one-fifth of the total. A third of the profit came from the major raffle, and the jumping castle raised $1100.
Determine the amount of money raised at the fete.

16. Lucy had half as much money as Mel, but since Grandma gave them each $20 she now has three-fifths as much. Determine the amount of money Lucy has.

Communicating, reasoning and problem solving

17. Answer the following question and justify your answer:
    a. Determine numbers smaller than 100 that have exactly 3 factors (including 1 and the number itself).
    b. Determine the two numbers smaller than 100 that have exactly 5 factors.
    c. Determine a number smaller than 100 that has exactly 7 factors.

18. To raise money for a charity, a Year 10 class has decided to organise a school lunch. Tickets will cost $6 each. The students have negotiated a special deal for delivery of drinks and pizzas, and they have budgeted $200 for drinks and $250 for pizzas. If they raise $1000 or more, they qualify for a special award.

    a. Write an equation to represent the minimum number of tickets required to be sold to qualify for the award.
    b. Solve the equation to determine the number of tickets they must sell to qualify for the award. Explain your answer.

19. If $\dfrac{x+7}{(x+2)(x+3)} \equiv \dfrac{a}{x+2} - \dfrac{4}{x+3}$, explain why a must be equal to 5.

    (*Note:* '$\equiv$' means identically equal to.)

20. Solve for x:
$$\frac{2}{9}(x-1) - \frac{5}{8}(x-2) = \frac{2}{5}(x-4) - \frac{7}{12}$$

21. If $\dfrac{2(4x+3)}{(x-3)(x+7)} \equiv \dfrac{a}{x-3} + \dfrac{b}{x+7}$, determine the values of a and b.

22. If $\dfrac{7x+20}{x^2+7x+12} = \dfrac{a}{x+3} + \dfrac{b}{x+4} + \dfrac{a+b}{x^2+7x+12}$, determine the values of a and b.

LESSON
4.4 Literal equations (Path)

LEARNING INTENTION

At the end of this lesson you should be able to:
- rearrange formulas with two or more pronumerals to make only one of the pronumerals the subject
- solve literal equations, which include multiple variables, by changing the subject of an equation to a particular pronumeral.

4.4.1 Literal equations

eles-4708

- **Literal equations** are equations that include several pronumerals or variables.
- Solving literal equations involves changing the subject of the equation to a particular pronumeral.
- A variable is the subject of an equation if it is expressed in terms of the other variables.
  For example, in the formula $v = u + at$, the subject of the equation is v as it is written in terms of the variables u, a and t.
  $v = u + at$ can be rearranged to make u, a or t the subject using our knowledge of solving linear equations.

WORKED EXAMPLE 7 Solving literal equations

Solve the following literal equations for x.

a. $y = kx + m$

b. $6(y + 1) = 7(x - 2)$

THINK

a. 1. Subtract m from both sides.

2. Divide both sides by k.

3. Rewrite the equation so that x is on the left-hand side.

b. 1. Expand the brackets.

2. Add 14 to both sides.

3. Divide both sides by 7.

4. Rewrite the equation so that x is on the left-hand side.

WRITE

a.
$$y = kx + m$$
$$y - m = kx$$

$$\frac{y - m}{k} = \frac{kx}{k}$$
$$\frac{y - m}{k} = x$$

$$x = \frac{y - m}{k}$$

b.
$$6(y + 1) = 7(x - 2)$$
$$6y + 6 = 7x - 14$$

$$6y + 20 = 7x$$

$$\frac{6y + 20}{7} = x$$

$$x = \frac{6y + 20}{7}$$

WORKED EXAMPLE 8 Solving literal equations for a given pronumeral

Solve the following literal equations.

a. If $g = 6d - 3$, solve for d.

b. Solve for v, given that $a = \dfrac{v - u}{t}$.

THINK	WRITE
a. 1. Add 3 to both sides.	**a.** $\quad g = 6d - 3$
	$g + 3 = 6d$
2. Divide both sides by 6.	$\dfrac{g + 3}{6} = d$
3. Rewrite the equation so that d is on the left-hand side.	$d = \dfrac{g + 3}{6}$
b. 1. Multiply both sides by t.	**b.** $\quad a = \dfrac{v - u}{t}$
	$at = v - u$
2. Add u to both sides.	$at + u = v$
3. Rewrite the equation so that v is on the left-hand side.	$v = at + u$

 Resources

 Interactivity Rearranging formulas (int-6040)

Exercise 4.4 Literal equations (Path)

learn **on**

4.4 Quick quiz **on**	4.4 Exercise

Individual pathways

■ PRACTISE	■ CONSOLIDATE	■ MASTER
1, 3, 6, 8, 11	2, 4, 7, 9, 12	5, 10, 13, 14, 15

Fluency

1. **WE7** Rearrange each formula to make x the subject.

   a. $y = ax$
   b. $y = ax + b$
   c. $y = 2ax - b$
   d. $y + 4 = 2x - 3$
   e. $6(y + 2) = 5(4 - x)$
   f. $x(y - 2) = 1$

2. Rearrange each formula to make x the subject.

   a. $x(y - 2) = y + 1$
   b. $5x - 4y = 1$
   c. $6(x + 2) = 5(x - y)$
   d. $7(x - a) = 6x + 5a$
   e. $5(a - 2x) = 9(x + 1)$
   f. $8(9x - 2) + 3 = 7(2a - 3x)$

3. **WE8** For each of the following, make the variable shown in brackets the subject of the formula.

  a. $g = 4P - 3$ (P)

  b. $f = \dfrac{9c}{5}$ (c)

  c. $f = \dfrac{9c}{5} + 32$ (c)

  d. $v = u + at$ (t)

4. For each of the following, make the variable shown in brackets the subject of the formula.

  a. $d = b^2 - 4ac$ (c)

  b. $m = \dfrac{y - k}{h}$ (y)

  c. $m = \dfrac{y - a}{x - b}$ (a)

  d. $m = \dfrac{y - a}{x - b}$ (x)

5. For each of the following, make the variable shown in brackets the subject of the formula.

  a. $C = \dfrac{2\pi}{r}$ (r)

  b. $f = ax + by$ (x)

  c. $s = ut + \dfrac{1}{2}at^2$ (a)

  d. $F = \dfrac{GMm}{r^2}$ (G)

Understanding

6. The cost to rent a car is given by the formula $C = 50d + 0.2k$, where $d =$ the number of days rented and $k =$ the number of kilometres driven. Lin has \$300 to spend on car rental for her four-day holiday.
   Calculate the distance that she can drive on her holiday.

7. A cyclist pumps up a bike tyre that has a slow leak. The volume of air (in cm^3) after t minutes is given by the formula $V = 24\,000 - 300t$.

   a. Determine the volume of air in the tyre when it is first filled.
   b. Write out the equation for the tyre's volume and solve it to work out how long it would take the tyre to go completely flat.

Communicating, reasoning and problem solving

8. The total surface area of a cylinder is given by the formula $T = 2\pi r^2 + 2\pi rh$, where $r =$ radius and $h =$ height. A car manufacturer wants its engines' cylinders to have a radius of 4 cm and a total surface area of 400 cm^2. Show that the height of the cylinder is approximately 11.92 cm, correct to 2 decimal places.
   (*Hint*: Express h in terms of T and r.)

9. If $B = 3x - 6xy$, write x as the subject. Explain the process you followed by showing your working.

10. The volume, V, of a sphere can be calculated using the formula $V = \dfrac{4}{3}\pi r^3$, where r is the radius of the sphere. Explain how you would work out the radius of a spherical ball that has the capacity to hold 5 litres of air.

11. The area of a trapezium is given by $A = \dfrac{1}{2}(a+b)h$, where a and b are the lengths of the top and the base and h is the height of the trapezium.

    a. Make b the subject of the formula.
    b. Determine the length of the base of a trapezium with a height of 4 cm and top of 5 cm and a total area of 32 cm^2.

12. T is the period of a pendulum whose length is l and g is the acceleration due to gravity. The formula relating these variables is $T = 2\pi\sqrt{\dfrac{l}{g}}$.

    a. Make l the subject of the formula.
    b. Determine the length of a pendulum that has a period of 3 seconds, given that $g = 9.8\,\text{m/s}^2$. Give your answer correct to one decimal place.

13. Use algebra to show that $\dfrac{1}{v} = \dfrac{1}{u} - \dfrac{1}{f}$ can also be written as $u = \dfrac{fv}{v+f}$.

14. Consider the formula $d = \sqrt{b^2 - 4ac}$. Rearrange the formula to make a the subject.

15. Evaluate the values for a and b such that:
$$\dfrac{4}{x+1} - \dfrac{3}{x+2} = \dfrac{ax+b}{(x+1)(x+2)}.$$

LESSON
4.5 Solving linear inequalities (Path)

LEARNING INTENTION

At the end of this lesson you should be able to:
- solve a linear inequality and represent the solution on a number line
- convert a worded statement to an inequality to solve a problem.

▶ 4.5.1 Inequalities between two expressions

eles-4774

- An equation, such as $y = 2x$, is a statement of *equality* as both sides are equal to each other.
- An **inequation**, such as $y < x + 3$, is a statement of **inequality** between two expressions.
- A linear equation such as $3x = 6$ will have a unique solution ($x = 2$), whereas an inequation such as $3x < 6$ will have an infinite number of solutions ($x = 1, 0, -1, -2 \ldots$ are all solutions.
  Note: Fractions and decimals can also be solutions).
- We use a number line to represent all possible solutions to a linear inequation. When representing an inequality on a number line, an open circle is used to represent that a value is not included, while a closed circle is used to indicate that a number is included.

- The table below shows four basic inequalities and their representation on a number line.

Mathematical statement	Worded statement	Number line diagram
$x > 2$	x is greater than 2	 −10 −8 −6 −4 −2 0 2 4 6 8 10
$x \geq 2$	x is greater than or equal to 2	 −10 −8 −6 −4 −2 0 2 4 6 8 10
$x < 2$	x is less than 2	 −10 −8 −6 −4 −2 0 2 4 6 8 10
$x \leq 2$	x is less than or equal to 2	 −10 −8 −6 −4 −2 0 2 4 6 8 10

Solving inequalities

- The following operations may be done to both sides of an inequality without affecting its truth.
    - A number can be added or subtracted from both sides of the inequality.

Adding or subtracting a number:	
For example: $6 > 2$ Add 3 to both sides: $9 > 5$ (True)	+3 +3 0 1 2 3 4 5 6 7 8 9 10
For example: $6 \geq 2$ Subtract 3 from both sides: $3 \geq -1$ (True)	−3 −3 −2 −1 0 1 2 3 4 5 6 7 8

Adding or subtracting moves both numbers the same distance along the number line.

- A number can be multiplied or divided by a positive number.

Multiplying or dividing by a positive number:	
For example: $6 > 2$ Multiply both sides by $\frac{1}{2}$: $3 > 1$ (True)	$\times \frac{1}{2}$ $\times \frac{1}{2}$ −1 0 1 2 3 4 5 6 7

The distance between the numbers has changed, but their relative position has not.

- Care must be taken when multiplying or dividing by a negative number.

Multiplying or dividing by a negative number:	
For example: $6 > 2$ Multiply both sides by -1: $-6 > -2$ (False)	$\times -1$ $\times -1$ −6 −5 −4 −3 −2 −1 0 1 2 3 4 5 6

Multiplying or dividing by a negative number reflects numbers about $x = 0$.
Their relative positions are reversed.

- When solving inequalities, if both sides are multiplied or divided by a negative number, then the inequality sign must be reversed.

  For example, $6 > 2$ implies that $-6 < -2$.

Solving a linear inequality

Solving a linear inequality is a similar process to solving a standard linear equation.

However, when multiplying or dividing by a negative number, the direction of the inequality sign reverses.

WORKED EXAMPLE 9 Solving linear inequalities

Solve each of the following linear inequalities and show the solution on a number line.

a. $4x - 1 < -2$

b. $6x - 7 \geq 3x + 5$

THINK	WRITE/DRAW
a. 1. Write the inequality.	a. $4x - 1 < -2$
2. Add 1 to both sides of the inequality.	$4x - 1 + 1 < -2 + 1$ $4x < -1$
3. Obtain x by dividing both sides of the inequality by 4.	$\dfrac{4x}{4} < -\dfrac{1}{4}$ $x < -\dfrac{1}{4}$
4. Show the solution on a number line. Use an open circle to show that the value of $-\dfrac{1}{4}$ is not included.	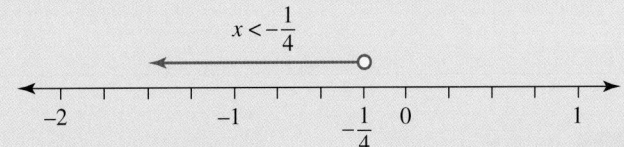
b. 1. Write the inequality.	b. $6x - 7 \geq 3x + 5$
2. Subtract $3x$ from both sides of the inequality.	$6x - 7 - 3x \geq 3x + 5 - 3x$ $3x - 7 \geq 5$
3. Add 7 to both sides of the inequality.	$3x - 7 + 7 \geq 5 + 7$ $3x \geq 12$
4. Obtain x by dividing both sides of the inequality by 3.	$3x \geq 12$ $\dfrac{3x}{3} \geq \dfrac{12}{3}$ $x \geq 4$
5. Show the solution on a number line. Use a closed circle to show that the value of 4 is included.	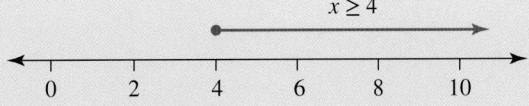

WORKED EXAMPLE 10 Solving complex linear inequalities

Solve each of the following linear inequalities.

a. $-3m + 5 < -7$

b. $5(x - 2) \geq 7(x + 3)$

THINK	WRITE
a. 1. Write the inequality.	**a.** $\qquad -3m + 5 < -7$
2. Subtract 5 from both sides of the inequality. (No change to the inequality sign.)	$-3m + 5 - 5 < -7 - 5$ $\qquad -3m < -12$
3. Obtain m by dividing both sides of the inequation by -3. Reverse the inequality sign, since we are dividing by a negative number.	$\qquad \dfrac{-3m}{-3} > \dfrac{-12}{-3}$ $\qquad m > 4$
b. 1. Write the inequality.	**b.** $\qquad 5(x - 2) \geq 7(x + 3)$
2. Expand both brackets.	$\qquad 5x - 10 \geq 7x + 21$
3. Subtract $7x$ from both sides of the inequality.	$5x - 10 - 7x \geq 7x + 21 - 7x$ $\qquad -2x - 10 \geq 21$
4. Add 10 to both sides of the inequation.	$-2x - 10 + 10 \geq 21 + 10$ $\qquad -2x \geq 31$
5. Divide both sides of the inequality by -2. Reverse the direction of the inequality sign as we are dividing by a negative number.	$\qquad \dfrac{-2x}{-2} \leq \dfrac{31}{-2}$ $\qquad x \leq \dfrac{-31}{2}$ $\qquad x \leq -15\dfrac{1}{2}$

 Resources

Interactivity Inequalities on the number line (int-6129)

DISCUSSION

In what real life context would you use an inequality or need to solve an inequality?

Discuss in pairs and share as a class.

Exercise 4.5 Solving linear inequalities (Path)

4.5 Quick quiz on	4.5 Exercise

Individual pathways

■ PRACTISE	■ CONSOLIDATE	■ MASTER
1, 4, 7, 10, 15, 18, 19, 22, 25, 28, 31	2, 5, 8, 11, 13, 16, 20, 23, 26, 29, 32	3, 6, 9, 12, 14, 17, 21, 24, 27, 30, 33

Fluency

WE9a For questions **1** to **3**, solve each of the following inequalities and show the solution on a number line.

1. a. $x + 1 > 3$ b. $a + 2 > 1$ c. $y - 3 \geq 4$ d. $m - 1 \geq 3$

2. a. $p + 4 < 5$ b. $x + 2 < 9$ c. $m - 5 \leq 4$ d. $a - 2 \leq 5$

3. a. $x - 4 > -1$ b. $5 + m \geq 7$ c. $6 + q \geq 2$ d. $5 + a > -3$

For questions **4** to **6**, solve each of the following inequalities. Check your solutions by substitution.

4. a. $3m > 9$ b. $5p \leq 10$ c. $2a < 8$ d. $4x \geq 20$

5. a. $5p > -25$ b. $3x \leq -21$ c. $2m \geq -1$ d. $4b > -2$

6. a. $\dfrac{m}{3} > 6$ b. $\dfrac{x}{2} < 4$ c. $\dfrac{a}{7} \leq -2$ d. $\dfrac{m}{5} \geq 5$

For questions **7** to **9**, solve each of the following inequalities.

7. a. $2m + 3 < 12$ b. $3x + 4 \geq 13$ c. $5p - 9 > 11$ d. $4n - 1 \leq 7$

8. a. $2b - 6 < 4$ b. $8y - 2 > 14$ c. $10m + 4 \leq -6$ d. $2a + 5 \geq -5$

9. a. $3b + 2 < -11$ b. $6c + 7 \leq 1$ c. $4p - 2 > -10$ d. $3a - 7 \geq -28$

WE9b For questions **10** to **14**, solve each of the following linear inequalities and show the solution on a number line.

10. a. $2m + 1 > m + 4$ b. $2a - 3 \geq a - 1$

    c. $5a - 3 < a - 7$ d. $3a + 4 \leq a - 2$

11. a. $5x - 2 > 40 - 2x$ b. $7x - 5 \leq 11 - x$

    c. $7b + 5 < 2b + 25$ d. $2(a + 4) > a + 13$

12. a. $3(m - 1) < m + 1$ b. $5(2m - 3) \leq 3m + 6$

    c. $3(5b + 2) \leq -10 + 4b$ d. $5(3m + 1) \geq 2(m + 9)$

13. a. $\dfrac{x + 1}{2} \leq 4$ b. $\dfrac{x - 2}{5} \geq -4$ c. $\dfrac{x + 7}{3} < -1$

14. a. $\dfrac{2x + 3}{4} > 6$ b. $\dfrac{3x - 1}{7} \geq 2$ c. $\dfrac{5x + 9}{6} < 0$

WE10 For questions **15** to **17**, solve each of the following inequalities.

15. a. $-2m > 4$ b. $-5p \leq 15$ c. $-2a \geq -10$

    d. $-p - 3 \leq 2$ e. $10 - y \geq 13$

16. **a.** $14 - x < 7$ **b.** $1 - 6p > 1$ **c.** $2 - 10a \leq 0$

     d. $2(3 - x) < 12$ **e.** $-4(a + 9) \geq 8$

17. **a.** $-15 \leq -3(2 + b)$ **b.** $2x - 3 > 5x + 6$ **c.** $k + 5 < 2k - 3$

     d. $3(x - 4) < 5(x + 5)$ **e.** $7(a + 4) \geq 4(2a - 3)$

18. **MC** When solving the inequality $-2x > -7$ we need to:

     A. change the sign to $\geq$ **B.** change the sign to $<$

     C. change the sign to $\leq$ **D.** keep the sign unchanged

For questions **19** to **24**, solve each of the following inequalities.

19. **a.** $\dfrac{2 - x}{3} > 1$ **b.** $\dfrac{5 - m}{4} \geq 2$

20. **a.** $\dfrac{-3 - x}{5} < -4$ **b.** $\dfrac{3 - 8a}{2} < -1$

21. **a.** $\dfrac{4 - 3m}{2} \leq 0$ **b.** $\dfrac{-2m + 6}{10} \leq 3$

22. **a.** $3k > 6$ **b.** $-a - 7 < -2$ **c.** $5 - 3m \geq 0$ **d.** $x + 4 > 9$

23. **a.** $10 - y \leq 3$ **b.** $5 + 3d < -1$ **c.** $\dfrac{7p}{3} \geq -2$ **d.** $\dfrac{1 - x}{3} \leq 2$

24. **a.** $\dfrac{-4 - 2m}{5} > 0$ **b.** $5a - 2 < 4a + 7$ **c.** $6p + 2 \leq 7p - 1$ **d.** $2(3x + 1) > 2x - 16$

Understanding

25. Write linear inequalities for the following statements, using x to represent the unknown. (Do not attempt to solve the equations.)

     a. The product of 5 and a certain number is greater than 10.
     b. When three is subtracted from a certain number the result is less than or equal to 5.
     c. The sum of seven and three times a certain number is less than 42.

26. Write linear inequalities for the following statements. Choose an appropriate letter to represent the unknown.

     a. Four more than triple a number is more than 19.
     b. Double the sum of six and a number is less than 10.
     c. Seven less the half the difference between a number and 8 is at least 9.

27. Write linear inequalities for the following situations. Choose an appropriate letter to represent the unknown.

     a. John makes $50 profit for each television he sells. Determine how many televisions John needs to sell to make at least $650 in profit.
     b. Determine what distances a person can travel with $60 if the cost of a taxi ride is $2.50 per km with a flagfall cost of $5.

Communicating, reasoning and problem solving

28. Tom is the youngest of 5 children. The five children were all born 1 year apart. If the sum of their ages is at most 150, set up an inequality and solve it to determine the possible ages of Tom.

29. Given the positive numbers a, b, c and d and the variable x, there is the following relationship:
$$-c < ax + b < -d.$$
    a. Determine the possible range of values of x if $a = 2$, $b = 3$, $c = 10$ and $d = 1$.
    b. Rewrite the original relationship in terms of x only (x by itself between the $<$ signs), using a, b, c and d.

30. Two speed boats are racing along a section of Lake Quikalong. The speed limit along this section of the lake is 50 km/h. Ella is travelling 6 km/h faster than Steven and the sum of the speeds at which they are travelling is greater than 100 km/h.

    a. Write an inequation and solve it to describe all possible speeds that Steven could be travelling at.
    b. At Steven's lowest possible speed, is he over the speed limit?
    c. The water police issue a warning to Ella for exceeding the speed limit on the lake. Show that the police were justified in issuing a warning to Ella.

31. Mick the painter has fixed costs (e.g. insurance, equipment, etc) of $3400 per year. His running cost to travel to jobs is based on $0.75 per kilometre. Last year Mick had costs that were less than $16 000.

    a. Write an inequality and solve it to determine how many kilometres Mick travelled for the year.
    b. Explain the information you have found.

32. A coffee store produces doughnuts and croissants to sell alongside its coffee. Each morning the bakery has to decide how many of each it will produce. The store has 240 minutes to produce food in the morning.
    It takes 20 minutes to make a batch of doughnuts and 10 minutes to make a batch of croissants. The store also has 36 kg of flour to use each day. A batch of doughnuts uses 2 kg of flour and a batch of croissants require 2 kg of flour.
    a. Set up an inequality around the amount of time available to produce doughnuts and croissants.
    b. Set up an inequality around the amount of flour available to produce doughnuts and croissants.
    c. Use technology to work out the possible number of each that can be made, taking into account both inequalities.

33. I have $40 000 to invest. Part of this I intend to invest in a stable 5% simple interest account. The remainder will be invested in my friend's IT business. She has said that she will pay me 7.5% interest on any money I give to her. I am saving for a European trip so want the best return for my money.
    Calculate the least amount of money I should invest with my friend so that I receive at least $2500 interest per year from my investments.

    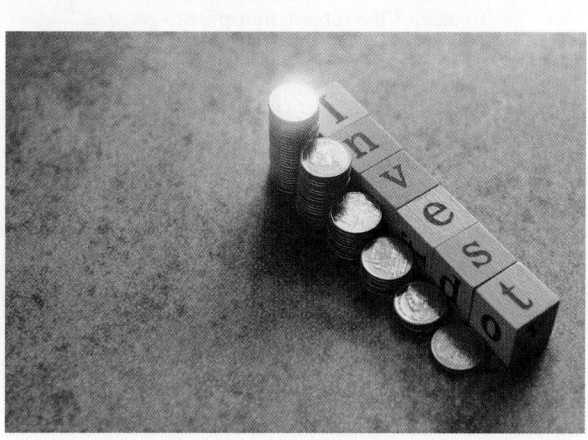

LESSON
4.6 Review

4.6.1 Topic summary

Solving equations

- Inverse operations are used to solve equations.
  - Add (+) and subtract (−) are inverses
  - Multiply (×) and divide (÷) are inverses
  - Squares (x^2) and square roots ($\sqrt{x}$) are inverses
- One-step equations can be solved using one inverse operation:
  e.g. $x + 5 = 12$
  $x + 5 - 5 = 12 - 5$
  $x = 7$

Solving linear inequalities (Path)

- Terms can be added to or subtracted from both sides.
- Each side can be multiplied or divided by a positive number.
- When multiplying or dividing by a negative number, switch the direction of the inequality sign.

Mathematical statement	Number line diagram
$x > 2$	
$x \geq 2$	
$x < 2$	
$x \leq 2$	

LINEAR EQUATIONS AND INEQUALITIES

Literal equations (Path)

- Literal equations are equations that involve multiple pronumerals or variables.
- The same processes (inverse operations etc.) are used to solve literal equations.
- Solving a literal equation is the same as making one variable the subject of the equation. This means it is expressed in terms of the other variables.

  e.g. P is the subject of the equation: $P = \dfrac{nRT}{V}$

  To make T the subject, transpose to get: $T = \dfrac{PV}{nR}$

Solving complex equations (Path)

- Solving two-step and multi-step equations will involving the following.
  - Using inverse operations
  - Expanding brackets
  - Collecting like terms
  - Finding the LCM of algebraic fractions, then multiplying by the LCM to remove all denominators

Algebraic basics

- We can only add and subtract like terms:
  $3x + 6y - 7x + 2z = 6y + 2z - 4x$
- Multiplying algebraic terms: $10x^3y^2 \times 4x^2z = 40x^5y^2z$
- Cancelling down fractions: only cancel what is common to all terms in both the numerator and denominator.

  $\dfrac{3ac + 5ab}{10abc} = \dfrac{a(3c + 5b)}{10abc} = \dfrac{3c + 5b}{10bc}$
- Expanding brackets: $x(a + b) = ax + bx$

4.6.2 Project

Checking for data entry errors

When entering numbers into an electronic device, or even writing numbers down, errors frequently occur. A common type of error is a transposition error, which occurs when two digits are written in the reverse order. Take the number 2869, for example. With this type of error, it could be written as 8269, 2689 or 2896. A common rule for checking these errors is as follows.

If the difference between the correct number and the recorded number is a multiple of 9, a transposition error has occurred.

We can use algebraic expressions to check this rule. Let the digit in the thousands position be represented by a, the digit in the hundreds position by b, the digit in the tens position by c and the digit in the ones position by d.
So the real number can be represented as $1000a + 100b + 10c + d$.

1. If the digits in the ones position and the tens position were written in the reverse order, the number would be $1000a + 100b + 10d + c$. The difference between the correct number and the incorrect one would then be:
   $1000a + 100b + 10c + d - (1000a + 100b + 10d + c)$.
   a. Simplify this expression.
   b. Is the expression a multiple of 9? Explain.

2. If a transposition error had occurred in the tens and hundreds position, the incorrect number would be $1000a + 100c + 10b + d$. Perform the procedure shown in question **1** to determine whether the difference between the correct number and the incorrect one is a multiple of 9.

3. Consider, lastly, a transposition error in the thousands and hundreds positions. Is the difference between the two numbers a multiple of 9?

4. Comment on the checking rule for transposition errors.

 Resources

 Interactivities Crossword (int-2830)
Sudoku puzzle (int-3589)

Exercise 4.6 Review questions

learn

Fluency

1. Solve the following equations.
   a. $p - 20 = 68$
   b. $s - 0.56 = 2.45$
   c. $3b = 48$

2. Solve the following equations.
   a. $\dfrac{r}{7} = -5$
   b. $2(x + 5) = -3$
   c. $\dfrac{y}{4} - 3 = 12$

3. Solve the following.
   a. $42 - 7b = 14$
   b. $12t - 11 = 4t + 5$
   c. $2(4p - 3) = 2(3p - 5)$

▶

4. Solve each of the following linear equations.
   a. $5(x-2)+3(x+2)=0$
   b. $7(5-2x)-3(1-3x)=1$

5. Solve each of the following linear equations.
   a. $5(x+1)-6(2x-1)=7(x+2)$
   b. $8(3x-2)+(4x-5)=7x$
   c. $7(2x-5)-4(x+20)=x-5$
   d. $3(x+1)+6(x+5)=3x+40$

6. **PATH** Solve the following linear equations for x.
   a. $3(5x-1)=4x-14$
   b. $\dfrac{4-x}{3}+\dfrac{3x-2}{4}=5$

7. Solve the following linear equations for x.
   a. $2x-5=11$
   b. $\dfrac{x}{2}+4=-1$
   c. $2x-4=5x+3$

   d. $5x+3=-4(1-x)$
   e. $\dfrac{3-2x}{6}=\dfrac{x+4}{4}$
   f. $\dfrac{4x-5}{6}-\dfrac{x+4}{12}=1$

8. **PATH** Solve the following linear equations for x.
   a. $7(2x-3)=5(3+2x)$
   b. $\dfrac{4x}{5}-9=7$
   c. $4-2(x-6)=\dfrac{2x}{3}$

   d. $\dfrac{3x+5}{9}=\dfrac{4-2x}{5}$
   e. $\dfrac{x+2}{3}+\dfrac{x}{2}-\dfrac{x+1}{4}=1$
   f. $\dfrac{7x}{5}-\dfrac{3x}{10}=2\left(x+\dfrac{9}{2}\right)$

9. **PATH** Solve for x: $\dfrac{d-x}{a}=\dfrac{a-x}{d}$.

Understanding

10. **PATH** For each of the following, make the variable shown in brackets the subject of the formula.
    a. $y=6x-4$ (x)
    b. $y=mx+c$ (x)
    c. $q=2(P-1)+2r$ (P)
    d. $P=2l+2w$ (w)

11. **PATH** For each of the following, make the variable shown in brackets the subject of the formula.

    a. $v=u+at$ (a)
    b. $s=\left(\dfrac{u+v}{2}\right)t$ (t)

    c. $v^2=u^2+2as$ (a)
    d. $2A=h(a+b)$ (b)

12. **PATH** Calculate the values of x for which $7-\dfrac{3x}{8}\le-2$ and show this set of values on a number line.

13. **PATH** Solve the following inequations for x.

    a. $4-2x\ge5$
    b. $\dfrac{x-6}{3}+4<1$
    c. $2x-3<4x+1$

    d. $-2(x-5)-x>3(x+4)$
    e. $1-\dfrac{4-x}{2}>-1$
    f. $\dfrac{5-x}{2}<-\dfrac{3x+2}{8}$

14. **PATH** Solve the following inequations for x.

    a. $3x-5\le-8$
    b. $4(x-6)+3(2-2x)<0$
    c. $1-\dfrac{2x}{3}\ge-11$
    d. $\dfrac{5x}{6}-\dfrac{4-x}{2}>2$

    e. $8x+7(1-4x)\le7x-3(x+3)$
    f. $\dfrac{2}{3}(x-6)-\dfrac{3}{2}(x+4)>1+x$

15. **PATH** Solve each of the following equations.

a. $\dfrac{x}{2} + \dfrac{x}{5} = \dfrac{3}{5}$

b. $\dfrac{x}{3} - \dfrac{x}{5} = 3$

c. $-\dfrac{1}{21} = \dfrac{x}{7} - \dfrac{x}{6}$

d. $\dfrac{3}{x} + \dfrac{2}{5} = \dfrac{5}{x}$

e. $\dfrac{2x-3}{2} - \dfrac{3}{5} = \dfrac{x+3}{5}$

f. $\dfrac{2(x+2)}{3} = \dfrac{3}{7} + \dfrac{5(x+1)}{3}$

16. **PATH** a. Make x the subject of $bx + cx = \dfrac{d}{2}$.

b. Make r the subject of $V = \dfrac{4}{3}\pi r^3$.

Communicating, reasoning and problem solving

17. A production is in town and many parents are taking their children. An adult ticket costs $15 and a child's ticket costs $8. Every child must be accompanied by an adult and each adult can have no more than 4 children with them. It costs the company $12 per adult and $3 per child to run the production. There is a seating limit of 300 people and all tickets are sold.

a. Determine how much profit the company makes on each adult ticket and on each child's ticket.

b. To maximise profit, the company should sell as many children's tickets as possible. Of the 300 available seats, determine how many should be allocated to children if there is a maximum of 4 children per adult.

c. Using your answer to part **b**, determine how many adults would make up the remaining seats.

d. Construct an equation to represent the profit that the company can make depending on the number of children and adults attending the production.

e. Substitute your values to calculate the maximum profit the company can make.

18. You are investigating prices for having business cards printed for your new games store. A local printing company charges a flat rate of $250 for the materials used and $40 per hour for labour.

a. If h is the number of hours of labour required to print the cards, construct an equation for the cost of the cards, C.

b. You have budgeted $1000 for the printing job. Determine the number of hours of labour you can afford. Give your answer to the nearest minute.

c. The printer estimates that it can print 1000 cards per hour of labour. Evaluate the number of cards that will be printed with your current budget.

d. An alternative to printing is photocopying. The company charges 15 cents per side for the first 10 000 cards and then 10 cents per side for the remaining cards. Justify which is the cheaper option for 18 750 single-sided cards and by how much.

19. A company manufactures a special batch of mobile phone covers. The fixed costs are $250 and materials cost $1.80 per cover. The company sells the covers for $12 each. Determine how many mobile phone covers the company will need to sell to cover all of their costs.

20. The local football club is raising funds for new equipment by setting up a takeaway stall during all of their games. The stall cost the club $65. They sell drinks for $4 each, but the club purchases the drinks for $1.20 each. The cost of the new equipment is $450.
Determine how many drinks the club will need to sell before it has the funds for this new equipment.

on To test your understanding and knowledge of this topic, go to your learnON title at www.jacplus.com.au and complete the **post-test**.

Answers

Topic 4 Linear equations and inequalities

4.1 Pre-test

1. C
2. $x = \dfrac{16}{3}$
3. $x = 7$
4. B
5. D
6. $x < -1$
7. $x < -7$
8. $x > -\dfrac{1}{6}$
9. $25\dfrac{1}{2}$
10. $x = -16$
11. $x = 8$
12. $\$24\,000$
13. C
14. A
15. D

4.2 Solving linear equations review

1. a. $a = 24$ b. $k = 121$ c. $g = 2.9$
2. a. $r = 3$ b. $h = 0.26$ c. $i = -2$
3. a. $t = 5$ b. $q = \dfrac{1}{6}$ c. $x = 0$
4. a. $f = 12$ b. $i = -60$ c. $z = -7$
5. a. $v = 7$ b. $w = -5\dfrac{1}{3}$ c. $k = 10$
6. a. $a = 0.425$ b. $m = 16\dfrac{5}{8}$ c. $y = 21\dfrac{1}{2}$
7. a. $a = 4$ b. $b = 6$ c. $i = 3$
8. a. $f = 9$ b. $q = 1\dfrac{1}{8}$ c. $r = 5\dfrac{2}{5}$
9. a. $s = 4\dfrac{5}{6}$ b. $t = 9\dfrac{4}{5}$ c. $a = -7\dfrac{1}{2}$
10. a. $f = 40$ b. $g = 30$ c. $r = -10$
11. a. $m = 18$ b. $n = 28$ c. $p = 62.4$
12. a. $x = 1\dfrac{1}{3}$ b. $y = 9$ c. $m = 4\dfrac{2}{5}$
13. a. $k = 1\dfrac{1}{2}$ b. $n = 5\dfrac{2}{3}$ c. $c = 1\dfrac{1}{3}$
14. a. $x = 10$ b. $x = 4$ c. $x = -7\dfrac{1}{5}$
15. a. $k = 25$ b. $m = 16$ c. $p = -11\dfrac{3}{7}$
16. a. $u = -4\dfrac{1}{8}$ b. $x = \dfrac{8}{11}$ c. $v = 3$
17. a. $x = 26$ b. $m = -5$ c. $w = \dfrac{25}{3}$
18. a. $t = 5$ b. $x = 9$ c. $n = -\dfrac{19}{3}$
19. a. A   b. D   c. C
20. a. $h = -2$ b. $d = -1$ c. $p = 7$ d. $x = -11$
21. a. $t = 5$ b. $v = -20$ c. $r = -3$ d. $g = -0.8$
22. a. $x = -1$ b. $v = 1$ c. $l = 2$ d. $g = -2$
23. a. $t = 3$ b. $e = -23\dfrac{1}{3}$ c. $k = -36$ d. $f = -12\dfrac{1}{4}$
24. a. $x = 2$ b. $b = 5$ c. $w = 2$
25. a. $f = 7$ b. $t = 3$ c. $r = 2\dfrac{1}{3}$
26. a. $g = -1\dfrac{1}{3}$ b. $h = -2\dfrac{1}{5}$ c. $a = 0$
27. a. $x = -1$ b. $c = 2$ c. $r = 2\dfrac{2}{3}$
28. a. $k = 1$ b. $y = -1\dfrac{1}{8}$ c. $g = 7$
29. a. $w = 1$ b. $m = \dfrac{1}{5}$ c. $p = 1\dfrac{2}{3}$
30. a. $x = -15$ b. $y = -4\dfrac{4}{5}$ c. $t = 21$
31. a. $u = -2\dfrac{5}{7}$ b. $f = 12\dfrac{1}{2}$ c. $r = 7\dfrac{1}{2}$
32. a. $d = -6$ b. $h = -12$ c. $x = 1$
33. a. A   b. D   c. B

4.3 Solving multi-step equations, including fractions (Path)

1. a. $x = \dfrac{20}{31}$ b. $x = 3\dfrac{5}{8}$
2. a. $x = \dfrac{29}{36}$ b. $x = -7$
3. a. $x = -2\dfrac{8}{11}$ b. $x = \dfrac{10}{43}$
4. a. $x = 3$ b. $x = 12$ c. $x = -2$
5. a. $x = \dfrac{3}{2}$ b. $x = -3\dfrac{2}{3}$ c. $x = \dfrac{2}{13}$
6. a. $x = 3$ b. $x = \dfrac{5}{7}$ c. $x = \dfrac{7}{18}$

7. a. $x = \dfrac{5}{17}$ **b.** $x = 15$

c. $x = -6\dfrac{2}{9}$ **d.** $x = -\dfrac{10}{19}$

8. a. $x = -1\dfrac{1}{2}$ **b.** $x = -192$

c. $x = \dfrac{4}{7}$ **d.** $x = 12$

9. a. $x = 3\dfrac{1}{4}$ **b.** $x = 3$

c. $x = 52$ **d.** $x = 1\dfrac{5}{8}$

10. a. $x = \dfrac{5}{19}$ **b.** $x = 1\dfrac{31}{58}$

c. $x = 4\dfrac{11}{14}$ **d.** $x = -3\dfrac{15}{17}$

11. a. $x = 5\dfrac{20}{43}$ **b.** $x = -1\dfrac{10}{13}$

c. $x = 1\dfrac{2}{61}$ **d.** $x = -4\dfrac{9}{26}$

12. a. $x = 1.5$ **b.** $x = -4\dfrac{1}{3}$

c. $x = 3$ **d.** $x = 1$

13. $180

14. 60 hours

15. $12 000

16. $60

17. a. $4, 9, 25, 49$ **b.** $16, 81$ **c.** 64

18. a. $6x - 450 = 1000$

b. $241\dfrac{1}{3}$ tickets. This means they need to sell 242 tickets to qualify, as the number of tickets must be a whole number.

19. Sample responses can be found in the worked solutions in the online resources.

20. 4

21. $a = 3, b = 5$

22. $a = -8$ and $b = 15$

4.4 Literal equations (Path)

1. a. $x = \dfrac{y}{a}$ **b.** $x = \dfrac{y-b}{a}$ **c.** $x = \dfrac{y+b}{2a}$

d. $x = \dfrac{y+7}{2}$ **e.** $x = \dfrac{8-6y}{5}$ **f.** $x = \dfrac{1}{y-2}$

2. a. $x = \dfrac{y+1}{y-2}$ **b.** $x = \dfrac{4y+1}{5}$ **c.** $x = -5y - 12$

d. $x = 12a$ **e.** $x = \dfrac{5a-9}{19}$ **f.** $x = \dfrac{14a+13}{93}$

3. a. $P = \dfrac{g+3}{4}$ **b.** $c = \dfrac{5f}{9}$

c. $c = \dfrac{5(f-32)}{9}$ **d.** $t = \dfrac{v-u}{a}$

4. a. $c = \dfrac{b^2 - d}{4a}$ **b.** $y = hm + k$

c. $a = y - m(x - b)$ **d.** $x = \dfrac{y - a + mb}{m}$

5. a. $r = \dfrac{2\pi}{C}$ **b.** $x = \dfrac{f - by}{a}$

c. $a = \dfrac{2(s - ut)}{t^2}$ **d.** $G = \dfrac{Fr^2}{Mm}$

6. 500 km

7. a. $24\,000\ \text{cm}^3$

b. $t = 80\ \text{min} = 1\,\text{h}\,20\ \text{min}$

8. Sample responses can be found in the worked solutions in the online resources.

9. $\dfrac{B}{3(1 - 2y)} = x$

10. You would rearrange the formula for volume to make r the subject. If you then substitute $V = 5$, you can calculate that the radius is approximately 10.6 cm.

11. a. $b = \dfrac{2A - ah}{h}$ **b.** 11 cm

12. a. $l = \dfrac{gT^2}{4\pi^2}$ **b.** 2.2 m

13. Sample responses can be found in the worked solutions in the online resources.

14. $a = \dfrac{b^2 - d^2}{4c}$

15. $a = 1; b = 5$

4.5 Solving linear inequalities (Path)

1. a. $x > 2$

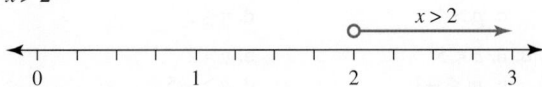

b. $a > -1$

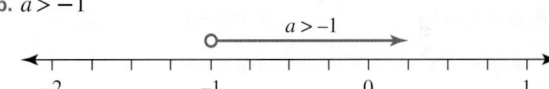

c. $y \geq 7$

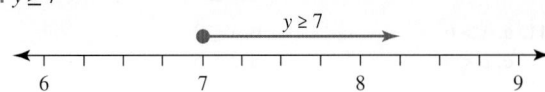

d. $m \geq 4$

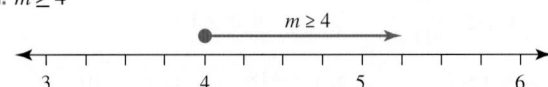

2. a. $p < 1$

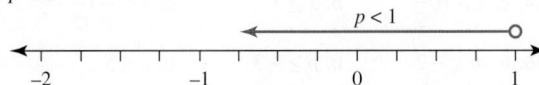

b. $x < 7$

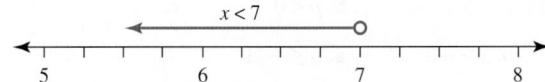

c. $m \leq 9$

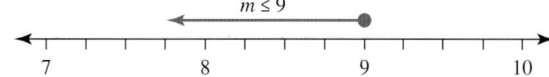

d. $a \leq 7$

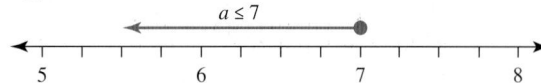

3. a. $x > 3$

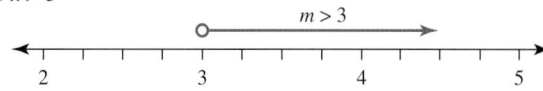

b. $m \geq 2$

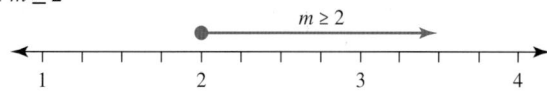

c. $q \geq -4$

d. $a > -8$

4. a. $m > 3$ **b.** $p \leq 2$
    c. $a < 4$ **d.** $x \geq 5$

5. a. $p > -5$ **b.** $x \leq -7$
    c. $m \geq -0.5$ **d.** $b > -0.5$

6. a. $m > 18$ **b.** $x < 8$
    c. $a \leq -14$ **d.** $m \geq 25$

7. a. $m < 4.5$ **b.** $x \geq 3$
    c. $p > 4$ **d.** $n \leq 2$

8. a. $b < 5$ **b.** $y > 2$
    c. $m \leq -1$ **d.** $a \geq -5$

9. a. $b < -4\frac{1}{3}$ **b.** $c \leq -1$
    c. $p > -2$ **d.** $a \geq -7$

10. a. $m > 3$ **b.** $a \geq 2$
    c. $a < -1$ **d.** $a \leq -3$

11. a. $x > 6$ **b.** $x \leq 2$
    c. $b < 4$ **d.** $a > 5$

12. a. $m < 2$ **b.** $m \leq 3$
    c. $b \leq -\frac{16}{11}$ **d.** $m \geq 1$

13. a. $x \leq 7$ **b.** $x \geq -18$ **c.** $x < -10$

14. a. $x > 10\frac{1}{2}$ **b.** $x \geq 5$ **c.** $x < -1\frac{4}{5}$

15. a. $m < -2$ **b.** $p \geq -3$ **c.** $a \leq 5$
    d. $p \geq -5$ **e.** $y \leq -3$

16. a. $x > 7$ **b.** $p < 0$ **c.** $a \geq \frac{1}{5}$

    d. $x > -3$ **e.** $a \leq -11$

17. a. $b \leq 3$ **b.** $x < -3$ **c.** $k > 8$

    d. $x > -18\frac{1}{2}$ **e.** $a \leq 40$

18. B

19. a. $x < -1$ **b.** $m \leq -3$

20. a. $x > 17$ **b.** $a > \frac{5}{8}$

21. a. $m \geq 1\frac{1}{3}$ **b.** $m \geq -12$

22. a. $k > 2$ **b.** $a > -5$

    c. $m \leq 1\frac{2}{3}$ **d.** $x > 5$

23. a. $y \geq 7$ **b.** $d < -2$

    c. $p \geq \frac{-6}{7}$ **d.** $x \geq -5$

24. a. $m < -2$ **b.** $a < 9$

    c. $p \geq 3$ **d.** $x > -4\frac{1}{2}$

25. a. $5x > 10$ **b.** $x - 3 \leq 5$ **c.** $7 + 3x < 42$

26. a. $4 + 3x > 19$ **b.** $2(x + 6) < 10$ **c.** $\frac{(x - 8)}{2} - 7 \geq 9$

27. a. $50x \geq 650$ **b.** $2.50d + 5 \leq 60$

28. Tom could be any age from 1 to 28.

29. a. $-6.5 < x < -2$ **b.** $\frac{-c - b}{a} < x < \frac{-d - b}{a}$

30. a. $S > 47$

    b. No

    c. Sample responses can be found in the worked solutions in the online resources.

31. a. $n < 16\,800$ km

    b. Mick travelled less than 16 800 km for the year and his costs stayed below $16 000.

32. a. $20d + 10c \leq 240$

    b. $2d + 2c \leq 36$

    c. $0 \leq d \leq 12$ and $0 \leq C \leq 18$

33. $20\,000

Project

1. a. $9(c - d)$

    b. Yes, this is a multiple of 9 as the number that multiples the brackets is 9.

2. $90(b - c)$; 90 is a multiple of 9 so the difference between the correct and incorrect one is a multiple of 9.

3. $900(a - b)$; again 900 is a multiple of 9.

4. If two adjacent digits are transposed, the difference between the correct number and the transposed number is a multiple of 9.

4.6 Review questions

1. a. $p = 88$ b. $s = 3.01$ c. $b = 16$

2. a. $r = -35$ b. $x = -\dfrac{13}{2}$ c. $y = 60$

3. a. $b = 4$ b. $t = 2$ c. $p = -2$

4. a. $x = \dfrac{1}{2}$ b. $x = 6\dfrac{1}{5}$

5. a. $x = -\dfrac{3}{14}$ b. $x = 1$

   c. $x = 12\dfrac{2}{9}$ d. $x = 1\dfrac{1}{6}$

6. a. $x = -1$ b. $x = 10$

7. a. $x = 8$ b. $x = -10$ c. $x = -\dfrac{7}{3}$

   d. $x = -7$ e. $x = -\dfrac{6}{7}$ f. $x = \dfrac{26}{7}$

8. a. $x = 9$ b. $x = 20$ c. $x = 6$

   d. $x = \dfrac{1}{3}$ e. $x = 1$ f. $x = -10$

9. $x = a + d$

10. a. $x = \dfrac{y + 4}{6}$ b. $x = \dfrac{y - c}{m}$

   c. $P = \dfrac{q - 2r}{2} + 1$ d. $w = \dfrac{P - 2l}{2}$

11. a. $a = \dfrac{v - u}{t}$ b. $t = \dfrac{2s}{u + v}$

   c. $a = \dfrac{v^2 - u^2}{2s}$ d. $b = \dfrac{2A - ah}{h}$

12.
$x \geq 24$

13. a. $x \leq -\dfrac{1}{2}$ b. $x < -3$ c. $x > -2$

   d. $x < -\dfrac{1}{3}$ e. $x > 0$ f. $x > 22$

14. a. $x \leq -1$ b. $x > -9$ c. $x \leq 18$

   d. $x > 3$ e. $x \geq \dfrac{2}{3}$ f. $x < -6$

15. a. $x = \dfrac{6}{7}$ b. $x = 22\dfrac{1}{2}$ c. $x = 2$

   d. $x = 5$ e. $x = 3\dfrac{3}{8}$ f. $x = -\dfrac{16}{21}$

16. a. $x = \dfrac{d}{2(b + c)}$ b. $r = \sqrt[3]{\dfrac{3V}{4\pi}}$

17. a. $3 per adult ticket; $5 per child's ticket
   b. 240
   c. 60
   d. $P = 3a + 5c$, where a = number of adults and c = number of children
   e. $1380

18. a. $C = 250 + 40h$
   b. 18 hours 45 minutes
   c. 18 750
   d. Printing is the cheaper option by $1375.

19. 25 phone covers

20. 184 drinks

5 Simultaneous linear equations (Path)

LESSON SEQUENCE

LESSON
5.1 Overview

Why learn this?

Often in life, we will be faced with a trade-off situation. This means that you are presented with multiple options and must decide on a combination of outcomes that provides you with the best result. Imagine a race with both swimming and running components, in which athletes start from a boat, swim to shore and then run along the beach to the finish line. Each athlete would have the following options:

- swim directly to shore and run a longer distance along the beach
- swim a longer distance diagonally through the ocean and reduce the distance required to run to reach the finish line
- swim directly through the ocean to the finish, covering the shortest possible distance.

Which option should an athlete take? This would depend on how far the athlete can swim or run, because reducing the swimming distance increases the running distance. To find the best combination of swimming and running, an athlete could form equations based on speed, time and distance and solve simultaneously to find the best combination.

Just like the athletes in the scenario above, businesses face trade-offs like these every day, where they have to decide how much of each product they should produce in order to produce the highest possible profit. As an example, a baker might make the most per-item profit from selling cakes, but if they don't produce muffins, bread and a range of other products, then they will attract fewer customers and miss out on sales, reducing overall profit. Thus, a baker could use simultaneous equations to find the best combination of baked goods to produce in order to maximise profit.

Hey students! Bring these pages to life online

▶ Watch videos

🧩 Engage with interactivities

A⁺ Answer questions and check solutions

Find all this and MORE in jacPLUS

Reading content and rich media, including interactivities and videos for every concept

Extra learning resources

Differentiated question sets

Questions with immediate feedback, and fully worked solutions to help students get unstuck

1. State whether the following is True or False. The point $\left(\dfrac{10}{3}, 5\right)$ is the solution to the simultaneous equations $3x + 2y = 10$ and $-x - 4y = 5$.

2. Identify the solution to the simultaneous equations shown in the graph.

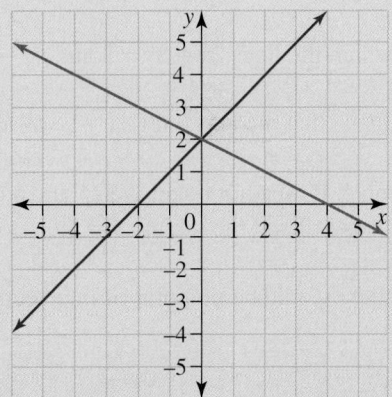

3. State the number of solutions to the pair of simultaneous equations $2x - y = 1$ and $-6x + 3y = -3$.

4. Use substitution to solve the simultaneous equations $y = 0.2x$ and $y = -0.3x + 0.5$.
   Give your answer as a coordinate pair.

5. Solve the simultaneous equations $2x + y = 6$ and $5x - 2y = 24$.

6. Find the value of a so that the line $ax - 7y = 8$ is perpendicular to the line $3y + 6x = 7$.

7. Dylan received a better result for his Maths test than for his English test. If the sum of his two test results is 159 and the difference is 25, determine Dylan's Maths test result.

8. **MC** Solve the pair of simultaneous equations $mx + ny = m$ and $x = y + n$ for x and y in terms of m and n.

   A. $x = \dfrac{m + n^2}{m + n}$ and $y = \dfrac{m(1 - n)}{m + n}$

   B. $x = 1$ and $y = \dfrac{m(1 - n)}{m + n}$

   C. $x = \dfrac{1}{n}$ and $y = \dfrac{m(1 - n)}{m + n}$

   D. $x = \dfrac{m + n^2}{m + n}$ and $y = \dfrac{1 - n}{n}$

9. **MC** Solve the pair of simultaneous equations $\dfrac{x}{3} - \dfrac{y}{2} = \dfrac{1}{6}$ and $\dfrac{x}{4} + \dfrac{y}{3} = \dfrac{1}{2}$.

   A. $x = \dfrac{13}{102}$ and $y = \dfrac{1}{34}$

   B. $x = \dfrac{81}{17}$ and $y = \dfrac{17}{9}$

   C. $x = \dfrac{17}{22}$ and $y = \dfrac{9}{22}$

   D. $x = \dfrac{22}{17}$ and $y = \dfrac{9}{17}$

10. The perimeter of both of the two rectangles shown is 22 cm. Find the smaller area.

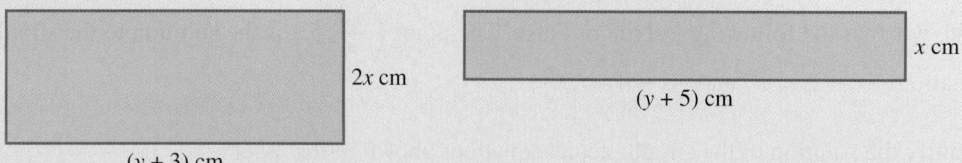

11. Determine the value of p for which the lines $2x + 3y = 3$ and $px - 5y = 17$ will not intersect.

12. Determine the values of a and b so that the two lines $ax + y = b$ and $3x - 2y = 4$ are coincidental.

13. A movie theatre holds 200 people. If an adult ticket is $25 and a child's ticket is $10, find the number of children present if the theatre was full and the takings for the session were $3200.

14. **MC** The value of a such that there would be no point of intersection between the two lines $ay + 3x = 4$ and $2y + 4x = 3$ is:

   A. 2   B. 1.5   C. $\dfrac{8}{3}$ D. -0.5

15. **MC** The point of intersection of the straight lines $x - 5 = 0$ and $y + 2 = 0$ is:

   A. $(5, 2)$ B. $(5, -2)$ C. $(-5, -2)$ D. $(-5, 2)$

LESSON
5.2 Solving simultaneous linear equations graphically

LEARNING INTENTION

At the end of this lesson you should be able to:
   • use the graph of two simultaneous equations to determine the point of intersection
   • determine whether two simultaneous equations will have 0, 1 or infinite solutions
   • determine whether two lines are parallel or perpendicular.

ⓘ 5.2.1 Simultaneous linear equations and graphical solutions
eles-4763

   • **Simultaneous** means occurring at the same time.
   • When a point lies on more than one line, the coordinates of that point are said to satisfy all equations of the lines it lies on. The equations of the lines are called **simultaneous equations**.
   • A **system of equations** is a set of two or more equations with the same variables.
   • Solving a system of simultaneous equations is to find the coordinates of any point/s that satisfy all equations in the system.
   • Any point or points that satisfy a system of simultaneous equations is said to be the solution. For the equations shown to the right, the solution is the point $(-1, 1)$.
   • Simultaneous equations can be solved by finding these points graphically or algebraically.

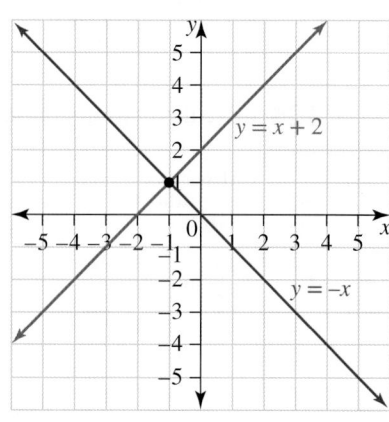

Graphical solution

- The solution to a pair of simultaneous equations can be found by graphing the two equations and identifying the coordinates of the point of intersection.
- The accuracy of the solution depends on having an accurate graph.

WORKED EXAMPLE 1 Solving simultaneous equations graphically

Use the graphs of the given simultaneous equations to determine the point of intersection and, hence, the solution of the simultaneous equations.

$$x + 2y = 4$$
$$y = 2x - 3$$

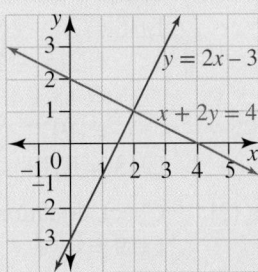

THINK	WRITE/DRAW
1. Write the equations, one under the other and number them.	$x + 2y = 4$ [1] $y = 2x - 3$ [2]
2. Locate the point of intersection of the two lines. This gives the solution.	Point of intersection $(2, 1)$ Solution: $x = 2$ and $y = 1$ 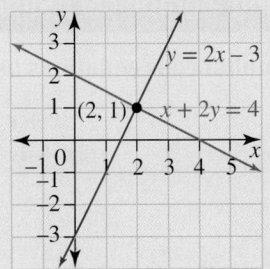

3. Check the solution by substituting $x = 2$ and $y = 1$ into the given equations. Comment on the results obtained.	Check equation [1]:
	LHS $= x + 2y$ RHS $= 4$
	    $= 2 + 2(1)$
	    $= 4$
	LHS $=$ RHS
	Check equation [2]:
	LHS $= y$ RHS $= 2x - 3$
	    $= 1$ $= 2(2) - 3$
	                   $= 4 - 3$
	                   $= 1$
	LHS $=$ RHS

4. State the solution.	In both cases, LHS $=$ RHS. Therefore, the solution set $(2, 1)$ is correct.

WORKED EXAMPLE 2 Verifying a solution using substitution

Verify whether the given pair of coordinates, $(5, -2)$, is the solution to the following pair of simultaneous equations.

$$3x - 2y = 19$$
$$4y + x = -3$$

THINK	WRITE
1. Write the equations and number them.	$3x - 2y = 19$ [1] $4y + x = -3$ [2]
2. Substitute $x = 5$ and $y = -2$ into equation [1].	Check equation [1]: LHS $= 3x - 2y$ RHS $= 19$ $= 3(5) - 2(-2)$ $= 15 + 4$ $= 19$ LHS $=$ RHS
3. Substitute $x = 5$ and $y = -2$ into equation [2].	Check equation [2]: LHS $= 4y + x$ RHS $= -3$ $= 4(-2) + 5$ $= -3$ LHS $=$ RHS
4. State the solution.	Therefore, the solution set $(5, -2)$ is a solution to both equations.

WORKED EXAMPLE 3 Using a graphical method to solve simultaneously

Solve the following pair of simultaneous equations using a graphical method.

$$x + y = 6$$
$$2x + 4y = 20$$

THINK	WRITE/DRAW
1. Write the equations, one under the other and number them.	$x + y = 6$ [1] $2x + 4y = 20$ [2]
2. Calculate the x- and y-intercepts for equation [1]. For the x-intercept, substitute $y = 0$ into equation [1].	Equation [1] x-intercept: when $y = 0$, $x + 0 = 6$ $x = 6$ The x-intercept is at $(6, 0)$.
For the y-intercept, substitute $x = 0$ into equation [1].	y-intercept: when $x = 0$, $0 + y = 6$ $y = 6$ The y-intercept is at $(0, 6)$.
3. Calculate the x- and y-intercepts for equation [2]. For the x-intercept, substitute $y = 0$ into equation [2]. Divide both sides by 2.	Equation [2] x-intercept: when $y = 0$, $2x + 0 = 20$ $2x = 20$ $x = 10$ The x-intercept is at $(10, 0)$.

For the y-intercept, substitute $y = 0$ into equation [2].
Divide both sides by 4.

y-intercept: when $x = 0$,
$$0 + 4y = 20$$
$$4y = 20$$
$$y = 5$$
The y-intercept is at $(0, 5)$.

4. Use graph paper to rule up a set of axes and label the x-axis from 0 to 10 and the y-axis from 0 to 6.

5. Plot the x- and y-intercepts for each equation.

6. Produce a graph of each equation by ruling a straight line through its intercepts.

7. Label each graph.

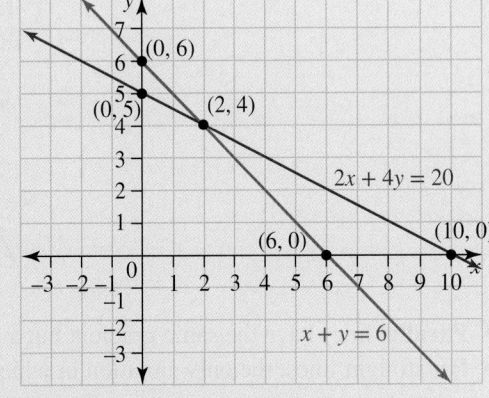

8. Locate the point of intersection of the lines.

The point of intersection is $(2, 4)$.

9. Check the solution by substituting $x = 2$ and $y = 4$ into each equation.

Check [1]:
$$\text{LHS} = x + y \qquad \text{RHS} = 6$$
$$= 2 + 4$$
$$= 6$$
$$\text{LHS} = \text{RHS}$$

Check [2]:
$$\text{LHS} = 2x + 4y \qquad \text{RHS} = 20$$
$$= 2(2) + 4(4)$$
$$= 4 + 16$$
$$= 20$$
$$\text{LHS} = \text{RHS}$$

10. State the solution.

In both cases, $\text{LHS} = \text{RHS}$. Therefore, the solution set $(2, 4)$ is correct.
The solution is $x = 2$, $y = 4$.

⏵ 5.2.2 Solutions to coincident, parallel and perpendicular lines

eles-4764

- The equation $3x - 1 = -1 + 3x$ is an **identity** as it is true for all values of x; therefore, the lines $y = 3x - 1$ and $y = -1 + 3x$ are identical.
- Two lines are **coincident** if they lie one on top of the other. For example, in the graph shown, the line in blue and line segment in pink are coincident.
- There are an infinite number of solutions to coincident equations. Every point where the lines coincide satisfies both equations and hence is a solution to the simultaneous equations.
- Coincident equations have the same equation, although the equations may have been transposed so they look different. For example, $y = 2x + 3$ and $2y = 4x + 6$ are coincident equations as the second equation is double the first.

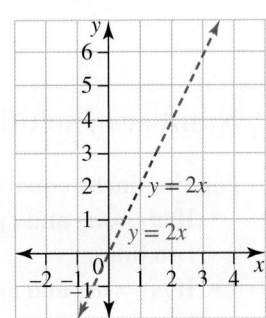

Parallel lines

- The equation $3x - 1 = -2 + 3x$ is a **contradiction** as it has no solutions; therefore, the lines $y = 3x - 1$ and $y = -2 + 3x$ do not intersect.
- If two lines do not intersect, there is no simultaneous solution to the equations. For example, the graph lines shown do not intersect, so there is no point that belongs to both lines.

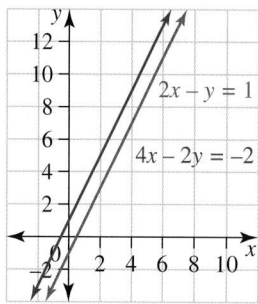

- **Parallel** lines have the same gradient but a different y-intercept.
- For straight lines, the only situation in which the lines do not cross is if the lines are parallel *and* not coincident.

$$2x - y = 1 \qquad [1] \qquad 4x - 2y = -2 \qquad [2]$$
$$-y = 1 - 2x \qquad\qquad -2y = -2 - 4x$$
$$-y = -2x - 1 \qquad\qquad -2y = -4x - 2$$
$$y = 2x - 1 \qquad\qquad y = 2x + 1$$
$$\text{Gradient } m = 2 \qquad\qquad \text{Gradient } m = 2$$

- Writing both equations in the form $y = mx + c$ confirms that the lines are parallel since the gradients are equal.

Perpendicular lines

- Two lines are **perpendicular** if they intersect at right angles (90°).
- The product of the gradients of two perpendicular lines is equal to -1:

$$m_1 \times m_2 = -1 \text{ or } m_1 = -\frac{1}{m_2}$$

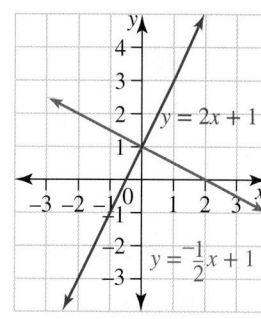

- The two lines in the graph shown are perpendicular as $m_1 \times m_2 = 2 \times -\dfrac{1}{2} = -1$.

Number of solutions for a pair of simultaneous linear equations

For two linear equations given by $y = m_1 x + c_1$ and $y = m_2 x + c_2$:
- If $m_1 = m_2$ and $c_1 \neq c_2$, then the two lines are parallel and there will be no solutions between the two lines.
- If $m_1 = m_2$ and $c_1 = c_2$, then the two lines are coincident and there will be infinite solutions between the two lines.
- If $m_1 \neq m_2$, then the lines will cross once, so there will be one solution.
- If $m_1 \times m_2 = -1$, then the lines are perpendicular and will intersect (once) at right angles (90°).

WORKED EXAMPLE 4 Determining the number of solutions between two lines

Determine the number of solutions between the following pairs of simultaneous equations. If there is only one solution, determine whether the lines are perpendicular.

a. $2y = 4x + 6$ and $-3y = -6x - 12$

b. $y = -3x + 2$ and $-3y = x + 15$

c. $5y = 25x - 30$ and $2y - 10x + 12 = 0$

THINK	WRITE
a. 1. Re-write both equations in the form $y = mx + c$.	**a.** $2y = 4x + 6$ $\quad y = 2x + 3$ [1] $-3y = -6x - 12$ $\quad y = 2x + 4$ [2]
2. Determine the gradient of both lines.	$m_1 = 2$ and $m_2 = 2$
3. Check if the lines are parallel, coincident or perpendicular.	The gradients are the same and the y-intercepts different. So, the two lines are parallel.
4. Write the answer.	There will be *no* solutions between this pair of simultaneous equations as the lines are parallel.
b. 1. Re-write both equations in the form $y = mx + c$.	**b.** $y = -3x + 2$ [1] $-3y = x + 15$ $\quad y = \dfrac{x}{-3} - 5$ [2]
2. Determine the gradient of both lines.	$m_1 = -3$ and $m_2 = -\dfrac{1}{3}$
3. Check if the lines are parallel, coincident or perpendicular and comment on the number of solutions.	The gradients are different, so there will be *one* solution.
4. Determine if the lines are perpendicular by calculating the product of the gradients.	$m_1 \times m_2 = -3 \times -\dfrac{1}{3} = 1$
5. Write the answer.	The lines have one solution but they are not perpendicular.
c. 1. Re-write both equations in the form $y = mx + c$.	**c.** $5y = 25x - 30$ $\quad y = 5x - 6$ [1] $2y - 10x + 12 = 0$ $\quad 2y = 10x - 12$ $\quad y = 5x - 6$ [2]
2. Determine the gradient of both lines.	$m_1 = 5$ and $m_2 = 5$
3. Check if the lines are parallel, coincident, or perpendicular.	The gradients are the same and the y-intercepts are also the same, so the two lines are coincident.
4. Write the answer.	The lines are coincident, so there are *infinite* solutions between the two lines.

 Resources

Interactivities Solving simultaneous equations graphically (int-6452)
Parallel lines (int-3841)
Perpendicular lines (int-6124)

Exercise 5.2 Solving simultaneous linear equations graphically

5.2 Quick quiz on	5.2 Exercise

Individual pathways

■ PRACTISE	■ CONSOLIDATE	■ MASTER
1, 4, 7, 10, 14, 15, 18	2, 5, 8, 11, 12, 16, 19	3, 6, 9, 13, 17, 20, 21

Fluency

WE1 For questions **1** to **3**, use the graphs to determine the point of intersection and hence the solution of the simultaneous equations.

1. **a.** $x + y = 3$
   $x - y = 1$

   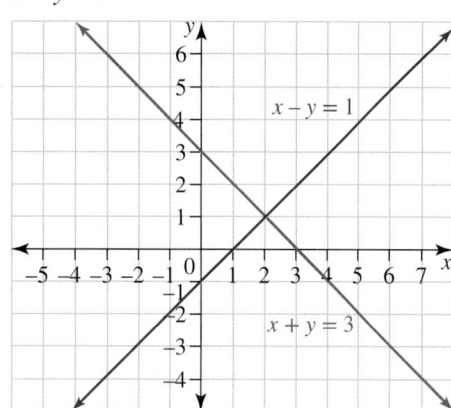

   b. $x + y = 2$
   $3x - y = 2$

   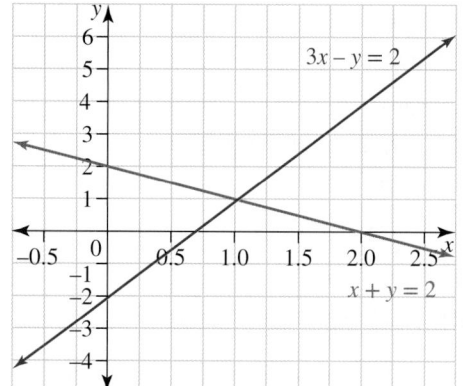

2. **a.** $y - x = 4$
   $3x + 2y = 8$

   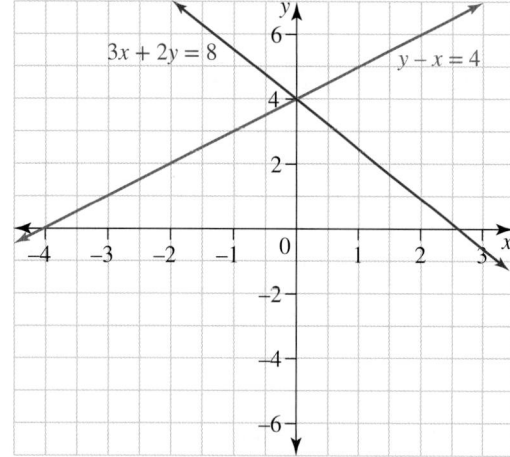

   b. $y + 2x = 3$
   $2y + x = 0$

   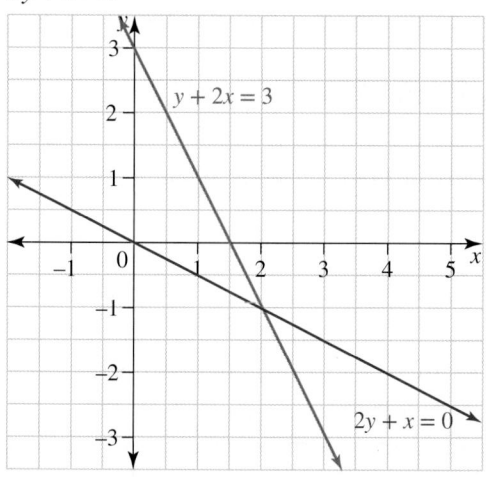

3. a. $y - 3x = 2$
$x - y = 2$

b. $2y - 4x = 5$
$4y + 2x = 5$

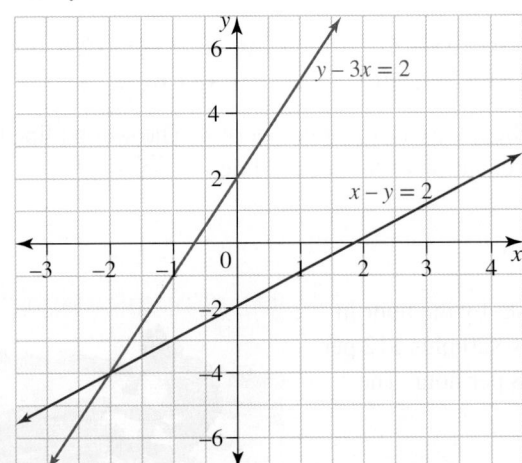

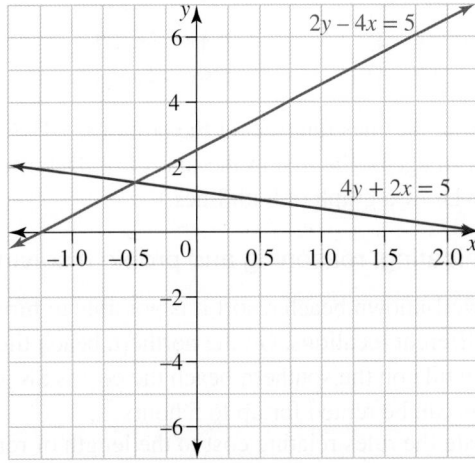

WE2 For questions **4** to **6**, use substitution to check if the given pair of coordinates is a solution.

4. a. $(7, 5)$ $\quad 3x + 2y = 31$
$2x + 3y = 28$

b. $(3, 7)$ $\quad y - x = 4$
$2y + x = 17$

c. $(9, 1)$ $\quad x + 3y = 12$
$5x - 2y = 43$

d. $(2, 5)$ $\quad x - y = 7$
$2x + 3y = 18$

5. a. $(4, -3)$ $\quad y = 3x - 15$
$4x + 7y = -5$

b. $(6, -2)$ $\quad x - y = 7$
$3x + y = 16$

c. $(4, -2)$ $\quad 2x + y = 6$
$x - 3y = 8$

d. $(5, 1)$ $\quad y - 5x = -24$
$3y + 4x = 23$

6. a. $(-2, -5)$ $\quad 3x - 2y = -4$
$2x - 3y = 11$

b. $(-3, -1)$ $\quad y - x = 2$
$2y - 3x = 7$

c. $\left(-\dfrac{1}{2}, 2\right)$ $\quad 6x + 4y = 5$
$20x - 5y = 0$

d. $\left(\dfrac{3}{2}, \dfrac{5}{3}\right)$ $\quad 8x + 6x = 22$
$10x - 9y = 0$

WE3 For questions **7** to **9**, solve each of the following pairs of simultaneous equations using a graphical method.

7. a. $x + y = 5$
$2x + y = 8$

b. $x + 2y = 10$
$3x + y = 15$

c. $2x + 3y = 6$
$2x - y = -10$

d. $x - 3y = -8$
$2x + y = -2$

8. a. $6x + 5y = 12$
$5x + 3y = 10$

b. $y + 2x = 6$
$2y + 3x = 9$

c. $y = 3x + 10$
$y = 2x + 8$

d. $y = 8$
$3x + y = 17$

9. a. $4x - 2y = -5$
$x + 3y = 4$

b. $3x + y = 11$
$4x - y = 3$

c. $3x + 4y = 27$
$x + 2y = 11$

d. $3y + 3x = 8$
$3y + 2x = 6$

Understanding

WE4 For questions **10** to **12**, using technology, determine which of the following pairs of simultaneous equations have no solutions. Confirm by finding the gradient of each line.

10. a. $y = 2x - 4$
$3y - 6x = 10$

b. $5x - 3y = 13$
$4x - 2y = 10$

c. $x + 2y = 8$
$5x + 10y = 45$

d. $y = 4x + 5$
$2y - 10x = 8$

11. **a.** $3y + 2x = 9$ **b.** $y = 5 - 3x$ **c.** $4y + 3x = 7$ **d.** $2y - x = 0$
       $6x + 4y = 22$ $3y = -9x + 18$ $12y + 9x = 22$ $14y - 6x = 2$

12. **a.** $y = 3x - 4$ **b.** $4x - 6y = 12$ **c.** $3y = 5x - 22$ **d.** $3x = 12 - 4y$
       $5y = 12 + 15x$ $6x - 4y = 12$ $5x = 3y + 26$ $8y + 6x = 14$

13. Two straight lines intersect at the point $(3, -4)$. One of the lines has a y-intercept of 8. The second line is a mirror image of the first in the line $x = 3$. Determine the equation of the second line.
    (*Hint:* Draw a graph of both lines.)

Communicating, reasoning and problem solving

14. At a well-known beach resort it is possible to hire a jet ski by the hour in two different locations. On the northern beach the cost is \$20 plus \$12 per hour, while on the southern beach the cost is \$8 plus \$18 per hour. The jet skis can be rented for up to 5 hours.
    a. Write the rules relating cost to the length of rental.
    b. On the same set of axes sketch a graph of cost (y-axis) against length of rental (x-axis) for 0–5 hours.
    c. For what rental times, if any, is the northern beach rental cheaper than the southern beach rental? Use your graph to justify your answer.
    d. For what length of rental time are the two rental schemes identical? Use the graph and your rules to justify your answer.

15. For each of the pairs of simultaneous equations below, determine whether they are the same line, parallel lines, perpendicular lines or intersecting lines. Show your working.
    a. $2x - y = -9$ **b.** $x - y = 7$ **c.** $x + 6 = y$ **d.** $x + y = -2$
          $-4x - 18 = -2y$ $x + y = 7$ $2x + y = 6$ $x + y = 7$

16. For each of the following, explain if the equations have one solution, an infinite number of solutions or no solution.
    a. $x - y = 1$ **b.** $2x - y = 5$ **c.** $x - 2y = -8$
          $2x - 3y = 2$ $4x - 2y = -6$ $4x - 8y = -16$

17. Determine whether the following pairs of equations will have one, infinite or no solutions. If there is only one solution, determine whether the lines are perpendicular.
    a. $3x + 4y = 14$ **b.** $2x + y = 5$ **c.** $3x - 5y = -6$ **d.** $2y - 4x = 6$
          $4x - 3y = 2$ $3y + 6x = 15$ $5x - 3y = 24$ $2x - y = -10$

18. Use the information below to determine the value of a in each of the following equations:
    a. $y = ax + 3$, which is parallel to $y = 3x - 2$
    b. $y = ax - 2$, which is perpendicular to $y = -4x + 6$
    c. $y = ax - 4$, which intersects the line $y = 3x + 6$ when $x = 2$.

19. Line A is parallel to the line with equation $y - 3x - 3 = 0$ and passes through the point $(1, 9)$. Line B is perpendicular to the line with equation $2y - x + 6 = 0$ and passes through the point $(2, -3)$.
    a. Determine the equation of line A.
    b. Determine the equation of line B.
    c. Sketch both lines on the one set of axes to find where they intersect.

20. Solve the system of three simultaneous equations graphically.
$$3x - y = 2$$
$$y + 3x = 4$$
$$2y - x = 1$$

21. A line with equation $4x + 5y = 4$ intersects a second line when $x = -4$. Determine the equation of the second line if it is perpendicular to the first line.

LESSON
5.3 Solving simultaneous linear equations using substitution

LEARNING INTENTION

At the end of this lesson you should be able to:
- identify when it is appropriate to solve using the substitution method
- solve a system of two linear simultaneous equations using the substitution method.

▶ 5.3.1 Solving simultaneous equations using the substitution method

eles-4766

- A variable is considered the **subject of an equation** if it is expressed in terms of the other variables. In the equation $y = 3x + 4$, the variable y is the subject.
- The **substitution method** is used when one (or both) of the equations is presented in a form where one of the two variables is the subject of the equation.
- When solving two linear simultaneous equation, the substitution method involves replacing a variable in one equation with the other equation. This produces a new third equation expressed in terms of a single variable.
- Consider the pair of simultaneous equations:

$$y = 2x - 4$$
$$3x + 2y = 6$$

- In the first equation, y is written as the subject and is equal to $(2x - 4)$. In this case, substitution is performed by replacing y in the second equation with the expression $(2x - 4)$.

$$y = (2x - 4)$$

$$3x + 2(y) = 10$$

$$3x + 2(2x - 4) = 6$$

- This produces a third equation, all in terms of x, so that the value of x can be found.
- Once a value for one variable is found, it can be substituted back into either equation to find the value of the other variable.
- It is often helpful to use brackets when substituting an expression into another equation.

WORKED EXAMPLE 5 Solving using the substitution method

Solve the simultaneous equations $y = 2x - 1$ and $3x + 4y = 29$ using the substitution method.

THINK

1. Write the equations, one under the other and number them.

2. y and $2x - 1$ are equal so substitute the expression $(2x - 1)$ for y into equation [2].

3. Solve for x.

   i. Expand the brackets on the LHS of the equation.

   ii. Collect like terms.

   iii. Add 4 to both sides of the equation.

   iv. Divide both sides by 11.

4. Substitute $x = 3$ into either of the equations, say [1], to find the value of y.

5. Write your answer.
6. Check the solution by substituting $(3, 5)$ into equation [2].

WRITE

$y = 2x - 1$ [1]
$3x + 4y = 29$ [2]

Substituting $(2x - 1)$ into [2]:
$3x + 4(2x - 1) = 29$

$3x + 8x - 4 = 29$

$11x - 4 = 29$

$11x = 33$

$x = 3$

Substituting $x = 3$ into [1]:
$y = 2(3) - 1$
$ = 6 - 1$
$ = 5$

Solution: $x = 3, y = 5$ or $(3, 5)$

Check: Substitute $(3, 5)$ into
$3x + 4y = 29$.

LHS $= 3(3) + 4(5)$ RHS $= 29$
$ = 9 + 20$
$ = 29$

As LHS $=$ RHS, the solution is correct.

⏵ 5.3.2 Equating equations

eles-4767

- To **equate** in mathematics is to take two expressions that have the same value and make them equal to each other.
- When both linear equations are written with the same variable as the subject, we can equate the equations to solve for the other variable. Consider the following simultaneous equations:

$$y = 4x - 3$$
$$y = 2x + 9$$

- In the first equation, y is equal to $(4x - 3)$, and in the second equation, y is equal to $(2x + 9)$. Since both expressions are equal to the same thing (y), they must also be equal to each other. Thus, equating the equations gives:

$$y = (4x - 3)$$
$$(y) = 2x + 9$$

$$4x - 3 = 2x + 9$$

- As can be seen above, equating equations is still a form of substitution. A third equation is produced, all in terms of x, allowing for a value of x to be solved.

WORKED EXAMPLE 6 Substitution by equating two equations

Solve the pair of simultaneous equations $y = 5x - 8$ and $y = -3x + 16$ by equating the equations.

THINK	WRITE
1. Write the equations, one under the other and number them.	$y = 5x - 8 \quad [1]$ $y = -3x + 16 \quad [2]$
2. Both equations are written with y as the subject, so equate them.	$5x - 8 = -3x + 16$
3. Solve for x.	

i. Add $3x$ to both sides of the equation.
    ii. Add 8 to both sides of the equation.
    iii. Divide both sides of the equation by 8.
	$8x - 8 = 16$
	 $8x = 24$
	 $x = 3$

4. Substitute the value of x into either of the original equations, say [1], and solve for y.	Substituting $x = 3$ into [1]: $y = 5(3) - 8$ $\quad = 15 - 8$ $\quad = 7$
5. Write your answer.	Solution: $x = 3, y = 7$ or $(3, 7)$
6. Check the answer by substituting the point of intersection into equation [2].	Check: Substitute into $y = -3x + 16$. LHS $= y$ $\quad = 7$ RHS $= -3x + 16$ $\quad = -3(3) + 16$ $\quad = -9 + 16$ $\quad = 7$ As LHS $=$ RHS, the solution is correct.

on Resources

 Video eLesson Solving simultaneous equations using substitution (eles-1932)

 Interactivity Solving simultaneous equations using substitution (int-6453)

Exercise 5.3 Solving simultaneous linear equations using substitution

learn on

5.3 Quick quiz **on**	5.3 Exercise

Individual pathways

■ PRACTISE	■ CONSOLIDATE	■ MASTER
1, 4, 7, 8, 14, 15, 18	2, 5, 9, 12, 16, 19	3, 6, 10, 11, 13, 17, 20

Fluency

WE5 For questions **1** to **3**, solve the following simultaneous equations using the substitution method. Check your solutions using technology.

1.   **a.** $x = -10 + 4y$
       $3x + 5y = 21$
   b. $3x + 4y = 2$
        $x = 7 + 5y$
   c. $3x + y = 7$
        $x = -3 - 3y$
   d. $3x + 2y = 33$
        $y = 41 - 5x$

2. a. $y = 3x - 3$
$-5x + 3y = 3$

b. $4x + y = 9$
$y = 11 - 5x$

c. $x = -5 - 2y$
$5y + x = -11$

d. $x = -4 - 3y$
$-3x - 4y = 12$

3. a. $x = 7 + 4y$
$2x + y = -4$

b. $x = 14 + 4y$
$-2x + 3y = -18$

c. $3x + 2y = 12$
$x = 9 - 4y$

d. $y = 2x + 1$
$-5x - 4y = 35$

WE6 For questions **4** to **6**, solve the following pairs of simultaneous equations by equating the equations. Check your solutions using technology.

4. a. $y = 2x - 11$ and $y = 4x + 1$
c. $y = 2x - 10$ and $y = -3x$

b. $y = 3x + 8$ and $y = 7x - 12$
d. $y = x - 9$ and $y = -5x$

5. a. $y = -4x - 3$ and $y = x - 8$
c. $y = -x - 2$ and $y = x + 1$

b. $y = -2x - 5$ and $y = 10x + 1$
d. $y = 6x + 2$ and $y = -4x$

6. a. $y = 0.5x$ and $y = 0.8x + 0.9$

b. $y = 0.3x$ and $y = 0.2x + 0.1$

c. $y = -x$ and $y = -\dfrac{2}{7}x + \dfrac{4}{7}$

d. $y = -x$ and $y = -\dfrac{3}{4}x - \dfrac{1}{4}$

Understanding

7. A small farm has sheep and chickens. There are twice as many chickens as sheep, and there are 104 legs between the sheep and the chickens. Calculate the total number of chickens.

For questions **8** to **10**, use substitution to solve each of the following pairs of simultaneous equations.

8. a. $5x + 2y = 17$

$$y = \frac{3x - 7}{2}$$

b. $2x + 7y = 17$

$$x = \frac{1 - 3y}{4}$$

9. a. $2x + 3y = 13$

$$y = \frac{4x - 15}{5}$$

b. $-2x - 3y = -14$

$$x = \frac{2 + 5y}{3}$$

10. a. $3x + 2y = 6$

$$y = 3 - \frac{5x}{3}$$

b. $-3x - 2y = -12$

$$y = \frac{5x - 20}{3}$$

11. Use substitution to solve each of the following pairs of simultaneous equations for x and y in terms of m and n.

a. $mx + y = n$
$y = mx$

b. $x + ny = m$
$y = nx$

c. $mx - y = n$
$y = nx$

d. $mx - ny = n$
$y = x$

e. $mx - ny = -m$
$x = y - n$

f. $mx + y = m$
$x = \dfrac{y + m}{n}$

12. Determine the values of a and b so that the pair of equations $ax + by = 17$ and $2ax - by = -11$ has a unique solution of $(-2, 3)$.

13. The earliest record of magic squares is from China in about 2200 BCE. In magic squares the sums of the numbers of each row, column and diagonal are all equal to a magic number. Let z be the magic number. By creating a set of equations, solve to find the magic number and the missing values in the magic square.

m	11	7
9		
n	5	10

Communicating, reasoning and problem solving

14. **a.** Consider the pair of simultaneous equations:

$$8x - 7y = 9$$
$$x + 2y = 4$$

Identify which equation is the logical choice to make x the subject.

b. Use the substitution method to solve the system of equations. Show all your working.

15. A particular chemistry book costs $6 less than a particular physics book, while two such chemistry books and three such physics books cost a total of $123. Construct two simultaneous equations and solve them using the substitution method. Show your working.

16. The two shorter sides of a right triangle are 1 cm and 8 cm shorter than the hypotenuse. If the area of the triangle is $30\,\text{cm}^2$, determine the perimeter of the triangle.

17. Andrew is currently ten years older than his sister Prue. In four years' time he will be twice as old as Prue. Determine how old Andrew and Prue are now.

18. Use the substitution method to solve the following.

$$2x + y - 9 = 0$$
$$4x + 5y + 3 = 0$$

19. Use the substitution method to solve the following.

$$\frac{y - x}{2} - \frac{x + y}{3} = \frac{1}{6}$$
$$\frac{x}{5} + \frac{y}{2} = \frac{1}{2}$$

20. Consider the following pair of equations:

$$kx - \frac{y}{k} = 2$$
$$27x - 3y = 12k - 18$$

Determine the values of k when they will have:

a. one solution   **b.** no solutions   **c.** infinite solutions.

LESSON
5.4 Solving simultaneous linear equations using elimination

LEARNING INTENTION

At the end of this lesson you should be able to:
- solve two simultaneous linear equations using the elimination method.

ⓘ 5.4.1 Solving simultaneous equations using the elimination method

eles-4768

- The **elimination method** is an algebraic method to solve simultaneous linear equations. It involves adding or subtracting equations in order to eliminate one of the variables.
- In order to eliminate a variable, the variable must be on the same side of the equal sign in both equations and must have the same coefficient.
- If the coefficients of the variable have the same sign, we subtract one equation from the other to eliminate the variable.
- If the coefficients of the variables have the opposite sign, we add the two equations together to eliminate the variable.

$$3x + 4y = 14 \qquad\qquad 6x - 2y = 12$$
$$5x - 4y = 2 \qquad\qquad 6x + 3y = 27$$
$$\text{(add equations to eliminate } y) \quad \text{(subtract equations to eliminate } x)$$

- The process of elimination is carried out by adding (or subtracting) the left-hand sides and the right-hand sides of each equation together. Consider the equations $2x + y = 5$ and $x + y = 3$. The process of subtracting each side of the equation from each other is visualised on the scales to the right.
- To represent this process algebraically, the setting out would look like:

$$\begin{array}{r} 2x + y = 5 \\ -(x + y = 3) \\ \hline x = 2 \end{array}$$

- Once the value of x has been found, it can be substituted into either original equation to find y.

$$2(2) + y = 5 \Rightarrow y = 1$$

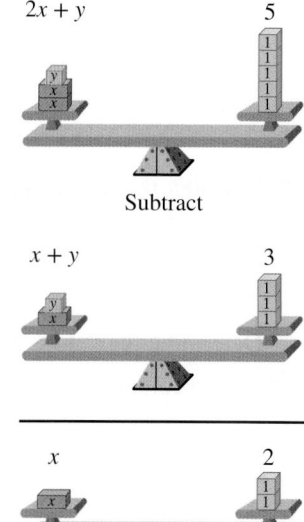

Solve the following pair of simultaneous equations using the elimination method.

$$-2x - 3y = -9$$
$$2x + y = 7$$

THINK	WRITE
1. Write the equations, one under the other and number them.	$-2x - 3y = -9$ [1] $2x + y = 7$ [2]
2. Look for an addition or subtraction that will eliminate either x or y. *Note:* Adding equations [1] and [2] in order will eliminate x.	[1] + [2]: $-2x - 3y + (2x + y) = -9 + 7$ $-2x - 3y + 2x + y = -2$ $-2y = -2$
3. Solve for y by dividing both sides of the equation by -2.	$y = 1$
4. Substitute the value of y into equation [2]. *Note:* $y = 1$ may be substituted into either equation.	Substituting $y = 1$ into [2]: $2x + 1 = 7$

5. Solve for x.
    i. Subtract 1 from both sides of the equation.
    ii. Divide both sides of the equation by 2.
	$2x = 6$
	
	 $x = 3$

6. Write the solution.	Solution: $x = 3$, $y = 1$ or $(3, 1)$

7. Check the solution by substituting $(3, 1)$ into equation [1] since equation [2] was used to find the value of x.	Check: Substitute into $-2x - 3y = -9$.
	 LHS $= -2(3) - 3(1)$
	        $= -6 - 3$
	        $= -9$
	 RHS $= -9$
	 LHS = RHS, so the solution is correct.

⊙ 5.4.2 Solving simultaneous equations by multiplying by a constant

eles-4769

- If neither variable in the two equations have the same coefficient, it will be necessary to multiply one or both equations by a constant so that a variable can be eliminated.
- The equals sign in an equation acts like a balance, so as long as both sides of equation are correctly multiplied by the same value, the new statement is still a valid equation.

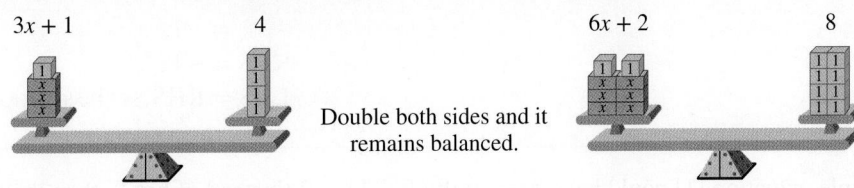

Double both sides and it remains balanced.

- Consider the following pairs of equations:

$$3x + 7y = 23 \qquad\qquad 4x + 5y = 22$$
$$6x + 2y = 22 \qquad\qquad 3x - 4y = -6$$

- For the first pair: the easiest starting point is to work towards eliminating x. This is done by first multiplying the top equation by 2 so that both equations have the same coefficient of x.

$$2(3x + 7y = 23) \Rightarrow 6x + 14y = 46$$

- For the second pair: in this case, both equations will need to be multiplied by a constant. Choosing to eliminate x would require the top equation to be multiplied by 3 and the bottom equation by 4 in order to produce two new equations with the same coefficient of x.

$$3(4x + 5y = 23) \Rightarrow 12x + 15y = 69$$

$$4(3x - 4y = -6) \Rightarrow 12x - 16y = -24$$

- Once the coefficient of one of the variables is the same, you can begin the elimination method.

WORKED EXAMPLE 8 Multiplying one equation by a constant to eliminate

Solve the following pair of simultaneous equations using the elimination method.

$$x - 5y = -17$$
$$2x + 3y = 5$$

THINK

1. Write the equations, one under the other and number them.

2. Look for a single multiplication that will create the same coefficient of either x or y. Multiply equation [1] by 2 and call the new equation [3].

3. Subtract equation [2] from [3] in order to eliminate x.

4. Solve for y by dividing both sides of the equation by -13.

5. Substitute the value of y into equation [2].

6. Solve for x.

   i. Subtract 9 from both sides of the equation.

   ii. Divide both sides of the equation by 2.

7. Write the solution.

8. Check the solution by substituting into equation [1].

WRITE

$x - 5y = -17$ [1]
$2x + 3y = 5$ [2]

$[1] \times 2 : 2x - 10y = -34$ [3]

$[3] - [2]:$
$2x - 10y - (2x + 3y) = -34 - 5$
$2x - 10y - 2x - 3y = -39$
$-13y = -39$

$y = 3$

Substituting $y = 3$ into [2]:
$2x + 3(3) = 5$
$2x + 9 = 5$

$2x = -4$

$x = -2$

Solution: $x = -2, y = 3$ or $(-2, 3)$

Check: Substitute into $x - 5y = -17$.
LHS $= (-2) - 5(3)$
       $= -2 - 15$
       $= -17$
RHS $= -17$
LHS = RHS, so the solution is correct.

Note: In this example, equation [1] could have been multiplied by -2 (instead of by 2), then the two equations added (instead of subtracted) to eliminate x.

Solve the following pair of simultaneous equations using the elimination method.

$$6x + 5y = 3$$
$$5x + 4y = 2$$

THINK	WRITE
1. Write the equations, one under the other and number them.	$6x + 5y = 3$ [1] $5x + 4y = 2$ [2]
2. Decide which variable to eliminate, say y. Multiply equation [1] by 4 and call the new equation [3]. Multiply equation [2] by 5 and call the new equation [4].	Eliminate y. [1] $\times 4$: $\ 24x + 20y = 12$ [3] [2] $\times 5$: $\ 25x + 20y = 10$ [4]
3. Subtract equation [3] from [4] in order to eliminate y.	[4] $-$ [3]: $25x + 20y - (24x + 20y) = 10 - 12$ $\quad 25x + 20y - 24x - 20y = -2$ $\qquad\qquad\qquad\qquad\qquad x = -2$
4. Substitute the value of x into equation [1].	Substituting $x = -2$ into [1]: $6(-2) + 5y = 3$ $\quad -12 + 5y = 3$

5. Solve for y.
   i. Add 12 to both sides of the equation.
   ii. Divide both sides of the equation by 5.
	
	$5y = 15$
	$\ y = 3$

6. Write your answer.	Solution $x = -2$, $y = 3$ or $(-2, 3)$
7. Check the answer by substituting the solution into equation [2].	Check: Substitute into $5x + 4y = 2$. LHS $= 5(-2) + 4(3)$ $\qquad\quad = -10 + 12$ $\qquad\quad = 2$ RHS $= 2$ LHS $=$ RHS, so the solution is correct.

Note: Equation [1] could have been multiplied by -4 (instead of by 4), then the two equations added (instead of subtracted) to eliminate y.

DISCUSSION

Discuss where in real life the elimination method could be used to solve a problem.

on Resources

▶ **Video eLesson** Solving simultaneous equations using elimination (eles-1931)

✦ **Interactivity**   Solving simultaneous equations using elimination (int-6127)

Exercise 5.4 Solving simultaneous linear equations using elimination

learn on

5.4 Quick quiz on	5.4 Exercise

Individual pathways

■ PRACTISE	■ CONSOLIDATE	■ MASTER
1, 3, 5, 10, 13, 18	2, 6, 8, 11, 14, 15, 19	4, 7, 9, 12, 16, 17, 20

Fluency

1. **WE7** Solve the following pairs of simultaneous equations by adding equations to eliminate either x or y.

   a. $x + 2y = 5$
   $-x + 4y = 1$

   b. $5x + 4y = 2$
   $5x - 4y = -22$

   c. $-2x + y = 10$
   $2x + 3y = 14$

2. Solve the following pairs of equations by subtracting equations to eliminate either x or y.

   a. $3x + 2y = 13$
   $5x + 2y = 23$

   b. $2x - 5y = -11$
   $2x + y = 7$

   c. $-3x - y = 8$
   $-3x + 4y = 13$

3. Solve each of the following equations using the elimination method.

   a. $6x - 5y = -43$
   $6x - y = -23$

   b. $x + 4y = 27$
   $3x - 4y = 17$

   c. $-4x + y = -10$
   $4x - 3y = 14$

4. Solve each of the following equations using the elimination method.

   a. $-5x + 3y = 3$
   $-5x + y = -4$

   b. $5x - 5y = 1$
   $2x - 5y = -5$

   c. $4x - 3y - 1 = 0$
   $4x + 7y - 11 = 0$

WE8 For questions **5** to **7**, solve the following pairs of simultaneous equations.

5. a. $6x + y = 9$
   $-3x + 2y = 3$

   b. $x + 3y = 14$
   $3x + y = 10$

   c. $5x + y = 27$
   $4x + 3y = 26$

6. a. $-6x + 5y = -14$
   $-2x + y = -6$

   b. $2x + 5y = 14$
   $3x + y = -5$

   c. $-3x + 2y = 6$
   $x + 4y = -9$

7. a. $3x - 5y = 7$
   $x + y = -11$

   b. $2x + 3y = 9$
   $4x + y = -7$

   c. $-x + 5y = 7$
   $5x + 5y = 19$

8. **WE9** Solve the following pairs of simultaneous equations.

   a. $-4x + 5y = -9$
   $2x + 3y = 21$

   b. $2x + 5y = -6$
   $3x + 2y = 2$

   c. $2x - 2y = -4$
   $5x + 4y = 17$

9. Solve the following pairs of simultaneous equations.

   a. $2x - 3y = 6$
   $4x - 5y = 9$

   b. $\dfrac{x}{2} + \dfrac{y}{3} = 2$
   $\dfrac{x}{4} + \dfrac{y}{3} = 4$

   c. $\dfrac{x}{3} + \dfrac{y}{2} = \dfrac{3}{2}$
   $\dfrac{x}{2} + \dfrac{y}{5} = -\dfrac{1}{2}$

Understanding

For questions **10** to **12**, solve the following simultaneous equations using an appropriate method. Check your answer using technology.

10. a. $7x + 3y = 16$
    $y = 4x - 1$

    b. $2x + y = 8$
    $4x + 3y = 16$

    c. $-3x + 2y = 19$
    $4x + 5y = 13$

11. a. $-3x + 7y = 9$
    $4x - 3y = 7$

    b. $-4x + 5y = -7$
    $x = 23 - 3y$

    c. $y = -x$
    $y = -\dfrac{2}{5}x - \dfrac{1}{5}$

12. a. $4x + 5y = 41$

$y = \dfrac{3x}{2} - 1$

b. $3x - 2y = 9$

$2x + 5y = -13$

c. $\dfrac{x}{3} + \dfrac{y}{4} = 7$

$3x - 2y = 12$

Communicating, reasoning and problem solving

13. The cost of a cup of coffee and croissant is $8.50 from a local bakery, and an order of 5 coffees and 3 croissants costs $35.70. Determine the cost of one croissant.

14. Celine notices that she only has 5-cent and 10-cent coins in her coin purse. She counts up how much she has and finds that from the 34 coins in the purse the total value is $2.80. Determine how many of each type of coin she has.

15. Abena, Bashir and Cecily wanted to weigh themselves, but the scales they had were broken and would only give readings over 100 kg. They decided to weigh themselves in pairs and calculate their weights from the results.
    * Abena and Bashir weighed 119 kg
    * Bashir and Cecily weighed 112 kg
    * Cecily and Abena weighed 115 kg
      Determine the weight of each student.

16. **a.** For the general case $ax + by = e$ [1]

    $cx + dy = f$ [2]

    y can be found by eliminating x.

    i. Multiply equation [1] by c to create equation 3.
    ii. Multiply equation [2] by a to create equation 4.
    iii. Use the elimination method to find a general solution for y.

    b. Use a similar process to that outlined above to find a general solution for x.
    c. Use the general solution for x and y to solve each of the following.

    i. $2x + 5y = 7$

    $7x + 2y = 24$

    ii. $3x - 5y = 4$

    $x + 3y = 5$

    Choose another method to check that your solutions are correct in each part.
    d. For y to exist, it is necessary to state that $bc - ad \neq 0$. Explain.
    e. Is there a necessary condition for x to exist? Explain.

17. A family of two parents and four children go to the movies and spend $95 on the tickets. Another family of one parent and two children go to see the same movie and spend $47.50 on the tickets. Determine if it is possible to work out the cost of an adult's ticket and child's ticket from this information.

18. The sum of two numbers is equal to k. The difference of the two numbers is given by $k - 20$. Determine the possible solutions for the two numbers.

19. Use the method of elimination to solve:

$$\dfrac{x-4}{3} + y = -2$$

$$\dfrac{2y-1}{7} + x = 6$$

20. Use an appropriate method to solve:

$$2x + 3y + 3z = -1$$
$$3x - 2y + z = 0$$
$$z + 2y = 0$$

LESSON
5.5 Applications of simultaneous linear equations

LEARNING INTENTION

At the end of this lesson you should be able to:
- define unknown quantities with appropriate variables
- form two simultaneous equations using the information presented in a problem
- choose an appropriate method to solve simultaneous equations in order to find the solution to a problem.

▶ 5.5.1 Applications of simultaneous linear equations

eles-4770

- When solving practical problems, the following steps can be useful:
  Step 1. Define the unknown quantities using appropriate pronumerals.
  Step 2. Use the information given in the problem to form two equations in terms of these pronumerals.
  Step 3. Solve these equations using an appropriate method.
  Step 4. Write the solution in words.
  Step 5. Check the solution.

Key language used in worded problems

To help set up equations from the information presented in a problem question, make sure you look out for the following key terms:
- *Addition:* sum, altogether, add, more than, and, in total
- *Subtraction:* difference, less than, take away, take off, fewer than
- *Multiplication:* product, groups of, times, of, for each, double, triple
- *Division:* quotient, split into, halve, thirds
- *Equals:* gives, is

WORKED EXAMPLE 10 Applying the elimination method to problem solving

Ashley received better results for his Mathematics test than for his English test. If the sum of the two marks is 164 and the difference is 22, calculate the mark he received for each subject.

THINK	WRITE
1. Define the two variables.	Let x = the Mathematics mark. Let y = the English mark.
2. Formulate two equations from the information given and number them. The sum of the two marks is $x + y$. The difference of the two marks is $x - y$.	$x + y = 164$ [1] $x - y = 22$ [2]
3. Use the elimination method by adding equations [1] and [2] to eliminate y.	$[1] + [2] : 2x = 186$
4. Solve for x by dividing both sides of the equation by 2.	$x = 93$

5.	Substitute the value of x into equation [1].	Substituting $x = 93$ into [1]:
		$x + y = 164$
		$93 + y = 164$
6.	Solve for y by subtracting 93 from both sides of the equation.	$y = 71$
7.	Write the solution.	Solution:
		Mathematics mark $(x) = 93$
		English mark $(y) = 71$
8.	Check the solution by substituting $x = 93$ and $y = 71$ into equation [1].	Check: Substitute into $x + y = 164$.
		LHS $= 93 + 71$ RHS $= 164$
		$\quad = 164$
		As LHS $=$ RHS, the solution is correct.

WORKED EXAMPLE 11 Applying the substitution method to problem solving

To finish a project, Genevieve bought a total of 25 nuts and bolts from a hardware store. If each nut costs 12 cents, each bolt costs 25 cents and the total purchase price is \$4.30, calculate how many nuts and how many bolts Genevieve bought.

THINK	WRITE
1. Define the two variables.	Let $x =$ the number of nuts.
	Let $y =$ the number of bolts.
2. Formulate two equations from the information given and number them.	$x + y = 25 \qquad$ [1]
	$12x + 25y = 430 \qquad$ [2]
Note: The total number of nuts and bolts is 25. Each nut cost 12 cents, each bolt cost 25 cents and the total cost is 430 cents (\$4.30).	
3. Solve simultaneously using the substitution method, since equation [1] is easy to rearrange.	Rearrange equation [1]:
	$x + y = 25$
Rearrange equation [1] to make x the subject by subtracting y from both sides of equation [1].	$x = 25 - y$
4. Substitute the expression $(25 - y)$ for x into equation [2].	Substituting $(25 - y)$ into [2]:
	$12(25 - y) + 25y = 430$
5. Solve for y.	$300 - 12y + 25y = 430$
	$300 + 13y = 430$
	$13y = 130$
	$y = 10$

6. Substitute the value of y into the rearranged equation $x = 25 - y$ from step **3.**

Substituting $y = 10$ into $x = 25 - y$:
$x = 25 - 10$
$x = 15$

7. Write the solution.

Solution:
The number of nuts $(x) = 15$.
The number of bolts $(y) = 10$.

8. Check the solution by substituting $x = 15$ and $y = 10$ into equation [1].

Check: Substitute into $x + y = 25$.
LHS $= 15 + 10$ RHS $= 25$
$\qquad = 25$
As LHS $=$ RHS, the solution is correct.

- It is also possible to determine solutions to worded problems using the graphical method by forming and then graphing equations.

WORKED EXAMPLE 12 Applying the graphical method to problem solving

Cecilia buys 2 pairs of shorts and 3 T-shirts for \$160. Ida buys 1 pair of shorts and 2 T-shirts for \$90. Develop two equations to describe the situation and solve them graphically to determine the cost of one pair of shorts and one T-shirt.

THINK

1. Define the two variables.

2. Formulate two equations from the information given and number them.

3. Calculate the x- and y-intercepts for both graphs.

4. Graph the two lines either by hand or using technology. Only the first quadrant of the graph is required, as cost cannot be negative.

5. Identify the point of intersection to solve the simultaneous equations.

6. Write the answer as a sentence.

WRITE

Let $x =$ cost of a pair of shorts.
Let $y =$ cost of a T-shirt.

$2x + 3y = 160$ [1]
$x + 2y = 90$ [2]

Equation [1]   Equation [2]
$\quad 2x + 3y = 160$ $\quad x + 2y = 90$
x-intercept, $y = 0$ x-intercept, $y = 0$
$\quad 2x + 3 \times 0 = 160$ $\quad x + 2 \times 0 = 90$
$\qquad 2x = 160$ $\qquad x = 90$
$\qquad\ x = 80$
y-intercept, $x = 0$ y-intercept, $x = 0$
$\quad 2 \times 0 + 3y = 160$ $\quad 0 + 2y = 90$
$\qquad 3y = 160$ $\qquad 2y = 90$
$\qquad\ y = 53\frac{1}{3}$ $\qquad\ y = 45$

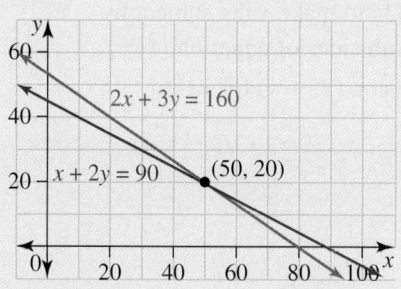

The point of intersection is (50, 20).

The cost of one pair of shorts is \$50 and the cost of one T-shirt is \$20.

5.5 Quick quiz on	5.5 Exercise

Individual pathways

■ PRACTISE	■ CONSOLIDATE	■ MASTER
1, 7, 8, 12, 13, 15, 17, 22	2, 5, 9, 10, 14, 18, 19, 23	3, 4, 6, 11, 16, 20, 21, 24, 25

Fluency

1. **WE10** Rick received better results for his Maths test than for his English test. If the sum of his two marks is 163 and the difference is 31, calculate the mark received for each subject.

2. **WE11** Rachael buys 30 nuts and bolts to finish a project. If each nut costs 10 cents, each bolt costs 20 cents and the total purchase price is $4.20, how many nuts and how many bolts does she buy?

3. Eloise has a farm that raises chickens and sheep. Altogether there are 1200 animals on the farm. If the total number of legs from all the animals is 4000, calculate how many of each type of animal there is on the farm.

Understanding

4. Determine the two numbers whose difference is 5 and whose sum is 11.

5. The difference between two numbers is 2. If three times the larger number minus twice the smaller number is 13, determine the values of the two numbers.

6. One number is 9 less than three times a second number. If the first number plus twice the second number is 16, determine the values of the two numbers.

7. A rectangular house has a perimeter of 40 metres and the length is 4 metres more than the width. Calculate the dimensions of the house.

8. **WE12** Mike has 5 lemons and 3 oranges in his shopping basket. The cost of the fruit is $3.50. Voula, with 2 lemons and 4 oranges, pays $2.10 for her fruit. Develop two equations to describe the situation and solve them graphically to determine the cost of each type of fruit.

9. A surveyor measuring the dimensions of a block of land finds that the length of the block is three times the width. If the perimeter is 160 metres, calculate the dimensions of the block.

10. Julie has $3.10 in change in her pocket. If she has only 50-cent and 20-cent pieces and the total number of coins is 11, calculate how many coins of each type she has.

11. Mr Yang's son has a total of twenty-one $1 and $2 coins in his moneybox. When he counts his money, he finds that its total value is $30. Determine how many coins of each type he has.

12. If three Magnums and two Paddlepops cost $8.70 and the difference in price between a Magnum and a Paddlepop is 90 cents, calculate how much each type of ice-cream costs.

13. If one Red Frog and four Killer Pythons cost $1.65, whereas two Red Frogs and three Killer Pythons cost $1.55, calculate how much each type of lolly costs.

14. A catering firm charges a fixed cost for overheads and a price per person. It is known that a party for 20 people costs $557, whereas a party for 35 people costs $909.50. Determine the fixed cost and the cost per person charged by the company.

15. The difference between Sally's PE mark and Science mark is 12, and the sum of the marks is 154. If the PE mark is the higher mark, calculate what mark Sally got for each subject.

16. Mozza's Cheese Supplies sells six Mozzarella cheeses and eight Swiss cheeses to Munga's deli for $83.60, and four Mozzarella cheeses and four Swiss cheeses to Mina's deli for $48. Calculate how much each type of cheese costs.

Communicating, reasoning and problem solving

17. If the perimeter of the triangle in the diagram is 12 cm and the length of the rectangle is 1 cm more than the width, determine the value of x and y.

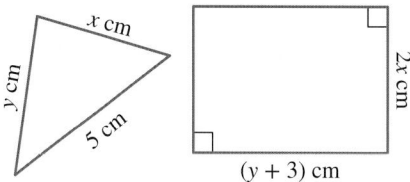

18. Mr and Mrs Waugh want to use a caterer for a birthday party for their twin sons. The manager says the cost for a family of four would be $160. However, the sons want to invite 8 friends, making 12 people in all. The cost for this would be $360. If the total cost in each case is made up of the same cost per person and the same fixed cost, calculate the cost per person and the fixed cost. Show your working.

19. Joel needs to buy some blank DVDs and USB sticks to back up a large amount of data that has been generated by an accounting firm. He buys 6 DVDs and 3 USB sticks for $96. He later realises these are not sufficient and so buys another 5 DVDs and 4 USB sticks for $116. Determine how much each DVD and each USB stick cost. (Assume the same rate per item was charged for each visit.) Show your working.

20. Four years ago Tim was 4 times older than his brother Matthew. In six years' time Tim will only be double his brother's age. Calculate how old the two brothers currently are.

21. A local cinema has different prices for movie tickets for children (under 12), adults and seniors (over 60). Consider the following scenarios:
    • For a senior couple (over 60) and their four grandchildren, the total cost is $80.
    • For two families with four adults and seven children, the total cost is $160.50.
    • For a son (under 12), his father and his grandfather (over 60), the total cost is $45.75.
    Determine the cost of each type of ticket.

22. Reika completes a biathlon (swimming and running) that has a total distance of 37 kilometres. Reika knows that it takes her 21.2 minutes to swim 1 kilometre and 4.4 minutes to run 1 kilometre. If her total time for the race was 6 hours and 39 minutes, calculate the length of the swimming component of the race.

23. At the football, hot chips are twice as popular as meat pies and three times as popular as hot dogs. Over the period of half an hour during half time, a fast-food outlet serves 121 people who each bought one item. Determine how many serves of each of the foods were sold during this half-hour period.

24. Three jet skis in a 300 kilometres handicap race leave at two-hour intervals. Jet ski 1 leaves first and has an average speed of 25 kilometres per hour for the entire race. Jet ski 2 leaves two hours later and has an average speed of 30 kilometres per hour for the entire race. Jet ski 3 leaves last, two hours after jet ski 2 and has an average speed of 40 kilometres per hour for the entire race.

    a. Sketch a graph to show each jet ski's journey on the one set of axes.
    b. Determine who wins the race.
    c. Check your findings algebraically and describe what happened to each jet ski during the course of the race.

25. Alice is competing in a cycling race on an extremely windy day. The race is an 'out and back again' course, so the wind is against Alice in one direction and assisting her in the other. For the first half of the race the wind is blowing against Alice, slowing her down by 4 km per hour. Given that on a normal day Alice could maintain a pace of 36 km per hour and that this race took her 4 hours and 57 minutes, calculate the total distance of the course.

LESSON
5.6 Review

5.6.1 Topic summary

Solving graphically

- Solving simultaneous equations involves finding the point (or points) of intersection between two lines.
- We can determine these points by accurately sketching both equations, or using technology to find these points.

Solving by substitution and elimination

- The substitution and elimination methods are two algebraic techniques used to solve simultaneous equations.
- We can use substitution when one (or both) of the equations have a variable as the subject. e.g. $y = 3x - 4$
- We use the elimination method when substitution is more tedious.
- Elimination method involves adding or subtracting equations to eliminate one of the variables. e.g.

$$\begin{array}{r} 3x + y = 5 \\ 4x - y = 2 \\ \hline 7x = 7 \end{array} +$$

Applications

- Choose appropriate variables to define the unknown quantities.
- Use the information in the question to form two or more equations.
- Pick an appropriate technique to solve simultaneously.
- Write out a statement that explicitly answers the question.

SIMULTANEOUS LINEAR EQUATIONS (PATH)

Parallel and perpendicular lines

- Two lines are parallel if they have the same gradient.
  e.g. $\quad y = \mathbf{3}x - 6$
  $\quad\quad\quad y = \mathbf{3}x + 1$
- Two lines are perpendicular if the product of their gradients is −1.
  e.g. $\quad y = 2x + 3$
  $\quad\quad\quad y = -\dfrac{x}{2} - 4$
  $m_1 \times m_2 = 2 \times -\dfrac{1}{2} = -1$

Number of solutions

- Parallel lines with *different* y-intercepts will never intersect.
- Parallel lines with the *same* y-intercept are called **coincident lines** and will intersect an infinite number of times.
- Perpendicular lines will intersect once and cross at right angles to each other.

5.6.2 Project

Documenting business expenses

In business, expenses can be represented graphically so that relevant features are clearly visible. The graph compares the costs of hiring cars from two different car rental companies. It will be cheaper to use Plan A when travelling distances less than 250 kilometres and Plan B when travelling more than 250 kilometres. Both plans cost the same when you are travelling exactly 250 kilometres.

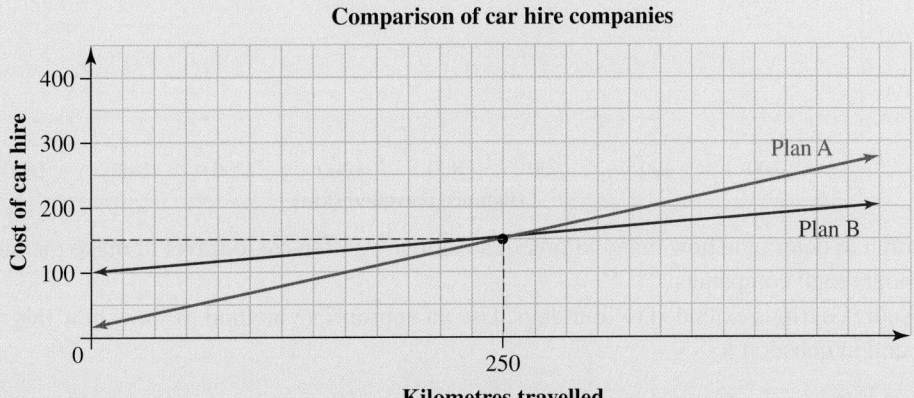

Andrea works as a travelling sales representative. She needs to plan her next business trip to Port Hedland, which she anticipates will take her away from the office for 3 or 4 days. Due to other work commitments, she is not sure whether she can make the trip by the end of this month or early next month.

She plans to fly to Port Hedland and use a hire car to travel when she arrives. Andrea's boss has asked her to supply documentation detailing the anticipated costs for the hire car, based on the following quotes received.

| A1 Rentals | $35 per day plus 28c per kilometre of travel |
| Cut Price Rentals | $28 per day plus 30c per kilometre of travel |

Andrea is aware that although the Cut Price Rentals deal looks cheaper, it could work out more expensive in the long run because of the higher cost per kilometre of travel; she intends to travel a considerable distance.

Andrea is advised by both rental companies that their daily hire charges are due to rise by $2 per day from the first day of next month.

Assuming that Andrea is able to travel this month and her trip will last 3 days, use the information given to answer questions **1** to **4**.

1. Write equations to represent the costs of hiring a car from A1 Rentals and Cut Price Rentals. Use the pronumeral C to represent the cost (in dollars) and d to represent the distance travelled (in kilometres).
2. Copy the following set of axes to plot the two equations from question **1** to show how the costs compare over 1500 km.

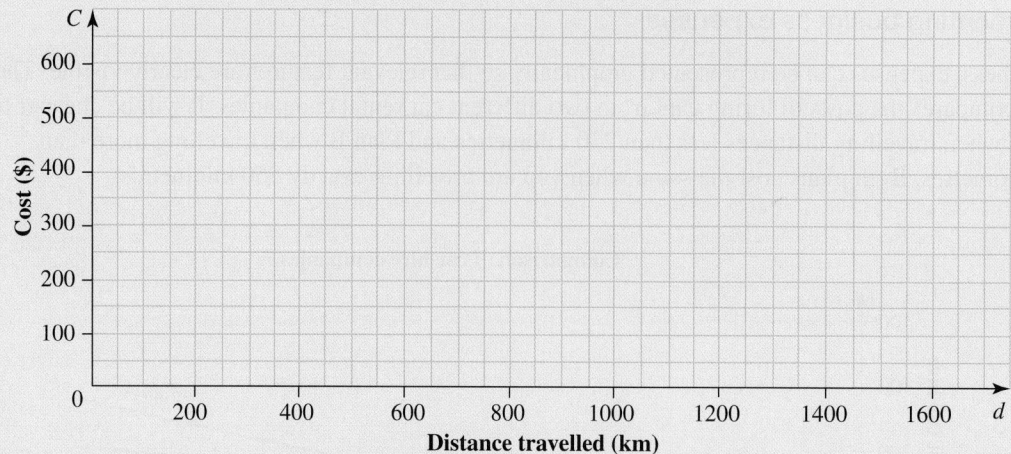

Comparison of cost of hiring a car from A1 Rentals and Cut Price Rentals

3. Use the graph to determine how many kilometres Andrea would have to travel to make the hire costs the same for both rental companies.

4. Assume Andrea's trip is extended to four days. Use an appropriate method to show how this changes the answer found in question **3**.

For questions **5** to **7**, assume that Andrea has delayed her trip until next month when the hire charges have increased.

5. Write equations to show the cost of hiring a car from both car rental companies for a trip lasting:
   a. 3 days
   b. 4 days.

6. Copy the following set of axes to plot the four equations from question **5** to show how the costs compare over 1500 km.

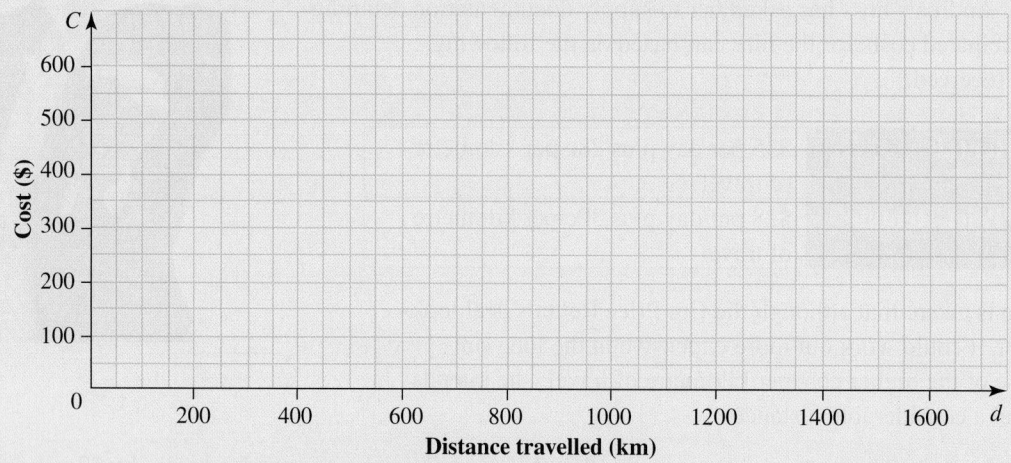

Comparison of cost of hiring a car from A1 Rentals and Cut Price Rentals

7. Comment on the results displayed in your graph.

8. Andrea needs to provide her boss with documentation of the hire car costs, catering for all options. Prepare a document for Andrea to hand to her boss.

 Resources

Interactivities Crossword (int-2836)
                 Sudoku puzzle (int-3591)

Fluency

1. Determine the value of m so that the set of simultaneous equations below has no solution.

$$mx - y = 2$$
$$3x + 4y = 12$$

2. **MC** A music shop charges a flat rate of $5 postage for 2 CDs and $11 for 5 CDs. Identify the equation that best represents this, if C is the cost and n is the number of CDs.
   - **A.** $C = n + 2$
   - **B.** $C = 2n + 1$
   - **C.** $C = 5n + 11$
   - **D.** $C = 6n + 5$

3. **MC** Identify which of the following pairs of coordinates is the solution to the simultaneous equations:

$$2x + 3y = 18$$
$$5x - y = 11$$

   - **A.** $(6, 2)$
   - **B.** $(3, -4)$
   - **C.** $(3, 9)$
   - **D.** $(3, 4)$

4. **MC** Identify the graphical solution to the following pair of simultaneous equations:

$$y = 5 - 2x$$
$$y = 3x - 10$$

A.

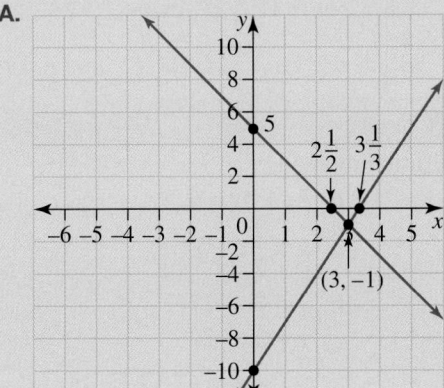

B.

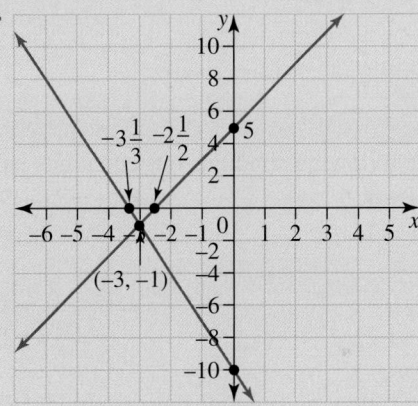

C.

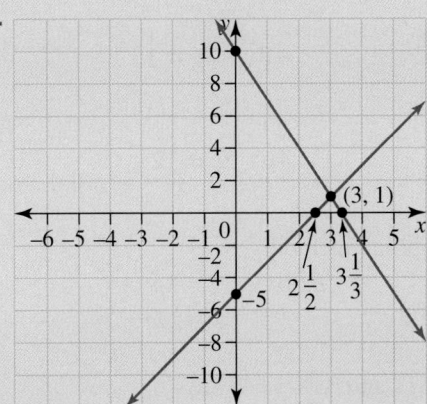

D.

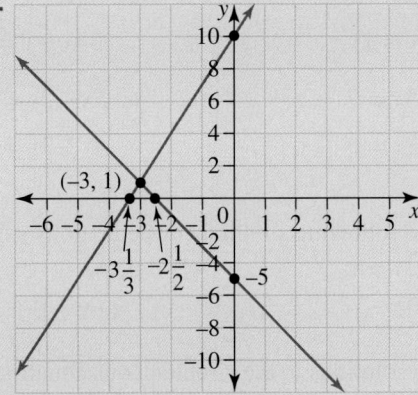

5. True or false?
   The three lines $5x + 3y = 1$, $4x + 7y = 10$ and $2x - y = -4$ have one point in common.

Understanding

6. Use substitution to check if the given pair of coordinates is a solution to the given simultaneous equations.

   a. $x - 2y = 5$ $(7, 1)$
   $5y + 2x = 18$

   b. $y = 7 - 3$ $(4, 3)$
   $5y - 2x = 7$

7. Solve each of the following pairs of simultaneous equations using a graphical method.

   a. $4y - 2x = 8$
   $x + 2y = 0$

   b. $y = 2x - 2$
   $x - 4y = 8$

   c. $2x + 5y = 20$
   $y = 7$

8. Use the graphs below, showing the given simultaneous equations, to write the point of intersection of the graphs and, hence, the solution of the simultaneous equations.

   a. $x + 3y = 6$
   $y = 2x - 5$

   b. $3x + 2y = 12$
   $2y = 3x$

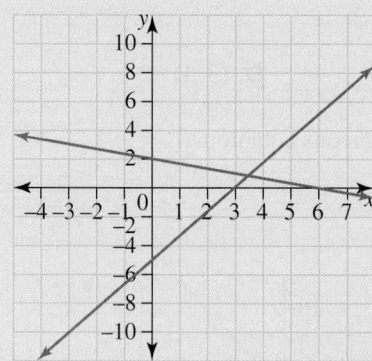
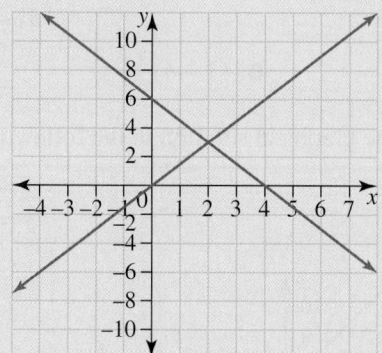

9. Solve the following simultaneous equations using the substitution method.

   a. $y = 3x + 1$
   $x + 2y = 16$

   b. $y = 2x + 7$
   $3y - 4x = 11$

   c. $2x + 5y = 6$
   $y = \dfrac{3}{2}x + 5$

   d. $y = -x$
   $y = 8x + 21$

   e. $y = 3x - 11$
   $y = 5x + 17$

   f. $y = 4x - 17$
   $y = 6x - 22$

10. Solve the following simultaneous equations using the elimination method.

   a. $3x + y = 17$
   $7x - y = 33$

   b. $4x + 3y = 1$
   $-4x + y = 11$

   c. $3x - 7y = -2$
   $-2x - 7y = 13$

   d. $4y - 3x = 9$
   $y + 3x = 6$

   e. $5x + 2y = 6$
   $4x + 3y = 2$

   f. $x - 4y = -4$
   $4x - 2y = 12$

11. Solve the following simultaneous equations using an appropriate method.

   a. $3x + 2y = 6$
   $3y + 5x = 9$

   b. $6x - 4y = -6$
   $7x + 3y = -30$

   c. $6x + 2y = 14$
   $x = -3 + 5y$

12. **MC** The solutions to the simultaneous equations $7x - 2y = 11$ and $3x + y = 1$ are:

   A. $x = 9, y = -26$

   B. $x = 5, y = 11$

   C. $x = -1, y = 2$

   D. $x = 1, y = -2$

13. Determine the value of p for which the lines $2x + 3y = 23$ and $7x + py = 8$ will not intersect.

Communicating, reasoning and problem solving

14. Write the following as a pair of simultaneous equations and solve.
    a. Determine which two numbers have a difference of 5, and their sum is 23.
    b. A rectangular house has a total perimeter of 34 metres and the width is 5 metres less than the length. Calculate the dimensions of the house.
    c. If two Chupa Chups and three Wizz Fizzes cost $2.55, but five Chupa Chups and seven Wizz Fizzes cost $6.10, determine the price of each type of lolly.

15. Laurie buys milk and bread for his family on the way home from school each day, paying with a $10 note. If he buys three cartons of milk and two loaves of bread, he receives 5 cents in change. If he buys two cartons of milk and one loaf of bread, he receives $4.15 in change. Calculate how much each item costs.

16. A paddock contains some cockatoos (2-legged) and kangaroos (4-legged). The total number of animals is 21 and they have 68 legs in total. Using simultaneous equations, determine how many cockatoos and kangaroos there are in the paddock.

17. Warwick was solving a pair of simultaneous equations using the elimination method and reached the result that $0 = -5$. Suggest a solution to the problem, giving a reason for your answer.

18. There are two sections to a concert hall. Seats in the 'Dress circle' are arranged in rows of 40 and cost $140 each. Seats in the 'Bleachers' are arranged in rows of 70 and cost $60 each. There are 10 more rows in the 'Dress circle' than in the 'Bleachers' and the capacity of the hall is 7000.
    a. If d represents the number of rows in the 'Dress circle' and b represents the number of rows in the 'Bleachers', write an equation in terms of these two variables based on the fact that there are 10 more rows in the 'Dress circle' than in the 'Bleachers'.
    b. Write an equation in terms of these two variables based on the fact that the capacity of the hall is 7000 seats.
    c. Solve the two equations from **a** and **b** simultaneously using the method of your choice to find the number of rows in each section.
    d. Now that you have the number of rows in each section, calculate the number of seats in each section.
    e. Hence, calculate the total receipts for a concert where all tickets are sold.

19. John is comparing two car rental companies, Golden Ace Rental Company and Silver Diamond Rental Company.
Golden Ace Rental Company charges a flat rate of $38 per day and $0.20 per kilometre. Silver Diamond Rental Company charges a flat rate of $30 per day plus $0.32 per kilometre.
   a. Write an algebraic equation for the cost of renting a car for three days from Golden Ace Rental Company in terms of the number of kilometres travelled, k.
   b. Write an algebraic equation for the cost of renting a car for three days from Silver Diamond Rental Company in terms of the number of kilometres travelled, k.
   c. Determine how many kilometres John would have to travel so that the cost of hiring from each company for three days is the same.

20. Frederika has $24 000 saved for a holiday and a new stereo. Her travel expenses are $5400 and her daily expenses are $260.
   a. Write an equation for the cost of her holiday if she stays for d days.
   b. Upon her return from holidays Frederika wants to purchase a new stereo system that will cost her $2500. Calculate how many days she can spend on her holiday if she wishes to purchase a new stereo upon her return.

on To test your understanding and knowledge of this topic, go to your learnON title at www.jacplus.com.au and complete the **post-test**.

Answers

Topic 5 Simultaneous linear equations (Path)

5.1 Pre-test

1. False
2. $(0, 2)$
3. An infinite number of solutions.
4. $(1, 0.2)$
5. $(4, -2)$
6. $a = \dfrac{7}{2}$
7. 92
8. A
9. D
10. $18\,\text{cm}^2$
11. $p = -\dfrac{10}{3}$
12. $a = -\dfrac{3}{2}, b = -2$
13. 120 children
14. B
15. B

5.2 Solving simultaneous linear equations graphically

1. a. $(2, 1)$ b. $(1, 1)$
2. a. $(0, 4)$ b. $(2, -1)$
3. a. $(-2, -4)$ b. $(-0.5, 1.5)$
4. a. No   b. Yes
   c. Yes   d. No
5. a. Yes   b. No
   c. No   d. Yes
6. a. No   b. Yes
   c. No   d. Yes
7. a. $(3, 2)$ b. $(4, 3)$
   c. $(-3, 4)$ d. $(-2, 2)$
8. a. $(2, 0)$ b. $(3, 0)$
   c. $(-2, 4)$ d. $(3, 8)$
9. a. $\left(-\dfrac{1}{2}, 1\dfrac{1}{2}\right)$ b. $(2, 5)$
   c. $(5, 3)$ d. $\left(2, \dfrac{2}{3}\right)$
10. a. No solution   b. $(2, -1)$
    c. No solution   d. $(1, 9)$
11. a. $(3, 1)$ b. No solution
    c. No solution   d. $(2, 1)$

12. a. No solution   b. $\left(\dfrac{6}{5}, -\dfrac{6}{5}\right)$
    c. $(2, -4)$ d. No solution
13. $y = 4x - 16$
14. a. Northern beach
    $C = 20 + 12t, 0 \le t \le 5$
    Southern beach
    $D = 8 + 18t, 0 \le t \le 5$
    b. Northern beaches in red, southern beaches in blue
    c. Time > 2 hours
    d. Time $= 2$ hours, cost $= \$44$

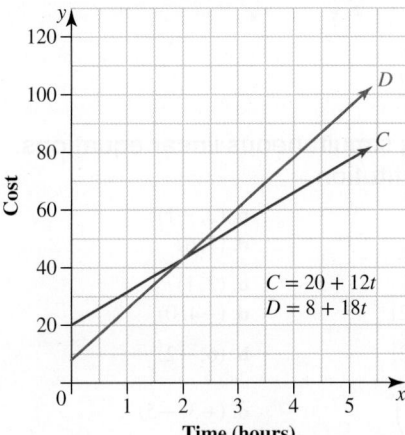

15. a. Same line   b. Perpendicular
    c. Intersecting   d. Parallel
16. a. 1 solution
    b. No solution (parallel lines)
    c. No solution (parallel lines)
17. a. 1 solution (perpendicular lines)
    b. Infinite solutions (coincident)
    c. 1 solution
    d. No solution (parallel lines)
18. a. $a = 3$ b. $a = \dfrac{1}{4}$ c. $a = 8$
19. a. $y = 3x + 6$
    b. $y = -2x + 1$
    c.

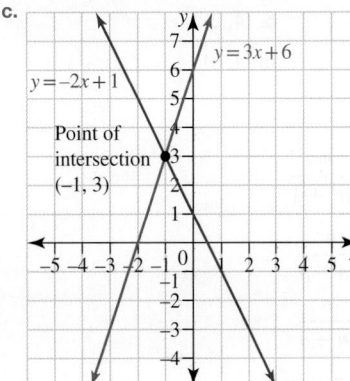

20.

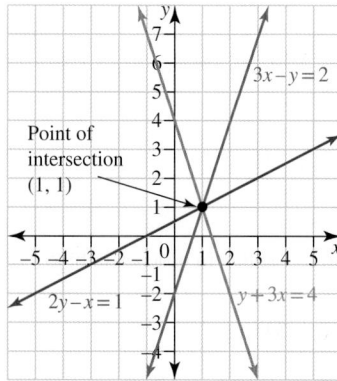

21. $y = \dfrac{5x}{4} + 9$

5.3 Solving simultaneous linear equations using substitution

1. a. $(2, 3)$ **b.** $(2, -1)$
   c. $(3, -2)$ **d.** $(7, 6)$

2. a. $(3, 6)$ **b.** $(2, 1)$
   c. $(-1, -2)$ **d.** $(-4, 0)$

3. a. $(-1, -2)$ **b.** $(6, -2)$
   c. $\left(3, 1\dfrac{1}{2}\right)$ **d.** $(-3, -5)$

4. a. $(-6, -23)$ **b.** $(5, 23)$
   c. $(2, -6)$ **d.** $\left(\dfrac{3}{2}, -\dfrac{15}{2}\right)$

5. a. $(1, -7)$ **b.** $\left(-\dfrac{1}{2}, -4\right)$
   c. $\left(-\dfrac{3}{2}, -\dfrac{1}{2}\right)$ **d.** $\left(-\dfrac{1}{5}, \dfrac{4}{5}\right)$

6. a. $(-3, -1.5)$ **b.** $(1, 0.3)$
   c. $\left(-\dfrac{4}{5}, \dfrac{4}{5}\right)$ **d.** $(1, -1)$

7. 26 chickens

8. a. $(3, 1)$ **b.** $(-2, 3)$

9. a. $(5, 1)$ **b.** $(4, 2)$

10. a. $(0, 3)$ **b.** $(4, 0)$

11. a. $x = \dfrac{n}{2m}, y = \dfrac{n}{2}$

   b. $x = \dfrac{m}{n^2 + 1}, y = \dfrac{mn}{n^2 + 1}$

   c. $x = \dfrac{n}{m - n}, y = \dfrac{n^2}{m - n}$

   d. $x = \dfrac{n}{m - n}, y = \dfrac{n}{m - n}$

   e. $x = \dfrac{n^2 - m}{m - n}, y = \dfrac{m(n - 1)}{m - n}$

   f. $x = \dfrac{2m}{m + n}, y = \dfrac{m(n - m)}{m + n}$

12. $a = -1, b = 5$

13. $z = 24, m = 6, n = 9$

m	11	7
9	8	7
n	5	10

14. a. $x + 2y = 4$ **b.** $x = 2, y = 1$

15. Chemistry \$21, physics \$27.

16. 30 cm

17. Andrew is 16, Prue is 6.

18. $x = 8, y = -7$

19. $x = 0, y = 1$

20. a. $k \neq \pm 3$ **b.** $k = 3$ **c.** $k = -3$

5.4 Solving simultaneous linear equations using elimination

1. a. $(3, 1)$ **b.** $(-2, 3)$ **c.** $(-2, 6)$

2. a. $(5, -1)$ **b.** $(2, 3)$ **c.** $(-3, 1)$

3. a. $(-3, 5)$ **b.** $(-5, -8)$ **c.** $(2, -2)$

4. a. $\left(1\dfrac{1}{2}, 3\dfrac{1}{2}\right)$ **b.** $\left(2, 1\dfrac{4}{5}\right)$ **c.** $(1, 1)$

5. a. $(1, 3)$ **b.** $(2, 4)$ **c.** $(5, 2)$

6. a. $(4, 2)$ **b.** $(-3, 4)$ **c.** $\left(-3, -1\dfrac{1}{2}\right)$

7. a. $(-6, -5)$ **b.** $(-3, 5)$ **c.** $(2, 1.8)$

8. a. $(6, 3)$ **b.** $(2, -2)$ **c.** $(1, 3)$

9. a. $(-1.5, -3)$ **b.** $(-8, 18)$ **c.** $(-3, 5)$

10. a. $(1, 3)$ **b.** $(4, 0)$ **c.** $(-3, 5)$

11. a. $(4, 3)$ **b.** $(8, 5)$ **c.** $\left(\dfrac{1}{3}, -\dfrac{1}{3}\right)$

12. a. $(4, 5)$ **b.** $(1, -3)$ **c.** $(12, 12)$

13. \$3.40 (coffee is \$5.10)

14. 12 5-cent coins and 22 10-cent coins.

15. Abena 61 kg, Bashir 58 kg, Cecily 54 kg.

16. a. i. $acx + bcy = ce$ (3)
      ii. $acx + ady = af$ (4)
      iii. $y = \dfrac{ce - af}{bc - ad}$
   b. $x = \dfrac{de - bf}{ad - bc}$
   c. i. $\left(\dfrac{106}{31}, \dfrac{1}{31}\right)$ ii. $\left(\dfrac{37}{14}, \dfrac{11}{14}\right)$
   d. Because you cannot divide by 0.
   e. $ad - bc \neq 0$

17. It is not possible. When the two equations are set up it is impossible to eliminate one variable without eliminating the other.

18. $k - 10, 10$

19. $x = 7, y = -3$

20. $x = 4, y = 3, z = -6$

5.5 Applications of simultaneous linear equations

1. Maths mark = 97, English mark = 66.
2. 18 nuts, 12 bolts.
3. 800 sheep, 400 chickens.
4. 8 and 3
5. 9 and 7
6. 6 and 5
7. Length = 12 m and width = 8 m.
8. Lemons cost 55 cents and oranges cost 25 cents.
9. Length 60 m and width 20 m.
10. Eight 20-cent coins and three 50-cent coins.
11. Twelve $1 coins and nine $2 coins.
12. Paddlepops cost $1.20 and a Magnum costs $2.10.
13. Cost of the Killer Python = 35 cents and cost of the Red frog = 25 cents.
14. Fixed costs = $87, cost per person = $23.50.
15. PE mark is 83 and Science mark is 71.
16. Mozzarella costs $6.20, Swiss cheese costs $5.80.
17. $x = 3$ and $y = 4$.
18. Fixed costs = $60, cost per person = $25.

19. $4 each for DVDs and $24 each for zip disks.
20. 9 and 24 years old.
21. Child $12.50, Adult $18.25, Elderly $15.
22. 6.5 km
23. 66 cups of hot chips, 33 meat pies and 22 hot dogs were sold during the half-hour period.
24. a. See the graph at the bottom of the page.*
    b. Jet ski 3 wins the race.
    c. Jet ski 1 and 2 reach the destination at the same time although jet ski 2 started 2 hours after jet ski 1. Jet ski 3 overtakes jet ski 1 6 hours and 40 minutes after its race begins or 10 hours and 40 minutes after jet ski 1 starts the race. Jet ski 3 overtakes jet ski 2 6 hours after it starts the race or 8 hours after jet ski 2 started the race.
25. 176 km

Project

1. A1 Rentals: $C = \$35 \times 3 + 0.28d$
   Cut Price Rentals: $C = \$28 \times 3 + 0.3d$
2. See the graph at the bottom of the page.*
3. 1050 km
4. 1400 km

*24.a.

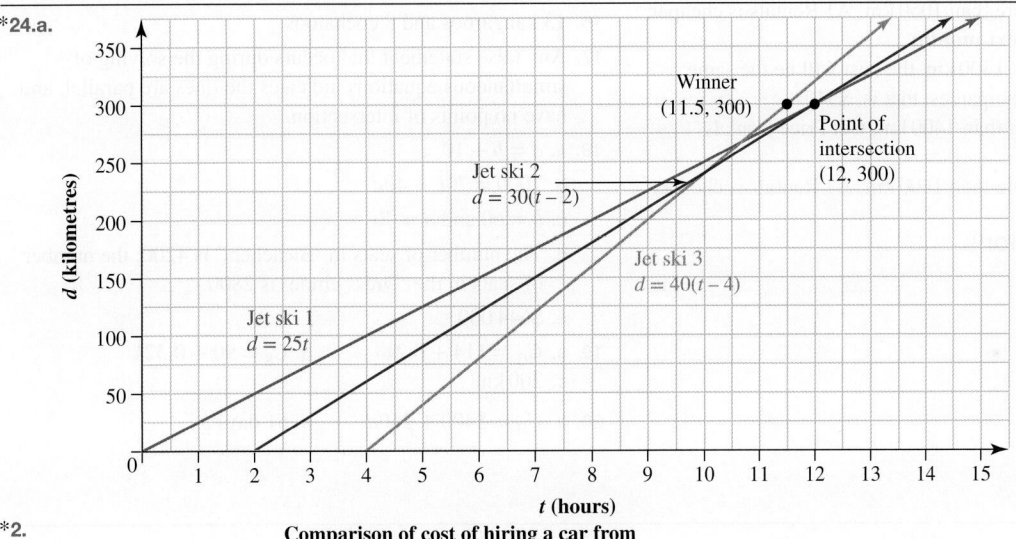

*2.

Comparison of cost of hiring a car from A1 Rentals and Cut Price Rentals

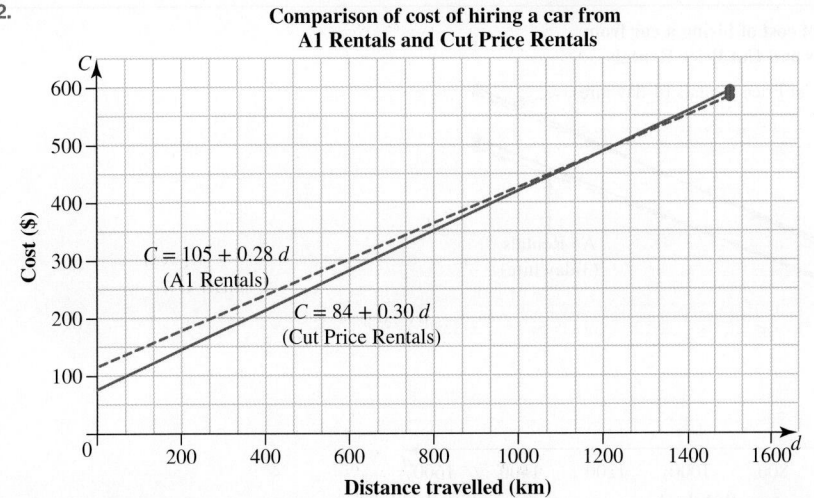

$C = 105 + 0.28\,d$
(A1 Rentals)

$C = 84 + 0.30\,d$
(Cut Price Rentals)

5. a. A1 Rentals: $C = \$37 \times 3 + 0.28d$
   Cut Price Rentals: $C = \$30 \times 3 + 0.3d$

   b. A1 Rentals: $C = \$37 \times 4 + 0.28d$
   Cut Price Rentals: $C = \$30 \times 4 + 0.3d$

6. See the graph at the bottom of the page.*

7. The extra cost of $2 per day for both rental companies has not affected the charges they make for the distances travelled. However, the overall costs have increased.

8. Presentation of the answers will vary. Answers will include:
   Travelling 3 days this month:
   - If Andrea travels 1050 km, the cost will be the same for both rental companies; that is, $399.
   - If she travels less than 1050 km, Cut Price Rentals is cheaper.
   - If she travels more than 1050 km, A1 Rentals is cheaper.
   Travelling 4 days this month:
   - If Andrea travels 1400 km, the cost will be the same for both rental companies; that is, $532.
   - If she travels less than 1400 km, Cut Price Rentals is cheaper.
   - If she travels more than 1400 km, A1 Rentals is cheaper.
   Travelling 3 days next month:
   - If Andrea travels 1050 km, the cost will be the same for both rental companies; that is, $405.
   - If she travels less than 1050 km, Cut Price Rentals is cheaper.
   - If she travels more than 1050 km, A1 Rentals is cheaper.
   Travelling 4 days next month:
   - If Andrea travels 1400 km, the cost will be the same for both rental companies; that is, $540.
   - If she travels less than 1400 km, Cut Price Rentals is cheaper.
   - If she travels more than 1400 km, A1 Rentals is cheaper.

5.6 Review questions

1. $m = -\dfrac{3}{4}$

2. B

3. D

4. A

5. True

6. a. No   b. Yes

7. a. $(-2, 1)$ b. $(0, -2)$
   c. $(-7.5, 7)$

8. a. $(3, 1)$ b. $(2, 3)$

9. a. $(2, 7)$ b. $(-5, -3)$
   c. $(-2, 2)$ d. $\left(-\dfrac{7}{3}, \dfrac{7}{3}\right)$
   e. $(-14, -53)$ f. $\left(\dfrac{5}{2}, -7\right)$

10. a. $(5, 2)$ b. $(-2, 3)$
    c. $(-3, -1)$ d. $(1, 3)$
    e. $(2, -2)$ f. $(4, 2)$

11. a. $(0, 3)$ b. $(-3, -3)$
    c. $(2, 1)$

12. D

13. $p = \dfrac{21}{2}$

14. a. The numbers are 9 and 14.
    b. Length $= 11$ m, width $= 6$ m.
    c. Chupa Chups cost 45 cents and Wizz Fizzes cost 55 cents.

15. Milk $1.75, bread $2.35.

16. 13 kangaroos and 8 cockatoos.

17. Any false statement that occurs during the solving of simultaneous equations indicates the lines are parallel, and have no points of intersection.

18. a. $d = b + 10$
    b. $7000 = 70b + 40d$
    c. $b = 60$ and $d = 70$
    d. The number of seats in 'Bleachers' is 4200; the number of seats in the 'Dress circle' is 2800.
    e. $644\,000

19. a. $C_G = 114 + 0.2k$ b. $C_S = 90 + 0.32k$
    c. 200 km

20. a. $C_H = 5400 + 260d$ b. 61 days

*6.

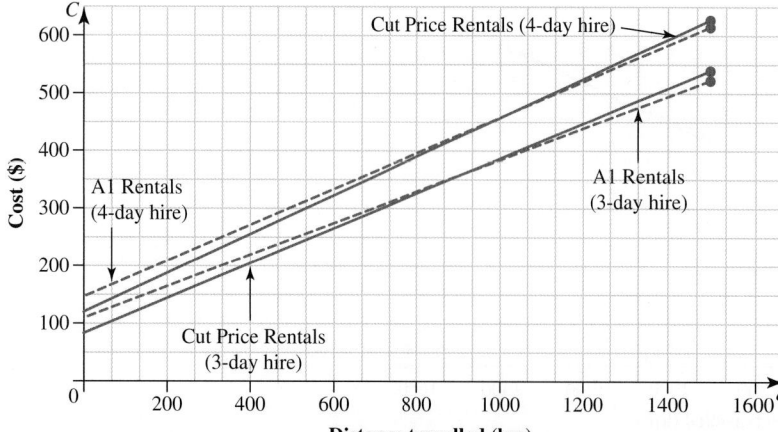

Comparison of cost of hiring a car from A1 Rentals and Cut Price Rentals

Cut Price Rentals (4-day hire)
A1 Rentals (4-day hire)
A1 Rentals (3-day hire)
Cut Price Rentals (3-day hire)

Cost ($) — vertical axis: 100, 200, 300, 400, 500, 600
Distance travelled (km) — horizontal axis: 200, 400, 600, 800, 1000, 1200, 1400, 1600

6 Quadratic equations (Path)

LEARNING SEQUENCE

LESSON
6.1 Overview

Why learn this?

Have you ever thought about the shape a ball makes as it flies through the air? Or have you noticed the shape that the stream of water from a drinking fountain makes? These are both examples of quadratic equations in the real world. When you learn about quadratic equations, you learn about the mathematics of these real-world shapes, but quadratic equations are so much more than interesting shapes.

Being able to use and understand quadratic equations lets you unlock incredible problem-solving skills. It allows you to quickly understand situations and solve complicated problems that would be close to impossible without these skills. Many professionals rely on their understanding of quadratic equations to make important decisions, designs and discoveries. Some examples include: sports people finding the perfect spot to intercept a ball, architects designing complex buildings such as the Sydney Opera House and scientists listening to sound waves coming from the furthest regions of space using satellite dishes. All of these professionals use the principles of quadratic equations.

By learning about quadratic equations you are also embarking on your first step on the path to learning about non-linear equations. When you learn about non-linear equations, a whole world of differently shaped curves and different problem-solving skills will open up.

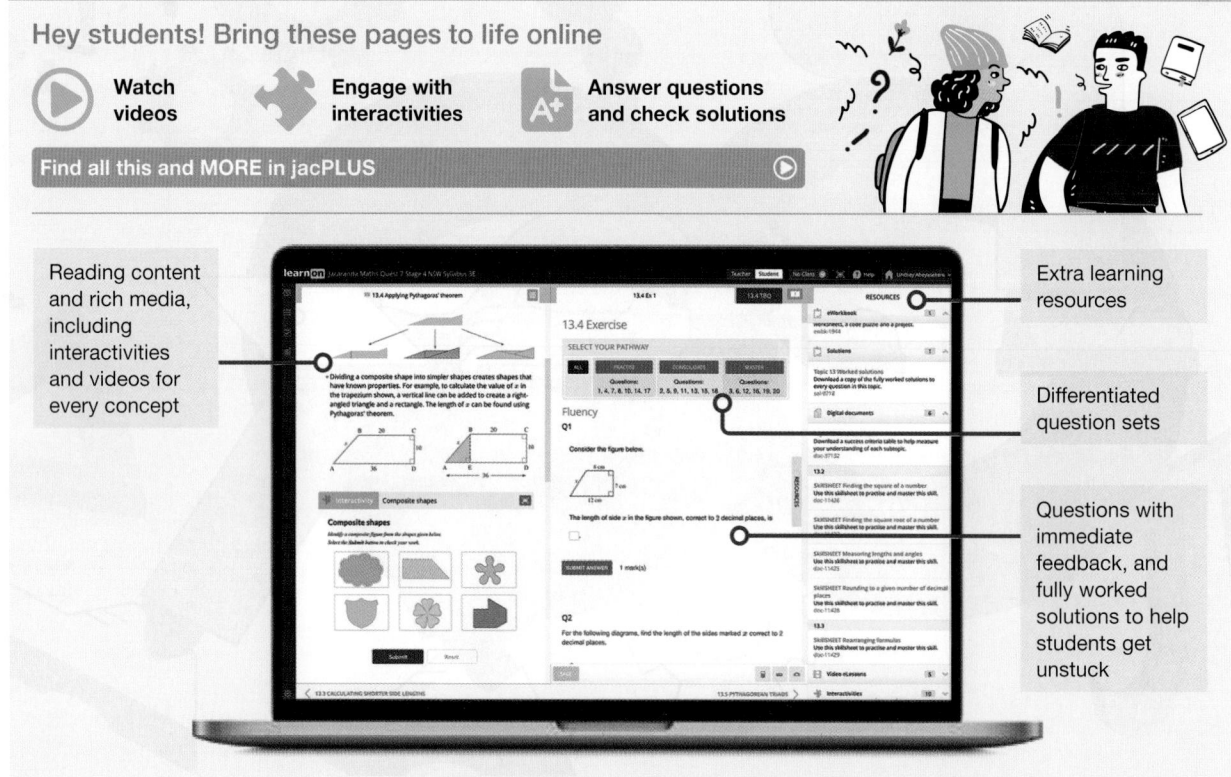

1. State whether the following statement is true or false.
   The Null Factor Law states that if the product of two numbers is zero then one or both of the numbers must equal zero.

2. Calculate the two solutions for the equation $(2x - 1)(x + 5) = 0$.

3. Solve $16x^2 - 9 = 0$ for x.

4. **MC** The solutions for $x(x - 3)(x + 2) = 0$ are:
   A. $x = 3$ or $x = -2$
   B. $x = 3$ or $x = -2$
   C. $x = 0$ or $x = -3$ or $x = 2$
   D. $x = 0$ or $x = 3$ or $x = -2$

5. **MC** The solutions to the equation $x^2 + x - 6 = 0$ are:
   A. $x = 3$ or $x = -2$
   B. $x = -3$ or $x = 2$
   C. $x = 3$ or $x = 2$
   D. $x = -6$ or $x = 1$

6. **MC** Four times a number is subtracted from 3 times its square. If the result is 4, the possible numbers are:
   A. $x = -4$
   B. $x = 0$ or $x = -4$
   C. $x = 0$ or $x = 4$
   D. $x = -\dfrac{2}{3}$ or $x = 2$

7. **MC** An exact solution to the equation $x^2 - 3x - 1 = 0$ is:

   A. $x = \dfrac{3 \pm \sqrt{13}}{2}$
   B. $x = \dfrac{-3 \pm \sqrt{13}}{2}$
   C. $x = \dfrac{3 \pm 2\sqrt{2}}{2}$
   D. $x = \dfrac{-3 \pm 2\sqrt{2}}{2}$

8. Calculate the discriminant for $5x^2 - 8x + 2 = 0$.

9. Match the discriminant value to the number of solutions for a quadratic equation.

Discriminant value	Number of solutions
a. $\Delta = 0$	No real solutions
b. $\Delta < 0$	Two solutions
c. $\Delta > 0$	Only one solution

10. **MC** If $\Delta = 0$, the graph of a quadratic:
    A. does not cross or touch the x-axis
    B. touches the x-axis
    C. intersects the x-axis twice
    D. intersects the x-axis three times.

11. **MC** The equation $3x^2 - 27 = 0$ has:
    A. two rational solutions
    B. two irrational solutions
    C. one solution
    D. one rational and one irrational solution.

12. A rectangle's length is 4 cm more than its width. Determine the dimensions of the rectangle if its area is $(4x + 9)\,\text{cm}^2$.

13. Calculate the value of m for which $mx^2 - 12x + 9 = 0$ has one solution.

14. ▆MC▆ Identify for what values of a the straight line $y = ax - 12$ intersects once with the parabola $y = x^2 - 2x - 8$.

    A. $a = -6$ or $a = 2$ **B.** $a = 6$ or $a = -2$

    C. $a = 4$ or $a = -2$ **D.** $a = -4$ or $a = 2$

15. ▆MC▆ Solve the following for x.

$$\frac{x}{x-2} - \frac{x+1}{x+4} = 1$$

    A. $x = -1$ or $x = -4$ **B.** $x = 0$ or $x = 2$

    C. $x = 2$ or $x = -4$ **D.** $x = 5$ or $x = -2$

LESSON
6.2 Solving quadratic equations algebraically

LEARNING INTENTION

At the end of this lesson you should be able to:
- solve quadratic equations using the Null Factor Law
- solve quadratic equations by completing the square
- solve equations that are reducible to quadratics
- solve worded questions algebraically.

▶ 6.2.1 Quadratic equations

eles-4843

- **Quadratic equations** are equations in the form:
$$ax^2 + bx + c = 0, \text{ where } a, b \text{ and } c \text{ are numbers.}$$
- For example, the equation $3x^2 + 5x - 7 = 0$ has $a = 3, b = 5$ and $c = -7$.

The Null Factor Law

- We know that any number multiplied by zero will equal zero.

> **The Null Factor Law**
>
> **If $a \times b = 0$ then:**
>
> $a = 0,\ b = 0,$ **or both a and b equal 0.**

- To apply the Null Factor Law, quadratic equations must be in a factorised form.
- To review factorising quadratic expressions, see topic 3, Algebraic techniques.

WORKED EXAMPLE 1 Applying the Null Factor Law to solve quadratic equations

Solve the equation $(x-7)(x+11) = 0$.

THINK	WRITE
1. Write the equation and check that the right-hand side equals zero. (The product of the two numbers is zero.)	$(x-7)(x+11) = 0$
2. The left-hand side is factorised, so apply the Null Factor Law.	$x-7 = 0$ or $x+11 = 0$
3. Solve for x.	$x = 7$ or $x = -11$

WORKED EXAMPLE 2 Factorising then applying the Null Factor Law

Solve each of the following equations.
a. $x^2 - 3x = 0$
b. $3x^2 - 27 = 0$
c. $x^2 - 13x + 42 = 0$
d. $36x^2 - 21x = 2$

THINK	WRITE
a. 1. Write the equation. Check that the right-hand side equals zero.	**a.** $x^2 - 3x = 0$
2. Factorise by taking out the common factor of x^2 and $3x$, which is x.	$x(x-3) = 0$
3. Apply the Null Factor Law.	$x = 0$ or $x - 3 = 0$
4. Solve for x.	$x = 0$ $\quad\quad x = 3$
b. 1. Write the equation. Check that the right-hand side equals zero.	**b.** $\quad 3x^2 - 27 = 0$
2. Factorise by taking out the common factor of $3x^2$ and 27, which is 3.	$3(x^2 - 9) = 0$
3. Factorise using the difference of two squares rule.	$3(x^2 - 3^2) = 0$ $\quad 3(x+3)(x-3) = 0$
4. Apply the Null Factor Law.	$x+3 = 0$ or $x - 3 = 0$
5. Solve for x.	$x = -3$ $\quad\quad x = 3$
	(Alternatively, $x = \pm 3$.)
c. 1. Write the equation. Check that the right-hand side equals zero.	**c.** $x^2 - 13x + 42 = 0$
2. Factorise by identifying a factor pair of 42 that adds to -13.	

Factors of 42	Sum of factors
-6 and -7	-13

THINK	WRITE
3. Use the Null Factor Law to write two linear equations.	$(x-6)(x-7) = 0$
4. Solve for x.	$x-6 = 0$ or $x-7 = 0$ $\quad x = 6 \quad\quad x = 7$

d. 1. Write the equation. Check that the right-hand side equals zero. (It does not.)

d. $36x^2 - 21x = 2$

2. Rearrange the equation so the right-hand side of the equation equals zero as a non-monic quadratic trinomial in the form $ax^2 + bx + c = 0$.

$36x^2 - 21x - 2 = 0$

3. Recognise that the expression to factorise is a non-monic quadratic trinomial. Calculate $ac = 36 \times -2 = -72$. Factorise by identifying a factor pair of -72 that adds to -21.

Factors of -72	Sum of factors
3 and -24	-21

$36x^2 - 24x + 3x - 2 = 0$

4. Factorise the expression.

$12x(3x - 2) + 1(3x - 2) = 0$
$(3x - 2)(12x + 1) = 0$

5. Use the Null Factor Law to write two linear equations.

$3x - 2 = 0$ or $12x + 1 = 0$
$3x = 2$ $12x = -1$

6. Solve for x.

$x = \dfrac{2}{3}$ $x = -\dfrac{1}{12}$

▶ 6.2.2 Solving quadratic equations by completing the square

eles-4844

- Completing the square is another technique that can be used for factorising quadratics.
- Use completing the square when other techniques will not work.
- To complete the square, the value of a must be 1. If a is not 1, divide everything by the coefficient of x^2 so that $a = 1$.
- Steps to complete the square:

Factorise $x^2 + 6x + 2 = 0$.

Step 1: $ax^2 + bx + c = 0$. Since $a = 1$ in this example, we can complete the square.

Step 2: Add and subtract $\left(\dfrac{1}{2}b\right)^2$. In this example, $b = 6$.

$$x^2 + 6x + \left(\dfrac{1}{2} \times 6\right)^2 + 2 - \left(\dfrac{1}{2} \times 6\right)^2 = 0$$
$$x^2 + 6x + 9 + 2 - 9 = 0$$

Step 3: Factorise the first three terms.

$$x^2 + 6x + 9 - 7 = 0$$
$$(x + 3)^2 - 7 = 0$$

Step 4: Factorise the quadratic by using difference of two squares

$$(x + 3)^2 - \left(\sqrt{7}\right)^2 = 0 \text{ because } \left(\sqrt{7}\right)^2 = 7$$
$$\left(x + 3 + \sqrt{7}\right)\left(x + 3 - \sqrt{7}\right) = 0$$

When $a \neq 1$:
- factorise the expression if possible before completing the square
- if a is not a factor of each term, divide each term by a to ensure the coefficient of x^2 is 1.

Completing the square

$$x^2 \pm px = \left[x^2 \pm px + \left(\frac{p}{2} \right)^2 \right] - \left(\frac{p}{2} \right)^2$$

$$= \left(x \pm \frac{p}{2} \right)^2 - \left(\frac{p}{2} \right)^2$$

The expression $x^2 \pm px$ becomes a difference of two squares.

WORKED EXAMPLE 3 Solving quadratic equations by completing the square

Solve the equation $x^2 + 2x - 4 = 0$ by completing the square. Give exact answers.

THINK	WRITE
1. Write the equation.	$x^2 + 2x - 4 = 0$
2. Identify the coefficient of x, halve it and square the result.	$\left(\frac{1}{2} \times 2 \right)^2$
3. Add the result of step **2** to the equation, placing it after the x-term. To balance the equation, we need to subtract the same amount as we have added.	$x^2 + 2x + \left(\frac{1}{2} \times 2 \right)^2 - 4 - \left(\frac{1}{2} \times 2 \right)^2 = 0$ $x^2 + 2x + (1)^2 - 4 - (1)^2 = 0$ $x^2 + 2x + 1 - 4 - 1 = 0$
4. Insert brackets around the first three terms to group them and then simplify the remaining terms.	$(x^2 + 2x + 1) - 5 = 0$
5. Factorise the first three terms to produce a perfect square.	$(x + 1)^2 - 5 = 0$
6. Express as the difference of two squares and then factorise.	$(x + 1)^2 - \left(\sqrt{5} \right)^2 = 0$ $\left(x + 1 + \sqrt{5} \right) \left(x + 1 - \sqrt{5} \right) = 0$
7. Apply the Null Factor Law to identify linear equations.	$x + 1 + \sqrt{5} = 0 \quad \text{or} \quad x + 1 - \sqrt{5} = 0$
8. Solve for x. Keep the answer in surd form to provide an exact answer.	$x = -1 - \sqrt{5} \quad \text{or} \quad x = -1 + \sqrt{5}$ (Alternatively, $x = -1 \pm \sqrt{5}$.)

▶ 6.2.3 Equations that reduce to quadratic form

eles-6265

- Substitution techniques can be applied to reduce equations such as $ax^4 + bx^2 + c = 0$ to quadratic form.
- The equation $ax^4 + bx^2 + c = 0$ can be expressed in the form

$$a(x^2)^2 + b(x^2) + c = 0$$

Letting $u = x^2$, it becomes $au^2 + bu + c = 0$, a quadratic equation in the variable u.
- Solve for u, then substitute $u = x^2$ and solve for x.
- Not all values of u will give values of x, since $x^2 \geq 0$.

Solve the following equations for x.

a. $x^4 - 10x^2 + 9 = 0$
b. $4x^4 - 35x^2 - 9 = 0$

THINK		WRITE
a. 1.	Use an appropriate substitution to reduce the given equation to quadratic form.	a. $x^4 - 10x^2 + 9 = 0$ Let $u = x^2$. $u^2 - 10u + 9 = 0$
2.	Solve for u by factorising and applying the Null Factor Law.	$(u-9)(u-1) = 0$ $\therefore u = 9 \quad \text{or} \quad u = 1$
3.	Substitute back, replacing u with x^2.	$x^2 = 9 \qquad x^2 = 1$
4.	Solve the equations for x.	$x = \pm 3 \qquad x = \pm 1$
b. 1.	Use an appropriate substitution to reduce the given equation to quadratic form.	b. $4x^4 - 35x^2 - 9 = 0$ Let $u = x^2$. $4u^2 - 35u - 9 = 0$
2.	Solve for u by factorising and applying the Null Factor Law.	$(4u + 1)(u - 9) = 0$ $\therefore u = -\dfrac{1}{4} \quad \text{or} \quad u = 9$
3.	Substitute back, replacing u with x^2.	$x^2 = -\dfrac{1}{4} \qquad x^2 = 9$
4.	Since x^2 cannot be negative, any negative value of u needs to be rejected.	Reject $x^2 = -\dfrac{1}{4}$, since there are no real solutions.
5.	Solve the remaining equation for x.	$x^2 = 9$ $x = \pm\sqrt{9}$ $x = \pm 3$

▶ 6.2.4 Solving worded questions

eles-4845

- For worded questions:
  - start by identifying the unknowns
  - then write the equation and solve it
  - give answers in a full sentence and include units if required.

When two consecutive numbers are multiplied together, the result is 20. Determine the numbers.

THINK	WRITE
1. Define the unknowns. First number $= x$, second number $= x + 1$.	Let the two numbers be x and $(x + 1)$.
2. Write an equation using the information given in the question.	$x(x + 1) = 20$
3. Transpose the equation so that the right-hand side equals zero.	$x(x + 1) - 20 = 0$

4. Expand to remove the brackets.	$x^2 + x - 20 = 0$
5. Factorise.	$(x + 5)(x - 4) = 0$
6. Apply the Null Factor Law to solve for x.	$x + 5 = 0$ or $x - 4 = 0$
	$x = -5$ $x = 4$
7. Use the answer to determine the second number.	If $x = -5, x + 1 = -4$.
	If $x = 4, x + 1 = 5$.
8. Check the solutions.	Check:
	$4 \times 5 = 20$ $(-5) \times (-4) = 20$
9. Write the answer in a sentence.	The numbers are 4 and 5 or -5 and -4.

WORKED EXAMPLE 6 Solving worded application questions

The height of a football after being kicked is determined by the formula $h = -0.1d^2 + 3d$, where d is the horizontal distance from the player in metres.

a. Calculate how far the ball is from the player when it hits the ground.
b. Calculate the horizontal distance the ball has travelled when it first reaches a height of 20 m.

THINK

a. 1. Write the formula.

2. The ball hits the ground when $h = 0$. Substitute $h = 0$ into the formula.

3. Factorise.

4. Apply the Null Factor Law and simplify.

5. Interpret the solutions.

6. Write the answer in a sentence.

b. 1. The height of the ball is 20 m, so, substitute $h = 20$ into the formula.

2. Transpose the equation so that zero is on the right-hand side.

3. Multiply both sides of the equation by 10 to remove the decimal from the coefficient.

WRITE

a. $h = -0.1d^2 + 3d$

$-0.1d^2 + 3d = 0$

$-0.1d^2 + 3d = 0$
$d(-0.1d + 3) = 0$

$d = 0$ or $-0.1d + 3 = 0$
$-0.1d = -3$
$d = \dfrac{-3}{-0.1}$
$= 30$

$d = 0$ is the origin of the kick.
$d = 30$ is the distance from the origin that the ball has travelled when it lands.

The ball is 30 m from the player when it hits the ground.

b. $h = -0.1d^2 + 3d$
$20 = -0.1d^2 + 3d$

$0.1d^2 - 3d + 20 = 0$

$d^2 - 30d + 200 = 0$

4. Factorise.

$(d-20)(d-10)=0$

5. Apply the Null Factor Law.

$d-20=0 \quad \text{or} \quad d-10=0$

6. Solve.

$d=20 \qquad\qquad d=10$

7. Interpret the solution. The ball reaches a height of 20 m on the way up and on the way down. The *first* time the ball reaches a height of 20 m is the smaller value of d. Write the answer in a sentence.

The ball first reaches a height of 20 m after it has travelled a distance of 10 m.

 Resources

 Video eLesson The Null Factor Law (eles-2312)

 Interactivity The Null Factor Law (int-6095)

Exercise 6.2 Solving quadratic equations algebraically

learnon

6.2 Quick quiz on	6.2 Exercise

Individual pathways

■ PRACTISE	■ CONSOLIDATE	■ MASTER
1, 4, 7, 10, 13, 16, 17, 18, 21, 27, 30, 33, 34, 38, 42, 47	2, 5, 8, 11, 14, 19, 22, 24, 25, 28, 31, 35, 36, 39, 43, 44, 48	3, 6, 9, 12, 15, 20, 23, 26, 29, 32, 37, 40, 41, 45, 46, 49, 50

Fluency

 For questions 1–6, solve each of the following equations.

1. a. $(x+7)(x-9)=0$
   b. $(x-3)(x+2)=0$
   c. $(x-2)(x-3)=0$
   d. $x(x-3)=0$

2. a. $x(x-1)=0$
   b. $x(x+5)=0$
   c. $2x(x-3)=0$
   d. $9x(x+2)=0$

3. a. $\left(x-\dfrac{1}{2}\right)\left(x+\dfrac{1}{2}\right)=0$
   b. $-(x+1.2)(x+0.5)=0$
   c. $2(x-0.1)(2x-1.5)=0$
   d. $\left(x+\sqrt{2}\right)\left(x-\sqrt{3}\right)=0$

4. a. $(2x-1)(x-1)=0$
   b. $(3x+2)(x+2)=0$
   c. $(4x-1)(x-7)=0$

5. a. $(7x+6)(2x-3)=0$
   b. $(5x-3)(3x-2)=0$
   c. $(8x+5)(3x-2)=0$

6. a. $x(x-3)(2x-1)=0$
   b. $x(2x-1)(5x+2)=0$
   c. $x(x+3)(5x-2)=0$

WE2a For questions 7–9, solve each of the following equations.

7. a. $x^2-2x=0$
   b. $x^2+5x=0$
   c. $x^2=7x$

8. a. $3x^2=-2x$
   b. $4x^2-6x=0$
   c. $6x^2-2x=0$

9. **a.** $4x^2 - 2\sqrt{7}x = 0$ **b.** $3x^2 + \sqrt{3}x = 0$ **c.** $15x - 12x^2 = 0$

WE2b For questions **10–12**, solve each of the following equations.

10. **a.** $x^2 - 4 = 0$ **b.** $x^2 - 25 = 0$
    c. $3x^2 - 12 = 0$ **d.** $4x^2 - 196 = 0$

11. **a.** $9x^2 - 16 = 0$ **b.** $4x^2 - 25 = 0$
    c. $9x^2 = 4$ **d.** $36x^2 = 9$

12. **a.** $x^2 - \dfrac{1}{25} = 0$ **b.** $\dfrac{1}{36}x^2 - \dfrac{4}{9} = 0$

    c. $x^2 - 5 = 0$ **d.** $9x^2 - 11 = 0$

WE2c For questions **13–15**, solve each of the following equations.

13. **a.** $x^2 - x - 6 = 0$ **b.** $x^2 + 6x + 8 = 0$
    c. $x^2 - 6x - 7 = 0$ **d.** $x^2 - 8x + 15 = 0$

14. **a.** $x^2 - 3x - 4 = 0$ **b.** $x^2 - 10x + 25 = 0$ **c.** $x^2 - 3x - 10 = 0$
    d. $x^2 - 8x + 12 = 0$ **e.** $x^2 - 4x - 21 = 0$

15. **a.** $x^2 - x - 30 = 0$ **b.** $x^2 - 7x + 12 = 0$ **c.** $x^2 - 8x + 16 = 0$
    d. $x^2 + 10x + 25 = 0$ **e.** $x^2 - 20x + 100 = 0$

16. **MC** The solutions to the equation $x^2 + 9x - 10 = 0$ are:
    A. $x = 1$ and $x = 10$ **B.** $x = 1$ and $x = -10$
    C. $x = -1$ and $x = 10$ **D.** $x = -1$ and $x = -10$

17. **MC** The solutions to the equation $x^2 - 100 = 0$ are:
    A. $x = 0$ and $x = 10$ **B.** $x = 0$ and $x = -10$
    C. $x = -10$ and $x = 10$ **D.** $x = 0$ and $x = 100$

WE2d For questions **18–20**, solve each of the following equations.

18. **a.** $2x^2 - 5x = 3$ **b.** $3x^2 + x - 2 = 0$
    c. $5x^2 + 9x = 2$ **d.** $6x^2 - 11x + 3 = 0$

19. **a.** $14x^2 - 11x = 3$ **b.** $12x^2 - 7x + 1 = 0$
    c. $6x^2 - 7x = 20$ **d.** $12x^2 + 37x + 28 = 0$

20. **a.** $10x^2 - x = 2$ **b.** $6x^2 - 25x + 24 = 0$
    c. $30x^2 + 7x - 2 = 0$ **d.** $3x^2 - 21x = -36$

WE3 For questions **21–26**, solve the following equations by completing the square. Give exact answers.

21. **a.** $x^2 - 4x + 2 = 0$ **b.** $x^2 + 2x - 2 = 0$ **c.** $x^2 + 6x - 1 = 0$

22. **a.** $x^2 - 8x + 4 = 0$ **b.** $x^2 - 10x + 1 = 0$ **c.** $x^2 - 2x - 2 = 0$

23. **a.** $x^2 + 2x - 5 = 0$ **b.** $x^2 + 4x - 6 = 0$ **c.** $x^2 + 4x - 11 = 0$

24. **a.** $x^2 - 3x + 1 = 0$ **b.** $x^2 + 5x - 1 = 0$ **c.** $x^2 - 7x + 4 = 0$

25. **a.** $x^2 - 5 = x$ **b.** $x^2 - 11x + 1 = 0$ **c.** $x^2 + x = 1$

26. **a.** $x^2 + 3x - 7 = 0$ **b.** $x^2 - 3 = 5x$ **c.** $x^2 - 9x + 4 = 0$

For questions **27–29**, solve each of the following equations, rounding answers to 2 decimal places.

27. **a.** $2x^2 + 4x - 6 = 0$ **b.** $3x^2 + 12x - 3 = 0$ **c.** $5x^2 - 10x - 15 = 0$

28. **a.** $4x^2 - 8x - 8 = 0$ **b.** $2x^2 - 6x + 2 = 0$ **c.** $3x^2 - 9x - 3 = 0$

29. **a.** $5x^2 - 15x - 25 = 0$ **b.** $7x^2 + 7x - 21 = 0$ **c.** $4x^2 + 8x - 2 = 0$

WE4 For questions **30–32**, solve each of the following equations using a suitable substitution.

30. **a.** $x^4 - 13x^2 + 36 = 0$ **b.** $x^4 - 5x^2 + 4 = 0$ **c.** $x^4 - x^2 - 12 = 0$

31. **a.** $x^4 - 17x^2 + 16 = 0$ **b.** $x^4 - 7x^2 - 18 = 0$ **c.** $x^4 + 5x^2 + 4 = 0$

32. **a.** $x^4 - 29x^2 + 100 = 0$ **b.** $4x^4 + 11x^2 - 3 = 0$ **c.** $(3x + 4)^2 + 9(3x + 4) - 10 = 0$

Understanding

33. **WE5** When two consecutive numbers are multiplied, the result is 72. Determine the numbers.

34. When two consecutive even numbers are multiplied, the result is 48. Determine the numbers.

35. When a number is added to its square the result is 90. Determine the number.

36. Twice a number is added to three times its square. If the result is 16, determine the number.

37. Five times a number is added to two times its square. If the result is 168, determine the number.

38. **WE6** A soccer ball is kicked. The height, h, in metres, of the soccer ball t seconds after it is kicked can be represented by the equation $h = -t(t - 6)$. Calculate how long it takes for the soccer ball to hit the ground again.

39. The length of an Australian flag is twice its width and the diagonal length is 45 cm.

     a. If x cm is the width of the flag, express its length in terms of x.
     b. Draw a diagram of the flag marking in the diagonal. Label the length and the width in terms of x.
     c. Use Pythagoras' theorem to write an equation relating the lengths of the sides to the length of the diagonal.
     d. Solve the equation to calculate the dimensions of the Australian flag. Round your answer to the nearest cm.

40. If the length of a paddock is 2 m more than its width and the area is 48 m^2, calculate the length and width of the paddock.

41. Solve for x.

     a. $x + 5 = \dfrac{6}{x}$ **b.** $x = \dfrac{24}{x - 5}$ **c.** $x = \dfrac{1}{x}$

Communicating, reasoning and problem solving

42. The sum of the first n numbers $1, 2, 3, 4 \ldots n$ is given by the formula $S = \dfrac{n(n+1)}{2}$.

     a. Use the formula to calculate the sum of the first 6 counting numbers.
     b. Determine how many numbers are added to give a sum of 153.

43. If these two rectangles have the same area, determine the value of x.

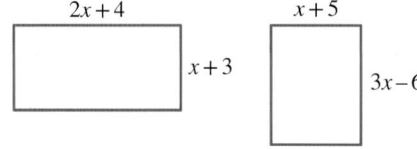

44. Henrietta is a pet rabbit who lives in an enclosure that is 2 m wide and 4 m long. Her human family has decided to purchase some more rabbits to keep her company and so the size of the enclosure must be increased.

　　a. Draw a diagram of Henrietta's enclosure, clearly marking the lengths of the sides.
　　b. If the length and width of the enclosure are increased by x m, express the new dimensions in terms of x.
　　c. If the new area is to be 24 m^2, write an equation relating the sides and the area of the enclosure (area = length × width).
　　d. Use the equation to calculate the value of x and, hence, the length of the sides of the new enclosure. Justify your answer.

45. The cost per hour, C, in thousands of dollars of running two cruise ships, *Annabel* and *Betty*, travelling at a speed of s knots is given by the following relationships.
$$C_{Annabel} = 0.3s^2 + 4.2s + 12 \text{ and } C_{Betty} = 0.4s^2 + 3.6s + 8$$
　　a. Determine the cost per hour for each ship if they are both travelling at 28 knots.
　　b. Calculate the speed in knots at which both ships must travel for them to have the same cost.
　　c. Explain why only one of the solutions obtained in your working for part **b** is valid.

46. Explain why the equation $x^2 + 4x + 10 = 0$ has no real solutions.

47. Solve $(x^2 - x)^2 - 32(x^2 - x) + 240 = 0$ for x.

48. Solve $\dfrac{3z^2 - 35}{16} - z = 0$ for z.

49. A garden measuring 12 metres by 16 metres is to have a pedestrian pathway installed all around it, increasing the total area to 285 square metres. Determine the width of the pathway.

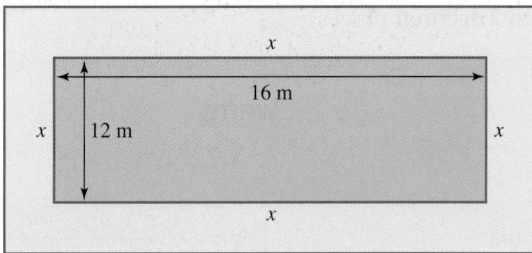

50. Solve $2\left(1 + \dfrac{3}{x}\right)^2 + 5\left(1 + \dfrac{3}{x}\right) + 3 = 0$ for x.

LESSON
6.3 The quadratic formula

LEARNING INTENTION

At the end of this lesson you should be able to:
- solve equations using the quadratic formula
- solve worded problems involving quadratic equations.

▶ 6.3.1 Using the quadratic formula

eles-4846

- The **quadratic formula** can be used to solve any equations in the form:

$$ax^2 + bx + c = 0$$

where a, b and c are numbers.

- To use the quadratic formula:
  - identify the values for a, b and c
  - substitute the values into the quadratic formula.

The quadratic formula

The solutions of quadratic equation $y = ax^2 + bx + c$ are:

$$x = \frac{-b \pm \sqrt{b^2 - 4ac}}{2a}$$

- Note that the derivation of this formula uses completing the square and is beyond the scope of this course.
- Note that the $\pm$ symbol is called the plus–minus sign. To use it, identify the first solution by treating it as a plus, then identify the second solution by treating it as a minus.
- There will be no real solutions if the value in the square root is negative; that is, there will be no solutions if $b^2 - 4ac < 0$.

WORKED EXAMPLE 7 Using the quadratic formula to solve equations

Use the quadratic formula to solve each of the following equations.
a. $3x^2 + 4x + 1 = 0$ (exact answer)

b. $-3x^2 - 6x - 1 = 0$ (round to 2 decimal places)

THINK	WRITE
a. 1. Write the equation.	a. $3x^2 + 4x + 1 = 0$
2. Write the quadratic formula.	$x = \dfrac{-b \pm \sqrt{b^2 - 4ac}}{2a}$
3. State the values for a, b and c.	where $a = 3, b = 4, c = 1$.
4. Substitute the values into the formula.	$x = \dfrac{-4 \pm \sqrt{4^2 - 4 \times 3 \times 1}}{2 \times 3}$

5. Simplify and solve for x.

$$= \frac{-4 \pm \sqrt{4}}{6}$$

$$= \frac{-4 \pm 2}{6}$$

$$x = \frac{-4 + 2}{6} \quad \text{or} \quad x = \frac{-4 - 2}{6}$$

6. Write the two solutions.

$$x = -\frac{1}{3} \qquad\qquad x = -1$$

b. 1. Write the equation.

b. $-3x^2 - 6x - 1 = 0$

2. Write the quadratic formula.

$$x = \frac{-b \pm \sqrt{b^2 - 4ac}}{2a}$$

3. State the values for a, b and c.

where $a = -3, b = -6, c = -1$.

4. Substitute the values into the formula.

$$x = \frac{-(-6) \pm \sqrt{(-6)^2 - 4 \times (-3) \times (-1)}}{2 \times -3}$$

5. Simplify the fraction.

$$= \frac{6 \pm \sqrt{24}}{-6}$$

$$= \frac{6 \pm 2\sqrt{6}}{-6}$$

$$= \frac{3 \pm \sqrt{6}}{-3}$$

$$x = \frac{3 + \sqrt{6}}{-3} \quad \text{or} \quad x = \frac{3 - \sqrt{6}}{-3}$$

6. Write the two solutions correct to 2 decimal places.

$$x = -1.82 \qquad\qquad x = -0.18$$

Note: When asked to give an answer in exact form, you should simplify any surds as necessary.

 Resources

 Video eLesson The quadratic formula (eles-2314)

Interactivity The quadratic formula (int-2561)

6.3 Quick quiz on	6.3 Exercise

Individual pathways

■ PRACTISE	■ CONSOLIDATE	■ MASTER
1, 3, 6, 9, 13, 18, 19, 22	2, 4, 7, 10, 11, 14, 17, 20, 23, 24	5, 8, 12, 15, 16, 21

Fluency

1. State the values for a, b and c in each of the following equations of the form $ax^2 + bx + c = 0$.

    a. $3x^2 - 4x + 1 = 0$ b. $7x^2 - 12x + 2 = 0$ c. $8x^2 - x - 3 = 0$ d. $x^2 - 5x + 7 = 0$

2. State the values for a, b and c in each of the following equations of the form $ax^2 + bx + c = 0$.

    a. $5x^2 - 5x - 1 = 0$ b. $4x^2 - 9x - 3 = 0$ c. $12x^2 - 29x + 103 = 0$ d. $43x^2 - 81x - 24 = 0$

WE7a For questions **3–5**, use the quadratic formula to solve each of the following equations. Give exact answers.

3. a. $x^2 + 5x + 1 = 0$ b. $x^2 + 3x - 1 = 0$ c. $x^2 - 5x + 2 = 0$ d. $x^2 - 4x - 9 = 0$

4. a. $x^2 + 2x - 11 = 0$ b. $x^2 - 7x + 1 = 0$ c. $x^2 - 9x + 2 = 0$ d. $x^2 - 6x - 3 = 0$

5. a. $x^2 + 8x - 15 = 0$ b. $-x^2 + x + 5 = 0$ c. $-x^2 + 5x + 2 = 0$ d. $-x^2 - 2x + 7 = 0$

WE7b For questions **6–8**, use the quadratic formula to solve each of the following equations. Give approximate answers rounded to 2 decimal places.

6. a. $3x^2 - 4x - 3 = 0$ b. $4x^2 - x - 7 = 0$ c. $2x^2 + 7x - 5 = 0$
   d. $7x^2 + x - 2 = 0$ e. $5x^2 - 8x + 1 = 0$

7. a. $2x^2 - 13x + 2 = 0$ b. $-3x^2 + 2x + 7 = 0$ c. $-7x^2 + x + 8 = 0$
   d. $-12x^2 + x + 9 = 0$ e. $-6x^2 + 4x + 5 = 0$

8. a. $-11x^2 - x + 1 = 0$ b. $-4x^2 - x + 7 = 0$ c. $-2x^2 + 12x - 1 = 0$ d. $-5x^2 + x + 3 = 0$

9. **MC** The solutions of the equation $3x^2 - 7x - 2 = 0$ are:

    A. 1, 2   **B.** 1, −2   **C.** −0.257, 2.59   **D.** −0.772, 7.772

10. **MC** In the expansion of $(6x - 5)(3x + 4)$, the coefficient of x is:

    A. 18   **B.** −15   **C.** 9   **D.** 6

11. **MC** In the expanded form of $(x - 2)(x + 4)$, identify which of the following statements is incorrect.

    A. The value of the constant is −8.
    B. The coefficient of the x term is −6.
    C. The coefficient of the x term is 2.
    D. The coefficient of the x^2 term is 1.

12. **MC** Identify an exact solution to the equation $x^2 + 2x - 5 = 0$.

    A. −3.449   **B.** $-1 + \sqrt{24}$ **C.** $-1 + \sqrt{6}$ **D.** $\dfrac{2 + \sqrt{-16}}{2}$

Understanding

For questions **13–15**, use each of the following equations using any suitable method. Round to 3 decimal places where appropriate.

13. a. $2x^2 - 7x + 3 = 0$ b. $x^2 - 5x = 0$ c. $x^2 - 2x - 3 = 0$ d. $x^2 - 3x + 1 = 0$

14. **a.** $x^2 - 6x + 8 = 0$ **b.** $x^2 - 5x + 8 = 0$ **c.** $x^2 - 7x - 8 = 0$ **d.** $x^2 + 2x - 9 = 0$

15. **a.** $2x^2 + 11x - 21 = 0$ **b.** $7x^2 - 2x + 1 = 0$ **c.** $-x^2 + 9x - 14 = 0$ **d.** $-6x^2 - x + 1 = 0$

16. The surface area of a closed cylinder is given by the formula $SA = 2\pi r(r + h)$, where r cm is the radius of the can and h cm is the height.
    The height of a can of wood finish is 7 cm and its surface area is 231 cm².

    a. Substitute values into the formula to form a quadratic equation using the pronumeral, r.
    b. Use the quadratic formula to solve the equation and, hence, determine the radius of the can correct to 1 decimal place.
    c. Calculate the area of the curved surface of the can, correct to the nearest square centimetre.

17. To satisfy lighting requirements, the window shown must have an area of 1500 cm².

    a. Write an expression for the area of the window in terms of x.
    b. Write an equation so that the window satisfies the lighting requirements.
    c. Use the quadratic formula to solve the equation and calculate x to the nearest mm.

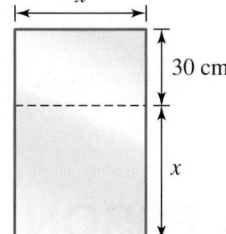

18. When using the quadratic formula, you are required to calculate b^2 (inside the square root sign). When $b = -3$, Breanne says that $b^2 = -(3)^2$ but Kelly says that $b^2 = (-3)^2$.

    a. What answer did Breanne calculate for b^2?
    b. What answer did Kelly calculate for b^2?
    c. Identify who is correct and explain why.

Communicating, reasoning and problem solving

19. There is one solution to the quadratic equation if $b^2 - 4ac = 0$. The following have one solution only. Determine the missing value.

    a. $a = 1$ and $b = 4$, $c = ?$
    b. $a = 2$ and $c = 8$, $b = ?$
    c. $2x^2 + 12x + c = 0$, $c = ?$

20. Two competitive neighbours build rectangular pools that cover the same area but are different shapes. Pool A has a width of $(x + 3)$ m and a length that is 3 m longer than its width. Pool B has a length that is double the width of Pool A. The width of Pool B is 4 m shorter than its length.

    a. Determine the exact dimensions of each pool if their areas are the same.
    b. Verify that the areas are the same.

21. A block of land is in the shape of a right-angled triangle with a perimeter of 150 m and a hypotenuse of 65 m. Determine the lengths of the other two sides. Show your working.

22. Solve $\left(x + \dfrac{1}{x}\right)^2 - 14\left(x + \dfrac{1}{x}\right) = 72$ for x.

23. Gunoor's tennis serve can be modelled using $y = -\dfrac{1}{16}x^2 + \dfrac{7}{8}x + 2$. The landing position of his serve will be one of the solutions to this equation when $y = 0$.

    a. For $-\dfrac{1}{16}x^2 + \dfrac{7}{8}x + 2 = 0$, determine the values of a, b and c in $ax^2 + bx + c = 0$.

    b. Using a calculator, calculate the solutions to this equation.
    c. Gunoor's serve will be 'in' if one solution is between 12 and 18, and a 'fault' if no solutions are between those values. Interpret whether Gunoor's serve was 'in' or if it was a 'fault'.

24. Triangle MNP is an isosceles triangle with sides $MN = MP = 3$ cm. Angle MPN is equal to $72°$. The line NQ bisects the angle MNP.

a. Prove that triangles MNP and NPQ are similar.
b. If $NP = m$ cm and $PQ = 3 - m$ cm, show that $m^2 + 3m - 9 = 0$.
c. Solve the equation $m^2 + 3m - 9 = 0$ and determine the side length of NP, giving your answer correct to 2 decimal places.

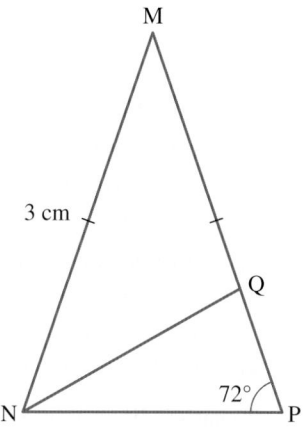

LESSON
6.4 Solving quadratic equations graphically

LEARNING INTENTION

At the end of this lesson you should be able to:
• identify the solutions to $ax^2 + bx + c = 0$ by inspecting the graph of $y = ax^2 + bx + c$
• recognise whether $ax^2 + bx + c = 0$ has two, one or no solutions by looking at the graph.

▶ 6.4.1 Solving quadratic equations graphically

eles-4847

• To solve quadratic equations, look at the graph of $y = ax^2 + bx + c$.
  • The solutions to $ax^2 + bx + c = 0$ are the x-axis intercepts, where the graph 'cuts' the x-axis.
• The number of x-axis intercepts will indicate the number of solutions.

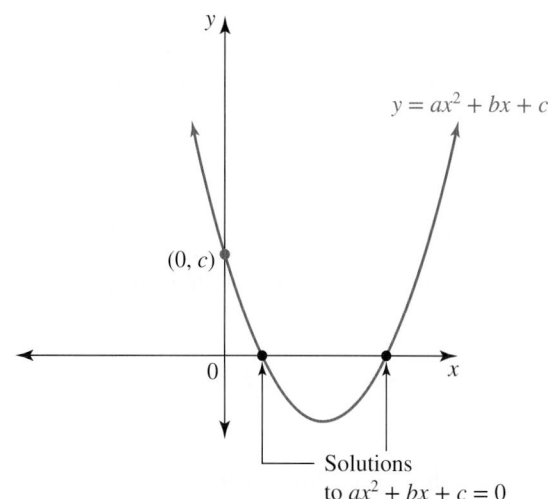

Number of solutions

Two solutions	One solution	No real solutions
The graph 'cuts' the x-axis twice.	The graph 'touches' the x-axis.	The graph does not reach the x-axis.
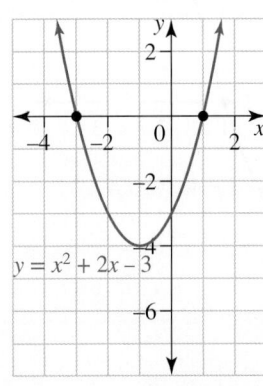	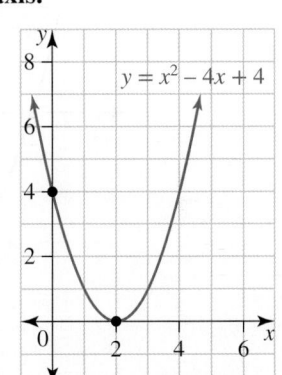	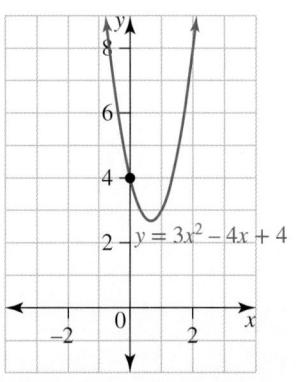

WORKED EXAMPLE 8 Determining solutions from the graph

Determine the solutions of each of the following quadratic equations by inspecting their corresponding graphs. Give answers to 1 decimal place where appropriate.

a. $x^2 + x - 2 = 0$

b. $2x^2 - 4x - 5 = 0$

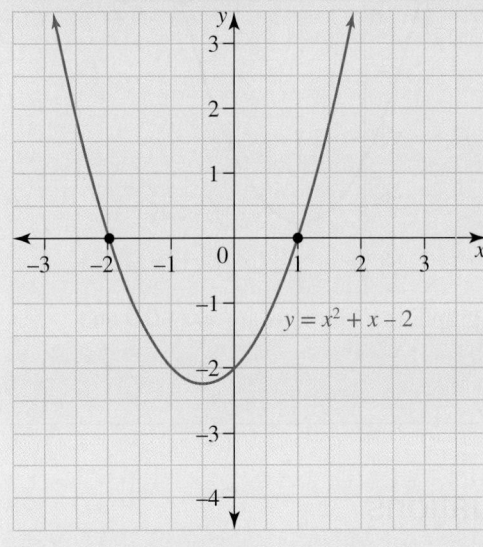

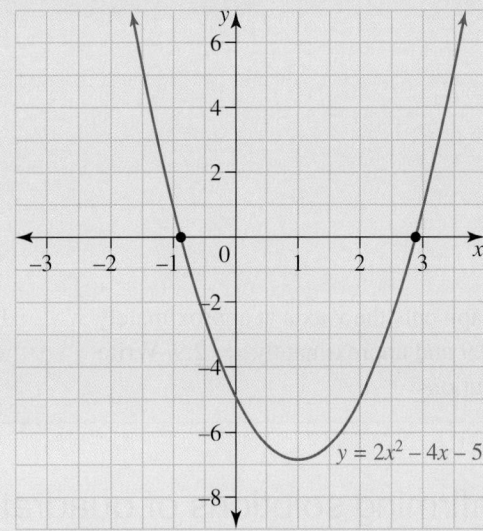

THINK	**WRITE/DRAW**

a. 1. Examine the graph of $y = x^2 + x - 2$ and locate the points where $y = 0$, that is, where the graph intersects the x-axis.

a.

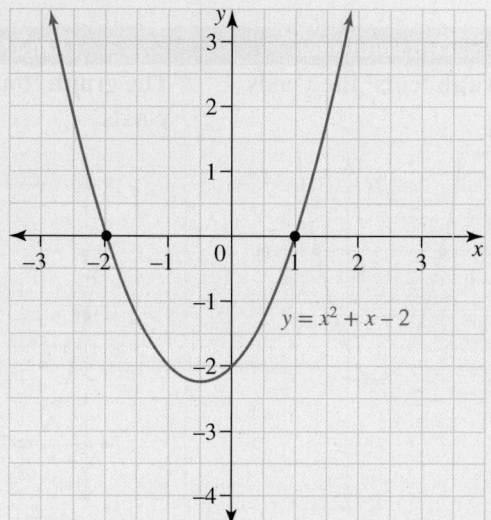

2. The graph cuts the x-axis $(y = 0)$ at $x = 1$ and $x = -2$. Write the solutions.

From the graph, the solutions are $x = 1$ and $x = -2$.

b. 1. The graph of $y = 2x^2 - 4x - 5$ is equal to zero when $y = 0$. Look at the graph to see where $y = 0$, that is, where it intersects the x-axis. By sight, we can only give estimates of the solutions.

b.

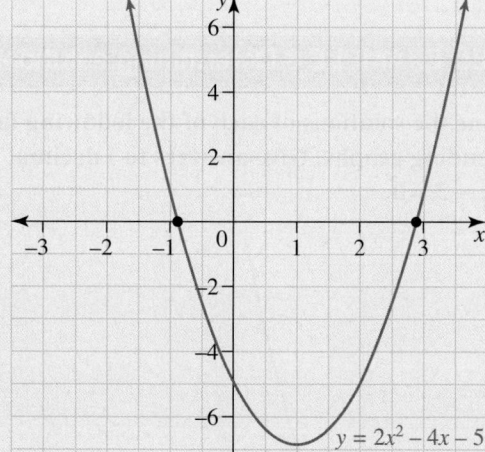

2. The graph cuts the x-axis at approximately $x = -0.9$ and approximately $x = 2.9$. Write the solutions.

From the graph, the solutions are $x \approx -0.9$ and $x \approx 2.9$.

▶ 6.4.2 Confirming solutions of quadratic equations

eles-4848

- To confirm a solution is correct:
  - substitute one solution for x into the quadratic equation
  - the solution is confirmed if the left-hand side (LHS) equals the right-hand side (RHS)
  - repeat for the second solution if applicable.

WORKED EXAMPLE 9 Confirming solutions by substitution

Confirm, by substitution, the solutions obtained in Worked example 8a:
$x^2 + x - 2 = 0$; solutions: $x = 1$ and $x = -2$.

THINK	WRITE
1. Write the left-hand side of the equation and substitute $x = 1$ into the expression.	When $x = 1$, LHS: $x^2 + x - 2 = 1^2 + 1 - 2$ $\qquad = 0$
2. Write the right-hand side.	RHS $= 0$
3. Confirm the solution.	LHS $=$ RHS $\Rightarrow$ The solution is confirmed.
4. Write the left-hand side and substitute $x = -2$.	When $x = -2$, LHS: $x^2 + x - 2 = (-2)^2 + -2 - 2$ $\qquad = 4 - 2 - 2$ $\qquad = 0$
5. Write the right-hand side.	RHS $= 0$
6. Confirm the solution.	LHS $=$ RHS $\Rightarrow$ The solution is confirmed

WORKED EXAMPLE 10 Solving application questions using a graph

A golf ball hit along a fairway follows the path shown in the
following graph. The height, h metres after it has travelled

x metres horizontally, follows the rule $h = -\dfrac{1}{270}\left(x^2 - 180x\right)$.

Use the graph to identify how far the ball landed from
the golfer.

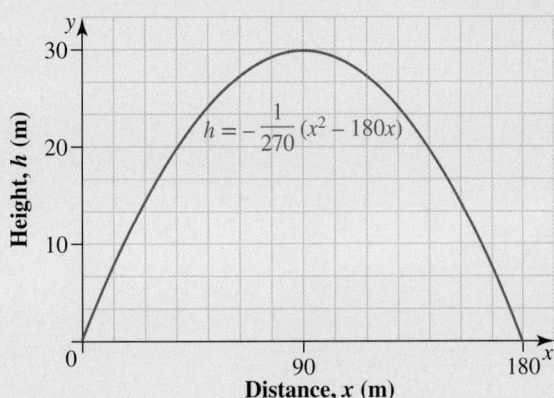

Distance, x (m)

THINK	WRITE
On the graph, the ground is represented by the x-axis since this is where $h = 0$. The golf ball lands when the graph intersects the x-axis.	The golf ball lands 180 m from the golfer.

Interactivity Solving quadratic equations graphically (int-6148)

Exercise 6.4 Solving quadratic equations graphically

learn on

| **6.4 Quick quiz** on | **6.4 Exercise** |

Individual pathways

■ PRACTISE	■ CONSOLIDATE	■ MASTER
1, 4, 7, 10	2, 5, 8, 11	3, 6, 9, 12

Fluency

1. **a.** WE8 Determine the solutions of each of the following quadratic equations by inspecting the corresponding graphs.

    i. $x^2 - x - 6 = 0$ **ii.** $x^2 - 11x + 10 = 0$

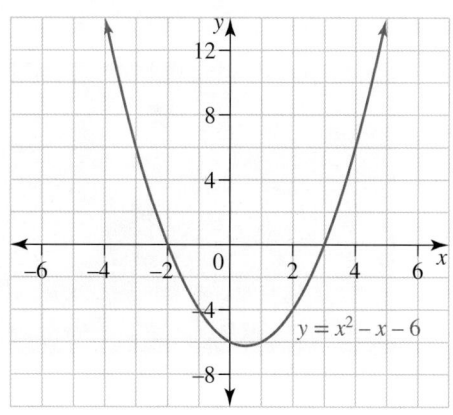   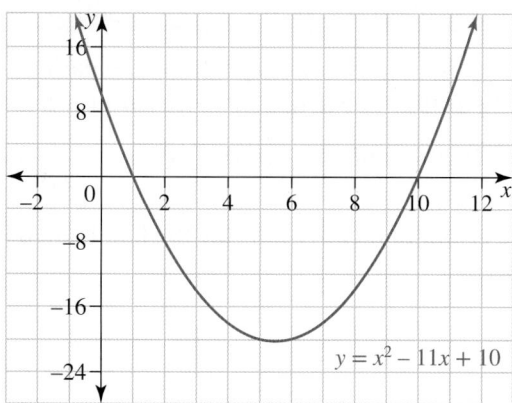

    iii. $-x^2 + 25 = 0$

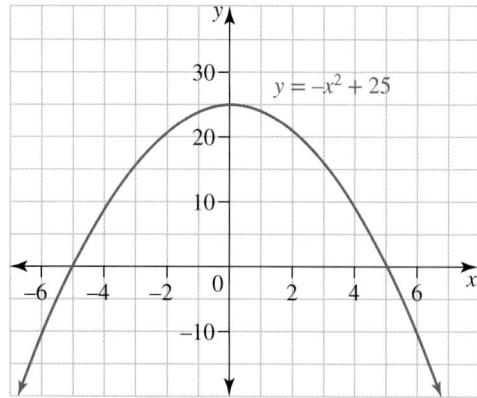

    b. WE9 Confirm, by substitution, the solutions obtained in part **a.**

2. a. Determine the solutions of each of the following quadratic equations by inspecting the corresponding graphs. Give answers correct to 1 decimal place where appropriate.

i. $2x^2 - 8x + 8 = 0$

ii. $x^2 - 3x - 4 = 0$

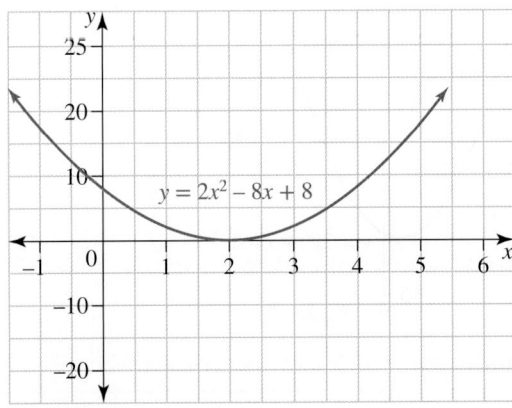

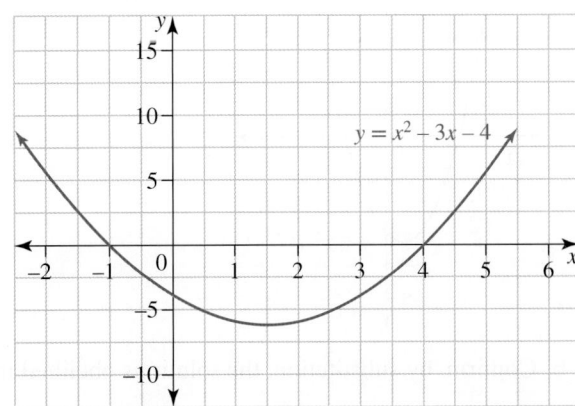

iii. $x^2 - 3x - 6 = 0$

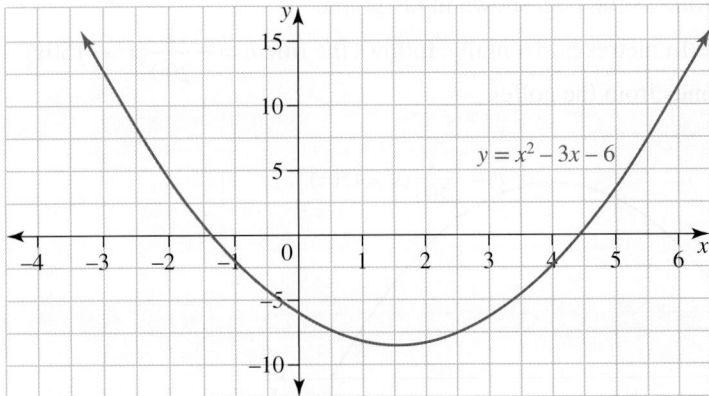

b. Confirm, by substitution, the solutions obtained in part **a.**

3. a. Determine the solutions of each of the following quadratic equations by inspecting the corresponding graphs.

i. $x^2 + 15x - 250 = 0$

ii. $-x^2 = 0$

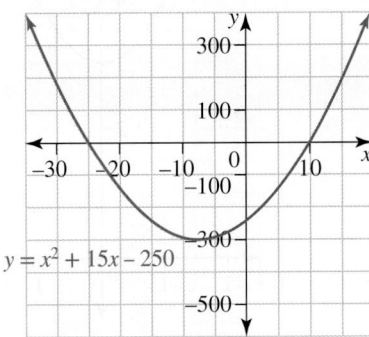

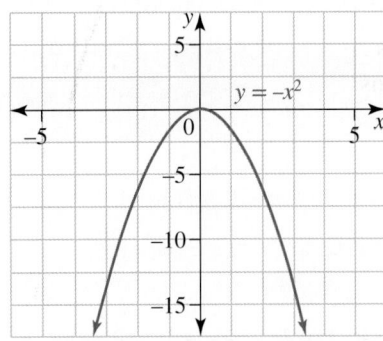

iii. $x^2 + x - 3 = 0$

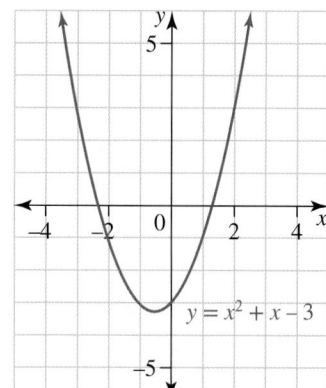

iv. $2x^2 + x - 3 = 0$

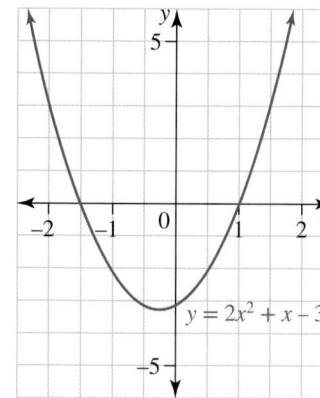

b. Confirm, by substitution, the solutions obtained in part **a**.

Understanding

4. **WE10** A golf ball hit along a fairway follows the path shown in the graph.

   The height, h metres after it has travelled x metres horizontally, follows the rule $h = -\dfrac{1}{200}(x^2 - 150x)$. Use the graph to identify how far the ball lands from the golfer.

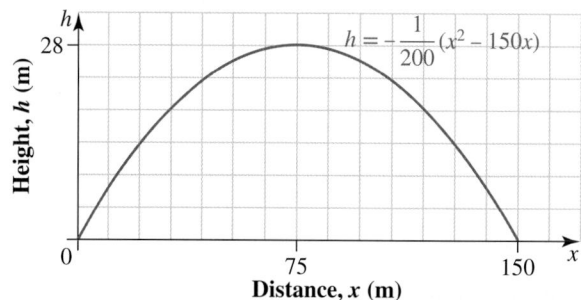

5. **MC** Use the graph to determine how many solutions the equation $x^2 + 3 = 0$ has.

   A. No real solutions
   B. One solution
   C. Two solutions
   D. Three solutions

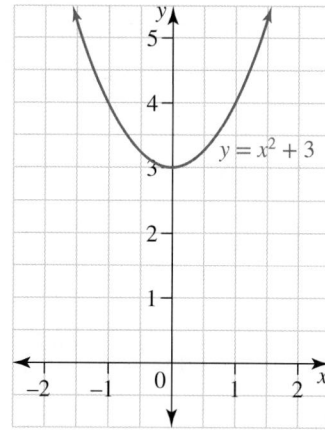

6. **MC** For the equation $x^2 + c = 0$, what values of c will give two solutions, one solution and no real solutions?

   A. Two solutions $c > 0$; one solution $c = 0$; no solutions $c < 0$
   B. Two solutions $c = 2$; one solution $c = 0$; no solutions $c = -2$
   C. Two solutions $c < 0$; one solution $c = 0$; no solutions $c > 0$
   D. Two solutions $c = 2$; one solution $c = 1$; no solutions $c = 0$

Communicating, reasoning and problem solving

7. Two graphs are shown.

   a. What are the solutions to both graphs?
   b. Look at the shapes of the graphs. Explain the similarities and differences.
   c. Look at the equations of the graphs. Explain the similarities and differences.

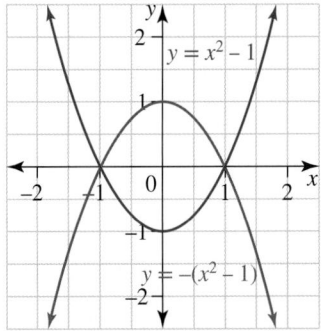

8. a. The x-intercepts of a particular equation are $x = 2$ and $x = 5$. Suggest a possible equation.
   b. If the y-intercept in part **a** is $(0, 4)$, give the exact equation.

9. a. The x-intercepts of a particular equation are $x = p$ and $x = q$. Suggest a possible equation.
   b. If the y-intercept in part **a** is $(0, r)$, give the exact equation.

10. A ball is thrown upwards from a building and follows the path shown in the graph until it lands on the ground.
   The ball is h metres above the ground when it is a horizontal distance of x metres from the building.
   The path of the ball follows the rule $h = -x^2 + 4x + 21$.

   a. Use the graph to identify how far from the building the ball lands.
   b. Use the graph to determine the height the ball was thrown from.

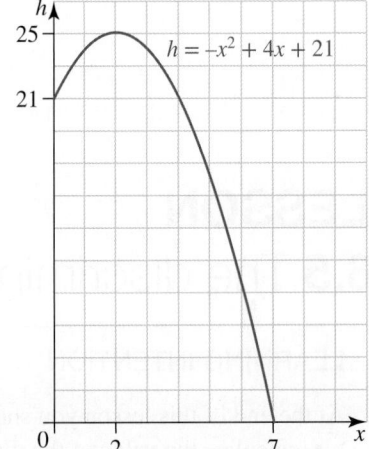

11. A platform diver follows a path determined by the equation $h = -0.5d^2 + 2d + 6$, where h represents the height of the diver above the water and d represents the distance from the diving board. Both pronumerals are measured in metres.

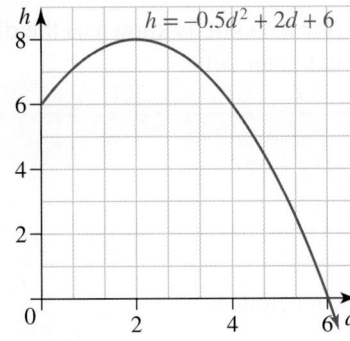

   Use the graph to determine:

   a. how far the diver landed from the edge of the diving board
   b. how high the diving board is above the water
   c. the maximum height reached by the diver when they are in the air.

12. Determine the equation of the given parabola. Give your answer in the form $y = ax^2 + bx + c$.

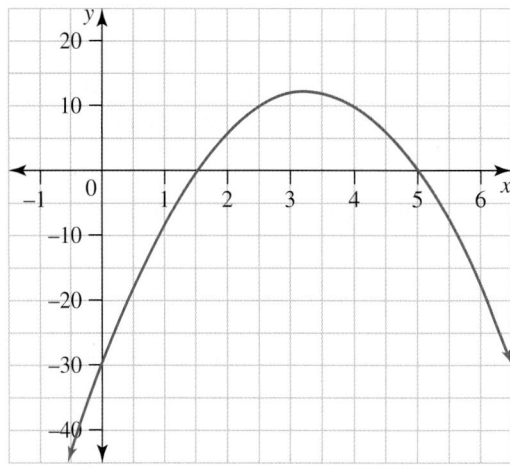

LESSON
6.5 The discriminant

LEARNING INTENTION

At the end of this lesson you should be able to:
- calculate the value of the discriminant
- use the value of the discriminant to determine the number of solutions and type of solutions
- calculate the discriminant for simultaneous equations to determine whether two graphs intersect.

6.5.1 Using the discriminant

eles-4849

- In the quadratic equation $ax^2 + bx + c = 0$, the expression $\Delta = b^2 - 4ac$ is known as the **discriminant**.
- The symbol used for the discriminant, Δ, is the Greek capital letter delta.

> **The discriminant**
>
> $$\Delta = b^2 - 4ac$$

- The discriminant is found in the quadratic formula, as shown below.

$$x = \frac{-b \pm \sqrt{b^2 - 4ac}}{2a} = \frac{-b \pm \sqrt{\Delta}}{2a}$$

- The discriminant is the value that determines the number of solutions to a quadratic equation.
  - If $\Delta > 0$ and a perfect square, there are *two rational solutions* and the quadratic can be factorised.
  - If $\Delta > 0$ and *not* a perfect square, there are *two irrational solutions* (surds).
  - If $\Delta = 0$, there is *one distinct solution*, it is a perfect square.
  - If $\Delta < 0$, there are *no real solutions*.

The discriminant and solutions

- If $\Delta > 0$, there are two distinct real solutions to the equation.
- If $\Delta = 0$, there is only one real solution to the equation.
- If $\Delta < 0$, there are no real solutions to the equation.

WORKED EXAMPLE 11 Using the discriminant to determine the number of solutions

Calculate the value of the discriminant for each of the following and use it to determine how many solutions the equation will have.

a. $2x^2 + 9x - 5 = 0$

b. $x^2 + 10 = 0$

THINK	WRITE
a. 1. Write the expression and determine the values of a, b and c given $ax^2 + bx + c = 0$.	a. $\quad 2x^2 + 9x - 5 = 0$ $2x^2 + 9x + -5 = 0$ $a = 2, b = 9, c = -5$
2. Write the formula for the discriminant and substitute values of a, b and c.	$\Delta = b^2 - 4ac$ $= 9^2 - 4 \times 2 \times -5$
3. Simplify the equation and solve.	$= 81 - (-40)$ $= 121$
4. State the number of solutions. In this case $\Delta > 0$, which means there are two solutions.	$\Delta > 0$, so there will be two solutions to the equation $2x^2 + 9x - 5 = 0$.
b. 1. Write the expression and determine the values of a, b and c given $ax^2 + bx + c = 0$.	b. $\qquad x^2 + 10 = 0$ $1x^2 + 0x + 10 = 0$ $a = 1, b = 0, c = 10$
2. Write the formula for the discriminant and substitute the values of a, b and c.	$\Delta = b^2 - 4ac$ $= 0^2 - 4 \times 1 \times 10$ $= 0 - 40$ $= -40$
3. State the number of solutions. In this case $\Delta < 0$, which means there are no solutions.	$\Delta < 0$, so there will be no solutions to the equation $x^2 + 10 = 0$.

▶ 6.5.2 Using the discriminant to determine if graphs intersect

- The table below summarises the number of points of intersection by the graph, indicated by the discriminant.

	$\Delta < 0$ (negative)	$\Delta = 0$ (zero)	$\Delta > 0$ (positive)	
			Perfect square	**Not a perfect square**
Number of solutions	No real solutions	1 rational solution	2 rational solutions	2 irrational (surd) solutions
Description	Graph does not cross or touch the x-axis	Graph touches the x-axis	Graph intersects the x-axis twice	
Graph				

WORKED EXAMPLE 12 Using the discriminant to determine the number and type of solutions

By using the discriminant, determine whether the following equations have:

i. two rational solutions   **ii. two irrational solutions**
iii. one solution   **iv. no real solutions.**

a. $x^2 - 9x - 10 = 0$ **b.** $x^2 - 2x - 14 = 0$
c. $x^2 - 2x + 14 = 0$ **d.** $x^2 + 14x = -49$

THINK	WRITE
a. 1. Write the equation.	**a.** $x^2 - 9x - 10 = 0$
2. Identify the coefficients a, b and c.	$a = 1, b = -9, c = -10$
3. Calculate the discriminant.	$\Delta = b^2 - 4ac$ $= (-9)^2 - 4 \times 1 \times (-10)$ $= 121$
4. Identify the number and type of solutions when $\Delta > 0$ and is a perfect square.	The equation has two rational solutions.
b. 1. Write the equation.	**b.** $x^2 - 2x - 14 = 0$
2. Identify the coefficients a, b and c.	$a = 1, b = -2, c = -14$
3. Calculate the discriminant.	$\Delta = b^2 - 4ac$ $= (-2)^2 - 4 \times 1 \times (-14)$ $= 60$
4. Identify the number and type of solutions when $\Delta > 0$ but not a perfect square.	The equation has two irrational solutions.

c. **1.** Write the equation.

c. $x^2 - 2x + 14 = 0$

2. Identify the coefficients a, b and c.

$a = 1, b = -2, c = 14$

3. Calculate the discriminant.

$\Delta = b^2 - 4ac$
$= (-2)^2 - 4 \times 1 \times 14$
$= -52$

4. Identify the number and type of solutions when $\Delta < 0$.

The equation has no real solutions.

d. **1.** Write the equation, then rewrite it so the right side equals zero.

d. $\quad x^2 + 14x = -49$
$x^2 + 14x + 49 = 0$

2. Identify the coefficients a, b and c.

$a = 1, b = 14, c = 49$

3. Calculate the discriminant.

$\Delta = b^2 - 4ac$
$= 14^2 - 4 \times 1 \times 49$
$= 0$

4. Identify the number and types of solutions when $\Delta = 0$.

The equation has one solution.

- In Topic 5 we learned that simultaneous equations can be solved graphically, where the intersection of the two graphs is the solution.
- The discriminant can be used to determine whether a solution exists for two equations and, hence, whether the graphs intersect.

WORKED EXAMPLE 13 Using the discriminant to determine if graphs intersect

Determine whether the parabola $y = x^2 - 2$ and the line $y = x - 3$ intersect.

THINK

1. If the parabola and the line intersect, there will be at least one solution to the simultaneous equations:

2. Collect all terms on one side and simplify.

3. Use the discriminant to check if any solutions exist.
If $\Delta < 0$, then no solutions exist.

4. Write the answer in a sentence.

WRITE

$y = x^2 - 2$
$y = x - 3$
$\therefore x^2 - 2 = x - 3$

$x^2 - 2 - x + 3 = 0$
$x^2 - x + 1 = 0$

$\Delta = b^2 - 4ac$
$a = 1, b = -1, c = 1$

$\Delta = (-1)^2 - 4 \times 1 \times 1$
$= 1 - 4$
$= -3$

$\Delta < 0$; therefore, no solutions exist.

The parabola and the line do not intersect.

DISCUSSION

What does the discriminant tell us?

 Resources

 Video eLesson The discriminant (eles-1946)

 Interactivity The discriminant (int-2560)

Exercise 6.5 The discriminant

learn

| 6.5 Quick quiz **on** | 6.5 Exercise |

Individual pathways

■ PRACTISE	■ CONSOLIDATE	■ MASTER
1, 4, 9, 11, 15, 19, 22	2, 5, 7, 12, 14, 18, 20, 23	3, 6, 8, 10, 13, 16, 17, 21, 24

Fluency

WE11 For questions **1–3**, calculate the value of the discriminant for each of the following and use it to determine how many solutions the equation will have.

1. **a.** $6x^2 + 13x - 5 = 0$ **b.** $x^2 + 9x - 90 = 0$ **c.** $x^2 + 4x - 2 = 0$
   d. $36x^2 - 1 = 0$ **e.** $x^2 + 2x + 8 = 0$

2. **a.** $x^2 - 5x - 14 = 0$ **b.** $36x^2 + 24x + 4 = 0$ **c.** $x^2 - 19x + 88 = 0$
   d. $x^2 - 10x + 17 = 0$ **e.** $30x^2 + 17x - 21 = 0$

3. **a.** $x^2 + 16x + 62 = 0$ **b.** $9x^2 - 36x + 36 = 0$ **c.** $2x^2 - 16x = 0$ **d.** $x^2 - 64 = 0$

4. **WE12** By using the discriminant, determine whether the equations below have:
   i. two rational solutions   **ii.** two irrational solutions
   iii. one solution   **iv.** no real solutions.
   a. $x^2 - 3x + 5$ **b.** $4x^2 - 20x + 25 = 0$
   c. $x^2 + 9x - 22 = 0$ **d.** $9x^2 + 12x + 4$

5. By using the discriminant, determine whether the equations below have:
   i. two rational solutions   **ii.** two irrational solutions
   iii. one solution   **iv.** no real solutions.
   a. $x^2 + 3x - 7 = 0$ **b.** $25x^2 - 10x + 1 = 0$
   c. $3x^2 - 2x - 4 = 0$ **d.** $2x^2 - 5x + 4 = 0$

6. By using the discriminant, determine whether the equations below have:
   i. two rational solutions   **ii.** two irrational solutions
   iii. one solution   **iv.** no real solutions.
   a. $x^2 - 10x + 26 = 0$ **b.** $3x^2 + 5x - 7 = 0$
   c. $2x^2 + 7x - 10 = 0$ **d.** $x^2 - 11x + 30 = 0$

7. **WE13** Determine whether the following graphs intersect.
   a. $y = -x^2 + 3x + 4$ and $y = x - 4$ **b.** $y = -x^2 + 3x + 4$ and $y = 2x + 5$
   c. $y = -(x + 1)^2 + 3$ and $y = -4x - 1$ **d.** $y = (x - 1)^2 + 5$ and $y = -4x - 1$

8. Consider the equation $3x^2 + 2x + 7 = 0$.
   a. Identify the values of a, b and c.
   b. Calculate the value of $b^2 - 4ac$.
   c. State how many real solutions, and hence x-intercepts, are there for this equation.

9. Consider the equation $-6x^2 + x + 3 = 0$.
   a. Identify the values of a, b and c.
   b. Calculate the value of $b^2 - 4ac$.
   c. State how many real solutions, and hence x-intercepts, there are for this equation.

10. Consider the equation $-6x^2 + x + 3 = 0$. With the information gained from the discriminant, use the most efficient method to solve the equation. Give an exact answer.

11. **MC** Identify the discriminant of the equation $x^2 - 4x - 5 = 0$.
   A. 36
   B. 11
   C. 4
   D. 0

12. **MC** Identify which of the following quadratic equations has two irrational solutions.
   A. $x^2 - 8x + 16 = 0$
   B. $2x^2 - 7x = 0$
   C. $x^2 + 8x + 9 = 0$
   D. $x^2 - 4 = 0$

13. **MC** The equation $x^2 = 2x - 3$ has:
   A. two rational solutions
   B. exactly one solution
   C. no solutions
   D. two irrational solutions

Understanding

14. Determine the value of k if $x^2 - 2x - k = 0$ has one solution.

15. Determine the value of m for which $mx^2 - 6x + 5 = 0$ has one solution.

16. Determine the values of n when $x^2 - 3x - n = 0$ has two solutions.

17. The path of a dolphin as it leaps out of the water can be modelled by the equation $h = -0.4d^2 + d$, where h is the dolphin's height above water and d is the horizontal distance from its starting point. Both h and d are in metres.

   a. Calculate how high above the water the dolphin is when it has travelled 2 m horizontally from its starting point.
   b. Determine the horizontal distance the dolphin has covered when it first reaches a height of 25 cm.
   c. Determine the horizontal distance the dolphin has covered when it next reaches a height of 25 cm. Explain your answer.
   d. Determine the horizontal distance the dolphin covers in one leap. (*Hint:* What is the value of h when the dolphin has completed its leap?)
   e. During a leap, determine whether this dolphin reaches a height of:
      i. 0.5 m   ii. 1 m.
      Explain how you can determine this without actually solving the equation.
   f. Determine the greatest height the dolphin reaches during a leap.

18. Determine how many times the parabolas $y = x^2 - 4$ and $y = 4 - x^2$ intersect.

Communicating, reasoning and problem solving

19. Show that $3x^2 + px - 2 = 0$ will have real solutions for all values of p.

20. Answer the following questions.

    a. Determine the values of a for which the straight line $y = ax + 1$ will have one intersection with the parabola $y = -x^2 - x - 8$.

    b. Determine the values of b for which the straight line $y = 2x + b$ will *not* intersect with the parabola $y = x^2 + 3x - 5$.

21. Answer the following questions.

    a. Identify how many points of intersection exist between the parabola $y = -2(x + 1)^2 - 5$, where $y = f(x), x \in R$, and the straight line $y = mx - 7$, where $y = f(x), x \in R$.

    b. Determine the value of m (where $m < 0$) such that $y = mx - 7$ has one intersection point with $y = -m(x + 1)^2 - 5$.

22. Answer the following questions.

    a. If $\Delta = 9$, $a = 1$ and $b = 5$, use the quadratic formula to determine the two solutions of x.

    b. If $\Delta = 9$, $a = 1$ and $b = 5$, calculate the value of c.

23. The parabola with the general equation $y = ax^2 + bx + 9$ where $0 < a < 0$ and $0 < b < 20$ touches the x-axis at one point only. The graph passes through the point $(1, 25)$. Determine the values of a and b.

24. The line with equation $kx + y = 3$ is a tangent to the curve with equation $y = kx^2 + kx - 1$. Determine the value of k.

LESSON
6.6 Review

6.6.1 Topic summary

Standard quadratic equation

- Quadratic equations are in the form:
  $ax^2 + bx + c = 0$
  where a, b and c are numbers.
  e.g. In the equation $-5x^2 + 2x - 4 = 0$,
  $a = -5$, $b = 2$ and $c = -4$.

Solving quadratic equations graphically

- Inspect the graph of $y = ax^2 + bx + c$.
- The solutions to $ax^2 + bx + c = 0$ are the x-axis intercepts.

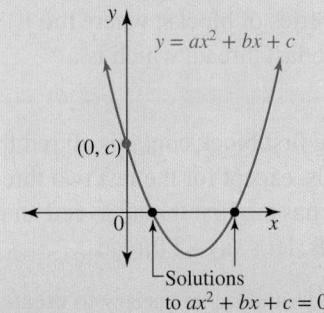

Solutions to $ax^2 + bx + c = 0$

The discriminant

- The discriminant indicates the number and type of solutions.
  $$\Delta = b^2 - 4ac$$
- $\Delta < 0$: no real solutions
- $\Delta = 0$: one solution
- $\Delta > 0$: two solutions (if perfect square then two rational solutions, if not a perfect square then two irrational solutions)
- A positive discriminant that is a perfect square (e.g. 16, 25, 144) gives two *rational* solutions.
- A positive discriminant that is not a perfect square (e.g. 7, 11, 15) gives two *irrational* (surd) solutions.

QUADRATIC EQUATIONS (PATH)

Solving quadratic equations algebraically

- Use algebra to determine the x-value solutions to $ax^2 + bc + c = 0$.
- There will be two, one or no solutions.

Factorising

Factorise using:
- common factor
- difference of two squares
- factor pairs
- completing the square.

Null Factor Law

- Once an equation is factorised, then apply the Null Factor Law.
  - Set the products equal to zero.
  - Solve for x.
- If the product of two numbers is zero, then one or both numbers must be zero.

Completing the square

- Completing the square is another technique to factorise quadratics when the factors are not whole numbers.
- To complete the square, the value of a must be 1. If a is not 1, divide everything by the coefficient of x^2 so that $a = 1$.
- The formula for completing the square is:
  $$x^2 \pm px = \left[x^2 \pm px + \left(\frac{p}{2}\right)^2\right] - \left(\frac{p}{2}\right)^2$$
  $$= \left(x \pm \frac{p}{2}\right)^2 - \left(\frac{p}{2}\right)^2$$
- The formula to complete the square can be remembered as 'add and subtract half the coefficient of x and square it'.

Quadratic formula

To use the quadratic formula:
- identify the values for a, b and c
- substitute the values into the quadratic formula.
  $$x = \frac{-b \pm \sqrt{b^2 - 4ac}}{2a}$$

Note: The $\pm$ symbol is called the plus–minus sign.

6.6.2 Project

Weaving

Many articles of clothing are sewn from materials that show designs and patterns made by weaving together threads of different colours. Intricate and complex designs can result. Let's investigate some very simple repetitive patterns. Knowledge of quadratic equations and the quadratic formula is helpful in creating these designs.

We need to understand the process of weaving. Weaving machines have parts called *warps*. Each warp is divided into a number of *blocks*. Consider a pattern that is made up of a series of blocks, where the first block is all one colour except for the last thread, which is a different colour.

Let's say our pattern is red and blue. The first block contains all red threads, except for the last one, which is blue. The next block has all red threads, except for the last two threads, which are blue. The pattern continues in this manner. The last block has the first thread as red and the remainder as blue. The warp consists of a particular number of threads, let's say 42 threads.

How many blocks and threads per block would be necessary to create a pattern of this type?

To produce this pattern, we need to divide the warp into equally sized blocks, if possible. What size block and how many threads per block would give us the 42-thread warp? We will need to look for a mathematical pattern. Look at the table (below), where we consider the smallest block consisting of 2 threads through to a block consisting of 7 threads.

Block 1

Block 2

Block n

Pattern	Number of threads per block	Number of blocks	Total threads in warp
RB	2	1	2
RRB RBB	3	2	6
RRRB RRBB RBBB	4		
	5		
	6		
	7		

1. Complete the entries in the table.
2. Consider a block consisting of n threads.
   a. How many blocks would be needed?
   b. What would be the total number of threads in the warp?

The 42-thread warp was chosen as a simple example to show the procedure involved in determining the number of blocks required and the number of threads per block. In this particular case, 6 blocks of 7 threads per block would give us our design for a 42-thread warp. In practice, you would not approach the problem by drawing up a table to determine the number of blocks and the size of each block.

3. Take your expression in question **2b** and let it equal 42. This should form a quadratic equation. Solve this equation to verify that you would need 6 blocks with 7 threads per block to fulfil the size of a 42-thread warp.
4. In reality, the size of each block is not always clearly defined. Also, the thread warp sizes are generally much larger, about 250. Let's determine the number of threads per block and the number of blocks required for a 250-thread warp.
   a. Form your quadratic equation with the thread warp size equal to 250.
   b. A solution to this equation can be found using the quadratic formula. Use the quadratic formula to determine a solution.
   c. The number of threads per block is represented by n and this obviously must be a whole number. Round your solution down to the nearest whole number.
   d. How many whole blocks are needed?
   e. Use your solutions to c and d to determine the total number of threads used for the pattern.
   f. How many more threads do you need to make the warp size equal to 250 threads?
   g. Distribute these threads by including them at the beginning of the first block and the end of the last block. Describe your overall pattern.
5. Investigate the number of blocks required and threads per block required for a 400-thread warp.
6. Investigate changing the pattern. Let the first block be all red. In the next block, change the colour of the first and last threads to blue. With each progressive block, change the colour of an extra thread at the top and bottom to blue until the last block is all blue. On a separate sheet of paper, draw a table to determine the thread warp size for a block size of n threads. Draw the pattern and describe the result for a particular warp size.

  Resources

Interactivities Crossword (int-2848)
Sudoku puzzle (int-3595)

Exercise 6.6 Review questions

learn on

Fluency

1. **MC** Identify the solutions to the equation $x^2 + 10x - 11 = 0$.
   A. $x = 1$ and $x = 11$
   B. $x = 1$ and $x = -11$
   C. $x = -1$ and $x = 11$
   D. $x = -1$ and $x = -11$

2. **MC** Identify the solutions to the equation $-5x^2 + x + 3 = 0$.
   A. $x = 1$ and $x = \dfrac{3}{5}$
   B. $x = -0.68$ and $x = 0.88$
   C. $x = 3$ and $x = -5$
   D. $x = 0.68$ and $x = -0.88$

3. **MC** Identify the discriminant of the equation $x^2 - 11x + 30 = 0$.
   A. 1
   B. 241
   C. 91
   D. 19

4. **MC** Choose from the following equations which has two irrational solutions.
   A. $x^2 - 6x + 9 = 0$
   B. $4x^2 - 11x = 0$
   C. $x^2 - 25 = 0$
   D. $x^2 + 8x + 2 = 0$

5. The area of a pool is $(6x^2 + 11x + 4)\,\text{m}^2$. Determine the length of the rectangular pool if its width is $(2x + 1)\,\text{m}$.

6. Determine the solutions of the following equations, by first factorising the left-hand side.
   a. $x^2 + 8x + 15 = 0$
   b. $x^2 + 7x + 6 = 0$
   c. $x^2 + 11x + 24 = 0$
   d. $x^2 + 4x - 12 = 0$
   e. $x^2 - 3x - 10 = 0$

7. Determine the solutions of the following equations, by first factorising the left-hand side.
   a. $x^2 + 3x - 28 = 0$
   b. $x^2 - 4x + 3 = 0$
   c. $x^2 - 11x + 30 = 0$
   d. $x^2 - 2x - 35 = 0$

8. Determine the solutions of the following equations, by first factorising the left-hand side.
   a. $2x^2 + 16x + 24 = 0$
   b. $3x^2 + 9x + 6 = 0$
   c. $4x^2 + 10x - 6 = 0$
   d. $5x^2 + 25x - 70 = 0$
   e. $2x^2 - 7x - 4 = 0$

9. Determine the solutions of the following equations, by first factorising the left-hand side.
   a. $6x^2 - 8x - 8 = 0$
   b. $2x^2 - 6x + 4 = 0$
   c. $6x^2 - 25x + 25 = 0$
   d. $2x^2 + 13x - 7 = 0$

10. Determine the solutions to the following equations by completing the square.
    a. $x^2 + 8x - 1 = 0$
    b. $3x^2 + 6x - 15 = 0$
    c. $-4x^2 - 3x + 1 = 0$

Understanding

11. Ten times an integer is added to seven times its square. If the result is 152, calculate the original number.

12. By using the quadratic formula, determine the solutions to the following equations. Give your answers correct to 3 decimal places.
    a. $4x^2 - 2x - 3 = 0$
    b. $7x^2 + 4x - 1 = 0$
    c. $-8x^2 - x + 2 = 0$

13. By using the quadratic formula, determine the solutions to the following equations. Give your answers correct to 3 decimal places.

   a. $18x^2 - 2x - 7 = 0$ b. $29x^2 - 105x - 24 = 0$ c. $-5x^2 + 2 = 0$

14. The graph of $y = x^2 - 4x - 21$ is shown.

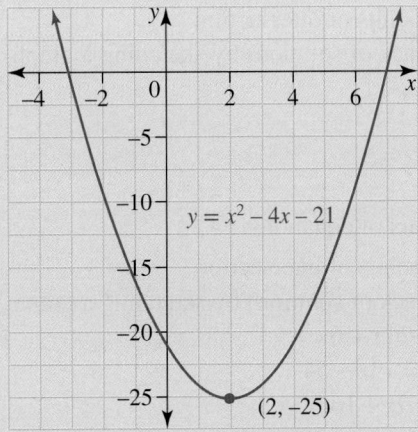

   Use the graph to identify the solutions to the quadratic equation $x^2 - 4x - 21 = 0$.

15. Determine the solutions to the equation $-2x^2 - 4x + 6 = 0$.

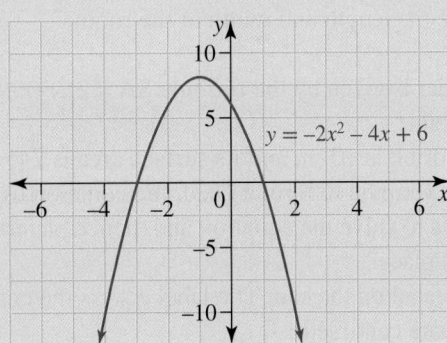

16. By using the discriminant, determine the number and nature of the solutions for the following equations.

   a. $x^2 + 11x + 9 = 0$ b. $3x^2 + 2x - 5 = 0$ c. $x^2 - 3x + 4 = 0$

17. Use the discriminant to determine if the following equations intersect.

$$y = x^2 + 4x - 10$$
$$y = 6 - 2x$$

18. Use the discriminant to determine if the following equations intersect.

$$y = x^2 - 7x + 20$$
$$y = 3x - 5$$

19. Use the discriminant to determine if the following equations intersect.

$$y = x^2 + 7x + 11$$
$$y = x$$

20. For each of the following pairs of equations:
    i. determine the number of points of intersection
    ii. illustrate the solution (or lack of solution) by sketching a graph.
    a. $y = x^2 + 6x + 5$ and $y = 11x - 1$
    b. $y = x^2 + 5x - 6$ and $y = 8x - 8$
    c. $y = x^2 + 9x + 14$ and $y = 3x + 5$

21. For each of the following pairs of equations:
    i. determine the number of points of intersection
    ii. illustrate the solution (or lack of solution) by sketching a graph.
    a. $y = x^2 - 7x + 10$ and $y = -11x + 6$
    b. $y = -x^2 + 14x - 48$ and $y = 13x - 54$
    c. $y = -x^2 + 4x + 12$ and $y = 9x + 16$

Communicating, reasoning and problem solving

22. When a number is added to its square, the result is 56. Determine the number.

23. Leroy measures his bedroom and finds that its length is 3 metres more than its width. If the area of the bedroom is $18\,\text{m}^2$, calculate the length and width of the room.

24. The surface area of a cylinder is given by the formula $SA = 2\pi r\,(r + h)$, where r cm is the radius of the cylinder and h cm is the height.
    The height of a can of soft drink is 10 cm and its surface area is $245\,\text{cm}^2$.
    a. Substitute values into the formula to form a quadratic equation using the pronumeral r.
    b. Use the quadratic formula to solve the equation and, hence, determine the radius of the can. Round your answer to 1 decimal place.
    c. Calculate the area of the label on the can. The label covers the entire curved surface. Round the answer to the nearest square centimetre.

25. Determine the value of d when $2x^2 - 5x - d = 0$ has one solution.

26. Determine the values of k where $(k - 1)\,x^2 - (k - 1)\,x + 2 = 0$ has two distinct solutions.

27. Let m and n be the solutions to the quadratic equation $x^2 - 2\sqrt{5}x - 2 = 0$. Determine the value of $m^2 + n^2$.

28. Although it requires a minimum of two points to determine the graph of a line, it requires a minimum of three points to determine the shape of a parabola. The general equation of a parabola is $y = ax^2 + bx + c$, where a, b and c are the constants to be determined.
    a. Determine the equation of the parabola that has a y-intercept of $(0, -2)$, and passes though the points $(1, -5)$ and $(-2, 16)$.
    b. Determine the equation of a parabola that goes through the points $(0, 0)$, $(2, 2)$ and $(5, 5)$. Show full working to justify your answer.

29. When the radius of a circle increases by 6 cm, its area increases by 25%. Use the quadratic formula to calculate the exact radius of the original circle.

30. A football player received a hand pass and ran directly towards goal. Right on the 50-metre line he kicked the ball and scored a goal. The graph shown represents the path of the ball. Using the graph, answer the following questions.

a. State the height of the ball from the ground when it was kicked.

b. Identify the greatest height the ball reached.

c. Identify the length of the kick.

d. If there were defenders in the goal square, explain if it would have been possible for one of them to mark the ball right on the goal line to prevent a goal. (*Hint:* What was the height of the ball when it crossed the goal line?)

e. As the footballer kicked the ball, a defender rushed at him to try to smother the kick. If the defender can reach a height of 3 m when he jumps, determine how close to the player kicking the ball he must be to just touch the football as it passes over his outstretched hands.

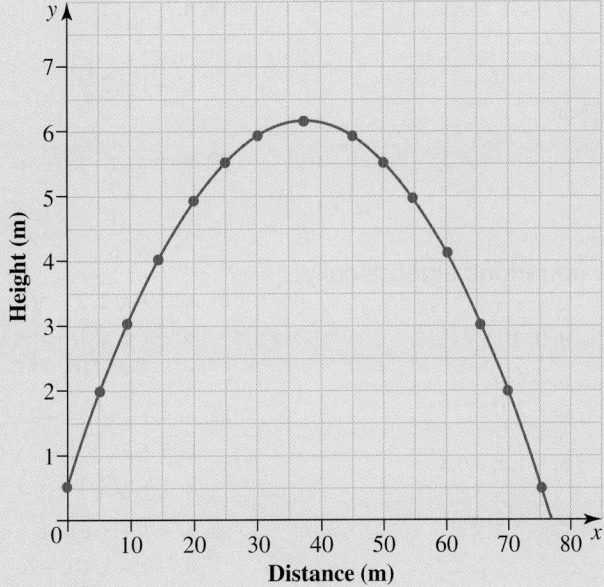

31. The quadratic formula is $x = \dfrac{-b \pm \sqrt{b^2 - 4ac}}{2a}$.

An alternative form of the quadratic formula is $x = \dfrac{2c}{-b \pm \sqrt{b^2 - 4ac}}$.

Choose a quadratic equation and show that the two formulas give the same answers.

on To test your understanding and knowledge of this topic, go to your learnON title at www.jacplus.com.au and complete the **post-test**.

Answers

Topic 6 Quadratic equations (Path)

6.1 Pre-test

1. True

2. $x = -5$ or $x = \dfrac{1}{2}$

3. $x = \pm\dfrac{3}{4}$

4. D

5. B

6. D

7. A

8. 24

9. a. Only one solution
   b. No real solutions
   c. Two solutions

10. B

11. A

12. Width $= 3\,\text{cm}$, length $= 7\,\text{cm}$

13. $m = 4$

14. A

15. D

6.2 Solving quadratic equations algebraically

1. a. $-7, 9$ b. $-2, 3$
   c. $2, 3$ d. $0, 3$

2. a. $0, 1$ b. $-5, 0$
   c. $0, 3$ d. $-2, 0$

3. a. $-\dfrac{1}{2}, \dfrac{1}{2}$ b. $-1.2, -0.5$
   c. $0.1, 0.75$ d. $-\sqrt{2}, \sqrt{3}$

4. a. $\dfrac{1}{2}, 1$ b. $-2, -\dfrac{2}{3}$ c. $\dfrac{1}{4}, 7$

5. a. $-\dfrac{6}{7}, 1\dfrac{1}{2}$ b. $\dfrac{3}{5}, \dfrac{2}{3}$ c. $-\dfrac{5}{8}, \dfrac{2}{3}$

6. a. $0, \dfrac{1}{2}, 3$ b. $0, \dfrac{1}{2}, -\dfrac{2}{5}$ c. $0, -3, \dfrac{2}{5}$

7. a. $0, 2$ b. $-5, 0$ c. $0, 7$

8. a. $-\dfrac{2}{3}, 0$ b. $0, 1\dfrac{1}{2}$ c. $0, \dfrac{1}{3}$

9. a. $0, \dfrac{\sqrt{7}}{2}$ b. $-\dfrac{\sqrt{3}}{3}, 0$ c. $0, 1\dfrac{1}{4}$

10. a. $-2, 2$ b. $-5, 5$
    c. $-2, 2$ d. $-7, 7$

11. a. $-1\dfrac{1}{3}, 1\dfrac{1}{3}$ b. $-2\dfrac{1}{2}, 2\dfrac{1}{2}$
    c. $-\dfrac{2}{3}, \dfrac{2}{3}$ d. $-\dfrac{1}{2}, \dfrac{1}{2}$

12. a. $-\dfrac{1}{5}, \dfrac{1}{5}$ b. $-4, 4$
    c. $-\sqrt{5}, \sqrt{5}$ d. $-\dfrac{\sqrt{11}}{3}, \dfrac{\sqrt{11}}{3}$

13. a. $-2, 3$ b. $-4, -2$
    c. $-1, 7$ d. $3, 5$

14. a. $-1, 4$ b. 5 c. $-2, 5$
    d. $2, 6$ e. $-3, 7$

15. a. $-5, 6$ b. $3, 4$ c. 4
    d. -5 e. 10

16. B

17. C

18. a. $-\dfrac{1}{2}, 3$ b. $\dfrac{2}{3}, -1$
    c. $-2, \dfrac{1}{5}$ d. $\dfrac{1}{3}, 1\dfrac{1}{2}$

19. a. $-\dfrac{3}{14}, 1$ b. $\dfrac{1}{4}, \dfrac{1}{3}$
    c. $-1\dfrac{1}{3}, 2\dfrac{1}{2}$ d. $-1\dfrac{3}{4}, -1\dfrac{1}{3}$

20. a. $-\dfrac{2}{5}, \dfrac{1}{2}$ b. $1\dfrac{1}{2}, 2\dfrac{2}{3}$
    c. $-\dfrac{2}{5}, \dfrac{1}{6}$ d. $3, 4$

21. a. $2 + \sqrt{2}, 2 - \sqrt{2}$
    b. $-1 + \sqrt{3}, -1 - \sqrt{3}$
    c. $-3 + \sqrt{10}, -3 - \sqrt{10}$

22. a. $4 + 2\sqrt{3}, 4 - 2\sqrt{3}$
    b. $5 + 2\sqrt{6}, 5 - 2\sqrt{6}$
    c. $1 + \sqrt{3}, 1 - \sqrt{3}$

23. a. $-1 + \sqrt{6}, -1 - \sqrt{6}$
    b. $-2 + \sqrt{10}, -2 - \sqrt{10}$
    c. $-2 + \sqrt{15}, -2 - \sqrt{15}$

24. a. $\dfrac{3}{2} + \dfrac{\sqrt{5}}{2}, \dfrac{3}{2} - \dfrac{\sqrt{5}}{2}$
    b. $-\dfrac{5}{2} + \dfrac{\sqrt{29}}{2}, -\dfrac{5}{2} - \dfrac{\sqrt{29}}{2}$
    c. $\dfrac{7}{2} + \dfrac{\sqrt{33}}{2}, \dfrac{7}{2} - \dfrac{\sqrt{33}}{2}$

25. a. $\dfrac{1}{2} + \dfrac{\sqrt{21}}{2}, \dfrac{1}{2} - \dfrac{\sqrt{21}}{2}$
    b. $\dfrac{11}{2} + \dfrac{\sqrt{117}}{2}, \dfrac{11}{2} - \dfrac{\sqrt{117}}{2}$
    c. $-\dfrac{1}{2} + \dfrac{\sqrt{5}}{2}, -\dfrac{1}{2} - \dfrac{\sqrt{5}}{2}$

26. a. $-\dfrac{3}{2}+\dfrac{\sqrt{37}}{2}, -\dfrac{3}{2}-\dfrac{\sqrt{37}}{2}$

    b. $\dfrac{5}{2}+\dfrac{\sqrt{37}}{2}, \dfrac{5}{2}-\dfrac{\sqrt{37}}{2}$

    c. $\dfrac{9}{2}+\dfrac{\sqrt{65}}{2}, \dfrac{9}{2}-\dfrac{\sqrt{65}}{2}$

27. a. $-3, 1$ b. $-4.24, 0.24$ c. $-1, 3$

28. a. $-0.73, 2.73$ b. $0.38, 2.62$ c. $-0.30, 3.30$

29. a. $-1.19, 4.19$ b. $-2.30, 1.30$ c. $-2.22, 0.22$

30. a. $x = \pm2, \pm3$ b. $x = \pm1, \pm2$ c. $x = \pm2$

31. a. $x = \pm1, \pm4$ b. $x = \pm3$ c. No real solutions

32. a. $x = \pm2, \pm5$ b. $x = \pm\dfrac{1}{2}$ c. $x = -\dfrac{14}{3}, -\dfrac{5}{3}$

33. 8 and 9 or -8 and -9

34. 6 and 8, -6 and -8

35. 9 or -10

36. 2 or $-2\dfrac{2}{3}$

37. 8 or $-10\dfrac{1}{2}$

38. 6 seconds

39. a. $l = 2x$

    b.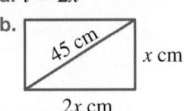

    c. $x^2 + (2x)^2 = 45^2, 5x^2 = 2025$

    d. Length 40 cm, width 20 cm

40. 8 m, 6 m

41. a. $-6, 1$ b. $8, -3$ c. $x = \pm1$

42. a. 21   b. 17

43. 7

44. a.

    b. $(2 + x)\,\text{m}, (4 + x)\,\text{m}$

    c. $(2 + x)(4 + x) = 24$

    d. $x = 2$, 4 m wide, 6 m long

45. a. $C_{\text{Annabel}}(28) = \$364\,800, C_{\text{Betty}}(28) = \$422\,400$

    b. 10 knots

    c. Speed can only be a positive quantity, so the negative solution is not valid.

46. No real solutions — when we complete the square we get the sum of two squares, not the difference of two squares and we cannot factorise the expression.

47. $x = 5, -4, 4, -3$

48. $z = -\dfrac{5}{3}, 7$

49. The width of the pathway is 1.5 m.

50. $x = -\dfrac{3}{2}$ or $x = -\dfrac{6}{5}$

6.3 The quadratic formula

1. a. $a = 3, b = -4, c = 1$
    b. $a = 7, b = -12, c = 2$
    c. $a = 8, b = -1, c = -3$
    d. $a = 1, b = -5, c = 7$

2. a. $a = 5, b = -5, c = -1$
    b. $a = 4, b = -9, c = -3$
    c. $a = 12, b = -29, c = 103$
    d. $a = 43, b = -81, c = -24$

3. a. $\dfrac{-5 \pm \sqrt{21}}{2}$ b. $\dfrac{-3 \pm \sqrt{13}}{2}$

    c. $\dfrac{5 \pm \sqrt{17}}{2}$ d. $2 \pm \sqrt{13}$

4. a. $-1 \pm 2\sqrt{3}$ b. $\dfrac{7 \pm 3\sqrt{5}}{2}$

    c. $\dfrac{9 \pm \sqrt{73}}{2}$ d. $3 \pm 2\sqrt{3}$

5. a. $-4 \pm \sqrt{31}$ b. $\dfrac{1 \pm \sqrt{21}}{2}$

    c. $\dfrac{5 \pm \sqrt{33}}{2}$ d. $-1 \pm 2\sqrt{2}$

6. a. $-0.54, 1.87$ b. $-1.20, 1.45$
    c. $-4.11, 0.61$ d. $-0.61, 0.47$
    e. $0.14, 1.46$

7. a. $0.16, 6.34$ b. $-1.23, 1.90$ c. $-1.00, 1.14$
    d. $-0.83, 0.91$ e. $-0.64, 1.31$

8. a. $-0.35, 0.26$ b. $-1.45, 1.20$ c. $0.08, 5.92$
    d. $-0.68, 0.88$

9. C

10. C

11. B

12. C

13. a. $0.5, 3$ b. $0, 5$
    c. $-1, 3$ d. $0.382, 2.618$

14. a. $2, 4$ b. No real solution
    c. $-1, 8$ d. $-4.162, 2.162$

15. a. $-7, 1.5$ b. No real solution
    c. $2, 7$ d. $-\dfrac{1}{2}, \dfrac{1}{3}$

16. a. $2\pi r^2 + 14\pi r - 231 = 0$
    b. 3.5 cm
    c. 154 cm^2

17. a. $x(x + 30)$
    b. $x(x + 30) = 1500$
    c. 265 mm

18. a. -9 b. 9   c. Kelly

19. a. 4   b. 8 or -8 c. 18

20. a. Pool A: $3\dfrac{2}{3}$ m by $6\dfrac{2}{3}$ m; Pool B: $3\dfrac{1}{3}$ m by $7\dfrac{1}{3}$ m

    b. The area of each is $24\dfrac{4}{9}$ m^2.

21. $25\,\text{m}, 60\,\text{m}$

22. $-2 \pm \sqrt{3}, 9 \pm 4\sqrt{5}$

23. a. $a = -\dfrac{1}{16}, b = \dfrac{7}{8}, c = 2$

   b. $x = -2$ and $x = 16$

   c. One solution is between 12 and 18 ($x = 16$), which means his serve was 'in'.

24. a. Sample responses can be found in the worked solutions in the online resources.

   b. Sample responses can be found in the worked solutions in the online resources.

   c. $m = 1.85$ so NP is $1.85\,\text{cm}$.

6.4 Solving quadratic equations graphically

1. a. i. $x = -2, x = 3$ ii. $x = 1, x = 10$
   iii. $x = -5, x = 5$

   b. Sample responses can be found in the worked solutions in the online resources.

2. a. i. $x = 2$ ii. $x = -1, x = 4$
   iii. $x \approx -1.4, x \approx 4.4$

   b. Sample responses can be found in the worked solutions in the online resources.

3. a. i. $x = -25, x = 10$ ii. $x = 0$
   iii. $x \approx -2.3, x \approx 1.3$ iv. $x \approx -1.5, x = 1$

   b. Sample responses can be found in the worked solutions in the online resources.

4. $150\,\text{m}$

5. A

6. C

7. a. $x = -1$ and $x = 1$

   b. Similarity: shapes; Difference: one is inverted

   c. Similarity: both have $x^2 - 1$; Difference: one has a negative and brackets

8. a. $y = a(x - 2)(x - 5)$

   b. $y = \dfrac{2}{5}(x - 2)(x - 5)$

9. a. $y = a(x - p)(x - q)$

   b. $y = \dfrac{r}{pq}(x - p)(x - q)$

10. a. $7\,\text{m}$ b. $21\,\text{m}$

11. a. $6\,\text{m}$ b. $6\,\text{m}$ c. $8\,\text{m}$

12. $y = -4x^2 + 26x - 30$

6.5 The discriminant

1. a. $\Delta = 289$, 2 solutions   b. $\Delta = 441$, 2 solutions
   c. $\Delta = 24$, 2 solutions   d. $\Delta = 144$, 2 solutions
   e. $\Delta = -28$, 0 solutions

2. a. $\Delta = 81$, 2 solutions   b. $\Delta = 0$, 1 solution
   c. $\Delta = 9$, 2 solutions   d. $\Delta = 32$, 2 solutions
   e. $\Delta = 2809$, 2 solutions

3. a. $\Delta = 8$, 2 solutions   b. $\Delta = 0$, 1 solution
   c. $\Delta = 256$, 2 solutions   d. $\Delta = 256$, 2 solutions

4. a. No real solutions   b. 1 solution
   c. 2 rational solutions   d. 1 solution

5. a. 2 irrational solutions   b. 1 solution
   c. 2 irrational solutions   d. No real solutions

6. a. No real solutions   b. 2 irrational solutions
   c. 2 irrational solutions   d. 2 rational solutions

7. a. Yes   b. No   c. Yes   d. No

8. a. $a = 3, b = 2, c = 7$ b. -80
   c. No real solutions

9. a. $a = -6, b = 1, c = 3$ b. 73
   c. 2 real solutions

10. $\dfrac{1 \pm \sqrt{73}}{12}$

11. A

12. C

13. C

14. $k = -1$

15. $m = 1.8$

16. $n > -\dfrac{9}{4}$

17. a. $0.4\,\text{m}$

   b. $0.28\,\text{m}$

   c. $2.22\,\text{m}$

   d. $2.5\,\text{m}$

   e. i. Yes

   ii. No
   Identify the halfway point between the beginning and the end of the leap, and substitute this value into the equation to determine the maximum height.

   f. $0.625\,\text{m}$

18. Two times

19. p^2 can only give a positive number, which, when added to 24, is always a positive solution.

20. a. $a = -7$ or 5 will give one intersection point.

   b. For values of $< -\dfrac{21}{4}$, there will be no intersection points.

21. a. The straight line crosses the parabola at $(0, -7)$, so no matter what value m takes, there will be at least one intersection point and a maximum of two.

   b. $m = -\dfrac{8}{5}$

22. a. $x = -4$ and $x = -1$

   b. 4

23. $a = 4, b = 12$

24. $k = -4$

Project

1. See table at the bottom of the page.*

2. a. $n - 1$ b. $n^2 - n$

3. Answers will vary. Students should form a quadratic equation and let it equal 42. Then solve this equation.

4. a. $n^2 - n = 250$

   b. $n = \dfrac{\sqrt{1001} + 1}{2}$

   c. $n = 16$

   d. 15

   e. 240

   f. 10

   g. Answers will vary. Students should show overall pattern.

5. Answers will vary. Students should investigate the number of blocks required and threads per block required for a 400-thread warp.

6. Answers will vary. Students should investigate changing the pattern. Students should draw a table to determine the thread warp size for a block size of n threads and also draw the pattern and describe the result for a particular warp size.

6.6 Review questions

1. B

2. B

3. A

4. D

5. $(3x + 4)$ m

6. a. $-5, -3$ b. $-6, -1$ c. $-8, -3$
   d. $2, -6$ e. $5, -2$

7. a. $4, -7$ b. $3, 1$ c. $5, 6$
   d. $7, -5$

8. a. $-2, -6$ b. $-2, -1$ c. $\dfrac{1}{2}, -3$
   d. $2, -7$ e. $-\dfrac{1}{2}, 4$

9. a. $-\dfrac{2}{3}, 2$ b. $2, 1$ c. $\dfrac{5}{3}, \dfrac{5}{2}$
   d. $-7, \dfrac{1}{2}$

10. a. $-4 \pm \sqrt{17}$ b. $-1 \pm \sqrt{6}$ c. $-1, \dfrac{1}{4}$

11. 4

12. a. $-0.651, 1.151$ b. $-0.760, 0.188$
    c. $0.441, -0.566$

13. a. $-0.571, 0.682$ b. $-0.216, 3.836$
    c. $-0.632, 0.632$

14. $-3, 7$

15. $-3, 1$

16. a. 2 irrational solutions
    b. 2 rational solutions
    c. No real solutions

17. 2 solutions

18. 1 solution

19. No solution

20. a.

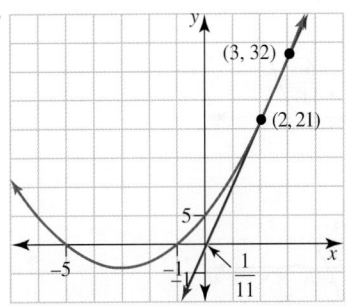

b.

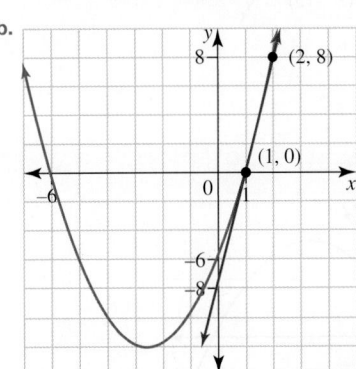

---

*1.

Pattern	Number of threads per block	Number of blocks	Total threads in warp
RB	2	1	2
RRB RRB	3	2	6
RRRB RRBB RBBB	4	3	12
RRRRB RRRBB RRBBB RBBBB	5	4	20
RRRRRB RRRRBB RRRBBB RRBBBB RBBBBB	6	5	30
RRRRRRB RRRRRBB RRRRBBB RRRBBBB RRBBBBB RBBBBBB	7	6	42

c.

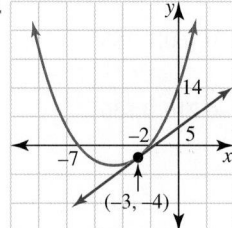

21. a.

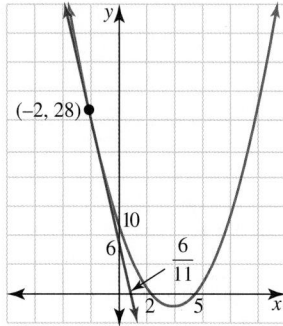

b.

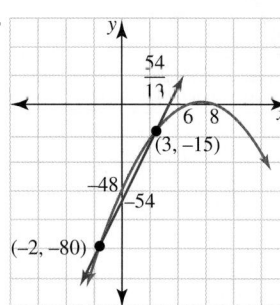

c.

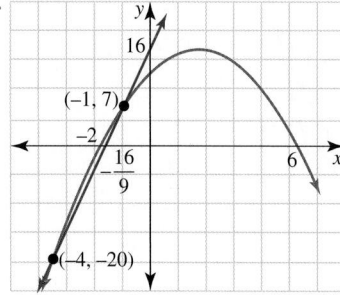

22. -8 and 7

23. Length $= 6\,\text{m}$, width $= 3\,\text{m}$

24. a. $2\pi r(r+10) = 245$
    b. $3.0\,\text{cm}$
    c. $188\,\text{cm}^2$

25. $-\dfrac{25}{8}$

26. $k > 9$ and $k < 1$

27. 24

28. a. $y = 2x^2 - 5x - 2$
    b. No parabola is possible. The points are on the same straight line.

29. $12\left(\sqrt{5}+2\right)\text{cm}$

30. a. $0.5\,\text{m}$
    b. $6.1\,\text{m}$
    c. $76.5\,\text{m}$
    d. No, the ball is 5.5 m off the ground and nobody can reach it.
    e. $9.5\,\text{m}$ away

31. Sample responses can be found in the worked solutions in the online resources.

7 Linear relationships

LESSON
7.1 Overview

Why learn this?

Linear relationships in many ways represent the foundation upon which your understanding in maths will rely on over the final years of your secondary schooling. The principles you learn in this topic will be applied to a variety of contexts you encounter as you learn about higher order polynomial functions and conic sections. Indeed, skills presented in this subject, such as determining the midpoint and length of a line segment, are regularly applicable to the study of differential calculus, which forms a large part of your study in the final years of high-school mathematics.

In the world beyond education, understanding the principles of linear relationships will help you model real-world data and behaviour, interpret the nature of market trends and population trends, and determine points of market equilibrium in the finance sector. A knowledge of algebra, linear quadratic and simultaneous equations is used to create the computer games. Establishing a relationship between variables is also fundamental to the study of science, and the principles learned in this topic will help inform your understanding of the world around us!

Exercise 7.1 Pre-test

learn on

1. **MC** Lines that have the same gradient are:

   A. parallel   **B.** collinear   **C.** perpendicular   **D.** of same lengths

2. Determine the x-intercept of the line $6x + y - 3 = 0$.

3. Sammy has \$35 credit from an App Store. She only buys apps that cost \$2.50 each.
   Calculate the number of apps Sammy can buy and still have \$27.50 credit.

4. Determine the equation of the line, in the form $y = mx + c$.

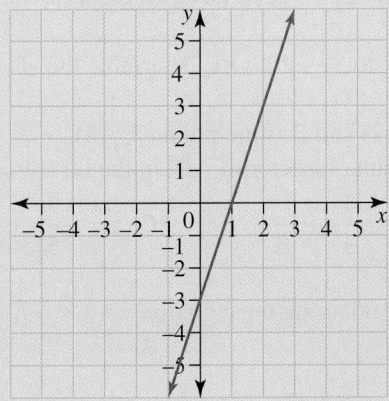

5. **PATH** **MC** The distance between the points $(-3a, 6b)$ and $(a, 2b)$ is:

   A. $\sqrt{4a^2 + 16b^2}$ **B.** $\sqrt{2a^2 + 4b^2}$ **C.** $\sqrt{2a^2 + 8b^2}$ **D.** $4\sqrt{a^2 + b^2}$

6. **MC** Identify the equation of the vertical line passing through the point $(-2, 3)$.

   A. $y = -2$ **B.** $x = -2$ **C.** $y = 3$ **D.** $x = 3$

7. **PATH** The distance between the points $(-3, 10)$ and $(6, a)$ is 15 units. Determine the possible values of a.
   Write the lowest value first.

8. **PATH** **MC** The midpoint of a line segment AB is $(3, -2)$. If the coordinates of A are $(10, 7)$, the coordinates of B are:

   A. $\left(\dfrac{13}{2}, \dfrac{5}{2}\right)$ **B.** $\left(\dfrac{7}{2}, \dfrac{9}{2}\right)$ **C.** $\left(\dfrac{7}{2}, \dfrac{5}{2}\right)$ **D.** $(-4, 11)$

9. **PATH** **MC** The equation of the straight line, in the form $y = mx + c$, passing through the point $(3, -1)$ with a gradient of -2 is:

   A. $y = -2x + 2$ **B.** $y = -2x + 5$ **C.** $y = -2x + 3$ **D.** $y = -2x - 1$

TOPIC 7 Linear relationships **269**

10. **PATH** **MC** The equation of the straight line, in the form $by + ax = k$, that passes through $\left(2, -\dfrac{1}{2}\right)$ and $\left(-6, \dfrac{3}{2}\right)$ is:

A. $2y - 4x = 15$ B. $2y + 4x = 7$ C. $2y + 8x = 7$ D. $2y + 8x = 15$

11. **PATH** ABCD is a parallelogram. The coordinates are A(3, 8), B(6, 1), C(4, −1) and D(1, a). Calculate the value of a.

12. **PATH** Determine the equation of the straight line, in the form $y = mx + c$, that passes through the midpoint of A(0.5, −3) and B(−2.5, 7) and has a gradient of −2.

13. If $2y + 5x = 7$ is perpendicular to $3y + 12 = nx$, determine the value of n.

14. **PATH** **MC** A is the point (−3, 2) and B is the point (7, −4).
The equation of the perpendicular bisector of AB, in the form $y = mx + c$, is:

A. $y = \dfrac{5}{3}x - \dfrac{13}{3}$ B. $y = \dfrac{5}{3}x + \dfrac{11}{3}$ C. $y = -\dfrac{3}{5}x + \dfrac{1}{5}$ D. $y = -\dfrac{3}{5}x + \dfrac{7}{5}$

15. **PATH** **MC** Identify the region satisfying the equation $y + x < -2$.

A.
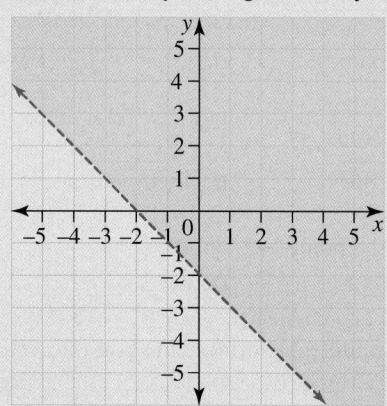

B.
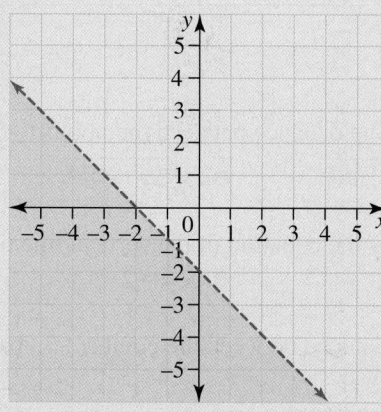

C.
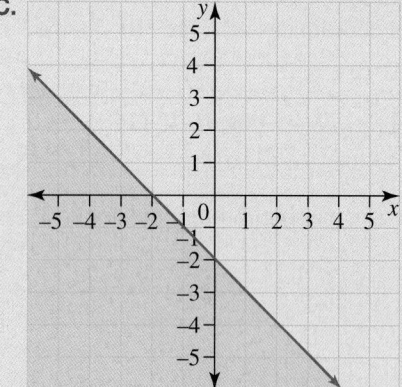

D.
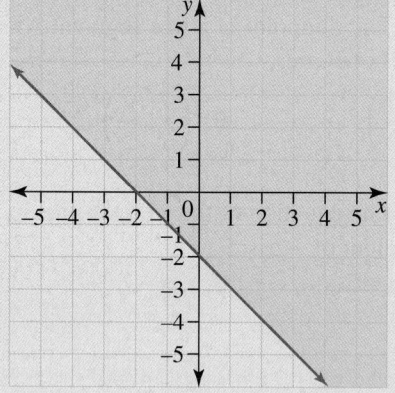

LESSON
7.2 Sketching linear graphs

LEARNING INTENTION

At the end of this lesson you should be able to:
- construct a table of values and use the coordinates to graph linear relationships
- graph linear equations by determining the *x*- and *y*-intercepts
- graph linear equations using the gradient and *y*-intercept
- sketch the graphs of horizontal and vertical lines
- model linear graphs from a real-life context.

▶ 7.2.1 Plotting linear graphs

eles-4736

- If a series of points (x, y) is plotted using the rule $y = mx + c$, then the points always lie in a straight line whose gradient equals m and whose *y*-intercept equals c.
- The rule $y = mx + c$ is called the equation of a straight line written in 'gradient–intercept' form.
- There are different ways to plot a **linear graph** which will be discussed in this lesson:
  - using the table of values
  - using the intercepts
  - using the gradient and the *y*-intercept.

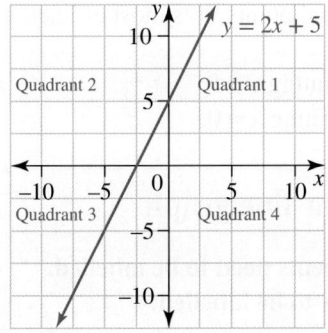

WORKED EXAMPLE 1 Graphing linear equations using the table of values

Plot the linear graph defined by the rule $y = 2x - 5$ for the *x*-values −3, −2, −1, 0, 1, 2 and 3.

THINK

1. Construct a table of values using the given *x*-values.

2. Determine the corresponding *y*-values by substituting each *x*-value into the rule.

WRITE/DRAW

x	−3	−2	−1	0	1	2	3
y							

x	−3	−2	−1	0	1	2	3
y	−11	−9	−7	−5	−3	−1	1

▶

3. Plot the points on a Cartesian plane and rule a straight line through them. Since the *x*-values have been specified, the line should only be drawn between the *x*-values of −3 and 3.

4. Label the graph.

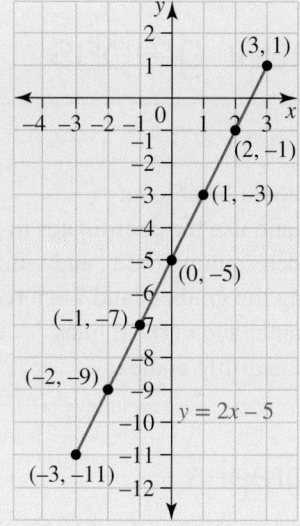

7.2.2 Sketching linear graphs

eles-4737

Sketching a straight line using the *x*- and *y*-intercepts

- Only two points are needed to sketch a straight-line (linear) graph.
- Since it is necessary to label all critical points, it is most efficient to plot these graphs by determining the *x*- and *y*-intercepts.
- Determine the **x-intercept** by substituting $y = 0$.
- Determine the **y-intercept** by substituting $x = 0$.

Sketching a straight-line graph

- **The *x*- and *y*-intercepts need to be labelled.**
- **The equation needs to be labelled.**

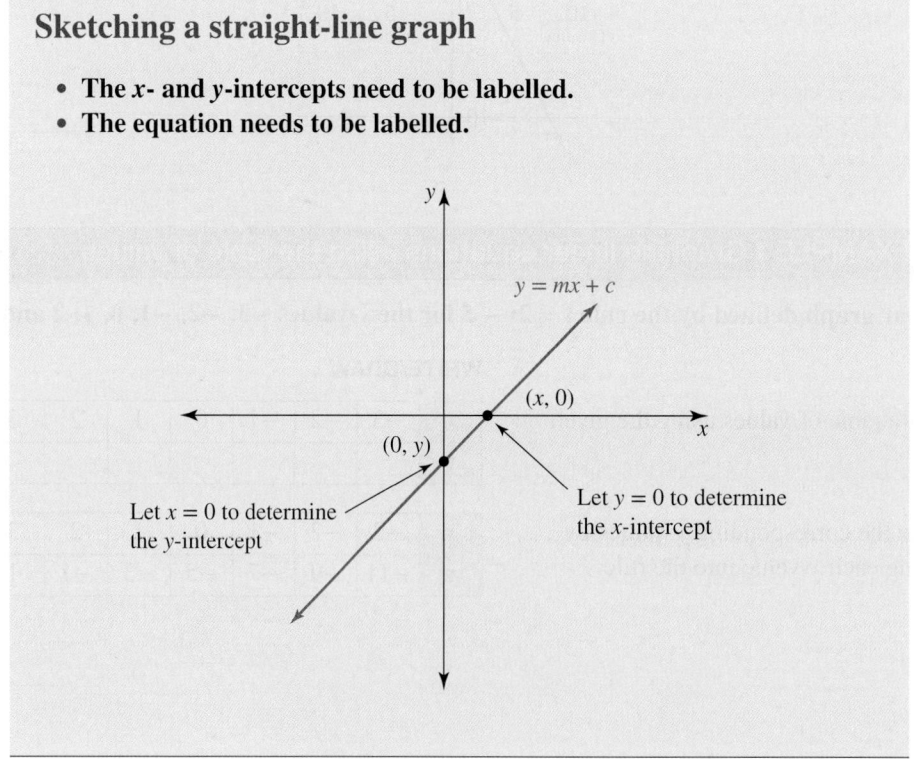

Sketch graphs of the following linear equations.

a. $2x + y = 6$ b. $y = -3x - 12$

THINK	WRITE/DRAW
a. 1. Write the equation.	a. $2x + y = 6$
2. Determine the x-intercept by substituting $y = 0$.	x-intercept: when $y = 0$, $2x + 0 = 6$ $2x = 6$ $x = 3$ The coordinates of the x-intercept are $(3, 0)$.
3. Determine the y-intercept by substituting $x = 0$.	y-intercept: when $x = 0$, $2(0) + y = 6$ $y = 6$ The coordinates of the y-intercept are $(0, 6)$.
4. Plot both points and rule the line. 5. Label the graph.	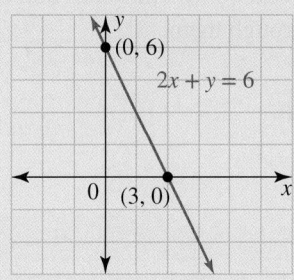
b. 1. Write the equation.	b. $y = -3x - 12$

2. Determine the x-intercept by substituting $y = 0$
   i. Add 12 to both sides of the equation.
   ii. Divide both sides of the equation by -3.
	x-intercept: when $y = 0$,
	$-3x - 12 = 0$
	$-3x = 12$
	$x = -4$
	The coordinates of the
	x-intercept are $(-4, 0)$.

3. Determine the y-intercept. The equation is in the form $y = mx + c$, so compare this with our equation to determine the y-intercept, c.	$c = -12$ The coordinates of the y-intercept are $(0, -12)$.
4. Plot both points and rule the line. 5. Label the graph.	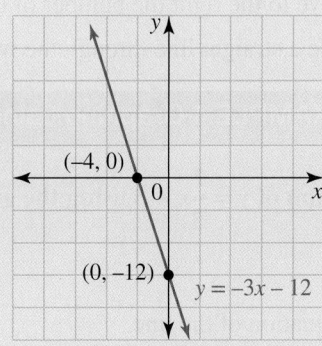

Sketching linear graphs of the form $y = mx$

- Graphs given by $y = mx$ pass through the origin $(0, 0)$, since $c = 0$.
- A second point may be determined using the rule $y = mx$ by substituting a value for x to determine y.

WORKED EXAMPLE 3 Sketching linear graphs of the form $y = mx$

Sketch the graph of $y = 3x$.

THINK	WRITE/DRAW
1. Write the equation.	$y = 3x$
2. Determine the x- and y-intercepts. *Note:* By recognising the form of this linear equation, $y = mx$ you can simply state that the graph passes through the origin, $(0, 0)$.	x-intercept: when $y = 0$, $0 = 3x$ $x = 0$ y-intercept: $(0, 0)$ Both the x- and y-intercepts are at $(0, 0)$.
3. Determine another point to plot by calculating the y-value when $x = 1$.	When $x = 1$, $\quad y = 3 \times 1$ $\qquad\qquad = 3$ Another point on the line is $(1, 3)$.
4. Plot the two points $(0, 0)$ and $(1, 3)$ and rule a straight line through them. 5. Label the graph.	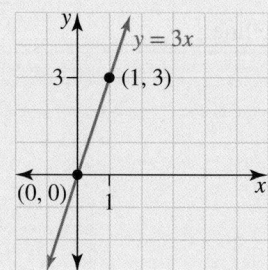

Sketching a straight line using the gradient–intercept method

- This method is often used if the equation is in the form $y = mx + c$, where m represents the gradient (slope) of the straight line, and c represents the y-intercept.
- The steps below outline how to use the gradient–intercept method to sketch a linear graph.

Step 1: Plot a point at the y-intercept.

Step 2: Write the gradient in the form $m = \dfrac{\text{rise}}{\text{run}}$. (To write a whole number as a fraction, place it over a denominator of 1.)

Step 3: Starting from the y-intercept, move up the number of units suggested by the rise (move down if the gradient is negative).

Step 4: Move to the right the number of units suggested by the run and plot the second point.

Step 5: Rule a straight line through the two points.

WORKED EXAMPLE 4 Graphing using the gradient and the y-intercept

Sketch the graph of $y = \dfrac{2}{5}x - 3$ using the gradient–intercept method.

THINK	WRITE
1. Write the equation of the line.	$y = \dfrac{2}{5}x - 3$

---

2. Identify the value of c (that is, the y-intercept) and plot this point.

$c = -3$, so y-intercept: $(0, -3)$

3. Write the gradient, m, as a fraction.

$m = \dfrac{2}{5}$

4. $m = \dfrac{\text{rise}}{\text{run}}$, note the rise and run.

So rise $= 2$; run $= 5$.

5. Starting from the y-intercept at $(0, -3)$, move 2 units up and 5 units to the right to find the second point $(5, -1)$. We have still not found the x-intercept.

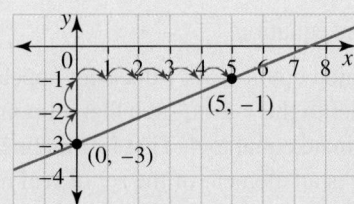

7.2.3 Sketching horizontal and vertical lines

eles-4739

- The line $y = c$ is parallel to the x-axis, has a gradient of zero, a y-intercept of c and forms a horizontal line.
- The line $x = a$ is parallel to the y-axis, has an undefined (infinite) gradient and forms a vertical line.

Horizontal and vertical lines

- **Horizontal lines are in the form $y = c$.**
- **Vertical lines are in the form $x = a$.**

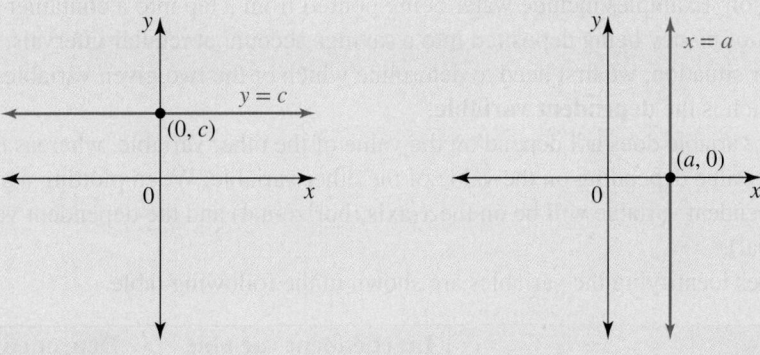

WORKED EXAMPLE 5 Sketching graphs of the form $y = c$ and $x = a$

Sketch graphs of the following linear equations.

a. $y = -3$

b. $x = 4$

THINK

a. 1. Write the equation.

2. The y-intercept is -3. As x does not appear in the equation, the line is parallel to the x-axis, such that all points on the line have a y-coordinate equal to -3. That is, this line is the set of points $(x, -3)$ where x is an element of the set of real numbers.

WRITE/DRAW

a. $y = -3$

y-intercept $= -3$, $(0, -3)$

3. Sketch a horizontal line through (0, −3).

4. Label the graph.

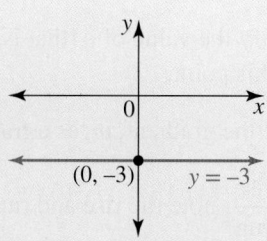

b. 1. Write the equation.

       2. The x-intercept is 4. As y does not appear in the equation, the line is parallel to the y-axis, such that all points on the line have an x-coordinate equal to 4. That is, this line is the set of points $(4, y)$ where y is an element of the set of real numbers.

       3. Sketch a vertical line through (4, 0).

       4. Label the graph.

b. $x = 4$

x-intercept $= 4$, $(4, 0)$

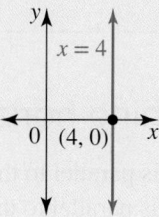

7.2.4 Using linear graphs to model real-life contexts

eles-4740

- If a real-life situation involves a constant increase or decrease at regular intervals, then it can be modelled by a linear equation. Examples include water being poured from a tap into a container at a constant rate, or the same amount of money being deposited into a savings account at regular intervals.
- To model a linear situation, we first need to determine which of the two given variables is the **independent variable** and which is the **dependent variable**.
- The independent variable does not depend on the value of the other variable, whereas the dependent variable takes its value depending on the value of the other variable. When plotting a graph of a linear model, the independent variable will be on the x-axis (horizontal) and the dependent variable will be on the y-axis (vertical).
- Real-life examples identifying the variables are shown in the following table.

Situation	Independent variable	Dependent variable
Money being deposited into a savings account at regular intervals	Time	Money in account
The age of a person in years and their height in cm	Age in years	Height in cm
The temperature at a snow resort and the depth of the snow	Temperature	Depth of snow
The number of workers building a house and the time taken to complete the project	Number of workers	Time

- Note that if time is one of the variables, it will usually be the independent variable. The final example above is a rare case of time being the dependent variable. Also, some of the above cases can't be modelled by linear graphs, as the increases or decreases aren't necessarily happening at constant rates.

WORKED EXAMPLE 6 Using linear graphs to model real-life situations

Water is leaking from a bucket at a constant rate. After 1 minute there is 45 litres in the bucket; after 3 minutes there is 35 litres in the bucket; after 5 minutes there is 25 litres in the bucket; and after 7 minutes there is 15 litres in the bucket.
a. Define two variables to represent the given information.
b. Determine which variable is the independent variable and which is the dependent variable.
c. Represent the given information in a table of values.
d. Plot a graph to represent how the amount of water in the bucket is changing.
e. Use your graph to determine how much water was in the bucket at the start and how long it will take for the bucket to be empty.

THINK

a. Determine which two values change in the relationship given.

b. The dependent variable takes its value depending on the value of the independent variable.
In this situation the amount of water depends on the amount of time elapsed, not the other way round.

c. The independent variable should appear in the top row of the table of values, with the dependent variable appearing in the second row.

d. The values in the top row of the table represent the values on the horizontal axis, and the values in the bottom row of the table represent the values on the vertical axis. As the value for time can't be negative and there can't be a negative amount of water in the bucket, only the first quadrant needs to be drawn for the graph. Plot the 4 points and rule a straight line through them. Extend the graph to meet the vertical and horizontal axes.

e. The amount of water in the bucket at the start is the value at which the line meets the vertical axis, and the time taken for the bucket to be empty is the value at which the line meets the horizontal axis.
Note: Determining the time when the bucket will be empty is an example of *extrapolation* as this time is determined by extending the graph beyond the known data points.

WRITE/DRAW

a. The two variables are 'time' and 'amount of water in bucket'.

b. Independent variable = time
Dependent variable = amount of water in bucket

c.

Time (minutes)	1	3	5	7
Amount of water in bucket (litres)	45	35	25	15

d.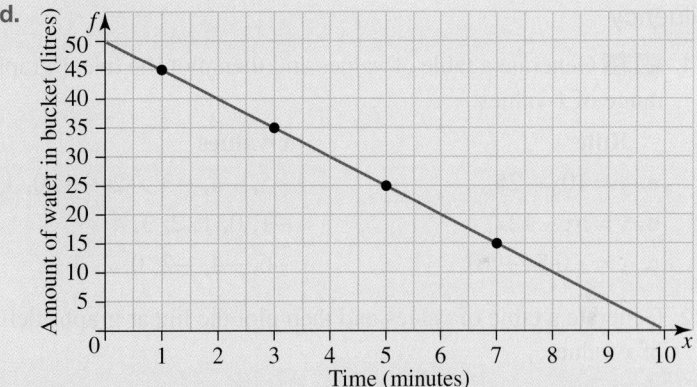

e. There was 50 litres of water in the bucket at the start, and it will take 10 minutes for the bucket to be empty.

What types of straight lines have an x- and y-intercept of the same value?

 Resources

Video eLessons Sketching linear graphs (eles-1919)
                Sketching linear graphs using the gradient-intercept method (eles-1920)
Interactivities Plotting linear graphs (int-3834)
                 The gradient-intercept method (int-3839)
                 The intercept method (int-3840)
                 Equations of straight lines (int-6485)

Exercise 7.2 Sketching linear graphs

learnon

7.2 Quick quiz on	7.2 Exercise

Individual pathways

■ PRACTISE	■ CONSOLIDATE	■ MASTER
1, 4, 7, 10, 13, 16, 21, 24	2, 5, 8, 11, 14, 17, 19, 22, 25	3, 6, 9, 12, 15, 18, 20, 23, 26

Fluency

1. **WE1** Generate a table of values and then plot the linear graphs defined by the following rules for the given range of x-values.

Rule	x-values
a. $y = 10x + 25$	$-5, -4, -3, -2, -1, 0, 1$
b. $y = 5x - 12$	$-1, 0, 1, 2, 3, 4$
c. $y = -0.5x + 10$	$-6, -4, -2, 0, 2, 4$

2. Generate a table of values and then plot the linear graphs defined by the following rules for the given range of x-values.

Rule	x-values
a. $y = 100x - 240$	$0, 1, 2, 3, 4, 5$
b. $y = -5x + 3$	$-3, -2, -1, 0, 1, 2$
c. $y = 7 - 4x$	$-3, -2, -1, 0, 1, 2$

3. Plot the linear graphs defined by the following rules for the given range of x-values.

   Rule   **x-values**

   a. $y = -3x + 2$

x	-6	-4	-2	0	2	4	6
y							

   b. $y = -x + 3$

x	-3	-2	-1	0	1	2	3
y							

c. $y = -2x + 3$

x	-6	-4	-2	0	2	4	6
y							

WE2 For questions **4** to **6**, sketch the graphs of the following linear equations by determining the x- and y-intercepts.

4. a. $5x - 3y = 10$ **b.** $5x + 3y = 10$ **c.** $-5x + 3y = 10$
   d. $-5x - 3y = 10$ **e.** $2x - 8y = 20$

5. a. $4x + 4y = 40$ **b.** $-x + 6y = 120$ **c.** $-2x + 8y = -20$
   d. $10x + 30y = -150$ **e.** $5x + 30y = -150$

6. a. $-9x + 4y = 36$ **b.** $6x - 4y = -24$ **c.** $y = 2x - 10$

   d. $y = -5x + 20$ **e.** $y = -\dfrac{1}{2}x - 4$

WE4 For questions **7** to **9**, sketch the graphs of the following using the gradient–intercept method.

7. a. $y = 4x + 1$ **b.** $y = 3x - 7$ **c.** $y = -2x + 3$

8. a. $y = -5x - 4$ **b.** $y = \dfrac{1}{2}x - 2$ **c.** $y = -\dfrac{2}{7}x + 3$

9. a. $y = 0.6x + 0.5$ **b.** $y = 8x$ **c.** $y = x - 7$

WE3 For questions **10** to **12**, sketch the graphs of the following linear equations on the same set of axes.

10. a. $y = 2x$ **b.** $y = \dfrac{1}{2}x$ **c.** $y = -2x$

11. a. $y = 5x$ **b.** $y = \dfrac{1}{3}x$ **c.** $y = -\dfrac{5}{2}x$

12. a. $y = \dfrac{2}{3}x$ **b.** $y = -3x$ **c.** $y = -\dfrac{3}{2}x$

WE5 For questions **13** to **15**, sketch the graphs of the following linear equations.

13. a. $y = 10$ **b.** $x = -10$ **c.** $x = 0$

14. a. $y = -10$ **b.** $y = 100$ **c.** $x = -100$

15. a. $x = 10$ **b.** $y = 0$ **c.** $y = -12$

Understanding

For questions **16** to **18**, transpose each of the equations to standard form (that is, $y = mx + c$). State the x- and y-intercept for each.

16. a. $5(y + 2) = 4(x + 3)$ **b.** $5(y - 2) = 4(x - 3)$ **c.** $2(y + 3) = 3(x + 2)$

17. a. $10(y - 20) = 40(x - 2)$ **b.** $4(y + 2) = -4(x + 2)$ **c.** $2(y - 2) = -(x + 5)$

18. a. $-5(y + 1) = 4(x - 4)$ **b.** $5(y + 2.5) = 2(x - 3.5)$ **c.** $2.5(y - 2) = -6.5(x - 1)$

19. Determine the x- and y-intercepts of the following lines.
   a. $-y = 8 - 4x$ **b.** $6x - y + 3 = 0$ **c.** $2y - 10x = 50$

20. Explain why the gradient of a horizontal line is equal to zero and the gradient of a vertical line is undefined.

Communicating, reasoning and problem solving

21. **WE6** Your friend loves to download music. She earns $50 and spends some of it buying music online at $1.75 per song. She saves the remainder. Her saving is given by the function $y = 50 - 1.75x$.

    a. Determine which variable is the independent variable and which is the dependent variable.
    b. Sketch the function.
    c. Determine the number of songs your friend can buy and still save $25.

22. Determine whether $\dfrac{x}{3} - \dfrac{y}{2} = \dfrac{7}{6}$ is the equation of a straight line by rearranging into an appropriate form and hence sketch the graph, showing all relevant features.

23. Nikita works a part-time job and is interested in sketching a graph of her weekly earnings. She knows that in a week where she does not work any hours, she will still earn $25.00 for being 'on call'. On top of this initial payment, Nikita earns $20.00 per hour for her regular work. Nikita can work a maximum of 8 hours per day as her employer is unwilling to pay her overtime.

    a. Write a linear equation that represents the amount of money Nikita could earn in a week.
       (*Hint:* You might want to consider the 'on call' amount as an amount of money earned for zero hours worked.)
    b. Sketch a graph of Nikita's weekly potential earnings.
    c. Determine the maximum amount of money that Nikita can earn in a single week.

24. The temperature in a room is rising at a constant rate. Initially (when time equals zero), the temperature of the room is 15 °C. After 1 hour, the temperature of the room has risen to 18 °C. After 3 hours, the temperature has risen to 24 °C.

    a. Using the variables t to represent the time in hours and T to represent the temperature of the room, identify the dependent and the independent variable in this linear relationship.
    b. i. Construct a table of values to represent this information.
       ii. Plot this relationship on a suitable axis.
    c. If the maximum temperature of the room was recorded to be 30 °C, evaluate after how many hours was this recording taken.

25. Water is flowing from a tank at a constant rate. The equation relating the volume of water in the tank, V litres , to the time the water has been flowing from the tank, t minutes, is given by $V = 80 - 4t$, $t \geq 0$.

    a. Determine which variable is the independent variable and which is the dependent variable.
    b. Calculate how much water is in the tank initially.
    c. Explain why it is important that $t \geq 0$.
    d. Determine the rate the water is flowing from the tank.
    e. Determine how long it takes for the tank to empty.
    f. Sketch the graph of V versus t.

26. A straight line has a general equation defined by $y = mx + c$. This line intersects the lines defined by the rules $y = 7$ and $x = 3$. The lines $y = mx + c$ and $y = 7$ have the same y-intercept while $y = mx + c$ and $x = 3$ have the same x-intercept.

    a. On the one set of axes, sketch all three graphs.
    b. Determine the y-axis intercept for $y = mx + c$.
    c. Determine the gradient for $y = mx + c$.
    d. **MC** The equation of the line defined by $y = mx + c$ is:

       A. $x + y = 3$ **B.** $7x + 3y = 21$ **C.** $3x + 7y = 21$ **D.** $x + y = 7$

LESSON
7.3 Determining the equation of a straight line

LEARNING INTENTION

At the end of this lesson you should be able to:
- determine the equation of a straight line when given its graph
- determine the equation of a straight line when given the gradient and the y-intercept
- determine the equation of a straight line passing through two points
- formulate the equation of a straight line from a written context.

▶ 7.3.1 Determining the equation of a straight line given two points

eles-4741

- The equation of a straight line can be found using the gradient-intercept form: $y = mx + c$, where c is the y-intercept and m is the gradient.
- When the graph of the straight line is given, you can determine the gradient by identifying the rise (vertical distance) and the run (horizontal distance).
  If the coordinates, (x_1, y_1) and (x_2, y_2) are given, calculate the gradient using the formula
  $m = \dfrac{\text{rise}}{\text{run}} = \dfrac{y_2 - y_1}{x_2 - x_1}$.
- Use the gradient and one of the points to determine the y-intercept, c.

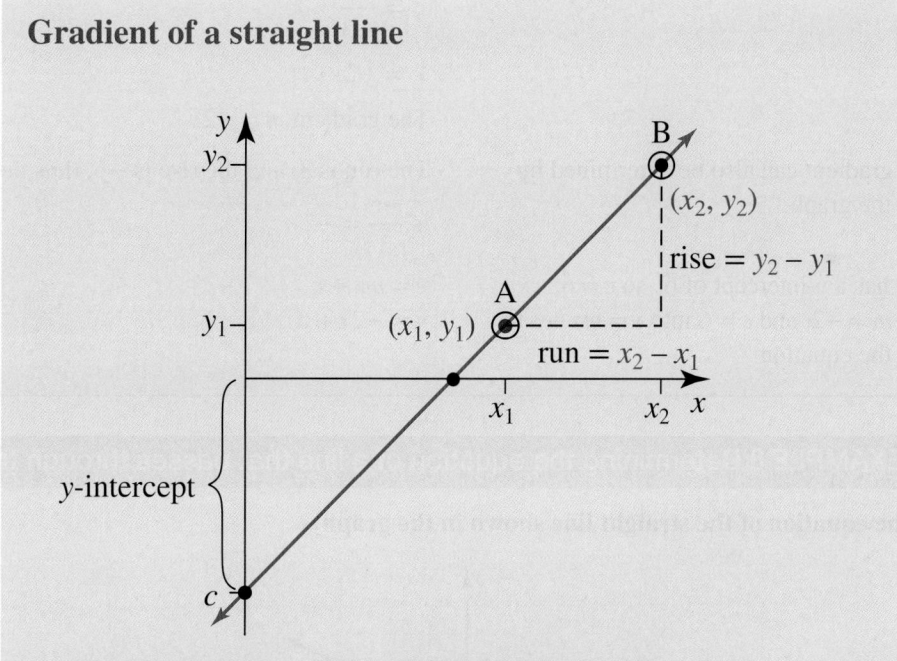

Gradient of a straight line

- **The equation of a straight line is given by $y = mx + c$.**
- **m is the value of the gradient and c is the value of the y-intercept.**

$$\text{Gradient} = m = \frac{\text{rise}}{\text{run}} = \frac{y_2 - y_1}{x_2 - x_1}$$

Determine the equation of the straight line shown in the graph.

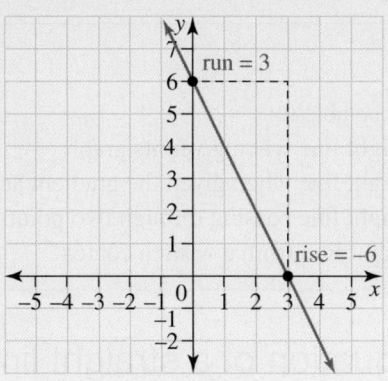

THINK

1. There are two points given on the straight line: the *x*-intercept $(3, 0)$ and the *y*-intercept $(0, 6)$.

2. Calculate the gradient of the line by applying the formula $m = \dfrac{\text{rise}}{\text{run}} = \dfrac{y_2 - y_1}{x_2 - x_1}$, where $(x_1, y_1) = (3, 0)$ and $(x_2, y_2) = (0, 6)$.

   Note: The gradient can also be determined by looking at the graph.

3. The graph has a *y*-intercept of 6, so $c = 6$. Substitute $m = -2$, and $c = 6$ into $y = mx + c$ to determine the equation.

WRITE

$(3, 0), (0, 6)$

$$m = \frac{\text{rise}}{\text{run}}$$

$$= \frac{y_2 - y_1}{x_2 - x_1}$$

$$= \frac{6 - 0}{0 - 3}$$

$$= \frac{6}{-3}$$

$$= -2$$

The gradient $m = -2$.

The run is 3, and the rise is -6, thus the gradient is $\dfrac{-6}{3} = -2$

$y = mx + c$
$y = -2x + 6$

Determine the equation of the straight line shown in the graph.

THINK	WRITE
1. There are two points given on the straight line: the x- and y-intercept $(0, 0)$ and another point $(2, 1)$.	$(0, 0)$, $(2, 1)$
2. Calculate the gradient of the line by applying the formula $m = \dfrac{\text{rise}}{\text{run}} = \dfrac{y_2 - y_1}{x_2 - x_1}$, where $(x_1, y_1) = (0, 0)$ and $(x_2, y_2) = (2, 1)$.	$m = \dfrac{\text{rise}}{\text{run}}$ $= \dfrac{y_2 - y_1}{x_2 - x_1}$ $= \dfrac{1 - 0}{2 - 0}$ $= \dfrac{1}{2}$ The gradient $m = \dfrac{1}{2}$.
Note: The gradient can also be determined looking at the graph.	The run is 2, and the rise is 1, thus the gradient is $\dfrac{1}{2}$
3. The y-intercept is 0, so $c = 0$. Substitute $m = \dfrac{1}{2}$ and $c = 0$ into $y = mx + c$ to determine the equation.	$y = mx + c$ $y = \dfrac{1}{2}x + 0$ $y = \dfrac{1}{2}x$

7.3.2 Determining the equation of a line given the gradient and one point

- If the gradient of a line is known, only one point is needed to determine the equation of the line.
- Follow similar steps.
  Step 1: Substitute the given value of the gradient, m.
  Step 2: Substitute the point and solve for the value of c.
  Step 3: State the equation of the line.

WORKED EXAMPLE 9 Determining equations given the gradient and one point

Determine the equation of the line passing through the point $(-1, 3)$ with a gradient of 2.

THINK	WRITE
1. State the general equation and substitute the value of m.	$y = mx + c$ $y = 2x + c$
2. Substitute the given point for x and y.	$3 = 2 \times (-1) + c$
3. Solve for c.	$3 = -2 + c$ $c = 5$
4. Substitute the value of c into the equation and state the equation of the line.	Equation of the line: $y = 2x + 5$

TOPIC 7 Linear relationships **283**

▶ 7.3.3 Point-gradient form of linear equations (Path)

eles-4742

- The diagram shows a line of gradient m passing through the point (x_1, y_1).
- If (x, y) is any other point on the line, then:

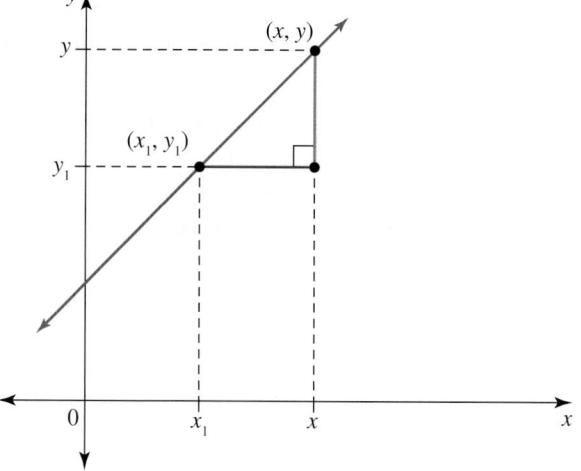

$$m = \frac{\text{rise}}{\text{run}}$$

$$m = \frac{y - y_1}{x - x_1}$$

$$m(x - x_1) = y - y_1$$

$$y - y_1 = m(x - x_1)$$

- The formula $y - y_1 = m(x - x_1)$ can be used to write down the equation of a line, given the gradient and the coordinates of one point.

The point-gradient form of a straight line

- **Determining the equation of a straight line with coordinates of one point (x_1, y_1) and the gradient (m):**

$$y - y_1 = m(x - x_1)$$

PATH

WORKED EXAMPLE 10 Determining the equation using the gradient and the x-intercept

Determine the equation of the straight line with a gradient of 2 and a x-intercept of -3.

THINK	WRITE
1. Write the point-gradient form of a straight line.	$y - y_1 = m(x - x_1)$
2. State the known variables.	$m = 2$, $(x_1, y_1) = (-3, 0)$
3. Substitute the values into the formula.	$y - (0) = 2(x - -3)$ $y = 2x + 6$
4. State the equation.	$y = 2x - 5$

PATH

WORKED EXAMPLE 11 Determining the equation using point-gradient form

Determine the equation of the straight line with a gradient of 3 and passing through the point $(5, -1)$.

THINK	WRITE
1. Write out the point-gradient formula.	$y - y_1 = m(x - x_1)$
2. State the known variables.	$m = 3$, $x_1 = 5$, $y_1 = -1$
3. Substitute the values $m = 3$, $x_1 = 5$, $y_1 = -1$ into the formula.	$y - (-1) = 3(x - 5)$ $y + 1 = 3x - 15$
4. Rearrange the formula to state the equation of the line in the form $y = mx + c$.	$y = 3x - 16$

- Follow these simple steps to determine the equation of a line given two points.
  Step 1: Find the gradient, m, of the line between the two points.
  Step 2: Substitute the gradient into the general equation $y = mx + c$.
  Step 3: Substitute either of the two points into the equation.
  Step 4: Solve for c.
  Step 5: State the equation of the line with the values of m and c.

PATH

WORKED EXAMPLE 12 Determining the equation of a line using two points

Determine the equation of the straight line passing through the points $(-2, 5)$ and $(1, -1)$.

THINK	WRITE
1. Write out the point-gradient formula.	$y - y_1 = m(x - x_1)$
2. State the known variables.	$(x_1, y_1) = (-2, -5)$ $(x_2, y_2) = (1, -1)$
3. Substitute the values $(x_1, y_1) = (-2, 5)$ and $(x_2, y_2) = (1, -1)$ to calculate the gradient from the given points.	$m = \dfrac{y_2 - y_1}{x_2 - x_1}$ $m = \dfrac{-1 - 5}{1 - (-2)}$ $m = \dfrac{-6}{3}$ $= -2$
4. Substitute the values $m = -2$, $(x_1, y_1) = (-2, 5)$ into the formula for the equation of a straight line.	$y - y_1 = m(x - x_1)$ $y - 5 = -2(x - (-2))$
5. Rearrange the formula to state the equation of the line in the form $y = mx + c$.	$y - 5 = -2(x + 2)$ $y = -2x - 4 + 5$ $y = -2x + 1$

PATH

WORKED EXAMPLE 13 Writing an equation in the form $ax + by + c = 0$

Determine the equation of the line with a gradient of -2 which passes through the point $(3, -4)$. Write the equation in general form, that is in the form $ax + by + c = 0$.

THINK	WRITE
1. Use the formula $y - y_1 = m(x - x_1)$. Write the values of x_1, y_1, and m.	$m = -2, \quad x_1 = 3, \quad y_1 = -4$ $y - y_1 = m(x - x_1)$
2. Substitute for x_1, y_1, and m into the equation.	$y - (-4) = -2(x - 3)$ $y + 4 = -2x + 6$
3. Transpose the equation into the form $ax + by + c = 0$.	$y + 4 + 2x - 6 = 0$ $2x + y - 2 = 0$

A printer prints pages at a constant rate. It can print 165 pages in 3 minutes and 275 pages in 5 minutes.
a. **Identify which variable is the independent variable (x) and which is the dependent variable (y).**
b. **Calculate the gradient of the equation and explain what this means in the context of the question.**
c. **Write an equation, in algebraic form, linking the independent and dependent variables.**
d. **Rewrite your equation in words.**
e. **Using the equation, determine how many pages can be printed in 11 minutes.**

THINK

a. The dependent variable takes its value depending on the value of the independent variable. In this situation the number of pages depends on the time elapsed, not the other way round.

b. 1. Determine the two points given by the information in the question.

2. Substitute the values of these two points into the formula to calculate the gradient.

3. The gradient states how much the dependent variable increases for each increase of 1 unit in the independent variable.

c. The graph travels through the origin, as the time elapsed for the printer to print 0 pages is 0 seconds. Therefore, the equation will be in the form $y = mx$. Substitute in the value of m.

d. Replace x and y in the equation with the independent and dependent variables.

e. 1. Substitute $x = 11$ into the equation.

2. Write the answer in words.

WRITE/DRAW

a. Independent variable = time
Dependent variable = number of pages

b. $(x_1, y_1) = (3, 165)$
$(x_2, y_2) = (5, 275)$

$$m = \frac{y_2 - y_1}{x_2 - x_1}$$

$$= \frac{275 - 165}{5 - 3}$$

$$= \frac{110}{2}$$

$$= 55$$

In the context of the question, this means that each minute 55 pages are printed.

c. $y = mx$
$y = 55x$

d. Number of pages = $55 \times$ time

e. $y = 55x$
$= 55 \times 11$
$= 605$

The printer can print 605 pages in 11 minutes.

DISCUSSION

What problems might you encounter when calculating the equation of a line whose graph is actually parallel to one of the axes?

 Resources

▶ **Video eLesson** The equation of a straight line (eles-2313)

✦ **Interactivity**   Linear graphs (int-6484)

Exercise 7.3 Determining the equation of a straight line   learn

7.3 Quick quiz on	7.3 Exercise

Individual pathways

■ PRACTISE	■ CONSOLIDATE	■ MASTER
1, 4, 7, 10, 13, 16, 19	2, 5, 8, 11, 14, 17, 20	3, 6, 9, 12, 15, 18, 21

Fluency

1. **WE7** Determine the equation for each of the straight lines shown.

a. 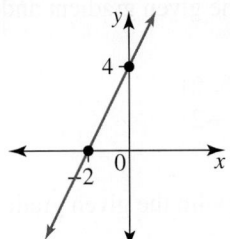   b. 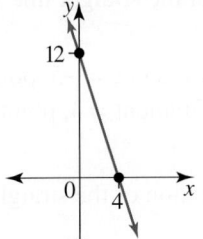   c. 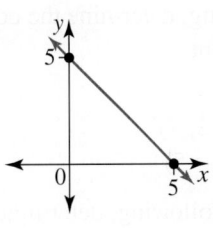   d.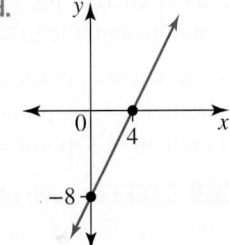

2. Determine the equation for each of the straight lines shown.

a. 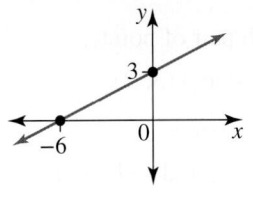   b. 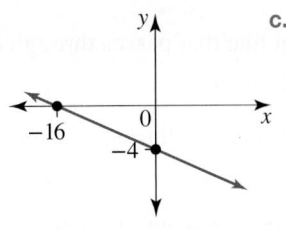   c. 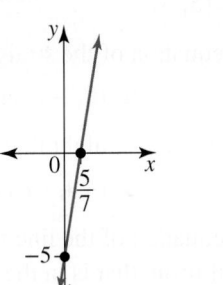   d.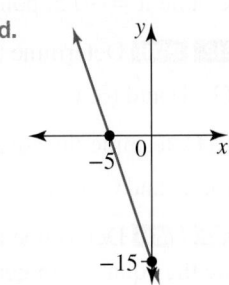

3. **WE8** Determine the equation of each of the straight lines shown.

a.
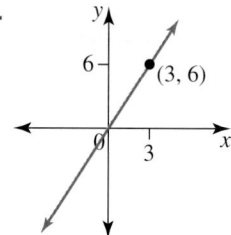

b.
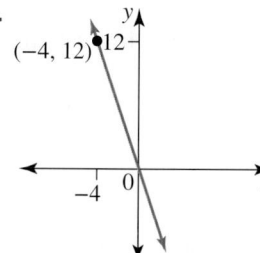

c.
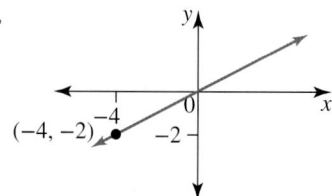

d.
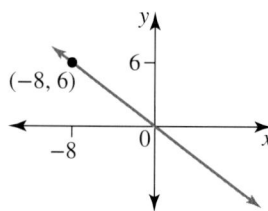

4. **WE10** Determine the linear equation given the information in each case below.

   a. Gradient $= 3$, y-intercept $= 3$
   b. Gradient $= -3$, y-intercept $= 4$
   c. Gradient $= -4$, y-intercept $= 2$
   d. Gradient $= 4$, y-intercept $= 2$
   e. Gradient $= -1$, y-intercept $= 4$

5. Determine the linear equation given the information in each case below.

   a. Gradient $= 0.5$, y-intercept $= -4$
   b. Gradient $= 5$, y-intercept $= 2.5$
   c. Gradient $= -6$, y-intercept $= 3$
   d. Gradient $= -2.5$, y-intercept $= 1.5$
   e. Gradient $= 3.5$, y-intercept $= 6.5$

6. **WE9** For each of the following, determine the equation of the straight line with the given gradient and passing through the given point.

   a. Gradient $= 5$, point $= (5, 6)$
   b. Gradient $= -5$, point $= (5, 6)$
   c. Gradient $= -4$, point $= (-2, 7)$
   d. Gradient $= 4$, point $= (8, -2)$
   e. Gradient $= 3$, point $= (10, -5)$

7. **WE11** **PATH** For each of the following, determine the equation of the straight line with the given gradient and passing through the given point.

   a. Gradient $= -3$, point $= (3, -3)$
   b. Gradient $= -2$, point $= (20, -10)$
   c. Gradient $= 2$, point $= (2, -0.5)$
   d. Gradient $= 0.5$, point $= (6, -16)$
   e. Gradient $= -0.5$, point $= (5, 3)$

8. **WE12** **PATH** Determine the equation of the straight line that passes through each pair of points.

   a. $(1, 4)$ and $(3, 6)$
   b. $(0, -1)$ and $(3, 5)$
   c. $(-1, 4)$ and $(3, 2)$

9. **PATH** Determine the equation of the straight line that passes through each pair of points.

   a. $(3, 2)$ and $(-1, 0)$
   b. $(-4, 6)$ and $(2, -6)$
   c. $(-3, -5)$ and $(-1, -7)$

10. **WE13** **PATH** Determine the equation of the line with a gradient of 2 which passes through the point $(2, -1)$. Write the equation in general form, that is in the form $ax + by + c = 0$.

11. **PATH** Determine the equation of the line with a gradient of -4 which passes through the point $(1, 3)$. Write the equation in general form, that is in the form $ax + by + c = 0$.

12. **PATH** Determine the equation of the line with a gradient of $-\dfrac{1}{2}$ which passes through the point $(-2, -4)$. Write the equation in general form, that is in the form $ax + by + c = 0$.

Understanding

13. **WE14** **PATH** a. Determine which variable (time or cost) is the independent variable and which is the dependent variable in the Supa-Bowl advertisement on the right.

    Save $$$ with Supa-Bowl!!!
    NEW Ten-Pin Bowling Alley
    Shoe rental just $2 (fixed fee)
    Rent a lane for ONLY $6/hour!

    b. If t represents the time in hours and C represents cost ($), construct a table of values for 0–3 hours for the cost of playing ten-pin bowling at the new alley.

    c. Use your table of values to plot a graph of time versus cost. (*Hint:* Ensure your time axis (horizontal axis) extends to 6 hours and your cost axis (vertical axis) extends to $40.)

    d. i. Identify the y-intercept.

       ii. Describe what the y-intercept represents in terms of the cost.

    e. Calculate the gradient and explain what this means in the context of the question.

    f. Write a linear equation to describe the relationship between cost and time.

    g. Use your linear equation from part f to calculate the cost of a 5-hour tournament.

    h. Use your graph to check your answer to part g.

14. **PATH** A local store has started renting out scooters to tour groups who pass through the city. Groups are charged based on the number of people hiring the equipment. There is a flat charge of $10.00 any time you book a day of rentals and it is known that the cost for 20 people to hire scooters is $310.00. The cost for 40 people to hire scooters is $610.00

    a. Label the cost in dollars for hiring scooters for a day as the variable C. Let the number of people hiring scooters be the variable n. Identify which is the dependent variable and which is the independent variable.

    b. Formulate a linear equation that models the cost of hiring scooters for a day.

    c. Calculate how much it will cost to hire 30 scooters.

    d. Sketch a graph of the cost function you created in part b.

15. **PATH** The Robinsons' water tank sprang a leak and has been losing water at a steady rate. Four days after the leak occurred, the tank contained 552 L of water, and ten days later it held only 312 L.

    a. Determine the rule linking the amount of water in the tank (w) and the number of days (t) since the leak occurred.

    b. Calculate how much water was in the tank initially.

    c. If water loss continues at the same rate, determine when the tank will be empty.

Communicating, reasoning and problem solving

16. When using the gradient to draw a line, does it matter if you rise before you run or run before you rise? Explain your answer.

17. **PATH** a. Using the graph shown, write a general formula for the gradient m in terms of x, y and c.

    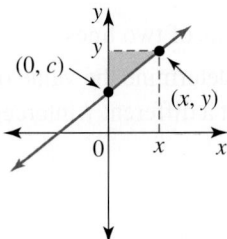

    b. Transpose your formula to make y the subject. Explain what you notice.

18. $A(x_1, y_1)$, $B(x_2, y_2)$ and $P(x, y)$ all lie on the same line. P is a general point that lies anywhere on the line. Given that the gradient from A to P must be equal to the gradient from P to B, show that an equation relating these three points is given by:

$$y - y_1 = \frac{y_2 - y_1}{x_2 - x_1}(x - x_1)$$

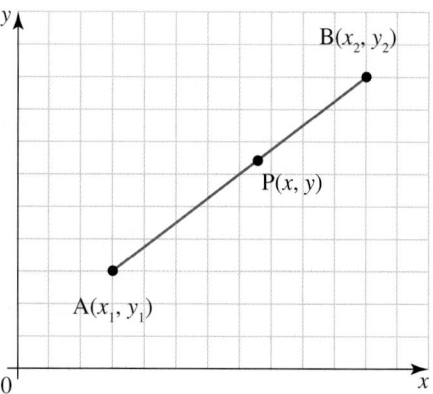

19. **PATH** ABCD is a parallelogram with coordinates A(2, 1), B(3, 6) and C(7, 10).

   a. Calculate the value of the gradient of the line AB.
   b. Determine the equation of the line AB.
   c. Calculate the value of the gradient of the line CD.
   d. Determine the coordinates of the point D.

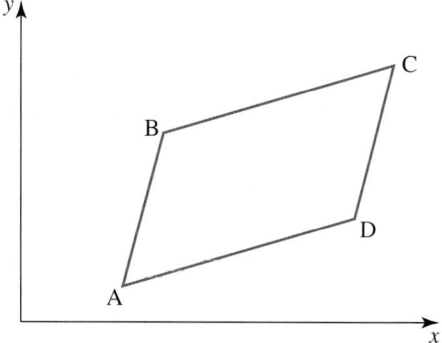

20. **PATH** Show that the quadrilateral ABCD is a parallelogram.

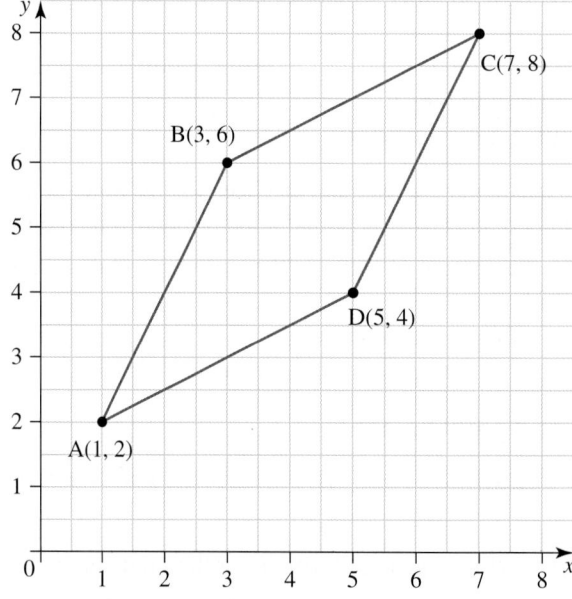

21. $2x + 3y = 5$ and $ax - 6y = b$ are the equations of two lines.

   a. If both lines have the same y intercept, determine the value of b.
   b. If both lines have the same gradient (but a different y intercept), determine the value of a.

LESSON
7.4 Parallel and perpendicular lines

LEARNING INTENTION

At the end of this lesson you should be able to:
- recognise that two lines are parallel, perpendicular or neither
- recognise that lines are perpendicular if the product of their gradients is -1
- determine the equation of a straight line that is parallel or perpendicular to a given line.

▶ 7.4.1 Parallel lines

eles-4743
- Lines that have equal gradients are **parallel**.
- The three lines (pink, green and blue) on the graph shown all have a gradient of 1 and are parallel to each other.
- Parallel lines will never intersect.

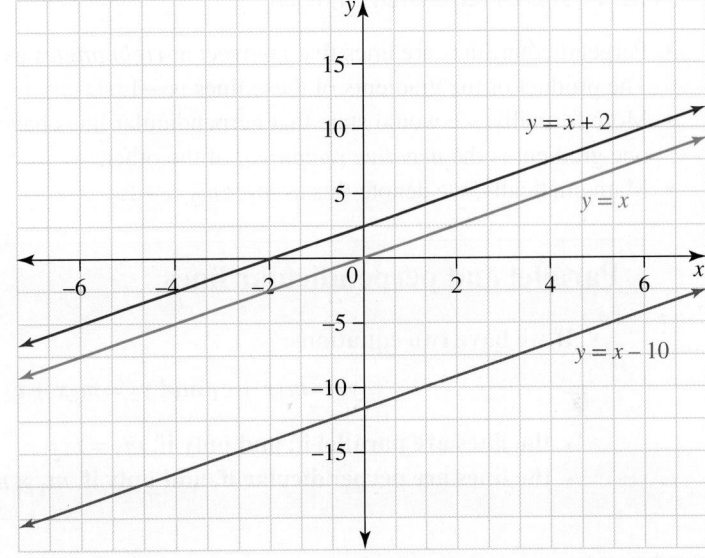

WORKED EXAMPLE 15 Proving that two lines are parallel

Show that AB is parallel to CD given that A has coordinates $(-1, -5)$, B has coordinates $(5, 7)$, C has coordinates $(-3, 1)$ and D has coordinates $(4, 15)$.

THINK

1. AB and CD are parallel if their gradients are equal.

2. Calculate the value of the gradient of AB by applying the formula $m = \dfrac{y_2 - y_1}{x_2 - x_1}$.

WRITE

AB $\parallel$ CD if $m_{AB} = m_{CD}$

Let $A(-1, -5) = (x_1, y_1)$ and $B(5, 7) = (x_2, y_2)$

Since $m = \dfrac{y_2 - y_1}{x_2 - x_1}$

$$m_{AB} = \frac{7 - (-5)}{5 - (-1)}$$

$$= \frac{12}{6}$$

$$= 2$$

3. Calculate the value of the gradient of CD.

Let $C(-3, 1) = (x_1, y_1)$ and $D(4, 15) = (x_2, y_2)$

$$m_{CD} = \frac{15 - 1}{4 - (-3)}$$

$$= \frac{14}{7}$$

$$= 2$$

4. Draw a conclusion.
(*Note*: ∥ means 'is parallel to'.)

Since $m_{AB} = m_{CD} = 2$, then AB ∥ CD.

▶ 7.4.2 Perpendicular lines

eles-4744

- Perpendicular lines are lines that intersect at *right angles* as seen in the diagram. The product of the gradients of these lines is -1.
- More formally, we would state that perpendicular lines have gradients in which one gradient is the *negative reciprocal* of the other.
- Mathematically, we denote this as $m_1 \times m_2 = -1$.

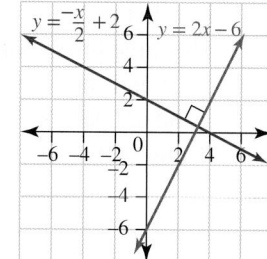

Parallel and perpendicular lines

- **If we have two equations:**

$$y_1 = m_1 x + c_1 \text{ and } y_2 = m_2 x + c_2$$

- the lines are parallel if, and only if, $m_1 = m_2$
- the lines are perpendicular if, and only if, $m_1 \times m_2 = -1$.

WORKED EXAMPLE 16 Proving two lines are perpendicular

Show that the lines $y = -5x + 2$ and $5y - x + 15 = 0$ are perpendicular.

THINK	WRITE
1. Identify the gradient of the first line $y_1 = -5x + 2$.	$y_1 = -5x + 2$ $m_1 = -5$
2. Identify the gradient of the second line $5y - x + 15 = 0$ by rearranging the equation in the form $y = mx + c$.	$5y - x + 15 = 0$ $5y = x - 15$ $y = \dfrac{x}{5} - 3$ $m_2 = \dfrac{1}{5}$
3. Test if the two points are perpendicular by checking whether the product of the two gradients is equal to -1.	$m_1 \times m_2 = -5 \times \dfrac{1}{5}$ $= -1$
4. Write the answer in a sentence.	As the product of the two gradients is equal to -1, therefore these two lines are perpendicular.

⊙ 7.4.3 Determining the equation of a line parallel or perpendicular to another line (Path)

eles-6259

- The gradient of the new line is found from the given parallel or perpendicular line.
- Since one point on the new line is known, determine the equation of the line by substituting the gradient and the point into the general equation $y = mx + c$
- The gradient properties of parallel and perpendicular lines can be used to solve many problems.

PATH

WORKED EXAMPLE 17 Determining the equation of a line parallel to another line

Determine the equation of the line that passes through the point $(3, -1)$ and is parallel to the straight line with equation $y = 2x + 1$.

THINK	WRITE
1. Write the general equation.	$y = mx + c$
2. Determine the gradient of the given line. The two lines are parallel, so they have the same gradient.	$y = 2x + 1$ has a gradient of 2. Hence, $m = 2$.
3. Substitute for m in the general equation.	$y = 2x + c$
4. Substitute the given point to find c.	$(x, y) = (3, -1)$ $-1 = 2(3) + c$ $-1 = 6 + c$ $c = -7$
5. Substitute for c in the general equation.	$y = 2x - 7$ or $2x - y - 7 = 0$

PATH

WORKED EXAMPLE 18 Determining the equation of a line perpendicular to another line

Determine the equation of the line that passes through the point $(2, 1)$ and is perpendicular to the line with a gradient of 5.

THINK	WRITE
1. To determine the equation of a line, we need both a gradient and a point. State the known values.	$(x, y) = (2, 1)$ $m_1 = 5$
2. As the lines are perpendicular so $m_1 \times m_2 = -1$. Calculate the value of the gradient of the other line using this formula.	$m_1 \times m_2 = -1$ $5 \times m_2 = -1$ $m_2 = \dfrac{-1}{5}$
3. Write the general formula and substitute the gradient, m	$y = mx + c$ $y = -\dfrac{1}{5}x + c$

▶

4. Substitute the given point $(x, y) = (2, 1)$ and solve for c.

$$1 = -\frac{1}{5} \times 2 + c$$

$$c = \frac{7}{5}$$

5. State the equation of the line.

$$y = -\frac{1}{5}x + \frac{7}{5}$$

Alternatively: Using the point-gradient formula (path)

Determine the line equation by using the gradient $= \dfrac{-1}{5}$ and the point $(2, 1)$.

$$y - y_1 = m(x - x_1)$$
$$y - 1 = \frac{-1}{5}(x - 2)$$
$$y - 1 = \frac{-x}{5} + \frac{2}{5}$$
$$y = \frac{-x}{5} + \frac{7}{5} \text{ or } 5y + x = 7$$

 Resources

 Interactivities Parallel lines (int-3841)
Perpendicular lines (int-6124)

Exercise 7.4 Parallel and perpendicular lines

learn on

7.4 Quick quiz on	7.4 Exercise

Individual pathways

■ PRACTISE	■ CONSOLIDATE	■ MASTER
1, 3, 7, 8, 13, 14, 18, 19, 22	2, 4, 9, 12, 17, 20, 23	5, 6, 10, 11, 15, 16, 21, 24

Fluency

WE15 For questions **1** to **4**, determine whether AB is parallel to CD given the following sets of points.

1. a. A(4, 13), B(2, 9), C(0, −10), D(15, 0)
   b. A(2, 4), B(8, 1), C(−6, −2), D(2, −6)
   c. A(−3, −10), B(1, 2), C(1, 10), D(8, 16)

2. a. A(1, −1), B(4, 11), C(2, 10), D(−1, −5)
   b. A(1, 0), B(2, 5), C(3, 15), D(7, 35)
   c. A(1, −6), B(−5, 0), C(0, 0), D(5, −4)

3. a. A(1, 6), B(3, 8), C(4, −6), D(−3, 1)
   b. A(2, 12), B(−1, −9), C(0, 2), D(7, 1)
   c. A(1, 3), B(4, 18), C(−5, 4), D(5, 0)

4. a. $A(1, -5), B(0, 0), C(5, 11), D(-10, 8)$
   b. $A(-4, 9), B(2, -6), C(-5, 8), D(10, 14)$
   c. $A(4, 4), B(-8, 5), C(-6, 2), D(3, 11)$

5. Determine which pairs of the following straight lines are parallel.

   a. $2x + y + 1 = 0$
   b. $y = 3x - 1$
   c. $2y - x = 3$
   d. $y = 4x + 3$
   e. $y = \dfrac{x}{2} - 1$
   f. $6x - 2y = 0$
   g. $3y = x + 4$
   h. $2y = 5 - x$

6. **WE16** Show that the lines $y = 6x - 3$ and $x + 6y - 6 = 0$ are perpendicular to one another.

7. **WE17** **PATH** Determine the equation of the line that passes through the point $(4, -1)$ and is parallel to the line with equation $y = 2x - 5$.

8. **WE18** **PATH** Determine the equation of the line that passes through the point $(-2, 7)$ and is perpendicular to a line with a gradient of $\dfrac{2}{3}$.

9. **PATH** Determine the equations of the following lines.

   a. Gradient 3 and passing through the point $(1, 5)$.
   b. Gradient -4 and passing through the point $(2, 1)$.
   c. Passing through the points $(2, -1)$ and $(4, 2)$.
   d. Passing through the points $(1, -3)$ and $(6, -5)$.
   e. Passing through the point $(5, -2)$ and parallel to $x + 5y + 15 = 0$.
   f. Passing through the point $(1, 6)$ and parallel to $x - 3y - 2 = 0$.
   g. Passing through the point $(-1, -5)$ and perpendicular to $3x + y + 2 = 0$.

10. **PATH** Determine the equation of the line that passes through the point $(-2, 1)$ and is:

    a. parallel to the line with equation $2x - y - 3 = 0$.
    b. perpendicular to the line with equation $2x - y - 3 = 0$.

11. **PATH** Determine the equation of the line that contains the point $(1, 1)$ and is:

    a. parallel to the line with equation $3x - 5y = 0$
    b. perpendicular to the line with equation $3x - 5y = 0$.

Understanding

12. **MC** a. The vertical line passing through the point $(3, -4)$ is given by:

    A. $y = -4$
    B. $x = 3$
    C. $y = 3x - 4$
    D. $y = -4x + 3$

   b. Select the point which passes through the horizontal line given by the equation $y = -5$.

    A. $(-5, 4)$
    B. $(4, 5)$
    C. $(3, -5)$
    D. $(5, -4)$

   c. Select which of the following statements is true.

    A. Vertical lines have a gradient of zero.
    B. The y-coordinates of all points on a vertical line are the same.
    C. Horizontal lines have an undefined gradient.
    D. The x-coordinates of all points on a vertical line are the same.

   d. Select which of the following statements is false.

    A. Horizontal lines have a gradient of zero.
    B. The line joining the points $(1, -1)$ and $(-7, -1)$ is vertical.
    C. Vertical lines have an undefined gradient.
    D. The line joining the points $(1, 1)$ and $(-7, 1)$ is horizontal.

13. **MC** **PATH** The point $(-1, 5)$ lies on a line parallel to $4x + y + 5 = 0$. Another point on the same line as $(-1, 5)$ is:

    A. $(2, 9)$ **B.** $(4, 2)$ **C.** $(4, 0)$ **D.** $(3, -11)$

14. **PATH** Determine the equation of the straight line given the following conditions.

    a. Passes through the point $(-1, 3)$ and parallel to $y = -2x + 5$.
    b. Passes through the point $(4, -3)$ and parallel to $3y + 2x = -3$.

15. Determine which pairs of the following lines are perpendicular.

    a. $x + 3y - 5 = 0$ **b.** $y = 4x - 7$ **c.** $y = x$ **d.** $2y = x + 1$
    e. $y = 3x + 2$ **f.** $x + 4y - 9 = 0$ **g.** $2x + y = 6$ **h.** $x + y = 0$

16. **PATH** Determine the equation of the straight line that cuts the x-axis at 3 and is perpendicular to the line with equation $3y - 6x = 12$.

17. Calculate the value of m for which lines with the following pairs of equations are perpendicular to each other.

    a. $2y - 5x = 7$ and $4y + 12 = mx$
    b. $5x - 6y = -27$ and $15 + mx = -3y$

18. **MC** **PATH** The gradient of the line perpendicular to the line with equation $3x - 6y = 2$ is:

    A. 3   **B.** -6 **C.** 2   **D.** -2

Communicating, reasoning and problem solving

19. **PATH** Determine the equation of a line, in the form of $ax + by + c = 0$, that is perpendicular to the line with equation $2x - y = 3$ and passes through the point $(2, 3)$.

20. **PATH** Form the equation of the line, in the form of $ax + by + c = 0$, that is perpendicular to the line with equation $-4x - 3y = 3$ and passes through the point $(-1, 4)$.

21. **MC** **PATH** Triangle ABC has a right angle at B. The vertices are $A(-2, 9)$, $B(2, 8)$ and $C(1, z)$. The value of z is:

    A. $8\dfrac{1}{4}$ **B.** 4   **C.** 12   **D.** $7\dfrac{3}{4}$

22. **PATH** Sketch the following.

    a. Sketch the graph of the equation $y = 2x - 4$.
    b. On the same set of axes, sketch the graph of the line parallel to $y = 2x - 4$ that has a y-intercept of -2.
    c. Sketch the graph of the line that is perpendicular to the lines found in part a and b that also passes through the origin.

23. **PATH** Determine the value(s) of a such that there would be no point of intersection between the lines $ay + 3x = 4a$ and $2x - y = 5$.

24. **PATH** A family of parallel lines has the equation $3x - 2y = k$ where k is a real number.

    a. Determine the gradient of each member of this family of lines.
    b. Show that all lines in the family contain the point (k, k).

LESSON
7.5 The distance between two points formula (Path)

LEARNING INTENTION

At the end of this lesson you should be able to:
- apply the formula to calculate the distance between 2 points on the Cartesian plane
- apply the distance formula.

▶ 7.5.1 The distance between two points

eles-4745

- The distance between two points can be calculated using Pythagoras' theorem.
- Consider two points $A(x_1, y_1)$ and $B(x_2, y_2)$ on the Cartesian plane as shown.

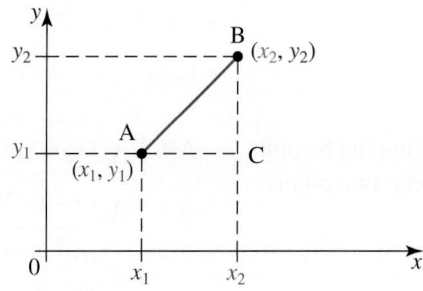

- If point C is placed as shown, ABC is a right-angled triangle and AB is the hypotenuse.

$$AC = x_2 - x_1$$
$$BC = y_2 - y_1$$

By Pythagoras' theorem:

$$AB^2 = AC^2 + BC^2$$
$$= (x_2 - x_1)^2 + (y_2 - y_1)^2$$

$$\text{Hence } AB = \sqrt{(x_2 - x_1)^2 + (y_2 - y_1)^2}$$

The distance between two points

The distance between two points $A(x_1, y_1)$ and $B(x_2, y_2)$ is:

$$AB = \sqrt{(x_2 - x_1)^2 + (y_2 - y_1)^2}$$

- This distance formula can be used to calculate the distance between any two points on the Cartesian plane.
- The distance formula has many geometric applications.

Note: If the coordinates were named in the reverse order, the formula would still give the same answer. Check this for yourself using $(x_1, y_1) = (3, 4)$ and $(x_2, y_2) = (-3, 1)$.

Determine the distance between the points A and B in the figure.

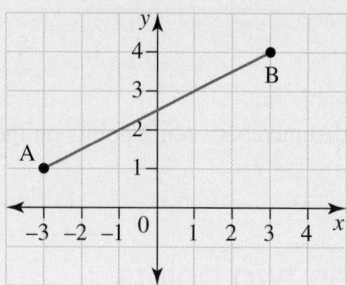

THINK	WRITE
1. From the graph, locate points A and B.	$A(-3, 1)$ and $B(3, 4)$
2. Let A have coordinates (x_1, y_1).	Let $(x_1, y_1) = (-3, 1)$
3. Let B have coordinates (x_2, y_2).	Let $(x_2, y_2) = (3, 4)$
4. Calculate the length AB by applying the formula for calculating the distance between two points.	$\begin{aligned} AB &= \sqrt{(x_2 - x_1)^2 + (y_2 - y_1)^2} \\ &= \sqrt{(3 - (-3))^2 + (4 - 1)^2} \\ &= \sqrt{(6)^2 + (3)^2} \\ &= \sqrt{36 + 9} \\ &= \sqrt{45} \\ &= 3\sqrt{5} \end{aligned}$
Note: You can also use Pythagoras' theorem, as AB is the hypotenuse of a right-angled triangle of side lengths 6 and 3.	$\begin{aligned} AB^2 &= 6^2 + 3^2 \\ AB &= \sqrt{45} \\ &= 3\sqrt{5} \end{aligned}$

Calculate the distance between the points $P(-1, 5)$ and $Q(3, -2)$.

THINK	WRITE
1. Let P have coordinates (x_1, y_1).	Let $(x_1, y_1) = (-1, 5)$
2. Let Q have coordinates (x_2, y_2).	Let $(x_2, y_2) = (3, -2)$
3. Calculate the length PQ by applying the formula for the distance between two points.	$\begin{aligned} PQ &= \sqrt{(x_2 - x_1)^2 + (y_2 - y_1)^2} \\ &= \sqrt{(3 - (-1))^2 + (-2 - 5)^2} \\ &= \sqrt{(4)^2 + (-7)^2} \\ &= \sqrt{16 + 49} \\ &= \sqrt{65} \end{aligned}$

WORKED EXAMPLE 21 Applying the distance formula

Prove that the points A(1, 1), B(3, −1) and C(−1, −3) are the vertices of an isosceles triangle.

THINK	WRITE/DRAW
1. Plot the points and draw the triangle. *Note*: For triangle ABC to be isosceles, two sides must have the same magnitude.	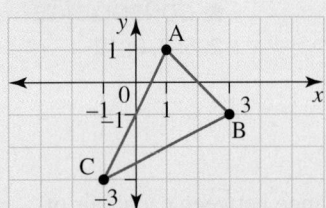
2. AC and BC seem to be equal. Calculate the length AC. $A(1, 1) = (x_2, y_2)$ $C(-1, -3) = (x_1, y_1)$	$AC = \sqrt{[1-(-1)]^2 + [1-(-3)]^2}$ $= \sqrt{(2)^2 + (4)^2}$ $= \sqrt{20}$ $= 2\sqrt{5}$
3. Calculate the length BC. $B(3, -1) = (x_2, y_2)$ $C(-1, -3) = (x_1, y_1)$	$BC = \sqrt{[3-(-1)]^2 + [-1-(-3)]^2}$ $= \sqrt{(4)^2 + (2)^2}$ $= \sqrt{20}$ $= 2\sqrt{5}$
4. Calculate the length AB. $A(1, 1) = (x_1, y_1)$ $B(3, -1) = (x_2, y_2)$	$AB = \sqrt{[3-(1)]^2 + [-1-(1)]^2}$ $= \sqrt{(2)^2 + (-2)^2}$ $= \sqrt{4+4}$ $= 2\sqrt{2}$
5. Write your conclusion.	Since $AC = BC \neq AB$, triangle ABC is an isosceles triangle.

DISCUSSION

How could you use the distance formula to show that a series of points lay on the circumference of a circle with centre C?

 Resources

⬥ **Interactivity** Distance between two points (int-6051)

Exercise 7.5 The distance between two points formula (Path) learn on

7.5 Quick quiz on   **7.5 Exercise**

Individual pathways

■ PRACTISE	■ CONSOLIDATE	■ MASTER
1, 4, 7, 11	2, 5, 8, 9, 12	3, 6, 10, 13

Fluency

1. **WE19** Determine the distance between each pair of points shown in the graph.

2. **WE20** Calculate the distance between the following pairs of points.

   a. $(2, 5)$, $(6, 8)$ b. $(-1, 2)$, $(4, 14)$
   c. $(-1, 3)$, $(-7, -5)$ d. $(5, -1)$, $(10, 4)$
   e. $(4, -5)$, $(1, 1)$

3. Calculate the distance between the following pairs of points.

   a. $(-3, 1)$, $(5, 13)$ b. $(5, 0)$, $(8, 0)$
   c. $(1, 7)$, $(1, -6)$ d. (a, b), $(2a, -b)$
   e. $(-a, 2b)$, $(2a, -b)$

4. The vertices of a quadrilateral are A$(1, 4)$, B$(-1, 8)$, C$(1, 9)$ and D$(3, 5)$.

   a. Determine the lengths of the sides.
   b. Determine the lengths of the diagonals.
   c. State the type of quadrilateral.

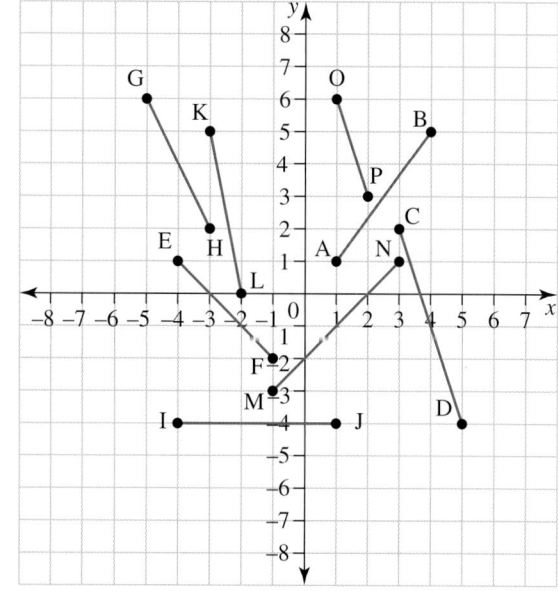

Understanding

5. **MC** If the distance between the points $(3, b)$ and $(-5, 2)$ is 10 units, then the value of b is:

   A. -8 B. -4 C. 4   D. 0

6. **MC** A rhombus has vertices A$(1, 6)$, B$(6, 6)$, C$(-2, 2)$ and D(x, y). The coordinates of D are:

   A. $(2, -3)$ B. $(2, 3)$ C. $(-2, 3)$ D. $(3, 2)$

Communicating, reasoning and problem solving

7. **WE21** Prove that the points A$(0, -3)$, B$(-2, -1)$ and C$(4, 3)$ are the vertices of an isosceles triangle.

8. The points P$(2, -1)$, Q$(-4, -1)$ and R$(-1, 3\sqrt{3} - 1)$ are joined to form a triangle. Prove that triangle PQR is equilateral.

9. Prove that the triangle with vertices D$(5, 6)$, E$(9, 3)$ and F$(5, 3)$ is a right-angled triangle.

10. A rectangle has vertices A$(1, 5)$, B$(10.6, z)$, C$(7.6, -6.2)$ and D$(-2, 1)$. Determine:

    a. the length of CD
    b. the length of AD
    c. the length of the diagonal AC
    d. the value of z.

11. Triangle ABC is an isosceles triangle where AB = AC, B is the point (−1, 2), C is the point (6, 3) and A is the point (a, 3a). Determine the value of the integer constant a.

12. Show that the triangle ABC with coordinates A(a, a), B(m, −a) and C(−a, m) is isosceles.

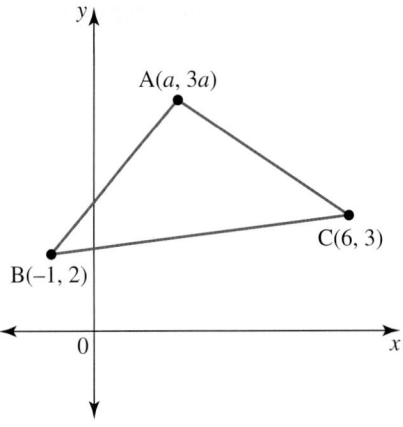

13. ABCD is a parallelogram.
   a. Evaluate the gradients of AB and BC.
   b. Determine the coordinates of the point D(x, y).
   c. Show that the diagonals AC and BD bisect each other.

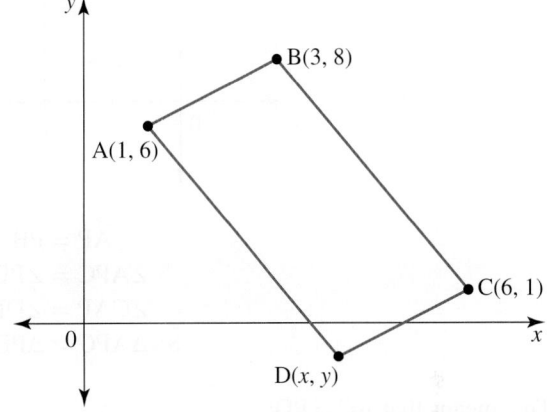

LESSON
7.6 The midpoint of a line segment formula (Path)

LEARNING INTENTION

At the end of this lesson you should be able to:
- apply the formula to determine the midpoint between 2 points on the Cartesian plane
- apply the formula for midpoint.

⏵ 7.6.1 Midpoint of a line segment

eles-4746

- The **midpoint** of a **line segment** is the halfway point.
- The x- and y-coordinates of the midpoint is calculated by determining the average of the x and y values of the endpoints.
- The following diagram shows the line interval AB joining points $A(x_1, y_1)$ and $B(x_2, y_2)$.

The midpoint of AB is P, so AP = PB.

Points $C(x, y_1)$ and $D(x_2, y)$ are added to the diagram and are used to make the two right-angled triangles $\triangle ABC$ and $\triangle PBD$.

The two triangles are congruent:

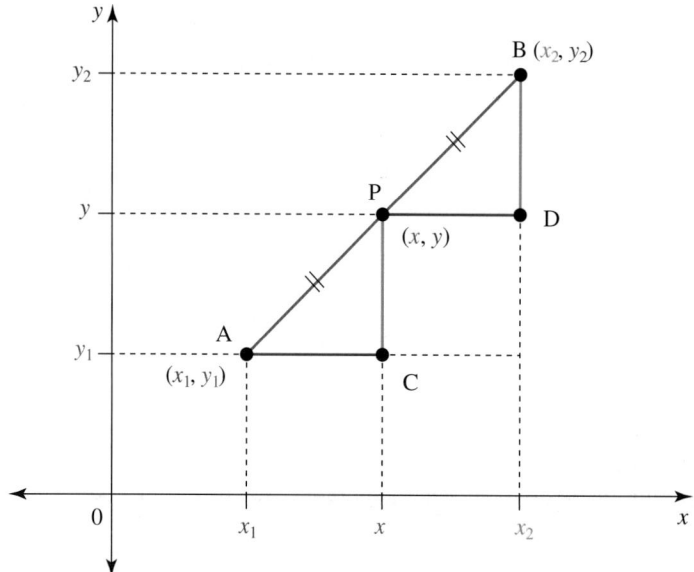

$$AP = PB \quad \text{(given)}$$
$$\angle APC = \angle PBD \quad \text{(corresponding angles)}$$
$$\angle CAP = \angle DPB \quad \text{(corresponding angles)}$$
$$\text{So } \triangle APC = \triangle PBD \quad \text{(ASA)}$$

This means that $AC = PD$;

$$\text{i.e. } x - x_1 = x_2 - x \quad \text{(solve for } x\text{)}$$
$$\text{i.e. } \quad 2x = x_1 + x_2$$
$$x = \frac{x_1 + x_2}{2}$$

In other words, x is simply the average x_1 and x_2.

Similarly, $y = \dfrac{y_1 + y_2}{2}$.

The midpoint formula

To calculate the midpoint (x, y) of the two points A (x_1, y_1) and B (x_2, y_2):
- **The x-value is the average of x_1 and x_2.**
- **The y-value is the average of y_1 and y_2.**

$$\textbf{Midpoint} = \left(\frac{x_1 + x_2}{2}, \frac{y_1 + y_2}{2} \right)$$

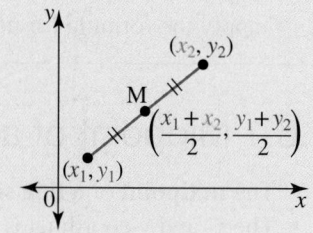

Calculate the coordinates of the midpoint of the line segment joining $(-2, 5)$ and $(7, 1)$.

THINK

1. Label the given points (x_1, y_1) and (x_2, y_2).

2. Determine the x-coordinate of the midpoint.

3. Determine the y-coordinate of the midpoint.

4. Write the coordinates of the midpoint.

WRITE

Let $(x_1, y_1) = (-2, 5)$ and $(x_2, y_2) = (7, 1)$

$$x = \frac{x_1 + x_2}{2}$$

$$= \frac{-2 + 7}{2}$$

$$= \frac{5}{2}$$

$$= 2\frac{1}{2}$$

$$y = \frac{y_1 + y_2}{2}$$

$$= \frac{5 + 1}{2}$$

$$= \frac{6}{2}$$

$$= 3$$

The midpoint is $\left(2\frac{1}{2}, 3\right)$.

The coordinates of the midpoint, M, of the line segment AB are $(7, 2)$. If the coordinates of A are $(1, -4)$, determine the coordinates of B.

THINK

1. Let the start of the line segment be (x_1, y_1) and the midpoint be (x, y).

2. The average of the x-coordinates is 7. Determine the x-coordinate of the end point.

3. The average of the y-coordinates is 2. Determine the y-coordinate of the end point.

WRITE/DRAW

Let $(x_1, y_1) = (1, -4)$ and $(x, y) = (7, 2)$

$$x = \frac{x_1 + x_2}{2}$$

$$7 = \frac{1 + x_2}{2}$$

$$14 = 1 + x_2$$

$$x_2 = 13$$

$$y = \frac{y_1 + y_2}{2}$$

$$2 = \frac{-4 + y_2}{2}$$

$$4 = -4 + y_2$$

$$y_2 = 8$$

4. Write the coordinates of the end point.

The coordinates of the point B are (13, 8).

5. Check that the coordinates are feasible by drawing a diagram.

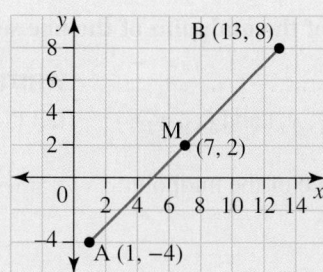

DISCUSSION

If the midpoint of a line segment is the origin, what are the possible values of the x- and y- coordinates of the end points?

Exercise 7.6 The midpoint of a line segment formula (Path)   learn on

7.6 Quick quiz on	7.6 Exercise

Individual pathways

■ PRACTISE	■ CONSOLIDATE	■ MASTER
1, 3, 6, 7, 12, 15	4, 5, 8, 9, 13, 16	2, 10, 11, 14, 17

Fluency

1. WE22 Calculate the coordinates of the midpoint of the line segment joining the following pairs of points.

   a. $(-5, 1), (-1, -8)$
   b. $(4, 2), (11, -2)$
   c. $(0, 4), (-2, -2)$

2. Calculate the coordinates of the midpoint of the line segment joining the following pairs of points.

   a. $(3, 4), (-3, -1)$
   b. $(a, 2b), (3a, -b)$
   c. $(a + 3b, b), (a - b, a - b)$

3. WE23 The coordinates of the midpoint, M, of the line segment AB are $(2, -3)$. If the coordinates of A are $(7, 4)$, determine the coordinates of B.

4. Determine the midpoint of the following sets of coordinates.

   a. $(1, 2)$ and $(3, -4)$
   b. $(7, -2)$ and $(-4, 13)$
   c. $(3, a)$ and $(1, 4a)$

5. If M$(2, -2)$ is the midpoint of the line segment joining the points X$(4, y)$ and Y$(x, -1)$, then calculate the value of $x + y$.

Understanding

6. A square has vertices A(0, 0), B(2, 4), C(6, 2) and D(4, −2). Determine:
   a. the coordinates of the centre   b. the length of a side   c. the length of a diagonal.

7. **MC** The midpoint of the line segment joining the points (−2, 1) and (8, −3) is:
   A. (6, −2)   B. (5, 2)   C. (6, 2)   D. (3, −1)

8. **MC** If the midpoint of AB is (−1, 5) and the coordinates of B are (3, 8), then A has coordinates:
   A. (1, 6.5)   B. (2, 13)   C. (−5, 2)   D. (4, 3)

9. a. The vertices of a triangle are A(2, 5), B(1, −3) and C(−4, 3). Determine:
      i. the coordinates of P, the midpoint of AC
      ii. the coordinates of Q, the midpoint of AB
      iii. the length of PQ.
   b. Show that BC = 2PQ.

10. a. A quadrilateral has vertices A(6, 2), B(4, −3), C(−4, −3) and D(−2, 2). Determine:
       i. the midpoint of the diagonal AC
       ii. the midpoint of the diagonal BD.
    b. State what you can infer about the quadrilateral.

11. a. The points A(−5, 3.5), B(1, 0.5) and C(−6, −6) are the vertices of a triangle. Determine:
       i. the midpoint, P, of AB   ii. the length of PC
       iii. the length of AC   iv. the length of BC.

    b. Describe the triangle. State what PC represents.

Communicating, reasoning and problem solving

12. a. Plot the following points on a Cartesian plane: A(−1, −4), B(2, 3), C(−3, 8) and D(4, −5).
    b. Show that the midpoint of the interval AC is (−2, 2).
    c. Calculate the exact distance between the points A and C.
    d. If B is the midpoint of an interval CM, determine the coordinates of point M.
    e. Show that the gradient of the line segment AB is $\dfrac{7}{3}$.
    f. Determine the equation of the line that passes through the points B and D.

13. Write down the coordinates of the midpoint of the line joining the points $(3k − 1, 4 − 5k)$ and $(5k − 1, 3 − 5k)$. Show that this point lies on the line with equation $5x + 4y = 9$.

14. The points A($2m$, $3m$), B($5m$, $−2m$) and C($−3m$, 0) are the vertices of a triangle. Show that this is aright-angled triangle.

15. Determine the equation of the straight line that passes through the midpoint of A(−2, 5) and B(−2, 3), and has a gradient of −3.

16. Determine the equation of the straight line that passes through the midpoint of A(−1, −3) and B(3, −5), and has a gradient of $\dfrac{2}{3}$.

17. Determine the equation of the straight line passing through the midpoint of (3, 2) and (5, −2) that is also perpendicular to the line $3x − 2y = 7$.

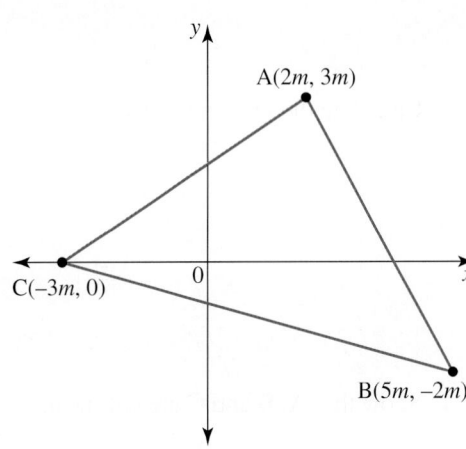

LESSON
7.7 Applications to solving problems (Path)

LEARNING INTENTION

At the end of this lesson you should be able to:
- determine whether a set of coordinates are collinear
- determine the equation of a perpendicular bisector of a line segment
- determine equations of horizontal and vertical lines.

⏵ 7.7.1 Collinear points

eles-4747

- **Collinear points** are points that all lie on the same straight line.

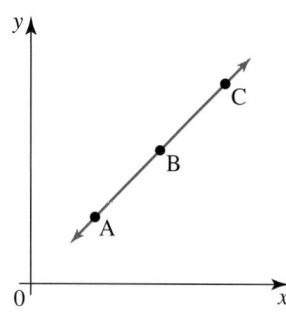

- If A, B and C are collinear, then $m_{AB} = m_{BC}$.

WORKED EXAMPLE 24 Proving points are collinear

Show that the points A(2, 0), B(4, 1) and C(10, 4) are collinear.

THINK	WRITE
1. Calculate the gradient of AB.	Let A(2, 0) $= (x_1, y_1)$ and B(4, 1) $= (x_2, y_2)$
	since $\quad m = \dfrac{y_2 - y_1}{x_2 - x_1}$
	$m_{AB} = \dfrac{1 - 0}{4 - 2}$
	$\quad\quad = \dfrac{1}{2}$
2. Calculate the gradient of BC.	Let B(4, 1) $= (x_1, y_1)$ and C(10, 4) $= (x_2, y_2)$
	$m_{BC} = \dfrac{4 - 1}{10 - 4}$
	$\quad\quad = \dfrac{3}{6}$
	$\quad\quad = \dfrac{1}{2}$
3. Show that A, B and C are collinear.	Since $m_{AB} = m_{BC} = \dfrac{1}{2}$ and B is common to both line segments, A, B and C are collinear.

⊙ 7.7.2 Equations of horizontal and vertical lines

eles-4748

- Horizontal lines are parallel to the x-axis, have a gradient of zero, are expressed in the form $y = c$ and have no x-intercept.
- Vertical lines are parallel to the y-axis, have an undefined (infinite) gradient, are expressed in the form $x = a$ and have no y-intercept.

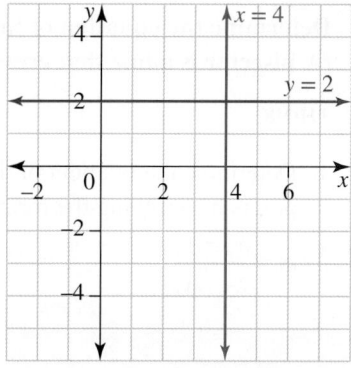

WORKED EXAMPLE 25 Determining the equation of vertical and horizontal lines

Determine the equation of:
a. **the vertical line that passes through the point (2, −3)**
b. **the horizontal line that passes through the point (−2, 6).**

THINK	WRITE
a. The equation of a vertical line is $x = a$. The x-coordinate of the given point is 2.	a. $x = 2$
b. The equation of a horizontal line is $y = c$. The y-coordinate of the given point is 6.	b. $y = 6$

⊙ 7.7.3 Perpendicular bisectors

eles-4749

- A perpendicular bisector is a line that intersects another line at a right angle and cuts it into two equal lengths.
- A perpendicular bisector passes through the midpoint of a line segment.

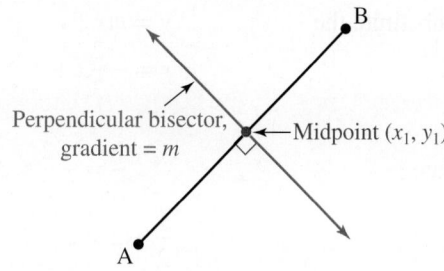

- The equation of a perpendicular bisector can be found by:
  Step 1: determining the midpoint of the interval joining the two points
  Step 2: determining the gradient of the interval joining the two points
  Step 3: state the perpendicular gradient
  Step 4: substitute values into the general equation $y = mx + c$

Determine the equation of the perpendicular bisector of the line joining the points $(0, -4)$ and $(6, 5)$. (A bisector is a line that crosses another line at right angles and cuts it into two equal lengths.)

THINK

WRITE

1. Determine the gradient of the line joining the given points by applying the formula.
$$m = \frac{y_2 - y_1}{x_2 - x_1}.$$

Let $(0, -4) = (x_1, y_1)$.
Let $(6, 5) = (x_2, y_2)$.
$$m_1 = \frac{y_2 - y_1}{x_2 - x_1}$$
$$m_1 = \frac{5 - (-4)}{6 - 0}$$
$$= \frac{9}{6}$$
$$= \frac{3}{2}$$

2. Calculate the gradient of the perpendicular line.
$$m_1 \times m_2 = -1$$

$$m_1 = \frac{3}{2}$$
$$m_2 = -\frac{2}{3}$$

3. Determine the midpoint of the line joining the given points.
$$M = \left(\frac{x_1 + x_2}{2}, \frac{y_1 + y_2}{2} \right) \text{ where } (x_1, y_1) = (0, -4)$$
and $(x_2, y_2) = (6, 5)$.

$$x = \frac{x_1 + x_2}{2} \qquad y = \frac{y_1 + y_2}{2}$$
$$= \frac{0 + 6}{2} \qquad = \frac{-4 + 5}{2}$$
$$= 3 \qquad = \frac{1}{2}$$

Hence $\left(3, \frac{1}{2} \right)$ are the coordinates of the midpoint.

4. Write the general formula and substitute the perpendicular gradient.

$$y = mx + c$$
$$y = -\frac{2}{3}x + c$$

5. Substitute the midpoint to evaluate c.

$$\frac{1}{2} = -\frac{2}{3} \times 3 + c$$
$$\frac{1}{2} + 2 = c$$
$$c = \frac{5}{2}$$

6. State the equation of the perpendicular bisector of the interval.

$$y = -\frac{2}{3}x + \frac{5}{2}$$

Alternatively: Using the simple formula (path)

4. Determine the equation of the line with gradient $-\dfrac{2}{3}$ that passes through $\left(3, \dfrac{1}{2}\right)$.

Since $y - y_1 = m(x - x_1)$,

then $y - \dfrac{1}{2} = -\dfrac{2}{3}(x - 3)$

5. Simplify by removing the fractions.
   Multiply both sides by 3.
   Multiply both sides by 2.

$$3\left(y - \dfrac{1}{2}\right) = -2(x - 3)$$

$$3y - \dfrac{3}{2} = -2x + 6$$

$$6y - 3 = -4x + 12$$

$$4x + 6y - 15 = 0$$

DISCUSSION

How could you use coordinate geometry to design a logo for an organisation?

 Resources

⬥ **Interactivity** Vertical and horizontal lines (int-6049)

Exercise 7.7 Applications to solving problems (Path)   learn on

7.7 Quick quiz on	7.7 Exercise

Individual pathways

■ PRACTISE	■ CONSOLIDATE	■ MASTER
1, 3, 4, 11, 12, 13, 18, 19	2, 7, 9, 14, 15, 20	5, 6, 8, 10, 16, 17, 21

Fluency

1. **WE24** Show that the points A$(0, -2)$, B$(5, 1)$ and C$(-5, -5)$ are collinear.

2. Show that the line that passes through the points $(-4, 9)$ and $(0, 3)$ also passes through the point $(6, -6)$.

3. **WE25** Determine the equation of:
   a. the vertical line that passes through the point $(1, -8)$
   b. the horizontal line that passes through the point $(-5, -7)$.

4. **WE26** Determine the equation of the perpendicular bisector of the line joining the points $(1, 2)$ and $(-5, -4)$.

5. a. Show that the following three points are collinear: $\left(1, \dfrac{7}{5}\right)$, $\left(\dfrac{5}{2}, 2\right)$ and $(5, 3)$.

   b. Determine the equation of the perpendicular bisector of the line joining the points $\left(1, \dfrac{7}{5}\right)$ and $(5, 3)$.

6. The triangle ABC has vertices A(9, −2), B(3, 6), and C(1, 4).

   a. Determine the midpoint, M, of BC.
   b. Determine the gradient of BC.
   c. Show that AM is the perpendicular bisector of BC.
   d. Describe triangle ABC.

7. Determine the equation of the perpendicular bisector of the line joining the points (−2, 9) and (4, 0).

8. ABCD is a parallelogram. The coordinates of A, B and C are (4, 1), (1, −2) and (−2, 1) respectively. Determine:

   a. the equation of AD
   b. the equation of DC
   c. the coordinates of D.

Understanding

9. In each of the following, show that ABCD is a parallelogram.
   a. A(2, 0), B(4, − 3), C (2, − 4), D(0, − 1)   **b.** A(2, 2), B(0, − 2), C(−2, − 3), D(0, 1)
   c. A(2.5, 3.5), B(10, − 4), C(2.5, − 2.5), D(−5, 5)

10. In each of the following, show that ABCD is a trapezium.
    a. A(0, 6), B(2, 2), C(0, − 4), D(−5, − 9)   **b.** A(26, 32), B(18, 16), C(1, − 1), D(−3, 3)
    c. A(2, 7), B(1, − 1), C(−0.6, − 2.6), D(−2, 3)

11. **MC** The line that passes through the points (0, −6) and (7, 8) also passes through:

    A. (4, 3)   **B.** (5, 4)   **C.** (−2, 10)   **D.** (1, −8)

Communicating, reasoning and problem solving

12. The map shows the proposed course for a yacht race. Buoys have been positioned at A(1, 5), B(8, 8), C(12, 6), and D(10, w).

    a. Calculate how far it is from the start, O, to buoy A.
    b. The race marshall boat, M, is situated halfway between buoys A and C. Determine the coordinates of the boat's position.
    c. Stage 4 of the race (from C to D) is perpendicular to stage 3 (from B to C). Evaluate the gradient of CD.
    d. Determine the linear equation that describes stage 4.
    e. Hence determine the exact position of buoy D.
    f. An emergency boat is to be placed at point E, (7, 3). Determine how far the emergency boat is from the hospital, located at H, 2 km north of the start.

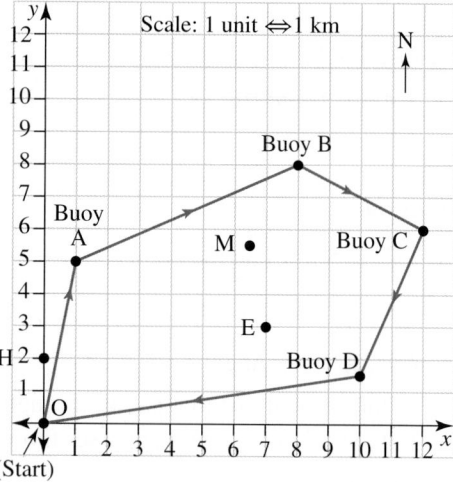

13. Prove that the quadrilateral ABCD is a rectangle with A(2, 5), B(6, 1), C (3, −2) and D(−1, 2).

14. Show that the following sets of points form the vertices of a right-angled triangle.
    a. A(1, − 4), B(2, − 3), C(4, − 7)   **b.** A(3, 13), B(1, 3), C(−4, 4)
    c. A(0, 5), B(9, 12), C(3, 14)

15. a. A square has vertices at (0, 0) and (2, 0). Determine where the other 2 vertices are. (There are 3 sets of answers.)

b. An equilateral triangle has vertices at (0, 0) and (2, 0). Determine where the other vertex is. (There are 2 answers.)

c. A parallelogram has vertices at (0, 0) and (2, 0) and (1, 1). Determine where the other vertex is. (There are 3 sets of answers.)

16. Prove that the quadrilateral ABCD is a rhombus, given A(2, 3), B(3, 5), C(5, 6) and D(4, 4).
Hint: A rhombus is a parallelogram with diagonals that intersect at right angles.

17. A is the point (0, 0) and B is the point (0, 2).

a. Determine the perpendicular bisector of AB.

b. Show that any point on this line is equidistant from A and B.

Questions **18** and **19** relate to the diagram.

M is the midpoint of OA.
N is the midpoint of AB.
P is the midpoint of OB.

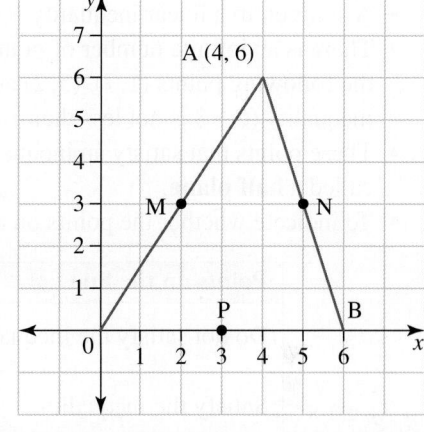

18. A simple investigation:

a. Show that MN is parallel to OB.

b. Is PN parallel to OA? Explain.

c. Is PM parallel to AB? Explain.

19. A difficult investigation:

a. Determine the perpendicular bisectors of OA and OB.

b. Determine the point W where the two bisectors intersect.

c. Show that the perpendicular bisector of AB also passes through W.

d. Explain why W is equidistant from O, A and B.

e. W is called the circumcentre of triangle OAB. Using W as the centre, draw a circle through O, A, and B.

20. Line A is parallel to the line with equation $2x - y = 7$ and passes through the point (2, 3). Line B is perpendicular to the line with equation $4x - 3y + 3 = 0$ and also passes through the point (2, 3). Line C intersects with line A where it cuts the y-axis and intersects with line B where it cuts the x-axis.

a. Determine the equations for all three lines. Give answers in the form $ax + by + c = 0$.

b. Sketch all three lines on the one set of axes.

c. Determine whether the triangle formed by the three lines is scalene, isosceles or equilateral.

21. The lines l_1 and l_2 are at right angles to each other. The line l_1 has the equation $px + py + r = 0$. Show that the distance from M to the origin is given by

$$\frac{r}{\sqrt{p^2 + p^2}}.$$

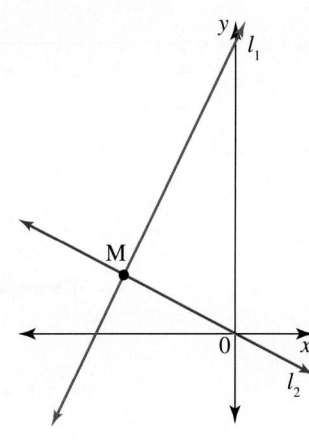

LESSON
7.8 Graphing linear inequalities (Path)

LEARNING INTENTION

At the end of this lesson you should be able to:
- sketch the graph of a half plane: the region represented by an inequality
- sketch inequalities using digital technology
- determine the required region by testing points.

▶ 7.8.1 Inequalities on the Cartesian plane

eles-4776

- A solution to a linear inequality is any ordered pair (coordinate) that makes the inequality true.
- There is an infinite number of points that can satisfy an inequality. If we consider the inequality $x + y < 10$, the following points $(1, 7)$, $(5, 2)$ and $(4, 3)$ are all solutions, whereas $(6, 8)$ is not as it does not satisfy the inequality ($6 + 8$ is not less than 10).
- These points that satisfy an inequality are represented by a region that is found on one side of a line and is called a **half plane**.
- To indicate whether the points on a line satisfy the inequality, a specific type of **boundary line** is used.

Points on the line	Symbol	Type of boundary line used
Do not satisfy the inequality	< or >	Dashed - - - - - - - - - - - - - - -
Satisfy the inequality	≤ or ≥	Solid _____

- The **required region** is the region that contains the points that satisfy the inequality.
- Shading or no shading is used to indicate which side of the line is the required region, and a key is shown to indicate the region.

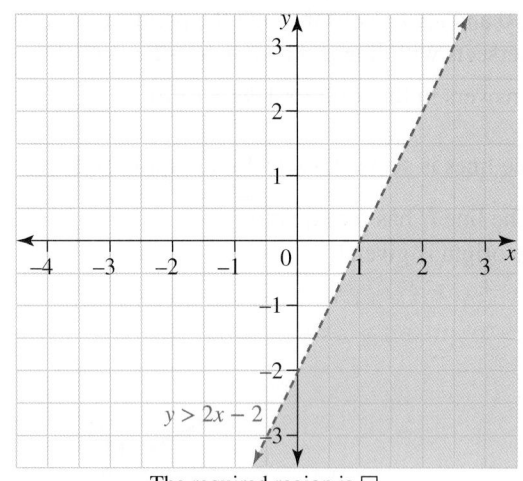

The required region is □.

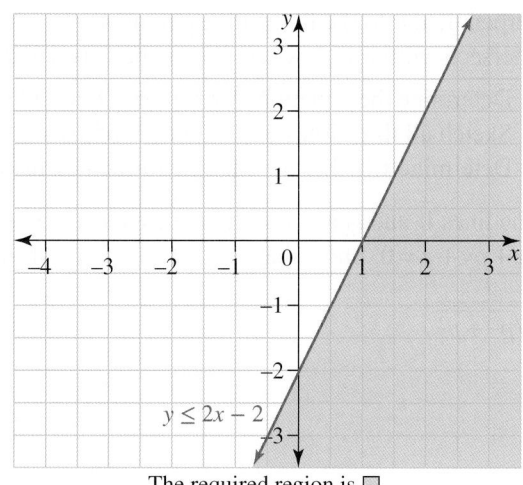

The required region is ▨.

- Consider the line $x = 2$. It divides the **Cartesian plane** into two distinct regions or half-planes.

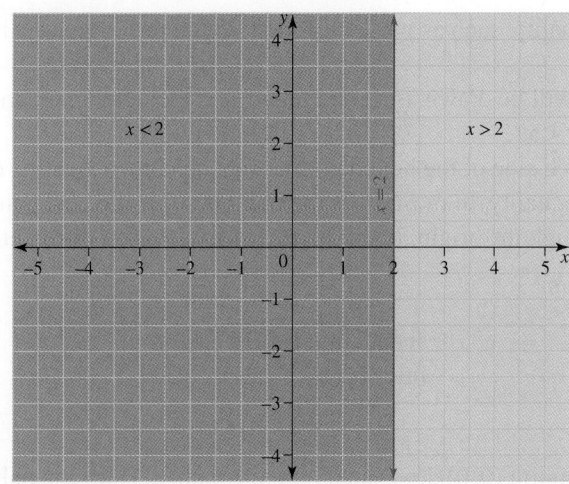

- The region on the left (shaded pink) contains all the points whose x-coordinate is less than 2, for example $(1, 3)$, so this region is given the name $x < 2$.
- The region on the right (shaded blue) contains all the points whose x-coordinate is greater than 2, for example $(3, -2)$, so this region is given the name $x > 2$.
- There are three distinct parts to the graph:
  - the boundary line, where $x = 2$
  - the pink region, where $x < 2$
  - the blue region, where $x > 2$.

WORKED EXAMPLE 27 Sketching graphs of simple inequalities

Sketch a graph of each of the following regions.
a. $x \geq -1$ b. $y < 3$

THINK

a. 1. $x \geq -1$ includes the line $x = -1$ and the region $x > -1$.
    2. On a neat Cartesian plane sketch the line $x = -1$. Because the line is required, it will be drawn as a continuous (unbroken) line.
    3. Identify a point where $x > -1$, say $(2, 1)$.
    4. Shade the region that includes this point. Label the region $x \geq -1$.

DRAW

a.
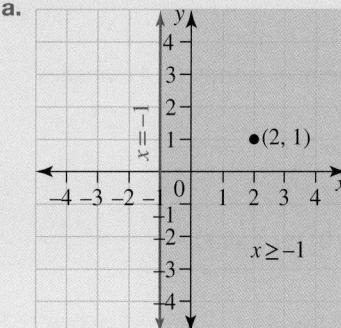

b. 1. The line $y = 3$ is not included.
    2. Sketch the line $y = 3$. Because the line is not included, show it as a dashed (broken) line.
    3. Identify a point where $y < 3$, say $(1, 2)$.
    4. Shade the region where $y < 3$.
    5. Label the region.

b.
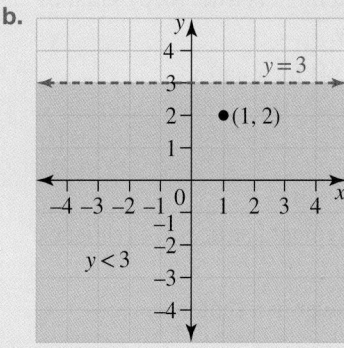

▶ 7.8.2 Determining the required region on the Cartesian plane

eles-4777

- For a more complex inequality, such as $y < 2x + 3$, first sketch the boundary line which is given by the equation $y = 2x + 3$.
  Note: The boundary line will be drawn as a solid line if it is included in the inequality ($y \le x$) or as a broken line if it is not included ($y < x$).
- In order to determine which side of the boundary line satisfies the inequality, choose a point and test whether it satisfies the inequality. In most cases the point $(0, 0)$ is the best point to choose, but if the boundary line passes through the origin, it will be necessary to test a different point such as $(0, 1)$.
  For example:

$$\text{Inequality: } y < 2x + 3:$$
$$\text{Test } (0, 0): \ 0 < 2(0) + 3$$
$$0 < 3 \qquad \text{True}$$

- Since 0 is less than 3, the point $(0, 0)$ does satisfy the inequality. Thus, the half plane containing $(0, 0)$ is the required region.

WORKED EXAMPLE 28 Verifying inequalities at points on the Cartesian plane

Determine whether the points $(0, 0)$ and $(3, 4)$ satisfy either of the following inequalities.
a. $x - 2y < 3$ **b. $y > 2x - 3$**

THINK	WRITE
a. 1. Substitute $(0, 0)$ for x and y.	a. $x - 2y < 3$ Substitute $(0, 0)$:
2. Since the statement is true, $(0, 0)$ satisfies the inequality.	$0 - 0 < 3$ $\quad 0 < 3 \quad$ True
3. Substitute $(3, 4)$ for x and y.	$x - 2y < 3$ Substitute $(3, 4)$: $3 - 2(4) < 3$ $\quad 3 - 8 < 3$
4. Since the statement is true, $(3, 4)$ satisfies the inequality.	$\quad -5 < 3 \quad$ True
5. Write the answer in a sentence.	The points $(0, 0)$ and $(3, 4)$ both satisfy the inequality
b. 1. Substitute $(0, 0)$ for x and y.	b. $y > 2x - 3$ Substitute $(0, 0)$: $0 > 0 - 3$
2. Since the statement is true, $(0, 0)$ satisfies the inequality.	$0 > -3 \qquad$ True
3. Substitute $(3, 4)$ for x and y.	$y > 2x - 3$ Substitute $(3, 4)$: $4 > 2(3) - 3$ $4 > 6 - 3$
4. Since the statement is true, $(3, 4)$ satisfies the inequality.	$4 > 3 \quad$ True
5. Write the answer in a sentence.	The points $(0, 0)$ and $(3, 4)$ both satisfy the inequality

WORKED EXAMPLE 29 Sketching a linear inequality

Sketch a graph of the region $2x + 3y < 6$.

THINK	WRITE/DRAW

1. Locate the boundary line $2x + 3y < 6$ by finding the x- and y-intercepts.	$x = 0$: $0 + 3y = 6$
	          $y = 2$
	$y = 0$: $2x + 0 = 6$
	          $x = 3$

2. The line is not required due to the $<$ inequality, so rule a broken line.	
3. Test with the point $(0, 0)$. Does $(0, 0)$ satisfy $2x + 3y < 6$?	Test $(0, 0)$: $2(0) + 3(0) = 0$ As $0 < 6$, $(0, 0)$ is in the required region.
4. Shade the region that includes $(0, 0)$.	
5. Label the region.	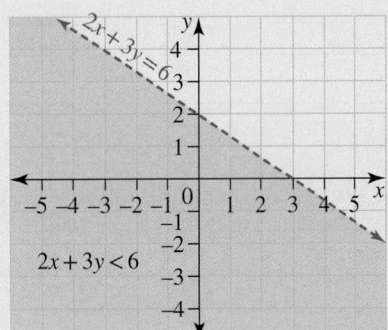

WORKED EXAMPLE 30 Modelling real-life situations

In the school holidays you have been given $160 to arrange some activities for your family. A ticket to the movies costs $10 and a ticket for the trampoline park costs $16.
a. If m represents the movie tickets and t represents the trampoline park tickets, write an inequality in terms of m and t that represents your entertainment budget.
b. Sketch the inequality from part a on the Cartesian plane.
c. Using the graph from part b explore the maximum number of movie and trampoline park tickets you can buy to use the maximum amount of your holiday budget.

THINK	WRITE
a. Each movie ticket, m, costs $12, and each trampoline ticket, t, costs $15. The maximum amount you have to spend is $160.	a. $10m + 16t \leq 160$
b. 1. To draw the boundary line $10m + 15t \leq 160$, identify two points on the line. Let m be the x-axis and t be the y-axis.	b. For the line $10m + 16t = 160$ x-intercept; let $t = 0$ $10m + 16 \times 0 = 160$ $10m = 160$ $m = 16$ x-intercept is $(16, 0)$ y-intercept; let $m = 0$ $10 \times 0 + 16t = 160$ $16t = 160$ $t = 10$ y-intercept is $(0, 10)$

2. Plot the two points and draw the line. As you can spend up to and including $160, the boundary line is solid.
Only the first quadrant of the graph is required, as the number of tickets cannot be negative.

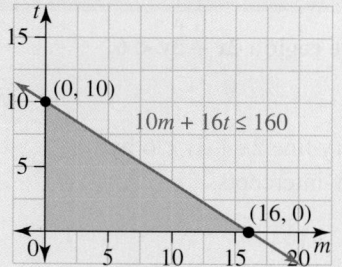

c. 1. To determine the maximum number of movie and trampoline park tickets, identify the nearest whole numbers of each to the graph line. These must be whole numbers as you cannot buy part of a ticket.

c. To spend the entire $160, only 16 movie tickets or 10 trampoline park tickets can be purchased. If less than $160 was spent you could purchase any whole number combinations, such as 6 movie tickets and 6 trampoline park tickets for $156.

DISCUSSION

Think of some real-life situations where inequalities could be used to help solve a problem.

on Resources

 Interactivity Linear inequalities in two variables (int-6488)

Exercise 7.8 Graphing linear inequalities (Path)

learn on

| 7.8 Quick quiz on | 7.8 Exercise |

Individual pathways

■ PRACTISE	■ CONSOLIDATE	■ MASTER
1, 4, 7, 10, 13, 16	2, 5, 8, 11, 14, 17, 19	3, 6, 9, 12, 15, 18, 20

Fluency

WE27 For questions **1** to **3**, sketch a graph of each of the following regions.

1. a. $x < 1$
   c. $x \geq 0$
   b. $y \geq -2$
   d. $y < 0$

2. a. $x > 2$
   c. $y \geq 3$
   b. $x \leq -6$
   d. $y \leq 2$

3. a. $x < \dfrac{1}{2}$
   c. $y \geq -4$
   b. $y < \dfrac{3}{2}$
   d. $x \leq \dfrac{3}{2}$

WE28 For questions **4** to **6**, determine which of the points A $(0, 0)$, B $(1, -2)$ and C $(4, 3)$ satisfy each of the following inequalities.

4. **a.** $x + y > 6$ **b.** $x - 3y < 2$

5. **a.** $y > 2x - 5$ **b.** $y < x + 3$

6. **a.** $3x + 2y < 0$ **b.** $x \geq 2y - 2$

WE29 For questions **7** to **9**, sketch the graphs for the regions given by each of the following inequations. Verify your solutions using technology.

7. **a.** $y \geq x + 1$ **b.** $y < x - 6$ **c.** $y > -x - 2$ **d.** $y < 3 - x$

8. **a.** $y > x - 2$ **b.** $y < 4$ **c.** $2x - y < 6$ **d.** $y \leq x - 7$

9. **a.** $x - y > 3$ **b.** $y < x + 7$ **c.** $x + 2y \leq 5$ **d.** $y \leq 3x$

10. **MC** The shaded region satisfying the inequality $y > 2x - 1$ is:

A.
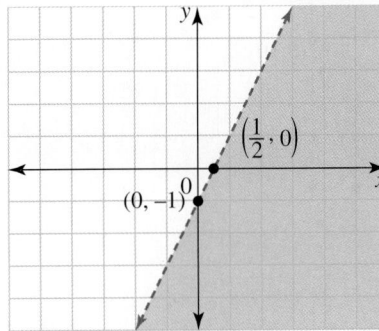

B.
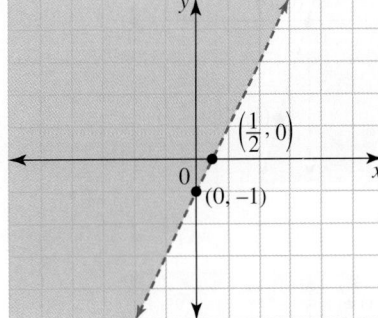

C.
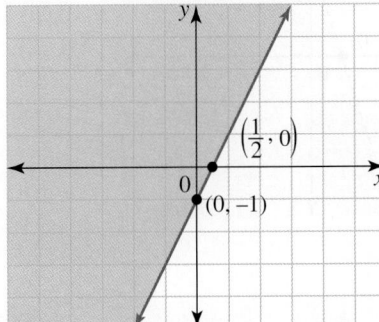

D.
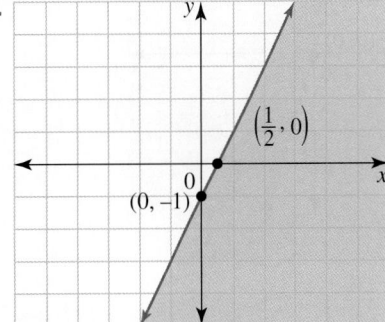

11. **MC** The shaded region satisfying the inequality $y \leq x + 4$ is:

A.
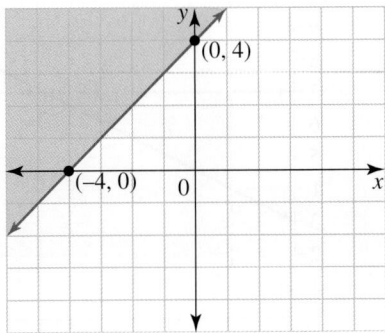

B.
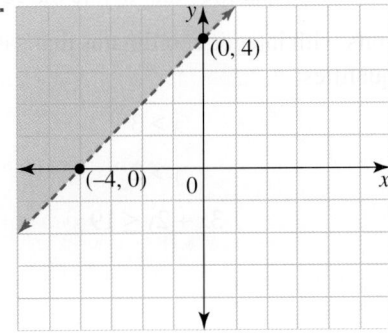

C.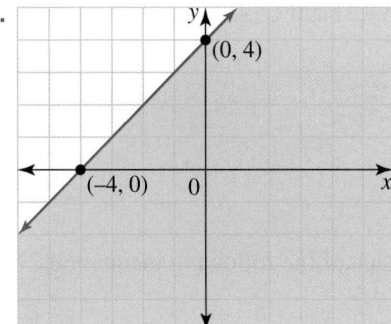

D.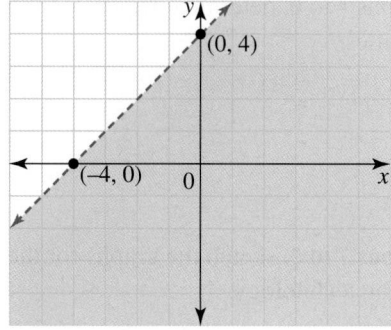

12. **MC** The region satisfying the inequality $y < -3x$ is:

A.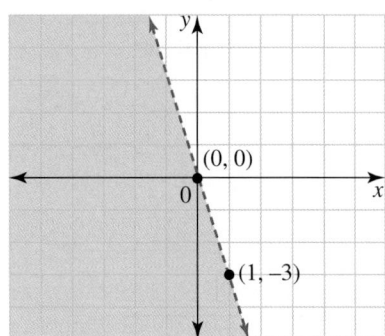

B.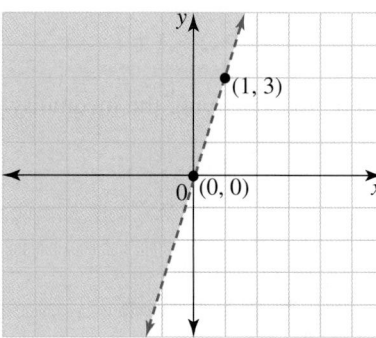

C.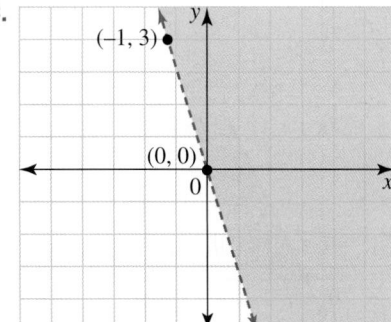

D.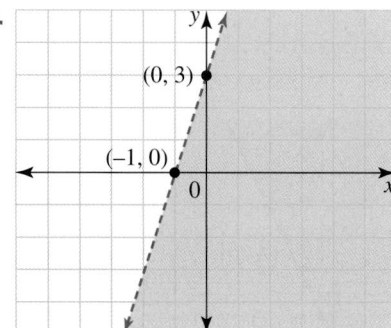

Understanding

13. **a.** Determine the equation of the line l shown in the diagram.
    b. Write down three inequalities that define the region R.

14. Identify all points with integer coordinates that satisfy the following inequalities:

$$x \geq 3$$
$$y > 2$$
$$3x + 2y \leq 19$$

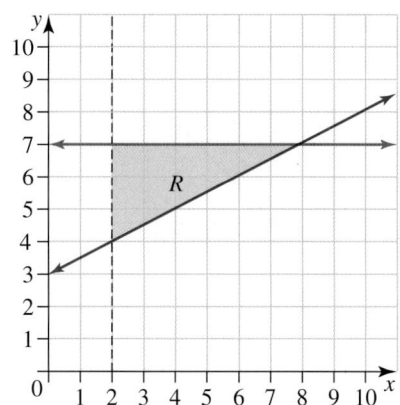

15. **WE30** Happy Yaps Dog Kennels charges $35 per day for large dogs (dogs over 20 kg) and $20 per day for small dogs (less than 20 kg). On any day, Happy Yaps Kennels can only accommodate a maximum of 30 dogs.

    a. If l represents the number of large dogs and s represents the number of small dogs, write an inequality in terms of l and s that represents the total number of dogs at Happy Yaps.

    b. Another inequality can be written as $s \geq 12$. In the context of this problem, write down what this inequality represents.

    c. The inequality $l \leq 15$ represents the number of large dogs that Happy Yaps can accommodate on any day. Draw a graph that represents this situation.

    d. Explore the maximum number of small and large dogs Happy Yaps Kennels can accommodate to receive the maximum amount in fees.

Communicating, reasoning and problem solving

16. Use technology to sketch and then find the area of the region formed by the following inequalities.

$$y \geq -4$$
$$y < 2x - 4$$
$$2y + x \leq 2$$

17. Given the following graph.

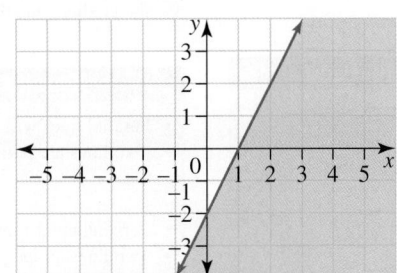

The required region is ☐.

    a. State the inequality it represented.

    b. Choose a point from each half plane and show how this point confirms your answer to part **a**.

18. Answer the following questions.

    a. Determine the equation of the line, l.

    b. Write an inequation to represent the unshaded region.

    c. Write an inequation to represent the shaded region.

    d. Rewrite the answer for part **b** if the line was not broken.

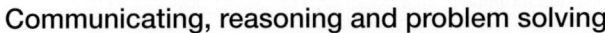

19. a. Sketch the graph of:

$$\frac{x+1}{2} - \frac{x+1}{3} = 2 - y$$

    b. Shade the region that represents:

$$\frac{x+1}{2} - \frac{x+1}{3} \leq 2 - y$$

20. Use your knowledge about linear inequations to sketch the region defined by $y \geq x^2 + 4x + 3$.

LESSON
7.9 Review

7.9.1 Topic summary

Sketching linear graphs

- The x- and y-intercept method involves calculating both axis intercepts, then drawing the line through them.
- Determine the x-intercept but substituting $y = 0$.
- Determine the y-intercept but substituting $x = 0$.
- Graphs given by $y = mx$ pass through the origin $(0, 0)$, since $c = 0$.
- The line $y = c$ is parallel to the x-axis, having a gradient of zero and a y-intercept of c.
- The line $x = a$ is parallel to the y-axis, having a undefined (infinite) gradient and a x-intercept of a.

Equation of a straight line

- The equation of a straight line is:
$$y = mx + c$$
Where: m is the gradient and c is the y-intercept
e.g. $y = ②x + ⑤$
- The rule $y = mx + c$ is called the equation of a straight line in the **gradient-intercept** form.
- The gradient of a straight line can be determined by the formulas:
$$m = \frac{\text{rise}}{\text{run}} = \frac{y_2 - y_1}{x_2 - x_1}$$

LINEAR RELATIONSHIPS

Determining linear equations (Path)

- The formula:
$$y - y_1 = m(x - x_1)$$
can be used to write the equation of a line, given the gradient and the coordinates (x_1, y_1) of one point.

The midpoint of a line segment (Path)

- The midpoint of two points, (x_1, y_1) and (x_2, y_2) is:
$$M = \left(\frac{x_1 + x_2}{2}, \frac{y_1 + y_2}{2} \right)$$

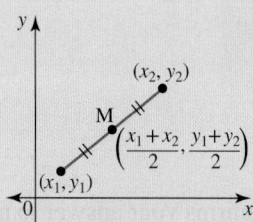

Parallel and perpendicular lines

- Parallel lines will never intersect with each other.
- Two lines are parallel if they have the same gradient.
e.g.
$$y = 3x - 6$$
$$y = 3x + 1$$
- Perpendicular lines are lines that intersect at *right angles*.
- Two lines are perpendicular if the product of their gradients is -1.

Perpendicular bisector (Path)

- A perpendicular bisector is a line that passes through the midpoint of a line segment, cutting it at right angles in two equal lengths.

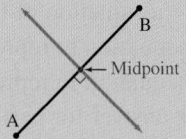

The distance between two points (Path)

- The distance between two points, (x_1, y_1) and (x_2, y_2) is:
$$d = \sqrt{(x_2 - x_1)^2 + (y_2 - y_1)^2}$$

Sketching inequalities (Path)

- When sketching a linear inequality such as $y < 3x + 4$, sketch the boundary line first which is given by $y = 3x + 4$.
- To determine which side of the boundary line is the required ragion test the point $(0, 0)$ and see if it satisfies the inequality.
Test $(0, 0) = 0 < 3(0) + 4$
$ 0 < 4$ which is true
- In this case, the region required is the region with the point $(0, 0)$.
- When sketching simultaneous inequalities, the required region is the overlap region of each individual inequality.

Inequalities and half planes (Path)

- The graph of a linear inequality is called a **half plane** and is the region above or below a boundary line.

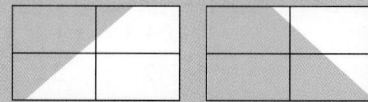

- If the inequality has $<$ or $>$ the boundary line is dotted as it is not included in the solution.
- If the inequality has $\leq$ or $\geq$ the boundary line is solid as it is included in the solution.

7.9.2 Project

What common computer symbol is this?

On computer hardware, and on many different software applications, a broad range of symbols is used. These symbols help us to identify where things need to be plugged into, what buttons we need to push, or what option needs to be selected.

The main focus of this task involves constructing a common symbol found on the computer. The instructions are given below. Use grid paper to construct the symbol.

The construction part of this task requires you to graph nine lines to reveal a common computer symbol. Draw the scale of your graph to accommodate x- and y-values in the following ranges: $-10 \leq x \leq 16$ and $-10 \leq y \leq 16$.

Centre the axes on the grid lines.
- Line 1 has an equation $y = x - 1$. Graph this line in the range $-7 \leq x \leq -2$.
- Line 2 is perpendicular to line 1 and has a y-intercept of -5. Determine the equation of this line, and then draw the line in the range $-5 \leq x \leq -1$.
- Line 3 is parallel to line 1, with a y-intercept of 3. Determine the equation of the line, and then graph the line in the range $-9 \leq x \leq -4$.
- Line 4 is parallel to line 1, with a y-intercept of -3. Determine the equation of the line, and then graph the line in the range $-1 \leq x \leq 2$.
- Line 5 has the same length as line 4 and is parallel to it. The point $(-2, \ 3)$ is the starting point of the line, which decreases in both x- and y-values from there.
- Line 6 commences at the same starting point as line 5, and then runs at right angles to line 5. It has an x-intercept of 1 and is the same length as line 2.
- Line 7 commences at the same starting point as both lines 5 and 6. Its equation is $y = 6x + 15$. The point $(-1, 9)$ lies at the midpoint.
- Line 8 has the equation $y = -x + 15$. Its midpoint is the point $(7, 8)$ and its extremities are the points where the line meets line 7 and line 9.
- Line 9 has the equation $6y - x + 8 = 0$. It runs from the intersection of lines 4 and 6 until it meets line 8.
1. Determine what common computer symbol you have drawn.
2. The top section of your figure is a familiar geometric shape. Use the coordinates on your graph, together with the distance formula to determine the necessary lengths to calculate the area of this figure.
3. Using any symbol of interest to you, draw your symbol on grid lines and provide instructions for your design. Ensure that your design involves aspects of coordinate geometry that have been used throughout this task.

on Resources

Interactivities Crossword (int-2833)

Sudoku puzzle (int-3590)

Exercise 7.9 Review questions

learnon

Fluency

1. **MC** The equation of the following line is:

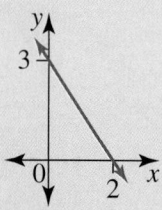

A. $3x + 2y = 6$
B. $3x - 2y = 6$
C. $2x + 3y = 6$
D. $2x - 3y = 6$

2. **MC** The equation of a linear graph with gradient -3 and x-intercept of 4 is:
A. $y = -3x - 12$
B. $y = -3x + 4$
C. $y = -3x - 4$
D. $y = -3x + 12$

3. **MC** **PATH** The equation of a linear graph which passes through $(2, -7)$ and $(-2, -2)$ is:
A. $4x - 5y + 18 = 0$
B. $5x + 4y + 18 = 0$
C. $5x + 4y - 18 = 0$
D. $5x - 4y - 18 = 0$

4. **MC** **PATH** The distance between the points $(1, 5)$ and $(6, -7)$ is:
A. $\sqrt{53}$
B. $\sqrt{29}$
C. 13
D. $\sqrt{193}$

5. **MC** **PATH** The midpoint of the line segment joining the points $(-4, 3)$ and $(2, 7)$ is:
A. $(-1, 5)$
B. $(-2, 10)$
C. $(-6, 4)$
D. $(-2, 4)$

6. **MC** **PATH** If the midpoint of the line segment joining the points A$(3, 7)$ and B(x, y) has coordinates $(6, 2)$, then the coordinates of B are:
A. $(15, 3)$
B. $(0, -6)$
C. $(9, -3)$
D. $(4.5, 4.5)$

7. **MC** **PATH** If the points $(-6, -11)$, $(2, 1)$ and $(x, 4)$ are collinear, then the value of x is:
A. 4
B. 3.2
C. $\dfrac{1}{4}$
D. $\dfrac{5}{16}$

8. **MC** The gradient of the line perpendicular to $3x - 4y + 7 = 0$ is:
A. $\dfrac{3}{4}$
B. $\dfrac{4}{3}$
C. $-\dfrac{4}{3}$
D. 3

9. **MC** **PATH** The equation of the line perpendicular to $2x + y - 1 = 0$ and passing through the point $(1, 4)$ is:
A. $2x + y - 6 = 0$
B. $2x + y - 2 = 0$
C. $x - 2y + 7 = 0$
D. $x + 2y + 9 = 0$

10. 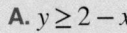 Identify the inequality that is represented by the following region.

   A. $y \geq 2 - x$
   B. $y \geq x - 2$
   C. $y \leq 2 - x$
   D. $y \leq x - 2$

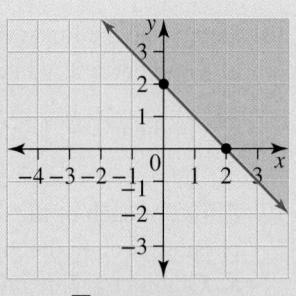

☐ Region required

Understanding

11. Produce a table of values, and sketch the graph of the equation $y = -5x + 15$ for values of x between -10 and $+10$.

12. Sketch the graph of the following linear equations, labelling the x- and y-intercepts.

   a. $y = 3x - 2$
   b. $y = -5x + 15$
   c. $y = -\dfrac{2}{3}x + 1$
   d. $y = \dfrac{7}{5}x - 3$

13. Determine the x- and y-intercepts of the following straight lines.

   a. $y = -7x + 6$
   b. $y = \dfrac{3}{8}x - 5$
   c. $y = \dfrac{4}{7}x - \dfrac{3}{4}$
   d. $y = 0.5x + 2.8$

14. Sketch graphs of the following linear equations by finding the x- and y-intercepts.
   a. $2x - 3y = 6$
   b. $3x + y = 0$
   c. $5x + y = -3$
   d. $x + y + 3 = 0$

15. Sketch the graph of each of the following.

   a. $y = \dfrac{1}{2}x$
   b. $y = -4x$
   c. $x = -2$
   d. $y = 7$

16. Sketch the graph of the equation $3(y - 5) = 6(x + 1)$.

17. Determine the equations of the straight lines in the following graphs.

   a.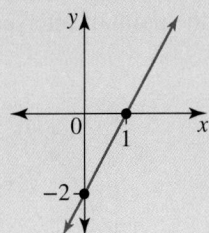
   b.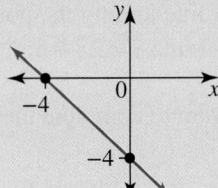
   c.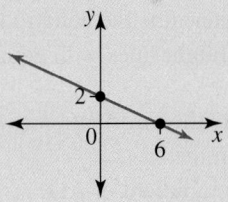

   d.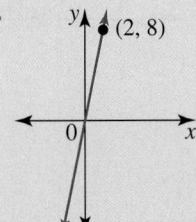
   e.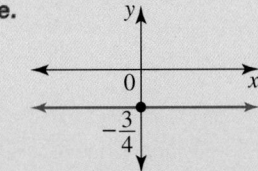
   f.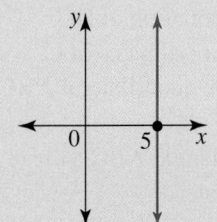

18. Determine the linear equation given the information in each case below.
    a. Gradient $= 3$, y-intercept $= -4$
    b. Gradient $= -2$, y-intercept $= -5$
    c. Gradient $= \dfrac{1}{2}$, y-intercept $= 5$
    d. Gradient $= 0$, y-intercept $= 6$

19. **PATH** For each of the following, determine the equation of the straight line with the given gradient and passing through the given point.
    a. Gradient $= 7$, point $(2, 1)$
    b. Gradient $= -3$, point $(1, 1)$
    c. Gradient $= \dfrac{1}{2}$, point $(-2, 5)$
    d. Gradient $= \dfrac{3}{5}$, point $(1, -3)$

20. **PATH** Determine the distance between the points $(1, 3)$ and $(7, -2)$ in exact form.

21. **PATH** Prove that triangle ABC is isosceles given A(3, 1), B(−3, 7) and C(−1, 3).

22. **PATH** Show that the points A(1, 1), B(2, 3) and C(8, 0) are the vertices of a right-angled triangle.

23. **PATH** The midpoint of the line segment AB is (6, −4). If B has coordinates (12, 10), determine the coordinates of A.

24. **PATH** Show that the points A(3, 1), B(5, 2) and C(11, 5) are collinear.

25. Show that the lines $y = 2x - 4$ and $x + 2y - 10 = 0$ are perpendicular to one another.

26. **PATH** Determine the equation of the straight line passing through the point (6, −2) and parallel to the line $x + 2y - 1 = 0$.

27. **PATH** Determine the equation of the line perpendicular to $3x - 2y + 6 = 0$ and that has the same y-intercept.

28. **PATH** Determine the equation of the perpendicular bisector of the line joining the points (−2, 7) and (4, 11).

29. **PATH** Determine the equation of the straight line joining the point (−2, 5) and the point of intersection of the straight lines with equations $y = 3x - 1$ and $y = 2x + 5$.

30. **PATH** Use the information given in the diagram to complete the following.
    a. Determine:
        i. the gradient of AD
        ii. the gradient of AB
        iii. the equation of BC
        iv. the equation of DC
        v. the coordinates of C.
    b. Describe quadrilateral ABCD.

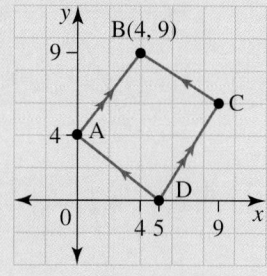

31. **PATH** In triangle ABC, A is (1, 5), B is (−2, −3) and C is (8, −2).
    a. Determine:
        i. the gradient of BC
        ii. the midpoint, P, of AB
        iii. the midpoint, Q, of AC.
    b. Hence show that:
        i. PQ is parallel to BC
        ii. PQ is half the length of BC.

32. PATH Sketch the half plane given by each of the following inequalities.
  a. $y \leq x + 1$ **b.** $y \geq 2x + 10$ **c.** $y > 3x - 12$ **d.** $y < 5x$

Communicating, reasoning and problem solving

33. John has a part-time job working as a gardener and is paid $13.50 per hour.
  a. Complete the following table of values relating the amount of money received to the number of hours worked.

Number of hours	0	2	4	6	8	10
Pay $						

  b. Determine a linear equation relating the amount of money received to the number of hours worked.
  c. Sketch the linear equation on a Cartesian plane over a suitable domain.
  d. Using algebra, calculate the pay that John will receive if he works for $6\frac{3}{4}$ hours.

34. A fun park charges a $12.50 entry fee and an additional $2.50 per ride.
  a. Complete the following table of values relating the total cost to the number of rides.

Number of rides	0	2	4	6	8	10
Cost $						

  b. Determine a linear equation relating total cost to the number of rides.
  c. Sketch the linear equation on a Cartesian plane over a suitable domain.
  d. Using algebra, calculate the cost for 7 rides.

35. The cost of hiring a boat is $160 plus $22.50 per hour.
  a. Sketch a graph showing the total cost for between 0 and 12 hours.
  b. State the equation relating cost to time rented.
  c. Predict the cost of hiring a boat for 12 hours and 15 minutes.

36. PATH ABCD is a quadrilateral with vertices A(4, 9), B(7, 4), C(1, 2) and D(a, 10).
  Given that the diagonals are perpendicular to each other, determine:
  a. the equation of the diagonal AC
  b. the equation of the diagonal BD
  c. the value of a.

37. **PATH** An architect decides to design a building with a 14-metre-square base such that the external walls are initially vertical to a height of 50 metres, but taper so that their separation is 8 metres at its peak height of 90 metres. A profile of the building is shown with the point (0, 0) marked as a reference at the centre of the base.

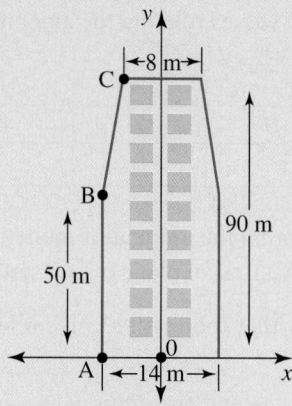

a. Write the equation of the vertical line connecting A and B.
b. Write the coordinates of B and C.
c. Determine the length of the tapered section of wall from B to C.

38. **PATH** In a game of lawn bowls, the object is to bowl a biased ball so that it gets as close as possible to a smaller white ball called a jack. During a game, a player will sometimes bowl a ball quite quickly so that it travels in a straight line in order to displace an opponents 'guard balls'. In a particular game, player x has two guard balls close to the jack. The coordinates of the jack are (0, 0) and the coordinates of the guard balls are $A\left(-1, \frac{4}{5}\right)$ and $B\left(-\frac{1}{2}, \frac{57}{40}\right)$. Player Y bowls a ball so that it travels in a straight line toward the jack. The ball is bowled from the position S, with the coordinates $(-30, 24)$.

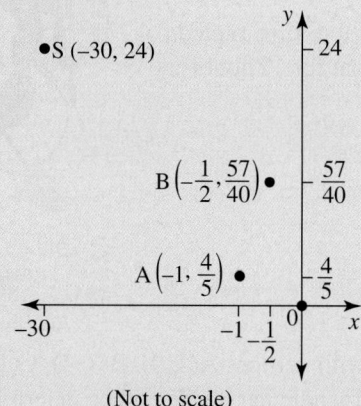

(Not to scale)

a. Will player Y displace one of the guard balls? If so, which one? Explain your answer.
b. Due to bias, the displaced guard ball is knocked so that it begins to travel in a straight line (at right angles to the path found in part **a**. Determine the equation of the line of the guard ball.
c. Show that guard ball A is initially heading directly toward guard ball B.
d. Given its initial velocity, guard ball A can travel in a straight line for 1 metre before its bias affects it path. Calculate and explain whether guard ball A will collide with guard ball B.

39. The graph shows the line p passing through the points A(-1, 1) and B(5, 5).

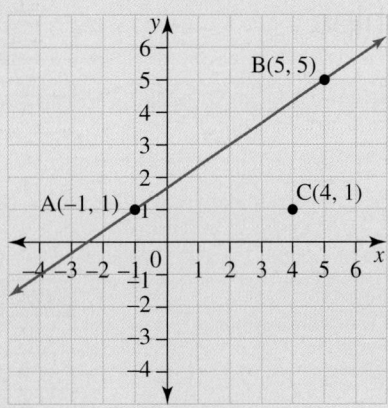

Given that C is the point (4, 1), determine:

a. the gradient of p

b. the equation of p

c. the area of $\triangle$ABC

d. the length BC, giving your answer correct to 2 decimal places.

40. The temperature of the air ($T \,°C$) is related to the height above sea level (h metres) by the formula $T = 18 - 0.005h$.

a. Evaluate the temperature at the heights of:

  i. 600 m

  ii. 1000 m

  iii. 3000 m

b. Draw a graph using the results from part **a**.

c. Use the graph to determine the temperature at 1200 m and 2500 m.

d. Predict the height at which the temperature is 9 °C.

41. An old theory is that the number of hours of sleep (h) that a child of c years of age should have each night is

$h = 8 + \dfrac{18 - c}{2}$.

a. Determine how many hours a 10-year-old should have.

b. Evaluate the age of a child that requires 10 hours sleep.

c. For every year, determine how much less sleep a child requires.

on To test your understanding and knowledge of this topic, go to your learnON title at www.jacplus.com.au and complete the **post-test**.

Answers

Topic 7 Linear relationships

7.1 Pre-test

1. A
2. $x = \dfrac{1}{2}$
3. 3 apps
4. $y = 3x - 3$
5. D
6. B
7. $a = -2$ or $a = 22$
8. D
9. B
10. D
11. $a = 6$
12. $y = -2x$
13. $n = \dfrac{6}{5}$
14. A
15. B

7.2 Sketching linear graphs

1. a.

x	y
−5	−25
−4	−15
−3	−5
−2	5
−1	15
0	25
1	35

b.

x	y
−1	−17
0	−12
1	−7
2	−2
3	3
4	8

c.

x	y
−6	13
−4	12
−2	11
0	10
2	9
4	8

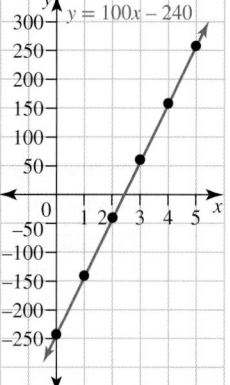

2. a.

x	y
0	−240
1	−140
2	−40
3	60
4	160
5	260

b.

x	y
−3	18
−2	13
−1	8
0	3
1	−2
2	−7

c.

x	y
−3	19
−2	15
−1	11
0	7
1	3
2	−1

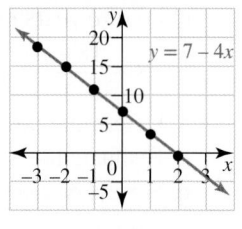

3. a.

x	y
−6	20
−4	14
−2	8
0	2
2	−4
4	−10
6	−16

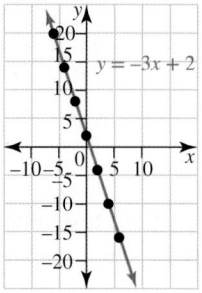

b.

x	y
−3	6
−2	5
−1	4
0	3
1	2
2	1
3	0

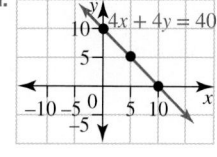

$y = -x + 3$

c.

x	y
−6	15
−4	11
−2	7
0	3
2	−1
4	−5
6	−9

$y = -2x + 3$

4. a.

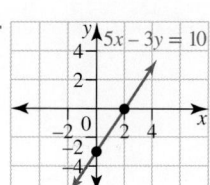

$5x - 3y = 10$

b.

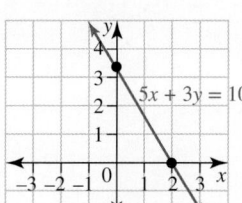

$5x + 3y = 10$

c.

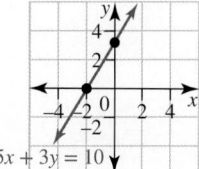

$-5x + 3y = 10$

d.

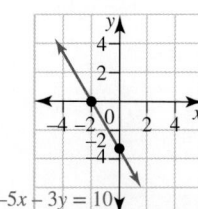

$-5x - 3y = 10$

e.

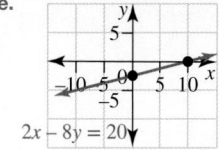

$2x - 8y = 20$

5. a.

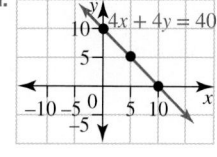

$4x + 4y = 40$

b.

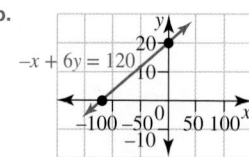

$-x + 6y = 120$

c.

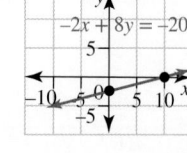

$-2x + 8y = -20$

d.

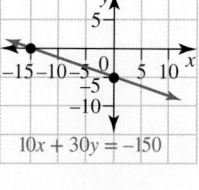

$10x + 30y = -150$

e.

$5x + 30y = -150$

6. a.

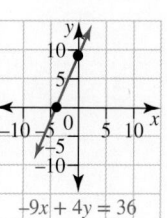

$-9x + 4y = 36$

b.

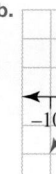

$6x - 4y = -24$

c.

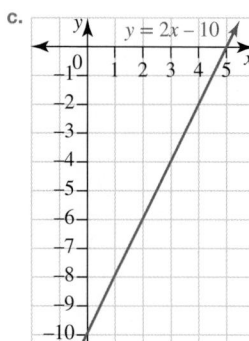

$y = 2x - 10$

d.

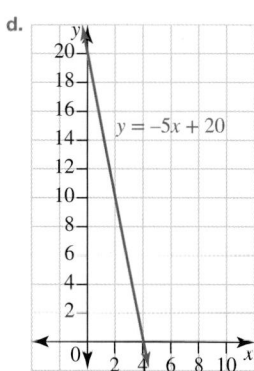

$y = -5x + 20$

e.

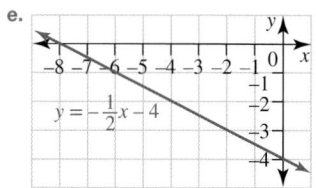

$y = -\frac{1}{2}x - 4$

7. a.

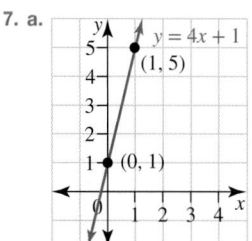

$y = 4x + 1$

(1, 5)

(0, 1)

b.

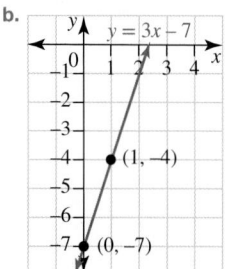

$y = 3x - 7$

(1, −4)

(0, −7)

c.

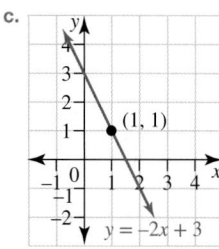

(1, 1)

$y = -2x + 3$

8. a.

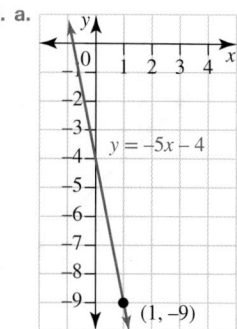

$y = -5x - 4$

(1, −9)

b.

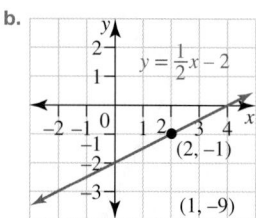

$y = \frac{1}{2}x - 2$

(2, −1)

(1, −9)

c.

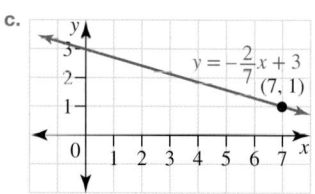

$y = -\frac{2}{7}x + 3$

(7, 1)

9. a.

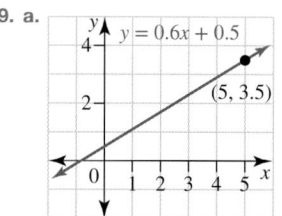

$y = 0.6x + 0.5$

(5, 3.5)

b.

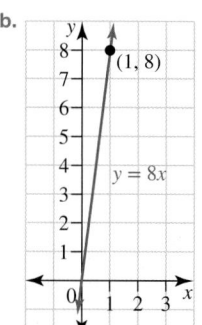

(1, 8)

$y = 8x$

c.

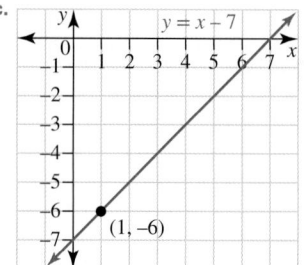

$y = x - 7$

(1, −6)

10.

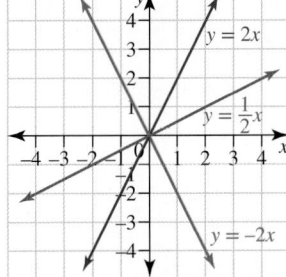

11.

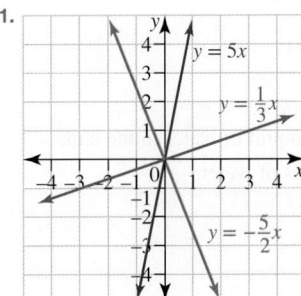

12.

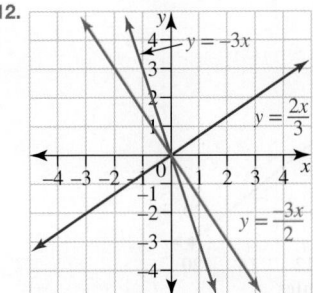

13. a.

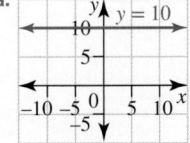

b.

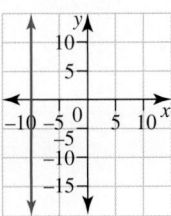

c.

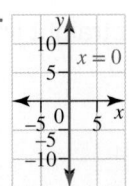

14. a.

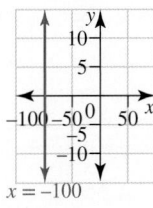

b.

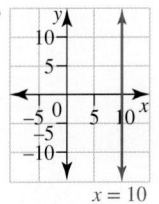

c.

$x = -100$

15. a.

$x = 10$

b.

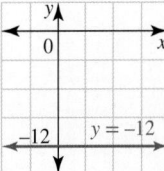

c.

$y = -12$

16. a. x-intercept: -0.5; y-intercept: 0.4

b. x-intercept: 0.5; y-intercept: -0.4

c. x-intercept: 0; y-intercept: 0

17. a. x-intercept: -3; y-intercept: 12

b. x-intercept: -4; y-intercept: -4

c. x-intercept: -1; y-intercept: -0.5

18. a. x-intercept: 2.75; y-intercept: 2.2

b. x-intercept: 9.75; y-intercept: -3.9

c. x-intercept: $-\dfrac{23}{13} \approx 1.77$; y-intercept: 4.6

19. a. $(2, 0)$, $(0, -8)$

b. $\left(-\dfrac{1}{2}, 0\right)$, $(0, 3)$

c. $(-5, 0)$, $(0, 25)$

20. Sample responses can be found in the worked solutions in the online resources.

21. a. Independent variable = number of songs bought, dependent variable = amount of money saved

b.
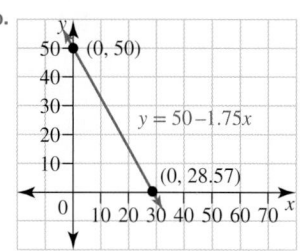

c. 14 songs

22. $y = \dfrac{2}{3}x - \dfrac{7}{3}$

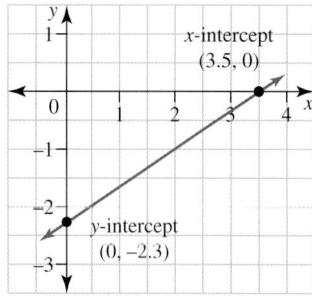

23. a. $y = 20x + 25$

b.
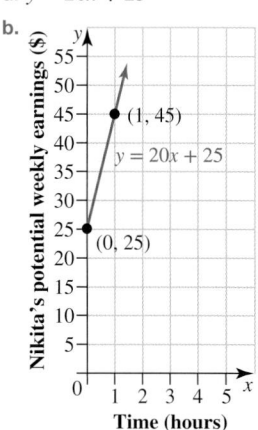

c. Nikita can earn a maximum of $1145.00 in a single week.

24. a. T is the dependent variable (temperature) and t is the independent variable (time).

b. i.
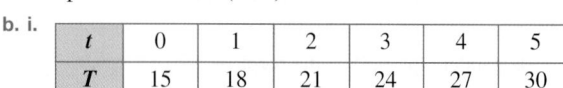

t	0	1	2	3	4	5
T	15	18	21	24	27	30

ii.
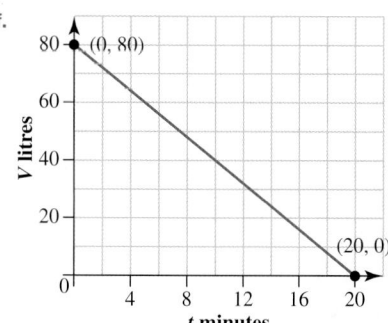

c. 5 hours

25. a. Independent variable = time, dependent variable = amount of water in the tank

b. Initially there are 80 litres of water.

c. Time cannot be negative.

d. 4 litres per minute

e. 20 minutes

f.
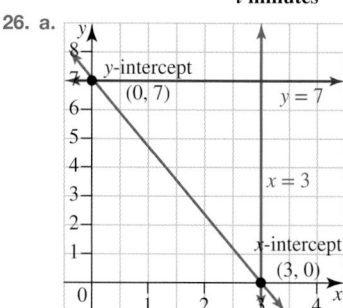

26. a.

b. 7

c. $-\dfrac{7}{3}$

d. B

7.3 Determining the equation of a straight line

1. a. $y = 2x + 4$ **b.** $y = -3x + 12$
c. $y = -x + 5$ **d.** $y = 2x - 8$

2. a. $y = \dfrac{1}{2}x + 3$ **b.** $y = -\dfrac{1}{4}x - 4$

c. $y = 7x - 5$ **d.** $y = -3x - 15$

3. a. $y = 2x$ **b.** $y = -3x$

c. $y = \dfrac{1}{2}x$ **d.** $y = -\dfrac{3}{4}x$

4. a. $y = 3x + 3$ **b.** $y = -3x + 4$ **c.** $y = -4x + 2$
  d. $y = 4x + 2$ **e.** $y = -x - 4$

5. a. $y = 0.5x - 4$ **b.** $y = 5x + 2.5$
  c. $y = -6x + 3$ **d.** $y = -2.5x + 1.5$
  e. $y = 3.5x + 6.5$

6. a. $y = 5x - 19$ **b.** $y = -5x + 31$
  c. $y = -4x - 1$ **d.** $y = 4x - 34$
  e. $y = 3x - 35$

7. a. $y = -3x + 6$ **b.** $y = -2x + 30$
  c. $y = 2x - 4.5$ **d.** $y = 0.5x - 19$
  e. $y = -0.5x + 5.5$

8. a. $y = x + 3$ **b.** $y = 2x - 1$ **c.** $y = -\frac{1}{2}x + \frac{7}{2}$

9. a. $y = \frac{1}{2}x + \frac{1}{2}$ **b.** $y = -2x - 2$ **c.** $y = -x - 8$

10. $2x - y - 5 = 0$

11. $4x + y - 7 = 0$

12. $\frac{1}{2}x + y + 5 = 0$

13. a. Independent variable = time (in hours), dependent variable = cost (in \$)

b.

t	0	1	2	3
c	2	8	14	20

c.

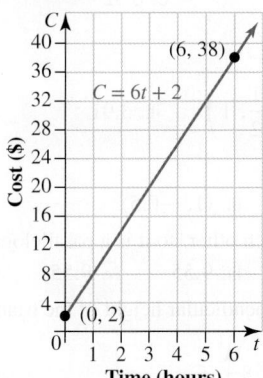

d. i. $(0, 2)$
  ii. The y-intercept represents the initial cost of bowling at the alley, which is the shoe rental.
  e. $m = 6$, which represents the cost to hire a lane for an additional hour.
  f. $C = 6t + 2$
  g. \$32
  h. Sample responses can be found in the worked solutions in the online resources.

14. a. C = dependent variable, n = independent variable
  b. $C = 15n + 10$
  c. \$460.00

d.

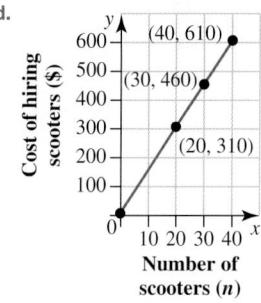

15. a. $W = -40t + 712$
  b. 712 L
  c. 18 days

16. It does not matter if you rise before you run or run before you rise, as long as you take into account whether the rise or run is negative.

17. a. $m = \dfrac{y - c}{x}$ **b.** $y = mx + c$

18. Sample responses can be found in the worked solutions in the online resources.

19. a. $m_{AB} = 5$ **b.** $y = 5x - 3$
  c. $m_{CD} = 5$ **d.** $D = (6, 5)$

20. $m_{AB} = m_{CD} = 2$ and $m_{BC} = m_{AD} = \dfrac{1}{2}$. As opposite sides have the same gradients, this quadrilateral is a parallelogram.

21. a. $b = -10$ **b.** $a = -4$

7.4 Parallel and perpendicular lines

1. a. No **b.** Yes **c.** No

2. a. No **b.** Yes **c.** No

3. a. No **b.** No **c.** No

4. a. No **b.** No **c.** No

5. b and f are parallel. c and e are parallel.

6. Sample responses can be found in the worked solutions in the online resources.

7. $y = 2x - 9$

8. $3x + 2y - 8 = 0$

9. a. $y = 3x + 2$ **b.** $y = -4x + 9$
  c. $y = \dfrac{3x}{2} - 4$ **d.** $y = \dfrac{-2x}{5} - \dfrac{13}{5}$
  e. $y = \dfrac{x}{5} - 1$ **f.** $y = \dfrac{x}{3} + \dfrac{17}{3}$
  g. $y = \dfrac{x}{3} - \dfrac{14}{3}$

10. a. $2x - y + 5 = 0$ **b.** $x + 2y = 0$

11. a. $3x - 5y + 2 = 0$ **b.** $5x + 3y - 8 = 0$

12. a. B **b.** C **c.** D **d.** B

13. D

14. a. $y = -2x + 1$ **b.** $y = \dfrac{-2x}{3} - \dfrac{1}{3}$

15. a and e are perpendicular; b and f are perpendicular; c and h are perpendicular; d and g are perpendicular.

16. $y = \dfrac{-x}{2} + \dfrac{3}{2}$

17. a. $m = \dfrac{-8}{5}$ **b.** $m = \dfrac{18}{5}$

18. D

19. $2y + x - 8 = 0$

20. $4y - 3x + 15 = 0$

21. B

22. a.

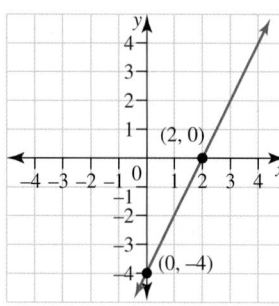

b.

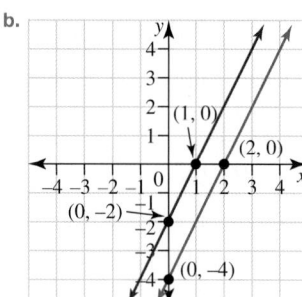

c.

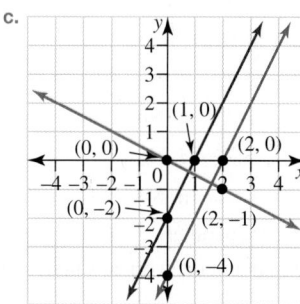

23. $a = \dfrac{-3}{2}$

24. a. $\dfrac{3}{2}$

 b. Sample responses can be found in the worked solutions in the online resources.

7.5 The distance between two points formula (Path)

1. $AB = 5$, $CD = 2\sqrt{10}$ or 6.32, $EF = 3\sqrt{2}$ or 4.24, $GH = 2\sqrt{5}$ or 4.47, $IJ = 5$, $KL = \sqrt{26}$ or 5.10, $MN = 4\sqrt{2}$ or 5.66, $OP = \sqrt{10}$ or 3.16

2. a. 5 **b.** 13 **c.** 10 **d.** 7.07 **e.** 6.71

3. a. 14.42 **b.** 13 **c.** 13

 d. $\sqrt{a^2 + 4b^2}$ **e.** $3\sqrt{a^2 + b^2}$

4. a. $AB = 4.47$, $BC = 2.24$, $CD = 4.47$, $DA = 2.24$

 b. $AC = 5$, $BD = 5$

 c. Rectangle

5. B

6. D

7. 8. and **9.** Sample responses can be found in the worked solutions in the online resources.

10. a. 12 **b.** 5 **c.** 13 **d.** -2.2

11. $a = 2$

12. Sample responses can be found in the worked solutions in the online resources.

13. a. $m_{AB} = 1$ and $m_{BC} = -\dfrac{7}{3}$

 b. $D(4, -1)$

 c. Sample responses can be found in the worked solutions in the online resources.

7.6 The midpoint of a line segment formula (Path)

1. a. $\left(-3, -3\dfrac{1}{2}\right)$ **b.** $\left(7\dfrac{1}{2}, 0\right)$ **c.** $(-1, 1)$

2. a. $\left(0, 1\dfrac{1}{2}\right)$ **b.** $\left(2a, \dfrac{1}{2}b\right)$ **c.** $\left(a + b, \dfrac{1}{2}a\right)$

3. $(-3, -10)$

4. a. $(2, -1)$ **b.** $\left(\dfrac{3}{2}, \dfrac{11}{2}\right)$ **c.** $\left(2, \dfrac{5a}{2}\right)$

5. $x + y = -3$

6. a. $(3, 1)$ **b.** 4.47 **c.** 6.32

7. D

8. C

9. a. i. $(-1, 4)$ **ii.** $\left(1\dfrac{1}{2}, 1\right)$ **iii.** 3.91

 b. $BC = 7.8 = 2PQ$

10. a. i. $(1, -0.5)$ **ii.** $(1, -0.5)$

 b. The diagonals bisect each other, so it is a parallelogram.

11. a. i. $(-2, 2)$ **ii.** 8.94 **iii.** 9.55 **iv.** 9.55

 b. Isosceles. PC is the perpendicular height of the triangle.

12. a.

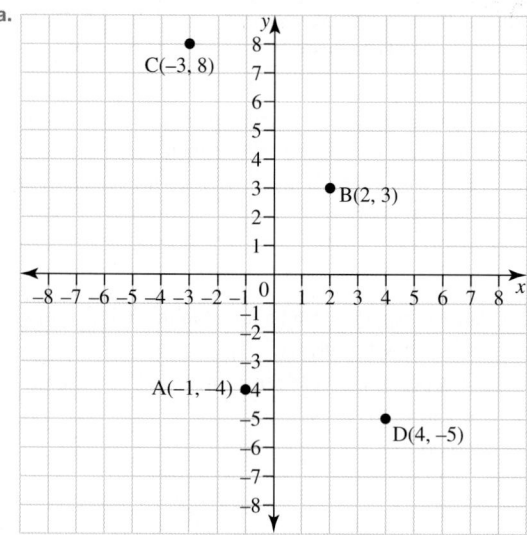

 b. $M = \left(\dfrac{-1 + -3}{2}, \dfrac{-4 + 8}{2}\right)$
 $= (-2, 2)$

c. $2\sqrt{37}$

d. $(7, -2)$

e. $\dfrac{3 - (-4)}{2 - (-1)} = \dfrac{7}{3}$

f. $y = -4x + 11$

13. $(4k - 1, 3.5 - 5k)$

14. Sample responses can be found in the worked solutions in the online resources.

15. $y = -3x - 2$

16. $3y - 2x + 14 = 0$

17. $y = \dfrac{-2x}{3} + \dfrac{8}{3}$

7.7 Applications to solving problems (Path)

1. and 2. Sample responses can be found in the worked solutions in the online resources.

3. a. $x = 1$ b. $y = -7$

4. $x + y + 3 = 0$

5. a. Since $m_1 = m_2 = \dfrac{2}{5}$ and $\left(\dfrac{5}{2}, 2\right)$ is common to both line segments, these three points are collinear.

b. $y = \dfrac{-5}{2}x + \dfrac{97}{10}$

6. a. $(2, 5)$

b. 1

c. Sample responses can be found in the worked solutions in the online resources.

d. Isosceles triangle

7. $4x - 6y + 23 = 0$

8. a. $y = -x + 5$ b. $y = x + 3$ c. $(1, 4)$

9. and 10. Sample responses can be found in the worked solutions in the online resources.

11. B

12. a. $5.10\,\text{km}$ b. $(6.5, 5.5)$ c. 2
d. $y = 2x - 18$ e. $(10, 2)$ f. $7.07\,\text{km}$

13. and 14. Sample responses can be found in the worked solutions in the online solutions.

15. a. $(0, 2)$, $(2, 2)$ or $(0, -2)$, $(-2, -2)$ or $(1, 1)$, $(1, -1)$

b. $\left(1, \sqrt{3}\right)$ or $\left(1, -\sqrt{3}\right)$

c. $(3, 1)$, $(-1, 1)$ or $(1, -1)$

16. Sample responses can be found in the worked solutions in the online resources.

17. Sample responses can be found in the worked solutions in the online resources.

a. $y = 1$

b. Sample responses can be found in the worked solutions in the online resources.

18. a. Sample responses can be found in the worked solutions in the online resources.

b. Yes

c. Yes

19. a. OA: $2x + 3y - 13 = 0$; OB: $x = 3$

b. $\left(3, \dfrac{7}{3}\right)$

c. d. and e. Sample responses can be found in the worked solutions in the online resources.

20. a. Line A: $2x - y - 1 = 0$, Line B: $3x + 4y - 18 = 0$, Line C: $x - 6y - 6 = 0$

b.

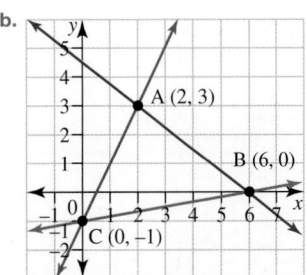

c. Scalene

21. Sample responses can be found in the worked solutions in the online resources.

7.8 Graphing linear inequalities (Path)

1. a.

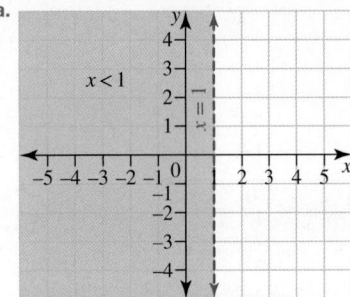

b.

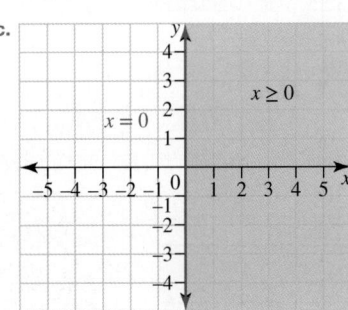

c.

d.
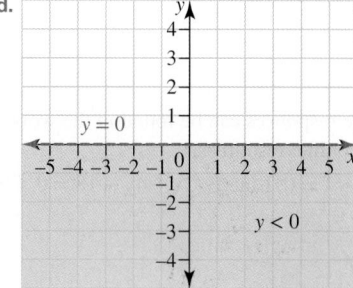

2. a.
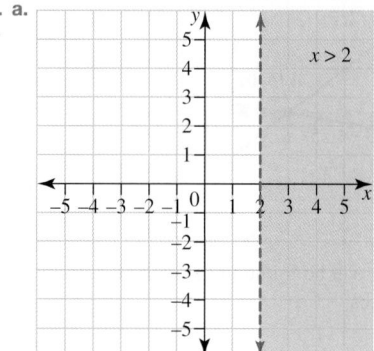

b.
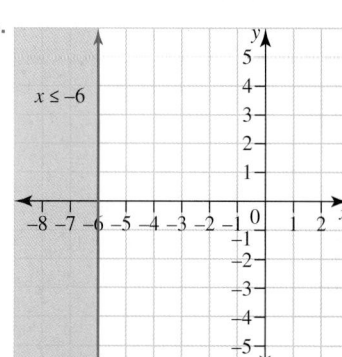

c.
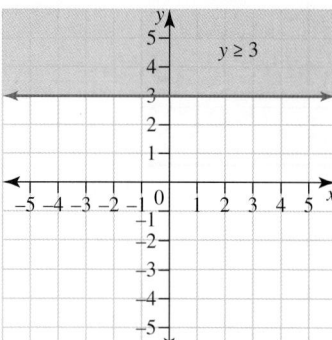

d.
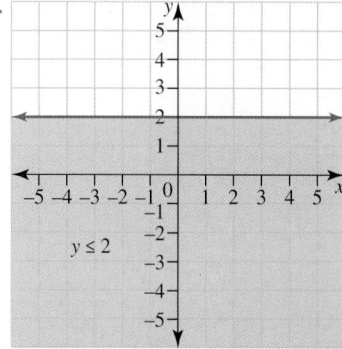

3. a.
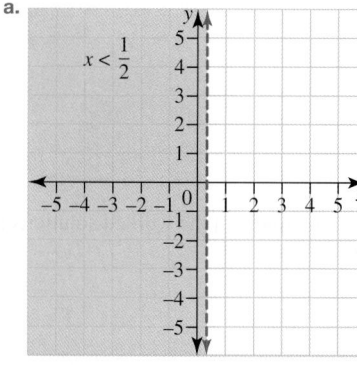

b.
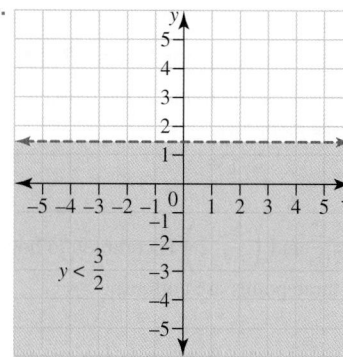

c.
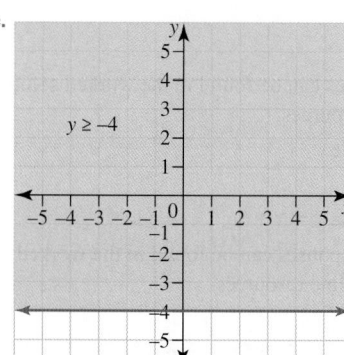

d.
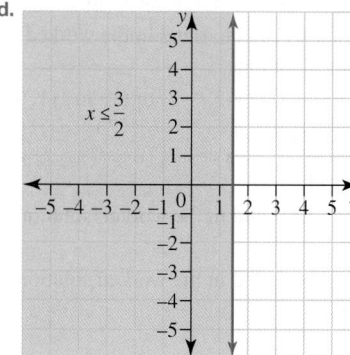

4. a. C   b. A, C
5. a. A, B   b. A, B, C
6. a. B   b. A, B, C

7. a. $y \geq x + 1$

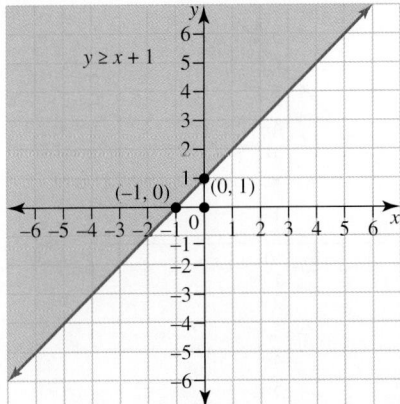

b. $y < x - 6$

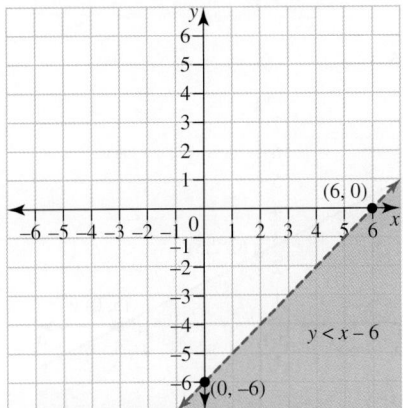

c. $y > -x - 2$

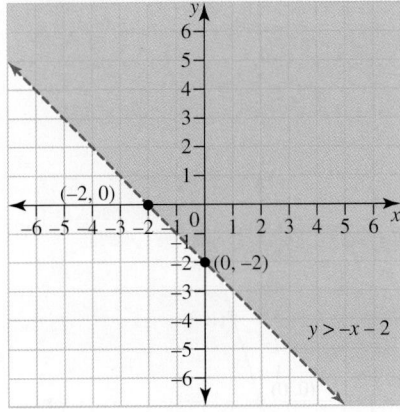

d. $y < 3 - x$

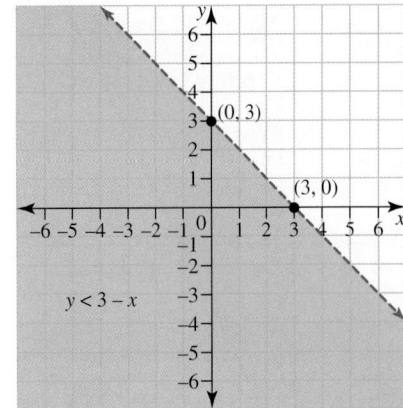

8. a. $y > x - 2$

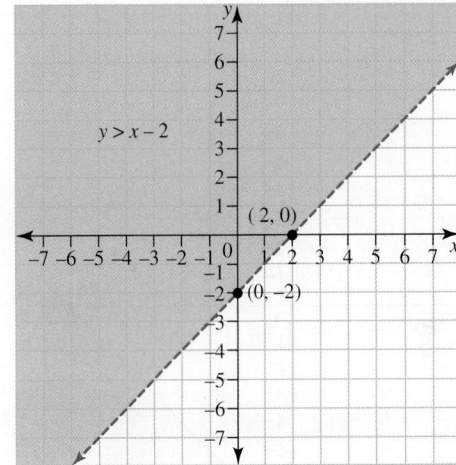

b. $y < 4$

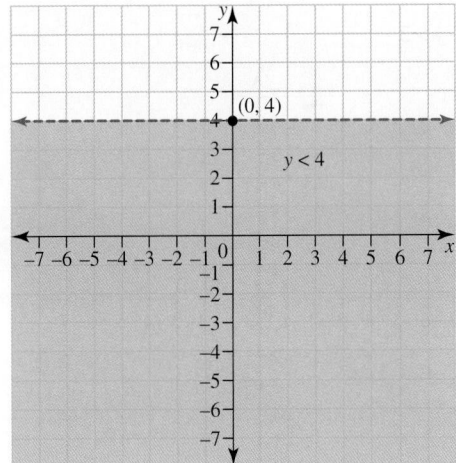

c. $2x - y < 6$

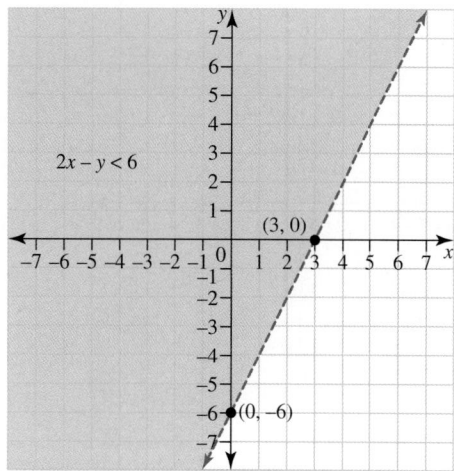

$2x - y < 6$
$(3, 0)$
$(0, -6)$

b. $y < x + 7$

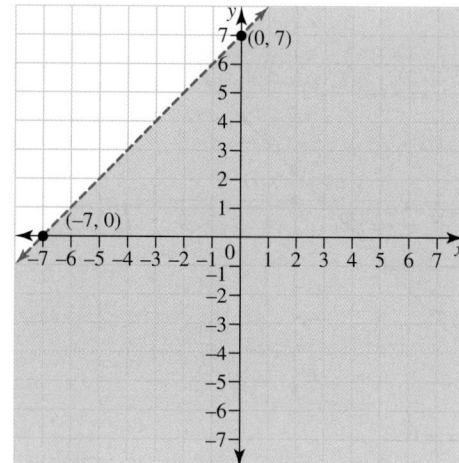

$(0, 7)$
$(-7, 0)$

d. $y \le x - 7$

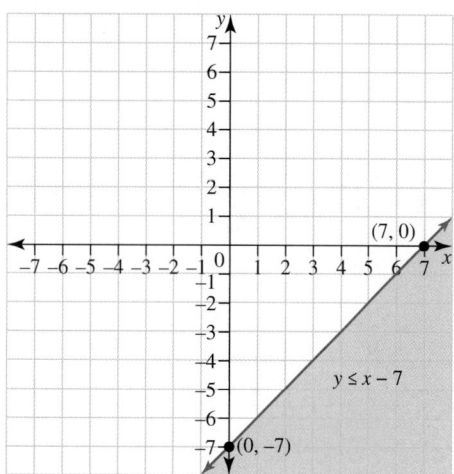

$(7, 0)$
$y \le x - 7$
$(0, -7)$

c. $x + 2y \le 5$

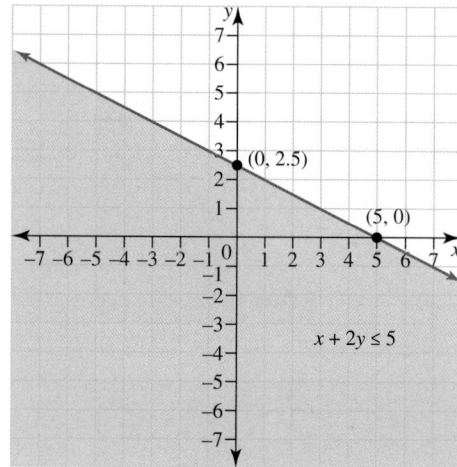

$(0, 2.5)$
$(5, 0)$
$x + 2y \le 5$

9. a. $x - y > 3$

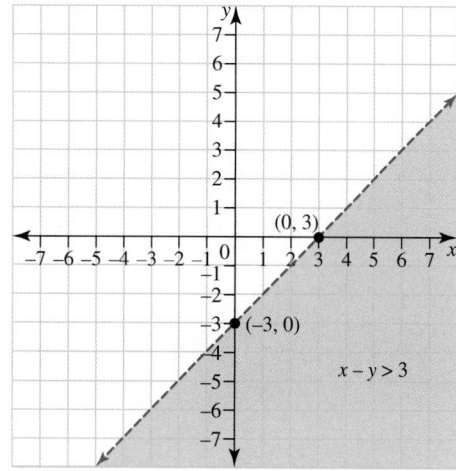

$(0, 3)$
$(-3, 0)$
$x - y > 3$

d. $y \le 3x$

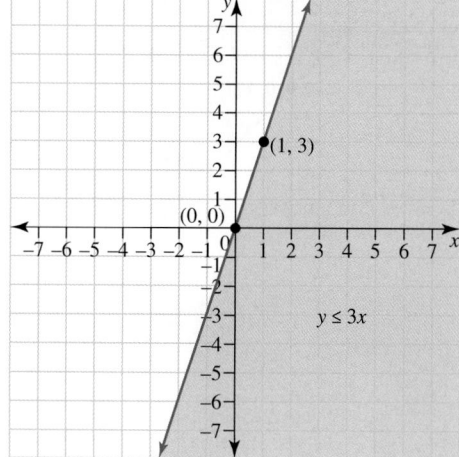

$(1, 3)$
$(0, 0)$
$y \le 3x$

10. B

11. C

12. A

13. a. $y = \dfrac{1}{2}x + 3$

b. $y \geq \dfrac{1}{2}x + 3, x > 2, y \leq 7$

14. $(3,3), (3,4), (3,5), (4,3)$

15. a. $l + s \leq 30$

b. At least 12 small dogs

c.

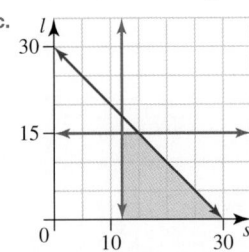

d. 15 large and 15 small dogs

16. $20\,\text{units}^2$

17. a. $y \leq 2x - 2$

b. Sample responses can be found in the worked solutions in the online resources.

18. a. $y = -\dfrac{2}{3}x + 3$ **b.** $y > -\dfrac{2}{3}x + 3$

c. $y < -\dfrac{2}{3}x + 3$ **d.** $y \geq -\dfrac{2}{3}x + 3$

19. a. $y = \dfrac{11}{6} - \dfrac{1}{6}x$

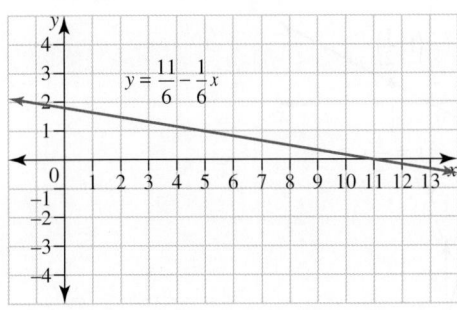

b. The unshaded region is the required region.

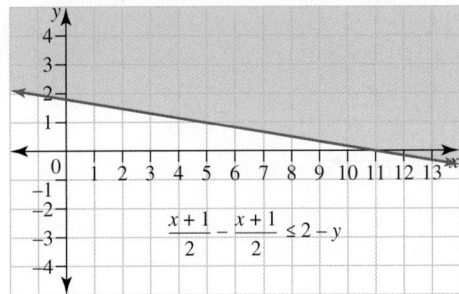

20. The unshaded region is the required region.

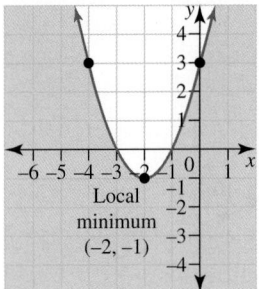

Project

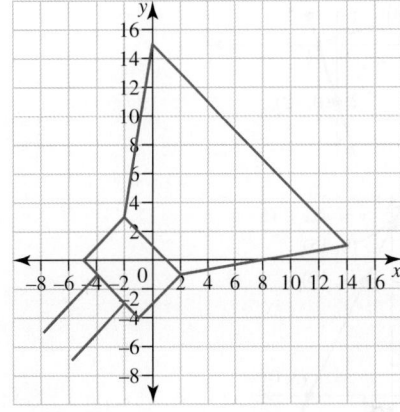

1. The symbol is the one used to represent a speaker.

2. The shape is a trapezium.

$$\text{Area} = \frac{1}{2}\left(\text{length line 6} + \text{length line 8}\right)$$

$$\times \text{Perpendicular distance between these lines.}$$

$$= \frac{1}{2}\left(4\sqrt{2} + 14\sqrt{2}\right) \times 7\sqrt{2}$$

$$= 126\,\text{units}^2$$

3. Sample responses can be found in the worked solutions in the online resources. You could use any symbol of interest and provide instructions for your design. Ensure that your design involves aspects of coordinate geometry that have been used throughout this task.

7.9 Review questions

1. A
2. D
3. B
4. C
5. A
6. C
7. A
8. C
9. C
10. A

11. See table at the bottom of the page.*

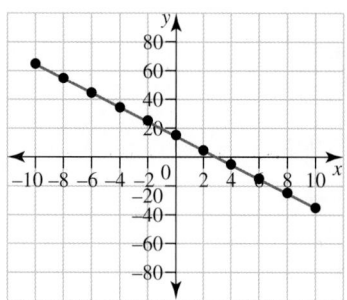

12. a.

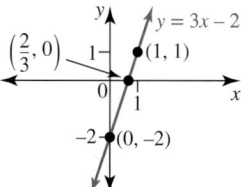

b.

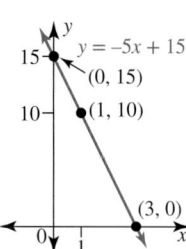

c.

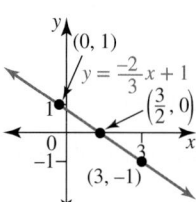

d.

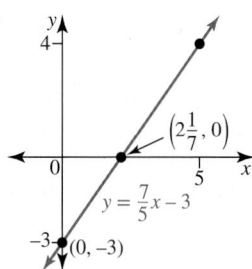

13. a. x-intercept: $= \dfrac{6}{7}$; y-intercept: $b = 6$

b. x-intercept: $= \dfrac{40}{3}$, $\left(= 13\dfrac{1}{3}\right)$, y-intercept: $= -5$

c. x-intercept: $= \dfrac{21}{16}$, $\left(= 1\dfrac{5}{16}\right)$, y-intercept: $= -\dfrac{3}{4}$

d. x-intercept: -5.6 y-intercept: $= 2.8$

14. a.

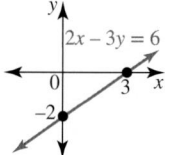

b.

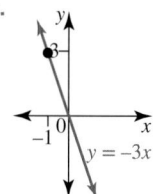

c.

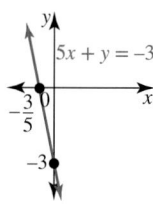

d.

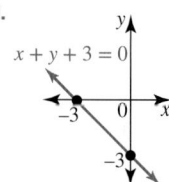

15. a.

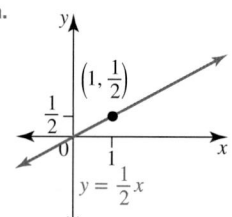

b.

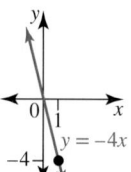

c.

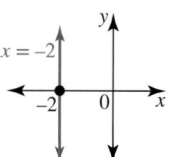

*11.

x	−10	−8	−6	−4	−2	0	2	4	6	8	10
y	65	55	45	35	25	15	5	−5	−15	−25	−35

d.

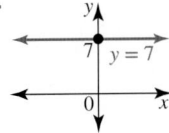

16.

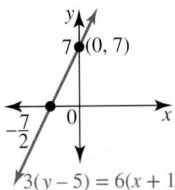

17. a. $y = 2x - 2$ b. $y = -x - 4$ c. $y = -\frac{1}{3}x + 2$

d. $y = 4x$ e. $y = -\frac{3}{4}$ f. $x = 5$

18. a. $y = 3x - 4$ b. $y = -2x - 5$

c. $y = \frac{1}{2}x + 5$ d. $y = 6$

19. a. $y = 7x - 13$ b. $y = -3x + 4$

c. $y = \frac{1}{2}x + 6$ d. $y = \frac{3}{5}x - \frac{18}{5}$

20. $\sqrt{61}$

21. and 22. Sample responses can be found in the worked solutions in the online resources.

23. $(0, -18)$

24. and 25. Sample responses can be found in the worked solutions in the online resources.

26. $x + 2y - 2 = 0$

27. $2x + 3y - 9 = 0$

28. $3x + 2y - 21 = 0$

29. $3x - 2y + 16 = 0$

30. a. i. $-\frac{4}{5}$ ii. $\frac{5}{4}$

iii. $4x + 5y - 61 = 0$ iv. $5x - 4y - 25 = 0$

v. $(9, 5)$

b. Square

31. a. i. $\frac{1}{10}$ ii. $\left(-\frac{1}{2}, 1\right)$ iii. $\left(4\frac{1}{2}, 1\frac{1}{2}\right)$

b. Sample responses can be found in the worked solutions in the online resources.

32. *Note:* The shaded region is the region required.

a.

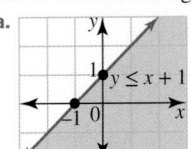

b.

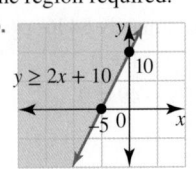

c.

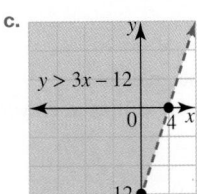

d.

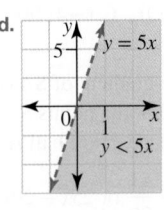

33. a. See table at the bottom of the page.*

b. Pay = $13.50 × (number of hours worked)

c.

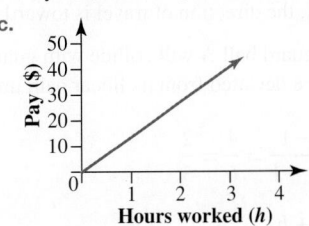

d. $91.13

34. a. See table at the bottom of the page.*

b. Cost = $2.50 × number of rides + $12.50

c.

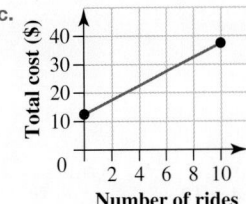

d. $30

35. a.

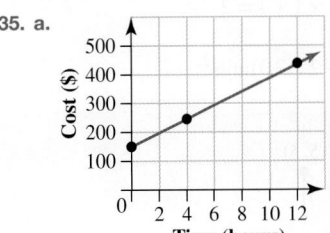

*33a.

Number of hours	0	2	4	6	8	10
Pay ($)	0	27	54	81	108	135

*34a.

Number of rides	0	2	4	6	8	10
Cost($)	12.50	17.50	22.50	27.50	32.50	37.50

b. $C = 22.50h + 160$

c. Approximately $436

36. a. $7x - 3y - 1 = 0$

b. $3x + 7y - 49 = 0$

c. -7

37. a. $x = -7$

b. B $(-7, 50)$, C $(-4, 90)$

c. $40.11\,\text{m}$

38. a. Since the gradient of SA equals the gradient of SO $= -0.8$, the points S, A and O are collinear. Player Y will displace guard ball A.

b. $y = \dfrac{5}{4}x + \dfrac{41}{20}$ or $25x - 20y + 41 = 0$

c. Since the gradient of the path AB is $\dfrac{5}{4}$, which is the same as the gradient of the known path of travel from the common point A, the direction of travel is toward B.

d. $d_{AB} = 0.80\,\text{m}$. Yes, guard ball A will collide with guard ball B as it will not be deviated from its linear path under 1 metre of travel.

39. a. Gradient $= m = \dfrac{5-1}{5--1} = \dfrac{4}{6} = \dfrac{2}{3}$

b. $y = mx + b$, $y = \dfrac{2}{3}x + b$

If $x = -1$ and $y = 1$, substitute in the question:

$1 = \dfrac{2}{3}(-1) + b$

$b = 1\dfrac{2}{3}$

$y = \dfrac{2}{3}x + 1\dfrac{2}{3}$

c. Plot the point (5, 1).

Area of large $\Delta = \dfrac{1}{2} \times 6 \times 4 = 12$

Area of small $\Delta = \dfrac{1}{2} \times 1 \times 4 = 2$

Area of $\Delta ABC = 12 - 2 = 10\,\text{units}^2$

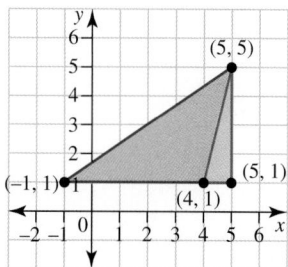

d. $BC^2 = 4^2 + 1^2$

$BC^2 = 16 + 1$

$BC^2 = 17$

$BC^2 = \sqrt{17} \approx 4.12\,\text{units}$

40. a. i. $T = 18 - 0.005(600) = 15\,°C$

ii. $T = 18 - 0.005(1000) = 13\,°C$

iii. $T = 18 - 0.005(3000) = 3\,°C$

b.

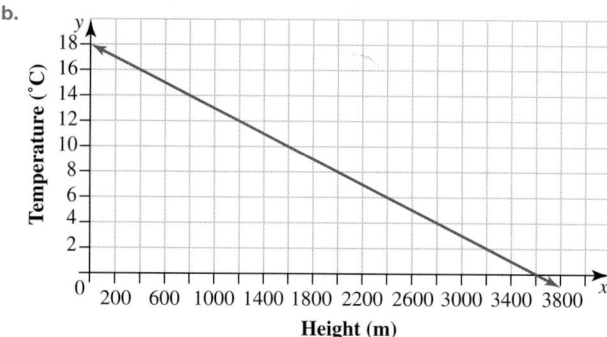

c. $1200\,\text{m} = 12\,°C$, $2500\,\text{m} = 5.5\,°C$

d. $1800\,\text{m}$

41. a. 12 hours

b. 14 years old

c. $h = 8 + \dfrac{18 - c}{2}$

$2h = 16 + 18 - c$

$3h = -c + 34$

$h = \dfrac{1}{2}c + 17$

For every year, the child requires half an hour less sleep.

8 Non-linear relationships

LESSON
8.1 Overview

Why learn this?

So far, throughout high school, much of the focus of algebraic sketching has been on linear graphs. A linear graph is the graph of a straight line; therefore, a non-linear graph is any graph of a curve that is not straight. This means that non-linear graphing is a huge field of mathematics, encompassing many topics and areas of study.

If we think of the purpose of sketching graphs, to some extent it is to model relationships between real-life variables. Yet how often do we actually come up across a linear relationship? Throwing a ball through the air, the speed of a car as it accelerates from rest, the path of a runner around a track, the temperature of coffee as it cools — none of these relationships are linear. Even the path of light from a star in the night sky may not be a straight line through space to Earth, due to the curvature of space itself based on gravity!

Non-linear graphs, whether they be parabolas, hyperbolas, exponentials or circles, are some of the more common types of graphs that are used to model phenomena in everyday life. Thus, it is important to study these relationships and their graphs so we can use them to help us model concepts, such as exponential growth of a colony of bacteria, or an inversely proportional relationship such as the decay of radioactive material over time.

1. Complete the table of values for the equation $y = \frac{1}{2}(x-2)^2 - 3$.

x	-3	-2	-1	0
y				

2. **PATH** For the graph of $y = -3(x+1)^2 - 2$, state the equation of the axis of symmetry.

3. **MC** **PATH** For the graph of $y = 2(x+5)^2 + 8$, the turning point is:
   A. $(-5, -8)$
   B. $(-5, 8)$
   C. $(5, 8)$
   D. $(8, 5)$

4. **MC** **PATH** The graph of $y = -(x-1)^2 - 3$ has:
   A. a maximum turning point
   B. a minimum turning point
   C. no turning point
   D. two maximum turning points

5. **MC** **PATH** The x-intercepts of $y = 2(x-2)^2 - 8$ are:
   A. $x = 2$ or $x = 8$
   B. $x = 0$ or $x = 8$
   C. $x = 10$ or $x = -6$
   D. $x = 0$ or $x = 4$

6. **PATH** Calculate the coordinates of the turning point for the graph of $y = (x+3)(x+5)$.
   Write your answer in the form (a, b).

7. The equation $y = 2x^2 + bx - 1200$ has x-intercepts of $(-30, 0)$ and $(20, 0)$. Determine the value of b.

8. **MC** **PATH** The radius r, of the circle $4x^2 + 4y^2 = 16$ is:
   A. $r = 16$
   B. $r = 8$
   C. $r = 4$
   D. $r = 2$

9. **MC** **PATH** The centre of the circle with equation $x^2 + 4x + y^2 - 6y + 9 = 0$ is:
   A. $(-2, 3)$
   B. $(4, -6)$
   C. $(-4, 6)$
   D. $(2, -3)$

10. **PATH** Calculate the points of intersection between the parabola $y = x^2$ and the circle $x^2 + y^2 = 1$, correct to two decimal places.

11. **MC** **PATH** The horizontal asymptote for $y = 3^{-x} + 1$ is:
    A. $y = 0$
    B. $y = 1$
    C. $x = -1$
    D. $x = 0$

12. **MC** **PATH** For the equation of a hyperbola $y = \dfrac{2}{x+3} - 1$, the vertical and horizontal asymptotes are:
    A. $x = \dfrac{2}{3}$ and $y = -1$
    B. $x = 3$ and $y = -1$
    C. $x = -3$ and $y = -1$
    D. $x = -1$ and $y = -3$

13. Match each graph with its correct equation.

a. $y = 2^x$

i.
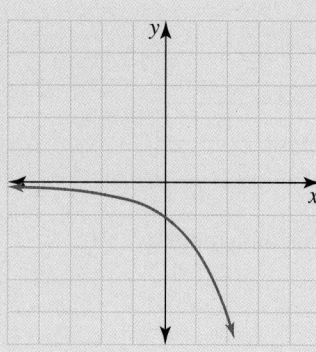

b. -2^x

ii.
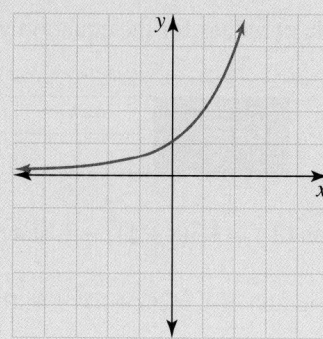

14. **MC** **PATH** From the graph of the hyperbola, the equation is:

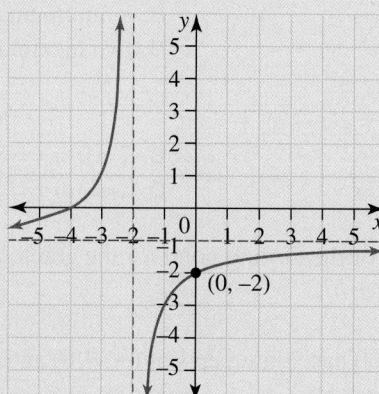

A. $y = \dfrac{1}{x+2} - 1$

B. $y = \dfrac{-1}{x+2} - 1$

C. $y = \dfrac{2}{x+2} - 1$

D. $y = \dfrac{-2}{x+2} - 1$

15. **MC** **PATH** Select which of the following is the equation of a parabola with a vertex (turning point) at $(-5, 2)$.

A. $y = -5x^2 + 2$

B. $y = 2 - (x - 5)^2$

C. $y = (x + 2)^2 - 5$

D. $y = -(x + 5)^2 + 2$

LESSON
8.2 The graph of a parabola

LEARNING INTENTION

At the end of this lesson you should be able to:
- use a table of values to sketch the graph of a parabola
- identify the axis of symmetry, turning point and y-intercept of a parabola.

▶ 8.2.1 Graphing parabolas

eles-4864

- The graphs of all quadratic relationships are called **parabolas**.
- If the equation of the parabola is given, a table of values can be produced by substituting x-values into the equation to obtain the corresponding y-values. These x-and y-values provide the coordinates for points that can be plotted and joined to form the shape of the graph.
- The graph of $y = x^2$ shown has been produced by generating a table of values.

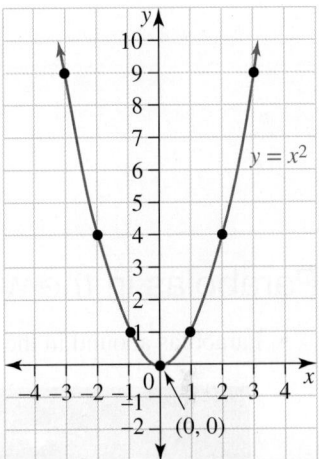

x	−3	−2	−1	0	1	2	3
y	9	4	1	0	1	4	9

- Parabolas are symmetrical; in other words, they have an **axis of symmetry**. In the parabola shown the axis of symmetry is the y-axis, also called the line $x = 0$.
- A parabola has a **vertex** or **turning point**. In this case, the vertex is at the origin and is called a 'minimum turning point'.
- The **y-intercept** of a quadratic is the y-coordinate where the parabola cuts the y-axis. This can be found from a table of values by looking for the point where $x = 0$.
- The **x-intercept** of a quadratic is the x-coordinate where the parabola cuts the x-axis. This can be found from a table of values by looking for the point where $y = 0$.
- Consider the key features of the equation $y = x^2 + 2x - 8$.

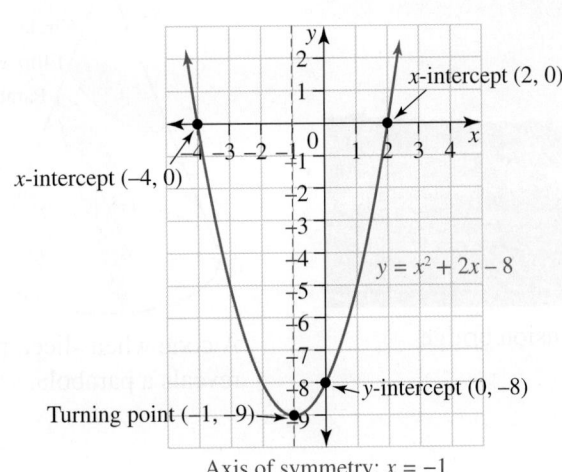

Shapes of parabolas

- **Parabolas with the ∪ shape are said to be 'concave up' and have a minimum turning point.**
- **Parabolas with the ∩ shape are said to be 'concave down' and have a maximum turning point.**

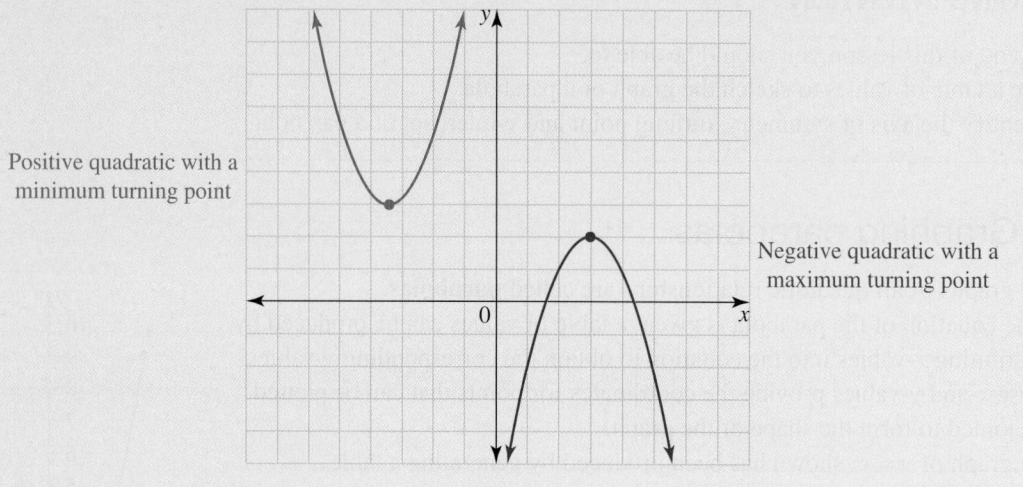

Positive quadratic with a minimum turning point

Negative quadratic with a maximum turning point

Parabolas in the world around us

- Parabolas abound in the world around us. Here are some examples.

Satellite dishes

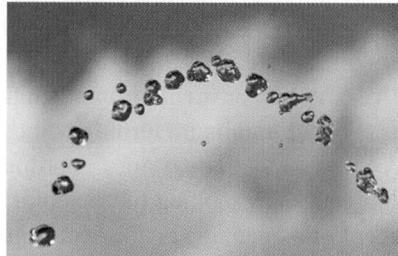

Water droplets from a hose

The cables from a suspension bridge

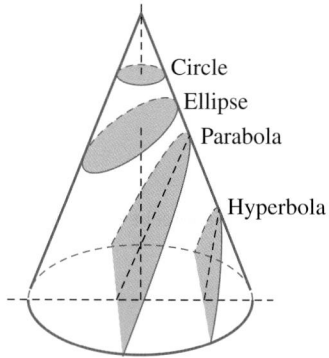

A cone when sliced parallel to its edge reveals a parabola.

Plot the graph of each of the following equations. In each case, use the values of x shown as the values in your table.

State the equation of the axis of symmetry and the coordinates of the turning point.

a. $y = 2x^2$ for $-3 \leq x \leq 3$

b. $y = \dfrac{1}{2}x^2$ for $-3 \leq x \leq 3$

THINK

a. 1. Write the equation.

2. Produce a table of values using x-values from -3 to 3.

3. Draw a set of clearly labelled axes, plot the points and join them with a smooth curve. The scale would be from -2 to 20 on the y-axis and -4 to 4 on the x-axis.

4. Label the graph.

WRITE/DRAW

a. $y = 2x^2$

x	-3	-2	-1	0	1	2	3
y	18	8	2	0	2	8	18

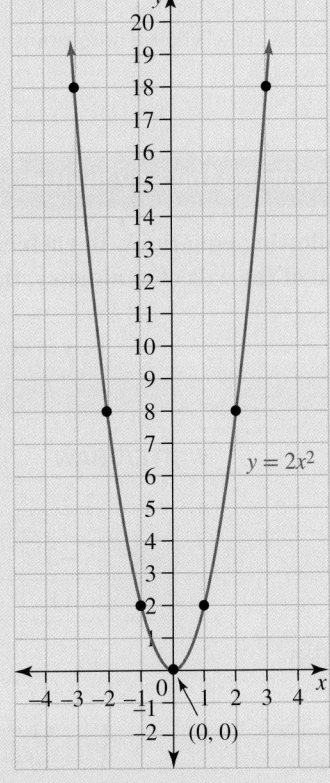

5. Write the equation of the axis of symmetry that divides the parabola exactly in half.

The equation of the axis of symmetry is $x = 0$.

6. Write the coordinates of the turning point.

The turning point is $(0, 0)$.

b. 1. Write the equation.

b. $y = \dfrac{1}{2}x^2$

2. Produce a table of values using x-values from -3 to 3.

x	-3	-2	-1	0	1	2	3
y	4.5	2	0.5	0	0.5	2	4.5

3. Draw a set of clearly labelled axes, plot the points and join them with a smooth curve. The scale would be from −2 to 7 on the y-axis and −4 to 4 on the x-axis.

4. Label the graph.

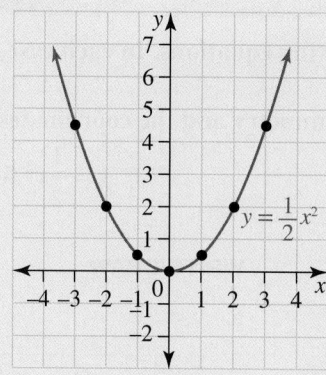

5. Write the equation of the line that divides the parabola exactly in half.

The equation of the axis of symmetry is $x = 0$.

6. Write the coordinates of the turning point.

The turning point is $(0, 0)$.

WORKED EXAMPLE 2 Determining the key features of a parabola

Plot the graph of each of the following equations. In each case, use the values of x shown as the values in your table. State the equation of the axis of symmetry, the coordinates of the turning point and the y-intercept for each one.

a. $y = x^2 + 2$ for $-3 \leq x \leq 3$

b. $y = (x + 3)^2$ for $-6 \leq x \leq 0$

c. $y = -x^2$ for $-3 \leq x \leq 3$

THINK

a. 1. Write the equation.

2. Produce a table of values.

WRITE/DRAW

a. $y = x^2 + 2$

x	−3	−2	−1	0	1	2	3
y	11	6	3	2	3	6	11

3. Draw a set of clearly labelled axes, plot the points and join them with a smooth curve. The scale on the y-axis would be from −2 to 12 and −4 to 4 on the x-axis.

4. Label the graph.

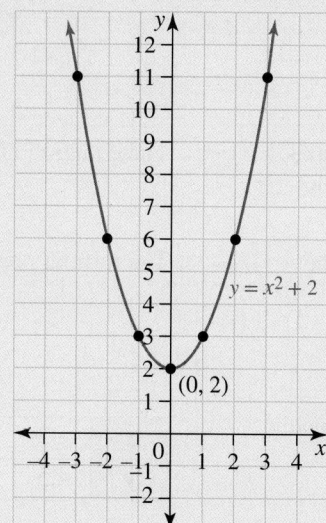

5. Write the equation of the line that divides the parabola exactly in half.

The equation of the axis of symmetry is $x = 0$.

6. Write the coordinates of the turning point.

The turning point is $(0, 2)$.

7. Determine the y-coordinate of the point where the graph crosses the y-axis.

The y-intercept is 2.

b. 1. Write the equation.

b. $y = (x + 3)^2$

2. Produce a table of values.

x	−6	−5	−4	−3	−2	−1	0
y	9	4	1	0	1	4	9

3. Draw a set of clearly labelled axes, plot the points and join them with a smooth curve. The scale on the y-axis would be from −2 to 10 and −7 to 1 on the x-axis.

4. Label the graph.

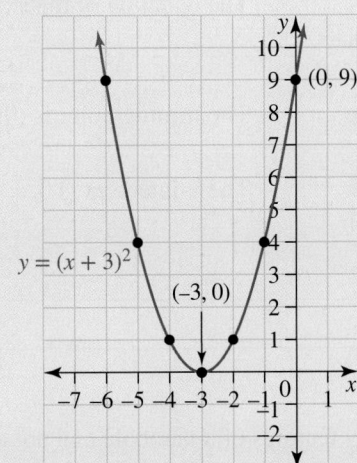

5. Write the equation of the line that divides the parabola exactly in half.

The equation of the axis of symmetry is $x = -3$.

6. Write the coordinates of the turning point.

The turning point is $(-3, 0)$.

7. Determine the y-coordinate of the point where the graph crosses the y-axis.

The y-intercept is 9.

c. 1. Write the equation.

c. $y = -x^2$

2. Produce a table of values.

x	−3	−2	−1	0	1	2	3
y	−9	−4	−1	0	−1	−4	−9

3. Draw a set of clearly labelled axes, plot the points and join them with a smooth curve. The scale on the y-axis would be from -9 to 1 and from -4 to 4 on the x-axis.

4. Label the graph.

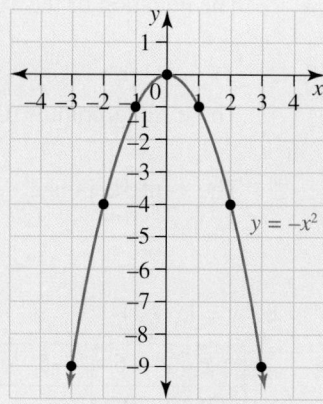

5. Write the equation of the line that divides the parabola exactly in half.

The equation of the axis of symmetry is $x = 0$.

6. Write the coordinates of the turning point.

The turning point is $(0, 0)$.

7. Determine the y-coordinate of the point where the graph crosses the y-axis.

The y-intercept is 0.

DISCUSSION

Why is it important to note the key features of a parabola and not always rely on the table of values when sketching its graph?

COMMUNICATING — COLLABORATIVE TASK: What's the same, what's different?

Equipment: pen, paper, graphing application

1. As a pair, select one non-zero integer a between -10 and $+10$.
2. Sketch the graphs of $y = (x + a)^2$ and $y = x^2 + a$. You can use a graphing application to do so.
3. State the equation of the axis of symmetry, the coordinates of the turning point and the y-intercept for $y = x^2 + a$.
4. Do the same for $y = (x - a)^2$.
5. Share your findings as a class. Can you generalise your results for all pairs of equations in the form $y = (x + a)^2$ and $y = x^2 + a$?

 Resources

Interactivities Plotting quadratic graphs (int-6150)
Parabolas in the world around us (int-7539)

Exercise 8.2 The graph of a parabola

8.2 Quick quiz on	8.2 Exercise

Individual pathways

■ PRACTISE	■ CONSOLIDATE	■ MASTER
1, 3, 5, 11, 13, 15, 19, 22, 26, 29	2, 6, 8, 10, 16, 17, 20, 23, 27, 30	4, 7, 9, 12, 14, 18, 21, 24, 25, 28, 31

You may wish to use a graphing calculator or graphing application for this exercise.

Fluency

1. **WE1** Plot the graph of each of the following equations. In each case, use the values of x shown as the values in your table. State the equation of the axis of symmetry and the coordinates of the turning point.

   a. $y = 3x^2$ for $-3 \leq x \leq 3$

   b. $y = \dfrac{1}{4}x^2$ for $-3 \leq x \leq 3$

2. Compare the graphs drawn for question **1** with that of $y = x^2$. Explain how placing a number in front of x^2 affects the graph obtained.

3. **WE2a** Plot the graph of each of the following for values of x between -3 and 3. State the equation of the axis of symmetry, the coordinates of the turning point and the y-intercept for each one.

   a. $y = x^2 + 1$ b. $y = x^2 + 3$ c. $y = x^2 - 3$ d. $y = x^2 - 1$

4. Compare the graphs drawn for question **3** with the graph of $y = x^2$. Explain how adding to or subtracting from x^2 affects the graph obtained.

   WE2b For questions **5** to **8**, plot the graph of each of the following equations. In each case, use the values of x shown as the values in your table. State the equation of the axis of symmetry and the coordinates of the turning point and the y-intercept for each one.

5. $y = (x + 1)^2$ $-5 \leq x \leq 3$

6. $y = (x - 2)^2$ $-1 \leq x \leq 5$

7. $y = (x - 1)^2$ $-2 \leq x \leq 4$

8. $y = (x + 2)^2$ $-6 \leq x \leq 2$

9. Compare the graphs drawn for questions **5** to **8** with that for $y = x^2$. Explain how adding to or subtracting from x before squaring affects the graph obtained.

   WE2c For questions **10** to **13**, plot the graph of each of the following equations. In each case, use the values of x shown as the values in your table. State the equation of the axis of symmetry, the coordinates of the turning point and the y-intercept for each one.

10. $y = -x^2 + 1$ $-3 \leq x \leq 3$

11. $y = -(x + 2)^2$ $-5 \leq x \leq 1$

12. $y = -x^2 - 3$ $-3 \leq x \leq 3$

13. $y = -(x - 1)^2$ $-2 \leq x \leq 4$

14. Compare the graphs drawn for questions **10** to **13** with that for $y = x^2$. Explain how a negative sign in front of x^2 affects the graph obtained. Also compare the graphs obtained in questions **10** to **13** with those in questions **3** and **5** to **8**.

    State which graphs have the same turning point. Describe how are they different.

Understanding

For questions **15** to **20**:

a. sketch the graph

b. state the equation of the axis of symmetry

c. state the coordinates of the turning point and whether it is a maximum or a minimum

d. state the y-intercept.

15. $y = (x-5)^2 + 1$ $0 \le x \le 6$

16. $y = 2(x+2)^2 - 3$ $-5 \le x \le 1$

17. $y = -(x-3)^2 + 4$ $0 \le x \le 6$

18. $y = -3(x-1)^2 + 2$ $-2 \le x \le 4$

19. $y = x^2 + 4x - 5$ $-6 \le x \le 2$

20. $y = -3x^2 - 6x + 24$ $-5 \le x \le 3$

21. Use the equation $y = a(x-b)^2 + c$ to answer the following.

    a. Explain how you can determine whether a parabola has a minimum or maximum turning point by looking only at its equation.

    b. Explain how you can determine the coordinates of the turning point of a parabola by looking only at the equation.

    c. Explain how you can obtain the equation of the axis of symmetry by looking only at the equation of the parabola.

22. **MC** For the graph of $y = (x-2)^2 + 5$, the turning point is:

    A. $(5, 2)$ **B.** $(2, -5)$

    C. $(2, 5)$ **D.** $(-2, -5)$

23. **MC** For the graph of $y = 3(x-1)^2 + 12$, the turning point is:

    A. $(3, 12)$ **B.** $(1, 12)$

    C. $(-1, 12)$ **D.** $(-3, 12)$

24. **MC** For the graph of $y = (x+2)^2 - 7$, the y-intercept is:

    A. -2 **B.** -7

    C. -3 **D.** -11

25. **MC** Select which of the following is true for the graph of $y = -(x-3)^2 + 4$.

    A. Turning point $(3, 4)$, y-intercept -5 **B.** Turning point $(3, 4)$, y-intercept 5

    C. Turning point $(-3, 4)$, y-intercept -5 **D.** Turning point $(-3, 4)$, y-intercept 5

Communicating, reasoning and problem solving

26. A ball is thrown into the air. The height, h metres, of the ball at any time, t seconds, can be found by using the equation $h = -(t-4)^2 + 16$.

    a. Plot the graph for values of t between 0 and 8.

    b. Use the graph to determine:

        i. the maximum height of the ball

        ii. how long it takes for the ball to fall back to the ground from the moment it is thrown.

27. From a crouching position in a ditch, an archer wants to fire an arrow over a horizontal tree branch, which is 15 metres above the ground. The height, in metres (h), of the arrow t seconds after it has been fired is given by the equation $h = -8t(t-3)$.

   a. Plot the graph for $t = 0, 1, 1.5, 2, 3$.
   b. From the graph, determine:

      i. the maximum height the arrow reaches
      ii. whether the arrow clears the branch and the distance by which it clears or falls short of the branch
      iii. the time it takes to reach maximum height
      iv. how long it takes for the arrow to hit the ground after it has been fired.

28. There are $0, 1, 2$ and infinite possible points of intersection for two parabolas.

   a. Illustrate these on separate graphs.
   b. Explain why infinite points of intersection are possible. Give an example.
   c. Determine how many points of intersection are possible for a parabola and a straight line. Illustrate these.

29. The area of a rectangle in cm² is given by the equation $A = \dfrac{-8w}{5}(w-6)$, where w is the width of the rectangle in centimetres.

   a. Complete a table of values for $-1 \le w \le 7$.
   b. Explain which of the values for w from part a should be discarded and why.
   c. Sketch the graph of A for suitable values of w.
   d. Evaluate the maximum possible area of the rectangle. Show your working.
   e. Determine the dimensions of the rectangle that produce the maximum area found in part d.

30. The path taken by a netball thrown by a rising Australian player is given by the quadratic equation $y = -x^2 + 3.2x + 1.8$, where y is the height of the ball and x is the horizontal distance from the player's upstretched hand.

   a. Complete a table of values for $-1 \le x \le 4$.
   b. Plot the graph.
   c. Explain what values of x are 'not reasonable'.
   d. Evaluate the maximum height reached by the netball.
   e. Assuming that nothing hits the netball, determine how far away from the player the netball will strike the ground.

31. The values of a, b and c in the equation $y = ax^2 + bx + c$ can be calculated using three points that lie on the parabola. This requires solving triple simultaneous equations by algebra. This can also be done using a CAS calculator.
   If the points $(0, 1)$, $(1, 0)$ and $(2, 3)$ all lie on one parabola, determine the equation of the parabola.

LESSON
8.3 Parabolas and transformations

LEARNING INTENTION

At the end of this lesson you should be able to:
- identify the key features of the basic graph of a quadratic, $y = x^2$ and $y = kx^2$
- recognise that dilation makes the parabola wider or narrower
- translate parabolas vertically and identify their respective equations
- translate parabolas horizontally and identify their respective equations (Path).

▶ 8.3.1 Parabola transformations

eles-6305

- Using the features of the parabola: the concavity, the turning point, the x- and y-intercepts, the axis of symmetry, helps us sketch the graph of a quadratic relationship.
- The basic quadratic graph has the equation $y = x^2$. Transformations or changes in the features of the graph can be observed when the equation changes. These transformations include:
  - dilation
  - vertical translation
  - horizontal translation.
  - reflection

Dilation

- A dilation stretches a graph away from an axis. A dilation of factor 3 from the x-axis triples the distance of each point from the x-axis. This means the point $(2, 4)$ would become $(2, 12)$.
- Compare the graph of $y = 2x^2$ with that of $y = x^2$. This graph is **narrower** or closer to the y-axis and has a dilation factor of 2 from the x-axis. As the magnitude (or size) of the coefficient of x^2 increases, the graph becomes narrower and closer to the y-axis.

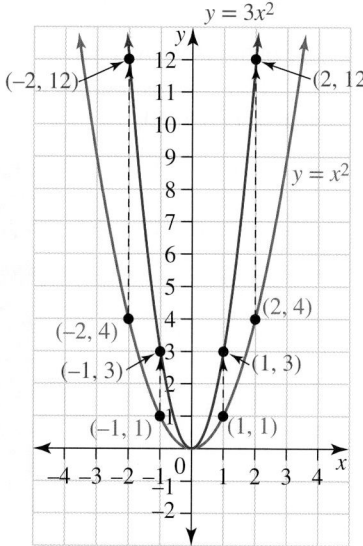

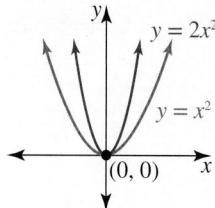

- The turning point has not changed under the transformation and is still $(0, 0)$.
- Compare the graph $y = \dfrac{1}{4}x^2$ with that of $y = x^2$.

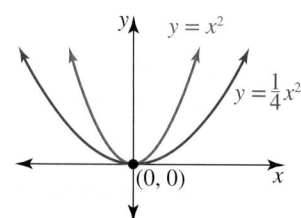

The graph is **wider** or closer to the x-axis and has a dilation factor of factor $\dfrac{1}{4}$.

The turning point has not changed and is still $(0, 0)$. As the coefficient of x^2 decreases (but remains positive), the graph becomes wider or closer to the x-axis.

WORKED EXAMPLE 3 Describing the dilation and determining the turning point

State whether each of the following graphs is wider or narrower than the graph of $y = x^2$ and state the coordinates of the turning point of each one.

a. $y = \dfrac{1}{5}x^2$

b. $y = 4x^2$

THINK	WRITE
a. 1. Write the equation.	a. $y = \dfrac{1}{5}x^2$
2. Look at the coefficient of x^2 and decide whether it is greater than or less than 1.	$\dfrac{1}{5} < 1$, so the graph is wider than that of $y = x^2$.
3. The dilation doesn't change the turning point.	The turning point is $(0, 0)$.
b. 1. Write the equation.	b. $y = 4x^2$
2. Look at the coefficient of x^2 and decide whether it is greater than or less than 1.	$4 > 1$, so the graph is narrower than that of $y = x^2$.
3. The dilation doesn't change the turning point.	The turning point is $(0, 0)$.

▶ 8.3.2 Vertical translation

eles-4866

- Compare the graph of $y = x^2 + 2$ with that of $y = x^2$.
  The whole graph has been moved or translated 2 units upwards. The turning point has become $(0, 2)$.

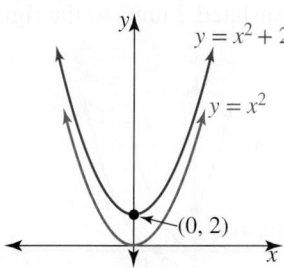

- Compare the graph of $y = x^2 - 3$ with that of $y = x^2$.
  The whole graph has been moved or translated 3 units downwards. The turning point has become $(0, -3)$.

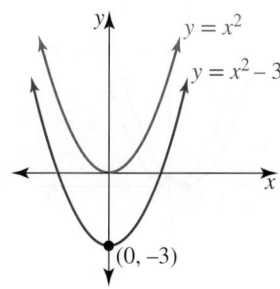

> **WORKED EXAMPLE 4 Describing the vertical translation and determining the turning point**

State the vertical translation and the coordinates of the turning point for the graphs of the following equations when compared to the graph of $y = x^2$.

a. $y = x^2 + 5$

b. $y = x^2 - 4$

THINK	WRITE
a. 1. Write the equation.	**a.** $y = x^2 + 5$
2. $+5$ means the graph is translated upwards 5 units.	Vertical translation of 5 units up.
3. Translate the turning point of $y = x^2$ which is $(0, 0)$. The x-coordinate of the turning point remains 0, and the y-coordinate has 5 added to it.	The turning point becomes $(0, 5)$.
b. 1. Write the equation.	**b.** $y = x^2 - 4$
2. -4 means the graph is translated downwards 4 units.	Vertical translation of 4 units down.
3. Translate the turning point of $y = x^2$, which is $(0, 0)$. The x-coordinate of the turning point remains 0, and the y-coordinate has 4 subtracted from it.	The turning point becomes $(0, -4)$.

⊙ 8.3.3 Horizontal translation (Path)

eles-4867

- Compare the graph of $y = (x - 2)^2$ with that of $y = x^2$.
  The whole graph has been moved or translated 2 units to the right. The turning point has become $(2, 0)$.

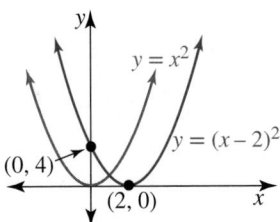

- Compare the graph of $y = (x + 1)^2$ with that of $y = x^2$.
  The whole graph has been moved or translated 1 unit left. The turning point has become $(-1, 0)$.

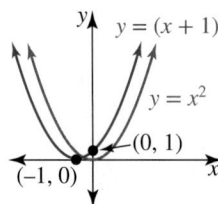

- *Note:* Horizontal translations appear to cause the graph to move in the opposite direction to the sign inside the brackets.

WORKED EXAMPLE 5 Describing the horizontal translation and determining the turning point

State the horizontal translation and the coordinates of the turning point for the graphs of the following equations when compared to the graph of $y = x^2$.

a. $y = (x - 3)^2$ b. $y = (x + 2)^2$

THINK	WRITE
a. 1. Write the equation.	**a.** $y = (x - 3)^2$
2. -3 means the graph is translated to the right 3 units.	Horizontal translation of 3 units to the right
3. Translate the turning point of $y = x^2$, which is $(0, 0)$. The y-coordinate of the turning point remains 0, and the x-coordinate has 3 added to it.	The turning point becomes $(3, 0)$.
b. 1. Write the equation.	**b.** $y = (x + 2)^2$
2. $+2$ means the graph is translated to the left 2 units.	Horizontal translation of 2 units to the left
3. Translate the turning point of $y = x^2$, which is $(0, 0)$. The y-coordinate of the turning point remains 0, and the x-coordinate has 2 subtracted from it.	The turning point becomes $(-2, 0)$.

⊳ 8.3.4 Reflection

eles-4868

- Compare the graph of $y = -x^2$ with that of $y = x^2$.

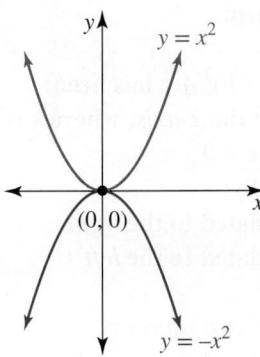

In each case the axis of symmetry is the line $x = 0$ and the turning point is $(0, 0)$. The only difference between the equations is the negative sign in $y = -x^2$, and the difference between the graphs is that $y = x^2$ 'sits' on the x-axis and $y = -x^2$ 'hangs' from the x-axis. (One is a reflection or mirror image of the other.) The graph of $y = x^2$ has a minimum turning point, and the graph of $y = -x^2$ has a maximum turning point.

WORKED EXAMPLE 6 Identifying the coordinates of the turning point

For each of the following graphs, identify the coordinates of the turning point and state whether it is a maximum or a minimum.

a. $y = 5 - x^2$

b. $y = -(x-7)^2$

THINK	WRITE
a. 1. Write the equation.	a. $y = 5 - x^2$
2. Rewrite the equation so that the x^2 term is first.	$y = -x^2 + 5$
3. The vertical translation is 5 units up, so 5 units is added to the y-coordinate of $(0,0)$.	The turning point is $(0,5)$.
4. The sign in front of the x^2 term is negative, so the graph is inverted.	Maximum turning point.
b. 1. Write the equation.	b. $y = -(x-7)^2$
2. It is a horizontal translation of 7 units to the right, so 7 units is added to the x-coordinate of $(0,0)$.	The turning point is $(7,0)$.
3. The sign in front of the x^2 term is negative, so it is inverted.	Maximum turning point.

▶ 8.3.5 Combining transformations (Path)

eles-4869

- Often, multiple transformations will be applied to the equation $y = x^2$ to produce a new graph.
- We can determine the transformations applied by looking at the equation of the resulting quadratic.

Combining transformations

A quadratic of the form $y = k(x-b)^2 + c$ has been:
- dilated by a factor of k from the *x-axis*, where k is the magnitude (or size) of k.
- reflected about the *x-axis* if $k < 0$
- translated b *units* horizontally:
  - if $b > 0$, the graph is translated to the *right*
  - if $b < 0$, the graph is translated to the *left*
- translated c *units* vertically:
  - if $c > 0$, the graph is translated *upwards*
  - if $c < 0$, the graph is translated *downwards*.

WORKED EXAMPLE 7 Determining transformations

For each of the following quadratic equations:
  i. state the appropriate dilation, reflection and translation of the graph of $y = x^2$ needed to obtain the graph
  ii. state the coordinates of the turning point
  iii. hence, sketch the graph.

a. $y = (x+3)^2$

b. $y = -2x^2$

THINK	**WRITE/DRAW**

a. 1. Write the quadratic equation.

a. $y = (x+3)^2$

2. Identify the transformation needed — horizontal translation only, no dilation or reflection.

i. Horizontal translation of 3 units to the left

3. State the turning point.

ii. The turning point is $(-3, 0)$.

4. Sketch the graph of $y = (x+3)^2$. You may find it helpful to lightly sketch the graph of $y = x^2$ on the same set of axes first.

iii.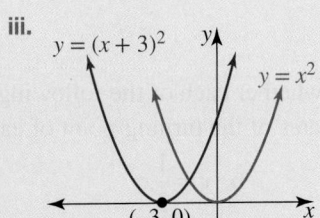

b. 1. Write the quadratic equation.

b. $y = -2x^2$

2. Identify the transformations needed — dilation (2 in front of x^2) and reflection (negative in front of x^2), no translation.

i. This is a reflection, so the graph is inverted. As $2 > 1$, the graph is narrower than that of $y = x^2$.

3. The turning point remains the same as there is no translation.

ii. The turning point is $(0, 0)$.

4. Sketch the graph of $y = -2x^2$. You may find it helpful to lightly sketch the graph of $y = x^2$ on the same set of axes first.

iii.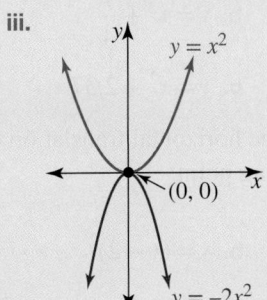

DISCUSSION

Determine the turning points of the graphs $y = x^2 + k$ and $y = (x - h)^2$.

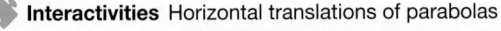

on Resources

Interactivities Horizontal translations of parabolas (int-6054)
Vertical translations of parabolas (int-6055)
Dilation of parabolas (int-6096)
Reflection of parabolas (int-6151)

8.3 Quick quiz on	**8.3 Exercise**

Individual pathways

■ PRACTISE	■ CONSOLIDATE	■ MASTER
1, 4, 7, 10, 13, 16, 17, 22, 25	2, 5, 8, 11, 14, 18, 19, 23, 26	3, 6, 9, 12, 15, 20, 21, 24, 27, 28

Fluency

WE3 For questions **1** to **3**, state whether each of the following graphs is wider or narrower than the graph of $y = x^2$ and determine the coordinates of the turning point of each one.

1. a. $y = 5x^2$
   b. $y = \dfrac{1}{3}x^2$

2. a. $y = 7x^2$
   b. $y = 10x^2$
   c. $y = \dfrac{2}{5}x^2$

3. a. $y = 0.25x^2$
   b. $y = 1.3x^2$
   c. $y = \sqrt{3}x^2$

WE4 For questions **4** to **6**, describe the vertical translation of each of the following equations from $y = x^2$ and determine the coordinates of the turning point.

4. a. $y = x^2 + 3$
   b. $y = x^2 - 1$

5. a. $y = x^2 - 7$
   b. $y = x^2 + \dfrac{1}{4}$
   c. $y = x^2 - \dfrac{1}{2}$

6. a. $y = x^2 - 0.14$
   b. $y = x^2 + 2.37$
   c. $y = x^2 + \sqrt{3}$

WE5 For questions **7** to **9**, describe the horizontal translation of each of the following equations from $y = x^2$ and determine the coordinates of the turning point.

7. **PATH**
   a. $y = (x - 1)^2$
   b. $y = (x - 2)^2$

8. **PATH**
   a. $y = (x + 10)^2$
   b. $y = (x + 4)^2$
   c. $y = \left(x - \dfrac{1}{2}\right)^2$

9. **PATH**
   a. $y = \left(x + \dfrac{1}{5}\right)^2$
   b. $y = (x + 0.25)^2$
   c. $y = (x + \sqrt{3})^2$

WE6 For questions **10** to **12**, for each of the following graphs identify the coordinates of the vertex and state whether it is the maximum or the minimum point.

10. a. $y = -x^2 + 1$
    b. $y = x^2 - 3$

11. **PATH**
    a. $y = -(x + 2)^2$
    b. $y = 3x^2$
    c. $y = 4 - x^2$

12. **PATH**
    a. $y = -2x^2$
    b. $y = (x - 5)^2$
    c. $y = 1 + x^2$

For questions **13** to **15**, in each of the following state whether the graph is wider or narrower than $y = x^2$ and whether it has a maximum or a minimum turning point.

13. **a.** $y = 3x^2$
    b. $y = -3x^2$

14. **a.** $y = \dfrac{1}{2}x^2$
    b. $y = -\dfrac{1}{5}x^2$
    c. $y = -\dfrac{4}{3}x^2$

15. **a.** $y = 0.25x^2$
    b. $y = \sqrt{3}x^2$
    c. $y = -0.16x^2$

Understanding

WE7 For questions **16** to **21**:
 i. describe the dilation, reflection and translation of the graph of $y = x^2$ needed to obtain the graph
 ii. state the coordinates of the turning point
 iii. hence, sketch the graph.

16. **PATH**
    a. $y = (x+1)^2$
    b. $y = -3x^2$
    c. $y = x^2 + 1$

17. **a.** $y = \dfrac{1}{3}x^2$
    b. $y = x^2 - 3$

18. **PATH**
    a. $y = (x-4)^2$
    b. $y = -\dfrac{2}{5}x^2$
    c. $y = 5x^2$

19. **a.** $y = -x^2 + 2$
    b. $y = -x^2 - 4$

20. **PATH**
    a. $y = -(x-6)^2$
    b. $y = 2(x+1)^2 - 4$
    c. $y = \dfrac{1}{2}(x-3)^2 + 2$

21. **PATH**
    a. $y = -\dfrac{1}{3}(x+2)^2 + \dfrac{1}{4}$
    b. $y = -\dfrac{7}{4}(x-1)^2 - \dfrac{3}{2}$

Communicating, reasoning and problem solving

22. **PATH** A vase 25 cm tall is positioned on a bench near a wall as shown.

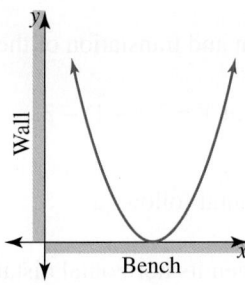

The shape of the vase follows the curve $y = (x - 10)^2$, where y cm is the height of the vase and x cm is the distance of the vase from the wall.

  a. Identify how far the base of the vase is from the wall.
  b. Determine the shortest distance from the top of the vase to the wall.
  c. If the vase is moved so that the top just touches the wall, determine the new distance from the wall to the base.
  d. Determine the new equation that follows the shape of the vase.

23. **PATH** Tom is standing at the start of a footpath at $(0, 0)$ that leads to the base of a hill. The height of the hill is modelled by the equation $h = -\dfrac{1}{10}(d - 25)^2 + 40$, where h is the height of the hill in metres and d is the horizontal distance from the start of the path.

   a. Calculate how tall the hill is.

   b. Determine how far the base of the hill is from the beginning of the footpath.

   c. If the footpath is to be extended so the lead in to the hill is 50 m, determine the new equation that models the height of the hill.

   d. The height of the hill has been incorrectly measured and should actually be 120 m. Adjust the equation from part c to correct this error and state the transformation applied.

24. A ball is thrown vertically upwards. Its height in metres after t seconds is given by $h = 7t - t^2$.

   a. Sketch the graph of the height of the ball against time.

   b. Evaluate the highest point reached by the ball. Show your full working.

A second ball is thrown vertically upwards. Its height in metres after t seconds is given by $h = 10t - t^2$.

   c. On the same set of axes used for part a, sketch the graph of the height of the second ball against time.

   d. State the difference in the highest point reached by the two balls.

25. Consider the quadratic equation $y = x^2 - 4x + 7$.

   a. Determine the equivalent inverted equation of the quadratic that just touches the one above at the turning point.

   b. Confirm your result graphically.

26. **PATH** Consider the equation $y = 3(x - 2)^2 - 7$.

   a. State the coordinates of the turning point and y-intercept.

   b. State a sequence of transformations that when applied to the graph of $y = \dfrac{-3}{2}(x - 2)^2 - 7$ will produce the graph of $y = x^2$.

27. **PATH** A parabola has the equation $y = -\dfrac{1}{2}(x - 3)^2 + 4$. A second parabola has an equation defined by $Y = 2(y - 1) - 3$.

   a. Determine the equation relating Y to x.

   b. State the appropriate dilation, reflection and translation of the graph of $Y = x^2$ required to obtain the graph of $Y = 2(y - 1) - 3$.

   c. State the coordinates of the turning point $Y = 2(y - 1) - 3$.

   d. Sketch the graph of $Y = 2(y - 1) - 3$.

28. A ball shot at a certain angle to the horizontal follows a parabolic path.
It reaches a maximum height of 200 m when its horizontal distance from its starting point is 10 m. When the ball's horizontal distance from the starting point was 1 m, the ball had reached a height of 38 m.
Determine an equation to model the ball's flight, clearly defining your chosen pronumerals.

LESSON
8.4 Parabolas in vertex form (Path)

LEARNING INTENTION

At the end of this lesson you should be able to:
- identify the axis of symmetry and turning point of a quadratic equation in vertex form
- calculate the y-intercept and any x-intercepts of a quadratic equation in vertex form.

▶ 8.4.1 Turning point form

eles-4870

- When a quadratic equation is expressed in the form $y = k(x - b)^2 + c$:
  - the turning point is the point (b, c)
  - the axis of symmetry is $x = b$
  - the x-intercepts are calculated by substituting $y = 0$ and solving $k(x - b)^2 + c = 0$.
  - the y-intercept is calculated by substituting $x = 0$.

$$y = k(x - b)^2 + c$$

| Reflects and dilates | Translates left and right | Translates up and down |

- Changing the values of k, b and c in the equation transforms the shape and position of the parabola when compared with the parabola $y = x^2$.

Turning point form

A quadratic equation of the form $y = k(x - b)^2 + c$ has:
- a turning point at the coordinate (b, c)
  - the turning point will be a minimum if $k > 0$
  - the turning point will be a maximum if $k < 0$
- an axis of symmetry with equation $x = b$

The number of x-intercepts depends on the values of k, b and c. Changing the value of k does not change the position of the turning point, only b and c.

WORKED EXAMPLE 8 Determining the coordinates of the vertex

For each of the following equations, state the coordinates of the turning point of the graph and whether it is a maximum or a minimum.

a. $y = (x - 6)^2 - 4$
b. $y = -(x + 3)^2 + 2$

THINK

a. 1. Write the equation.

2. Identify the transformations — horizontal translation of 6 units to the right and a vertical translation of 4 units down. State the turning point.

3. As k is positive ($k = 1$), the graph is upright with a minimum turning point.

WRITE

a. $y = (x - 6)^2 - 4$

The turning point is $(6, -4)$.

Minimum turning point.

b. 1. Write the equation		**b.** $y = -(x+3)^2 + 2$	

2. Identify the transformations — horizontal translation of 3 units to the left and a vertical translation of 2 units up. State the turning point.

The turning point is $(-3, 2)$.

3. As k is negative ($k = -1$), the graph is inverted with a maximum turning point.

Maximum turning point.

▶ 8.4.2 x- and y-intercepts of parabolic graphs

eles-4871

- The x- and y-intercepts can also be determined from the equation of a parabola.
- The point(s) where the graph cuts or touches the x-axis are called the x-intercept(s). At these points, $y = 0$.
- The point where the graph cuts the y-axis is called the y-intercept. At this point, $x = 0$.

WORKED EXAMPLE 9 Determining the x- and y-intercepts

For the parabolas with the following equations:
i. determine the y-intercept
ii. determine the x-intercepts (where they exist).

a. $y = (x+3)^2 - 4$ **b.** $y = 2(x-1)^2$ **c.** $y = -(x+2)^2 - 1$

THINK

WRITE

a. 1. Write the equation.

a. $y = (x+3)^2 - 4$

2. Calculate the y-intercept by substituting $x = 0$ into the equation.

y-intercept: when $x = 0$,
$y = (0+3)^2 - 4$
  $= 9 - 4$
  $= 5$
The y-intercept is 5.

3. Calculate the x-intercepts by substituting $y = 0$ into the equation and solving for x. Add 4 to both sides of the equation. Take the square root of both sides of the equation.

x-intercepts: when $y = 0$,
$(x+3)^2 - 4 = 0$
    $(x+3)^2 = 4$
    $(x+3) = +2$ or -2

4. Solve for x.

$x = 2 - 3$ or $x = -2 - 3$
$x = -1$ $x = -5$

The x-intercepts are -5 and -1.

b. 1. Write the equation.

b. $y = 2(x-1)^2$

2. Calculate the y-intercept by substituting $x = 0$ into the equation.

y-intercept: when $x = 0$,
$y = 2(0-1)^2$
  $= 2 \times 1$
  $= 2$
The y-intercept is 2.

3. Calculate the x-intercepts by substituting $y = 0$ into the equation and solving for x.
Note that there is only one solution for x and so there is only one x-intercept. (The graph touches the x-axis.)

x-intercepts: when, $y = 0$,
$$2(x-1)^2 = 0$$
$$(x-1)^2 = 0$$
$$x - 1 = 0$$
$$x = 0 + 1$$
$$x = 1$$
The x-intercept is 1.

c. 1. Write the equation.

c. $y = -(x+2)^2 - 1$

2. Calculate the y-intercept by substituting $x = 0$ into the equation.

y-intercept: when $x = 0$,
$$y = -(0+2)^2 - 1$$
$$= -4 - 1$$
$$= -5$$
The y-intercept is -5.

3. Calculate the x-intercepts by substituting $y = 0$ into the equation and solving for x. We cannot take the square root of -1 to obtain real solutions; therefore, there are no x-intercepts.

x-intercepts: when $y = 0$,
$$-(x+2)^2 - 1 = 0$$
$$(x+2)^2 = -1$$
There are no real solutions, so there are no x-intercepts.

WORKED EXAMPLE 10 Sketching the graph of a quadratic equation

For each of the following:
  i. write the coordinates of the turning point
 ii. state whether the graph has a maximum or a minimum turning point
iii. state whether the graph is wider, narrower or the same width as the graph of $y = x^2$
 iv. calculate the y-intercept
  v. calculate the x-intercepts
 vi. sketch the graph.

a. $y = (x-2)^2 + 3$

b. $y = -2(x+1)^2 + 6$

THINK

WRITE/DRAW

a. 1. Write the equation.

a. $y = (x-2)^2 + 3$

2. State the coordinates of the turning point from the equation. Use (b, c) as the equation is in the turning point form of $y = k(x-b)^2 + c$ where $k = 1, b = 2$ and $c = 3$.

The turning point is $(2, 3)$.

3. State the nature of the turning point by considering the sign of k.

The graph has a minimum turning point as the sign of k is positive.

4. Specify the width of the graph by considering the magnitude of k.

The graph has the same width as $y = x^2$ since $k = 1$.

5. Calculate the y-intercept by substituting $x = 0$ into the equation.

y-intercept: when $x = 0$,
$$y = (0-2)^2 + 3$$
$$= 4 + 3$$
$$= 7$$
y-intercept is 7.

6. Calculate the x-intercepts by substituting $y = 0$ into the equation and solving for x. As we have to take the square root of a negative number, we cannot solve for x.

x-intercepts: when $y = 0$,
$$(x-2)^2 + 3 = 0$$
$$(x-2)^2 = -3$$
There are no real solutions, and hence no x-intercepts.

7. Sketch the graph, clearly showing the turning point and the y-intercept.

8. Label the graph.

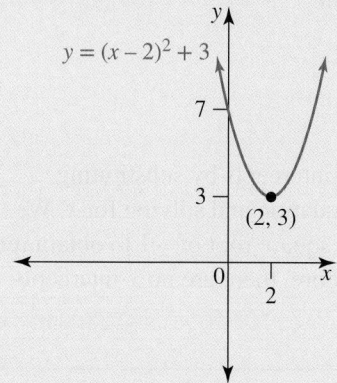

b. 1. Write the equation.

b. $y = -2(x+1)^2 + 6$

2. State the coordinates of the turning point from the equation. Use (b, c) as the equation is in the turning point form of $y = k(x-b)^2 + c$ where $k = -2, b = -1$ and $c = 6$.

The turning point is $(-1, 6)$.

3. State the nature of the turning point by considering the sign of k.

The graph has a maximum turning point as the sign of k is negative.

4. Specify the width of the graph by considering the magnitude of k.

The graph is narrower than $y = x^2$ since $|k| > 1$.

5. Calculate the y-intercept by substituting $x = 0$ into the equation.

y-intercept: when $x = 0$,
$$y = -2(0+1)^2 + 6$$
$$= -2 \times 1 + 6$$
$$= 4$$
The y-intercept is 4.

6. Calculate the x-intercepts by substituting $y = 0$ into the equation and solving for x.

x-intercepts: when $y = 0$,
$$-2(x+1)^2 + 6 = 0$$
$$2(x+1)^2 = 6$$
$$(x+1)^2 = 3$$
$$x+1 = \sqrt{3} \text{ or } x+1 = -\sqrt{3}$$
$$x = -1 + \sqrt{3} \text{ or } x = -1 - \sqrt{3}$$
The x-intercepts are $-1 - \sqrt{3}$ and $-1 + \sqrt{3}$ (or approximately -2.73 and 0.73).

7. Sketch the graph, clearly showing the turning point and the x- and y-intercepts.
8. Label the graph.

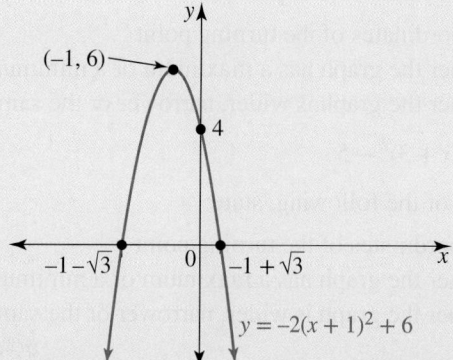

Note: Unless otherwise stated, exact values for the intercepts should be shown on sketch graphs.

DISCUSSION

Does k in the equation $y = k(x - b)^2 + c$ have any impact on the turning point? Discuss the strengths and weaknesses of having the quadratic equation in the vertex form: $y = k(x - b)^2 + c$.

on Resources

 Video eLessons Sketching quadratics in turning point form (eles-1926)
Solving quadratics in turning point form (eles-1941)

 Interactivity Quadratic functions (int-2562)

Exercise 8.4 Parabolas in vertex form (Path)

learn on

8.4 Quick quiz **on**	8.4 Exercise

Individual pathways

■ PRACTISE	■ CONSOLIDATE	■ MASTER
1, 4, 7, 9, 12, 14, 17, 20, 23, 27	2, 5, 8, 10, 13, 15, 18, 21, 24, 28	3, 6, 11, 16, 19, 22, 25, 26, 29

Fluency

WE8 For questions **1** to **3**, for each of the following equations, state the coordinates of the turning point of the graph and whether it is a maximum or a minimum.

1. a. $y = (x - 1)^2 + 2$
   b. $y = (x + 2)^2 - 1$
   c. $y = (x + 1)^2 + 1$

2. a. $y = -(x - 2)^2 + 3$
   b. $y = -(x - 5)^2 + 3$
   c. $y = (x + 2)^2 - 6$

3. a. $y = \left(x - \dfrac{1}{2}\right)^2 - \dfrac{3}{4}$
   b. $y = \left(x - \dfrac{1}{3}\right)^2 + \dfrac{2}{3}$
   c. $y = (x + 0.3)^2 - 0.4$

4. For each of the following, state:
   i. the coordinates of the turning point
   ii. whether the graph has a maximum or a minimum turning point
   iii. whether the graph is wider, narrower or the same width as that of $y = x^2$.

   a. $y = 2(x+3)^2 - 5$

   b. $y = -(x-1)^2 + 1$

5. For each of the following, state:
   i. the coordinates of the turning point
   ii. whether the graph has a maximum or a minimum turning point
   iii. whether the graph is wider, narrower or the same width as that of $y = x^2$.

   a. $y = -5(x+2)^2 - 4$

   b. $y = \frac{1}{4}(x-3)^2 + 2$

6. For each of the following, state:
   i. the coordinates of the turning point
   ii. whether the graph has a maximum or a minimum turning point
   iii. whether the graph is wider, narrower or the same width as that of $y = x^2$.

   a. $y = -\frac{1}{2}(x+1)^2 + 7$

   b. $y = 0.2\left(x - \frac{1}{5}\right)^2 - \frac{1}{2}$

7. Select the equation that best suits each of the following graphs.

   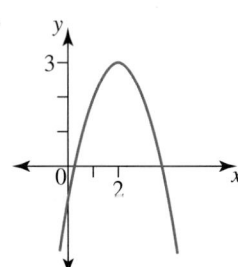 i.

   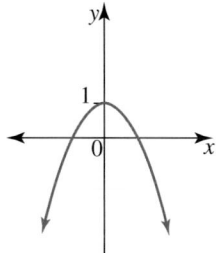 ii.

   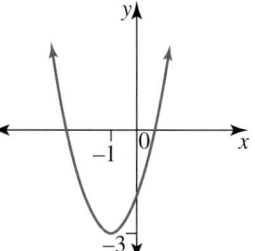 iii.

   a. $y = (x+1)^2 - 3$

   b. $y = -(x-2)^2 + 3$

   c. $y = -x^2 + 1$

8. Select the equation that best suits each of the following graphs.

   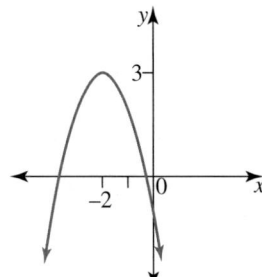 i.

   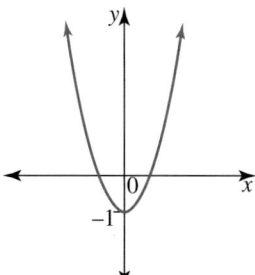 ii.

   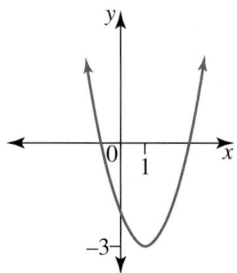 iii.

   a. $y = (x-1)^2 - 3$

   b. $y = -(x+2)^2 + 3$

   c. $y = x^2 - 1$

9. **MC** The translations required to change $y = x^2$ into $y = \left(x - \frac{1}{2}\right)^2 + \frac{1}{3}$ are:

   A. right $\frac{1}{2}$, up $\frac{1}{3}$

   B. left $\frac{1}{2}$, down $\frac{1}{3}$

   C. right $\frac{1}{2}$, down $\frac{1}{3}$

   D. left $\frac{1}{2}$, up $\frac{1}{3}$

10. **MC** For the graph $\frac{1}{4}\left(x - \frac{1}{2}\right)^2 + \frac{1}{3}$, the effect of the $\frac{1}{4}$ on the graph is:

    A. no effect   **B.** to make the graph narrower
    C. to make the graph wider   **D.** to invert the graph

11. **MC** Compared to the graph of $y = x^2$, $y = -2(x + 1)^2 - 4$ is:

    A. inverted and wider   **B.** inverted and narrower
    C. upright and wider   **D.** upright and narrower

12. **MC** A graph that has a minimum turning point $(1, 5)$ and that is narrower than the graph of $y = x^2$ is:

    A. $y = (x - 1)^2 + 5$ **B.** $y = \frac{1}{2}(x + 1)^2 + 5$

    C. $y = 2(x - 1)^2 + 5$ **D.** $y = 2(x + 1)^2 + 5$

13. **MC** Compared to the graph of $y = x^2$, the graph of $y = -3(x - 1)^2 - 2$ has the following features.

    A. Maximum TP at $(-1, -2)$, narrower
    B. Maximum TP at $(1, -2)$, narrower
    C. Maximum TP at $(1, 2)$, wider
    D. Minimum TP at $(1, -2)$, narrower

WE9 For questions **14** to **16**, for the parabolas with the following equations:
i. determine the y-intercept
ii. determine the x-intercepts (where they exist).

14. **a.** $y = (x + 1)^2 - 4$ **b.** $y = 3(x - 2)^2$

15. **a.** $y = -(x + 4)^2 - 2$ **b.** $y = (x - 2)^2 - 9$

16. **a.** $y = 2x^2 + 4$ **b.** $y = (x + 3)^2 - 5$

Understanding

17. **WE10** For each of the following:

    i. write the coordinates of the turning point
    ii. state whether the graph has a maximum or a minimum turning point
    iii. state whether the graph is wider, narrower or the same width as the graph of $y = x^2$
    iv. calculate the y-intercept
    v. calculate the x-intercepts
    vi. sketch the graph.

    a. $y = (x - 4)^2 + 2$ **b.** $y = (x - 3)^2 - 4$ **c.** $y = (x + 1)^2 + 2$

18. For each of the following:

    i. write the coordinates of the turning point
    ii. state whether the graph has a maximum or a minimum turning point
    iii. state whether the graph is wider, narrower or the same width as the graph of $y = x^2$
    iv. calculate the y-intercept
    v. calculate the x-intercepts
    vi. sketch the graph.

    a. $y = (x + 5)^2 - 3$ **b.** $y = -(x - 1)^2 + 2$ **c.** $y = -(x + 2)^2 - 3$

19. For each of the following:

   i. write the coordinates of the turning point
   ii. state whether the graph has a maximum or a minimum turning point
   iii. state whether the graph is wider, narrower or the same width as the graph of $y = x^2$
   iv. calculate the y-intercept
   v. calculate the x-intercepts
   vi. sketch the graph.

   a. $y = -(x+3)^2 - 2$
   b. $y = 2(x-1)^2 + 3$
   c. $y = -3(x+2)^2 + 1$

20. Consider the equation $2x^2 - 3x - 8 = 0$.

   a. Complete the square.
   b. Use the result to determine the exact solutions to the original equation.
   c. Determine the turning point of $y = 2x^2 - 3x - 8$ and indicate its type.

21. Answer the following questions.

   a. Determine the equation of a quadratic that has a turning point of $(-4, 6)$ and has an x-intercept at $(-1, 0)$.
   b. State the other x-intercept (if any).

22. Write the new equation for the parabola $y = x^2$ that has been:

   a. reflected in the x-axis
   b. dilated by a factor of 7 away from the x-axis
   c. translated 3 units in the negative direction of the x-axis
   d. translated 6 units in the positive direction of the y-axis
   e. dilated by a factor of $\dfrac{1}{4}$ from the x-axis, reflected in the x-axis, and translated 5 units in the positive direction of the x-axis and 3 units in the negative direction of the y-axis.

Communicating, reasoning and problem solving

23. The price of shares in fledgling company 'Lollies'r'us' plunged dramatically one afternoon, following the breakout of a small fire on the premises.
However, Ms Sarah Sayva of Lollies Anonymous agreed to back the company, and share prices began to rise.
Sarah noted at the close of trade that afternoon that the company's share price followed the curve:
$P = 0.1(t-3)^2 + 1$ where $\$P$ is the price of shares t hours after noon.

   a. Sketch a graph of the relationship between time and share price to represent the situation.
   b. Determine the initial share price.
   c. Determine the lowest price of shares that afternoon.
   d. Evaluate the time when the price was at its lowest.
   e. Determine the final price of 'Lollies'r'us' shares as trade closed at 5 pm.

24. Rocky is practising for a football kicking competition. After being kicked, the path that the ball follows can be modelled by the quadratic relationship:

$$h = -\frac{1}{30}(d - 15)^2 + 8$$

where h is the vertical distance the ball reaches (in metres), and d is the horizontal distance (in metres).

   a. Determine the initial vertical height of the ball.
   b. Determine the exact maximum horizontal distance the ball travels.
   c. Write down both the maximum height and the horizontal distance when the maximum height is reached.

25. Answer the following questions.

   a. If the turning point of a particular parabola is $(2, 6)$, suggest a possible equation for the parabola.
   b. If the y-intercept in part **a** is $(0, 4)$, determine the exact equation for the parabola.

26. Answer the following questions.

   a. If the turning point of a particular parabola is (p, q), suggest a possible equation for the parabola.
   b. If the y-intercept in part **a** is $(0, r)$, determine the exact equation for the parabola.

27. Use the completing the square method to write each of the following in turning point form and sketch the parabola for each part.

   a. $y = x^2 - 8x + 1$ b. $y = x^2 + 4x - 5$ c. $y = x^2 + 3x + 2$

28. Use the information given in the graph shown to answer the following questions.

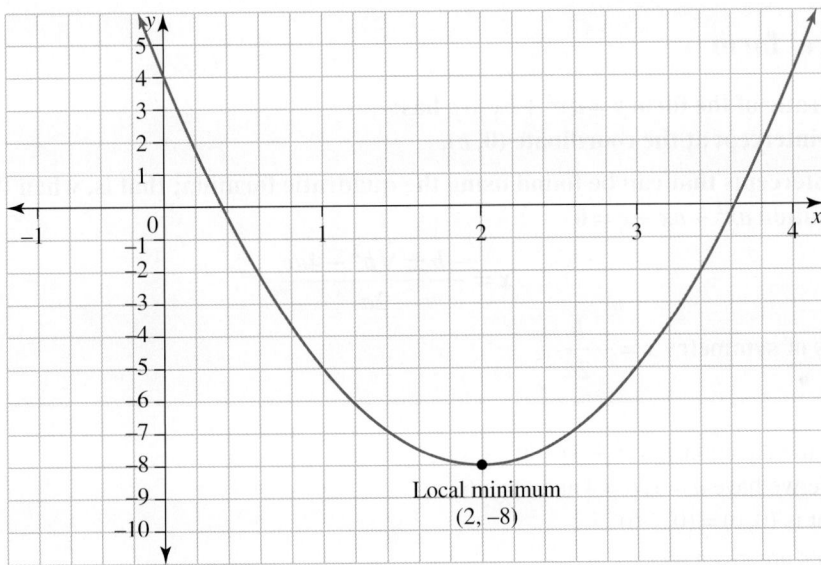

Local minimum
$(2, -8)$

   a. Determine the equation of the parabola shown.
   b. State the dilation and translation transformations that have been applied to $y = x^2$ to achieve this parabola.
   c. This graph is reflected in the x-axis. Determine the equation of the reflected graph.
   d. Sketch the graph of the reflected parabola.

29. The graph of a quadratic equation has a turning point at $(-3, 8)$ and passes through the point $(-1, 6)$.

   a. Determine the equation of this parabola.
   b. State the transformations that have been applied to $y = x^2$ to produce this parabola.
   c. Calculate the x and y-intercepts of this parabola.
   d. The graph is reflected in the x-axis, dilated by a factor of 2 from the x-axis, and then reflected in the y-axis. Sketch the graph of this new parabola.

LESSON
8.5 Parabolas in general form (Path)

LEARNING INTENTION

At the end of this lesson you should be able to:
- graph parabolas by determining and using features such as the x-intercept/s, y-intercept, the turning point and the axis of symmetry
- use the quadratic formula when needed to calculate the x-intercept(s).

▶ 8.5.1 Parabolas of the form $y = ax^2 + bx + c$

eles-4872

- The general form of a quadratic equation is $y = ax^2 + bx + c$ where a, b and c are constants.
- The x-intercepts can be found by letting $y = 0$, factorising and using the Null Factor Law to solve for x.
- The y-intercept can be found by letting $x = 0$ and solving for y.
- The x-coordinate of the turning point lies midway between the x-intercepts.
- The axis of symmetry gives the x-coordinate of the turning point, substituting it into the original equation gives the y-coordinate of the turning point.
- If an equation is not written in turning point form, and cannot be readily factorised, then we will need to use the quadratic formula to help find all key points.
- Graph the parabola showing all its key features.

> **General form**
>
> A quadratic of the form $y = ax^2 + bx + c$ has:
> - a y-intercept at the coordinate $(0, c)$.
> - x-intercepts that can be found using the quadratic formula; that is, when the equation $ax^2 + bx + c = 0$
>
> $$x = \frac{-b \pm \sqrt{b^2 - 4ac}}{2a}$$
>
> - axis of symmetry $x = -\dfrac{b}{2a}$

For example, given the equation $y = x^2 + 4x - 6$:
- In this example we have $a = 1, b = 4$ and $c = -6$.
- The y-intercept is $(0, c) = (0, -6)$
- The x-intercepts are given by:

$$x = \frac{-b \pm \sqrt{b^2 - 4ac}}{2a} = \frac{-(4) \pm \sqrt{(4)^2 - 4 \times (1) \times (-6)}}{2 \times (1)} = \frac{-4 \pm \sqrt{40}}{2} = \frac{-4 \pm 2\sqrt{10}}{2} = -2 \pm \sqrt{10}$$

- Therefore, the x-intercepts are $\left(-2 + \sqrt{10}, 0\right)$ and $\left(-2 - \sqrt{10}, 0\right)$.
- The axis of symmetry is $x = -\dfrac{b}{2a} = -\dfrac{4}{2 \times 1} = -2$
- Substituting $x = -2$ in the original equation:
  $y = (-2)^2 + 4 \times (-2) - 6 = -10$.
- Therefore the turning point is $(-2, -10)$.

Note: Do not convert answer to decimals unless specified by the question. It is always best practice to leave coordinates in exact form.

WORKED EXAMPLE 11 Sketching the graph of a quadratic equation in factored form

Sketch the graph of $y = (x - 3)(x + 2)$.

THINK

1. The equation is in factorised form. To calculate the x-intercepts, let $y = 0$ and use the Null Factor Law.

2. The x-coordinate of the turning point is midway between the x-intercepts. Calculate the average of the two x-intercepts to determine the midpoint between them.

3. To calculate the y-coordinate of the turning point, substitute x_{TP} into the equation.

4. State the turning point.

5. To calculate the y-intercept, let $x = 0$ and substitute.

6. State the y-intercept.

7. Sketch the graph, showing all the important features.
8. Label the graph.

WRITE/DRAW

$y = (x - 3)(x + 2)$
$0 = (x - 3)(x + 2)$
$x - 3 = 0$ or $x + 2 = 0$ (NFL)
$\quad x = 3$ or $\quad x = -2$
x-intercepts: $(3, 0)$ or $(-2, 0)$

$x_{TP} = \dfrac{3 + (-2)}{2}$
$\quad = 0.5$

$y = (x - 3)(x + 2)$
$y_{TP} = (0.5 - 3)(0.5 + 2)$
$\quad = -6.25$

Turning point: $(0.5, -6.25)$

$y = (0 - 3)(0 + 2)$
$\quad = -6$

y-intercept: $(0, -6)$

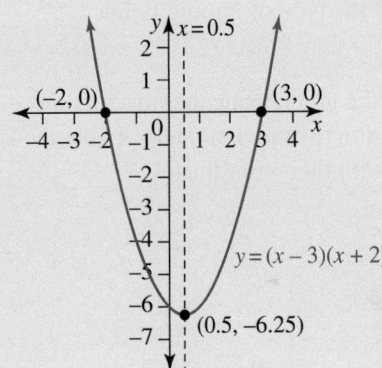

WORKED EXAMPLE 12 Sketching the graph of a quadratic equation in standard form

Sketch the graph of $y = \dfrac{1}{2}x^2 - x - 4$ and label the key points with their coordinates.

THINK

1. Write down the y-intercept.

WRITE

$y = \dfrac{1}{2}x^2 - x - 4$

y-intercept: if $x = 0$, then $y = -4 \Rightarrow (0, -4)$

2. Obtain any x-intercepts.

x-intercepts: let $y = 0$.
$$\frac{1}{2}x^2 - x - 4 = 0$$
$$x^2 - 2x - 8 = 0$$
$$(x + 2)(x - 4) = 0$$
$$\therefore x = -2, 4$$
$$\Rightarrow (-2, 0) \text{ or } (4, 0)$$

3. Determine the equation of the axis of symmetry.

Axis of symmetry formula: $x = -\dfrac{b}{2a}$

$a = \dfrac{1}{2}, b = -1$

$x = -\dfrac{-1}{\left(2 \times \frac{1}{2}\right)}$

$= 1$

4. Determine the coordinates of the turning point.

Turning point: when $x = 1$,
$$y = \frac{1}{2} - 1 - 4$$
$$= -4\frac{1}{2}$$
$$\Rightarrow \left(1, -4\frac{1}{2}\right) \text{ is the turning point.}$$

5. Identify the type of turning point.

Since $a > 0$, the turning point is a minimum turning point.

6. Sketch the graph using the information obtained in the previous steps. Label the key points with their coordinates.

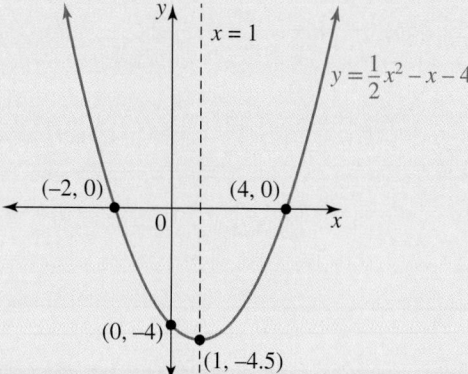

 Resources

Video eLessons Sketching quadratics in factorised form (eles-1927)
Sketching parabolas using the quadratic formula (eles-1945)

Exercise 8.5 Parabolas in general form (Path)

8.5 Quick quiz on	8.5 Exercise

Individual pathways

■ PRACTISE	■ CONSOLIDATE	■ MASTER
1, 2, 5, 8, 11, 14	3, 6, 9, 12, 15	4, 7, 10, 13, 16

Fluency

1. What information is necessary to be able to sketch a parabola?

WE11 For questions **2** to **4**, sketch the graph of each of the following.

2. **a.** $y = (x-5)(x-2)$ **b.** $y = (x+4)(x-7)$

3. **a.** $y = (x+3)(x+5)$ **b.** $y = (2x+3)(x+5)$

4. **a.** $y = (4-x)(x+2)$ **b.** $y = \left(\dfrac{x}{2}+3\right)(5-x)$

WE12 For questions **5** to **7**, sketch the graph of each of the following.

5. **a.** $y = x^2 + 4x + 2$ **b.** $y = x^2 - 4x - 5$ **c.** $y = 2x^2 - 4x - 3$

6. **a.** $y = -2x^2 + 11x + 5$ **b.** $y = -2x^2 + 12x$ **c.** $y = 3x^2 + 6x + 1$

7. **a.** $y = -3x^2 - 5x + 2$ **b.** $y = 2x^2 + 8x - 10$ **c.** $y = -3x^2 + 7x + 3$

Understanding

8. The path of a soccer ball kicked by the goal keeper can be modelled by the equation $y = -\dfrac{1}{144}(x^2 - 24x)$ where y is the height of the soccer ball and x is the horizontal distance from the goalie, both in metres.

   a. Sketch the graph.
   b. Calculate how far away from the player does the ball first bounce.
   c. Calculate the maximum height of the ball.

9. The monthly profit or loss, p, (in thousands of dollars) for a new brand of chicken loaf is given by $p = 3x^2 - 15x - 18$, where x is the number of months after its introduction (when $x = 0$).

   a. Sketch the graph.
   b. Determine during which month a profit was first made.
   c. Calculate the month in which the profit is $54\,000.

10. The height, h metres, of a model rocket above the ground t seconds after launch is given by the equation $h = 4t(50 - t)$, where $0 \le t \le 50$.

   a. Sketch the graph of the rocket's flight.
   b. State the height of the rocket above the ground when it is launched.
   c. Calculate the greatest height reached by the rocket.
   d. Determine how long the rocket takes to reach its greatest height.
   e. Determine how long the rocket is in the air.

Communicating, reasoning and problem solving

11. The equation $y = x^2 + bx + 7500$ has x-intercepts of $(-150, 0)$ and $(-50, 0)$. Determine the value of b in the equation. Justify your answer.

12. The equation $y = x^2 + bx + c$ has x-intercepts of m and n. Determine the value of b in the equation in terms of m and n. Justify your answer.

13. A ball thrown from a cliff follows a parabolic path of the form $y = ax^2 + bx + c$. The ball is released at the point $(0, 9)$, reaches a maximum height at $(2, 11)$ and passes through the point $(6, 3)$ on its descent.
   Determine the equation of the ball's path. Show full working.

14. A ball is thrown upwards from a building and follows the path given by the formula $h = -x^2 + 4x + 21$. The ball is h metres above the ground when it is a horizontal distance of x metres from the building.

   a. Sketch the graph of the path of the ball.
   b. Determine the maximum height of the ball.
   c. Determine how far the ball is from the wall when it reaches the maximum height.
   d. Determine how far from the building the ball lands.

15. During an 8-hour period, an experiment is done in which the temperature of a room follows the relationship $T = h^2 - 8h + 21$, where T is the temperature in degrees Celsius h hours after starting the experiment.

   a. Sketch the graph of this quadratic.
   b. Identify the initial temperature.
   c. Determine if the temperature is increasing or decreasing after 3 hours.
   d. Determine if the temperature is increasing or decreasing after 5 hours.
   e. Determine the minimum temperature and when it occurred.
   f. Determine the temperature after 8 hours.

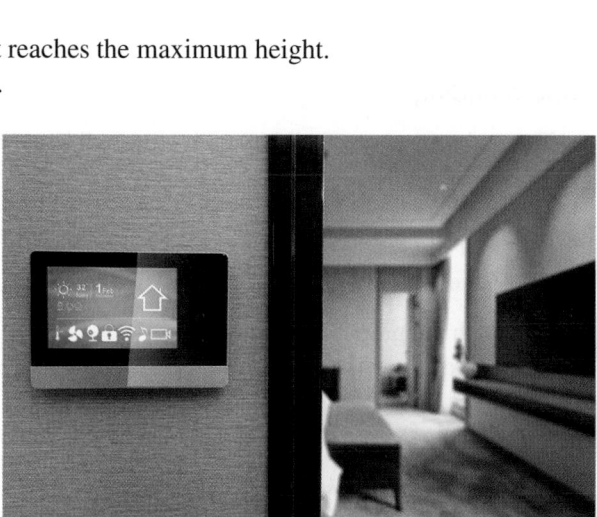

16. A ball is thrown out of a window, passing through the points $(0, 7)$, $(6, 18)$ and $(18, 28)$. A rule for the height of the ball in metres is given by $h = ax^2 + bx + c$ where x is the horizontal distance, in metres, covered by the ball.

   a. Determine the values of a, b and c in the rule for the height of the ball.
   b. Calculate the height that the ball was thrown from.
   c. Evaluate the maximum height reached by the ball.
   d. Determine horizontal distance covered by the ball when it hits the ground.
   e. Sketch the flight path of the ball, making sure you show all key points.

LESSON
8.6 Exponential graphs (Path)

LEARNING INTENTION

At the end of this lesson you should be able to:
- identify the features of the graph of an exponential equation and recognise its basic shape
- sketch the graph of an exponential equation using transformations.

▶ 8.6.1 Exponential equations

eles-4873

- Relationships of the form $y = a^x$ are called **exponential equations** with base a, where a is a real number not equal to 1, and x is the index power or exponent.

- The term 'exponential' is used, as x is an exponent (or index). For example, the graph of the exponential function $y = 2^x$ can be plotted by completing a table of values.

  Remember that $2^{-3} = \dfrac{1}{2^3}$

  $$= \frac{1}{8}, \text{ and so on.}$$

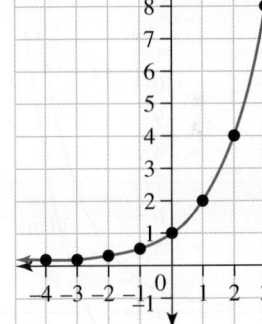

x	-4	-3	-2	-1	0	1	2	3	4
y	$\dfrac{1}{16}$	$\dfrac{1}{8}$	$\dfrac{1}{4}$	$\dfrac{1}{2}$	1	2	4	8	16

- The graph has many significant features.
  - The y-intercept is 1.
  - The value of y is always greater than zero.
  - As x decreases, y gets closer to but never reaches zero. So the graph gets closer to but never reaches the x-axis. The x-axis (or the line $y = 0$) is called an **asymptote.**
  - As x increases, y becomes very large.
- The basic shapes of exponential graphs are shown below.

$y = a^x, a > 1$	$y = a^{-x}, a > 1$	$y = a^x, 0 < a < 1$
![graph with point (0, 1), asymptote y = 0, increasing curve]	![graph with point (0, 1), asymptote y = 0, decreasing curve]	![graph with point (0, 1), asymptote y = 0, decreasing curve]

Comparing exponential graphs

- The diagram at right shows the graphs of $y = 2^x$ and $y = 3^x$.
- The graphs both pass through the point $(0, 1)$.
- The graph of $y = 3^x$ climbs more steeply than the graph of $y = 2^x$.
- $y = 0$ is an asymptote for both graphs.

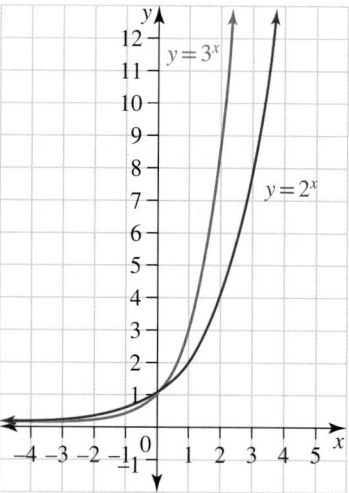

⏵ 8.6.2 Exponential transformations

eles-6308

Translation and reflection

Vertical translation	Reflection about the x-axis	Reflection about the y-axis
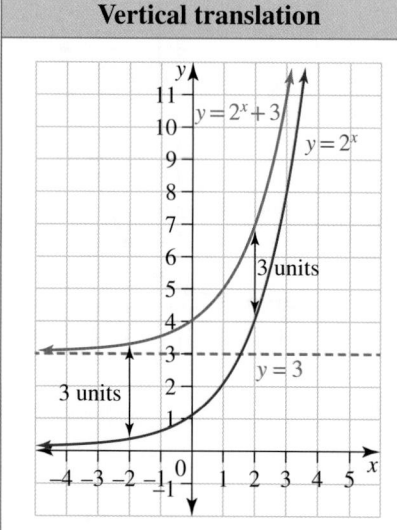	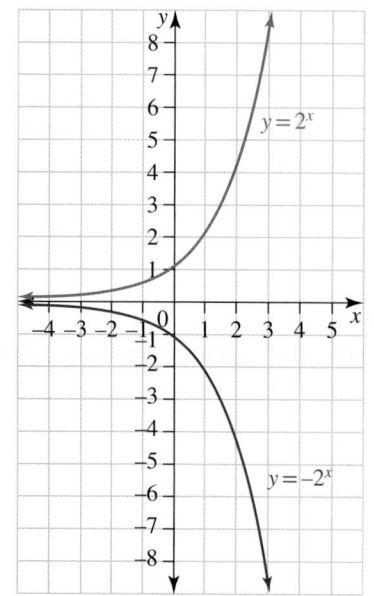	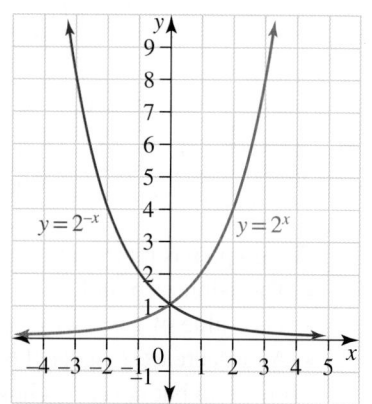

Vertical translation

- The diagram shows the graphs of $y = 2^x$ and $y = 2^x + 3$.
- The graphs have identical shape.
- Although they appear to get closer to each other, the graphs are constantly 3 vertical units apart.
- As x becomes very small, the graph of $y = 2^x + 3$ approaches but never reaches the line $y = 3$, so $y = 3$ is the horizontal asymptote.
- When the graph of $y = 2^x$ is translated 3 units upward, it becomes the graph of $y = 2^x + 3$.

Reflection about the x-axis

- The diagram shows the graphs of $y = 2^x$ and $y = -2^x$.
- The graphs have identical shape.
- The graph of $y = -2^x$ is a reflection about the x-axis of the graph of $y = 2^x$.
- The x-axis $(y = 0)$ is an asymptote for both graphs.
- In general, the graph of $y = -a^x$ is a reflection about the x-axis of the graph of $y = a^x$.

Reflection about the y-axis

- The diagram shows the graphs of $y = 2^x$ and $y = 2^{-x}$.
- The graphs have identical shape.
- The graph of $y = 2^{-x}$ is a reflection about the y-axis of the graph of $y = 2^x$.
- Both graphs pass through the point $(0, 1)$.
- The x-axis $(y = 0)$ is an asymptote for both graphs.
- In general, the graph of $y = a^{-x}$ is a reflection about the y-axis of the graph of $y = a^x$.

WORKED EXAMPLE 13 Sketching exponential graphs

Given the graph of $y = 4^x$, sketch on the same axes the graphs of:
a. $y = 4^x - 2$
b. $y = -4^x$
c. $y = 4^{-x}$.

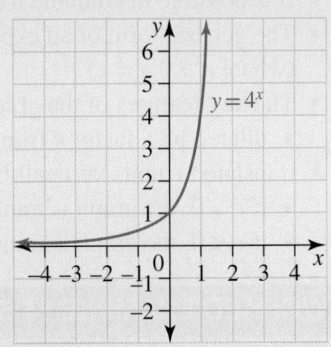

THINK

a. The graph of $y = 4^x$ has already been drawn. It has a y-intercept of 1 and a horizontal asymptote at $y = 0$. The graph of $y = 4^x - 2$ has the same shape as $y = 4^x$ but is translated 2 units vertically down. It has a y-intercept of -1 and a horizontal asymptote at $y = -2$.

b. $y = -4^x$ has the same shape as $y = 4^x$ but is reflected about the x-axis. It has a y-intercept of -1 and a horizontal asymptote at $y = 0$.

c. $y = 4^{-x}$ has the same shape as $y = 4^x$ but is reflected about the y-axis. The graphs have the same y-intercept and the same horizontal asymptote ($y = 0$).

DRAW

a.

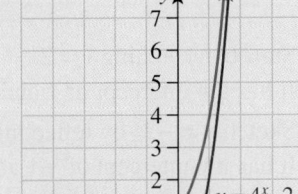

b.

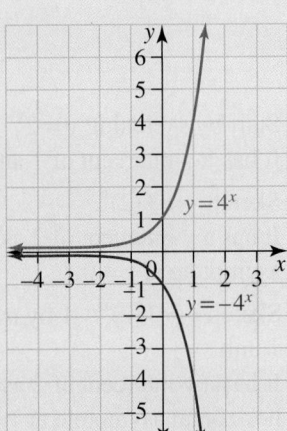

c.

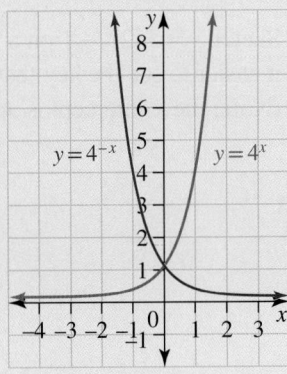

▶ 8.6.3 Combining transformations of exponential graphs

eles-4874

- It is possible to combine translations, dilations and reflections in one graph.
- The general form of an exponential graph is $y = k(a)^x + c$ or $y = k(a)^{-x} + c$, with integer values of k, a and c (where $a > 0$ $a \neq 1$)
- The key features of the graphs of are:
  - dilated by a factor k from the x-axis
- translated c units vertically:
  - if $c > 0$, the graph is translated *upwards*
  - if $c < 0$, the graph is translated *downwards*

WORKED EXAMPLE 14 Sketching exponential graphs with multiple transformations

By considering transformations to the graph of $y = 2^x$, sketch the graph of
a. $y = -2^x + 1$ **b.** $y = 3(2^x) - 1$

THINK

a. 1. Start by sketching $y = 2^x$.
It has a y-intercept of 1 and a horizontal asymptote at $y = 0$.

2. Sketch $y = -2^x$ by reflecting $y = 2^x$ about the x-axis.
It has a y-intercept of -1 and a horizontal asymptote at $y = 0$.

3. Sketch $y = -2^x + 1$ by translating $y = -2^x$ upwards by 1 unit.
The graph has a y-intercept of 0 and a horizontal asymptote at $y = 1$.

WRITE/DRAW

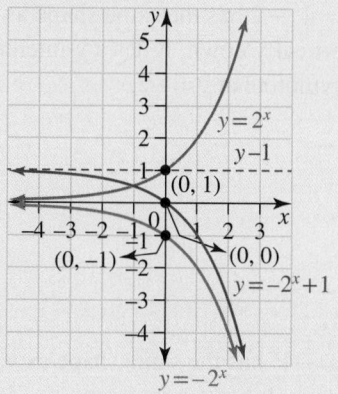

b. 1. Start by sketching $y = 2^x$
It has a y-intercept of 1 and a horizontal asymptote at $y = 0$.

2. Sketch $y = 3(2)^x$.
It has a y-intercept of 3 as $x = 0$, $y = 3 \times 1$ and a horizontal asymptote at $y = 0$.

3. Sketch $y = 3(2)^x - 1$ by translating $y = 3(2)^x$ downwards by 1 unit.
It has a y-intercept of 2 as $x = 0$, $y = 3 \times 1 - 1 = 2$ and a horizontal asymptote at $y = -1$.

4. Sketch $y = 3(2)^x - 1$ on a separate graph if necessary to make it easier to see the key features.
(Note: the x-intercept is approximately $(-1.58, 0)$)

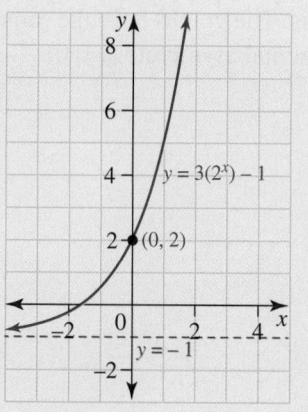

DISCUSSION

Explain why the graph of an exponential function will always have a horizontal asymptote.

 Resources

Interactivity Exponential functions (int-5959)

Exercise 8.6 Exponential graphs (Path)

learn

8.6 Quick quiz on	8.6 Exercise

Individual pathways

■ PRACTISE	■ CONSOLIDATE	■ MASTER
1, 6, 8, 10, 13, 15, 18	2, 4, 7, 11, 14, 16, 19	3, 5, 9, 12, 17, 20

Fluency

1. Complete the following table and use it to plot the graph of $y = 3^x$, for $-3 \leq x \leq 3$.

x	-3	-2	-1	0	1	2	3
y							

2. If $x = 1$, calculate the value of y when:
   a. $y = 2^x$
   b. $y = 3^x$
   c. $y = 4^x$

3. If $x = 1$, calculate the value of y when:
   a. $y = 10^x$
   b. $y = a^x$.

4. Using graphing technology, sketch the graphs of $y = 2^x$, $y = 3^x$ and $y = 4^x$ on the same set of axes.
   a. Describe the common features among the graphs.
   b. Describe how the value of the base $(2, 3, 4)$ affects the graph.
   c. Predict where the graph $y = 8^x$ would lie and sketch it in.

5. Using graphing technology, sketch the following graphs on one set of axes.
   $y = 3^x$, $y = 3^x + 2$, $y = 3^x + 5$ and $y = 3^x - 3$
   a. State what remains the same in all of these graphs.
   b. State what is changed.
   c. For the graph of $y = 3^x + 10$, write down:
      i. the y-intercept
      ii. the equation of the horizontal asymptote.

6. Using graphing technology, sketch the graphs of:
   a. $y = 2^x$ and $y = -2^x$
   b. $y = 3^x$ and $y = -3^x$
   c. $y = 6^x$ and $y = -6^x$.
   d. State the relationship between these pairs of graphs.

7. Using graphing technology, sketch the graphs of:
   a. $y = 2^x$ and $y = 2^{-x}$
   b. $y = 3^x$ and $y = 3^{-x}$
   c. $y = 6^x$ and $y = 6^{-x}$
   d. State the relationship between these pairs of graphs.

8. **WE13** Given the graph of $y = 2^x$, sketch on the same axes the graphs of:
   a. $y = 2^x + 6$
   b. $y = -2^x$
   c. $y = 2^{-x}$.

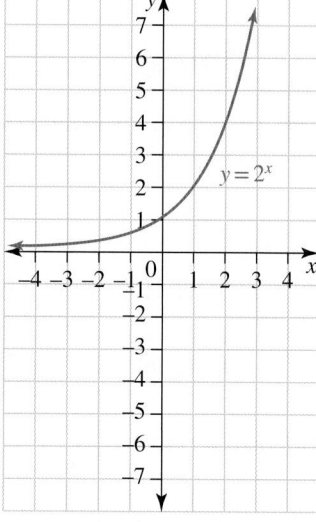

9. Given the graph of $y = 3^x$, sketch on the same axes the graphs of:
   a. $y = 3^x + 2$
   b. $y = -3^x$.

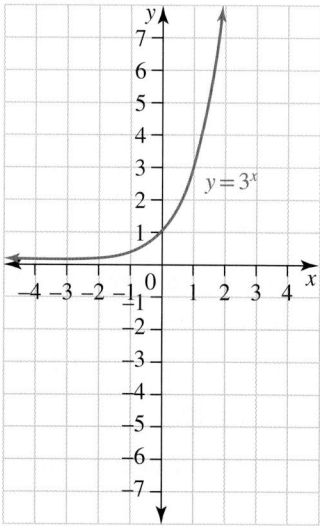

10. Given the graph of $y = 4^x$, sketch on the same axes the graphs of:
    a. $y = 4^x - 3$
    b. $y = 4^{-x}$.

Understanding

11. **WE14** By considering transformations of the graph of $y = 2^x$, sketch the following graphs on the same set of axes.
    a. $y = 2^{-x} + 2$
    b. $y = -2^x + 3$

12. By considering transformations of the graph of $y = 5^x$, sketch the following graphs on the same set of axes.
    a. $y = -5^x + 10$
    b. $y = 5^{-x} + 10$

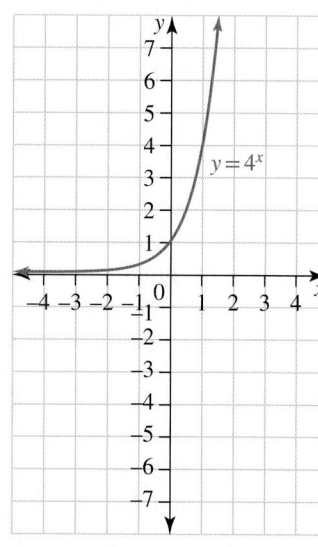

13. Match each equation with its correct graph.

   i. $y = 2^x$ **ii.** $y = 3^x$ **iii.** $y = -4^x$ **iv.** $y = 5^{-x}$

a.

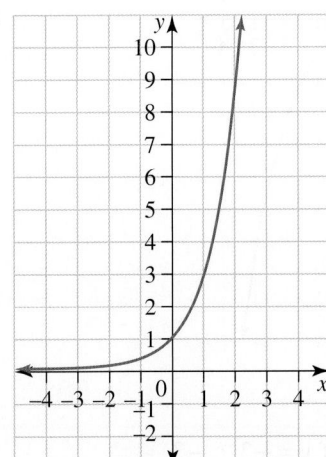

b.

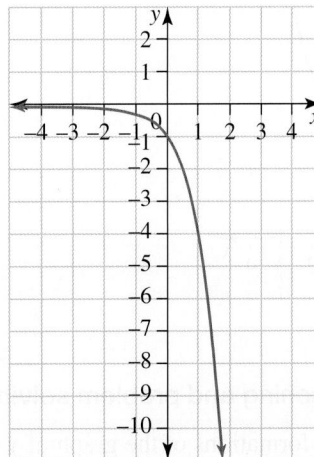

c.

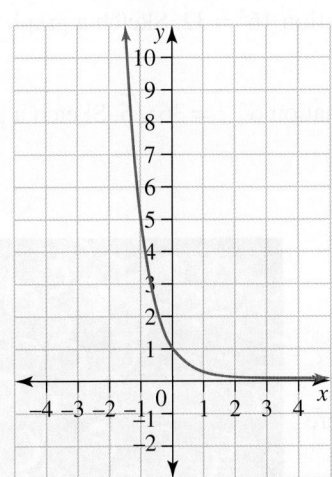

d.

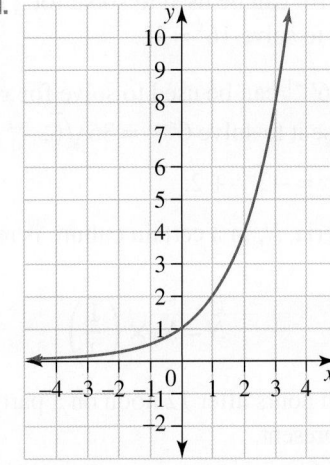

14. Match each equation with its correct graph. Explain your answer.

   i. $y = 2^x + 1$ **ii.** $y = 3^x + 1$ **iii.** $y = -2^x + 1$ **iv.** $y = 2^{-x} + 1$

a.

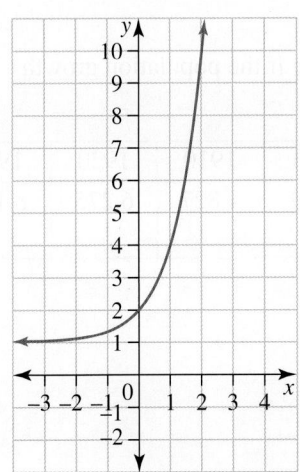

b.

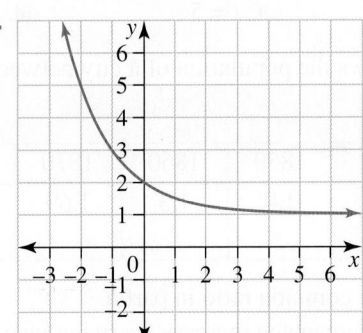

c.

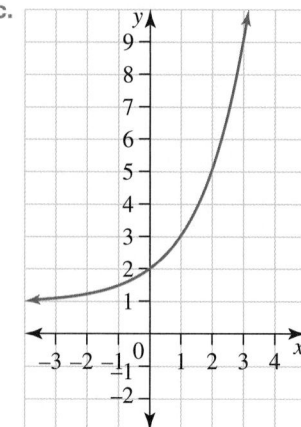

d.

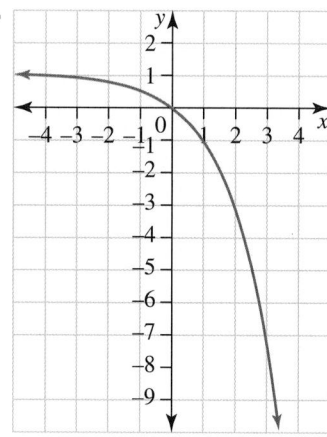

Communicating, reasoning and problem solving

15. By considering transformations of the graph of $y = 3^x$, sketch the graph of $y = -3^{-x} - 3$.

16. The graph of $f(x) = 16^x$ can be used to solve for x in the exponential equation $16^x = 32$. Sketch a graph of $f(x) = 16^x$ and use it to solve $16^x = 32$.

17. The graph of $f(x) = 6^{x-1}$ can be used to solve for x in the exponential equation $6^{x-1} = 36\sqrt{6}$. Sketch a graph of $f(x) = 6^{x-1}$ and use it to solve $6^{x-1} = 36\sqrt{6}$.

18. Sketch the graph of $y = -2^{-x} + 2$.

19. The number of bacteria, N, in a certain culture is reduced by a third every hour so

$$N = N_0 \times \left(\frac{1}{3}\right)^t$$

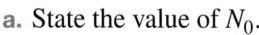

where t is the time in hours after 12 noon on a particular day. Initially there are 10 000 bacteria present.

a. State the value of N_0.
b. Calculate the number of bacteria, correct to the nearest whole number, in the culture when:

   i. $t = 2$ **ii.** $t = 5$ **iii.** $t = 10$.

20. a. The table shows the population of a city between 1850 and 1930. Explain if the population growth is exponential.

Year	1850	1860	1870	1880	1890	1900	1910	1920	1930
Population (million)	1.0	1.3	1.69	2.197	2.856	3.713	4.827	6.275	8.157

b. Determine the common ratio in part **a**.
c. Evaluate the percentage increase every ten years.
d. Estimate the population in 1895.
e. Estimate the population in 1980.

LESSON
8.7 The hyperbola (Path)

LEARNING INTENTION

At the end of this lesson you should be able to:
- identify the key features of the hyperbola
- determine the equation of the vertical and horizontal asymptote of a hyperbola
- sketch the graph of a hyperbola, and understand the effect dilations and reflections have on the shape of the graph.

▶ 8.7.1 Hyperbolas

eles-4876

- A hyperbola is a function of the form $y = \dfrac{k}{x}$ or $xy = k$.

WORKED EXAMPLE 15 Sketch the graph of a hyperbola using the table of values

Complete the table of values below and use it to plot the graph of $y = \dfrac{1}{x}$.

x	-3	-2	-1	$-\dfrac{1}{2}$	0	$\dfrac{1}{2}$	1	2	3
y									

THINK

1. Substitute each x-value into the function $y = \dfrac{1}{x}$ to obtain the corresponding y-value.

WRITE/DRAW

x	-3	-2	-1	$-\dfrac{1}{2}$	0	$\dfrac{1}{2}$	1	2	3
y	$-\dfrac{1}{3}$	$-\dfrac{1}{2}$	-1	-2	Undef.	2	1	$\dfrac{1}{2}$	$\dfrac{1}{3}$

2. Draw a set of axes and plot the points from the table. Join them with a smooth curve.

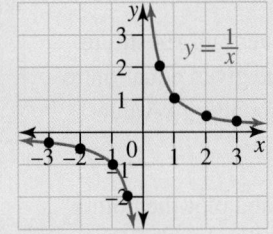

- The graph in Worked example 15 has several important features.
  1. There is no function value (y-value) when $x = 0$. At this point the hyperbola is undefined. When this occurs, the line that the graph approaches ($x = 0$) is called a vertical asymptote.
  2. As x becomes larger and larger, the graph gets very close to but will never touch the x-axis. The same is true as x becomes smaller and smaller. The hyperbola also has a horizontal asymptote at $y = 0$.
  3. The hyperbola has two separate branches. It cannot be drawn without lifting your pen from the page and is an example of a discontinuous graph.
- Graphs of the form $y = \dfrac{k}{x}$ are the same basic shape as $y = \dfrac{1}{x}$ with y-values dilated by a factor of k.

WORKED EXAMPLE 16 Determining the asymptotes of a hyperbola

a. **Plot the graph of** $y = \dfrac{4}{x}$ **for** $-2 \le x \le 2.$

b. **Write down the equation of each asymptote.**

THINK

a. 1. Prepare a table of values taking x-values from -2 to 2. Fill in the table by substituting each x-value into the given equation to calculate the corresponding y-value.

2. Draw a set of axes and plot the points from the table. Join them with a smooth curve.

WRITE/DRAW

a.

x	-2	-1	$-\dfrac{1}{2}$	0	$\dfrac{1}{2}$	1	2
y	-2	-4	-8	Undef.	8	4	2

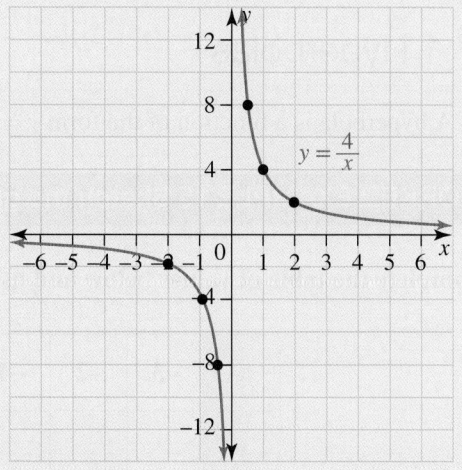

b. Consider any lines that the curve approaches but does not cross.

b. Vertical asymptote is $x = 0$.
Horizontal asymptote is $y = 0$.

WORKED EXAMPLE 17 Sketching hyperbolas

Plot the graph of $y = \dfrac{-3}{x}$ **for** $-3 \le x \le 3.$

THINK

1. Draw a table of values and substitute each x-value into the given equation to calculate the corresponding y-value.

WRITE/DRAW

x	-3	-2	-1	$-\dfrac{1}{2}$	0	$\dfrac{1}{2}$	1	2	3
y	1	1.5	3	6	Undef.	-6	-3	-1.5	-1

2. Draw a set of axes and plot the points from the table. Join them with a smooth curve.

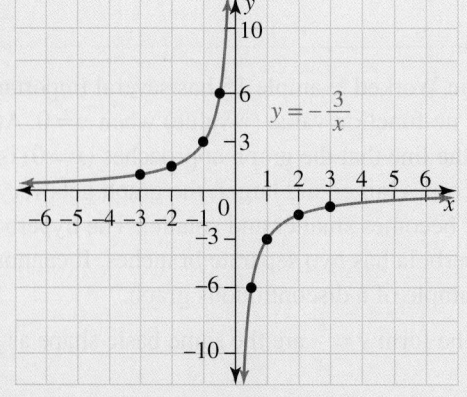

8.7.2 General form of a hyperbola

eles-6303

- The general form of a hyperbola with multiple transformations is given by the equation

$$y = \frac{k}{x-b} + c.$$

- This graph has a horizontal asymptote with equation $y = c$. This occurs because the fraction $\dfrac{k}{x-b}$ cannot be made to equal 0, which means y can never be equal to c, $y \neq c$.
- The graph has a vertical asymptote with equation $x = b$. This is because the denominator of a fraction cannot be equal to 0, so x can never be equal to b.
- The graph will be reflected in the x-axis if $k < 0$.

Note: Not all hyperbolas will have an x- or y-intercept. This will depend on the equation of the asymptotes.

Key features of the hyperbola

The key features of the hyperbola $y = \dfrac{k}{x-b} + c$ are:

- horizontal asymptote $y = c$
- vertical asymptote $x = b$
- reflected about the x-axis if $k < 0$

WORKED EXAMPLE 18 Determining the vertical and horizontal asymptotes of a hyperbola

Write the equation of the asymptotes of the following hyperbolas:

a. $y = \dfrac{3}{x+2}$

b. $y = -\dfrac{1}{x} + 4$

c. $y = \dfrac{5}{10-x} - 1$

THINK	WRITE
a. Consider what value of x would make the denominator equal to 0, as well as any value added to the fraction.	a. Vertical asymptote is $x = -2$ Horizontal asymptote is $y = 0$
b. Consider what value of x would make the denominator equal to 0, as well as any value added to the fraction.	b. Vertical asymptote is $x = 0$ Horizontal asymptote is $y = 4$
c. Consider what value of x would make the denominator equal to 0, as well as any value added to the fraction.	c. Vertical asymptote is $x = 10$ Horizontal asymptote is $y = -1$

WORKED EXAMPLE 19 Sketching the graph of a hyperbola involving transformations

Plot the graph of $y = -\dfrac{2}{x+1} + 3$.

THINK	WRITE
1. Consider what value of x would make the denominator equal to 0, as well as any value added to the fraction.	Vertical asymptote is $x = -1$ Horizontal asymptote is $y = 3$
2. Calculate the y-intercept (when $x = 0$).	$y = -\dfrac{2}{0+1} + 3 = 1$ y-intercept is $(0, 1)$

3. Calculate the x-intercept (when $y = 0$).

$$0 = -\frac{2}{x+1} + 3$$

$$-3 = -\frac{2}{x+1}$$

$$-3(x+1) = -2$$

$$x+1 = \frac{-2}{-3}$$

$$x = \frac{2}{3} - 1 = -\frac{1}{3}$$

The x-intercept is $\left(-\frac{1}{3}, 0\right)$ $x = -1$

4. Sketch the graph with coordinates of both intercepts and both asymptotes labelled.

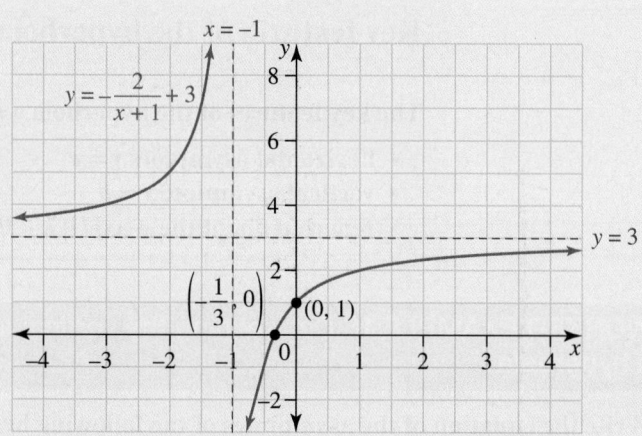

DISCUSSION

How could you summarise the effect of the transformations dealt with in this lesson on the shape of the basic hyperbola $y = \frac{1}{x}$?

 Resources

Interactivity Hyperbolas (int-6155)

Individual pathways

■ PRACTISE	■ CONSOLIDATE	■ MASTER
1, 4, 7, 10, 14, 17	2, 5, 8, 11, 15, 18, 20	3, 6, 9, 12, 13, 16, 19

Fluency

1. **WE15** Complete the table of values below and use it to sketch the graph of $y = \dfrac{10}{x}$.

x	−5	−4	−3	−2	−1	0	1	2	3	4	5
y											

WE16 For questions **2** to **4**, answer the following.
a. Sketch the graph of each hyperbola.
b. Write down the equation of each asymptote.

2. $y = \dfrac{5}{x}$

3. $y = \dfrac{20}{x}$

4. $y = \dfrac{100}{x}$

5. Draw the graphs of $y = \dfrac{2}{x}$, $y = \dfrac{3}{x}$ and $y = \dfrac{4}{x}$, on the same set of axes.

6. Describe the effect of increasing the value of k on the graph of $y = \dfrac{k}{x}$.

7. **WE17** Sketch the graph of $y = \dfrac{-10}{x}$ for $-5 \le x \le 5$.

8. Draw the graphs of $y = \dfrac{6}{x}$ and $y = \dfrac{-6}{x}$, on the same set of axes.

9. Describe the effect of the negative in $y = \dfrac{-k}{x}$.

10. Complete the table of values below and use the points to plot $y = \dfrac{1}{x-1}$. State the equation of the vertical asymptote.

x	−3	−2	−1	0	1	2	3	4
y								

Understanding

11. Sketch the graph of each hyperbola and label the vertical asymptote.

    a. $y = \dfrac{1}{x-2}$

    b. $y = \dfrac{1}{x-3}$

12. Plot the graph of $y = \dfrac{1}{x+1}$ and label the vertical asymptote.

13. Describe the effect of b in $y = \dfrac{1}{x-b}$.

For questions 14 to 16:
 i. write the equations of the asymptotes of the following hyperbolas
 ii. sketch the graphs of the following hyperbolas.

14. a. **WE18** $y = \dfrac{-4}{x+1}$

    b. $y = \dfrac{2}{x-1}$

15. a. $y = \dfrac{5}{x+2}$

    b. $y = \dfrac{3}{x+2} - 2$

16. **WE19**

    a. $y = -\dfrac{4}{x+2} + 1$

    b. $y = \dfrac{7}{3-x} + 5$

Communicating, reasoning and problem solving

17. Give an example of the equation of a hyperbola that has a vertical asymptote of:

    a. $x = 3$

    b. $x = -10$.

18. Give an example of a hyperbola that has the following key features.
    a. Asymptotes of $x = 2$ and $y = 3$
    b. Asymptotes of $x = -2$ and $y = 4$ and a y-intercept of -3

19. The graph of $y = \dfrac{1}{x}$ is reflected in the x-axis, dilated by a factor of 2 parallel to the y-axis or from the x-axis and translated 3 units to the left and down 1 unit. Determine the equation of the resulting hyperbola and give the equations of any asymptotes.

20. The temperature of a cup of coffee as it cools is modelled by the equation $T = \dfrac{780}{t+10} + 22$, where T represents the temperature in °C and t is the time in minutes since the coffee was first made.

    a. State the initial temperature of the cup of coffee.
    b. Calculate the temperature, to 1 decimal place, of the coffee after it has been left to cool for an hour.
    c. A coffee will be too hot to drink unless its temperature has dropped below 50°C. Determine how long someone would have to wait, to the nearest minute, before drinking the coffee.
    d. Explain whether the coffee will ever cool to 0°C. Justify your answer.

LESSON
8.8 The circle (Path)

LEARNING INTENTION

At the end of this lesson you should be able to:
- graph circles of the form $x^2 + y^2 = r^2$
- establish the equation of the circle with centre (a, b) and radius r
- determine the centre and radius of a circle by completing the square.

▶ 8.8.1 Circles

eles-5352

- A circle is the path traced out by a point at a constant distance (the radius) from a fixed point (the centre).
- Consider the circles shown. The first circle has its centre at the origin and radius r. Let $P(x, y)$ be a point on the circle.
  By Pythagoras: $x^2 + y^2 = r^2$.
  This relationship is true for all points, P, on the circle.
  The equation of a circle, with centre $(0, 0)$ and radius r is:

$$x^2 + y^2 = r^2$$

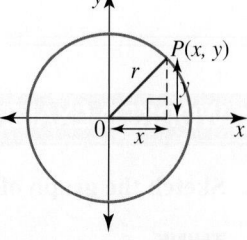

- If the circle is translated a units to the right, parallel to the x-axis, and b units upwards, parallel to the y-axis, then the centre of the circle will become (a, b). The radius will remain unchanged.

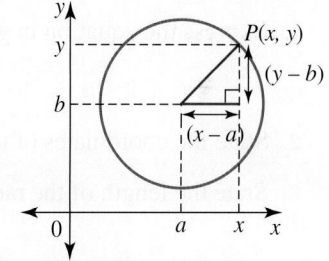

Equation of a circle

- **The equation of a circle, with centre $(0, 0)$ and radius r, is:**

$$x^2 + y^2 = r^2$$

- **The equation of a circle, with centre (a, b) and radius r, is:**

$$(x - a)^2 + (y - b)^2 = r^2$$

WORKED EXAMPLE 20 Identifying the centre and radius of a circle from its equation

Sketch the graph of $4x^2 + 4y^2 = 25$, stating the centre and radius.

THINK	WRITE/DRAW
1. Express the equation in general form by dividing both sides by 4.	$x^2 + y^2 = r^2$ $4x^2 + 4y^2 = 25$ $x^2 + y^2 = \dfrac{25}{4}$

▶

2. State the coordinates of the centre. Centre $(0, 0)$

3. Calculate the length of the radius by taking the square root of both sides. (Ignore the negative results.)

$$r^2 = \frac{25}{4}$$

$$r = \frac{5}{2}$$

Radius $= 2.5$ units

4. Sketch the graph.

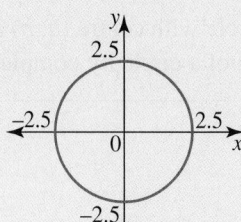

WORKED EXAMPLE 21 Sketching the graph of a circle

Sketch the graph of $(x-2)^2 + (y+3)^2 = 16$, clearly showing the centre and radius.

THINK	WRITE/DRAW
1. Express the equation in general form.	$(x-a)^2 + (y-b)^2 = r^2$ $(x-2)^2 + (y+3)^2 = 16$
2. State the coordinates of the centre.	Centre $(2, -3)$
3. State the length of the radius.	$r^2 = 16$ $r = 4$ Radius $= 4$ units
4. Sketch the graph.	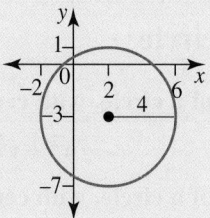

WORKED EXAMPLE 22 Determining the centre and radius of a circle by completing the square

Sketch the graph of the circle $x^2 + 2x + y^2 - 6y + 6 = 0$.

THINK	WRITE/DRAW
1. Express the equation in general form by completing the square on the x terms and again on the y terms.	$(x-h)^2 + (y-k)^2 = r^2$ $x^2 + 2x + y^2 - 6y + 6 = 0$ $(x^2 + 2x + 1) - 1 + (y^2 - 6y + 9) - 9 + 6 = 0$ $(x+1)^2 + (y-3)^2 - 4 = 0$ $(x+1)^2 + (y-3)^2 = 4$

2. State the coordinates of the centre. Centre $(-1, 3)$

3. State the length of the radius.
$$r^2 = 4$$
$$r = 2$$
Radius $= 2$ units

4. Sketch the graph.

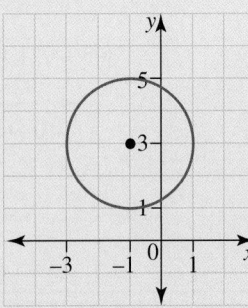

WORKED EXAMPLE 23 Sketching the graph of a circle with centre (*a*, *b*)

Sketch the graph of the circle $(x + 3)^2 + (y - 2)^2 = 25$. Make sure to show all axial intercepts.

THINK	WRITE
1. State the coordinate of the centre.	Centre $(-3, 2)$
2. State the length of the radius.	$r^2 = 25$ $r = 5$ Radius $= 5$ units
3. To determine the *y*-intercepts let $x = 0$.	$(0 + 3)^2 + (y - 2)^2 = 25$ $(y - 2)^2 = 25 - 9$ $(y - 2)^2 = 16$ $y = \pm 4$ $y = 4 + 2 \text{ or } y = 4 + 2$ $y = 6, \ y = -2$ The *y*-intercepts are $(0, 6)$ and $(0, -2)$
4. To determine the *x*-intercepts let $y = 0$.	$(x + 3)^2 + (0 - 2)^2 = 25$ $(x + 3)^2 = 25 - 4$ $x + 3 = \pm\sqrt{21}$ $x = -3 \pm \sqrt{21}$ The *x*-intercepts are $\left(-3 + \sqrt{21}, 0\right)$ and $\left(-3 - \sqrt{21}, 0\right)$.

5. Sketch the graph.

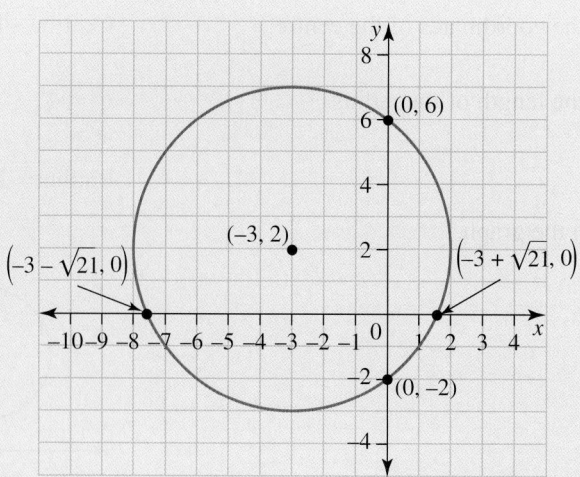

DISCUSSION

What effect does a and b have on the graph of a circle in the form $(x-a)^2 + (y-b)^2 = r^2$?

How would you derive the equation of a circle $x^2 + y^2 = r^2$ with centre $(0, 0)$ and radius r using the distance formula?

---

on Resources

 Interactivity Graphs of circles (int-6156)

Exercise 8.8 The circle (Path)

learn on

8.8 Quick quiz **on**	8.8 Exercise

Individual pathways

■ PRACTISE	■ CONSOLIDATE	■ MASTER
1, 4, 7, 10, 13, 17	2, 5, 8, 11, 14, 15, 18	3, 6, 9, 12, 16, 19, 20

Fluency

WE20 For questions **1** to **3**, sketch the graphs of the following, stating the centre and radius of each.

1. a. $x^2 + y^2 = 49$ **b.** $x^2 + y^2 = 4^2$

2. a. $x^2 + y^2 = 36$ **b.** $x^2 + y^2 = 81$

3. a. $2x^2 + 2y^2 = 50$ **b.** $9x^2 + 9y^2 = 100$

WE21 For questions **4** to **6**, sketch the graphs of the following, clearly showing the centre and the radius.

4. a. $(x-1)^2 + (y-2)^2 = 5^2$ **b.** $(x+2)^2 + (y+3)^2 = 6^2$

5. a. $(x+3)^2 + (y-1)^2 = 49$ **b.** $(x-4)^2 + (y+5)^2 = 64$

6. a. $x^2 + (y+3)^2 = 4$ **b.** $(x-5)^2 + y^2 = 100$

WE22 For questions **7** to **9**, sketch the graphs of the following circles.

7. **a.** $x^2 + 4x + y^2 + 8y + 16 = 0$ **b.** $x^2 - 10x + y^2 - 2y + 10 = 0$

8. **a.** $x^2 - 14x + y^2 + 6y + 9 = 0$ **b.** $x^2 + 8x + y^2 - 12y - 12 = 0$

9. **a.** $x^2 + y^2 - 18y - 19 = 0$ **b.** $2x^2 - 4x + 2y^2 + 8y - 8 = 0$

Understanding

10. **MC** The graph of $(x - 2)^2 + (y + 5)^2 = 4$ is:

A.
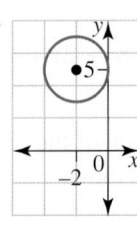
B.
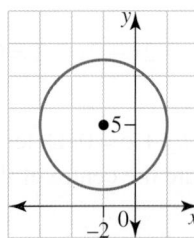
C.
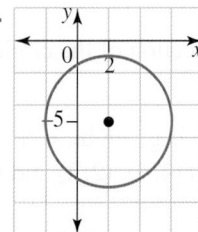
D.
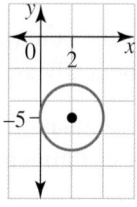

11. **MC** The centre and radius of the circle $(x + 1)^2 + (y - 3)^2 = 4$ is:

A. $(1, -3), 4$ **B.** $(-1, 3), 2$ **C.** $(3, -1), 4$ **D.** $(1, -3), 2$

12. **MC** The centre and radius of the circle with equation $x^2 + y^2 + 8x - 10y = 0$ is:

A. $(4, 5), \sqrt{41}$ **B.** $(-4, 5), 9$ **C.** $(4, -5), 3$ **D.** $(-4, 5), \sqrt{41}$

Communicating, reasoning and problem solving

13. Determine the equation representing the outer edge of the galaxy as shown in the photo below, using the astronomical units provided.

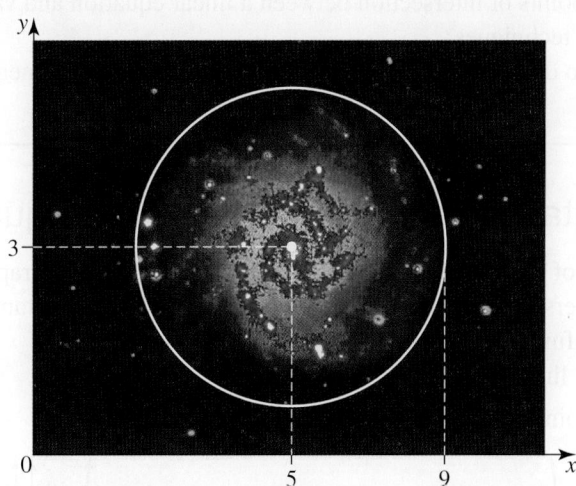

14. Circular ripples are formed when a water drop hits the surface of a pond.
    If one ripple is represented by the equation $x^2 + y^2 = 4$ and then 3 seconds later by $x^2 + y^2 = 190$, where the length of measurements are in centimetres:

    a. identify the radius (in cm) of the ripple in each case
    b. calculate how fast the ripple is moving outwards.
    (State your answers to 1 decimal place.)

15. Two circles with equations $x^2 + y^2 = 4$ and $(x-1)^2 + y^2 = 9$ intersect. Determine the point(s) of intersection. Show your working.

16. **a.** Graph the line $y = x$, the parabola $y = x^2$ and the circle $x^2 + y^2 = 1$ on the one set of axes.
    b. Evaluate algebraically the points of intersection of:
       i. the line and the circle   **ii.** the line and the parabola   **iii.** the parabola and the circle.

17. **WE23** Sketch the graph of $(x + 6)^2 + (y - 3)^2 = 100$ showing all axial intercepts.

18. Determine the points of intersection between the quadratic equation $y = x^2 - 5$ and the circle given by $x^2 + y^2 = 25$.

19. Determine the point(s) of intersection of the circles $x^2 + y^2 - 2x - 2y - 2 = 0$ and $x^2 + y^2 - 8x - 2y + 16 = 0$ both algebraically and graphically.

20. The general equation of a circle is given by $x^2 + y^2 + ax + by + c = 0$. Determine the equation of the circle which passes through the points $(4, 5)$, $(2, 3)$ and $(0, 5)$. State the centre of the circle and its radius.

LESSON
8.9 Simultaneous linear and non-linear equations (Path)

LEARNING INTENTION

At the end of this lesson you should be able to:
- determine the point or points of intersection between a linear equation and various non-linear equations using various techniques
- use digital technology to determine the points of intersection between a linear equation and a non-linear equation.

▶ 8.9.1 Solving simultaneous linear and quadratic equations

eles-4771
- The solution to a system of simultaneous equations can be determined by graphing.
- The point or points of intersection of the graphs is the solution to the simultaneous equations.
- The graph of a quadratic function is called a parabola.
- A parabola and a straight line may:

  - intersect at only one point

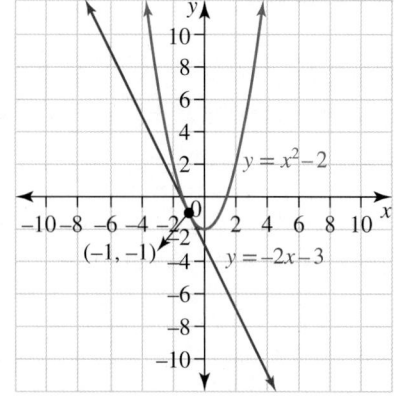

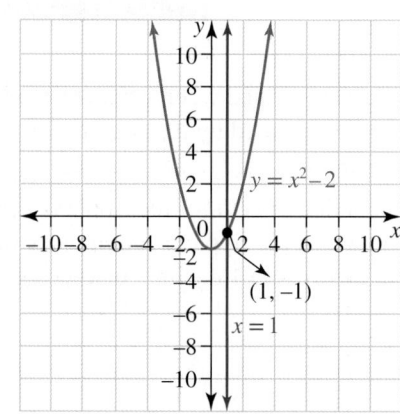

- intersect at two points

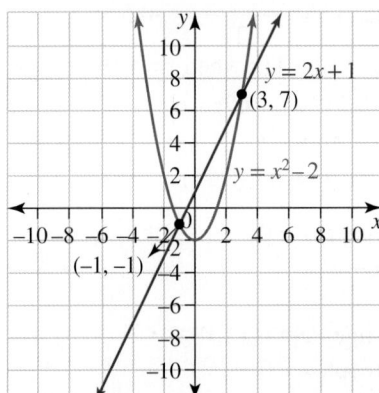

- not intersect at all.

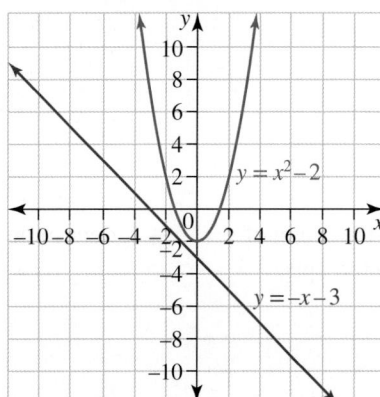

WORKED EXAMPLE 24 Solving linear and quadratic simultaneous equations

Determine the points of intersection of $y = x^2 + x - 6$ and $y = 2x - 4$:
a. **algebraically** b. **graphically.**

THINK **WRITE/DRAW**

a. 1. The points of intersections will be the (x, y) coordinates which satisfy the simultaneous equations.
Number the equations. Equate [1] and [2].

a.
$$y = x^2 + x - 6 \quad [1]$$
$$y = 2x - 4 \quad [2]$$
$$x^2 + x - 6 = 2x - 4$$

2. Collect all the terms on one side and simplify.

$$x^2 + x - 6 - 2x + 4 = 2x - 4 - 2x + 4$$
$$x^2 + x - 6 - 2x + 4 = 0$$
$$x^2 - x - 2 = 0$$

3. Factorise and solve the quadratic equation, using the Null Factor Law.

$$(x - 2)(x + 1) = 0$$
$$x - 2 = 0 \quad \text{or} \quad x + 1 = 0$$
$$x = 2 \qquad\qquad x = -1$$

4. Identify the y-coordinate for each point of intersection by substituting each x-value into one of the equations.

When $x = 2$,
$$y = 2(2) - 4$$
$$= 4 - 4$$
$$= 0$$
Intersection point $(2, 0)$
When $x = -1$
$$y = 2(-1) - 4$$
$$= -2 - 4$$
$$= -6$$

5. Write the solution.

Intersection point $(-1, -6)$

b. 1. To sketch the graph of $y = x^2 + x - 6$, determine the x- and y-intercepts and the turning point (TP). The x-value of the TP is the average of the x-axis intercepts. The y-value of the TP is calculated by substituting the x-value into the equation of the parabola.

b. x-intercepts: $y = 0$
$$0 = x^2 + x - 6$$
$$0 = (x + 3)(x - 2)$$
$$x = -3, x = 2$$
The x-intercepts are $(-3, 0)$ and $(2, 0)$.
y-intercept: $x = 0$
$$y = -6$$
The y-intercept is $(0, -6)$

x-value of TP: $\dfrac{-3+2}{2} = -0.5$

y-value of the TP; when $x = -0.5$:

$y = (-0.5)^2 + (-0.5) - 6$

$y = -6.25$

The TP is $(-0.5, -6.25)$

2. To sketch the graph of $y = 2x - 4$, determine the x- and y-intercepts.

x-intercept: $y = 0$

$0 = 2x - 4$

$x = 2$

The x-intercept is $(2, 0)$

y-intercept: $x = 0$

$y = -4$

The y-intercept is $(0, -4)$

3. On the same set of axes, sketch the graphs of $y = x^2 + x - 6$ and $y = 2x - 4$, labelling both.

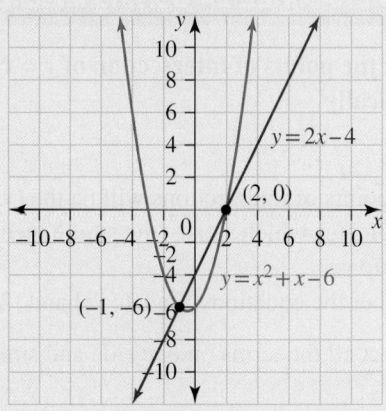

4. On the graph, locate the points of intersection and write the solutions.

The points of intersection are $(2, 0)$ and $(-1, -6)$.

8.9.2 Solving simultaneous equations involving lines and hyperbolas

eles-4772

- A hyperbola and a straight line may:
  - intersect at only one point. In the first case, the line is a tangent to the curve.

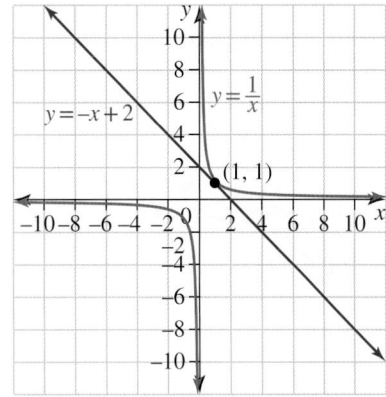

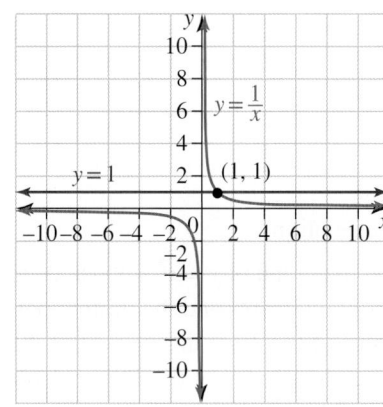

- intersect at two points

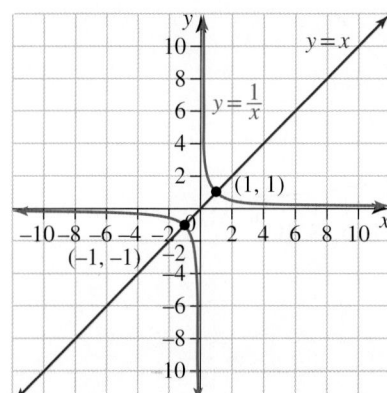

- not intersect at all.

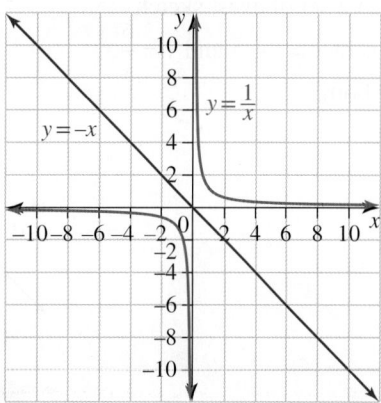

WORKED EXAMPLE 25 Solving linear and hyperbolic simultaneous equations

Determine the point(s) of intersection between $y = x + 5$ and $y = \dfrac{6}{x}$:

a. algebraically

b. graphically.

THINK

a. 1. Number the equations.

2. Equate [1] and [2]. Collect all terms on one side, factorise and simplify to solve for x.

3. To determine the y-coordinates of the points of intersection, substitute the values of x into [1].

4. Write the solutions.

b. 1. To sketch the graph of $y = \dfrac{6}{x}$, draw a table of values.

2. To sketch the graph of $y = x + 5$, determine the x- and y-intercepts.

WRITE/DRAW

a. $y = x + 5$ [1]

$y = \dfrac{6}{x}$ [2]

$x + 5 = \dfrac{6}{x}$

$x(x + 5) = 6$

$x^2 + 5x - 6 = 0$

$(x + 6)(x - 1) = 0$

$\qquad\qquad x = -6, \; x = 1$

$x = -6 \qquad x = 1$

$y = -6 + 5 \quad y = 1 + 5$

$y = -1 \qquad y = 6$

The points of intersection are $(-6, -1)$ and $(1, 6)$.

b.

x	-6	-5	-4	-3	-2	-1	-0	1	2
y	-1	$-1\frac{1}{5}$	$-1\frac{1}{2}$	-2	-3	-6	Undef.	6	3

x-intercept: $y = 0$

$\qquad\qquad 0 = x + 5$

$\qquad\qquad x = -5$

The x-intercept is $(-5, 0)$.

y-intercept: $x = 0$

$\qquad\qquad\qquad y = 5$

The y-intercept is $(0, 5)$.

3. On the same set of axes, sketch the graphs of $y = x + 5$ and $y = \dfrac{6}{x}$, labelling both.

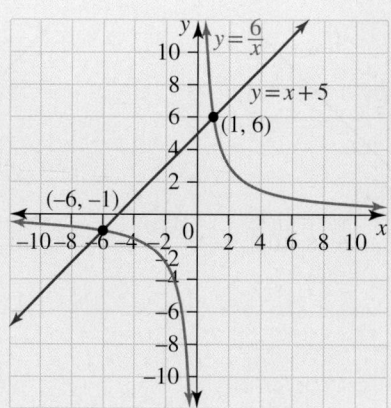

4. On the graph, locate the points of intersection and write the solutions.

The points of intersection are $(1, 6)$ and $(-6, -1)$.

8.9.3 Solving simultaneous equations involving lines and circles

eles-6304

- A circle and a straight line may:

  - intersect at only one point.
  - Here, the line is a tangent to the curve.

  - intersect at two points.
  - Here, the line is a secant to the curve.

  - not intersect at all.

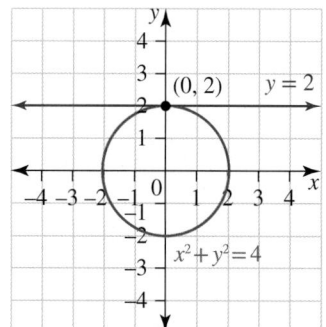

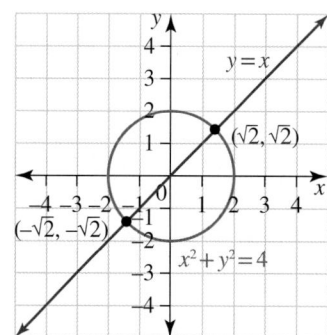

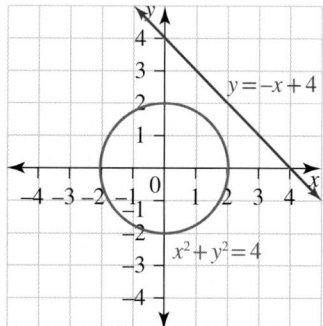

Solutions of a linear and non-linear equation

Depending on the equations, a linear equation and a non-linear equation can have a different number of solutions. For a linear equation and any of the following:
- **quadratic equations**
- **hyperbolic equations**
- **circles**

the number of possible solutions (points of intersections) is **0, 1 or 2.**

 Resources

Interactivity Solving simultaneous linear and non-linear equations (int-6128)

Exercise 8.9 Simultaneous linear and non-linear equations (Path)

8.9 Quick quiz on	8.9 Exercise

Individual pathways

■ PRACTISE	■ CONSOLIDATE	■ MASTER
1, 2, 5, 8, 9, 12, 15	3, 6, 10, 13, 16, 17	4, 7, 11, 14, 18, 19, 20

Fluency

1. Describe how a parabola and straight line may intersect. Use diagrams to illustrate your explanation.

2. **WE24** Determine the points of intersection of the following:
   i. algebraically
   ii. algebraically using a calculator
   iii. graphically using a calculator.
   a. $y = x^2 + 5x + 4$ and $y = -x - 1$ b. $y = -x^2 + 2x + 3$ and $y = -2x + 7$
   c. $y = -x^2 + 2x + 3$ and $y = -6$

3. Determine the points of intersection of the following.
   a. $y = -x^2 + 2x + 3$ and $y = 3x - 8$ b. $y = -(x-1)^2 + 2$ and $y = x - 1$
   c. $y = x^2 + 3x - 7$ and $y = 4x + 2$

4. Determine the points of intersection of the following.
   a. $y = 6 - x^2$ and $y = 4$

   b. $y = 4 + x - x^2$ and $y = \dfrac{3 - x}{2}$

   c. $x = 3$ and $y = 2x^2 + 7x - 2$

5. **MC** Identify which of the following graphs shows the parabola $y = x^2 + 3x + 2$, and the straight line $y = x + 3$.

A.
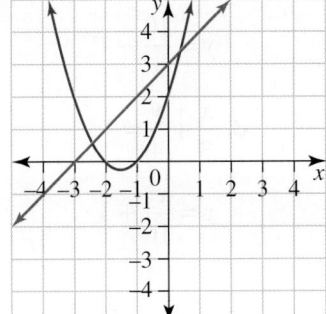

B.
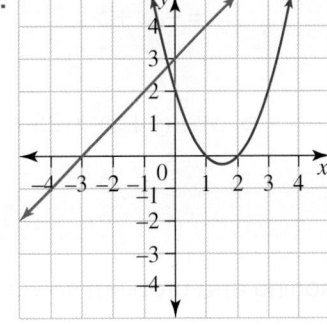

C.
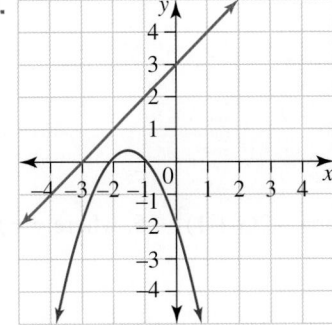

D.
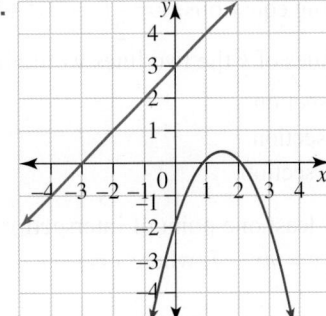

6. **MC** Identify which of the following equations are represented by the graph shown.

A. $y = 0.5(x + 1.5)^2 + 4$ and $y = -\dfrac{1}{3}x + 1$

B. $y = -0.5(x + 1.5)^2 - 4$ and $y = -\dfrac{1}{3}x + 1$

C. $y = -0.5(x - 1.5)^2 + 4$ and $y = \dfrac{1}{3}x + 1$

D. $y = 0.5(x - 1.5)^2 + 4$ and $y = -\dfrac{1}{3}x + 1$

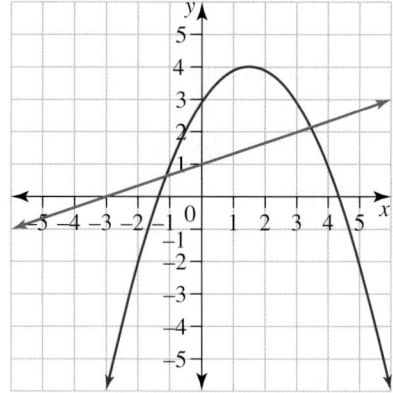

7. Determine whether the following graphs intersect.

a. $y = -x^2 + 3x + 4$ and $y = x - 4$
b. $y = -x^2 + 3x + 4$ and $y = 2x + 5$
c. $y = -(x + 1)^2 + 3$ and $y = -4x - 1$
d. $y = (x - 1)^2 + 5$ and $y = -4x - 1$

Understanding

WE25 For questions **8** to **11**, determine the point(s) of intersection between the following.

8. a. $y = x$
$y = \dfrac{1}{x}$

b. $y = x - 2$
$y = \dfrac{1}{x}$

c. $y = 3x$
$y = \dfrac{5}{x}$

d. $y = \dfrac{6}{x}$
$y = \dfrac{x}{2} + 2$

9. a. $y = 3x$
$x^2 + y^2 = 10$

b. $x^2 + y^2 = 25$
$3x + 4y = 0$

c. $x^2 + y^2 = 50$
$y = 5 - 2x$

d. $x^2 + y^2 = 9$
$y = 2 - x$

10. a. $y = \dfrac{1}{x}$
$y = 4x$

b. $x^2 + y^2 = 25$
$y = -2x + 5$

c. $y = 2x + 3$
$y = -4x^2 + 3$

d. $3x + 4y = 7$
$y = \dfrac{10}{x} - 4$

11. a. $y = x^2$
$y = 2x - 1$

b. $x^2 + (y + 1)^2 = 25$
$y = 3$

c. $y = -4x - 5$
$y = x^2 + 2x + 3$

d. $\dfrac{x}{3} + \dfrac{y}{4} = 7$
$y = \dfrac{x^2}{16} + 3$

Communicating, reasoning and problem solving

12. Consider the following equations: $y = \dfrac{1}{2}(x - 3)^2 + \dfrac{5}{2}$ and $y = x + k$.

Identify for what values of k the two lines would have:

a. no points of intersection
b. one point of intersection
c. two points of intersection.

13. Show that there is at least one point of intersection between the parabola $y = -2(x + 1)^2 - 5$, where $y = f(x)$, and the straight line $y = mx - 7$, where $y = f(x)$.

14. **a.** Using technology, sketch the following graphs and state how many ways a straight line could intersect with the equation.
   i. $y = x^3 - 4x$.
   ii. $y = x^4 - 8x^2 + 16$.
   iii. $y = x^5 - 8x^3 + 16x$.

   b. Comment on the connection between the highest power of x and the number of possible points of intersection.

15. If two consecutive numbers have a product of 306, calculate the possible values for these numbers.

16. The perimeter of a rectangular paddock is 200 m and the area is 1275 m². Determine the length and width of the paddock.

17. **a.** Determine the point(s) of intersection between the circle $x^2 + y^2 = 50$ and the linear equation $y = 2x - 5$.
   b. Confirm your solution to part **a** by plotting the equation of the circle and the linear equation on the same graph.

18. The sum of two positive numbers is 21. Twice the square of the larger number minus three times the square of the smaller number is 45. Determine the value of the two numbers.

19. **a.** Omar is running laps around a circular park with equation $x^2 + y^2 = 32$. Chae-won is running along another track where the path is given by $y = \left(\sqrt{2} - 1\right)x + \left(8 - 4\sqrt{2}\right)$. Determine the point(s) where the two paths intersect.

   b. Omar and Chae-won both start from the same point. If Chae-won gets between the two points in two hours, calculate the possible speeds Omar could run at along his circle in order to collide with
   Chae-won at the other point of intersection. Assume all distances are in kilometres and give your answer to 2 decimal places.

20. Adam and Eve are trying to model the temperature of a cup of coffee as it cools.
   • Adam's model: Temperature (°C) = $100 - 5 \times$ time
   • Eve's model: Temperature (°C) = $\dfrac{800}{10 + \text{time}} + 20$

   Time is measured in minutes.

   a. Using either model, identify the initial temperature of the cup of coffee.
   b. Determine at what times the two models predict the same temperature for the cup of coffee.
   c. Evaluate whose model is more realistic. Justify your answer.

LESSON
8.10 Review

8.10.1 Topic summary

Transformations of the parabola

- The parabola with the equation $y = x^2$ has a turning point at (0,0).
- A **dilation** of factor k from the x-axis produces the equation $y = kx^2$ and:
  - will produce a narrow graph for $k > 0$
  - will produce a wider graph for $0 < k < 1$
  - For $k < 0$ (k is negative), the graph is **reflected** in the x-axis.

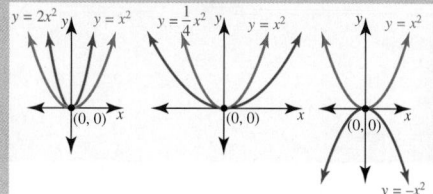

- The graph of $y = (x - b)^2$ has been translated left/right b units.
- The graph of $y = x^2 + c$ has been translated up/down c units.

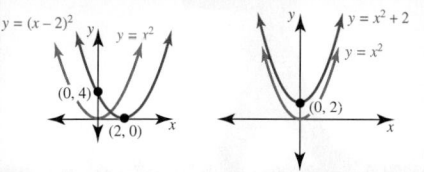

Hyperbolas (Path)

- The graph of an equation in the form $y = \dfrac{k}{x}$ is called a hyperbola.
- The general form of a hyperbola is $y = \dfrac{k}{x - b} + c$ and has:
  - vertical asymptote at $x = b$
  - horizontal asymptote at $y = c$
  - y-intercept at $\left(0, \dfrac{-k}{b + c}\right)$
  - x-intercept at $\left(\dfrac{-k}{b + c}, 0\right)$
- If $a < 0$ the graph is reflected in the x-axis.

Simultaneous linear and non-linear equations (Path)

- A system of equations which contains a linear equation and a non-linear equation can have 0, 1 or 2 solutions (points of intersection).
- The number of solutions will depend on the equations of both lines.
- Non-linear equations include:
  - quadratic equations (parabolas)
  - exponential equations
  - hyperbolic equations
  - circles.

The parabola

- The graph of a quadratic equation is called a parabola.

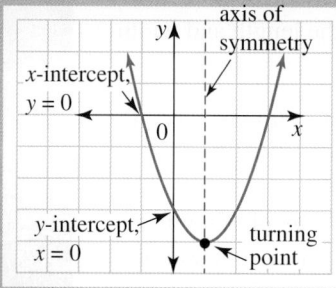

NON-LINEAR RELATIONSHIPS

Exponential graphs (Path)

- Equations $y = a^x$ are called exponential functions.
- The general form $y = k(a)^x + c$ has:
  - a horizontal asymptote at $y = c$
  - a y-intercept at $(0, k + c)$
  - a dilation of a factor of k from the x-axis
  - If $k < 0$ the graph is reflected in the x-axis.

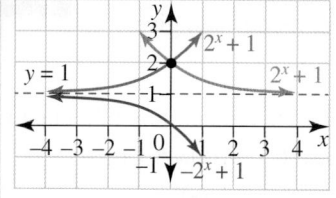

Circles (Path)

- $x^2 + y^2 = r^2$ is a circle with centre (0, 0) and radius r.
- $(x - a)^2 + (y - b)^2 = r^2$ is a circle with centre (a, b) and radius r.
- Use completing the square to determine the equation of a circle:
$$x^2 + 6x + y^2 + 8x = 0$$
$$x^2 + 6x + 9 + y^2 + 8x + 16 = 9 + 16$$
$$(x + 3)^2 + (y + 4)^2 = 25$$

Turning point form (Path)

- Completing the square allows us to write a quadratic in turning point form.
- Turning point form:
$$y = k(x - b)^2 + c$$
- Turning point at (b, c)
- The y-intercept is determined by setting $x = 0$ and solving for y.
- x-intercept/s is determined by setting $y = 0$ and solving for x.

Expanded form (Path)

- The expanded form of a quadratic is $y = ax^2 + bx + c$.
- The axis of symmetry is $x = -\dfrac{b}{2a}$.
- There are two methods for sketching: factorising or using the quadratic formula.
  e.g. Consider $y = x^2 - 2x - 3$.
- To sketch, first factorise:
  $y = (x - 3)(x + 1)$
- Use the null-factor law to determine the x-intercepts:
  when $y = 0$, $0 = (x - 3)(x + 1)$
  so $x = 3$ or -1.
- Axis of symmetry: $x = \dfrac{3 + (-1)}{2} = 1$
- y-coordinate of the turning point:
  $x = 1$, $y = (1 - 3)(1 + 1) = -4$.
  Turning point is $(1, -4)$. The y-intercept when $x = 0$ is $(0, -3)$.

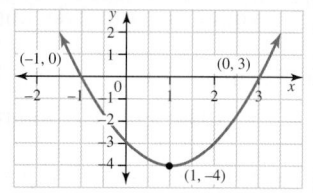

8.10.2 Project

Parametric equations

You are familiar with the quadratic equation $y = x^2$ and its resulting graph. Let us consider an application of this equation by forming a relationship between x and y through a third variable, say, t.

$$x = t \quad \text{and} \quad y = t^2$$

It is obvious that these two equations are equivalent to the equation $y = x^2$. This third variable t is known as a parameter, and the two equations are now called parametric equations. We cannot automatically assume that the resulting graph of these two parametric equations is the same as that of $y = x^2$ for all real values of x. It is dependent on the range of values of t.

Consider the parametric equations $x = t$ and $y = t^2$ for values of the parameter $t = 0$ for questions **1** to **3**.

1. Complete the following table by calculating x- and y-values from the corresponding t-value.

t	x	y
0		
1		
2		
3		
4		
5		

2. Graph the x-values and corresponding y-values on this Cartesian plane. Join the points with a smooth curve and place an arrow on the curve to indicate the direction of increasing t-values.
3. Is there any difference between this graph and that of $y = x^2$? Explain your answer.
4. Consider now the parametric equations $x = 1 - t$ and $y = (1 - t)^2$. These clearly are also equivalent to the equation $y = x^2$. Complete the table and draw the graph of these two equations for values of the parameter $t = 0$. Draw an arrow on the curve in the direction of increasing t-values.

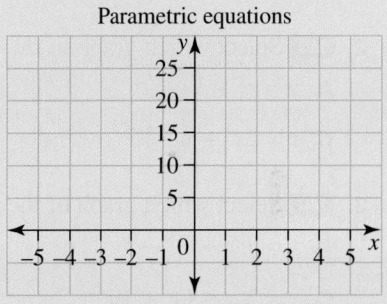

Parametric equations

t	x	y
0		
1		
2		
3		
4		
5		

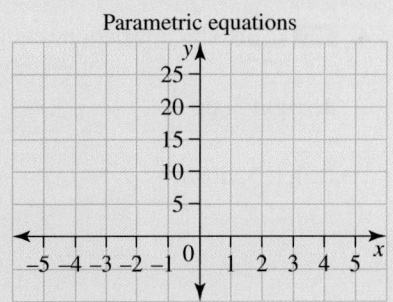

Parametric equations

Describe the shape of your resulting graph. What values of the parameter t would produce the same curve as that obtained in question **2**?

5. The graph of $y = -x^2$ is a reflection of $y = x^2$ in the x-axis. Construct a table and draw the graph of the parametric equations $x = t$ and $y = -t^2$ for parameter values $t = 0$. Remember to place an arrow on the curve in the direction of increasing t-values.

6. Without constructing a table, predict the shape of the graph of the parametric equations $x = 1 - t$ and $y = -(1 - t)^2$ for parameter values $t = 0$. Draw a sketch of the shape.

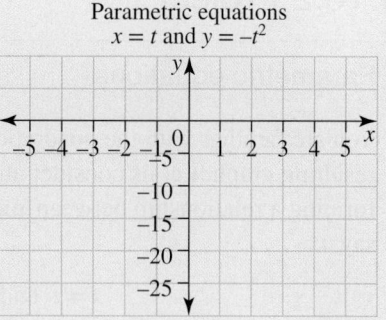

Parametric equations
$x = t$ and $y = -t^2$

 Resources

 Interactivities Crossword (int-2851)
Sudoku puzzle (int-3596)

Exercise 8.10 Review questions

learn on

Fluency

1. **MC** The turning point for the graph $y = 3x^2 - 4x + 9$ is:

A. $\left(\dfrac{1}{3}, 1\dfrac{2}{3} \right)$

B. $\left(\dfrac{1}{3}, \dfrac{2}{3} \right)$

C. $\left(\dfrac{1}{6}, 1\dfrac{1}{6} \right)$

D. $\left(\dfrac{2}{3}, 7\dfrac{2}{3} \right)$

2. **MC** Select which graph of the following equations has the x-intercepts closest together.

A. $y = x^2 + 3x + 2$

B. $y = x^2 + x - 2$

C. $y = 2x^2 + x - 15$

D. $y = 4x^2 + 27x - 7$

3. **MC** Select which graph of the equations below has the largest y-intercept.

A. $y = 3(x - 2)^2 + 9$

B. $y = 5(x - 1)^2 + 8$

C. $y = 2(x - 1)^2 + 19$

D. $y = 2(x - 5)^2 + 4$

4. **MC** **PATH** The translation required to change $y = x^2$ into $y = (x - 3)^2 + \dfrac{1}{4}$ is:

A. right 3, up $\dfrac{1}{4}$

B. right 3, down $\dfrac{1}{4}$

C. left 3, down $\dfrac{1}{4}$

D. left 3, up $\dfrac{1}{4}$

5. **MC** **PATH** The graph of $y = -3 \times 2^x$ is best represented by:

A.
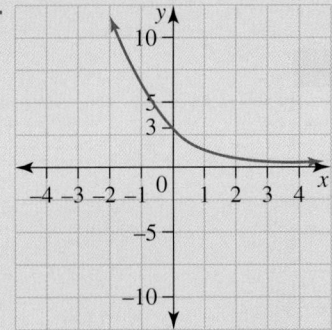

B.
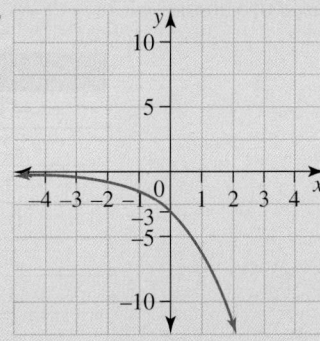

C.
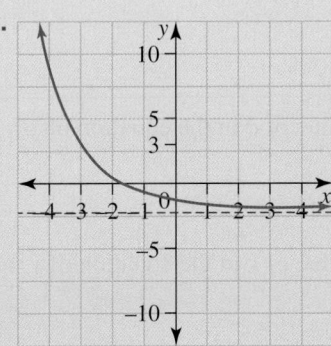

D.
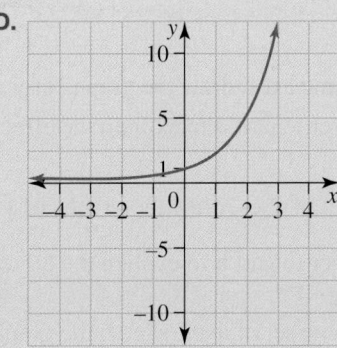

Understanding

6. **PATH** Use the completing the square method to determine the turning point for each of the following graphs.

   a. $y = x^2 - 8x + 1$ **b.** $y = x^2 + 4x - 5$

7. For the graph of the equation $y = x^2 + 8x + 7$, produce a table of values for the x-values between -9 and 1, and then plot the graph. Show the y-intercept and turning point. From your graph, state the x-intercepts.

8. **PATH** For each of the following, determine the coordinates of the turning point and the x- and y-intercepts and sketch the graph.

   a. $y = (x - 3)^2 + 1$ **b.** $y = 2(x + 1)^2 - 5$

9. For the equation $y = -x^2 - 2x + 15$, sketch the graph and determine the x- and y-intercepts and the coordinates of the turning point.

10. **PATH** Draw the graph of $y = 10 \times 3^x$ for $-2 \leq x \leq 4$.

11. **PATH** Identify the type of graph that can be drawn from each of following the equations. (Straight line, parabola, exponential, hyperbola, circle)

   a. $xy = 3$ **b.** $(x - 1)^2 + y^2 = 9$

   c. $y - 8 + 2x - x^2 = 0$ **d.** $\dfrac{y}{3} = 2^{-x}$

   e. $4 - 3x + 2y = 0$ **f.** $x^2 - y = 2$

12. For the exponential function $y = 5^x$:

a. complete the table of values

x	y
−3	
−2	
−1	
0	
1	
2	
3	

b. plot the graph.

13. PATH **a** On the same axes, draw the graphs of $y = (1.2)^x$ and $y = (1.5)^x$.

b. Use your answer to part **a** to explain the effect of changing the value of a in the equation of $y = a^x$.

14. PATH **a** On one set of axes, draw the graphs of $y = 2 \times 3^x$, $y = 5 \times 3^x$ and $y = \dfrac{1}{2} \times 3^x$

b. Use your answer to part **a** to explain the effect of changing the value of k in the equation of $y = ka^x$.

15. PATH **a** On the same set of axes, sketch the graphs of $y = (2.5)^x$ and $y = (2.5)^{-x}$.

b. Use your answer to part **a** to explain the effect of a negative index on the equation $y = a^x$.

16. PATH Sketch each of the following.

a. $y = \dfrac{4}{x}$
                                     b. $y = -\dfrac{2}{x}$

17. PATH Sketch $y = \dfrac{-3}{x - 2}$.

18. PATH Give an example of an equation of a hyperbola that has a vertical asymptote at $x = -3$.

19. PATH Sketch each of these circles. Clearly show the centre and the radius.

a. $x^2 + y^2 = 16$
b. $(x - 5)^2 + (y + 3)^2 = 64$

20. PATH Sketch the following circles. Remember to first complete the square.

a. $x^2 + 4x + y^2 - 2y = 4$
                              b. $x^2 + 8x + y^2 + 8y = 32$

21. PATH Determine the equation of this circle.

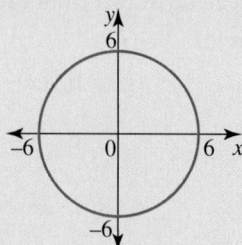

22. **PATH** Determine the point(s) of intersection for each of the following pairs of lines.

   a. $y = x^2 - 6$
      $y = 5x - 3$

   b. $y = \dfrac{2}{x}$
      $y = 5x - 3$

   c. $x^2 + y^2 = 2$
      $y = 5x - 3$

Communicating, reasoning and problem solving

23. The height, h, in metres of a golf ball t seconds after it is hit is given by the formula $h = 4t - t^2$.
   a. Sketch the graph of the path of the ball.
   b. Calculate the maximum height the golf ball reaches.
   c. Determine how long it takes for the ball to reach the maximum height.
   d. Determine how long is it before the ball lands on the ground after it has been hit.

24. A soccer ball is kicked upwards in the air. The height, h, in metres, t seconds after the kick is modelled by the quadratic equation $h = -5t^2 + 20t$.
   a. Sketch the graph of this relationship.
   b. Determine how many seconds the ball is in the air for.
   c. Determine how many seconds the ball is above a height of 15 m. That is, solve the quadratic inequation $-5t^2 + 20t > 15$.
   d. Calculate how many seconds the ball is above a height of 20 m.

25. The height of the water level in a cave is determined by the tides. At any time, t, in hours after 9 am, the height, $h(t)$, in metres, can be modelled by the function $h(t) = t^2 - 12t + 32, 0 \le t \le 12$.
   a. Determine the values of t for which the model is valid. Write your answer in interval notation.
   b. State the initial height of the water.
   c. Bertha has dropped her keys onto a ledge which is 7 metres from the bottom of the cave. By using a graphics calculator, determine the times in which she would be able to climb down to retrieve her keys. Write your answers correct to the nearest minute.

26. **PATH** When a drop of water hits the flat surface of a pool, circular ripples are made. One ripple is represented by the equation $x^2 + y^2 = 9$ and 5 seconds later, the ripple is represented by the equation $x^2 + y^2 = 225$, where the lengths of the radii are in cm.
   a. State the radius of each of the ripples.
   b. Sketch these graphs.
   c. Evaluate how fast the ripple are moving outwards.
   d. If the ripple continues to move at the same rate, determine when it will hit the edge of the pool which is 2 m from its centre.

27. Identify the type and a possible equation for each of the following graphs and verify using a graphing application.

   a.

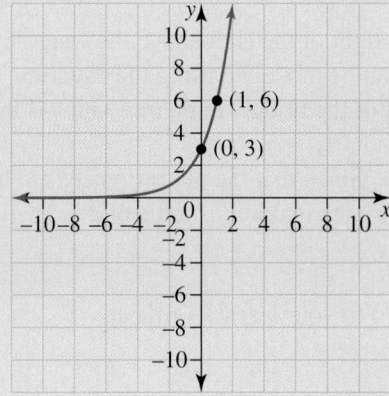

   b.

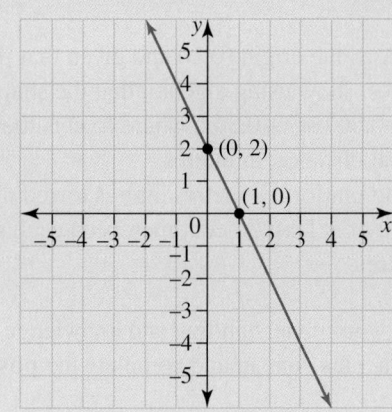

c.

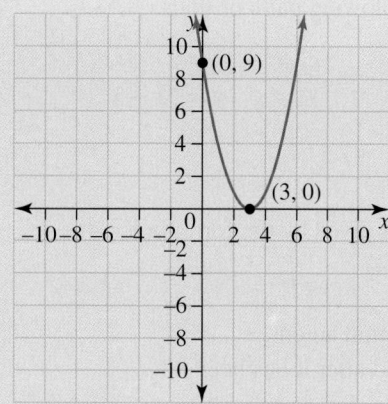

d.

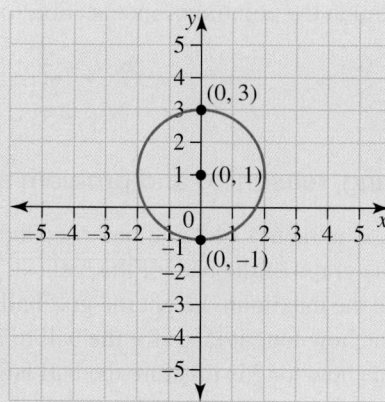

e.

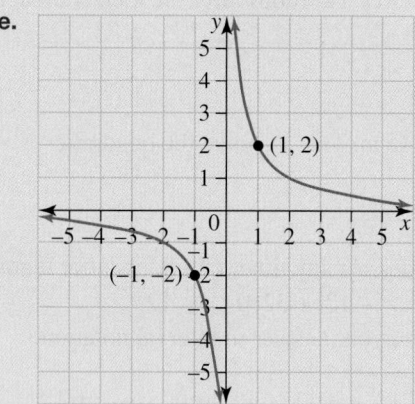

28. A stone arch bridge has a span of 50 metres. The shape of the curve AB can be modelled using a quadratic equation.

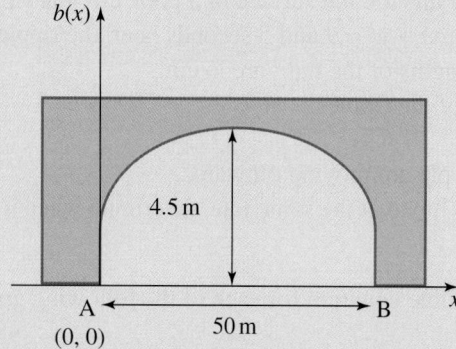

a. Taking A as the origin $(0, 0)$ and given that the maximum height of the arch above the water level is 4.5 metres, show using algebra, that the shape of the arch can be modelled using the equation $b(x) = -0.0072x^2 + 0.36x$, where $b(x)$ is the vertical height of the bridge, in metres, and x is the horizontal distance, in metres.

b. A floating platform p metres high is towed under the bridge. Given that the platform needs to have a clearance of at least 30 centimetres on each side, explain why the maximum value of p is 10.7 centimetres.

on To test your understanding and knowledge of this topic, go to your learnON title at www.jacplus.com.au and complete the **post-test**.

Answers

Topic 8 Non-linear relationships

8.1 Pre-test

1.

x	-3	-2	-1	0
y	9.5	5	1.5	-1

2. $x = -1$

3. B

4. A

5. D

6. $(-4, -1)$

7. $b = 20$

8. D

9. A

10. $(-0.79, 0.62)$ and $(0.79, 0.62)$

11. B

12. C

13. a. ii   **b.** i

14. D

15. D

8.2 The graph of a parabola

1. a.

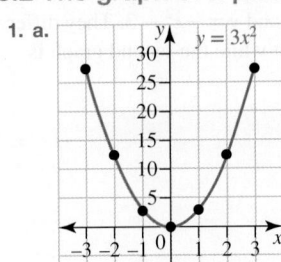

$x = 0, (0, 0)$

b.

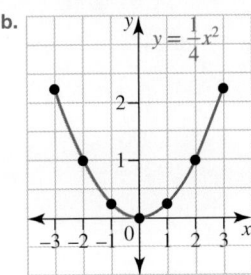

$x = 0, (0, 0)$

2. Placing a number greater than 1 in front of x^2 makes the graph thinner. Placing a number greater than 0 but less than 1 in front of x^2 makes the graph wider.

3. a.

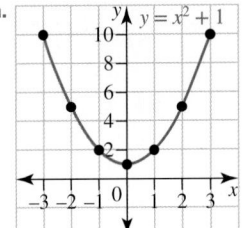

$x = 0, (0, 1), 1$

b.

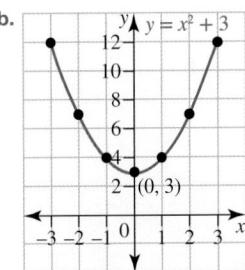

$x = 0, (0, 3), 3$

c.

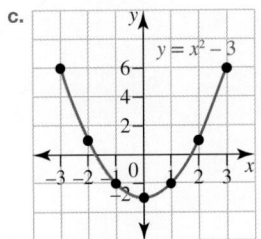

$x = 0, (0, -3), -3$

d.

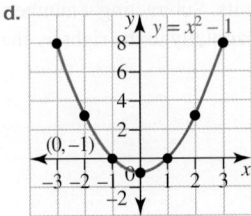

$x = 0, (0, -1), -1$

4. Adding a number raises the graph of $y = x^2$ vertically that number of units. Subtracting a number lowers the graph of $y = x^2$ vertically that number of units.

5.

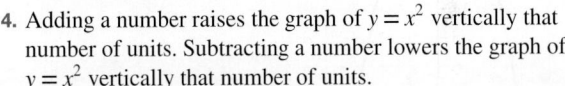
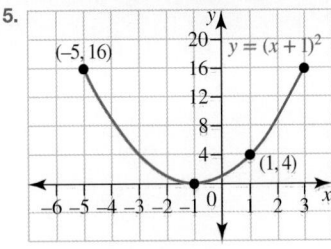

$x = -1, (-1, 0), 1$

6.

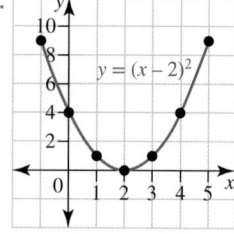

$x = 2, (2, 0), 4$

7.

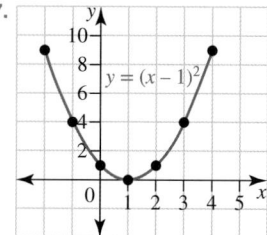

$x = 1, (1, 0), 1$

8.

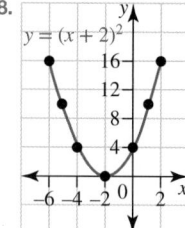

$x = -2, (-2, 0), 4$

9. Adding a number moves the graph of $y = x^2$ horizontally to the left by that number of units. Subtracting a number moves the graph of $y = x^2$ horizontally to the right by that number of units.

10.

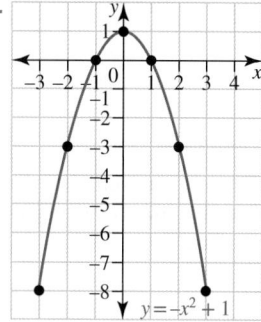

$x = 0, (0, 1), 1$

11.

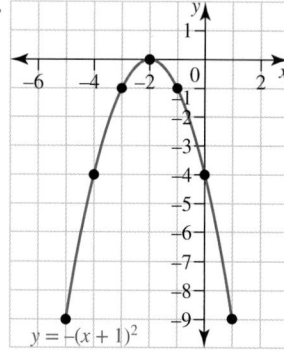

$x = -2, (-2, 0), -4$

12.

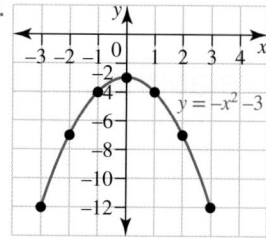

$x = 0, (0, -3), -3$

13.

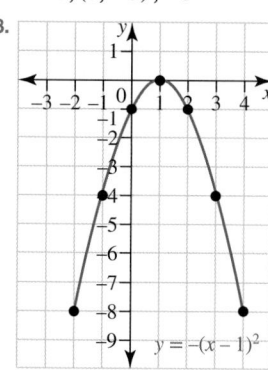

$x = 1, (1, 0), -1$

14. The negative sign inverts the graph of $y = x^2$. The graphs with the same turning points are: $y = x^2 + 1$ and $y = -x^2 + 1$; $y = (x-1)^2$ and $y = -(x-1)^2$; $y = (x+2)$ and $y = -(x+2)^2$; $y = x^2 - 3$ and $y = -x^2 - 3$. They differ in that the first graph is upright while the second graph is inverted.

15. a.

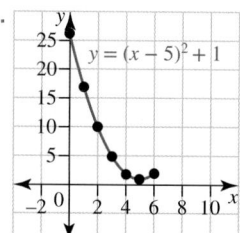

b. $x = 5$ **c.** $(5, 1)$, min   **d.** 26

16. a.

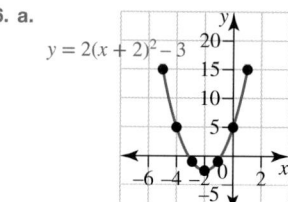

b. $x = -2$ **c.** $(-2, -3)$, min

d. 5

17. a.

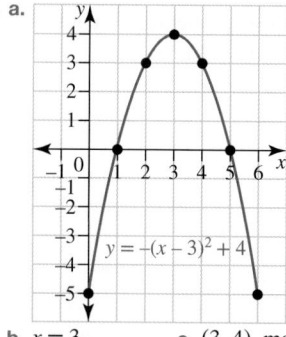

b. $x = 3$ **c.** $(3, 4)$, max   **d.** -5

18. a.

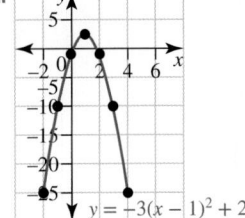

$y = -3(x-1)^2 + 2$

b. $x = 1$　　　**c.** $(1, 2)$, max　　**d.** -1

19. a.

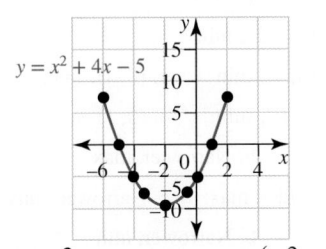

$y = x^2 + 4x - 5$

b. $x = -2$　　　　**c.** $(-2, -9)$, min
d. -5

20. a.

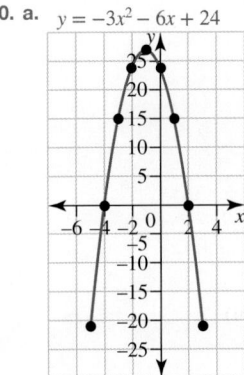

$y = -3x^2 - 6x + 24$

b. $x = -1$　　　　**c.** $(-1, 27)$, max
d. 24

21. a. If the x^2 term is positive, the parabola has a minimum turning point. If the x^2 term is negative, the parabola has a maximum turning point.

b. If the equation is of the form $y = a(x-b)^2 + c$, the turning point has coordinates (b, c).

c. The equation of the axis of symmetry can be found from the x-coordinate of the turning point. That is, $x = b$.

22. C

23. B

24. C

25. A

26. a.

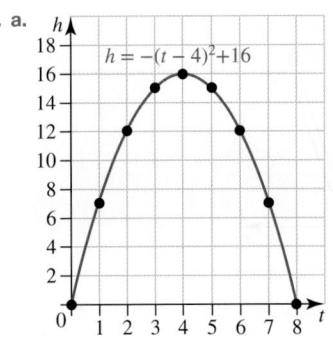

$h = -(t-4)^2 + 16$

b. i. 16 m　　　　**ii.** 8 s

27. a.

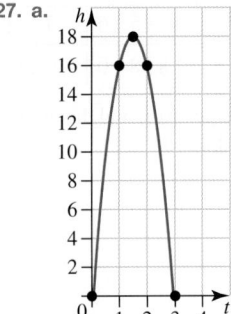

b.　i. 18 m　　　　　**ii.** Yes, by 3 m
iii. 1.5 s　　　　　**iv.** 3 s

28. a.

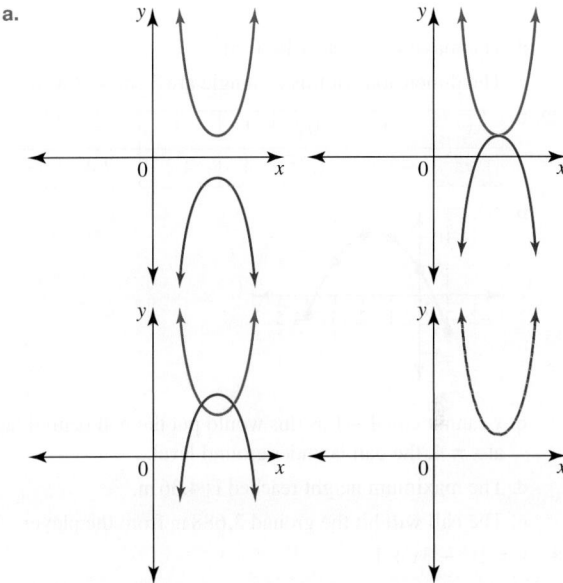

b. An infinite number of points of intersection occur when the two equations represent the same parabola, with the effect that the two parabolas superimpose. For example $y = x^2 + 4x + 3$ and $2y = 2x^2 + 8x + 6$.

c. It is possible to have 0, 1 or 2 points of intersection.

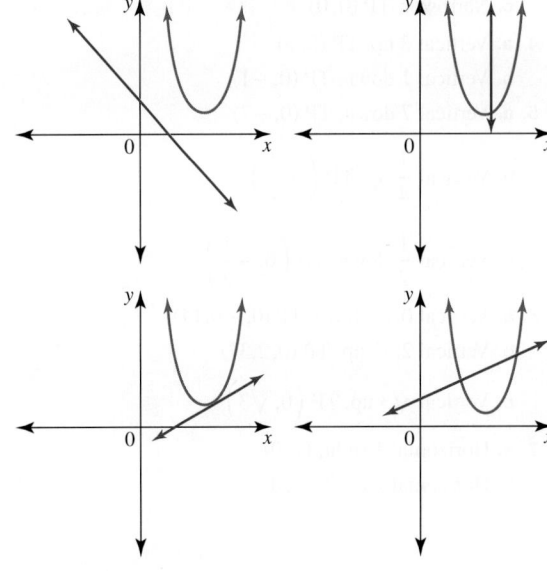

29. a.

w cm	−1	0	1	2	3	4	5	6	7
A cm²	−11.2	0	8	12.8	14.4	12.8	8	0	−11.2

b. The length of a rectangle must be positive, and the area must also be positive, so we can discard $w = -1, 0, 6$ and 7.

c.

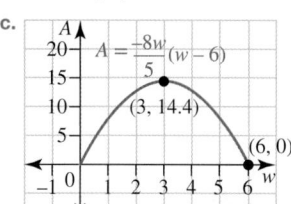

d. The maximum area is 14.4 cm^2.

e. The dimensions of this rectangle are $3 \text{ cm} \times 4.8 \text{ cm}$.

30. a.

x	−1	0	1	2	3	4
y	−2.4	1.8	4	4.2	2.4	−1.4

b.

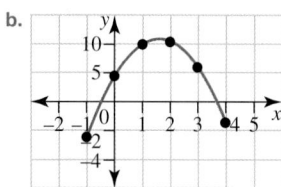

c. x cannot equal -1 as this would put the ball behind her; at $x = 4$, the ball is under ground level.

d. The maximum height reached is 4.36 m.

e. The ball will hit the ground 3.688 m from the player.

31. $y = 2x^2 - 3x + 1$

8.3 Parabolas and transformations

1. a. Narrower, TP $(0, 0)$ **b.** Wider, $(0, 0)$

2. a. Narrower, TP $(0, 0)$ **b.** Narrower, TP $(0, 0)$
c. Wider, TP $(0, 0)$

3. a. Wider, TP $(0, 0)$ **b.** Narrower, TP $(0, 0)$
c. Narrower, TP $(0, 0)$

4. a. Vertical 3 up, TP $(0, 3)$
b. Vertical 1 down, TP $(0, -1)$

5. a. Vertical 7 down, TP $(0, -7)$

b. Vertical $\frac{1}{4}$ up, TP $\left(0, \frac{1}{4}\right)$

c. Vertical $\frac{1}{2}$ down, TP $\left(0, -\frac{1}{2}\right)$

6. a. Vertical 0.14 down, TP $(0, -0.14)$
b. Vertical 2.37 up, TP $(0, 2.37)$

c. Vertical $\sqrt{3}$ up, TP $\left(0, \sqrt{3}\right)$

7. a. Horizontal 1 right, $(1, 0)$
b. Horizontal 2 right, $(2, 0)$

8. a. Horizontal 10 left, $(-10, 0)$
b. Horizontal 4 left, $(-4, 0)$

c. Horizontal $\frac{1}{2}$ right, $\left(\frac{1}{2}, 0\right)$

9. a. Horizontal $\frac{1}{5}$ left, $\left(-\frac{1}{5}, 0\right)$

b. Horizontal 0.25 left, $(-0.25, 0)$
c. Horizontal $\sqrt{3}$ left, $\left(-\sqrt{3}, 0\right)$

10. a. $(0, 1)$, max   **b.** $(0, -3)$, min

11. a. $(-2, 0)$, max   **b.** $(0, 0)$, min   **c.** $(0, 4)$, max

12. a. $(0, 0)$, max   **b.** $(5, 0)$, min   **c.** $(0, 1)$, min

13. a. Narrower, min   **b.** Narrower, max

14. a. Wider, min   **b.** Wider, max   **c.** Narrower, max

15. a. Wider, min   **b.** Narrower, min
c. Wider, max

16. a. **i.** Horizontal translation 1 left
ii. $(-1, 0)$
iii.

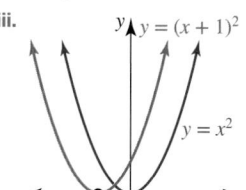

b. **i.** Reflected, narrower (dilation)
ii. $(0, 0)$
iii.

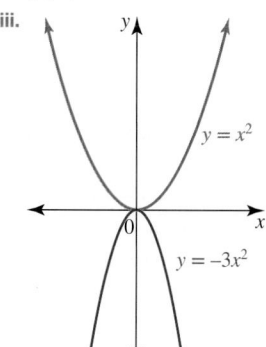

c. **i.** Vertical translation 1 up
ii. $(0, 1)$
iii.

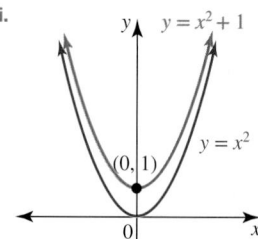

17. a. **i.** Wider (dilation)
ii. $(0, 0)$

iii.

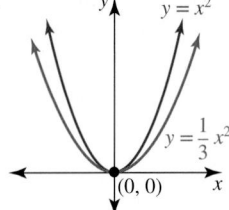

b. i. Vertical translation 3 down

ii. $(0, -3)$

iii.

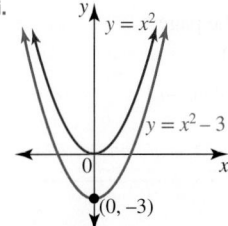

18. a. i. Horizontal translation 4 right

ii. $(4, 0)$

iii.

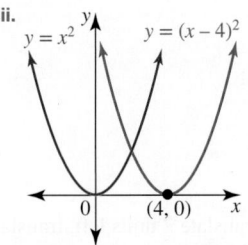

b. i. Reflected, wider (dilation)

ii. $(0, 0)$

iii.

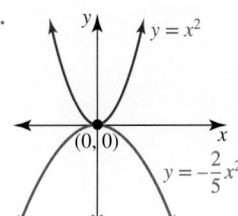

c. i. Narrower (dilation)

ii. $(0, 0)$

iii.

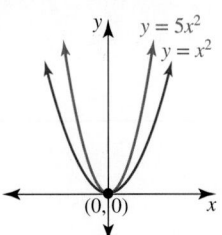

19. a. i. Reflected, vertical translation 2 up

ii. $(0, 2)$

iii.

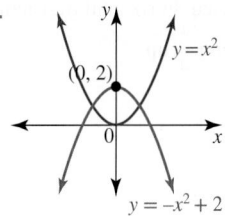

b. i. Reflected, vertical translation 4 down

ii. $(0, -4)$

iii.

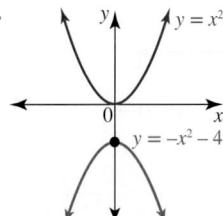

20. a. i. Reflected, horizontal translation 6 right

ii. $(6, 0)$

iii.

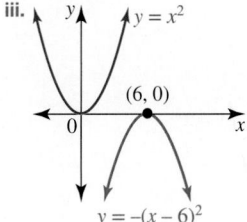

b. i. Narrower (dilation), horizontal translation 1 left, vertical translation 4 down

ii. $(-1, -4)$

iii.

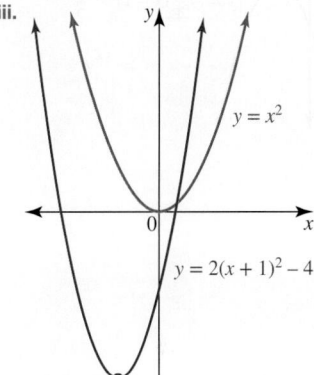

c. i. Wider (dilation), horizontal translation 3 right, vertical translation 2 up

ii. $(3, 2)$

iii.

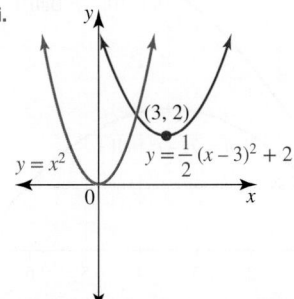

21. a. i. Wider (dilation), reflected, horizontal translation 2 left, vertical translation $\frac{1}{4}$ up

ii. $\left(-2, \frac{1}{4}\right)$

iii.

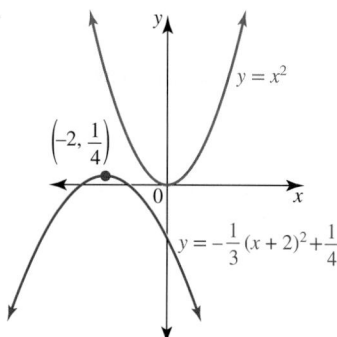

b. i. Narrower (dilation), reflected, horizontal translation 1 right, vertical translation $\frac{3}{2}$ down

ii. $\left(1, -\frac{3}{2}\right)$

iii.

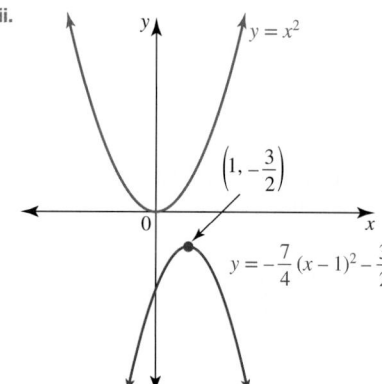

22. a. 10 cm   **b.** 5 cm   **c.** 5 cm

d. $y = (x - 5)^2$

23. a. 40 m

b. 5 m

c. $h = -\frac{1}{10}(d - 70)^2 + 40$

d. $h = -\frac{3}{10}(d - 70)^2 + 120$, this is a dilation by a factor of 3 from the d axis.

24. a. and c.
See figure at the bottom of the page.*

b. 12.25 m   **d.** 12.75 m

25. a. $y = -(x - 2)^2 + 3 = -x^2 + 4x - 1$

b.

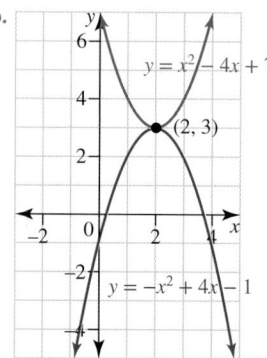

26. a. $(2, -7)$ and $(0, 5)$

b. In the following order: translate 2 units left, translate 7 units up, reflect in the x-axis then dilate by a factor of $\frac{2}{3}$ from the x-axis.

*24. a. and c.

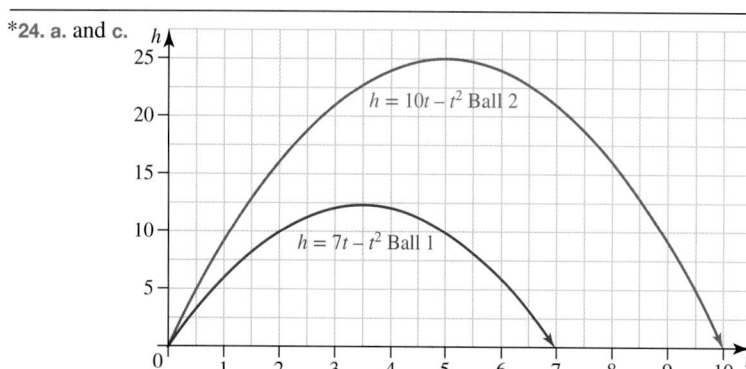

27. a. $Y = -(x-3)^2 + 3$

  b. Reflected in x-axis, translated 3 units to the right and up 3 units. No dilation.

  c. $(3, 3)$

  d. See figure at the bottom of the page.*

28. $v = -2h^2 + 40h$, where v is the vertical distance h is the horizontal distance.

8.4 Parabolas in vertex form (Path)

1. a. $(1, 2)$, min

  b. $(-2, -1)$, min

  c. $(-1, 1)$, min

2. a. $(2, 3)$, max   **b.** $(5, 3)$, max   **c.** $(-2, -6)$, min

3. a. $\left(\dfrac{1}{2}, -\dfrac{3}{4}\right)$, min   **b.** $\left(\dfrac{1}{3}, \dfrac{2}{3}\right)$, min

  c. $(-0.3, -0.4)$, min

4. a. i. $(-3, -5)$ **ii.** Min   **iii.** Narrower

  b. i. $(1, 1)$ **ii.** Max   **iii.** Same

5. a. i. $(-2, -4)$ **ii.** Max   **iii.** Narrower

  b. i. $(3, 2)$ **ii.** Min   **iii.** Wider

6. a. i. $(-1, 7)$ **ii.** Max   **iii.** Wider

  b. i. $\left(\dfrac{1}{5}, -\dfrac{1}{2}\right)$ **ii.** Min   **iii.** Wider

7. a. $y = (x+1)^2 - 3$, graph iii

  b. $y = -(x-2)^2 + 3$, graph i

  c. $y = -x^2 + 1$, graph ii

8. a. $y = (x-1)^2 - 3$, graph iii

  b. $y = -(x+2)^2 + 3$, graph i

  c. $y = x^2 - 1$, graph ii

9. A

10. C

11. B

12. C

13. B

14. a. i. -3 **ii.** $-3, 1$

  b. i. 12 **ii.** 2

15. a. i. -18 **ii.** No x-intercepts

  b. i. -5 **ii.** $-1, 5$

16. a. i. 4 **ii.** No x-intercepts

  b. i. 4

  ii. $-3 - \sqrt{5}, -3 + \sqrt{5}$ (approx. $-5.24, -0.76$)

17. a. i. $(4, 2)$

  ii. Min

  iii. Same width

  iv. 18

  v. No x-intercepts

  vi.

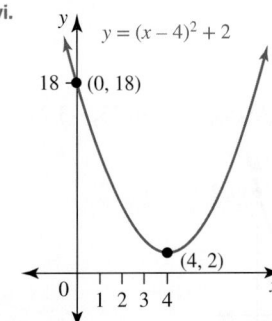

  b. i. $(3, -4)$

  ii. Min

  iii. Same width

  iv. 5

  v. $1, 5$

  vi.

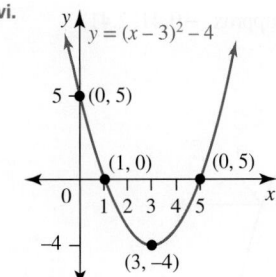

  c. i. $(-1, 2)$

  ii. Min

  iii. Same width

  iv. 3

  v. No x-intercepts

*27. d.

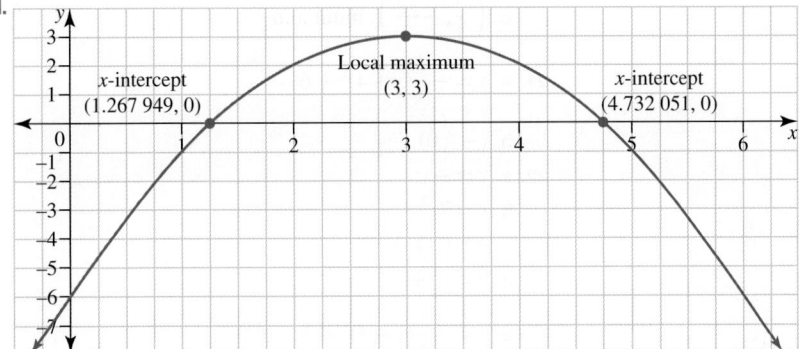

vi.

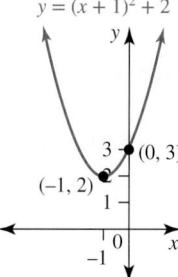
$y = (x + 1)^2 + 2$

18. a. i. $(-5, -3)$
    ii. Min
    iii. Same width
    iv. 22
    v. $-5 - \sqrt{3}, -5 + \sqrt{3}$ (approx. $-6.73, -3.27$)
    vi.

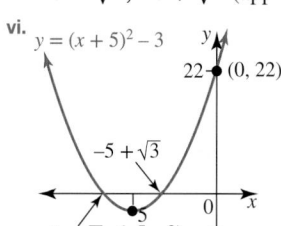
$y = (x + 5)^2 - 3$

   b. i. $(1, 2)$
    ii. Max
    iii. Same width
    iv. 1
    v. $1 - \sqrt{2}, 1 + \sqrt{2}$ (approx. $-0.41, 2.41$)
    vi.

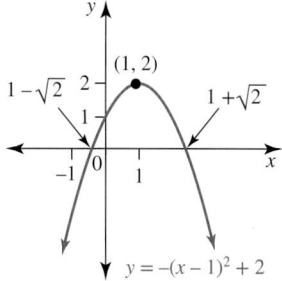
$y = -(x - 1)^2 + 2$

   c. i. $(-2, -3)$
    ii. Max
    iii. Same width
    iv. -7
    v. No x-intercepts
    vi.

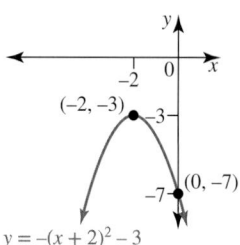
$y = -(x + 2)^2 - 3$

19. a. i. $(-3, -2)$
    ii. Max
    iii. Same width
    iv. -11

v. No x-intercepts
vi.

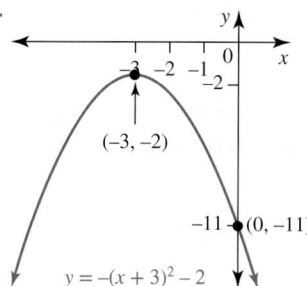
$y = -(x + 3)^2 - 2$

   b. i. $(1, 3)$
    ii. Min
    iii. Narrower
    iv. 5
    v. No x-intercepts
    vi.

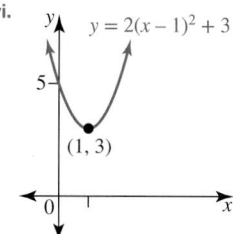
$y = 2(x - 1)^2 + 3$

   c. i. $(-2, 1)$
    ii. Max
    iii. Narrower
    iv. -11
    v. $-2 - \dfrac{1}{\sqrt{3}}, -2 + \dfrac{1}{\sqrt{3}}$ (approx. $-2.58, -1.42$)
    vi.

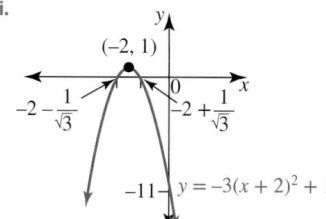
$y = -3(x + 2)^2 + 1$

20. a. $2\left(x - \dfrac{3}{4}\right)^2 - \dfrac{73}{8} = 0$

   b. $x = \dfrac{3}{4} \pm \dfrac{\sqrt{73}}{4}$

   c. $\left(\dfrac{3}{4}, -\dfrac{73}{8}\right)$, minimum

21. a. $y = -\dfrac{2}{3}(x + 4)^2 + 6$

   b. $(-7, 0)$

22. a. $y = -x^2$
   b. $y = 7x^2$
   c. $y = (x + 3)^2$
   d. $y = x^2 + 6$
   e. $y = -\dfrac{1}{4}(x - 5)^2 - 3$

23. a.

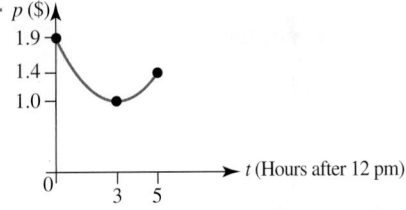

p ($), t (Hours after 12 pm)

b. $1.90   **c.** $1   **d.** 3 pm   **e.** $1.40

24. a. 0.5 m

b. $\left(15 + 4\sqrt{15}\right)$ m

c. Maximum height is 8 metres when horizontal distance is 15 metres.

25. a. Sample responses can be found in the worked solutions in the online resources.
An example is $y = (x - 2)^2 + 6$.

b. $y = -\dfrac{1}{2}(x - 2)^2 + 6$

26. a. Sample responses can be found in the worked solutions in the online resources.
An example is $y = (x - p)^2 + q$.

b. $y = \left(\dfrac{r - q}{p^2}\right)(x - p)^2 + q$

27. a. $y = (x - 4)^2 - 15$

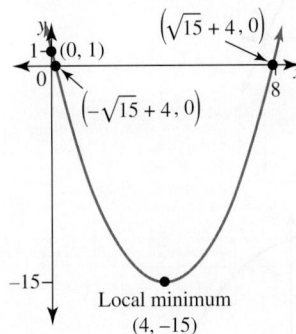

Local minimum (4, –15)

b. $y = (x + 2)^2 - 9$

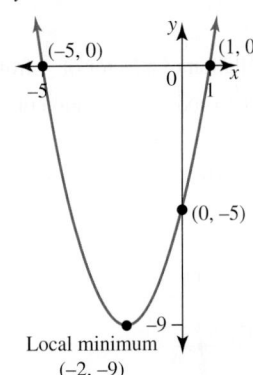

Local minimum (–2, –9)

c. $y = \left(x + \dfrac{3}{2}\right)^2 - \dfrac{1}{4}$

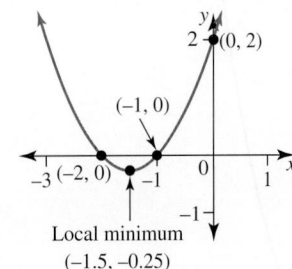

Local minimum (–1.5, –0.25)

28. a. $y = 3(x - 2)^2 - 8$

b. Dilated by a factor of 3 parallel to the y-axis or from the x-axis as well as being translated 2 units to the right and down 8 units.

c. $y = -3(x - 2)^2 + 8$

d. See figure at the bottom of the page.*

---

*28. d.

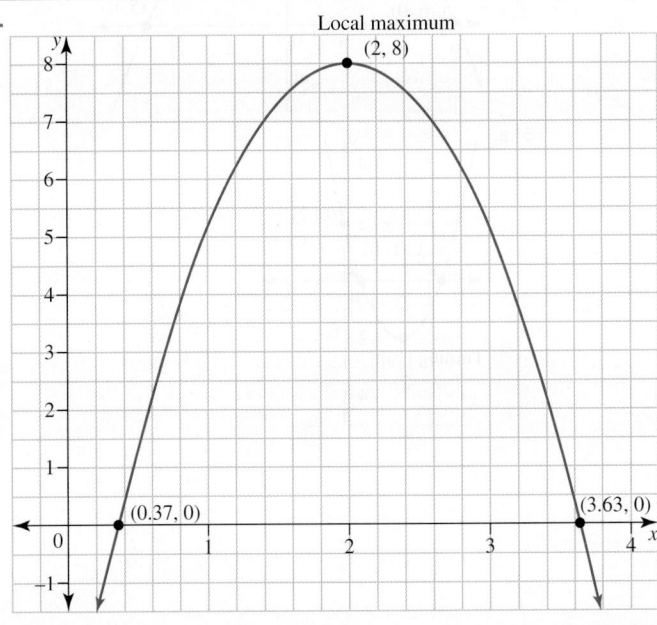

Local maximum (2, 8)

(0.37, 0)   (3.63, 0)

29. a. $y = -\dfrac{1}{2}(x+3)^2 + 8$

b. In the following order: a dilation of a factor of $\dfrac{1}{2}$ from the x-axis, a reflection in the x-axis, a translation of 3 units left and 8 units up.

c. x- intercepts: $(-7, 0)$ and $(1, 0)$

  y- intercept: $\left(0, \dfrac{7}{2}\right)$

d. The new rule is $y = (x-3)^2 - 16$

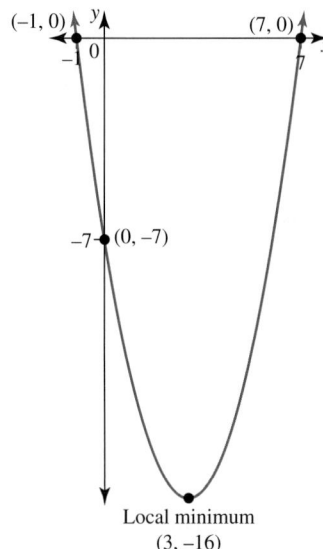

8.5 Parabolas in general form (Path)

1. You need the x-intercepts, the y-intercept and the turning point to sketch a parabola.

2. a.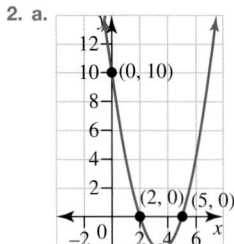

b.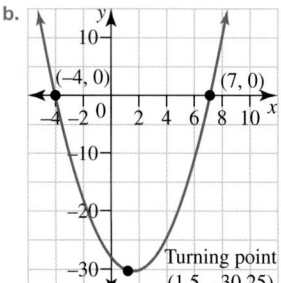

3. a.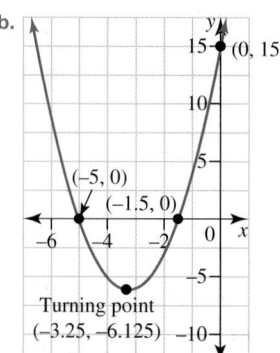

b.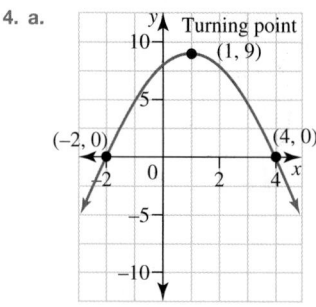

4. a.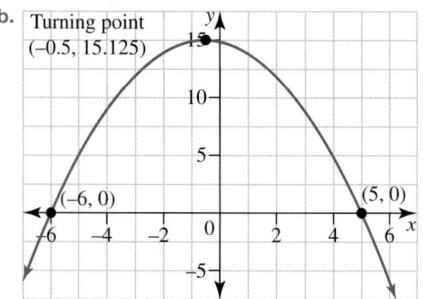

b.

5. a.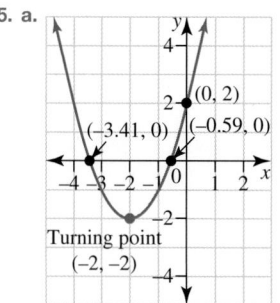

b.

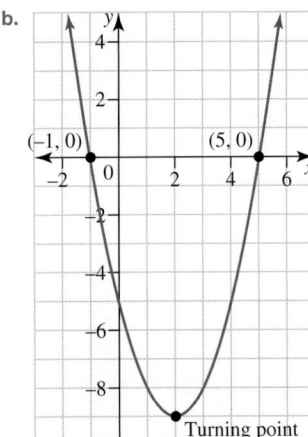

(−1, 0)   (5, 0)

Turning point
(2, −9)

c.

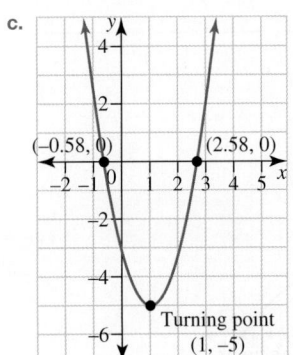

(−0.58, 0)   (2.58, 0)

Turning point
(1, −5)

6. a.

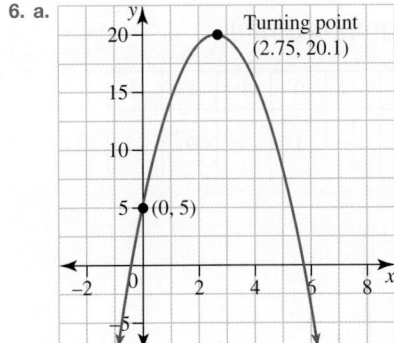

Turning point
(2.75, 20.1)

(0, 5)

b.

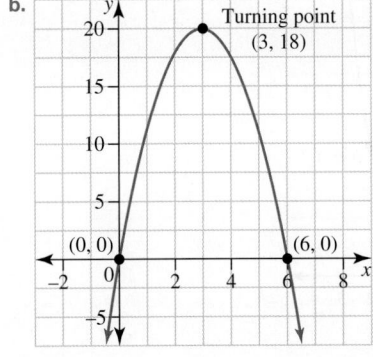

Turning point
(3, 18)

(0, 0)   (6, 0)

c.

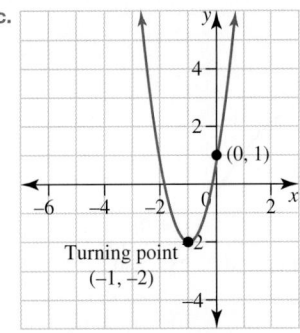

(0, 1)

Turning point
(−1, −2)

7. a.

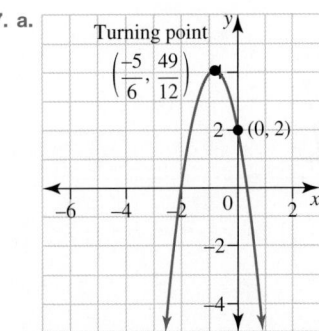

Turning point
$\left(\dfrac{-5}{6}, \dfrac{49}{12}\right)$

(0, 2)

b.

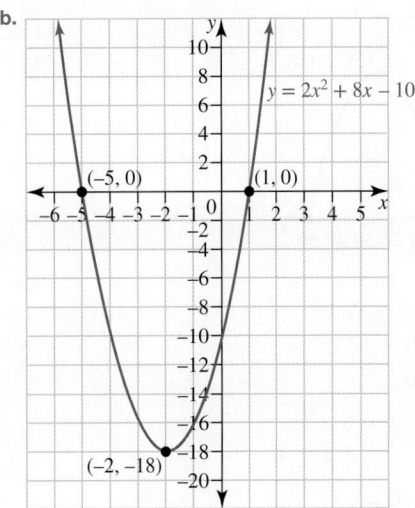

$y = 2x^2 + 8x - 10$

(−5, 0)   (1, 0)

(−2, −18)

c.

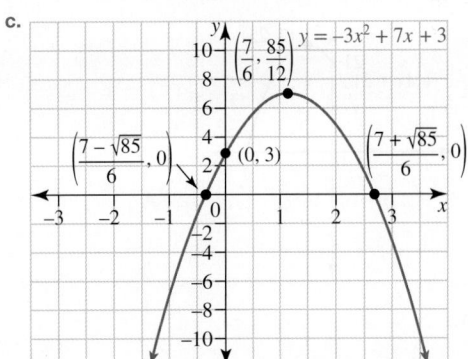

$\left(\dfrac{7}{6}, \dfrac{85}{12}\right)$ $y = -3x^2 + 7x + 3$

$\left(\dfrac{7 - \sqrt{85}}{6}, 0\right)$ (0, 3) $\left(\dfrac{7 + \sqrt{85}}{6}, 0\right)$

8. a.

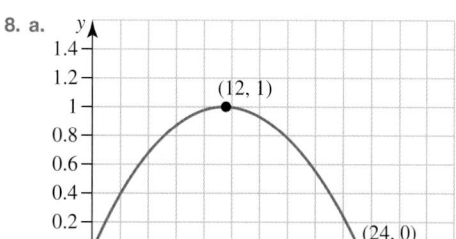

b. 24 m

c. 1 m

9. a.

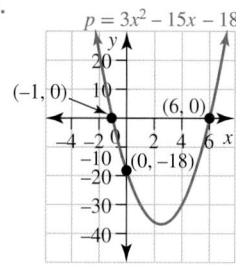

b. 6th month

c. 8th month

10. a.

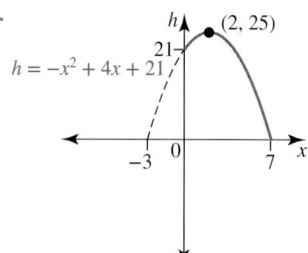

b. 0

c. 2500 m

d. 25 seconds

e. 50 seconds

11. 200

12. $-(m+n)$

13. $y = -\dfrac{1}{2}x^2 + 2x + 9$

14. a.

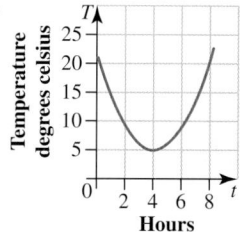

$h = -x^2 + 4x + 21$

b. 25 m

c. 2 m

d. 7 m

15. a.

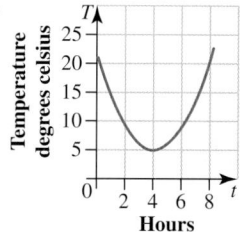

b. 21 °C

c. Decreasing

d. Increasing

e. 5 °C after 4 hours

f. 21 °C

16. a. $a = -\dfrac{1}{18}, b = \dfrac{13}{6}, c = 7$

b. 7 metres

c. $\dfrac{225}{8}$ metres

d. 42 metres

e.

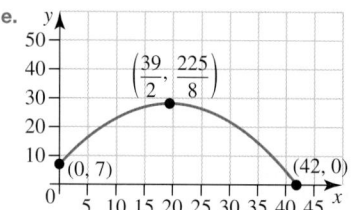

8.6 Exponential graphs (Path)

1.

x	-3	-2	-1	0	1	2	3
y	$\dfrac{1}{27}$	$\dfrac{1}{9}$	$\dfrac{1}{3}$	1	3	9	27

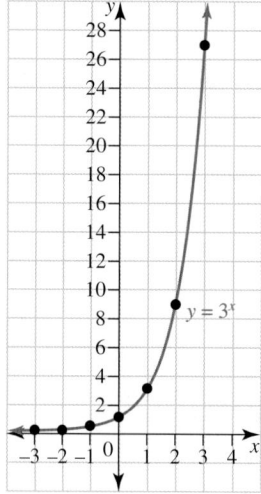

2. a. 2   **b.** 3   **c.** 4

3. a. 10   **b.** a

4.

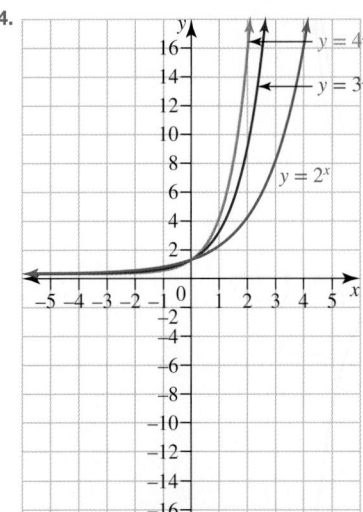

a. The graphs all pass through $(0, 1)$. The graphs have the same horizontal asymptote, $(y = 0)$. The graphs are all very steep.

b. As the base grows larger, the graphs become steeper.

c.

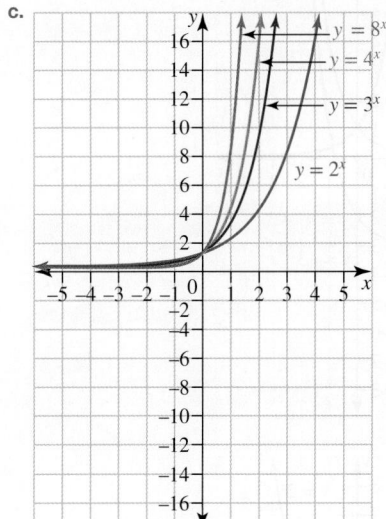

5.

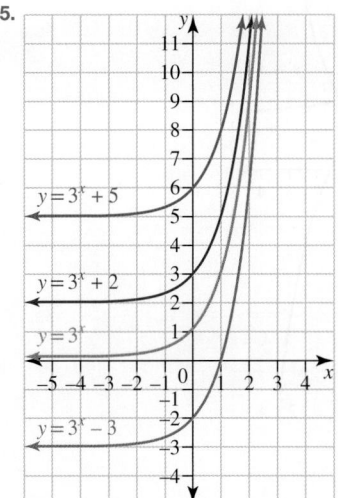

a. The shape of each graph is the same.

b. Each graph has a different y-intercept and a different horizontal asymptote.

c. i. $(0, 11)$

   ii. $y = 10$

6. a.

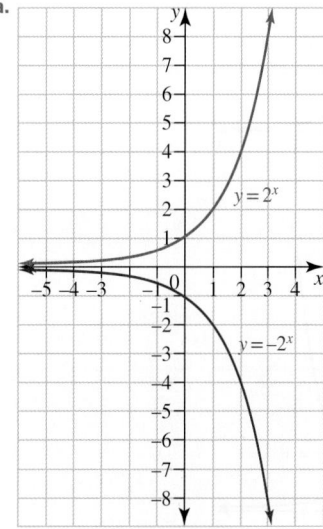

b.

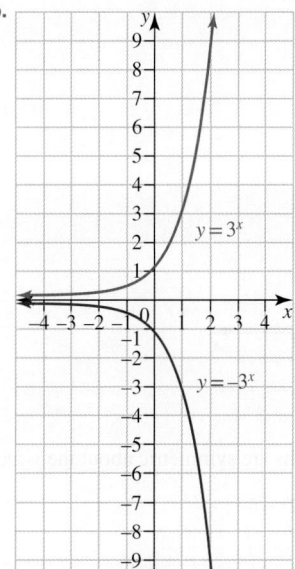

c.

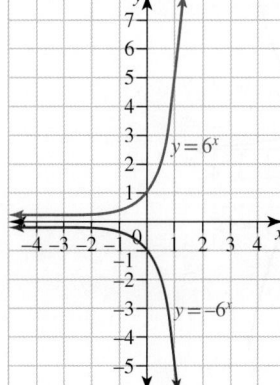

d. In each case the graphs are symmetric about the x-axis.

7. a.

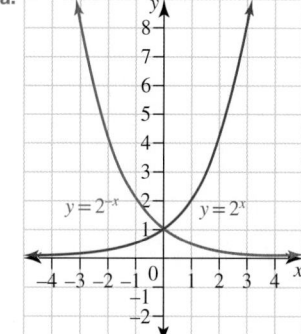

b.

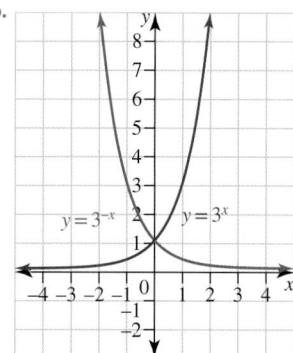

c.

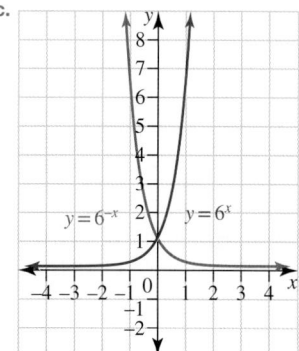

d. In each case the graphs are symmetric about the y-axis.

8. a–c.

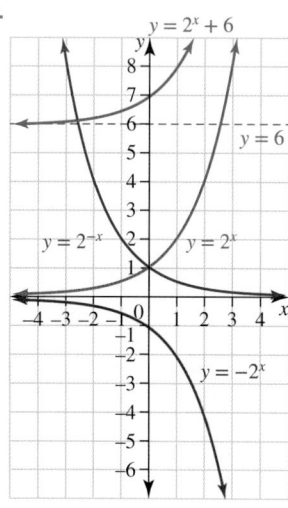

9. a, b

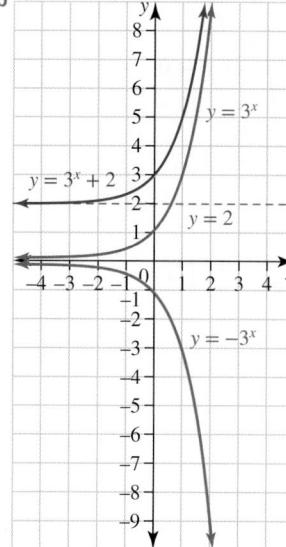

10. a, b

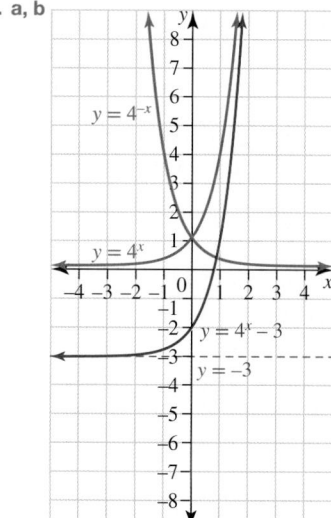

11. a, b

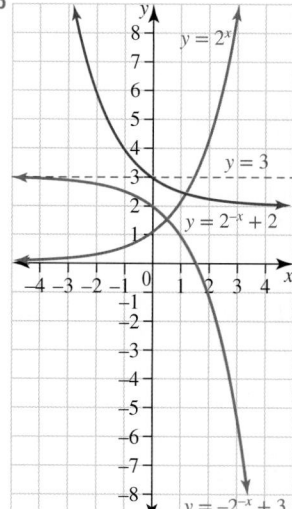

12. a, b

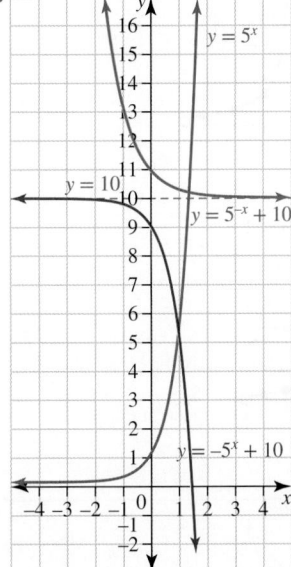

13. a. ii   **b.** iii   **c.** iv   **d.** i

14. a. ii   **b.** iv   **c.** i   **d.** iii

15.

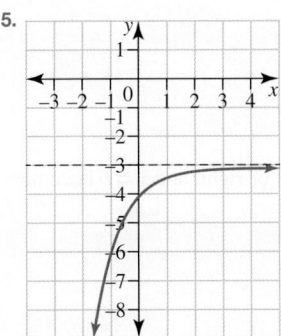

16.

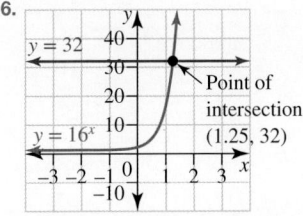

$x = 1.25$

17.

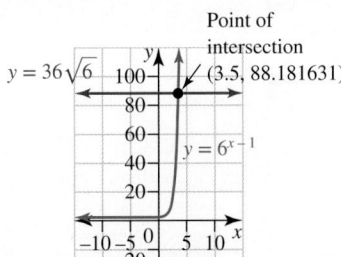

$x = 3.5$

18.

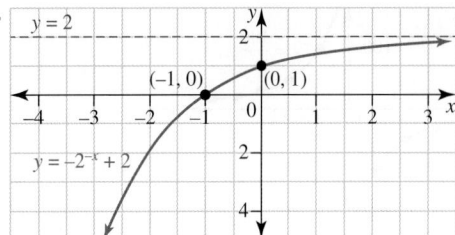

19. a. 10 000

    b. i. 1111

      ii. 41

      iii. 0

20. a. Yes

    b. There is a constant ratio of 1.3.

    c. 30%

    d. 3.26 million

    e. 30.26 million

8.7 The hyperbola (Path)

1. See table at the bottom of the page.*

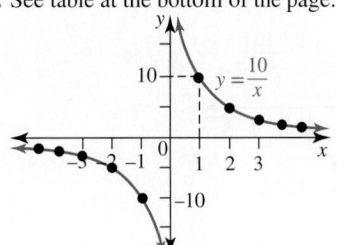

2. a.

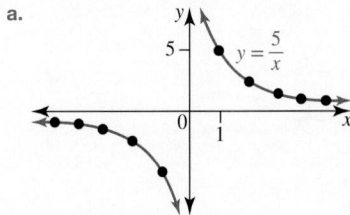

    b. $x = 0, y = 0$

3. a.

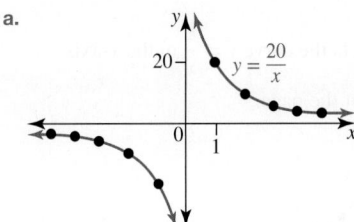

    b. $x = 0, y = 0$

*1.

x	−5	−4	−3	−2	−1	0	1	2	3	4	5
y	−2	−2.5	−3.3	−5	−10	Undef.	10	5	3.3	2.5	2

4. a.

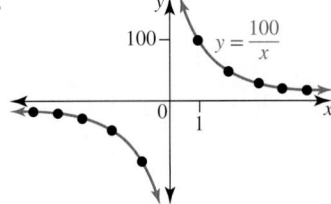

b. $x = 0, y = 0$

5.

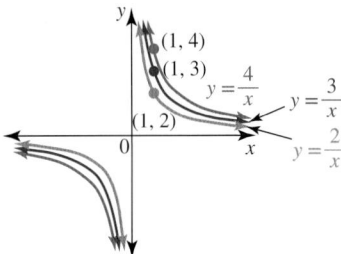

6. It increases the y-values by a factor of k and hence dilates the curve by a factor of k.

7.

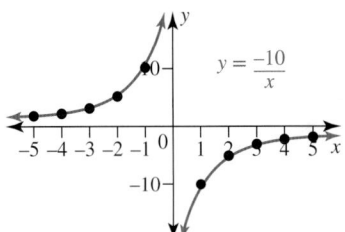

8.

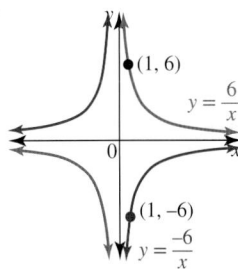

9. The negative reflects the curve $y = \dfrac{k}{x}$ in the x-axis.

10. See table bottom of the page.*

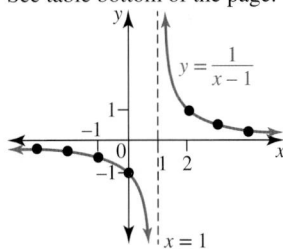

Equation of vertical asymptote is $x = 1$.

11. a.

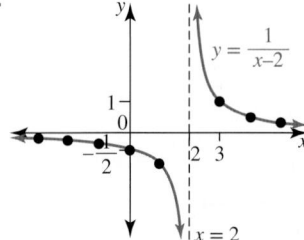

b.

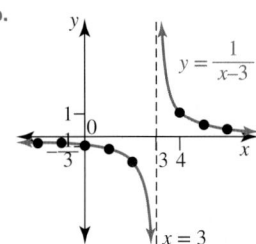

12.

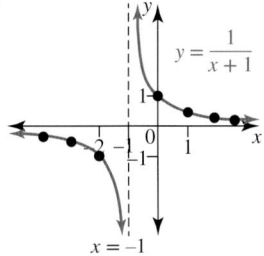

13. The b translates the graph left or right, and $x = b$ becomes the vertical asymptote.

14. a. i. $x = -1, y = 0$

ii.

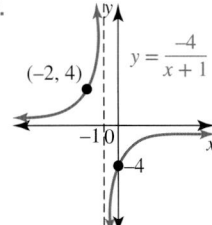

b. i. $x = 1, y = 0$

ii.

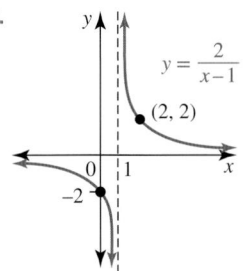

*10. a.

x	−3	−2	−1	0	1	2	3	4
y	−0.25	−0.33	−0.5	−1	Undef.	1	0.5	0.33

15. a. i. $x = -2, y = 0$

ii.

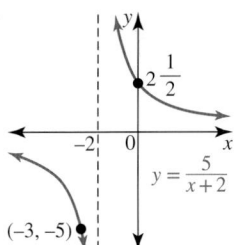

b. i. $x = -2, y = -2$

ii.

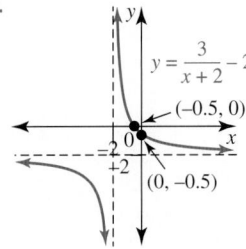

16. a. i. $x = -2, y = 1$

ii.

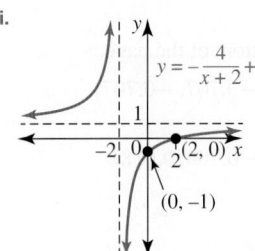

b. i. $x = 3, y = 5$

ii.

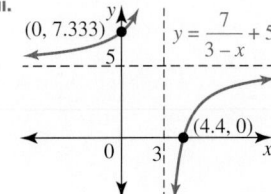

17. Sample responses can be found in the worked solutions in the online resources. Possible answers:

a. $y = \dfrac{1}{x - 3}$

b. $y = \dfrac{1}{x + 10}$

18. a. Any equation of the form $y = \dfrac{k}{x - 2} + 3$.

b. $y = \dfrac{-14}{x + 2} + 4$

19. $y = \dfrac{-2}{x + 3} - 1, x = -3, y = -1$

20. a. $100\,°C$

b. $33.1\,°C$

c. 17.85 minutes ≈ 18 minutes

d. No. The horizontal asymptote is $22\,°C$, so the temperature will never drop below this value.

8.8 The circle (Path)

1. a.

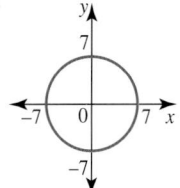

Centre $(0, 0)$, radius 7

b.

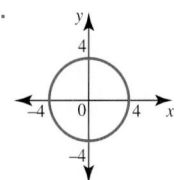

Centre $(0, 0)$, radius 4

2. a.

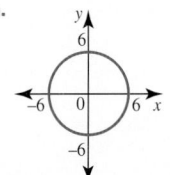

Centre $(0, 0)$, radius 6

b.

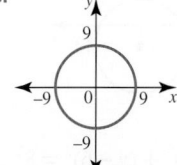

Centre $(0, 0)$, radius 9

3. a.

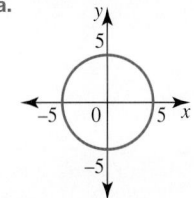

Centre $(0, 0)$, radius 5

b.

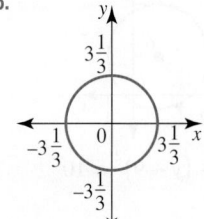

Centre $(0, 0)$, radius $\dfrac{10}{3}$

4. a.

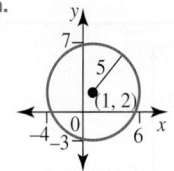

b.

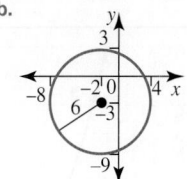

5. a.

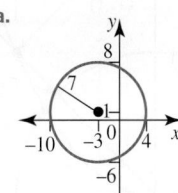

b.

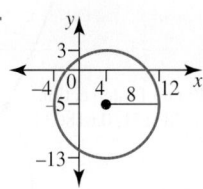

6. a.

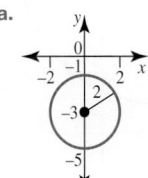

b.

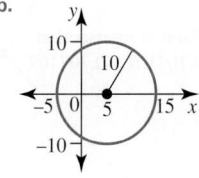

7. a. $(x+2)^2 + \left(y+4\right)^2 = 2^2$

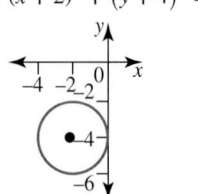

b. $(x-5)^2 + \left(y-1\right)^2 = 4^2$

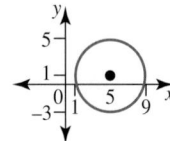

8. a. $(x-7)^2 + (y+3)^2 = 7^2$

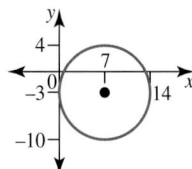

b. $(x+4)^2 + (y-6)^2 = 8^2$

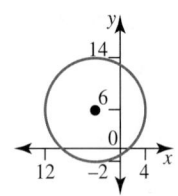

9. a. $x^2 + \left(y-9\right)^2 = 10^2$

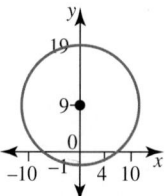

b. $(x-1)^2 + \left(y+2\right)^2 = 3^2$

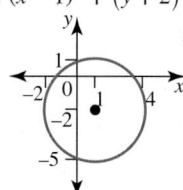

10. D

11. B

12. D

13. $(x-5)^2 + \left(y-3\right)^2 = 16$

14. a. 2 cm, 13.8 cm

  b. 3.9 cm/s

15. $(-2, 0)$

16. a. See the figure at the bottom of the page.*

  b. **i.** $(0.707, 0.707)$ and $(-0.707, -0.707)$

  ii. $(0, 0)$ and $(1, 1)$

  iii. $(0.786, 0.618)$ and $(-0.786, 0.618)$

---

*16. a.

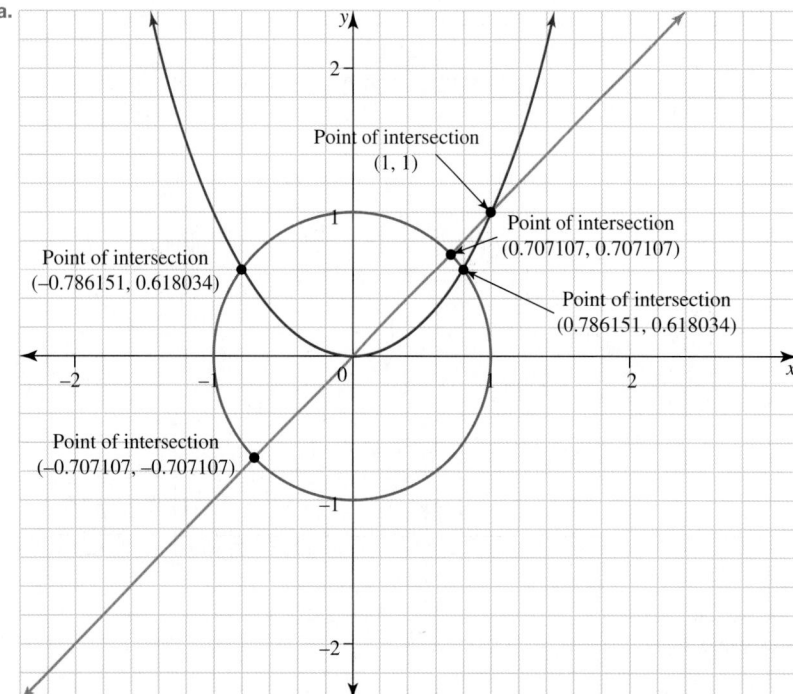

Point of intersection (1, 1)

Point of intersection (0.707107, 0.707107)

Point of intersection (-0.786151, 0.618034)

Point of intersection (0.786151, 0.618034)

Point of intersection (-0.707107, -0.707107)

17.

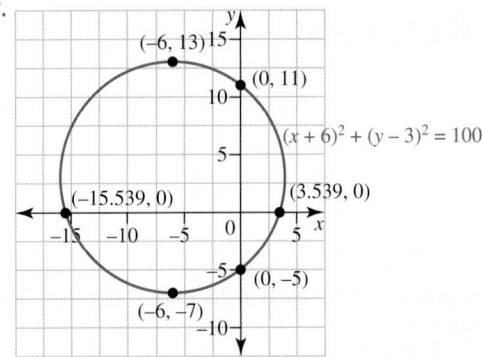

The graph shows the circle $(x + 6)^2 + (y - 3)^2 = 100$ with points $(-6, 13)$, $(0, 11)$, $(-15.539, 0)$, $(3.539, 0)$, $(0, -5)$ and $(-6, -7)$.

18. $(-3, 4), (3, 4), (0, 5)$ and $(0, -5)$

19. $(x - 1)^2 + (y - 1)^2 = 4$ centre at $(1, 1)$ and radius of 2 units.
$(x - 4)^2 + (y - 1)^2 = 1$ centre at $(4, 1)$ and radius of 1 unit.
The circles intersect (touch) at $(3, 1)$.
See figure at the bottom of the page.*

20. $(x - 2)^2 + (y - 5)^2 = 4$ centre at $(2, 5)$ and radius of 2 units.

8.9 Simultaneous linear and non-linear equations (Path)

1. A parabola may intersect with a straight line twice, once or not at all.

2. a. $(-5, 4)$ and $(-1, 0)$

b. $(2, 3)$

c. $\left(1 - \sqrt{10}, -6\right)$ and $\left(1 + \sqrt{10}, -6\right)$

3. a. $\left(\dfrac{-1}{2} - \dfrac{3\sqrt{5}}{2}, \dfrac{-19}{2} - \dfrac{9\sqrt{5}}{2}\right)$ and
$\left(\dfrac{-1}{2} + \dfrac{3\sqrt{5}}{2}, \dfrac{-19}{2} + \dfrac{9\sqrt{5}}{2}\right)$

b. $(-1, -2)$ and $(2, 1)$

c. $(-2.54, -8.17)$ and $(3.54, 16.17)$

4. a. $(-1.41, 4)$ and $(1.41, 4)$

b. $(-1, 2)$ and $\left(\dfrac{5}{2}, \dfrac{1}{4}\right)$

c. $(3, 37)$

5. B

6. C

7. a. Yes   **b.** No   **c.** Yes   **d.** No

8. a. $(1, 1), (-1, -1)$

b. $\left(1 + \sqrt{2}, -1 + \sqrt{2}\right), \left(1 - \sqrt{2}, -1 - \sqrt{2}\right)$

c. $\left(\dfrac{-\sqrt{15}}{3}, -\sqrt{15}\right), \left(\dfrac{\sqrt{15}}{3}, \sqrt{15}\right)$

d. $(-6, -1), (2, 3)$

9. a. $(-1, -3), (1, 3)$

b. $(-4, 3), (4, -3)$

c. $(-1, 7), (5, -5)$

d. $\left(\dfrac{(2 - \sqrt{14})}{2}, \dfrac{(\sqrt{14} + 2)}{2}\right), \left(\dfrac{(\sqrt{14} + 2)}{2}, \dfrac{(2 - \sqrt{14})}{2}\right)$

10. a. $\left(-\dfrac{1}{2}, -2\right), \left(\dfrac{1}{2}, 2\right)$ **b.** $(0, 5), (4, -3)$

c. $\left(-\dfrac{1}{2}, 2\right), (0, 3)$ **d.** $\left(\dfrac{8}{3}, -\dfrac{1}{4}\right), (5, -2)$

11. a. $(1, 1)$ **b.** $(-3, 3), (3, 3)$

c. $(-4, 11), (-2, 3)$ **d.** $\left(-\dfrac{100}{3}, \dfrac{652}{9}\right), (12, 12)$

12. a. $k < -1$ **b.** $k = -1$ **c.** $k > -1$

13. The straight line crosses the parabola at $(0, -7)$ so no matter what value m takes, there will be at least one intersection point.

*19.

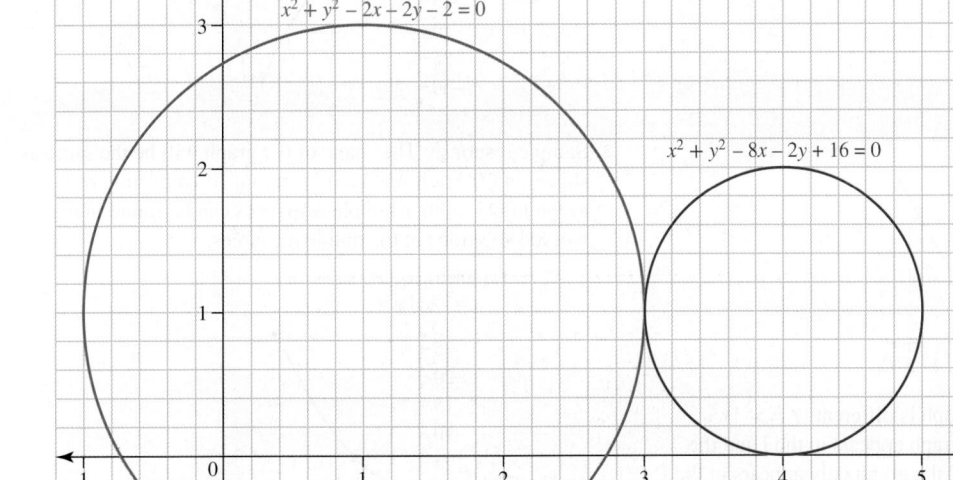

The graph shows two circles: $x^2 + y^2 - 2x - 2y - 2 = 0$ and $x^2 + y^2 - 8x - 2y + 16 = 0$.

14. a. i. $1, 2, 3$

ii. $0, 1, 2, 4$

iii. $1, 2, 3, 4, 5$

b. The number of possible intersections between an equation and a straight line is equal to the highest power of x.

15. $17, 18$ and $-17, -18$

16. Length $15\,\text{m}$, width $85\,\text{m}$.

17. a. $(5, 5), (-1, -7)$

b.

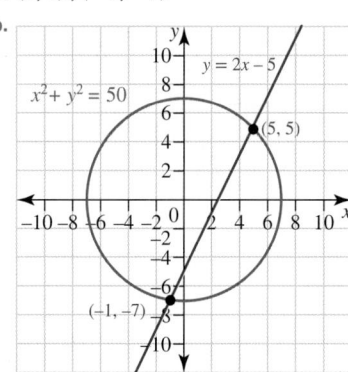

18. $9, 12$

19. a. $(4, 4), \left(-\sqrt{32}, 0\right)$

b. $6.66\,\text{km per hour}$ or $11.11\,\text{km per hour}$

20. a. $100\,°\text{C}$

b. 0 minutes, 6 minutes

c. Eve's model. This model flattens out at $20\,°\text{C}$, whereas Adam's becomes negative which would not occur.

Project

1.

t	x	y
0	0	0
1	1	1
2	2	4
3	3	9
4	4	16
5	5	25

2.

Parametric equations
$x = t$ and $y = t^2$

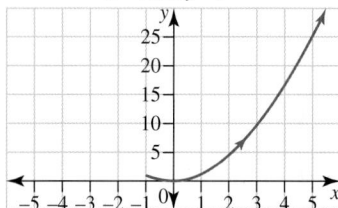

3. Sample response: Yes the graph is different. $Y = x^2$ is a parabola and therefore, the graph appears in the I and the II quadrants. Where as in Q2 the graph only appears in the I quadrant. Other sample responses can be found in the worked solutions in the online resources.

4.

t	x	y
0	1	1
1	0	0
2	-1	1
3	-2	4
4	-3	9
5	-4	16

Sample responses can be found in the worked solutions in the online resources.

$t = 1, 0, -1, -2, -3, -4$.

Parametric equations
$x = 1 - t$ and $y = (1 - t)^2$

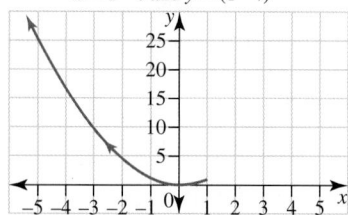

5.

t	x	y
0	0	0
1	1	-1
2	2	-4
3	3	-9
4	4	-16
5	5	-25

Parametric equations
$x = t$ and $y = -t^2$

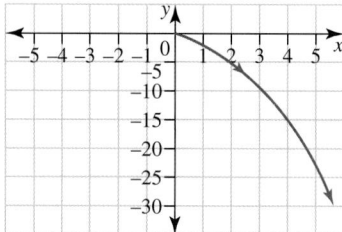

6. Sample response: The shape of the graph will be the same as shown in Q5 but it will be in quadrant III (reflection of the graph in Q5). Other sample responses can be found in the worked solutions in the online resources.

Parametric equations
$x = t$ and $y = t^2$

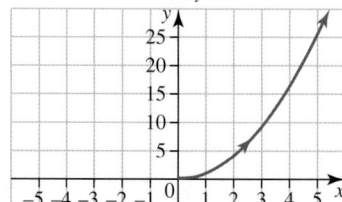

8.10 Review questions

1. D
2. A
3. D
4. A
5. B
6. a. $(4, -15)$ b. $(-2, -9)$
7. See table at the bottom of the page.*

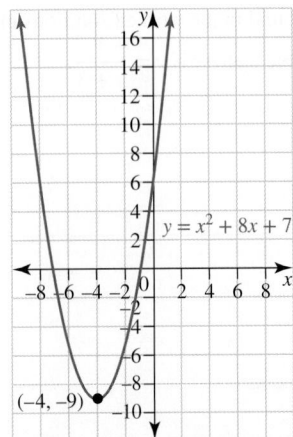

$$y = x^2 + 8x + 7$$

TP $(-4, -9)$; x-intercepts: -7 and -1

8. a. TP $(3, 1)$; no x-intercepts; y-intercept: $(0, 10)$

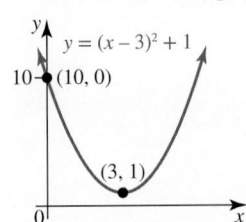

$$y = (x - 3)^2 + 1$$

b. TP $(-1, -5)$; x-intercepts: $-1 - \sqrt{\dfrac{5}{2}}, -1 + \sqrt{\dfrac{5}{2}}$;
y-intercept: $(0, -3)$

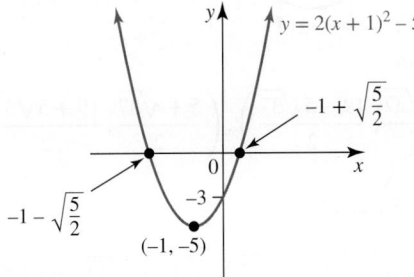

$$y = 2(x + 1)^2 - 5$$

9. TP $(-1, 16)$; x-intercepts: -5 and 3; y-intercept: $(0, 15)$

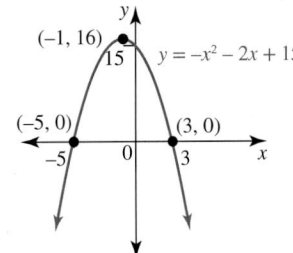

$$y = -x^2 - 2x + 15$$

10.
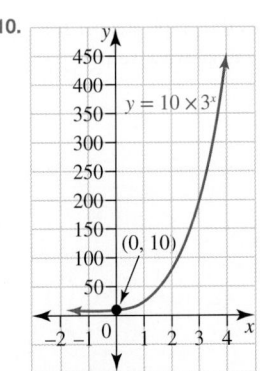

$$y = 10 \times 3^x$$

11. a. Hyperbola
    b. Circle
    c. Parabola
    d. Exponential
    e. Straight line
    f. Parabola

12. a. See table at the bottom of the page.*
    b.
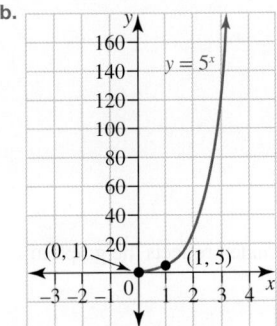

$$y = 5^x$$

*7.

x	-9	-8	-7	-6	-5	-4	-3	-2	-1	0	1
y	16	7	0	-5	-8	-9	-8	-5	0	7	16

*12. a.

x	-3	-2	-1	0	1	2	3
y	0.008	0.04	0.2	1	5	25	125

13. a.

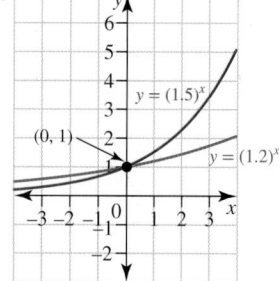

b. Increasing the value of a makes the graph steeper for positive x-values and flatter for negative x-values.

14. a.

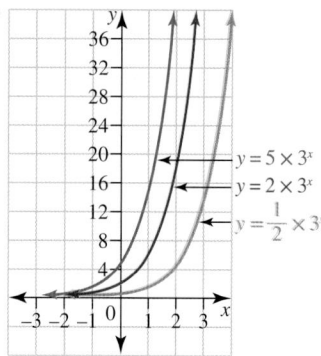

b. Increasing the value of k makes the graph steeper.

15. a.

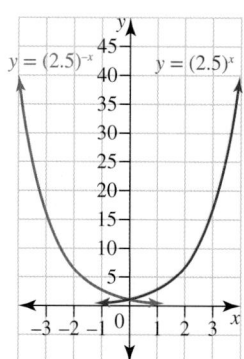

b. Changing the sign of the index reflects the graph in the y-axis.

16. a.

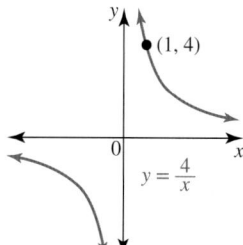

b.

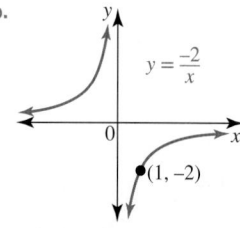

17.

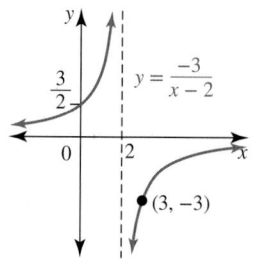

18. Sample responses can be found in the worked solutions in the online resources. Possible answer is $y = \dfrac{1}{x+3}$.

19. a.

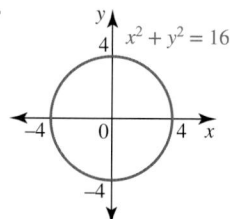

b.

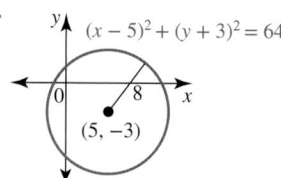

20. a.

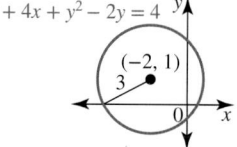

b.

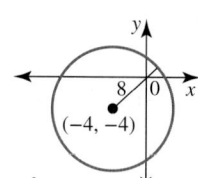

21. $x^2 + y^2 = 36$

22. a. $\left(\dfrac{5 - \sqrt{37}}{2}, \dfrac{19 - 5\sqrt{37}}{2}\right), \left(\dfrac{5 + \sqrt{37}}{2}, \dfrac{19 + 5\sqrt{37}}{2}\right)$

b. $\left(-\dfrac{2}{5}, -5\right)$ and $(1, 2)$

c. $\left(\dfrac{15 - \sqrt{43}}{26}, \dfrac{-3 - 5\sqrt{43}}{26}\right), \left(\dfrac{15 + \sqrt{43}}{26}, \dfrac{-3 + 5\sqrt{43}}{26}\right)$

23. a.

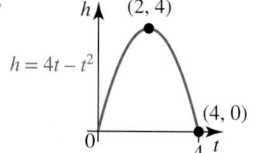

b. 4 m

c. 2 seconds

d. 4 seconds

24. a.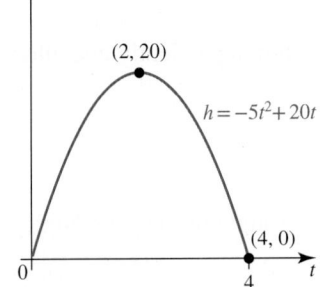

b. 4 seconds

c. 2 seconds $(1 < t < 3)$

d. The ball is never above a height of 20 m.

25. a. $[0, 12]$

b. 32 m

c. 11:41 am to 6:19 pm.

26. a. First ripple's radius is 3 cm, second ripple's radius is 15 cm.

b.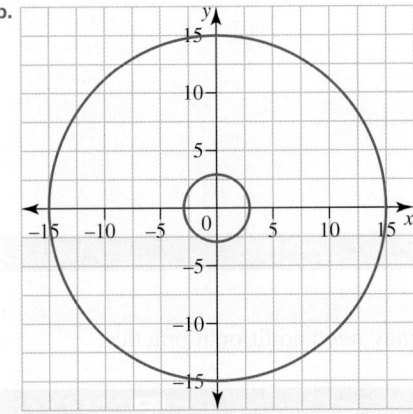

c. 2.4 cm/s

d. 1 minute 22.1 seconds after it is dropped.

27. a. Exponential, $y = 3 \times 2^x$

b. Straight line, $y = 2 - 2x$

c. Parabola, $y = (x - 3)^2$

d. Circle, $x^2 + (y - 1)^2 = 4$

e. Hyperbola, $y = \dfrac{2}{x}$

28. a. Sample responses can be found in the worked solutions in the online resources.

b. When $x = 0.3$, $b = 10.7$. Therefore if p is greater than 10.7 cm the platform would hit the bridge.

Semester review 1

The learnON platform is a powerful tool that enables students to complete revision independently and allows teachers to set mixed and spaced practice with ease.

Student self-study

Review the **Course Content** to determine which topics and lessons you studied throughout the year. Notice the green bubbles showing which elements were covered.

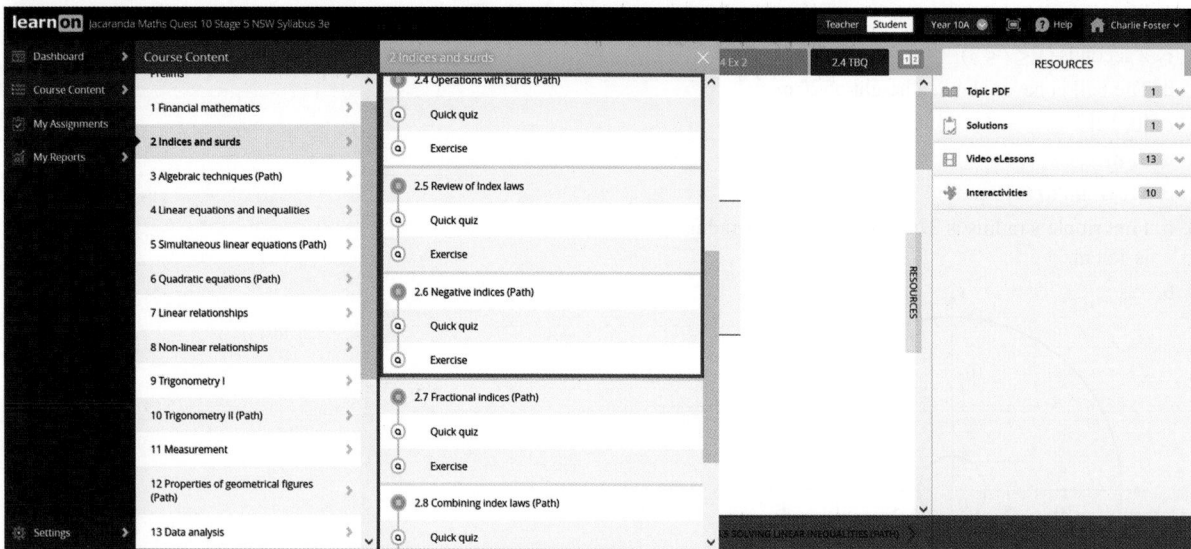

Review your results in **My Reports** and highlight the areas where you may need additional practice.

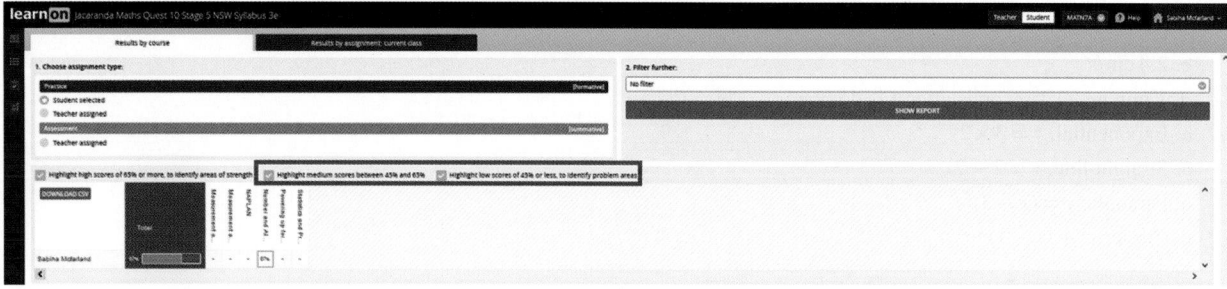

Use these and other tools to help identify areas of strengths and weakness and target those areas for improvement.

Teachers

It is possible to set questions that span multiple topics. These assignments can be given to individual students, to groups or to the whole class in a few easy steps.

Go to **Menu** and select **Assignments** and then **Create Assignment**. You can select questions from one or many topics simply by ticking the boxes as shown below.

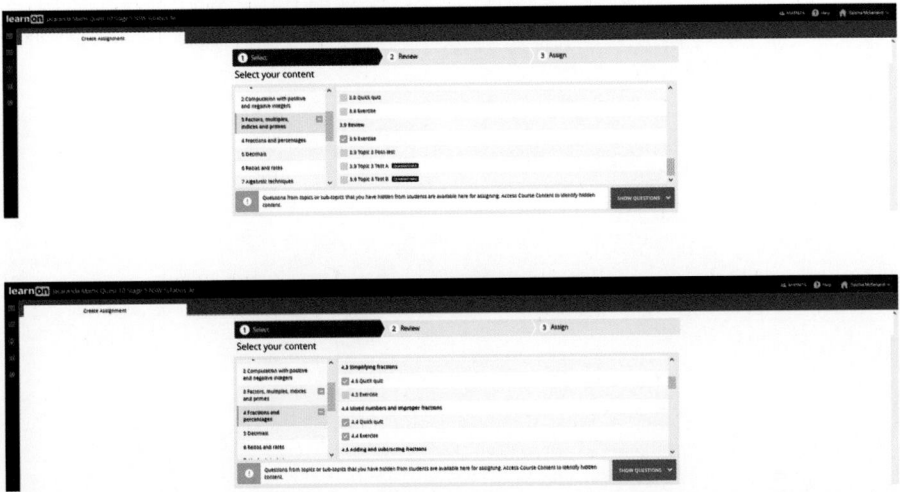

Once your selections are made, you can assign to your whole class or subsets of your class, with individualised start and finish times. You can also share with other teachers.

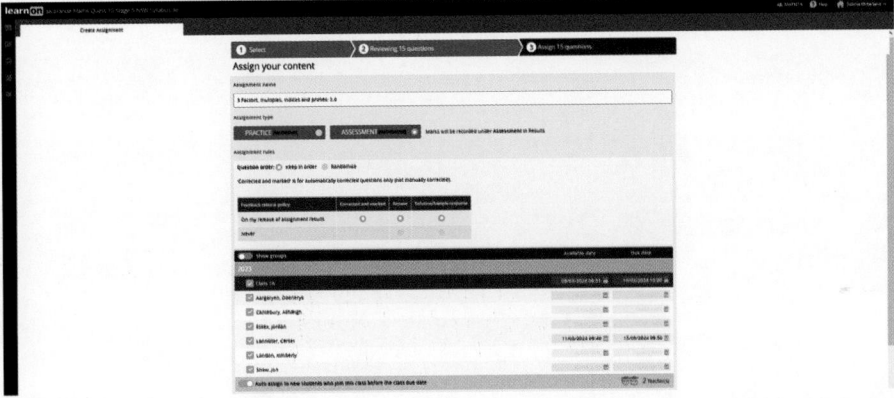

More instructions and helpful hints are available at www.jacplus.com.au.

9 Trigonometry I

LESSON SEQUENCE

LESSON
9.1 Overview

Why learn this?

Nearly 2000 years ago Ptolemy of Alexandria published the first book of trigonometric tables, which he used to chart the heavens and plot the courses of the Moon, stars and planets. He also created geographical charts and provided instructions on how to create maps.

The word trigonometry is derived from Greek words 'trigonon' and 'metron' meaning *triangles* and *measure* respectively, and reportedly has been studied since the third century BCE. This field of mathematics was studied across the world, with major discoveries made in India, China, Greece and Persia, to name a few. The works ranged from developing relationships, axioms and proofs to its application to everyday use and life.

Trigonometry is the branch of mathematics that makes the whole universe more easily understood. The role and use of trigonometry in navigation in the early years were crucial and its application and study grew from there. Today, it is used in architecture, surveying, astronomy and, as previously mentioned, navigation. It also provides the foundation of the study of sound and light waves, resulting in its application in the areas of music manufacturing and composition, study of tides, radiology and many other fields.

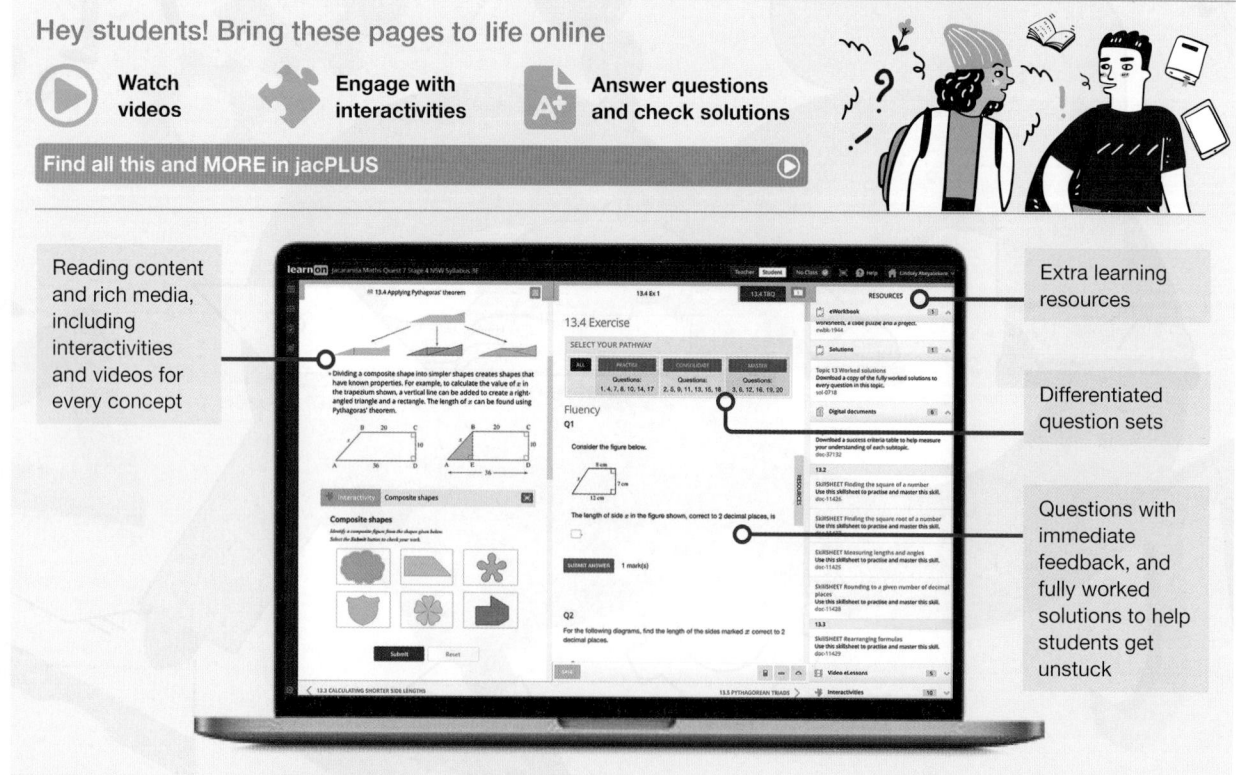

1. Determine the value of the pronumeral w, correct to 2 decimal places.

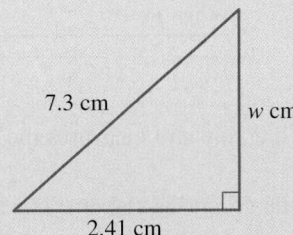

2. Determine the value of x, correct to 2 decimal places.

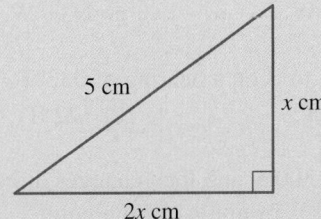

3. **PATH** A square-based pyramid is 16 cm high. Each sloping edge is 20 cm long. Calculate the length of the sides of the base, in cm correct to 2 decimal places.

4. **PATH** A cork is in the shape of a truncated cone; both the top and the base of the cork are circular.

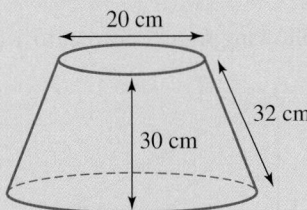

Calculate the sum of diameters of the top and the base. Give your answer in cm to 2 decimal places.

5. Evaluate $\sin(20°37')$ correct to 4 decimal places.

6. Calculate the size of the angle θ, correct to the nearest minute, given that $\cos(\theta) = 0.5712$. Give your answer in degrees and minutes.

7. Determine the size of the angle θ, correct to the nearest minute.

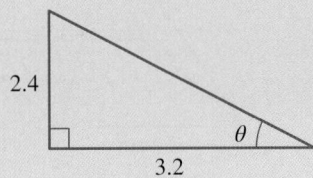

8. Calculate y, correct to 1 decimal place.

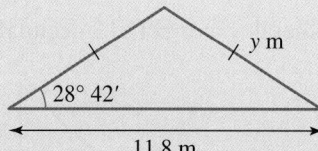

9. **MC** Tyler is standing 12 m from a flagpole and measures the angle of elevation from his eye level to the top of the pole as 62°.
The distance from Tyler's eyes to the ground is 185 cm. The height of the flagpole correct to 2 decimal places is:
A. 20.71 m   B. 22.56 m   C. 24.42 m   D. 207.56 m

10. Change each of the following compass bearings to true bearings.
a. N20°E   b. S47°W   c. N33°W   d. S17°E

11. **MC** A boat travels 15 km from A to B on a bearing of 032°T. The bearing from B to A is:
A. 032°T   B. 058°T   C. 122°T   D. 212°T

12. A bushwalker travels N50°W for 300 m and then changes direction 220°T for 0.5 km. Determine how many metres west the bushwalker is from his starting point.
Give your answer in km correct to 1 decimal place.

13. **MC** P and Q are two points on a horizontal line that are 120 metres apart. If the angles of elevation from P and Q to the top of the mountain are 34°5′ and 41°16′ respectively, the height of the mountain correct to 1 decimal place is:
A. 81.2 metres   B. 105.3 metres   C. 120.5 metres   D. 354.7 metres

14. Determine the value of x in the following figure, correct to 1 decimal place.

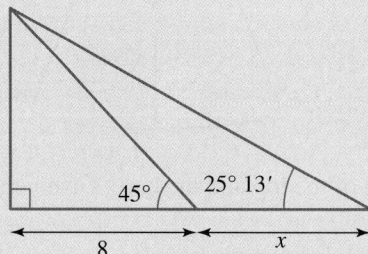

15. **PATH** **MC** In a right square-based pyramid, the square base has a length of 7.2 cm. If the angle between the triangular face and the base is 55°, the angle the sloping edge makes with the base is:

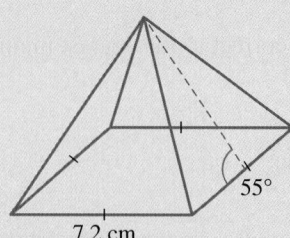

A. 45.3°   B. 55°   C. 63.4°   D. 47.7°

LESSON
9.2 Pythagoras' theorem

LEARNING INTENTION

At the end of this lesson you should be able to:
- identify similar right-angled triangles when corresponding sides are in the same ratio and corresponding angles are congruent
- apply Pythagoras' theorem to calculate the third side of a right-angled triangle when two other sides are known.

▶ 9.2.1 Similar right-angled triangles

eles-4799

- Two similar right-angled triangles have the same angles when the corresponding sides are in the same ratio.
- The **hypotenuse** is the longest side of a right-angled triangle and is always the side that is opposite the right angle.
- The corresponding sides are in the same ratio.

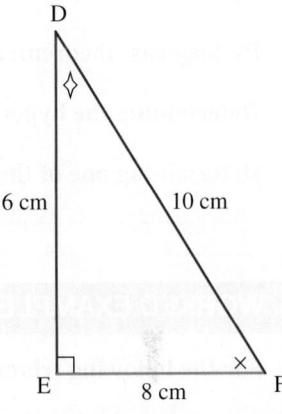

$$\frac{AB}{DE} = \frac{AC}{DF} = \frac{BC}{EF}$$

- To write this using the side lengths of the triangles gives:

$$\frac{AB}{DE} = \frac{3}{6} = \frac{1}{2}$$

$$\frac{AC}{DF} = \frac{5}{10} = \frac{1}{2}$$

$$\frac{BC}{EF} = \frac{4}{8} = \frac{1}{2}$$

This means that for right-angled triangles, when the angles are fixed, the ratios of the sides in the triangle are constant.

- We can examine this idea further by completing the following activity.
  Using a protractor and ruler, draw an angle of 70° measuring horizontal distances of 3 cm, 7 cm and 10 cm as demonstrated in the diagram below.
  Note: Diagram not drawn to scale.

  Measure the perpendicular heights a, b and c.

  $$a \approx 8.24 \text{ cm}, \ b \approx 19.23 \text{ cm}, \ c \approx 27.47 \text{ cm}$$

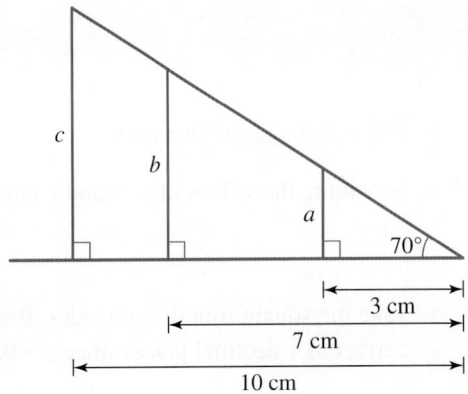

- To test if the theory for right-angled triangles, that when the angles are fixed the ratios of the sides in the triangle are constant, is correct, calculate the ratios of the side lengths.

$$\frac{a}{3} \approx \frac{8.24}{3} \approx 2.75, \quad \frac{b}{7} \approx \frac{19.23}{7} \approx 2.75, \quad \frac{c}{10} \approx \frac{27.47}{10} \approx 2.75$$

- The ratios are the same because the triangles are similar. This important concept forms the basis of trigonometry.

▶ 9.2.2 Review of Pythagoras' theorem

eles-4800

- Pythagoras' theorem gives us a way of finding the length of the third side in a right angle triangle, if we know the lengths of the two other sides.

Pythagoras' theorem

Pythagoras' theorem: $c^2 = a^2 + b^2$

Determining the hypotenuse: $c = \sqrt{a^2 + b^2}$

Determining one of the two shorter sides: $a = \sqrt{c^2 - b^2}$ or $b = \sqrt{c^2 - a^2}$

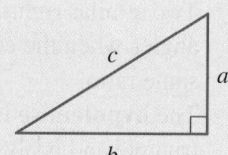

WORKED EXAMPLE 1 Calculating the hypotenuse

For the following triangle, calculate the length of the hypotenuse x, correct to 1 decimal place.

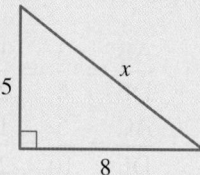

THINK	WRITE/DRAW
1. Copy the diagram and label the sides a, b and c. Remember to label the hypotenuse as c.	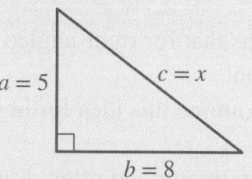
2. Write Pythagoras' theorem.	$c^2 = a^2 + b^2$
3. Substitute the values of a, b and c into this rule and simplify.	$x^2 = 5^2 + 8^2$ $= 25 + 64$ $= 89$
4. Take the square root of both sides. Round the positive answer correct to 1 decimal place, since $x > 0$.	$x = \pm\sqrt{89}$ $x \approx 9.4$

WORKED EXAMPLE 2 Calculating the length of the shorter side

Calculate the length, correct to 1 decimal place, of the unmarked side of the
following triangle.

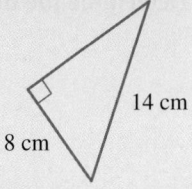

THINK

1. Copy the diagram and label the sides a, b and c. Remember
   to label the hypotenuse as c; it does not matter which side is
   a and which side is b.

2. Write Pythagoras' theorem.

3. Substitute the values of a, b and c into this rule and solve
   for a.

4. Evaluate a by taking the square root of both sides and
   round to 1 decimal place ($a > 0$).

WRITE/DRAW

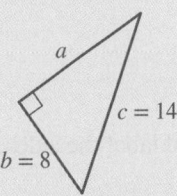

$$c^2 = a^2 + b^2$$

$$14^2 = a^2 + 8^2$$
$$196 = a^2 + 64$$
$$a^2 = 196 - 64$$
$$= 132$$

$$a = \pm\sqrt{132}$$
$$\approx 11.5 \text{ cm}$$

WORKED EXAMPLE 3 Solving a practical problem using Pythagoras' theorem

A ladder that is 5.5 m long leans up against a vertical wall. The foot of the ladder is 1.5 m from the
wall. Determine how far up the wall the ladder reaches. Give your answer in metres correct to
1 decimal place.

THINK

1. Draw a diagram and label the sides a, b and c. Remember to
   label the hypotenuse as c.

2. Write Pythagoras' theorem.

3. Substitute the values of a, b and c into this rule
   and simplify.

4. Evaluate a by taking the square root of 28. Round to
   1 decimal place, $a > 0$.

5. Write the answer in a sentence.

WRITE/DRAW

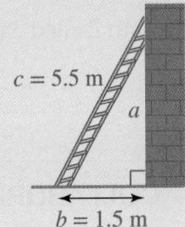

$$c^2 = a^2 + b^2$$

$$5.5^2 = a^2 + 1.5^2$$
$$30.25 = a^2 + 2.25$$
$$a^2 = 30.25 - 2.25$$
$$= 28$$

$$a = \pm\sqrt{28}$$
$$\approx 5.3$$

The ladder reaches 5.3 m up the wall.

WORKED EXAMPLE 4 Determining the unknown sides

Determine the unknown side lengths of the triangle, correct to 2 decimal places.

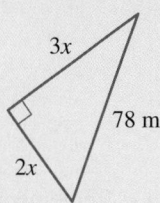

$3x$

78 m

$2x$

THINK	WRITE/DRAW
1. Copy the diagram and label the sides a, b and c.	$b = 3x$ $c = 78$ m $a = 2x$
2. Write Pythagoras' theorem.	$c^2 = a^2 + b^2$
3. Substitute the values of a, b and c into this rule and simplify.	$78^2 = (3x)^2 + (2x)^2$ $6084 = 9x^2 + 4x^2$ $6084 = 13x^2$
4. Rearrange the equation so that the pronumeral is on the left-hand side of the equation.	$13x^2 = 6084$
5. Divide both sides of the equation by 13.	$\dfrac{13x^2}{13} = \dfrac{6084}{13}$ $x^2 = 468$
6. Evaluate x by taking the square root of both sides. Round the answer correct to 2 decimal places.	$x = \pm\sqrt{468}$ ≈ 21.6333
7. Substitute the value of x into $2x$ and $3x$ to determine the lengths of the unknown sides.	$2x \approx 43.27$ m $3x \approx 64.90$ m

---

COMMUNICATING — COLLABORATIVE TASK: History of Pythagoras' theorem

Pythagoras' theorem was known about before the age of Pythagoras. In small groups, research which other civilisations knew about the theory and construct a timeline as a class for its history.

 Resources

Interactivities Finding a shorter side (int-3845)
Finding the hypotenuse (int-3844)

9.2 Quick quiz on	9.2 Exercise

Individual pathways

■ PRACTISE	■ CONSOLIDATE	■ MASTER
1, 5, 7, 10, 11, 14, 17, 21, 22, 25	2, 3, 8, 12, 15, 19, 20, 23, 26	4, 6, 9, 13, 16, 18, 24, 27

Fluency

1. **WE1** For each of the following triangles, calculate the length of the hypotenuse, giving answers correct to 2 decimal places.

   a.
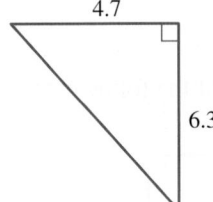

   b.
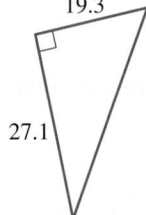

   c.
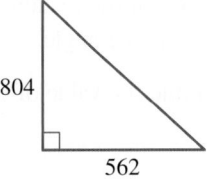

2. For each of the following triangles, calculate the length of the hypotenuse, giving answers correct to 2 decimal places.

   a.
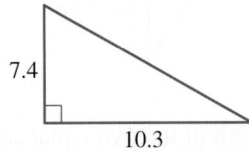

   b.
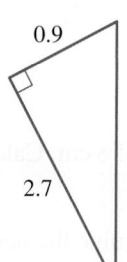

   c.
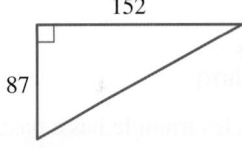

3. **WE2** Determine the value of the pronumeral, correct to 2 decimal places.

   a.
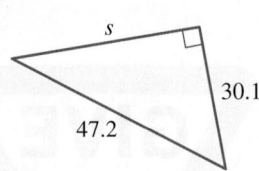

   b.
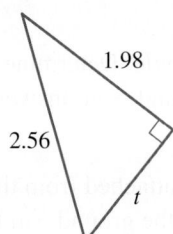

   c.
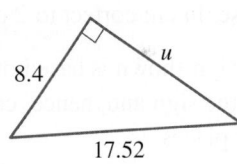

4. Determine the value of the pronumeral, correct to 2 decimal places.

   a.
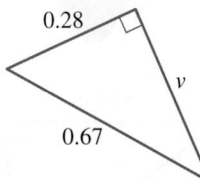

   b.
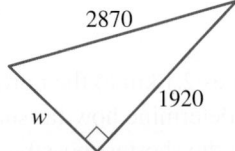

   c.
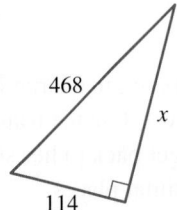

5. **WE3** The diagonal of a rectangular NO SMOKING sign is 34 cm. If the height of this sign is 25 cm, calculate the width of the sign, in cm correct to 2 decimal places.

6. A right-angled triangle has a base of 4 cm and a height of 12 cm. Calculate the length of the hypotenuse in cm correct to 2 decimal places.

7. Calculate the lengths of the diagonals (in cm to 2 decimal places) of squares that have side lengths of:

   a. 10 cm
   b. 17 cm
   c. 3.2 cm.

8. The diagonal of a rectangle is 90 cm. One side has a length of 50 cm. Determine, correct to 2 decimal places:

   a. the length of the other side
   b. the perimeter of the rectangle
   c. the area of the rectangle.

9. **WE4** Determine the value of the pronumeral, correct to 2 decimal places for each of the following.

a.
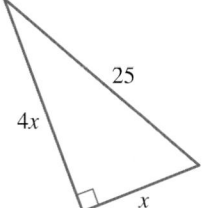

b.
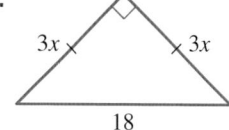

c.
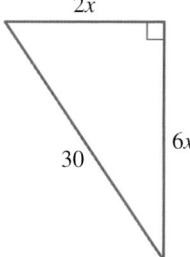

Understanding

10. An isosceles triangle has a base of 25 cm and a height of 8 cm. Calculate the length of the two equal sides, in cm correct to 2 decimal places.

11. An equilateral triangle has sides of length 18 cm. Determine the height of the triangle, in cm correct to 2 decimal places.

12. A right-angled triangle has a height of 17.2 cm, and a base that is half the height. Calculate the length of the hypotenuse, in cm correct to 2 decimal places.

13. The road sign shown is based on an equilateral triangle. Determine the height of the sign and, hence, calculate its area. Round your answers to 2 decimal places.

14. A flagpole, 12 m high, is supported by three wires, attached from the top of the pole to the ground. Each wire is pegged into the ground 5 m from the pole. Determine how much wire is needed to support the pole, correct to the nearest metre.

15. Sarah goes canoeing in a large lake. She paddles 2.1 km to the north, then 3.8 km to the west. Use the triangle shown to determine how far she must then paddle to get back to her starting point in the shortest possible way, in km correct to 2 decimal places.

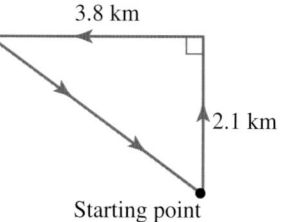

16. A baseball diamond is a square of side length 27 m. When a runner on first base tries to steal second base, the catcher has to throw the ball from home base to second base. Calculate the distance of the throw, in metres correct to 1 decimal place.

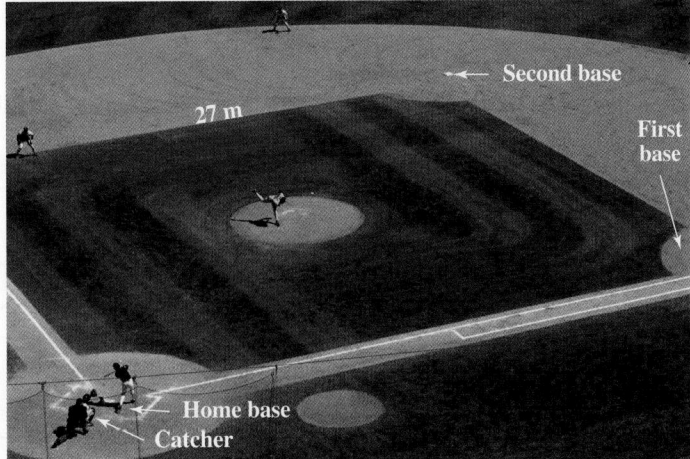

17. A rectangle measures 56 mm by 2.9 cm. Calculate the length of its diagonal in millimetres correct to 2 decimal places.

18. A rectangular envelope has a length of 24 cm and a diagonal measuring 40 cm. Calculate:

   a. the width of the envelope, correct to the nearest cm
   b. the area of the envelope, correct to the nearest cm^2.

19. A swimming pool is 50 m by 25 m. Peter is bored by his usual training routine, and decides to swim the diagonal of the pool. Determine how many diagonals he must swim to complete his normal distance of 1500 m.

20. A hiker walks 2.9 km north, then 3.7 km east. Determine how far in metres she is from her starting point. Give your answer in metres to 2 decimal places.

21. A square has a diagonal of 14 cm. Calculate the length of each side, in cm correct to 2 decimal places.

Communicating, reasoning and problem solving

22. The triangles below are right-angled triangles. Two possible measurements have been suggested for the hypotenuse in each case. For each triangle, complete calculations to determine which of the lengths is correct for the hypotenuse in each case. Show your working.

a.

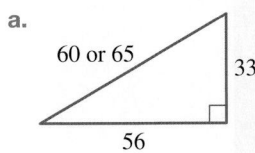

60 or 65
33
56

b.

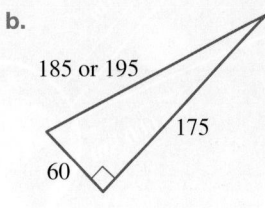

185 or 195
175
60

c.

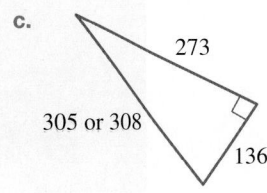

273
305 or 308
136

23. The square root of a number usually gives us both a positive and negative answer. Explain why we take only the positive answer when using Pythagoras' theorem.

24. Four possible side length measurements are 105, 208, 230 and 233. Three of them together produce a right-angled triangle.

 a. Explain which of the measurements could not be the hypotenuse of the triangle.
 b. Complete as few calculations as possible to calculate which combination of side lengths will produce a right-angled triangle.

25. The area of the rectangle MNPQ is 588 cm^2. Angles MRQ and NSP are right angles.

 a. Determine the integer value of x.
 b. Determine the length of MP.
 c. Calculate the value of y and hence determine the length of RS, in cm correct to 1 decimal place.

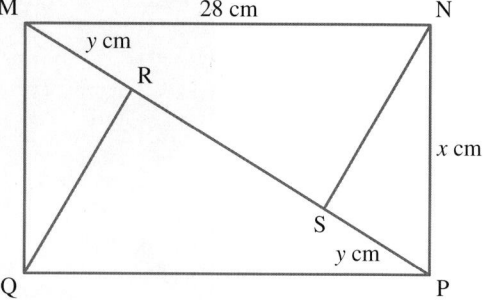

26. Triangle ABC is an equilateral triangle of side length x cm. Angles ADB and DBE are right angles. Determine the value of x in cm, correct to 2 decimal places.

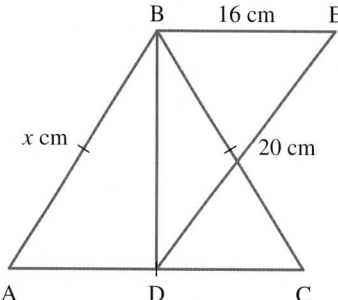

27. The distance from Earth to the Moon is approximately 385 000 km and the distance from Earth to the Sun is approximately 147 million kilometres.
In a total eclipse of the Sun, the moon moves between the Sun and Earth, thus blocking the light of the Sun from reaching Earth and causing a total eclipse of the Sun.
If the diameter of the Moon is approximately 3474 km, evaluate the diameter of the Sun. Express your answer to the nearest 10 000 km.

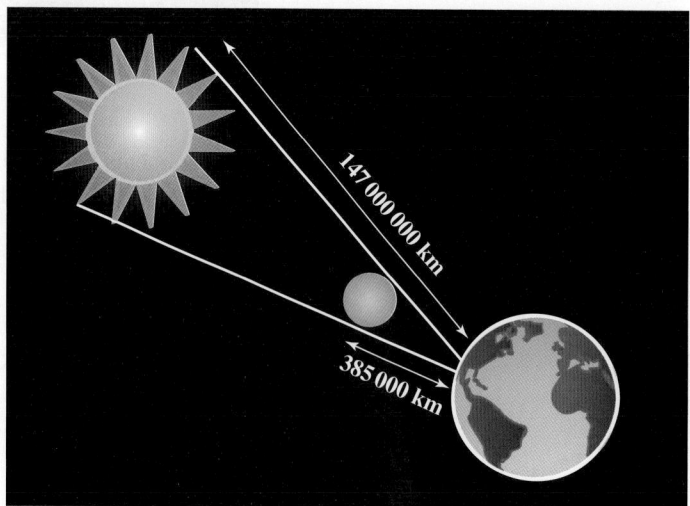

LESSON
9.3 Pythagoras' theorem in three dimensions (Path)

LEARNING INTENTION

At the end of this lesson you should be able to apply Pythagoras' theorem to:
- determine unknown lengths when a 3D diagram is given
- determine unknown lengths in situations by first drawing a diagram.

▶ 9.3.1 Applying Pythagoras' theorem in three dimensions

eles-4801

- Many real-life situations involve **3-dimensional** (3-D) objects: objects with length, width and height. Some common 3-D objects used in this section include cuboids, pyramids and right-angled wedges.

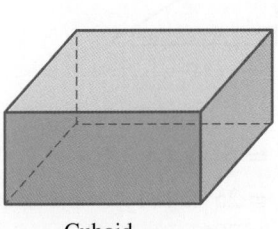

Cuboid

Pyramid

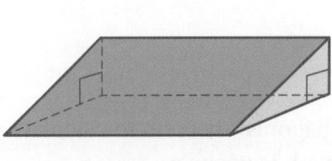

Right-angled wedge

- In diagrams of 3-D objects, right angles may not look like right angles, so it is important to redraw sections of the diagram in two dimensions, where the right angles can be seen accurately.

WORKED EXAMPLE 5 Applying Pythagoras' theorem to 3D objects

Determine the length AG in this rectangular prism (cuboid), in cm correct to 2 decimal places.

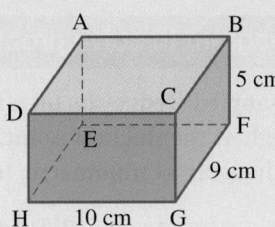

THINK	WRITE/DRAW
1. Draw the diagram in three dimensions. Draw the lines AG and EG. ∠AEG is a right angle.	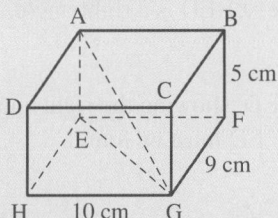
2. Draw △AEG, showing the right angle. Only 1 side is known, so EG must be found.	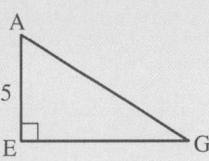

3. Draw EFGH in two dimensions and label the diagonal EG as x.

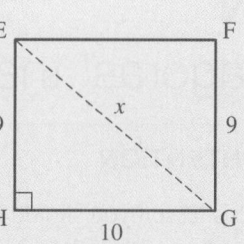

4. Use Pythagoras' theorem to calculate x where $x > 0$ since length is positive.

$$x^2 = 9^2 + 10^2$$
$$= 81 + 100$$
$$= 181$$
$$x = \sqrt{181}$$

5. Place this information on triangle AEG. Label the side AG as y.

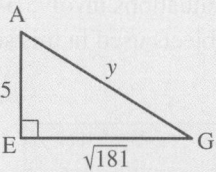

6. Use Pythagoras' theorem to calculate y where $y > 0$ since length is positive.

$$y^2 = 5^2 + \left(\sqrt{181}\right)^2$$
$$= 25 + 181$$
$$= 206$$
$$y = \pm\sqrt{206}$$
$$\approx 14.35$$

7. Write the answer in a sentence.

The length of AG is 14.35 cm.

WORKED EXAMPLE 6 Drawing a diagram to solve problems

A piece of cheese in the shape of a right-angled wedge sits on a table. It has a rectangular base measuring 14 cm by 8 cm, and is 4 cm high at the thickest point. An ant crawls diagonally across the sloping face. Determine how far, to the nearest millimetre, the ant walks.

THINK

1. Draw a diagram in three dimensions and label the vertices. Mark BD, the path taken by the ant, with a dotted line. $\angle BED$ is a right angle.

2. Draw $\triangle BED$, showing the right angle. Only one side is known, so ED must be found.

WRITE/DRAW

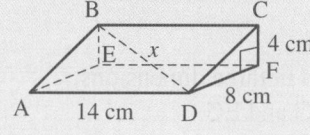

3. Draw EFDA in two dimensions, and label the diagonal ED. Label the side ED as x.

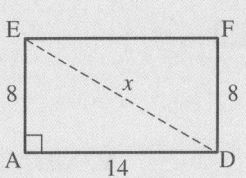

4. Use Pythagoras' theorem to calculate x where $x > 0$ since length is positive.

$$c^2 = a^2 + b^2$$
$$x^2 = 8^2 + 14^2$$
$$= 64 + 196$$
$$= 260$$
$$x = \sqrt{260}$$

5. Place this information on triangle BED. Label the side BD as y.

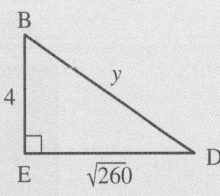

6. Use Pythagoras' theorem to calculate y where $y > 0$ since length is positive.

$$y^2 = 4^2 + \left(\sqrt{260}\right)^2$$
$$= 16 + 260$$
$$= 276$$
$$y = \sqrt{276}$$
$$\approx 16.61 \text{ cm}$$
$$\approx 166.1 \text{ mm}$$

7. Write the answer in a sentence.

The ant walks 166 mm, correct to the nearest millimetre.

DISCUSSION

Look around the room you are in. How many right angles can you spot in three-dimensional objects? Make a list of them and compare your list to that of another student.

---

on Resources

▶ **Video eLesson** Pythagoras' theorem in three dimensions (eles-1913)

✦ **Interactivity**   Right angles in 3-dimensional objects (int-6132)

Exercise 9.3 Pythagoras' theorem in three dimensions (Path) learn on

| 9.3 Quick quiz on | 9.3 Exercise |

Individual pathways

■ PRACTISE	■ CONSOLIDATE	■ MASTER
1, 4, 7, 11, 12, 15	2, 5, 8, 13, 16	3, 6, 9, 10, 14, 17

Where appropriate in this exercise, give answers correct to 2 decimal places.

Fluency

1. **WE5** Calculate the length of AG in each of the following figures.

a.

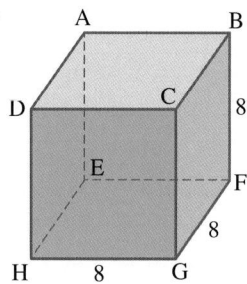

b.

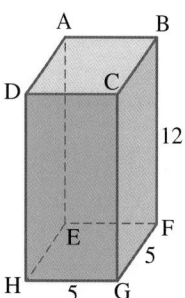

c.

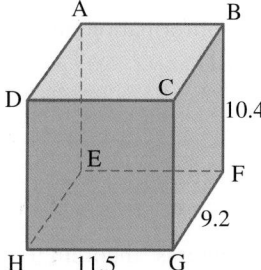

2. Consider the wedge shown. Calculate the length of CE in the wedge and, hence, obtain the length of AC .

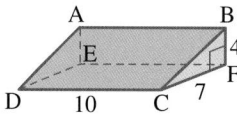

3. If $DC = 3.2\,m$, $AC = 5.8\,m$, and $CF = 4.5\,m$ in the figure, calculate the length of AD and BF.

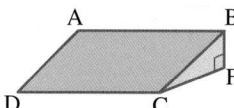

4. Consider the pyramid shown. Calculate the length of BD and, hence, the height of the pyramid.

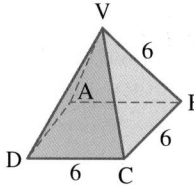

5. The sloping side of a cone is 16 cm and the height is 12 cm. Determine the length of the radius of the base.

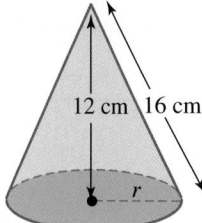

6. The pyramid ABCDE has a square base. The pyramid is 20 cm high. Each sloping edge measures 30 cm. Calculate the length of the sides of the base.

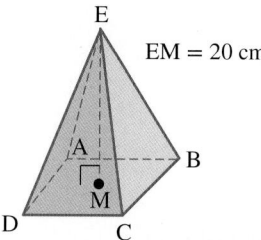

EM = 20 cm

Understanding

7. **WE6** A piece of cheese in the shape of a right-angled wedge sits on a table. It has a base measuring 20 mm by 10 mm, and is 4 mm high at the thickest point, as shown in the figure. A fly crawls diagonally across the sloping face. Determine how far, to the nearest millimetre, the fly walks.

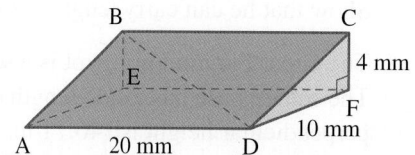

8. A 7 m high flagpole is in the corner of a rectangular park that measures 200 m by 120 m. Give your answers to the following questions correct to 2 decimal places.

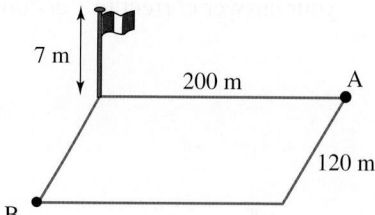

a. Calculate:
   i. the length of the diagonal of the park
   ii. the distance from A to the top of the pole
   iii. the distance from B to the top of the pole.

b. A bird flies from the top of the pole to the centre of the park. Calculate how far it flies.

9. A candlestick is in the shape of two cones, joined at the vertices as shown. The smaller cone has a diameter and sloping side of 7 cm, and the larger one has a diameter and sloping side of 10 cm. Calculate the total height of the candlestick.

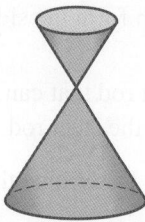

10. The total height of the shape below is 15 cm. Calculate the length of the sloping side of the pyramid.

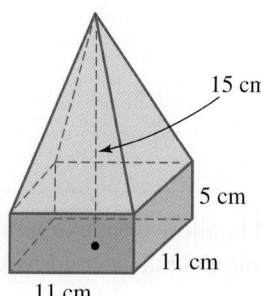

15 cm

5 cm

11 cm

11 cm

11. A sandcastle is in the shape of a truncated cone as shown. Calculate the length of the diameter of the base.

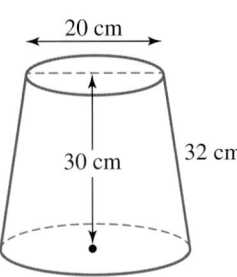

Communicating, reasoning and problem solving

12. Stephano is renovating his apartment, which he accesses through two corridors. The corridors of the apartment building are 2 m wide with 2-m-high ceilings, and the first corridor is at right angles to the second. Show that he can carry lengths of timber up to 6 m long to his apartment.

13. The Great Pyramid in Egypt is a square-based pyramid.
    The square base has a side length of 230.35 m and the perpendicular height is 146.71 m.
    Determine the slant height, s, of the great pyramid. Give your answer correct to 1 decimal place.

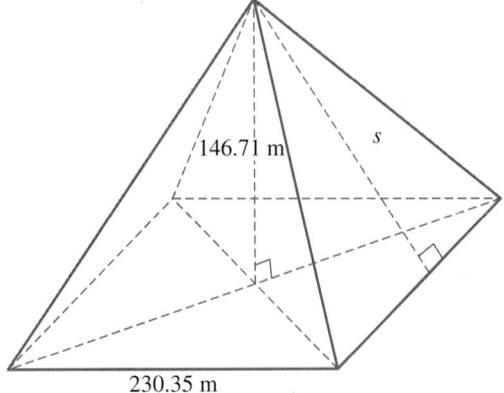

14. A tent is in the shape of a triangular prism, with a height of 140 cm as shown in the diagram. The width across the base of the door is 1 m and the tent is 2.5 m long.

    a. Calculate the length of each sloping side, in metres.
    b. Using your answer from part **a** calculate the area of fabric used in the construction of the sloping rectangles which form the sides. Show full working out.

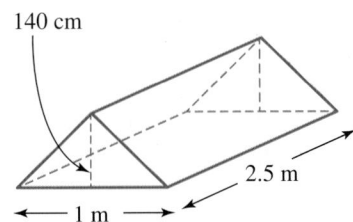

15. Determine the exact length of the longest steel rod that can sit inside a cuboid with dimensions 32 cm × 15 cm × 4 cm. Ignore the thickness of the steel rod.

16. Angles ABD, CBD and ABC are right angles. Determine the value of h, correct to 3 decimal places.

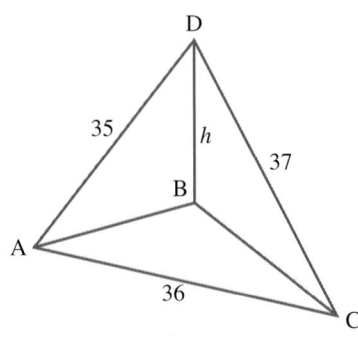

17. The roof of a squash centre is constructed to allow for maximum use of sunlight. Determine the value of h, giving your answer correct to 1 decimal place.

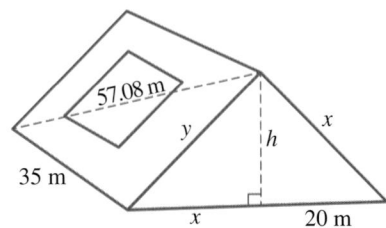

LESSON
9.4 Trigonometric ratios

LEARNING INTENTION

At the end of this lesson you should be able to:
- identify and define the trigonometric ratios for right-angled triangles and use trigonometric notation
- calculate the values of $\sin(\theta)$, $\cos(\theta)$ and $\tan(\theta)$ for different values of θ
- express angles in degrees, minutes and seconds.

▶ 9.4.1 Trigonometric ratios

eles-6274

- In a right-angled triangle, the longest side is called the hypotenuse.
- If one of the two acute angles is named (for example, θ), then the other two sides can also be given names, as shown in the following diagram.

Trigonometric ratios

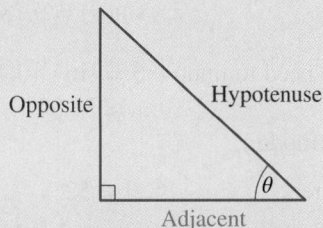

- **Using the diagram, the following three trigonometric ratios can be defined.**

  - **The sine ratio:**
  $$\text{sine}(\theta) = \frac{\text{length of Opposite}}{\text{length of Hypotenuse}}$$

  - **The cosine ratio:**
  $$\text{cosine}(\theta) = \frac{\text{length of Adjacent}}{\text{length of Hypotenuse}}$$

  - **The tangent ratio:**
  $$\text{tangent}(\theta) = \frac{\text{length of Opposite}}{\text{length of Adjacent}}$$

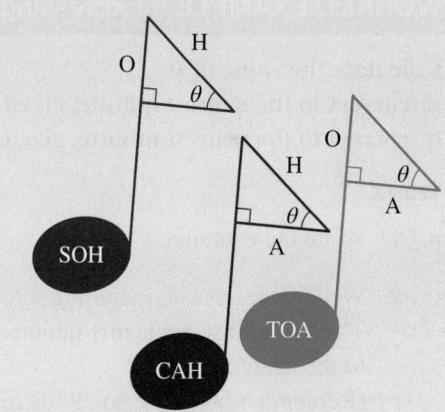

- **The names of the three ratios are usually shortened to $\sin(\theta)$, $\cos(\theta)$ and $\tan(\theta)$.**
- **The three ratios are often remembered using the mnemonic SOHCAHTOA, where SOH means $\sin(\theta)$ = Opposite over Hypotenuse and so on.**

Calculating trigonometric values using a calculator

- The sine, cosine and tangent of an angle have numerical values that can be found using a calculator.
- Traditionally angles were measured in **degrees**, minutes and seconds, where 60 seconds = 1 minute and 60 minutes = 1 degree. This is known as a sexagesimal system as the division are based on 60.
  For example, $50°33'48''$ means 50 degrees, 33 minutes and 48 seconds.
- In this topic, you will be asked to give answers in degrees and minutes only.

WORKED EXAMPLE 7 Calculating values (ratios) from angles

Calculate the value of each of the following, correct to **4 decimal places**, using a calculator.
(Remember to first work to 5 decimal places before rounding.)

a. $\cos(65°57')$

b. $\tan(56°45')$

THINK

a. Write your answer to the required number of decimal places.

b. Write your answer to the correct number of decimal places.

WRITE

a. $\cos(65°57') \approx 0.407\,53$
≈ 0.4075

b. $\tan(56°45') \approx 1.525\,253\,5$
≈ 1.5253

WORKED EXAMPLE 8 Calculating angles from ratios

Calculate the size of angle θ, correct to the nearest degree, given $\sin(\theta) = 0.7854$.

THINK

1. Write the given equation.

2. To calculate the size of the angle, we need to undo sine with its inverse, $\sin^{-1}$.
(Ensure your calculator is in degrees mode.)

3. Write your answer to the nearest degree.

WRITE

$\sin(\theta) = 0.7854$

$\theta = \sin^{-1}(0.7854)$
$\approx 51.8°$

$\theta \approx 52°$

WORKED EXAMPLE 9 Expressing angles in degrees and minutes

Calculate the value of θ:
a. correct to the nearest minute, given that $\cos(\theta) = 0.2547$
b. correct to the nearest minute, given that $\tan(\theta) = 2.364$.

THINK

a. 1. Write the equation.

2. Write your answer, including seconds.
There are 60 seconds in 1 minute. Round to the nearest minute.
(Remember $60'' = 1'$, so $39''$ is rounded up.)

b. 1. Write the equation.

2. Write the answer, rounding to the nearest minute.

WRITE

a. $\cos(\theta) = 0.2547$

$\cos^{-1}(0.2547) \approx 75°14'39''$
$\approx 75°15'$

b. $\tan(\theta) = 2.364$

$\tan^{-1} 2.364 \approx 67°4'15.8''$
$\approx 67°4'$

Write the equation that relates the two marked sides and the marked angle.

a.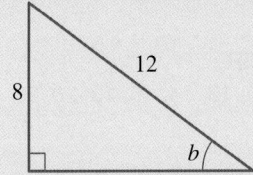

b.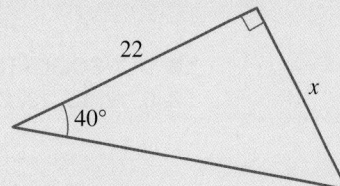

THINK	**WRITE/DRAW**
a. 1. Label the given sides of the triangle.	a. 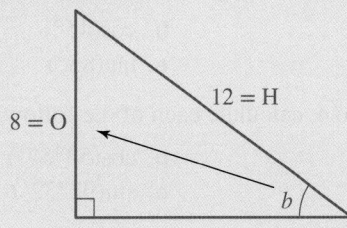
2. Write the ratio that contains O and H.	$\sin(\theta) = \dfrac{O}{H}$
3. Identify the values of the pronumerals.	$O = 8, H = 12$
4. Substitute the values of the pronumerals into the ratio and simplify the fraction. (Since the given angle is denoted with the letter b, replace θ with b.)	$\sin(b) = \dfrac{8}{12} = \dfrac{2}{3}$
b. 1. Label the given sides of the triangle.	b. 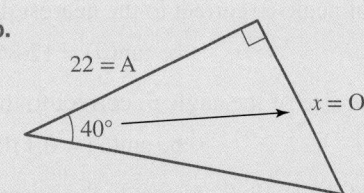
2. Write the ratio that contains O and A.	$\tan(\theta) = \dfrac{O}{A}$
3. Identify the values of the pronumerals.	$O = x, A = 22, \theta = 40°$
4. Substitute the values of the pronumerals into the ratio.	$\tan(40°) = \dfrac{x}{22}$

DISCUSSION

Can you think of a real-world situation where you could use trigonometric ratios to solve a problem?

 Resources

✦ **Interactivity** Trigonometric ratios (int-2577)

Exercise 9.4 Trigonometric ratios

| 9.4 Quick quiz on | 9.4 Exercise |

Individual pathways

■ PRACTISE	■ CONSOLIDATE	■ MASTER
1, 5, 7, 9, 11, 17, 22	2, 6, 12, 15, 18, 21	3, 4, 8, 10, 13, 14, 16, 19, 20, 23

Fluency

1. Calculate each of the following, correct to 4 decimal places.

    a. $\sin(30°)$
    b. $\cos(45°)$
    c. $\tan(25°)$
    d. $\sin(57°)$
    e. $\tan(83°)$
    f. $\cos(44°)$

WE7 For questions **2** to **4**, calculate each of the following, correct to 4 decimal places.

2. a. $\sin(40°30')$
   b. $\cos(53°57')$
   c. $\tan(27°34')$
   d. $\tan(123°40')$
   e. $\sin(92°32')$
   f. $\sin(42°8')$

3. a. $\cos(35°42')$
   b. $\tan(27°42')$
   c. $\cos(143°25')$
   d. $\sin(23°58')$
   e. $\cos(8°54')$

4. a. $\sin(286)°$
   b. $\tan(420°)$
   c. $\cos(845°)$
   d. $\sin(367°35')$

5. **WE8** Calculate the size of angle θ, correct to the nearest degree, for each of the following.

    a. $\sin(\theta) = 0.763$
    b. $\cos(\theta) = 0.912$
    c. $\tan(\theta) = 1.351$

6. Calculate the size of angle θ, correct to the nearest degree, for each of the following.

    a. $\cos(\theta) = 0.321$
    b. $\tan(\theta) = 12.86$
    c. $\cos(\theta) = 0.756$

7. **WE9a** Calculate the size of the angle θ, correct to the nearest minute.

    a. $\sin(\theta) = 0.814$
    b. $\sin(\theta) = 0.110$
    c. $\tan(\theta) = 0.015$

8. Calculate the size of the angle θ, correct to the nearest minute.

    a. $\cos(\theta) = 0.296$
    b. $\tan(\theta) = 0.993$
    c. $\sin(\theta) = 0.450$

9. **WE9b** Calculate the size of the angle θ, correct to the nearest minute.

    a. $\tan(\theta) = 0.5$
    b. $\cos(\theta) = 0.438$
    c. $\sin(\theta) = 0.9047$

10. Calculate the size of the angle θ, correct to the nearest minute.

    a. $\tan(\theta) = 1.1141$
    b. $\cos(\theta) = 0.8$
    c. $\tan(\theta) = 43.76.$

For questions **11** to **13**, calculate the value of each expression, correct to 3 decimal places.

11. a. $3.8 \cos(42°)$
    b. $118 \sin(37°)$
    c. $2.5 \tan(83°)$
    d. $\dfrac{2}{\sin(45°)}$

12. a. $\dfrac{220}{\cos(14°)}$
    b. $\dfrac{2 \cos(23°)}{5 \sin(18°)}$
    c. $\dfrac{12.8}{\tan(60°32')}$
    d. $\dfrac{18.7}{\sin(35°25')}$

13. a. $\dfrac{55.7}{\cos(89°21')}$
    b. $\dfrac{3.8 \tan(1°51')}{4.5 \sin(25°45')}$
    c. $\dfrac{2.5 \sin(27°8')}{10.4 \cos(83°2')}$
    d. $\dfrac{3.2 \cos(34°52')}{0.8 \sin(12°48')}$

14. For parts **a** to **f**, write an expression for:

   i. sine   **ii.** cosine   **iii.** tangent.

a.
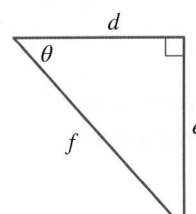

b.
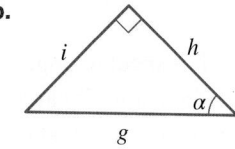

c.
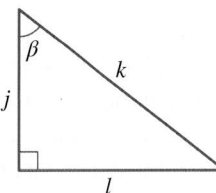

d.
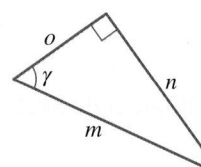

e.
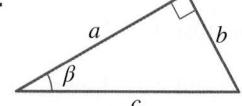

f.
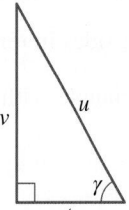

Understanding

15. [WE10] Write the equation that relates the two marked sides and the marked angle in each of the following triangles.

a.
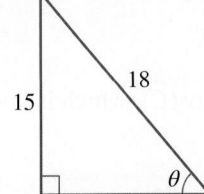

b.
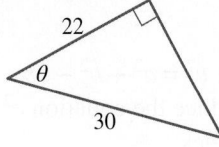

c.
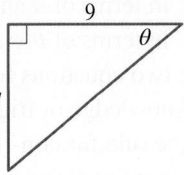

16. Write the equation that relates the two marked sides and the marked angle in each of the following triangles.

a.
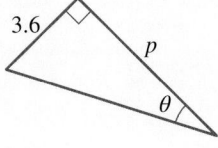

b.
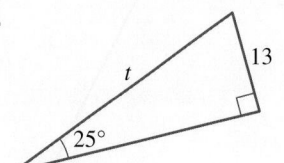

c.
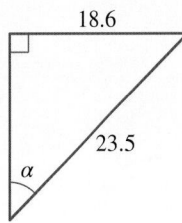

Communicating, reasoning and problem solving

17. Consider the right-angled triangle shown.

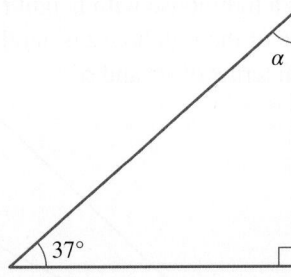

   a. Label each of the sides using the letters O, A and H with respect to the 37° angle.
   b. Determine the value of each trigonometric ratio. (Where applicable, answers should be given correct to 2 decimal places.)

   i. sin(37°)   **ii.** cos(37°)   **iii.** tan(37°)

18. Consider the right-angled triangle shown in question **17**.

    a. Determine the value of each of these trigonometric ratios, correct to 2 decimal places.

      i. $\sin(\alpha)$ **ii.** $\cos(\alpha)$ **iii.** $\tan(\alpha)$

    (*Hint:* First relabel the sides of the triangle with respect to angle α)

    b. What do you notice about the relationship between $\sin(37°)$ and $\cos(\alpha)$? Explain your answer.

    c. What do you notice about the relationship between $\sin(\alpha)$ and $\cos(37°)$? Explain your answer.

    d. Make a general statement about the two angles.

19. Using a triangle labelled with a, h and o, algebraically show that $\tan(\theta) = \dfrac{\sin(\theta)}{\cos(\theta)}$.

    (*Hint:* Write all the sides in terms of the hypotenuse.)

20. ABC is a scalene triangle with side lengths a, b and c as shown. Angles BDA and BDC are right angles.

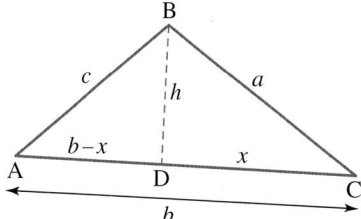

    a. Express h^2 in terms of a and x.

    b. Express h^2 in terms of b, c and x.

    c. Equate the two equations for h^2 to show that $c^2 = a^2 + b^2 - 2bx$.

    d. Use your knowledge of trigonometry to produce the equation $c^2 = a^2 + b^2 - 2ab\cos(C)$, which is known as the cosine rule for non-right-angled triangles.

21. Determine the length of the side DC in terms of x, y and θ.

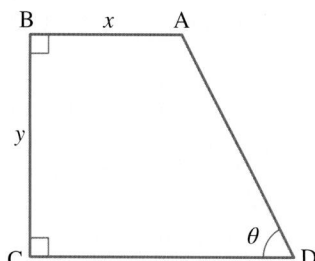

22. Explain how we determine whether to use sin, cos or tan in trigonometry questions.

23. From an observer on a boat 110 m away from a vertical cliff with height c, the angle from the base of the cliff to the top of the cliff is $\alpha°$. There is a lighthouse with height t on the cliff. From the observer, the angle from the base of the lighthouse to the top of the lighthouse is another $\theta°$ more than $\alpha°$.
Express the height of the lighthouse, t, in terms of $\theta°$ and $\alpha°$.

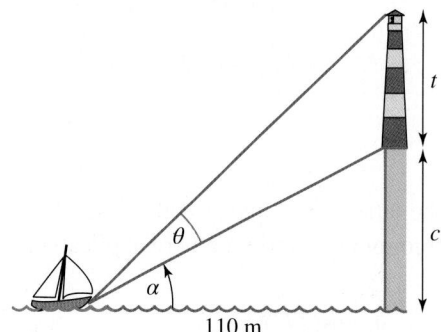

LESSON
9.5 Using trigonometry to calculate side lengths

LEARNING INTENTION

At the end of this lesson you should be able to:
- select and use appropriate trigonometric ratios to find the length of an unknown side when the length of one other side and an acute angle is known.

▶ 9.5.1 Using trigonometry to calculate side lengths

eles-4804

- When one acute angle and one side length are known in a right-angled triangle, this information can be used to find all other unknown sides or angles.

WORKED EXAMPLE 11 Using trigonometry to calculate side lengths

Calculate the value of each pronumeral, giving answers correct to 3 decimal places.

a.

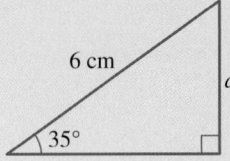

b.

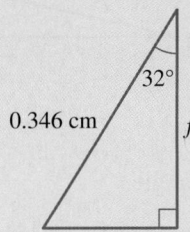

THINK	WRITE/DRAW

a. 1. Label the marked sides of the triangle.

a.

2. Identify the appropriate trigonometric ratio to use.

$$\sin(\theta) = \frac{O}{H}$$

3. Substitute $O = a$, $H = 6$ and $\theta = 35°$.

$$\sin(35°) = \frac{a}{6}$$

4. Make a the subject of the equation.

$$6\sin(35°) = a$$
$$a = 6\sin(35°)$$

5. Calculate and round the answer, correct to 3 decimal places.

$$a \approx 3.441 \text{ cm}$$

b. 1. Label the marked sides of the triangle.

b.

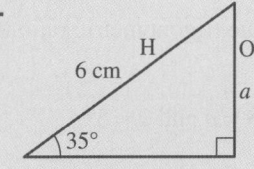

2. Identify the appropriate trigonometric ratio to use.

$$\cos(\theta) = \frac{A}{H}$$

3. Substitute $A = f$, $H = 0.346$ and $\theta = 32°$.

$$\cos(32°) = \frac{f}{0.346}$$

4. Make f the subject of the equation.

$$0.346 \cos(32°) = f$$
$$f = 0.346 \cos(32°)$$

5. Calculate and round the answer, correct to 3 decimal places.

$$f \approx 0.293 \text{ cm}$$

WORKED EXAMPLE 12 Using trigonometry to calculate side lengths

Calculate the value of the pronumeral in the triangle shown. Give the answer correct to 2 decimal places.

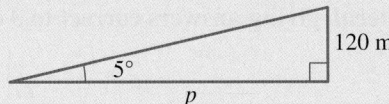

THINK

1. Label the marked sides of the triangle.

WRITE/DRAW

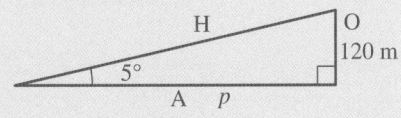

2. Identify the appropriate trigonometric ratio to use.

$$\tan(\theta) = \frac{O}{A}$$

3. Substitute $O = 120$, $A = p$ and $\theta = 5°$.

$$\tan(5°) = \frac{120}{p}$$

4. Make p the subject of the equation.
   i. Multiply both sides of the equation by p.
   ii. Divide both sides of the equation by $\tan(5°)$.

$$p \times \tan(5°) = 120$$
$$p = \frac{120}{\tan 5°}$$

5. Calculate and round the answer, correct to 2 decimal places.

$$p \approx 1371.61 \text{ m}$$

DISCUSSION

How does solving a trigonometric equation differ when we are finding the length of the hypotenuse side compared to when finding the length of a shorter side?

 Resources

 Interactivity Using trigonometry to calculate side lengths (int-6133)

9.5 Quick quiz on	9.5 Exercise

Individual pathways

■ PRACTISE	■ CONSOLIDATE	■ MASTER
1, 3, 7, 9, 12	2, 5, 8, 10, 13	4, 6, 11, 14, 15

Fluency

1. **WE11** Calculate the value of each pronumeral in each of the following, correct to 3 decimal places.

   a.
   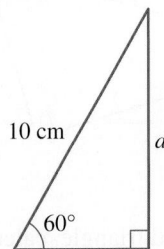

   b.
   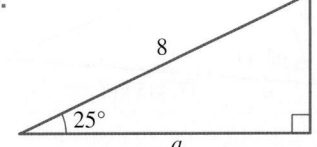

   c.
   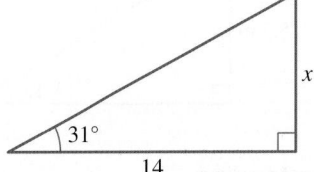

2. **WE12** Calculate the value of each pronumeral in each of the following, correct to 2 decimal places.

   a.
   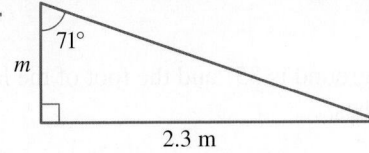

   b.
   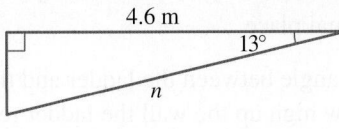

   c.
   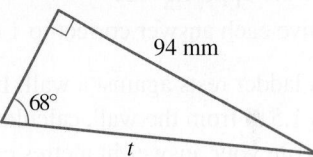

3. Determine the length of the unknown side in each of the following, correct to 2 decimal places.

   a.
   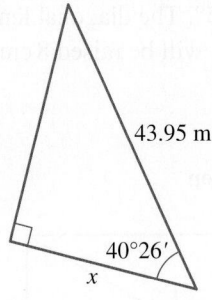

   b.
   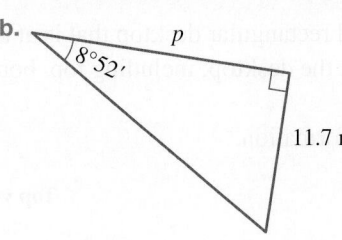

   c.
   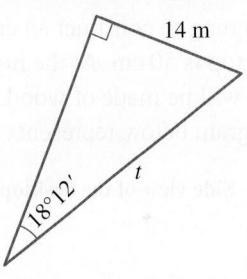

4. Determine the length of the unknown side in each of the following, correct to 2 decimal places.

   a.
   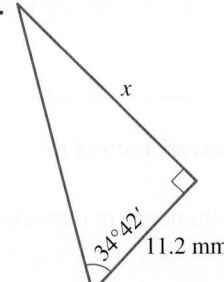

   b.
   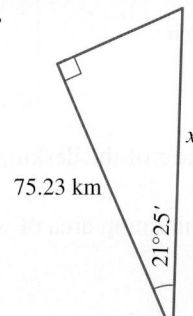

   c.
   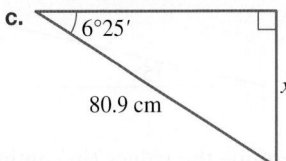

5. Calculate the value of the pronumeral in each of the following, correct to 2 decimal places.

a.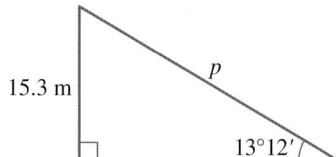
43.9 cm
x
46°

b.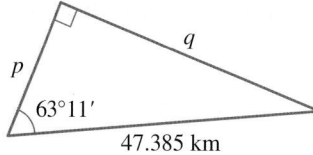
23.7 m
36°42′
y

c.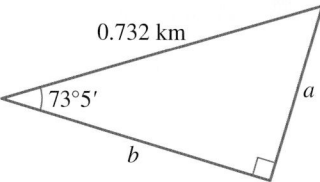
34°12′
z
12.3 m

6. Calculate the value of the pronumeral in each of the following, correct to 2 decimal places.

a.
15.3 m
p
13°12′

b.
p
q
63°11′
47.385 km

c.
0.732 km
73°5′
a
b

Understanding

7. Given that the angle θ is 42° and the length of the hypotenuse is 8.95 m in a right-angled triangle, calculate the length of:

   a. the opposite side

   b. the adjacent side.

   Give each answer correct to 1 decimal place.

8. A ladder rests against a wall. If the angle between the ladder and the ground is 35° and the foot of the ladder is 1.5 m from the wall, calculate how high up the wall the ladder reaches.
   Write your answer in metres correct to 2 decimal places.

Communicating, reasoning and problem solving

9. Tran is going to construct an enclosed rectangular desktop that is at an incline of 15°. The diagonal length of the desktop is 50 cm. At the high end, the desktop, including top, bottom and sides, will be raised 8 cm. The desktop will be made of wood.
   The diagram below represents this information.

Side view of the desktop

x
8 cm
15°
y

Top view of the desktop

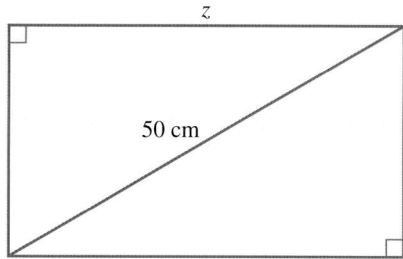
z
50 cm

   a. Determine the values (in centimetres) of x, y and z of the desktop. Write your answers correct to 2 decimal places.

   b. Using your answer from part **a** determine the minimum area of wood, in cm², Tran needs to construct his desktop including top, bottom and sides.
   Write your answer correct to 2 decimal places.

10. a. In a right-angled triangle, under what circumstances will the opposite side and the adjacent side have the same length?

    b. In a right-angled triangle, for what values of θ (the reference angle) will the adjacent side be longer than the opposite side?

11. In triangle ABC shown, the length of x correct to 2 decimal places is 8.41 cm. AM is perpendicular to BC. Jack found the length of x to be 5.11 cm. Below is his working. Identify his error and what he should have done instead.

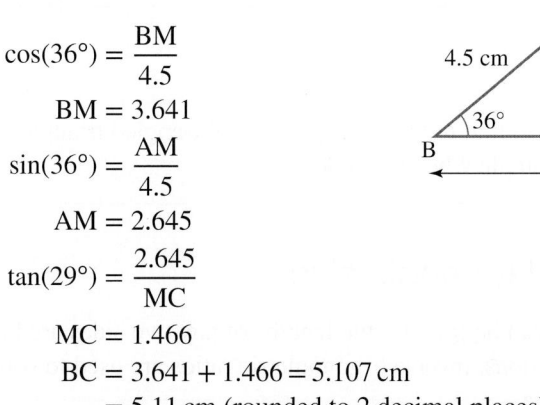

$$\cos(36°) = \frac{BM}{4.5}$$
$$BM = 3.641$$
$$\sin(36°) = \frac{AM}{4.5}$$
$$AM = 2.645$$
$$\tan(29°) = \frac{2.645}{MC}$$
$$MC = 1.466$$
$$BC = 3.641 + 1.466 = 5.107\,cm$$
$$= 5.11\,cm \text{ (rounded to 2 decimal places)}$$

12. A surveyor needs to determine the height of a building. She measures the angle of elevation of the top of the building from two points, 64 m apart. The surveyor's eye level is 195 cm above the ground.

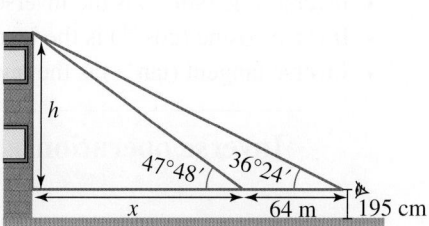

    a. Determine the expressions for the height of the building, h, in terms of x using the two angles.
    b. Solve for x by equating the two expressions obtained in part a. Give your answer to 2 decimal places.
    c. Determine the height of the building correct to 2 decimal places.

13. Building A and Building B are 110 m apart. From the base of Building A to the top of Building B, the angle is 15°. From the top of Building A looking down to the top of Building B, the angle is 22°. Evaluate the heights of each of the two buildings correct to 1 decimal place.

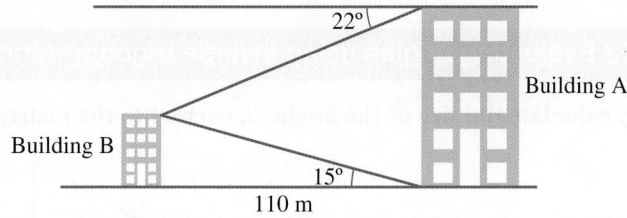

14. **PATH** If angles QNM, QNP and MNP are right angles, determine the length of NQ.

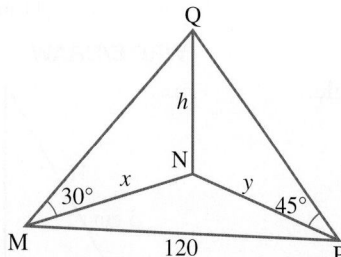

15. Determine how solving a trigonometric equation differs when we are calculating the length of the hypotenuse side compared to when determining the length of a shorter side.

LESSON
9.6 Using trigonometry to calculate angle size

LEARNING INTENTION

At the end of this lesson you should be able to:
- select and use appropriate trigonometric ratios to find unknown angles in right-angled triangles
- apply inverse operations to calculate the acute angle when two sides are given.

▶ 9.6.1 Using trigonometry to calculate angle size

eles-4805

- The size of any angle in a right-angled triangle can be found if the lengths of any two sides are known.
- Just as inverse operations are used to solve equations, inverse trigonometric ratios are used to solve trigonometric equations for the value of the angle.
  - Inverse sine $(\sin^{-1})$ is the inverse of sine.
  - Inverse cosine $(\cos^{-1})$ is the inverse of cosine.
  - Inverse tangent $(\tan^{-1})$ is the inverse of tangent.

Inverse operations

$$\text{If } \sin(\theta) = a, \text{ then } \sin^{-1}(a) = \theta.$$

$$\text{If } \cos(\theta) = a, \text{ then } \cos^{-1}(a) = \theta.$$

$$\text{If } \tan(\theta) = a, \text{ then } \tan^{-1}(a) = \theta.$$

For example, since $\sin(30°) = 0.5$, then $\sin^{-1}(0.5) = 30°$; this is read as 'inverse sine of 0.5 is 30 degrees'.

- A calculator can be used to calculate the values of inverse trigonometric ratios.

WORKED EXAMPLE 13 Evaluating angles using inverse trigonometric ratios

For each of the following, calculate the size of the angle, θ, correct to the nearest degree.

a.

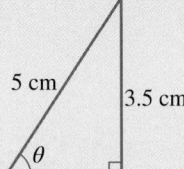

b.

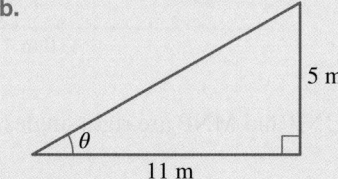

THINK	WRITE/DRAW
a. 1. Label the given sides of the triangle.	a.

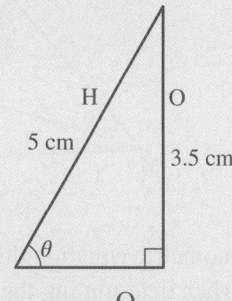

a. 2. Identify the appropriate trigonometric ratio to use. We are given O and H.

$$\sin(\theta) = \frac{O}{H}$$

---

3. Substitute $O = 3.5$ and $H = 5$ and evaluate the expression.

$$\sin(\theta) = \frac{3.5}{5}$$
$$= 0.7$$

4. Make θ the subject of the equation using inverse sine.

$$\theta = \sin^{-1}(0.7)$$
$$= 44.427\,004°$$

5. Evaluate θ and round the answer, correct to the nearest degree.

$$\theta \approx 44°$$

b. 1. Label the given sides of the triangle.

b.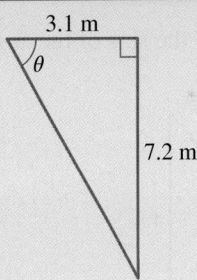

2. Identify the appropriate trigonometric ratio to use. Given O and A.

$$\tan(\theta) = \frac{O}{A}$$

3. Substitute $O = 5$ and $A = 11$.

$$\tan(\theta) = \frac{5}{11}$$

4. Make θ the subject of the equation using inverse tangent.

$$\theta = \tan^{-1}\left(\frac{5}{11}\right)$$
$$= 24.443\,954\,78°$$

5. Evaluate θ and round the answer, correct to the nearest degree.

$$\theta \approx 24°$$

WORKED EXAMPLE 14 Evaluating angles in degrees and minutes

Calculate the size of angle θ, correct to the nearest minute.

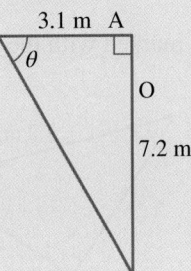

THINK

WRITE/DRAW

1. Label the given sides of the triangle.

2. Identify the appropriate trigonometric ratio to use.

$$\tan(\theta) = \frac{O}{A}$$

3. Substitute O = 7.2 and A = 3.1.

$$\tan(\theta) = \frac{7.2}{3.1}$$

4. Make θ the subject of the equation using inverse tangent.

$$\theta = \tan^{-1}\left(\frac{17.2}{3.1}\right)$$

5. Evaluate θ and write the calculator display.

$$\theta = 66.70543675°$$

6. Use the calculator to convert the answer to degrees, minutes and seconds.

$$\theta = 66°42'19.572''$$

7. Round the answer to the nearest minute.

$$\theta \approx 66°42'$$

 Resources

 Interactivity Finding the angle when two sides are known (int-6046)

Exercise 9.6 Using trigonometry to calculate angle size   **learn**

| 9.6 Quick quiz on | 9.6 Exercise |

Individual pathways

■ PRACTISE	■ CONSOLIDATE	■ MASTER
1, 4, 7, 9, 12	2, 5, 8, 10, 13	3, 6, 11, 14

Fluency

1. **WE13** Calculate the size of the angle, θ, in each of the following. Give your answer correct to the nearest degree.

a.
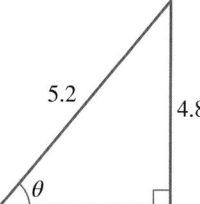

b.
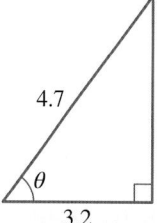

c.
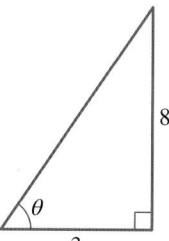

2. **WE14** Calculate the size of the angle marked with the pronumeral in each of the following. Give your answer correct to the nearest minute.

a.
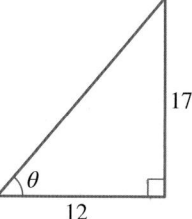

b.
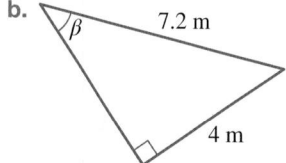

c.
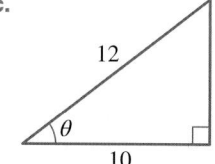

3. Calculate the size of the angle marked with the pronumeral in each of the following. Give your answer correct to the nearest minute.

a.
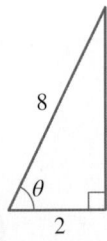

b.
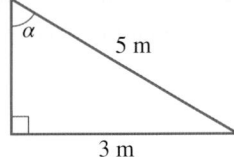

c.
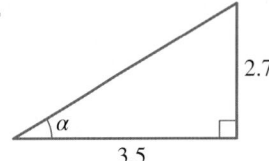

4. Calculate the size of the angle marked with the pronumeral in each of the following, giving your answer correct to the nearest degree.

a.
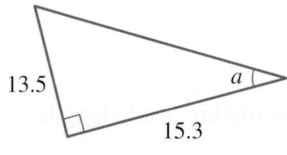

b.
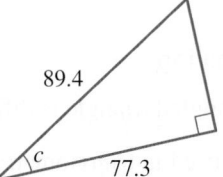

c.
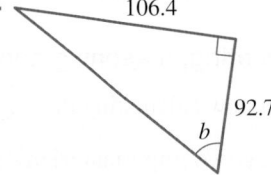

5. Calculate the size of the angle marked with the pronumeral in each of the following, giving your answer correct to the nearest degree.

a.
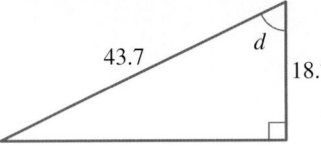

b.
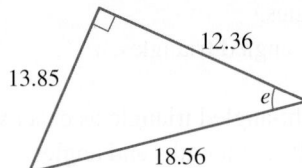

c.
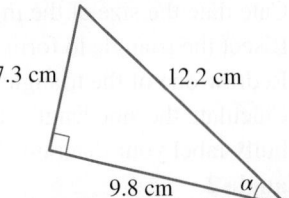

6. Calculate the size of each of the angles in the following, giving your answers correct to the nearest minute.

a.
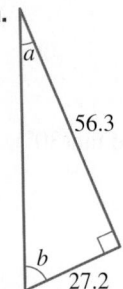

b.
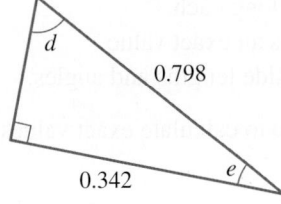

c.
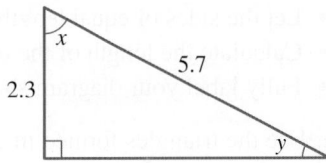

Understanding

7. Answer the following questions for the triangle shown.

a. Calculate the length of the sides r, l and h. Write your answers correct to 2 decimal places.

b. Calculate the area of ABC, correct to the nearest square centimetre.

c. Determine the size of $\angle$BCA.

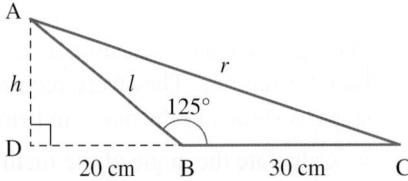

8. In the sport of air racing, small aeroplanes have to travel between two large towers (or pylons). The gap between a pair of pylons is smaller than the wing-span of the plane, so the plane has to go through on an angle with one wing higher than the other. The wing-span of a competition plane is 8 metres.

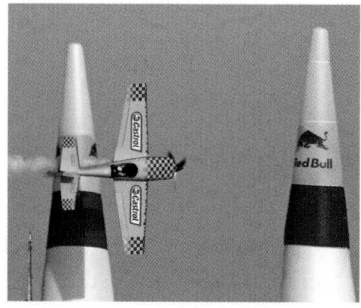

a. Determine the angle, correct to 1 decimal place, that the plane has to tilt if the gap between pylons is:

   i. 7 metres   ii. 6 metres   iii. 5 metres.

b. Because the plane has rolled away from the horizontal as it travels between the pylons it loses speed. If the plane's speed is below 96 km/h it will stall and possibly crash. For each degree of 'tilt' the speed of the plane is reduced by 0.98 km/h. Calculate the minimum speed the plane must go through each of the pylons in part **a**. Write your answer correct to 2 decimal places.

Communicating, reasoning and problem solving

9. Explain how calculating the angle of a right-angled triangle is different to calculating a side length.

10. There are two important triangles commonly used in trigonometry. Complete the following steps and answer the questions to create these triangles.
Triangle 1
 • Sketch an equilateral triangle with side length 2 units.
 • Calculate the size of the internal angles.
 • Bisect the triangle to form two right-angled triangles.
 • Redraw one of the triangles formed.
 • Calculate the side lengths of this right-angled triangle as exact values.
 • Fully label your diagram showing all side lengths and angles.
Triangle 2
 • Draw a right-angled isosceles triangle.
 • Calculate the sizes of the internal angles.
 • Let the sides of equal length be 1 unit long each.
 • Calculate the length of the third side as an exact value.
 • Fully label your diagram showing all side lengths and angles.

11. a. Use the triangles formed in question **10** to calculate exact values for $\sin(30°)$, $\cos(30°)$ and $\tan(30°)$. Justify your answers.

   b. Use the exact values for $\sin(30°)$, $\cos(30°)$ and $\tan(30°)$ to show that $\tan(30°) = \dfrac{\sin(30°)}{\cos(30°)}$.

   c. Use the formulas $\sin(\theta) = \dfrac{O}{H}$ and $\cos(\theta) = \dfrac{A}{H}$ to prove that $\tan(\theta) = \dfrac{\sin(\theta)}{\cos(\theta)}$.

12. During a Science excursion, a class visited an underground cave to observe rock formations. They were required to walk along a series of paths and steps as shown in the diagram below.

   a. Calculate the angle of the incline (slope) required to travel down between each site. Give your answers to the nearest whole number.

   b. Determine which path would have been the most challenging; that is, which path had the steepest slope.

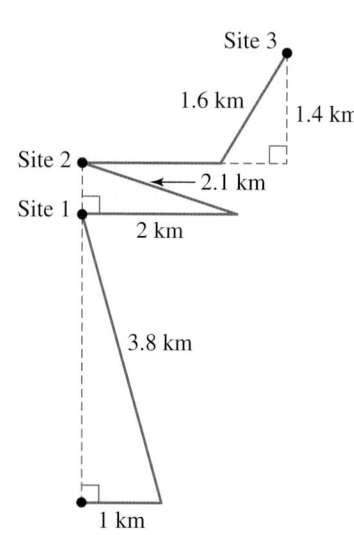

13. Determine the angle θ in degrees and minutes.

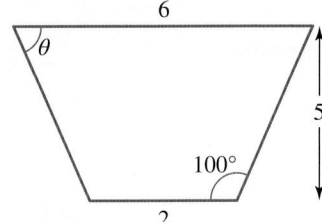

14. At midday, the hour hand and the minute hand on a standard clock are both pointing at the twelve. Calculate the angles the minute hand and the hour hand have moved 24.5 minutes later. Express both answers in degrees and minutes.

LESSON
9.7 Angles of elevation and depression

LEARNING INTENTION

At the end of this lesson you should be able to:
- identify angles of elevation and depression and solve for unknown side lengths and angles.

▶ 9.7.1 Angles of elevation and depression

- Solving real-life problems usually involves the person measuring angles or lengths from their position using trigonometry.
- They may have to either look up at the object or look down to it; hence the terms 'angle of elevation' and 'angle of depression' respectively.

Angle of elevation

- **If a horizontal line is drawn from A as shown, forming the angle θ, then θ is called the angle of elevation of B *from* A.**

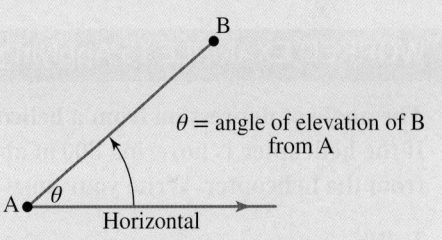

Angle of depression

- **If a horizontal line is drawn from B, forming the angle α, then α is called the angle of depression of A *from* B.**

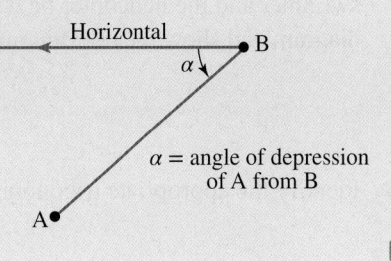

TOPIC 9 Trigonometry I **473**

Alternate angle rule

- **Because the horizontal lines are parallel, θ and α are the same size angle (alternate angles).**

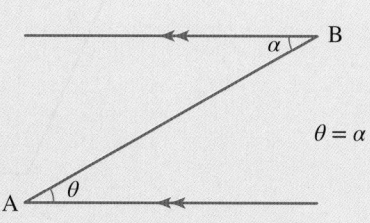

$\theta = \alpha$

WORKED EXAMPLE 15 Applying angles of elevation to solve problems

From a point P, on the ground, the angle of elevation of the top of a tree is 50°. If P is 8 metres from the tree, determine the height of the tree correct to 2 decimal places.

THINK	WRITE/DRAW
1. Let the height of the tree be h. Sketch a diagram and show the relevant information.	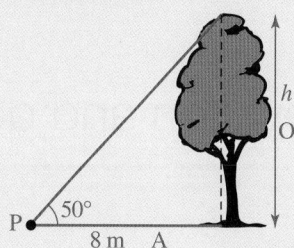
2. Identify the appropriate trigonometric ratio.	$\tan(\theta) = \dfrac{O}{A}$
3. Substitute $O = h$, $A = 8$ and $\theta = 50°$.	$\tan(50°) = \dfrac{h}{8}$
4. Rearrange to make h the subject.	$h = 8\tan(50°)$
5. Calculate and round the answer to 2 decimal places.	≈ 9.53
6. Write the answer in a sentence.	The height of the tree is 9.53 m.

WORKED EXAMPLE 16 Applying angles of depression to solve problems

The angle of depression from a helicopter, at point H, to a swimmer in distress in the water is 60°. If the helicopter is hovering 800 m above sea level, determine how far horizontally the swimmer is from the helicopter. Write your answer in metres correct to 2 decimal places.

THINK	WRITE/DRAW
1. Let the horizontal distance between the swimmer and the helicopter be d. Sketch a diagram and show the relevant information.	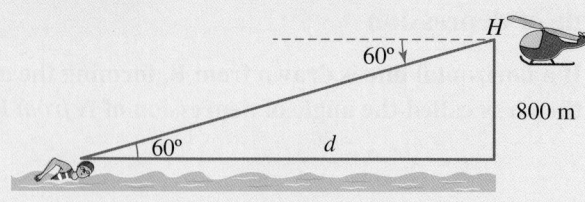
2. Identify the appropriate trigonometric ratio.	$\tan(\theta) = \dfrac{O}{A}$

3. Substitute O = 800, θ = 60° and A = d.	$\tan(60°) = \dfrac{800}{d}$
4. Rearrange to make d the subject.	$d = \dfrac{800}{\tan(60°)}$
5. Calculate and round to 2 decimal places.	$d \approx 461.88$ m
6. Write the answer in a sentence.	The horizontal distance between the swimmer and the helicopter is 461.88 m.

on Resources

❖ **Interactivity** Finding the angle of elevation and angle of depression (int-6047)

Exercise 9.7 Angles of elevation and depression   learn on

9.7 Quick quiz on   **9.7 Exercise**

Individual pathways

■ PRACTISE	■ CONSOLIDATE	■ MASTER
1, 4, 6, 9, 12, 15	2, 5, 7, 10, 13, 16	3, 8, 11, 14, 17

Fluency

1. **WE15** From a point P on the ground the angle of elevation from an observer to the top of a tree is 54°22'. If the tree is known to be 12.19 m high, determine how far P is from the tree (measured horizontally). Write your answer in metres correct to 2 decimal places.

2. **WE16** From the top of a cliff 112 m high, the angle of depression to a boat is 9°15'. Determine how far the boat is from the foot of the cliff. Write your answer in metres correct to 1 decimal place.

3. A person on a ship observes a lighthouse on the cliff, which is 830 metres away from the ship. The angle of elevation of the top of the lighthouse is 12°.

   a. Determine how far above sea level the top of the lighthouse is, correct to 2 decimal places.
   b. If the height of the lighthouse is 24 m, calculate the height of the cliff, correct to 2 decimal places.

4. At a certain time of the day a post, 4 m tall, casts a shadow of 1.8 m. Calculate the angle of elevation of the sun at that time. Write your answer correct to the nearest minute.

5. An observer who is standing 47 m from a building measures the angle of elevation of the top of the building as 17°. If the observer's eye is 167 cm from the ground, determine the height of the building. Write your answer in metres correct to 2 decimal places.

Understanding

6. A surveyor needs to determine the height of a building. She measures the angle of elevation of the top of the building from two points, 38 m apart. The surveyor's eye level is 180 cm above the ground.

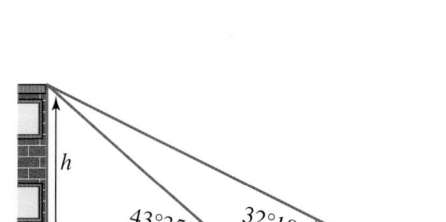

a. Determine two expressions for the height of the building, h, in terms of x using the two angles.

b. Solve for x by equating the two expressions obtained in **a**. Write your answer in metres correct to 2 decimal places

c. Determine the height of the building, in metres correct to 2 decimal places.

7. The height of another building needs to be determined but cannot be found directly. The surveyor decides to measure the angle of elevation of the top of the building from different sites, which are 75 m apart. The surveyor's eye level is 189 cm above the ground.

a. Determine two expressions for the height of the building above the surveyor's eye level, h, in terms of x using the two angles.

b. Using these two expressions, solve for x. Write your answer in metres correct to 2 decimal places.

c. Determine the height of the building, in metres correct to 2 decimal places.

8. A lookout tower has been erected on top of a cliff. At a distance of 5.8 km from the foot of the cliff, the angle of elevation to the base of the tower is 15.7° and to the observation deck at the top of the tower is 16° respectively, as shown in the figure below. Determine how high from the top of the cliff the observation deck is, to the nearest metre.

9. Elena and Sonja were on a camping trip to the Grampians, where they spent their first day hiking. They first walked 1.5 km along a path inclined at an angle of 10° to the horizontal. Then they had to follow another path, which was at an angle of 20° to the horizontal. They walked along this path for 1.3 km, which brought them to the edge of the cliff. Here Elena spotted a large gum tree 1.4 km away. If the gum tree is 150 m high, calculate the angle of depression from the top of the cliff to the top of the gum tree. Express your answer correct to the nearest degree.

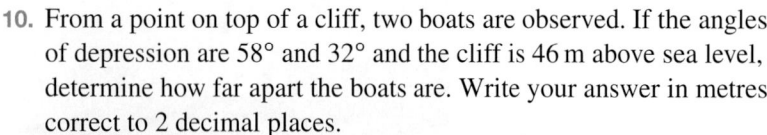

10. From a point on top of a cliff, two boats are observed. If the angles of depression are 58° and 32° and the cliff is 46 m above sea level, determine how far apart the boats are. Write your answer in metres correct to 2 decimal places.

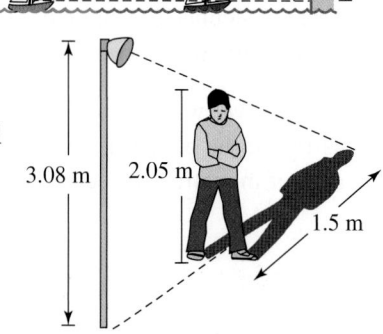

11. A 2.05 m tall man, standing in front of a street light 3.08 m high, casts a 1.5 m shadow.

a. Calculate the angle of elevation, to the nearest degree, from the ground to the source of light.

b. Determine how far the man is from the bottom of the light pole, in metres correct to 2 decimal places.

Communicating, reasoning and problem solving

12. Explain the difference between an angle of elevation and an angle of depression, and how they are related.

13. Joseph is asked to obtain an estimate of the height of his house using any mathematical technique. He decides to use an inclinometer and basic trigonometry. Using the inclinometer, Joseph determines the angle of elevation, θ, from his eye level to the top of his house to be 42°. The point from which Joseph measures the angle of elevation is 15 m away from his house and the distance from Joseph's eyes to the ground is 1.76 m.

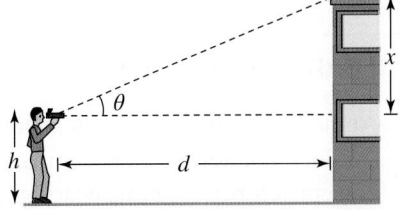

   a. Determine the values for the pronumerals h, d and θ.
   b. Determine the height of Joseph's house, in metres correct to 2 decimal places.

14. The angle of elevation of a vertically rising hot air balloon changes from 27° at 7:00 am to 61° at 7:03 am, according to an observer who is 300 m away from the take-off point.

   a. Assuming a constant speed, calculate that speed (in m/s and km/h) at which the balloon is rising, correct to 2 decimal places.
   b. The balloon then falls 120 metres. Determine the angle of elevation now. Write your answer in degrees correct to 1 decimal place.

15. The competitors of a cross-country run are nearing the finish line. From a lookout 100 m above the track, the angles of depression to the two leaders, Nathan and Rachel, are 40° and 62° respectively. Evaluate how far apart, to the nearest metre, the two competitors are.

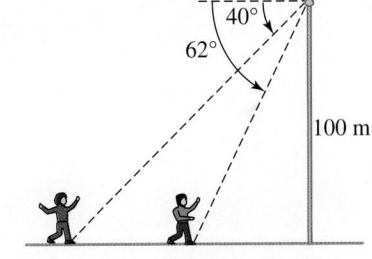

16. The angle of depression from the top of one building to the foot of another building across the same street and 45 metres horizontally away is 65°. The angle of depression to the roof of the same building is 30°. Evaluate the height of the shorter building. Write your answer in metres correct to 3 decimal places.

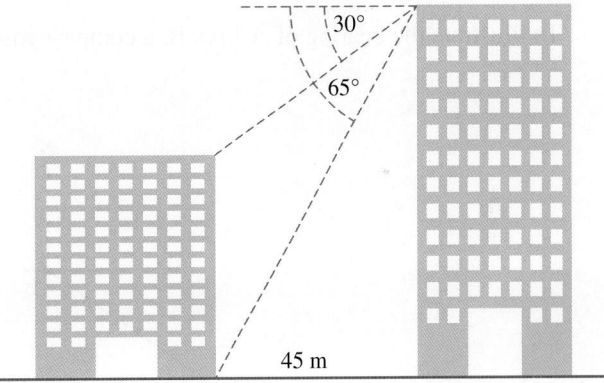

17. P and Q are two points on a horizontal line that are 120 metres apart. The angles of elevation from P and Q to the top of a mountain are 36° and 42° respectively. Determine the height of the mountain, in metres, correct to 1 decimal place.

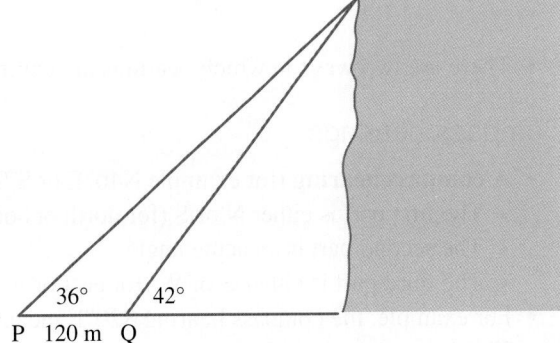

LESSON
9.8 Bearings

LEARNING INTENTION

At the end of this lesson you should be able to:
- identify, describe and compare true bearings and compass bearings
- convert between true bearings and compass bearings
- draw diagrams with correct angles to represent information to help solve triangles
- apply trigonometry to solve bearing problems involving compass and true bearings.

▶ 9.8.1 Using bearings

eles-4807

- A bearing gives the direction of travel from one point or object to another.
- The bearing of B from A tells how to get to B *from* A. A compass rose would be drawn at A.

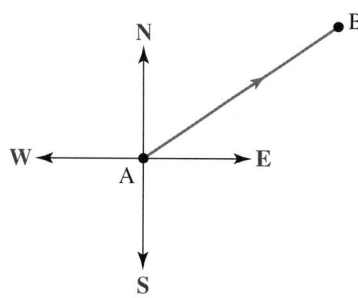

To illustrate the bearing of A *from* B, a compass rose would be drawn at B.

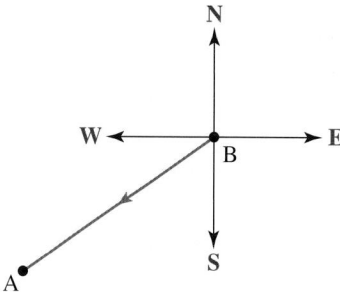

- There are two ways in which bearings are commonly written. They are compass bearings and true bearings.

Compass bearings

- A **compass bearing** (for example N40°E or S72°W) has three parts.
  - The first part is either N or S (for north or south).
  - The second part is an acute angle.
  - The third part is either E or W (for east or west).
- For example, the compass bearing S20°E means start by facing south and then turn 20° towards the east. This is the direction of travel.
  N40°W means start by facing north and then turn 40° towards the west.

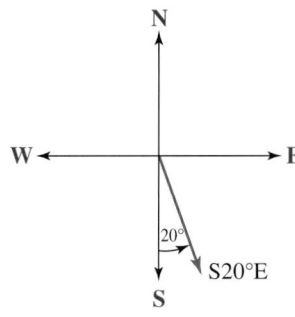

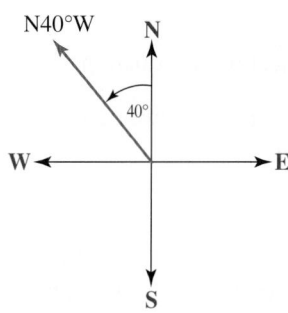

True bearings

- **True bearings** are measured from north in a clockwise direction and are expressed in 3 digits.
- The diagrams below show the bearings of 025° true and 250° true respectively. (These true bearings are more commonly written as 025°T and 250°T.)

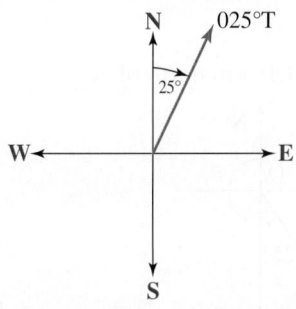

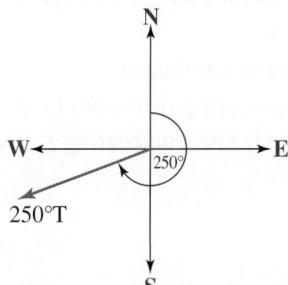

WORKED EXAMPLE 17 Solving trigonometric problems involving bearings

A boat travels a distance of 5 km from P to Q in a direction of 035°T.
a. Calculate how far Q is east of P, correct to 2 decimal places.
b. Calculate how far Q is north of P, correct to 2 decimal places.
c. Calculate the true bearing of P from Q.

THINK	WRITE/DRAW
a. 1. Draw a diagram showing the distance and bearing of Q from P. Complete a right-angled triangle travelling x km due east from P and then y km due north to Q.	a. 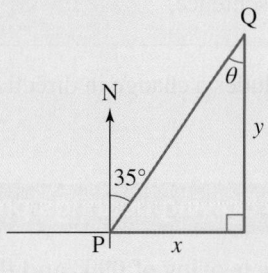
2. To determine how far Q is east of P, we need to determine the value of x. We are given the length of the hypotenuse (H) and need to find the length of the opposite side (O). Write the sine ratio.	$\sin(\theta) = \dfrac{O}{H}$
3. Substitute O $= x$, H $= 5$ and $\theta = 35°$.	$\sin(35°) = \dfrac{x}{5}$

4. Make x the subject of the equation. $x = 5 \sin(35°)$

5. Evaluate and round the answer, correct to ≈ 2.87
   2 decimal places.

6. Write the answer in a sentence.   Point Q is 2.87 km east of P.

b. 1. To determine how far Q is north of P, we   **b.** $\cos(\theta) = \dfrac{A}{H}$
      need to find the value of y. This can be
      done in several ways, namely: using the
      cosine ratio, the tangent ratio, or Pythagoras'
      theorem. Write the cosine ratio.

2. Substitute $A = y$, $H = 5$ and $\theta = 35°$. $\cos(35°) = \dfrac{y}{5}$

3. Make y the subject of the equation. $y = 5 \cos(35°)$

4. Evaluate and round the answer, correct to $y \approx 4.10$
   2 decimal places.

5. Write the answer in a sentence.   Point B is 4.10 km north of A.

c. 1. To determine the bearing of P from Q, draw a   **c.**
      compass rose at Q. The true bearing is given
      by $\angle\theta$.

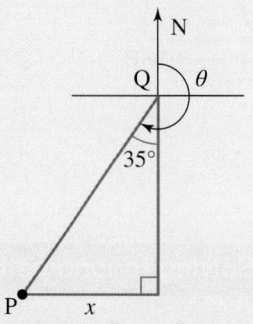

2. The value of θ is the sum of 180° (from north   True bearing $= 180° + \alpha$
   to south) and 35°. Write the value of θ. $\alpha = 35°$
                                                            True bearing $= 180° + 35°$
                                                                          $= 215°$

3. Write the answer in a sentence.   The bearing of P from Q is 215°T.

- Sometimes a journey includes a change in directions. In such cases, each section of the journey should be dealt with separately.

WORKED EXAMPLE 18 Solving bearings problems with 2 stages

A boy walks 2 km on a true bearing of 090° and then 3 km on a true bearing of 130°.
a. Calculate how far east of the starting point the boy is at the completion of his walk, correct to 1 decimal place.
b. Calculate how far south of the starting point the boy is at the completion of his walk, correct to 1 decimal place.
c. To return directly to his starting point, calculate how far the boy must walk and on what bearing. Write your answers in km correct to 2 decimal places and in degrees and minutes correct to the nearest minute.

THINK	WRITE/DRAW

a. 1. Draw a diagram of the boy's journey. The first leg of the journey is due east. Label the easterly component x and the southerly component y.

a.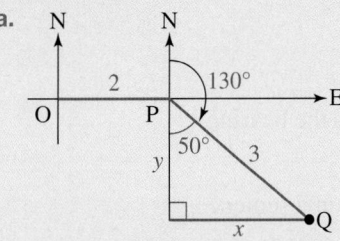

2. Write the ratio to determine the value of x.

$$\sin(\theta) = \frac{O}{H}$$

3. Substitute $O = x$, $H = 3$ and $\theta = 50°$.

$$\sin(50°) = \frac{x}{3}$$

4. Make x the subject of the equation.

$$x = 3\,\sin(50°)$$

5. Evaluate and round correct to 1 decimal place.

$$\approx 2.3\,\text{km}$$

6. Add to this the 2 km east that was walked in the first leg of the journey and write the answer in a sentence.

Total distance east = 2 + 2.3
= 4.3 km
The boy is 4.3 km east of the starting point.

b. 1. To determine the value of y (see the diagram in part **a**) we can use Pythagoras' theorem, as we know the lengths of two out of three sides in the right-angled triangle. Round the answer correct to 1 decimal place. *Note:* Alternatively, the cosine ratio could have been used.

b. Distance south = y km
$$a^2 = c^2 - b^2$$
$$y^2 = 3^2 - 2.3^2$$
$$= 9 - 5.29$$
$$= 3.71$$
$$y = \sqrt{3.71}$$
$$\approx 1.9\,\text{km}$$

2. Write the answer in a sentence.

The boy is 1.9 km south of the starting point.

c. 1. Draw a diagram of the journey and write in the results found in parts **a** and **b**. Draw a compass rose at Q.

c.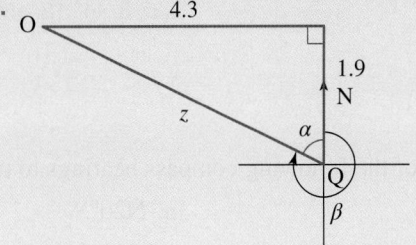

2. Determine the value of z using Pythagoras' theorem.

$$z^2 = 1.9^2 + 4.3^2$$
$$= 22.1$$
$$z = 22.1$$
$$\approx 4.70$$

3. Determine the value of α using trigonometry.

$$\tan(\alpha) = \frac{4.3}{1.9}$$

4. Make α the subject of the equation using the inverse tangent function.

$$\alpha = \tan^{-1}\left(\frac{4.3}{1.9}\right)$$

5. Evaluate and round to the nearest minute.

$$= 66.16125982°$$
$$= 66°9'40.535''$$
$$= 66°10'$$

6. The angle β gives the bearing.

$$\beta = 360° - 66°10'$$
$$= 293°50'$$

7. Write the answer in a sentence.

The boy travels 4.70 km on a bearing of 293°50' T.

DISCUSSION

Explain the difference between true bearings and compass directions.

 Resources

 Video eLesson Bearings (eles-1935)

Interactivity Bearings (int-6481)

Exercise 9.8 Bearings

learn on

9.8 Quick quiz	9.8 Exercise

Individual pathways

■ PRACTISE	■ CONSOLIDATE	■ MASTER
1, 3, 7, 10, 14, 17	2, 5, 8, 11, 12, 15, 18	4, 6, 9, 13, 16, 19

Fluency

1. Change each of the following compass bearings to true bearings.
   a. N20°E
   b. N20°W
   c. S35°W

2. Change each of the following compass bearings to true bearings.
   a. S28°E
   b. N34°E
   c. S42°W

3. Change each of the following true bearings to compass bearings.
   a. 049°T
   b. 132°T
   c. 267°T

4. Change each of the following true bearings to compass bearings.
   a. 330°T
   b. 086°T
   c. 234°T

5. Describe the following paths using true bearings.

a.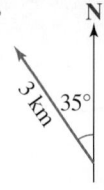

b.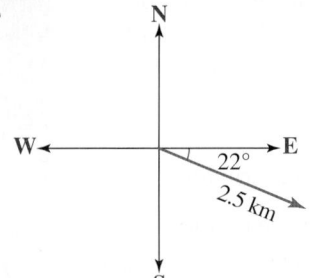

c.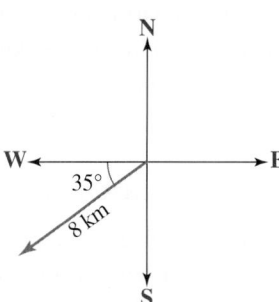

6. Describe the following paths using true bearings.

a.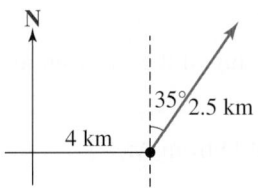

b.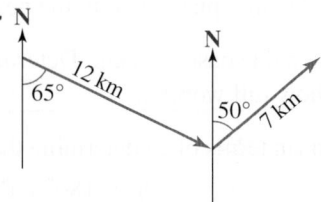

c.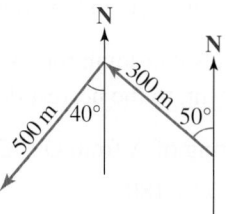

7. Show each of the following journeys as a diagram.

   a. A ship travels 040°T for 40 km and then 100°T for 30 km.
   b. A plane flies for 230 km in a direction 135°T and a further 140 km in a direction 240°T.

8. Show each of the following journeys as a diagram.

   a. A bushwalker travels in a direction 260°T for 0.8 km, then changes direction to 120°T for 1.3 km, and finally travels in a direction of 32° for 2.1 km.
   b. A boat travels N40°W for 8 km, then changes direction to S30°W for 5 km and then S50°E for 7 km.
   c. A plane travels N20°E for 320 km, N70°E for 180 km and S30°E for 220 km.

9. **WE17** A yacht travels 20 km from A to B on a bearing of 042°T.

   a. Calculate how far east of A is B, in km correct to 2 decimal places.
   b. Calculate how far north of A is B, in km correct to 2 decimal places.
   c. Calculate the bearing of A from B.

   The yacht then sails 80 km from B to C on a bearing of 130°T.

   d. Show the journey using a diagram.
   e. Calculate how far south of B is C, in km correct to 2 decimal places.
   f. Calculate how far east of B is C, in km correct to 2 decimal places.
   g. Calculate the bearing of B from C.

10. If a farmhouse is situated 220 m N35°E from a shed, calculate the true bearing of the shed from the house.

Understanding

11. A pair of hikers travel 0.7 km on a true bearing of 240° and then 1.3 km on a true bearing of 300°. Calculate how far west have they travelled from their starting point, in km correct to 3 decimal places.

12. **WE18** A boat travels 6 km on a true bearing of 120° and then 4 km on a true bearing of 080°

   a. Calculate how far east the boat is from the starting point on the completion of its journey, in km correct to 3 decimal places.
   b. Calculate how far south the boat is from the starting point on the completion of its journey, in km correct to 3 decimal places.
   c. Calculate the bearing of the boat from the starting point on the completion of its journey, correct to the nearest minute.

13. A plane flies on a true bearing of 320° for 450 km. It then flies on a true bearing of 350° for 130 km and finally on a true bearing of 050° for 330 km. Calculate how far north of its starting point the plane is. Write your answer in km correct to 2 decimal places.

Communicating, reasoning and problem solving

14. A bushwalker leaves her tent and walks due east for 4.12 km, then walks a further 3.31 km on a bearing of N20°E. If she wishes to return directly to her tent, determine how far she must walk and what bearing she should take.
Write your answers in km correct to 2 decimal places and to the nearest degree.

15. A car travels due south for 3 km and then due east for 8 km. Determine the bearing of the car from its starting point, to the nearest degree. Show full working.

16. If the bearing of A from O is θ°T, then (in terms of θ) determine the bearing of O from A:
   a. if 0° < θ° < 180°
   b. if 180° < θ° < 360°.

17. A boat sails on a compass direction of E12°S for 10 km then changes direction to S27°E for another 20 km. The boat then decides to return to its starting point.
   a. Determine how far, correct to 2 decimal places, the boat is from its starting point.
   b. Determine on what bearing should the boat travel to return to its starting point. Write the angle correct to the nearest degree.

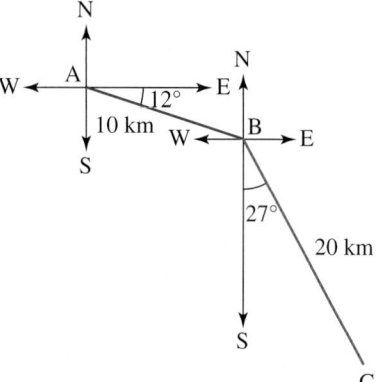

18. Samira and Tim set off early from the car park of a national park to hike for the day. Initially they walk N60°E for 12 km to see a spectacular waterfall. They then change direction and walk in a south-easterly direction for 6 km, then stop for lunch. Give all answers correct to 2 decimal places.

   a. Make a scale diagram of the hiking path they completed.
   b. Determine how far north of the car park they are at the lunch stop.
   c. Determine how far east of the car park they are at the lunch stop.
   d. Determine the bearing of the lunch stop from the car park.
   e. If Samira and Tim then walk directly back to the car park, calculate the distance they have covered after lunch.

19. Starting from their base in the national park, a group of bushwalkers travel 1.5 km at a true bearing of 030°, then 3.5 km at a true bearing of 160°, and then 6.25 km at a true bearing of 300°. Evaluate how far, and at what true bearing, the group should walk to return to its base.
Write your answers in km correct to 2 decimal places and to the nearest degree.

LESSON
9.9 Applications of trigonometry

LEARNING INTENTION

At the end of this lesson you should be able to:
- draw well-labelled diagrams to represent information
- apply trigonometry to solve various problems involving triangles.

▶ 9.9.1 Applications of trigonometry

eles-4808

- When applying trigonometry to practical situations, it is essential to draw good mathematical diagrams using points, lines and angles.
- Several diagrams may be required to show all the necessary right-angled triangles.

WORKED EXAMPLE 19 Applying trigonometry to solve problems

A ladder of length 3 m makes an angle of 32° with the wall.
a. Calculate how far the foot of the ladder is from the wall, in metres, correct to 2 decimal places.
b. Calculate how far up the wall the ladder reaches, in metres, correct to 2 decimal places.
c. Calculate the value of the angle the ladder makes with the ground.

THINK	WRITE/DRAW
Sketch a diagram and label the sides of the right-angled triangle with respect to the given angle.	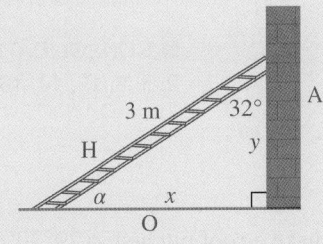
a. 1. We need to calculate the distance of the foot of the ladder from the wall (O) and are given the length of the ladder (H). Write the sine ratio.	**a.** $\sin(\theta) = \dfrac{O}{H}$
2. Substitute $O = x$, $H = 3$ and $\theta = 32°$.	$\sin(32°) = \dfrac{x}{3}$
3. Make x the subject of the equation.	$x = 3\sin(32°)$
4. Evaluate and round the answer to 2 decimal places.	$x \approx 1.59\,\text{m}$
5. Write the answer in a sentence.	The foot of the ladder is 1.59 m from the wall.
b. 1. We need to calculate the height the ladder reaches up the wall (A) and are given the hypotenuse (H). Write the cosine ratio.	**b.** $\cos(\theta) = \dfrac{A}{H}$
2. Substitute $A = y$, $H = 3$ and $\theta = 32°$.	$\cos(32°) = \dfrac{y}{3}$
3. Make y the subject of the equation.	$y = 3\cos(32°)$

4. Evaluate and round the answer to 2 decimal places. $\qquad$ $y \approx 2.54\,\text{m}$

5. Write the answer in a sentence. $\qquad$ The ladder reaches 2.54 m up the wall.

c. 1. To calculate the angle that the ladder makes with the ground, we could use any of the trigonometric ratios, as the lengths of all three sides are known. However, it is quicker to use the angle sum of a triangle.

$$
\begin{aligned}
\text{c. } \alpha + 90° + 32° &= 180° \\
\alpha + 122° &= 180° \\
\alpha &= 180° - 122° \\
\alpha &= 58°
\end{aligned}
$$

2. Write the answer in a sentence. $\qquad$ The ladder makes a 58° angle with the ground.

DISCUSSION

What are some real-life applications of trigonometry?

Exercise 9.9 Applications of trigonometry

learn on

| 9.9 Quick quiz on | 9.9 Exercise |

Individual pathways

■ PRACTISE	■ CONSOLIDATE	■ MASTER
1, 3, 4, 10, 11, 17	2, 5, 8, 12, 14, 16, 18	6, 7, 9, 13, 15, 19

Fluency

1. A carpenter wants to make a roof pitched at 29°30′, as shown in the diagram. Calculate how long, in metres correct to 2 decimal places, he should cut the beam PR.

2. The mast of a boat is 7.7 m high. A guy wire from the top of the mast is fixed to the deck 4 m from the base of the mast. Determine the angle, correct to the nearest minute, the wire makes with the horizontal.

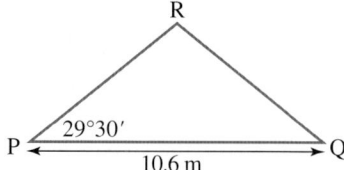

Understanding

3. A steel roof truss is to be made to the following design. Write your answers in metres correct to 2 decimal places.

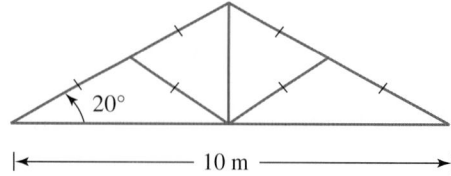

a. Calculate how high the truss is.

b. Determine the total length of steel required to make the truss.

4. **WE19** A ladder that is 2.7 m long is leaning against a wall at an angle of 20° as shown. If the base of the ladder is moved 50 cm further away from the wall, determine what angle the ladder will make with the wall. Write your answer correct to the nearest minute.

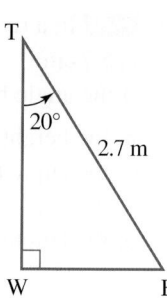

5. A wooden framework is built as shown.
   Bella plans to reinforce the framework by adding a strut from C to the midpoint of AB. Calculate the length of the strut, in metres correct to the 2 decimal places.

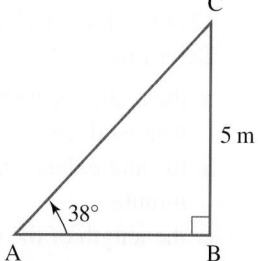

6. Atlanta is standing due south of a 20 m flagpole at a point where the angle of elevation of the top of the pole is 35°. Ginger is standing due east of the flagpole at a point where the angle of elevation of the top of the pole is 27°. Calculate how far, to the nearest metre, Ginger is from Atlanta.

7. From a point at ground level, Henry measures the angle of elevation of the top of a tall building to be 41°. After walking directly towards the building, he finds the angle of elevation to be 75°. If the building is 220 m tall, determine how far Henry walked between measurements. Write your answer correct to the nearest metre.

8. Sailing in the direction of a mountain peak of height 893 m, Imogen measured the angle of elevation to be 14°. A short time later the angle of elevation was 27°. Calculate how far, in km correct to 3 decimal places, Imogen had sailed in that time.

9. **PATH** A desk top of length 1.2 m and width 0.5 m rises to 10 cm.

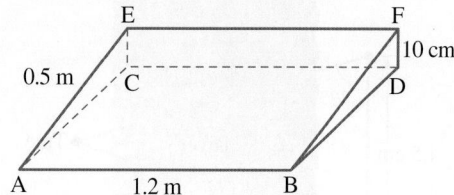

   Calculate, correct to the nearest minute:
   a. ∠DBF
   b. ∠CBE.

10. **PATH** A cuboid has a square end. If the length of the cuboid is 45 cm and its height and width are 25 cm each, calculate:

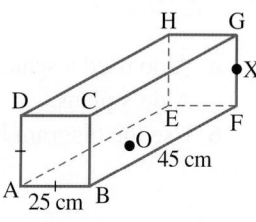

   a. the length of BD, correct to 2 decimal places
   b. the length of BG, correct to 2 decimal places
   c. the length of BE, correct to 2 decimal places
   d. the length of BH, correct to 2 decimal places
   e. ∠FBG, correct to the nearest minute
   f. ∠EBH, correct to the nearest minute.

   If the midpoint of FG is X and the centre of the rectangle ABFE is O calculate:

   g. the length OF, correct to 2 decimal places
   h. the length FX, correct to 1 decimal place
   i. ∠FOX, correct to the nearest minute
   j. the length OX, correct to 2 decimal places.

11. **PATH** In a right square-based pyramid, the length of the side of the square base is 5.7 cm.

If the angle between the triangular face and the base is 68°, calculate:

   a. the height of the pyramid, in cm correct to 2 decimal places

   b. the angle the sloping edge makes with the base, correct to the nearest minute

   c. the length of the sloping edge, in cm correct to 2 decimal places.

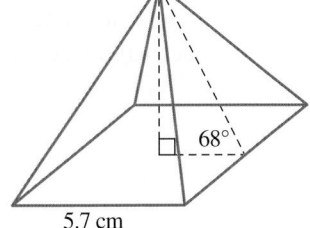

12. **PATH** In a right square-based pyramid, the length of the side of the base is 12 cm and the height is 26 cm.

Determine:

   a. the angle the triangular face makes with the base, correct to the nearest degree

   b. the angle the sloping edge makes with the base, correct to the nearest minute

   c. the length of the sloping edge, in cm correct to 2 decimal places.

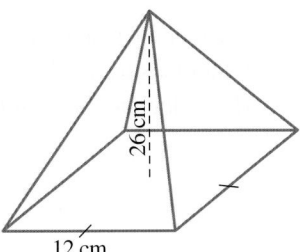

13. **PATH** In a right square-based pyramid, the height is 47 cm. If the angle between a triangular face and the base is 73°, calculate:

   a. the length of the side of the square base, in cm correct to 2 decimal places

   b. the length of the diagonal of the base, in cm correct to 2 decimal places

   c. the angle the sloping edge makes with the base, correct to the nearest minute.

Communicating, reasoning and problem solving

14. Explain whether sine of an acute angle can be 1 or greater.

15. Aldo the carpenter is lost in a rainforest. He comes across a large river and he knows that he can not swim across it. Aldo intends to build a bridge across the river. He draws some plans to calculate the distance across the river as shown in the diagram below.

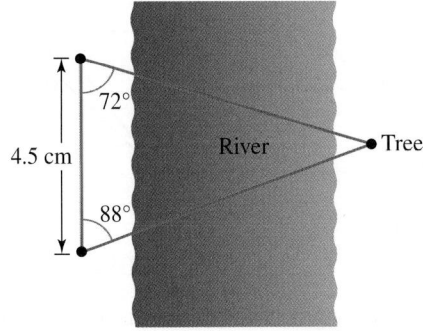

   a. Aldo used a scale of 1 cm to represent 20 m. Determine the real-life distance represented by 4.5 cm in Aldo's plans.

   b. Use the diagram below to write an equation for h in terms of d and the two angles.

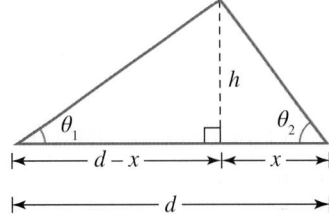

   c. Use your equation from part **b** to find the distance across the river, correct to the nearest metre.

16. **PATH** A block of cheese is in the shape of a rectangular prism as shown. The cheese is to be sliced with a wide blade that can slice it in one go. Calculate the angle (to the vertical correct to 2 decimal places) that the blade must be inclined if:

a. the block is to be sliced diagonally into two identical triangular wedges

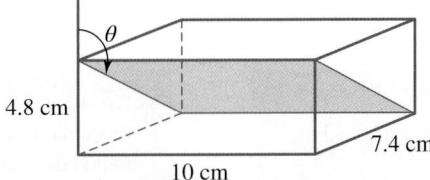

b. the blade is to be placed in the middle of the block and sliced through to the bottom corner, as shown.

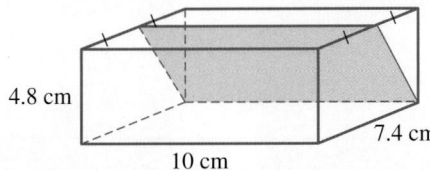

17. A ship travels north for 7 km, then on a true bearing of 140° for another 13 km.

a. Draw a sketch of the situation.
b. Determine how far south the ship is from its starting point, in km correct to 2 decimal places.
c. Evaluate the bearing, correct to the nearest degree, the ship is now from its starting point.

18. The ninth hole on a municipal golf course is 630 m from the tee. A golfer drives a ball from the tee a distance of 315 m at a 10° angle off the direct line as shown.
Determine how far the ball is from the hole and state the angle of the direct line that the ball must be hit along to go directly to the hole.
Give your answers correct to 1 decimal place.

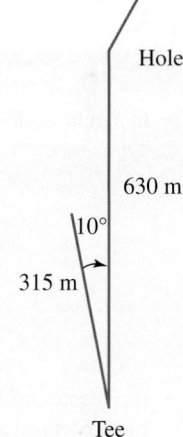

19. **PATH** A sphere of radius length 2.5 cm rests in a hollow inverted cone as shown. The height of the cone is 12.5 cm and its vertical angle is equal to 36°.

a. Evaluate the distance, d, from the tip of the cone to the point of contact with the sphere, correct to 2 decimal places.
b. Determine the distance, h, from the open end of the cone to the bottom of the ball, correct to 2 decimal places.

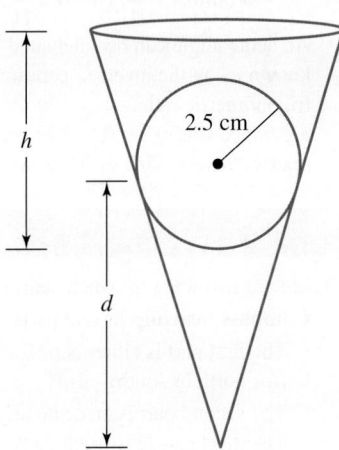

LESSON
9.10 Review

9.10.1 Topic summary

Similar triangles

- When triangles have common angles, they are said to be similar.
  e.g. Triangles OAH, OBG and ODE are similar.

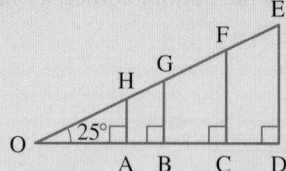

- Corresponding sides of similar triangles will have the same ratio.

  e.g. $\dfrac{FC}{OC} = \dfrac{ED}{OD} = \dfrac{HA}{OA}$

Pythagoras' theorem

- When the length of two sides are known in a right-angled triangle, the third side can be found using the rule $a^2 + b^2 = c^2$.

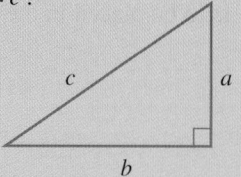

- Length of the longest side $c = \sqrt{a^2 + b^2}$
- Length of the shorter sides $a = \sqrt{c^2 - b^2}$ or $b = \sqrt{c^2 - a^2}$

TRIGONOMETRY I

Trigonometric ratios (SOHCAHTOA)

- In a right-angled triangle, the longest side is called the hypotenuse.

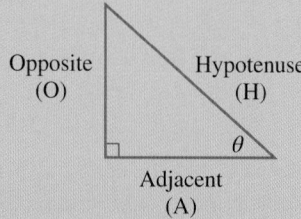

- If an acute angle is known, then the trigonometric ratios can be defined as:

  $\sin(\theta) = \dfrac{O}{H}, \cos(\theta) = \dfrac{A}{H}, \tan(\theta) = \dfrac{O}{A}$

- An acute angle can be calculated when two sides are known using the inverse operation of the correct trigonometric ratio.
  e.g. Since $\sin(30°) = 0.5$, then $\sin^{-1}(0.5) = 30°$; this is read as 'inverse sine of 0.5 is 30°'.

Angles of elevation and depression

- If a horizontal line is drawn from A as shown, forming the angle θ, then θ is called the **angle of elevation** of B *from* A.

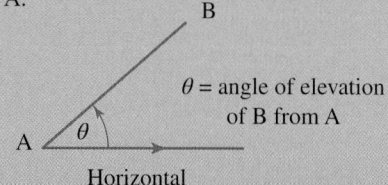

- If a horizontal line is drawn from B as shown, forming the angle α, then α is called the **angle of depression** of A *from* B.

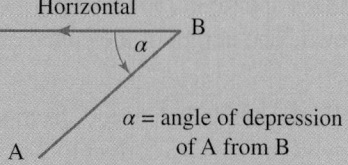

Bearings

There are two ways in which bearings can be written:
- **Compass bearings** have 3 parts:
  - The first part is either N or S (for north or south).
  - The second part is an acute angle.
  - The third part is either E or W (for east or west).

  e.g. S20°E means start by facing south and then turn 20° towards the east.

- **True bearings** are measured from north in a clockwise direction and are expressed in 3 digits.

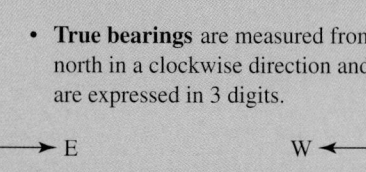

9.10.2 Project

How steep is the land?

When buying a block of land on which to build a house, the slope of the land is often not very obvious. The slab of a house built on the ground must be level, so it is frequently necessary to remove or build up soil to obtain a flat area. The gradient of the land can be determined from a contour map of the area.

Consider the building block shown. The contour lines join points having the same height above sea level. Their measurements are in metres. The plan clearly shows that the land rises from A to B. The task is to determine the angle of this slope.

1. A cross-section shows a profile of the surface of the ground. Let us look at the cross-section of the ground between A and B. The technique used is as follows.
   • Place the edge of a piece of paper on the line joining A and B.
   • Mark the edge of the paper at the points where the contour lines intersect the paper.
   • Transfer this paper edge to the horizontal scale of the profile and mark these points.
   • Choose a vertical scale within the range of the heights of the contour lines.
   • Plot the height at each point where a contour line crosses the paper.
   • Join the points with a smooth curve.

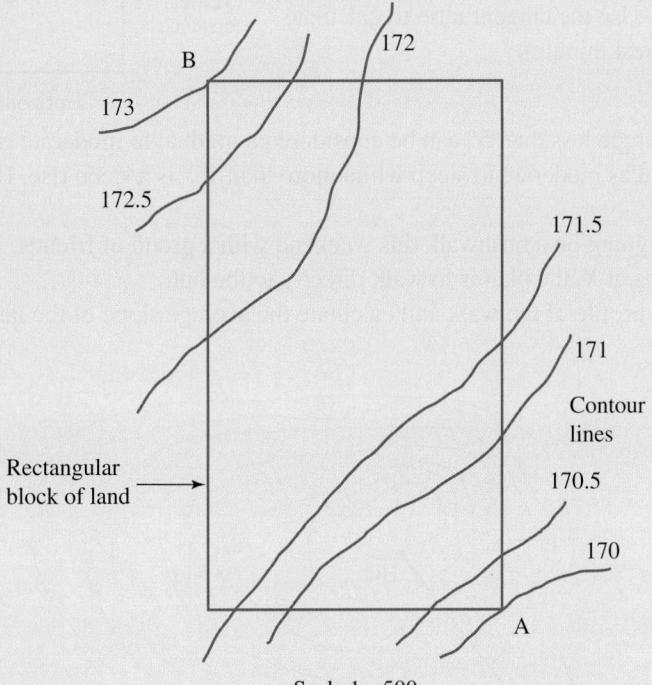

Scale 1 : 500

The cross-section has been started for you. Complete the profile of the line A B. You can now see a visual picture of the profile of the soil between A and B.

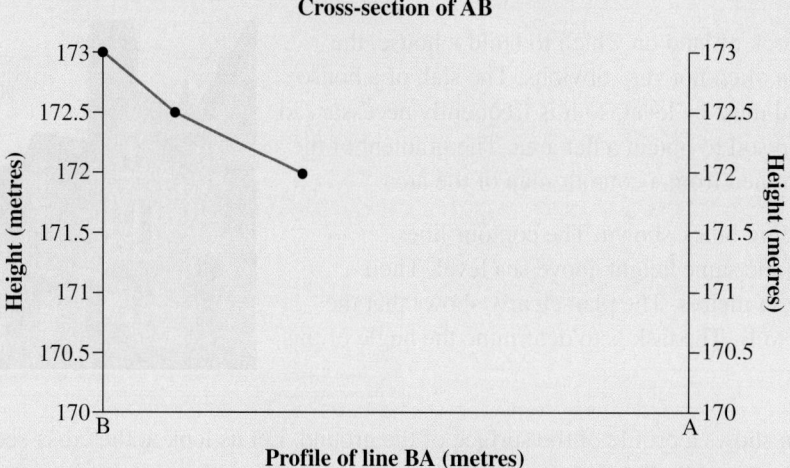

Cross-section of AB

Profile of line BA (metres)

2. We now need to determine the horizontal distance between A and B.
   a. Measure the map distance between A and B using a ruler. What is the map length?
   b. Using the scale of 1 : 500, calculate the actual horizontal distance AB (in metres).
3. The vertical difference in height between A and B is indicated by the contour lines. Calculate this vertical distance.
4. Complete the measurements on this diagram.
5. The angle *a* represents the angle of the average slope of the land from A to B. Use the tangent ratio to calculate this angle (to the nearest minute).

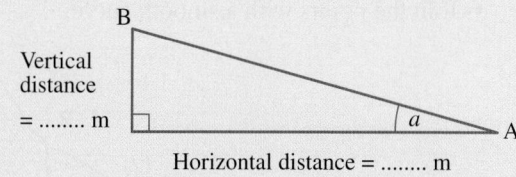

6. In general terms, an angle less than 5° can be considered a gradual to moderate rise. An angle between 5° and 15° is regarded as moderate to steep while more than 15° is a steep rise. How would you describe this block of land?
7. Imagine that you are going on a bushwalk this weekend with a group of friends. A contour map of the area is shown. Starting at *X*, the plan is to walk directly to the hut.
   Draw a cross-section profile of the walk and calculate the average slope of the land. How would you describe the walk?

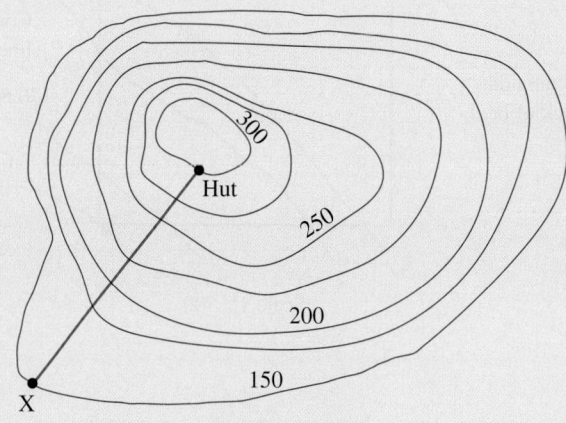

Scale 1 : 20 000

Resources ──

🧩 **Interactivities** Crossword (int-2869)
             Sudoku puzzle (int-3592)

Exercise 9.10 Review questions

learnon

Fluency

1. **MC** The most accurate measurement for the length of the third side in the triangle is:
   A. 483 m
   B. 3.94 m
   C. 2330 mm
   D. 4826 mm

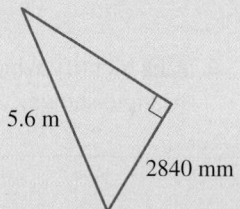

2. **MC** The value of x in this figure is:
   A. 5.4
   B. 7.5
   C. 10.1
   D. 10.3

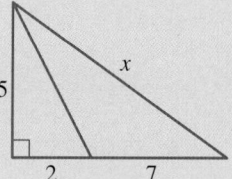

3. **PATH** **MC** Select the closest length of AG of the cube.
   A. 10
   B. 17
   C. 20
   D. 14

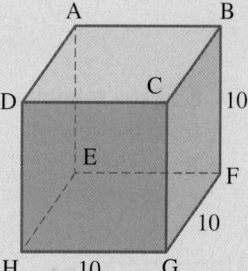

4. **MC** If $\sin(38°) = 0.6157$, identify which of the following will also give this result.
   A. $\sin(218°)$ **B.** $\sin(322°)$ **C.** $\sin(142°)$ **D.** $\sin(682°)$

5. **MC** If $\tan(\theta) = 1.0499$, the value of θ to the nearest minute is:
   A. $46°39'$ **B.** $47°$ **C.** $46°$ **D.** $46°24'$

6. **MC** Identify which trigonometric ratio for the triangle shown is incorrect.

   A. $\sin(\alpha) = \dfrac{b}{c}$ **B.** $\sin(\alpha) = \dfrac{a}{c}$

   C. $\cos(\alpha) = \dfrac{a}{c}$ **D.** $\tan(\alpha) = \dfrac{b}{a}$

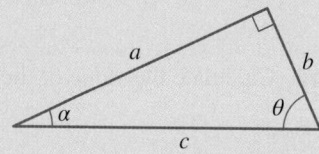

7. `MC` Identify which of the following statements is correct.

    A. $\sin(55°) = \cos(55°)$ **B.** $\sin(45°) = \cos(35°)$

    C. $\cos(15°) = \sin(85°)$ **D.** $\sin(42°) = \cos(48°)$

8. `MC` Identify which of the following can be used to determine the value of x in the diagram.

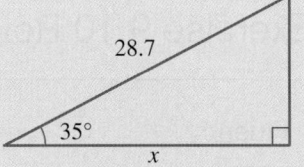

    A. $28.7\sin(35°)$ **B.** $28.7\cos(35°)$

    C. $28.7\tan(35°)$ **D.** $\dfrac{28.7}{\sin(35°)}$

9. `MC` Identify which of the following expressions can be used to determine the value of a in the triangle shown.

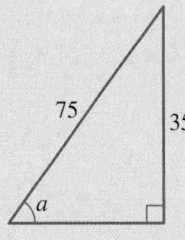

    A. $35\sin(75°)$ **B.** $\sin^{-1}\left(\dfrac{35}{75}\right)$

    C. $\sin^{-1}\left(\dfrac{75}{35}\right)$ **D.** $\cos^{-1}\left(\dfrac{35}{75}\right)$

10. `MC` If a school is 320 m S42°W from the police station, calculate the true bearing of the police station from the school.

    A. 042°T   **B.** 048°T

    C. 222°T   **D.** 228°T

Understanding

11. Calculate x, correct to 2 decimal places.

    a.

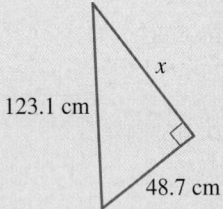

    b.

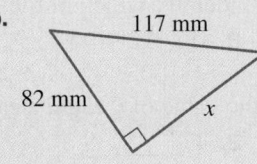

12. Calculate the value of the pronumeral, correct to 2 decimal places.

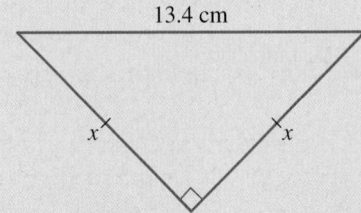

13. **PATH** Calculate the height of this pyramid, in mm correct to 2 decimal places.

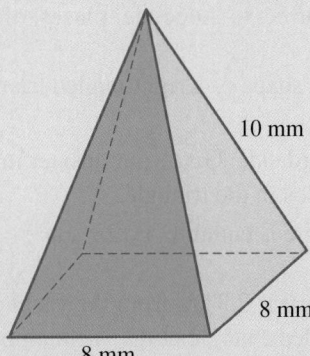

14. A person standing 23 m away from a tree observes the top of the tree at an angle of elevation of 35°. If the person's eye level is 1.5 m from the ground, calculate the height of the tree, in metres correct to 1 decimal place.

15. A man with an eye level height of 1.8 m stands at the window of a tall building. He observes his young daughter in the playground below. If the angle of depression from the man to the girl is 47° and the floor on which the man stands is 27 m above the ground, determine how far from the bottom of the building the child is, in metres correct to 2 decimal places.

16. A plane flies 780 km in a direction of 185°T. Evaluate how far west it has travelled from the starting point, in km correct to 2 decimal places.

17. A hiker travels 3.2 km on a bearing of 250°T and then 1.8 km on a bearing of 320°T. Calculate far west she has travelled from the starting point, in km correct to 2 decimal places.

18. If a 4 m ladder is placed against a wall and the foot of the ladder is 2.6 m from the wall, determine the angle (in degrees and minutes, correct to the nearest minute) the ladder makes with the wall.

Communicating, reasoning and problem solving

19. **PATH** The height of a right square-based pyramid is 13 cm. If the angle the face makes with the base is 67°, determine:
    a. the length of the edge of the square base, in cm correct to 2 decimal places
    b. the length of the diagonal of the base, in cm correct to 2 decimal places
    c. the angle the slanted edge makes with the base in degrees and minutes, correct to the nearest minute.

20. A car is travelling northwards on an elevated expressway 6 m above ground at a speed of 72 km/h. At noon another car passes under the expressway, at ground level, travelling west, at a speed of 90 km/h.
    a. Determine how far apart, in metres, the two cars are 40 seconds after noon, in metres correct to 2 decimal places.
    b. At this time the first car stops, while the second car keeps going. Determine the time when they will be 3.5 km apart. Write your answer correct to the nearest tenth of a second.

21. Two towers face each other separated by a distance, d, of 20 metres. As seen from the top of the first tower, the angle of depression of the second tower's base is $59°$ and that of the top is $31°$. Calculate the height, in metres correct to 2 decimal places, of each of the towers.

22. A piece of flat pastry is cut in the shape of a right-angled triangle. The longest side is $6b$ cm and the shortest is $2b$ cm.

    a. Determine the length of the third side. Give your answer in exact form.
    b. Determine the sizes of the angles in the triangle.
    c. Show that the area of the triangle is equal to $4\sqrt{2}b^2$ cm^2.

23. A yacht is anchored off an island. It is 2.3 km from the yacht club and 4.6 km from a weather station. The three points form a right angled triangle at the yacht club.

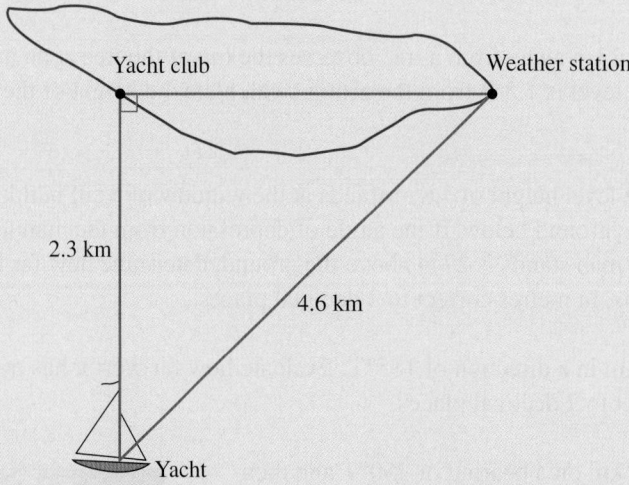

    a. Calculate the angle at the yacht between the yacht club and the weather station.
    b. Evaluate the distance between the yacht club and the weather station, in km correct to 2 decimal places.

    The next day the yacht travels directly towards the yacht club, but is prevented from reaching the club because of dense fog. The weather station notifies the yacht that it is now 4.2 km from the station.

    c. Calculate the new angle at the yacht between the yacht club and the weather station, in degrees correct to 1 decimal place.
    d. Determine how far the yacht is now from the yacht club, correct to 2 decimal places.

on To test your understanding and knowledge of this topic, go to your learnON title at www.jacplus.com.au and complete the **post-test**.

Answers

Topic 9 Trigonometry I

9.1 Pre-test

1. $w = 6.89\,\text{cm}$
2. $x = 2.24\,\text{cm}$
3. $16.97\,\text{cm}$
4. $62.28\,\text{cm}$
5. 0.3521
6. $\theta = 55°10'$
7. $\theta = 36°52'$
8. $y = 6.7\,\text{m}$
9. C
10. a. $020°T$ b. $227°T$ c. $327°T$ d. $163°T$
11. D
12. $551.2\,\text{m}$
13. E
14. $x = 9.0$
15. A

9.2 Pythagoras' theorem

1. a. 7.86 b. 33.27 c. 980.95
2. a. 12.68 b. 2.85 c. 175.14
3. a. 36.36 b. 1.62 c. 15.37
4. a. 0.61 b. 2133.19 c. 453.90
5. $23.04\,\text{cm}$
6. $12.65\,\text{cm}$
7. a. $14.14\,\text{cm}$ b. $24.04\,\text{cm}$ c. $4.53\,\text{cm}$
8. a. $74.83\,\text{cm}$ b. $249.67\,\text{cm}$ c. $3741.66\,\text{cm}^2$
9. a. 6.06 b. 4.24 c. 4.74
10. $14.84\,\text{cm}$
11. $15.59\,\text{cm}$
12. $19.23\,\text{cm}$
13. $72.75\,\text{cm}$; $3055.34\,\text{cm}^2$
14. $39\,\text{m}$
15. $4.34\,\text{km}$
16. $38.2\,\text{m}$
17. $63.06\,\text{mm}$
18. a. $32\,\text{cm}$ b. $768\,\text{cm}^2$
19. 26.83 diagonals, so would need to complete 27.
20. $4701.06\,\text{m}$
21. $9.90\,\text{cm}$
22. a. 65 b. 185 c. 305
23. The value found using Pythagoras' theorem represents length and therefore cannot be negative.
24. a. Neither 105 nor 208 can be the hypotenuse of the triangle, because they are the two smallest values. The other two values could be the hypotenuse if they enable the creation of a right-angled triangle.
    b. $105, 208, 233$

25. a. $21\,\text{cm}$
    b. $35\,\text{cm}$
    c. $y = 12.6\,\text{cm}$ and $RS = 9.8\,\text{cm}$
26. $13.86\,\text{cm}$
27. 1.33 million km

9.3 Pythagoras' theorem in three dimensions (Path)

1. a. 13.86 b. 13.93 c. 18.03
2. $12.21, 12.85$
3. $4.84\,\text{m}, 1.77\,\text{m}$
4. $8.49, 4.24$
5. $10.58\,\text{cm}$
6. $31.62\,\text{cm}$
7. $23\,\text{mm}$
8. a. i. $233.24\,\text{m}$ ii. $200.12\,\text{m}$ iii. $120.20\,\text{m}$
    b. $116.83\,\text{m}$
9. $14.72\,\text{cm}$
10. $12.67\,\text{cm}$
11. $42.27\,\text{cm}$
12. Sample responses can be found in the worked solutions in the online resources.
13. $186.5\,\text{m}$
14. a. $1.49\,\text{m}$ b. $7.43\,\text{m}^2$
15. $\sqrt{1265}\,\text{cm}$
16. 25.475
17. $28.6\,\text{m}$

9.4 Trigonometric ratios

1. a. 0.5000 b. 0.7071 c. 0.4663
    d. 0.8387 e. 8.1443 f. 0.7193
2. a. 0.6494 b. 0.5885 c. 0.5220
    d. -1.5013 e. 0.9990 f. 0.6709
3. a. 0.8121 b. 0.5250 c. -0.8030
    d. 0.4062 e. 0.9880
4. a. -0.9613 b. 1.7321 c. -0.5736 d. 0.1320
5. a. $50°$ b. $24°$ c. $53°$
6. a. $71°$ b. $86°$ c. $41°$
7. a. $54°29'$ b. $6°19'$ c. $0°52'$
8. a. $72°47'$ b. $44°48'$ c. $26°45'$
9. a. $26°34'$ b. $64°1'$ c. $64°47'$
10. a. $48°5'$ b. $36°52'$ c. $88°41'$
11. a. 2.824 b. 71.014 c. 20.361 d. 2.828
12. a. 226.735 b. 1.192 c. 7.232 d. 32.268
13. a. 4909.913 b. 0.063 c. 0.904 d. 14.814
14. a. i. $\sin(\theta) = \dfrac{e}{f}$ ii. $\cos(\theta) = \dfrac{d}{f}$ iii. $\tan(\theta) = \dfrac{e}{d}$

    b. i. $\sin(\alpha) = \dfrac{i}{g}$ ii. $\cos(\alpha) = \dfrac{h}{g}$ iii. $\tan(\alpha) = \dfrac{i}{h}$

c. i. $\sin(\beta) = \dfrac{1}{k}$ ii. $\cos(\beta) = \dfrac{j}{k}$ iii. $\tan(\beta) = \dfrac{1}{j}$

d. i. $\sin(\gamma) = \dfrac{n}{m}$ ii. $\cos(\gamma) = \dfrac{o}{m}$ iii. $\tan(\gamma) = \dfrac{n}{o}$

e. i. $\sin(\beta) = \dfrac{b}{c}$ ii. $\cos(\beta) = \dfrac{a}{c}$ iii. $\tan(\beta) = \dfrac{b}{a}$

f. i. $\sin(\gamma) = \dfrac{v}{u}$ ii. $\cos(\gamma) = \dfrac{t}{u}$ iii. $\tan(\gamma) = \dfrac{v}{t}$

15. a. $\sin(\theta) = \dfrac{15}{18}$ b. $\cos(\theta) = \dfrac{22}{30}$ c. $\tan(\theta) = \dfrac{7}{9}$

16. a. $\tan(\theta) = \dfrac{3.6}{p}$ b. $\sin(25°) = \dfrac{13}{t}$ c. $\sin(\alpha) = \dfrac{18.6}{23.5}$

17. a.

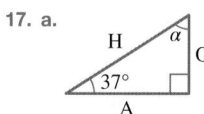

b. i. $\sin(37°) = 0.60$

ii. $\cos(37°) = 0.80$

iii. $\tan(37°) = 0.75$

18. a. i. $\sin(53°) = 0.80$

ii. $\cos(53°) = 0.60$

iii. $\tan(53°) = 1.33$

b. They are equal.

c. They are equal.

d. The sin of an angle is equal to the cos of its complement angle.

19. $\sin(\theta) = \dfrac{\text{opp}}{\text{hyp}}, \cos(\theta) = \dfrac{\text{adj}}{\text{hyp}} \Rightarrow \dfrac{\sin(\theta)}{\cos(\theta)} = \dfrac{\text{opp}}{\text{adj}} = \tan(\theta)$

20. a. $h^2 = a^2 - x^2$

b. $h^2 = c^2 - b^2 + 2bx - x^2$

c. Sample responses can be found in the worked solutions in the online resources.

d. Sample responses can be found in the worked solutions in the online resources.

21. $DC = x + \dfrac{y}{\tan(\theta)}$

22. To determine which trigonometric ratio to apply, the sides in relation to the angle relevant in the question need to be identified and named.

23. $110\tan(\theta + \alpha) - 110\tan(\alpha)$

9.5 Using trigonometry to calculate side lengths

1. a. 8.660   b. 7.250   c. 8.412

2. a. 0.79   b. 4.72   c. 101.38

3. a. 33.45 m   b. 75.00 m   c. 44.82 m

4. a. 7.76 mm   b. 80.81 km   c. 9.04 cm

5. a. $x = 31.58$ cm   b. $y = 17.67$ m   c. $z = 14.87$ m

6. a. $p = 67.00$ m

b. $p = 21.38$ km, $q = 42.29$ km

c. $a = 0.70$ km, $b = 0.21$ km

7. a. 6.0 m   b. 6.7 m

8. 1.05 m

9. a. $x = 30.91$ cm, $y = 29.86$ cm, $z = 39.30$ cm

b. 2941.54 cm^2

10. a. In an isosceles right-angled triangle

b. $\theta < 45°$

11. Sample responses can be found in the worked solutions in the online resources.

12. a. $h = \tan(47°48')x$ m

$h = \tan(36°24')(x + 64)$ m

b. 129.07 m

c. 144.29 m

13. Building A is 73.9 m and Building B is 29.5 m.

14. 60

15. Sample responses can be found in the worked solutions in the online resources.

9.6 Using trigonometry to calculate angle size

1. a. 67°   b. 47°   c. 69°

2. a. 54°47′   b. 33°45′   c. 33°33′

3. a. 75°31′   b. 36°52′   c. 37°39′

4. a. 41°   b. 30°   c. 49°

5. a. 65°   b. 48°   c. 37°

6. a. $a = 25°47'$, $b = 64°13'$

b. $d = 25°23'$, $e = 64°37'$

c. $x = 66°12'$, $y = 23°48'$

7. a. $r = 57.58$, $l = 34.87$, $h = 28.56$

b. 428 cm^2

c. 29.7°

8. a. i. 29.0°   ii. 41.4°   iii. 51.3°

b. i. 124.42 km/h   ii. 136.57 km/h   iii. 146.27 km/h

9. To find the size of acute angles, use inverse operations.

10. Sample responses can be found in the worked solutions in the online resources.

11. a. $\sin(30°) = \dfrac{1}{2}$, $\cos(30°) = \dfrac{\sqrt{3}}{2}$, $\tan(30°) = \dfrac{\sqrt{3}}{3}$

b. Sample responses can be found in the worked solutions in the online resources.

c. Sample responses can be found in the worked solutions in the online resources.

12. a. Between site 3 and site 2: 61°
Between site 2 and site 1: 18°
Between site 1 and bottom: 75°

b. Between site 1 and bottom: 75° slope

13. 58°3′

14. 147°0′; 12°15′

9.7 Angles of elevation and depression

1. 8.74 m

2. 687.7 m

3. a. 176.42 m   b. 152.42 m

4. 65°46′

5. 16.04 m

6. a. $h = x\tan(47°12')$ m; $h = (x + 38)\tan(35°50')$ m

   b. $x = 76.69$ m

   c. 84.62 m

7. a. $h = x\tan(43°35')$ m; $h = (x + 75)\tan(32°18')$ m

   b. 148.37 m

   c. 143.10 m

8. 0.033 km or 33 m

9. 21°

10. 44.88 m

11. a. 54°   b. 0.75 m

12. Angle of elevation is an angle measured upwards from the horizontal. Angle of depression is measured from the horizontal downwards.

13. a.
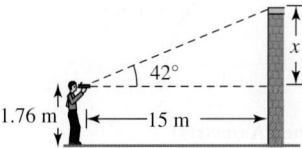

    b. 15.27 m

14. a. 2.16 m/s, 7.77 km/h

    b. 54.5°

15. 66 m

16. 70.522 m

17. 451.5 m

9.8 Bearings

1. a. 020°T   b. 340°T   c. 215°T

2. a. 152°T   b. 034°T   c. 222°T

3. a. N49°E   b. S48°E   c. S87°W

4. a. N30°W   b. N86°E   c. S54°W

5. a. 3 km 325°T   b. 2.5 km 112°T   c. 8 km 235°T

6. a. 4 km 090°T, then 2.5 km 035°T

   b. 12 km 115°T, then 7 km 050°T

   c. 300 m 310°T, then 500 m 220°T

7. a.
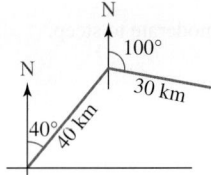

   b.
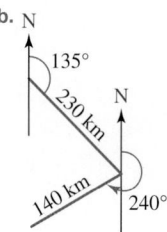

8. a.
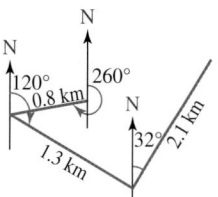

   b.
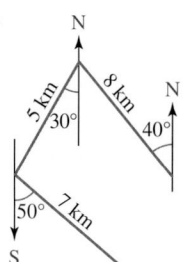

   c.
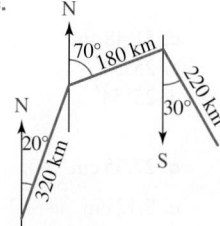

9. a. 13.38 km

   b. 14.86 km

   c. 222°T

   d.
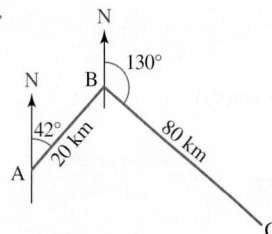

   e. 51.42 km

   f. 61.28 km

   g. 310°T

10. 215°T

11. 1.732 km

12. a. 9.135 km   b. 2.305 km   c. 104°10'T

13. 684.86 km

14. 6.10 km and 239°T

15. 111°T

16. a. $(\theta + 180)$ °T   b. $(\theta - 180)$ °T

17. a. 27.42 km   b. N43°W or 317°T

18. a.
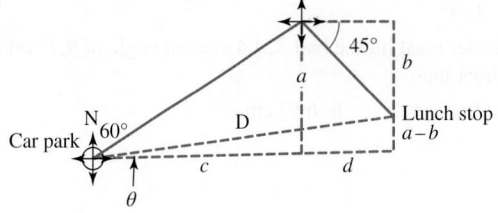

b. 1.76 km North

c. 14.63 km East

d. N83.15°E

e. D = 14.74 km

19. 3.65 km on a bearing of 108°T

9.9 Applications of trigonometry

1. 6.09 m

2. 62°33′

3. a. 1.82 m   b. 27.78 m

4. 31°49′

5. 5.94 m

6. 49 m

7. 194 m

8. 1.829 km

9. a. 11°32′   b. 4°25′

10. a. 35.36 cm   b. 51.48 cm   c. 51.48 cm
     d. 57.23 cm   e. 29°3′   f. 25°54′
     g. 25.74 cm   h. 12.5 cm   i. 25°54′
     j. 28.61 cm

11. a. 77°   b. 71°56′   c. 27.35 cm

12. a. 7.05 cm   b. 60°15′   c. 8.12 cm

13. a. 28.74 cm   b. 40.64 cm   c. 66°37′

14. $\sin(\theta) = \dfrac{O}{H}$. Since the hypotenuse H is the longest side in the right-angled triangle, when dividing O by H the value will be between 0 and 1.

15. a. 90 m

     b. $h = \dfrac{d\tan(\theta_1)}{\tan(\theta_1) + \tan(\theta_2)} \times \tan(\theta_2)$

     c. 250 m

16. a. 122.97°   b. 142.37°

17. a.

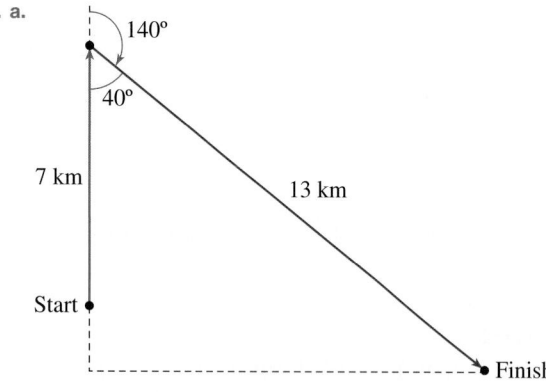

     b. 2.96 km

     c. 110°

18. Golfer must hit the ball 324.4 m at an angle of 9.7° off the direct line.

19. a. 7.69 cm   b. 6.91 cm

Project

1.

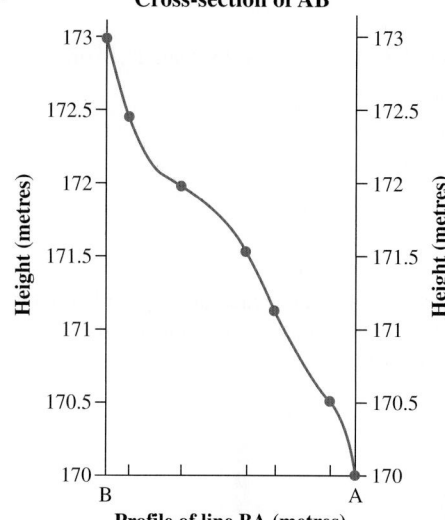

Cross-section of AB

Profile of line BA (metres)

2. a. 8 cm   b. 40 m

3. 3 m

4.

B
Vertical
distance
= 3 m
Horizontal distance = 40 m
a
A

5. $a = 4°17′$

6. Gradual to moderate.

7.

Cross-section X to hut

Profile of X to hut

The average slope is 11.46° — moderate to steep.

9.10 Review questions

1. D

2. D

3. B

4. C

5. D

6. B

7. D

8. B

9. B

10. A

11. a. $x = 113.06$ cm   b. $x = 83.46$ mm
12. 9.48 cm
13. 8.25 mm
14. 17.6 m
15. 26.86 m
16. 67.98 km
17. 4.16 km
18. $40°32'$
19. a. 11.04 cm   b. 15.61 cm   c. $59°1'$
20. a. 1280.64 m   b. $12:02:16.3$ pm
21. 33.29 m, 21.27 m
22. a. $4\sqrt{2}b$

    b. $19.5°$, $70.5°$, $90°$.

    c. Area $= \dfrac{1}{2}$ base $\times$ height

    $= \dfrac{1}{2} \times 2b \times 4\sqrt{2}b$

    $= 4\sqrt{2}b^2$ cm^2.
23. a. $60°$ b. 3.98 km   c. $71.5°$ d. 1.33 km

10 Trigonometry II (Path)

LESSON SEQUENCE

LESSON
10.1 Overview

Why learn this?

Trigonometry is the branch of mathematics that describes the relationship between the angles and side lengths in triangles. The ability to calculate distances using angles has long been critical. As early as the third century BCE, trigonometry was being used in the study of astronomy. Early explorers, using rudimentary calculations and the stars, were able to navigate their way around the world. They were even able to map coastlines along the way. Cartographers use trigonometry when they are making maps. It is essential to be able to calculate distances that can't be physically measured. Astronomers use trigonometry to calculate distances such as that from a particular planet to Earth. Our explorations have now turned towards the skies and outer space. Scientists design and launch space shuttles and rockets to explore our universe. By applying trigonometry, they can approximate the distances to other planets. As well as in astronomy and space exploration, trigonometry is widely used in many other areas. Surveyors use trigonometry in setting out a land subdivision.

Builders, architects and engineers use angles, lengths and forces in the design and construction of all types of buildings, both domestic and industrial. In music, a single note is a sine wave. Sound engineers manipulate sine waves to create the desired effect. Trigonometry has many real-life applications.

1. **MC** From the following options select the exact value of sin(30°).

    A. $\dfrac{1}{\sqrt{2}}$ **B.** $\dfrac{1}{2}$ **C.** $\dfrac{\sqrt{3}}{2}$ **D.** 1

2. Solve for x, correct to 2 decimal places.

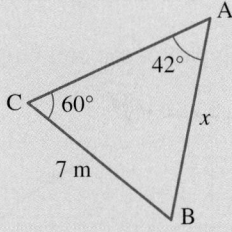

3. Solve for y, correct to 2 decimal places.

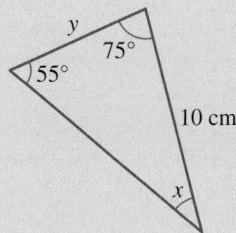

4. **MC** Choose the values of the angles B and B' in the below triangle, to the nearest degree. (Assume $BC = B'C$.)

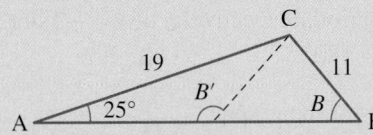

    A. 47° and 133°   **B.** 46° and 134°   **C.** 47° and 153°   **D.** 25° and 155°

5. Calculate the perimeter of the following triangle, correct to 2 decimal places.

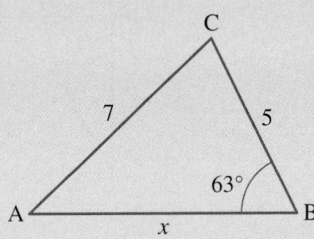

6. Solve for x, correct to 1 decimal place.

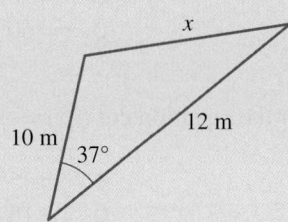

7. **MC** Calculate the area of the triangle shown.
   A. 13.05 cm^2
   B. 18.64 cm^2
   C. 22.75 cm^2
   D. 26.1 cm^2

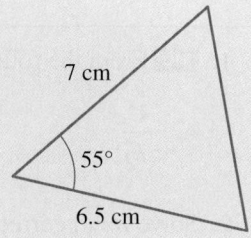

8. State in which quadrant of the unit circle is the angle $203°$ located.

9. Determine the value of $\cos(60°)$ using part of the unit circle.

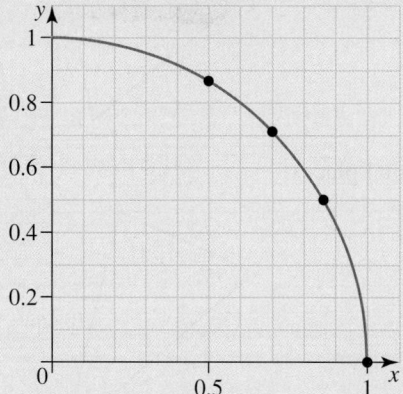

10. **MC** If $\cos(x°) = p$ for $0 \le x \le 90°$, then $\sin(180 - x°)$ in terms of p is:
    A. p B. $180 - p$ C. $1 - p$ D. $\sqrt{1 - p^2}$

11. **MC** Select the amplitude and period, respectively, of $y = -2\sin(2x)$ from the following options.
    A. $-2, 360°$ B. $-2, 180°$ C. $2, 2$ D. $2, 180°$

12. **MC** Select the correct equation for the graph shown.
    A. $y = 4\cos(2x)$
    B. $y = -4\cos(2x)$
    C. $y = -4\sin(2x)$
    D. $y = 4\sin(2x)$

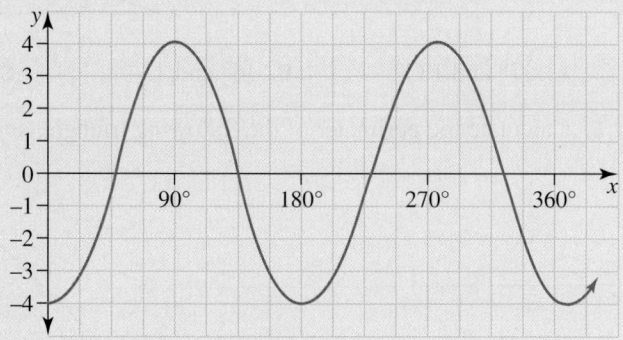

13. **MC** Select the correct solutions for the equation $\sin(x) = \dfrac{1}{2}$ for x over the domain $0 \le x \le 360°$.

    A. $x = 30°$ and $x = 150°$ B. $x = 30°$ and $x = 210°$
    C. $x = 60°$ and $x = 120°$ D. $x = 60°$ and $x = 240°$

14. **MC** Select the correct solutions for the equation $\cos(2x) = -\dfrac{\sqrt{2}}{2}$ for x over the domain $0 \le x \le 360°$.

    A. $x = 22.5°$ and $x = 337.5°$ B. $x = 45°$ and $x = 315°$
    C. $x = 67.5°$ and $x = 112.5°$ D. $x = 157.5°$ and $x = 202.5°$

15. Using the graph shown, solve the equation $7 \sin(x) = -7$ for $0 \le x \le 360°$.

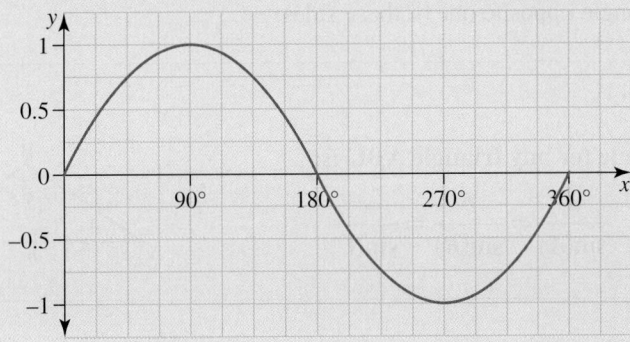

LESSON
10.2 The sine rule

LEARNING INTENTION

At the end of this lesson you should be able to:
- apply the sine rule to evaluate angles and sides of triangles
- recognise when the ambiguous case of the sine rule exists
- apply the sine rule to evaluate angles involving the ambiguous case.

 10.2.1 The sine rule

eles-6287

- In any triangle, the angles are named by the vertices A, B and C and the corresponding opposite sides as a, b and c as shown in the diagram at right.
- Let BD be the perpendicular line from B to AC, of length h, giving two right-angled triangles, $\triangle$ADB and $\triangle$CDB.

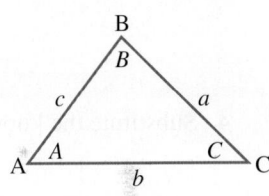

Using $\triangle$ADB:

$$\sin(A) = \frac{h}{c}$$

$$h = c \sin(A)$$

Using $\triangle$CDB:

$$\sin(C) = \frac{h}{a}$$

$$h = a \sin(C)$$

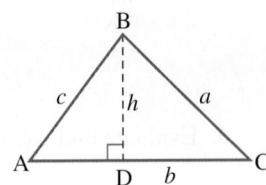

Equating the values of h:

$$c \sin(A) = a \sin(C)$$

$\frac{h}{c} = \sin(A)$ and $\frac{h}{a} = \sin C$)

giving:

$$\frac{c}{\sin(C)} = \frac{a}{\sin(A)}$$

- Similarly, if a perpendicular line is drawn from vertex A to BC, then:

$$\frac{c}{\sin(C)} = \frac{b}{\sin(B)}$$

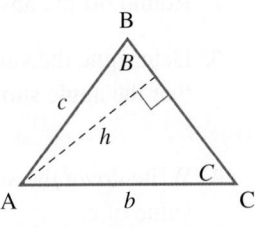

$h = c \sin(B)$ and $h = b \sin(C)$

- The **sine rule** can be used to solve non-right-angled triangles if we are given:
  1. two angles and one side
  2. two sides and an angle opposite one of these sides.

Sine rule

The sine rule for any triangle ABC is:

$$\frac{a}{\sin(A)} = \frac{b}{\sin(B)} = \frac{c}{\sin(C)}$$

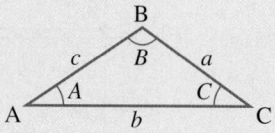

WORKED EXAMPLE 1 Determining unknown angles and sides of a given triangle

In the triangle ABC, $a = 4\,\text{m}$, $b = 7\,\text{m}$ and $B = 80°$. Calculate the values of A, C and c. Round angles to the nearest minute and lengths to 2 decimal places.

THINK	WRITE/DRAW
1. Draw a labelled diagram of the triangle ABC and fill in the given information.	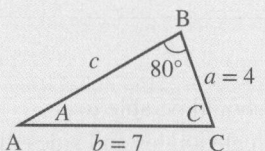
2. Check that one of the criteria for the sine rule has been satisfied.	The sine rule can be used since two side lengths and an angle opposite one of these side lengths have been given.
3. Write down the sine rule to calculate A.	To calculate angle A: $$\frac{a}{\sin(A)} = \frac{b}{\sin(B)}$$
4. Substitute the known values into the rule.	$$\frac{4}{\sin(A)} = \frac{7}{\sin(80°)}$$
5. Transpose the equation to make $\sin(A)$ the subject.	$$4\sin(80°) = 7\sin(A)$$ $$\sin(A) = \frac{4\sin(80°)}{7}$$
6. Evaluate and write your answer.	$$A = \sin^{-1}\left(\frac{4\sin(80°)}{7}\right)$$ $$\approx 34.246\,004\,71°$$
7. Round off the answer to degrees and minutes.	$$A = 34°15'$$
8. Determine the value of angle C using the fact that the angle sum of any triangle is 180°.	$$C \approx 180° - (80° + 34°15')$$ $$= 65°45'$$
9. Write down the sine rule to calculate the value of c.	To calculate side length c: $$\frac{c}{\sin(C)} = \frac{b}{\sin(B)}$$

10. Substitute the known values into the rule.		$\dfrac{c}{\sin(65°45')} = \dfrac{7}{\sin(80°)}$
11. Transpose the equation to make c the subject.		$c = \dfrac{7\sin(5°45')}{\sin(80°)}$
12. Evaluate. Round off the answer to 2 decimal places and include the appropriate unit.		$\approx 6.48\,\text{m}$

▶ 10.2.2 The ambiguous case

eles-5005

- If two side lengths and an angle opposite one of these side lengths are given, then two different triangles may be drawn.
- For example, if $a = 10$, $c = 6$ and $C = 30°$, two possible triangles could be created.
- In the first case angle A is an acute angle, while in the second case angle A is an obtuse angle.
- When using the sine rule to determine an angle, the inverse sine function is used.
- In lesson 10.7, you will see that the sine of an angle between 0° and 90° has the same value as the sine of its supplement.
  For example, $\sin(40°) \approx 0.6427$ and $\sin(140°) \approx 0.6427$.

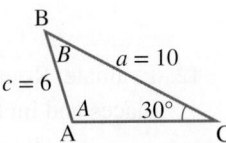

WORKED EXAMPLE 2 Solving triangles and checking for the ambiguous case

In the triangle ABC, $a = 10\,\text{m}$, $c = 6\,\text{m}$ and $C = 30°$. Determine two possible values of A, and hence two possible values of B and b.
Round angles to the nearest minute and lengths to 2 decimal places.

Case 1

THINK	WRITE/DRAW
1. Draw a labelled diagram of the triangle ABC and fill in the given information.	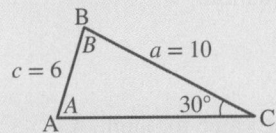
2. Check that one of the criteria for the sine rule has been satisfied.	The sine rule can be used since two side lengths and an angle opposite one of these side lengths have been given.
3. Write down the sine rule to determine A.	To determine angle A: $\dfrac{a}{\sin(A)} = \dfrac{c}{\sin(C)}$
4. Substitute the known values into the rule.	$\dfrac{10}{\sin(A)} = \dfrac{6}{\sin(30°)}$ $10\sin(30°) = 6\sin(A)$
5. Transpose the equation to make $\sin(A)$ the subject.	$\sin(A) = \dfrac{10\sin(30°)}{6}$

6. Evaluate angle A.

$$A = \sin^{-1}\left(\frac{10\sin(30°)}{6}\right)$$

$$\approx 56.442\,690\,24°$$

7. Round off the answer to degrees and minutes.

$$A = 56°27'$$

8. Determine the value of angle B, using the fact that the angle sum of any triangle is 180°.

$$B \approx 180° - (30° + 56°27')$$

$$= 93°33'$$

9. Write down the sine rule to calculate b.

To calculate side length b:

$$\frac{b}{\sin(B)} = \frac{c}{\sin(C)}$$

10. Substitute the known values into the rule.

$$\frac{b}{\sin(93°33')} = \frac{6}{\sin(30°)}$$

11. Transpose the equation to make b the subject.

$$b = \frac{6\sin(93°33')}{\sin(30°)}$$

12. Evaluate. Round off the answer to 2 decimal places and include the appropriate unit.

$$\approx 11.98\text{ m}$$

Note: The values we have just obtained are only one set of possible answers for the given dimensions of the triangle ABC.

We are told that $a = 10$ m, $c = 6$ m and $C = 30°$. Since side a is larger than side c, it follows that angle A will be larger than angle C. Angle A must be larger than 30°; therefore it may be an acute angle or an obtuse angle.

Case 2

THINK

1. Draw a labelled diagram of the triangle ABC and fill in the given information.

2. Write down the alternative value for angle A. Simply subtract the value obtained for A in case 1 from 180°.

3. Determine the alternative value of angle B, using the fact that the angle sum of any triangle is 180°.

4. Write down the sine rule to determine the alternative b.

5. Substitute the known values into the rule.

WRITE/DRAW

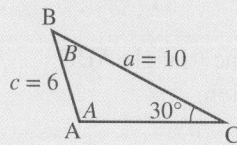

To determine the alternative angle A:
If $\sin A = 0.8333$, then A could also be:
$A \approx 180° - 56°27'$

$$= 123°33'$$

$B \approx 180° - (30° + 123°33')$

$$= 26°27'$$

To calculate side length b:

$$\frac{b}{\sin(B)} = \frac{c}{\sin(C)}$$

$$\frac{b}{\sin(26°27')} = \frac{6}{\sin(30°)}$$

6. Transpose the equation to make b the subject.	$b = \dfrac{6 \sin(26°27')}{\sin(30°)}$
7. Evaluate. Round off the answer to 2 decimal places and include the appropriate unit.	$\approx 5.34 \text{ m}$

- In Worked example 2 there were two possible solutions, as shown by the diagrams below.

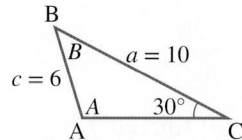 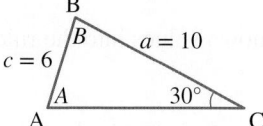

- The ambiguous case does not apply to every question.

  Consider Worked example 1.
  - Since $\angle A = 34°15'$, then it could also have been $\angle A = 145°45'$, the supplementary angle.
  - If $\angle A = 34°15'$ and $\angle B = 80°$, then
    $\angle C = 65°45'$ (angle sum of triangle).
  - If $\angle A = 145°45'$ and $\angle B = 80°$, then
    $\angle C = 180° - (145°45' + 80°)$ which is not possible.
    $\angle C = -45°45'$

  Hence, for Worked example 1, only one possible solution exists.
- The ambiguous case may exist if the angle found is opposite the larger given side.

COMMUNICATING – COLLABORATIVE TASK: Using graphing applications to verify the sine rule

Equipment: Graphing software.
1. Use a graphing application (such as Geogebra), to construct a triangle of any shape.
2. Select two angles and their opposite sides to substitute into the sine rule. What do you notice about the two fractions?
3. Modify the shape of your triangle and repeat step 2.
4. Compare your results with the rest of the class.

WORKED EXAMPLE 3 Calculating heights using given angles of elevation

To calculate the height of a building, Kevin measures the angle of elevation to the top as $52°$. He then walks 20 m closer to the building and measures the angle of elevation as $60°$.
Calculate the height of the building, correct to 2 decimal places.

THINK

1. Draw a labelled diagram of the situation and fill in the given information.

WRITE/DRAW

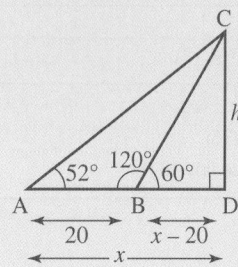

2. Check that one of the criteria for the sine rule has been satisfied for triangle ABC.

The sine rule can be used for triangle ABC since two angles and one side length have been given.

3. Calculate the value of angle ACB, using the fact that the angle sum of any triangle is 180°.

$\angle ACB = 180° - (52° + 120°)$
$= 8°$

4. Write down the sine rule to calculate b (or AC).

To calculate side length b of triangle ABC:

$$\frac{b}{\sin(B)} = \frac{c}{\sin(C)}$$

5. Substitute the known values into the rule.

$$\frac{b}{\sin(120°)} = \frac{20}{\sin(8°)}$$

6. Transpose the equation to make b the subject.

$$b = \frac{20 \sin(120°)}{\sin(8°)}$$

7. Evaluate. Round off the answer to 2 decimal places and include the appropriate unit.

$\approx 124.45 \, \text{m}$

8. Draw a diagram of the situation, that is, triangle ADC, labelling the required information. *Note*: There is no need to solve the rest of the triangle in this case as the values will not assist in calculating the height of the building.

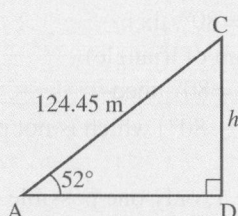

9. Write down what is given for the triangle.

Have: angle and hypotenuse

10. Write down what is needed for the triangle.

Need: opposite side

11. Determine which of the trigonometric ratios is required (SOH − CAH − TOA).

$$\sin(\theta) = \frac{O}{H}$$

12. Substitute the given values into the appropriate ratio.

$$\sin(52°) = \frac{h}{124.45}$$

13. Transpose the equation and solve for h.

$124.45 \sin(52°) = h$
$h = 124.45 \sin(52°)$

14. Round off the answer to 2 decimal places.

≈ 98.07

15. Write the answer.

The height of the building is 98.07 m.

DISCUSSION

Discuss a real-life situation in which the sine rule can be used.

 Resources

 Interactivities The sine rule (int-6275)
                 The ambiguous case (int-4818)

Exercise 10.2 The sine rule

10.2 Quick quiz on

10.2 Exercise

Individual pathways

■ PRACTISE	■ CONSOLIDATE	■ MASTER
1, 2, 6, 7, 11, 13, 16, 18, 21	3, 4, 8, 9, 12, 14, 17, 19, 22	5, 10, 15, 20, 23, 24, 25

Where appropriate in this exercise, write your angles correct to the nearest minute and side lengths correct to 2 decimal places.

Fluency

1. **WE1** In the triangle ABC, $a = 10$, $b = 12$ and $B = 58°$. Calculate A, C and c.

2. In the triangle ABC, $c = 17.35$, $a = 26.82$ and $A = 101°47'$. Calculate C, B and b.

3. In the triangle ABC, $a = 5$, $A = 30°$ and $B = 80°$. Calculate C, b and c.

4. In the triangle ABC, $c = 27$, $C = 42°$ and $A = 105°$. Calculate B, a and b.

5. In the triangle ABC, $a = 7$, $c = 5$ and $A = 68°$. Determine the perimeter of the triangle.

6. Calculate all unknown sides and angles for the triangle ABC, given $A = 57°$, $B = 72°$ and $a = 48.2$.

7. Calculate all unknown sides and angles for the triangle ABC, given $a = 105$, $B = 105°$ and $C = 15°$.

8. Calculate all unknown sides and angles for the triangle ABC, given $a = 32$, $b = 51$ and $A = 28°$.

9. Calculate the perimeter of the triangle ABC if $a = 7.8$, $b = 6.2$ and $A = 50°$.

10. **MC** In a triangle ABC, $B = 40°$, $b = 2.6$ and $c = 3$. Identify the approximate value of C.
    Note: There may be more than one correct answer.
    - **A.** $47°$
    - **B.** $48°$
    - **C.** $132°$
    - **D.** $133°$

Understanding

11. **WE2** In the triangle ABC, $a = 10$, $c = 8$ and $C = 50°$. Determine two possible values of A, and hence two possible values of b.

12. In the triangle ABC, $a = 20$, $b = 12$ and $B = 35°$. Determine two possible values for the perimeter of the triangle.

13. Calculate all unknown sides and angles for the triangle ABC, given $A = 27°$, $B = 43°$ and $c = 6.4$.

14. Calculate all unknown sides and angles for the triangle ABC, given $A = 100°$, $b = 2.1$ and $C = 42°$.

15. Calculate all unknown sides and angles for the triangle ABC, given $A = 25°$, $b = 17$ and $a = 13$.

Communicating, reasoning and problem solving

16. Calculate the value of h, correct to 1 decimal place. Show the full working.

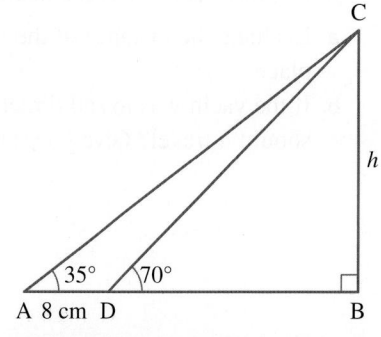

17. **WE3** To calculate the height of a building, Kevin measures the angle of elevation to the top as 48°. He then walks 18 m closer to the building and measures the angle of elevation as 64°. Calculate the height of the building.

18. A boat sails on a bearing of N15°E for 10 km and then on a bearing of S85°E until it is due east of the starting point. Determine the distance from the starting point to the nearest kilometre. Show all your working.

19. A hill slopes at an angle of 30° to the horizontal. A tree that is 8 m tall and leaning downhill is growing at an angle of 10° m to the vertical and is part-way up the slope. Evaluate the vertical height of the top of the tree above the slope. Show all your working.

20. A cliff is 37 m high. The rock slopes outward at an angle of 50° to the horizontal and then cuts back at an angle of 25° to the vertical, meeting the ground directly below the top of the cliff.

    Carol wishes to abseil from the top of the cliff to the ground as shown in the diagram. Her climbing rope is 45 m long, and she needs 2 m to secure it to a tree at the top of the cliff. Determine if the rope will be long enough to allow her to reach the ground.

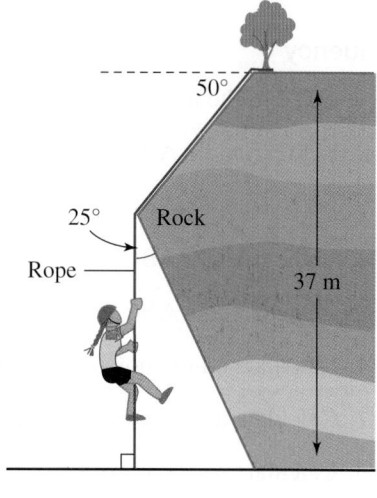

21. A river has parallel banks that run directly east–west. From the south bank, Kylie takes a bearing to a tree on the north side. The bearing is 047°T. She then walks 10 m due east, and takes a second bearing to the tree. This is 305°T. Determine:

    a. her distance from the second measuring point to the tree
    b. the width of the river, to the nearest metre.

22. A ship sails on a bearing of S20°W for 14 km; then it changes direction and sails for 20 km and drops anchor. Its bearing from the starting point is now N65°W.

    a. Determine the distance of the ship from the starting point of it.
    b. Calculate the bearing on which the ship sails for the 20 km leg.

23. A cross-country runner runs at 8 km/h on a bearing of 150°T for 45 mins; then she changes direction to a bearing of 053°T and runs for 80 mins at a different speed until she is due east of the starting point.

    a. Calculate the distance of the second part of the run.
    b. Calculate her speed for this section, correct to 1 decimal place.
    c. Evaluate how far she needs to run to get back to the starting point.

24. From a fire tower, A, a fire is spotted on a bearing of N42°E. From a second tower, B, the fire is on a bearing of N12°W. The two fire towers are 23 km apart, and A is N63°W of B. Determine how far the fire is from each tower.

25. A yacht sets sail from a marina and sails on a bearing of 065°T for 3.5 km. It then turns and sails on a bearing of 127°T for another 5 km.

    a. Evaluate the distance of the yacht from the marina, correct to 1 decimal place.
    b. If the yacht was to sail directly back to the marina, on what bearing should it travel? Give your answer rounded to the nearest minute.

LESSON
10.3 The cosine rule

LEARNING INTENTION

At the end of this lesson you should be able to:
- apply the cosine rule to calculate a side of a triangle
- apply the cosine rule to calculate the angles of a triangle.

▶ 10.3.1 The cosine rule

eles-5006

- In triangle ABC, let BD be the perpendicular line from B to AC, of length h, giving two right-angled triangles, $\triangle$ADB and $\triangle$CDB.
- Let the length of AD $= x$, then DC $= (b–x)$.
- Using triangle ADB and Pythagoras' theorem, we obtain:

$$c^2 = h^2 + x^2 \qquad [1]$$

Using triangle CDB and Pythagoras' theorem, we obtain:

$$a^2 = h^2 + (b - x)^2 \qquad [2]$$

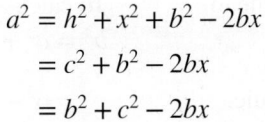

Expanding the brackets in equation [2]:

$$a^2 = h^2 + b^2 - 2bx + x^2$$

Rearranging equation [2] and using $c^2 = h^2 + x^2$ from equation [1]:

$$a^2 = h^2 + x^2 + b^2 - 2bx$$
$$= c^2 + b^2 - 2bx$$
$$= b^2 + c^2 - 2bx$$

From triangle ABD, $x = c\cos(A)$; therefore $a^2 = b^2 + c^2 - 2bx$ becomes

$$a^2 = b^2 + c^2 - 2bc\cos(A).$$

- The **cosine rule** can be used to solve non-right-angled triangles if we are given:
  1. three sides

  or
  2. two sides and the included angle.

Note: Once the third side has been calculated, the sine rule could be used to determine other angles if necessary.

Cosine rule — calculating an unknown side

The cosine rule to calculate an unknown side for any triangle ABC is:

$$a^2 = b^2 + c^2 - 2bc\cos(A)$$
$$b^2 = a^2 + c^2 - 2ac\cos(B)$$
$$c^2 = a^2 + b^2 - 2ab\cos(C)$$

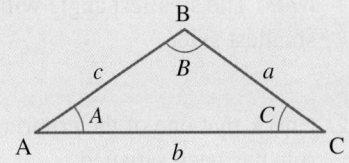

- If three sides of a triangle are known, an angle could be found by transposing the cosine rule to make $\cos(A)$, $\cos(B)$ or $\cos(C)$ the subject.

Cosine rule — calculating an unknown angle

- The cosine rule to calculate an unknown angle for any triangle ABC is:

$$a^2 = b^2 + c^2 - 2bc\cos(A) \Rightarrow \cos(A) = \frac{b^2 + c^2 - a^2}{2bc}$$

$$b^2 = a^2 + c^2 - 2ac\cos(B) \Rightarrow \cos(B) = \frac{a^2 + c^2 - b^2}{2ac}$$

$$c^2 = a^2 + b^2 - 2ab\cos(C) \Rightarrow \cos(C) = \frac{a^2 + b^2 - c^2}{2ab}$$

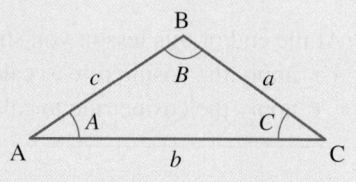

WORKED EXAMPLE 4 Calculating the length of a side using the cosine rule

Calculate the third side of triangle ABC given $a = 6$, $c = 10$ and $B = 76°$ correct to 2 decimal places.

THINK	WRITE/DRAW
1. Draw a labelled diagram of the triangle ABC and fill in the given information.	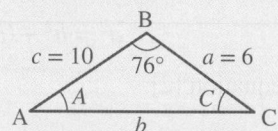
2. Check that one of the criteria for the cosine rule has been satisfied.	Yes, the cosine rule can be used since two side lengths and the included angle have been given.
3. Write down the appropriate cosine rule to calculate side b.	To calculate side b: $b^2 = a^2 + c^2 - 2ac\cos(B)$
4. Substitute the given values into the rule.	$= 6^2 + 10^2 - 2 \times 6 \times 10 \times \cos(76°)$
5. Evaluate.	$\approx 106.969\,372\,5$ $b \approx \sqrt{106.969\,372\,5}$
6. Round off the answer to 2 decimal places.	≈ 10.34

WORKED EXAMPLE 5 Calculating the size of an angle using the cosine rule

Calculate the smallest angle in the triangle with sides 4 cm, 7 cm and 9 cm correct to the nearest minute.

THINK	WRITE/DRAW
1. Draw a labelled diagram of the triangle, call it ABC and fill in the given information. *Note:* The smallest angle will correspond to the smallest side.	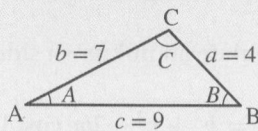 Let $a = 4$, $b = 7$, $c = 9$
2. Check that one of the criteria for the cosine rule has been satisfied.	The cosine rule can be used since three side lengths have been given.

3. Write down the appropriate cosine rule to calculate angle A.

$$\cos(A) = \frac{b^2 + c^2 - a^2}{2bc}$$

4. Substitute the given values into the rearranged rule.

$$= \frac{7^2 + 9^2 - 4^2}{2 \times 7 \times 9}$$

5. Evaluate.

$$= \frac{114}{126}$$

6. Transpose the equation to make A the subject by taking the inverse cos of both sides.

$$A = \cos^{-1}\left(\frac{114}{126}\right)$$
$$\approx 25.208\,765\,3°$$

7. Round off the answer to degrees and minutes.

$$\approx 25°13'$$

COMMUNICATING – COLLABORATIVE TASK: Using graphing applications to verify the cosine rule

Equipment: Graphing software.
1. Use a graphing application (such as Geogebra), to construct a triangle of any shape.
2. Select two sides and the angle between to substitute into the cosine rule.
3. Repeat the process for the other two angles in the triangle.
3. Modify the shape of your triangle and repeat steps 2 and 3.
4. Compare your results with the rest of the class.

WORKED EXAMPLE 6 Applying the cosine rule to solve problems

Two rowers, Harriet and Kate, set out from the same point. Harriet rows N70°E for 2000 m and Kate rows S15°W for 1800 m.
Calculate the distance between the two rowers, correct to 2 decimal places.

THINK

1. Draw a labelled diagram of the triangle, call it ABC and fill in the given information.

WRITE/DRAW

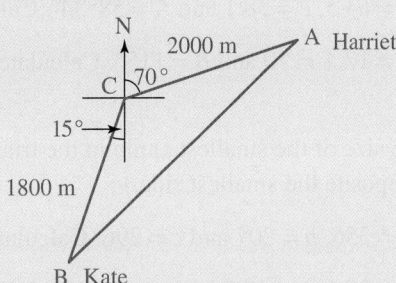

2. Check that one of the criteria for the cosine rule has been satisfied.

The cosine rule can be used since two side lengths and the included angle have been given.

3. Write down the appropriate cosine rule to calculate side c.

To calculate side c:
$$c^2 = a^2 + b^2 - 2ab\cos(C)$$

4. Substitute the given values into the rule.

$$= 2000^2 + 1800^2 - 2 \times 2000 \times 1800\,\cos(125°)$$

5. Evaluate.

$$\approx 11\,369\,750.342$$

$$c \approx \sqrt{11\,369\,750.342}$$

6. Round off the answer to 2 decimal places.

$$\approx 3371.91$$

7. Write the answer.

The rowers are 3371.91 m apart.

DISCUSSION

In what situations would you use the sine rule rather than the cosine rule?

 Resources

 Interactivity The cosine rule (int-6276)

Exercise 10.3 The cosine rule

learnon

10.3 Quick quiz on	10.3 Exercise

Individual pathways

■ PRACTISE	■ CONSOLIDATE	■ MASTER
1, 4, 7, 9, 14, 17	2, 5, 8, 10, 15, 18	3, 6, 11, 12, 13, 16, 19

Where appropriate in this exercise, write your angles correct to the nearest minute and side lengths correct to 2 decimal places.

Fluency

1. **WE4** Calculate the third side of triangle ABC given $a = 3.4$, $b = 7.8$ and $C = 80°$.

2. In triangle ABC, $b = 64.5$, $c = 38.1$ and $A = 58°34'$. Calculate the value of a.

3. In triangle ABC, $a = 17$, $c = 10$ and $B = 115°$. Calculate the value of b, and hence calculate the values of A and C.

4. **WE5** Calculate the size of the smallest angle in the triangle with sides 6 cm, 4 cm and 8 cm. (*Hint:* The smallest angle is opposite the smallest side.)

5. In triangle ABC, $a = 356$, $b = 207$ and $c = 296$. Calculate the size of the largest angle.

6. In triangle ABC, $a = 23.6$, $b = 17.3$ and $c = 26.4$. Calculate the size of all the angles.

7. **WE6** Two rowers set out from the same point. One rows N30°E for 1500 m and the other rows S40°E for 1200 m. Calculate the distance between the two rowers, correct to the nearest metre.

8. Maria cycles 12 km in a direction N68°W and then 7 km in a direction of N34°E.
   a. Calculate her distance from the starting point.
   b. Determine the bearing of the starting point from her finishing point.

Understanding

9. A garden bed is in the shape of a triangle, with sides of length 3 m, 4.5 m and 5.2 m.

   a. Calculate the size of the smallest angle.
   b. Hence, calculate the area of the garden, correct to 2 decimal places. (*Hint:* Draw a diagram, with the longest length as the base of the triangle.)

10. A hockey goal is 3 m wide. When Sophie is 7 m from one post and 5.2 m from the other, she shoots for goal. Determine within what angle, to the nearest degree, the shot must be made if it is to score a goal.

11. An advertising balloon is attached to two ropes 120 m and 100 m long. The ropes are anchored to level ground 35 m apart. Calculate the height of the balloon when both ropes are taut.

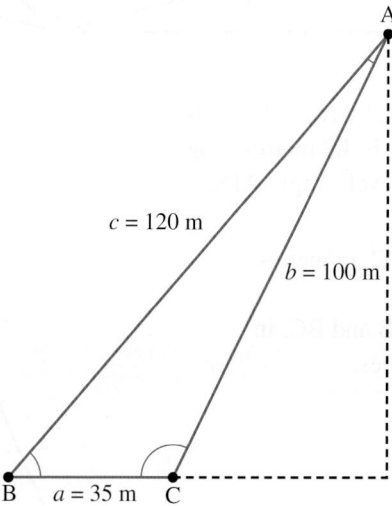

12. A plane flies in a direction of N70°E for 80 km and then on a bearing of S10°W for 150 km.

    a. Calculate the plane's distance from its starting point, correct to the nearest km.
    b. Calculate the plane's direction from its starting point.

13. Ship A is 16.2 km from port on a bearing of 053°T and ship B is 31.6 km from the same port on a bearing of 117°T. Calculate the distance between the two ships, in km correct to 1 decimal place.

Communicating, reasoning and problem solving

14. A plane takes off at 10.00 am from an airfield and flies at 120 km/h on a bearing of N35°W. A second plane takes off at 10.05 am from the same airfield and flies on a bearing of S80°E at a speed of 90 km/h. Determine how far apart the planes are at 10.25 am, in km correct to 1 decimal place.

15. Three circles of radii 5 cm, 6 cm and 8 cm are positioned so that they just touch one another. Their centres form the vertices of a triangle. Determine the largest angle in the triangle. Show your working.

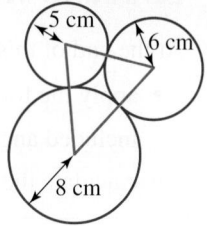

16. For the shape shown, determine:

    a. the length of the diagonal
    b. the magnitude (size) of angle B
    c. the length of x.

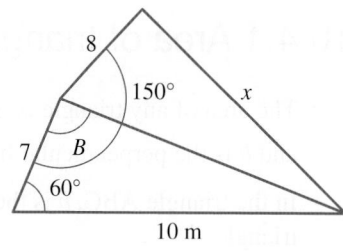

17. From the top of a vertical cliff 68 m high, an observer notices a yacht at sea. The angle of depression to the yacht is 47°. The yacht sails directly away from the cliff, and after 10 minutes the angle of depression is 15°. Determine the speed of the yacht, in km/h correct to 2 decimal places.

18. Determine the size of angles CAB, ABC and BCA.
Give your answers in degrees correct to 2 decimal places.

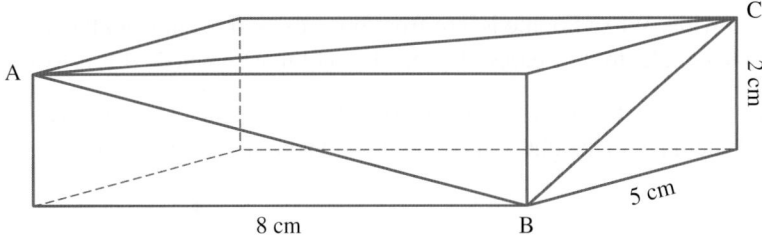

19. A vertical flag pole DB is supported by two wires AB and BC. AB is 5.2 metres long, BC is 4.7 metres long and B is 3.7 metres above ground level. Angle ADC is a right angle.

a. Evaluate the distance from A to C, in metres correct to 4 decimal places.
b. Determine the angle between AB and BC, in degrees correct to 2 decimal places.

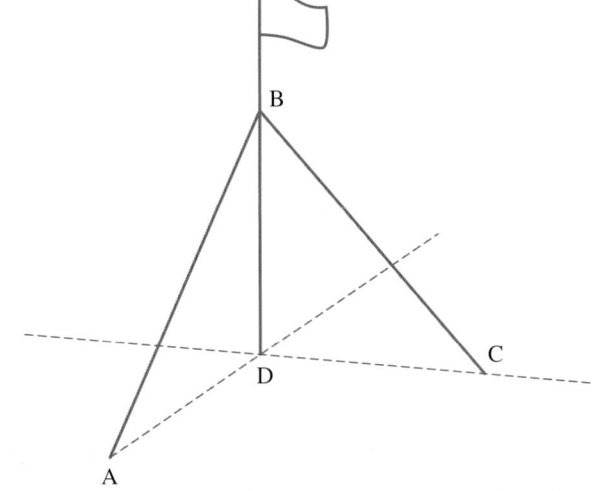

LESSON
10.4 Area of triangles

LEARNING INTENTION

At the end of this lesson you should be able to:
- apply the formula, area $= \frac{1}{2}ab\sin(C)$ to calculate the area of a triangle, given two sides and the included angle
- calculate the area of a triangle, given the three sides.

ⓟ 10.4.1 Area of triangles

eles-5007

- The area of any triangle is given by the formula area $= \frac{1}{2}bh$, where b is the base and h is the perpendicular height of the triangle.

- In the triangle ABC, b is the base and h is the perpendicular height of the triangle.

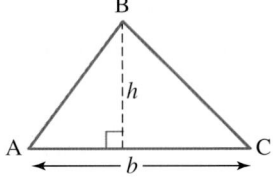

- Using the trigonometric ratio for sine:

$$\sin(A) = \frac{h}{c}$$

Transposing the equation to make h the subject, we obtain:

$$h = c\sin(A)$$

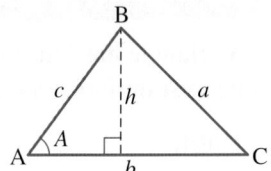

If we substitute h into the formula, area $= \frac{1}{2}bh$ we obtain the formula:

$$\text{Area} = \frac{1}{2}bc\sin(A).$$

Area of triangle

The area of triangle ABC using the sine ratio:

$$\text{Area} = \frac{1}{2}bc\sin(A)$$

- Depending on how the triangle is labelled, the formula could read:

$$\text{Area} = \frac{1}{2}ab\,\sin(C) \qquad \text{Area} = \frac{1}{2}ac\,\sin(B) \qquad \text{Area} = \frac{1}{2}bc\,\sin(A)$$

- The area formula may be used on any triangle provided that two sides of the triangle and the included angle (that is, the angle between the two given sides) are known.

WORKED EXAMPLE 7 Calculating the area of a triangle

Calculate the area of the triangle shown, in cm^2 correct to 2 decimal places.

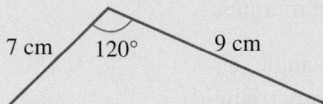

THINK	WRITE/DRAW
1. Draw a labelled diagram of the triangle, label it ABC and fill in the given information.	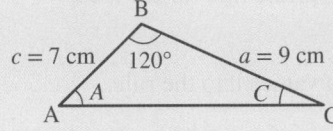 Let $a = 9$ cm, $c = 7$ cm, $B = 120°$.
2. Check that the criterion for the area rule has been satisfied.	The area rule can be used since two side lengths and the included angle have been given.
3. Write down the appropriate rule for the area.	$\text{Area} = \frac{1}{2}ac\sin(B)$
4. Substitute the known values into the rule.	$= \frac{1}{2} \times 9 \times 7 \times \sin 120°$
5. Evaluate. Round off the answer to 2 decimal places and include the appropriate unit.	$\approx 27.28 \text{ cm}^2$

A triangle has known dimensions of $a = 5$ cm, $b = 7$ cm and $B = 52°$. Determine A and C, correct to the nearest minute, and hence the area, in cm^2 correct to 2 decimal places.

THINK	WRITE/DRAW
1. Draw a labelled diagram of the triangle, label it ABC and fill in the given information.	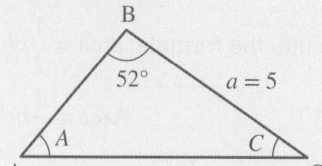
	Let $a = 5$, $b = 7$, $B = 52°$.
2. Check whether the criterion for the area rule has been satisfied.	The area rule cannot be used since the included angle has not been given.
3. Write down the sine rule to calculate A.	To calculate angle A: $$\frac{a}{\sin(A)} = \frac{b}{\sin(B)}$$
4. Substitute the known values into the rule.	$$\frac{5}{\sin(A)} = \frac{7}{\sin(52°)}$$
5. Transpose the equation to make sin A the subject.	$5\sin(52°) = 7\sin(A)$ $$\sin(A) = \frac{5\sin(52°)}{7}$$
6. Evaluate.	$$A = \sin^{-1}\left(\frac{5\sin(52°)}{7}\right)$$ $\approx 34.254\,15187°$
7. Round off the answer to degrees and minutes.	$\approx 34°15'$
8. Determine the value of the included angle, C, using the fact that the angle sum of any triangle is 180°.	$C \approx 180° - (52° + 34°15')$ $= 93°45'$
9. Write down the appropriate rule for the area.	$\text{Area} = \dfrac{1}{2}\,ab\sin(C)$
10. Substitute the known values into the rule.	$\approx \dfrac{1}{2} \times 5 \times 7 \times \sin(93°45')$
11. Evaluate. Round off the answer to 2 decimal places and include the appropriate unit.	$\approx 17.46\,cm^2$

▶ 10.4.2 Areas of triangles when three sides are known

eles-6288

- If the lengths of all the sides of the triangle are known but none of the angles, apply the cosine rule to calculate one of the angles.
- Substitute the values of the angle and the two adjacent sides into the area of a triangle formula
  $\text{Area} = \dfrac{1}{2}\,bc\sin(A)$

Calculate the area of the triangle with sides of **4 cm, 6 cm and 8 cm**, in cm² correct to 2 decimal places.

THINK	WRITE/DRAW
1. Draw a labelled diagram of the triangle, call it ABC and fill in the given information.

Let $a = 4$, $b = 6$, $c = 8$. |
2. Choose an angle and apply the cosine rule to calculate angle A.	$\cos(A) = \dfrac{b^2 + c^2 - a^2}{2bc}$ $= \dfrac{6^2 + 8^2 - 4^2}{2 \times 6 \times 8}$ $= \dfrac{7}{8}$ $A = \cos^{-1}\left(\dfrac{7}{8}\right)$ $\approx 28.95502437°$
3. Write down the appropriate rule for the area.	$\text{Area} = \dfrac{1}{2}\, bc \sin(A)$
4. Substitute the known values into the rule.	$\text{Area} = \dfrac{1}{2} \times 6 \times 8 \times \sin(28.95502437°)$
5. Evaluate. Round off the answer to 2 decimal places and include the appropriate unit.	≈ 11.61895 Area is approximately 11.62 cm².

COMMUNICATING — COLLABORATIVE TASK: Using graphing applications to verify the area rule

Equipment: Graphing software.

1. Use a graphing application (such as Geogebra), to construct a triangle of any shape.
2. Measure the base and height and calculate the area of the triangle using $A = \dfrac{1}{2}bh$.
3. Use the area rule $A = \dfrac{1}{2}ab\sin(C)$ to calculate the area of the triangle.
4. What did you notice about both calculations? Compare your results with the rest of the class.

Resources

Interactivity Area of triangles (int-6483)

Exercise 10.4 Area of triangles

10.4 Quick quiz on	10.4 Exercise

Individual pathways

■ PRACTISE	■ CONSOLIDATE	■ MASTER
1, 4, 7, 10, 12, 15, 18, 21	2, 5, 8, 11, 13, 16, 19, 22	3, 6, 9, 14, 17, 20, 23, 24

Where appropriate in this exercise, write your angles correct to the nearest minute and other measurements correct to 2 decimal places.

Fluency

1. **WE7** Calculate the area of the triangle ABC with $a = 7$, $b = 4$ and $C = 68°$.

2. Calculate the area of the triangle ABC with $a = 7.3$, $c = 10.8$ and $B = 104°40'$.

3. Calculate the area of the triangle ABC with $b = 23.1$, $c = 18.6$ and $A = 82°17'$.

4. **WE8** A triangle has $a = 10$ cm, $c = 14$ cm and $C = 48°$. Determine A and B and hence the area.

5. A triangle has $a = 17$ m, $c = 22$ m and $C = 56°$. Determine A and B and hence the area.

6. A triangle has $b = 32$ mm, $c = 15$ mm and $B = 38°$. Determine A and C and hence the area.

7. **MC** In a triangle, $a = 15$ m, $b = 20$ m and $B = 50°$. The area of the triangle is:
   - **A.** 86.2 m^2
   - **B.** 114.9 m^2
   - **C.** 149.4 m^2
   - **D.** 172.4 m^2

8. **WE9** Calculate the area of the triangle with sides of 5 cm, 6 cm and 8 cm.

9. Calculate the area of the triangle with sides of 40 mm, 30 mm and 5.7 cm.

10. Calculate the area of the triangle with sides of 16 mm, 3 cm and 2.7 cm.

11. **MC** A triangle has sides of length 10 cm, 14 cm and 20 cm. The area of the triangle is:
    - **A.** 41 cm^2
    - **B.** 65 cm^2
    - **C.** 106 cm^2
    - **D.** 137 cm^2

Understanding

12. A piece of metal is in the shape of a triangle with sides of length 114 mm, 72 mm and 87 mm. Calculate its area.

13. A triangle has the largest angle of 115°. The longest side is 62 cm and another side is 35 cm. Calculate the area of the triangle to the nearest whole number.

14. A triangle has two sides of 25 cm and 30 cm. The angle between the two sides is 30°. Determine:
    - **a.** its area
    - **b.** the length of its third side

15. The surface of a fish pond has the shape shown in the diagram.

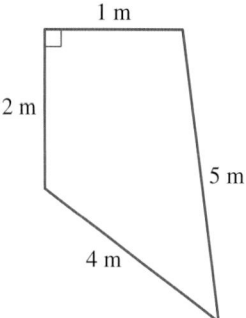

Calculate how many goldfish can the pond support if each fish requires 0.3 m^2 surface area of water.

16. **MC** A parallelogram has sides of 14 cm and 18 cm and an angle between them of 72°. The area of the parallelogram is:

A. 118.4 cm^2 B. 172.4 cm^2 C. 239.7 cm^2 D. 252 cm^2

17. **MC** An advertising hoarding is in the shape of an isosceles triangle, with sides of length 15 m, 15 m and 18 m. It is to be painted with two coats of purple paint. If the paint covers 12 m^2 per litre, the amount of paint needed, to the nearest litre, would be:

A. 9 L   B. 18 L   C. 24 L   D. 36 L

Communicating, reasoning and problem solving

18. A parallelogram has diagonals of length 10 cm and 17 cm. An angle between them is 125°. Determine:
    a. the area of the parallelogram
    b. the dimensions of the parallelogram.

19. A lawn is to be made in the shape of a triangle, with sides of length 11 m, 15 m and 17.2 m. Determine how much grass seed, to the nearest kilogram, needs to be purchased if it is sown at the rate of 1 kg per 5 m^2.

20. A bushfire burns out an area of level grassland shown in the diagram. (*Note:* This is a sketch of the area and is not drawn to scale.) Evaluate the area, in hectares correct to 1 decimal place, of the land that is burned.

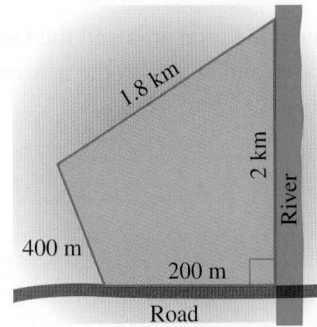

21. An earth embankment is 27 m long and has a vertical cross-section shown in the diagram. Determine the volume of earth needed to build the embankment, correct to the nearest cubic metre.

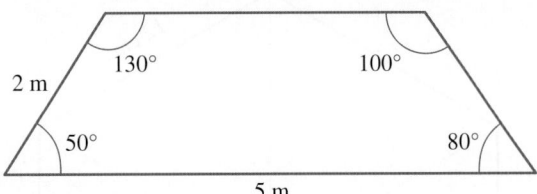

22. Evaluate the area of this quadrilateral.

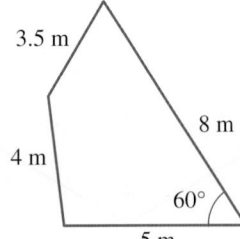

23. A surveyor measured the boundaries of a property as shown. The side AB could not be measured because it crossed through a marsh.

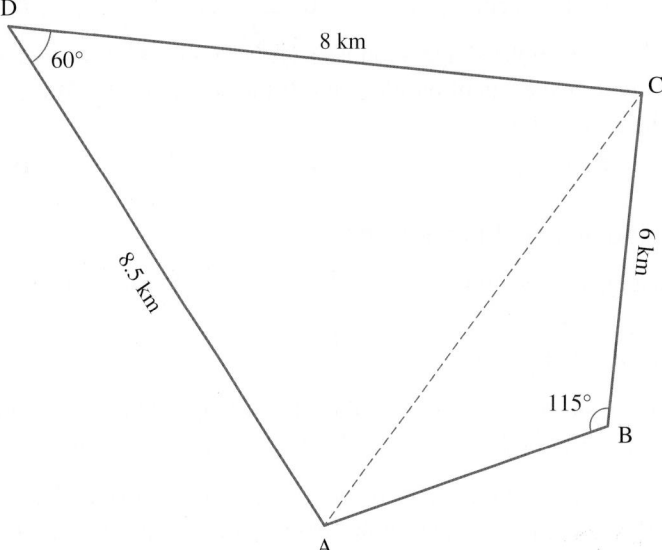

The owner of the property wanted to know the total area and the length of the side AB.
Give all lengths correct to 2 decimal places and angles to the nearest degree.

a. Calculate the area of the triangle ACD.
b. Calculate the distance AC.
c. Calculate the angle CAB.
d. Calculate the angle ACB.
e. Calculate the length AB.
f. Determine the area of the triangle ABC.
g. Determine the area of the property.

24. A regular hexagon has sides of length 12 centimetres. It is divided into six smaller equilateral triangles. Evaluate the area of the hexagon, giving your answer correct to 2 decimal places.

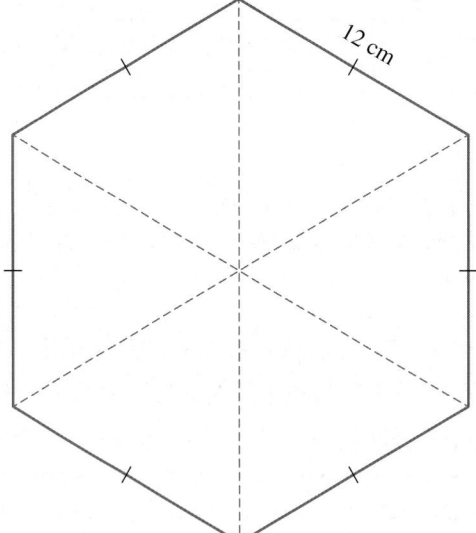

LESSON
10.5 The unit circle

LEARNING INTENTION

At the end of this lesson you should be able to:
- determine in which quadrant an angle lies
- apply and interpret the relationship between a point on the unit circle and the angle made with the positive x-axis
- use the unit circle to determine approximate trigonometric ratios for angles greater than 90°
- apply the relationships between supplementary and complementary angles
- relate the gradient of a line to its angle of inclination with the x-axis on the Cartesian plane.

▶ 10.5.1 The unit circle

eles-5009

- A **unit circle** is a circle with a radius of 1 unit.
- The unit circle is divided into 4 quadrants, numbered in an anticlockwise direction, as shown in the diagram.

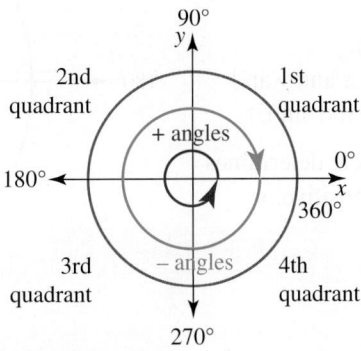

- *Positive angles* are measured anticlockwise from 0°.
- *Negative angles* are measured clockwise from 0°.

WORKED EXAMPLE 10 Identifying where an angle lies on the unit circle

State the quadrant of the unit circle in which each of the following angles is found.
a. 145°   **b. 282°**

THINK   **WRITE**

a. The given angle is between 90° and 180°.   **a.** 145° is in quadrant 2.
State the appropriate quadrant.

b. The given angle is between 270° and 360°.   **b.** 282° is in quadrant 4.
State the appropriate quadrant.

- Consider the unit circle with point $P(x, y)$ making the right-angled triangle OPN as shown in the diagram.
- Using the trigonometric ratios:

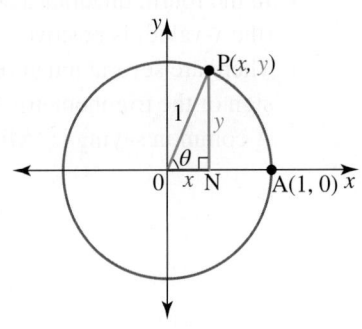

$$\frac{x}{1} = \cos(\theta), \quad \frac{y}{1} = \sin(\theta), \quad \frac{y}{x} = \tan(\theta)$$

where θ is measured anticlockwise from the positive x-axis.

> ## Calculate value of sine, cosine and tangent
>
> **To calculate the value of sine, cosine or tangent of any angle θ from the unit circle:**
>
> $$\cos(\theta) = x$$
>
> $$\sin(\theta) = y$$
>
> $$\tan(\theta) = \frac{y}{x} = \frac{\sin(\theta)}{\cos(\theta)}$$

⊙ 10.5.2 The four quadrants of the unit circle

eles-5010

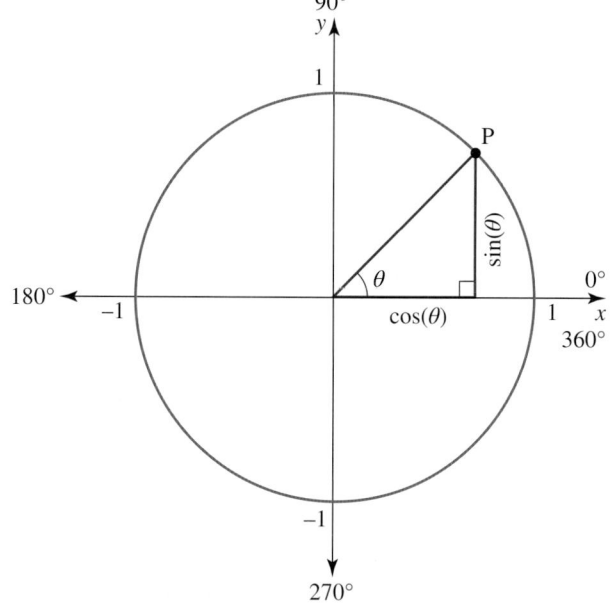

- Approximate values for sine, cosine and tangent of an angle can be found from the unit circle using the following steps, as shown in the diagram.

  Step 1: Draw a unit circle, label the x- and y-axes.
  Step 2: Mark the angles 0°, 90°, 180°, 270°, and 360°.
  Step 3: Draw the given angle θ.
  Step 4: Mark $x = \cos(\theta)$, $y = \sin(\theta)$.
  Step 5: Approximate the values of x and y and equate to give the values of $\cos(\theta)$ and $\sin(\theta)$.

- Where the angle lies in the unit circle determines whether the trigonometric ratio is positive or negative.

Sign of the trigonometric functions

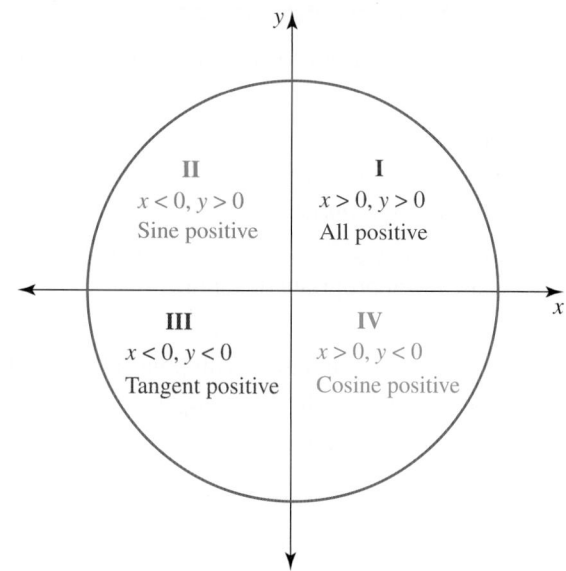

- Consider the following.
  - In the first quadrant $x > 0$, $y > 0$; therefore **All** trig ratios are positive.
  - In the second quadrant $x < 0$, $y > 0$; therefore **Sine** (the y-value) is positive.
  - In the third quadrant $x < 0$, $y < 0$; therefore **Tangent** $\left(\dfrac{y}{x} \text{ values} \right)$ is positive.
  - In the fourth quadrant $x > 0$, $y < 0$; therefore **Cosine** (the x-value) is positive.
  - There are several mnemonics for remembering the sign of the trigonometric functions.
    A common saying is '**A**ll **S**tations **T**o **C**entral'

WORKED EXAMPLE 11 Using the unit circle to approximate trigonometric ratios of an angle

Determine the approximate value of each of the following using the unit circle.
a. sin(200°)   b. cos(200°)   c. tan(200°)

THINK	WRITE/DRAW
Draw a unit circle and construct an angle of 200°. Label the point corresponding to the angle of 200° on the circle P. Highlight the lengths, representing the *x*- and *y*-coordinates of point P.	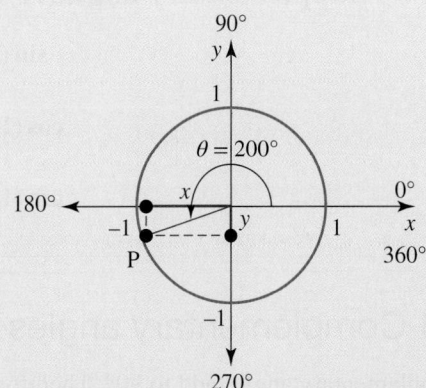

a. The sine of the angle is given by the *y*-coordinate of P. Determine the *y*-coordinate of P by measuring the distance along the *y*-axis. State the value of sin(200°). (*Note:* The sine value will be negative as the *y*-coordinate is negative.)

a. $\sin(200°) = -0.3$

b. The cosine of the angle is given by the *x*-coordinate of P. Determine the *x*-coordinate of P by measuring the distance along the *x*-axis. State the value of cos(200°). (*Note:* Cosine is also negative in quadrant 3, as the *x*-coordinate is negative.)

b. $\cos(200°) = -0.9$

c. $\tan(200°) = \dfrac{\sin(200°)}{\cos(200°)}$

c. $\dfrac{-0.3}{-0.9} = \dfrac{1}{3} = 0.3333$

- The approximate results obtained in Worked example 11 can be verified with the aid of a calculator:

$$\sin(200°) = -0.342\,020\,143, \cos(200°) = -0.939\,692\,62 \text{ and } \tan(200°) = 0.3640.$$

Rounding these values to 1 decimal place would give −0.3, −0.9 and 0.4 respectively, which approximately match the values obtained from the unit circle.

▶ 10.5.3 Supplementary angles

eles-6284

- Consider the special relationship between the sine, cosine and tangent of supplementary angles, say $A°$ and $(180 - A)°$.

In the diagram, the *y*-axis is an axis of symmetry.
  - The *y*-values of points C and E are the same.
    That is, $\sin(A°) = \sin(180 - A)°$
  - The *x*-values of points C and E are opposites in value.
    That is, $\cos(A°) = -\cos(180 - A)°$

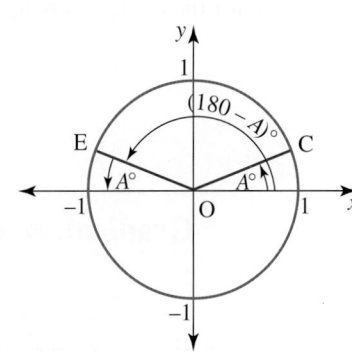

Thus: $\sin(180 - A)° = \sin(A°)$

$\cos(180 - A)° = -\cos(A°)$

$\tan(180 - A)° = \dfrac{\sin(180-A)°}{\cos(180-A)°} = \dfrac{\sin(A°)}{-\cos(A°)} = -\tan(A°)$

Supplementary angles $A°$ and $(180° - A)$ where $0° \leq A \leq 90°$

$$\sin(180° - A) = \sin(A)$$

$$\cos(180° - A) = -\cos(A)$$

$$\tan(180° - A) = -\tan(A)$$

▶ 10.5.4 Complementary angles

eles-6285

- Complementary angles add to 90°. Therefore, 30° and 60° are complementary angles, and θ and $(90° - \theta)$ are also complementary angles.
- The sine of an angle is equal to the cosine of its complement. Therefore, $\sin(60°) = \cos(30°)$.
- We say that sine and cosine are complementary functions.

$$\sin(A) = \cos(90° - A)$$
$$\cos(A) = \sin(90° - A)$$

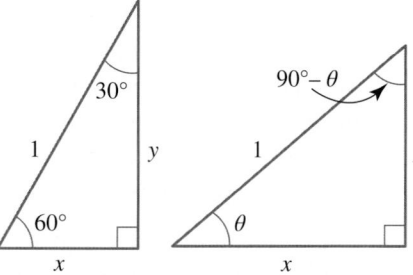

Complementary angles where $0° \leq A \leq 90°$

$$\sin(A) = \cos(90° - A)$$

$$\cos(A) = \sin(90° - A)$$

▶ 10.5.5 Gradient of a line

eles-6286

- The gradient of a line may be found using trigonometry if the angle the line makes with the positive direction of the x-axis is known.
- In the triangle shown, the gradient of the line is $\dfrac{\text{rise}}{\text{run}}$ which is the tangent ratio for the angle θ, giving the relationship $m = \tan(\theta)$.

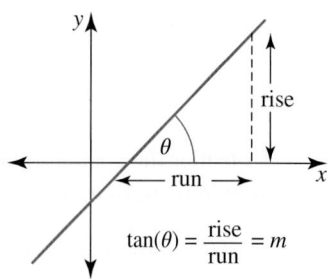

Gradient, m, of a line

$$m = \tan(\theta)$$

WORKED EXAMPLE 12 Gradients of lines given the angles made with the x-axis

a. Determine the gradient (accurate to 3 decimal places) of a line making an angle of 40° to the positive x-axis, as shown in the figure at right.

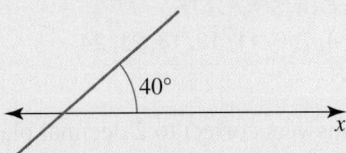

b. Determine the gradient of the line shown at right. Express your answer to 2 decimal places.

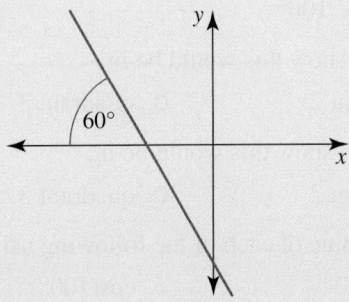

THINK	WRITE/DRAW
a. Since the angle the line makes with the positive x-axis is given, the formula $m = \tan(\theta)$ can be used.	**a.** $m = \tan(\theta)$ $= \tan 40°$ $= 0.839$
b. 1. The angle given is not the one between the graph and the positive direction of the x-axis. Calculate the required angle θ.	**b.** $\theta = 180° - 60°$ $= 120°$
2. Use $m = \tan(\theta)$ to calculate m to 2 decimal places.	$m = \tan(\theta)$ $= \tan(120°)$ $= -1.73$

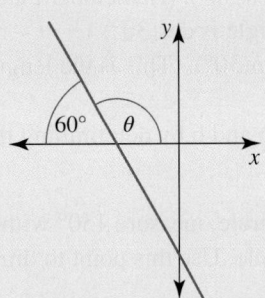

| 10.5 Quick quiz on | 10.5 Exercise |

Individual pathways

■ PRACTISE	■ CONSOLIDATE	■ MASTER
1, 2, 6, 8, 12, 17, 20, 23	3, 4, 7, 9, 11, 13, 18, 21, 24	5, 10, 14, 15, 16, 19, 22, 25

Where appropriate in this exercise, give answers correct to 2 decimal places.

Fluency

1. **WE10** State which quadrant of the unit circle each of the following angles is in.
    a. $60°$
    b. $130°$
    c. $310°$
    d. $260°$
    e. $100°$
    f. $185°$

2. **MC** If $\theta = 43°$, the triangle drawn to show this would be in:
    A. quadrant 1
    B. quadrant 2
    C. quadrant 3
    D. quadrant 4

3. **MC** If $\theta = 295°$, the triangle drawn to show this would be in:
    A. quadrant 1
    B. quadrant 2
    C. quadrant 3
    D. quadrant 4

4. **WE11** Determine the approximate value of each of the following using the unit circle.
    a. $\sin(20°)$
    b. $\cos(20°)$
    c. $\cos(100°)$
    d. $\sin(100°)$

5. Determine the approximate value of each of the following using the unit circle.
    a. $\sin(320°)$
    b. $\cos(320°)$
    c. $\sin(215°)$
    d. $\cos(215°)$

6. Use the unit circle to determine the approximate value of each of the following.
    a. $\sin(90°)$
    b. $\cos(90°)$
    c. $\sin(180°)$
    d. $\cos(180°)$

7. Use the unit circle to determine the approximate value of each of the following.
    a. $\sin(270°)$
    b. $\cos(270°)$
    c. $\sin(360°)$
    d. $\cos(360°)$

Understanding

8. On the unit circle, use a protractor to measure an angle of $30°$ from the positive x-axis. Mark the point P on the circle. Use this point to construct a triangle in quadrant 1 as shown.

    a. Calculate the value of $\cos(30°)$. (Remember that the length of the adjacent side of the triangle is $\cos(30°)$.)
    b. Calculate the value of $\sin(30°)$. (This is the length of the opposite side of the triangle.)
    c. Check your answers in a and b by determining these values with a calculator.

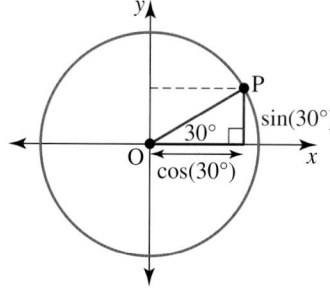

9. Using a graph of the unit circle, measure $150°$ with a protractor and mark the point P on the circle. Use this point to draw a triangle in quadrant 2 as shown.

    a. Determine the angle the radius OP makes with the negative x-axis.
    b. Remembering that $x = \cos(\theta)$, use your circle to determine the value of $\cos(150°)$.
    c. Comment on how $\cos(150°)$ compares to $\cos(30°)$.
    d. Remembering that $y = \sin(\theta)$, use your circle to determine the value of $\sin(150°)$.
    e. Comment on how $\sin(150°)$ compares with $\sin(30°)$.

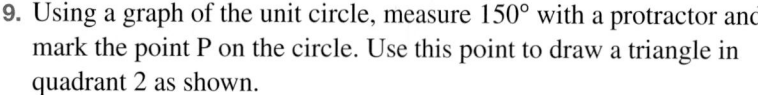

10. On a unit circle, measure 210° with a protractor and mark the point P on the circle. Use this point to draw a triangle in quadrant 3 as shown.

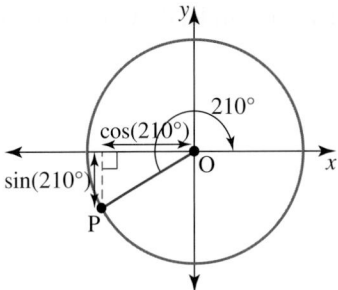

   a. Determine the angle the radius OP makes with the negative *x*-axis.

   b. Use your circle to determine the value of cos(210°).

   c. Comment on how cos(210°) compares to cos(30°).

   d. Use your circle to determine the value of sin(210°).

   e. Comment on how sin(210°) compares with sin(30°).

11. On a unit circle, measure 330° with a protractor and mark the point P on the circle. Use this point to draw a triangle in quadrant 4 as shown.

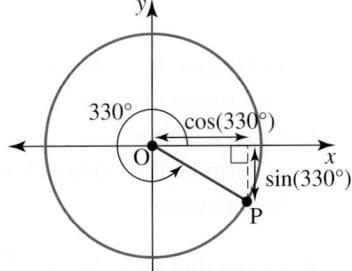

   a. Determine the angle the radius OP makes with the positive *x*-axis.

   b. Use your circle to determine the value of cos(330°).

   c. Comment on how cos(330°) compares to cos(30°).

   d. Use your circle to determine the value of sin(330°).

   e. Comment on how sin(330°) compares with sin(30°).

12. On a unit circle, draw an appropriate triangle for the angle of 20° in quadrant 1.

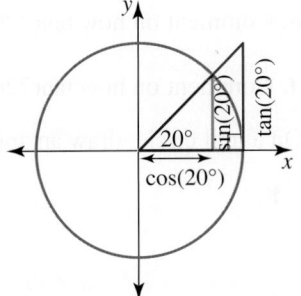

   a. Determine the value of sin(20°).

   b. Determine the value of cos(20°).

   c. Draw a tangent line and extend the hypotenuse of the triangle to meet the tangent as shown.
Accurately measure the length of the tangent between the *x*-axis and the point where it meets the hypotenuse and, hence, state the value of tan(20°).

   d. Determine the value of $\dfrac{\sin(20°)}{\cos(20°)}$.

   e. Comment on how tan(20°) compares with $\dfrac{\sin(20°)}{\cos(20°)}$.

13. On a unit circle, draw an appropriate triangle for the angle of 135° in quadrant 2.

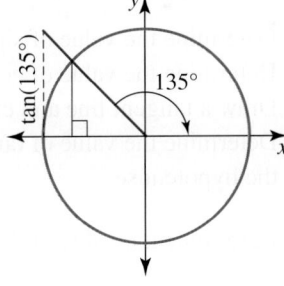

   a. Determine the value of sin(135°), using sin(45°).

   b. Determine the value of cos(135°), using cos(45°).

   c. Draw a tangent line and extend the hypotenuse of the triangle to meet the tangent as shown.
Accurately measure the length of the tangent to where it meets the hypotenuse to calculate the value of tan(135°).

   d. Determine the value of $\dfrac{\sin(135°)}{\cos(135°)}$.

   e. Comment on how tan(135°) compares with $\dfrac{\sin(135°)}{\cos(135°)}$.

   f. Comment on how tan(135°) compares with tan(45°).

14. On a unit circle, draw an appropriate triangle for the angle of 220° in quadrant 3.

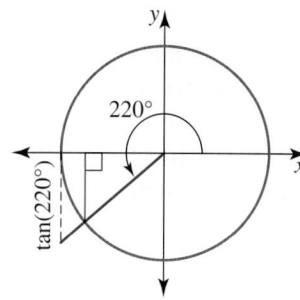

    a. Determine the value of sin(220°).

    b. Determine the value of cos(220°).

    c. Draw a tangent line and extend the hypotenuse of the triangle to meet the tangent as shown. Determine the value of tan(220°) by accurately measuring the length of the tangent to where it meets the hypotenuse.

    d. Determine the value of $\dfrac{\sin(220°)}{\cos(220°)}$.

    e. Comment on how tan(220°) compares with $\dfrac{\sin(220°)}{\cos(220°)}$.

    f. Comment on how tan(220°) compares with tan(40°). (Use a calculator.)

15. On a unit circle, draw an appropriate triangle for the angle of 300° in quadrant 4.

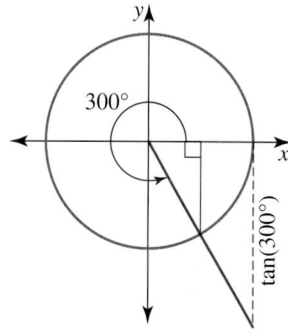

    a. Determine the value of sin(300°).

    b. Determine the value of cos(300°).

    c. Draw a tangent line and extend the hypotenuse of the triangle to meet the tangent as shown. Determine the value of tan(300°) by accurately measuring the length of the tangent to where it meets the hypotenuse.

    d. Determine the value of the value of $\dfrac{\sin(300°)}{\cos(300°)}$.

    e. Comment on how tan(300°) compares with $\dfrac{\sin(300°)}{\cos(300°)}$.

    f. Comment on how tan(300°) compares with tan(60°). (Use a calculator.)

16. **MC** In a unit circle, the length of the radius is equal to:

    A. sin(θ)   **B.** cos(θ)   **C.** tan(θ)   **D.** 1

17. **WE12a** Determine the gradients (accurate to 3 decimal places) of lines making the following angles with the positive x-axis.

   a. 50°
   b. 72°
   c. 10°
   d. −30°

18. Determine the gradients (accurate to 3 decimal places) of lines making the following angles with the positive x-axis.

   a. 150°
   b. 0°
   c. 45°
   d. 89°

19. **WE12b** Determine the gradient of each line in the figures below. Give your answers accurate to 2 decimal places.

   a.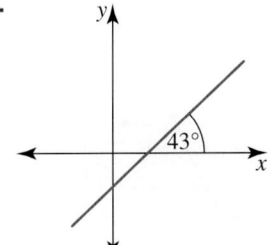

   b.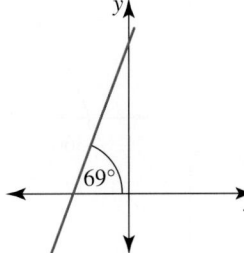

   c.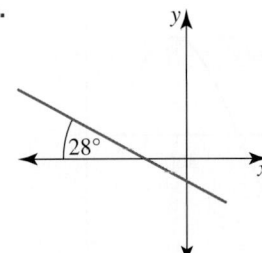

   d.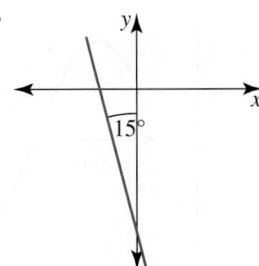

Communicating, reasoning and problem solving

20. Show that $\sin^2(\alpha°) + \cos^2(\alpha°) = 1$.

21. Show that $1 - \sin^2(180 - \alpha)° = \cos^2(180 - \alpha)°$.

22. Show that $1 + \tan^2(\alpha°) = \sec^2(\alpha°)$, where $\sec(\alpha°) = \dfrac{1}{\cos(\alpha°)}$.

23. If $\sin(x°) = p$, $0 \le x \le 90°$, write each of the following in terms of p.

   a. $\cos(x°)$
   b. $\sin(180 - x)°$
   c. $\cos(180 - x)°$
   d. $\cos(90 - x)°$

24. Simplify $\sin(180 - x)° - \sin(x°)$.

25. Simplify $\cos(180 - x)° + \cos(x°)$, where $0 < x° < 90°$.

LESSON
10.6 Trigonometric functions

LEARNING INTENTION

At the end of this lesson you should be able to:
- sketch the graphs of the sine, cosine and tangent graphs using graphing applications
- compare the features of sine, cosine and tangent graphs using graphing applications
- examine the periodic nature of trigonometric graphs.

▶ 10.6.1 Sine, cosine and tangent graphs

eles-5011

- The graphs of $y = \sin(x)$, $y = \cos(x)$ and $y = \tan(x)$ are shown below.

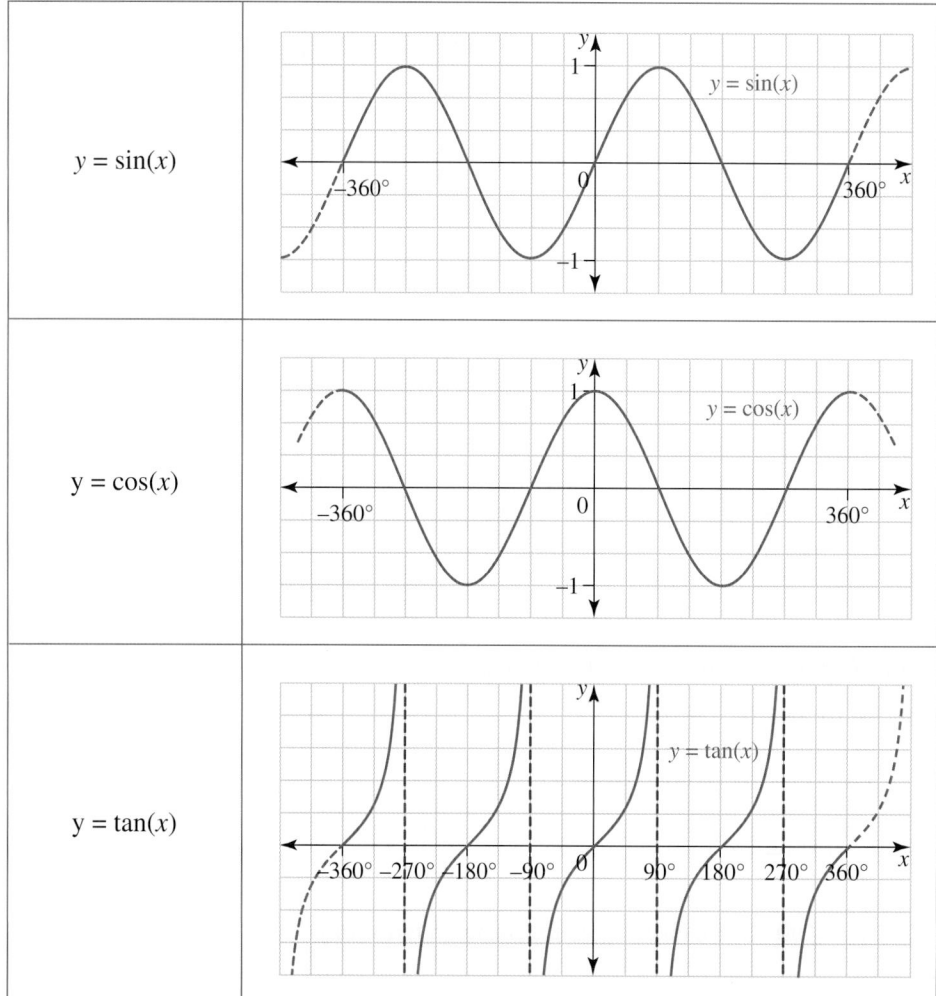

- Trigonometric graphs repeat themselves continuously in cycles, and hence they are called **periodic functions**.
- The **period** of the graph is the horizontal distance between repeating peaks or troughs. The period between the repeating peaks for $y = \sin(x)$ and $y = \cos(x)$ is 360°. The period of the graph $y = \tan(x)$ is 180°, and **asymptotes** occur at $x = 90°$ and intervals of 180°.
- The **amplitude** of a periodic graph is half the distance between the maximum and minimum values of the function. Amplitude can also be described as the amount by which the graph goes above and below its mean value, which is the x-axis for the graphs of $y = \sin(x)$, $y = \cos(x)$ and $y = \tan(x)$.

- The following can be summarised from the graphs of the trigonometric functions.

Graph	Period	Amplitude
$y = \sin(x)$	360°	1
$y = \cos(x)$	360°	1
$y = \tan(x)$	180°	Undefined

- A scientific calculator can be used to check significant points such as 0°, 90°, 180°, 270° and 360° for the shape of a sin, cos or tan graph

Transformations of trigonometric graphs

- **The sine, cosine and tangent graphs can be dilated, translated and reflected in the same way as other functions.**
- **These transformations are summarised in the table below.**

Graph	Period	Amplitude
$y = a \sin(nx)$	$\dfrac{360°}{n}$	a
$y = a \cos(nx)$	$\dfrac{360°}{n}$	a
$y = a \tan(nx)$	$\dfrac{180°}{n}$	Undefined

- **If $a < 0$, the graph is reflected in the x-axis. The amplitude is always the positive value of a.**

WORKED EXAMPLE 13 Sketching periodic functions

Sketch the following graphs for $0° \le x \le 360°$ using appropriate technology.
a. $y = 2\sin(x)$ b. $y = \cos(2x)$

THINK

a. 1. The graph must be drawn from 0° to 360°.
   2. Compared to the graph of $y = \sin(x)$ each value of $\sin(x)$ has been multiplied by 2, therefore the amplitude of the graph must be 2.
   3. Label the graph $y = 2\sin(x)$.

b. 1. The graph must be drawn from 0° to 360°.
   2. Compared to the graph of $y = \cos(x)$, each value of x has been multiplied by 2, therefore the period of the graph must become 180°.
   3. Label the graph $y = \cos(2x)$.

WRITE/DRAW

a.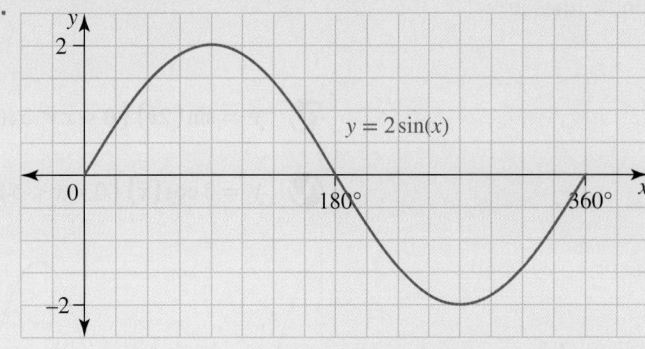

b.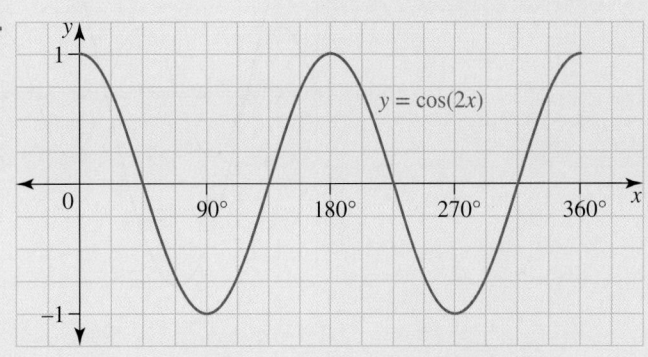

DIGITAL TECHNOLOGY

Digital tools allow trigonometric functions to be easily graphed. There are several applications available on the internet to graph trigonometric functions. The examples below use the Desmos Graphing Calculator.

- Ensure the mode is on degrees as angles may be in either degrees or radians. (Radian measure will be studied in the future.)
- Include the restrictions on the values of x when entering the graph.

The example below is the graph of $y = 2\sin(x)$ for $-180° \le x \le 360°$

The entry line is:

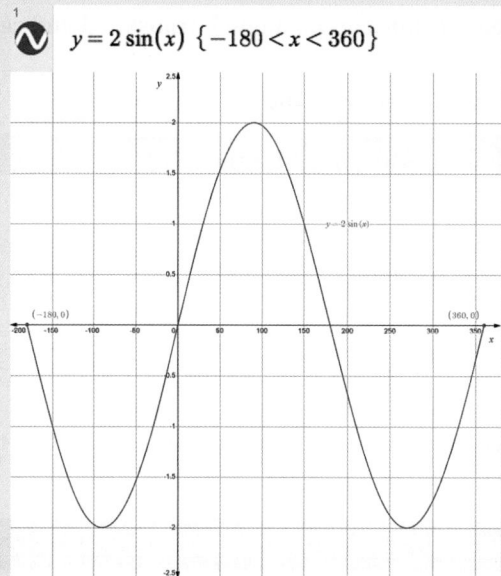

This graph shows the curve has an amplitude of 2 and a period of $360°$.

The graph below shows the curves, $y = \sin(2x)$ and $y = 3\cos(x)$ for $0° \le x \le 540°$.

The entry lines are:

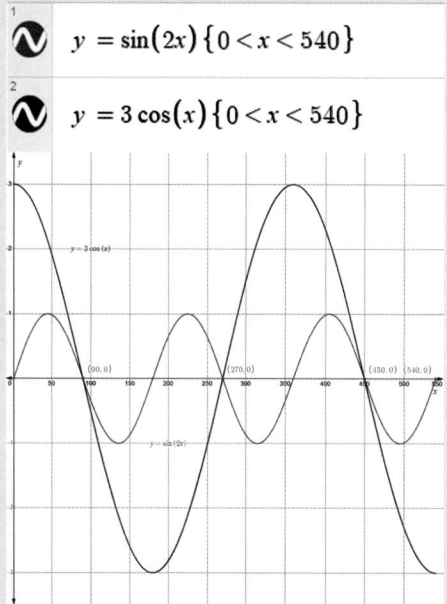

The curves can be compared easily from the graphs.

The purple curve, $y = \sin(2x)$, has an amplitude of 1 and a period of 180° as there are 3 complete cycles in the 540°.

The black curve, $y = 3\cos(x)$, has an amplitude of 3 and a period of 360° as there are 1.5 cycles in the 540°.

Points of intersection and axis intercepts can be observed from the graphs, or obtained from the application.

WORKED EXAMPLE 14 Stating the amplitude and period of given periodic functions

For each of the following graphs, state:
 i. the amplitude
 ii. the period.

a. $y = 2\sin(3x)$

b. $y = \cos\left(\dfrac{x}{3}\right)$

c. $y = \tan(2x)$

THINK	WRITE
a. The value of a is 2.	a. i. Amplitude $= 2$
The periods is $\dfrac{360°}{n}$.	ii. Period $= \dfrac{360}{3} = 120°$
b. The value of a is 1.	b. i. Amplitude $= 1$
The period is $\dfrac{360°}{n}$.	ii. Period $= \dfrac{360}{\frac{1}{3}} = 1080°$
c. The tangent curve has an undefined amplitude.	c. i. Amplitude $=$ undefined
The period is $\dfrac{180°}{2}$.	ii. Period $= \dfrac{180°}{2} = 90°$

---

on Resources

Interactivity Graphs of trigonometric functions (int-4821)

Exercise 10.6 Trigonometric functions

| 10.6 Quick quiz | on | | 10.6 Exercise |

Individual pathways

PRACTISE	CONSOLIDATE	MASTER
1, 4, 8, 12, 16, 20, 25, 26, 28, 29, 32	3, 5, 7, 10, 13, 14, 17, 21, 23, 27, 30, 33	2, 6, 9, 11, 15, 18, 19, 22, 24, 31, 34

Fluency

1. Using your calculator (or the unit circle if you wish), complete the following table.

x	0°	30°	60°	90°	120°	150°	180°	210°	240°	270°	300°	330°	360°
$\sin(x)$													
x	390°	420°	450°	480°	510°	540°	570°	600°	630°	660°	690°	720°	
$\sin(x)$													

For questions 2 to 7, using graph paper, rule x- and y-axes and carefully mark a scale along each axis.

2. Use 1 cm = 30° on the x-axis to show x-values from 0° to 720°.
   Use 2 cm = 1 unit along the y-axis to show y-values from −1 to 1.
   Carefully plot the graph of $y = \sin(x)$ using the values from the table in question 1.

3. State how long it takes for the graph of $y = \sin(x)$ to complete one full cycle.

4. From your graph of $y = \sin(x)$, estimate to 1 decimal place the value of y for each of the following.
   a. $x = 42°$ b. $x = 130°$ c. $x = 160°$ d. $x = 200°$

5. From your graph of $y = \sin(x)$, estimate to 1 decimal place the value of y for each of the following.
   a. $x = 180$ b. $x = 70°$ c. $x = 350°$ d. $x = 290°$

6. From your graph of $y = \sin(x)$, estimate to the nearest degree a value of x for each of the following.
   a. $y = 0.9$ b. $y = -0.9$ c. $y = 0.7$

7. From your graph of $y = \sin(x)$, estimate to the nearest degree a value of x for each of the following.
   a. $y = -0.5$ b. $y = -0.8$ c. $y = 0.4$

8. Using your calculator (or the unit circle if you wish), complete the following table.

x	0°	30°	60°	90°	120°	150°	180°	210°	240°	270°	300°	330°	360°
$\cos(x)$													
x	390°	420°	450°	480°	510°	540°	570°	600°	630°	660°	690°	720°	
$\cos(x)$													

For questions 9 to 14, using graph paper, rule x- and y-axes and carefully mark a scale along each axis.

9. Use 1 cm = 30° on the x-axis to show x-values from 0° to 720°.
   Use 2 cm = 1 unit along the y-axis to show y-values from −1 to 1.
   Carefully plot the graph of $y = \cos(x)$ using the values from the table in question 8.

10. If you were to continue the graph of $y = \cos(x)$, state what shape you would expect it to take.

11. State whether the graph of $y = \cos(x)$ is the same as the graph of $y = \sin(x)$. Explain how it differs. State what features are the same.

12. Using the graph of $y = \cos(x)$, estimate to 1 decimal place the value of y for each of the following.

   a. $x = 48°$ b. $x = 155°$ c. $x = 180°$ d. $x = 340°$

13. Using the graph of $y = \cos(x)$, estimate to 1 decimal place the value of y for each of the following.

   a. $x = 240°$ b. $x = 140°$ c. $x = 40°$ d. $x = 200°$

14. Using the graph of $y = \cos(x)$, estimate to the nearest degree a value of x for each of the following.

   a. $y = -0.5$ b. $y = 0.8$ c. $y = 0.7$

15. Using the graph of $y = \cos(x)$, estimate to the nearest degree a value of x for each of the following.

   a. $y = -0.6$ b. $y = 0.9$ c. $y = -0.9$

16. Using your calculator (or the unit circle if you wish), complete the following table.

x	0°	30°	60°	90°	120°	150°	180°	210°	240°	270°	300°	330°	360°
$\tan(x)$													
x	390°	420°	450°	480°	510°	540°	570°	600°	630°	660°	690°	720°	
$\tan(x)$													

For questions 17 to 22, using graph paper, rule x- and y-axes and carefully mark a scale along each axis.

17. Use 1 cm = 30° on the x-axis to show x-values from 0° to 720°.
   Use 2 cm = 1 unit along the y-axis to show y-values from -2 to 2.
   Carefully plot the graph of $y = \tan(x)$ using the values from the table in question 16.

18. If you were to continue the graph of $y = \tan(x)$, state what shape would you expect it to take.

19. State whether the graph of $y = \tan(x)$ is the same as the graphs of $y = \sin(x)$ and $y = \cos(x)$. Explain how it differs. State what features are the same.

20. Using the graph of $y = \tan(x)$, estimate to 1 decimal place the value of y for each of the following.

   a. $x = 60°$ b. $x = 135°$ c. $x = 310°$ d. $x = 220°$

21. Using the graph of $y = \tan(x)$, determine the value of y for each of the following.

   a. $x = 500°$ b. $x = 590°$ c. $x = 710°$ d. $x = 585°$

22. Using the graph of $y = \tan(x)$, estimate to the nearest degree a value of x for each of the following.

   a. $y = 1$ b. $y = 1.5$ c. $y = -0.4$
   d. $y = -2$ e. $y = 0.2$ f. $y = -1$

WE13&14 For each of the graphs in questions 23 and 24:
   i. state the period
  ii. state the amplitude
 iii. sketch the graph.

23. a. $y = \cos(x)$, for $x \in [-180°, 180°]$ b. $y = \sin(x)$, for $x \in [0°, 720°]$

24. a. $y = \sin(2x)$, for $x \in [0°, 360°]$ b. $y = 2\cos(x)$, for $x \in [-360°, 0°]$

25. For each of the following, state:
   i. the period
  ii. the amplitude.

   a. $y = 3\cos(2x)$ b. $y = 4\sin(3x)$ c. $y = 2\cos\left(\dfrac{x}{2}\right)$

   d. $y = \dfrac{1}{2}\sin\left(\dfrac{x}{4}\right)$ e. $y = -\sin(x)$ f. $y = -\cos(2x)$

Understanding

26. **MC** Use the graph shown to answer the following.

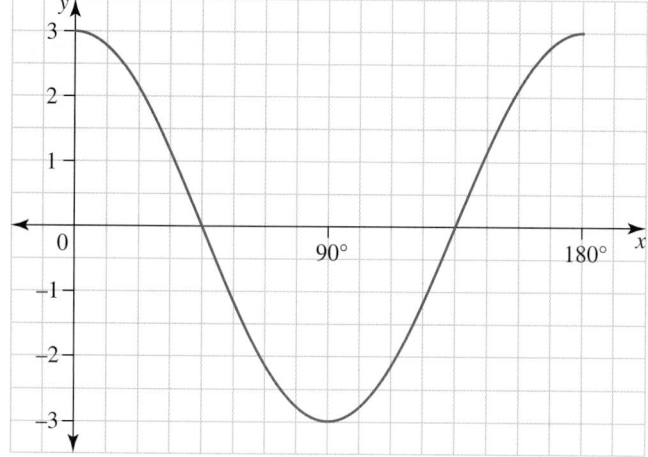

a. The amplitude of the graph is:

   A. 180°   **B.** 90°

   C. 3   **D.** −3

b. The period of the graph is:

   A. 180°   **B.** 360°

   C. 90°   **D.** 3

c. The equation of the graph could be:

   A. $y = \cos(x)$ **B.** $y = \sin(x)$

   C. $y = 3\cos\left(\dfrac{x}{3}\right)$ **D.** $y = 3\cos(2x)$

27. Sketch using graphing applications each of the following graphs, stating the period and amplitude of each.

a. $y = 2\cos\left(\dfrac{x}{3}\right)$, for $x \in [0°, 1080°]$

b. $y = -3\sin(2x)$, for $x \in [0°, 360°]$

c. $y = 3\sin\left(\dfrac{x}{2}\right)$, for $x \in [-180°, 180°]$

d. $y = -\cos(3x)$, for $x \in [0°, 360°]$

e. $y = 5\cos(2x)$, for $x \in [0°, 180°]$

f. $y = -\sin(4x)$, for $x \in [0°, 180°]$

28. Use technology to sketch the graphs of each of the following for $0° \le x \le 360°$.

a. $y = \cos(x) + 1$

b. $y = \sin(2x) - 2$

c. $y = \cos\left(\dfrac{\pi}{180}(x - 60)\right)$

d. $y = 2\sin(4x) + 3$

Communicating, reasoning and problem solving

29. a. Sketch the graph of $y = \cos(2x)$ for $x \in [0°, 360°]$ using graphing applications.

   i. State the minimum value of y for this graph.

   ii. State the maximum value of y for this graph.

b. Using the answers obtained in part a write down the maximum and minimum values of $y = \cos(2x) + 2$.

c. Determine what would be the maximum and minimum values of the graph of $y = 2\sin(x) + 3$. Explain how you obtained these values.

30. a. Complete the table below by filling in the exact values of $y = \tan(x)$.

x	0°	30°	60°	90°	120°	150°	180°
$y = \tan(x)$							

b. Sketch the graph of $y = \tan(x)$ for $[0°, 180°]$.

c. Determine what happens at $x = 90°$.

d. For the graph of $y = \tan(x)$, $x = 90°$ is called an asymptote. Write down when the next asymptote would occur.

e. Determine the period and amplitude of $y = \tan(x)$.

31. a. Sketch the graph of $y = \tan(2x)$ for $[0°, 180°]$ using graphing applications.

b. Determine when the asymptotes occur.

c. State the period and amplitude of $y = \tan(2x)$.

32. The height of the tide above the mean sea level on the first day of the month is given by the rule
$$h = 3\sin(30t°)$$
where t is the time in hours since midnight.
   a. Sketch the graph of h versus t.
   b. Determine the height of the high tide.
   c. Calculate the height of the tide at 8 am.

33. The height, h metres, of the tide on the first day of January at Trig Cove is given by the rule
$$h = 6 + 4\sin(30t°)$$
where t is the time in hours since midnight.
   a. Sketch the graph of h versus t, for $0 \leq t \leq 24$.
   b. Determine the height of the high tide.
   c. Determine the height of the low tide.
   d. Calculate the height of the tide at 10 am, correct to the nearest centimetre.

34. The temperature, T, inside a house t hours after 3 am is given by the rule
$$T = 22 - 2\cos(15t°) \text{ for } 0 \leq t \leq 24$$
where T is the temperature in degrees Celsius.
   a. Determine the temperature inside the house at 9 am.
   b. Sketch the graph of T versus t.
   c. Determine the warmest and coolest temperatures that it gets inside the house over the 24-hour period.

LESSON
10.7 Solving trigonometric equations

LEARNING INTENTION

At the end of this lesson you should be able to:
 • derive and apply the exact values of sin, cos and tan for 30°, 45°, 60° angles
 • solve trigonometric equations graphically for a given domain
 • solve trigonometric equations algebraically, using exact values, for a given domain.

▶ 10.7.1 Exact values of trigonometric functions

eles-6307

 • Most of the trigonometric values that we will deal with in this topic are approximations.
 • However, angles of 30° 45° and 60° have exact values of sine, cosine and tangent.
 • Consider an equilateral triangle, ABC, of side length 2 cm.
   Let BD be the perpendicular bisector of AC, then:
   $\triangle ABD \cong \triangle CBD$ (using RHS)
   giving:
   $AD = CD = 1$ cm
   $\angle ABD = \angle CBD = 30°$
   and:
   $(AB)^2 = (AD)^2 + (BD)^2$ (using Pythagoras' theorem)
   $2^2 = 1^2 + (BD)^2$
   $BD = \sqrt{3}$

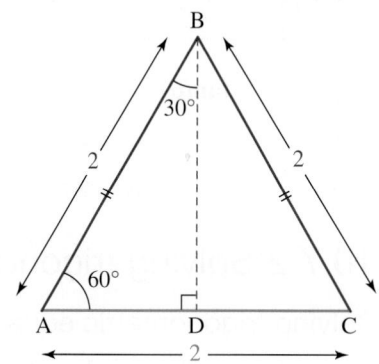

- Using $\triangle ABD$, the following exact values are obtained:

$$\sin(A) = \frac{\text{opp}}{\text{hyp}} \Rightarrow \sin(60°) = \frac{\sqrt{3}}{2} \qquad \sin(B) = \frac{\text{opp}}{\text{hyp}} \Rightarrow \sin(30°) = \frac{1}{2}$$

$$\cos(A) = \frac{\text{adj}}{\text{hyp}} \Rightarrow \cos(60°) = \frac{1}{2} \qquad \cos(B) = \frac{\text{adj}}{\text{hyp}} \Rightarrow \cos(30°) = \frac{\sqrt{3}}{2}$$

$$\tan(A) = \frac{\text{opp}}{\text{adj}} \Rightarrow \tan(60°) = \frac{\sqrt{3}}{1} \text{ or } \sqrt{3} \qquad \tan(B) = \frac{\text{opp}}{\text{adj}} \Rightarrow \tan(30°) = \frac{1}{\sqrt{3}} \text{ or } \frac{\sqrt{3}}{3}$$

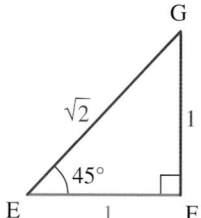

- Consider a right-angled isosceles $\triangle EFG$ with equal sides of 1 unit.

$$(EG)^2 = (EF)^2 + (FG)^2 \quad \text{(using Pythagoras' theorem)}$$
$$(EG)^2 = 1^2 + 1^2$$
$$EG = \sqrt{2}$$

- Using $\triangle EFG$, the following exact values are obtained:

$$\sin(E) = \frac{\text{opp}}{\text{hyp}} \Rightarrow \sin(45°) = \frac{1}{\sqrt{2}} \text{ or } \frac{\sqrt{2}}{2}$$

$$\cos(E) = \frac{\text{adj}}{\text{hyp}} \Rightarrow \cos(45°) = \frac{1}{\sqrt{2}} \text{ or } \frac{\sqrt{2}}{2}$$

$$\tan(E) = \frac{\text{opp}}{\text{adj}} \Rightarrow \tan(45°) = \frac{1}{1} \text{ or } 1$$

Summary of exact values

θ	30°	45°	60°
$\sin(\theta)$	$\dfrac{1}{2}$	$\dfrac{1}{\sqrt{2}} = \dfrac{\sqrt{2}}{2}$	$\dfrac{\sqrt{3}}{2}$
$\cos(\theta)$	$\dfrac{\sqrt{3}}{2}$	$\dfrac{1}{\sqrt{2}} = \dfrac{\sqrt{2}}{2}$	$\dfrac{1}{2}$
$\tan(\theta)$	$\dfrac{1}{\sqrt{3}} = \dfrac{\sqrt{3}}{3}$	1	$\sqrt{3}$

10.7.2 Solving trigonometric equations

eles-5012

Solving trigonometric equations graphically

- Because of the periodic nature of circular functions, there are infinitely many solutions to unrestricted trigonometric equations.
- Equations are usually solved within a particular domain (x-values), to restrict the number of solutions.

- The sine graph below shows the solutions between $0°$ and $360°$ for the equation $\sin(x) = 0.6$.

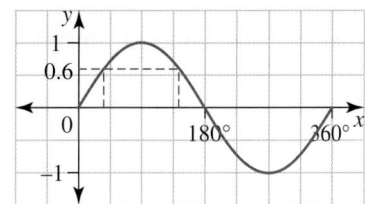

- In the example it can clearly be seen that there are two solutions to this equation, which are approximately $x = 37°$ and $x = 143°$.
- It is difficult to obtain accurate answers from a graph. More accurate answers can be obtained using technology.

Solving trigonometric equations algebraically

Exact values can be found for some trigonometric equations using the table in section 10.7.1.

WORKED EXAMPLE 15 Solving trigonometric equations using exact values

Solve the following equations.

a. $\sin(x) = \dfrac{\sqrt{3}}{2}, \ x \in [0°, 360°]$

b. $\cos(2x) = -\dfrac{1}{\sqrt{2}}, \ x \in [0°, 360°]$

THINK	WRITE
a. 1. The inverse operation of sine is $\sin^{-1}$.	a. $x = \sin^{-1}\left(\dfrac{\sqrt{3}}{2}\right)$
2. The first solution in the given domain from the table in section 10.7.1 is $x = 60°$. Since sine is positive in the first and second quadrants, another solution must be $x = 180° - 60° = 120°$.	There are two solutions in the given domain, $x = 60°$ and $x = 120°$.
b. 1. The inverse operation of cosine is $\cos^{-1}$.	b. $2x = \cos^{-1}\left(\dfrac{-1}{\sqrt{2}}\right)$
2. From the table of values, $\cos^{-1}\left(\dfrac{1}{\sqrt{2}}\right) = 45°$. Cosine is negative in the second and third quadrants, which gives the first two solutions to the equation as: $180° - 45°$ and $180° + 45°$.	$2x = 135°, 225°$
3. Solve for x by dividing by 2.	$x = 67.5°, 112.5°$
4. Since the domain in this case is $[0°, 360°]$ and the period has been halved, there must be 4 solutions altogether. The other 2 solutions can be found by adding the period onto each solution.	The period $= \dfrac{360°}{2} = 180°$ $x = 67.5° + 180°, 112.5° + 180°$ $x = 67.5°, 112.5°, 247.5°, 292.5°$

DISCUSSION

Explain why sine and cosine functions can be used to model situations that occur in nature such as tide heights and sound waves.

 Resources

 Interactivities Solving trigonometric equations graphically (int-4822)
Exact values of trigonometric functions (int-4816)

Exercise 10.7 Solving trigonometric equations

learn**on**

| 10.7 Quick quiz **on** | 10.7 Exercise |

Individual pathways

■ PRACTISE	■ CONSOLIDATE	■ MASTER
1, 3, 6, 9, 13, 16	2, 4, 7, 10, 14, 17	5, 8, 11, 12, 15

Fluency

For questions **1** and **2**, use the graph to determine approximate answers to the equations for the domain $0 \leq x \leq 360°$. Check your answers using a calculator.

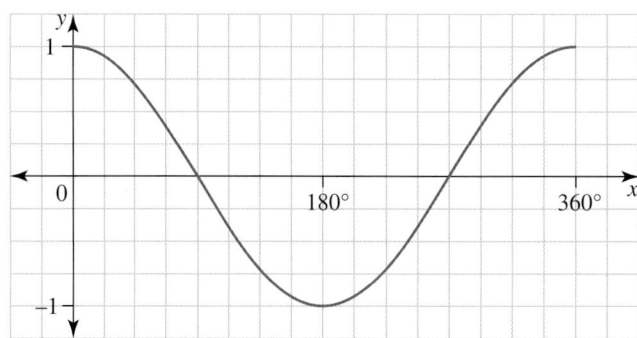

1. a. $\cos(x) = 0.9$
    b. $\cos(x) = 0.3$

2. a. $\cos(x) = -0.2$
    b. $\cos(x) = -0.6$

For questions **3** to **8**, solve the equations for the domain $0° \leq x \leq 360°$.

3. a. $\sin(x) = \dfrac{1}{2}$
    b. $\sin(x) = \dfrac{\sqrt{3}}{2}$

4. a. $\cos(x) = -\dfrac{1}{2}$
    b. $\cos(x) = -\dfrac{1}{\sqrt{2}}$

5. a. $\sin(x) = 1$
    b. $\cos(x) = -1$

6. a. $\sin(x) = -\dfrac{1}{2}$
    b. $\sin(x) = -\dfrac{1}{\sqrt{2}}$

7. **a.** $\cos(x) = \dfrac{\sqrt{3}}{2}$ **b.** $\cos(x) = -\dfrac{\sqrt{3}}{2}$

8. **a.** $\sin(x) = 1$ **b.** $\cos(x) = 0$

Understanding

WE15 For questions 9 and 12, solve the following equations for the given values of x.

9. **a.** $\sin(2x) = \dfrac{\sqrt{3}}{2}, x \in [0°, 360°]$ **b.** $\cos(2x) = -\dfrac{\sqrt{3}}{2}, x \in [0°, 360°]$

10. **a.** $\tan(2x) = \dfrac{1}{\sqrt{3}}, x \in [0°, 360°]$ **b.** $\sin(3x) = -\dfrac{1}{2}, x \in [0°, 180°]$

11. **a.** $\sin(4x) = -\dfrac{1}{2}, x \in [0°, 180°]$ **b.** $\sin(3x) = -\dfrac{1}{\sqrt{2}}, x \in [-180°, 180°]$

12. **a.** $\tan(3x) = -1, x \in [0°, 90°]$ **b.** $\cos(3x) = 0, x \in [0°, 360°]$

13. Solve the following equations for $x \in [0°, 360°]$.

    a. $2\sin(x) - 1 = 0$ **b.** $2\cos(x) = \sqrt{3}$

14. Solve the following equations for $x \in [0°, 360°]$.

    a. $\sqrt{2}\cos(x) - 1 = 0$ **b.** $\tan(x) + 1 = 0$

Communicating, reasoning and problem solving

15. Sam measured the depth of water at the end of the Intergate jetty at various times on Thursday 13 August 2020. The table below provides her results.

Time	6 am	7	8	9	10	11	12 pm	1	2	3	4	5	6	7	8	9
Depth	1.5	1.8	2.3	2.6	2.5	2.2	1.8	1.2	0.8	0.5	0.6	1.0	1.3	1.8	2.2	2.5

    a. Plot the data.

    b. Determine:

      i. the period

      ii. the amplitude.

    c. Sam fishes from the jetty when the depth is a maximum. Specify these times for the next 3 days.

    d. Sam's mother can moor her yacht when the depth is above 1.5 m. Determine during what periods she can moor the yacht on Sunday 16 January.

16. Solve:

    a. $\sqrt{3}\sin(x°) = \cos(x°)$ for $0° \le x \le 360°$

    b. $2\sin(x°) + \cos(x°) = 0$ for $0° \le x \le 360°$.

17. Solve $2\sin^2(x°) + 3\sin(x°) - 2 = 0$ for $0° \le x \le 360°$.

LESSON
10.8 Review

10.8.1 Topic summary

Sine rule
- Connects two sides with the two opposite angles in a triangle.
- $\dfrac{a}{\sin(A)} = \dfrac{b}{\sin(B)} = \dfrac{c}{\sin(C)}$
- Used to solve triangles:
  - Two angles and one side.
  - Two sides and an angle opposite one of these sides.

Area of a triangle
- Given two sides and the included angle:
$$\text{Area} = \frac{1}{2}\,ab\,\sin(C)$$

Cosine rule
- Connects three sides and one angle of a triangle.
- $a^2 = b^2 + c^2 - 2bc\,\cos(A)$
- Used to solve triangles given:
  - three sides, or
  - two sides and the included angle.

Ambiguous case
- When using the sine rule to calculate an angle, there may be two answers.
- The ambiguous case may occur when determining the angle opposite the larger side.
- Always check that the three angles add to 180°.
- e.g. In the triangle ABC, $a = 10$, $b = 6$ and $B = 30°$, using the sine rule, $A = 56°$ or $(180 - 56)°$.
  Angles in triangle would be:
  $A = 56°$, $B = 30°$ giving $C = 94°$
  $A = 124°$, $B = 30°$ giving $C = 26°$
  Two triangles are possible, so the ambiguous case exists.

TRIGONOMETRY II (PATH)

Trigonometric equations
- Infinite number of solutions, so domain is normally restricted:
  e.g. $0° \leq x \leq 180°$
- Equations can be solved:
  - graphically — not very accurate
  - using technology
  - algebraically, using the exact values.
  e.g. $\sin \alpha = -\dfrac{1}{2}$, $0° \leq x \leq 360°$
  from exact values: $\sin(30°) = \dfrac{1}{2}$
  sine is negative in 3rd and 4th quadrants and the angle is from the x-axis
  $\alpha = (180 + 30)°$ or $(360 - 30)°$
  $\alpha = 210°$ or $330°$

Exact values
- Exact trig ratios can be found using triangles.
- For 30° and 60° use the equilateral triangle.
- For 45°, use a right-angled isosceles triangle.
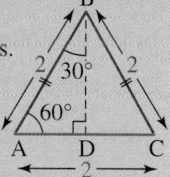
e.g. $\sin(30°) = \dfrac{1}{2}$
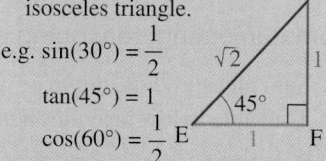
$\tan(45°) = 1$
$\cos(60°) = \dfrac{1}{2}$

Trigonometric graphs
- Trigonometric graphs repeat themselves continuously in cycles.
- Period: horizontal distance between repeating peaks or troughs.
- Amplitude: half the distance between the maximum and minimum values.

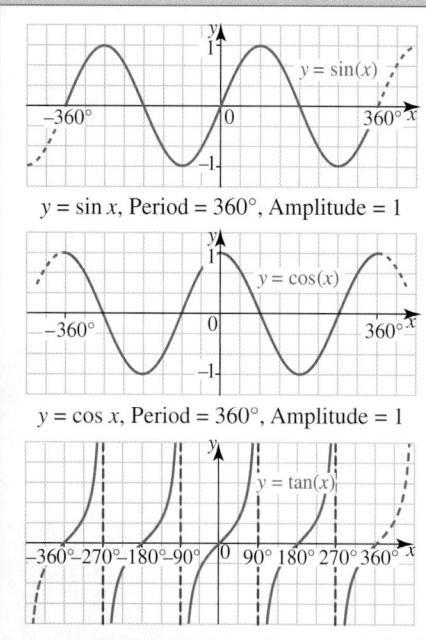

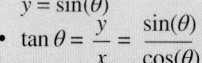
$y = \sin x$, Period = 360°, Amplitude = 1

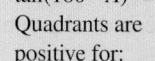
$y = \cos x$, Period = 360°, Amplitude = 1

$y = \tan x$, Period = 180°, Amplitude = undefined.

Unit circle
- Equation of the unit circle: $x^2 + y^2 = 1$
- Radius of length 1 unit.
- For any point on circumference:
  $x = \cos(\theta)$
  $y = \sin(\theta)$
- $\tan \theta = \dfrac{y}{x} = \dfrac{\sin(\theta)}{\cos(\theta)}$
- $\sin(180 - A)° = \sin(A°)$
  $\cos(180 - A)° = -\cos(A°)$
  $\tan(180 - A)° = -\tan(A°)$
- Quadrants are positive for:

II	**I**
$x < 0$, $y > 0$	$x > 0$, $y > 0$
Sin positive	All positive
III	**IV**
$x < 0$, $y < 0$	$x > 0$, $y < 0$
Tan positive	Cos positive

10.8.2 Project

What's an arbelos?

As an introduction to this task, you are required to complete the following construction. The questions that follow require the application of measurement formulas, and an understanding of semicircles related to this construction.

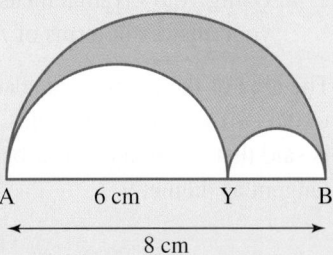

A   6 cm   Y   B

8 cm

1. **Constructing an arbelos**
   - Rule a horizontal line AB 8 cm long.
   - Determine the midpoint of the line and construct a semicircle on top of the line with AB as the diameter.
   - Mark Y as a point on AB such that AY = 6 cm.
   - Determine the midpoint of AY and draw a small semicircle inside the larger semicircle with AY as the diameter.
   - Determine the midpoint of YB and construct a semicircle (also inside the larger semicircle) with a diameter YB.

The shape enclosed by the three semicircles is known as an arbelos. The word, in Greek, means *shoemaker's knife* as it resembles the blade of a knife used by cobblers. The point Y is not fixed and can be located anywhere along the diameter of the larger semicircle, which can also vary in size.

2. **Perimeter of an arbelos**
   The perimeter of an arbelos is the sum of the arc length of the three semicircles. Perform the following calculations, leaving each answer in terms of π.
   a. Calculate the arc length of the semicircle with diameter AB.
   b. Calculate the arc length of the semicircle with diameter AY.
   c. Calculate the arc length of the semicircle on diameter YB.
   d. Compare the largest arc length with the two smaller arc lengths. What do you conclude?
3. We can generalise the arc length of an arbelos. The point Y can be located anywhere on the line AB, which can also vary in length. Let the diameter AB be d cm, AY be d_1 cm and YB be d_2 cm. Prove that your conclusion from question **2d** holds true for any value of d, where $d_1 + d_2 = d$.

4. Area of an arbelos

The area of an arbelos may be treated as the area of a composite shape.

a. Using your original measurements, calculate the area of the arbelos you drew in question **1**. Leave your answer in terms of π.

The area of the arbelos can also be calculated using another method. We can draw the common tangent to the two smaller semicircles at their point of contact and extend this tangent to the larger semicircle. It is said that the area of the arbelos is the same as the area of the circle constructed on this common tangent as diameter.

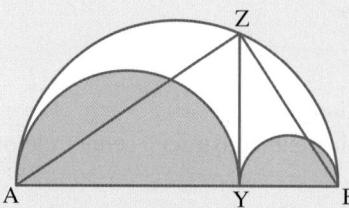

YZ is the common tangent.

Triangles AYZ, BYZ and AZB are all right-angled triangles. We can use Pythagoras' theorem, together with a set of simultaneous equations, to determine the length of the tangent YZ.

b. Complete the following.

In $\triangle AYZ$,
$$AZ^2 = AY^2 + YZ^2$$
$$= 6^2 + YZ^2$$

In $\triangle BYZ$,
$$BZ^2 = BY^2 + YZ^2$$
$$= \ldots\ldots\ldots\ldots\ldots + YZ^2$$

Adding these two equations,
$$AZ^2 + BZ^2 = \ldots\ldots\ldots\ldots\ldots \; + \; \ldots\ldots\ldots\ldots\ldots$$
$$AZ^2 + BZ^2 = AB^2$$

But, in $\triangle AZB$
$$= \ldots\ldots\ldots\ldots\ldots$$

$$\ldots\ldots\ldots\ldots\ldots \; + \; \ldots\ldots\ldots\ldots\ldots$$

So,
$$YZ = \ldots\ldots\ldots\ldots\ldots \text{ (Leave your answer in surd form.)}$$

c. Now calculate the area of the circle with diameter YZ. Is your answer the same as that calculated in question **4a**?

The area of an arbelos can be generalised.
Let the radii of the two smaller semicircles be r_1 and r_2.

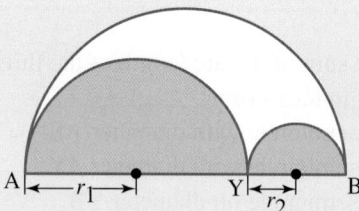

5. Develop a formula for the area of the arbelos in terms of r_1 and r_2. Demonstrate the use of your formula by checking your answer to question **4a**.

on Resources

Interactivities Crossword (int-2884)
Sudoku puzzle (int-3895)

Exercise 10.8 Review questions

Fluency

1. Calculate the value of x, correct to 2 decimal places.

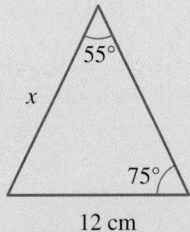

2. Calculate the value of θ, correct to the nearest minute.

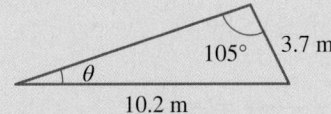

3. Determine all unknown sides (correct to 2 decimal places) and angles (correct to the nearest degree) of triangle ABC, given $a = 25\,\text{m}$, $A = 120°$ and $B = 50°$.

4. Calculate the value of x, correct to 2 decimal places.

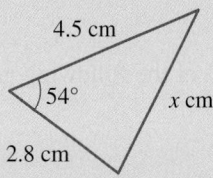

5. Calculate the value of θ, correct to the nearest degree.

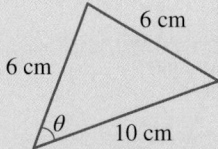

6. A triangle has sides of length 12 m, 15 m and 20 m. Calculate the magnitude (size) of the largest angle, correct to the nearest minute.

Understanding

7. A triangle has two sides of 18 cm and 25 cm. The angle between the two sides is 45°. Calculate, correct to 2 decimal places:
   a. its area
   b. the length of its third side

8. If an angle of $\theta = 290°$ was represented on the unit circle, state which quadrant the triangle to show this would be drawn in.

9. On the unit circle, draw an appropriate triangle for the angle 110° in quadrant 2.
   a. Determine the value of $\sin(110°)$ and $\cos(110°)$, correct to 2 decimal places.
   b. Determine the value of $\tan(110°)$, correct to 2 decimal places.

10. **MC** The value of $\sin(53°)$ is equal to:
    A. $\cos(53°)$　　　　**B.** $\cos(37°)$　　　　**C.** $\sin(37°)$　　　　**D.** $\tan(53°)$

11. Simplify $\dfrac{\sin(53°)}{\sin(37°)}$.

12. Draw a sketch of $y = \sin(x)$ from $0° \le x \le 360°$.

13. Draw a sketch of $y = \cos(x)$ from $0° \le x \le 360°$.

14. Draw a sketch of $y = \tan(x)$ from $0° \le x \le 360°$.

15. Label the shown triangle so that $\dfrac{x}{\sin(46°)} = \dfrac{y}{\sin(68°)}$.

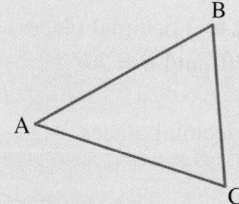

16. State the period and amplitude of each of the following graphs.
    a. $y = 2\sin(3x)$
    b. $y = -3\cos(2x)$
    c.

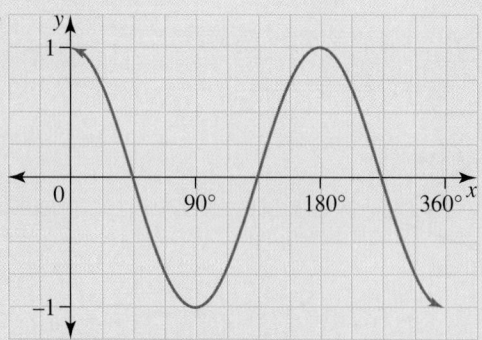

17. Sketch the following graphs.
    a. $y = 2\sin(x), x \in [0°, 360°]$　　　　**b.** $y = \cos(2x), x \in [-180°, 180°]$

18. Use technology to write down the solutions to the following equations for the domain $0° \leq x \leq 360°$ to 2 decimal places.

   a. $\sin(x) = -0.2$ b. $\cos(2x) = 0.7$ c. $3\cos(x) = 0.1$ d. $2\tan(2x) = 0.5$

19. Solve each of the following equations.

   a. $\sin(x) = \dfrac{1}{2}, x \in [0°, 360°]$

   b. $\cos(x) = \dfrac{\sqrt{3}}{2}, x \in [0°, 360°]$

   c. $\cos(x) = \dfrac{1}{\sqrt{2}}, x \in [0°, 360°]$

   d. $\sin(x) = \dfrac{1}{\sqrt{2}}, x \in [0°, 360°]$

20. **MC** The equation that represents the graph shown could be:

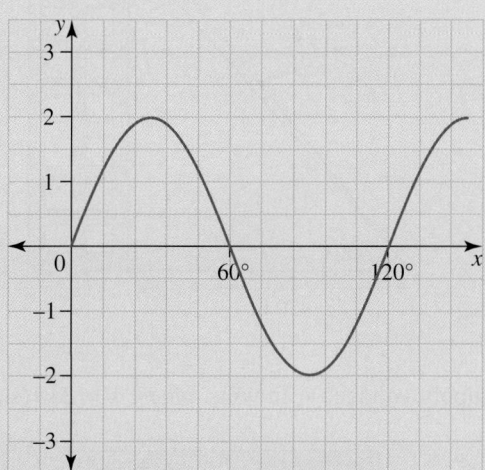

   A. $y = 3\sin(2x)$ B. $y = 2\cos(3x)$ C. $y = 2\sin(2x)$ D. $y = 2\sin(3x)$

21. a. Use technology to help sketch the graph of $y = 2\sin(2x) - 3$.
   b. Write down the period and the amplitude of the graph in part **a**.

Communicating, reasoning and problem solving

22. Sketch the graphs of each of the following, stating:
   i. the period
   ii. the amplitude.
   a. $y = 2\cos(2x), x \in [0°, 360°]$
   b. $y = 3\sin(4x), x \in [0°, 180°]$
   c. $y = -2\cos(3x), x \in [-60°, 60°]$
   d. $y = 4\sin(2x), x \in [-90°, 90°]$

23. Solve each of the following equations for the given values of x.

   a. $\cos(2x) = \dfrac{\sqrt{3}}{2}, x \in [0°, 360°]$

   b. $\sin(3x) = \dfrac{1}{2}, x \in [-90°, 90°]$

   c. $\sin(2x) = \dfrac{1}{\sqrt{2}}, x \in [0°, 360°]$

   d. $\cos(3x) = -\dfrac{1}{\sqrt{2}}, x \in [0°, 360°]$

   e. $\sin(4x) = 0, x \in [0°, 180°]$

   f. $\tan(4x) = -1, x \in [0°, 180°]$

24. Solve the following for $x \in [0°, 360°]$.
   a. $2\cos(x) - 1 = 0$
   b. $2\sin(x) = -\sqrt{3}$
   c. $-\sqrt{2}\cos(x) + 1 = 0$
   d. $\sqrt{2}\sin(x) + 1 = 0$

25. Sketch the graph of $y = \tan(2x), x \in [0°, 180°]$. Write down the period, amplitude and the equations of any asymptotes.

26. A satellite dish is placed on top of an apartment building as shown in the diagram. Determine the height of the satellite dish, in metres correct to 2 decimal places.

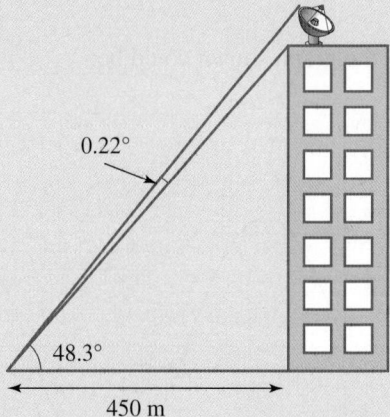

27. Australian power points supply voltage, V, in volts, where $V = 240\,(\sin 18\,000t)$ and t is measured in seconds.

t	V
0.000	
0.005	
0.010	
0.015	
0.020	
0.025	
0.030	
0.035	
0.040	

a. Copy and complete the table and sketch the graph, showing the fluctuations in voltage over time.
b. State the times at which the maximum voltage output occurs.
c. Determine how many seconds there are between times of maximum voltage output.
d. Determine how many periods (or cycles) are there per second.

on   To test your understanding and knowledge of this topic, go to your learnON title at www.jacplus.com.au and complete the **post-test**.

Answers

Topic 10 Trigonometry II (Path)

10.1 Pre-test

1. B
2. 9.06 m
3. 9.35 cm
4. A
5. 19.67
6. 7.2 m
7. B
8. 3rd quadrant
9. 0.5
10. D
11. D
12. B
13. A
14. C
15. 270°

10.2 The sine rule

1. $44°58', 77°2', 13.79$
2. $39°18', 38°55', 17.21$
3. $70°, 9.85, 9.40$
4. $33°, 38.98, 21.98$
5. 19.12
6. $C = 51°, b = 54.66, c = 44.66$
7. $A = 60°, b = 117.11, c = 31.38$
8. $B = 48°26', C = 103°34', c = 66.26$; or $B = 131°34'$, $C = 20°26', c = 23.8$
9. 24.17
10. B, C
11. $A = 73°15', b = 8.73$; or $A = 106°45', b = 4.12$
12. 51.90 or 44.86
13. $C = 110°, a = 3.09, b = 4.64$
14. $B = 38°, a = 3.36, c = 2.28$
15. $B = 33°33', C = 121°27', c = 26.24$; or $B = 146°27'$, $C = 8°33', c = 4.57$
16. 43.62 m
17. $h = 7.5$ cm
18. 113 km
19. 8.68 m
20. Yes, she needs 43 m altogether.
21. a. 6.97 m   b. 4 m
22. a. 13.11 km   b. N20°47'W
23. a. 8.63 km   b. 6.5 km/h   c. 9.90 km
24. 22.09 km from A and 27.46 km from B.
25. a. 7.3 km   b. 282°3'

10.3 The cosine rule

1. 7.95
2. 55.22

3. $23.08, 41°53', 23°7'$
4. $28°57'$
5. $88°15'$
6. $A = 61°15', B = 40°, C = 78°45'$
7. 2218 m
8. a. 12.57 km   b. S35°1'E
9. a. 35°6'   b. 6.73 m^2
10. 23°
11. 89.12 m
12. a. 130 km   b. S22°12'E
13. 28.5 km
14. 74.3 km
15. 70°49'
16. a. 8.89 m   b. 77°0'   c. $x = 10.07$m
17. 1.14 km/h
18. $\angle CAB = 34.65°, \angle ABC = 84.83°$ and $\angle BCA = 60.52°$
19. a. 4.6637 m   b. 55.93°

10.4 Area of triangles

1. 12.98
2. 38.14
3. 212.88
4. $A = 32°4', B = 99°56'$, area = 68.95 cm^2
5. $A = 39°50', B = 84°10'$, area = 186.03 m^2
6. $A = 125°14', B = 16°46'$, area = 196.03 mm^2
7. C
8. 14.98 cm^2
9. 570.03 mm^2
10. 2.15 cm^2
11. B
12. 3131.41 mm^2
13. 610 cm^2
14. a. 187.5 cm^2 b. 15.03 cm
15. 17 goldfish
16. C
17. B
18. a. Area = 69.63 cm^2
    b. Dimensions are 12.08 cm and 6.96 cm.
19. 17 kg
20. 52.2 hectares
21. 175 m^3
22. 22.02 m^2
23. a. 29.44 km^2 b. 8.26 km   c. 41°
    d. 24°   e. 3.72 km   f. 10.11 km^2
    g. 39.55 km^2
24. 374.12 cm^2

10.5 The unit circle

1. a. 1st   b. 2nd   c. 4th
   d. 3rd   e. 2nd   f. 3rd
2. A
3. D

4. a. 0.34   b. 0.94   c. −0.17   d. 0.98

5. a. −0.64   b. 0.77   c. −0.57   d. −0.82

6. a. 1   b. 0   c. 0   d. −1

7. a. −1   b. 0   c. 0   d. 1

8. a. 0.87   b. 0.50

9. a. 30°

    b. −0.87

    c. $\cos(150°) = -\cos(30°)$

    d. 0.5

    e. $\sin(150°) = \sin(30°)$

10. a. 30°

    b. −0.87

    c. $\cos(210°) = -\cos(30°)$

    d. −0.50

    e. $\sin(210°) = -\sin(30°)$

11. a. 30°

    b. 0.87

    c. $\cos(330°) = -\cos(30°)$

    d. −0.50

    e. $\sin(330°) = -\sin(30°)$

12. a. 0.34   b. 0.94

    c. 0.36   d. 0.36

    e. They are equal.

13. a. 0.71   b. −0.71

    c. −1   d. −1

    e. They are equal.   f. $\tan(135°) = -\tan(45°)$

14. a. −0.64

    b. −0.77

    c. 0.84

    d. 0.83

    e. They are approximately equal.

    f. $\tan(220°) = \tan(40°)$

15. a. −0.87

    b. 0.5

    c. −1.73

    d. −1.74

    e. They are approximately equal.

    f. $\tan(300°) = -\tan(60°)$

16. D

17. a. 1.192   b. 3.078   c. 0.176   d. −0.577

18. a. −0.577   b. 0   c. 1   d. 57.290

19. a. 0.93   b. 2.61   c. −0.53   d. −3.73

20, 21, 22. Sample responses can be found in the worked solutions in the online resources.

23. a. $\sqrt{1 - p^2}$ b. p

    c. $-\sqrt{1 - p^2}$ d. p

24. 0

25. 0

10.6 Trigonometric functions

1. See table at the bottom of the page.*

2.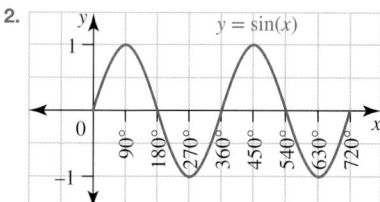

3. 360°

4. a. 0.7   b. 0.8   c. 0.3   d. −0.3

5. a. 0   b. 0.9   c. −0.2   d. −0.9

6. a. 64°, 116°, 424°, 476°

    b. 244°, 296°, 604°, 656°

    c. 44°, 136°, 404°, 496°

7. a. 210°, 330°, 570°, 690°

    b. 233°, 307°, 593°, 667°

    c. 24°, 156°, 384°, 516°

8. See table at the bottom of the page.*

9.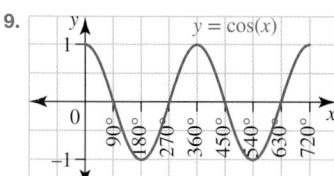

10. The graph would continue with the cycle.

11. It is a very similar graph with the same shape; however, the sine graph starts at (0, 0), whereas the cosine graph starts at (0, 1).

*1.

x	0°	30°	60°	90°	120°	150°	180°	210°	240°	270°	300°	330°	360°
sin(x)	0	0.5	0.87	1	0.87	0.5	0	−0.5°	−0.87	−1	−0.87	−0.5	0
x	390°	420°	450°	480°	510°	540°	570°	600°	630°	660°	690°	720°	
sin(x)	0.5	0.87	1	0.87	0.5	0	−0.5	−0.87	−1	−0.87	−0.5	0	

*8.

x	0°	30°	60°	90°	120°	150°	180°	210°	240°	270°	300°	330°	360°
cos(x)	1	0.87	0.5	0	−0.5	−0.87	−1	−0.87	−0.5	0	0.5	0.87	1
x	390°	420°	450°	480°	510°	540°	570°	600°	630°	660°	690°	720°	
cos(x)	0.87	0.5	0	−0.5	−0.87	−1	−0.87	−0.5	0	0.5	0.87	1	

12. a. 0.7   **b.** −0.9   **c.** −1   **d.** 0.9

13. a. −0.5   **b.** −0.8   **c.** 0.8   **d.** −0.9

14. a. 120°, 240°, 480°, 600°

   b. 37°, 323°, 397°, 683°

   c. 46°, 314°, 406°, 674°

15. a. 127°, 233°, 487°, 593°

   b. 26°, 334°, 386°, 694°

   c. 154°, 206°, 514°, 566°

16. See table at the bottom of the page.*

17. $y = \tan(x)$

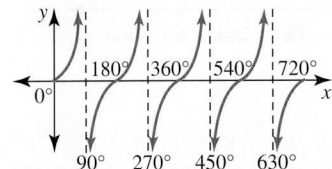

18. The graph would continue repeating every 180° as above.

19. Quite different. $y = \tan(x)$ has undefined values (asymptotes) and repeats every 180° rather than 360°. It also gives all y-values, rather than just values between −1 and 1.

20. a. 1.7   **b.** −1   **c.** −1.2   **d.** 0.8

21. a. −0.8   **b.** 1.2   **c.** −0.2   **d.** 1

22. a. 45°, 225°, 405°, 585°

   b. 56°, 236°, 416°, 596°

   c. 158°, 338°, 518°, 698°

   d. 117°, 297°, 477°, 657°

   e. 11°, 191°, 371°, 551°

   f. 135°, 315°, 495°, 675°

23. a. i. 360°

   ii. 1

   iii.

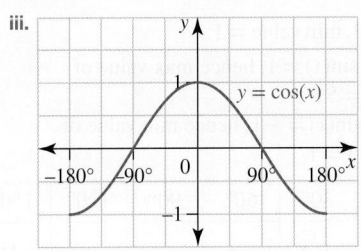

   b. i. 360°

   ii. 1

iii.

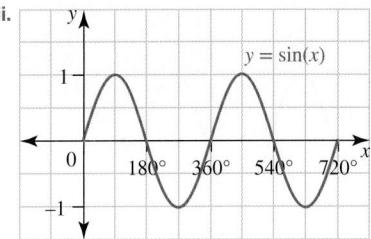

24. a. i. 180°

   ii. 1

iii.

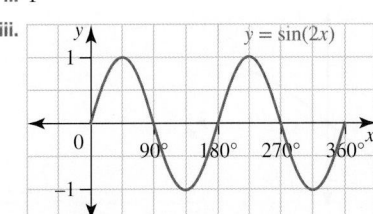

   b. i. 360°

   ii. 2

iii.

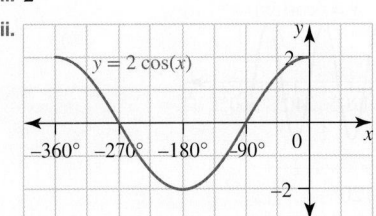

25. a. i. 180°   **ii.** 3

   b. i. 120°   **ii.** 4

   c. i. 720°   **ii.** 2

   d. i. 1440°   **ii.** $\dfrac{1}{2}$

   e. i. 360°   **ii.** 1

   f. i. 180°   **ii.** 1

26. a. C   **b.** A   **c.** D

27. a.

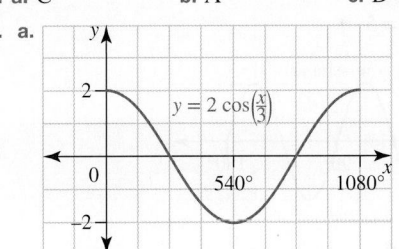

Period = 1080°
Amplitude = 2

---

*16.

x	0°	30°	60°	90°	120°	150°	180°	210°	240°	270°	300°	330°	360°
$\tan(x)$	0	0.58	1.73	undef.	−1.73	−0.58	0	0.58	1.73	undef.	−1.73	−0.58	0
x	390°	420°	450°	480°	510°	540°	570°	600°	630°	660°	690°	720°	
$\tan(x)$	0.58	1.73	undef.	−1.73	−0.58	0	0.58	1.73	undef.	−1.73	−0.58	0	

b.

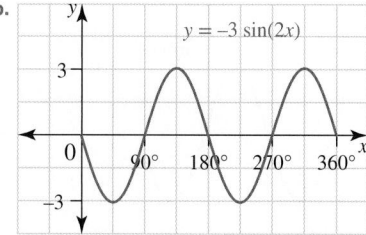

Period = 180°
Amplitude = 3

c.

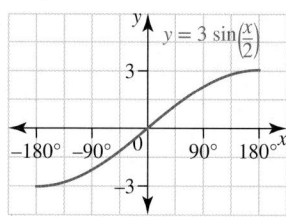

Period = 720°
Amplitude = 3

d.

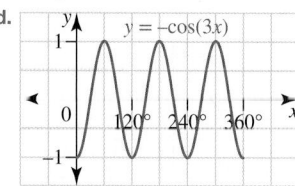

Period = 120°
Amplitude = 1

e.

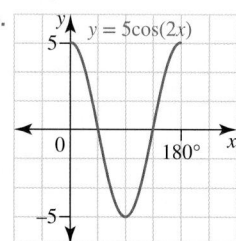

Period = 180°
Amplitude = 5

f.

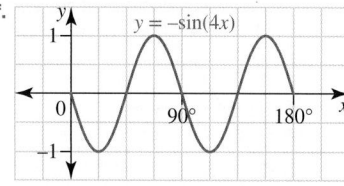

Period = 90°
Amplitude = 1

28. a.

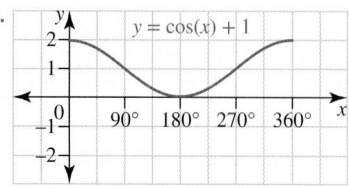

b.

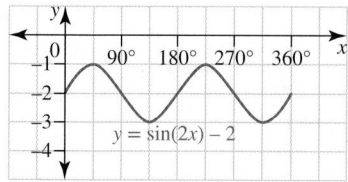

c.

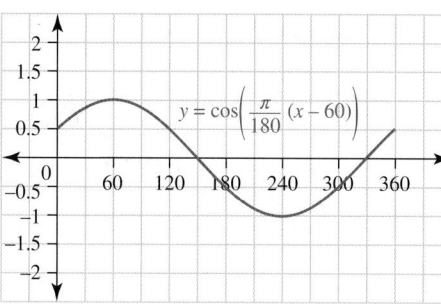

d.

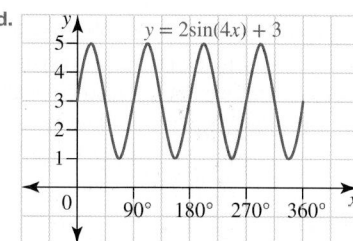

29. a.

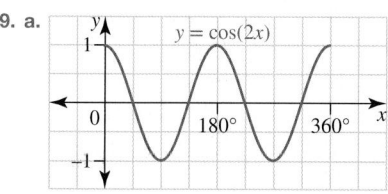

  i. −1   ii. 1

b. Max value = 3, min value = 1

c. Max value of $\sin(x) = 1$, hence max value of
$y = 2 \times 1 + 3 = 5$
Min value of $\sin(x) = -1$, hence min value of
$y = 2 \times -1 + 3 = 1$

30. a.

x	0	30°	60°	90°	120°	150°	180°
y	0	$\dfrac{\sqrt{3}}{3}$	$\sqrt{3}$	undef	$-\sqrt{3}$	$-\dfrac{\sqrt{3}}{3}$	0

b.

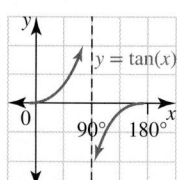

c. At $x = 90°$, y is undefined.

d. $x = 270°$

e. The period = 180°, amplitude is undefined.

31. a.

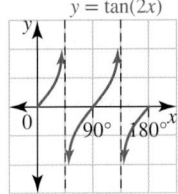

$y = \tan(2x)$

b. $x = 45°$ and $x = 135°$

c. The period $= 90°$ and amplitude is undefined.

32. a.

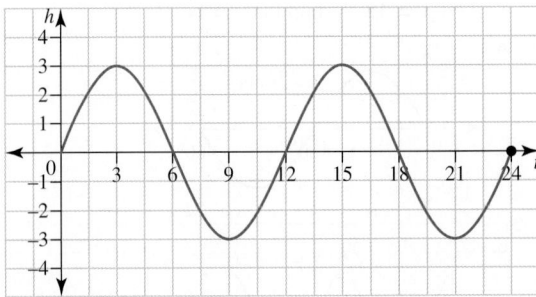

b. 3 metres

c. −2.6 metres

33. a.

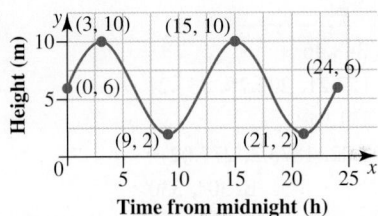

b. 10 metres

c. 2 metres

d. 2.54 metres

34. a. 22 °C

b.

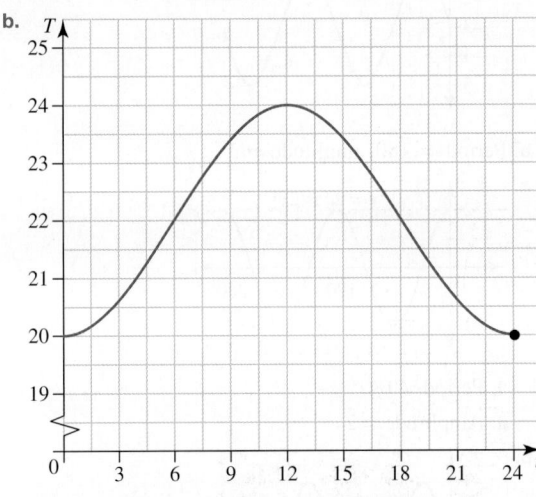

c. Coolest 20 °C, warmest 24 °C

10.7 Solving trigonometric equations

1. Calculator answers
  a. 25.84°, 334.16°   **b.** 72.54°, 287.46°

2. a. 101.54°, 258.46°   **b.** 126.87°, 233.13°

3. a. 30°, 150°   **b.** 60°, 120°

4. a. 120°, 240°   **b.** 135°, 225°

5. a. 90°   **b.** 180°

6. a. 210°, 330°   **b.** 225°, 315°

7. a. 30°, 330°   **b.** 150°, 210°

8. a. 90°   **b.** 90°, 270°

9. a. 30°, 60°, 210°, 240°   **b.** 75°, 105°, 255°, 285°

10. a. 15°, 105°, 195°, 285°
  b. 70°, 110°

11. a. 52.5°, 82.5°, 142.5°, 172.5°
  b. −165°, −135°, −45°, −15°, 75°, 105°

12. a. 45°
  b. 30°, 90°, 150°, 210°, 270°, 330°

13. a. 30°, 150°   **b.** 30°, 330°

14. a. 45°, 315°   **b.** 135°, 315°

15. a.

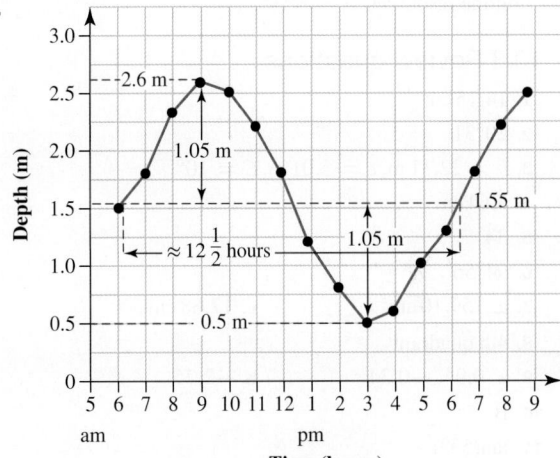

b. i. $12\frac{1}{2}$ hours   **ii.** 1.05 m

c. 10.00 am, 10.30 pm, 11.00 am, 11.30 pm, noon.

d. Until 1.45 am Sunday, 8 am to 2.15 pm and after 8.30 pm.

16. a. $x = 30°, 210°$ **b.** $x = 153.43°, 333.43°$

17. $x = 30°, 150°$

Project

1. Follow given instructions.

2. a. 4π cm
  b. 3π cm
  c. π cm
  d. The largest arc length equals the sum of the two smaller arc lengths.

3. Sample responses can be found in the worked solutions in the online resources.

4. a. 3π cm^2

b. In $\triangle$AYZ:

$$AZ^2 = AY^2 + YZ^2$$
$$= 6^2 + YZ^2$$

In $\triangle$BYZ:

$$BZ^2 = BY^2 + YZ^2$$
$$= 2^2 + YZ^2$$

Adding these equations: $AZ^2 + BZ^2 = 6^2 + YZ^2 + 2^2 + YZ^2$

But in $\triangle$AZB: $AZ^2 + BZ^2 = AB^2$

$$6^2 + YZ^2 + 2^2 + YZ^2 = 8^2$$
$$2\,YZ^2 = 64\text{–}36\text{–}4$$
$$YZ^2 = 12$$
$$YZ = \pm\sqrt{12}$$

But YZ > 0 as it is a length: YZ $= 2\sqrt{3}$

c. 3π cm^2

Yes, same area

5. Area of the arbelos $= \pi r_1 r_2$

Sample responses can be found in the worked solutions in the online resources.

10.8 Review questions

1. 14.15 cm

2. 20°31′

3. $b = 22.11\,\text{m}, c = 5.01\,\text{m}, C = 10°$

4. 3.64 cm

5. 34°

6. 94°56′

7. **a.** 159.10 cm^2 **b.** 17.68 cm

8. 4th quadrant

9. **a.** 0.94, -0.34 **b.** -2.75

10. B

11. tan(53°)

12.
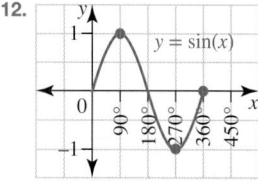

13.
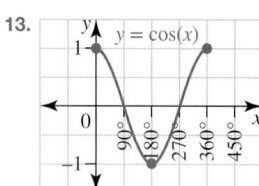

14.
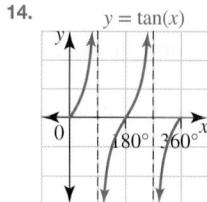

15.
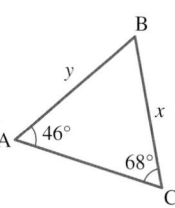

16. **a.** Period $= 120°$, amplitude $= 2$
    b. Period $= 180°$, amplitude $= 3$
    c. Period $= 180°$, amplitude $= 1$

17. **a.**
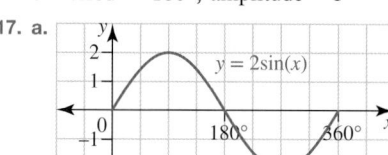

b.
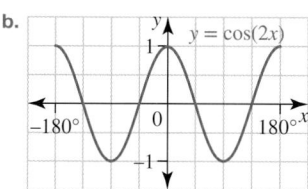

18. **a.** $x = 191.54, 348.46$
    b. $x = 22.79, 157.21, 202.79, 337.21$
    c. $x = 88.09, 271.91$
    d. $x = 7.02, 97.02, 187.02, 277.02$

19. **a.** 30°, 150°   **b.** 30°, 330°
    c. 45°, 315°   **d.** 45°, 135°

20. D

21. **a.**
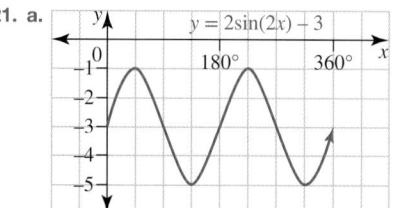

b. Period $= 180$, amplitude $= 2$

22. **a.**
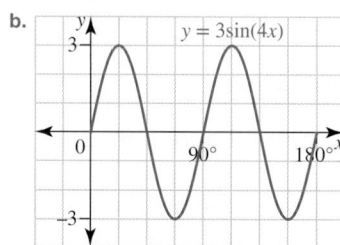

    i. Period $= 180°$
    ii. Amplitude $= 2$

b.

    i. Period $= 90°$
    ii. Amplitude $= 3$

c. $y = -2\cos(3x)$

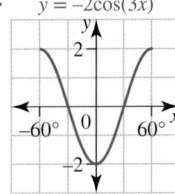

    i. Period = 120°

    ii. Amplitude = 2

d. $y = 4\sin(2x)$

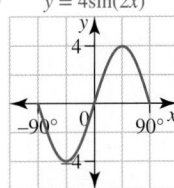

    i. Period = 180°

    ii. Amplitude = 4

23. a. 15°, 165°, 195°, 345°

    b. −70°, 10°, 50°

    c. 22.5°, 67.5°, 202.5°, 247.5°

    d. 45°, 75°, 165°, 195°, 285°, 315°

    e. 0°, 45°, 90°, 135°, 180°

    f. 33.75°, 78.75°, 123.75°, 168.75°

24. a. 60°, 300°   **b.** 240°, 300°

    c. 45°, 315°   **d.** 225°, 315°

25.

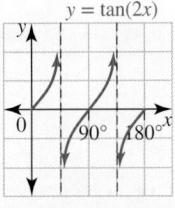

Period = 90°, amplitude is undefined.
Asymptotes are at $x = 45°$ and $x = 135°$.

26. 3.92 m

27. a.

t	V
0.000	0
0.005	240
0.010	0
0.015	−240
0.020	0
0.025	240
0.030	0
0.035	−240
0.040	0

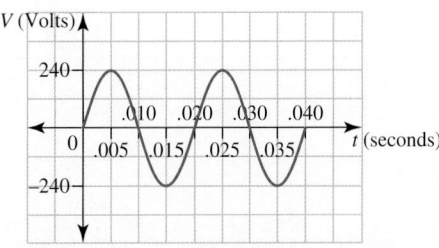

b. Maximum voltage occurs at $t = 0.005$ s, 0.025 s

c. 0.02 s

d. 50 cycles per second

11 Measurement

LESSON
11.1 Overview

Why learn this?

People must measure! How much paint or carpet will you need to redecorate your bedroom? How many litres of water will it take to fill the new pool? How many tiles do you need to order to retile the bathroom walls? How far is it from the North Pole to the South Pole? These are just a few examples where measurement skills are needed.

Measuring tools have advanced significantly in their capability to measure extremely small and extremely large amounts and objects, leading to many breakthroughs in medicine, engineering, science, architecture and astronomy.

In architecture, not all buildings are simple rectangular prisms. In our cities and towns, you will see buildings that are cylindrical in shape, buildings with domes and even buildings that are hexagonal or octagonal in shape. Architects, engineers and builders all understand the relationships between these various shapes and how they are connected. Industrial and interior designers use the properties of plane figures, prisms, pyramids and spheres in various aspects of their work.

Have you ever wondered why tennis balls are sold in cylindrical containers? This is an example of manufacturers wanting to minimise the amount of waste in packaging. Understanding the concepts involved in calculating the surface area and volume of common shapes we see around us is beneficial in many real-life situations.

Hey students! Bring these pages to life online

- ▶ Watch videos
- 🧩 Engage with interactivities
- A+ Answer questions and check solutions

Find all this and MORE in jacPLUS ▶

Reading content and rich media, including interactivities and videos for every concept

Extra learning resources

Differentiated question sets

Questions with immediate feedback, and fully worked solutions to help students get unstuck

1. Calculate the area of the shape, correct to 2 decimal places.

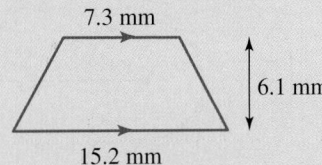

7.3 mm

6.1 mm

15.2 mm

2. Calculate the area of the sector, correct to 1 decimal place.

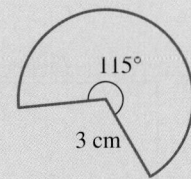

115°

3 cm

3. **MC** Select the total surface area of the rectangular prism from the following.

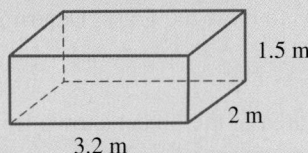

1.5 m

2 m

3.2 m

    A. $9.6 \, \text{m}^2$ **B.** $14.2 \, \text{m}^2$ **C.** $22.0 \, \text{m}^2$ **D.** $28.4 \, \text{m}^2$

4. **PATH** Calculate the total surface area of the sphere, correct to 1 decimal place.

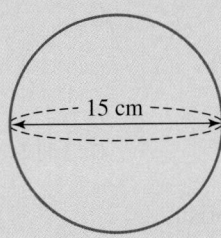

15 cm

5. **PATH** Calculate the volume of the solid.

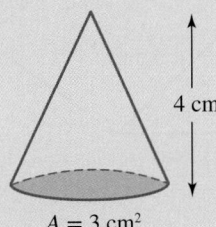

4 cm

$A = 3 \, \text{cm}^2$

6. Calculate the area of the shape, correct to 1 decimal place.

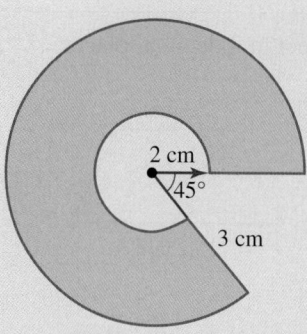

7. A council park is shown below.

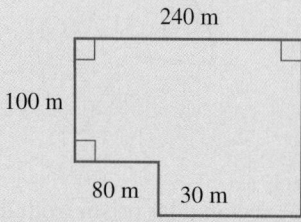

A worker charges $30 per 1000 m² to mow the grass. Determine how much it will cost the council to have the grass mown.

8. **MC** **PATH** Select the total surface area of the object shown from the following.

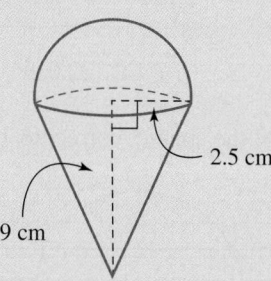

A. 109.96 cm²   **B.** 112.63 cm²   **C.** 151.9 cm²   **D.** 124.36 cm²

9. Determine the volume of the triangular prism.

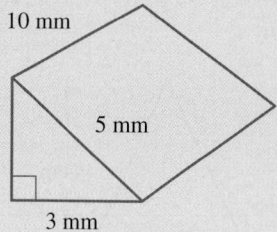

10. **MC** **PATH** Select the volume of the object from the following.

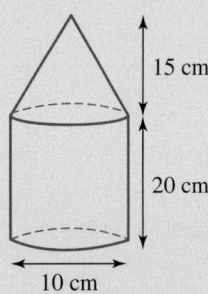

A. 2748.9 cm³ B. 1701.7 cm³ C. 1963.5 cm³ D. 7854 cm³

11. **PATH** If a square-based pyramid has a height of 12 cm and a volume of 128 cm³, then calculate the length of the square base, correct to 2 decimal places.

12. **PATH** The volume of a ball is given by the formula $V = \frac{4}{3}\pi r^3$. Evaluate the radius of a ball with a volume of 384.66 cm³.

Give your answer correct to 1 decimal place.

13. **MC** **PATH** Determine what effect doubling the radius and halving the height of a cone will have on its volume.
A. The volume will be the same.
B. The volume will be halved.
C. The volume will be doubled.
D. The volume will be quadrupled.

14. **PATH** Calculate the volume of the shape shown, correct to two decimal places.

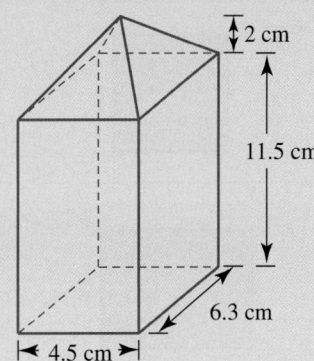

15. A cylindrical soft drink can has a diameter of 6.4 cm and a height of 14.3 cm.
If the can is only half full, determine what capacity of soft drink remains, to the nearest millilitre.

LESSON
11.2 Area

LEARNING INTENTION

At the end of this lesson you should be able to:
- convert between units of area
- calculate the area of plane figures using area formulas.

▶ 11.2.1 Area

eles-4809

- The **area** of a figure is the amount of surface covered by the figure.
- The units used for area are mm^2, cm^2, m^2, km^2 and ha (hectares).
- One unit that is often used when measuring land is the hectare. It is equal to $10\,000\,m^2$.
- The following diagram can be used to convert between units of area.

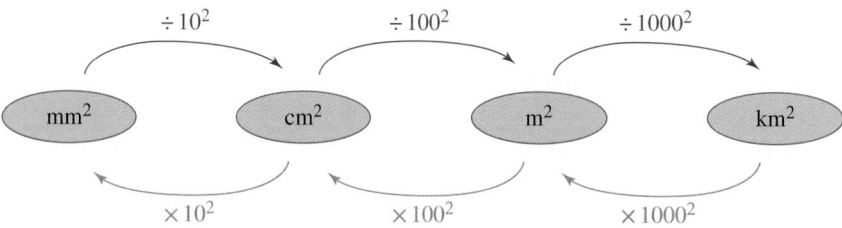

Area formulas

- The table below shows the formula for the area of some common shapes.

Shape	Diagram	Formula
Square		$A = l^2$
Rectangle		$A = lw$
Triangle		$A = \dfrac{1}{2}bh$
Parallelogram		$A = bh$

Shape	Diagram	Formula
Trapezium		$A = \dfrac{1}{2}(a+b)h$
Kite (including rhombus)		$A = \dfrac{1}{2}xy$
Circle		$A = \pi r^2$
Sector		$A = \dfrac{\theta°}{360°} \times \pi r^2$

WORKED EXAMPLE 1 Calculating areas of plane figures

Calculate the areas of the following plane figures, correct to 2 decimal places.

a.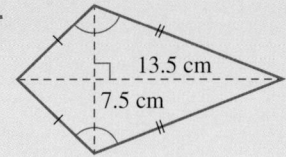

b.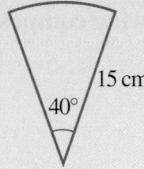

THINK

a. 1. The shape shown is an kite. Write the appropriate area formula.

2. Identify the values x and y

3. Substitute the values of x and y into the formula and evaluate, rounding your answer correct to 2 decimal places.

b. 1. The shape shown is a sector. Write the formula for finding the area of a sector.

WRITE

a. $A = \dfrac{1}{2}xy$

$x = 13.5\,\text{cm}, \ y = 7.5\,\text{cm}$

$A = \dfrac{1}{2} \times 13.5 \times 7.5$

$= 50.63\,\text{cm}^2$

b. $A = \dfrac{\theta}{360°} \times \pi r^2$

2. Write the value of θ and r.	$\theta = 40°$, $r = 15$
3. Substitute and evaluate the expression, correct to 2 decimal places.	$A = \dfrac{40°}{360°} \times \pi \times 15^2$
	$= 78.54\,\text{cm}^2$

⏵ 11.2.2 Areas of composite figures

eles-4810

- A **composite figure** is a figure made up of a combination of simple figures.
- The area of a composite figure can be calculated by:
  - calculating the sum of the areas of the simple figures that make up the composite figure

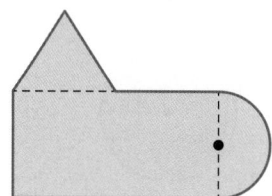

  - calculating the area of a larger shape and then subtracting the extra area involved.

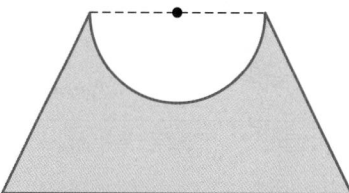

WORKED EXAMPLE 2 Calculating areas of composite shapes

Calculate the area of each of the following composite shapes.

a.
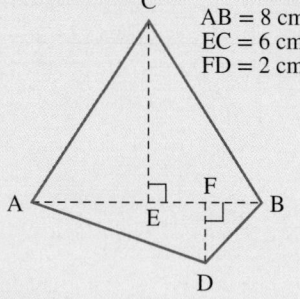

AB = 8 cm
EC = 6 cm
FD = 2 cm

b.
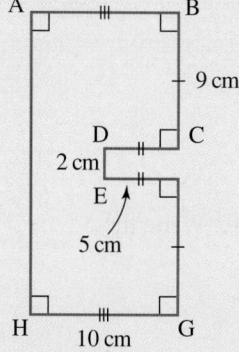

9 cm

2 cm

5 cm

10 cm

THINK

a. 1. ACBD is a quadrilateral that can be split into two triangles: △ABC and △ABD.

2. Write the formula for the area of a triangle containing base and height.

3. Identify the values of b and h for △ABC.

WRITE

a. Area ACBD = Area △ABC + Area △ABD

$A_{\text{triangle}} = \dfrac{1}{2}\,bh$

△ABC: $b = \text{AB} = 8$, $h = \text{EC} = 6$

4. Substitute the values of the pronumerals into the formula and calculate the area of $\triangle ABC$.

$$\text{Area of } \triangle ABC = \frac{1}{2} \times AB \times EC$$
$$= \frac{1}{2} \times 8 \times 6$$
$$= 24 \text{ cm}^2$$

5. Identify the values of b and h for $\triangle ABD$.

$\triangle ABD: b = AB = 8, h = FD = 2$

6. Calculate the area of $\triangle ABD$.

$$\text{Area of } \triangle ABD = \frac{1}{2} AB \times FD$$
$$= \frac{1}{2} \times 8 \times 2$$
$$= 8 \text{ cm}^2$$

7. Add the areas of the two triangles together to calculate the area of the quadrilateral ACBD.

$$\text{Area of } ACBD = 24 \text{ cm}^2 + 8 \text{ cm}^2$$
$$= 32 \text{ cm}^2$$

b. 1. One way to calculate the area of the shape shown is to calculate the total area of the rectangle ABGH and then subtract the area of the smaller rectangle DEFC.

b. Area = Area ABGH − Area DEFC

2. Write the formula for the area of a rectangle.

$A_{\text{rectangle}} = l \times w$

3. Identify the values of the pronumerals for the rectangle ABGH.

Rectangle ABGH: $l = 9 + 2 + 9$
$= 20$
$w = 10$

4. Substitute the values of the pronumerals into the formula to calculate the area of the rectangle ABGH.

$$\text{Area of } ABGH = 20 \times 10$$
$$= 200 \text{ cm}^2$$

5. Identify the values of the pronumerals for the rectangle DEFC.

Rectangle DEFC: $l = 5, w = 2$

6. Substitute the values of the pronumerals into the formula to calculate the area of the rectangle DEFC.

$$\text{Area of } DEFC = 5 \times 2$$
$$= 10 \text{ cm}^2$$

7. Subtract the area of the rectangle DEFC from the area of the rectangle ABGH to calculate the area of the given shape.

$$\text{Area} = 200 - 10$$
$$= 190 \text{ cm}^2$$

on Resources

Video eLesson Composite area (eles-1886)

Interactivities Conversion chart for area (int-3783)
Area of rectangles (int-3784)
Area of parallelograms (int-3786)
Area of trapeziums (int-3790)
Area of circles (int-3788)
Area of a sector (int-6076)
Area of a kite (int-6136)
Area of an ellipse (int-6137)

Exercise 11.2 Area

11.2 Quick quiz on	11.2 Exercise

Individual pathways

■ PRACTISE	■ CONSOLIDATE	■ MASTER
1, 6, 8, 10, 13, 14, 18, 21	2, 4, 7, 11, 15, 16, 19, 22	3, 5, 9, 12, 17, 20, 23

Unless told otherwise, where appropriate, give answers correct to 2 decimal places.

Fluency

1. Calculate the areas of the following shapes.

   a.
   4 cm

   b.
   4 cm
   12 cm

   c.
   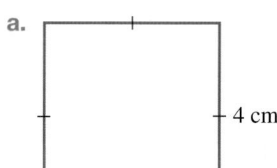
   15 cm
   10 cm

2. Calculate the areas of the following shapes.

   a.
   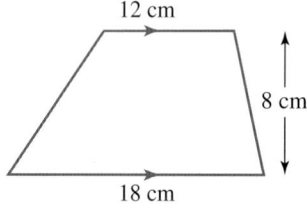
   12 cm
   8 cm
   18 cm

   b.
   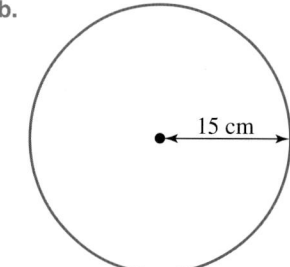
   15 cm

   c.
   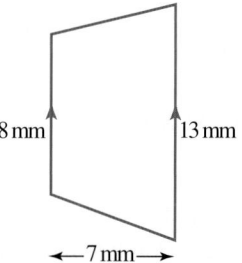
   8 mm   13 mm
   ←7 mm→

3. **WE1a** Calculate the areas of the following rhombuses. Answer correct to 1 decimal place.

   a.
   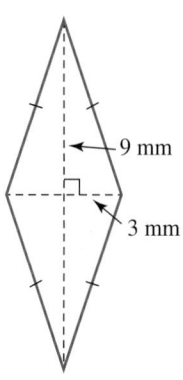
   9 mm
   3 mm

   b.
   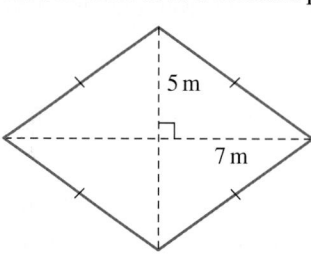
   5 m
   7 m

4. Calculate the areas of the following shapes.

   a.
   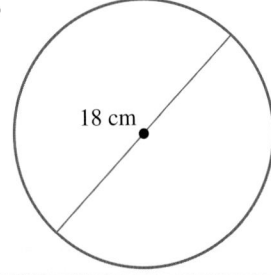
   18 cm

   b.
   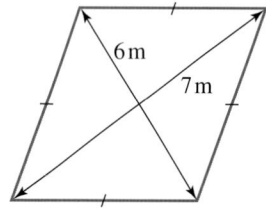
   6 m
   7 m

   c.
   15 cm
   10 cm

5. **WE1b** Calculate the area of each of the following shapes:

   i. stating the answer exactly, that is, in terms of π

   ii. correct to 2 decimal places.

a.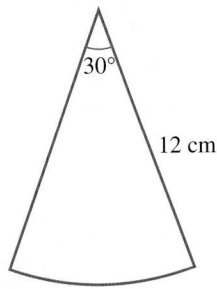

b.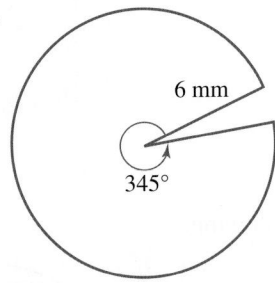

c.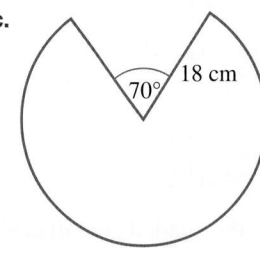

6. **MC** A figure has an area of about $64\,\text{cm}^2$. Identify which of the following *cannot* possibly represent the figure.

   A. A triangle with base length $16\,\text{cm}$ and height $8\,\text{cm}$
   B. A circle with radius $4.51\,\text{cm}$
   C. A rectangle with dimensions $16\,\text{cm}$ and $4\,\text{cm}$
   D. A rhombus with diagonals $16\,\text{cm}$ and $4\,\text{cm}$

7. **MC** Identify from the following list, all the lengths required to calculate the area of the quadrilateral shown.

   A. AC, BE and FD
   B. AB, BE, AC and CD
   C. BC, BE, AD and CD
   D. AC, CD and AB

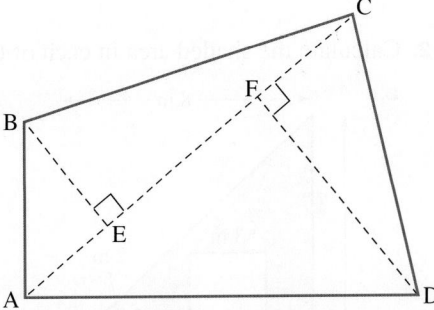

8. **WE2** Calculate the areas of the following composite shapes.

a.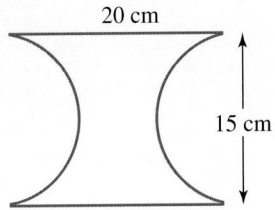

b.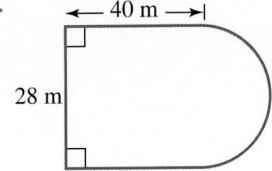

c.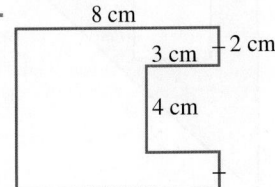

9. Calculate the areas of the following composite shapes.

a.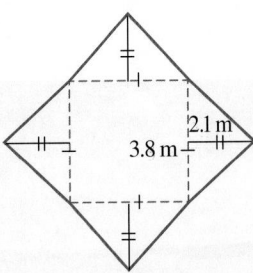

b.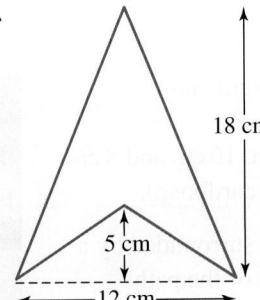

c.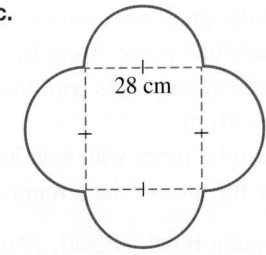

10. Calculate the shaded area in each of the following.

a.

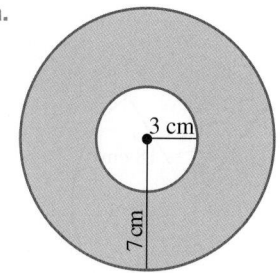

b.

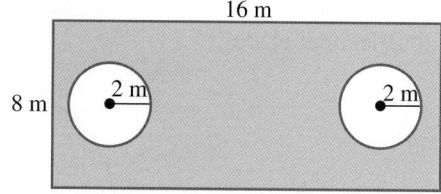

11. Calculate the shaded area in each of the following.

a.

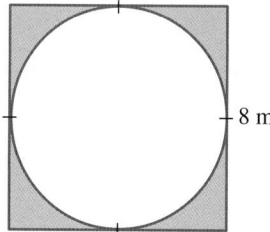

b.

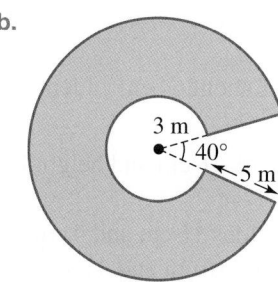

12. Calculate the shaded area in each of the following.

a.

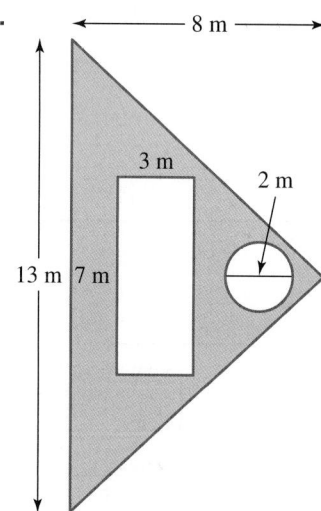

b.

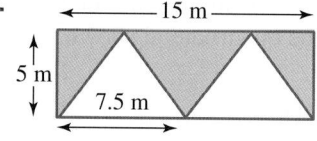

Understanding

13. A sheet of cardboard is 1.6 m by 0.8 m. The following shapes are cut from the cardboard:
    - a circular piece with radius 12 cm
    - a rectangular piece 20 cm by 15 cm
    - two triangular pieces with base length 30 cm and height 10 cm
    - a triangular piece with side lengths 12 cm, 10 cm and 8 cm.
    Calculate the area of the remaining piece of cardboard.

14. A rectangular block of land, 12 m by 8 m, is surrounded by a concrete path 0.5 m wide. Calculate the area of the path.

15. Concrete slabs 1 m by 0.5 m are used to cover a footpath 20 m by 1.5 m. Determine how many slabs are needed.

16. A city council builds a 0.5 m wide concrete path around the garden as shown below.

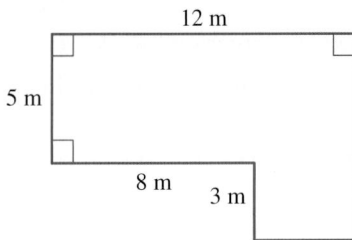

Determine the cost of the job if the worker charges $40.00 per m².

17. A tennis court used for doubles is 10.97 m wide, but a singles court is only 8.23 m wide, as shown in the diagram.

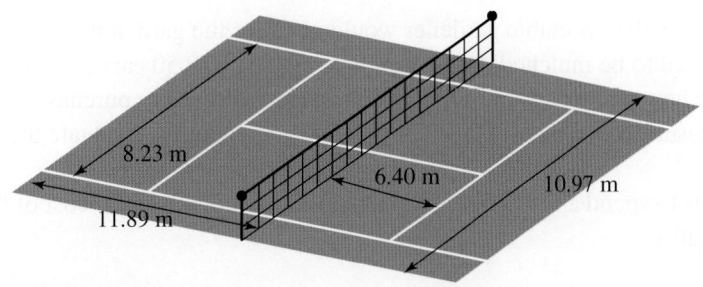

a. Calculate the area of the doubles tennis court.
b. Calculate the area of the singles court.
c. Determine the percentage of the doubles court that is used for singles. Give your answer to the nearest whole number.

Communicating, reasoning and problem solving

18. Dan has purchased a country property with layout and dimensions as shown in the diagram.

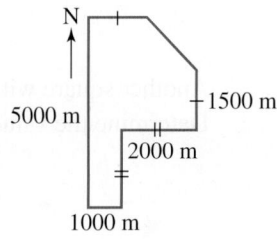

a. Show that the property has a total area of 987.5 ha.
b. Dan wants to split the property in half (in terms of area) by building a straight-lined fence running either north–south or east–west through the property. Assuming the cost of the fencing is a fixed amount per linear metre, justify where the fence should be built (that is, how many metres from the top left-hand corner and in which direction) to minimise the cost.

19. Romesh the excavator operator has 100 metres of barricade mesh and needs to enclose an area to safely work in. He chooses to make a rectangular region with dimensions x m and $y = (50 - x)$ m. Show your working when required.

a. Write an equation for the area of the region in terms of x.
b. Fill in the table for different values of x.

x	0	5	10	15	20	25	30	35	40	45	50
Area (m²)											

c. Sketch a graph of area against x.
d. Determine the values of x and y that makes the area a maximum and the maximum area.

Romesh decides to choose to make a circular area with the barricade mesh.

e. Calculate the radius of this circular region.
f. Determine how much extra area Romesh now has compared to his rectangular region.

20. In question **19**, Romesh the excavator operator could choose to enclose a rectangular or circular area with 100 m of barricade mesh.
In this case, the circular region resulted in a larger safe work area.

a. Show that for 150 m of barricade mesh, a circular region again results in a larger safe work area as opposed to a rectangular region.

b. Show that for *n* metres of barricade mesh, a circular region will result in a larger safe work area as opposed to a rectangular region.

21. A vegetable gardener is going to build four new rectangular garden beds side by side. Each garden bed measures 12.5 metres long and 3.2 metres wide.
To access the garden beds, the gardener requires a path 1 metre wide between each garden bed and around the outside of the beds.

a. Evaluate the total area the vegetable gardener would need for the garden beds and paths.

b. The garden beds need to be mulched. Bags of mulch, costing $29.50 each, cover an area of 25 square metres. Determine how many bags of mulch the gardener will need to purchase.

c. The path is to be resurfaced at a cost of $39.50 per 50 square metres. Evaluate the cost of resurfacing the path.

d. The gardener needs to spend a further $150 on plants. Determine the total cost of building these new garden beds and paths.

22. The diagram shows one smaller square drawn inside a larger square on grid paper.

a. Determine what fraction of the area of the larger square is the area of the smaller square.

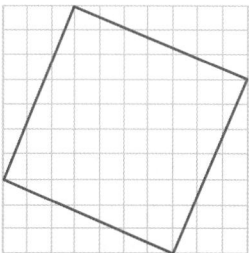

b. Another square with side lengths of 10 cm has a smaller square drawn inside. Determine the values of *x* and *y* if the smaller square is half the larger square.

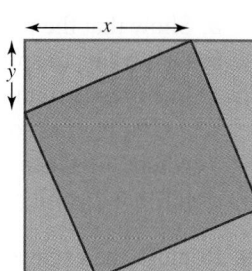

23. The shaded area in the diagram is called a segment of a circle.
A circle with a radius of 10 cm has ∠AOB equal to 90°.
A second circle, also with a radius of 10 cm, has ∠AOB equal to 120°.
Evaluate the difference in the areas of the segments of these two circles, correct to 2 decimal places.

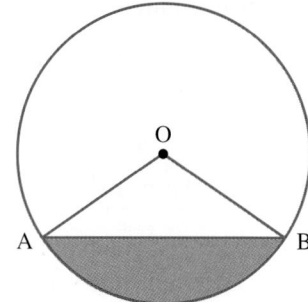

LESSON
11.3 Total surface area

LEARNING INTENTION

At the end of this lesson you should be able to:
- calculate the total surface area of rectangular prisms and pyramids
- calculate the total surface area of cylinders and spheres
- calculate the total surface area of cones
- calculate the surface area of composite solids
- solve worded problems involving surface area.

⊙ 11.3.1 Total surface area of solids

eles-4811

- The **total surface area (TSA)** of a solid is the sum of the areas of all the faces of that solid.
- TSA can be found by summing the areas of each face.
- Check the total number of faces to ensure that none are left out.
- For right prisms, drawing a net, such as with the examples below, can be useful to ensure all faces are accounted for.

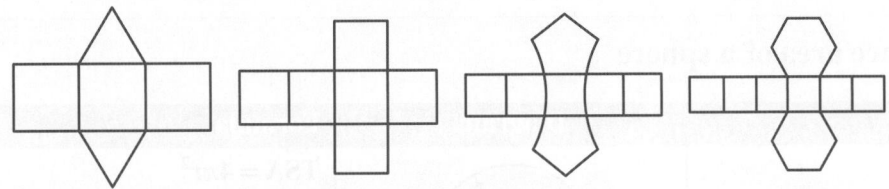

TSA of rectangular prisms, cubes and cylinders

Shape	Diagram	Formula
Rectangular prism (cuboid)		$\text{TSA} = 2(lh + lw + wh)$
Cube		$\text{TSA} = 6l^2$
Cylinder		$\text{TSA} = A_{\text{curved surface}} + A_{\text{circular ends}}$ $= 2\pi rh + 2\pi r^2$ $= 2\pi r(h + r)$

WORKED EXAMPLE 3 Calculating the TSA of a cylinder

WORKED EXAMPLE 3 Calculating the TSA of a cylinder

Calculate the total surface area of the cylinder, correct to the nearest cm².

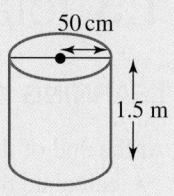

50 cm

1.5 m

THINK	WRITE
1. Write the formula for the TSA of a cylinder.	$\text{TSA} = 2\pi r(r+h)$
2. Identify the values for r and h. Note that the units will need to be the same.	$r = 50\,\text{cm}, \ h = 1.5\,\text{m}$ $\qquad = 150\,\text{cm}$
3. Substitute and evaluate.	$\text{TSA} = 2 \times \pi \times 50 \times (50 + 150)$ $\qquad = 62\,831.9\,\text{cm}^2$
4. Write the answer to correct to the nearest cm², with units.	$\approx 62\,832\,\text{cm}^2$

11.3.2 Total surface area of spheres, cones and pyramids (Path)

eles-4812

Total surface area of a sphere

Shape	Diagram	Formula
Sphere	Radius	$\textbf{TSA} = 4\pi r^2$

PATH

WORKED EXAMPLE 4 Calculating the TSA of a sphere

Calculate the total surface area of the sphere, correct to the nearest cm².

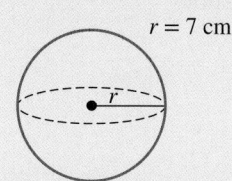

$r = 7$ cm

r

THINK	WRITE
1. Write the formula for the TSA of a sphere.	$\text{TSA} = 4\pi r^2$
2. Identify the value for r.	$r = 7$
3. Substitute and evaluate.	$\text{TSA} = 4 \times \pi \times 7^2$ $\qquad \approx 615.8\,\text{cm}^2$
4. Write the answer to correct to the nearest cm², with units.	$\approx 616\,\text{cm}^2$

- The total surface area of a cone can be found by considering its net, which is comprised of a small circle and a sector of a larger circle.

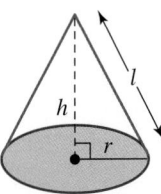 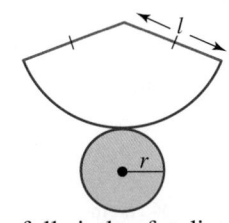

r = radius of the cone
l = slant height of the cone
h = perpendicular height of the cone

- The sector is a fraction of the full circle of radius l with circumference $2\pi l$.
- The sector has an arc length equivalent to the circumference of the base of the cone, $2\pi r$.
- The fraction of the full circle represented by the sector can be found by writing the arc length as a fraction of the circumference of the full circle, $\dfrac{2\pi r}{2\pi l} = \dfrac{r}{l}$.

$$\text{Area of a sector} = \text{fraction of the circle} \times \pi l^2$$
$$= \frac{r}{l} \times \pi l^2$$
$$= \pi r l$$

- *Note*: the perpendicular height h is not used in the formula to calculate the surface area of a cone, the relation between h, l and r can be determined using Pythagoras' theorem: $h = \sqrt{l^2 - r^2}$.

Total surface area of a cone

Shape	Diagram	Formula
Cone	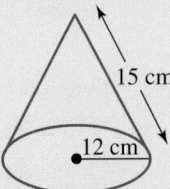	$\text{TSA} = A_{\text{circular end}} + A_{\text{curved surface}}$ $= \pi r^2 + \pi r l$ $= \pi r (r + l)$

PATH

WORKED EXAMPLE 5 Calculating the TSA of a cone

Calculate the total surface area of the cone shown, correct to one decimal place.

15 cm

12 cm

THINK

1. Write the formula for the TSA of a cone.

2. State the values of r and l.

3. Substitute and evaluate to obtain the answer.

WRITE

$\text{TSA} = \pi r (r + l)$

$r = 12, l = 15$

$\text{TSA} = \pi \times 12 \times (12 + 15)$

$= 1017.9 \text{ cm}^2$

Calculate the total surface area of the square-based pyramid shown.

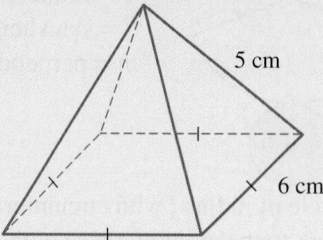

THINK

WRITE/DRAW

1. There are five faces: The square base and four identical triangles.

TSA = Area of square base + area of four triangular faces

2. Calculate the area of the square base.

Area of base = l^2, where $l = 6$

Area of base = 6^2

$= 36 \text{ cm}^2$

3. Draw and label one triangular face and write the formula for determining its area.

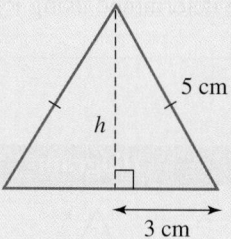

Area of a triangular face = $\dfrac{1}{2} bh$; $b = 6$

4. Calculate the height of the triangle, h, using Pythagoras' theorem.

$a^2 = c^2 - b^2$, where $a = h$, $b = 3$, $c = 5$

$h^2 = 5^2 - 3^2$

$h^2 = 25 - 9$

$h^2 = 16$

$h = 4 \text{ cm}$

5. Calculate the area of the triangular face by substituting $b = 6$ and $h = 4$.

Area of triangular face = $\dfrac{1}{2} \times 6 \times 4$

$= 12 \text{ cm}^2$

6. Calculate the TSA by adding the area of the square base and the area of four identical triangular faces together.

TSA = $36 + 4 \times 12$

$= 36 + 48$

$= 84 \text{ cm}^2$

WORKED EXAMPLE 7 Calculating the TSA of a composite solid

Calculate the total surface area of the solid shown correct to 1 decimal place.

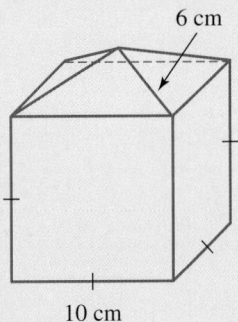

6 cm

10 cm

THINK	**WRITE/DRAW**
1. The solid shown has nine faces — five identical squares and four identical triangles.	$\text{TSA} = 5 \times \text{area of a square}$ $+ 4 \times \text{area of a triangle}$
2. Calculate the area of one square face with the side length 10 cm.	$A_{\text{square}} = l^2$, where $l = 10$ $A = 10^2$ $A = 100 \, \text{cm}^2$
3. Draw a triangular face and label the three sides. Use Pythagoras's theorem to calculate the height of the triangle, and substitute this value in the formula for the area of a triangle.	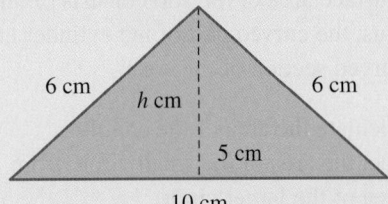 6 cm h cm 6 cm 5 cm 10 cm
4. Use Pythagoras's theorem in the triangle with sides 5 cm and h cm and hypothenuse 6 cm. Evaluate the value of h.	$5^2 + h^2 = 6^2$ $25 + h^2 = 36$ $h^2 = 11$ $h^2 = \sqrt{11}$
5. State the formula for the area of a triangle of base b and height h. Substitute and evaluate.	$A = \dfrac{b \times h}{2}$ $A = \dfrac{10 \times \sqrt{11}}{2}$ $A = 5\sqrt{11}$ $A = 16.583\,124 \dots \text{cm}^2$
6. Determine the TSA of the solid by adding the area of the five squares and four triangles.	$\text{TSA} = 5 \times 100 + 4 \times 16.583\,124 \dots$ $= 566.3325 \dots$ $= 566.3 \, \text{cm}^2$ (to 1 decimal place)

Note: Rounding is not done until the final step. It is important to realise that rounding too early can affect the accuracy of results.

WORKED EXAMPLE 8 Applying surface area in worded problems

The silo shown is to be built from metal. The top portion of the silo is a cylinder of diameter 4 m and height 8 m. The bottom part of the silo is a cone of slant height 3 m. The silo has a circular opening of radius 30 cm on the top.

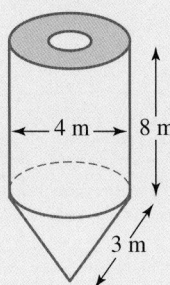

a. Calculate the area of metal (to the nearest m²) that is required to build the silo.
b. If it costs $12.50 per m² to cover the surface with an anti-rust material, determine how much will it cost to cover the silo completely.

THINK

a. 1. The surface area of the silo consists of an annulus, the curved part of the cylinder and the curved section of the cone.

2. To calculate the area of the annulus, subtract the area of the small circle from the area of the larger circle.
Let R = radius of small circle. Remember to convert all measurements to the same units.

3. The middle part of the silo is the curved part of a cylinder. Determine its area.
(Note that in the formula $\text{TSA}_{\text{cylinder}} = 2\pi r^2 + 2\pi rh$, the curved part is represented by $2\pi rh$.)

4. The bottom part of the silo is the curved section of a cone. Determine its area. (Note that in the formula $\text{TSA}_{\text{cone}} = \pi r^2 + \pi rl$, the curved part is given by πrl.)

5. Calculate the total surface area of the silo by finding the sum of the surface areas calculated above.

6. Write the answer in words.

b. To determine the total cost, multiply the total surface area of the silo by the cost of the anti-rust material per m² ($12.50).

WRITE

a. TSA = area of annulus
          + area of curved section of a cylinder
          + area of curved section of a cone

Area of annulus = $A_{\text{large circle}} - A_{\text{small circle}}$
            = $\pi r^2 - \pi R^2$
where $r = \dfrac{4}{2} = 2$ m and $R = 30$ cm $= 0.3$ m.
Area of annulus = $\pi \times 2^2 - \pi \times 0.3^2$
               = 12.28 m

Area of curved section of cylinder = $2\pi rh$
where $r = 2$, $h = 8$.
Area of curved section of cylinder = $2 \times \pi \times 2 \times 8$
                          = 100.53 m²

Area of curved section of cone = πrl
where $r = 2$, $l = 3$.
Area of curved section of cone = $\pi \times 2 \times 3$
                        = 18.85 m²

TSA = $12.28 + 100.53 + 18.85$
      = 131.66 m²

The area of metal required is 132 m², correct to the nearest square metre.

b. Cost = $132 \times \$12.50$
       = \$1650.00

on Resources

Video eLesson Total surface area of prisms (eles-1909)

Interactivities Surface area of a prism (int-6079)
Surface area of a cylinder (int-6080)
Surface area (int-6477)

Exercise 11.3 Total surface area

learn on

11.3 Quick quiz **on**	**11.3 Exercise**

Individual pathways

■ PRACTISE	■ CONSOLIDATE	■ MASTER
1, 5, 7, 11, 12, 14, 19, 22	2, 6, 9, 10, 13, 16, 17, 20, 23	3, 4, 8, 15, 18, 21, 24

Unless told otherwise, where appropriate, give answers correct to 1 decimal place.

Fluency

1. Calculate the total surface areas of the solids shown.

a.
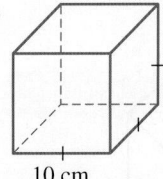
10 cm

b.
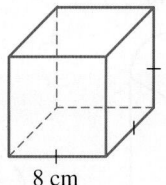
8 cm

c.
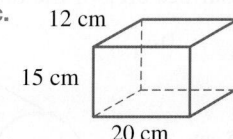
12 cm
15 cm
20 cm

d.
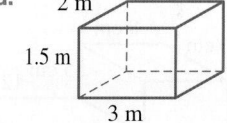
2 m
1.5 m
3 m

2. **WE3** Calculate the total surface area of the solids shown below.

a.
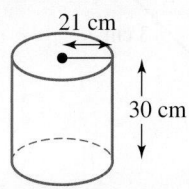
21 cm
30 cm

b.
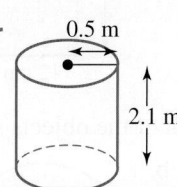
0.5 m
2.1 m

3. **WE4&5** **PATH** Calculate the total surface area of the spheres and cones below.

a.
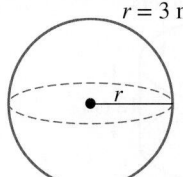
r = 3 m
r

b.
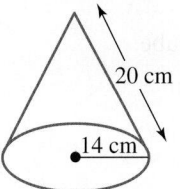
12 cm

c.
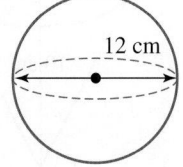
20 cm
14 cm

d.
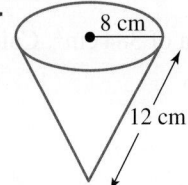
8 cm
12 cm

4. WE6 PATH Calculate the total surface area of the solids below.

a.

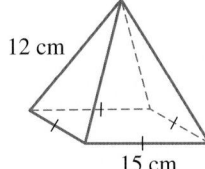

12 cm

15 cm

b.

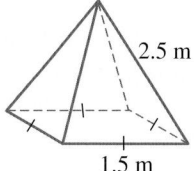

2.5 m

1.5 m

c.

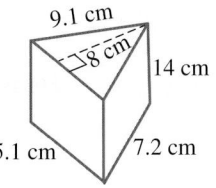

9.1 cm

8 cm

14 cm

5.1 cm   7.2 cm

d.

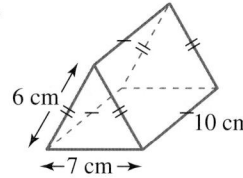

6 cm

7 cm

10 cm

5. Calculate the surface areas of the following.

   a. A cube of side length 1.5 m.

   b. A rectangular prism 6 m × 4 m × 2.1 m.

   c. A cylinder of radius 30 cm and height 45 cm, open at one end.

6. PATH Calculate the surface areas of the following.

   a. A sphere of radius 28 mm.

   b. An open cone of radius 4 cm and slant height 10 cm.

   c. A square pyramid of base length 20 cm and slant edge 30 cm.

7. WE7 PATH Calculate the total surface area of the objects shown.

a.

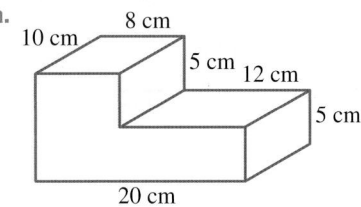

10 cm   8 cm

5 cm   12 cm

5 cm

20 cm

b.

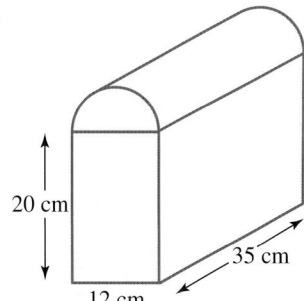

20 cm

35 cm

12 cm

c.

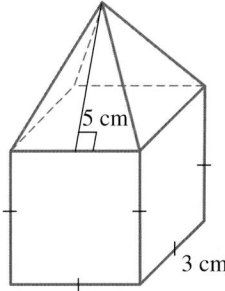

5 cm

3 cm

8. PATH Calculate the total surface area of the objects shown.

a.

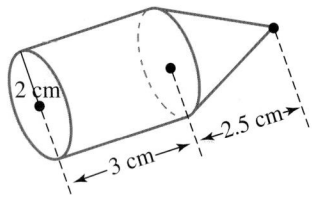

2 cm

3 cm   2.5 cm

b.

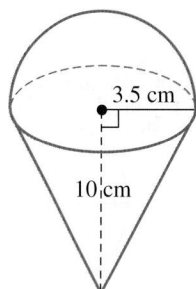

3.5 cm

10 cm

c.

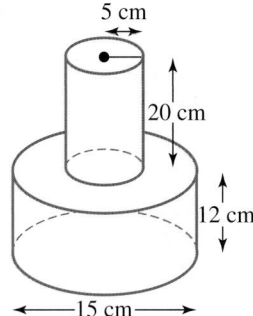

5 cm

20 cm

12 cm

15 cm

9. MC A cube has a total surface area of 384 cm². Calculate the length of the edge of the cube.

   A. 9 cm   B. 8 cm   C. 7 cm   D. 6 cm

Understanding

10. **PATH** In each of these diagrams name the dimensions given.

a.

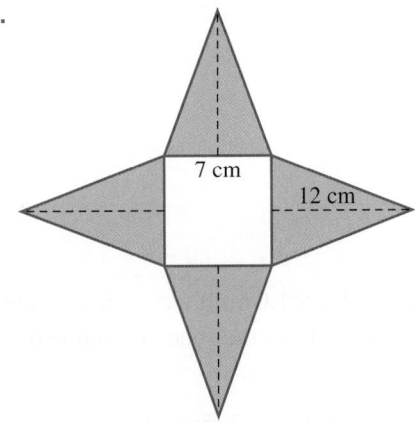

7 cm

12 cm

b.

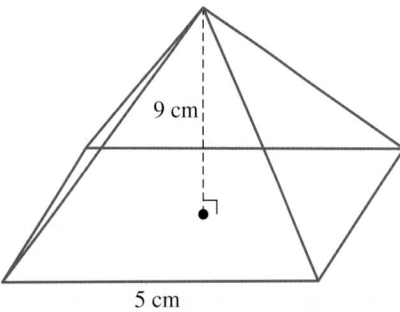

9 cm

5 cm

c.

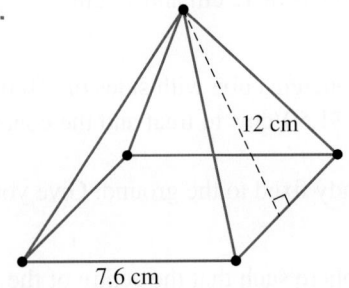

12 cm

7.6 cm

11. **PATH** Calculate the surface area of the following.

a.

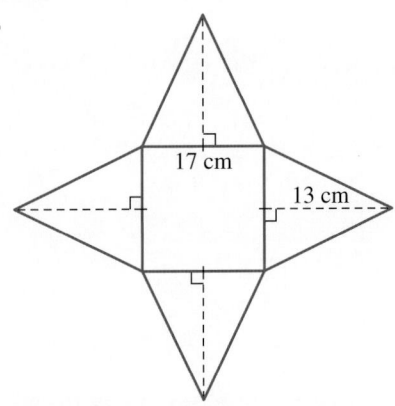

17 cm

13 cm

b.

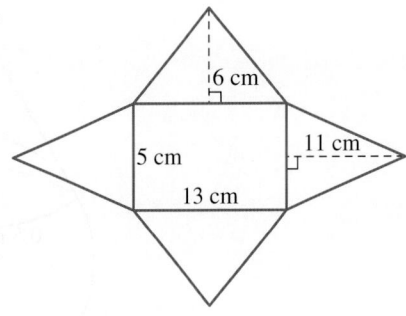

6 cm

5 cm

11 cm

13 cm

12. **WE8** **PATH** The greenhouse shown is to be built using shade cloth. It has a wooden door of dimensions $1.2\,\text{m} \times 0.5\,\text{m}$.

a. Calculate the total area of shade cloth needed to complete the greenhouse.

b. Determine the cost of the shade cloth at $6.50 per m^2.

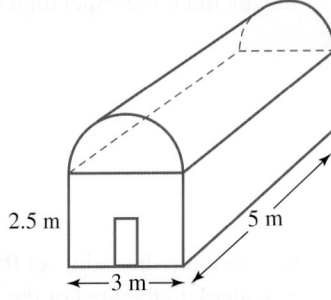

2.5 m

5 m

3 m

13. **PATH** A cylinder is joined to a hemisphere to make a cake holder, as shown. The surface of the cake holder is to be chromed at 5.5 cents per cm^2.

a. Calculate the total surface area to be chromed.

b. Determine the cost of chroming the cake holder.

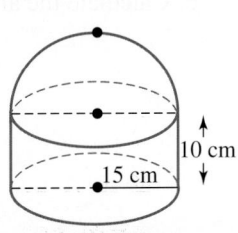

10 cm

15 cm

14. A steel girder is to be painted. Calculate the area of the surface to be painted.

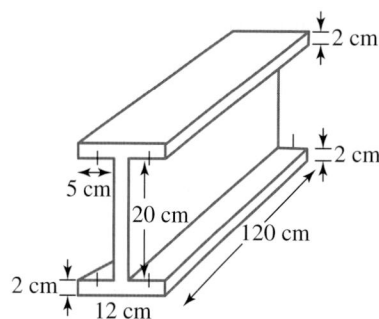

15. PATH Open cones are made from nets cut from a large sheet of paper 1.2 m × 1.0 m. If a cone has a radius of 6 cm and a slant height of 10 cm, determine how many cones can be made from the sheet. (Assume there is 5% wastage of paper.)

16. A prism of height 25 cm has a base in the shape of a rhombus with diagonals of 12 cm and 16 cm. Calculate the total surface area of the prism.

17. PATH A hemispherical glass dome, with a diameter of 24 cm, sits on a concrete cube with sides of 50 cm. To protect the structure, all exposed sides are to be treated. The glass costs $1.50/cm² to treat and the concrete costs 5 c/cm².
Calculate the cost in treating the structure if the base of the cube is already fixed to the ground. Give your answer to the nearest dollar.

18. PATH An inverted cone with side length 4 metres is placed on top of a sphere such that the centre of the cone's base is 0.5 metres above the centre of the sphere. The radius of the sphere is $\sqrt{2}$ metres.

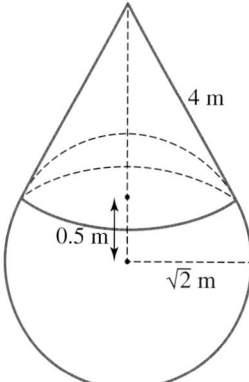

a. Calculate the exact total surface area of the sphere.

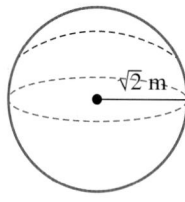

b. Calculate the radius of the cone exactly.
c. Calculate the area of the curved surface of the cone exactly.

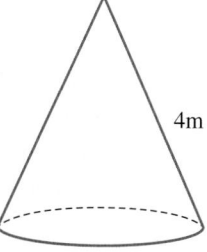

Communicating, reasoning and problem solving

19. A shower recess with dimensions 1500 mm (back wall) by 900 mm (side wall) needs to have the back and two side walls tiled to a height of 2 m.

    a. Calculate the area to be tiled in m².
    b. Justify that 180 tiles (including those that need to be cut) of dimension 20 cm by 20 cm will be required. Disregard the grout and assume that once a tile is cut, only one piece of the tile can be used.
    c. Evaluate the cheapest option of tiling; $1.50/tile or $39.50/box, where a box covers 1 m², or tiles of dimension 30 cm by 30 cm costing $3.50/tile.

20. The table shown below is to be varnished (including the base of each leg). The tabletop has a thickness of 180 mm and the cross-sectional dimensions of the legs are 50 mm by 50 mm. A friend completes the calculation without a calculator as shown. Assume there are no simple calculating errors.
    Analyse the working presented and justify if the TSA calculated is correct.

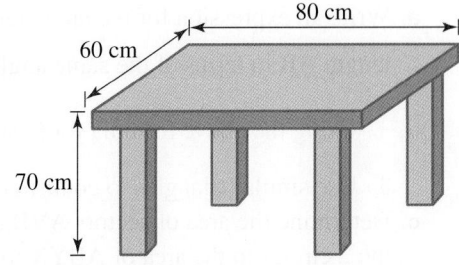

Tabletop (inc. leg bases)	0.96	$2 \times (0.8 \times 0.6)$
Legs	0.416	$16 \times (0.52 \times 0.05)$
Tabletop edging	0.504	$0.18 \times (2(0.8 + 0.6))$
TSA	1.88 m²	

21. A soccer ball is made up of a number of hexagons sewn together on its surface. Each hexagon can be considered to have dimensions as shown in the diagram.

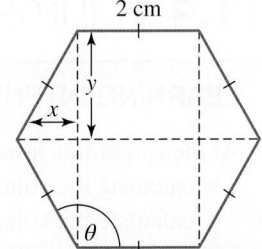

    a. Calculate $\theta°$.
    b. Calculate the values of x and y exactly.
    c. Calculate the area of the trapezium in the diagram.
    d. Hence, determine the area of the hexagon.
    e. If the total surface area of the soccer ball is $192\sqrt{3}$ cm², determine how many hexagons are on its surface.

22. Phuong is re-covering a footstool in the shape of a cylinder with diameter 50 cm and height 30 cm. She also intends to cover the base of the cushion. She has 1 m² of fabric to make this footstool.
    When calculating the area of fabric required, allow an extra 20% of the total surface area to cater for seams and pattern placings.
    Explain whether Phuong has enough material to cover the footstool.

23. If the surface area of a sphere to that of a cylinder is in the ratio 4 : 3 and the sphere has a radius of $3a$, show that if the radius of the cylinder is equal to its height, then the radius of the cylinder is $\dfrac{3\sqrt{3}a}{2}$.

24. **PATH** A frustum of a cone is a cone with the top sliced off, as shown.

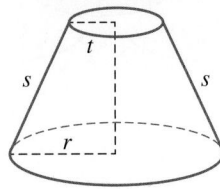

When the curved side is 'opened up', it creates a shape, ABYX, as shown in the diagram.

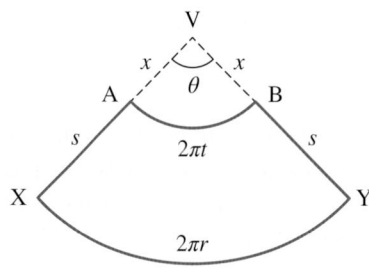

a. Write an expression for the arc length XY in terms of the angle θ. Write another expression for the arc length AB in terms of the same angle θ. Show that, in radians, $\theta = \dfrac{2\pi(r-t)}{s}$.

b. i. Using the above formula for θ, show that $x = \dfrac{st}{(r-t)}$.

   ii. Use similar triangles to confirm this formula.

c. Determine the area of sectors AVB and XVY and hence determine the area of ABYX. Add the areas of the 2 circles to the area of ABYX to determine the TSA of a frustum.

LESSON
11.4 Volume

LEARNING INTENTION

At the end of this lesson you should be able to:
- calculate the volume of prisms, including cylinders
- calculate the volume of spheres
- calculate the volume of pyramids
- calculate the volume of composite solids
- solve worded problems involving volume.

▶ 11.4.1 Volume

eles-4814

- The **volume** of a 3-dimensional object is the amount of space it takes up.
- Volume is measured in units of mm^3, cm^3 and m^3.
- The following diagram can be used to convert between units of volume.

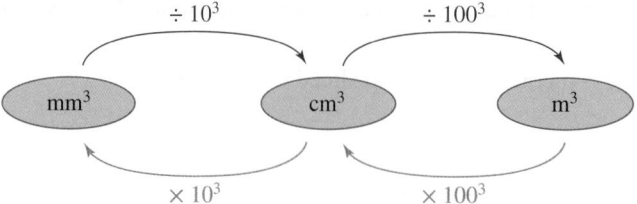

Volume of a solid with a uniform cross-sectional area

- The volume of any solid with a uniform cross-sectional area is given by the formula shown below.

> ## Volume of a solid with uniform cross-sectional area
>
> $$V = AH$$
>
> **where A is the area of the cross-section and H is the height of the solid.**

Shape	Diagram	Formula
Cube		Volume $= AH$ $\quad = $ area of a square $\times$ height $\quad = l^2 \times l$ $\quad = l^3$
Rectangular prism		Volume $= AH$ $\quad = $ area of a rectangle $\times$ height $\quad = lwh$
Cylinder		Volume $= AH$ $\quad = $ area of a circle $\times$ height $\quad = \pi r^2 h$
Triangular prism		Volume $= AH$ $\quad = $ area of a triangle $\times$ height $\quad = \dfrac{1}{2}bh \times H$

WORKED EXAMPLE 9 Calculating volumes of prisms

Calculate the volumes of the following shapes when necessary, correctly rounding to two decimal places.

a.
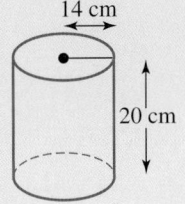
14 cm
20 cm

b.
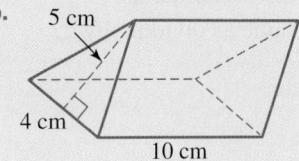
5 cm
4 cm
10 cm

THINK	WRITE
a. 1. Write the formula for the volume of the cylinder (prism).	**a.** $V = AH$ $\quad = \pi r^2 h$
2. Identify the value of the pronumerals.	$r = 14, h = 20$
3. Substitute and evaluate the answer, express answer with units.	$V = \pi \times 14^2 \times 20$ $\quad \approx 12\,315.04 \text{ cm}^3$
b. 1. Write the formula for the volume of a triangular prism.	**b.** $V = \dfrac{1}{2}bh \times H$
2. Identify the value of the pronumerals. (*Note:* h is the height of the triangle and H is the depth of the prism.)	$b = 4, h = 5, H = 10$
3. Substitute and evaluate the answer, express answer with units.	$V = \dfrac{1}{2} \times 4 \times 5 \times 10$ $\quad = 100 \text{ cm}^3$

WORKED EXAMPLE 10 Changing the dimensions of a prism

a. If each of the side lengths of a cube are doubled, then determine the effect on its volume.
b. If the radius is halved and the height of a cylinder is doubled, then determine the effect on its volume.

THINK	WRITE
a. 1. Write the formula for the volume of the cube.	**a.** $V = l^3$
2. Identify the value of the pronumeral. *Note:* Doubling is the same as multiplying by 2.	$l_{\text{new}} = 2l$
3. Substitute and evaluate.	$V_{\text{new}} = (2l)^3$
4. Compare the answer obtained in step 3 with the volume of the original shape.	$\quad = 8l^3$
5. Write your answer.	Doubling each side length of a cube increases the volume by a factor of 8; that is, the new volume will be 8 times as large as the original volume.
b. 1. Write the formula for the volume of the cylinder.	**b.** $V = \pi r^2 h$
2. Identify the value of the pronumerals. *Note:* Halving is the same as dividing by 2.	$r_{\text{new}} = \dfrac{r}{2}, h_{\text{new}} = 2h$
3. Substitute and evaluate.	$V_{\text{new}} = \pi \left(\dfrac{r}{2}\right)^2 2h$ $\quad = \pi \times \dfrac{r^2}{2\cancel{4}} \times \cancel{2}h$ $\quad = \dfrac{\pi r^2 h}{2}$

4. Compare the answer obtained in step 3 with the volume of the original shape.

$$= \frac{1}{2}\pi r^2 h$$

5. Write your answer.

Halving the radius and doubling the height of a cylinder decreases the volume by a factor of 2; that is, the new volume will be half the original volume.

DISCUSSION

The volumes of two solids were being compared.

The first solid is a rectangular prism, with a square base (length 6 cm) and a height of 20 cm. It has a cylindrical cut out straight through.

The second is a cylinder with diameter of 6 cm and a height of 20 cm. It has a square prism cut straight through.

Predict, giving reasons, which would have the greater volume.

Calculate both solids volume.

Do these calculations support your prediction?

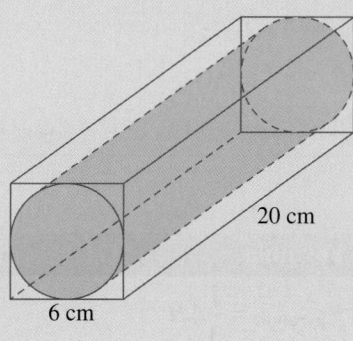

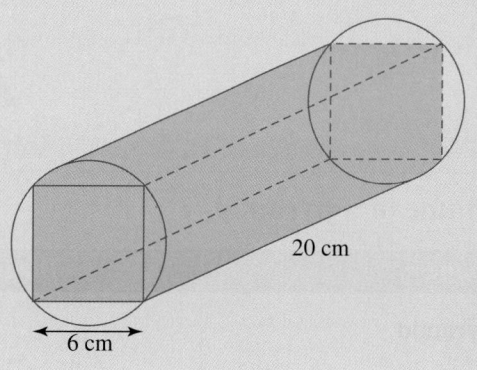

⊙ 11.4.2 Volumes of common shapes
eles-4815

Volume of a sphere (Path)

- The volume of a sphere of radius r is given by the following formula.

Volume of a sphere

Shape	Diagram	Formula
Sphere		$V = \dfrac{4}{3}\pi r^3$

PATH

WORKED EXAMPLE 11 Calculating the volume of a sphere

Calculate the volume of a sphere of radius 9 cm. Answer correct to 1 decimal place.

THINK	WRITE
1. Write the formula for the volume of a sphere.	$V = \dfrac{4}{3}\pi r^3$
2. Identify the value of r.	$r = 9$
3. Substitute and evaluate, express answer with units.	$V = \dfrac{4}{3} \times \pi \times 9^3$
	$= 3053.6 \text{ cm}^3$

Volume of a pyramid (Path)

- Pyramids are not prisms, as the cross-section changes from the base upwards.
- The volume of a pyramid is one-third the volume of the prism with the same base and height.

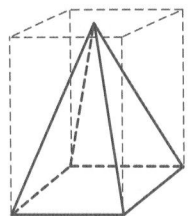

Volume of a pyramid

Shape	Diagram	Formula
Pyramid	Area of base $= A$ Base	$V_{\text{pyramid}} = \dfrac{1}{3}AH$

Volume of a cone

- The cone is a pyramid with a circular base.

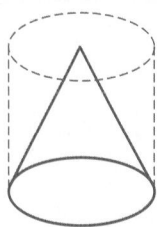

Volume of a cone

Shape	Diagram	Formula
Cone		$V_{\text{cone}} = \dfrac{1}{3}\pi r^2 h$

PATH

WORKED EXAMPLE 12 Calculating the volume of pyramids and cones

Calculate the volume of each of the following solids, rounding to two decimal places where necessary.

a.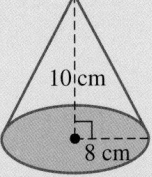

b.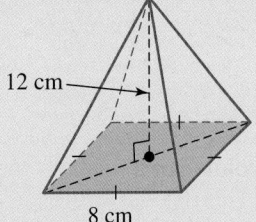

THINK	WRITE
a. 1. Write the formula for the volume of a cone.	a. $V = \dfrac{1}{3}\pi r^2 h$
2. Identify the values of r and h.	$r = 8,\ h = 10$
3. Substitute and evaluate, express answer with units.	$V = \dfrac{1}{3} \times \pi \times 8^2 \times 10$ $= 670.21\ \text{cm}^3$
b. 1. Write the formula for the volume of a pyramid.	b. $V = \dfrac{1}{3}AH$
2. Calculate the area of the square base.	$A = l^2$ where $l = 8$ $A = 8^2$ $\quad = 64\ \text{cm}^2$
3. Identify the value of H.	$H = 12$
4. Substitute and evaluate, express answer with units.	$V = \dfrac{1}{3} \times 64 \times 12$ $= 256\ \text{cm}^3$

▶ 11.4.3 Volume of composite solids (Path)

eles-4816

- A composite solid is a combination of a number of solids.
- Calculate the volume of each solid separately.
- Sum these volumes to give the volume of the composite solid.

WORKED EXAMPLE 13 Calculating the volume of a composite solid

Calculate the volume of the composite solid shown.

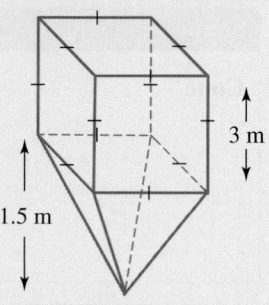

3 m

1.5 m

THINK	WRITE
1. The given solid is a composite figure, made up of a cube and a square-based pyramid.	$V = $ Volume of cube $+$ Volume of pyramid
2. Calculate the volume of the cube.	$V_{\text{cube}} = l^3$ where $l = 3$ $V_{\text{cube}} = 3^3$ $\quad\quad = 27 \,\text{m}^3$
3. Write the formula for the volume of a square-based pyramid.	$V_{\text{square-based pyramid}} = \dfrac{1}{3}AH$
4. Calculate the area of the square base.	$A = l^2$ $\quad = 3^2$ $\quad = 9 \,\text{m}^2$
5. Identify the value of H.	$H = 1.5$
6. Substitute and evaluate the volume of the pyramid.	$V_{\text{square-based pyramid}} = \dfrac{1}{3} \times 9 \times 1.5$ $\quad\quad\quad = 4.5 \,\text{m}^3$
7. Calculate the total volume by adding the volume of the cube and pyramid.	$V = 27 + 4.5$ $\quad = 31.5 \,\text{m}^3$

⏵ 11.4.4 Capacity

eles-4817

- Some 3-dimensional objects are hollow and can be filled with liquid or some other substance.
- The amount of substance that a container can hold is called its capacity.
- **Capacity** is essentially the same as volume but is usually measured in mL, L, kL and ML (megalitres) where $1 \,\text{mL} = 1 \,\text{cm}^3$

  $1 \,\text{L} = 1000 \,\text{cm}^3$

  $1 \,\text{kL} = 1 \,\text{m}^3$.
- The following diagram can be used to convert between units of capacity.

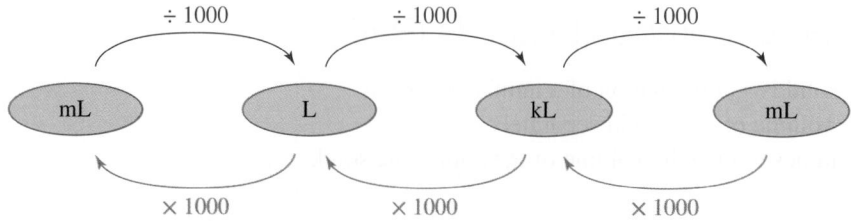

WORKED EXAMPLE 14 Calculating the capacity of a prism

Determine the capacity (in litres) of a cuboidal aquarium that is
50 cm long, 30 cm wide and 40 cm high.

THINK	WRITE
1. Write the formula for the volume of a rectangular prism.	$V = lwh$
2. Identify the values of the pronumerals.	$l = 50$, $w = 30$, $h = 40$
3. Substitute and evaluate.	$V = 50 \times 30 \times 40$ $= 60\,000\,\text{cm}^3$
4. State the capacity of the container in millilitres, using $1\,\text{cm}^3 = 1\,\text{mL}$.	$= 60\,000\,\text{mL}$
5. Since $1\,\text{L} = 1000\,\text{mL}$, to convert millilitres to litres divide by 1000.	$= 60\,\text{L}$
6. Write the answer in a sentence.	The capacity of the fish tank is $60\,\text{L}$.

on Resources

Interactivities   Volume 1 (int-3791)
Volume 2 (int-6476)
Volume of solids (int-3794)

Exercise 11.4 Volume

learn**on**

11.4 Quick quiz   **on**

11.4 Exercise

Individual pathways

■ PRACTISE	■ CONSOLIDATE	■ MASTER
1, 3, 5, 7, 9, 12, 15, 18, 23, 27	2, 6, 8, 10, 13, 16, 19, 21, 24, 28	4, 11, 14, 17, 20, 22, 25, 26, 29, 30

Fluency

1. Calculate the volumes of the following prisms.

a.
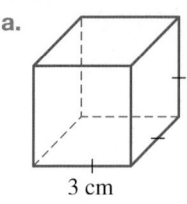
3 cm

b.
4.2 m

c.
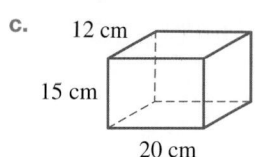
12 cm
15 cm
20 cm

d.
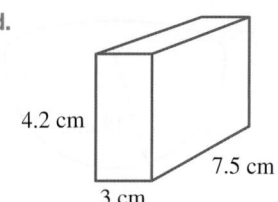
4.2 cm
7.5 cm
3 cm

2. Calculate the volume of each of these solids.

a.
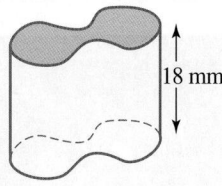
18 mm

[Base area: 25 mm²]

b.
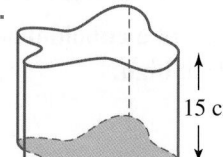
15 cm

[Base area: 24 cm²]

3. **WE9** Calculate the volume of each of the following. Give each answer correct to 1 decimal place where appropriate.

a.
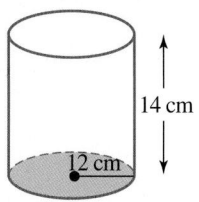
14 cm
12 cm

b.
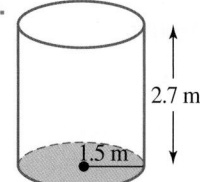
2.7 m
1.5 m

c.
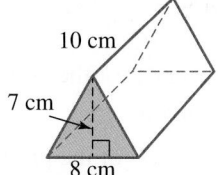
10 cm
7 cm
8 cm

4. Calculate the volume of each of the following. Give each answer correct to 1 decimal place where appropriate.

a.
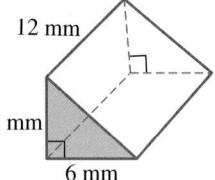
12 mm
8 mm
6 mm

b.
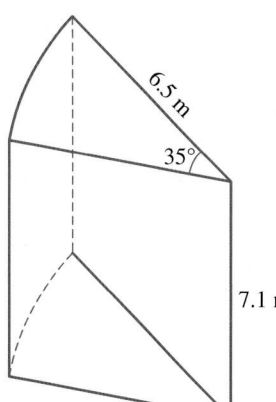
6.5 m
35°
7.1 m

c.
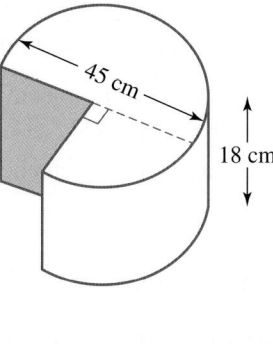
45 cm
18 cm

5. **WE11** **PATH** Determine the volume of a sphere (correct to 1 decimal place) with a radius of:

a. 1.2 m   b. 15 cm   c. 7 mm   d. 50 cm

6. **PATH** Calculate the volume of each of these figures, correct to 2 decimal places.

a.
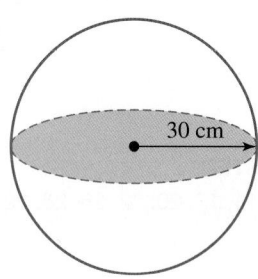
30 cm

b.
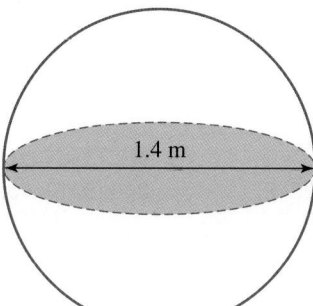
1.4 m

c.
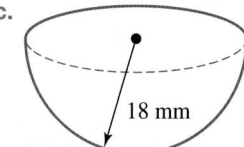
18 mm

d.
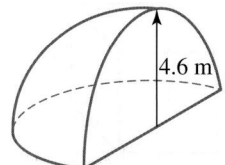
4.6 m

7. WE12a PATH Determine the volume of each of the following cones, correct to 1 decimal place.

a.
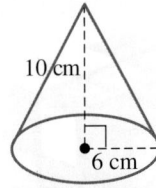
10 cm

6 cm

b.
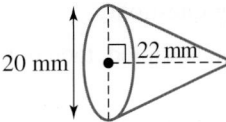
20 mm   22 mm

8. WE12b PATH Calculate the volume of each of the following pyramids.

a.
12 cm
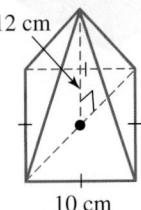
10 cm

b.
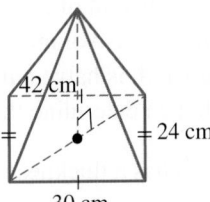
42 cm   24 cm
30 cm

9. WE13 PATH Calculate the volume of each of the following composite solids correct to 2 decimal places where appropriate.

a.
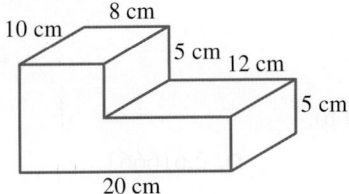
10 cm   8 cm
5 cm   12 cm
5 cm
20 cm

b.
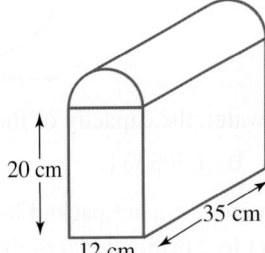
20 cm
12 cm   35 cm

10. PATH Calculate the volume of each of the following composite solids correct to 2 decimal places where appropriate.

a.
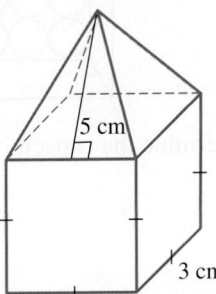
5 cm
3 cm

b.
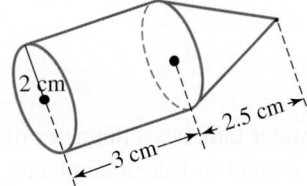
2 cm
3 cm   2.5 cm

11. PATH Calculate the volume of each of the following composite solids correct to 2 decimal places where appropriate.

a.
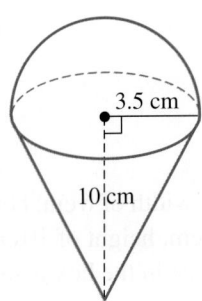
3.5 cm
10 cm

b.
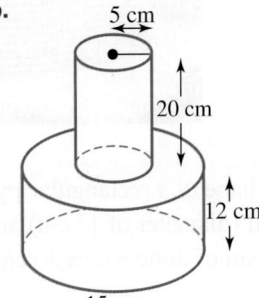
5 cm
20 cm
12 cm
15 cm

Understanding

12. **WE10** Answer the following questions.

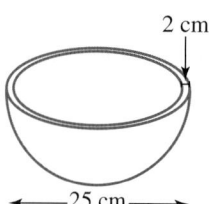

a. If the side length of a cube is tripled, then determine the effect on its volume.

b. If the side length of a cube is halved, then determine the effect on its volume.

c. If the radius is doubled and the height of a cylinder is halved, then determine the effect on its volume.

d. If the radius is doubled and the height of a cylinder is divided by four, then determine the effect on its volume.

e. If the length is doubled, the width is halved and the height of a rectangular prism is tripled, then determine the effect on its volume.

13. **MC** **PATH** A hemispherical bowl has a thickness of 2 cm and an outer diameter of 25 cm.

2 cm

25 cm

If the bowl is filled with water, the capacity of the water will be closest to:

A. 2.42452 L   B. 1.30833 L   C. 3.05208 L   D. 2.61666 L

14. **PATH** Tennis balls of diameter 8 cm are packed in a box 40 cm × 32 cm × 10 cm, as shown. Determine, correct to 2 decimal places, how much space is left unfilled.

15. **WE13** A cylindrical water tank has a diameter of 1.5 m and a height of 2.5 m. Determine the capacity (in litres) of the tank, correct to 1 decimal place.

16. **PATH** A monument in the shape of a rectangular pyramid (base length of 10 cm, base width of 6 cm, height of 8 cm), a spherical glass ball (diameter of 17 cm) and conical glassware (radius of 14 cm, height of 10 cm) are packed in a rectangular prism of dimensions 30 cm by 25 cm by 20 cm. The extra space in the box is filled up by a packing material.

Determine, correct to 2 decimal places, the volume of packing material that is required.

17. **PATH** A swimming pool is being constructed so that it is the upper part of an inverted square-based pyramid.

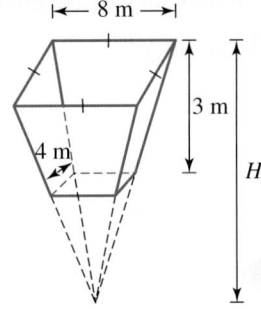

a. Calculate H.
b. Calculate the volume of the pool.
c. Determine how many $6\,m^3$ bins will be required to take the dirt away.
d. Determine how many litres of water are required to fill this pool.
e. Determine how deep the pool is when it is half-filled.

18. **PATH** A soft drink manufacturer is looking to repackage cans of soft drink to minimise the cost of packaging while keeping the volume constant.
Consider a can of soft drink with a capacity of 400 mL.

a. If the soft drink was packaged in a spherical can:
    i. calculate the radius of the sphere, correct to 2 decimal places
    ii. determine the total surface area of this can, correct to 1 decimal place.
b. If the soft drink was packaged in a cylindrical can with a radius of 3 cm:
    i. calculate the height of the cylinder, correct to 2 decimal places
    ii. determine the total surface area of this can, correct to 2 decimal places.
c. If the soft drink was packaged in a square-based pyramid with a base side length of 6 cm:
    i. calculate the height of the pyramid, correct to 2 decimal places
    ii. determine the total surface area of this can, correct to 2 decimal places.
d. Explain which can you would recommend the soft drink manufacturer use for its repackaging.

19. The volume of a cylinder is given by the formula $V = \pi r^2 h$.

a. Transpose the formula to make h the subject.
b. A given cylinder has a volume of $1600\,cm^3$.
   Calculate its height, correct to 1 decimal place, if it has a radius of:
    i. 4 cm   ii. 8 cm.
c. Transpose the formula to make r the subject.
d. Explain what restrictions must be placed on r.
e. A given cylinder has a volume of $1800\,cm^3$. Determine its radius, correct to 1 decimal place, if it has a height of:
    i. 10 cm   ii. 15 cm.

20. **PATH** A toy maker has enough rubber to make one super-ball of radius 30 cm. Determine how many balls of radius 3 cm he can make from this rubber.

21. A manufacturer plans to make a cylindrical water tank to hold 2000 L of water.

a. Calculate the height, correct to 2 decimal places, if he uses a radius of 500 cm.
b. Calculate the radius, correct to 2 decimal places if he uses a height of 500 cm.
c. Determine the surface area of each of the two tanks. Assume the tank is a closed cylinder and give your answer in square metres correct to 2 decimal places.

22. **PATH** The ancient Egyptians knew that the volume of the frustum of a square-based pyramid was given by the formula $V = \frac{1}{3}h\left(x^2 + xy + y^2\right)$, although how they discovered this is unclear. (A frustum is the part of a cone or pyramid that is left when the top is cut off.)

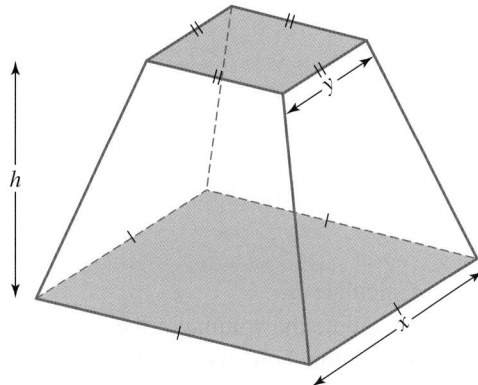

   a. Calculate the volume of the frustum below, correct to 2 decimal places.
   b. Determine the volume of the missing portion of the square-based pyramid shown, correct to 2 decimal places.

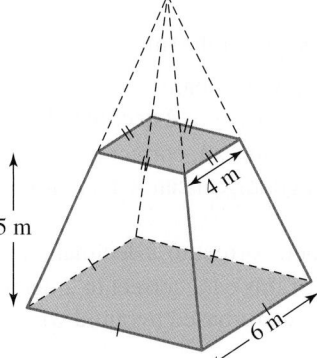

Communicating, reasoning and problem solving

23. The Hastings' family house has a rectangular roof with dimensions $17\,\text{m} \times 10\,\text{m}$ providing water to three cylindrical water tanks, each with a radius of $1.25\,\text{m}$ and a height of $2.1\,\text{m}$.
   Show that approximately 182 millimetres of rain must fall on the roof to fill the tanks.

24. **PATH** Archimedes is considered to be one of the greatest mathematicians of all time. He discovered several of the formulas used in this chapter. Inscribed on his tombstone was a diagram of his proudest discovery. It shows a sphere inscribed (fitting exactly) into a cylinder.
   Show that:

$$\frac{\text{volume of the cylinder}}{\text{volume of the sphere}} = \frac{\text{surface area of the cylinder}}{\text{surface area of the sphere}}$$

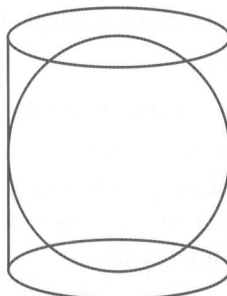

25. PATH Marion has mixed together ingredients for a cake. The recipe requires a baking tin that is cylindrical in shape with a diameter of 20 cm and a height of 5 cm.

Marion only has a tin in the shape of a trapezoidal prism and a muffin tray consisting of 24 muffin cups. Each of the muffin cups in the tray is a portion of a cone. Both the tin and muffin cup are shown in the diagrams.

Explain whether Marion should use the tin or the muffin tray.

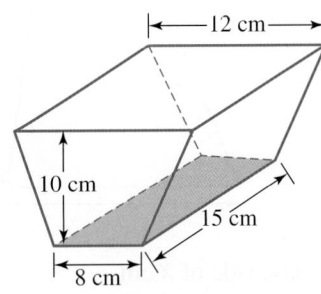

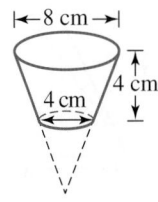

26. Sam is having his 16th birthday party and wants to make an ice trough to keep drinks cold. He has found a square piece of sheet metal with a side length of 2 metres.

He cuts squares of side length x metres from each corner, then bends the sides of the remaining sheet.

When four squares of the appropriate side length are cut from the corners, the capacity of the trough can be maximised at 588 litres.

Explain how Sam should proceed to maximise the capacity of the trough.

27. PATH Nathaniel and Reiko are going to the snow for survival camp. They plan to construct an igloo, consisting of an entrance and a hemispherical living section as shown.

Nathaniel and Annie are asked to redraw their plans and increase the height of the liveable region (hemispherical structure) so that the total volume (including entrance) is doubled.

Determine what must the new height of the hemisphere be to achieve this so that the total volume (including entrance) is doubled. Write your answer in metres correct to 2 decimal places.

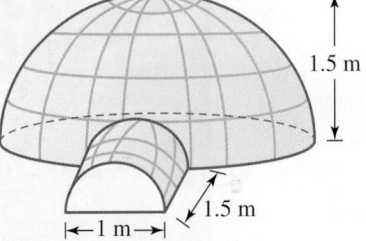

28. PATH Six tennis balls are just contained in a cylinder as the balls touch the sides and the end sections of the cylinder.

Each tennis ball has a radius of R cm.

a. Express the height of the cylinder in terms of R.
b. Evaluate the total volume of the tennis balls.
c. Determine the volume of the cylinder in terms of R.
d. Show that the ratio of the volume of the tennis balls to the volume of the cylinder is 2 : 3.

29. **PATH** A frustum of a square-based pyramid is a square pyramid with the top sliced off. H is the height of the full pyramid and h is the height of the frustum.

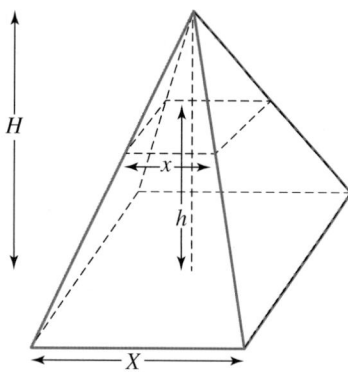 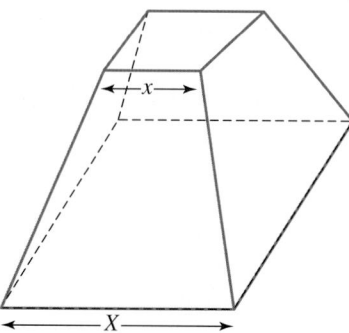

   a. Determine the volume of the large pyramid that has a square base side of X cm.
   b. Evaluate the volume of the small pyramid that has a square base side of x cm.

   c. Show that the relationship between H and h is given by $H = \dfrac{Xh}{X-x}$.

   d. Show that the volume of the frustum is given by $\dfrac{1}{3}h\left(X^2 + x^2 + Xx\right)$.

30. **PATH** A large container is five-eighths full of ice-cream. After removing 27 identical scoops, it is one-quarter full. Determine how many scoops of ice-cream are left in the container.

LESSON
11.5 Review

11.5.1 Topic summary

Prisms and cylinders

- Prisms are 3D objects that have a uniform cross section and all flat surfaces.
- The surface area of a prism is calculated by adding the areas of its faces.
- The volume of a prism is $V = AH$, where A is the cross-sectional area of the prism, and H is the perpendicular height.
- A cylinder is a 3D object that has a circular cross-section.
- The curved surface area of a cylinder is $2\pi rh$, and the total surface area is $2\pi r^2 + 2\pi rh = 2\pi r(r + h)$.
- The volume of a cylinder is $V = \pi r^2 h$.

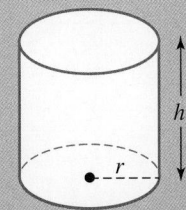

MEASUREMENT

Pyramids (Path)

- The surface area of a pyramid can be calculated by adding the surface areas of its faces.
- The volume of a pyramid is $V = \frac{1}{3}AH$, where A is the area of the base and H is the height.

Cones (Path)

- The curved surface area of a cone is $SA_{\text{curved}} = \pi rl$, where l is the slant height.
- The total surface area is $SA = \pi rl + \pi r^2 = \pi r(l + r)$.
- The volume of a cone is $V = \frac{1}{3}\pi r^2 h$.

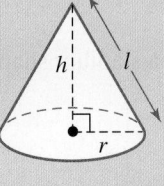

Spheres (Path)

- The surface area of a sphere is $A = 4\pi r^2$.
- The volume of a sphere is $V = \frac{4}{3}\pi r^3$.

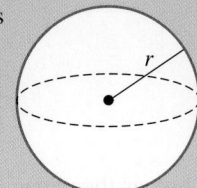

Units of area, volume and capacity

Area:

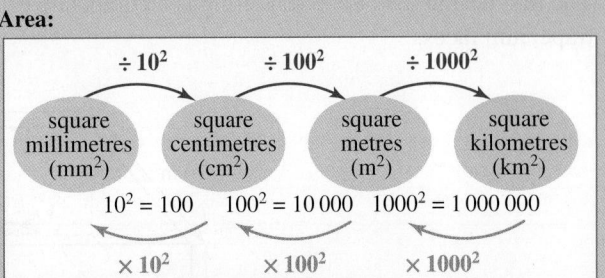

1 hectare (ha) = 10 000 m^2

Volume:

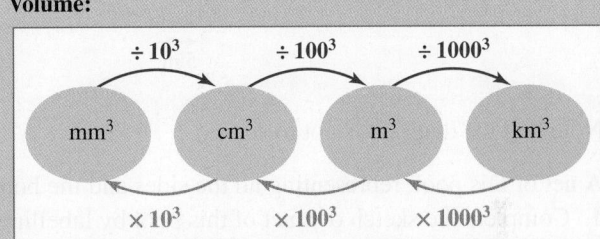

Capacity:

$$\xrightarrow{\div 1000} \text{mL} \quad \text{L} \quad \text{kL} \quad \text{ML}$$

$$\times 1000 \quad \times 1000 \quad \times 1000$$

1 cm^3 = 1 mL
1 L = 1 000 cm^3

Area formulas

- Square: $A = l^2$
- Rectangle: $A = lw$
- Triangle: $A = \frac{1}{2}bh$
- Parallelogram: $A = bh$
- Trapezium: $A = \frac{1}{2}(a + b)h$
- Kite: $A = \frac{1}{2}xy$
- Circle: $A = \pi r^2$
- Sector: $A = \frac{\theta}{360°} \times \pi r^2$

11.5.2 Project

Total surface area and capacity of a pool

Swimming pools come in a wide variety of shapes, not just the popular rectangular prisms or cylinders.

The Kaur family is planning to get a swimming pool constructed in the backyard of their Newcastle family home.

The first design they are considering is a composite prism, represented below, with rectangles and trapezium faces.

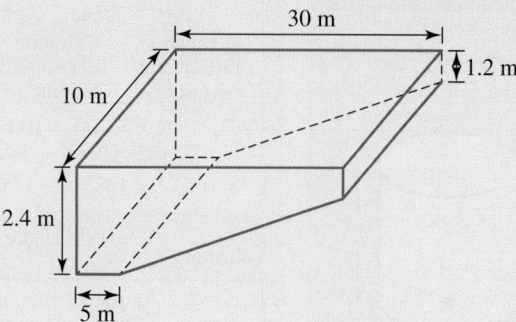

Note that the diagram is not to scale.

A net of this pool, representing all the sides and the bottom, has been drawn.

1. Complete the sketch of a net of this pool by labelling the dimensions you would need to calculate the total surface area of the pool.
2. Calculate the total surface area if this first pool.
3. To paint the surface of the pool, the Kaur family would use 3.5- litre cans of paint sufficient to cover 12 square metres each. Determine the minimum number of cans they would need for one coat of paint.

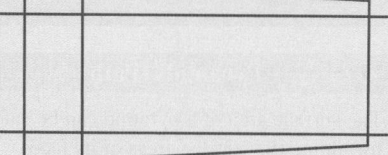

4. Each can of paint costs $37. Calculate how much it would cost to apply two coats of paint.
5. Calculate the capacity of this pool. Give your answer in m^3 and in L.
6. Calculate how many hours it would take to fill this pool using a garden hose with a flow rate of 1.2 L per second. Give your answer to the nearest hour.

The Kaur family is also considering a second, simpler design, with a stadium shape, represented below, and a constant depth d.

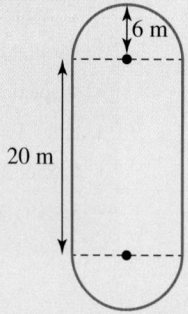

7. Determine the value of d so that the capacity of this pool is the same as the capacity of the first pool. Give your answer to the nearest centimetre.

The Kaur family prefers this second shape. However, they have young children and believe that the depth found in part 7 is too much, and decide to go with this second design but with a constant depth of 1.3 m.

8. Calculate how much it would cost them to paint the total surface area of this pool, with two coats of paint, using the same paint as in part **3**. Give your answer to the nearest dollar.

  Resources

Interactivities Crossword (int-2842)
Sudoku puzzle (int-3593)

Exercise 11.5 Review questions

learn on

Unless told otherwise, where appropriate, give answers correct to 2 decimal places.

Fluency

1. **MC** If all measurements are in cm, the area of the figure is:
   A. 16.49 cm^2
   B. 39.25 cm^2
   C. 9.81 cm^2
   D. 23.56 cm^2

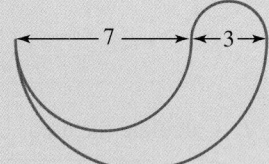

2. **MC** If all measurements are in centimetres, the area of the figure is:
   A. 50.73 cm^2
   B. 99.82 cm^2
   C. 80.18 cm^2
   D. 90 cm^2

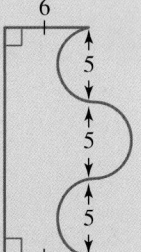

3. **MC** If all measurements are in centimetres, the shaded area of the figure is:
   A. 3.93 cm^2
   B. 129.59 cm^2
   C. 388.77 cm^2
   D. 141.11 cm^2

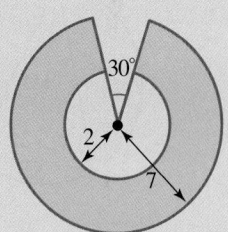

4. **MC** The total surface area of the solid is:
   A. 8444.6 mm^2
   B. 9221 mm^2
   C. $14\,146.5 \text{ mm}^2$
   D. $50\,271.1 \text{ mm}^2$

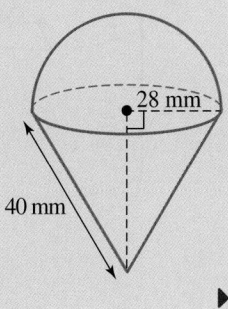

Understanding

5. Calculate the areas of the following plane figures. All measurements are in cm.

a.
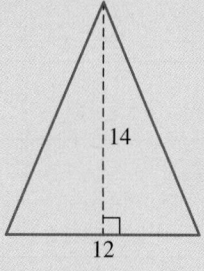

b.
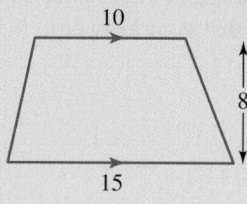

c.
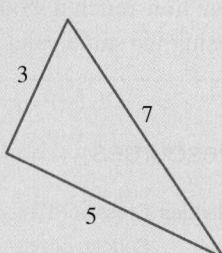

6. Calculate the areas of the following plane figures. All measurements are in cm.

a.
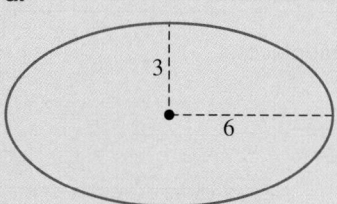

b.
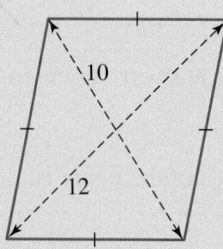

c.
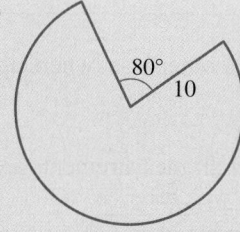

7. Calculate the areas of the following figures. All measurements are in cm.

a.
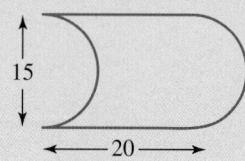

b.
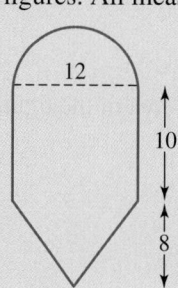

c.
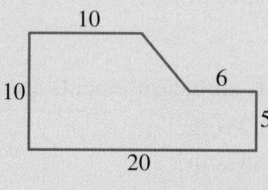

8. Calculate the blue shaded area in each of the following. All measurements are in cm.

a.
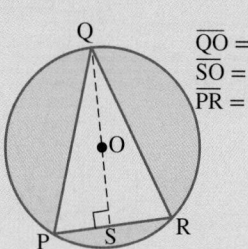

$\overline{QO} = 15$ cm
$\overline{SO} = 8$ cm
$\overline{PR} = 18$ cm

b.
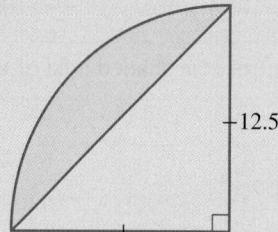

c.
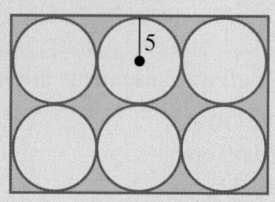

9. Calculate the total surface area of each of the following solids.

a.
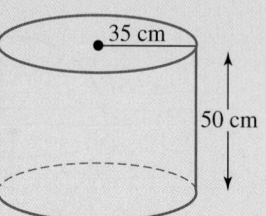

b.
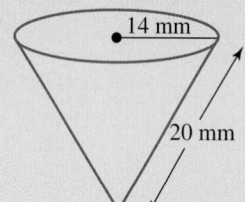

c.
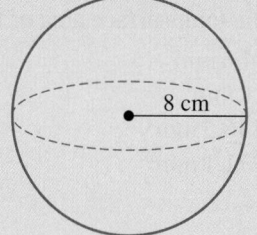

10. Calculate the total surface area of each of the following solids.

a.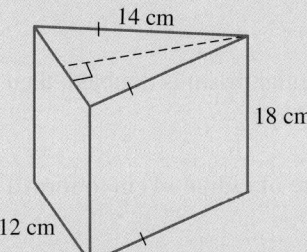
14 cm
18 cm
12 cm

b.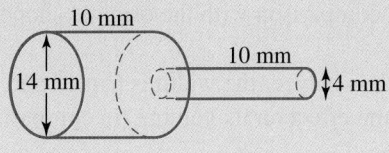
10 mm
10 mm
14 mm
4 mm

[closed at both ends]

c.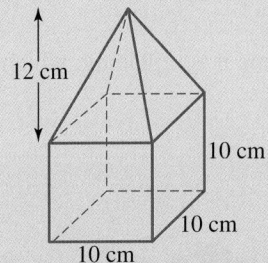
12 cm
10 cm
10 cm
10 cm

11. Calculate the volume of each of the following.

a.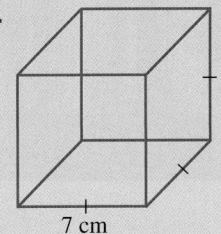
7 cm

b.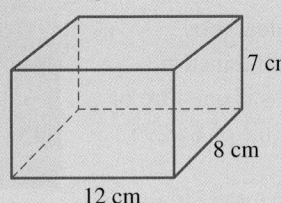
7 cm
8 cm
12 cm

c.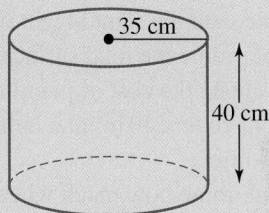
35 cm
40 cm

12. Determine the volume of each of the following.

a.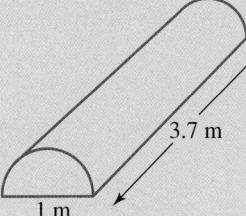
3.7 m
1 m

b.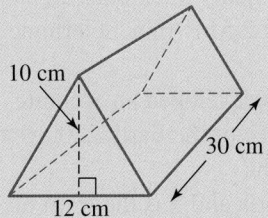
10 cm
30 cm
12 cm

c.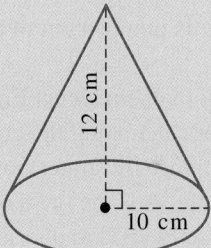
12 cm
10 cm

13. Determine the volume of each of the following.

a.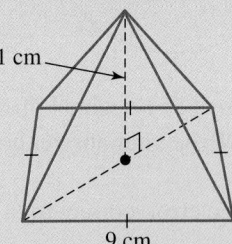
11 cm
9 cm

b.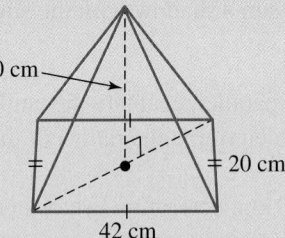
30 cm
20 cm
42 cm

c.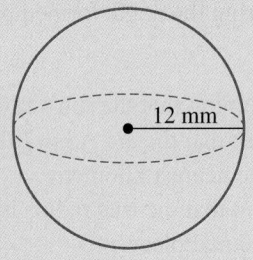
12 mm

Communicating, reasoning and problem solving

14. A rectangular block of land 4 m × 25 m is surrounded by a concrete path 1 m wide.
  a. Calculate the area of the path.
  b. Determine the cost of concreting at $45 per square metre.

15. If the radius is tripled and the height of a cylinder is divided by six, then determine the effect on its volume (in comparison with the original shape).

16. If the length is halved, the width is tripled and the height of a rectangular prism is doubled, then determine the effect on its volume (in comparison with the original shape).

17. A cylinder of radius 14 cm and height 20 cm is joined to a hemisphere of radius 14 cm to form a bread holder.

   a. Calculate the total surface area.
   b. Determine the cost of chroming the bread holder on the outside at \$0.05 per cm^2.
   c. Calculate the storage volume of the bread holder.
   d. Determine how much more space is in this new bread holder than the one it is replacing, which had a quarter circle end with a radius of 18 cm and a length of 35 cm.

18. Bella Silos has two rows of silos for storing wheat. Each row has 16 silos and all the silos are identical, with a cylindrical base (height of 5 m, diameter of 1.5 m) and conical top (diameter of 1.5 m, height of 1.1 m).

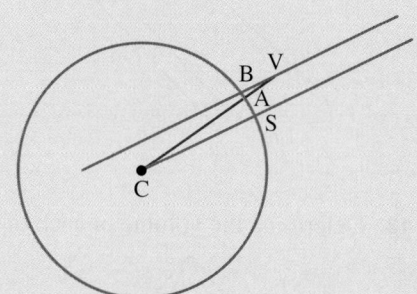

   a. Calculate the slant height of the conical tops.
   b. Determine the total surface area of all the silos.
   c. Evaluate the cost of painting the silos if one litre of paint covers 40 m^2 at a bulk order price of \$28.95 per litre.
   d. Determine how much wheat can be stored altogether in these silos.
   e. Wheat is pumped from these silos into cartage trucks with rectangular containers 2.4 m wide, 5 m long and 2.5 m high. Determine how many truckloads are necessary to empty all the silos.
   f. If wheat is pumped out of the silos at 2.5 m^3/min, determine how long it will take to fill one truck.

19. The Greek mathematician Eratosthenes developed an accurate method for calculating the circumference of the Earth 2200 years ago! The figure illustrates how he did this.
In this figure, A is the town of Alexandria and S is the town of Syene, exactly 787 km due south. When the sun's rays (blue lines) were vertical at Syene, they formed an angle of 7.2° at Alexandria ($\angle BVA = 7.2°$), obtained by placing a stick at A and measuring the angle formed by the sun's shadow with the stick.

   a. Assuming that the sun's rays are parallel, evaluate the angle $\angle SCA$, correct to 1 decimal place.
   b. Given that the arc AS = 787 km, determine the radius of the Earth, SC. Write your answer correct to the nearest kilometre.
   c. Given that the true radius is 6380 km, determine Eratosthenes' percentage error, correct to 1 decimal place.

To test your understanding and knowledge of this topic, go to your learnON title at www.jacplus.com.au and complete the **post-test**.

Answers

Topic 11 Measurement

11.1 Pre-test

1. 68.63 mm^2
2. 9.0 cm^2
3. D
4. 706.9 cm^2
5. 4 cm^3
6. 57.7 cm^2
7. $864
8. B
9. 60 mm^3
10. C
11. 5.66 cm
12. 4.5 cm
13. C
14. 344.93 cm^2
15. 230 mL

11.2 Area

1. a. 16 cm^2 b. 48 cm^2 c. 75 cm^2
2. a. 120 cm^2 b. 706.86 cm^2 c. 73.5 mm^2
3. a. 13.5 mm^2 b. 17.5 mm^2
4. a. 254.47 cm^2 b. 21 m^2 c. 75 cm^2
5. a. i. 12π cm^2 ii. 37.70 cm^2
   b. i. $\dfrac{69\pi}{2}$ mm^2 ii. 108.38 mm^2
   c. i. 261π cm^2 ii. 819.96 cm^2
6. D
7. A
8. a. 123.29 cm^2 b. 1427.88 m^2 c. 52 cm^2
9. a. 30.4 m^2 b. 78 cm^2 c. 2015.50 cm^2
10. a. 125.66 cm^2 b. 102.87 m^2
11. a. 13.73 m^2 b. 153.59 m^2
12. a. 27.86 m^2 b. 37.5 m^2
13. 11 707.92 cm^2
14. 21 m^2
15. 60
16. $840
17. a. 260.87 m^2 b. 195.71 m^2 c. 75%
18. a. Sample responses can be found in the worked solutions in the online resources.
    b. 2020.83 m; horizontal. If vertical split 987.5 m.

19. a. Area $= 50x - x^2$
    b. See the table at bottom of the page.*
    c.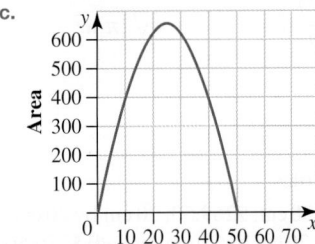
    d. $x = 25$, $y = 25$, Area $= 625$ m^2
    e. $r = 15.92$ m
    f. 170.77 m^2
20. a. Circular area, 1790.49 m^2; rectangular area, 1406.25 m^2
    b. Circular area, $\left(\dfrac{1}{4\pi}n^2\right)$ m^2; rectangular (square) area, $\left(\dfrac{1}{16}n^2\right)$ m^2. Circular area is always $\dfrac{4}{\pi}$ or 1.27 times larger.
21. a. 258.1 m^2 b. 7 bags   c. $79
    d. $435.50
22. a. $\dfrac{29}{50}$ b. $x = 5, y = 5$
23. 32.88 cm^2

11.3 Total surface area

1. a. 600 cm^2 b. 384 cm^2 c. 1440 cm^2
   d. 27 m^2
2. a. 6729.3 cm^2 b. 8.2 m^2
3. a. 113.1 m^2 b. 452.4 cm^2 c. 1495.4 cm^2
   d. 502.7 cm^2
4. a. 506.0 cm^2 b. 9.4 m^2 c. 340.4 cm^2
   d. 224.1 cm^2
5. a. 13.5 m^2 b. 90 m^2 c. 11 309.7 cm^2
6. a. 9852.0 mm^2 b. 125.7 cm^2 c. 1531.4 cm^2
7. a. 880 cm^2 b. 3072.8 cm^2 c. 75 cm^2
8. a. 70.4 cm^2 b. 193.5 cm^2 c. 1547.2 cm^2
9. B
10. a. Slant height and base length.
    b. Perpendicular height and base length.
    c. Slant height and base length.
11. a. 731 m^2 b. 198 m^2
12. a. 70.0 m^2 b. $455
13. a. 3063.1 cm^2 b. $168.47
14. 11 216 cm^2
15. 60

*19. b.

x	0	5	10	15	20	25	30	35	40	45	50
Area (m^2)	0	225	400	525	600	625	600	525	400	225	0

16. $1192\,\text{cm}^2$

17. $\$1960$

18. a. $8\pi\,\text{m}^2$ b. $\dfrac{\sqrt{7}}{2}\,\text{m}$ c. $4\sqrt{2}\pi\,\text{m}^2$

19. a. $6.6\,\text{m}^2$
 b. Back wall = 80 tiles
 Side wall = 50 tiles
 $80 + 50 + 50 = 180$ tiles
 c. Cheapest: 30 cm by 30 cm, \$269.50; 20 cm by 20 cm (individually) \$270; 20 cm by 20 cm (boxed) \$276.50

20. The calculation is correct.

21. a. $\theta = 120°$ b. $x = 1; y = \sqrt{3}$
 c. $3\sqrt{3}\,\text{cm}^2$ d. $6\sqrt{3}\,\text{cm}^2$
 e. 32

22. The area of material required is $1.04\,\text{m}^2$. If Phuong is careful in placing the pattern pieces, she may be able to cover the footstool.

23. $r = \dfrac{3\sqrt{3}a}{2}$

24. a. Arc length $XY = (x + s)\theta$
 Arc length $AB = x\theta$
 b. i. $x = \dfrac{2\pi t}{\theta} = \dfrac{st}{r - t}$
 ii. $\dfrac{x}{x + s} = \dfrac{t}{r}$
 c. Area of sector $AVB = \dfrac{x^2\theta}{2}$
 Area of sector $XVY = \dfrac{(s + x)^2\theta}{2}$
 Area of $ABYX = \dfrac{s\theta\,(s + 2x)}{2}$
 TSA of frustum $= \pi\left(t^2 + r^2\right) + \dfrac{s\theta\,(s + 2x)}{2}$

11.4 Volume

1. a. $27\,\text{cm}^3$ b. $74.088\,\text{m}^3$ c. $3600\,\text{cm}^3$
 d. $94.5\,\text{cm}^3$

2. a. $450\,\text{mm}^3$ b. $360\,\text{cm}^2$

3. a. $6333.5\,\text{cm}^3$ b. $19.1\,\text{m}^3$ c. $280\,\text{cm}^3$

4. a. $288\,\text{mm}^3$ b. $91.6\,\text{m}^3$ c. $21\,470.8\,\text{cm}^3$

5. a. $7.2\,\text{m}^3$ b. $14\,137.2\,\text{cm}^3$ c. $1436.8\,\text{mm}^3$
 d. $523\,598.8\,\text{cm}^3$

6. a. $113\,097.34\,\text{cm}^3$ b. $1.44\,\text{m}^3$
 c. $12\,214.51\,\text{mm}^3$ d. 101.93

7. a. $377.0\,\text{cm}^3$ b. $2303.8\,\text{mm}^3$

8. a. $400\,\text{cm}^3$ b. $10\,080\,\text{cm}^3$

9. a. $1400\,\text{cm}^3$ b. $10\,379.20\,\text{cm}^3$

10. a. $41.31\,\text{cm}^3$ b. $48.17\,\text{cm}^3$

11. a. $218.08\,\text{cm}^3$ b. $3691.37\,\text{cm}^3$

12. a. $V_{\text{new}} = 27l^3$, the volume will be 27 times as large as the original volume.
 b. $V_{\text{new}} = \dfrac{1}{8}l^2$, the volume will be $\dfrac{1}{8}$ of the original volume.
 c. $V_{\text{new}} = 2\pi r^2 h$, the volume will be twice as large as the original volume.
 d. $V_{\text{new}} = \pi r^2 h$, the volume will remain the same.
 e. $V_{\text{new}} = 3lwh$, the volume will be 3 times as large as the original value.

13. A

14. $7438.35\,\text{cm}^3$

15. $4417.9\,\text{L}$

16. $10\,215.05\,\text{cm}^3$

17. a. $H = 6\,\text{m}$ b. $112\,\text{m}^3$ c. 19 bins
 d. $112\,000\,\text{L}$ e. $1.95\,\text{m}$ from floor

18. a. i. $4.57\,\text{cm}$
 ii. $262.5\,\text{cm}^2$
 b. i. $14.15\,\text{cm}$
 ii. $323.27\,\text{cm}^2$
 c. i. $33.33\,\text{cm}$
 ii. $437.62\,\text{cm}^2$
 d. Sphere. Costs less for a smaller surface area.

19. a. $h = \dfrac{V}{\pi r^2}$
 b. i. $31.8\,\text{cm}$
 ii. $8.0\,\text{cm}$
 c. $\sqrt{\dfrac{V}{\pi h}}$
 d. $r \geq 0$, since r is a length
 e. i. $7.6\,\text{cm}$
 ii. $6.2\,\text{cm}$

20. 1000

21. a. $2.55\,\text{cm}$
 b. $35.68\,\text{cm}$
 c. $A_a = 157.88\,\text{m}^2$, $A_b = 12.01\,\text{m}^2$

22. a. $126.67\,\text{m}^3$ b. $53.33\,\text{m}^3$

23. Sample responses can be found in the worked solutions in the online resources.

24. Sample responses can be found in the worked solutions in the online resources.

25. Required volume $= 1570.80\,\text{cm}^3$; tin volume $= 1500\,\text{cm}^3$; muffin tray volume $= 2814.72\,\text{cm}^3$. Marion could fill the tin and have a small amount of mixture left over, or she could almost fill 14 of the muffin cups and leave the remaining cups empty.

26. Cut squares of side length $s = 0.3\,\text{m}$ or $0.368\,\text{m}$ from the corners.

27. $1.94\,\text{m}$.

28. a. $H = 12R$ b. $8\pi R^3$ c. $12\pi R^3$
 d. $8 : 12 = 2 : 3$

29. a. $\frac{1}{3}X^2H$

   b. $\frac{1}{3}x^2(H-h)$

   c. Sample responses can be found in the worked solutions in the online resources.

   d. Sample responses can be found in the worked solutions in the online resources.

30. 18 scoops

Project

1.

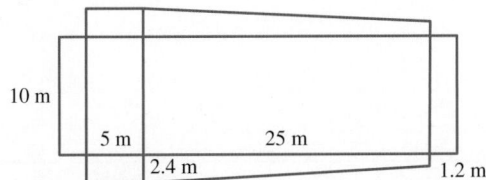

2. TSA = 450 m^2

3. 38 cans

4. $2812

5. Capacity = 570 m^3 = 570 000 L

6. 132 hours

7. $d = 1.61$ m

8. $2516

11.5 Review questions

1. D

2. C

3. B

4. A

5. a. 84 cm^2 b. 100 cm^2 c. 6.50 cm^2

6. a. 56.52 cm^2 b. 60 cm^2 c. 244.35 cm^2

7. a. 300 cm^2 b. 224.55 cm^2 c. 160 cm^2

8. a. 499.86 cm^2 b. 44.59 cm^2 c. 128.76 cm^2

9. a. 18 692.48 cm^2 b. 1495.40 mm^2 c. 804.25 cm^2

10. a. 871.79 cm^2 b. 873.36 mm^2 c. 760 cm^2

11. a. 343 cm^3 b. 672 cm^3 c. 153 938.04 cm^3

12. a. 1.45 m^3 b. 1800 cm^3 c. 1256.64 cm^3

13. a. 297 cm^3 b. 8400 cm^3 c. 7238.23 mm^3

14. a. 62 m^2 b. $2790

15. $V = \frac{3}{2}\pi r^2 h$, the volume will be 1.5 times as large as the original volume.

16. $V = 3lwh$, the volume will be 3 times as large as (or triple) the original volume.

17. a. 3606.55 cm^2 b. $180.33 c. 18 062.06 cm^3

   d. 9155.65 cm^3

18. a. 1.33 m

   b. 910.91 m^2

   c. $618.35 or $636.90 assuming you have to buy full litres (i.e. not 0.7 of a litre)

   d. 303.48 m^3

   e. 11 trucks

   f. 12 minutes

19. a. 7.2°   b. 6263 km   c. 1.8% error

12 Properties of geometrical figures (Path)

LESSON SEQUENCE

LESSON
12.1 Overview

Why learn this?

Geometry is an area of mathematics that has an abundance of real-life applications. The first important skill that geometry teaches is the ability to reason deductively and prove logically that certain mathematical statements are true.

Euclid (c. 300 BCE) was the mathematician who developed a systematic approach to geometry, now referred to as Euclidean geometry, which relied on mathematical proofs. Mathematicians and research scientists today spend a large part of their time trying to prove new theories, and they rely heavily on all the proofs that have gone before.

Geometry is used extensively in professions such as navigation and surveying. Planes circle our world, land needs to be surveyed before any construction can commence, and architects, designers and engineers all use geometry in their drawings and plans. Geometry is also used extensively in software and computing. Computer-aided design programs, computer imaging, animations, video games and 3D printers all rely greatly on built-in geometry packages.

Just about every sport involves geometry heavily. In cricket alone there are many examples; bowlers adjust the angle at which they release the ball to make the ball bounce towards the batsmen at different heights; fielders are positioned so they cover as much of the ground as efficiently as possible and batsmen angle their bat as they hit the ball to ensure the ball rolls along the ground instead of in the air. Netballers must consider the angle at which they shoot the ball to ensure it arcs into the ring and cyclists must consider the curved path of their turns that will allow them to corner in the quickest and most efficient way.

1. Calculate the value of the pronumeral x.

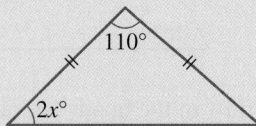

2. Calculate the value of the pronumeral x.

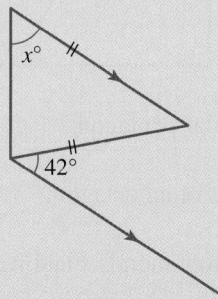

3. State what type of triangles have the same size and shape.

4. Determine the values of the pronumerals x and y.

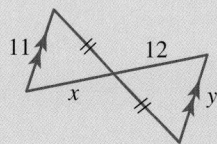

5. Triangles ABC and DEF are congruent; calculate the values of the pronumerals x, y and z.

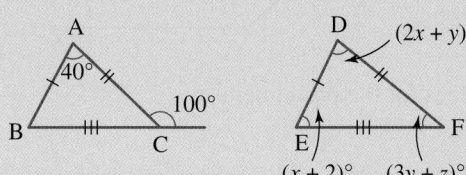

6. **MC** Choose which congruency test will prove these triangles are congruent.

   A. SAS   B. SSS   C. AAS   D. RHS

7. State what the AAA test checks about two triangles.

8. Calculate the value of the pronumeral x in the quadrilateral shown.

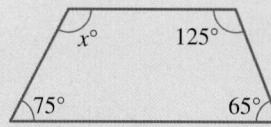

9. Determine the value of the pronumeral x in the quadrilateral shown.

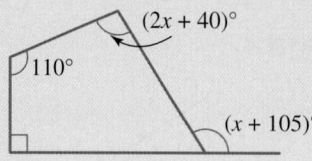

10. Determine the exterior angle of a regular pentagon.

11. Evaluate the sum of the interior angles of an octagon.

12. **MC** Select the correct values of the pronumerals x and y.

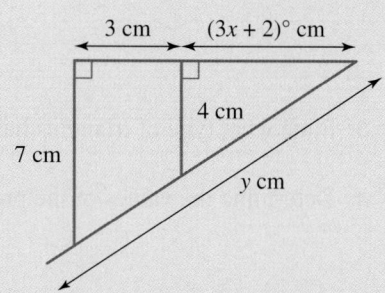

A. $x = \dfrac{1}{3}, y = 7\sqrt{2}$

B. $x = \dfrac{2}{3}, y = 5\sqrt{2}$

C. $x = \dfrac{2}{3}, y = 7\sqrt{2}$

D. $x = 1\dfrac{1}{3}, y = 7\sqrt{2}$

13. Calculate the value of the pronumeral x.

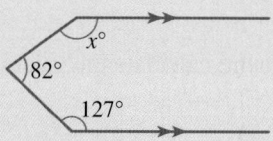

14. **MC** Choose the correct values for the pronumeral x.
A. $x = 6$ or $x = 5$
B. $x = 4$ or $x = 4$
C. $x = 5$ or $x = 4$
D. $x = 6$ or $x = 4$

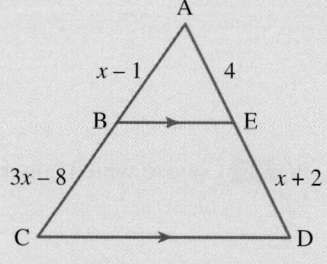

15. **MC** Select the correct values of the pronumerals x and y.
A. $x = 1.25$ and $y = 6.5$
B. $x = 0.8$ and $y = 5.6$
C. $x = 0.25$ and $y = 1.125$
D. $x = 5$ and $y = 3.5$

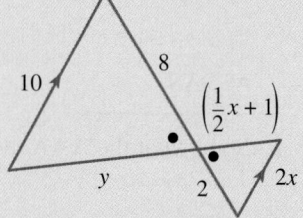

LESSON
12.2 Angles, triangles and congruence

LEARNING INTENTION

At the end of this lesson you should be able to:
- apply properties of straight lines and triangles to determine the value of an unknown angle
- construct simple geometric proofs for angles in triangles or around intersecting lines
- prove that triangles are congruent by applying the appropriate congruency test.

12.2.1 Proofs and theorems of angles

eles-4892

- Euclid (c. 300 BC) was the mathematician who developed a systematic approach to geometry, now referred to as Euclidean geometry, which relied on mathematical proofs.
- A **proof** is an argument that shows why a statement is true.
- A **theorem** is a statement that can be demonstrated to be true. It is conventional to use the following structure when setting out a theorem.
  - **Given:** a summary of the information given
  - **To prove:** a statement that needs to be proven
  - **Construction:** a description of any additions to the diagram given
  - **Proof:** a sequence of steps that can be justified and form part of a formal mathematical proof.

Sums of angles

Angles at a point
- The sum of the angles at a point is 360°.

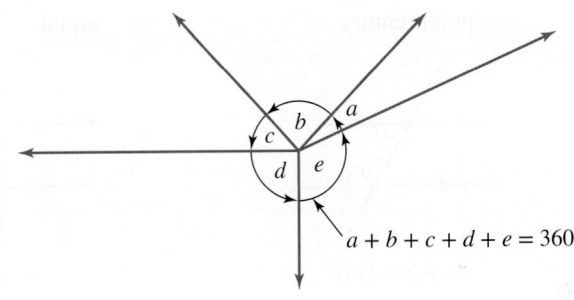

$$a + b + c + d + e = 360°$$

Supplementary angles
- The sum of the angles on a straight line is 180°.
- Angles that add up to 180° are called **supplementary angles**.
- In the diagram angles a, b and c are supplementary.

$$a + b + c = 180°$$

Complementary angles
- The sum of the angles in a right angle is 90°.
- Angles that add up to 90° are called **complementary angles**.
- In the diagram angles a, b and c are complementary.

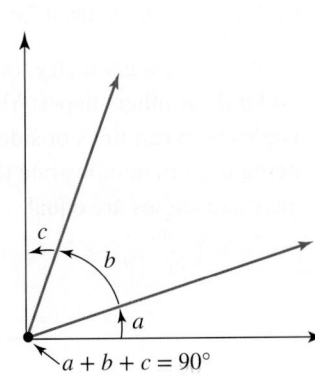

$$a + b + c = 90°$$

Vertically opposite angles

Theorem 1

- **Vertically opposite angles** are equal.

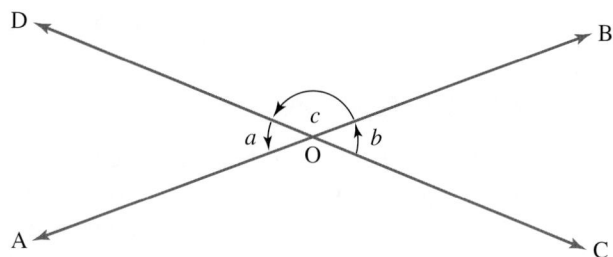

Given:	Straight lines AB and CD, which intersect at O.
To prove:	$\angle AOD = \angle BOC$ and $\angle BOD = \angle AOC$
Construction:	Label $\angle AOD$ as a, $\angle BOC$ as b and $\angle BOD$ as c.
Proof:	Let $\angle AOD = a°$, $\angle BOC = b°$ and $\angle BOD = c°$.

$$a + c = 180° \quad \text{(supplementary angles)}$$
$$b + c = 180° \quad \text{(supplementary angles)}$$
$$\therefore a + c = b + c$$
$$\therefore a = b$$

So, $\angle AOD = \angle BOC$.
Similarly, $\angle BOD = \angle AOC$.

Parallel lines

- If two lines are parallel and cut by a **transversal**, then:

  - co-interior angles are supplementary

  - corresponding angles are equal

  - alternate angles are equal.

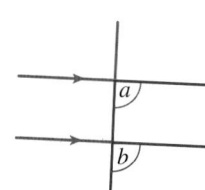

$$a + b = 180°$$

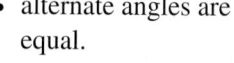 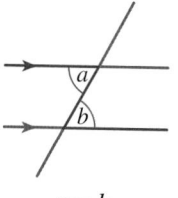

$$a = b$$

$$a = b$$

Digital technology

There are many online tools that can be used to play around with lines, shapes and angles. One good tool to explore is the Desmos geometry tool, which can be used for free at www.desmos.com/geometry.

In the Desmos geometry tool, you can draw lines, circles, polygons and all kinds of other shapes. You can then use the angle tool to explore the angles between lines or sides. The figure at right shows the angle tool being used to demonstrate that co-interior angles are supplementary and alternate angles are equal.

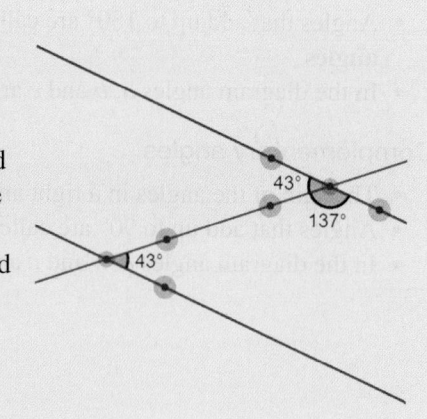

12.2.2 Angle properties of triangles

eles-5353

Theorem 2

• The sum of the interior angles of a triangle is 180°.

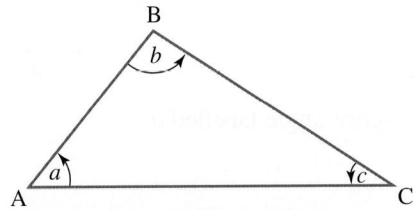

Given: $\triangle ABC$ with interior angles a, b and c

To prove: $a + b + c = 180°$

Construction: Draw a line parallel to AC, passing through B and label it DE as shown. Label $\angle ABD$ as x and $\angle CBE$ as y.

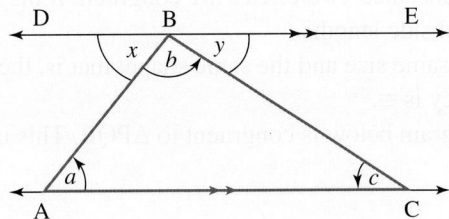

Proof:

$a = x$	(alternate angles)
$c = y$	(alternate angles)
$x + b + y = 180°$	(supplementary angles)
$\therefore\ a + b + c = 180°$	

Equiangular triangles

• It follows from Theorem 2 that each interior angle of an **equiangular triangle** is 60°.

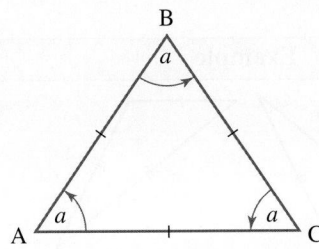

$$a + a + a = 180° \qquad \text{(sum of interior angles in a triangle is } 180°\text{)}$$
$$3a = 180°$$
$$a = 60°$$

Note that an equiangular triangle is also an **equilateral triangle**.

Theorem 3

- The exterior angle of a triangle is equal to the sum of the opposite interior angles.

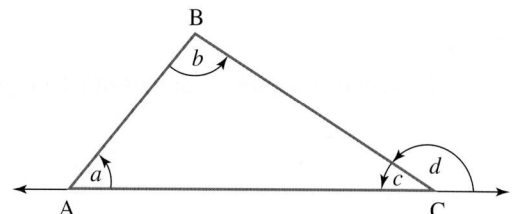

Given: $\triangle ABC$ with the exterior angle labelled d

To prove: $d = a + b$

Proof: $c + d = 180°$ (supplementary angles)

$a + b + c = 180°$ (sum of interior angles in a triangle is $180°$)

$\therefore d = a + b$

12.2.3 Congruent triangles

eles-4897

- Two objects or figures are said to be congruent if they have the same shape and size or if they are the image of one another in a mirror. For instance, two circles are congruent if they have equal radii, and two squares are congruent if they have equal side lengths.
- **Congruent triangles** have the same size and the same shape; that is, they are identical in all respects.
- The symbol used for congruency is $\equiv$.
- For example, $\triangle ABC$ in the diagram below is congruent to $\triangle PQR$. This is written as $\triangle ABC \equiv \triangle PQR$.

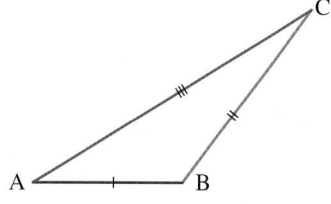   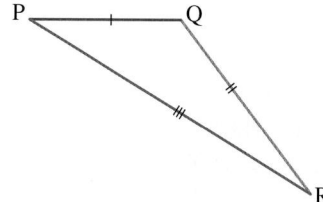

- Note that the vertices of the two triangles are written in corresponding order.
- There are four tests designed to check whether triangles are congruent. The tests are summarised in the table below.

Congruence test	Example	Description
Side-side-side (SSS)		The three corresponding sides are the same lengths.
Side-angle-side (SAS)		Two corresponding sides are the same length and the angle in between these sides is equal.
Angle-angle-side (AAS)		A pair of corresponding angles and a non-contained side are equal.

Congruence test	Example	Description
Right angle-hypotenuse-side (RHS)	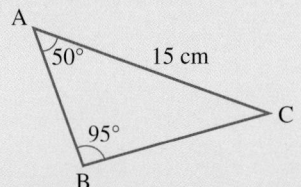	The triangles are right-angled, and the hypotenuse and one other side of one triangle are equal to the hypotenuse and a side of the other triangle.

- In each of the tests we need to show three equal measurements about a pair of triangles in order to show they are congruent.

WORKED EXAMPLE 1 Determining pairs of congruent triangles

Select a pair of congruent triangles from the diagrams shown, giving a reason for your answer.

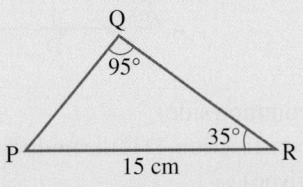
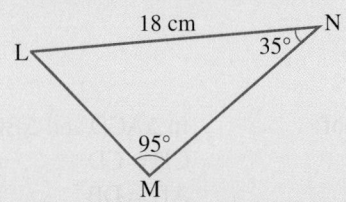

THINK

1. In each triangle the length of the side opposite the 95° angle is given. If triangles are to be congruent, the sides opposite the angles of equal size must be equal in length. Draw your conclusion.

2. To test whether ΔABC is congruent to ΔPQR, first evaluate the angle C.

3. Apply a test for congruence. Triangles ABC and PQR have a pair of corresponding sides equal in length and 2 pairs of angles the same, so draw your conclusion.

WRITE

All three triangles have equal angles, but the sides opposite the angle 95° are not equal.
AC = PR = 15 cm and LN = 18 cm

ΔABC: ∠A = 50°, ∠B = 95°,
   ∠C = 180° − 50° − 95°
     = 35°

A pair of corresponding angles
(∠B = ∠Q and ∠C = ∠R) and a corresponding side (AP = PR) are equal.
ΔABC ≡ ΔPQR (AAS)

12.2.4 Isosceles triangles

eles-4898

- A triangle is isosceles if the lengths of two sides are equal but the third side is not equal.

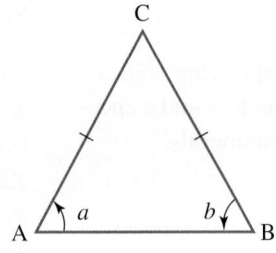

Theorem 4

- The angles at the base of an **isosceles triangle** are equal.

Given: $AC = CB$

To prove: $\angle BAC = \angle CBA$

Construction: Draw a line from the vertex C to the midpoint of the base AB and label the midpoint D. CD is the bisector of $\angle ACB$.

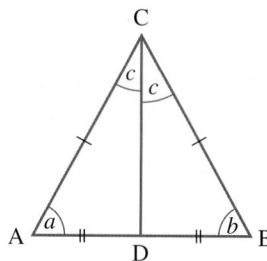

Proof: In $\triangle ACD$ and $\triangle BCD$, (common side)
              $CD = CD$ (construction, D is the midpoint of AB)
              $AD = DB$ (given)
              $AC = CB$ (SSS)
            $\Rightarrow \triangle ACD \equiv \triangle BCD$
              $\therefore \angle BAC = \angle CBA$

- Conversely, if two angles of a triangle are equal, then the sides opposite those angles are equal.

WORKED EXAMPLE 2 Determining values in congruent triangles

Given that $\triangle ABD \equiv \triangle CBD$, determine the values of the pronumerals in the figure shown.

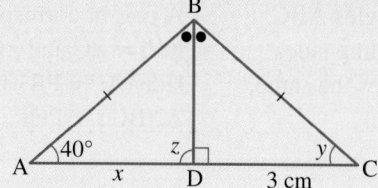

THINK	WRITE
1. In congruent triangles corresponding sides are equal in length. Side AD (marked x) corresponds to side DC, so state the value of x.	$\triangle ABD \equiv \triangle CBD$ $AD = CD, \quad AD = x, \quad CD = 3$ So $x = 3$ cm.
2. Since the triangles are congruent, corresponding angles are equal. State the angles corresponding to y and z and hence determine the values of these pronumerals.	$\angle BAD = \angle BCD$ $\angle BAD = 40°, \quad \angle BCD = y$ So $y = 40°$ $\angle BDA = \angle BDC$ $\angle BDA = z, \quad \angle BDC = 90°$ So $z = 90°$.

Prove that ΔPQS is congruent to ΔRSQ.

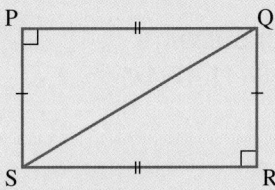

THINK	WRITE
1. Write the information given.	Given: Rectangle PQRS with diagonal QS.
2. Write what needs to be proved.	To prove: that ΔPQS is congruent to ΔRSQ.
	QP = SR (opposite sides of a rectangle)
	∠SPQ = ∠SRQ = 90° (given)
	QS is common.
3. Select the appropriate congruency test for proof. (In this case, it is RHS because the triangles have an equal side, a right angle and a common hypotenuse.)	So ΔPQS ≡ ΔRSQ (RHS)

DISCUSSION

Tessellation, or tiling, describes a repeating pattern of shapes over a surface. Repeating geometric patterns are often used in art, notably in Islamic art. Observe the tessellation shown. Can you identify any congruent figures?

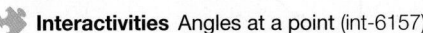 Resources

Interactivities Angles at a point (int-6157)
Supplementary angles (int-6158)
Angles in a triangle (int-3965)
Interior and exterior angles of a triangle (int-3966)
Vertically opposite and adjacent angles (int-3968)
Corresponding angles (int-3969)
Co-interior angles (int-3970)
Alternate angles (int-3971)
Congruency tests (int-3755)
Congruent triangles (int-3754)
Angles in an isosceles triangle (int-6159)

12.2 Quick quiz on	12.2 Exercise

Individual pathways

■ PRACTISE	■ CONSOLIDATE	■ MASTER
1, 4, 7, 8, 9, 16	2, 5, 10, 11, 12, 17	3, 6, 13, 14, 15, 18

Fluency

1. Determine the values of the unknown in each of the following.

a.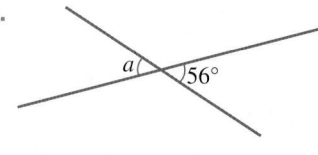

b.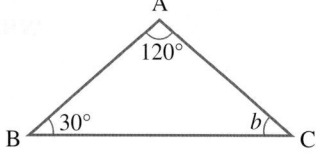

c.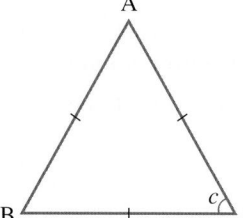

d.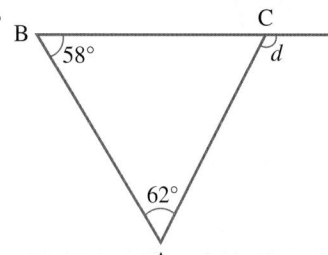

e.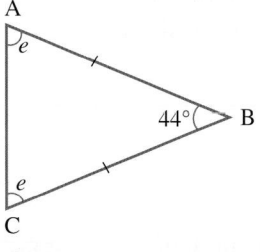

2. Determine the values of the pronumerals in the following diagrams.

a.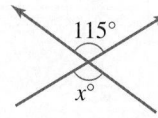

b.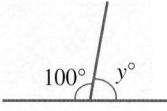

c.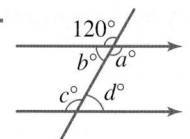

d.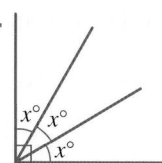

3. **WE1** Select a pair of congruent triangles in each of the following, giving a reason for your answer. All side lengths are in cm.

a.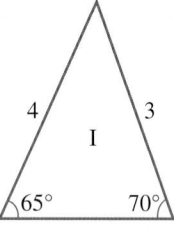

b.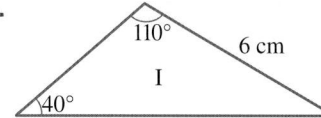

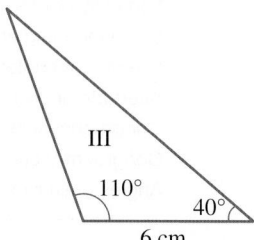

c.

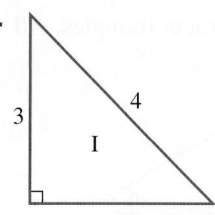

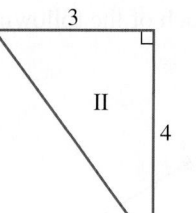

d.

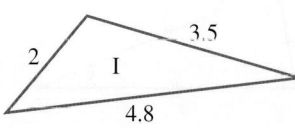

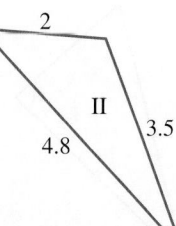

Understanding

4. Determine the missing values of x and y in each of the following diagrams. Give reasons for your answers.

a.

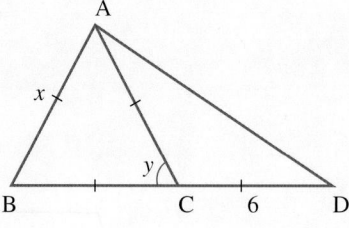

b.

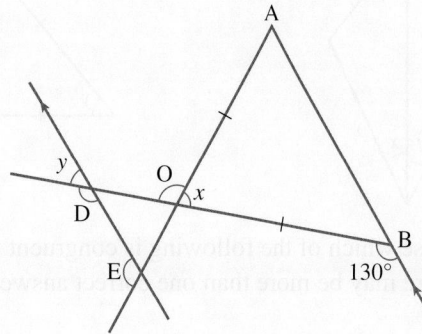

c.

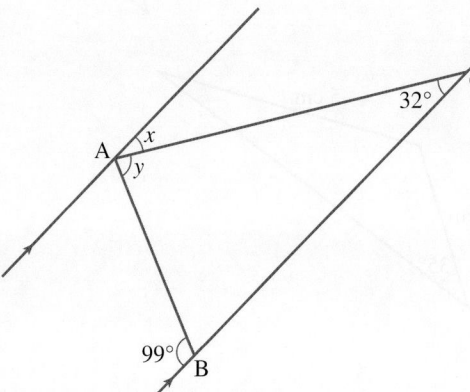

d.

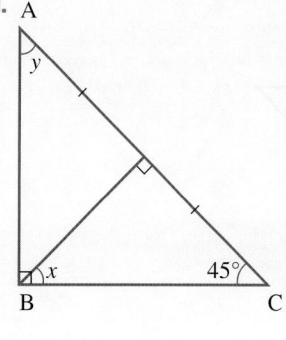

5. Determine the values of the pronumerals. Give reasons for your answers.

a.

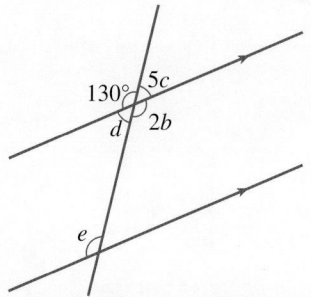

b.

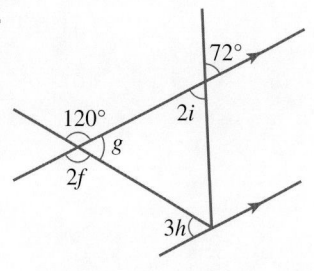

6. **WE2** Determine the value of the pronumeral in each of the following pairs of congruent triangles. All side lengths are in cm.

a.

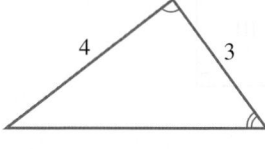

b.

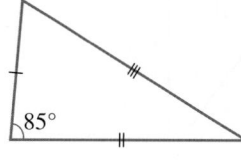

c.

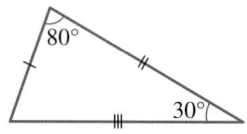

d.

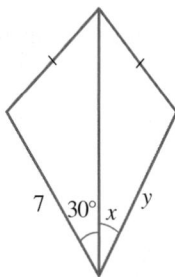

e.

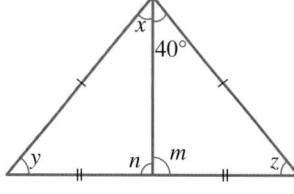

7. **MC** Choose which of the following is congruent to the triangle shown.
Note: There may be more than one correct answer.

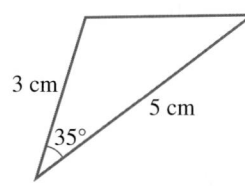

A.

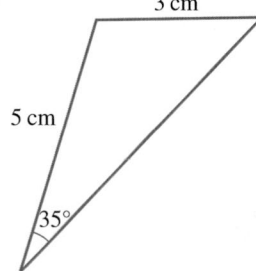

B.

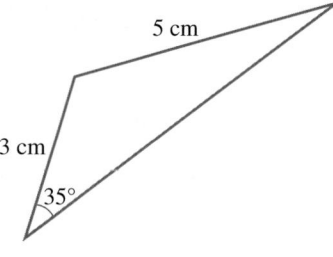

C.

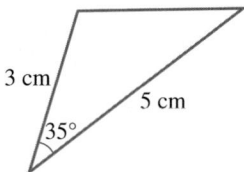

D.

Communicating, reasoning and problem solving

8. Prove that $\triangle ABC \equiv \triangle ADC$ and hence determine the value of the pronumerals.

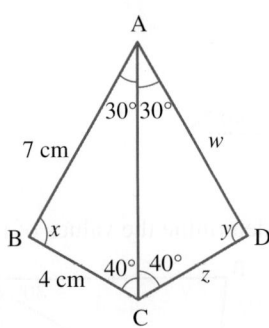

9. If $DA = DB = DC$, prove that $\angle ABC$ is a right angle.

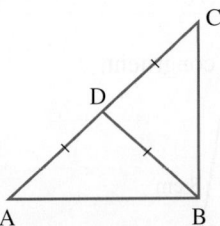

10. **WE3** Prove that each of the following pairs of triangles are congruent.

a.

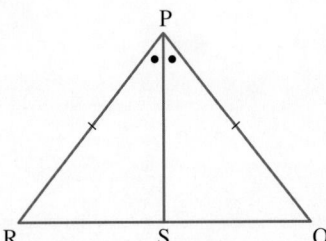

b.

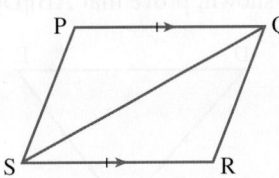

c.

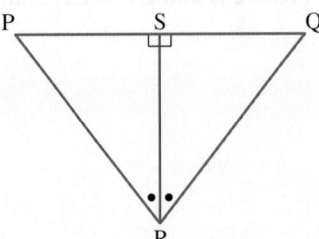

d.

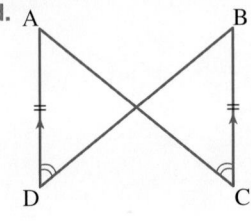

e.

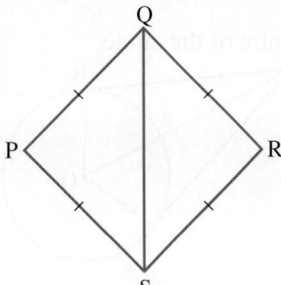

11. Prove that $\triangle ABC \equiv \triangle ADC$ and hence determine the value of x.

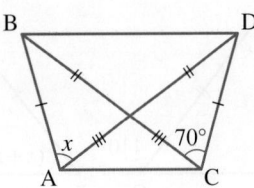

12. Explain why the triangles shown are not necessarily congruent.

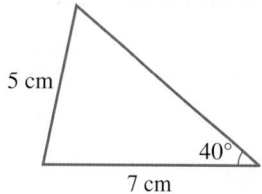
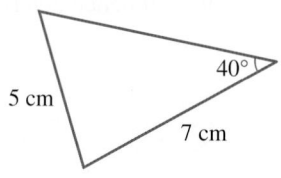

13. Prove that $\triangle ABC \equiv \triangle ADC$ and hence determine the values of the pronumerals.

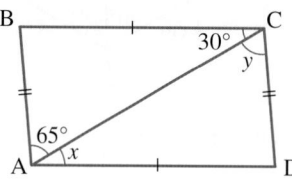

14. Explain why the triangles shown are not congruent.

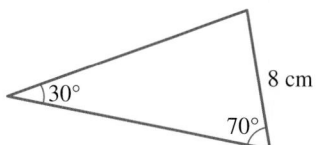
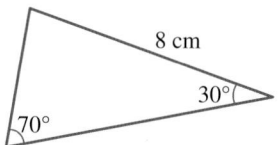

15. If $AC = CB$ and $DC = CE$ in the diagram shown, prove that $AB \| DE$.

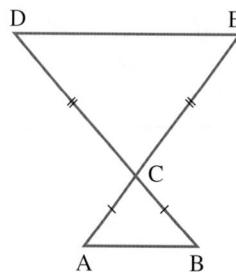

16. Show that $\triangle ABO \equiv \triangle ACO$, if O is the centre of the circle.

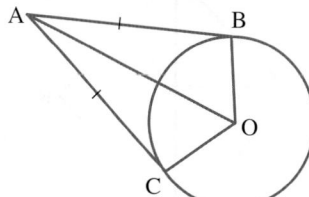

17. Triangles ABC and DEF are congruent.

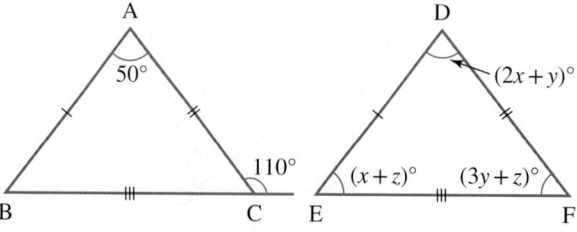

Determine the values of the pronumerals x, y and z.

18. ABC is an isosceles triangle in which AB and AC are equal in length. BDF is a right-angled triangle.

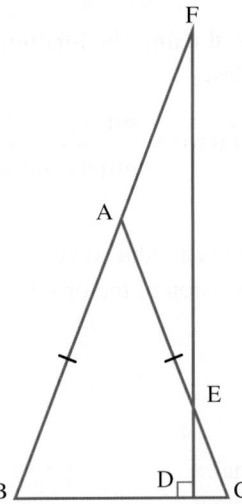

Show that triangle AEF is an isosceles triangle.

LESSON
12.3 Similar triangles

eles-4899

LEARNING INTENTION

At the end of this lesson you should be able to:
- identify similar figures
- calculate the scale factor in similar figures
- show that two triangles are similar using the appropriate similarity test.

12.3.1 Similar figures

- Two geometric shapes are **similar** when one is an **enlargement** or reduction of the other shape.
  - An enlargement increases the length of each side of a figure in all directions by the same factor. For example, in the diagram shown, triangle A'B'C' is an enlargement of triangle ABC by a factor of 3 from its **centre of enlargement** at O.

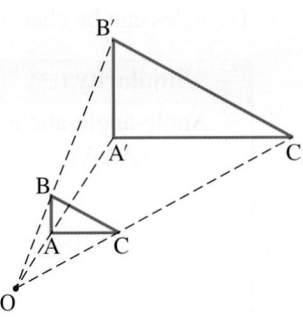

- Similar figures have the same shape. The corresponding angles are the same and each pair of corresponding sides is in the same ratio.
- The symbol for similarity is ~ and is read as 'is similar to'.
- The **image** of the original object is the enlarged or reduced shape.
- To create a similar shape, use a **scale factor** to enlarge or reduce the original shape called the object.

> **Calculating scale factor**
>
> The scale factor can be found using the formula below and the lengths of a pair of corresponding sides.
>
> $$\text{Scale factor} = \frac{\text{image side length}}{\text{object side length}}$$

- If the scale factor is less than 1, the image is a reduced version of the original shape. If the scale factor is greater than 1, the image is an enlarged version of the original shape.

Similar triangles

- Two triangles are similar if:
  - the angles are equal, or
  - the corresponding sides are proportional.
- Consider the pair of **similar triangles** below.

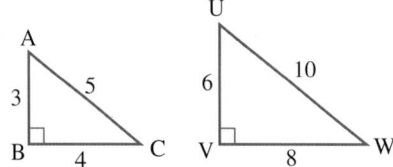

- The following statements are true for these triangles.
  - Triangle UVW is similar to triangle ABC or, using symbols, $\Delta UVW \sim \Delta ABC$.
  - The corresponding angles of the two triangles are equal in size:

  $$\angle CAB = \angle WUV, \angle ABC = \angle UVW \text{ and } \angle ACB = \angle UWV$$

  - The corresponding sides of the two triangles are in the same ratio. $\dfrac{UV}{AB} = \dfrac{VW}{BC} = \dfrac{UW}{AC} = 2$; that is, ΔUVW has each of its sides twice as long as the corresponding sides in ΔABC.
  - The scale factor is 2.

▶ 12.3.2 Testing triangles for similarity

eles-4900

- Triangles can be checked for similarity using one of the tests described in the table below.

Similarity test	Example	Description
Angle-angle-angle (AAA)		The three corresponding angles are equal.
Side-side-side (SSS)		The three sides of one triangle are proportional to the three sides of the other triangle.

Similarity test	Example	Description
Side-angle-side (SAS)	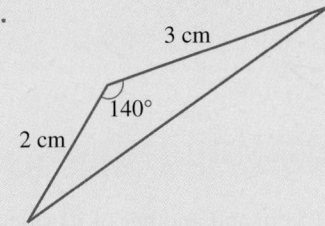	Two sides of one triangle are proportional to two sides of the other triangle, and the included angle is equal.
Right angle-hypotenuse-side (RHS)		The hypotenuse and a second side of a right-angled triangle are proportional to the hypotenuse and a second side of another right-angled triangle.

- *Note:* When using the equiangular test, only two corresponding angles have to be checked. Since the sum of the interior angles in any triangle is a constant number (180°), the third pair of corresponding angles will automatically be equal, provided that the first two pairs match exactly.

WORKED EXAMPLE 4 Determining pairs of similar triangles

Determine a pair of similar triangles among those shown. Give a reason for your answer.

a.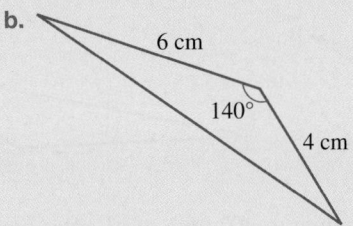

THINK

1. In each triangle the lengths of two sides and the included angle are known, so the SAS test can be applied.
   Since all included angles are equal (140°), we need to the calculate ratios of corresponding sides, taking two triangles at a time.

2. Only triangles **a** and **b** have corresponding sides in the same ratio (and the included angle is equal). State your conclusion, specifying the similarity test you used.

WRITE

For triangles **a** and **b**: $\dfrac{6}{3} = \dfrac{4}{2} = 2$

For triangles **a** and **c**: $\dfrac{5}{3} = 1.6, \dfrac{3}{2} = 1.5$

For triangles **b** and **c**: $\dfrac{5}{6} = 0.83, \dfrac{3}{4} = 0.75$

Triangle **a** ~ triangle **b** (SAS)

WORKED EXAMPLE 5 Proving two triangles are similar

Prove that $\triangle ABC$ is similar to $\triangle EDC$.

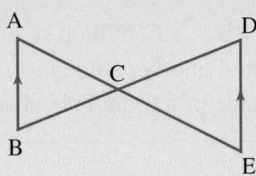

THINK	WRITE
1. Write the information given. AB is parallel to DE. Transversal BD forms two alternate angles: ∠ABC and ∠EDC.	Given: △ABC and △DCE AB‖DE C is common.
2. Write what is to be proved.	To prove: △ABC ~ △EDC
3. Write the proof.	Proof: ∠ABC = ∠EDC (alternate angles) ∠BAC = ∠DEC (alternate angles) ∠BCA = ∠DCE (vertically opposite angles) ∴ △ABC ~ △EDC (equiangular, AAA)

COMMUNICATING — COLLABORATIVE TASK: are SSA and AAA valid congruency tests?

Equipment: pen, paper, ruler, protractor

In this lesson, you have seen four ways of testing triangles for similarity. The object of this task is to determine whether AAA (Angle-angle-angle) and SSA (side-side-angle) can be used to test for congruence.
1. Observe the similar triangles △ABC and △DEF, with the same angles, below. In pairs, discuss whether they are congruent.

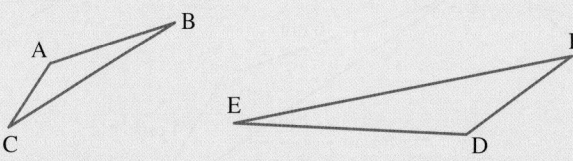

2. Draw a triangle with one of its angles equal to 30°, one of its sides equal to 5 cm and another of its sides equal to 7 cm.
3. In small groups, compare the triangles you have drawn. Are they always congruent?
4. As a class, discuss whether SSA and AAA are valid congruency tests.

DISCUSSION

Can you draw a triangle with sides 3 cm, 5 cm and 7 cm?

What about a triangle with sides 6 cm, 8 cm and 13 cm?

Can you prove that for a triangle to exist, the sum of the lengths of any two of its sides must be greater than or equal to the length of its third side?

COMMUNICATING — COLLABORATIVE TASK: why the longest side of a triangle is always opposite the largest angle?

Equipment: pen, paper, ruler

1. Draw a triangle △ABC such that ∠A > ∠B > ∠C. Which side is the longest?
2. Exchange your triangle with the triangle of another student. Is their answer the same?
3. As a class, discuss whether the longest side of a triangle is always the side opposite to the largest angle.

On Resources

▶ **Video eLesson** Similar triangles (eles-1925)

✦ **Interactivities** Scale factors (int-6041)
Angle-angle-angle condition of similarity (AAA) (int-6042)
Side-angle-side condition of similarity (SAS) (int-6447)
Side-side-side condition of similarity (SSS) (int-6448)

Exercise 12.3 Similar triangles

learn on

12.3 Quick quiz **on**	12.3 Exercise

Individual pathways

■ PRACTISE	■ CONSOLIDATE	■ MASTER
1, 3, 6, 11, 14	2, 4, 7, 9, 12, 15, 16	5, 8, 10, 13, 17, 18

Fluency

1. **WE4** Select a pair of similar triangles among those shown in each part. Give a reason for your answer.

a. i.
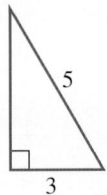

ii.
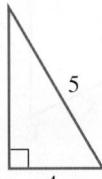

iii.
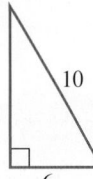

b. i.
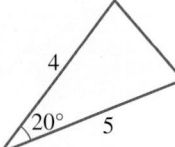

ii.
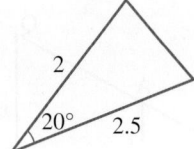

iii.
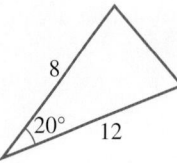

c. i.
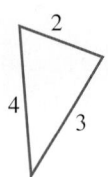

ii.
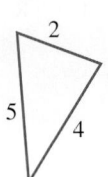

iii.
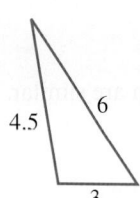

d. i.
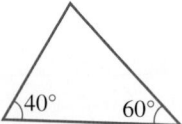

ii.
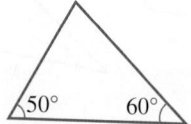

iii.
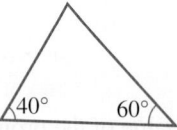

e. i.
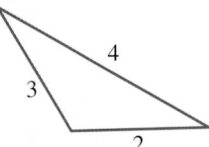

ii.
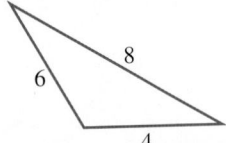

iii.
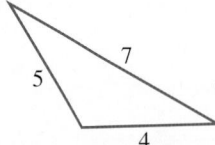

2. Name two similar triangles in each of the following figures.

a.

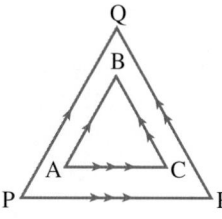

b.

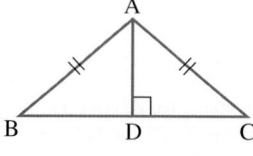

c.

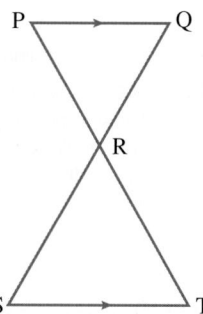

d.

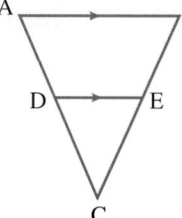

e.

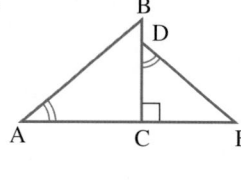

3. a. Complete this statement: $\dfrac{AB}{AD} = \dfrac{BC}{} = \dfrac{}{AE}$.

b. Determine the value of the pronumerals.

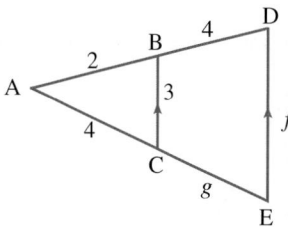

4. Determine the value of the pronumeral in the diagram shown.

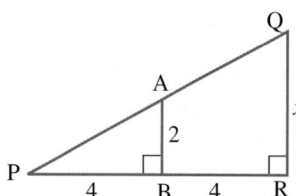

5. The triangles shown are similar.

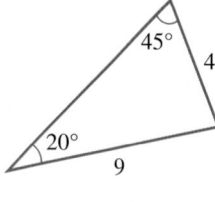

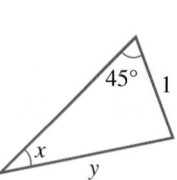

Determine the value of the pronumerals x and y.

Understanding

6. **a.** State why the two triangles shown are similar.

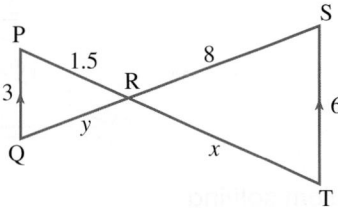

   b. Determine the values of the pronumerals x and y in the diagram.

7. Calculate the values of the pronumerals in the following diagrams.

a.

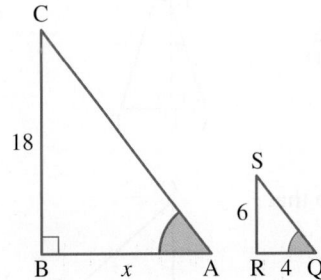

b.

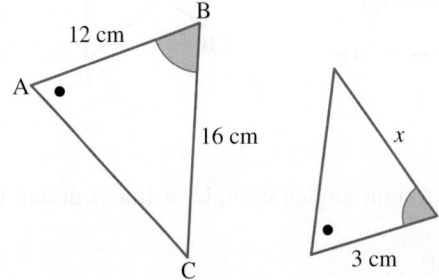

c.

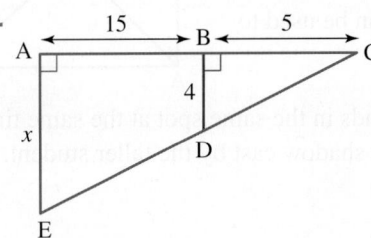

d.

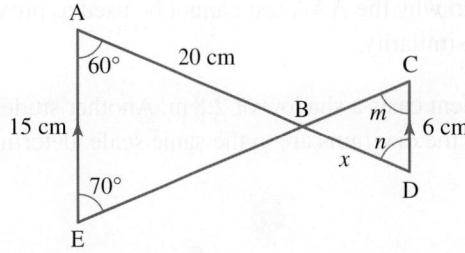

8. Calculate the values of the pronumerals in the following diagrams.

a.

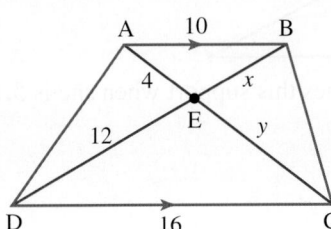

b.

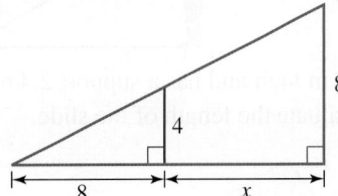

9. Determine the value of x in the diagram.

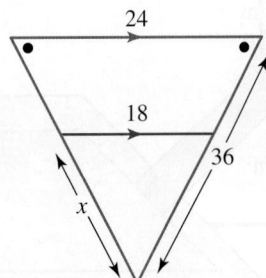

10. Calculate the values of the pronumerals.

a.
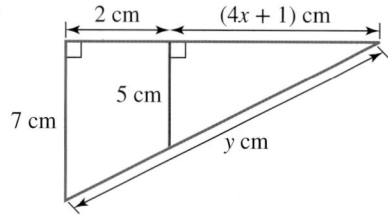

b.
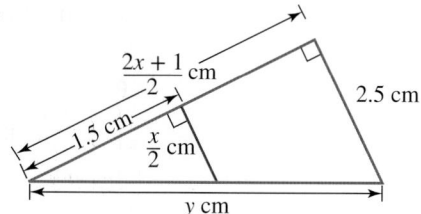

Communicating, reasoning and problem solving

11. WE5 Prove that △ABC is similar to △EDC in each of the following.

a.
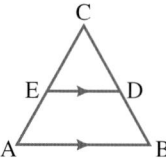

b.
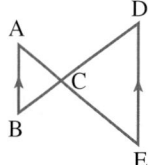

c.
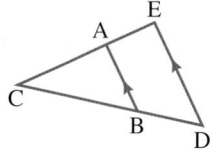

d.
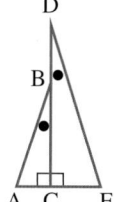

12. △ABC is a right-angled triangle. A line is drawn from A to D as shown so that AD⊥BC.
Prove that:
a. △ABD ~ △ACB
b. △ACD ~ △ACB.

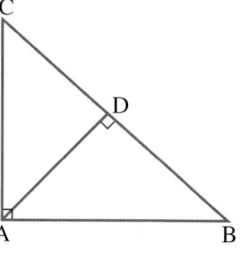

13. Explain why the AAA test cannot be used to prove congruence but can be used to prove similarity.

14. A student casts a shadow of 2.8 m. Another student, who is taller, stands in the same spot at the same time of day. If the diagrams are to the same scale, determine the length of the shadow cast by the taller student.

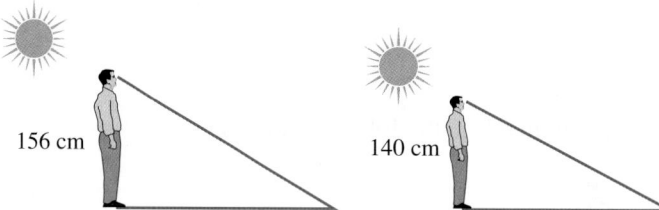

15. A waterslide is 4.2 m high and has a support 2.4 m tall. If a student reaches this support when she is 3.1 m down the slide, evaluate the length of the slide.

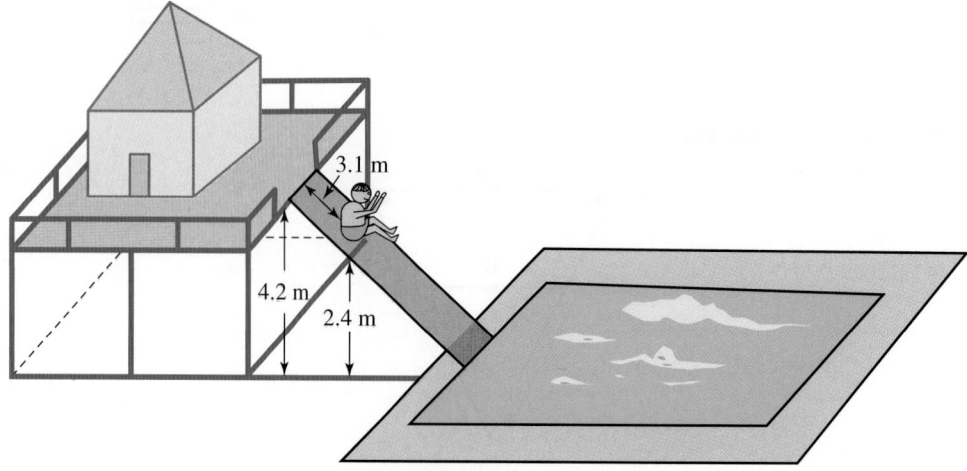

16. Prove that $\triangle EFO \sim \triangle GHO$.

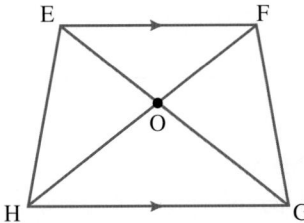

17. A storage tank as shown in the diagram is made of a 4-m-tall cylinder joined by a 3-m-tall cone.

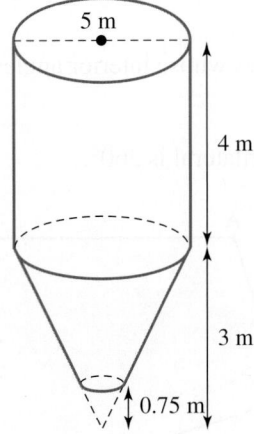

If the diameter of the cylinder is 5 m, evaluate the radius of the end of the cone if 0.75 m has been cut off the tip.

18. Determine the value of x in the diagram shown.

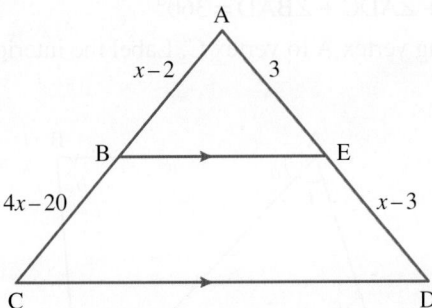

LESSON
12.4 Quadrilaterals

LEARNING INTENTION

At the end of this lesson you should be able to:
- identify the different types of quadrilaterals
- construct simple geometric proofs for angles, sides and diagonals in quadrilaterals.

12.4.1 Quadrilaterals

eles-4901

- Quadrilaterals are four-sided plane shapes whose interior angles sum to 360°.

Theorem 5

- The sum of the interior angles in a quadrilateral is 360°.

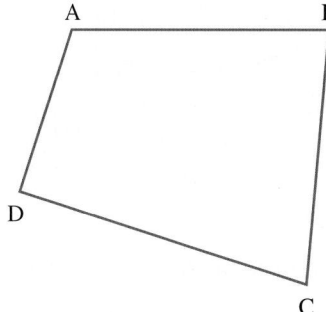

Given:	A quadrilateral ABCD
To prove:	$\angle ABC + \angle BCD + \angle ADC + \angle BAD = 360°$
Construction:	Draw a line joining vertex A to vertex C. Label the interior angles of the triangles formed.

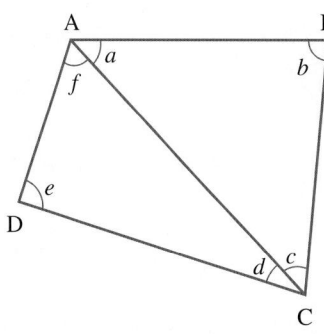

Proof:

$a + b + c = 180°$ (sum of interior angles in a triangle is 180°)

$d + e + f = 180°$ (sum of interior angles in a triangle is 180°)

$\Rightarrow a + b + c + d + e + f = 360°$

$\therefore \angle ABC + \angle BCD + \angle ADC + \angle BAD = 360°$

▶ 12.4.2 Parallelograms

eles-5354

- A **parallelogram** is a quadrilateral with two pairs of parallel sides.

Theorem 6

- Opposite angles of a parallelogram are equal.

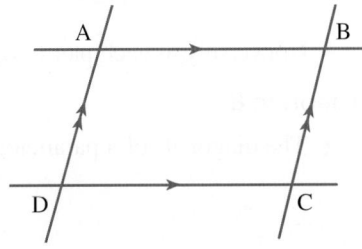

Given:	AB ∥ DC and AD ∥ BC
To prove:	∠ABC = ∠ADC
Construction:	Draw a diagonal from B to D.

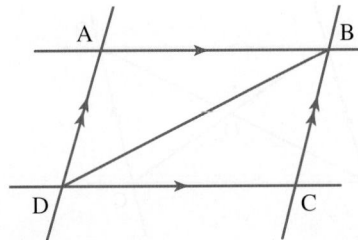

Proof:	∠ABD = ∠BDC	(alternate angles)
	∠ADB = ∠CBD	(alternate angles)
	∠ABC = ∠ABD + ∠CBD	(by construction)
	∠ADC = ∠BDC + ∠ADB	(by construction)
	∴ ∠ABC = ∠ADC	

- Conversely, if each pair of opposite angles of a quadrilateral is equal, then it is a parallelogram.

Theorem 7

- Opposite sides of a parallelogram are equal.

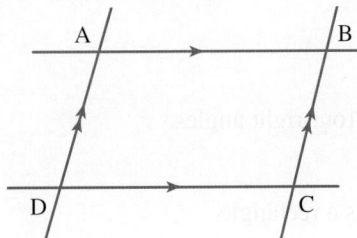

Given:	AB ∥ DC and AD ∥ BC
To prove:	AB = DC
Construction:	Draw a diagonal from B to D.

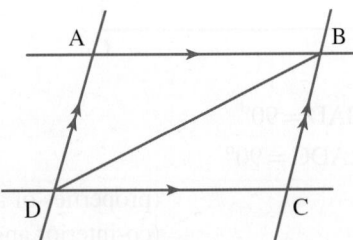

Proof: ∠ABD = ∠BDC   (alternate angles)
                    ∠ADB = ∠CBD   (alternate angles)
                    BD is common to ΔABD and ΔBCD.
                    ⇒ ΔABD ≡ ΔBCD   (ASA)
                    ∴ AB = DC

- Conversely, if each pair of opposite sides of a quadrilateral is equal, then it is a parallelogram.

Theorem 8

- The diagonals of a parallelogram bisect each other.

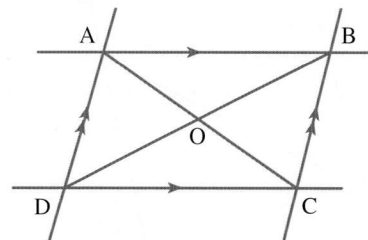

Given: AB ∥ DC and AD ∥ BC with diagonals AC and BD
To prove: AO = OC and BO = OD
Proof: In ΔAOB and ΔCOD,
                ∠OAB = ∠OCD   (alternate angles)
                ∠OBA = ∠ODC   (alternate angles)
                AB = CD   (opposite sides of a parallelogram)
                ⇒ ΔAOB ≡ ΔCOD   (ASA)
                ⇒ AO = OC   (corresponding sides in congruent triangles)
                and BO = OD   (corresponding sides in congruent triangles)

▶ 12.4.3 Rectangles

eles-5355

- A rectangle is a parallelogram with four right angles.

Theorem 9

- A parallelogram with a right angle is a rectangle.

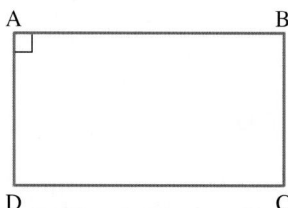

Given: Parallelogram ABCD with ∠BAD = 90°
To prove: ∠BAD = ∠ABC = ∠BCD = ∠ADC = 90°
Proof: AB ∥ CD   (properties of a parallelogram)
                ⇒ ∠BAD + ∠ADC = 180°   (co-interior angles)
                But ∠BAD = 90°   (given)
                ⇒ ∠ADC = 90°
                Similarly, ∠BCD = ∠ADC = 90°
                ∴ ∠BAD = ∠ABC = ∠BCD = ∠ADC = 90°

Theorem 10

• The diagonals of a rectangle are equal.

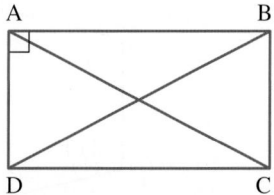

Given: Rectangle ABCD with diagonals AC and BD

To prove: $AC = BD$

Proof: In $\triangle ADC$ and $\triangle BCD$,

$AD = BC$	(opposite sides equal in a rectangle)
$DC = CD$	(common)
$\angle ADC = \angle BCD = 90°$	(right angles in a rectangle)
$\Rightarrow \triangle ADC \equiv \triangle BCD$	(SAS)
$\therefore AC = BD$	

12.4.4 Rhombuses

eles-5356

• A **rhombus** is a parallelogram with four equal sides.

Theorem 11

• The diagonals of a rhombus are perpendicular.

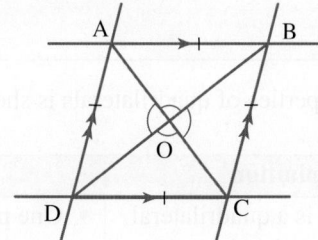

Given: Rhombus ABCD with diagonals AC and BD

To prove: $AC \perp BD$

Proof: In $\triangle AOB$ and $\triangle BOC$,

$AO = OC$	(property of parallelogram)
$AB = BC$	(property of rhombus)
$BO = OB$	(common)
$\Rightarrow \triangle AOB \equiv \triangle BOC$	(SSS)
$\Rightarrow \angle AOB = \angle BOC$	
But $\angle AOB + \angle BOC = 180°$	(supplementary angles)
$\Rightarrow \angle AOB = \angle BOC = 90°$	

Similarly, $\angle AOD = \angle DOC = 90°$.

Hence, $AC \perp BD$

a.

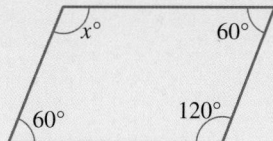

b.

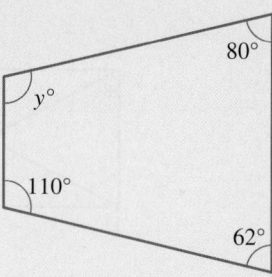

THINK

a. 1. Identify the shape.

2. To determine the values of $x°$, apply theorem 6, which states that opposite angles of a parallelogram are equal.

b. 1. Identify the shape.

2. To determine the values of $y°$, apply theorem 5, which states the sum of interior angles in a quadrilateral is 360°.

3. Simplify and solve for $y°$.

WRITE

a. The shape is a parallelogram as the shape has two pairs of parallel sides.

$x° = 120°$

b. The shape is a trapezium, as one pair of opposite sides is parallel but not equal in length.

The sum of all the angles = 360°
$$y° + 110° + 80° + 62° = 360°$$

$$y° + 110° + 80° + 62° = 360°$$
$$y° + 252° = 360°$$
$$y° = 360° - 252°$$
$$y° = 108°$$

• A summary of the definitions and properties of quadrilaterals is shown in the following table.

Shape	Definition	Properties
Trapezium	A trapezium is a quadrilateral with one pair of opposite sides parallel.	• One pair of opposite sides is parallel but not equal in length.
Parallelogram	A parallelogram is a quadrilateral with both pairs of opposite sides parallel.	• Opposite angles are equal. • Opposite sides are equal. • Diagonals bisect each other.
Rhombus	A rhombus is a parallelogram with four equal sides.	• Diagonals bisect each other at right angles. • Diagonals bisect the angles at the vertex through which they pass.

(continued)

(continued)

Shape	Definition	Properties
Rectangle *(figure)*	A rectangle is a parallelogram whose interior angles are right angles.	• Diagonals are equal. • Diagonals bisect each other.
Square *(figure)*	A square is a parallelogram whose interior angles are right angles with four equal sides.	• All angles are right angles. • All side lengths are equal. • Diagonals are equal in length and bisect each other at right angles. • Diagonals bisect the vertex through which they pass (45°).

Relationships between quadrilaterals

• The flowchart below shows the relationships between quadrilaterals.

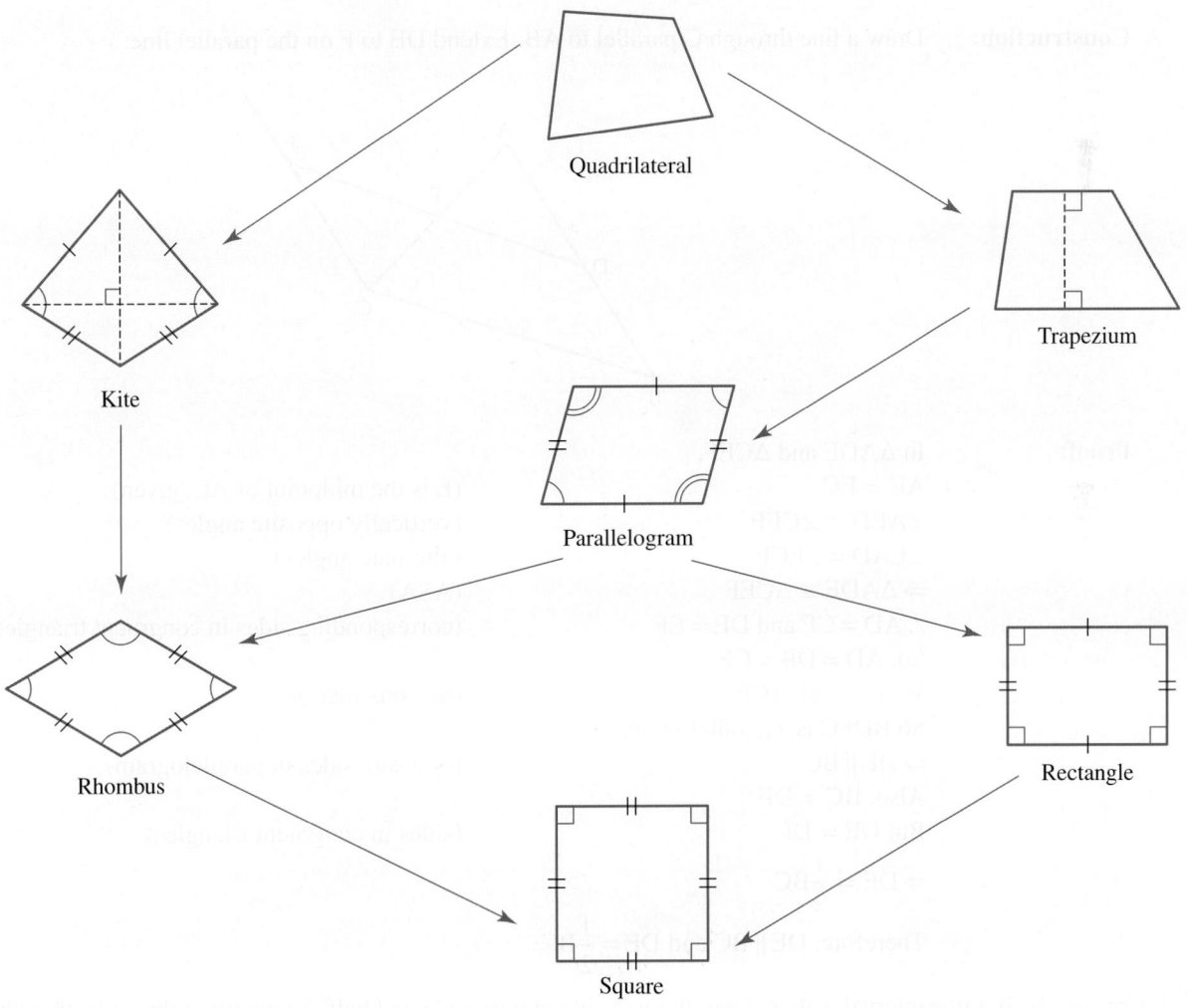

12.4.5 The midpoint theorem

eles-4905

• Now that the properties of quadrilaterals have been explored, the midpoint theorem can be tackled.

Theorem 12

• The interval joining the midpoints of two sides of a triangle is parallel to the third side and half its length.

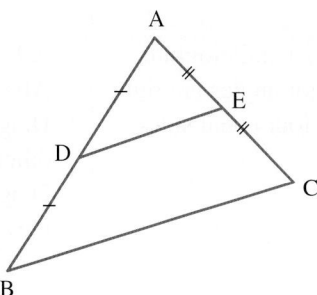

Given: $\triangle ABC$ in which $AD = DB$ and $AE = EC$

To prove: $DE \parallel BC$ and $DE = \dfrac{1}{2}BC$

Construction: Draw a line through C parallel to AB. Extend DE to F on the parallel line.

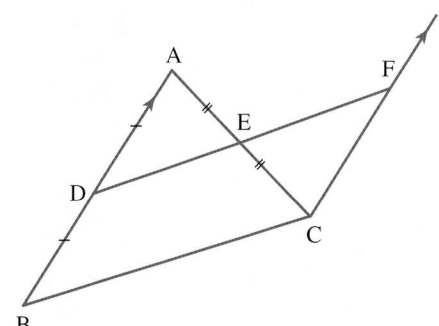

Proof:
In $\triangle ADE$ and $\triangle CEF$,
AE = EC   (E is the midpoint of AC, given)
$\angle AED = \angle CEF$ (vertically opposite angles)
$\angle EAD = \angle ECF$ (alternate angles)
$\Rightarrow \triangle ADE \equiv \triangle CEF$ (ASA)
$\therefore AD = CF$ and $DE = EF$ (corresponding sides in congruent triangles)
So, AD = DB = CF
We have $AB \parallel CF$ (by construction)
So BDFC is a parallelogram.
$\Rightarrow DE \parallel BC$ (opposite sides in parallelogram)
Also, BC = DF
But DE = DF   (sides in congruent triangles)
$\Rightarrow DE = \dfrac{1}{2}BC$

Therefore, $DE \parallel BC$ and $DE = \dfrac{1}{2}BC$.

• Conversely, if a line interval is drawn parallel to a side of a triangle and half the length of that side, then the line interval bisects each of the other two sides of the triangle.

In triangle ABC, the midpoints of AC and AB are D and E respectively.
Determine the value of DE, if BC = 18 cm.

THINK	WRITE
1. Determine the midpoints on the line AB and AC.	D is midpoint of AB and E is midpoint of AC.
2. Apply the midpoint theorem to determine the length of DE.	$DE = \dfrac{1}{2}BC$
3. Substitute the value of BC = 18 cm into the formula.	$DE = \dfrac{1}{2} \times 18$
4. Simplify and determine the length of DE.	$DE = 9\,\text{cm}$

on Resources

Interactivities Quadrilateral definitions (int-2786)
Angles in a quadrilateral (int-3967)
Opposite angles of a parallelogram (int-6160)
Opposite sides of a parallelogram (int-6161)
Diagonals of a parallelogram (int-6162)
Diagonals of a rectangle (int-6163)
Diagonals of a rhombus (int-6164)
The midpoint theorem (int-6165)
Quadrilaterals (int-3756)

Exercise 12.4 Quadrilaterals

learn on

12.4 Quick quiz on	12.4 Exercise

Individual pathways

■ PRACTISE	■ CONSOLIDATE	■ MASTER
1, 3, 7, 9, 12, 13, 18	2, 4, 8, 10, 14, 15, 19, 20	5, 6, 11, 16, 17, 21, 22

Fluency

1. Use the definitions of the five special quadrilaterals to decide if the following statements are true or false.

   a. A square is a rectangle.
   b. A rhombus is a parallelogram.
   c. A square is a rhombus.
   d. A rhombus is a square.

2. Use the definitions of the five special quadrilaterals to decide if the following statements are true or false.

   a. A square is a trapezium.
   b. A parallelogram is a rectangle.
   c. A trapezium is a rhombus.
   d. A rectangle is a square.

3. **WE6** Determine the value of the pronumeral in each of the following quadrilaterals.

a.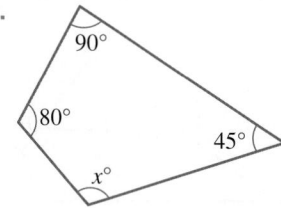

b.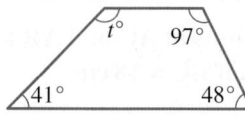

c.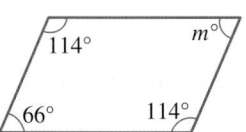

d.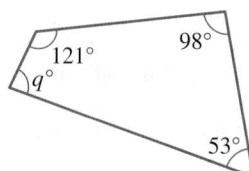

4. Determine the values of the pronumerals in the following diagrams.

a.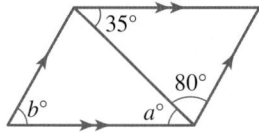

b.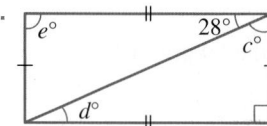

c.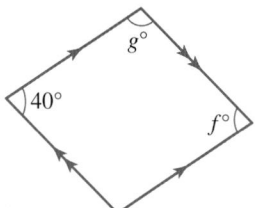

d.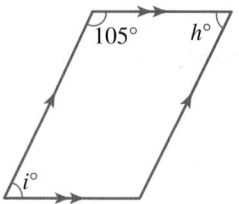

5. Determine the values of the pronumerals in each of the following figures.

a.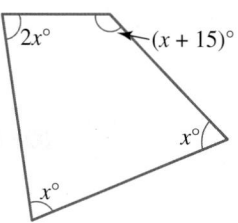

b.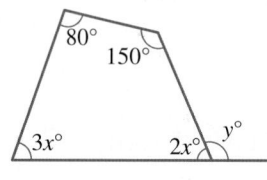

6. Determine the values of x and y in each of the following figures.

a.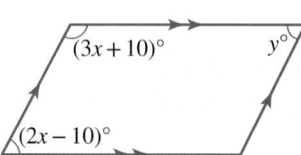

b.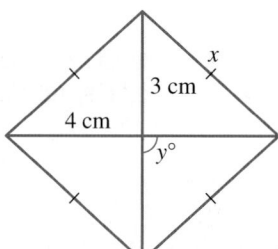

c.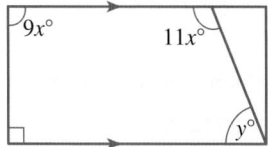

d.

Understanding

7. Draw three different trapeziums. Using your ruler, compass and protractor, decide which of the following properties are true in a trapezium.

  a. Opposite sides are equal.
  b. All sides are equal.
  c. Opposite angles are equal.
  d. All angles are equal.
  e. Diagonals are equal in length.
  f. Diagonals bisect each other.
  g. Diagonals are perpendicular.
  h. Diagonals bisect the angles they pass through.

8. Draw three different parallelograms. Using your ruler and protractor to measure, decide which of the following properties are true in a parallelogram.

  a. Opposite sides are equal.
  b. All sides are equal.
  c. Opposite angles are equal.
  d. All angles are equal.
  e. Diagonals are equal in length.
  f. Diagonals bisect each other.
  g. Diagonals are perpendicular.
  h. Diagonals bisect the angles they pass through.

9. Choose which of the following properties are true in a rectangle.

  a. Opposite sides are equal.
  b. All sides are equal.
  c. Opposite angles are equal.
  d. All angles are equal.
  e. Diagonals are equal in length.
  f. Diagonals bisect each other.
  g. Diagonals are perpendicular.
  h. Diagonals bisect the angles they pass through.

10. Name four quadrilaterals that have at least one pair of opposite sides that are parallel and equal.

11. Name a quadrilateral that has equal diagonals that bisect each other and bisect the angles they pass through.

Communicating, reasoning and problem solving

12. Prove that the diagonals of a rhombus bisect each other.

13. Give reasons why a square is a rhombus, but a rhombus is not necessarily a square.

14. **WE7** ABCD is a parallelogram. X is the midpoint of AB and Y is the midpoint of DC.

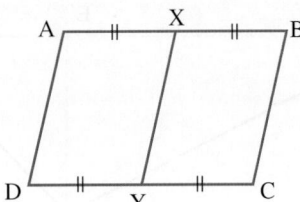

Prove that AXYD is also a parallelogram.

15. The diagonals of a parallelogram meet at right angles. Prove that the parallelogram is a rhombus.

16. ABCD is a parallelogram. P, Q, R and S are all midpoints of their respective sides of ABCD.

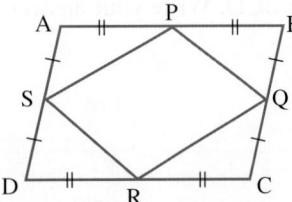

  a. Prove ΔPAS ≡ ΔRCQ.
  b. Prove ΔSDR ≡ ΔPBQ.
  c. Hence, prove that PQRS is also a parallelogram.

17. Two congruent right-angled triangles are arranged as shown.

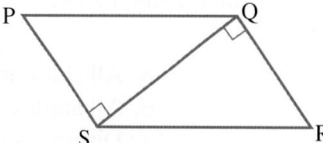

Show that PQRS is a parallelogram.

18. ABCD is a trapezium.

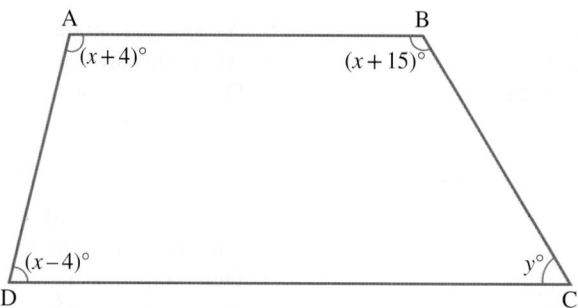

a. Describe a fact about a trapezium.

b. Determine the values of x and y.

19. ABCD is a kite where $AC = 8$ cm, $BE = 5$ cm and $ED = 9$ cm.

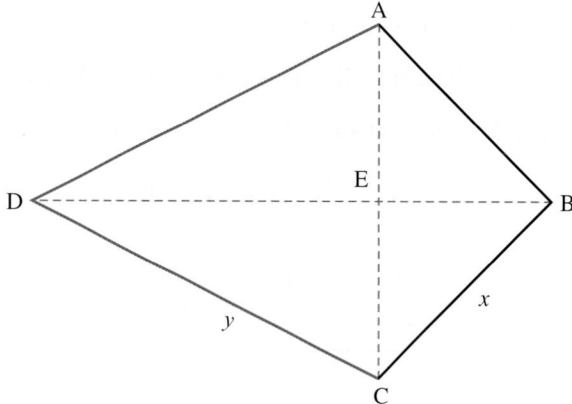

a. Determine the exact values of:

 i. x

 ii. y.

b. Evaluate angle BAD and hence angle BCD. Write your answer in degrees and minutes, correct to the nearest minute.

20. ABCDE is a regular pentagon whose side lengths are 2 cm. Each diagonal is x cm long.

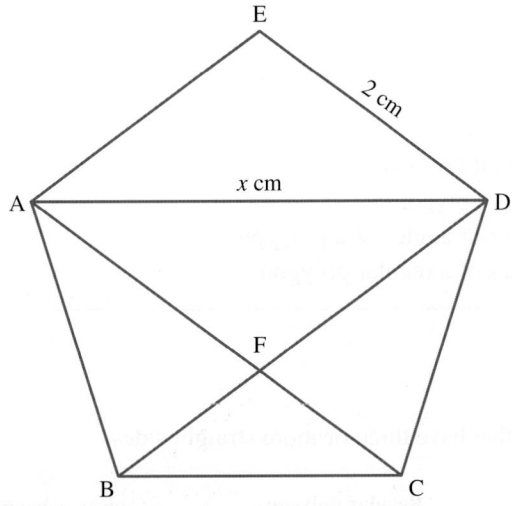

a. What kind of shape is AEDF and what is the length of FD?
b. What kind of shape is ABCD?
c. If ∠EDA is 40°, determine the value of ∠ACB, giving reasons for your findings.
d. Which triangle is similar to AED?
e. Explain why FB = $(x - 2)$ cm.
f. Show that $x^2 - 2x - 4 = 0$.
g. Solve the equation $x^2 - 2x - 4 = 0$, giving your answer as an exact value.

21. ABCD is called a cyclic quadrilateral because it is inscribed inside a circle.

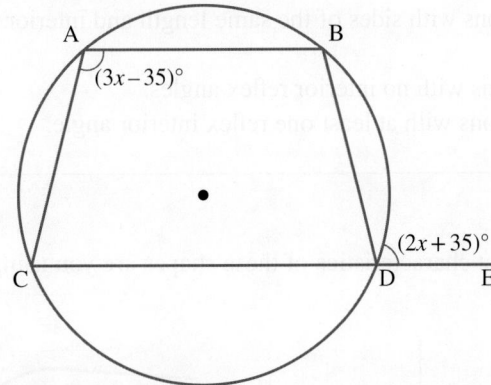

A characteristic of a cyclic quadrilateral is that the opposite angles are supplementary.
Determine the value of x.

22. The perimeter of this kite is 80 cm. Determine the exact value of x.

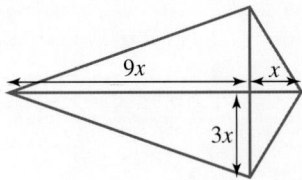

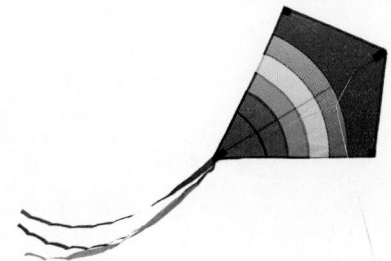

LESSON
12.5 Polygons

LEARNING INTENTION

At the end of this lesson you should be able to:
- identify regular and irregular polygons
- calculate the sum of the interior angles of a polygon
- determine the exterior angles of a regular polygon.

▶ 12.5.1 Polygons

eles-4906

- **Polygons** are closed shapes that have three or more straight sides.

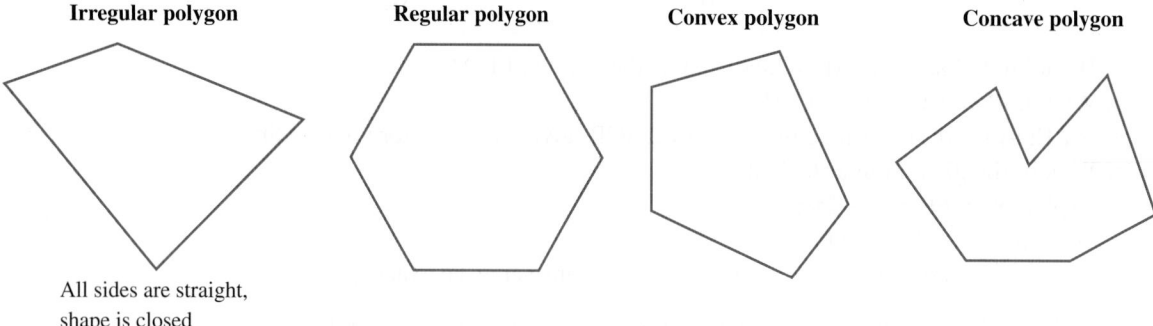

Irregular polygon   Regular polygon   Convex polygon   Concave polygon

All sides are straight,
shape is closed

- **Regular polygons** are polygons with sides of the same length and interior angles of the same size, like the hexagon shown below.
- **Convex polygons** are polygons with no interior reflex angles.
- **Concave polygons** are polygons with at least one reflex interior angle.

DISCUSSION

Are these shapes polygons? What characteristics of these shapes are you using to answer this question?

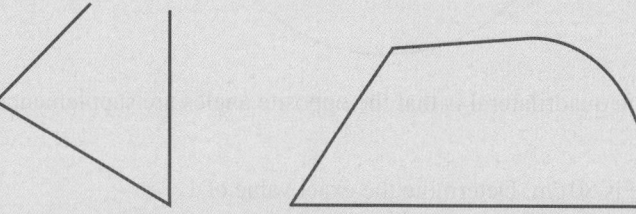

▶ 12.5.2 Interior angles of a polygon

eles-6266

- The interior angles of a polygon are the angles inside the polygon at each vertex.
- The sum of the interior angles of a polygon is given by the formula shown below.

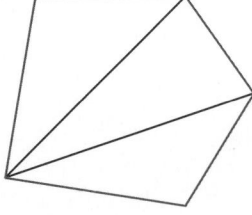

> **Sum of interior angles of a polygon**
>
> **Sum of interior angles $= 180° \times (n-2)$**
>
> **where $n =$ the number of sides of the polygon.**

This formula is illustrated in the diagram shown.

This polygon has five sides. Two diagonals can be drawn from a vertex, dividing the polygon into three triangles.

The angle sum of this polygon $= 3 \times 180° = 540°$

Using the formula, where $n = 5 = 180° \times (5-2) = 180° \times 3 = 540°$

WORKED EXAMPLE 8 Calculating the values of angles in a given diagram

Calculate the value of the pronumerals in the figure shown.

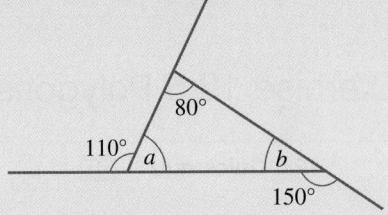

THINK	WRITE
1. Angles a and $110°$ form a straight line and so are supplementary (add to $180°$).	$a + 110° = 180°$ $a + 110° - 110° = 180° - 110°$ $a = 70°$
2. The interior angles of a triangle sum to $180°$.	$b + a + 80° = 180$
3. Substitute $70°$ for a and solve for b.	$b + 70° + 80° = 180°$ $b + 150° = 180°$ $b = 30°$
4. Write the value of the pronumerals.	$a = 70°, b = 30°$

▶ 12.5.3 Exterior angles of a polygon

eles-6267

- The exterior angles of a polygon are formed by the side of the polygon and an extension of its adjacent side.
  For example, x, y and z are exterior angles for the polygon (triangle) below and q, r, s and t are the exterior angles of the quadrilateral.

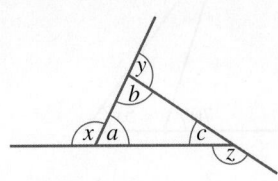 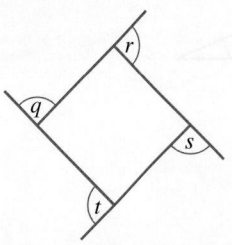

- The exterior angle and interior angle at that vertex are supplementary (add to 180°). For example, in the triangle above, $x + a = 180°$.
- Exterior angles of polygons can be measured in a clockwise or anticlockwise direction.
- In a regular polygon, the size of the exterior angle can be found by dividing 360° by the number of sides.

> **Exterior angles of a regular polygon**
>
> $$\text{Exterior angles of a regular polygon} = \frac{360°}{n}$$
>
> where n = the number of sides of the regular polygon.

- The sum of the exterior angles of a polygon equals 360°.
- The exterior angle of a triangle is equal to the sum of the opposite interior angles.

 Resources

Interactivities Interior angles of a polygon (int-6166)
Exterior angles of a polygon (int-6167)

Exercise 12.5 Polygons

learnon

12.5 Quick quiz on	12.5 Exercise

Individual pathways

■ PRACTISE	■ CONSOLIDATE	■ MASTER
1, 3, 6, 7, 11, 14	2, 4, 8, 12, 15, 17	5, 9, 10, 13, 16, 18

Fluency

1. Calculate the values of the pronumerals in the diagrams shown.

a.
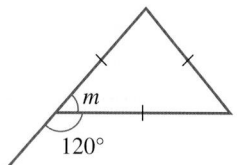

b.
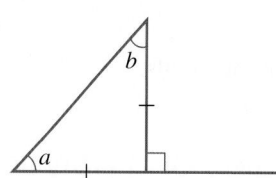

c.
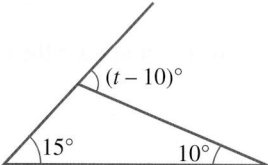

d.
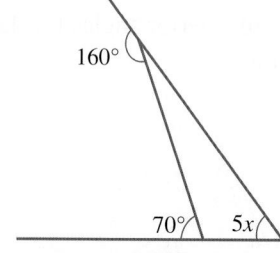

2. Determine the value of the pronumeral in each of the following polygons.

a.

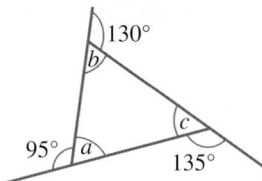

130°
b
c
95° a
135°

b.

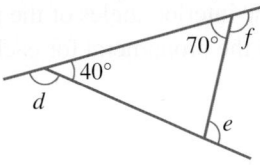

70° f
40°
d
e

c.

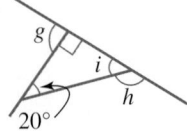

g
i
h
20°

d.

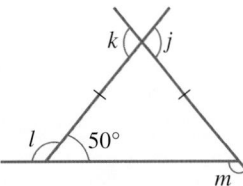

k j
l 50°
m

3. For the triangles shown, evaluate the pronumerals and determine the size of the interior angles.

a.

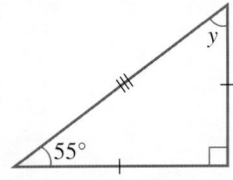

y
55°

b.

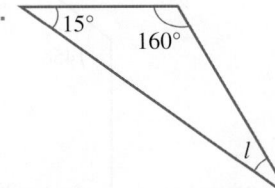

15°
160°
l

c.

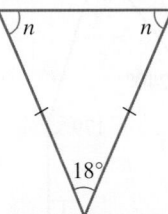

n   n
18°

d.

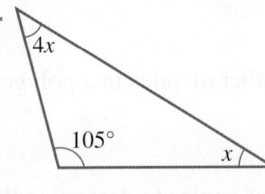

4x
105°
x

e.

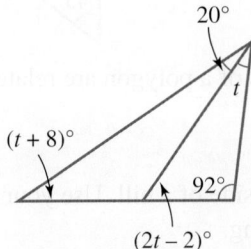

20°
t
$(t + 8)°$
92°
$(2t − 2)°$

4. For the five quadrilaterals shown:
   i. label the quadrilaterals as regular or irregular
   ii. determine the value of the pronumeral for each shape.

a.

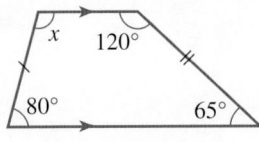

x   120°
80°
65°

b.

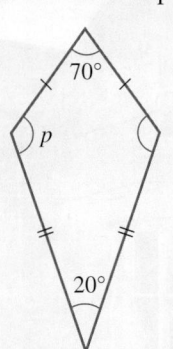

70°
p
20°

c.

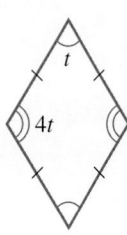

t
4t

d.

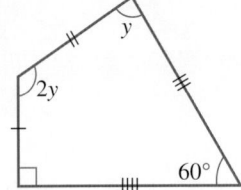

y
2y
60°

e.

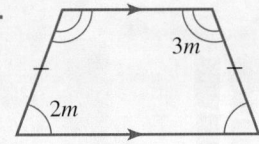

3m
2m

5. For the four polygons:
   i. calculate the sum of the interior angles of the polygon
   ii. determine the value of the pronumeral for each shape.

a.

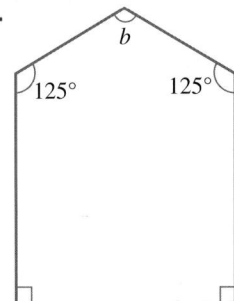

b.

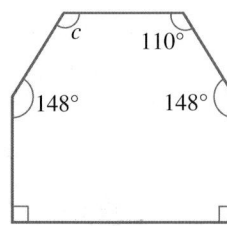

c.

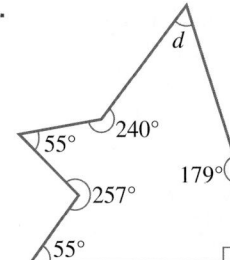

d.

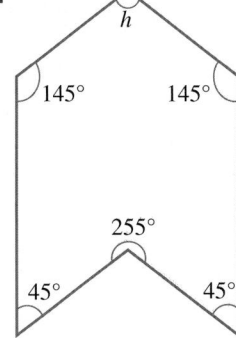

6. Explain how the interior and exterior angles of a polygon are related to the number of sides in a polygon.

Understanding

7. The photograph shows a house built on the side of a hill. Use your knowledge of angles to determine the values of the pronumerals. Show full working.

8. Determine the values of the four interior angles of the front face of the building in the photograph shown. Show full working.

9. Determine the values of the pronumerals for the irregular polygons shown. Show full working.

a.

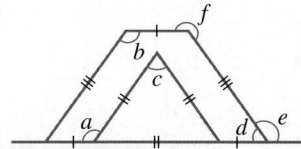

b.

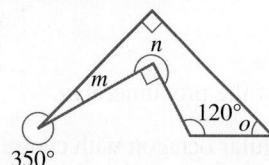

10. Calculate the size of the exterior angle of a regular hexagon (six sides).

Communicating, reasoning and problem solving

11. State whether the following polygons are regular or irregular. Give a reason for your answer.

a.

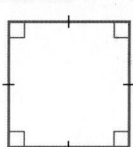

b.

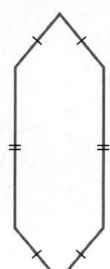

c.

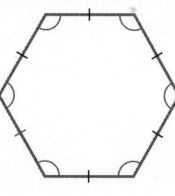

d.

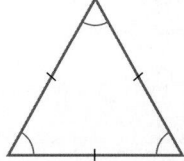

e.

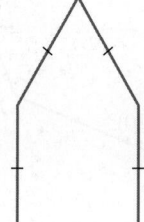

f.

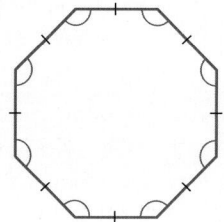

12. A diagonal of a polygon joins two vertices.

    a. Calculate the number of diagonals in a regular polygon with:

       i. 4 sides   ii. 5 sides   iii. 6 sides   iv. 7 sides.

    b. Using your results from part a, show that the number of diagonals for an n-sided polygon is $\frac{1}{2}n(n-3)$.

13. The exterior angle of a polygon can be calculated using the formula:

$$\text{Exterior angle} = \frac{360°}{n}$$

Use the relationship between interior and exterior angles of a polygon to write a formula for the internal angle of a regular polygon.

14. a. Name the polygon that best describes the road sign shown.

    b. Determine the value of the pronumeral m.

15. The diagram shows a regular octagon with centre O.

    a. Calculate the size of $\angle CBD$.

    b. Calculate the size of $\angle CBO$.

    c. Calculate the size of the exterior angle of the octagon, $\angle ABD$.

    d. Calculate the size of $\angle BOC$.

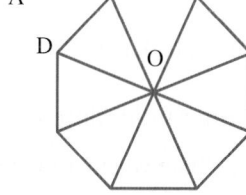

16. ABCDEFGH is an eight-sided polygon.

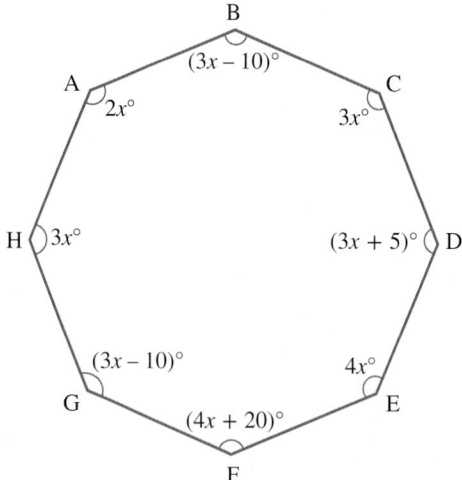

    a. Evaluate the sum of the interior angles of an eight-sided polygon.

    b. Determine the value of the pronumeral x.

17. Answer the following questions for the given shape.

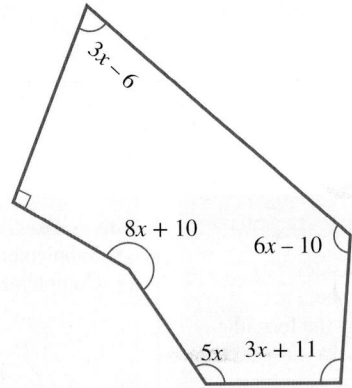

a. Evaluate the sum of the interior angles of this shape.
b. Determine the value of the pronumeral x.

18. Answer the following questions for the given shape.

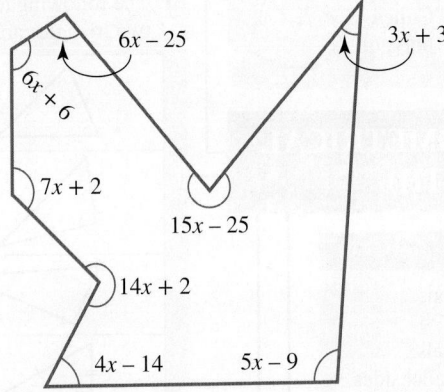

a. Evaluate the sum of the interior angles of this shape.
b. Determine the value of the pronumeral x.

LESSON
12.6 Review

12.6.1 Topic summary

Polygons

- Polygons are closed shapes with straight sides.
- The number of sides a polygon has is denoted n.
- The sum of the interior angles is given by the formula:
  Interior angle sum $= 180°(n-2)$
- **Regular polygons** have all sides the same length and all interior angles equal.
- **Convex polygons** have no internal reflex angles.
- **Concave polygons** have at least one internal reflex angle.
- The exterior angles of a regular polygon are given by the formula:
  Exterior angles $= \dfrac{360°}{n}$

Regular polygon

Convex polygon

Concave polygon

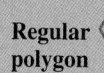

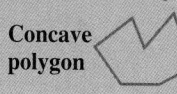

PROPERTIES OF GEOMETRICAL FIGURES (PATH)

Quadrilaterals

- Quadrilaterals are four sided polygons.
- The interior angles sum to 360°.
- There are many types of quadrilaterals:
  - **Trapeziums** have one pair of parallel sides.
  - **Parallelograms** have two pairs of parallel sides.
  - **Rhombuses** are parallelograms that have four equal sides.
  - A **rectangle** is a parallelogram whose interior angles are right angles.
  - A **square** is a rectangle with four equal sides.

Parallel lines

- If parallel lines are cut by a transversal, then:

Alternate angles are equal: $a = b$	
Corresponding angles are equal: $a = b$	
Co-interior angles are supplementary: $a + b = 180°$	

Supplementary and complementary angles

- Supplementary angles are angles that add up to 180°.
- Complementary angles are angles that add up to 90°.

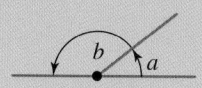
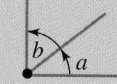

Supplementary angles   **Complementary angles**

Congruent figures

- Congruent figures have the same size and shape.
- The symbol for congruence is $\cong$.
- The following tests can be used to determine whether two triangles are congruent:

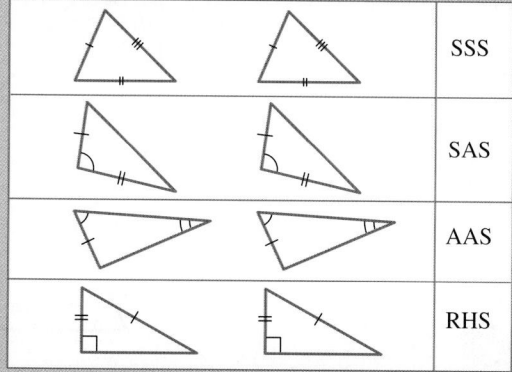

	SSS
	SAS
	AAS
	RHS

Similar figures

- Similar figures have the same shape but different size.
- The symbol for similarity is ~.
- Scale factor $= \dfrac{\text{image side length}}{\text{object side length}}$
- The following tests can be used to determine whether two triangles are similar:

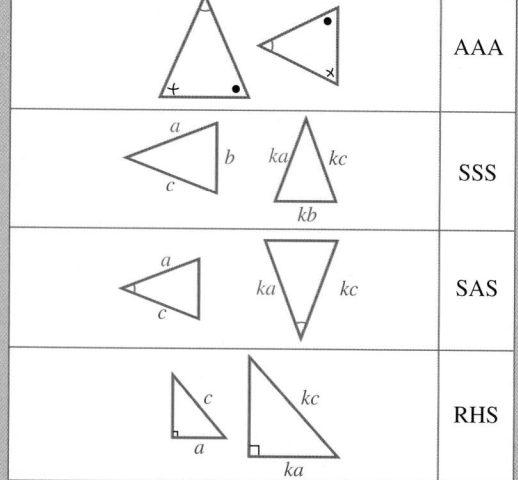

	AAA
	SSS
	SAS
	RHS

12.6.2 Project

What's this object?

When using geometrical tools to construct shapes, we have to make sure that our measurements are precise. Even a small error in one single step of the measuring process can result in an incorrect shape. The object you will be making in this task is a very interesting one. This object is made by combining three congruent shapes.

The instructions for making the congruent shape that you will use to then make the final object are given below.

Part 1: Making the congruent shape

1. Using a ruler, a protractor, a pencil and two compasses, draw a shape in your workbook by following the instructions below.
   - Measure a horizontal line (AB) that is 9.5 cm long. To make sure there is enough space for the whole of the final object, draw this first line close to the bottom of the page you are working on.
   - At point B, measure an angle of 120° above the line AB. Draw a line that follows this angle for 2 cm. Mark the end of this line as point C, creating the line BC.
   - At point C, measure an angle of 60° on the same side of line BC as point A. Draw a line that follows this angle for 7.5 cm. Mark the end of this line as point D, creating the line CD.
   - At point D, measure an angle of 60° above the line CD. Draw a line that follows this angle for 3.5 cm. Mark the end of this line as point E, creating the line DE.
   - At point E. measure an angle of 120° above the line DE. Draw a line that follows this angle for 2 cm. Mark the end of this line as point F, creating the line EF.
   - Join point F to point A.
     Colour or shade this shape using any colour you wish.
2. Determine the length of the line joining point A to point F.
3. Describe what you notice about the size of the angles FAB and AFE.

Part 2: Making the final object

You have now constructed the shape that will be used three limes to make the final object. To make this object, follow the instructions below.
   - Trace the original shape onto a piece of tracing paper twice. Cut around the edges of both of the shapes you have just drawn. Label the shapes with the same letters you used to mark the points A–F when making the original shape.
   - Place line AF of the first traced copy of the original shape so that it is covering up line CD of the shape you first drew. Use the tracing paper to transfer this shape onto the original shape. Colour or shade the traced shape using a colour that is different to the colour you used for the first shape.
   - Place the line DC of your second traced copy so that it is covering up line FA of the shape you first drew. Use the tracing paper to transfer this shape onto the original shape. Colour or shade this third section with a third colour.

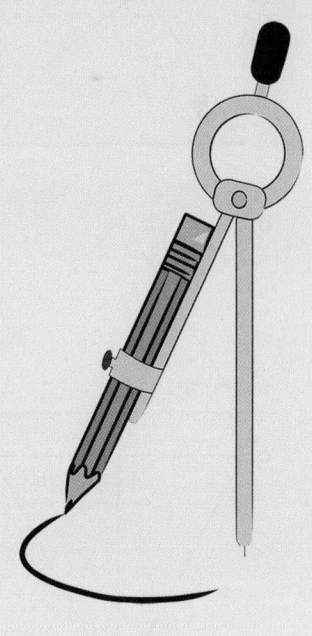

4. Describe the object you have drawn.
5. Using the internet, your school library or other references, investigate other 'impossible objects' that can be drawn as two-dimensional shapes. Recreate these shapes on a separate sheet of paper. Besides these shapes, briefly write some reasons for these shapes being referred to as 'impossible'.

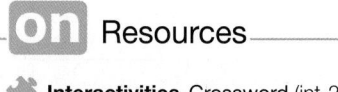

Resources

Interactivities Crossword (int-2854)
Sudoku puzzle (int-3597)

Exercise 12.6 Review questions

learn on

Fluency

1. Select a pair of congruent triangles in each of the following sets of triangles, giving a reason for your answer. All angles are in degrees and side lengths in cm. (The figures are not drawn to scale.)

   a.

   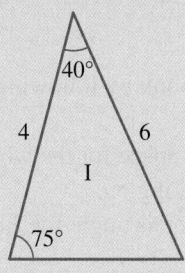
   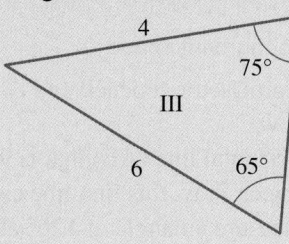

   b.

   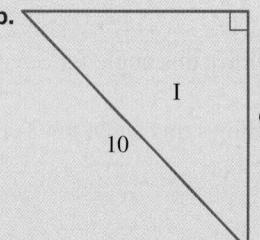
   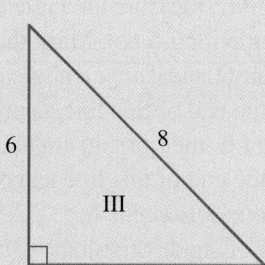

2. Determine the value of the pronumeral in each pair of congruent triangles. All angles are given in degrees and side lengths in cm.

   a.

   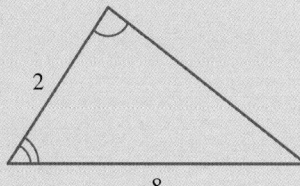

   b.

   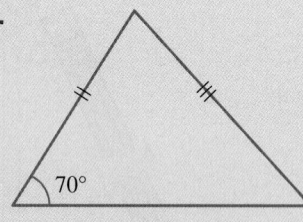
   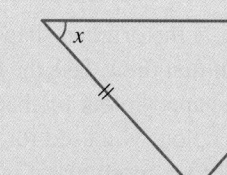

   c.

3. a. Prove that the two triangles shown in the diagram are congruent.

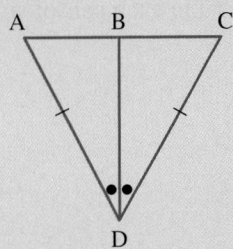

b. Prove that ΔPQR is congruent to ΔQPS.

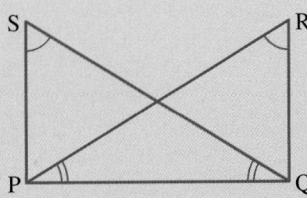

4. Test whether the following pairs of triangles are similar. For similar triangles, determine the scale factor. All angles are in degrees and side lengths in cm.

a.

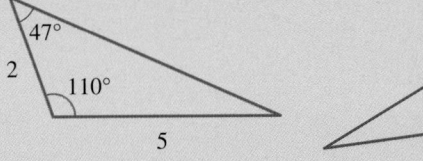

b.

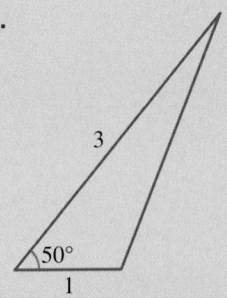

c.

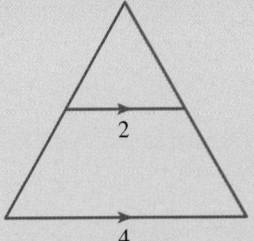

Understanding

5. Determine the value of the pronumeral in each pair of similar triangles. All angles are given in degrees and side lengths in cm.

a.

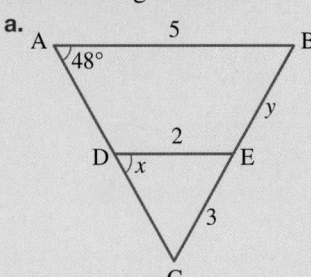

b.

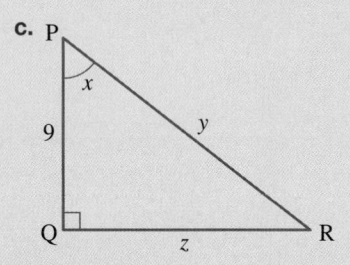

c.

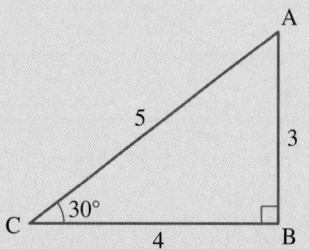

6. Prove that $\triangle ABC \sim \triangle EDC$.

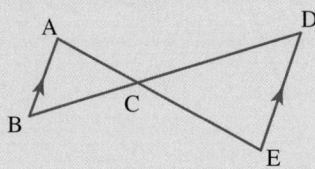

7. Prove that $\triangle PST \sim \triangle PRQ$.

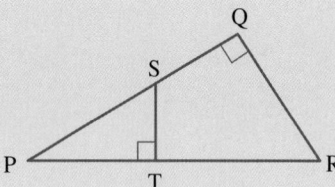

8. State the definition of a rhombus.

9. MC Two corresponding sides in a pair of similar octagons have lengths of 4 cm and 60 mm. The respective scale factor in length is:

    A. 1 : 15   **B.** 3 : 20   **C.** 2 : 3   **D.** 3 : 2

10. A regular nonagon has side length x cm. Use a scale factor of $\dfrac{x+1}{x}$ to calculate the side length of a similar nonagon.

Communicating, reasoning and problem solving

11. ABC is a triangle. D is the midpoint of AB, E is the midpoint of AC and F is the midpoint of BC
DG $\perp$ AB, EG $\perp$ AC and FG $\perp$ BC.
  a. Prove that $\triangle$GDA $\equiv$ $\triangle$GDB.
  b. Prove that $\triangle$GDE $\equiv$ $\triangle$GCE.
  c. Prove that $\triangle$GBF $\equiv$ $\triangle$GCF.
  d. State what this means about AG, BG and CG.
  e. A circle centred at G is drawn through A.
     Determine what other points it must pass through.

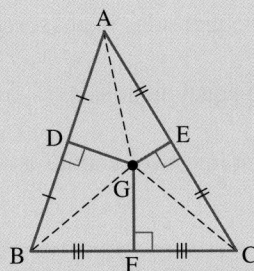

12. PR is the perpendicular bisector of QS. Prove that $\triangle$PQS is isosceles.

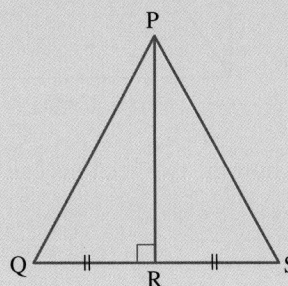

13. Name any quadrilaterals that have diagonals that bisect the angles they pass through.

14. State three tests that can be used to show that a quadrilateral is a rhombus.

15. Prove that WXYZ is a parallelogram.

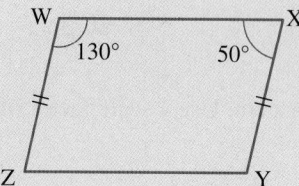

16. Prove that the diagonals in a rhombus bisect the angles they pass through.

17. Explain why the triangles shown are not congruent.

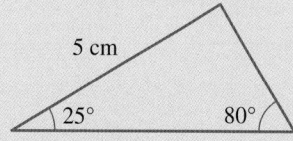 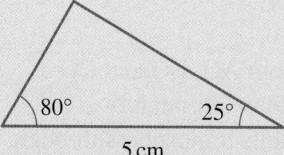

18. Prove that the angles opposite the equal sides in an isosceles triangle are equal.

19. Name any quadrilaterals that have equal diagonals.

20. This 8 cm by 12 cm rectangle is cut into two sections as shown.

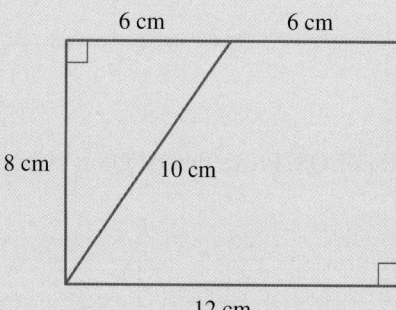

    a. Draw labelled diagrams to show how the two sections can be rearranged to form:
      i. a parallelogram
      ii. a right-angled triangle
      iii. a trapezium.
    b. Comment on the perimeters of the figures.

on To test your understanding and knowledge of this topic, go to your learnON title at www.jacplus.com.au and complete the **post-test**.

Answers

Topic 12 Properties of geometrical figures (Path)

12.1 Pre-test

1. $17.5°$
2. $69°$
3. Congruent triangles
4. $x = 12$, $y = 11$
5. $x = 58$, $y = -76$, $z = 308$
6. A
7. Whether they are similar or not
8. $95°$
9. $45°$
10. $72°$
11. $1080°$
12. C
13. $151°$
14. A
15. A

12.2 Angles, triangles and congruence

1. a. $a = 56°$ b. $b = 30°$ c. $c = 60°$
   d. $d = 120°$ e. $e = 68°$
2. a. $x = 115°$
   b. $y = 80°$
   c. $a = 120°$, $b = 60°$, $c = 120°$, $d = 60°$
   d. $x = 30°$
3. a. I and III, SAS b. I and II, AAS
   c. II and III, RHS d. I and II, SSS
4. a. $x = 6$, $y = 60°$ b. $x = 80°$, $y = 50°$
   c. $x = 32°$, $y = 67°$ d. $x = 45°$, $y = 45°$
5. a. $b = 65°$, $c = 10°$, $d = 50°$, $e = 130°$
   b. $f = 60°$, $g = 60°$, $h = 20°$, $i = 36°$
6. a. $x = 3$ cm
   b. $x = 85°$
   c. $x = 80°$, $y = 30°$, $z = 70°$
   d. $x = 30°$, $y = 7$ cm
   e. $x = 40°$, $y = 50°$, $z = 50°$, $m = 90°$, $n = 90°$
7. C, D
8. $x = 110°$, $y = 110°$, $z = 4$ cm, $w = 7$ cm
9. Sample responses can be found in the worked solutions in the online resources.
10. Sample responses can be found in the worked solutions in the online resources.
    a. Use SAS. b. Use SAS. c. Use ASA.
    d. Use ASA. e. Use SSS.
11. $x = 70°$
12. The third sides are not necessarily the same.
13. $x = 30°$, $y = 65°$
14. Corresponding sides are not the same.
15. Sample responses can be found in the worked solutions in the online resources.

16. Use SSS. Sample responses can be found in the worked solutions in the online resources.
17. $x = 20°$, $y = 10°$ and $z = 40°$
18. Sample responses can be found in the worked solutions in the online resources.

12.3 Similar triangles

1. a. i and iii, RHS b. i and ii, SAS
   c. i and iii, SSS d. i and iii, AAA
   e. i and ii, SSS
2. a. Triangles PQR and ABC
   b. Triangles ADB and ADC.
   c. Triangles PQR and TSR.
   d. Triangles ABC and DEC.
   e. Triangles ABC and DEC.
3. a. $\dfrac{AB}{AD} = \dfrac{BC}{DE} = \dfrac{AC}{AE}$ b. $f = 9$, $g = 8$
4. $x = 4$
5. $x = 20°$, $y = 2\dfrac{1}{4}$
6. a. AAA b. $x = 3$, $y = 4$
7. a. $x = 12$
   b. $x = 4$ cm
   c. $x = 16$
   d. $x = 8$ cm, $n = 60°$, $m = 70°$
8. a. $x = 7.5$, $y = 6.4$ b. $x = 8$
9. $x = 27$
10. a. $x = 1$, $y = 7\sqrt{2}$ b. $x = 2.5$, $y = 3.91$
11. Sample responses can be found in the worked solutions in the online resources.
12. a. $\angle ABD = \angle ABC$ (common angle)
    $\angle ADB = \angle BAC = 90°$
    $\triangle ABD \sim \triangle ACB$ (AAA)
    b. $\angle ACD = \angle BCA$ (common angle)
    $\angle ADC = \angle CAB = 90°$
    $\triangle ACD \sim \triangle ACB$ (AAA)
13. Congruent triangles must be identical; that is, the angles must be equal and the side lengths must be equal. Therefore, it is not enough just to prove that the angles are equal.
14. The taller student's shadow is 3.12 metres long.
15. The slide is 7.23 m long.
16. $\angle FEO = \angle OGH$ (alternate angles equal as EF ∥ HG)
    $\angle EFO = \angle OHG$ (alternate angles equal as EF ∥ HG)
    $\angle EOF = \angle HOG$ (vertically opposite angles equal)
    ∴ $\triangle EFO \sim \triangle GHO$ (equiangular)
17. Radius $= 0.625$ m
18. $x = 6$ or $x = 11$

12.4 Quadrilaterals

1. a. True b. True
   c. True d. False
2. a. False b. False
   c. False d. False
3. a. $x = 145°$ b. $t = 174°$
   c. $m = 66°$ d. $q = 88°$

4. a. $a = 35°, b = 65°$
   b. $c = 62°, d = 28°, e = 90°$
   c. $f = 40°, g = 140°$
   d. $h = 75°, i = 75°$

5. a. $x = 69°$ b. $x = 26°, y = 128°$

6. a. $x = 36°, y = 62°$ b. $x = 5\,\text{cm}, y = 90°$
   c. $x = 10°, y = 70°$ d. $x = 40°, y = 60°$

7. None are true, unless the trapezium is a regular trapezium, then **e** is true.

8. a, c, f

9. a, c, d, e, f

10. Parallelogram, rhombus, rectangle, square

11. Square

12. Use AAS. Sample responses can be found in the worked solutions in the online resources.

13. All the sides of a square are equal, so a square is a special rhombus. But the angles of a rhombus are not equal, so can't be a square.

14. AX || DY because ABCD is a parallelogram
    AX = DY (given)
    ∴ AXYD is a parallelogram since opposite sides are equal and parallel.

15. Use SAS. Sample responses can be found in the worked solutions in the online resources.

16. a. Use SAS. Sample responses can be found in the worked solutions in the online resources.
    b. Use SAS. Sample responses can be found in the worked solutions in the online resources.
    c. Opposite sides are equal. Sample responses can be found in the worked solutions in the online resources.

17. PS = QR (corresponding sides in congruent triangles are equal)
    PS || QR (alternate angles are equal)
    ∴ PQRS is a parallelogram since one pair of opposite sides are parallel and equal.

18. a. One pair of opposite sides are parallel.
    b. $x = 90°, y = 75°$

19. a. i. $x = \sqrt{41}$ ii. $y = \sqrt{97}$
    b. $\angle BAD = \angle BCD = 117°23'$

20. a. Rhombus, 2 cm   b. Trapezium
    c. 40°   d. Triangle BFC
    e. See worked solutions.   f. See worked solutions.
    g. $x = \left(1 + \sqrt{5}\right)$ cm

21. 70°

22. $x = \sqrt{10}\,\text{cm}$

12.5 Polygons

1. a. $m = 60°$ b. $a = 45°, b = 45°$
   c. $t = 35°$ d. $x = 10°$

2. a. $a = 85°, b = 50°, c = 45°$
   b. $d = 140°, e = 110°, f = 110°$
   c. $g = 90°, h = 110°, i = 70°$
   d. $j = 100°, k = 100°, l = 130°, m = 130°$

3. a. $y = 35°$ b. $t = 5°$ c. $n = 81°$
   d. $x = 15°$ e. $t = 30°$

4. a. i. Irregular   ii. $x = 95°$
   b. i. Irregular   ii. $p = 135°$
   c. i. Irregular   ii. $t = 36°$
   d. i. Irregular   ii. $y = 70°$
   e. i. Irregular   ii. $p = 36°$

5. a. i. 540°   ii. $b = 110°$
   b. i. 720°   ii. $c = 134°$
   c. i. 900°   ii. $d = 24°$
   d. i. 720°   ii. $h = 85°$

6. The sum of the interior angles is based on the number of sides of the polygon. The size of the exterior angle can be found by dividing 360° by the number of sides.

7. $w = 75°, x = 105°, y = 94°, z = 133°$

8. 82.5°, 82.5°, 97.5°, 97.5°

9. a. $a = 120°, b = 120°, c = 60°, d = 60°, e = 120°, f = 240°$
   b. $m = 10°, n = 270°, o = 50°$

10. 60°

11. a. Regular: all sides and interior angles are equal.
    b. Irregular: all sides and interior angles are not equal.
    c. Regular: all sides and interior angles are equal.
    d. Regular: all sides and interior angles are equal.
    e. Irregular: the sides are all equal, but the interior angles are not equal.
    f. Regular: all sides and interior angles are equal.

12. a. i. 2   ii. 5   iii. 9   iv. 14
    b. Sample responses can be found in the worked solutions in the online resources.

13. Internal angle $= 180° - \dfrac{360°}{n}$

14. a. Equilateral triangle
    b. $m = 150°$

15. a. 135°   b. 67.5°   c. 45°   d. 45°

16. a. 1080°   b. 43°

17. a. 720°   b. $x = 25°$

18. a. 1080°   b. $x = 17°$

Project

1.

2. 7.5 cm

3. $\angle FAB = 60°, \angle AFE = 60°$

4. The impossible triangle

5. Students are required to research and investigate other 'impossible' object drawn as two-dimensional shapes. These need to be created on a separate sheet of paper with reasons as to why they are termed 'impossible'.

12.6 Review questions

1. a. I and III, ASA or SAS
   b. I and II, RHS
2. a. $x = 8$ cm
   b. $x = 70°$
   c. $x = 30°, y = 60°, z = 90°$
3. Sample responses can be found in the worked solutions in the online resources.
   a. Use SAS.   b. Use ASA.
4. a. Similar, scale factor $= 1.5$
   b. Not similar
   c. Similar, scale factor $= 2$
5. a. $x = 48°, y = 4.5$ cm
   b. $x = 86°, y = 50°, z = 12$ cm
   c. $x = 60°, y = 15$ cm, $z = 12$ cm
6. Use the equiangular test. Sample responses can be found in the worked solutions in the online resources.
7. Use the equiangular test. Sample responses can be found in the worked solutions in the online resources.
8. A rhombus is a parallelogram with two adjacent sides equal in length.
9. C
10. $x + 1$
11. a. Use SAS. Sample responses can be found in the worked solutions in the online resources.
    b. Use SAS. Sample responses can be found in the worked solutions in the online resources.
    c. Use SAS. Sample responses can be found in the worked solutions in the online resources.
    d. They are all the same length.
    e. B and C
12. Use SAS. Sample responses can be found in the worked solutions in the online resources.
    PQ = PS (corresponding sides in congruent triangles are equal)
13. Rhombus, square.
14. A quadrilateral is a rhombus if:
    1. All sides are equal.
    2. The diagonals bisect each other at right angles.
    3. The diagonals bisect the angles they pass through.
15. WZ ∥ XY (co-interior angles are supplementary) and WZ = XY (given)
    ∴ WXYZ is a parallelogram since one pair of sides is parallel and equal.

16. Sample responses can be found in the worked solutions in the online resources.
17. Corresponding sides are not the same.
18.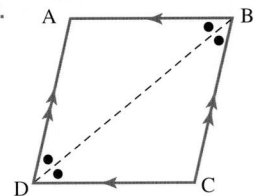

Bisect ∠BAC
AB = AC (given)
∠BAD = ∠DAC
AD is common.
∴ ΔABD ≡ ΔACD (SAS)
∴ ∠ABD = ∠ACD (corresponding sides in congruent triangles are equal)
19. Rectangle, square.
20. a. i.
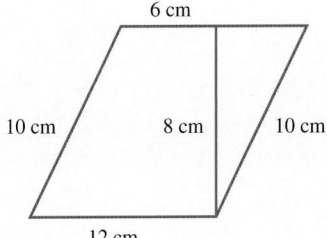

ii.
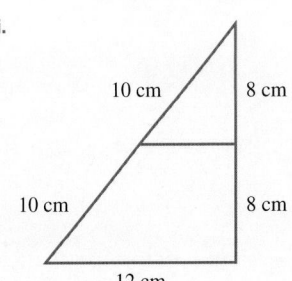

iii.
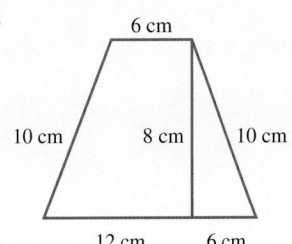

b. Perimeter of rectangle = 40 cm
   Perimeter of parallelogram = 44 cm
   Perimeter of triangle = 48 cm
   Perimeter of trapezium = 44 cm
   The triangle has the largest perimeter, while the rectangle has the smallest.

13 Data analysis

LESSON
13.1 Overview

Why learn this?

Bivariate data can be collected from all kinds of places. This includes data about the weather, data about athletic performance and data about the profitability of a business. By learning the tools you need to analyse bivariate data, you will be gaining skills that help you turn numbers (data) into powerful information that can be used to make predictions (plots).

The use of bivariate data is not limited to the classroom; in fact, many professionals rely on bivariate data to help make decisions. Some examples of bivariate data in the real world are:

- when a new drug is created, scientists will run drug trials in which they collect bivariate data about how the drug works. When the drug is approved for use, the results of the scientific analysis help guide doctors, nurses, pharmacists and patients as to how much of the drug to use and how often.
- manufacturers of products can use bivariate data about sales to help make decisions about when to make products and how many to make. For example, a beach towel manufacturer would know that they need to produce more towels for summer, and analysis of the data would help them decide how many to make.

By studying bivariate data you can learn how to use data to make predictions. By studying and understanding how these predictions work, you will be able to understand the strengths and limitations of these types of predictions.

Hey students! Bring these pages to life online

▶ Watch videos

🧩 Engage with interactivities

A+ Answer questions and check solutions

Find all this and MORE in jacPLUS ▶

Reading content and rich media, including interactivities and videos for every concept

Extra learning resources

Differentiated question sets

Questions with immediate feedback, and fully worked solutions to help students get unstuck

Exercise 13.1 Pre-test

learn**on**

1. **MC** Choose the following graphs that shows whether there is a relationship between two variables and each data value is shown as a point on a Cartesian plane.
   A. Box plot
   B. Scatterplot
   C. Dot plot
   D. Ogive

2. **MC** Select which of the following statements is incorrect.
   A. Bivariate data are data with two variables.
   B. Correlation describes the strength, the direction and the form of the relationship between two variables.
   C. The independent variable is placed on the y-axis and the dependent variable on the x-axis.
   D. The dependent variable is the one whose value depends on the other variable.

3. Data is compared from 20 students on the number of hours spent studying for an examination and the result of the examination. State if the number of hours spent studying is the independent or dependent variable.

4. Match the type of correlation with the data shown on the scatterplots.

Scatterplot	Type of correlation
a. 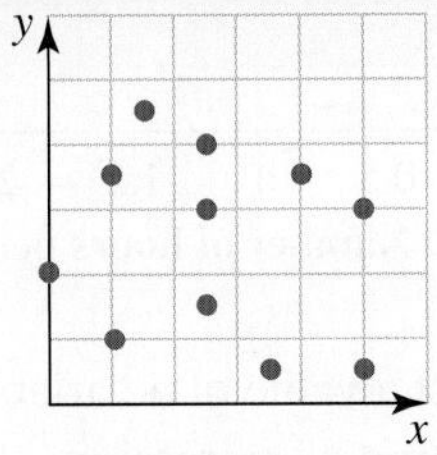	A. Strong negative linear correlation
b. 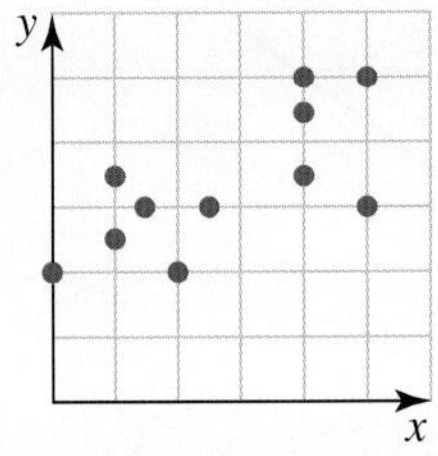	B. No correlation
c. 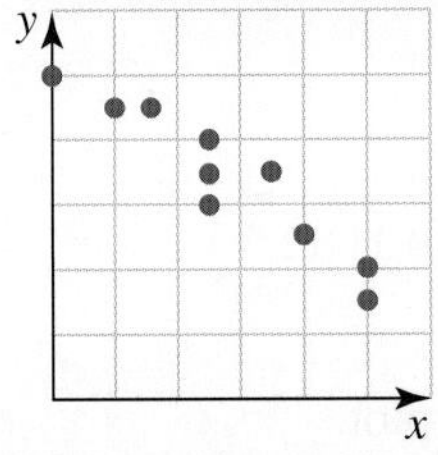	C. Weak positive linear correlation

▶

5. **MC** The table below shows the number of hours spent doing a problem-solving task for a subject and the corresponding total score for the task.

Number of hours spent on task	0	1.5	2	1	2	1.5	2.5	3	2	2.5
Task score (%)	20	50	60	45	80	70	75	97	85	20

Choose which data point is a possible outlier.

A. (0, 20) **B.** (1.5, 50) **C.** (1.5, 70) **D.** (2.5, 20)

6. **MC** Each point on the scatterplot shows the number of hours per week spent exercising by a person and their fitness level.

Choose the statement that best describes the scatterplot.

A. The more time exercising the worse the fitness level.

B. The number of hours per week spent exercising is the independent variable.

C. The correlation between the number of hours per week exercising and the fitness levels is a weak positive non-linear correlation.

D. Information has been collected from six people.

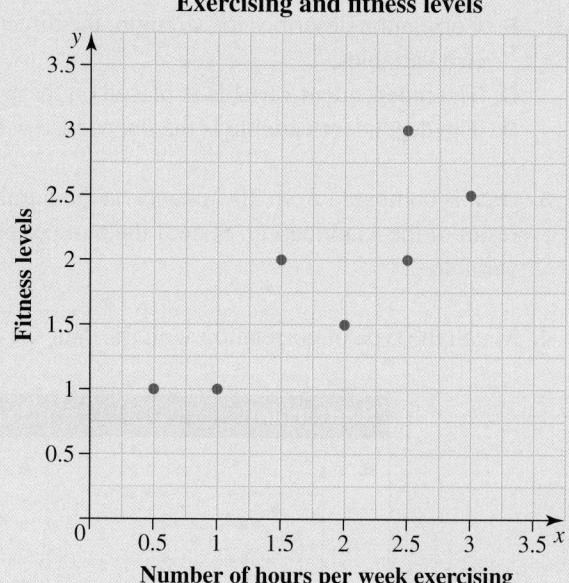

Exercising and fitness levels

7. Select a term that describes a line of best fit being used to predict a value of a variable from within a given range from the following options: extrapolation, interpolation or regression.

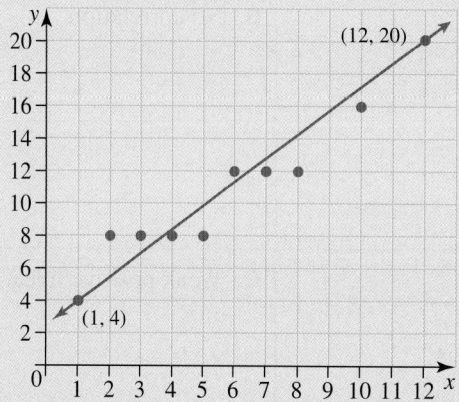

8. Determine the gradient of the line of best fit shown in the scatterplot.

9. In time series data, explain whether time is the independent or dependent variable.

10. Select another term for an independent variable from the following options: a response variable or an explanatory variable.

11. As preparation for a Mathematics test, a group of 12 students was given a revision sheet containing 60 questions. The table below shows the number of questions from the revision sheet successfully completed by each student and the mark, out of 100, of that student on the test.

Number of questions	9	12	40	25	35	29	19	44	49	20	16	58
Test result	18	21	67	50	62	54	30	70	82	37	28	99

   a. State which of the variables is dependent and which is independent.

   b. Construct a scatterplot of the data.

   c. State the type of correlation between the two variables suggested by the scatterplot and draw a corresponding conclusion.

   d. Suggest why the relationship is not perfectly linear.

12. Use the line of best fit shown on the graph to answer the following questions.

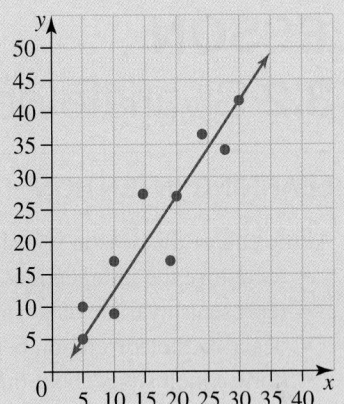

   a. Predict the value of y, when the value of x is:

      i. 10   **ii.** 35.

   b. Predict the value of x, when the value of y is:

      i. 15   **ii.** 30.

   c. Determine the equation of a line of best fit if it is known that it passes through the points $(5, 5)$ and $(20, 27)$.

   d. Use the equation of the line to algebraically verify the values obtained from the graph in parts **a** and **b**.

13. Use the given scatterplot and the line of best fit to determine the value of x when $y = 2$.

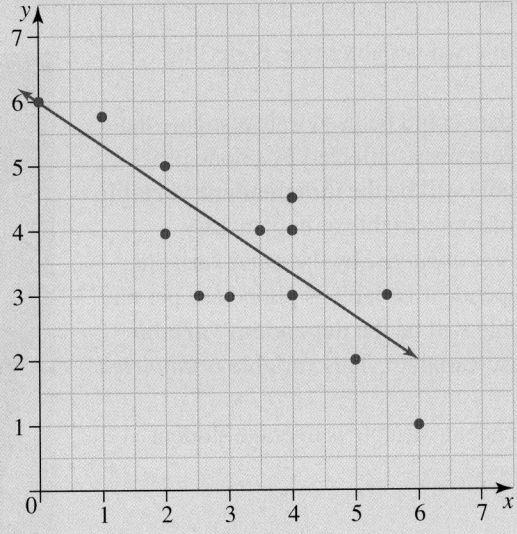

14. Use the information in the data table to answer the following questions.

Age in years (x)	7	11	8	16	9	8	14	19	17	10	20	15
Hours of television watched in a week (y)	20	19	25	55	46	50	53	67	59	25	70	58

   a. Use technology to determine the equation of the line of best fit for the following data.

   b. Use technology to predict the value of the number of hours of television watched by a person aged 15.

15. Use the given scatterplot and line of best fit to predict:
    a. the value of y when $x = 4$
    b. the value of x when $y = 1$.

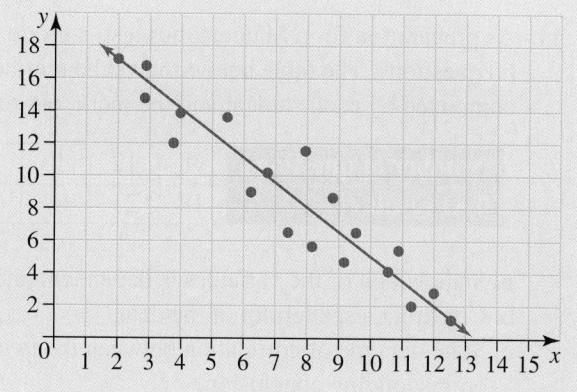

LESSON
13.2 Bivariate data

LEARNING INTENTION

At the end of this lesson you should be able to:
- recognise the independent and dependent variables in bivariate data
- represent bivariate data using a scatterplot
- describe the correlation between two variables in a bivariate data set
- draw conclusions about the correlation between two variables in a bivariate data set.

▶ 13.2.1 Bivariate data

eles-4965

- **Bivariate data** are data with two variables (the prefix 'bi' means 'two').
  - For example, bivariate data could be used to investigate the question: 'How are student marks affected by phone use?'
- In bivariate data, one variable will be the **independent variable** (also known as the experimental variable or explanatory variable). This variable *is not impacted* by the other variable.
  - In the example the independent variable is phone use per day.
- In bivariate data one variable will be the **dependent variable** (also known as the response variable). This variable *is impacted* by the other variable.
  - In the example the dependent variable is average student marks.

Scatterplots

- A **scatterplot** is a way of displaying bivariate data.
- A scatterplot will have:
  - the independent variable placed on the x-axis with a label and scale
  - the dependent variable placed on the y-axis with a label and scale
  - the data points shown on the plot (each dot is one data point).

Features of a scatterplot

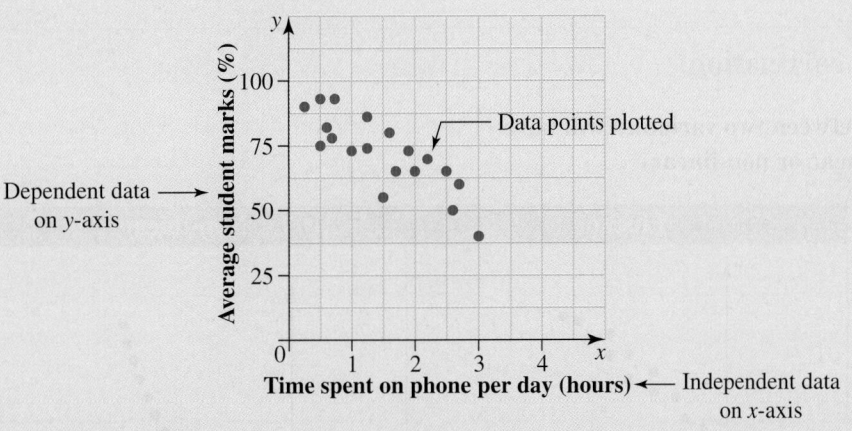

Dependent data on y-axis →

Data points plotted

Time spent on phone per day (hours) ← Independent data on x-axis

---

WORKED EXAMPLE 1 Representing bivariate data on a scatterplot

The table shows the total revenue from selling tickets for a number of different chamber music concerts. Represent the given data on a scatterplot.

Number of tickets sold	400	200	450	350	250	300	500	400	350	250
Total revenue ($)	8000	3600	8500	7700	5800	6000	11 000	7500	6600	5600

THINK

1. Determine which is the dependent variable and which is the independent variable.

2. Draw a set of axes. Label the title of the graph. Label the horizontal axis 'Number of tickets sold' and the vertical axis 'Total revenue ($)'.

3. Use an appropriate scale on the horizontal and vertical axes.

4. Plot the points on the scatterplot.

WRITE/DRAW

The total revenue depends on the number of tickets being sold, so the total revenue is the dependent variable and the number of tickets in the independent variable.

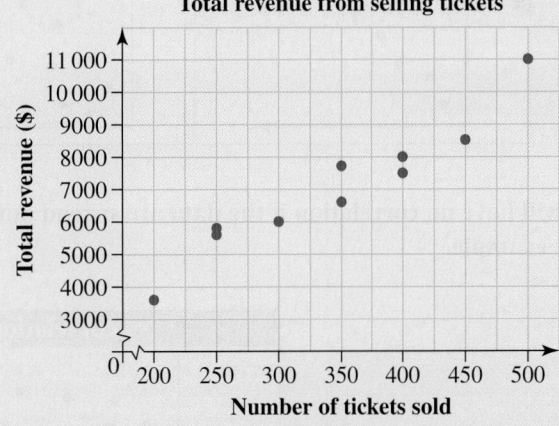

Total revenue from selling tickets

▶ 13.2.2 Correlation

eles-4966

- **Correlation** is a way of describing a connection between variables in a bivariate data set.

Describing correlation

Correlation between two variables will have:
- **a type (linear or non-linear)**

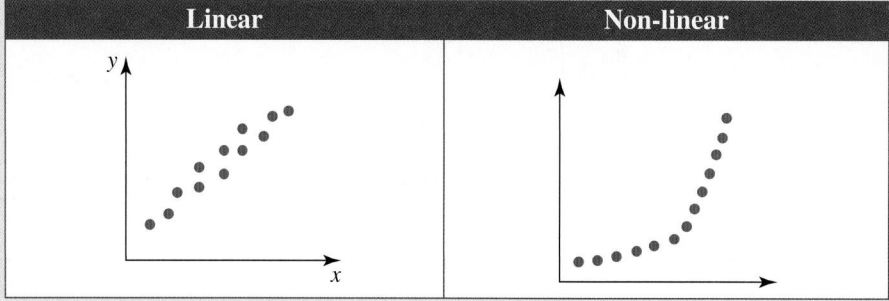

- **a direction (positive or negative)**

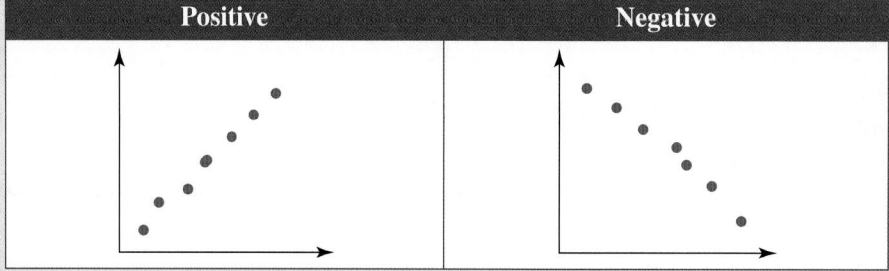

- **a strength (strong, moderate or weak).**

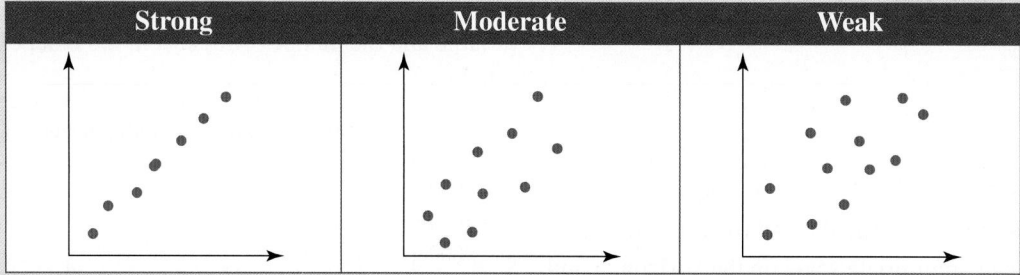

Data will have no correlation if the data are spread out across the plot with no clear pattern, as shown in this example.

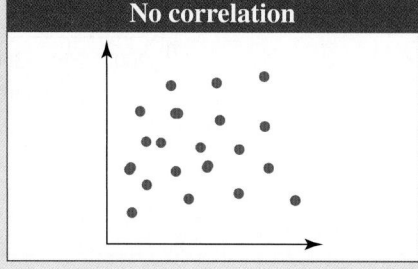

676 Jacaranda Maths Quest 10 Stage 5 NSW Syllabus Third Edition

State the type of correlation between the variables *x* and *y* shown on the scatterplot.

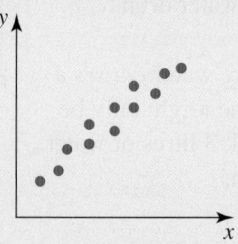

THINK	WRITE
Carefully analyse the scatterplot and comment on its form, direction and strength.	The points on the scatterplot are close together and constantly increasing therefore the relationship is linear.
	The path is directed from the bottom left corner to the top right corner and the value of *y* increases as *x* increases. Therefore the correlation is positive.
	The points are close together so the correlation can be classified as strong.
	There is a linear, positive and strong relationship between *x* and *y*.

13.2.3 Drawing conclusions from correlation

eles-4967

- When drawing a conclusion from a scatterplot, state how the independent variable appears to affect the dependent variable and explain what that means.
  - For the example here of comparing time spent on phone to average marks, a good conclusion would be:

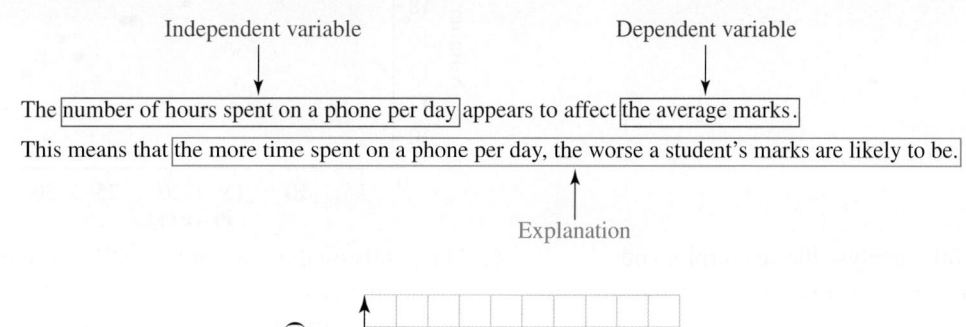

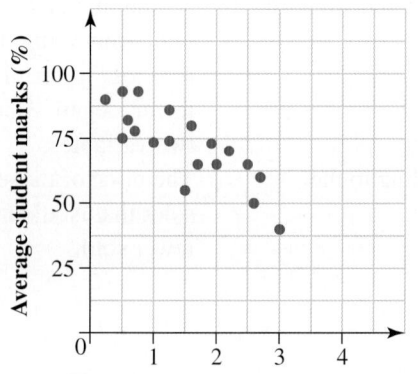

- Based on scatterplots, it is possible to draw conclusions about *correlation* but not *causation*.
  For example, for this graph:

  - it is correct to say that the amount of water drunk *appears to affect* the distance run
  - it is incorrect to say that drinking more water *causes* a person to run further, because someone might only be able to run 1 km and even if they drink 3 litres of water, they will still only be able to run 1 km.

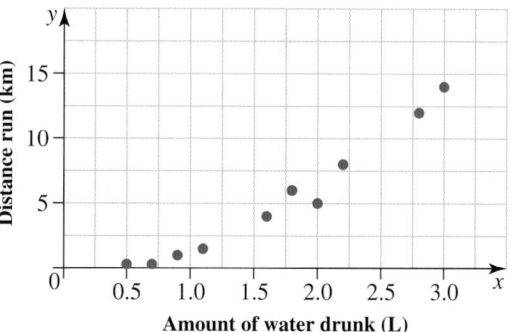

WORKED EXAMPLE 3 Stating conclusions from bivariate data

Mary sells business shirts in a department store. She always records the number of different styles of shirt sold during the day.
The table below shows her sales over one week.

Price ($)	14	18	20	21	24	25	28	30	32	35
Number of shirts sold	21	22	18	19	17	17	15	16	14	11

a. **Construct a scatterplot of the data.**
b. **State the type of correlation between the two variables and, hence, draw a corresponding conclusion.**

THINK

a. Draw the scatterplot showing 'Price ($)' (independent variable) on the horizontal axis and 'Number of shirts sold' (dependent variable) on the vertical axis.

WRITE/DRAW

a.

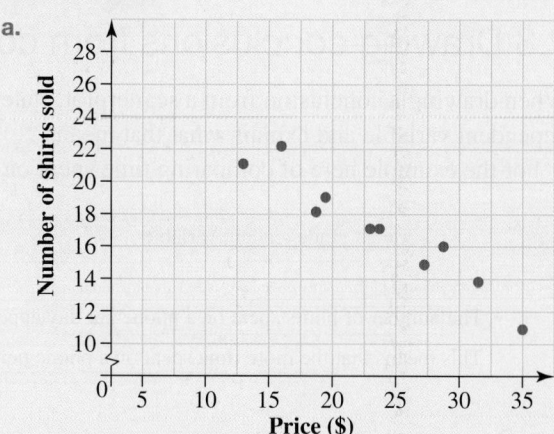

b. 1. Carefully analyse the scatterplot and comment on its form, direction and strength.

b. The points on the plot form a path that resembles a straight, narrow band, directed from the top left corner to the bottom right corner. The points are close to forming a straight line. There is a linear, negative and strong correlation between the two variables.

2. Draw a conclusion corresponding to the analysis of the scatterplot.

The price of the shirt appears to affect the number sold; that is, the more expensive the shirt the fewer sold.

DIGITAL TECHNOLOGY

Bivariate data can be displayed in a scatterplot using graphical applications and spreadsheets.

To display a scatterplot using technology:
1. First input the data with the independent variable in the first column and the dependent variable in the second column.

	A	B
1	Time	Distance
2	1	3.11
3	2	4.73
4	3	6.08
5	4	7.54
6		

2. Use the graphical application or spreadsheet to create a scatterplot. Highlight all of your data (including headings).

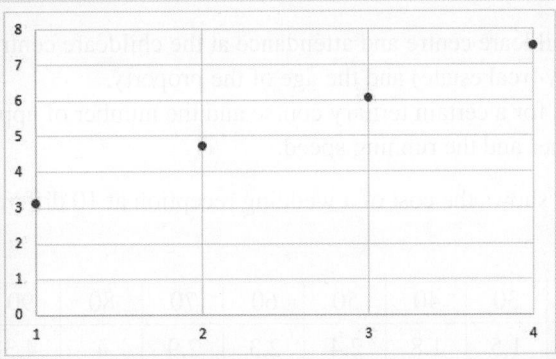

For Google Sheets:
Go to **Insert** and select **Chart**.
For Excel:
Go to **Insert**, select the scatterplot icon and choose the **Scatter** option.

DISCUSSION

How could you determine whether the change in one variable causes the change in another variable?

Individual pathways

■ PRACTISE	■ CONSOLIDATE	■ MASTER
1, 4, 5, 7, 9, 11, 13	2, 6, 8, 12, 14	3, 10, 15, 16

Fluency

For questions **1** and **2**, decide which of the variables is independent and which is dependent.

1. a. Number of hours spent studying for a Mathematics test and the score on that test.
   b. Daily amount of rainfall (in mm) and daily attendance at the Botanical Gardens.
   c. Number of hours per week spent in a gym and the annual number of visits to the doctor.
   d. The amount of computer memory taken by an essay and the length of the essay (in words).

2. a. The cost of care in a childcare centre and attendance at the childcare centre.
   b. The cost of the property (real estate) and the age of the property.
   c. The entry requirements for a certain tertiary course and the number of applications for that course.
   d. The heart rate of a runner and the running speed.

3. **WE1** The following table shows the cost of a wedding reception at 10 different venues. Represent the data on a scatterplot.

No of guests	30	40	50	60	70	80	90	100	110	120
Total cost ($\times$ $1000)	1.5	1.8	2.4	2.3	2.9	4	4.3	4.5	4.6	4.6

4. **WE2** State the type of relationship between x and y for each of the following scatterplots.

a.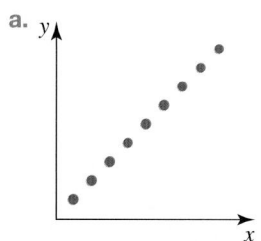

b.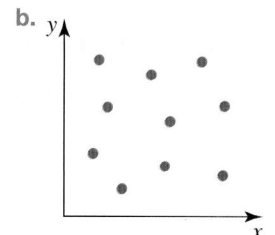

c.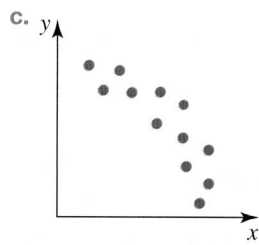

d.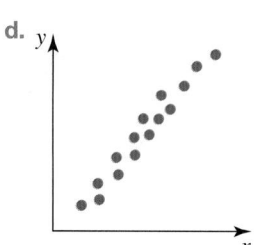

e.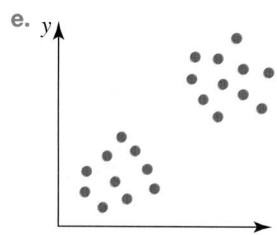

5. State the type of relationship between x and y for each of the following scatterplots.

a.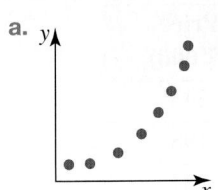

b.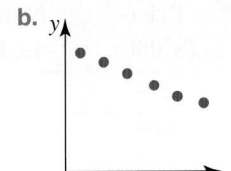

c.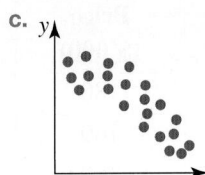

d.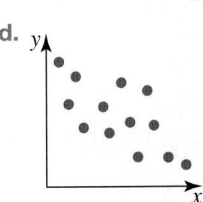

e.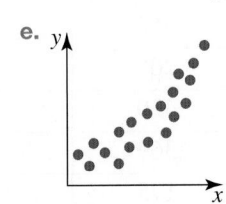

6. State the type of relationship between x and y for each of the following scatterplots.

a.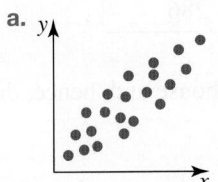

b.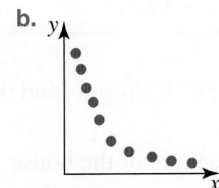

c.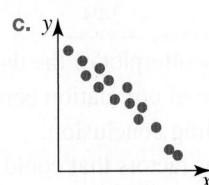

d.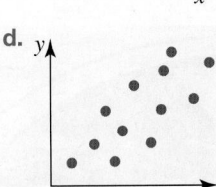

e.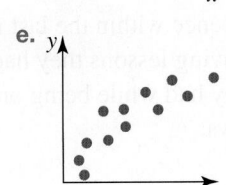

Understanding

7. **WE3** Eugene is selling leather bags at the local market. During the day he keeps records of his sales. The table below shows the number of bags sold over one weekend and their corresponding prices (to the nearest dollar).

Price ($) of a bag	30	35	40	45	50	55	60	65	70	75	80
Number of bags sold	10	12	8	6	4	3	4	2	2	1	1

a. Construct a scatterplot of the data.

b. State the type of correlation between the two variables and, hence, draw a corresponding conclusion.

8. The table below shows the number of questions solved by each student on a test, and the corresponding total score on that test.

Number of questions	2	0	7	10	5	2	6	3	9	4	8	3	6
Total score (%)	22	39	69	100	56	18	60	36	87	45	84	32	63

a. Construct a scatterplot of the data.

b. Suggest the type of correlation shown in the scatterplot.

c. Give a possible explanation as to why the scatterplot is not perfectly linear.

9. The table below shows the number of bedrooms and the price of each of 30 houses.

Number of bedrooms	Price ($'000)	Number of bedrooms	Price ($'000)	Number of bedrooms	Price ($'000)
2	180	3	279	3	243
2	160	2	195	3	198
3	240	6	408	3	237
2	200	4	362	2	226
2	155	2	205	4	359
4	306	7	420	4	316
3	297	5	369	2	200
5	383	1	195	2	158
2	212	3	265	1	149
4	349	2	174	3	286

a. Construct a scatterplot of the data.

b. State the type of correlation between the number of bedrooms and the price of the house and, hence, draw a corresponding conclusion.

c. Suggest other factors that could contribute to the price of the house.

10. A sample of 25 drivers who had obtained a full licence within the last month was asked to recall the approximate number of driving lessons they had taken (to the nearest 5), and the number of accidents they had while being on P plates. The results are summarised in the table that follows.

a. Represent these data on a scatterplot.
b. Specify the relationship suggested by the scatterplot.
c. Suggest some reasons why this scatterplot is not perfectly linear.

Number of lessons	5	20	15	25	10	35	5	15	10	20	40	25	10
Number of accidents	6	2	3	3	4	0	5	1	3	1	2	2	5
Number of lessons	5	20	40	25	30	15	35	5	30	15	20	10	
Number of accidents	5	3	0	4	1	4	1	4	0	2	3	4	

Communicating, reasoning and problem solving

11. **MC** The scatterplot that best represents the relationship between the amount of water consumed daily by a certain household for a number of days in summer and the daily temperature is:

A.
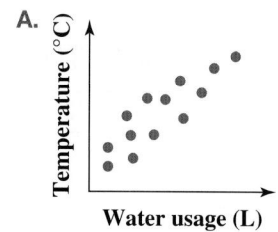

B.
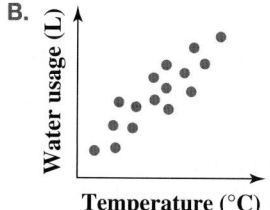

C.
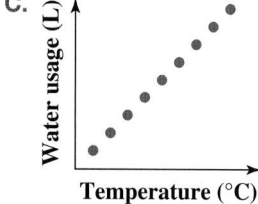

D.
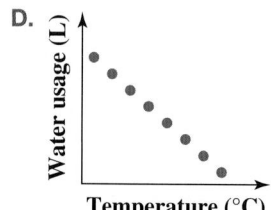

12. **MC** The scatterplot shows the number of sides and the sum of interior angles for a number of polygons.
Select the statement that is NOT true of the following statements.

A. The correlation between the number of sides and the angle sum of the polygon is perfectly linear.
B. The increase in the number of sides causes the increase in the size of the angle sum.
C. The number of sides depends on the sum of the angles.
D. The correlation between the two variables is positive.

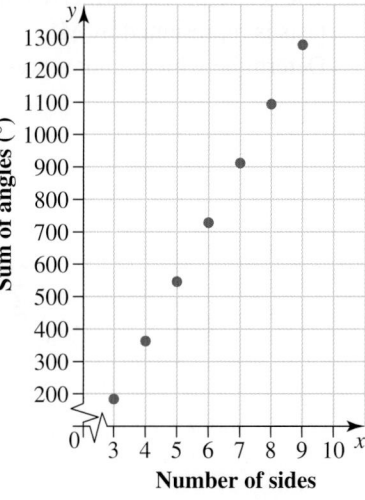

13. **MC** After studying a scatterplot, it was concluded that there was evidence that the greater the level of one variable, the smaller the level of the other variable. The scatterplot must have shown a:

A. strong, positive correlation
B. strong, negative correlation
C. moderate, positive correlation
D. moderate, negative correlation

14. The table below gives the number of kicks and handballs obtained by the top 8 players in an AFL game.

Player	A	B	C	D	E	F	G	H
Number of kicks	20	27	21	19	17	18	21	22
Number of handballs	11	3	11	6	5	1	9	7

a. Represent this information on a scatterplot by using the x-axis as the number of kicks and the y-axis as the number of handballs.
b. State whether the scatterplot supports the claim that the more kicks a player obtains, the more handballs they give.

15. Each point on the scatterplot shows the time (in weeks) spent by a person on a healthy diet and the corresponding mass lost (in kg).
Study the scatterplot and state whether each of the following statements is true or false.

a. The number of weeks that the person stays on a diet is the independent variable.
b. The y-coordinates of the points represent the time spent by a person on a diet.
c. There is evidence to suggest that the longer the person stays on a diet, the greater the loss in mass.
d. The time spent on a diet is the only factor that contributes to the loss in mass.
e. The correlation between the number of weeks on a diet and the number of kilograms lost is positive.

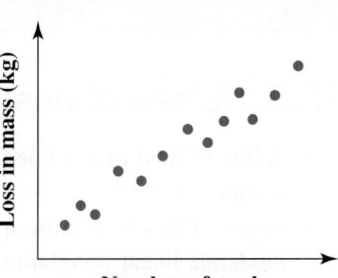

16. The scatterplot shown gives the marks obtained by students in two mathematics tests. Mardi's score in the tests is represented by M.
Determine which point represents each of the following students.

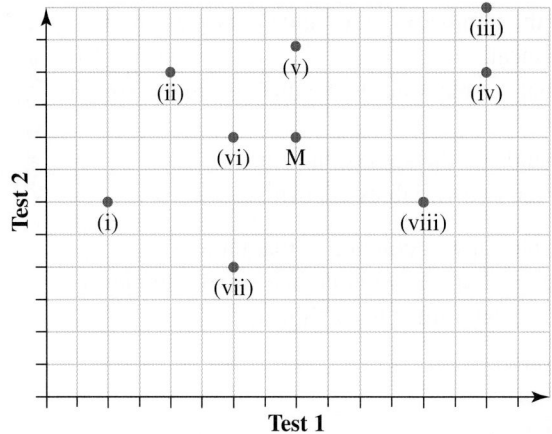

a. Saanvi, who got the highest mark in both tests.
b. Duong, who got the top mark in test 1 but not in test 2.
c. Charlotte, who did better on test 1 than Mardi but not as well in test 2.
d. Dario, who did not do as well as Charlotte in both tests.
e. Edward, who got the same mark as Mardi in test 2 but did not do so well in test 1.
f. Rory, who got the same mark as Mardi for test 1 but did better than her for test 2.
g. Amelia, who was the lowest in test 1.
h. Harrison, who had the greatest discrepancy between his two marks.

LESSON
13.3 Lines of best fit by eye

LEARNING INTENTION

At the end of this lesson you should be able to:
- draw a line of best fit by eye
- determine the equation of the line of best fit
- use the line of best fit to make interpolation or extrapolation predictions.

▶ 13.3.1 Lines of best fit by eye

eles-4968

- A **line of best fit** is a line that follows the trend of the data in a scatterplot.
- A line of best fit is most appropriate for data with strong or moderate linear correlation.
- Drawing lines of best fit by eye is done by placing a line that:
  - represents the data trend
  - usually has an equal number of points above and below the line.

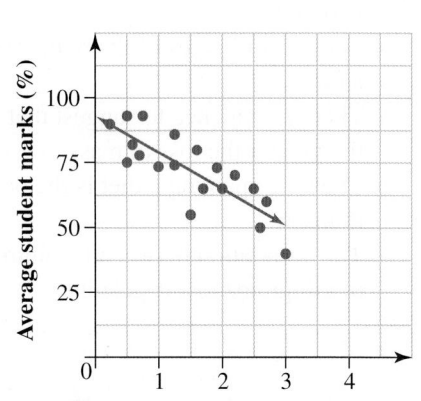

Determine the equation of a line of best fit by eye

To determine the equation of the line of best fit, follow these steps:
1. Choose two points on the line.
   Note: It is best to use two data points on the line if possible.
2. Write the points in the coordinate form (x_1, y_1) and (x_2, y_2).
3. Calculate the gradient using $m = \dfrac{y_2 - y_1}{x_2 - x_1}$.
4. Write the equation in the form $y = mx + c$ using the m found in step 3.
5. Substitute one coordinate into the equation and rearrange to determine the value of c.
6. Write the final equation, replacing x and y if needed.

WORKED EXAMPLE 4 Determining the equation for a line of best fit

The data in the table shows the cost of mobile data roaming for a number of different mobile providers, based on hours used per month.

Hours used per month	10	12	20	18	10	13	15	17	14	11
Total monthly cost ($)	15	18	30	32	18	20	22	23	22	18

a. Construct a scatterplot of the data.
b. Draw the line of best fit by eye.
c. Determine the equation of the line of best fit in terms of the variables n (number of hours) and C (monthly cost).

THINK

a. Draw the scatterplot placing the independent variable (hours used per month) on the horizontal axis and the dependent variable (total monthly cost) on the vertical axis. Label the axes.

WRITE/DRAW

a.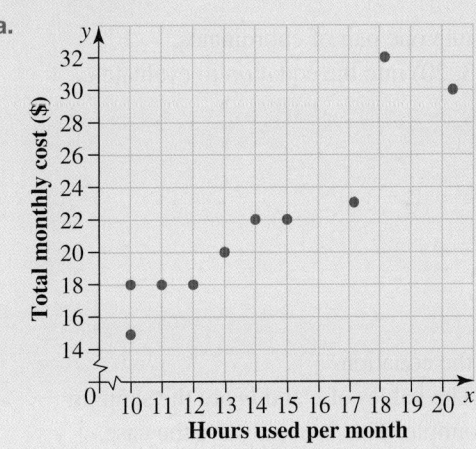

b. 1. Carefully analyse the scatterplot.

2. Position the line of best fit so there is approximately an equal number of data points on either side of the line and so that all points are close to the line.
Note: With the line of best fit, there is no single definite solution.

b.

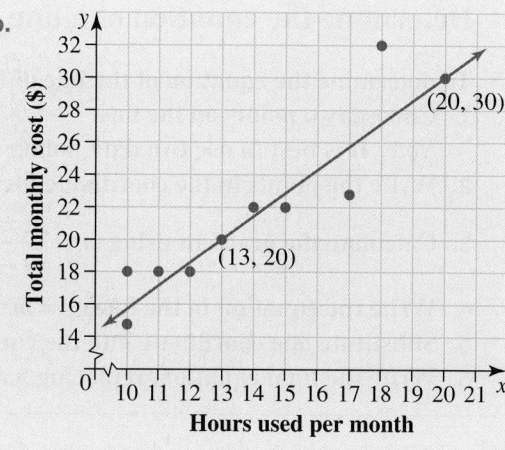

c. 1. Select two points on the line that are not too close to each other.

2. Calculate the gradient of the line.

3. Write the rule for the equation of a straight line.

4. Substitute the known values into the equation.

5. Substitute one pair of coordinates, say (13, 20) into the equation to evaluate c.

6. Write the equation.
Note: The values of c and m are the same in this example. This is not always the case.

7. Replace x with n (number of hours used) and y with C (the total monthly cost) as required.

c. Let $(x_1, y_1) = (13, 20)$ and $(x_2, y_2) = (20, 30)$.

$$m = \frac{y_2 - y_1}{x_2 - x_1}$$

$$m = \frac{30 - 20}{20 - 13}$$

$$= \frac{10}{7}$$

$$y = mx + c$$

$$y = \frac{10}{7}x + c$$

$$20 = \frac{10}{7}(13) + c$$

$$c = 20 - \frac{130}{7}$$

$$= \frac{140 - 130}{7}$$

$$= \frac{10}{7}$$

$$y = \frac{10}{7}x + \frac{10}{7}$$

$$C = \frac{10}{7}n + \frac{10}{7}$$

⏵ 13.3.2 Predictions using lines of best fit

eles-4969

- Predictions can be made by using the line of best fit.
- To make a prediction, use one coordinate then determine the other coordinate by using:
  - the line of best fit
  - the equation of the line of best fit.
- Predictions will be made using:
  - **interpolation** if the prediction sits within the given data
  - **extrapolation** if the prediction sits outside the given data.

Interpolation vs extrapolation

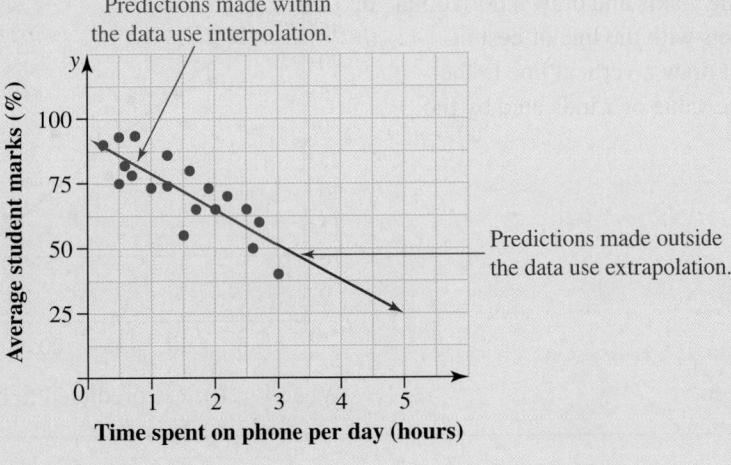

Predictions made within the data use interpolation.

Predictions made outside the data use extrapolation.

- Predictions will be more reliable if they are made:
  - using interpolation
  - from data with a strong correlation
  - from a large number of data.

WORKED EXAMPLE 5 Making predictions using the line of best fit

Use the given scatterplot and line of best fit to predict:
a. the value of y when $x = 10$
b. the value of x when $y = 10$.

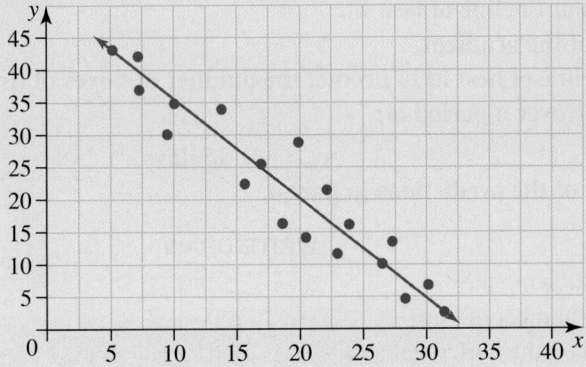

THINK

a. 1. Locate 10 on the x-axis and draw a vertical line until it meets with the line of best fit. From that point, draw a horizontal line to the y-axis. Read the value of y indicated by the horizontal line.

WRITE/DRAW

a.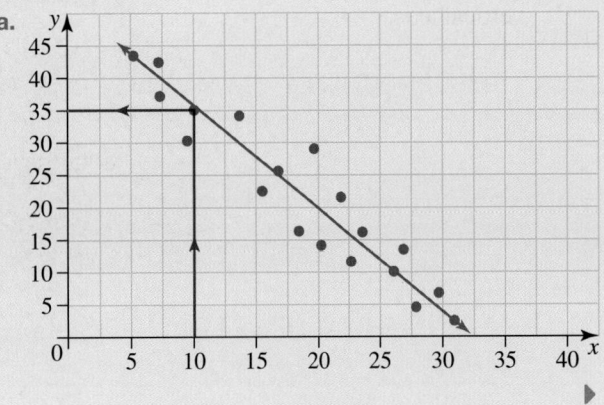

2. Write the answer.   When $x = 10$, y is predicted to be 35.

b. 1. Locate 10 on the y-axis and draw a horizontal line until it meets with the line of best fit. From that point draw a vertical line to the x-axis. Read the value of x indicated by the vertical line.

b.

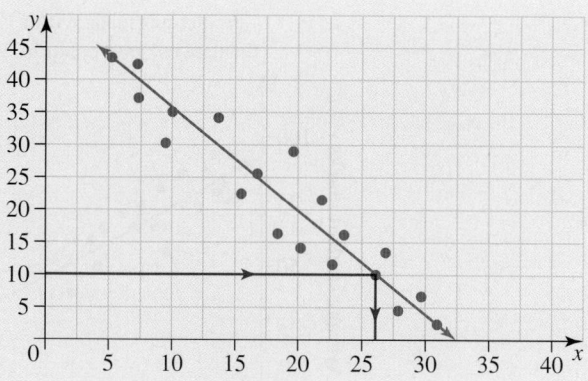

2. Write the answer.   When $y = 10$, x is predicted to be 27.

WORKED EXAMPLE 6 Interpreting meaning and making predictions

The table below shows the number of boxes of tissues purchased by hay fever sufferers and the number of days affected by hay fever during the blooming season in spring.

Number of days affected by hay fever (d)	3	12	14	7	9	5	6	4	10	8
Total number of boxes of tissues purchased (T)	1	4	5	2	3	2	2	2	3	3

a. Construct a scatterplot of the data and draw a line of best fit.
b. Determine the equation of the line of best fit.
c. Interpret the meaning of the gradient.
d. Use the equation of the line of best fit to predict the number of boxes of tissues purchased by people suffering from hay fever over a period of:

    i. 11 days   ii. 15 days.

e. Compare the reliability of the predictions in part d.

THINK

a. 1. Draw the scatterplot showing the independent variable (number of days affected by hay fever) on the horizontal axis and the dependent variable (total number of boxes of tissues purchased) on the vertical axis.

WRITE/DRAW

a.

[Scatterplot: Total no. boxes of tissues purchased (T, vertical axis, 0–5) vs No. days affected by hay fever (d, horizontal axis, 3–14)]

2. Position the line of best fit on the scatterplot so there is approximately an equal number of data points on either side of the line.

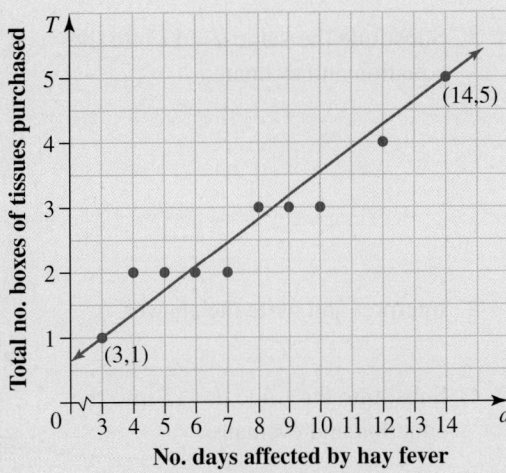

b. 1. Select two points on the line that are not too close to each other.

b. Let $(x_1, y_1) = (3, 1)$ and $(x_2, y_2) = (14, 5)$.

2. Calculate the gradient of the line.

$$m = \frac{y_2 - y_1}{x_2 - x_1}$$

$$m = \frac{5 - 1}{14 - 3} = \frac{4}{11}$$

3. Write the rule for the equation of a straight line.

$$y = mx + c$$

4. Substitute the known values into the equation, say (3, 1), into the equation to calculate c.

$$y = \frac{4}{11}x + c$$

$$1 = \frac{4}{11}(3) + c$$

$$c = 1 - \frac{12}{11}$$

$$= \frac{-1}{11}$$

5. Write the equation.

$$y = \frac{4}{11}x - \frac{1}{11}$$

6. Replace x with d (number of days with hay fever) and y with T (total number of boxes of tissues used) as required.

$$T = \frac{4}{11}d - \frac{1}{11}$$

c. Interpret the meaning of the gradient of the line of best fit.

c. The gradient indicates an increase in sales of tissues as the number of days affected by hay fever increases. A hay fever sufferer is using on average $\frac{4}{11}$ (or about 0.36) of a box of tissues per day.

d. i. 1. Substitute the value $d = 11$ into the equation and evaluate.

d. i. When $d = 11$,

$$T = \frac{4}{11} \times 11 - \frac{1}{11}$$

$$= 4 - \frac{1}{11}$$

$$= 3\frac{10}{11}$$

2. Interpret and write the answer.

In 11 days the hay fever sufferer will need 4 boxes of tissues.

ii. 1. Substitute the value $d = 15$ into the equation and evaluate.

ii. When $d = 15$,

$$T = \frac{4}{11} \times 15 - \frac{1}{11}$$

$$= \frac{60}{11} - \frac{1}{11}$$

$$= 5\frac{4}{11}$$

2. Interpret and write the answer.

In 15 days the hay fever sufferer will need about 6 boxes of tissues.

e. 1. Consider how the predictions have been made.

e. 1. The predictions have been made using a line of best fit for data with strong correlation and a large number of data. These are criteria for more reliable predictions. The prediction for part **i** used interpolation, and the one for part **ii** used extrapolation.

2. Interpret and write the answer.

2. The prediction made for 11 days, using interpolation, is more reliable than the one made for 15 days, using extrapolation.

DISCUSSION

In your own words, explain whether extrapolated values can be considered reliable.

on Resources

Interactivities Lines of best fit (int-6180)

Interpolation and extrapolation (int-6181)

Exercise 13.3 Lines of best fit by eye

13.3 Quick quiz **on**	13.3 Exercise

Individual pathways

■ PRACTISE	■ CONSOLIDATE	■ MASTER
1, 2, 5, 8, 11	3, 6, 9, 13	4, 7, 10, 12

Fluency

1. **WE4** The data in the table shows the distances travelled by 10 cars and the amount of petrol used for their journeys (to the nearest litre).

Distance travelled, d (km)	52	36	83	12	44	67	74	23	56	95
Petrol used, P (L)	7	5	9	2	7	9	12	3	8	14

   a. Construct a scatterplot of the data and draw the line of best fit.
   b. Determine the equation of the line of best fit in terms of the variables d (distance travelled) and P (petrol used).

2. **WE5** Use the given scatterplot and line of best fit to predict:

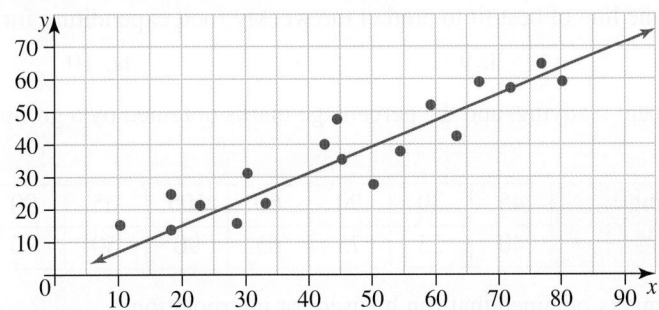

   a. the value of y when $x = 45$
   b. the value of x when $y = 15$.

3. Analyse the following graph.

   a. Use the line of best fit to estimate the value of y when the value of x is:

      i. 7   ii. 22   iii. 36.

   b. Use the line of best fit to estimate the value of x when the value of y is:

      i. 120   ii. 260   iii. 480.

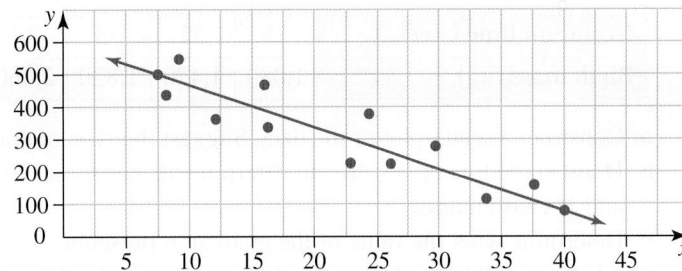

   c. Determine the equation of the line of best fit, if it is known that it passes through the points $(5, 530)$ and $(40, 75)$.
   d. Use the equation of the line to verify the values obtained from the graph in parts **a** and **b**.

4. A sample of ten Year 10 students who have part-time jobs was randomly selected. Each student was asked to state the average number of hours they work per week and their average weekly earnings (to the nearest dollar). The results are summarised in the table below.

Hours worked, h	4	8	15	18	10	5	12	16	14	6
Weekly earnings, E ($)	23	47	93	122	56	33	74	110	78	35

a. Construct a scatterplot of the data and draw the line of best fit.
b. Write the equation of the line of best fit, in terms of variables h (hours worked) and E (weekly earnings).
c. Interpret the meaning of the gradient.

Understanding

5. **WE6** The following table shows the average weekly expenditure on food for households of various sizes.

Number of people in a household	1	2	4	7	5	4	3	5
Cost of food ($ per week)	70	100	150	165	150	140	120	155
Number of people in a household	2	4	6	5	3	1	4	
Cost of food ($ per week)	90	160	160	160	125	75	135	

a. Construct a scatterplot of the data and draw in the line of best fit.
b. Determine the equation of the line of best fit. Write it in terms of variables n (for the number of people in a household) and C (weekly cost of food).
c. Interpret the meaning of the gradient.
d. Use the equation of the line of best fit to predict the weekly food expenditure for a family of:

   i. 8   ii. 9   iii. 10.

6. The number of hours spent studying, and the percentage marks obtained by a group of students on a test are shown in this table.

Hours spent studying	45	30	90	60	105	65	90	80	55	75
Marks obtained	40	35	75	65	90	50	90	80	45	65

a. State the values for marks obtained that can be used for interpolation.
b. State the values for hours spent studying that can be used for interpolation.

7. The following table shows the gestation time and the birth mass of 10 babies.

Gestation time (weeks)	31	32	33	34	35	36	37	38	39	40
Birth mass (kg)	1.080	1.470	1.820	2.060	2.230	2.540	2.750	3.110	3.080	3.370

a. Construct a scatterplot of the data. Suggest the type of correlation shown by the scatterplot.
b. Draw in the line of best fit and determine its equation. Write it in terms of the variables t (gestation time) and M (birth mass).
c. Determine what the value of the gradient represents.
d. Although full term of gestation is considered to be 40 weeks, some pregnancies last longer. Use the equation obtained in part **b** to predict the birth mass of babies born after 41 and 42 weeks of gestation.
e. Many babies are born prematurely. Using the equation obtained in part **b**, predict the birth mass of a baby whose gestation time was 30 weeks.
f. Calculate their gestation time (to the nearest week), if the birth mass of the baby was 2.390 kg.

Communicating, reasoning and problem solving

8. **MC** Consider the scatterplot shown.

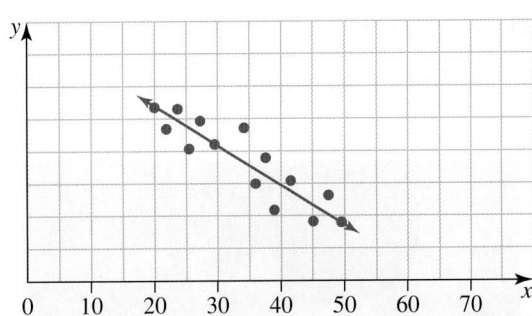

The line of best fit on the scatterplot is used to predict the values of y when $x = 15$, $x = 40$ and $x = 60$.

a. Interpolation would be used to predict the value of y when the value of x is:

   A. 15 and 40   B. 15 and 60   C. 15 only   D. 40 only

b. The prediction of the y-value(s) can be considered reliable when:

   A. $x = 15$ and $x = 40$ B. $x = 15$, $x = 40$ and $x = 60$
   C. $x = 40$ D. $x = 40$ and $x = 60$

9. **MC** The scatterplot below is used to predict the value of y when $x = 300$. This prediction is:

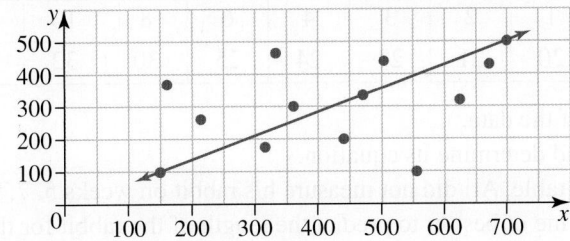

   A. reliable, because it is obtained using interpolation
   B. not reliable, because only x-values can be predicted with confidence
   C. reliable because the scatterplot contains a large number of points
   D. not reliable, because there is no correlation between x and y

10. As a part of her project, Rachel is growing a crystal. Every day she measures the crystal's mass using special laboratory scales and records it.
The table below shows the results of her experiment.

Day number	1	2	3	4	5	8	9	10	11	12	15	16
Mass (g)	2.5	3.7	4.2	5.0	6.1	8.4	9.9	11.2	11.6	12.8	16.1	17.3

Measurements on days 6, 7, 13 and 14 are missing, since these were 2 consecutive weekends and, hence, Rachel did not have a chance to measure her crystal, which is kept in the school laboratory.

a. Construct the scatterplot of the data and draw in the line of best fit.
b. Determine the equation of the line of best fit. Write the equation, using variables d (day of the experiment) and M (mass of the crystal).
c. Interpret the meaning of the gradient.
d. For her report, Rachel would like to fill in the missing measurements (that is, the mass of the crystal on days 6, 7, 13 and 14). Use the equation of the line of best fit to help Rachel determine these measurements.
Explain whether this is an example of interpolation or extrapolation.

e. Rachel needed to continue her experiment for 2 more days, but she fell ill and had to miss school. Help Rachel to predict the mass of the crystal on those two days (that is, days 17 and 18), using the equation of the line of best fit.
Explain whether these predictions are reliable.

11. Ari was given a baby rabbit for his birthday. To monitor the rabbit's growth, Ari decided to measure it once a week.

The table below shows the length of the rabbit for various weeks.

Week number, n	1	2	3	4	6	8	10	13	14	17	20
Length, l (cm)	20	21	23	24	25	30	32	35	36	37	39

a. Construct a scatterplot of the data.
b. Draw a line of best fit and determine its equation.
c. As can be seen from the table, Ari did not measure his rabbit on weeks 5, 7, 9, 11, 12, 15, 16, 18 and 19. Use the equation of the line of best fit to predict the length of the rabbit for those weeks.
d. Explain whether the predictions made in part **c** were an example of interpolation or extrapolation.
e. Predict the length of the rabbit in the next three weeks (that is, weeks 21–23), using the line of best fit from part **c**.
f. Explain whether the predictions that have been made in part **e** are reliable.

12. Sam has a mean score of 88 per cent for his first nine tests of the semester. In order to receive an A$^+$ his score must be 90 per cent or higher. There is one test remaining for the semester.
Explain whether it is possible for him to receive an A$^+$.

13. Laurie is training for the long jump, hoping to make the Australian Olympic team. Their best jump each year is shown in the table below.

Age, a (years)	Best jump, B (metres)
8	4.31
9	4.85
10	5.29
11	5.74
12	6.05
13	6.21
14	—
15	6.88
16	7.24
17	7.35
18	7.57

a. Plot the points generated by the table on a scatterplot.
b. Join the points generated with straight line segments.
c. Draw a line of best fit and determine its equation.
d. The next Olympic Games will occur when Laurie is 20 years old. Use the equation of the line of best fit to estimate Laurie's best jump that year and whether it will pass the qualifying mark of 8.1 metres.
e. Explain whether a line of best fit is a good way to predict future improvement in this situation. State the possible problems are there with using a line of best fit.
f. Olympic Games will also be held when Laurie is 24 years old and 28 years old. Using extrapolation, what length would you predict Laurie could jump at these two ages? Discuss whether this is realistic.
g. When Laurie was 14, they twisted a knee in training and did not compete for the whole season. In that year, a national junior championship was held. The winner of that championship jumped 6.5 metres. Use your line of best fit to predict whether Laurie would have won that championship.

LESSON
13.4 Evaluating statistical reports (Path)

LEARNING INTENTION

At the end of this lesson you should be able to:
- identify key factors in deciding which type of data collection method is appropriate
- evaluate whether or not the data has been represented accurately and fairly
- interpret a range of graphs for the purpose of detailed analysis.

▶ 13.4.1 Data collection methods

eles-6295

- Statistical investigations involve collecting data, recording the data, analysing the data and reporting the results.
- Collection methods involve gathering primary data, or using secondary data from stored records.
- Primary data can be collected by observation, digital footprint, measurement, survey, experiment or simulation.
- Secondary data can be collected electronically or by reading records.
- It is important to be able to justify the particular method chosen for each of these processes.
- Sometimes alternative methods are just as appropriate.

WORKED EXAMPLE 7 Choosing an appropriate data collection method

You have been given an assignment to investigate which year level uses the school library, after school, the most. The school does not keep records on library use.

a. Explain whether it is more appropriate to use primary or secondary data in this case. Justify your choice.

b. Describe how the data could be collected. Discuss any problems which might be encountered.

c. Explain whether an alternative method would be just as appropriate.

THINK	WRITE
a. No records have been kept on library use.	a. Since records are not kept on library use, secondary data is not an option.
b. The data can be collected via a questionnaire	b. A questionnaire could be designed online and sent out to all students. This would be a way of collecting data that could be easily collated. The difficulty would be how to ensure enough people responded.
c. The data could be collected in person by standing at the school gate on a designated day.	c. Observation could be used to personally interview students as they left the school. This would take more time, and could not ensure all students were included.

WORKED EXAMPLE 8 Selecting data collection methods

Which method would be the most appropriate to collect the following data? Suggest an alternative method in each case.
a. The number of cars parked in the staff car park each day.
b. The mass of books students carry to school each day.
c. The length a spring stretches when weights are added to it.
d. The cost of mobile phone plans with various network providers.

THINK	WRITE
a. Observation	**a.** The best way would probably be observation of the staff car park to count the number of cars there. An alternative method would be to conduct a census of all workers to ask if they parked in the staff car park. This is probably not as accurate as the direct observation method.
b. Measurement	**b.** The mass of the books could be measured by weighing each student's pack on scales. A random sample would probably yield a reasonably accurate result.
c. Experiment	**c.** Conduct an experiment and measure the extension of the spring with various weights. There is probably no alternative to this method.
d. Internet search	**d.** An internet search would enable data to be collected. Alternatively, a visit to multiple mobile phone outlets would yield similar results.

13.4.2 Analysing the data

eles-6296

- Once the data have been collected and collated, a decision must be made with regard to the best methods for analysing the data. Possible methods include:
    - a measure of central tendency — mean, median or mode
    - a measure of spread — range, interquartile range or standard deviation
    - a measure of skewness
    - an appropriate graph.

Statistical graphs

- Data can be graphed in a variety of ways, such as line graphs, bar graphs, histograms, stem plots and box plots. These have all been discussed in detail previously.
- In media reports it is common to see line and bar graphs.
- Graphs can summarise a lot of data into one visual representation. It can therefore be tempting to not study them in great detail. However, graphs can be misleading.
- It is easy to manipulate a graph to create a biased impression. This is achieved by careful choice of scale on the horizontal and vertical axes.
    - Shortening the horizontal axis tends to highlight the increasing/decreasing nature of the trend of the graph. Lengthening the vertical axis tends to have the same effect.
    - Lengthening the horizontal and shortening the vertical axes tends to level out the trends.

DISCUSSION

Observe this data representation of a company's profit over the 2019–2023 period. Do you think it is misleading?

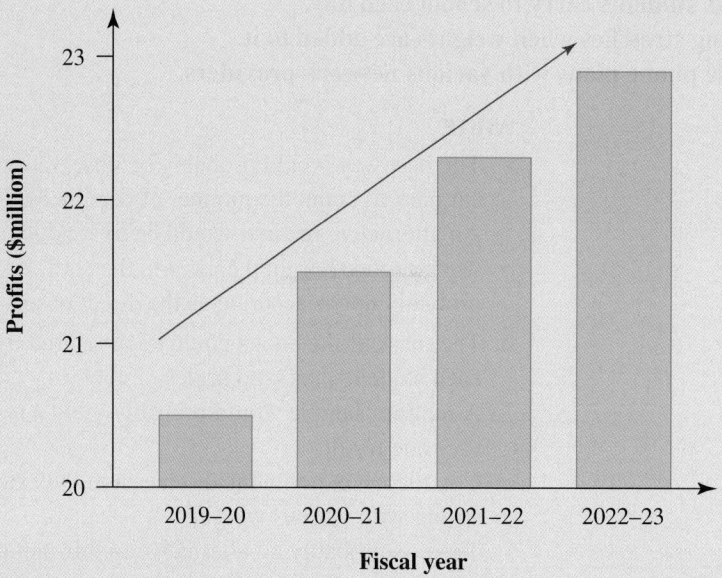

Explain your reasoning.

WORKED EXAMPLE 9 Manipulating graphs

This report shows the annual median unit prices in a number of suburbs from 2021–2022 to 2022–2023. Note that the percentages are rounded to 1 decimal place.

a. Draw a bar graph that would give the impression that the percentage annual change was much the same throughout all of the suburbs.

b. Draw a bar graph to give the impression that the percentage annual change in Rutherford was far smaller than that in the other suburbs.

Suburb or locality	Median price in 2021–22	Median price in 2022–23	Change in median price (%)
Orange, NSW	$380 000	$476 000	+25.2
Rutherford, NSW	$431 500	$489 000	+13.4
Cessnock, NSW	$390 000	$460 000	+17.9
Flora Hill, VIC	$379 400	$444 500	+17.0
Bundall, QLD	$451 000	$535 000	+18.6
Merrimac, QLD	$507 000	$590 000	+16.4
Christies Beach, SA	$398 000	$474 500	+19.0
Ascot Park, SA	$415 000	$511 000	+23.1
Port Hedland, WA	$343 000	$421 000	+22.7
Bunbury, WA	$348 500	$430 000	+23.3

<table>
<tr><td>

THINK

a. To flatten out trends, lengthen the horizontal axis and shorten the vertical axis.

</td><td>

DRAW

a.

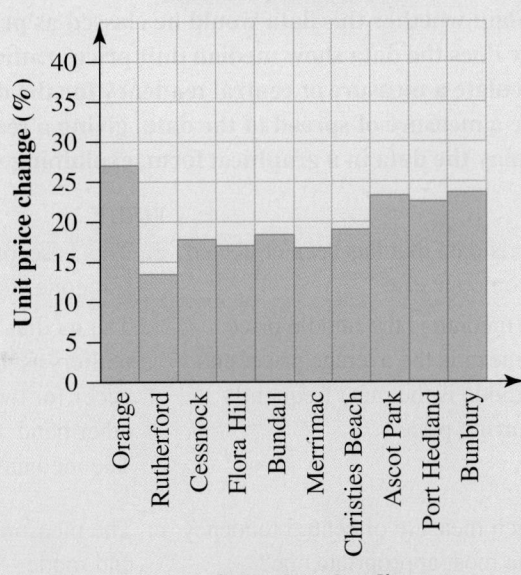

</td></tr>
<tr><td>

b. To accentuate trends, shorten the horizontal axis and lengthen the vertical axis.

</td><td>

b.

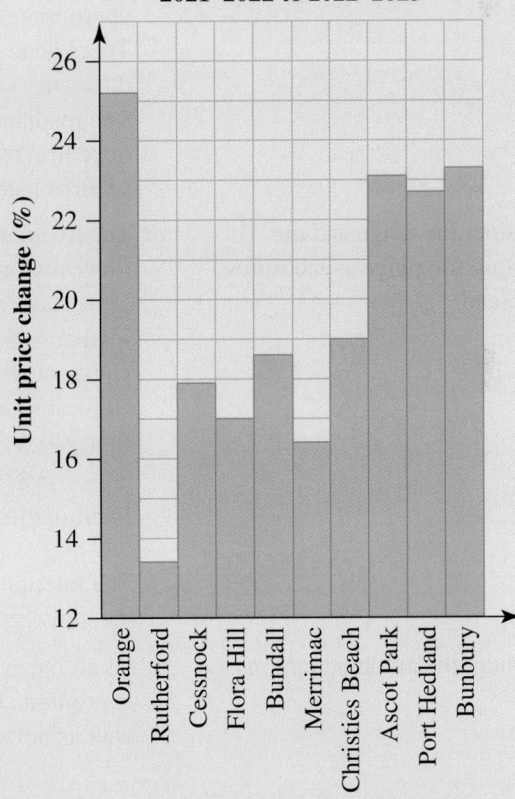

</td></tr>
</table>

Consider the data displayed in the table of Worked example 9. Use the data collected for the median house prices in 2022–2023.
a. Explain whether this data would be classed as primary or secondary data.
b. Why does the data show median unit prices rather than the mean unit price?
c. Calculate a measure of central tendency for the data. Explain the reason for this choice.
d. Give a measure of spread of the data, giving a reason for the particular choice.
e. Display the data in a graphical form, explaining why this particular form was chosen.

THINK

a. This is data that has been collected by someone else.

b. The median is the middle price, the mean is the average price, and the mode is the most frequently occurring price.

c. Which measure of central tendency is the most appropriate one?

d. Consider the range and the interquartile range as measures of spread.

e. Consider the graphing options.

WRITE

a. This is secondary data because it has been collected by someone else.

b. The median price is the middle one. It is not affected by outliers as the mean is. The modal house price may only occur for two house sales with the same value. On the other hand, there may not be any mode.
The median price is the most appropriate in this case.

c. The measures of central tendency are the mean, median and mode.
The mean ($483 000 here) is affected by high values (e.g. $590 000) and low values (e.g. $421 000). These might not be typical values, so the mean would not be appropriate if there are outliers.
There is no modal value, as all the house prices are different.
The median house price is the most suitable measure of central tendency to represent the house prices in the suburbs listed. The median value is $475 000.

d. The five-number summary values are:
Lowest score = $421 000
Lowest quartile = $448 375
Median = $475 000
Upper quartile = $505 500
Highest score = $590 000
Range = $590 000 − $421 000
          = $169 000
Interquartile range = $505 500 − $448 375
          = $57 125
The interquartile range is a better measure for the range, as the house prices form a cluster in this region.

e. Of all the graphing options, the boxplot seems the most appropriate because it shows the spread of the prices, as well as how they are grouped around the median price.

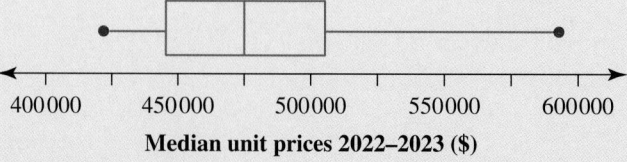

Median unit prices 2022–2023 ($)

The Australian women's national basketball team, the Opals, has 12 team members. Their heights (in metres) have been recorded:

$$1.85, \ 1.72, \ 1.78, \ 1.93, \ 1.65, \ 1.96, \ 1.85, \ 1.73, \ 1.85, \ 1.88, \ 2.03, \ 1.94$$

Source: http://www.basketball.net.au/ba_player_team/opals/

Provide calculations and explanations as evidence to verify or refute the following statements.

a. The mean height of the team is greater than their median height.
b. The range of the heights of the 12 players is almost 3 times their interquartile range.
c. Given that only 5 players are on the court at any one time, a team of 5 players can be chosen such that their mean, median and modal heights are all the same.

THINK	WRITE
a. 1. Calculate the mean height of the 12 players.	a. $\text{Mean} = \dfrac{\sum x}{n} = \dfrac{22.7}{12} = 1.85 \text{ m}$
2. Order the heights to determine the median.	The heights of the players, in order, are: 11.65, 1.72, 1.73, 1.78, 1.85, 1.85, 1.88, 1.93, 1.94, 1.96, 2.03 There are 12 scores, so the median is the average of the 6th and 7th scores. $\text{Median} = \dfrac{1.85 + 1.85}{2} = 1.85 \text{ m}$
3. Comment on the statement	Both the mean and the median are 1.85 m. This means that the statement is not true.
b. 1. Determine the range and the interquartile range of the 12 heights.	b. $\text{Range} = 2.03 - 1.65 = 0.38 \text{ m}$ The lower quartile is the average of 3rd and 4th scores. $\text{Lower quartile} = \dfrac{1.73 + 1.78}{2} = 1.755 \text{ m}$ The upper quartile is average of 3rd and 4th scores from the end. $\text{Upper quartile} = \dfrac{1.93 + 1.94}{2} = 1.935 \text{ m}$ $\text{Interquartile range} = 1.935 - 1.755 = 0.18 \text{ m}$
2. Compare the two values.	$\text{Range} = 0.38 \text{ m}$ $\text{Interquartile range} = 0.18 \text{ m}$ $\dfrac{\text{Range}}{\text{Interquartile range}} = \dfrac{0.38}{0.18} = 2.1$
3. Comment on the statement.	$\text{Range} = 2.11 \times \text{interquartile range}$ This is only 2.11 times, so the statement is not true.
c. 1. Choose 5 players whose mean, median and modal heights are all equal. Trial and error is appropriate here. There may be more than one answer.	c. Three players have a height of 1.85 m. If a player shorter and one taller are chosen, both the same measurement from 1.85 m, this would make the mean, median and mode all the same. Choose players with heights: 1.78, 1.85, 1.85, 1.85, 1.93 $\text{Mean} = \dfrac{9.26}{5} = 1.852 \text{ m}$ $\text{Median} = 3\text{rd score} = 1.85 \text{ m}$ $\text{Mode} = \text{most frequent score} = 1.85 \text{ m}$

If alternatively the players with heights as follow are chosen:
1.73, 1.85, 1.85, 1.85, 1.96

$$\text{Mean} = \frac{9.24}{5} = 1.848\,\text{m}$$

Median = 3rd score = 1.85 m

Mode = most frequent score = 1.85 m

2. Comment on the statement.

In both cases, if the players with heights as listed are chosen and if the mean values are rounded to 2 decimal places, then we will have mean, median and modal heights of 1.85 m. It is true that a team of 5 such players can be chosen.

Statistical reports

- Reported data must not be simply taken at face value. All reports should be examined with a critical eye.

WORKED EXAMPLE 12 Reading statistical reports

This article appeared in a newspaper. Read the article, then answer the following questions.

Sydney Metro surveys likened to 'push polling'

Sydney Metro is being accused of 'push polling' the public over the Sydenham to Bankstown section of the $20 billion rail line in telephone surveys critics say omit crucial information about the impact of the project.

Two companies, The Knowledge Warehouse and Newgate Research, have been conducting the surveys to test public attitudes towards the metro, which will replace the Bankstown heavy rail line.

Sydney Metro to deliver new stations

New rail stations will be built at Crows Nest, Victoria Cross, Central, Waterloo, Martin Place, Pitt St and Barangaroo under the Sydney Metro rail project.

The surveys cost $1 22 500 and involved more than 2800 people who live along the 66 kilometre Sydney Metro route from Rouse Hill in Sydney's northwest, through the city to Bankstown.

After establishing a respondent's attitude, the surveys present a series of statements about the benefits of the Sydney Metro.

They include that Sydney Metro 'will help increase rail capacity across Sydney, allowing the number of trains entering the CBD to go from 120 an hour to 200 during peak times' and highlight 24 hour security at stations and extra car parking in the north-west.

The survey notes that during construction of the city to Bankstown section 'stations between Sydenham and Bankstown will be closed so the line can be converted to Sydney Metro – alternative transport arrangements will be in place during this time'.

However, a community group opposing the Sydenham to Bankstown section argues it fails to mention many negative aspects of the project.

They include a reduction in the number of seats on each train from 896 to 378, the loss of direct access to St Peters, Erskineville, Town Hall, Circular Quay and other City Circle stations and demolition of houses.

'The state government appear to be push-polling the community through this questionnaire, presenting many statements that are at best half-truths,' said Peter Olive, spokesman for the Sydenham to Bankstown Corridor Alliance.

'Shutting down the existing train service between Sydenham and Bankstown and replacing it with another train service, in the same corridor, is a waste of taxpayers' money and just plain stupid. The state government's questionnaire is trying to put spin on this dodgy plan.'

'For commuters between Sydenham and Bankstown the metro will mean fewer seats, loss of access to City Circle Stations, major disruption during construction and the loss of the existing quality service.'

A spokesman for Sydney Metro said community feedback 'is crucial to getting this once-in-a-century project right and shaping it for the future'.

'This questionnaire used industry-standard methods and guidelines to ensure the views of the community were accurately captured to allow planners and designers to take on board the community's feedback,' he said.

'The questionnaire specifically addressed people's concerns including inconvenience caused by construction, quality of public transport, changes to suburbs and the upgrade of two existing railway lines to metro standards.'

Source: The Sydney Morning Herald, 6 February 2017.

a. **Comment on the sample used in this survey.**
b. **Comment on the claims of the survey.**
c. **Is the heading of the article appropriate?**

THINK	WRITE
a. Look at sample size and selection of sample.	a. The report claims that the sample size was more than 2800 and was selected from people who live along the 66-kilometre Sydney Metro route, from Rouse Hill, in Sydney's northwest, through the city to Bankstown. Two points to keep in mind are whether this sample is truly representative of the population consisting of all people who use the train service, not just those who live along the route, and how many of those sampled gave recordable and worthwhile responses. We have no way of knowing.
b. What are the results of the survey?	b. The stated purpose of the surveys was to test public attitudes towards the metro, which will replace the Bankstown heavy rail line, but it would appear that the writer of the article focused on the report's statements about the benefits of the project. From this article, we cannot tell what the actual results of the survey were. We can only read that 'After establishing a respondent's attitude, the surveys present a series of statements about the benefits of the Sydney Metro.'
c. Examine the heading in the light of the contents of the article.	c. The heading is sensational, designed to catch the attention of readers. The article focuses on statements made by the spokesman for the Sydenham and Bankstown Corridor Alliance, who alleges that 'The state government appear to be push-polling the community through this questionnaire, presenting many statements that are at best half-truths.' The article details a number of negative aspects of the project that the spokesman says are not mentioned in the report.
d. From reading the article, what do you understand by the term 'push polling'?	d. In broad terms, one meaning could be that the companies have used the poll to achieve a pre-determined outcome that will focus on only the positives of the project.

DISCUSSION

There is much discussion in the media about 'fake news'. How do we
determine what is 'fake news' and what is 'authentic news'?

on Resources

Interactivity Compare statistical reports (int-2790)

Exercise 13.4 Evaluating statistical reports (Path)

learn on

13.4 Quick quiz on	13.4 Exercise

Individual pathways

■ PRACTISE	■ CONSOLIDATE	■ MASTER
1, 4, 7, 9, 12	2, 6, 10, 13	3, 5, 8, 11

Fluency

1. **WE7&8** You have been given an assignment to investigate which year level has the greatest number of students who are driven to school each day by car.

   a. Explain whether it is more appropriate to use primary or secondary data in this case. Justify your choice.
   b. Describe how the data could be collected. Discuss any problems which might be encountered.
   c. Explain whether an alternative method would be just as appropriate.

2. **WE9** You run a small company that is listed on the Australian Stock Exchange (ASX). During the past year you have given substantial rises in salary to all your staff. However, profits have not been as spectacular as in the year before.
   This table gives the figures for the salary and profits for each quarter.

	1st quarter	2nd quarter	3rd quarter	4th quarter
Profits ($' 000 000)	6	5.9	6	6.5
Salaries ($' 000 000)	4	5	6	7

   Draw two graphs, one showing profits, the other showing salaries, which will show you in the best possible light to your shareholders.

3. **WE10** The data below were collected from a real estate agent and show the sale prices of ten blocks of land in a new estate.
$150\,000$, $190\,000$, $175\,000$, $150\,000$, $650\,000$
$150\,000$, $165\,000$, $180\,000$, $160\,000$, $180\,000$

   a. Calculate a measure of central tendency for the data. Explain the reason for this choice.
   b. Give a measure of spread of the data, giving a reason for the particular choice.
   c. Display the data in a graphical form, explaining why this particular form was chosen.
   d. The real estate agent advertises the new estate land as:
   Own one of these amazing blocks of land for only $150\,000$ (average)!
   Comment on the agent's claims.

4. **WE11** Using the data for the heights of the Opal players in Worked example 11, provide calculations and explanations as evidence to verify or refute the following statements.

   a. The mean height of the team is closer to the lower quartile than it is to the median.
   b. Half the players have a height within the interquartile range.
   c. Which 5 players could be chosen to have the minimum range in heights?

5. The table below shows the number of shoes of each size that were sold over a week at a shoe store.

Size	Number sold
4	5
5	7
6	19
7	24
8	16
9	8
10	7

   a. Calculate the mean shoe size sold.
   b. Determine the median shoe size sold.
   c. Determine the modal shoe size sold.
   d. Explain which measure of central tendency has the most meaning to the store proprietor.

Understanding

6. A small manufacturing plant employs 80 workers.
   The table below shows the structure of the plant's workforce.

Position	Salary $	Number of employees
Machine operator	55 000	50
Machine mechanic	65 000	15
Floor steward	38 000	10
Manager	95 000	4
Chief Executive Officer	120 000	1

   a. Workers are arguing for a pay rise, but the management of the factory claims that workers are well paid because the mean salary of the factory is $57\,563$. Explain whether this is a sound argument.
   b. Suppose that you were representing the factory workers and had to write a short submission in support of the pay rise. How could you explain the management's claim?
   Provide some other statistics to support your case.

7. The batting scores for two cricket players over six innings were recorded as follows.
   Player A: 31, 34, 42, 28, 30, 41
   Player B: 0, 0, 1, 0, 250, 0
   Player B was hailed as a hero for his score of 250. Comment on the performance of the two players.

8. **WE12** Read the information below and look carefully at the graphs, then answer the following questions.

PROFIT SLIP

On 25 August 2009, Woolworths announced a plan to enter the Australian hardware sector via a joint venture with US-based hardware chain Lowe's. The joint venture was to be called Masters Home Improvement. The Masters chain was operated by Woolworths between September 2011 and December 2016.

On 18 January 2016, Woolworths announced that it was exiting the home improvement business. During 2016, Woolworths incurred costs of $ 2988.2 million (after tax) relating to its decision to close the Masters chain.

The Woolworths share price at 2:23 pm on 28 April 2017 was $ 26.88.

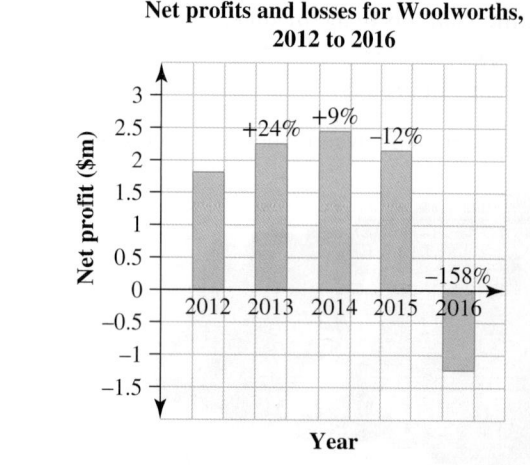

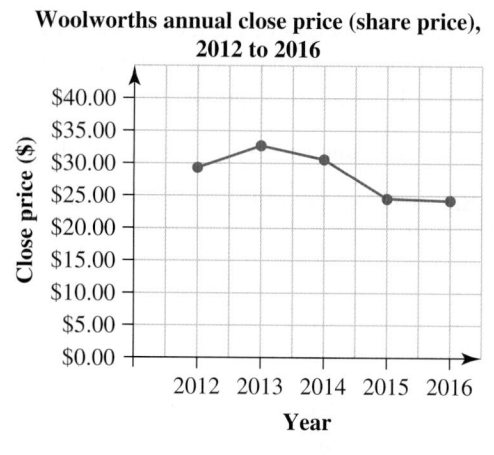

The information provided above paints a picture that could indicate a direct relationship between the profit and share price of Woolworths and the operation of the Woolworths-run home improvement business, Masters.

a. Comment on the above statement and detail any other information that is provided by the graphs.
b. Provide your thoughts as to why the Woolworths share price did not rise sharply after Masters was closed in December 2016.

Communicating, reasoning and problem solving

9. Look at the following bar charts and discuss why the one on the left is misleading. What characteristics does the graph on the right possess that make it acceptable?

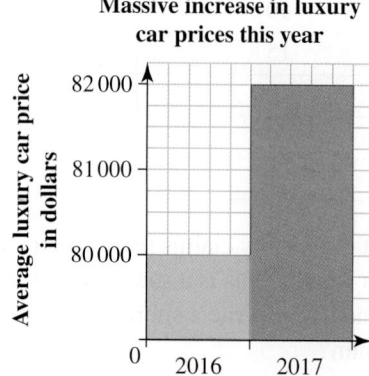

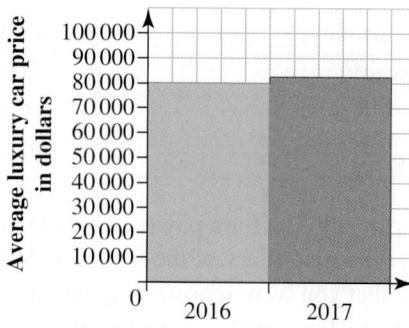

10. Two brands of light globes were tested by a consumer organisation. They obtained the following results.

    Brand A (Hours lasted)

    385 390 425 426 570
    640 645 730 735 760

    Brand B (Hours lasted)

    500 555 560 630 720
    735 742 770 820 860

    a. Complete a back-to-back stem plot for the data.
    b. Which brand had the shortest lifetime?
    c. Which brand had the longest lifetime?
    d. If you wanted to be certain that a globe you bought would last at least 500 hours, which brand would you buy?

11. The following graph shows the fluctuation in the Australian dollar in terms of the US dollar during the period 1 March to 1 May 2016. The lower the Australian dollar, the more expensive it is for Australian companies to import goods from overseas, and the more they should be able to charge the Australian public for their goods.

    The board of Company XYZ wanted to raise the price of the goods they sold in line with their understanding of the change in exchange rates as shown in the graph. However, the manager of the company produced another graph to support his claim that, because there hadn't been much change in the Australian dollar over that period, there shouldn't be any change in the price he charged for his imported goods to the Australian public. Draw a graph that would support his claim. Explain how you were able to achieve this effect.

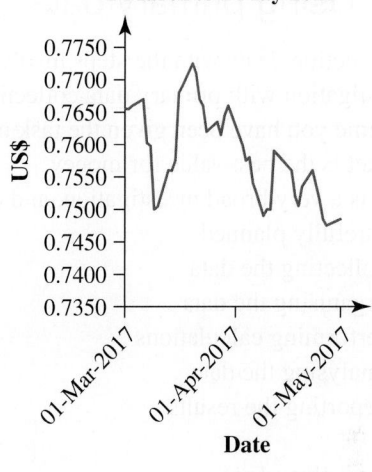

Australian dollar exchange rate with the US dollar, 1 March to 1 May 2017

12. a. What is wrong with this pie graph?

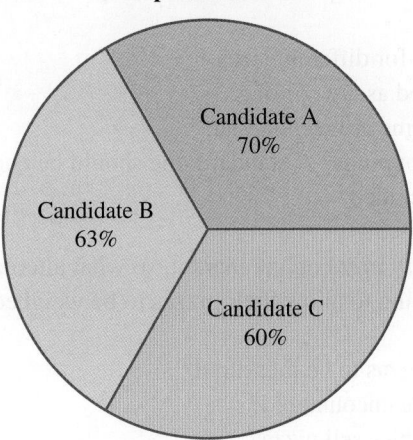

2016 presidential run

Candidate A 70%

Candidate B 63%

Candidate C 60%

    b. Why is the following information misleading?

    > Did scientists falsify research to support their own theories on global warming?
    > 59% somewhat likely
    > 35% very likely
    > 26% not very likely

    c. Discuss the implications of this falsification by statistics.

13. What is the point of drawing a misleading graph in a report?

LESSON
13.5 Statistical investigations (Path)

LEARNING INTENTION

At the end of this lesson you should be able to:
- plan, conduct and review a statistical inquiry into a question of interest
- examine reports of studies in digital media to understand how they are planned and implemented.

▶ 13.5.1 Using primary data

eles-6297

- This section deals with the steps involved in carrying out a statistical investigation with primary data collection.
- Assume you have been given the task of describing which pizza on the market is the best value for money.
- This is a very broad investigation, and each stage of the investigation must be carefully planned.
  - Collecting the data
  - Organising the data
  - Performing calculations
  - Analysing the data
  - Reporting the results

Collecting the data

- At this initial stage, questions should be posed with regard to the data.
  - What data should be collected?
    - Best value for money involves the price and size of the pizza. Data on both of these need to be collected.
    - Stores have different prices for different sizes.
    - Would size best be measured as area or mass?
    - Not all pizzas are round; some are rectangular.
    - What about the variety of toppings? A standard one should be chosen.
    - Should frozen pizzas be included?
  - How should the data be collected?
    - It is not possible to buy every pizza on the market, so what alternatives are there?
    - A store is probably not willing to allow their pizzas to be weighed, so mass is most likely out of the question.
    - Will the store allow their pizzas to be measured?
  - What problems are likely to be encountered?
    - How many different companies sell pizza?

Organising the data

- The data should be organised into some sort of table format.
  - What format is appropriate for this investigation?
  - A table with column headings 'Price' and 'Measurements' will organise the data.

Price	Measurements	Area	Value for money

- Take time to design the table so figures required for calculation are readily visible.
    - What calculations are required at this stage?
    - Measurements are required to calculate the area of each pizza.
- Think forward and add extra columns for future calculations.
    - What further calculations are needed?
    - 'Area' and 'Value for money' need to be calculated. Provide two extra columns for these.

Performing calculations

- What calculations need to be performed?
- Area and value for money are required in this case. How should these be calculated? Dividing price by area gives results in $/cm^2$, whereas dividing area by price gives results in $cm^2/$.

Analysing the data

- Are there any anomalies or obvious calculation errors?
- Do the calculated results make sense?
- In this case, if value for money is calculated in units of $/cm^2$, the pizza with the smallest of these values is the best value for money. Using units of $cm^2/$, the pizza with the highest of these values gives the best value for money.
- Would the inclusion of graphs be appropriate?

Reporting the results

- The results should be reported in a clear, concise manner.
- Justify any conclusions.
- Are there any anomalies or exceptions to mention?

▶ 13.5.2 Using secondary data

eles-6298

- The procedure for undertaking a statistical investigation using secondary data is similar to that for primary data, the difference being that you sometimes have to search for data in several areas before you find the appropriate source.
  Suppose you were given this assignment.

There have been _____ prime ministers of Australia since 1901 until this day.
There have been _____ elections.
_____ prime ministers have been defeated at a general election.
There have been _____ changes of prime minister without an election.
The average length these prime ministers served in office is _____.
Undertake a statistical investigation to complete the details.

Collecting the data

- What data should be collected?
- Where can the data be found? The internet is probably a good starting point, but not all sites are reliable.
- If there are multiple sources for the data, are they all in agreement?
- How many of these statements require calculations?

Organising the data

- Design a table to record all the data.
- Consider how many columns are necessary.
- Leave columns for calculations.

Performing calculations

- There is at least one calculation here — to determine the average length of time served in office. Are there any more?

Analysing the data

- Do all the calculated values make sense?
- Would a graph be appropriate?

Reporting the results

- Complete the details.
- Acknowledge the source(s) of the secondary data.

▶ 13.5.3 Investigating media reports

eles-6299

- Reports in the media often provide a good starting point for an interesting investigation. Here are a few extracts from media articles.

Media report 1: Australia's women's cricket team

"In the next five years, women's cricket will see a massive 62.5% increase in the pool from which the players are paid. The top players' earnings will reach \$800 000 per year, potentially stretching to \$1 million if they also play in the Women's Premier League and UK's Hundred."

Source: Adapted from Aussies add four to contract list, Carey turns down deal | cricket.com.au

TABLE Nation's seven top earning cricketers for 2022–2023

Squad	Retainer fees
Pat Cummins	\$2 million
Josh Hazlewood	\$1.6 million
David Warner	\$1.5 million
Mitchell Starc	\$1.4 million
Steve Smith	\$1.3 million
Marnus Labuschagne	\$1.2 million
Nathan Lyon	\$1.1 million

Source: https://www.sportyreport.com/australia-cricket-players-salary/

- The table above is a list of the top 7 men's cricketers in Australia in 2022. Compare this to the top women's cricketers. Is there a difference and, if so, why? Discuss in pairs or as a class.

Media report 2: How much money do social media companies make from advertising?

"Social media companies with free-to-use services are generally able to turn a tremendous profit through advertising, as most companies are more than willing to pay them to display their ads to their users. Social media companies made \$153 billion in 2021 and this number is expected to grow to \$252 billion by 2026."

Source: Adapted from How much money do social media companies make from advertising? - Zippia

Do you pay for using social media?
- Is it fair that companies make money from placing advertising in social media?
- How do companies know which ads to put on your social media?

Media report 3: How Much Time Do Children Spend on TikTok? New Report Reveals Staggering Stats

"The CDC reports that children and teens spend anywhere from six to nine hours on screens each day, and new data reveals what apps children scroll through the most.

Cribbage Challenge found that kids spend 113 minutes on TikTok, followed by 90 minutes on Snapchat, 20 minutes on Pinterest and 18 minutes on Reddit each day.

On streaming services, children and teens spend 77 minutes on YouTube, 52 minutes on Netflix, 42 minutes on Disney+ and 35 minutes on Amazon Prime every day."

Source: How Much Time Do Children Spend on TikTok? New Report Reveals Staggering Stats (movieguide.org)

- With permission from your teacher, look at Settings in your phone, find the statistics on device usage (how much time you spend on your phone). Write down the key statistics of your own usage in your book. Compare to your partner. Discuss any differences. Discuss as a class.

ACTIVITY: SOCIAL MEDIA

Look at the following statistics on the average monthly usage of the most popular social media apps.

Platform	Monthly active users
Facebook	2.9 billion
Youtube	2.2 billion
Instagram	1.4 billion
TikTok	1 billion
Snapchat	500 million
Pinterest	480 million
Twitter	397 million

1. Identify which age groups would use the different apps.
2. Why do you think that Facebook is still the most popular app?
3. How do you think these statistics will change in 5 year's time?
4. Write an estimate for each app in terms of monthly users in 5 year's time.
5. Discuss in pairs or as a class how the use of apps will change over time.

Exercise 13.5 Statistical investigations (Path)

learn on

13.5 Quick quiz on	13.5 Exercise

Individual pathways

■ PRACTISE	■ CONSOLIDATE	■ MASTER
1, 2, 5, 8	3, 6, 9	4, 7, 10, 11

Fluency

1. a. Write a plan detailing how you would collect primary data to undertake an investigation to determine which pizza on the market is the best value for money.
   b. Undertake your investigation.
   c. Report on the results of your findings.

2. What is the most popular social media app in the world as of 2022?

3. How many minutes per day do teenagers spend on TikTok on average in 2022?

4. How do social media companies make money from their users?

Understanding

5. On average, how many hours of screen-time do teenagers spend per day?

6. List 2 reasons why you think there is still a gap between the pay for men and women in sport in 2022.

7. An investigation is to be conducted to find the two most popular television programs in Australia. For each of the following samples, explain why they would be biased.

   a. A sample of 100 students from a city secondary school.
   b. A sample of 100 people passing a certain point in a busy city street at lunchtime.
   c. How would you go about selecting a sample of people?

Communicating, reasoning and problem solving

8. There has been a rise in supermarket-owned brands in Australia. These are commonly available in supermarkets such as Woolworths, Coles and Aldi. It has been said that these brands account for almost one-quarter of all grocery sales. It has also been claimed that the quality of supermarket-owned brands is comparable with the equivalent market-leading brands, although the supermarket-owned brands are much cheaper.
   Assume you are planning to undertake a study of a particular grocery line (e.g. baked beans or breakfast cereal). Write a plan of how you would undertake this study.

9. You wish to assess the opinion of a local population on the possible closure of their hospital.

   a. What target population would you use?
   b. What resources would you use?
   c. What are the possible biases in conducting such an experiment?

10. What would you consider to be the most important factor in reporting the results of a statistical investigation?

11. List five positive even integers that have a mean of 8 and a median of 10. How many possible solutions are there?

LESSON
13.6 Selecting a sample size (Path)

LEARNING INTENTION

At the end of this lesson you should be able to:
- understand how to choose an appropriate sample size.

▶ 13.6.1 Populations

eles-6300

- The term **population** refers to a complete set of individuals, objects or events belonging to some category.
- When data are collected from a whole population, the process is known as a **census**.
  - It is often not possible, nor cost-effective, to conduct a census.
  - For this reason, **samples** have to be selected carefully from the population. A sample is a subset of its population.
  - Samples represent all units in a population of interest. They can be selected using different methods, such as simple random sampling, cluster sampling, convenience sampling or systematic sampling.

WORKED EXAMPLE 13 Sampling populations

List some of the problems you might encounter in trying to collect data on the following populations.
a. The life of a mobile phone battery.
b. The number of possums in a local area.
c. The number of males in Australia.
d. The average cost of a loaf of white bread.

THINK

For each of these scenarios, consider how the data might be collected, and the problems in obtaining these data.

a. The life of a mobile phone battery.

b. The number of possums in a local area.

c. The number of males in Australia.

d. The average cost of a loaf of white bread.

WRITE

a. The life of a mobile phone battery cannot be measured until it is dead. The battery life also depends on how the phone is used, and how many times it has been recharged.

b. It would be almost impossible to find all the possums in a local area in order to count them. The possums also may stray into other areas.

c. The number of males in Australia is constantly changing. There are births and deaths every second.

d. The price of one particular loaf of white bread varies widely from one location to another. Sometimes the bread is on 'special' and this would affect the calculations.

13.6.2 Why the deviations are squared

eles-6301

- Surveys are conducted using samples. Ideally the sample should reveal generalisations about the population.
- A random sample is generally accepted as being an ideal representation of the population from which it was drawn. However, it must be remembered that different random samples from the same population can produce different results. This means that we must be cautious about making predictions about a population, as results of surveys conducted using random samples may vary.
- A sample size must be sufficiently large. As a general rule, the sample size should be about $\sqrt{N}$, where N is the size of the population.

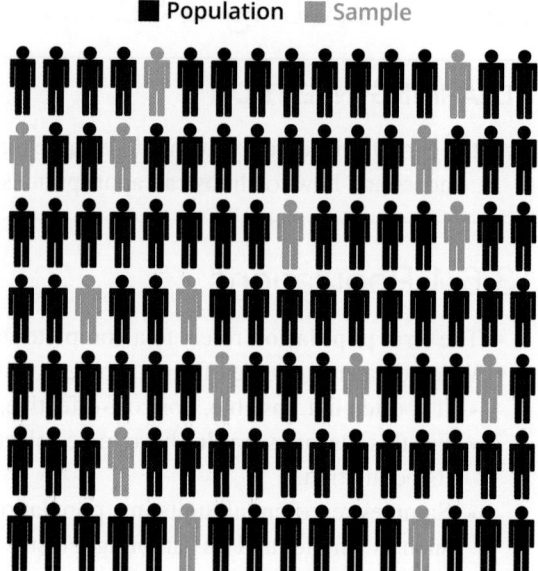

■ **Population** ■ **Sample**

WORKED EXAMPLE 14 Analysing the impact of sample size

A die was rolled 50 times and the following results were obtained.

$$6, 5, 3, 1, 6, 2, 3, 6, 2, 5, 3, 4, 1, 3, 2, 6, 4, 5, 5, 4, 3, 1, 2, 1, 6,$$
$$4, 5, 2, 3, 6, 1, 5, 3, 3, 2, 4, 1, 4, 2, 3, 2, 6, 3, 4, 6, 2, 1, 2, 4, 2$$

a. Determine the mean of the population (to 1 decimal place).

b. A suitable sample size for this population would be 7 $\left(\sqrt{50} \approx 7.1\right)$.
   i. Select a random sample of 7 scores and determine the mean of these scores.
   ii. Select a second random sample of 7 scores and determine the mean of these.
   iii. Select a third random sample of 20 scores and determine the mean of these.

c. Comment on your answers to parts a and b.

THINK	WRITE
a. Calculate the mean by first finding the sum of all the scores, then dividing by the number of scores (50).	a. Population mean $$= \frac{\sum x}{n}$$ $$= \frac{169}{50}$$ $$= 3.4$$
b. i. Use a calculator to randomly generate 7 scores from 1 to 50. Relate these numbers back to the scores, then calculate the mean.	b. i. The 7 scores randomly selected are numbers 17, 50, 11, 40, 48, 12, 19 in the set of 50 scores. These correspond to the scores 4, 2, 3, 3, 2, 4, 5. The mean of these scores $= \frac{23}{7} = 3.3$.

---

ii. Repeat the previous step to obtain a second set of 7 randomly selected scores. This second set of random numbers produced the number 1 twice. Try again. Another attempt produced the number 14 twice. Try again. A third attempt produced 7 different numbers. This set of 7 random numbers will then be used to, again, calculate the mean of the scores.

iii. Repeat for 20 randomly selected scores.

c. Comment on the results.

ii. Ignore the second and third attempts to select 7 random numbers because of repeated numbers. The second set of 7 scores randomly selected is numbers 16, 49, 2, 42, 31, 11, 50 of the set of 50. These correspond to the scores 6, 4, 5, 6, 1, 3, 2.

The mean of these scores $= \dfrac{27}{7} = 3.9$.

iii. The set of 20 randomly selected numbers produced a total of 68.

The mean of 20 random scores $= \dfrac{68}{20} = 3.4$.

c. The population mean is 3.4. The means of the two samples of 7 are 3.3 and 3.9. This shows that, even though the samples are randomly selected, their calculated means may be different. The mean of the sample of 20 scores is 3.4. This indicates that by using a bigger sample the result is more accurate than those obtained with the smaller samples.

ACTIVITY: SAMPLING ERRORS, WAY TO BE RANDOM!

Equipment: pen and paper

1. In groups of 3, think of 3 different ways that you could randomly select 10 people from a group of 50 people (25 males and 25 females ranged in age from 20 to 70) to survey their thoughts on global warming.
2. What factors do you need to consider? What would be considered a random sample? List any possible bias that could come from this group.
3. Using the information in the table below, trial your 3 methods of random sampling and comment on the groups you obtain.

Male	45	68	21	36	70	22	23	37	41	55	59	66	29	33	65	51	24	70	48	26	32	21	57	42	61
Female	33	32	57	52	44	68	70	61	23	25	57	36	41	49	26	47	55	68	70	20	34	49	51	67	42

4. As a class, discuss why it is important that randomly chosen data values have to have an equally likely chance of being selected.

13.6.3 To sample or to conduct a census?

eles-6302

- The particular circumstances determine whether data are collected from a population, or from a sample of the population. For example, suppose you collected data on the height of every Year 10 student in your class. If your class was the only Year 10 class in the school, your class would be the population. If, however, there were several Year 10 classes in your school, your class would be a sample of the Year 10 population.
- Worked example 14 showed that different random samples can produce different results. For this reason, it is important to acknowledge that there could be some uncertainty when using sample results to make predictions about the population.

ACTIVITY: DIGITAL LIVES OF YOUNG AUSTRALIANS

TABLE Top social networking sites or apps used by young people (%)

91%	83%	81%	55%	55%	31%	26%	26%
Facebook	Youtube	Instagram	WhatsApp	Snapchat	Reddit	TikTok	Twitter

Source: Australian Communications and Media Authority, The digital lives of Younger Australians, 2021

There are many studies into the use of social media by different age groups. The above statistics were part of a larger report by the Australian Communications and Media Authority in 2021. In this report, 'young' refers to people from 18 to 34 years old .

1. Is it useful to report on different age groups in terms of their social media usage? Write a paragraph to justify your answer.

It is important to be able to find out the methodology of official reports in order to check the validity of what is being reported on. Part of the checking the validity of statistics is to know information about the sample size. If the sample size is too small, the validity of the statistics is reduced. The report on digital lives of young Australians had a sample size of 2009 people in 2020.

2. Is this a suitable sample size to represent young Australians and their social media habits? Discuss in pairs or as a class.

WORKED EXAMPLE 15 Collecting data by census or survey

For each of the following situations, state whether the information was obtained by census or survey. Justify why that particular method was used.
a. A roll call is conducted each morning at school to determine which students are absent.
b. TV ratings are collected from a selection of viewers to discover the popular TV shows.
c. Every hundredth light bulb off an assembly production line is tested to determine the life of that type of light bulb.
d. A teacher records the examination results of her class.

THINK

a. Every student is recorded as being present or absent at the roll call.

b. Only a selection of the TV audience contributed to these data.

c. Only 1 bulb in every 100 is tested.

d. Every student's result is recorded.

WRITE

a. This is a census. If the roll call only applied to a sample of the students, there would not be an accurate record of attendance at school. A census is essential in this case.

b. This is a survey. To collect data from the whole viewer population would be time-consuming and expensive. For this reason, it is appropriate to select a sample to conduct the survey.

c. This is a survey. Light bulbs are tested to destruction (burn-out) to determine their life. If every bulb was tested in this way, there would be none left to sell! A survey on a sample is essential.

d. This is a census. It is essential to record the result of every student.

on Resources

Interactivity   Sample sizes (int-6183)

Exercise 13.6 Selecting a sample size (Path)

learn on

13.6 Quick quiz on	13.6 Exercise

Individual pathways

■ PRACTISE	■ CONSOLIDATE	■ MASTER
1, 2, 3, 5, 7, 14	4, 6, 8, 10	9, 11, 12, 13, 15

Fluency

1. **WE13** List some of the problems you might encounter in trying to collect data from the following populations.
   a. The life of a laptop computer battery.
   b. The number of dogs in your neighbourhood.
   c. The number of fish for sale at the fish markets.
   d. The average number of pieces of popcorn in a bag of popcorn.

2. **WE14** A die was rolled 50 times and the following results were obtained.

   6, 5, 3, 1, 6, 2, 3, 6, 2, 5, 3, 4, 1, 3, 2, 6, 4, 5, 5, 4, 3, 1, 2, 1, 6,
   4, 5, 2, 3, 6, 1, 5, 3, 3, 2, 4, 1, 4, 2, 3, 2, 6, 3, 4, 6, 2, 1, 2, 4, 2

   The mean of the population is 3.4. Select your own samples for the following questions.

   a. Select a random sample of 7 scores and determine the mean of these scores.
   b. Select a second random sample of 7 scores and determine the mean of these.
   c. Select a third random sample of 20 scores and determine the mean of these.
   d. Comment on your answers to parts a, b and c.

3. **WE15** In each of the following scenarios, state whether the information was obtained by census or survey. Justify why that particular method was used.

   a. Seating for all passengers is recorded for each aeroplane flight.
   b. Movie ratings are collected from a selection of viewers to discover the best movies for the week.
   c. Every hundredth soft drink bottle on an assembly production line is measured to determine the volume of its contents.
   d. A car driving instructor records the number of hours each learner driver has spent driving.

4. For each of the following, state whether a census or a survey has been used.

   a. Two hundred people in a shopping centre are asked to nominate the supermarket where they do most of their grocery shopping.
   b. To find the most popular new car on the road, 500 new car buyers are asked what make and model they purchased.
   c. To find the most popular new car on the road, data are obtained from the transport department.
   d. Your Year 10 Maths class completed a series of questions on the amount of maths homework for Year 10 students.

Understanding

5. To conduct a statistical investigation, Gloria needs to obtain information from 630 students.
   a. What size sample would be appropriate?
   b. Describe a method of generating a set of random numbers for this sample.

6. A local council wants the opinions of its residents regarding its endeavours to establish a new sporting facility for the community. It has specifically requested all residents over 10 years of age to respond to a set of on-line questions.
   a. Is this a census or a survey?
   b. What problems could you encounter collecting data this way?

7. A poll was conducted at a school a few days before the election for Head Boy and Head Girl. After the election, it was discovered that the polls were completely misleading. Explain how this could have happened.

8. A sampling error is said to occur when results of a sample are different from those of the population from which the sample was drawn. Discuss some factors which could introduce sampling errors.

Communicating, reasoning and problem solving

9. Since 1961, a census has been conducted in Australia every 5 years. Some people object to the census on the basis that their privacy is being invaded. Others say that the expense involved could be directed to a better cause. Others say that a sample could obtain statistics that are just as accurate.
   What are your views on this? Justify your statements.

10. Australia has a very small population compared with other countries such as China and India. These are the world's most populous nations, so the problems we encounter in conducting a census in Australia would be insignificant compared with those encountered in those countries.
    What different problems would authorities come across when conducting a census there?

11. The game of Lotto involves picking the same 6 numbers in the range 1 to 45 as have been randomly selected by a machine containing 45 numbered balls. The balls are mixed thoroughly, then 8 balls are selected representing the 6 main numbers, plus 2 extra numbers, called supplementary numbers.
    The following two lists show the number of times each number had been drawn over a period of time, and the number of weeks since each particular number has been drawn.

NUMBER OF WEEKS SINCE EACH NUMBER DRAWN

1	2	3	4	5	6	7	8
1	5	2	1	1	7	-	4
9	10	11	12	13	14	15	16
3	3	1	5	5	7	-	4
17	18	19	20	21	22	23	24
9	-	9	2	2	12	10	8
25	26	27	28	29	30	31	32
5	11	17	2	3	3	-	22
33	34	35	36	37	38	39	40
4	3	-	1	12	-	6	-
41	42	43	44	45			
6	1	7	-	31			

NUMBER OF TIMES EACH NUMBER DRAWN SINCE DRAW 413

1	2	3	4	5	6	7	8
246	238	244	227	249	241	253	266
9	10	11	12	13	14	15	16
228	213	250	233	224	221	240	223
17	18	19	20	21	22	23	24
217	233	240	226	238	240	253	228
25	26	27	28	29	30	31	32
252	239	198	229	227	204	230	226
33	34	35	36	37	38	39	40
246	233	232	251	222	221	219	259
41	42	43	44	45			
245	242	237	221	224			

If these numbers are randomly chosen, explain the differences shown in the tables.

12. A sample of 30 people was selected at random from those attending a local swimming pool. Their ages (in years) were recorded as follows:

$$19, \; 7, \; 58, \; 41, \; 17, \; 23, \; 62, \; 55, \; 40, \; 37, \; 32, \; 29, \; 21, \; 18, \; 16,$$
$$10, \; 40, \; 36, \; 33, \; 59, \; 65, \; 68, \; 15, \; 9, \; 20, \; 29, \; 38, \; 24, \; 10, \; 30$$

    a. Determine the mean and the median age of the people in this sample.

    b. Group the data into class intervals of 10 (0–9 etc.) and complete the frequency distribution table.

    c. Use the frequency distribution table to calculate an estimate of the mean age.

    d. Calculate the cumulative frequency and, hence, plot the ogive.

    e. Estimate the median age from the ogive.

    f. Compare the mean and median of the original data in part **a** with the estimates of the mean and the median obtained for the grouped data in parts **c** and **e**.

    g. Were the estimates good enough? Explain your answer.

13. The typing speed (words per minute) was recorded for a group of Year 8 and Year 10 students. The results are displayed in this back-to-back stem plot.

Key: $2\,|\,6 = 26$ wpm

Leaf: Year 8	Stem	Leaf: Year 10
9 9	0	
9 8 6 5 4 2 0	1	7 9
9 8 8 6 4 2 1 0 0	2	2 3 6 8 9
9 7 7 6 4 1 0	3	0 2 4 5 5 7 8 8
8 6 5 2 0	4	1 2 5 8 8 9 9
	5	0 3 5 7 8
	6	0 0 3

Write a report comparing the typing speeds of the two groups.

14. A well-known saying about statistics is: *Statistics means never having to say you're certain.* What does this saying mean?

15. A fisheries and wildlife officer released 200 tagged trout into a lake. A week later, the officer took a sample of 50 trout and found that 8 of them were tagged. The officer can use this information to estimate the population of trout in the lake.
How many trout are now in the lake?

LESSON
13.7 Review

13.7.1 Topic summary

DATA ANALYSIS

Properties of bivariate data

- **Bivariate data** can be displayed, analysed and used to make predictions.
- **Types of variables:**
  - Independent (experimental or explanatory variable): not impacted by the other variable.
  - Dependent (response variable): impacted by the other variable.

Statistics

- Data can be collected as a census (whole population) or a survey (a sample is small percentage of the population)
- The graphical display of statistics can be used to help understand the data. However, sometimes statistics are used to misrepresent the data, so we need to evaluate carefully.
- Estimating population means and medians can be helpful to analyse trends in data and compare different sets of data.

Representing the data

- **Scatterplots** can be created by hand or using technology (CAS or Excel).
- The independent variable is placed on the x-axis and the dependent variable on the y-axis.

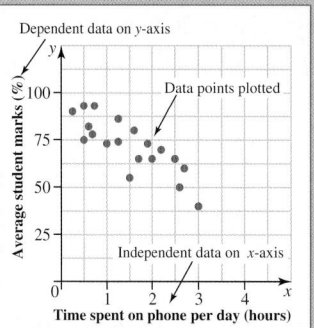

Line of best fit

- A line that follows the trend of the data in a scatterplot.
- It is most appropriate for data with strong or moderate linear correlation.
- Can be sketched as a line of best fit by eye or a regression line using technology.
  - The equation for the line can be found by using the gradient and equation of the straght line.
  - The line can be used to make predictions.
- Regression lines are only valid if the independent and dependent variables have a connection.

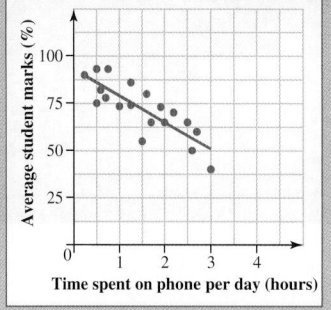

Correlations from scatterplots

- Correlation is a way of describing a connection between variables in a bivariate data set.
- Correlation between the two variables will have:
  - a type (linear or non-linear)
  - a direction (positive or negative)
  - a strength (strong, moderate or weak).
- Correlations can be used to make conclusions.
- There is no correlation if the data are spread out across the plot with no clear pattern.

Interpolation and extrapolation

- Interpolation and extrapolation can be used to make predictions.
- **Interpolation:**
  - is more reliable from a large number of data
  - is used if the prediction sits within the given data.
- **Extrapolation:**
  - assumes the trend will continue
  - is used if the prediction sits outside the given data.

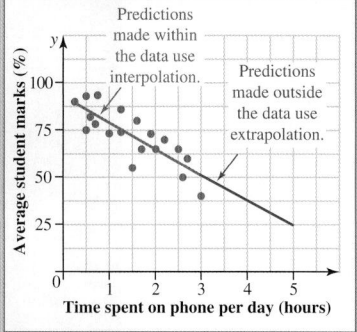

13.7.2 Project

Populations and samples

In this task we will see how closely a sample resembles the population. (Digital technology should be used to answer the following questions.)

The following table gives information about literacy rates (as percentages) for the entire population in 100 countries and for males and females in each country. This task will only use the first column of literacy rates for the country. Other conclusions may be drawn from the male and female literacy rates and possibly compared to the overall literacy rates in this investigation.

1. Determine the five-figure summary for the 100 countries' literacy rates and draw a boxplot.
2. Determine the mean and standard deviation.
3. Take a sample of 20 from the set of 100 countries using a random number generator and make a list of the countries and their literacy rates.
4. Determine the five-number summary for your sample.
5. Determine the mean and standard deviation for your sample.
6. Draw a boxplot that compares your sample with the whole population.
7. Describe the sample data obtained in terms of the population as a whole. In particular, concentrate on the middle 50 per cent of the data.
8. Take another two random samples of 20 countries and draw boxplots of all three on the same scale. Comment on these three samples.
9. Determine the mean and standard deviation for the three samples. How do they compare?
10. From the results you have obtained, how would you describe the reliability of a sample compared with using the whole population?

	Country	Literacy (%)	Male literacy (%)	Female literacy (%)
1	Afghanistan	29	44	14
2	Argentina	98	99	98
3	Armenia	98	100	100
4	Australia	100	100	100
5	Azerbaijan	98	100	100
6	Bahrain	77	55	55
7	Bangladesh	35	47	22
8	Barbados	99	99	99
9	Belarus	99	100	100
10	Bolivia	78	85	71
11	Botswana	72	32	16
12	Brazil	81	82	80
13	Bulgaria	98	99	98
14	Burkina Faso	24	34	14
15	Burundi	50	61	40
16	Cambodia	35	48	22
17	Cameroon	54	66	45
18	Canada	97	97	97
19	Cent. Afri. R	27	33	15

	Country	Literacy (%)	Male literacy (%)	Female literacy (%)
20	Chile	93	94	93
21	China	78	87	68
22	Colombia	87	88	86
23	Costa Rica	93	93	93
24	Croatia	98	99	97
25	Cuba	94	95	93
26	Dominican R.	83	85	82
27	Ecuador	88	90	86
28	Egypt	48	63	34
29	El Salvador	73	76	70
30	Estonia	99	100	100
31	Ethiopia	24	32	16
32	Finland	100	100	100
33	France	99	99	98
34	Gabon	61	74	48
35	Gambia	27	39	16
36	Georgia	99	100	100
37	Germany	99	99	98
38	Greece	93	98	89
39	Guatemala	55	63	47
40	Haiti	53	59	47
41	Honduras	73	76	71
42	Hungary	99	99	98
43	Iceland	100	100	100
44	India	52	64	39
45	Indonesia	77	84	68
46	Iran	54	64	43
47	Iraq	60	70	49
48	Ireland	98	99	97
49	Israel	92	95	89
50	Italy	97	98	96
51	Japan	99	99	99
52	Jordan	80	89	70
53	Kenya	69	80	58
54	Kuwait	73	77	67
55	Latvia	99	100	100
56	Lebanon	80	88	73
57	Liberia	40	50	29
58	Libya	64	75	50
59	Lithuania	99	99	98
60	Malaysia	78	86	70
61	Mexico	87	90	85
62	Morocco	50	61	38

	Country	Literacy (%)	Male literacy (%)	Female literacy (%)
63	N. Korea	99	99	99
64	Netherlands	100	100	100
65	New Zealand	99	99	99
66	Nicaragua	57	57	57
67	Nigeria	51	62	40
68	Norway	99	99	99
69	Oman	71	80	62
70	Pakistan	35	47	21
71	Panama	88	88	88
72	Paraguay	90	92	88
73	Peru	85	92	79
74	Philippines	90	90	90
75	Poland	99	99	98
76	Portugal	85	89	82
77	Russia	99	100	100
78	Rwanda	50	64	37
79	S. Korea	96	99	99
80	Saudi Arabia	62	73	48
81	Senegal	38	52	25
82	Singapore	88	93	84
83	Somalia	24	36	14
84	South Africa	85	86	86
85	Spain	95	97	93
86	Sweden	99	99	99
87	Switzerland	99	99	99
88	Syria	64	78	51
89	Thailand	93	96	90
90	Turkey	81	90	71
91	U. Arab Em.	68	70	63
92	UK	99	99	99
93	USA	97	97	97
94	Uganda	48	62	35
95	Ukraine	97	100	100
96	Uruguay	96	97	96
97	Uzbekistan	97	100	100
98	Venezuela	88	90	87
99	Vietnam	88	93	83
100	Zambia	73	81	65

 Resources

Interactivities Crossword Topic 13 (int-2866)
              Sudoku Topic 13 (int-3601)

Fluency

1. List some problems you might encounter in trying to collect data from the following populations.
   a. The average number of mL in a can of soft drink.
   b. The number of fish in a dam.
   c. The number of workers who catch public transport to work each weekday morning.

2. a. Calculate the mean of the integers 1 to 100.
   b. i. Randomly select 10 numbers in the range 1 to 100.
      ii. Calculate the mean of these numbers.
   c. i. Randomly select 20 numbers in the range 1 to 100.
      ii. Calculate the mean of these numbers.
   d. Comment on the similarities/differences between your means calculated in parts a, b and c.

3. For each of the following investigations, state whether a census or a survey has been used.
   a. The average price of petrol in Canberra was estimated by averaging the price at 30 petrol stations in the area.
   b. The performance of a cricketer is measured by looking at their performance in every match they have played.
   c. Public opinion on an issue is sought by a telephone poll of 2000 homes.

4. Traffic lights (red, amber, green) are set so that each colour shows for a set amount of time. Describe how you could use a spinner to simulate the situation so that you could determine (on average) how many sets of lights you must encounter in order to get two green lights in succession.

5. **MC** John and Bill play squash each week. In any given game they are evenly matched. A device that could not be used to represent the outcomes of the situation is:
   A. a die
   B. a coin
   C. a circular spinner divided into 2 equal sectors
   D. a circular spinner divided into 5 equal sectors

Understanding

6. The table shows the number of students in each year level from Years 7 to 12.

Year	Number of students
7	230
8	200
9	189
10	175
11	133
12	124

Draw two separate graphs to illustrate the following.
   a. The principal of the school claims a high retention rate in Years 11 and 12 (that is, most of the students from Year 10 continue on to complete Years 11 and 12).
   b. The parents claim that the retention rate of students in Years 11 and 12 is low (that is, a large number of students leave at the end of Year 10).

7. Records from a school were examined to determine the
    number of absent days of both boys and girls over the two
    years of Year 9 and Year 10. The result is shown in this
    stem-and-leaf plot.
    a. Calculate the median number of days absent for both
       boys and girls.
    b. Calculate the range for both boys and girls.
    c. Comment on the distribution of days absent for
       each group.

Key: 2 | 1 = 21 days

Leaf: Boys	Stem	Leaf: Girls
	0	1 7
7 4 1 0	1	2 4 7 9 9
9 9 7 6 6 5 3 1 1 0	2	1 3 3 4 6 6
8 7 7 5 2	3	4 4 4 8
2	4	3 6
	5	4

8. Fifteen boys and fifteen girls were randomly chosen from a group of 900 students. Their heights
    (in metres) were measured as shown below.
    Boys: 1.65, 1.71, 1.59, 1.74, 1.66, 1.69, 1.72, 1.66, 1.65, 1.64, 1.68, 1.74, 1.57, 1.59, 1.60
    Girls: 1.66, 1.69, 1.58, 1.55, 1.51, 1.56, 1.64, 1.69, 1.70, 1.57, 1.52, 1.58, 1.64, 1.68, 1.67
    a. Comment on the size of the sample.
    b. Display the data as a back-to-back stem plot.
    c. Compare the heights of the boys and girls.

9. The stem plot shown displays the number of vehicles sold by
    the Ford and Holden dealerships in a Sydney suburb each week
    for a three-month period.
    a. State the medians of both distributions.
    b. Calculate the ranges of both distributions.
    c. Calculate the interquartile ranges of both distributions.
    d. Show both distributions on a parallel box plot.

Key: 1 | 5 = 15 vehicles

Leaf: Ford	Stem	Leaf: Holden
7 4	0	3 9
9 2 2 1 0	1	1 1 1 6 6 8
8 5 4 4	2	2 2 7 9
0	3	5

10. The box plots shown display statistical data for two AFL teams over a season.

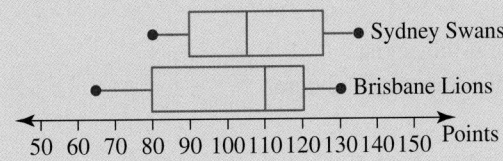

    a. Which team had the higher median score?
    b. What was the range of scores for each team?
    c. For each team calculate the interquartile range.

11. Tanya measures the heights (in m) of a group of Year 10 boys and girls and produces the following
    five-point summaries for each data set.
    Boys: 1.45, 1.56, 1.62, 1.70, 1.81
    Girls: 1.50, 1.55, 1.62, 1.66, 1.73
    a. Draw a box plot for both sets of data and display them on the same scale.
    b. What is the median of each distribution?
    c. What is the range of each distribution?
    d. What is the interquartile range for each distribution?
    e. Comment on the spread of the heights among the boys and the girls.

12. The box plots shown display the average daily sales of cold drinks at the school canteen in summer and winter.

    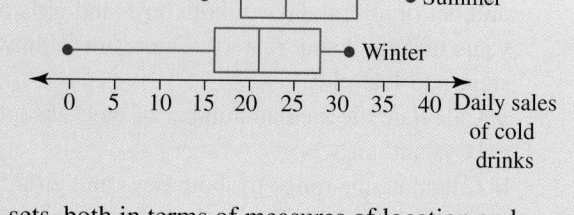

    a. Calculate the range of sales in both summer and winter.
    b. Calculate the interquartile range of the sales in both summer and winter.
    c. Comment on the relationship between the two data sets, both in terms of measures of location and measures of spread.

13. **MC** A movie theatre has taken a survey of the ages of people at a showing of two of their movies. The results are shown in these box plots.
    Which of the following conclusions could be drawn based on the preceding information?
    A. Movie A attracts an older audience than Movie B.
    B. Movie B attracts an older audience than Movie A.
    C. Movie A appeals to a wider age group than Movie B.
    D. Movie B appeals to a wider age group than Movie A.

Communicating, reasoning and problem solving

14. The stem plot shown displays the ages of a group of 30 males and 30 females as they enter hospital for the first time.

    a. Construct a pair of parallel boxplots to represent the two sets of data. Show your working for the median and the 1st and 3rd quartiles.
    b. Calculate the mean, range and IQR for both sets of data.
    c. Determine any outliers, if they exist.
    d. Write a short paragraph comparing the data.

Leaf: Male	Stem	Leaf: Female
9 8	0	5
9 9 8 8 8 6 3 2 1	1	7 7 8 9 9
8 7 7 6 4 3 2 0	2	0 0 1 2 4 5 5 6 7 9
8 6 3 1 0	3	0 1 3 3 5 8
7 5 2	4	2 3 6 8
5 3	5	1 3 4
	6	2
8	7	

15. The test scores out of a total score of 50 for two classes, A and B, are shown in the stem plot.

    a. Ms Vinculum teaches both classes. She made the statement that 'Class A's performance on the test showed that the students' ability was more closely matched than the students' ability in Class B.' By finding the median, the first and third quartiles, and the interquartile range for the test scores for each class, explain if Ms Vinculum's statement was correct.
    b. Would it be correct to say that Class A performed better on the test than Class B? Justify your answer by comparing the quartiles and median for each class.

Leaf: Class A	Stem	Leaf: Class B
5	0	1 2 4
9 7 5 3	1	1 4 5
9 7 7 5 4	2	0 0 5
8 8 6 5 5 1	3	1 5 5
3 2 0	4	1 5 7 7 8 9
0	5	0 0

16. The times, in seconds, of the duration of 20 TV advertisements shown in the 6–8 pm time slot are recorded below.

    16, 60, 35, 23, 45, 15, 25, 55, 33, 20, 22, 30, 28, 38, 40, 18, 29, 19, 35, 75

a. From the data, determine the:

    i. mode

    ii. median

    iii. mean (write your answer correct to 2 decimal places)

    iv. range

    v. lower quartile

    vi. upper quartile

    vii. interquartile range.

b. Using your results from part **a**, construct a boxplot for the time, in seconds, for the 20 TV advertisements in the 6–8 pm time slot.

c. From your boxplot, determine:

    i. the percentage of advertisements that are more than 39 seconds in length

    ii. the percentage of advertisements that last between 21 and 39 seconds

    iii. the percentage of advertisements that are more than 21 seconds in length.

    The types of TV advertisements during the 6–8 pm time slot were categorised as fast food, supermarkets, program information, and retail (clothing, sporting goods, furniture). A frequency table for the frequency of these advertisements being shown during this time slot is shown.

Type	Frequency
Fast food	7
Supermarkets	5
Program information	3
Retail	5

d. What type of data has been collected in the table?

e. What percentage of advertisements are advertisements for fast food outlets?

f. What would be good options for a graphical representation of this type of data?

17. The speeds, in km/h, of 55 cars travelling along a major road were recorded. The results are shown in the table.

a. By finding the midpoint for each class interval, determine the mean speed, in km/h, of the cars travelling along the road. Write your answer correct to 2 decimal places.

b. The speed limit along the road is 75 km/h. A speed camera is set to photograph the license plates of cars travelling 7% more than the speed limit. A speeding fine is automatically sent to the owners of the cars photographed. Based on the 55 cars recorded, how many speeding fines were issued?

c. Drivers of cars travelling 5 km/h up to 15 km/h over the speed limit are fined $135. Drivers of cars travelling more than 15 km/h and up to 25 km/h over the speed limit are fined $165, and drivers of cars recorded travelling more than 25 km/h and up to 35 km/h are fined $250. Drivers travelling more than 35 km/h pay a $250 fine in addition to having their driver's license suspended.

Assume that this data is representative of the speeding habits of drivers along a major road and that there are 30 000 cars travelling along this road on any given month.

Speed	Frequency
60–64	1
65–69	1
70–74	10
75–79	13
80–84	9
85–89	8
90–94	6
95–99	3
100–104	2
105–109	1
110–114	1
Total	**55**

    i. Determine the amount, in dollars, collected in fines throughout the month. Write your answer correct to the nearest cent.

    ii. How many drivers would expect to have their licenses suspended throughout the month?

on To test your understanding and knowledge of this topic, go to your learnON title at www.jacplus.com.au and complete the **post-test**.

Answers

Topic 13 Data analysis

13.1 Pre-test

1. B
2. C
3. Independent variable
4. a. B
   b. C
   c. A
5. D
6. B
7. Interpolation
8. $\dfrac{16}{11}$
9. Independent variable
10. Explanatory variable
11. a. Number of questions: independent;
    test result: dependent
    b.
    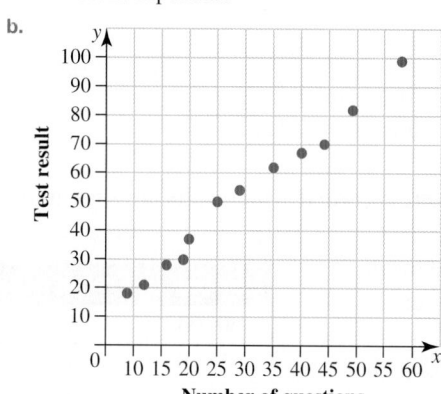
    c. Strong, positive, linear correlation; the larger the number of completed revision questions, the higher the mark on the test.
    d. Different abilities of the students
12. a. i. 12.5   ii. 49
    b. i. 12   ii. 22.5
    c. $y = \dfrac{22}{15}x - \dfrac{7}{3}$
    d. i. 12.33
       ii. 49
       iii. 11.82
       iv. 22.05
13. $x = 6$
14. a. $y = 3.31x + 3.05$
    b. Approximately 53 hours.
15. a. $y = 14$
    b. $x = 12.5$

13.2 Bivariate data

1.

	Independent	Dependent
a.	Number of hours	Test results
b.	Rainfall	Attendance
c.	Hours in gym	Visits to the doctor
d.	Lengths of essay	Memory taken

2.

	Independent	Dependent
a.	Cost of care	Attendance
b.	Age of property	Cost of property
c.	Number of applicants	Entry requirements
d.	Running speed	Heart rate

3.
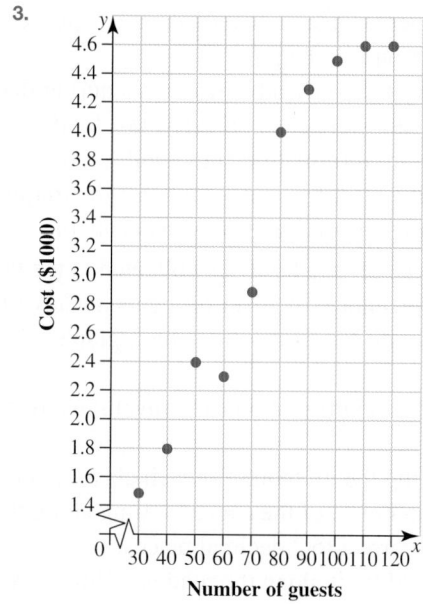

4. a. Perfectly linear, positive
   b. No correlation
   c. Non-linear, negative, moderate
   d. Strong, positive, linear
   e. No correlation
5. a. Non-linear, positive, strong
   b. Strong, negative, linear
   c. Non-linear, moderate, negative
   d. Weak, negative, linear
   e. Non-linear, moderate, positive
6. a. Positive, moderate, linear
   b. Non-linear, strong, negative
   c. Strong, negative, linear
   d. Weak, positive, linear
   e. Non-linear, moderate, positive

7. a.

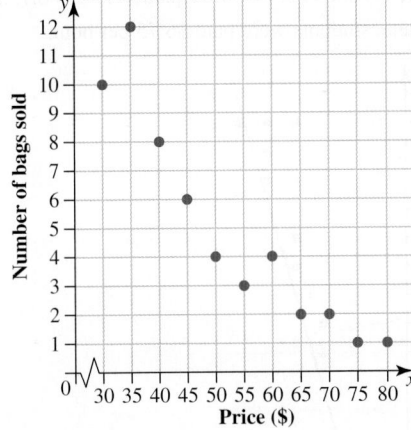

b. Negative, linear, moderate. The price of the bag appeared to affect the numbers sold; that is, the more expensive the bag, the fewer sold.

8. a.

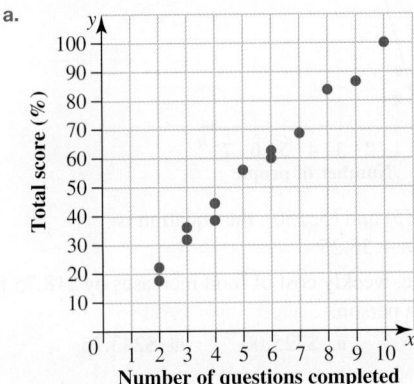

b. Strong, positive, linear correlation

c. Various answers; some students are of different ability levels and they may have attempted the questions but had incorrect answers.

9. a.

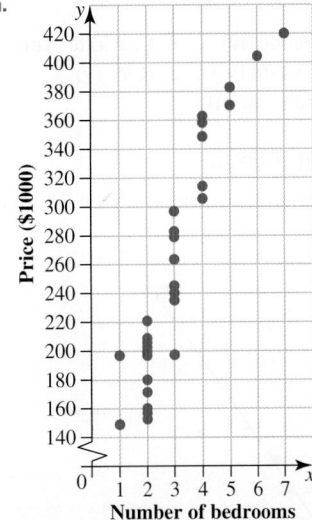

b. Moderate positive linear correlation. There is evidence to show that the larger the number of bedrooms, the higher the price of the house.

c. Various answers; location, age, number of people interested in the house, and so on.

10. a.

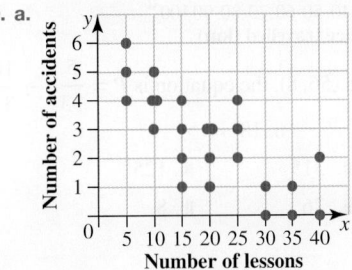

b. Weak, negative, linear relation

c. Various answers; some drivers are better than others, live in lower traffic areas, traffic conditions etc.

11. B

12. C

13. D

14. a. See the figure at the bottom of the page.*

b. This scatterplot does not support the claim.

*14. a.

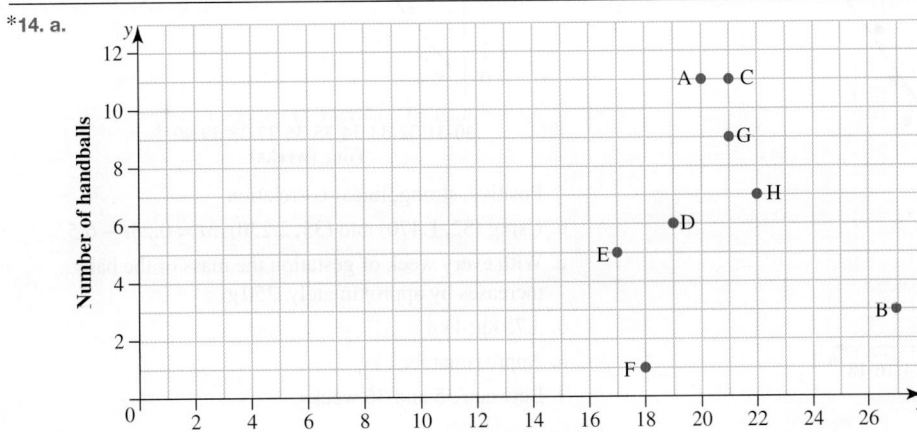

15. a. T   **b.** F   **c.** T   **d.** F   **e.** T

16. a. Saanvi (iii)   **b.** Duong (iv)   **c.** Charlotte (viii)
    d. Dario (vii)   **e.** Edward (vi)   **f.** Rory (v)
    g. Amelia (i)   **h.** Harrison (ii)

13.3 Lines of best fit by eye

Note: Answers may vary slightly depending on the line of best fit drawn.

1. a

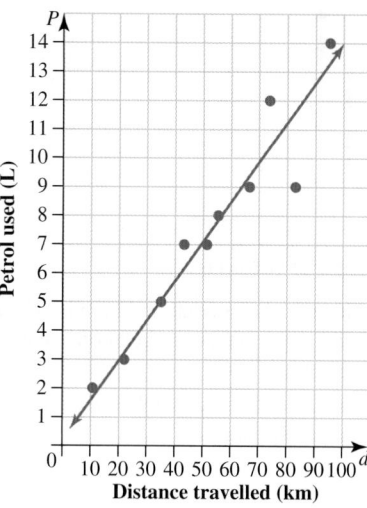

    b. Using, $(23, 3)$ and $(56, 8)$, the equation is $P = \dfrac{5}{33}d - \dfrac{16}{33}$.

2. a. 38   **b.** 18

3. a. i. 510   **ii.** 315   **iii.** 125

    b. i. 36.5   **ii.** 26   **iii.** 8

    c. $y = -13x + 595$

    d. *y*-values (**a**):

      i. 504

      ii. 309

      iii. 127

      x-values (**b**):

      i. 36.54

      ii. 25.77

      iii. 8.85

4. a

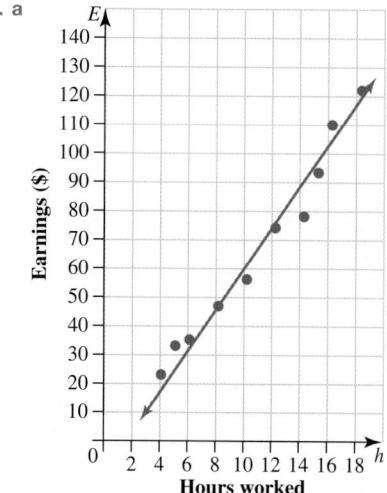

b. Using $(8, 47)$ and $(12, 74)$, the equation is $E = 6.75h - 7$.

c. On average, students were paid \$6.75 per hour.

5. a.

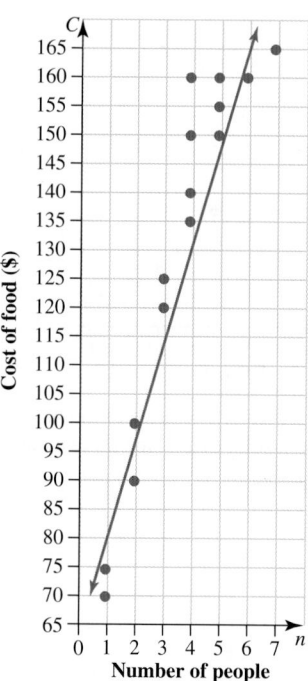

    b. Using $(1, 75)$ and $(5, 150)$, the equation is
      $C = 18.75n + 56.25$

    c. On average, weekly cost of food increases by \$18.75 for
      every extra person.

    d. i. \$206.25   **ii.** \$225.00   **iii.** \$243.75

6. a. 35 to 90

    b. 30 to 105

7. a.

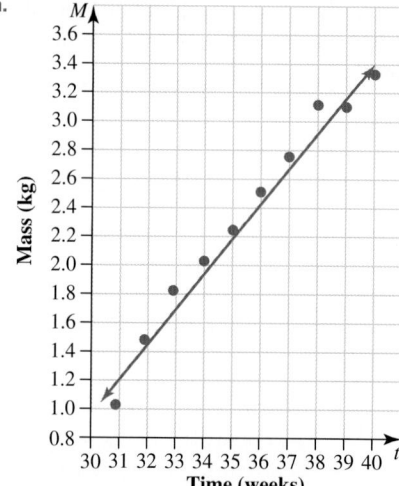

    Positive, strong, linear correlation

    b. Using $(32, 1.470)$ and $(35, 2.230)$, $M = 0.25t - 6.5$.

    c. With every week of gestation the mass of the baby
      increases by approximately. 250 g.

    d. 3.75 kg; 4 kg

    e. Approximately 1 kg

    f. Between 35 and 36 weeks

8. a. D   b. C

9. A

10. a.

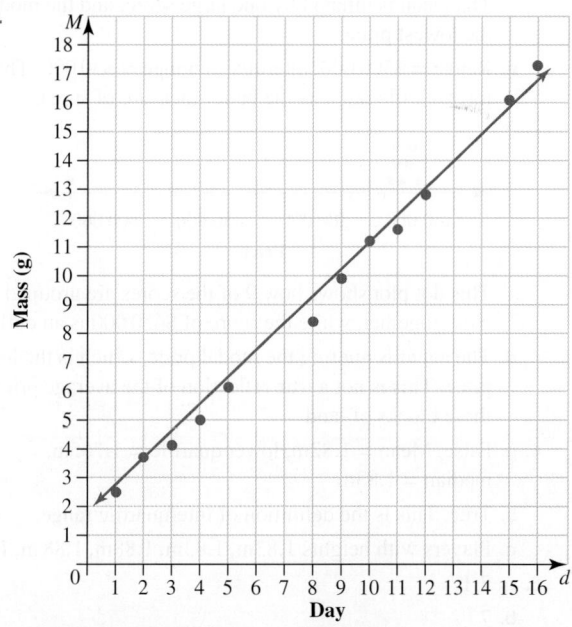

b. Using (2, 3.7) and (10, 11.2), $M = 0.88d + 1.94$.

c. Each day Rachel's crystal gains $0.88\,g$ in mass. Line of best fit appears appropriate.

d. $7.22\,g$; $8.10\,g$; $13.38\,g$ and $14.26\,g$; interpolation (within the given range of 1–16).

e. $16.9\,g$ and $17.78\,g$; predictions are not reliable, since they were obtained using extrapolation.

11. a. See the figure at the bottom of the page.*

b. $L = 1.07n + 18.9$

c. 24.25 cm; 26.39 cm; 28.53 cm; 30.67 cm; 31.74 cm; 34.95 cm; 36.02 cm; 38.16 cm; 39.23 cm

d. Interpolation (within the given range of 1–20).

e. 41.37 cm; 42.44 cm; 43.51 cm

f. Not reliable, because extrapolation has been used.

12. No. He would have to get 108% which would be impossible on a test.

13. a.

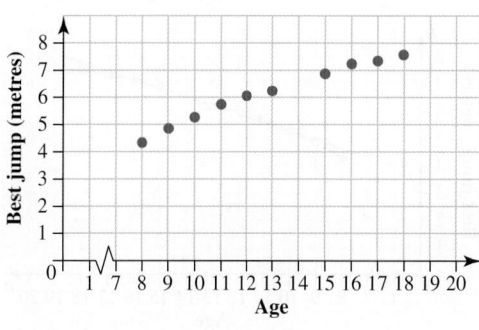

*11. a.

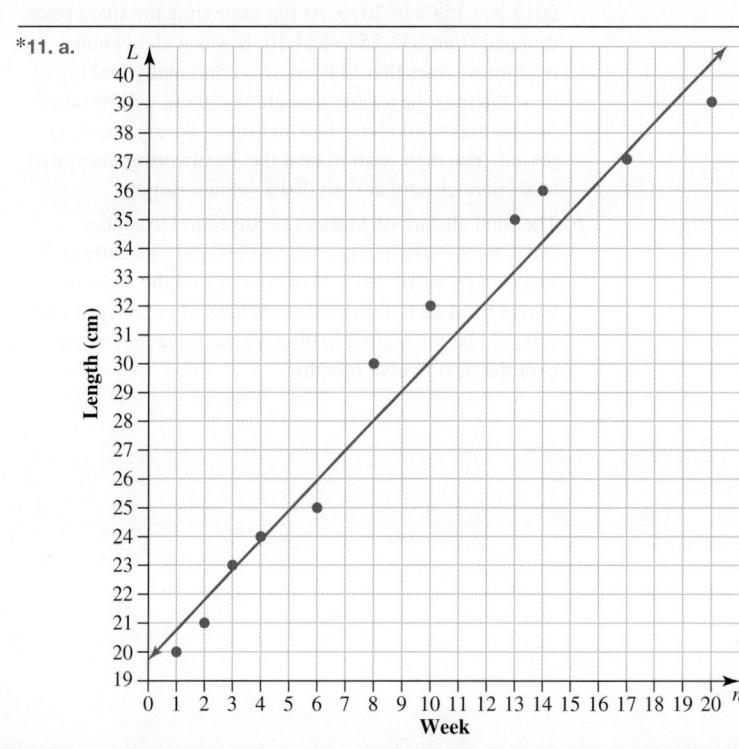

b.

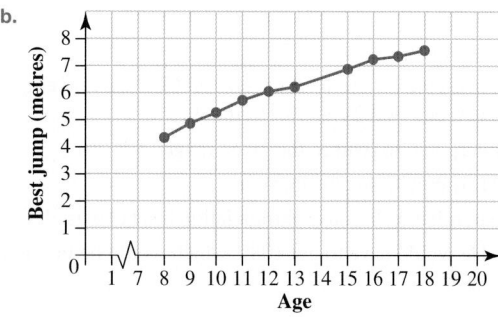

c.

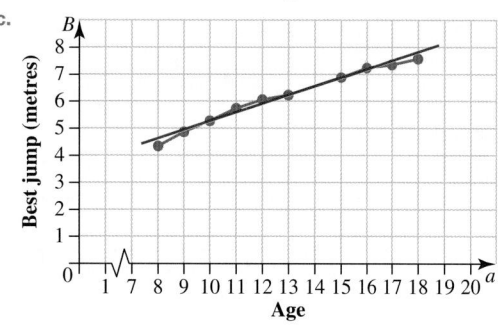

d. Yes. Using points $(9, 4.85)$ and $(16, 7.24)$, $B = 0.34a + 1.8$; estimated best jump $= 8.6$ m.

e. No, trends work well over the short term but in the long term are affected by other variables.

f. 24 years old: 9.97 m; 28 years old: 11.33 m. It is unrealistic to expect his jumping distance to increase indefinitely.

g. Equal first.

13.4 Evaluating statistical reports (Path)

1. a. Primary. There is probably no secondary data available.

b. Answers will vary. Check with your teacher.

c. Sample responses can be found in the worked solutions in the online resources.

2. Company profits

Company profits

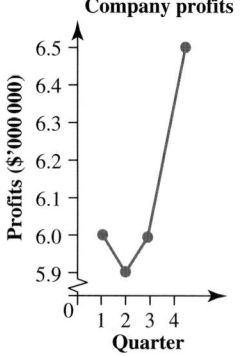

Mean salaries

Company salaries

3. a. Mean $= \$215\,000$, median $= \$170\,000$, mode $= \$150\,000$. The median best represents these land prices. The mean is inflated by one large score, and the mode is the lowest price.

b. Range $= \$500\,000$, interquartilerange $= \$30\,000$. The interquartile range is the better measure of spread.

c.

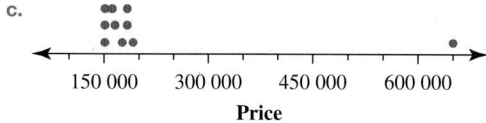

This dot plot shows how 9 of the scores are grouped close together, while the score of $\$650\,000$ is an outlier.

d. The agent is quoting the modal price, which is the lowest price. This is not a true reflection of the average price of these blocks of land.

4. a. False. Mean $= 1.82$m, lower quartile $= 1.765$m, median $= 1.83$m

b. True. This is the definition of interquartile range.

c. Players with heights 1.83m, 1.83m, 1.88m, 1.88 m, 1.83m

5. a. 7.1

b. 7

c. 7

d. The mode has the most meaning as this size sells the most.

6. a. The statement is true but misleading, as most of the employees earn $\$55\,000$.

b. The median and modal salary is $\$57\,563$, and only 20 out of 80 (25%) earn more than the mean.

7. Player B appears to be the better player if the mean result is used. However, Player A is the more consistent player.

8. a. On face value the statement is true, as the profit and share price from 2013 have both dropped remarkably. The profit rose by only 9% in 2014, then dropped by 12% in 2015 and 158% in 2016. At the same time the share price dropped from $\$33.85$ to $\$24.10$. It was widely known over those years that Masters was continually making a loss. We must be aware, though, that there will be other factors that would have had an impact during the same period. You may wish to visit the Woolworths website to look more closely at their 2016 Annual Report.

b. The final closure of Masters occurred in December 2016, so the sale had not been or had only recently been finalised by April 2017. Investors caused the share price to rise from $\$24.10$ in 2016 to $\$26.88$ as at 28 April 207. It would be interesting to observe the price movement over the next twelve months.

9. The bar chart on the left suggests that prices have tripled in one year; this is because the vertical axis does not start at zero. The bar chart on the right is truly indicative of situation.

10. a. Key: 3|85 = 385 hours

Leaf: Brand B	Stem	Leaf: Brand A
	3	85 90
	4	25 26
60 55 00	5	70
30	6	40 45
70 42 35 20	7	30 35
60 20	8	60

Brand A: mean = 570.6, median = 605
Brand B: mean = 689.2, median = 727.5

b. Brand A had the shortest mean lifetime.

c. Brand B had the longest mean lifetime.

d. Brand B

11. Shorten the *y*-axis and expand the *x*-axis. See the figure at the bottom of the page.*

12. a, b. Percentages do not add to 100%.

c. Such representation allows multiple choices to have greater percentages than really exist.

13. To support an idea presented by the creator of the graph.

13.5 Statistical investigations (Path)

1. Sample responses can be found in the worked solutions in the online resources.

2. Facebook (2.9 billion users)

3. 113 minutes per day

4. Advertising

5. 6 to 9 hours per day

6. Answers will vary. Discuss with your teacher. A sample answer is:
   – Men have been paid more in the past and the pay gap is decreasing.
   – Women's sport has recently significantly increased in prominence, bringing in much larger advertising revenue and ticket sales.

7. a. The sample would be biased since respondents would all be from the same area and the same age group, and would not be representative of the total population.

b. The sample would be biased since it is from only one city, so for example the respondents might not be representative of different socio-economic groups and ethnic backgrounds. The survey should include people from the country as well as the city.

c. Sample responses can be found in the worked solutions in the online resources.

8. Student's plan for an investigation — check with your teacher.

9. Answers will vary but could include:
   a. All households in the area serviced by the hospital
   b. Contacts obtained from the local electoral rolls
   c. Bias could be introduced in the wording of the questions; only people who feel passionately about the issue might respond.

10. Sample responses can be found in the worked solutions in the online resources.

11. 2, 4, 10, 10, 14; 8 solutions

13.6 Selecting a sample size (Path)

1. a. When was it first put into the machine? How old was the battery before being purchased? How frequently has the computer been used on battery?

b. Can't always see if a residence has a dog; a census is very time-consuming; perhaps could approach council for dog registrations.

c. This number is never constant with ongoing purchases, and continuously replenishing stock.

d. Would have to sample in this case as a census would involve opening every packet.

2. Answers will vary with the samples chosen. Sample responses can be found in the worked solutions in the online resources.

3. a. Census. The airline must have a record of every passenger on every flight.

b. Survey. It would be impossible to interview everyone.

c. Survey. A census would involve opening every bottle.

d. Census. The instructor must have an accurate record of each learner driver's progress.

*9.

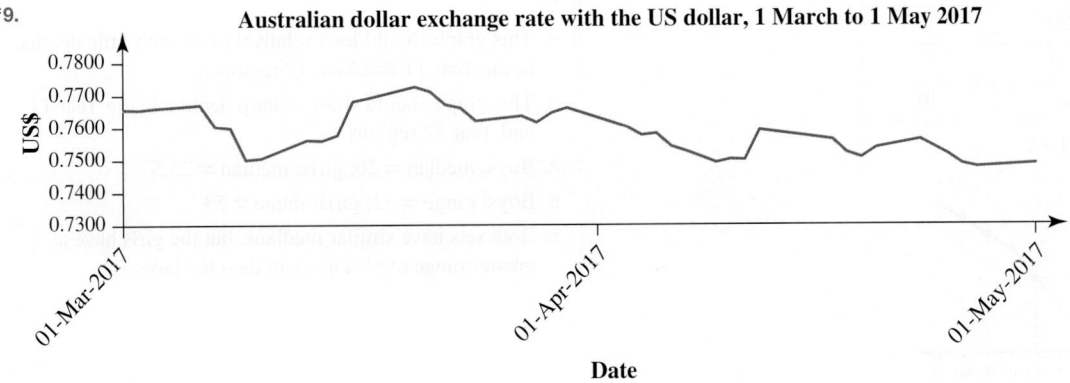

Australian dollar exchange rate with the US dollar, 1 March to 1 May 2017

4. a. Survey   **b.** Survey
  c. Census   **d.** Survey

5. a. About 25

  b. Answers will vary. Possible answers include drawing numbers from a hat or using the random number generator on a calculator.

6. a. The council is probably hoping it is a census, but it will probably be a survey because not all those over 10 will respond.

  b. Residents may not all have internet access. Only those who are highly motivated are likely to respond.

7. The sample could have been biased. The questionnaire may have been unclear.

8. Sample size, randomness of sample.

9. Sample responses can be found in the worked solutions in the online resources.

10. Answers will vary. Possible answers include:
The populations of China and India are growing very rapidly.
Many expatriate workers in China have different backgrounds, and forms need to be modified for them (such as people from Hong Kong working in mainland China).
There is a large migrant population in New Delhi, and often migrants don't have residency permits (so the truth of their answers is questionable).
Many people live in remote areas that are difficult to access.
Some families in China have more than 1 child and do not disclose this (due to social pressures relating to the Chinese government's former one-child policy).

11. There is quite a variation in the frequency of particular numbers drawn. For example, the number 45 has not been drawn for 31 weeks, while most have been drawn within the last 10 weeks. In the long term, one should find the frequency of drawing each number is roughly the same. It may take a long time for this to happen, as only 8 numbers are drawn each week.

12. a. Mean = 32.03; median = 29

  b.

Class interval	Frequency
0–9	2
10–19	7
20–29	6
30–39	6
40–49	3
50–59	3
60–69	3
Total	**30**

  c. Mean = 31.83

  d.

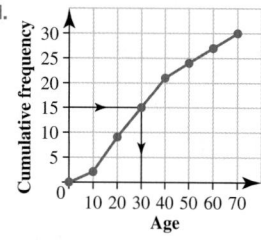

e. Median = 30

 f. Estimates from parts **c** and **e** were fairly accurate.

 g. Yes, they were fairly close to the mean and median of the raw data.

13. Year 8: mean = 26.83, median = 27, range = 39, IQR = 19
Year 10: mean = 40.7, median = 39.5, range = 46, IQR = 20
The typing speed of Year 10 students is about 13 to 14 wpm faster than that of Year 8 students. The spread of data in Year 8 is slightly less than the spread in Year 10.

14. Statistics quantifies uncertainty.

15. 1250 trout

Project

1. Data for the whole population: minimum = 24, $Q_1 = 61.5$, median = 87, $Q_3 = 98$, maximum = 100

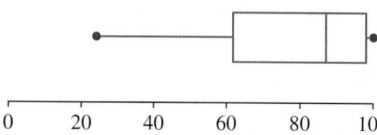

2. Mean = 77.85, standard deviation = 23

3–10. Answers will vary depending on the sample. Sample responses can be found in the worked solutions in the online resources.

13.7 Review questions

1. a. You would need to open every can to determine this.

  b. Fish are continuously dying, being born and being caught.

  c. Approaching work places and public transport offices

2. a. 50.5

  b, c, d. Sample responses can be found in the worked solutions in the online resources.

3. a. Survey

  b. Census

  c. Survey

4. Use a spinner with the green and red sectors being an equal size (say 174°) and the amber section being a smaller size (say 12°). Twirl the spinner until a green-green combination has been obtained. This is defined as one experiment. Count the number of trials required for this experiment. Repeat this procedure a number of times and determine an average.

5. D

6. a. This graph should look relatively flat, with little decline in the Year 11 and Year 12 regions.

  b. This graph should show a sharp decline in the Year 11 and Year 12 regions.

7. a. Boys: median = 26; girls: median = 23.5

  b. Boys: range = 32; girls: range = 53

  c. Both sets have similar medians, but the girls have a greater range of absenteeism than the boys.

8. a. The sample is an appropriate size as $\sqrt{900} = 30$.

b. Key: 16|1 = 1.61

Leaf: Boys	Stem	Leaf: Girls
9 9 7	15	1 2 5 6 7 8 8
9 8 6 6 5 5 4 0	16	4 4 6 7 8 9 9
4 4 2 1	17	0

c. The boys are generally taller than the girls, with the mean of the boys being 1.66 m and that of the girls being 1.62m. The five-number summaries are:
Boys: 1.57 m, 1.60 m, 1.66 m, 1.71 m, 1.74 m
Girls: 1.51 m, 1.56 m, 1.64 m, 1.68 m, 1.70 m

9. a. Ford: median = 15; Holden: median = 16

b. Ford: range = 26; Holden: range = 32

c. Ford: IQR = 14; Holden: IQR = 13.5

d.

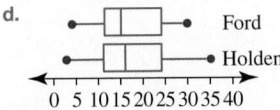

10. a. Brisbane Lions

b. Brisbane Lions: range = 65; Sydney Swans: range = 55

c. Brisbane Lions: IQR = 40; Sydney Swans: IQR = 35

11. a.

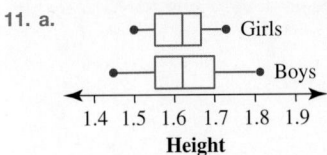

b. Boys: median = 1.62 m; girls: median = 1.62 m

c. Boys: range = 0.36 m; girls: range = 0.23 m

d. Boys: IQR = 0.14 m; girls: IQR = 0.11 m

e. Although the boys and girls have the same median height, the spread of heights is greater among boys as shown by the greater range and interquartile range.

12. a. Summer: range = 23; winter: range = 31

b. Summer: IQR = 13; winter: IQR = 12

c. There are generally more cold drinks sold in summer as shown by the higher median. The spread of data is similar as shown by the IQR although the range in winter is greater.

13. A

14. a. See the figure at the bottom of the page.*

b.

	Males	Females
Mean	28.2	31.1
Range	70	57
IQR	18	22

c. There is one outlier — a male aged 78.

d. Typically males seem to enter hospital for the first time at a younger age than females.

15. a. Class A: $Q_1 - 21.5$, Median $- 30$, $Q_3 - 38$, IQR $- 16.5$
Class B: $Q_1 - 14.5$, Median $- 33$, $Q_3 - 47$, IQR $- 32.5$
Based on the comparison between Class A's IQR (16.5) and Class B's IQR (32.5), Ms Vinculum was correct in her statement.

b. No, Class B has a higher median and upper quartile score than Class A, whereas Class A has a higher lower quartile. You can't confidently say that either class did better in the test than the other.

16. a. i. 35 s

　ii. 29.5 s

　iii. 33.05 s

　iv. 60 s

　v. 21 s

　vi. 39 s

　vii. 18 s

b.

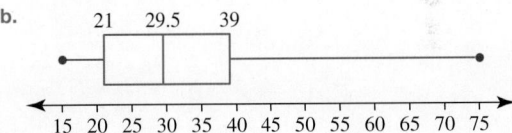

c. i. 25%

　ii. 50%

　iii. 75%

d. Categorical

e. 35%

f. Pictogram, pie chart or bar chart

17. a. 82.73 km/h

b. 30 cars

c. i. $2607272.73

　ii. About 545

*14. a.

Females

Males

14 Probability

LESSON
14.1 Overview

Why learn this?

Probability is a broad and interesting area of mathematics that affects our day-to-day lives far more than we can imagine. Here is a fun fact: did you know there are so many possible arrangements of the 52 cards in a deck $\left(52! = 8.0658 \times 10^{67}\right)$ that the probability of ever getting the same arrangement after shuffling is virtually zero? This means every time you shuffle a deck of cards, you are almost certainly producing an arrangement that has never been seen before. Probability is also a big part of computer and board games; letters X and Q in Scrabble are worth more points because you are less likely to be able to form a word using those letters. It goes without saying that probability is a big part of any casino game and of the odds and payouts when gambling on the outcome of racing or sports.

While it is handy to know probability factoids and understand gambling, this isn't the reason we spend time learning probability. Probability helps us build critical thinking skills, which are required for success in almost any career and even just for navigating our own lives. For example, if you were told your chance of catching a rare disease had doubled you probably wouldn't need to worry, as a 1-in-a-million-chance becoming a 2-in-a-million chance isn't a significant increase in the probability of you developing the disease. On the other hand, if a disease has a 1% mortality rate that may seem fairly low, but it means that if a billion people developed that disease, then 10 million would die. Using probability to understand risk helps us steer clear of manipulation by advertising, politicians and the media. Building on this understanding helps us as individuals make wise decisions in our day-to-day life, whether it be investing in the stock market, avoiding habits that increase our risk of sickness, or building our career.

1. PATH Calculate $P(A \cap B)$ if $P(A) = 0.4$, $P(B) = 0.3$ and $P(A \cup B) = 0.5$.

2. PATH If events A and B are mutually exclusive, and $P(B) = 0.38$ and $P(A \cup B) = 0.89$, calculate $P(A)$.

3. State whether the events $A = \{$drawing a red marble from a bag$\}$ and $B = \{$rolling a 1 on a die$\}$ are independent or dependent.

4. PATH The Venn diagram shows the number of university students in a group of 25 who own a computer and/or tablet.

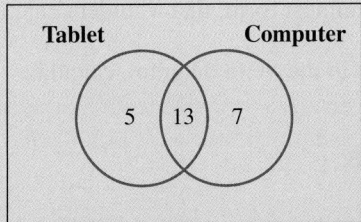

   In simplest form, calculate the probability that a university student selected at random will own only a tablet.

5. MC Two unbiased four-sided dice are rolled. Determine the probability that the total sum of two face-down numbers obtained is 6.

   A. $\dfrac{1}{8}$ B. $\dfrac{1}{16}$ C. $\dfrac{3}{16}$ D. $\dfrac{1}{4}$

6. PATH MC Identify which Venn diagram best illustrates $P(A \cup B)'$.

   A.
   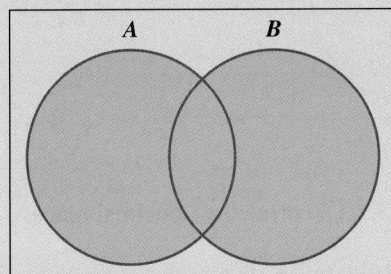

   B.
   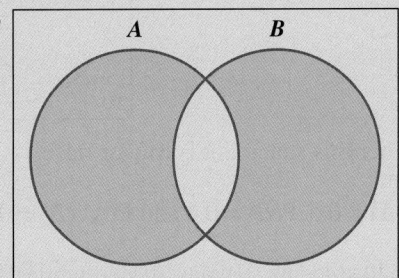

   C.
   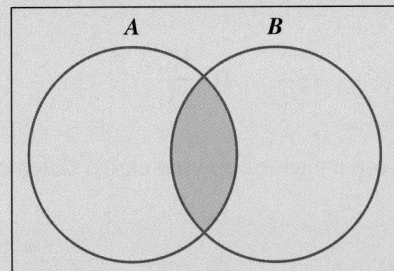

   D.
   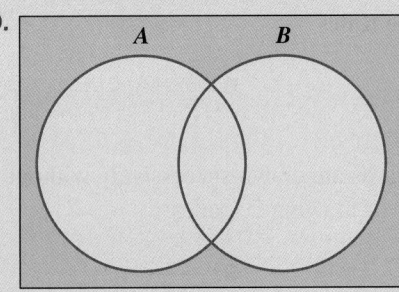

7. `PATH` `MC` From a group of 25 people, 12 use Instagram (I), 14 use Snapchat (S), and 6 use both Instagram and Snapchat applications.
Determine the probability that a person selected at random will use neither application.

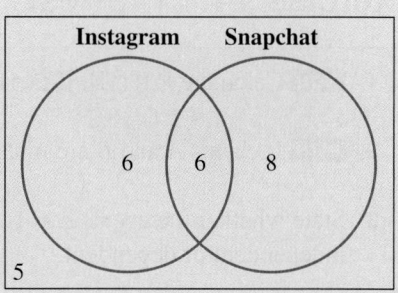

A. $\dfrac{1}{5}$ B. $\dfrac{6}{25}$ C. $\dfrac{8}{25}$ D. $\dfrac{12}{25}$

8. The probability that a student will catch a bus to school is 0.7 and the independent probability that a student will be late to school is 0.2.
Determine the probability, in simplest form, that a student catches a bus and is not late to school.

9. `PATH` `MC` From events A and B in the Venn diagram, calculate P(A|B).

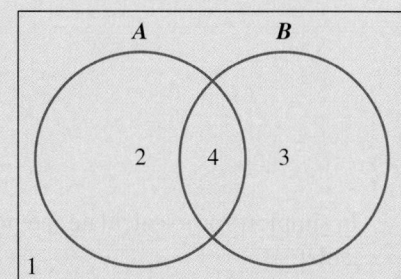

A. $\dfrac{4}{9}$ B. $\dfrac{4}{7}$

C. $\dfrac{2}{5}$ D. $\dfrac{1}{3}$

10. `PATH` In the Venn diagram, events A and B are independent.

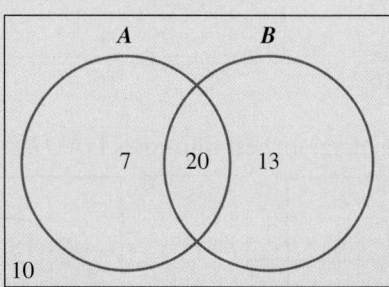

State whether this statement is true or false.

11. `PATH` If $P(A) = 0.5$, $P(B) = 0.4$ and $P(A \cup B) = 0.8$, calculate $P(B \mid A)$, correct to 1 decimal place.

12. `MC` From 20 students, 10 play soccer, while 15 play Aussie Rules and 8 play both soccer and Aussie Rules. Calculate the probability that a student randomly selected plays soccer given that they play Aussie Rules.

A. $\dfrac{2}{3}$ B. $\dfrac{2}{5}$ C. $\dfrac{8}{15}$ D. $\dfrac{10}{23}$

13. `MC` Two cards are drawn successively without replacement from a pack of playing cards. Calculate the probability of drawing 2 spades.

A. $\dfrac{1}{17}$ B. $\dfrac{2}{2652}$ C. $\dfrac{1}{2652}$ D. $\dfrac{1}{2704}$

14. A survey of a school of 800 students found that 100 used a bus (B) to get to school, 75 used a train (T) and 650 used neither.
In simplest form, determine the probability that a student uses both a bus and a train to get to school.

15. **PATH** **MC** On the first day at school, students are asked to tell the class about their holidays. There are 30 students in the class and all have spent part or all of their holidays at one of the following: a coastal resort, interstate, or overseas.

The teacher finds that:
- 5 students went to a coastal resort only
- 2 students went interstate only
- 2 students holidayed in all three ways
- 8 students went to a coastal resort and travelled overseas only
- 20 students went to a coastal resort
- no less than 4 students went overseas only
- no less than 13 students travelled interstate

Determine the probability that a student travelled overseas and interstate only.

A. $\dfrac{2}{15}$ B. $\dfrac{5}{6}$ C. $\dfrac{3}{15}$ D. $\dfrac{1}{3}$

LESSON
14.2 Review of probability and simulations

LEARNING INTENTION

At the end of this lesson you should be able to:
- use key probability terminology such as: trials, frequency, sample space, likely and unlikely events
- compare theoretical and experimental probabilities
- determine probabilities for complementary events
- design and use simulations to model and examine events involving probability.

▶ ### 14.2.1 The language of probability

eles-4922

- **Probability** measures the chance of an event taking place and ranges from 0 for an impossible event to 1 for a certain event.

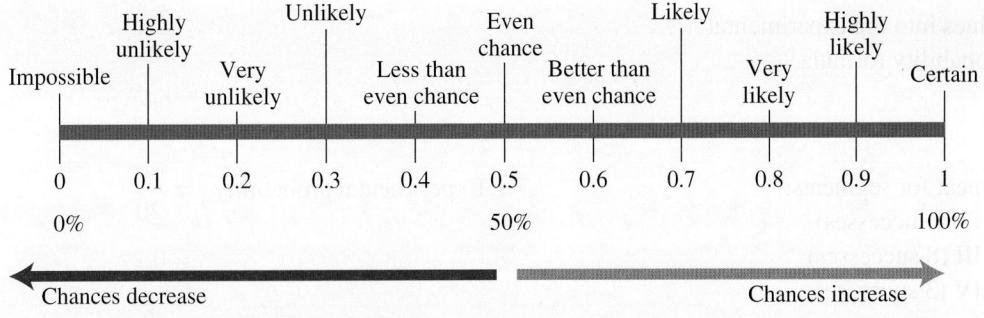

- The **experimental probability** of an event is based on the outcomes of experiments, simulations or surveys.
- A **trial** is a single experiment, for example, a single flip of a coin.

Experimental probability

$$\text{Experimental probability} = \frac{\text{number of successful trials}}{\text{total number of trials}}$$

- The experimental probability of an event is also known as the **relative frequency**.
- The list of all possible outcomes of an experiment is known as the **event space** or **sample space**.
  For example, when flipping a coin there are two possible outcomes: Heads or Tails. The event space can be written, using set notation, as {H, T}.

WORKED EXAMPLE 1 Sample space and calculating experimental probability

The spinner shown here is made up of 4 equal-sized segments. It is known that the probability that the spinner will land on any one of the 4 segments from one spin is $\frac{1}{4}$. To test if the spinner shown here is fair, a student spun the spinner 20 times and each time recorded the segment in which the spinner stopped. The spinner landed as follows.

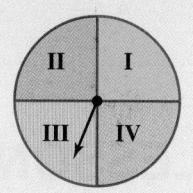

Segment	I	II	III	IV
Tally	5	4	8	3

a. List the sample space.
b. Given the experimental results, determine the experimental probability of each segment.
c. Compare the experimental probabilities with the known probabilities and suggest how the experiment could be changed to ensure that the results give a better estimate of the true probability.

THINK

a. The sample space lists all possible outcomes from one spin of the spinner. There are 4 possible outcomes.

b. 1. For segment I there were 5 successful trials out of the 20. Substitute these values into the experimental probability formula.

2. Repeat for segments:
   - II (4 successes)
   - III (8 successes)
   - IV (3 successes).

WRITE

a. Sample space = {I, II, III, IV}

b. $\text{Experimental probability}_{\text{I}} = \dfrac{\text{number of successful trials}}{\text{total number of trials}}$

$\phantom{\text{Experimental probability}_{\text{I}}} = \dfrac{5}{20}$

$\phantom{\text{Experimental probability}_{\text{I}}} = 0.25$

$\text{Experimental probability}_{\text{II}} = \dfrac{4}{20}$

$\phantom{\text{Experimental probability}_{\text{II}}} = 0.2$

$\text{Experimental probability}_{\text{III}} = \dfrac{8}{20}$

$\phantom{\text{Experimental probability}_{\text{III}}} = 0.4$

$\text{Experimental probability}_{\text{IV}} = \dfrac{3}{20}$

$\phantom{\text{Experimental probability}_{\text{IV}}} = 0.15$

c. Compare the experimental frequency values with the known value of $\frac{1}{4}$ (0.25). Answer the question.	c. The experimental probability of segment I was the only segment that mirrored the known value. To ensure that experimental probability gives a better estimate of the true probability, the spinner should be spun many more times.

Theoretical probability

- **Theoretical probability** is the probability of an event occurring, based on the number of possible favourable outcomes, $n(E)$, and the total number of possible outcomes, $n(S)$.

Theoretical probability

When all outcomes are equally likely, the theoretical probability of an event can be calculated using the formula:

$$P(\text{event}) = \frac{\text{number of favourable outcomes}}{\text{total number of possible outcomes}} \quad \text{or} \quad P(\text{event}) = \frac{n(E)}{n(S)}$$

where $n(E)$ is the number of favourable events and $n(S)$ is the total number of possible outcomes.

WORKED EXAMPLE 2 Calculating theoretical probability

A fair die is rolled and the value of the uppermost side is recorded. Calculate the theoretical probability that a 4 is uppermost.

THINK	WRITE
1. Write the number of favourable outcomes and the total number of possible outcomes. The number of 4s on a fair die is 1. There are 6 possible outcomes.	$n(E) = 1$ $n(S) = 6$
2. Substitute the values found in part **1** to calculate the probability of the event that a 4 is uppermost when a die is rolled.	$P(\text{a } 4) = \dfrac{n(E)}{n(S)}$ $= \dfrac{1}{6}$
3. Write the answer in a sentence.	The probability that a 4 is uppermost when a fair die is rolled is $\dfrac{1}{6}$.

14.2.2 Complementary events

eles-4923

- The **complement** of the set A is the set of all elements that belong to the universal set (S) but that do *not* belong to A.
- The complement of A is written as A' and is read as 'A dashed' or 'A prime'.
- The probability of the complement of an event is the probability that the event will not happen.
  For example, if an event is 'drawing a diamond' from a deck of cards, then the complement of this is 'not drawing a diamond' from a deck of cards, which can also be described as 'drawing a heart, club or spade' from a deck of cards.

Complementary events

Since complementary events fill the entire sample space:

$$P(A) + P(A') = 1$$

- This equation can be expressed in terms of the complement.

Probability of a complementary event

$$P(A') = 1 - P(A)$$

WORKED EXAMPLE 3 Determining complementary events

A player is chosen from a cricket team. Are the events 'selecting a batter' and 'selecting a bowler' complementary events if a player can have more than one role? Give a reason for your answer.

THINK	WRITE
Explain the composition of a cricket team. Players who can bat or bowl are not necessarily the only players in a cricket team. There is a wicket-keeper as well. Some players (all rounders) can both bat and bowl.	No, the events 'selecting a batter' and 'selecting a bowler' are not complementary events. These events may have common elements, that is, the all rounders in the team who can bat and bowl. The cricket team also includes a wicket-keeper.

14.2.3 Simulations

eles-6289

- Sometimes it is not possible to conduct trials of a real experiment because it is too expensive, too difficult, or impracticable.
- In situations like this, outcomes of events can be modelled using devices such as spinners, dice or coins to simulate or represent what happens in real life.
- These simulated experiments are called **simulations**.

WORKED EXAMPLE 4 Modelling real life situations using simulations

To simulate whether a baby is born male or female, a coin is flipped. If the coin lands Heads up, the baby is a boy. If the coin lands Tails up, the baby is a girl. Student 1 flips a coin 10 times and obtains 3 Heads and 7 Tails, while student 2 flips a coin 100 times and obtains 43 Heads and 57 Tails.

a. Determine the relative frequency of female babies in both cases. Give your answers as decimals.

b. Compare the results from the two simulations and determine which is a better estimate of the true probability.

THINK	WRITE
a. 1. A baby born female is a successful trial. For student 1: in the simulation, females are Tails. There were 7 Tails, so 7 females were born and there were 7 successful trials out of 10 trials. Write the formula for relative frequency and substitute the results.	a. Student 1: $$\text{Relative frequency} = \frac{\text{number of successful trials}}{\text{total number of trials}}$$ $$= \frac{7}{10}$$ $$= 0.7$$

2. For student 2: there were 57 Tails, so 57 females were born and there were 57 successful trials out of 100 trials. Write the formula for relative frequency and substitute the results.

Student 2:

$$\text{Relative frequency} = \frac{\text{number of successful trials}}{\text{total number of trials}}$$

$$= \frac{57}{100}$$

$$= 0.57$$

b. 1. Compare the results of the two simulations. Theoretically, there should be an equal number of males and females born, so we would expect the results to be close to 0.5.

2. Give your answer.

b.
Student	Relative frequency
1	0.7
2	0.57

The true probability is likely to be close to 0.5, so student 2 achieved the better estimate of probability.

▶ 14.2.4 Using technology to perform simulations

eles-6293

- Simulating experiments using manual devices such as dice and spinners can take a lot of time.
- A more efficient method of collecting results is to use a list of randomly generated numbers.
- **Random number generators** can generate a series of numbers between two given values.

Digital technology

Scientific calculators have the ability to be random number generators in two different features.

- **Ran#** (SHIFT +·):
  Ran# generates a 3-decimal random number between 0 and 1. Each time you press = another random number will be displayed.

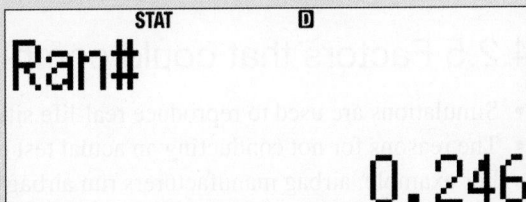

- **RanInt** (ALPHA +·):
  RanInt generates a random number between a certain interval. For example, RanInt#(1,6) will randomly generate a number from 1 to 6 (which can simulate a die). Each time you press = another random number in the interval will be displayed.

- The following table shows some Excel formulas that generate random numbers.

Formula	Output
= RAND()	A random decimal from 0 to 0.9999
= INT(6*RAND() + 1)	Integers between 1 and 6 that can be used to simulate rolling a die. Both of these formulas can be modified to generate the type of numbers that you require.
= RANDBETWEEN(1, 6)	Random numbers between 1 and 6

WORKED EXAMPLE 5 Using technology for simulation and to calculate probabilities

a. **Use a random number generator to simulate the number of chocolate chips in 50 biscuits, with a maximum of 80 chocolate chips in each.**

b. **Calculate the experimental probability that there will be more than 30 chocolate chips in a randomly chosen biscuit.**

THINK

a. Use a random number generator to generate 50 numbers between 0 and 80.
Use either a scientific calculator or an Excel spreadsheet.

WRITE

a.

8	55	8	62	80	50	42	80	57	39
15	7	64	73	47	12	74	74	16	42
41	22	50	33	68	72	64	16	6	72
70	72	52	48	14	22	59	48	65	34
67	62	72	59	10	30	13	7	40	18

b. 1. Count the number of biscuits with more than 30 chocolate chips. There are 34 in the example shown.

b.

8	55	8	62	80	50	42	80	57	39
15	7	64	73	47	12	74	74	16	42
41	22	50	33	68	72	64	16	6	34
70	72	52	48	14	22	59	48	65	34
62	62	72	59	10	30	13	7	40	18

2. The experimental probability is the number of successful outcomes (34) divided by the number of trials (50).

$$P(\text{more than 30 chocolate chips}) = \frac{34}{50}$$
$$= \frac{17}{25}$$

▶ 14.2.5 Factors that could complicate simulations

eles-6294

- Simulations are used to reproduce real-life situations when it is not feasible to implement the actual test.
- The reasons for not conducting an actual test can include financial or safety constraints.
  For example, airbag manufacturers run airbag tests using cars, but instead of using real humans, they use mannequins. These simulations are fairly accurate, but they have their limitations.

WORKED EXAMPLE 6 Identifying limitations of simulations

Six runners are competing in a race.

a. **Explain how simulation could be used to determine the winner.**

b. **Determine if there are any limitations in your answer.**

THINK

a. There are six runners, so six numbers are needed to represent them. A die is the obvious choice. Assign numbers 1 to 6 to each runner.

b. In this simulation, each runner had an equally likely chance of winning each game.

WRITE

a. Each runner could be assigned a number on a die, which could be rolled to determine the winner of the race. When the die is rolled, it simulates a race. The number that lands uppermost is the winner.

b. The answer depends on each runner having an equally likely chance of winning the race. This is rarely the case.

 Resources

Interactivities Experimental probability (int-3825)
Theoretical probability (int-6081)

Exercise 14.2 Review of probability and simulations   **learn** on

14.2 Quick quiz on	**14.2 Exercise**

Individual pathways

■ PRACTISE	■ CONSOLIDATE	■ MASTER
1, 2, 4, 7, 8, 9, 13, 16	3, 5, 10, 14, 15, 17, 19	6, 11, 12, 18, 20

Fluency

1. Explain the difference between experimental and theoretical probability.

2. **WE1** The spinner shown was spun 50 times and the outcome each time was recorded in the table.

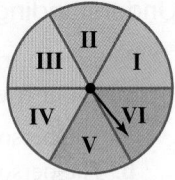

Segment	I	II	III	IV	V	VI
Tally	10	6	8	7	12	7

   a. List the sample space.
   b. Given the experimental results, determine the relative frequency for each segment.
   c. The theoretical probability of the spinner landing on any particular segment with one spin is $\frac{1}{6}$. State how the experiment could be changed to give a better estimate of the true probabilities.

3. A laptop company conducted a survey to see what were the most appealing colours for laptop computers among 15–18-year-old students. The results were as follows:

Colour	Black Black	Sizzling Silver	Power Pink	Blazing Blue	Gooey Green	Glamour Gold
Number	102	80	52	140	56	70

   a. Calculate the number of students who were surveyed.
   b. Calculate the relative frequency of students who found silver the most appealing laptop colour.
   c. Calculate the relative frequency of students who found black and green to be their most appealing colours.
   d. State which colour was found to be most appealing.

4. **WE2** A die is rolled. Calculate the probability that the outcome is an even number or a 5.

5. A die is rolled. Calculate the probability of not getting a 4.

6. **WE3** For each of the following pairs of events:

   i. state, giving justification, if the pair are complementary events

   ii. alter the statements, where applicable, so that the events become complementary events.

     a. Having Weet Bix or having Strawberry Pops for breakfast

     b. Walking to a friend's place or driving there

     c. Watching TV or reading as a leisure activity

     d. Rolling a number less than 5 or rolling a number greater than 5 with a ten-sided die with faces numbered 1 to 10

     e. Passing a maths test or failing a maths test

Understanding

7. You and a friend are playing a dice game. You have an eight-sided die (with faces numbered 1 to 8 inclusive) and your friend has a six-sided die (with faces numbered 1 to 6 inclusive). You each roll your own die.

   a. The person who rolls the number 4 wins. Determine if this game is fair.

   b. The person who rolls an odd number wins. Determine if this game is fair.

8. A six-sided die has three faces numbered 5; the other faces are numbered 6. Determine if the events 'rolling a 5' and 'rolling a 6' are equally likely.

9. A card is drawn from a shuffled pack of 52 cards. Calculate the probability that the card drawn is:

   a. an ace   **b.** a club

   c. a red card   **d.** not a jack

   e. a green card   **f.** not a red card.

10. A bag contains 4 blue marbles, 7 red marbles and 9 yellow marbles. All marbles are the same size. A marble is selected at random. Calculate the probability that the marble is:

   a. blue   **b.** red

   c. not yellow   **d.** black.

11. A six-sided die has three faces numbered 1 and the other three faces numbered 2. Determine if the events 'rolling a 1' and 'rolling a 2' are equally likely.

12. A drawer contains purple socks and red socks. The chance of obtaining a red sock is 2 in 9. There are 10 red socks in the drawer.

   Determine the smallest number of socks that need to be added to the drawer so that the probability of drawing a red sock increases to 3 in 7.

13. **WE4** To simulate whether or not the weather will be suitable for sailing, a coin is flipped. If the coin lands Heads up, the weather is perfect. If the coin lands Tails up, the weather is not suitable.

Mandy flips a coin 20 times and obtains 13 Heads and 7 Tails. Sophia flips a coin 100 times and obtains 47 Heads and 53 Tails.

   a. Determine the relative frequency of perfect weather days in both cases.
   b. Compare the results from the two simulations and determine which is a better estimate of the true probability of the experiment.

14. a. **WE5** Use a random number generator to simulate the number of walnuts in 40 different carrot cakes, with a maximum of 60 in each cake.
    b. Calculate the probability that there will be more than 40 walnuts in a randomly chosen cake.

15. The gender of babies in a set of triplets is simulated by flipping 3 coins. If a coin lands Tails up, the baby is a boy. If a coin lands Heads up, the baby is a girl. In the simulation, the trial is repeated 40 times and the following results show the number of Heads obtained in each trial.

   0, 3, 2, 1, 1, 0, 1, 2, 1, 0, 1, 0, 2, 0, 1, 0, 1, 2, 3, 2,
   1, 3, 0, 2, 1, 2, 0, 3, 1, 3, 0, 1, 0, 1, 3, 2, 2, 1, 2, 1

   a. Calculate the experimental probability, as a percentage, that exactly one of the babies in a set of triplets is female.
   b. Calculate the experimental probability, as a percentage, that more than one of the babies in the set of triplets is female.

Communicating, reasoning and problem solving

16. **WE6** There are 24 drivers lining up at the start of a Formula 1 race.

   a. Explain how simulation could be used to determine the winner.
   b. Consider if there are any limitations in your answer.

17. If a computer manufacturer wanted to simulate the probability of its laptops having a fault, determine if it would be appropriate to use a random number generator to select random computers on the assembly line. Explain your reasoning.

18. A random number generator was used to select how many students will receive a free lunch on any given day. The maximum number of students who can receive a free lunch is 10 and the minimum is 0.

   a. If the test is run for 30 days, calculate the theoretical probability that:

      i. 8 or more students are selected on any given day
      ii. 0 students are selected on any given day
      iii. 10 students are selected on any given day.

   b. Use a random number generator to simulate the free-lunch program over 30 days, and compare the results to those from part **a**.
   c. State if the results from the random number generator are the same as your predictions. Explain your answer.

19. You are about to sit a Mathematics examination that contains 40 multiple-choice questions. You didn't have time to study so you are going to choose the answers completely randomly. There are four choices for each question.
Explain how you could use random numbers to select your answers for you.

20. A pair of dice is rolled and the sum of the numbers shown is noted.

   a. List the sample space.
   b. Determine how many different ways the sum of 7 can be obtained.
   c. Determine if all outcomes are equally likely.
   d. Complete the given table.

Sum	2	3	4	5	6	7	8	9	11	12
Frequency										

   e. Determine the relative frequencies of the following sums.

      i. 2
      ii. 7
      iii. 11

   f. Determine the probability of obtaining the following sums.

      i. 2
      ii. 7
      iii. 11

   g. If a pair of dice is rolled 300 times, calculate how many times you would expect the sum of 7.

LESSON
14.3 Tree diagrams and multistage chance experiments

LEARNING INTENTION

At the end of this lesson you should be able to:
• create tree diagrams to represent multistage chance experiments
• use tree diagrams to solve probability problems involving multistage trials or events.

14.3.1 Two-step chance experiments

eles-4924

• In **two-step chance experiments** the result is obtained after performing two trials. Two-step chance experiments are often represented using tree diagrams.
• **Tree diagrams** are used to list all possible outcomes of two or more events that are not necessarily equally likely.
• The probability of obtaining the result for a particular event is listed on the branches.

- The probability for each outcome in the sample space is the product of the probabilities associated with the respective branches.
  For example, the tree diagram shown here represents the sample space for flipping a coin, then choosing a marble from a bag containing three red marbles and one black marble.

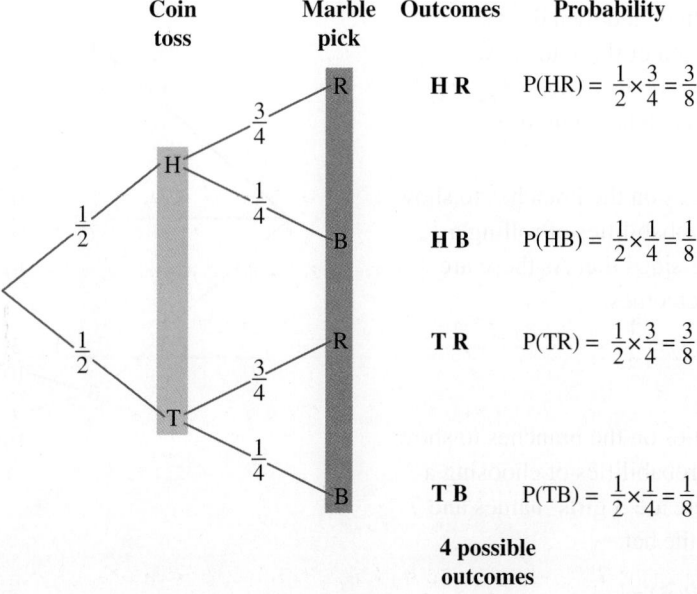

- When added together, all the probabilities for the outcomes should sum to 1. They are complementary events.
  For example,

$$P(HR) + P(HB) + P(TR) + P(TB) = \frac{3}{8} + \frac{1}{8} + \frac{3}{8} + \frac{1}{8}$$
$$= 1$$

- Other probabilities can also be calculated from the tree diagram.
  For example, the probability of getting an outcome that contains a red marble can be calculated by summing the probabilities of each of the possible outcomes that include a red marble.
  Outcomes that contain a red marble are HR and TR, therefore:

$$P(\text{red marble}) = P(HR) + P(TR)$$

$$= \frac{3}{8} + \frac{3}{8}$$
$$= \frac{6}{8}$$
$$= \frac{3}{4}$$

WORKED EXAMPLE 7 Using a tree diagrams for a two-step chance experiment

A three-sided die is rolled and a name is picked out of a hat that contains 3 girls' names and 7 boys' names.
a. Construct a tree diagram to display the sample space.
b. Calculate the probability of:
   i. rolling a 3, then choosing a boy's name
   ii. choosing a boy's name after rolling an odd number.

a. 1. Draw 3 branches from the starting point to show the 3 possible outcomes of rolling a three-sided die (shown in blue), and then draw 2 branches off each of these to show the 2 possible outcomes of choosing a name out of a hat (shown in red).

a.

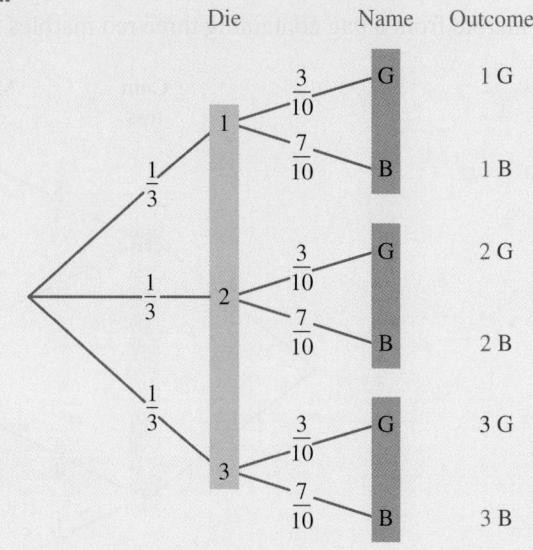

2. Write probabilities on the branches to show the individual probabilities of rolling a 1, 2 or 3 on a three-sided die. As these are equally likely outcomes,

$$P(1) = P(2) = P(3) = \frac{1}{3}.$$

3. Write probabilities on the branches to show the individual probabilities of choosing a name. Since there are 3 girls' names and 7 boys' names in the hat,

$$P(G) = \frac{3}{10} \text{ and } P(B) = \frac{7}{10}.$$

b. i. 1. Follow the pathway of rolling a 3 $\left[P(3) = \frac{1}{3}\right]$ and choosing a boy's name $\left[P(B) = \frac{7}{10}\right]$, and multiply the probabilities.

b. i. $P(3B) = P(3) \times P(B)$

$$= \frac{1}{3} \times \frac{7}{10}$$

$$= \frac{7}{30}$$

2. Write the answer.

The probability of rolling a 3, then choosing a boy's name is $\frac{7}{30}$.

ii. 1. To roll an odd number (1 or 3) then choose a boy's name:
- roll a 1, then choose a boy's name or
- roll a 3, then choose a boy's name.

Calculate the probability of each of these and add them together to calculate the total probability. Simplify the result if possible.

ii. $P(\text{odd } B) = P(1B) + P(3B)$

$$= P(1) \times P(B) + P(3) \times P(B)$$

$$= \frac{1}{3} \times \frac{7}{10} + \frac{1}{3} \times \frac{7}{10}$$

$$= \frac{7}{30} + \frac{7}{30}$$

$$= \frac{14}{30}$$

$$= \frac{7}{15}$$

2. Write the answer.

The probability of choosing a boy's name after rolling an odd number is $\frac{7}{15}$.

⊙ 14.3.2 Three-step chance experiments

eles-4925

- Outcomes are often made up of combinations of events. For example, when a coin is flipped three times, three of the possible outcomes are HHT, HTH and THH. These outcomes all contain 2 Heads and 1 Tail.
- The probability of an outcome with a particular order is written such that the order required is shown. For example, P(HHT) is the probability of H on the first coin, H on the second coin and T on the third coin.
- The probability of an outcome with a particular combination of events in which the order is not important is written describing the particular combination required. For example, P(2 Heads and 1 Tail).

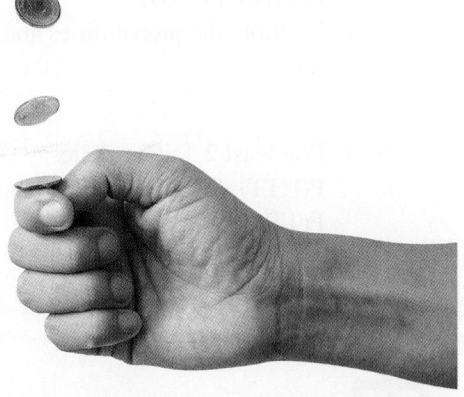

WORKED EXAMPLE 8 Using a tree diagram for a three-step chance experiment

A coin is biased so that the chance of it falling as a Head when flipped is 0.75.
a. Construct a tree diagram to represent the coin being flipped three times.
b. Calculate the following probabilities, correct to 3 decimal places:

  i. P(HTT)
  ii. P(1H and 2T)
  iii. P(at least 2 Tails).

THINK

a. 1. Tossing a coin has two outcomes. Draw 2 branches from the starting point to show the first toss, 2 branches off each of these to show the second toss and then 2 branches off each of these to show the third toss.

2. Write probabilities on the branches to show the individual probabilities of tossing a Head (0.75) and a Tail. Because tossing a Head and tossing a Tail are mutually exclusive,
$P(T) = 1 - P(H) = 0.25$.

WRITE

a.

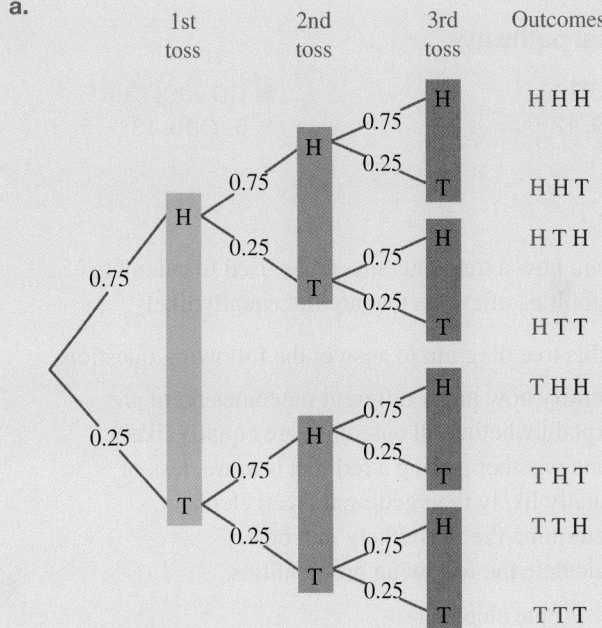

b. i. 1. P(HTT) implies the order:
H (0.75),
T (0.25), T (0.25).

2. Multiply the probabilities and round.

b. i. $P(HTT) = P(H) \times P(T) \times P(T)$

$= (0.75) \times (0.25)^2$

$= 0.046875$

≈ 0.047

ii. 1. P(1H and 2T) implies: P(HTT), P(THT), P(TTH).
   2. Multiply the probabilities and round.

ii. $P(1H \text{ and } 2T) = P(HTT) + P(THT) + P(TTH)$
   $= 3(0.75 \times 0.25^2)$
   $= 0.140625$
   ≈ 0.141

iii. 1. P(at least 2 Tails) implies: P(HTT), P(THT), P(TTH) and P(TTT).
   2. Add these probabilities and round.

iii. $P(\text{at least } 2T) = P(HTT) + P(THT) + P(TTH) + P(TTT)$
   $= 3(0.75 \times 0.25^2) + 0.25^3$
   $= 0.15625$
   ≈ 0.156

on Resources

▶ **Video eLesson** Tree diagrams (eles-1894)

✦ **Interactivity** Tree diagrams (int-6171)

Exercise 14.3 Tree diagrams and multistage chance experiments

learn on

14.3 Quick quiz on	**14.3 Exercise**

Individual pathways

■ PRACTISE	■ CONSOLIDATE	■ MASTER
1, 2, 6, 9, 12	3, 5, 7, 10, 13	4, 8, 11, 14

Fluency

1. Explain how a tree diagram can be used to calculate probabilities of events that are not equally likely.

2. Use this tree diagram to answer the following questions.

   a. Identify how many different outcomes there are.
   b. Explain whether all outcomes are equally likely.
   c. State whether getting a red fish is more, less or equally likely than getting a green elephant.
   d. Determine the most likely outcome.
   e. Calculate the following probabilities.

      i. P(blue elephant)
      ii. P(indigo elephant)
      iii. P(donkey)

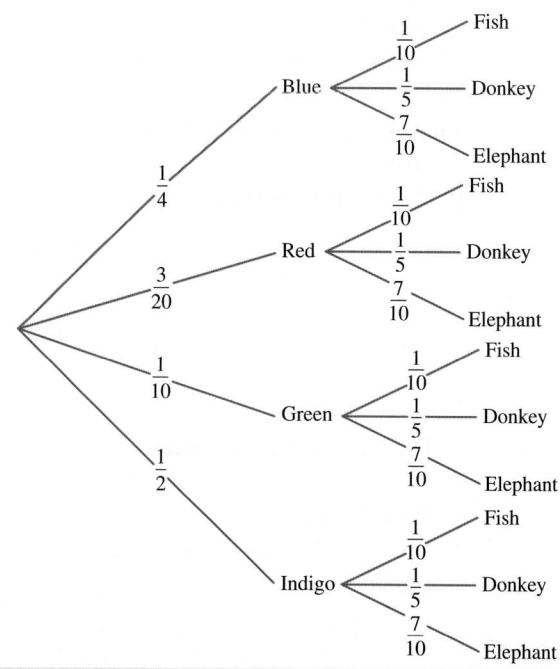

3. a. Copy the tree diagram shown here and complete the labelling for tossing a biased coin three times when the chance of tossing one Head in one toss is 0.7.

b. Calculate the probability of tossing three Heads.

c. Determine the probability of getting at least one Tail.

d. Calculate the probability of getting exactly two Tails.

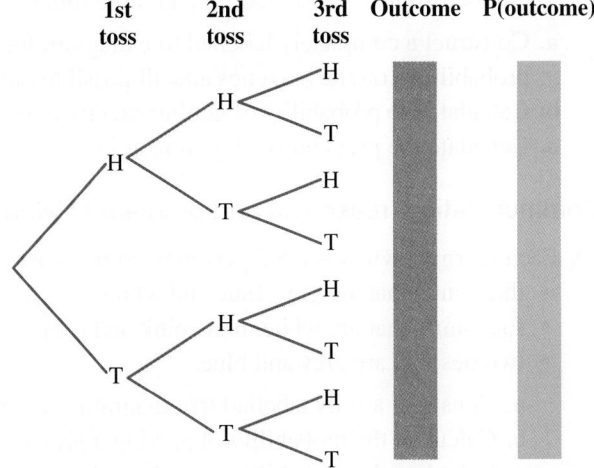

4. The questions below relate to rolling a fair die.

a. Calculate the probability of each of the following outcomes from one roll of a die.

   i. P(rolling number < 4)

   ii. P(rolling a 4)

   iii. P(rolling a number other than a 6)

b. The tree diagram shown has been condensed to depict rolling a die twice, noting the number relative to 4 on the first roll and 6 on the second. Complete a labelled tree diagram, showing probabilities on the branches and all outcomes, similar to that shown.

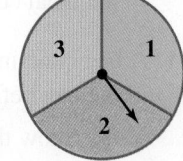

c. Determine the probability of rolling the following with 2 rolls of the die.

   i. P(a 4 then a 6)

   ii. P(a number less than 4 then a 6)

   iii. P(a 4 then 6′)

   iv. P(a number > 4 and then a number < 6)

5. **WE7** The spinner shown is divided into 3 equal-sized wedges labelled 1, 2 and 3. It is spun three times, and it is noted whether the spinner lands on a prime number, P = {2, 3} = 'prime', or not a prime number, P = {1} = 'not prime'.

a. Construct a labelled tree diagram for 3 spins of the spinner, showing probabilities on the branches and all possible outcomes.

b. Calculate the following probabilities.

   i. P(3 prime numbers)   **ii.** P(PPP′ in this order)   **iii.** P(PPP′ in any order)

Understanding

6. **WE8** A coin is biased so that the chance of it falling as a Tail when tossed is 0.2.

a. Construct a tree diagram to represent the coin being tossed 3 times.

b. Determine the probability of getting the same outcome on each toss.

7. A die is tossed twice and each time it is recorded whether or not the number is a multiple of 3. If M = the event of getting a multiple of 3 on any one toss and M' = the event of not getting a multiple of 3 on any one toss:

a. construct a tree diagram to represent the 2 tosses

b. calculate the probability of getting two multiples of 3.

8. The biased spinner illustrated is spun three times.

a. Construct a completely labelled tree diagram for 3 spins of the spinner, showing probabilities on the branches and all possible outcomes and associated probabilities.

b. Calculate the probability of getting exactly two 1s.

c. Calculate the probability of getting at most two 1s.

Communicating, reasoning and problem solving

9. Each morning when Ahmed gets dressed for work he has the following choices:
- three suits that are grey, blue and white
- four shirts that are white, blue, pink and grey
- two ties that are grey and blue.

a. Construct a fully labelled tree diagram showing all possible clothing choices.

b. Calculate the probability of picking a grey suit, a pink shirt and a blue tie.

c. Calculate the probability of picking the same colour for all three options.

d. Ahmed works five days a week for 48 weeks of the year. Determine how many times each combination would get repeated over the course of one year at work.

e. Determine how many more combinations of clothing he would have if he bought another tie and another shirt.

10. A restaurant offers its customers a three-course dinner, where they choose between two entrées, three main meals and two desserts.

The managers find that 30% choose soup and 70% choose prawn cocktail for the entrée; 20% choose vegetarian, 50% chicken, and the rest have beef for their main meal; and 75% have sticky date pudding while the rest have apple crumble for dessert.

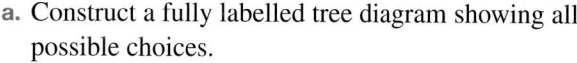

a. Construct a fully labelled tree diagram showing all possible choices.

b. Determine the probability that a customer will choose the soup, chicken and sticky date pudding.

c. If there are 210 people booked for the following week at the restaurant, determine how many you would expect to have the meal combination referred to in part **b**.

11. A bag contains 7 red and 3 white balls. A ball is taken at random, its colour noted and it is then placed back in the bag before a second ball is chosen at random and its colour noted.

a. **i.** Show the possible outcomes with a fully labelled tree diagram.

  ii. As the first ball was chosen, determine how many balls were in the bag.

  iii. As the second ball was chosen, determine how many balls were in the bag.

  iv. Explain whether the probability of choosing a red or white ball changes from the first selection to the second.

  v. Calculate the probability of choosing a red ball twice.

b. Suppose that after the first ball had been chosen it was not placed back in the bag.

  i. As the second ball is chosen, determine how many balls are in the bag.

  ii. Explain if the probability of choosing a red or white ball changes from the first selection to the second.

  iii. Construct a fully labelled tree diagram to show all possible outcomes.

  iv. Evaluate the probability of choosing two red balls.

12. An eight-sided die is rolled three times to see whether 5 occurs.

   a. Construct a tree diagram to show the sample space.
   b. Calculate:

   i. P(three 5s)
   ii. P(no 5s)
   iii. P(two 5s)
   iv. P(at least two 5s).

13. A tetrahedral die (four faces labelled 1, 2, 3 and 4) is rolled and a coin is tossed simultaneously.

   a. Construct a tree diagram and list all outcomes and their respective probabilities.
   b. Determine the probability of getting a Head on the coin and an even number on the die.

14. A biased coin that has an 80% chance of getting a Head is flipped four times. Use a tree diagram to answer to the following.

   a. Calculate the probability of getting 4 Heads.
   b. Determine the probability of getting 2 Heads then 2 Tails in that order.
   c. Calculate the probability of getting 2 Heads and 2 Tails in any order.
   d. Determine the probability of getting more Tails than Heads.

LESSON
14.4 Independent and dependent events

LEARNING INTENTION

At the end of this lesson you should be able to:
- describe independent and dependent events in the context of chance experiments involving 2 stages
- record all possible outcomes for multistage chance experiments
- determine the probabilities of outcomes for multistage experiments involving independent events
- determine the probabilities of outcomes for multistage experiments involving dependent events.

14.4.1 Independent events for multistage chance experiments

eles-4926

- **Independent events** are events that have no effect on each other. The outcome of the first event does not influence the outcome of the second.
- An example of independent events are successive coin tosses. The outcome of the first toss has no effect on the second coin toss. Similarly, the outcome of the roll on a die will not affect the outcome of the next roll.
- If events A and B are independent, then the **Multiplication Law of probability** states that:
  $P(A \text{ and } B) = P(A) \times P(B)$ or $P(A \cap B) = P(A) \times P(B)$
- The reverse is also true. If:
  $P(A \text{ and } B) = P(A) \times P(B)$ or $P(A \cap B) = P(A) \times P(B)$
  then event A and event B are independent events.

Independent events

For independent events A and B:
$$P(A \text{ and } B) = P(A \cap B) = P(A) \times P(B)$$

Adam is one of the 10 young golfers to represent his state. Baz is one of the 12 netball players to represent her state. All the players in their respective teams have an equal chance of being nominated as captains.

a. Explain whether the events 'Adam is nominated as captain' and 'Baz is nominated as captain' are independent.
b. Calculate:
   i. P(Adam is nominated as captain)
   ii. P(Baz is nominated as captain).
c. Determine the probability that both Adam and Baz are nominated as captains of their respective teams.

THINK

a. Determine whether the given events are independent and write your answer.

b. i. 1. Determine the probability of Adam being nominated as captain. He is one of 10 players.

   2. Write your answer.

   ii. 1. Determine the probability of Baz being nominated as captain. She is one of 12 players.

   2. Write your answer.

c. 1. Write the Multiplication Law of probability for independent events.

   2. Substitute the known values into the rule.

   3. Evaluate.

   4. Write your answer.

WRITE

a. Adam's nomination has nothing to do with Baz's nomination and vice versa. Therefore, the events are independent.

b. i. $P(\text{Adam is nominated}) = P(A)$
$$= \frac{n(\text{Adam is nominated})}{n(S)}$$
$$P(\text{Adam is nominated}) = \frac{1}{10}$$
The probability that Adam is nominated as captain is $\frac{1}{10}$.

ii. $P(\text{Baz is nominated}) = P(B)$
$$= \frac{n(\text{Baz is nominated})}{n(S)}$$
$$P(\text{Baz is nominated}) = \frac{1}{12}$$
The probability that Baz is nominated as captain is $\frac{1}{12}$.

c. $P(A \text{ and } B)$
$= P(A \cap B)$
$= P(A) \times P(B)$

$P(\text{Adam and Baz are nominated})$
$= P(\text{Adam is nominated}) \times P(\text{Baz is nominated})$

$= \frac{1}{10} \times \frac{1}{12}$

$= \frac{1}{120}$

The probability that both Adam and Baz are nominated as captains is $\frac{1}{120}$.

14.4.2 Dependent events for multistage chance experiments

- **Dependent events** are events where the outcome of one event affects the outcome of the other event.
- An example of dependent events is drawing a card from a deck of playing cards. The probability that the first card drawn is a heart, P(hearts), is $\frac{13}{52}$ (or $\frac{1}{4}$). If this card is a heart and is not replaced, then this will affect the probability of subsequent draws. The probability that the second card drawn is a heart will be $\frac{12}{51}$, while the probability that the second card is not a heart will be $\frac{39}{51}$.
- If two events are dependent, then the probability of occurrence of one event affects that of the subsequent event.

WORKED EXAMPLE 10 Calculating the probability of outcomes of dependent events

A bag contains 5 blue, 6 green and 4 yellow marbles. The marbles are identical in all respects except their colours. Two marbles are picked in succession without replacement. Calculate the probability of picking 2 blue marbles.

THINK	WRITE
1. Determine the probability of picking the first blue marble.	$P(\text{picking a blue marble}) = \dfrac{n(B)}{n(S)}$ $P(\text{picking a blue marble}) = \dfrac{5}{15}$ $= \dfrac{1}{3}$
2. Determine the probability of picking the second blue marble. *Note:* The two events are dependent since marbles are not being replaced. Since we have picked a blue marble this leaves 4 blue marbles remaining out of a total of 14 marbles.	$P(\text{picking second blue marble}) = \dfrac{n(B)}{n(S)}$ $P(\text{picking second blue marble}) = \dfrac{4}{14}$ $= \dfrac{2}{7}$
3. Calculate the probability of obtaining 2 blue marbles.	$P(\text{2 blue marbles}) = P(\text{1st blue}) \times P(\text{2nd blue})$ $= \dfrac{1}{3} \times \dfrac{2}{7}$ $= \dfrac{2}{21}$
4. Write your answer.	The probability of obtaining 2 blue marbles is $\dfrac{2}{21}$.

Note: Alternatively, a tree diagram could be used to solve this question.

The probability of selecting 2 blue marbles successively can be read directly from the first branch of the tree diagram.

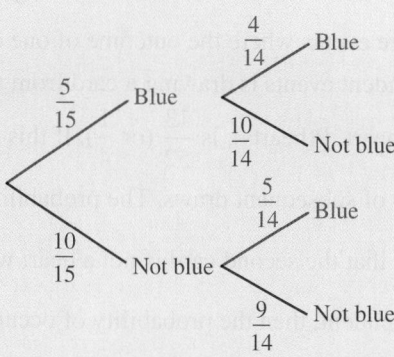

$$P(2 \text{ blue marbles}) = \frac{5}{15} \times \frac{4}{14}$$

$$= \frac{1}{3} \times \frac{2}{7}$$

$$= \frac{2}{21}$$

---

on Resources

🧩 **Interactivities** Independent and dependent events (int-2787)
Multiplication Law of probability (int-6172)
Dependent events (int-6173)

Exercise 14.4 Independent and dependent events

learn on

14.4 Quick quiz **on**	14.4 Exercise

Individual pathways

■ PRACTISE	■ CONSOLIDATE	■ MASTER
1, 4, 8, 11, 14, 17, 21	2, 5, 9, 12, 15, 18, 22	3, 6, 7, 10, 13, 16, 19, 20, 23

Fluency

1. If A and B are independent events and $P(A) = 0.7$ and $P(B) = 0.4$, calculate:
   a. $P(A \text{ and } B)$
   b. $P(A' \text{ and } B)$ where A' is the complement of A
   c. $P(A \text{ and } B')$ where B' is the complement of B
   d. $P(A' \text{ and } B')$.

2. Determine whether two events A and B with $P(A) = 0.6$, $P(B') = 0.84$ and $P(A \cap B) = 0.96$ are independent or dependent.

3. Determine whether two events A and B with $P(A) = 0.25$, $P(B) = 0.72$ and $P(A \cup B) = 0.79$ are independent or dependent.

4. **WE9** A die is rolled and a coin is tossed.
   a. Explain whether the outcomes are independent.
   b. Calculate:
      i. P(Head) on the coin
      ii. P(6) on the die.
   c. Calculate P(6 on the die and Head on the coin).

5. A tetrahedron (4-faced) die and a 10-sided die are rolled simultaneously. Calculate the probability of getting a 3 on the tetrahedral die and an 8 on the 10-sided die.

6. A blue die and a green die are rolled. Calculate the probability of getting a 5 on the blue die and not a 5 on the green die.

7. Dean is an archer. The experimental probability that Dean will hit the target is $\frac{4}{5}$.

   a. Calculate the probability that Dean will hit the target on two successive attempts.
   b. Calculate the probability that Dean will hit the target on three successive attempts.
   c. Calculate the probability that Dean will not hit the target on two successive attempts.
   d. Calculate the probability that Dean will hit the target on the first attempt but miss on the second attempt.

Understanding

8. **MC** A bag contains 20 apples, of which 5 are bruised. Peter picks an apple and realises that it is bruised. He puts the apple back in the bag and picks another one.

   a. The probability that Peter picks 2 bruised apples is:
      A. $\frac{1}{4}$ B. $\frac{1}{2}$ C. $\frac{1}{16}$ D. $\frac{3}{4}$
   b. The probability that Peter picks a bruised apple first but a good one on his second attempt is:
      A. $\frac{1}{4}$ B. $\frac{1}{2}$ C. $\frac{3}{4}$ D. $\frac{3}{16}$

9. The probability that John will be late for a meeting is $\frac{1}{7}$ and the probability that Phil will be late for a meeting is $\frac{3}{11}$.
   Calculate the probability that:

   a. John and Phil are both late
   b. neither of them is late
   c. John is late but Phil is not late
   d. Phil is late but John is not late.

10. On the roulette wheel at the casino there are 37 numbers, 0 to 36 inclusive.
    Bidesi puts his chip on number 8 in game 20 and on number 13 in game 21.

    a. Calculate the probability that he will win in game 20.
    b. Calculate the probability that he will win in both games.
    c. Calculate the probability that he wins at least one of the games.

11. Based on her progress through the year, Karen was given a probability of 0.8 of passing the Physics exam. If the probability of passing both Maths and Physics is 0.72, determine her probability of passing the Maths exam.

12. Suresh found that, on average, he is delayed 2 times out of 7 at Melbourne airport. Rakesh made similar observations at Brisbane airport, but found he was delayed 1 out of every 4 times.
    Determine the probability that both Suresh and Rakesh will be delayed if they are flying out of their respective airports.

13. Bronwyn has 3 pairs of Reebok and 2 pairs of Adidas running shoes. She has 2 pairs of Reebok, 3 pairs of Rio and a pair of Red Robin socks.
    Preparing for an early morning run, she grabs at random for a pair of socks and a pair of shoes. Calculate the probability that she chooses:

    a. Reebok shoes and Reebok socks
    b. Rio socks and Adidas shoes
    c. Reebok shoes and Red Robin socks
    d. Adidas shoes and socks that are not Red Robin.

14. **WE10** Two cards are drawn successively and without replacement from a pack of playing cards. Calculate the probability of drawing:

    a. 2 hearts
    b. 2 kings
    c. 2 red cards.

15. In a class of 30 students there are 17 students who study Music. Two students are picked randomly to represent the class in the Student Representative Council. Calculate the probability that:

    a. both students don't study Music
    b. both students do study Music
    c. one of the students doesn't study Music.

16. In a box, there are 4 red balls, 5 blue balls and 3 green balls. A player will draw balls from the box in the following manner.

    a. The player draws 2 balls without replacement. Determine the probability of drawing 2 blue balls consecutively.
    b. After part **a**, the player puts back the 2 balls and then draws 3 balls without replacement. Determine the probability of drawing exactly 1 red ball, 1 blue ball, and 1 green ball in that order.

Communicating, reasoning and problem solving

17. Greg has tossed a Tail on each of 9 successive coin tosses. He believes that his chances of tossing a Head on his next toss must be very high. Explain whether Greg is correct.

18. The Multiplication Law of probability relates to *independent events*. Tree diagrams can illustrate the sample space of successive *dependent events* and the probability of any one combination of events can be calculated by *multiplying* the stated probabilities along the branches.
    Explain whether this a contradiction to the Multiplication Law of probability.

19. Explain whether it is possible for two events, *A* and *B*, to be mutually exclusive and independent.

20. Consider the following sets:
$$S = \{1, 2, 3, 4, 5, 6, 7, 8, 9, 10, 11, 12, 13, 14, 15, 16, 17, 18, 19, 20\}$$

$$A = \{\text{evens in } S\}$$

$$B = \{\text{multiples of 3 in } S\}$$

    a. Explain whether the events *A* and *B* are independent.
    b. If *S* is changed to the integers from 1 to 10 only, explain whether the result from part **a** changes.

21. There are three coins in a box. One coin is a fair coin, one coin is biased with an 80% chance of landing Heads, and the third is a biased coin with a 40% chance of landing Heads. A coin is selected at random and flipped.
    If the result is a Head, determine the probability that the fair coin was selected.

22. A game at a carnival requires blindfolded contestants to throw balls at numbered ducks sitting on 3 shelves. The game ends when 3 ducks have been knocked off the shelves.
    Assume that the probability of hitting each duck is equal.

    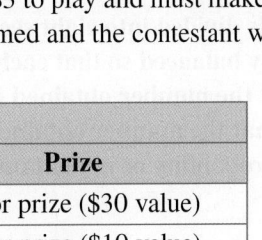

    a. Explain whether the events described in the game are dependent or independent.
    b. Determine the initial probabilities of hitting each number.
    c. Draw a labelled tree diagram to show the possible outcomes for a contestant.
    d. Calculate the probabilities of hitting the following:

    i. P(1, 1, 1)   ii. P(2, 2, 2)   iii. P(3, 3, 3)   iv. P(at least one 3).

23. Question **22** described a game at a carnival. A contestant pays $3 to play and must make 3 direct hits to be eligible for a prize. The numbers on the ducks hit are then summed and the contestant wins a prize according to the winners' table.

Winners' table	
Total score	**Prize**
9	Major prize ($30 value)
7, 8	Minor prize ($10 value)
5, 6	$2 prize
3, 4	No prize

    a. Determine the probability of winning each prize listed.
    b. Suppose 1000 games are played on an average show day. Evaluate the profit that could be expected to be made by the sideshow owner on any average show day.

LESSON
14.5 Venn diagrams, 2-way tables and mutually exclusive events (Path)

LEARNING INTENTION

At the end of this lesson you should be able to:
- represent and interpret data in Venn diagrams for mutually exclusive and non-mutually exclusive events
- construct Venn diagrams and 2-way tables to represent relationships between attributes
- convert between Venn diagrams and 2-way tables
- use set language and notation for events, including $A \cup B$, $A \cap B$ and A'
- identify and describe mutually and non-mutually exclusive events
- identify and shade regions on a Venn diagram.

▶ 14.5.1 Probabilities and Venn diagrams

eles-6290

Complementary events

- The **complement** of the set A is the set of all elements that belong to the universal or sample set but that do *not* belong to A.
- On a Venn diagram, **complementary events** appear as separate regions that together occupy the whole universal or sample set.
- As an example, the complement of {drawing a diamond} from a deck of cards is {not drawing a diamond}, which can also be described as {drawing a heart, spade or club}. This is shown in the Venn diagram below.
- For complementary events, since they fill the entire sample set, the relationship is:

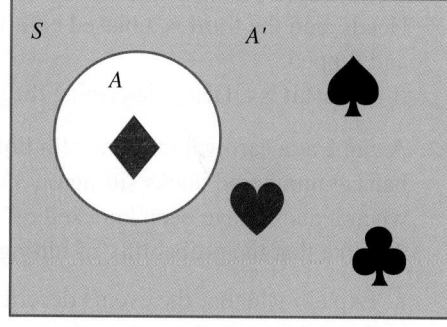

$$P(A) + P(A') = 1$$

WORKED EXAMPLE 11 Probability of complementary events

A spinning wheel is divided into eight sectors, each of which is marked with one of the numbers 1 to 8. The wheel is evenly balanced so that each sector has the same chance of being chosen.
a. If A is the event the number obtained is less than 4, calculate $P(A)$ and $P(A')$.
b. B is the event that the number obtained is either an odd number or a multiple of 3. Calculate the probability of not obtaining an element of B.

THINK	WRITE
a. 1. List the sample set, S.	a. $S = \{1, 2, 3, 4, 5, 6, 7, 8\}$
2. List the elements of set A.	$A = \{1, 2, 3\}$
3. State the probability of spinning set A.	$P(A) = \dfrac{3}{8}$

4. State the relationship for complementary events, substitute and solve for P(A').

$$P(A) + P(A') = 1$$
$$\frac{3}{8} + P(A') = 1$$
$$P(A') = \frac{5}{8}$$

5. Write the answer.

$$P(A) = \frac{3}{8} \text{ and } P(A') = \frac{5}{8}.$$

b. 1. List the sample set, S.

b. $S = \{1, 2, 3, 4, 5, 6, 7, 8\}$

2. List the elements of set B.

$B = \{1, 3, 5, 7, 6\}$

3. State the probability of spinning set B.

$$P(B) = \frac{5}{8}$$

4. State the relationship for complementary events, substitute and solve for P(B').

$$P(B) + P(B') = 1$$
$$\frac{5}{8} + P(B') = 1$$

5. Write the answer.

$$P(B') = \frac{3}{8}$$

The probability of NOT obtaining an odd number or a multiple of 3 is $\frac{3}{8}$.

The intersection and union of A and B

- The intersection of two events A and B is written $A \cap B$. These are the outcomes that are in A 'and' in B and so the intersection is often referred to as the event 'A and B'.
- The union of two events A and B is written $A \cup B$. These are the outcomes that are in A 'or' in B and so the union is often referred to as the event 'A or B'.

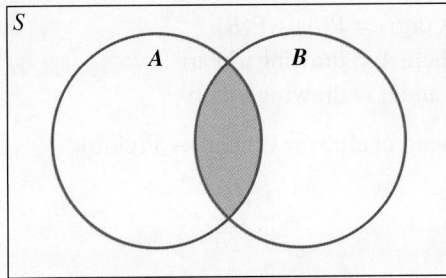

The region $A \cap B$

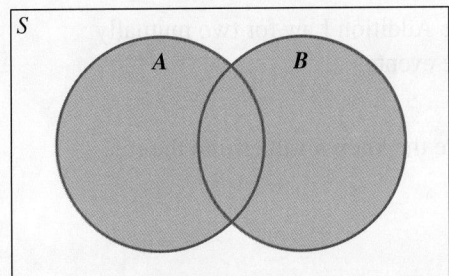

The region $A \cup B$

- When calculating the probability, $P(A \cup B)$, we cannot simply add the probabilities of A and B.
- The formula for the probability, $P(A \cup B)$, is therefore given by the following equation, which is known as the **Addition Law of probability**.

The Addition Law of probability

For intersecting events A and B:

$$P(A \cup B) = P(A) + P(B) - P(A \cap B)$$

Mutually exclusive events

- Two events are **mutually exclusive** if one event happening excludes the other from happening.
  For example, when rolling a die, the events {getting a 1} and {getting a 5} are mutually exclusive.
- On a Venn diagram, mutually exclusive events appear as disjointed sets within the universal or sample set.
- For mutually exclusive events A and B, $P(A \cap B) = 0$.

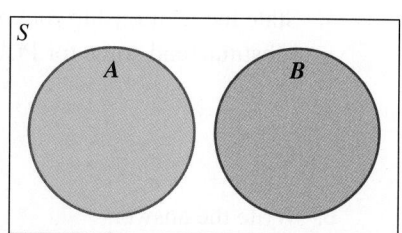

Mutually exclusive probabilities

For mutually exclusive events A and B:

$$P(A \cap B) = 0$$

and

$$P(A \cup B) = P(A) + P(B)$$

WORKED EXAMPLE 12 Determining the probability of mutually exclusive events

A card is drawn from a pack of 52 playing cards. Determine the probability that the card is a heart or a club.

THINK	WRITE
1. Determine whether the given events are mutually exclusive.	The two events are mutually exclusive as they have no common elements.
2. Determine the probability of drawing a heart and of drawing a club.	$P(\text{heart}) = \dfrac{13}{52}$ $\quad P(\text{club}) = \dfrac{13}{52}$ $\qquad\qquad = \dfrac{1}{4} \qquad\qquad = \dfrac{1}{4}$
3. Write the Addition Law for two mutually exclusive events.	$P(A \text{ or } B) = P(A) + P(B)$ where A = drawing a heart and B = drawing a club
4. Substitute the known values into the rule.	$P(\text{heart or club}) = P(\text{heart}) + P(\text{club})$ $= \dfrac{1}{4} + \dfrac{1}{4}$ $= \dfrac{2}{4}$ $= \dfrac{1}{2}$
5. Evaluate and simplify.	$= \dfrac{1}{2}$
6. Write your answer.	The probability of drawing a heart or a club is $\dfrac{1}{2}$.
Note: Alternatively, we can use the formula for theoretical probability.	$P(\text{heart or club}) = \dfrac{n(\text{heart or club})}{n(S)}$ $= \dfrac{26}{52}$ $= \dfrac{1}{2}$

A die is rolled. Determine:
a. P(an odd number)
b. P(a number less than 4)
c. P(an odd number or a number less than 4).

THINK	WRITE
a. 1. Determine the probability of obtaining an odd number, that is, $\{1, 3, 5\}$.	**a.** $P(\text{odd}) = \dfrac{3}{6}$ $= \dfrac{1}{2}$
2. Write your answer.	The probability of obtaining an odd number is $\dfrac{1}{2}$.
b. 1. Determine the probability of obtaining a number less than 4, that is, $\{1, 2, 3\}$.	**b.** $P(\text{less than 4}) = \dfrac{3}{6}$ $= \dfrac{1}{2}$
2. Write your answer.	The probability of obtaining a number less than 4 is $\dfrac{1}{2}$.
c. 1. Determine which numbers are odd or less than 4.	**c.** Less than $4 = \{1, 2, 3\}$ Odd $= \{1, 3, 5\}$ The numbers $\{1, 2, 3, 5\}$ are odd or less than 4.
2. Determine the probability of obtaining a number that is odd or less than 4.	$P(\text{odd or less than 4}) = \dfrac{4}{6}$ $= \dfrac{2}{3}$
3. Write your answer.	The probability of obtaining an odd number or a number less than 4 is $\dfrac{2}{3}$.

14.5.2 Venn diagrams and 2-way tables

eles-6291

- The information in a Venn diagram can also be represented using a **2-way table**.
- A 2-way table represents the outcomes of two events simultaneously in a two-dimensional table.
- The relationship between the two is shown below.

	Event B	**Event B′**	**Total**
Event A	$A \cap B$	$A \cap B'$	A
Event A′	$A' \cap B$	$A' \cap B'$	A'
Total	B	B'	

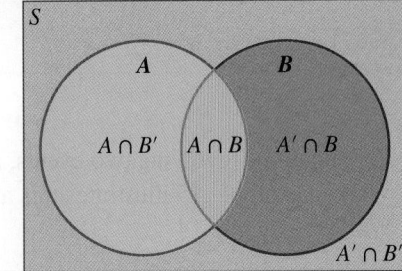

- The four distinct regions in the Venn diagram and 2-way table for two events A and B are:
  $A \cap B$ contains the outcomes in event A and in event B
  $A \cap B'$ contains the outcomes in event A but not in event B
  $A' \cap B$ contains the outcomes not in event A but in event B
  $A' \cap B'$ contains the outcomes not in event A and not in event B
- Venn diagrams and 2-way tables can show either the number of elements in the outcome or the probability of the outcome occurring.

Given P(A) = 0.6, P(B) = 0.4 and P(A ∪ B) = 0.9:
a. **use the Addition Law of probability to calculate the value of P(A ∩ B)**
b. **draw a Venn diagram to represent the universal set**
c. **calculate P(A ∩ B′).**

THINK

a. 1. Write the Addition Law of probability and substitute given values.

2. Collect like terms and rearrange to make P(A ∩ B) the subject. Solve the equation.

b. 1. Draw intersecting sets A and B within the universal set and write P(A ∩ B) = 0.1 inside the overlapping section, as shown.

2. • As P(A) = 0.6, 0.1 of this belongs in the overlap, the remainder of set A is 0.5 (0.6 − 0.1).
   • Since P(B) = 0.4, 0.1 of this belongs in the overlap, the remainder of set B is 0.3 (0.4 − 0.1).

3. The total probability for the universal set is 1. That means P(A ∪ B)′ = 0.1. Write 0.1 outside sets A and B to form the remainder of the universal set.

c. P(A ∩ B′) is the overlapping region of P(A) and P(B′). Shade the region and write down the corresponding probability value for this area.

WRITE

a. $P(A \cup B) = P(A) + P(B) - P(A \cap B)$
   $0.9 = 0.6 + 0.4 - P(A \cap B)$

   $0.9 = 1.0 - P(A \cap B)$
   $P(A \cap B) = 1.0 - 0.9$
   $= 0.1$

b.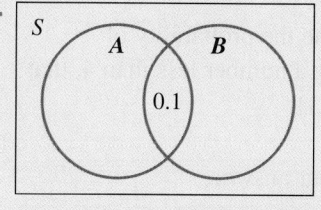

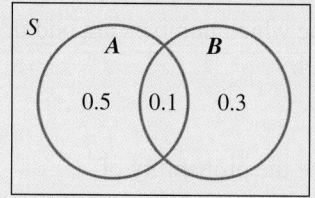

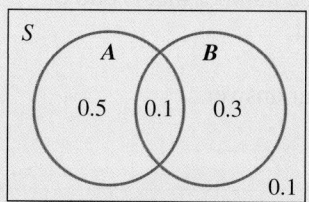

c.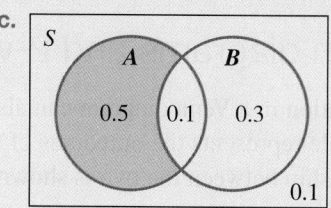

$P(A \cap B') = 0.5$

- For situations involving two events, a 2-way table can provide an alternative to a Venn diagram.
- Worked example 15 illustrates this alternative method to using a Venn diagram in a question similar to Worked example 14.

Given P(A) = 0.5, P(B) = 0.6 and P(A ∪ B) = 0.9:
a. **use the Addition Law of probability to calculate the value of P(A ∩ B)**
b. **draw a 2-way table to represent the sample set**
c. **determine P(A′ ∩ B′).**

a. 1. Write the Addition Law of probability and substitute given values.

2. Collect like terms and solve the equation.

b. 1. Construct a 2-way table and write $P(A \cap B) = 0.2$ in the intersection of the A row with the B column.

2. As $P(A) = 0.5$ and $P(A \cap B) = 0.2$, calculate $P(A \cap B')$.
Similarly, as $P(B) = 0.6$ and $P(A \cap B) = 0.2$, calculate $P(A' \cap B)$

3. Complete these two cells of the table

4. The four regions or cells of the table add to 1 for the sample set.
Calculate the remaining cells and complete the table.

c. 1. Shade in the required region.

2. Write the answer.

a. $P(A \cup B) = P(A) + P(B) - P(A \cap B)$
$0.9 = 0.5 + 0.6 - P(A \cap B)$
$0.9 = 1.1 - P(A \cap B)$
$P(A \cap B) = 0.2$

b. 1.

	B	B'	
A	0.2		0.5
A'			
	0.6		1

2. $P(A) = P(A \cap B) + P(A \cap B')$
$0.5 = 0.2 + P(A \cap B')$
$P(A \cap B') = 0.3$
$P(B) = P(B \cap A) + P(B \cap A')$
$0.6 = 0.2 + P(A' \cap B)$
$P(A' \cap B) = 0.4$

3.

	B	B'	
A	0.2	0.3	**0.5**
A'	0.4		
	0.6		**1**

4. $P(A' \cap B') = 1 - (0.2 + 0.3 + 0.4)$
$P(A' \cap B') = 1 - 0.9 = 0.1$

	B	B'	
A	0.2	0.3	**0.5**
A'	0.4	0.1	0.5
	0.6	0.4	**1**

c.

	B	B'	
A	0.2	0.3	**0.5**
A'	0.4	0.1	0.5
	0.6	0.4	**1**

$P(A' \cap B') = 0.1.$

WORKED EXAMPLE 16 Using a Venn diagram to represent sets and find probabilities

a. **Draw a Venn diagram representing the relationship between the following sets. Show the position of all the elements in the Venn diagram.**

$S = \{1, 2, 3, 4, 5, 6, 7, 8, 9, 10, 11, 12, 13, 14, 15, 16, 17, 18, 19, 20\}$
$A = \{3, 6, 9, 12, 15, 18\}$
$B = \{2, 4, 6, 8, 10, 12, 14, 16, 18, 20\}$

b. **Determine:**

   i. $P(A)$ ii. $P(B)$ iii. $P(A \cap B)$ iv. $P(A \cup B)$ v. $P(A' \cap B')$

THINK

a. **1.** Draw a rectangle with two partly intersecting circles labelled A and B.

2. Analyse sets A and B and place any common elements in the central overlap.

3. Place the remaining elements of set A in circle A.

4. Place the remaining elements of set B in circle B.

5. Place the remaining elements of the universal set S in the rectangle.

WRITE/DRAW

a.

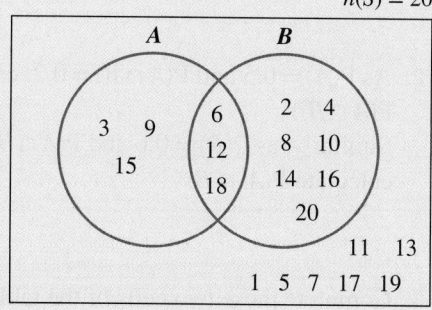

b. **i.** **1.** Write the number of elements that belong to set A and the total number of elements.

2. Write the rule for probability.

3. Substitute the known values into the rule.

4. Evaluate and simplify.

ii. **1.** Write the number of elements that belong to set B and the total number of elements.

2. Repeat steps 2 to 4 of part **b i.**

iii. **1.** Write the number of elements that belong to set $(A \cap B)$ and the total number of elements.

2. Repeat steps 2 to 4 of part **b i.**

b. **i.** $n(A) = 6, n(S) = 20$

$$P(A) = \frac{n(A)}{n(S)}$$

$$P(A) = \frac{6}{20}$$

$$= \frac{3}{10}$$

ii. $n(B) = 10, n(S) = 20$

$$P(B) = \frac{n(B)}{n(S)}$$
$$P(B) = \frac{10}{20}$$
$$= \frac{1}{2}$$

iii. $n(A \cap B) = 3, n(S) = 20$

$$P(A \cap B) = \frac{n(A \cap B)}{n(S)}$$
$$P(A \cap B) = \frac{3}{20}$$

iv. 1. Write the number of elements that belong to set $(A \cup B)$ and the total number of elements.	**iv.** $n(A \cup B) = 13, n(S) = 20$
2. Repeat steps 2 to 4 of part **b i.**	$n(A \cup B) = \dfrac{n(A \cup B)}{n(S)}$ $n(A \cup B) = \dfrac{13}{20}$
v. 1. Write the number of elements that belong to set $(A' \cap B')$ and the total number of elements.	**v.** $n(A' \cap B') = 7, n(S) = 20$
2. Repeat steps 2 to 4 of part **b i.**	$P(A' \cap B') = \dfrac{n(A' \cap B')}{n(S)}$ $P(A' \cap B') = \dfrac{7}{20}$

14.5.3 Venn diagrams with three intersecting circles

eles-6292

- Venn diagrams may represent three different events with three intersecting circles.

WORKED EXAMPLE 17 Creating a Venn diagram for three intersecting events

In a class of 35 students, 6 students like all three subjects: PE, Science and Music. Eight of the students like PE and Science, 10 students like PE and Music, and 12 students like Science and Music. Also, 22 students like PE, 18 students like Science and 17 like Music.

Two students don't like any of the subjects.

a. Display this information on a Venn diagram.
b. Determine the probability of selecting a student who:
   i. likes PE only
   ii. does not like Music.
c. Calculate $P[(\text{Science} \cup \text{Music}) \cap \text{PE}']$.

THINK

a. 1. Draw a rectangle with three partly intersecting circles, labelled PE, Science and Music.

WRITE/DRAW

a.

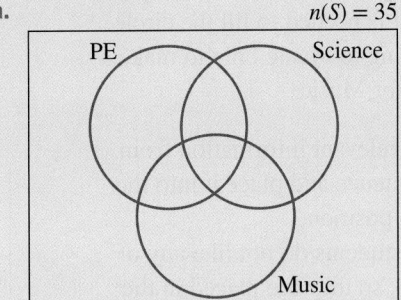

2. Extract the information relating to students liking all three subjects.
   Note: The central overlap is the key to solving these problems. Six students like all three subjects, so place the number 6 into the section corresponding to the intersection of the three circles.

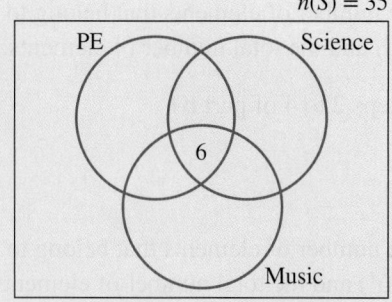

3. Extract the relevant information from the second sentence and place it into the appropriate position.
   Note: Eight students like PE and Science; however, 6 of these students have already been accounted for in step 2. Therefore, 2 will fill the intersection of only PE and Science. Similarly, 4 of the 10 who like PE and Music will fill the intersection of only PE and Music, and 6 of the 12 students will fill the intersection of only Science and Music.

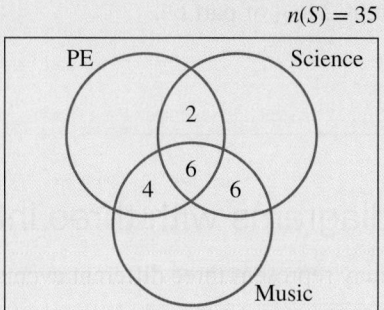

4. Extract the relevant information from the third sentence and place it into the appropriate position.
   Note: Twenty-two students like PE and 12 have already been accounted for in the set. Therefore, 10 students are needed to fill the circle corresponding to PE only. Similarly, 4 students are needed to fill the circle corresponding to Science only to make a total of 18 for Science. One student is needed to fill the circle corresponding to Music only to make a total of 17 for Music.

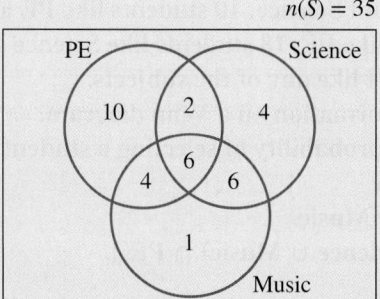

5. Extract the relevant information from the final sentence and place it into the appropriate position.
   Note: Two students do not like any of the subjects, so they are placed in the rectangle outside the three circles.

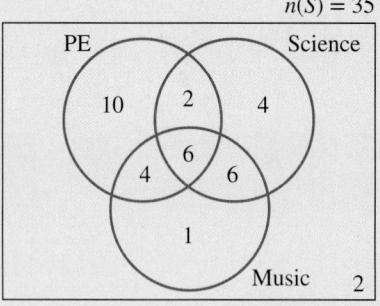

6. Check that the total number in all positions is equal to the number in the universal set.

$$10 + 2 + 4 + 4 + 6 + 6 + 1 + 2 = 35$$

b.	**i.**	**1.**	Write the number of students who like PE only and the total number of students in the class.

b. **i.** n(students who like PE only) $= 10$

$$n(S) = 35$$

		2.	Write the rule for probability.

$$P(\text{likes PE only}) = \frac{n(\text{likes PE only})}{n(S)}$$

		3.	Substitute the known values into the rule.

$$P(\text{likes PE only}) = \frac{10}{35}$$

		4.	Evaluate and simplify.

$$= \frac{2}{7}$$

		5.	Write your answer.

The probability of selecting a student who likes PE only is $\frac{2}{7}$.

	ii.	**1.**	Write the number of students who do not like Music and the total number of students in the class. *Note:* Add all the values that do not appear in the Music circle as well as the two that sit in the rectangle outside the circles.

ii. n(students who do not like Music) $= 18$

$$n(S) = 35$$

		2.	Write the rule for probability.

$$P(\text{does not like Music}) = \frac{n(\text{does not like Music})}{n(S)}$$

		3.	Substitute the known values into the rule.

$$P(\text{does not like Music}) = \frac{18}{35}$$

		4.	Write your answer.

The probability of selecting a student who does not like Music is $\frac{18}{35}$.

c.		**1.**	Write the number of students who like Science and Music but not PE. *Note:* Add the values that appear in the Science and Music circles but do not overlap with the PE circle.

c. $n[(\text{Science} \cup \text{Music}) \cap \text{PE}'] = 11$

$$n(S) = 35$$

		2.	Repeat steps 2 to 4 of part **b. ii.**

$$P[(\text{Science} \cup \text{Music}) \cap \text{PE}']$$
$$= \frac{n[(\text{Science} \cup \text{Music}) \cap \text{PE}']}{n(S)}$$

$$P[(\text{Science} \cup \text{Music}) \cap \text{PE}'] = \frac{11}{35}.$$

The probability of selecting a student who likes Science or Music but not PE is $\frac{11}{35}$.

- Shading a region on a Venn diagram can be used to represent an event.
- In shading a required region, careful attention must be given to the notation.

Draw a Venn diagram to represent the region defined by $(A \cup B) \cap C$

THINK

WRITE

1. Draw a Venn diagram with three intersecting circles and label the circles A, B and C.

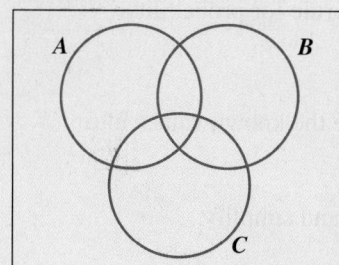

2. $A \cup B$ is the union of A and B. Shade in the two circles labelled A and B.

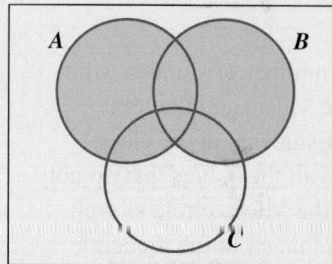

3. On the same set of intersecting circles, shade in C in a different colour.

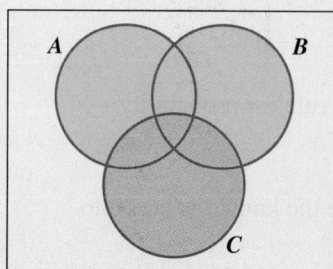

4. $(A \cup B) \cap C$ is the overlap between the two different shaded regions. Draw a new set of intersecting circles and shade the overlapping region.

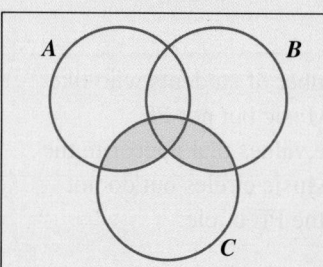

---

on Resources

▶ **Video eLesson** Venn diagrams (eles-1934)

✦ **Interactivities** Venn diagrams (int-3828)
2-way tables (int-6082)
Addition Law of probability (int-6168)

14.5 Quick quiz on	14.5 Exercise

Individual pathways

■ PRACTISE	■ CONSOLIDATE	■ MASTER
1, 2, 4, 8, 9, 10, 16	3, 5, 6, 11, 12, 14, 17	7, 13, 15, 18, 19, 20

Fluency

1. **WE11** A spinning wheel is divided into 10 equal sectors, each of which is marked with one of the numbers 1 to 10 and equally likely to be chosen.

   a. If A is the event of spinning a number greater than 7, calculate $P(A)$ and $P(A')$.
   b. B is the event that the number obtained is either odd or a prime number. Calculate the probability of not obtaining an element of B.

2. A bag contains 5 green, 6 pink, 4 orange and 8 blue counters. A counter is selected at random. Find the probability that the counter is:

   a. green
   c. not blue

   b. orange or blue
   d. black.

3. An unbiased six-sided die is thrown onto a table. State the probability that the number that is uppermost is

   a. 4
   d. smaller than 5

   b. not 4
   e. at least 5

   c. even
   f. greater than 12.

4. **WE12** A card is drawn from a shuffled pack of 52 cards. Determine the probability that the card is:

   a. a heart or a spade
   b. an ace or a jack
   c. a black king or a red queen

5. A bag contains 4 blue marbles, 7 red marbles and 9 yellow marbles. All of the marbles are the same size. A marble is selected at random. Determine the probability that the marble is:

   a. a blue or a red marble
   b. a blue or a red or a yellow marble
   c. not a yellow marble.

6. **WE13** A card is drawn from a well-shuffled pack of 52 playing cards. Calculate:

   a. P(a king is drawn)
   b. P(a heart is drawn)
   c. P(a king or a heart is drawn).

7. **WE14** Given $P(A) = 0.5, P(B) = 0.4$ and $P(A \cup B) = 0.8$:

   a. use the Addition Law of probability to calculate the value of $P(A \cap B)$
   b. draw a Venn diagram to represent the universal set
   c. calculate $P(A \cap B')$.

8. **WE15** Given $P(A) = 0.4$, $P(B) = 0.7$ and $P(A \cup B) = 0.9$:

   a. use the Addition Law of probability to calculate the value of $P(A \cap B)$
   b. draw a 2-way table to represent the sample set
   c. determine $P(A' \cap B)$.

Understanding

9. Let $P(A) = 0.25$, $P(B) = 0.65$ and $P(A \cap B) = 0.05$.
Calculate:

   a. $P(A \cup B)$
   b. $P(A \cap B)'$.

10. **MC** Choose which Venn diagram below best illustrates $P(A \cap B)'$.

A.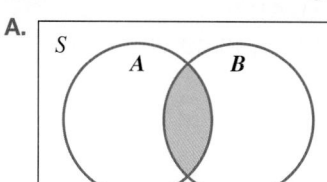

B.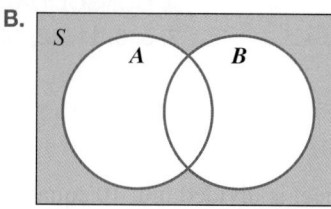

C.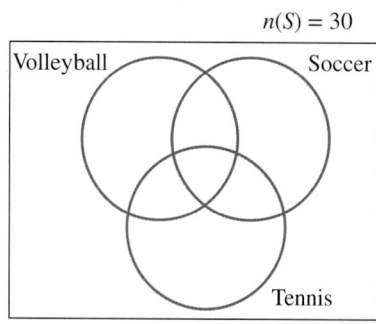

D.

11. a. **WE16** Draw a Venn diagram representing the relationship between the following sets. Show the position of all the elements in the Venn diagram.

$$S = \{1, 2, 3, 4, 5, 6, 7, 8, 9, 10, 11, 12, 13, 14, 15, 16, 17, 18, 19, 20\}$$
$$A = \{1, 3, 5, 7, 9, 11, 13, 15, 17, 19\}$$
$$B = \{1, 4, 9, 16\}$$

   b. Calculate:

      i. $P(A)$
      ii. $P(B)$
      iii. $P(A \cap B)$
      iv. $P(A \cup B)$
      v. $P(A' \cap B')$.

12. **WE17** Thirty students were asked which lunchtime sports they enjoyed — volleyball, soccer or tennis. Five students chose all three sports. Six students chose volleyball and soccer, 7 students chose volleyball and tennis, and 9 chose soccer and tennis. Fifteen students chose volleyball, 14 students chose soccer and 18 students chose tennis.

   a. Copy the Venn diagram shown and enter the given information.

   $n(S) = 30$

   Volleyball   Soccer

   Tennis

   b. If a student is selected at random, determine the probability of selecting a student who:

      i. chose volleyball
      ii. chose all three sports
      iii. chose both volleyball and soccer but not tennis
      iv. did not choose tennis
      v. chose soccer.

   c. Determine:

      i. $P[(\text{soccer} \cup \text{tennis}) \cap \text{volleyball}']$
      ii. $P[(\text{volleyball} \cup \text{tennis}) \cap \text{soccer}']$.

13. Show that $P(A' \cap B') = P(A \cup B)'$ by using:

   a. Venn diagrams   **b.** 2-way tables.

14. `WE18` Draw a Venn diagram to represent the region $(A \cap B) \cup C$.

15. Draw Venn diagrams to represent the following regions:

   a. $A \cap B \cap C$ **b.** $A \cap (B \cup C)'$ **c.** $A \cup (B \cup C)'$ **d.** $(A \cap B') \cap C$

Communicating, reasoning and problem solving

16. Ninety students were asked which lunchtime sports on offer, basketball, netball and soccer, they had participated in on at least one occasion in the last week.
The results are shown in the following table.

Sport	Basketball	Netball	Soccer	Basketball and netball	Basketball and soccer	Netball and soccer	All three
Number of students	35	25	39	5	18	8	3

   a. Copy and complete the Venn diagram shown below to illustrate the sample space.

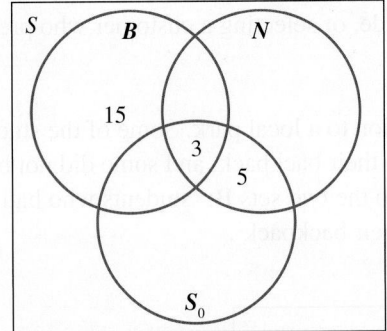

   b. Determine how many students did not play basketball, netball or soccer at lunchtime.
   c. Determine how many students played basketball and/or netball but not soccer.
   d. Determine how many students are represented by the region (basketball ∩ not netball ∩ soccer).
   e. Calculate the relative frequency of the region described in part **d** above.
   f. Estimate the probability that a student will play three of the sports offered.

17. The Venn diagram shows the results of a survey completed by a Chinese restaurateur to find out the food preferences of his regular customers.

   a. Determine the number of customers:
      i. surveyed
      ii. showing a preference for fried rice only
      iii. showing a preference for fried rice
      iv. showing a preference for chicken wings and dim sims.

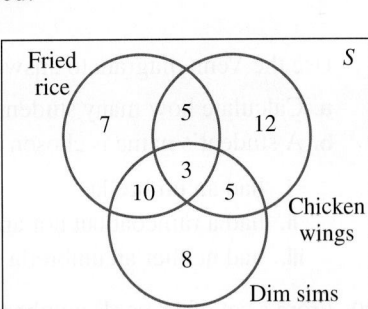

   b. A customer from this group won the draw for a lucky door prize.
   Determine the probability that this customer:
      i. likes fried rice
      ii. likes all three — fried rice, chicken wings and dim sims
      iii. prefers chicken wings only.

18. A survey was conducted with a group of 50 customers. This survey yielded the following results: 2 customers liked all three foods; 6 preferred fried rice and chicken wings; 7 preferred chicken wings and dim sims; 8 preferred fried rice and dim sims; 22 preferred fried rice; 23 preferred chicken wings; and 24 preferred dim sims.

a. Display this information on a Venn diagram.
b. Determine the probability of selecting a customer who prefers all three foods, if a random selection is made.
c. Determine the probability, if a random selection is made, of selecting a customer who prefers at least two of these foods.
d. Determine the probability, if a random selection is made, of selecting a customer who prefers only fried rice.

19. A class of 20 primary school students went on an excursion to a local park. Some of the students carried an umbrella in their backpack, some carried a raincoat in their backpack, and some did not bring either an umbrella or a raincoat. The Venn diagram shown refers to the two sets B: 'students who had an umbrella in their backpack' and C: 'students who had a raincoat in their backpack'.

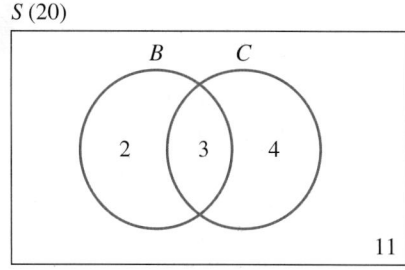

Use the Venn diagram to answer the following.

a. Calculate how many students had umbrellas.
b. A student's name is chosen at random. Calculate the probability that this student

   i. had an umbrella
  ii. had a raincoat but not an umbrella.
 iii. had neither an umbrella nor a raincoat.

20. From a set of 18 cards numbered $1, 2, 3, \ldots, 18$, one card is drawn at random.
Let A be the event of obtaining a multiple of 3, B be the event of obtaining a multiple of 4 and let C be the event of obtaining a multiple of 5.

a. List the elements of each event and then illustrate the three events as sets on a Venn diagram.
b. State which events are mutually exclusive.
c. State the value of $P(A)$.
d. Calculate the following.

   i. $P(A \cup C)$ ii. $P(A \cap B')$ iii. $P((A \cup B \cup C)')$

LESSON
14.6 Conditional probability (Path)

LEARNING INTENTION

At the end of this lesson you should be able to:
- use language such as 'if … then', to examine conditional probability statements
- calculate the probabilities of events where a condition restricts the sample space
- use a Venn diagram or the rule for conditional probability to calculate probabilities.

14.6.1 Recognising conditional probability

eles-4928

- **Conditional probability** is when the probability of an event is conditional (depends) on another event occurring first.
- The effect of conditional probability is to reduce the sample space and thus increase the probability of the desired outcome.
- For two events, A and B, the conditional probability of event B, given that event A occurs, is denoted by $P(B\,|\,A)$ and can be calculated using the following formula.

> **Probability of B given A**
>
> $$P(B\,|\,A) = \frac{P(A \cap B)}{P(A)}, P(A) \neq 0$$

- Conditional probability can be expressed using a variety of language. Some examples of conditional probability statements follow. The key words to look for in a conditional probability statement have been highlighted in each instance.
  - **If** a student receives a B+ or better in their first Maths test, **then** the chance of them receiving a B+ or better in their second Maths test is 75%.
  - **Given that** a red marble was picked out of the bag with the first pick, the probability of a blue marble being picked out with the second pick is 0.35.
  - **Knowing that** the favourite food of a student is hot chips, the probability of their favourite condiment being tomato sauce is 68%.

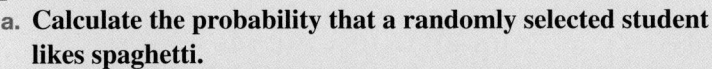

WORKED EXAMPLE 19 Using a Venn diagram to find conditional probabilities

A group of students was asked whether they like spaghetti (*Sp*) or lasagne (*L*). The results are illustrated in the Venn diagram shown. Use the Venn diagram to calculate the following probabilities relating to a student's preferred food.

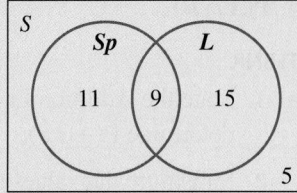

a. Calculate the probability that a randomly selected student likes spaghetti.
b. Determine the probability that a randomly selected student likes lasagne given that they also like spaghetti.

THINK	WRITE/DRAW

a. 1. Determine how many students were surveyed to identify the total number of possible outcomes. Add each of the numbers shown on the Venn diagram.

a. Total number of students $= 11 + 9 + 15 + 5 = 40$

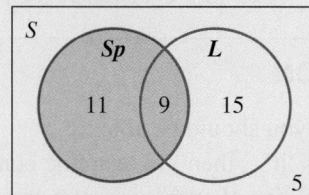

2. There are 20 students that like 'spaghetti' or 'spaghetti and lasagne', as shown in pink.

3. The probability that a randomly selected student likes spaghetti is found by substituting these values into the probability formula.

$$P(\text{event}) = \frac{\text{number of favourable outcomes}}{\text{total number of possible outcomes}}$$

$$P(\text{spaghetti}) = \frac{20}{40}$$

$$= \frac{1}{2}$$

b. 1. The condition imposed 'given they also like spaghetti' alters the sample space to the 20 students described in part **a**, as shaded in blue. Of these 20 students, 9 stated that they liked lasagne and spaghetti, as shown in pink.

b.

2. State the probability.

$$P(L \,|\, Sp) = \frac{9}{20}$$

3. Alternatively, the probability that a randomly selected student likes lasagne, given that they like spaghetti, is found by substituting these values into the probability formula for conditional probability.

$$P(B \,|\, A) = \frac{P(A \cap B)}{P(A)}$$

$$P(L \,|\, Sp) = \frac{\frac{9}{40}}{\frac{1}{2}}$$

$$= \frac{9}{20}$$

WORKED EXAMPLE 20 Using the rule for conditional probability

If $P(A) = 0.3$, $P(B) = 0.5$ and $P(A \cup B) = 0.6$, calculate:

a. $P(A \cap B)$ **b.** $P(B \,|\, A)$

THINK	WRITE

a. 1. State the Addition Law for probability to determine $P(A \cup B)$.

a. $P(A \cup B) = P(A) + P(B) - P(A \cap B)$

2. Substitute the values given in the question into this formula and simplify.

$$0.6 = 0.3 + 0.5 - P(A \cap B)$$
$$P(A \cap B) = 0.3 + 0.5 - 0.6$$
$$= 0.2$$

b. 1. State the formula for conditional probability. **b.** $P(B|A) = \dfrac{P(A \cap B)}{P(A)}, P(A) \neq 0$

2. Substitute the values given in the question into this formula and simplify.

$$P(B|A) = \dfrac{0.2}{0.3}$$
$$= \dfrac{2}{3}$$

- It is possible to transpose the formula for conditional probability to calculate $P(A \cap B)$:

$$P(B|A) = \dfrac{P(A \cap B)}{P(A)}, P(A) \neq 0$$
$$P(A \cap B) = P(A) \times P(B|A)$$

 Resources

▶ **Video eLesson** Conditional probability (eles-1928)

✦ **Interactivity** Conditional probability (int-6085)

Exercise 14.6 Conditional probability (Path)   learn

14.6 Quick quiz on	14.6 Exercise

Individual pathways

■ PRACTISE	■ CONSOLIDATE	■ MASTER
1, 3, 5, 9, 12, 15	2, 6, 10, 13, 16	4, 7, 8, 11, 14, 17

Fluency

1. **WE19** A group of students was asked whether they liked the following forms of dance: hip hop (H) or jazz (J). The results are illustrated in the Venn diagram. Use the Venn diagram to calculate the following probabilities relating to a student's favourite form of dance.

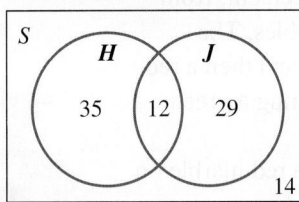

a. Calculate the probability that a randomly selected student likes jazz.
b. Determine the probability that a randomly selected student likes hip hop, given that they like jazz.

2. A group of students was asked which seats they liked: the seats in the computer lab or the science lab. The results are illustrated in the Venn diagram.

Use the Venn diagram to calculate the following probabilities relating to the most comfortable seats.

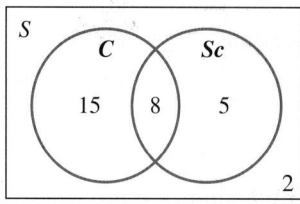

a. Calculate the probability that a randomly selected student likes the seats in the science lab.
b. Determine the probability that a randomly selected student likes the seats in the science lab, given that they like the seats in the computer lab or the science lab.

3. **WE20** If $P(A) = 0.7, P(B) = 0.5$ and $P(A \cup B) = 0.9$, calculate:

   a. $P(A \cap B)$ b. $P(B \mid A)$.

4. If $P(A) = 0.65, P(B) = 0.75$ and $P(A \cap B) = 0.45$, calculate:

   a. $P(B \mid A)$ b. $P(A \mid B)$.

Understanding

5. A medical degree requires applicants to participate in two tests: an aptitude test and an emotional maturity test. This year 52% passed the aptitude test and 30% passed both tests.

   Use the conditional probability formula to calculate the probability that a student who passed the aptitude test also passed the emotional maturity test.

6. At a school classified as a 'Music school for excellence', the probability that a student elects to study Music and Physics is 0.2. The probability that a student takes Music is 0.92. Determine the probability that a student takes Physics, given that the student is taking Music.

7. The probability that a student is well and misses a work shift the night before an exam is 0.045, and the probability that a student misses a work shift is 0.05.

   Determine the probability that a student is well, given they miss a work shift the night before an exam.

8. Two marbles are chosen, without replacement, from a jar containing only red and green marbles. The probability of selecting a green marble and then a red marble is 0.67. The probability of selecting a green marble on the first draw is 0.8.

   Determine the probability of selecting a red marble on the second draw, given the first marble drawn was green.

9. Consider rolling a red and a black die and the probabilities of the following events:

> Event A the red die lands on 5
> Event B the black die lands on 2
> Event C the sum of the dice is 10

a. **MC** The initial probability of each event described is:

A. $P(A) = \dfrac{1}{6}$ **B.** $P(A) = \dfrac{5}{6}$ **C.** $P(A) = \dfrac{5}{6}$ **D.** $P(A) = \dfrac{1}{6}$

 $P(B) = \dfrac{1}{6}$ $P(B) = \dfrac{2}{6}$ $P(B) = \dfrac{2}{6}$ $P(B) = \dfrac{1}{6}$

 $P(C) = \dfrac{1}{6}$ $P(C) = \dfrac{7}{36}$ $P(C) = \dfrac{5}{18}$ $P(C) = \dfrac{1}{12}$

b. Calculate the following probabilities.

   i. $P(A \mid B)$ ii. $P(B \mid A)$ iii. $P(C \mid A)$ iv. $P(C \mid B)$

10. **MC** A group of 80 schoolgirls consists of 54 dancers and 35 singers. Each member of the group is either a dancer, a singer, or both.
The probability that a randomly selected student is a singer given that she is a dancer is:

A. 0.17   **B.** 0.44   **C.** 0.68   **D.** 0.11

11. The following is the blood pressure data from 232 adult patients admitted to a hospital over a week. The results are displayed in a 2-way frequency table.

Age	Blood pressure			Total
	Low	**Medium**	**High**	
Under 60 years	92	44	10	146
60 years or above	17	46	23	86
Total	109	90	33	**232**

a. Calculate the probability that a randomly chosen patient has low blood pressure.
b. Determine the probability that a randomly chosen patient is under 60 years of age.
c. Calculate the probability that a randomly chosen patient has high blood pressure, given they are aged 60 years or above.
d. Determine the probability that a randomly chosen patient is under the age of 60, given they have medium blood pressure.

Communicating, reasoning and problem solving

12. Explain how imposing a condition alters probability calculations.

13. At a school, 65% of the students are male and 35% are female. Of the male students, 10% report that dancing is their favourite activity; of the female students, 25% report that dancing is their favourite activity.
Determine the probability that:
a. a student selected at random prefers dancing and is female
b. a student selected at random prefers dancing and is male.
c. Construct a tree diagram to present the information given, and use it to calculate:
   i. the probability that a student is male and does not prefer dancing
   ii. the overall percentage of students who prefer dancing.

14. Consider the following sets of numbers.

$$S = \{1, 2, 3, 4, 5, 6, 7, 8, 9, 10, 11, 12, 13, 14, 15, 16, 17, 18, 19, 20\}$$

$$A = \{1, 2, 3, 4, 5, 6, 7, 8, 9, 10\}$$

$$B = \{2, 3, 5, 7, 11, 13, 17, 19\}$$

a. Calculate the following.
   i. $P(A)$
   ii. $P(B)$
   iii. $P(A \cap B)$
   iv. $P(A \mid B)$
   v. $P(B \mid A)$

b. Explain whether the events A and B are independent.
c. Write a statement that connects $P(A), P(A \cap B), P(A \mid B)$ and independent events.

15. The rapid test used to determine whether a person is infected with COVID-19 is not perfect. For one type of rapid test, the probability of a person with the disease returning a positive result is 0.98, while the probability of a person without the disease returning a positive result is 0.04.

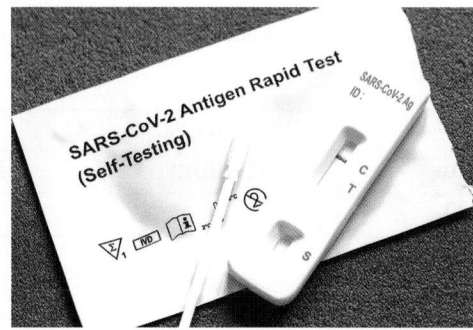

At its peak in a certain country, the probability that a randomly selected person has COVID-19 is 0.05. Determine the probability that a randomly selected person will return a positive result.

16. Two marbles are chosen, without replacement, from a jar containing only red and green marbles. The probability of selecting a green marble and then a red marble is 0.72. The probability of selecting a green marble on the first draw is 0.85.
Determine the probability of selecting a red marble on the second draw if the first marble drawn was green.

17. When walking home from school during the summer months, Harold buys either an ice-cream or a drink from the corner shop. If Harold bought an ice-cream the previous day, there is a 30% chance that he will buy a drink the next day. If he bought a drink the previous day, there is a 40% chance that he will buy an ice-cream the next day. On Monday, Harold bought an ice-cream. Determine the probability that he buys an ice-cream on Wednesday.

LESSON
14.7 Review

14.7.1 Topic summary

Review of probability

- The sample space is the set of all possible outcomes. It is denoted S. For example, the sample space of rolling a die is $S = \{1, 2, 3, 4, 5, 6\}$.
- An event is a set of favourable outcomes. For example, if A is the event {rolling an even number on a die}, $A = \{2, 4, 6\}$.
- The probability of an outcome or event is always between 0 and 1.
- The sum of all probabilities in a sample space is 1.
- An event that is **certain** has a probability of 1.
- An event that is **impossible** has a probability of 0.
- When all outcomes are equally likely the theoretical probability of an event is given by the following formula:

$$P(\text{event}) = \frac{\text{number of favourable outcomes}}{\text{total number of outcomes}} = \frac{n(E)}{n(S)}$$

- The experimental probability is given by the following formula:

$$\text{Experimental } P = \frac{n(\text{successful trials})}{n(\text{trials})}$$

- The complement of an event A is denoted A' and is the set of outcomes that are not in A.
  - $n(A) + n(A') = n(S)$
  - $P(A) + P(A') = 1$

PROBABILITY

Intersection and union

- The intersection of two events A and B is written $A \cap B$. These are the outcomes that are in A '**and**' in B.

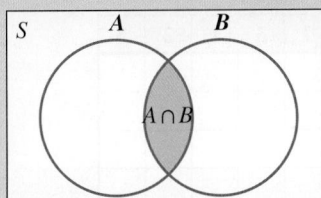

- The union of two events A and B is written $A \cup B$. These are the outcomes that are in A '**or**' in B.

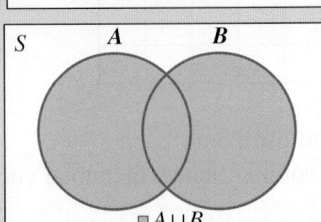

- A Venn diagram can be split into four distinct sections as shown at right.

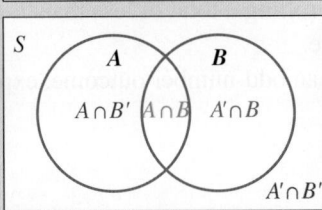

- The Addition Law of probability states that:
$$P(A \cup B) = P(A) + P(B) - P(A \cap B)$$

Independent and dependent events and conditional probability

- Two events are **independent** if the outcome of one event does not affect the outcome of the other event.
- If A and B are independent, then:
$$P(A \cap B) = P(A) \times P(B)$$
- Two events are **dependent** if the outcome of one event affects the outcome of the other event.
- **Conditional probability** applies to dependent events.
- The probability of A given that B has already occurred is given by:
$$P(A \mid B) = \frac{P(A \cap B)}{P(B)}$$
- This can be rearranged into:
$$P(A \cap B) = P(A \mid B) \times P(B)$$

Mutually exclusive events (Path)

- Mutually exclusive events have no elements in common.
- Venn diagrams show mutually exclusive events having no intersection. In the Venn diagram below A and B are mutually exclusive.

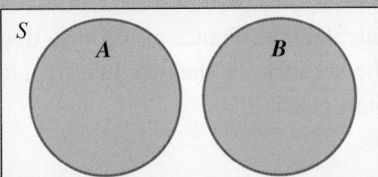

- For mutually exclusive events A and B:
  - $P(A \cap B) = 0$
  - $P(A \cup B) = P(A) + P(B)$

Multistage chance experiments

- Some chance experiments involve multiple trials performed separately.
- 2-way tables can be used to show the sample space of two-step experiments.
- Tree diagrams can be used to show the sample space and probabilities of multistage experiments.

		Coin 1	
		H	T
Coin 2	H	HH	TH
	T	HT	TT

First selection	Second selection	Sample space
B	R	BR
	G	BG
	Y	BY
W	R	WR
	G	WG
	Y	WY

Tricky dice

Dice games have been played throughout the world for many years. Professional gamblers resort to all types of devious measures in order to win. Often the other players are unaware of the tricks employed.

Imagine you are playing a game that involves rolling two dice. Instead of having each die marked with the numbers 1 to 6, let the first die have only the numbers 1, 2 and 3 (two of each) and the second die the numbers 4, 5 and 6 (two of each). If you were an observer to this game, you would see the numbers 1 to 6 occurring and probably not realise that the dice were not the regular type.

1. Complete the grid below to show the sample space on rolling these two dice.
2. Identify how many different outcomes there are. Compare this with the number of different outcomes using two regular dice.

	Die 1					
	1	2	3	1	2	3
Die 2 4						
5						
6						
4						
5						
6						

3. Calculate the chance of rolling a double using these dice.
4. The numbers on the two dice are added after rolling. Complete the table below to show the totals possible.

	Die 1					
	1	2	3	1	2	3
Die 2 4						
5						
6						
4						
5						
6						

5. Identify how many different totals are possible and list them.
6. State which total you have the greatest chance of rolling. State which total you have the least chance of rolling.
7. If you played a game in which you had to bet on rolling a total of less than 7, equal to 7 or greater than 7, explain which option would you be best to take.
8. If you had to bet on an even-number outcome or an odd-number outcome, explain which would be the better option.

9. The rules are changed to subtracting the numbers on the two dice instead of adding them. Complete the following table to show the outcomes possible.

		Die 1					
		1	2	3	1	2	3
Die 2	4						
	5						
	6						
	4						
	5						
	6						

10. Identify how many different outcomes are possible in this case and list them.
11. State the most frequently occurring outcome and how many times it occurs.
12. Devise a game of your own using these dice. On a separate sheet of paper, write out rules for your game and provide a solution, indicating the best options for winning.

 Resources

 Interactivities Crossword (int-2857)
Sudoku puzzle (int-3598)

Exercise 14.7 Review questions

learnon

Fluency

1. **MC** Choose which of the following is always true for an event, M, and its complementary event, M'.
   A. $P(M) + P(M') = 1$
   B. $P(M) - P(M') = 1$
   C. $P(M) + P(M') = 0$
   D. $P(M) - P(M') = 0$

2. **PATH** **MC** A number is chosen from the set $\{0, 1, 2, 3, 4, 5, 6, 7, 8, 9, 10\}$. Select which of the following pairs of events is mutually exclusive.
   A. $\{2, 4, 6\}$ and $\{4, 6, 7, 8\}$
   B. $\{1, 2, 3, 5\}$ and $\{4, 6, 7, 8\}$
   C. {multiples of 2} and {factors of 8}
   D. {even numbers} and {multiples of 3}

3. **PATH** **MC** Choose which of the following states the Multiplication Law of probability correctly.
   A. $P(A \cap B) = P(A) + P(B)$
   B. $P(A \cap B) = P(A) \times P(B)$
   C. $P(A \cup B) = P(A) \times P(B)$
   D. $P(A \cup B) = P(A) + P(B)$

4. **PATH** **MC** Given $S = \{1, 2, 3, 4, 5, 6, 7, 8, 9, 10\}$ and $A = \{2, 3, 4\}$ and $B = \{3, 4, 5, 8\}$, $A \cap B$ is:
   A. $\{3, 4\}$
   B. $\{2, 3, 4\}$
   C. $\{2, 3, 4, 5, 8\}$
   D. $\{2, 5, 8\}$

5. **PATH** **MC** Given $S = \{1, 2, 3, 4, 5, 6, 7, 8, 9, 10\}$ and $A = \{2, 3, 4\}$ and $B = \{3, 4, 5, 8\}$, $A \cap B'$ is:
   A. $\{3, 4\}$
   B. $\{2\}$
   C. $\{2, 3, 4\}$
   D. $\{1, 2, 6, 7, 9, 10\}$

▶

6. **PATH** Shade the region stated for each of the following Venn diagrams.
   a. $A' \cup B$
   b. $A' \cap B'$
   c. $A' \cap B' \cap C$

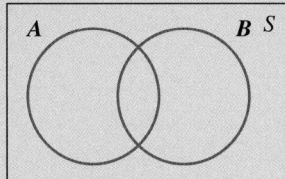
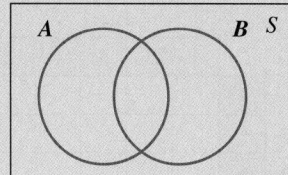
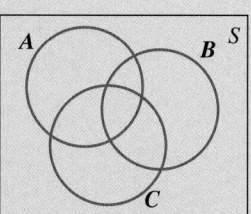

Understanding

7. **MC** From past experience, it is concluded that there is a 99% probability that July will be a wet month in Launceston (it has an average rainfall of approximately 80 mm). The probability that July will not be a wet month next year in Launceston is:

   A. 99%
   B. 0.99
   C. $\dfrac{1}{100}$
   D. 1

8. **MC** A card is drawn from a well-shuffled deck of 52 cards. Select the theoretical probability of not selecting a red card.

   A. $\dfrac{3}{4}$
   B. $\dfrac{1}{4}$
   C. $\dfrac{1}{13}$
   D. $\dfrac{1}{2}$

9. **MC** Choose which of the following events is not equally likely.
   A. Obtaining a 5 or obtaining a 1 when a die is rolled
   B. Obtaining a club or obtaining a diamond when a card is drawn from a pack of cards
   C. Obtaining 2 Heads or obtaining 2 Tails when a coin is tossed twice
   D. Obtaining 2 Heads or obtaining 1 Head when a coin is tossed twice
   E. Obtaining a 3 or obtaining a 6 when a die is rolled

10. **MC** The Australian cricket team has won 12 of the last 15 Test matches. Select the experimental probability of Australia not winning its next Test match.

    A. $\dfrac{4}{5}$
    B. $\dfrac{1}{5}$
    C. $\dfrac{1}{4}$
    D. $\dfrac{3}{4}$

11. A card is drawn from a well-shuffled pack of 52 cards. Calculate the theoretical probability of drawing:
    a. an ace
    b. a spade
    c. a queen or a king
    d. not a heart.

12. A die is rolled five times.
    a. Calculate the probability of rolling five 6s.
    b. Calculate the probability of not rolling five 6s.

13. Alan and Mary own 3 of the 8 dogs in a race. Evaluate the probability that:
    a. one of Alan's or Mary's dogs will win
    b. none of Alan's or Mary's dogs will win.

Communicating, reasoning and problem solving

14. **PATH** A die is rolled. Event A is obtaining an even number. Event B is obtaining a 3.
    a. Explain if events A and B are mutually exclusive.
    b. Calculate P(A) and P(B).
    c. Calculate P($A \cup B$).

15. **PATH** A card is drawn from a shuffled pack of 52 playing cards. Event A is drawing a club and event B is drawing an ace.
    a. Explain if events A and B are mutually exclusive.
    b. Calculate $P(A)$, $P(B)$ and $P(A \cap B)$.
    c. Calculate $P(A \cup B)$.

16. A tetrahedral die is numbered 0, 1, 2 and 3. Two of these dice are rolled and the sum of the numbers (the number on the face that the die sits on) is taken.
    a. Show the possible outcomes in a 2-way table.
    b. Determine if all the outcomes are equally likely.
    c. Determine which total has the least chance of being rolled.
    d. Determine which total has the best chance of being rolled.
    e. Determine which sums have the same chance of being rolled.

17. A bag contains 20 pears, of which 5 are bad. Cathy picks 2 pears (without replacement) from the bag. Evaluate the probability that:
    a. both pears are bad   b. both pears are good   c. one of the two pears is good.

18. Determine the probability of drawing 2 aces from a pack of cards if:
    a. the first card is replaced before the second one is drawn
    b. the first card drawn is not replaced.

19. **PATH** On grandparents day at a school, a group of grandparents was asked where they most like to take their grandchildren — the beach (B) or shopping (Sh). The results are illustrated in the Venn diagram. Use the Venn diagram to calculate the following probabilities relating to the place grandparents most like to take their grandchildren.

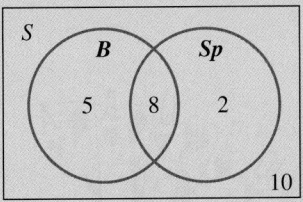

    a. Determine the probability that a randomly selected grandparent preferred to take their grandchildren to the beach or shopping.
    b. Determine the probability that a randomly selected grandparent liked to take their grandchildren to the beach, given that they liked to take their grandchildren shopping.

20. Andrew does not know the answer to two questions on a multiple-choice exam. The first question has four choices and the second question he does not know has five choices.
    a. Determine the probability that he will get both questions wrong.
    b. If he is certain that one of the choices cannot be the answer in the first question, determine how this will change the probability that he will get both questions wrong.

21. When all of Saphron's team players turn up for their twice weekly netball training the chance that they then win their Saturday game is 0.65. If not all players are at the training session, then the chance of winning their Saturday game is 0.40. Over a four-week period, Saphron's players all turn up for training three times.

    a. Using a tree diagram, with T to represent all players training and W to represent a win, represent the winning chance of Saphron's netball team.

    b. Using the tree diagram constructed in part **a**, determine the probability of Saphron's team winning their Saturday game. Write your answer correct to 4 decimal places.

    c. Determine the exact probability that Saphron's team did not train given that they won their Saturday game.

22. Mariah the Mathematics teacher wanted to give her students a chance to win a reward at the end of the term. She placed 20 cards into a box, and wrote the word ON on 16 cards, and OFF on 4 cards. After a student chooses a card, that card is replaced into the box for the next student to draw.

If a student chooses an OFF card, then they do not have to attend school on a specified day. If they choose an ON card, then they do not receive a day off.

    a. Mick, a student, chose a random card from the box. Calculate the probability he received a day off.

    b. Juanita, a student, chose a random card from the box after Mick. Calculate the probability that she did not receive a day off.

    c. Determine the probability that Mick and Juanita both received a day off.

23. In the game of draw poker, a player is dealt 5 cards from a deck of 52. To obtain a flush, all 5 cards must be of the same suit.

    a. Determine the probability of getting a diamond flush.

    b. Determine the probability of getting any flush.

24. a. A Year 10 boy is talking with a Year 10 girl and asks her if she has any brothers or sisters. She says, 'Yes, I have one.' Determine the probability that she has at least one sister.

    b. A Year 10 boy is talking with a Year 10 girl and asks her if she has any brothers or sisters. She says, 'Yes, I have three.' Determine the probability that she has at least one sister.

on To test your understanding and knowledge of this topic, go to your learnON title at www.jacplus.com.au and complete the **post-test**.

Answers

Topic 14 Probability

14.1 Pre-test

1. $P(A \cap B) = 0.2$
2. $P(A) = 0.51$
3. Independent
4. $\dfrac{1}{5}$
5. C
6. D
7. A
8. 0.56
9. B
10. False
11. $P(B|A) = 0.2$
12. C
13. A
14. 0.03125
15. A

14.2 Review of probability and simulations

1. Experimental probability is based on the outcomes of experiments, simulations or surveys. Theoretical probability is based on the number of possible favourable outcomes and the total possible outcomes.

2. a. $\{I, II, III, IV, V, VI\}$
   b. Relative frequency for I = 0.2
      Relative frequency for II = 0.12
      Relative frequency for III = 0.16
      Relative frequency for IV = 0.14
      Relative frequency for V = 0.24
      Relative frequency for VI = 0.14
   c. The spinner should be spun a larger number of times.

3. a. 500 students
   b. Frequency for silver = 0.16
   c. Frequency for black and green = 0.316
   d. Blazing Blue

4. $\dfrac{2}{3}$

5. a. $\dfrac{1}{13}$
   b. $\dfrac{1}{4}$
   c. $\dfrac{4}{13}$

6. Sample responses are given for part ii.
   a. i. No. There are many others foods one could have.
      ii. Having Weet Bix and not having Weet Bix

b. i. No. There are other means of transport, for example, catching a bus.
   ii. Walking to a friend's place and not walking to a friend's place
c. i. No. There are other possible leisure activities.
   ii. Watching TV and not watching TV
d. i. No. The number 5 can be rolled too.
   ii. Rolling a number less than 5 and rolling a number 5 or greater
e. i. Yes. There are only two possible outcomes: passing or failing.
   ii. No change is required to make these events complementary.

7. a. No. For a 6-sided die, $P(4) = \dfrac{1}{6}$; for an 8-sided die, $P(4) = \dfrac{1}{8}$.

   b. Yes; $P(\text{odd}) = \dfrac{1}{2}$ for both 6-sided and 8-sided dice

8. Yes; $P(5) = \dfrac{1}{2}, P(6) = \dfrac{1}{2}$.

9. a. $\dfrac{1}{13}$ b. $\dfrac{1}{4}$ c. $\dfrac{1}{2}$ d. $\dfrac{12}{13}$
   e. 0   f. $\dfrac{1}{2}$

10. a. $\dfrac{1}{5}$ b. $\dfrac{7}{20}$ c. $\dfrac{11}{20}$ d. 0

11. Yes. Both have a probability of $\dfrac{1}{2}$.

12. 17 red and 1 purple, i.e. 18 more socks.

13. a. Mandy's simulation: $\dfrac{13}{20}$

    Sophia's simulation: $\dfrac{47}{100}$

    b. Mandy's simulation shows a higher probability of perfect weather days. Sophia's simulation is likely to be more reliable as there is a higher number of coin tosses.

14. a. Answers will vary. Sample responses can be found in the worked solutions in the online resources. You can use =RANDBETWEEEN(1,60) in Excel to generate a random number.
    b. Answers will vary. Sample responses can be found in the worked solutions in the online resources. You can use =RANDBETWEEEN(1,60) in Excel to generate a random number.

15. a. 35%   b. 40%

16. Sample responses can be found in the worked solutions in the online resources.

17. Sample responses can be found in the worked solutions in the online resources.

18. a. i. $\dfrac{3}{11}$ ii. $\dfrac{1}{11}$ iii. $\dfrac{1}{11}$

b. Sample responses can be found in the worked solutions in the online resources.

c. Sample responses can be found in the worked solutions in the online resources.

19. Sample responses can be found in the worked solutions in the online resources.

20. a.

Die 2 outcomes

	1	2	3	4	5	6
1	2	3	4	5	6	7
2	3	4	5	6	7	8
3	4	5	6	7	8	9
4	5	6	7	8	9	10
5	6	7	8	9	10	11
6	7	8	9	10	11	12

(Die 1 outcomes — row labels)

b. 6

c. No. The frequency of the numbers is different.

d.

Sum	2	3	4	5	6	7	8	9	10	11	12
Frequency	1	2	3	4	5	6	5	4	3	2	1

e. i. $\dfrac{1}{36}$ ii. $\dfrac{1}{6}$ iii. $\dfrac{1}{18}$

f. i. $\dfrac{1}{36}$ ii. $\dfrac{1}{6}$ iii. $\dfrac{1}{6}$

g. 50

14.3 Tree diagrams and multistage chance experiments

1. If the probabilities of two events are different, the first column of branches indicates the probabilities for the first event and the second column of branches indicates the probabilities for the second event. The product of each branch gives the probability. All probabilities add to 1.

2. a. 12 different outcomes

b. No. Each branch is a product of different probabilities.

c. Less likely

d. Indigo elephant

e. i. $P(\text{Blue elephant}) = \dfrac{7}{40}$

ii. $P(\text{Indigo elephant}) = \dfrac{7}{20}$

iii. $P(\text{Donkey}) = \dfrac{1}{5}$

3. a.

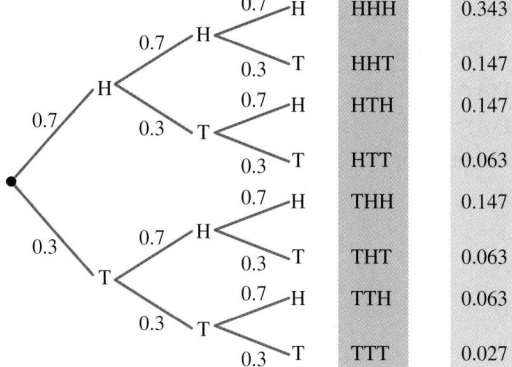

b. $P(\text{HHH}) = 0.343$

c. $P(\text{at least 1 Tail}) = 0.657$

d. $P(\text{exactly 2 Tails}) = 0.189$

4. a. i. $\dfrac{1}{2}$ ii. $\dfrac{1}{6}$ iii. $\dfrac{5}{6}$

b.

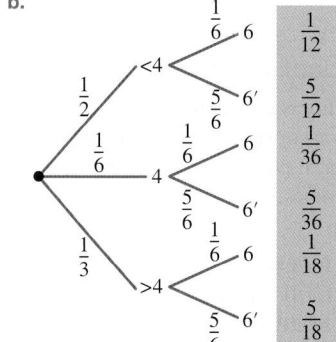

c. i. $\dfrac{1}{36}$ ii. $\dfrac{1}{12}$ iii. $\dfrac{5}{36}$ iv. $\dfrac{5}{18}$

5. a.

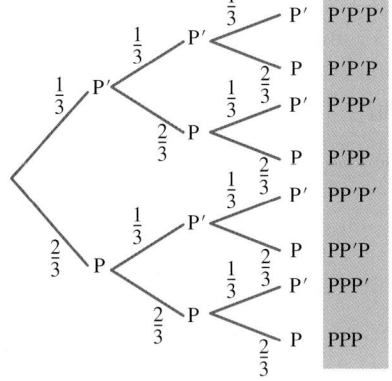

b. i. $\dfrac{8}{27}$ ii. $\dfrac{4}{27}$ iii. $\dfrac{12}{27}$

6. a.

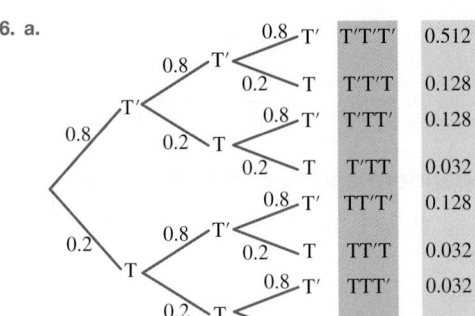

b. 0.520

7. a.

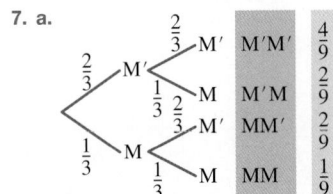

b. $\dfrac{1}{9}$

8. a.

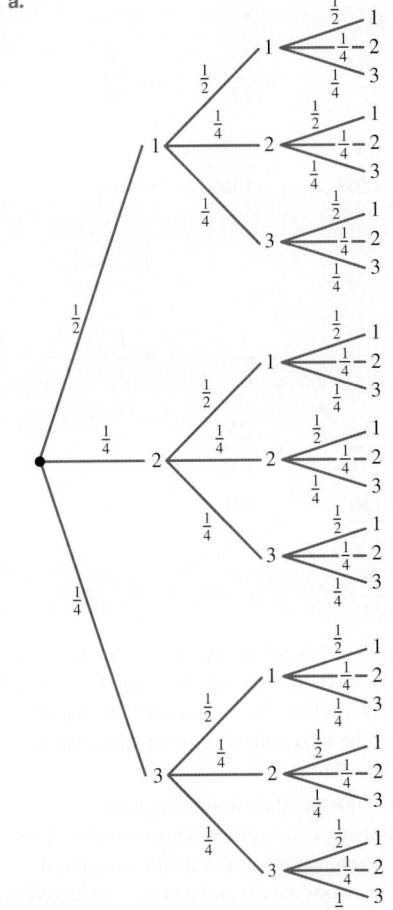

b. $\dfrac{3}{8}$

c. $\dfrac{7}{8}$

9. a.

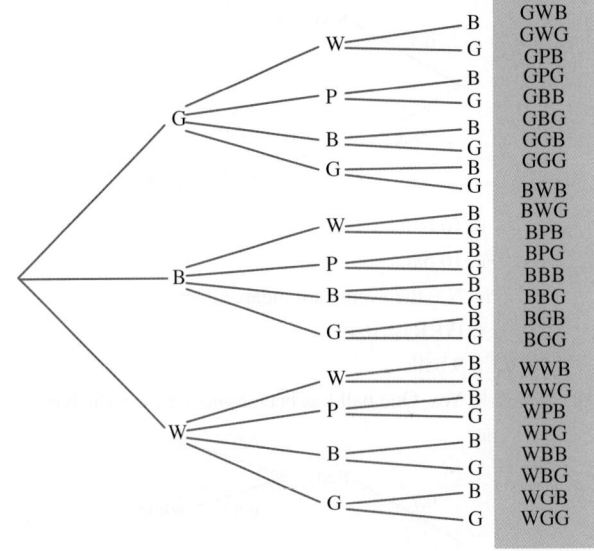

b. $\dfrac{1}{24}$

c. $\dfrac{1}{12}$

d. Each combination would be worn 10 times in a year.

e. 21 new combinations.

10. a.

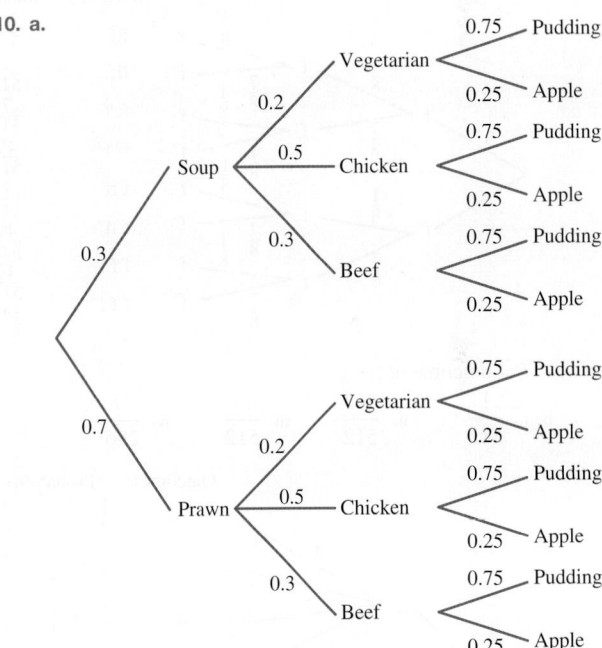

b. 0.1125

c. 24 people

11. a. i.

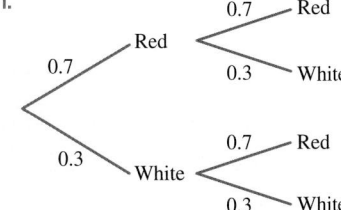

ii. 10 balls

iii. 10 balls

iv. No; the events are independent.

v. P(RR) = 0.49

b. i. 9 balls

ii. Yes. One ball has been removed from the bag.

iii.

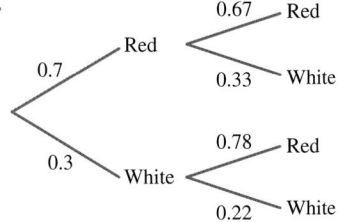

iv. P(RR) = $\frac{7}{15}$ or 0.469 using the rounded values from iii.

12. a.

	Outcomes	Probability
f	fff	$\frac{1}{512}$
f'	fff'	$\frac{7}{512}$
f	ff'f	$\frac{7}{512}$
f'	ff'f'	$\frac{49}{512}$
f	f'ff	$\frac{7}{512}$
f'	f'ff'	$\frac{49}{512}$
f	f'f'f	$\frac{49}{512}$
f'	f'f'f'	$\frac{343}{512}$

f = outcome of 5

b. i. $\frac{1}{512}$ **ii.** $\frac{343}{512}$ **iii.** $\frac{21}{512}$ **iv.** $\frac{11}{256}$

13. a.

	Outcomes	Probability
1	H1	$\frac{1}{2} \times \frac{1}{4} = \frac{1}{8}$
2	H2	$\frac{1}{2} \times \frac{1}{4} = \frac{1}{8}$
3	H3	$\frac{1}{2} \times \frac{1}{4} = \frac{1}{8}$
4	H4	$\frac{1}{2} \times \frac{1}{4} = \frac{1}{8}$
1	T1	$\frac{1}{2} \times \frac{1}{4} = \frac{1}{8}$
2	T2	$\frac{1}{2} \times \frac{1}{4} = \frac{1}{8}$
3	T3	$\frac{1}{2} \times \frac{1}{4} = \frac{1}{8}$
4	T4	$\frac{1}{2} \times \frac{1}{4} = \frac{1}{8}$
		1

b. $\frac{1}{4}$

14. a. $\frac{256}{625}$ **b.** $\frac{16}{625}$ **c.** $\frac{96}{625}$ **d.** $\frac{17}{625}$

14.4 Independent and dependent events

1. a. 0.28 **b.** 0.12 **c.** 0.42 **d.** 0.18

2. Dependent

3. Independent

4. a. Yes, the outcomes are independent.

b. i. $\frac{1}{2}$ **ii.** $\frac{1}{6}$

c. $\frac{1}{12}$

5. $\frac{1}{40}$

6. $\frac{5}{36}$

7. a. $\frac{16}{25}$ **b.** $\frac{64}{125}$ **c.** $\frac{1}{25}$ **d.** $\frac{4}{25}$

8. a. C **b.** D

9. a. $\frac{3}{77}$ **b.** $\frac{48}{77}$ **c.** $\frac{8}{77}$ **d.** $\frac{18}{77}$

10. a. $\frac{1}{37}$ **b.** $\frac{1}{1369}$ **c.** $\frac{73}{1369}$

11. 0.9

12. $\frac{1}{14}$

13. a. $\frac{1}{5}$ **b.** $\frac{1}{5}$ **c.** $\frac{1}{10}$ **d.** $\frac{1}{3}$

14. a. $\frac{1}{17}$ **b.** $\frac{1}{221}$ **c.** $\frac{25}{102}$

15. a. $\frac{26}{145}$ **b.** $\frac{136}{435}$ **c.** $\frac{221}{435}$

16. a. $\frac{5}{33}$ **b.** $\frac{1}{22}$

17. No. Coin tosses are independent events. No one toss affects the outcome of the next. The probability of a Head or Tail on a fair coin is always 0.5. Greg has a 50% chance of tossing a Head on the next coin toss as was the chance in each of the previous 9 tosses.

18. No. As events are illustrated on a tree diagram, the individual probability of each outcome is recorded. The probability of a dependent event is calculated (altered according to the previous event) and can be considered as if it was an independent event. As such, the multiplication law of probability can be applied along the branches to calculate the probability of successive events.

19. Only if P(A) = 0 or P(B) = 0 can two events be independent and mutually exclusive. For an event to have a probability of 0 means that it is impossible, so it is a trivial scenario.

20. a. A and B are independent.

b. A and B are not independent in this situation.

21. $\dfrac{5}{17}$

22. a. Dependent

b. $P(1) = \dfrac{7}{15}$

$P(2) = \dfrac{5}{15} = \dfrac{1}{3}$

$P(3) = \dfrac{3}{15} = \dfrac{1}{5}$

c.

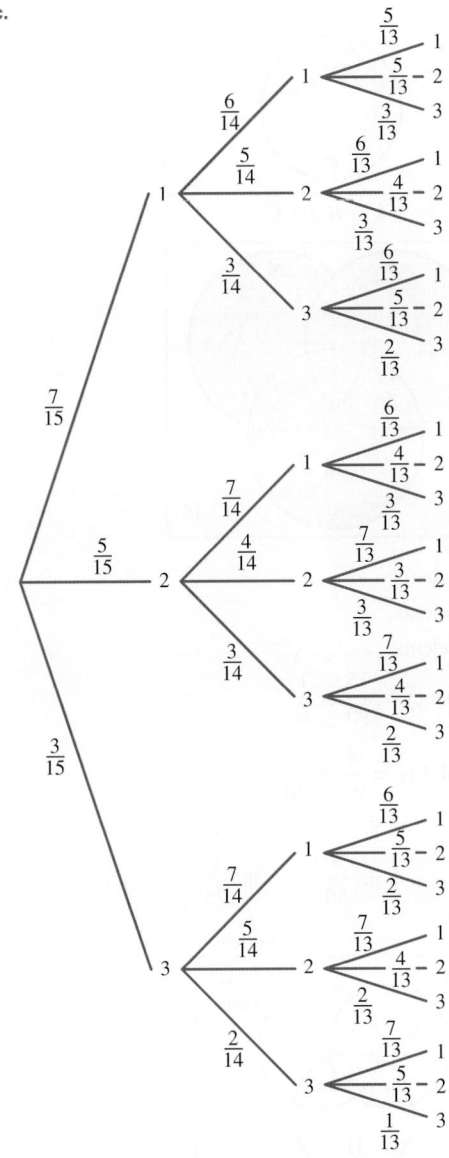

d. i. $P(1,1,1) = \dfrac{1}{13}$ **ii.** $P(2,2,2) = \dfrac{2}{91}$

iii. $P(3,3,3) = \dfrac{1}{455}$ **iv.** $P(\text{at least one } 3) = \dfrac{47}{91}$

23. a. $P(9) = \dfrac{1}{455}$

$P(7 \text{ or } 8) = \dfrac{66}{455}$

$P(5 \text{ or } 6) = \dfrac{248}{455}$

$P(3 \text{ or } 4) = \dfrac{4}{13}$

b. \$393.40

14.5 Venn diagrams, 2-way tables and mutually exclusive events (Path)

1. a. $P(A) = \dfrac{3}{10}$ $P(A') = \dfrac{7}{10}$

b. $\dfrac{2}{5}$

2. a. $\dfrac{5}{23}$ **b.** $\dfrac{12}{23}$ **c.** $\dfrac{15}{23}$ **d.** 0

3. a. $\dfrac{1}{6}$ **b.** $\dfrac{5}{6}$ **c.** $\dfrac{1}{2}$

d. $\dfrac{2}{3}$ **e.** $\dfrac{1}{3}$ **f.** 0

4. a. $\dfrac{1}{2}$ **b.** $\dfrac{2}{13}$ **c.** $\dfrac{1}{13}$

5. a. $\dfrac{11}{20}$ **b.** 1   **c.** $\dfrac{11}{20}$

6. a. $\dfrac{1}{13}$ **b.** $\dfrac{1}{4}$ **c.** $\dfrac{4}{13}$

7. a. $P(A \cap B) = 0.1$

b.

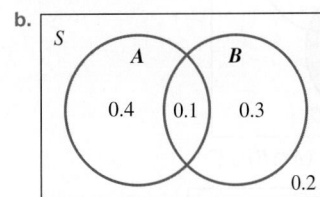

c. $P(A \cap B') = 0.4$

8. a. 0.2

b.

	B	B$'$	
A	0.2	0.2	0.4
A$'$	0.5	0.1	0.6
	0.7	0.3	1

c. 0.5

9. a. $P(A \cup B) = 0.85$ **b.** $P(A \cap B)' = 0.95$

10. C

11. a.

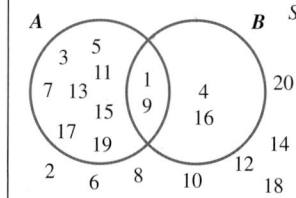

b. i. $\dfrac{10}{20} = \dfrac{1}{2}$ **ii.** $\dfrac{4}{20} = \dfrac{1}{5}$ **iii.** $\dfrac{2}{20} = \dfrac{1}{10}$

iv. $\dfrac{12}{20} = \dfrac{3}{5}$ **v.** $\dfrac{8}{20} = \dfrac{2}{5}$

12. a.

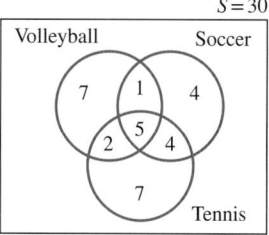

b. i. $\dfrac{1}{2}$ **ii.** $\dfrac{1}{6}$ **iii.** $\dfrac{1}{30}$ **iv.** $\dfrac{2}{5}$ **v.** $\dfrac{7}{15}$

c. i. $\dfrac{1}{2}$ **ii.** $\dfrac{8}{15}$

13. Sample responses can be found in the worked solutions in the online resources.

14.

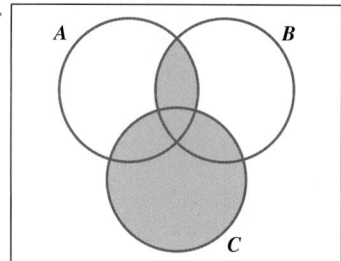

$(A \cap B) \cup C$

15. a.

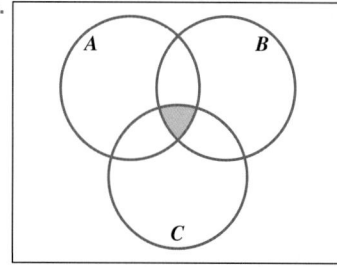

$A \cap B \cap C$

b.

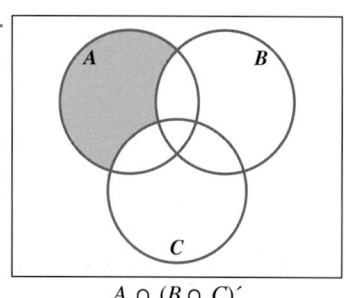

$A \cap (B \cap C)'$

c.

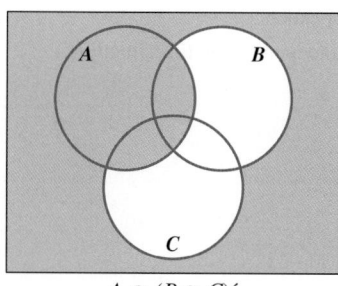

$A \cap (B \cap C)'$

d.

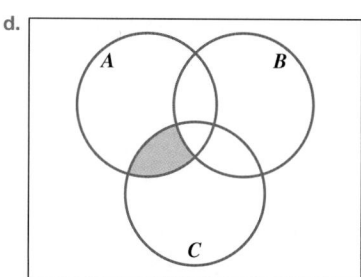

$(A \cap B') \cap C$

16. a.

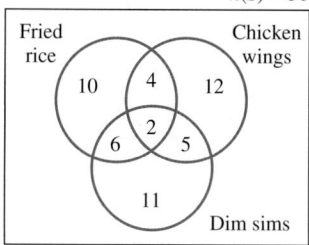

b. 19 students

c. 32 students

d. 15 students

e. Frequency $= \dfrac{15}{90}$ or $\dfrac{1}{6}$

f. Probability $= \dfrac{3}{90} = \dfrac{1}{30}$

17. a. i. 50   **ii.** 7   **iii.** 25   **iv.** 8

b. i. $\dfrac{1}{2}$ **ii.** $\dfrac{3}{50}$ **iii.** $\dfrac{6}{25}$

18. a.

b. $\dfrac{1}{25}$

c. $\dfrac{17}{50}$

d. $\dfrac{1}{5}$

19. a. 5

b. i. $\frac{1}{4}$ ii. $\frac{1}{5}$ iii. $\frac{11}{20}$

20. a. $S = \{1, 2, 3, ..., 18\}$, $A = \{3, 6, 9, 12, 15, 18\}$, $B = \{4, 8, 12, 16\}$ and $C = \{5, 10, 15\}$

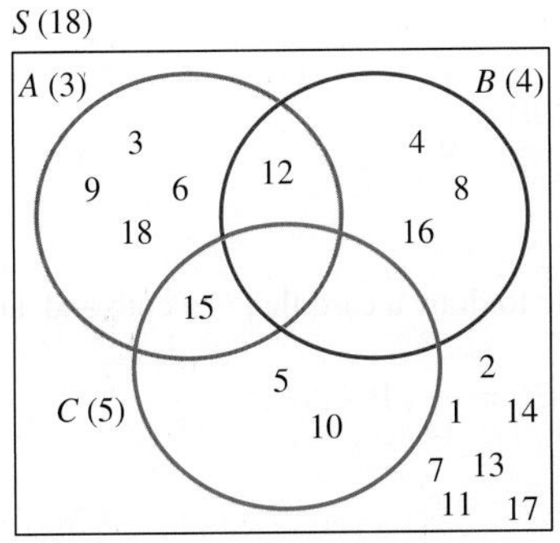

b. B and C

c. $\frac{1}{3}$

d. i. $\frac{4}{9}$ ii. $\frac{5}{18}$ iii. $\frac{7}{18}$

14.6 Conditional probability (Path)

1. a. $P(J) = \frac{41}{90}$ b. $P(z \mid J) = \frac{12}{41}$

2. a. $P(Sc) = \frac{13}{30}$ b. $P\left(Sc \mid (C \cup Sc)\right) = \frac{13}{28}$

3. a. 0.3 b. $\frac{3}{7}$

4. a. $\frac{9}{13}$ b. $\frac{3}{5}$

5. 0.58 or $\frac{15}{26}$

6. 0.22 or $\frac{5}{23}$

7. 0.9

8. 0.8375

9. a. D

b. i. $P(A \mid B) = \frac{1}{6}$ ii. $P(B \mid A) = \frac{1}{6}$

iii. $P(C \mid A) = \frac{1}{6}$ iv. $P(C \mid B) = 0$

10. A

11. a. $\frac{109}{232}$ b. $\frac{73}{116}$ c. $\frac{23}{86}$ d. $\frac{22}{45}$

12. Conditional probability is when the probability of one event depends on the outcome of another event.

13. a. 0.0875

b. 0.065

c. i. 0.585

ii. 15.25%

14. a. i. $\frac{1}{2}$ ii. $\frac{2}{5}$ iii. $\frac{1}{5}$ iv. $\frac{1}{2}$ v. $\frac{2}{5}$

b. Yes, A and B are independent.

c. If $P(A \mid B) = \frac{P(A \cap B)}{P(B)} = P(A)$ then A and B will be independent events. This is because the probability of A given B occurs is the same as the probability of A, meaning the probability of A is independent of B occurring.

15. 0.087 or 8.7% chance

16. 0.847

17. 0.61

Project

1.

Die 2 \ Die 1	1	2	3	1	2	3
4	(1, 4)	(2, 4)	(3, 4)	(1, 4)	(2, 4)	(3, 4)
5	(1, 5)	(2, 5)	(3, 5)	(1, 5)	(2, 5)	(3, 5)
6	(1, 6)	(2, 6)	(3, 6)	(1, 6)	(2, 6)	(3, 6)
4	(1, 4)	(2, 4)	(3, 4)	(1, 4)	(2, 4)	(3, 4)
5	(1, 5)	(2, 5)	(3, 5)	(1, 5)	(2, 5)	(3, 5)
6	(1, 6)	(2, 6)	(3, 6)	(1, 6)	(2, 6)	(3, 6)

2. 9

3. 0

4.

Die 2 \ Die 1	1	2	3	1	2	3
4	5	6	7	5	6	7
5	6	7	8	6	7	8
6	7	8	9	7	8	9
4	5	6	7	5	6	7
5	6	7	8	6	7	8
6	7	8	9	7	8	9

5. 5; 5, 6, 7, 8, 9

6. 7; 5, 9

7. Equal to 7; probability is the highest.

8. Odd-number outcome; probability is higher.

Coin \ Die	1	2	3	4	5	6	7	8
Head	(H, 1)	(H, 2)	(H, 3)	(H, 4)	(H, 5)	(H, 6)	(H, 7)	(H, 8)
Tail	(T, 1)	(T, 2)	(T, 3)	(T, 4)	(T, 5)	(T, 6)	(T, 7)	(T, 8)

9.

Die 1

	1	2	3	1	2	3
4	3	2	1	3	2	1
5	4	3	2	4	3	2
6	5	4	3	5	4	3
4	3	2	1	3	2	1
5	4	3	2	4	3	2
6	5	4	3	5	4	3

Die 2 (left vertical label)

10. 5; 1, 2, 3, 4, 5,

11. 3; 12

12. Students need to apply the knowledge of probability and create a new game using dice given. They also need to provide rules for the game and solution, indicating the best options for winning.

14.7 Review questions

1. A

2. B

3. B

4. A

5. B

6. a.

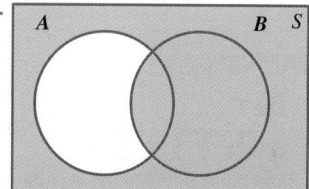

b.

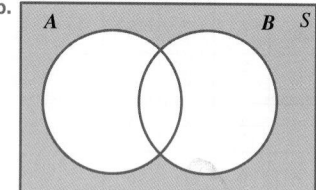

c.

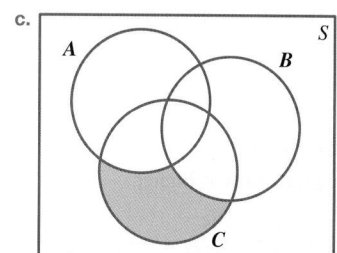

7. C

8. D

9. D

10. B

11. a. $\dfrac{1}{13}$ **b.** $\dfrac{1}{4}$ **c.** $\dfrac{2}{13}$ **d.** $\dfrac{3}{4}$

12. a. $\dfrac{1}{7776}$ **b.** $\dfrac{7775}{7776}$

13. a. $\dfrac{3}{8}$ **b.** $\dfrac{5}{8}$

14. a. Yes. It is not possible to roll an even number and for that number to be a 3.

   b. $P(A) = \dfrac{1}{2}$ and $P(B) = \dfrac{1}{6}$

   c. $\dfrac{2}{3}$

15. a. No. It is possible to draw a card that is a club and an ace.

   b. $P(A) = \dfrac{1}{4}$ and $P(B) = \dfrac{1}{13}$, $P(A \cap B) = \dfrac{1}{52}$

   c. $\dfrac{4}{13}$

16. a.

Die 2 outcomes

	0	**1**	**2**	**3**
0	0	1	2	3
1	1	2	3	4
2	2	3	4	5
3	3	4	5	6

Die 1 outcomes (left vertical label)

   b. No

   c. 0 and 6

   d. 3

   e. 0 and 6, 1 and 5, 2 and 4

17. a. $\dfrac{1}{19}$ **b.** $\dfrac{21}{38}$ **c.** $\dfrac{15}{38}$

18. a. $\dfrac{1}{169}$ **b.** $\dfrac{1}{221}$

19. a. $\dfrac{15}{25} = \dfrac{3}{5}$ **b.** $\dfrac{8}{10} = \dfrac{4}{5}$

20. a. $\dfrac{3}{5}$ **b.** $\dfrac{8}{15}$

21. a.

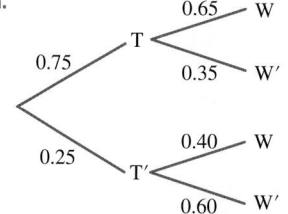

   b. 0.5875

   c. $\dfrac{8}{47}$

22. a. $\dfrac{1}{5}$ **b.** $\dfrac{4}{5}$ **c.** $\dfrac{1}{25}$

23. a. 0.000 495   **b.** 0.001 981

24. a. $\dfrac{1}{2}$ **b.** $\dfrac{7}{8}$

15 Polynomials (Path)

LESSON SEQUENCE

LESSON
15.1 Overview

Why learn this?

Just as your knowledge of numbers is learned in stages, so too are graphs. You have been building your knowledge of graphs and functions over time. First, you encountered linear functions. You saw how straight lines are everywhere in our daily lives. Then you learned about quadratic functions, or parabolas. Again, you saw, in everyday situations, how bridges and arches can be based on quadratic or parabolic functions. Circles and hyperbolas are other functions that you have studied. A polynomial is an algebraic expression with integer powers that are greater than or equal to zero, such as a parabola. Polynomial functions are represented by smooth and continuous curves. They can be used to model situations in many different fields, including business, science, architecture, design and engineering. An engineer and designer would use polynomials to model the curves on a rollercoaster. Economists use polynomials to model changes and fluctuations in the stock market. Scientists and researchers use polynomials when looking at changes in the behaviour of objects in different circumstances. Designers and architects incorporate polynomial functions in many areas of their designs in buildings and in landscaping. This topic introduces the building blocks of polynomials.

Hey students! Bring these pages to life online

Watch videos

Engage with interactivities

Answer questions and check solutions

Find all this and MORE in jacPLUS

Reading content and rich media, including interactivities and videos for every concept

Extra learning resources

Differentiated question sets

Questions with immediate feedback, and fully worked solutions to help students get unstuck

Exercise 15.1 Pre-test

1. State the degree of the following polynomials.

    a. $3x^2 - 5x + 1$ **b.** $2x^3 - 4x^2 + 3x$ **c.** $\dfrac{1}{2}x + 7$ **d.** 4

2. **MC** Choose which of the following is a polynomial.

    A. $\dfrac{x^3 + x}{2}$ **B.** $x^{-2} - \sqrt{7}$ **C.** $-x^2 + 3\sqrt{x}$ **D.** $6x^4 - \dfrac{1}{x}$

3. If $x^2 + 4x - 2 \equiv (x+1)^2 + a(x-1) + b$, determine the values of a and b.

4. **MC** Select the expanded and simplified expression of $(2x)(-3x)(x+1)$.

    A. $-6x^3 + 1$ **B.** $-6x^2 + 1$ **C.** $-6x^2 - 6x$ **D.** $-6x^3 - 6x^2$

5. **MC** Select the expanded and simplified expression for $(1 - 2x)(3 - x)(4x + 1)(x - 5)$.

    A. $8x^4 - 66x^3 + 135x^2 - 22x - 15$ **B.** $8x^4 - 66x^3 + 135x^2 - 12x - 10$
    C. $9x^4 - 66x^2 + 135x^2 - 8x - 10$ **D.** $8x^3 - 66x^2 + 135x^2 - 22x - 15$

6. Consider the polynomial $-3x^3 + 2x^2 + 4x - 1$. State:

    a. the degree of the polynomial   **b.** the leading coefficient
    c. the value of the constant term   **d.** the coefficient of x^2.

7. **MC** Select the simplified expression for $(x^6 + 2x^3 + 3x + 1) - (x^5 - 2x^2 + 4x - 2)$.

    A. $x^6 - x^5 + 2x^3 + 2x^2 - x + 3$ **B.** $2x^3 + 2x^2 - 2x + 3$
    C. $x^6 - x^5 + 4x^3 - x + 3$ **D.** $x^6 - x^5 + 2x^3 + 2x^2 - 3x$

8. Calculate the quotient when $x^3 + 2x^2 + x - 1$ is divided by $x + 1$.

9. Determine the remainder when $x^3 - x^2 + 4x - 3$ is divided by $x - 2$.

10. If $P(x) = 3x^4 - 2x^3 + x^2 - 4x + 8$, calculate $P(-1)$.

11. **MC** If $P(x) = x^3 - x^2 + x + 1$, select the value of $P(x + 1)$:

    A. $x^3 + 2x^2 + 5x + 4$ **B.** $x^3 + 4x^2 + 5x + 4$
    C. $x^3 + 3x^2 + 3x + 1$ **D.** $x^3 + 2x^2 + 2x + 2$

12. When $x^3 - 2x^2 + bx + 3$ is divided by $x - 1$ the remainder is 4. Calculate the value of b.

13. **MC** Select the correct factor for $x^3 - x^2 - 5x - 3$:

    A. x **B.** $x + 1$ **C.** $x - 1$ **D.** $x + 3$

14. **MC** Select the factorised from of $x^3 + 4x^2 + x - 6$.

    A. $(x + 1)(x - 2)(x + 3)$ **B.** $(x + 1)(x - 2)(x - 3)$
    C. $(x - 1)(x + 2)(x + 3)$ **D.** $(x - 1)(x - 2)(x - 3)$

15. Evaluate the correct value of x, for $2x^3 + 15x^2 + 19x + 6 = 0$.

LESSON
15.2 Polynomials

LEARNING INTENTION

At the end of this lesson you should be able to:
- identify polynomial expressions
- state the degree, leading term and leading coefficient of a polynomial.

⊙ 15.2.1 Polynomials

eles-4975

A **polynomial** in x, sometimes denoted $P(x)$, is an expression containing only non-negative integer powers of x.

- The **degree** of a polynomial in x is the highest power of x in the expression.
  For example:

$3x + 1$	is a polynomial of degree 1, or linear polynomial.
$x^2 + 4x - 7$	is a polynomial of degree 2, or quadratic polynomial.
$-5x^3 + \dfrac{x}{2}$	is a polynomial of degree 3, or cubic polynomial.
10	is a polynomial of degree 0 (think of 10 as $10x^0$).

- Expressions containing a term similar to any of the following terms are **not** polynomials:

$$\frac{1}{x}, \qquad x^{-2}, \qquad \sqrt{x}, \qquad 2^x, \qquad \sin(x), \qquad \text{and so on.}$$

  For example, the following are not polynomials.

$$3x^2 - 4x + \frac{2}{x} \qquad -5x^4 + x^3 - 2\sqrt{x} \qquad x^2 + \sin(x) + 1$$

- In the expression $6x^3 + 13x^2 - x + 1$
  - x is the *variable*.
  - 6 is the *coefficient* of x^3.
  - 13 is the *coefficient* of x^2.
  - -1 is the *coefficient* of x.
  - $6x^3$, $13x^2$, $-x$ and $+1$ are all *terms*.
  - The *constant* term is $+1$.
  - The *degree* of the polynomial is 3.

- The **leading term** is $6x^3$ because it is the term that contains the highest power of x.
- The **leading coefficient** is 6.
- Any polynomial with a leading coefficient of 1 is called **monic**.

General form of a polynomial

A polynomial of degree n has the form $ax^n + a_{n-1}x^{n-1} + \ldots + a_1x + a_0$, where n is a natural number and the coefficients $a_n, a_{n-1}, \ldots, a_1, a_0$ are real numbers.

The leading term is $a_n x^n$ and the constant term is a_0.

Answer the following questions.
 i. State the degree of each of the following polynomials.
 ii. State the variable for each of the following polynomials.

a. $x^3 + 3x^2 + x - 5$ **b.** $y^4 + 4y^3 - 8y^2 + 2y - 8$ **c.** $a^3 + 34a^6 - 12a - 72$

THINK

a. i. Determine the highest power of x in the expression.

ii. Determine the variable (unknown quantity) in the expression.

b. i. Determine the highest power of y in the expression.

ii. Determine the variable (unknown quantity) in the expression.

c. i. Determine the highest power of a in the expression.

ii. Determine the variable (unknown quantity) in the expression.

WRITE

a. i. In the expression $x^3 + 3x^2 + x - 5$, the highest power of x is 3. Therefore, this polynomial is of degree 3.

ii. x is the variable in this expression.

b. i. In the expression $y^4 + 4y^3 - 8y^2 + 2y - 8$ the highest power of y is 4. Therefore, this polynomial is of degree 4.

ii. y is the variable in this expression.

c. i. In the expression $a^3 + 34a^6 - 12a - 72$, the highest power of a is 6. Therefore, this polynomial is of degree 6.

ii. a is the variable in this expression.

Consider the polynomial $P(x) = 3x^4 - 5x^3 + 2x^2 + x - 12$.
a. State the degree and variable of the polynomial.
b. State the coefficient of x^3.
c. State the value of the constant term.
d. Determine the term that has a coefficient of 2.
e. Determine the leading term.

THINK

a. Determine the highest power of x and the variable (unknown quantity) in the expression.

b. Determine the term with x^3 in the expression.

c. Determine the term without variable in the expression.

d. Determine the term that has a coefficient of 2 in the expression.

e. Determine the term that contains the highest power of x in the expression.

WRITE

a. The highest power of x is 4 and therefore, the degree is 4.
x is the variable in this expression.

b. The coefficient in the term $-5x^3$ is -5.

c. The value of the constant term is -12.

d. The term that has a coefficient of 2 is $2x^2$.

e. The leading term is $3x^4$.

DISCUSSION

An example where polynomials are useful is in the construction of a greenhouse. The surface area, S, of a greenhouse of length L and height x can be approximated by the polynomial $S(x) = \pi x^2 + L\pi x - 4$.

For a 10-metre-long greenhouse, can you determine the minimal height of the greenhouse so that its surface area is greater than $70\,\text{m}^2$?

on Resources

Interactivity Degrees of polynomials (int-6203)

Exercise 15.2 Polynomials

learn on

15.2 Quick quiz **on**	15.2 Exercise

Individual pathways

■ PRACTISE	■ CONSOLIDATE	■ MASTER
1, 4, 5, 8, 11, 14	2, 6, 9, 12, 15	3, 7, 10, 13, 16

Fluency

WE1 For questions **1** to **3**, answer the following questions.
   i. State the degree of each of the following polynomials.
   ii. State the variable for each of the following polynomials.

1. **a.** $x^3 - 9x^2 + 19x + 7$ **b.** $65 + 2x^7$ **c.** $3x^2 - 8 + 2x$

2. **a.** $x^6 - 3x^5 + 2x^4 + 6x + 1$ **b.** $y^8 + 7y^3 - 5$ **c.** $\dfrac{1}{2}u^5 - \dfrac{u^4}{3} + 2u - 6$

3. **a.** $18 - \dfrac{e^5}{6}$ **b.** $2g - 3$ **c.** $1.5f^6 - 800f$

4. Identify the polynomials in questions **1** to **3** that are:
   a. linear   **b.** quadratic   **c.** cubic   **d.** monic.

For questions **5** to **7**, state whether each of the following is a polynomial (P) or not (N).

5. **a.** $7x + 6x^2 + \dfrac{5}{x}$ **b.** $33 - 4p$ **c.** $\dfrac{x^2}{9} + x$

6. **a.** $3x^4 - 2x^3 - 3\sqrt{x} - 4$ **b.** $k^{-2} + k - 3k^3 + 7$ **c.** $5r - r^9 + \dfrac{1}{3}$

7. **a.** $\dfrac{4c^6 - 3c^3 + 1}{2}$ **b.** $2^x - 8x + 1$ **c.** $\sin(x) + x^2$

Understanding

8. **WE2** Consider the polynomial $P(x) = -2x^3 + 4x^2 + 3x + 5$.

a. State the degree of the polynomial.
b. State the variable.
c. State the coefficient of x^2.
d. State the value of the constant term.
e. State the term that has a coefficient of 3.
f. Determine the leading term.

9. Consider the polynomial $P(w) = 6w^7 + 7w^6 - 9$.

a. State the degree of the polynomial.
b. State the variable.
c. State the coefficient of w^6.
d. Determine the coefficient of w.
e. State the value of the constant term.
f. State the term that has a coefficient of 6.

10. Consider the polynomial $f(x) = 4 - x^2 + x^4$.

a. State the degree of the polynomial.
b. State the coefficient of x^4.
c. Determine the leading term.
d. State the leading coefficient.

Communicating, reasoning and problem solving

11. Write the following polynomials as simply as possible, arranging terms in descending powers of x.

a. $7x + 2x^2 - 8x + 15 + 4x^3 - 9x + 3$
b. $x^2 - 8x^3 + 3x^4 - 2x^2 + 7x + 5x^3 - 7$
c. $x^3 - 5x^2 - 11x - 1 + 4x^3 - 2x + x^2 - 5$

12. A sports scientist determines the following equation for the velocity of a breaststroke swimmer during one complete stroke:

$$v(t) = 63.876t^6 - 247.65t^5 + 360.39t^4 - 219.41t^3 + 53.816t^2 + 0.4746t.$$

a. Determine the degree of the polynomial.
b. State the variable.
c. State the number of terms that are there.
d. Use a graphics calculator or graphing software to draw the graph of this polynomial.
e. Match what happens during one complete stroke with points on the graph.

13. The distance travelled by a body after t seconds is given by $d(t) = t^3 + 2t^2 - 4t + 5$.
Using a graphing calculator or suitable computer software, draw a graph of the above motion for $0 \le t \le 3$.
Use the graph to help you answer the following:

a. State what information the constant term gives.
b. Evaluate the position of the body after 1 second.
c. Describe in words the motion in the first 2 second.

14. If $x^2 - 3x + 5 = x^2 + (a + b)x + (a - b)$, determine the values of a and b.

15. If $x^2 + 2x - 1 \equiv (x - 1)^2 + a(x + 1) + b$, evaluate a and b.

16. If $x^3 + 9x^2 + 12x + 7 \equiv x^3 + (ax + b)^2 + 3$, evaluate a and b.

LESSON
15.3 Adding, subtracting and multiplying polynomials

LEARNING INTENTION

At the end of this lesson you should be able to:
- add, subtract and multiply polynomial expressions
- simplify polynomial expressions.

▶ 15.3.1 Operations with polynomials

eles-4976

- To add or subtract polynomials, simply add or subtract any like terms in the expressions.
- Polynomials are normally expressed with descending powers of x.

WORKED EXAMPLE 3 Simplifying polynomial expressions

Simplify each of the following.
a. $(5x^3 + 3x^2 - 2x - 1) + (x^4 + 5x^2 - 4)$ b. $(5x^3 + 3x^2 - 2x - 1) - (x^4 + 5x^2 - 4)$

THINK	WRITE
a. 1. Write the expression.	a. $(5x^3 + 3x^2 - 2x - 1) + (x^4 + 5x^2 - 4)$
2. Remove any grouping symbols, watching any signs.	$= 5x^3 + 3x^2 - 2x - 1 + x^4 + 5x^2 - 4$
3. Re-order the terms with descending degrees of x.	$= x^4 + 5x^3 + 3x^2 + 5x^2 - 2x - 1 - 4$
4. Simplify by collecting like terms.	$= x^4 + 5x^3 + 8x^2 - 2x - 5$
b. 1. Write the expression.	b. $(5x^3 + 3x^2 - 2x - 1) - (x^4 + 5x^2 - 4)$
2. Remove any grouping symbols, watching any signs.	$= 5x^3 + 3x^2 - 2x - 1 - x^4 - 5x^2 + 4$
3. Re-order the terms with descending degrees of x.	$= -x^4 + 5x^3 + 3x^2 - 5x^2 - 2x - 1 + 4$
4. Simplify by collecting like terms.	$= -x^4 + 5x^3 - 2x^2 - 2x + 3$

- To expand linear factors, for example $(x + 1)(x + 2)(x - 7)$, use FOIL from quadratic expansions.

WORKED EXAMPLE 4 Expanding polynomial expressions

Expand and simplify:
a. $x(x + 2)(x - 3)$ b. $(x - 1)(x + 5)(x + 2)$.

THINK	WRITE
a. 1. Write the expression.	a. $x(x + 2)(x - 3)$
2. Expand the last two linear factors, using FOIL and simplify	$= x(x^2 - 3x + 2x - 6)$
	$= x(x^2 - x - 6)$
3. Multiply the expression in brackets by x.	$= x^3 - x^2 - 6x$

---

b. 1. Write the expression.

2. Expand the last two linear factors, using FOIL, and simplify

3. Multiply the expression in the second bracket by x and then by -1.

4. Collect like terms.

b. $(x-1)(x+5)(x+2)$

$= (x-1)(x^2+2x+5x+10)$

$= (x-1)(x^2+7x+10)$
$= x^3+7x^2+10x-x^2-7x-10$

$= x^3+6x^2+3x-10$

---

on Resources

✦ **Interactivity** Adding and subtracting polynomials (int-6204)

Exercise 15.3 Adding, subtracting and multiplying polynomials

learn on

15.3 Quick quiz   **on**

15.3 Exercise

Individual pathways

■ PRACTISE	■ CONSOLIDATE	■ MASTER
1, 3, 5, 7, 9, 13, 16	2, 4, 6, 10, 14, 17	8, 11, 12, 15, 18

Fluency

1. **WE3a** Simplify each of the following.
   a. $(x^4+x^3-x^2+4)+(x^3-14)$
   b. $(x^6+x^4-3x^3+6x^2)+(x^4+3x^2+5)$
   c. $(x^3+x^2+2x-4)+(4x^3-6x^2+5x-9)$
   d. $(2x^4-3x^3+7x^2+9)+(6x^3+5x^2-4x+5)$
   e. $(15x^4-3x^2+4x-7)+(x^5-2x^4+3x^2-4x-3)$

2. **WE3b** Simplify each of the following.
   a. $(x^4+x^3+4x^2+5x+5)-(x^3+2x^2+3x+1)$
   b. $(x^6+x^3+1)-(x^5-x^2-1)$
   c. $(5x^7+6x^5-4x^3+8x^2+5x-3)-(6x^5+8x^2-3)$
   d. $(10x^4-5x^2+16x+11)-(2x^2-4x+6)$
   e. $(6x^3+5x^2-7x+12)-(4x^3-x^2+3x-3)$

3. **WE4a** Expand and simplify each of the following.
   a. $x(x+6)(x+1)$
   b. $x(x-9)(x+2)$
   c. $x(x-3)(x+11)$
   d. $2x(x+2)(x+3)$
   e. $-3x(x-4)(x+4)$

4. Expand and simplify each of the following.
   a. $5x(x+8)(x+2)$
   b. $x^2(x+4)$
   c. $-2x^2(7-x)$
   d. $(5x)(-6x)(x+9)$
   e. $-7x(x+4)^2$

WE4b For questions **5** to **10**, expand and simplify each of the following.

5. a. $(x+7)(x+2)(x+3)$
   b. $(x-2)(x+4)(x-5)$
   c. $(x-1)(x-4)(x+8)$
   d. $(x-1)(x-2)(x-3)$
   e. $(x+6)(x-1)(x+1)$

6. **a.** $(x-7)(x+7)(x+5)$ **b.** $(x+11)(x+5)(x-12)$ **c.** $(x+5)(x-1)^2$
   d. $(x+2)(x-7)^2$ **e.** $(x+1)(x-1)(x+1)$

7. **a.** $(x-2)(x+7)(x+8)$ **b.** $(x+5)(3x-1)(x+4)$ **c.** $(4x-1)(x+3)(x-3)(x+1)$
   d. $(5x+3)(2x-3)(x-4)$ **e.** $(1-6x)(x+7)(x+5)$

8. **a.** $3x(7x-4)(x-4)(x+2)$ **b.** $-9x(1-2x)(3x+8)$ **c.** $(6x+5)(2x-7)^2$
   d. $(3-4x)(2-x)(5x+9)(x-1)$ **e.** $2(7+2x)(x+3)(x+4)$

Understanding

9. **a.** $(x+2)^3$ **b.** $(x+5)^3$ **c.** $(x-1)^3$

10. **a.** $(x-3)^4$ **b.** $(2x-6)^3$ **c.** $(3x+4)^4$

11. Simplify the expression $2(ax+b)-5(c-bx)$.

12. Expand and simplify the expression $(x+a)(x-b)(x^2-3bx+2a)$.

Communicating, reasoning and problem solving

13. If $(x-3)^4 = ax^4 + bx^3 + cx^2 + dx + e$, determine the values of a, b, c, d and e. Show your working.

14. Simplify the expression $(2x-3)^3 - (4-3x)^2$.

15. Determine the difference in volume between a cube of side $\dfrac{3(x-1)}{2}$ and a cuboid whose sides are x, $(x+1)$ and $(2x+1)$. Show your working.

16. Determine the values of the pronumerals a and b if: $\dfrac{5x+1}{(x-1)(x+2)} \equiv \dfrac{a}{(x-1)} + \dfrac{b}{(x+2)}$

17. Evaluate the constants a, b and c if: $\dfrac{5x-7}{(x-1)(x+1)(x-2)} \equiv \dfrac{a}{(x-1)} + \dfrac{b}{(x+1)} + \dfrac{c}{(x-2)}$.

18. Write $\dfrac{3x-5}{(x^2+1)(x-1)}$ in the form $\dfrac{ax+b}{(x^2+1)} + \dfrac{c}{(x-1)}$ and hence determine the values of a, b and c.

LESSON
15.4 Long division of polynomials

LEARNING INTENTION

At the end of this lesson you should be able to:
- divide polynomials by linear expressions using long division
- determine the quotient and remainder when dividing polynomials by linear expressions.

▶ 15.4.1 Long division of polynomials

eles-4977

- The reverse of expanding is factorising (expressing a polynomial as a product of its linear factors).
- To factorise polynomials we need to use a form of long division.
- The following steps show how to divide a polynomial by a linear factor using long division.

Consider $\left(x^3 + 2x^2 - 13x + 10\right) \div (x - 3)$.		
Step 1	Write the division out using long division notation.	$x - 3 \overline{)x^3 + 2x^2 - 13x + 10}$
Step 2	Consider the leading terms only. Determine how many times x goes into x^3.	$x - 3 \overline{)x^3 + 2x^2 - 13x + 10}$
Step 3	x into x^3 goes x^2 times. Write x^2 above the x^2 term of the polynomial.	$\begin{array}{r} x^2 \\ x - 3 \overline{)x^3 + 2x^2 - 13x + 10} \end{array}$
Step 4	Multiply the term at the top $\left(x^2\right)$ by the linear factor $(x - 3)$: $x^2 \times (x - 3) = x^3 - 3x^2$. Write the result beneath the first two terms of the polynomial.	$\begin{array}{r} x^2 \\ x - 3 \overline{)x^3 + 2x^2 - 13x + 10} \\ x^3 - 3x^2 \end{array}$
Step 5	Subtract the first two terms of the polynomial by the terms written below them. $x^3 - x^3 = 0$ and $2x^2 - \left(-3x^2\right) = 5x^2$	$\begin{array}{r} x^2 \\ x - 3 \overline{)x^3 + 2x^2 - 13x + 10} \\ \underline{x^3 - 3x^2} \\ 5x^2 \end{array}$
Step 6	Bring the next term of the polynomial down to sit next to $5x^2$.	$\begin{array}{r} x^2 \\ x - 3 \overline{)x^3 + 2x^2 - 13x + 10} \\ \underline{x^3 - 3x^2} \quad \downarrow \\ 5x^2 - 13x \end{array}$

- The process now restarts, looking at the newly created $5x^2 - 13x$ expression.

Step 7	Consider the leading terms only. Determine how many times x goes into $5x^2$.	$\begin{array}{r} x^2 \\ x - 3 \overline{)x^3 + 2x^2 - 13x + 10} \\ \underline{x^3 - 3x^2} \\ 5x^2 - 13x \end{array}$
Step 8	x into $5x^2$ goes $5x$ times. Write $5x$ above the x term of the polynomial.	$\begin{array}{r} x^2 + 5x \\ x - 3 \overline{)x^3 + 2x^2 - 13x + 10} \\ \underline{x^3 - 3x^2} \\ 5x^2 - 13x \end{array}$
Step 9	Multiply the term at the top $(5x)$ by the linear factor $(x - 3)$: $5x \times (x - 3) = 5x^2 - 15x$. Write the result beneath the two terms written in step 6.	$\begin{array}{r} x^2 + 5x \\ x - 3 \overline{)x^3 + 2x^2 - 13x + 10} \\ \underline{x^3 - 3x^2} \\ 5x^2 - 13x \\ 5x^2 - 15x \end{array}$
Step 10	Subtract the two terms of the polynomial by the terms written below them. $5x^2 - 5x^2 = 0$ and $(-13x) - (-15x) = 2x$	$\begin{array}{r} x^2 + 5x \\ x - 3 \overline{)x^3 + 2x^2 - 13x + 10} \\ \underline{x^3 - 3x^2} \\ 5x^2 - 13x \\ \underline{5x^2 - 15x} \\ 2x \end{array}$

Step 11	Bring the next term of the polynomial down to sit next to 2x.	$$\begin{array}{r} x^2 + 5x \\ x - 3 \overline{)\,x^3 + 2x^2 - 13x + 10} \\ \underline{x^3 - 3x^2 } \\ 5x^2 - 13x \downarrow \\ \underline{5x^2 - 15x } \downarrow \\ 2x + 10 \end{array}$$

• Once again the process restarts, looking at the newly created $2x + 10$ expression.

Step 12	Consider the leading terms only. Determine how many times x goes into $2x$.	$$\begin{array}{r} x^2 + 5x \\ x - 3 \overline{)\,x^3 + 2x^2 - 13x + 10} \\ \underline{x^3 - 3x^2 } \\ 5x^2 - 13x \downarrow \\ \underline{5x^2 - 15x } \downarrow \\ 2x + 10 \end{array}$$
Step 13	x into $2x$ goes 2 times. Write 2 above the constant term of the polynomial.	$$\begin{array}{r} x^2 + 5x + 2 \\ x - 3 \overline{)\,x^3 + 2x^2 - 13x + 10} \\ \underline{x^3 - 3x^2 } \\ 5x^2 - 13x \\ \underline{5x^2 - 15x } \\ 2x + 10 \end{array}$$
Step 14	Multiply the term at the top (2) by the linear factor $(x - 3)$: $2 \times (x - 3) = 2x - 6$. Write the result beneath the two terms written in step 11.	$$\begin{array}{r} x^2 + 5x + 2 \\ x - 3 \overline{)\,x^3 + 2x^2 - 13x + 10} \\ \underline{x^3 - 3x^2 } \\ 5x^2 - 13x \\ \underline{5x^2 - 15x } \\ 2x + 10 \\ 2x - 6 \end{array}$$
Step 15	Subtract the two terms of the polynomial by the terms written below them. $2x - 2x = 0$ and $10 - (-6) = 16$	$$\begin{array}{r} x^2 + 5x + 2 \\ x - 3 \overline{)\,x^3 + 2x^2 - 13x + 10} \\ \underline{x^3 - 3x^2 } \\ 5x^2 - 13x \\ \underline{5x^2 - 15x } \\ 2x + 10 \\ \underline{2x - 6} \\ 16 \end{array}$$
Step 16	The division is now complete! The top line is the quotient ($Q(x)$), and the bottom number is the remainder ($R(x)$).	$$\begin{array}{r} x^2 + 5x + 2 \quad \longleftarrow \text{Quotient} \\ x - 3 \overline{)\,x^3 + 2x^2 - 13x + 10} \\ \underline{x^3 - 3x^2 } \\ 5x^2 - 13x \\ \underline{5x^2 - 15x } \\ 2x + 10 \\ \underline{2x - 6} \\ 16 \quad \longleftarrow \text{Remainder} \end{array}$$

Write the answer: $(x^3 + 2x^2 - 13x + 10) \div (x - 3) = x^2 + 5x + 2$ remainder 16

Note: $P(x)$, in the form $P(x) = D(x)Q(x) + R(x)$, is $x^3 + 2x^2 - 13x + 10 = (x - 3)(x^2 + 5x + 2) + 16$

WORKED EXAMPLE 5 Performing long division of cubic polynomials

Perform the following long divisions and state the quotient and remainder.

a. $\left(x^3 + 3x^2 + x + 9\right) \div (x + 2)$

b. $\left(x^3 - 4x^2 - 7x - 5\right) \div (x - 1)$

c. $\left(2x^3 + 6x^2 - 3x + 2\right) \div (x - 6)$

THINK	WRITE
a. 1. Write the question in long division format. **2.** Perform the long division process.	**a.** $$\begin{array}{r} x^2 + x - 1 \leftarrow Q(x) \\ x+2\overline{)x^3 + 3x^2 + x + 9} \\ \underline{x^3 + 2x^2} \\ x^2 + x \\ \underline{x^2 + 2x} \\ -x + 9 \\ \underline{-x - 2} \\ 11 \leftarrow R(x) \end{array}$$
3. Write the quotient and remainder.	Quotient is $x^2 + x - 1$; remainder is 11.
b. 1. Write the question in long division format. **2.** Perform the long division process.	**b.** $$\begin{array}{r} x^2 - 3x - 10 \leftarrow Q(x) \\ x-1\overline{)x^3 - 4x^2 - 7x - 5} \\ \underline{x^3 - x^2} \\ -3x^2 - 7x \\ \underline{-3x^2 + 3x} \\ -10x - 5 \\ \underline{-10x + 10} \\ -15 \leftarrow R(x) \end{array}$$
3. Write the quotient and remainder.	Quotient is $x^2 - 3x - 10$; remainder is -15.
c. 1. Write the question in long division format. **2.** Perform the long division process.	**c.** $$\begin{array}{r} 2x^2 + 18x + 105 \leftarrow Q(x) \\ x-6\overline{)2x^3 + 6x^2 - 3x + 2} \\ \underline{2x^3 - 12x^2} \\ 18x^2 - 3x \\ \underline{18x^2 - 108x} \\ 105x + 2 \\ \underline{105x - 630} \\ 632 \leftarrow R(x) \end{array}$$
3. Write the quotient and remainder.	Quotient is $2x^2 + 18x + 105$; remainder is 632.

WORKED EXAMPLE 6 Determining the quotient and remainder of a degree 4 polynomial

Determine the quotient and the remainder when $x^4 - 3x^3 + 2x^2 - 8$ is divided by the linear expression $x + 2$.

THINK	WRITE
1. Set out the long division with each polynomial in descending powers of x. If one of the powers of x is missing, include it with 0 as the coefficient.	$$x+2\overline{)x^4 - 3x^3 + 2x^2 + 0x - 8}$$

2. Divide x into x^4 and write the result above.

$$x + 2 \overline{)x^4 - 3x^3 + 2x^2 + 0x - 8} \quad \begin{array}{c} x^3 \end{array}$$

3. Multiply the result x^3 by $x + 2$ and write the result underneath.

$$\begin{array}{r} x^3 \\ x + 2 \overline{)x^4 - 3x^3 + 2x^2 + 0x - 8} \\ x^4 + 2x^3 \end{array}$$

4. Subtract and bring down the remaining terms to complete the expression.

$$\begin{array}{r} x^3 \\ x + 2 \overline{)x^4 - 3x^3 + 2x^2 + 0x - 8} \\ \underline{x^4 + 2x^3} \\ -5x^3 + 2x^2 + 0x - 8 \end{array}$$

5. Divide x into $-5x^3$ and write the result above.
6. Continue this process to complete the long division.

$$\begin{array}{r} x^3 - 5x^2 + 12x - 24 \\ x + 2 \overline{)x^4 - 3x^3 + 2x^2 + 0x - 8} \\ \underline{x^4 + 2x^3} \\ -5x^3 + 2x^2 + 0x - 8 \\ \underline{-5x^3 - 10x^2} \\ 12x^2 + 0x - 8 \\ \underline{12x^2 + 24x} \\ -24x - 8 \\ \underline{-24x - 48} \\ 40 \end{array}$$

7. The polynomial $x^3 - 5x^2 + 12x - 24$, at the top, is the quotient.

The quotient is $x^3 - 5x^2 + 12x - 24$.

8. The result of the final subtraction, 40, is the remainder.

The remainder is 40.

 Resources

 Interactivity Long division of polynomials (int-2793)

Exercise 15.4 Long division of polynomials

learn on

| 15.4 Quick quiz | 15.4 Exercise |

Individual pathways

■ PRACTISE	■ CONSOLIDATE	■ MASTER
1, 3, 5, 9, 14, 17	2, 4, 7, 10, 12, 15, 18	6, 8, 11, 13, 16, 19

Fluency

1. **WE5a** Perform the following long divisions and state the quotient and remainder.

   a. $(x^3 + 4x^2 + 4x + 9) \div (x + 2)$
   b. $(x^3 + 2x^2 + 4x + 1) \div (x + 1)$
   c. $(x^3 + 6x^2 + 3x + 1) \div (x + 3)$
   d. $(x^3 + 3x^2 + x + 3) \div (x + 4)$

2. Perform the following long divisions and state the quotient and remainder.

   a. $(x^3 + 6x^2 + 2x + 2) \div (x + 2)$ b. $(x^3 + x^2 + x + 3) \div (x + 1)$
   c. $(x^3 + 8x^2 + 5x + 4) \div (x + 8)$ d. $(x^3 + x^2 + 4x + 1) \div (x + 2)$

3. **WE5b** State the quotient and remainder for each of the following.

   a. $(x^3 + 2x^2 - 5x - 9) \div (x - 2)$ b. $(x^3 + x^2 + x + 9) \div (x - 3)$
   c. $(x^3 + x^2 - 9x - 5) \div (x - 2)$ d. $(x^3 - 4x^2 + 10x - 2) \div (x - 1)$

4. State the quotient and remainder for each of the following.

   a. $(x^3 - 5x^2 + 3x - 8) \div (x - 3)$ b. $(x^3 - 7x^2 + 9x - 7) \div (x - 1)$
   c. $(x^3 + 9x^2 + 2x - 1) \div (x - 5)$ d. $(x^3 + 4x^2 - 5x - 4) \div (x - 4)$

WE5c For questions 5 to 8, divide the first polynomial by the second and state the quotient and remainder.

5. a. $3x^3 - x^2 + 6x + 5, \ (x + 2)$ b. $4x^3 - 4x^2 + 10x - 4, \ (x + 1)$
   c. $2x^3 - 7x^2 + 9x + 1, \ (x - 2)$

6. a. $2x^3 + 8x^2 - 9x - 1, \ (x + 4)$ b. $4x^3 - 10x^2 - 9x + 8, \ (x - 3)$
   c. $3x^3 + 16x^2 + 4x - 7, \ (x + 5)$

7. a. $6x^3 - 7x^2 + 4x + 4, \ (2x - 1)$ b. $6x^3 + 23x^2 + 2x - 31, \ (3x + 4)$
   c. $8x^3 + 6x^2 - 39x - 13, \ (2x + 5)$

8. a. $2x^3 - 15x^2 + 34x - 13, \ (2x - 7)$ b. $3x^3 + 5x^2 - 16x - 23, \ (3x + 2)$
   c. $9x^3 - 6x^2 - 5x + 9, \ (3x - 4)$

Understanding

For questions 9 to 11, state the quotient and remainder for each of the following.

9. a. $\dfrac{-x^3 - 6x^2 - 7x - 16}{x + 1}$ b. $\dfrac{-3x^3 + 7x^2 + 10x - 15}{x - 3}$

   c. $\dfrac{-2x^3 + 9x^2 + 17x + 15}{2x + 1}$ d. $\dfrac{4x^3 - 20x^2 + 23x - 2}{-2x + 3}$

10. a. $(x^3 - 3x + 1) \div (x + 1)$ b. $(x^3 + 2x^2 - 7) \div (x + 2)$
    c. $(x^3 - 5x^2 + 2x) \div (x - 4)$ d. $(-x^3 - 7x + 8) \div (x - 1)$

11. a. $(5x^2 + 13x + 1) \div (x + 3)$ b. $(2x^3 + 8x^2 - 4) \div (x + 5)$
    c. $(-2x^3 - x + 2) \div (x - 2)$ d. $(-4x^3 + 6x^2 + 2x) \div (2x + 1)$

12. **WE6** Determine the quotient and the remainder when each polynomial is divided by the linear expression given.

    a. $x^4 + x^3 + 3x^2 - 7x, \ (x - 1)$ b. $x^4 - 13x^2 + 36, \ (x - 2)$
    c. $x^5 - 3x^3 + 4x + 3, \ (x + 3)$

13. Determine the quotient and the remainder when each polynomial is divided by the linear expression given.

    a. $2x^6 - x^4 + x^3 + 6x^2 - 5x, \ (x + 2)$ b. $6x^4 - x^3 + 2x^2 - 4x, \ (x - 3)$
    c. $3x^4 - 6x^3 + 12x, \ (3x + 1)$

Communicating, reasoning and problem solving

14. Determine the quotient and remainder when $3x^4 - 6x^3 + 12x + a$ is divided by $(3x + 6)$. Show your working.

15. Determine the quotient and remainder when $ax^2 + bx + c$ is divided by $(x - d)$. Show your working.

16. A birthday cake in the shape of a cube had side length $(x+p)$ cm. The cake was divided between $(x-p)$ guests. The left-over cake was used for lunch the next day. There were q^3 guests for lunch the next day and each received c^3 cm^3 of cake, which was then all finished. Determine an expression for q in terms of p and c. Show your working.

17. When $x^3 - 2x^2 + 4x + a$ is divided by $x - 1$ the remainder is zero. Use long division to determine the value of a.

18. When $x^3 + 3x^2 + a$ is divided by $x + 1$, the remainder is 8. Use long division to determine the value of a.

19. When $2x^2 + ax + b$ is divided by $x - 1$ the remainder is zero but when $2x^2 + ax + b$ is divided by $x - 2$ the remainder is 9. Use long division to determine the value of the pronumerals a and b.

LESSON
15.5 Polynomial values

LEARNING INTENTION

At the end of this lesson you should be able to:
- determine the value of a polynomial for a given value.

15.5.1 Polynomial values

eles-4978

- Consider the polynomial $P(x) = x^3 - 5x^2 + x + 1$.
- The value of the polynomial when $x = 3$ is denoted by $P(3)$ and is found by substituting $x = 3$ into the equation in place of x, as shown.

Substitution to determine polynomial values

When $x = 3$:
$$P(x) = x^3 - 5x^2 + x + 1$$
$$P(3) = (3)^3 - 5(3)^2 + (3) + 1$$
$$P(3) = 27 - 5(9) + 3 + 1$$
$$P(3) = 27 - 45 + 4$$
$$P(3) = -14$$

WORKED EXAMPLE 7 Evaluating polynomials for values of x

If $P(x) = 2x^3 + x^2 - 3x - 4$, determine the value of:
a. $P(1)$ b. $P(-2)$ c. $P(a)$ d. $P(2b)$ e. $P(x+1)$.

THINK	WRITE
a. 1. Write the expression.	a. $P(x) = 2x^3 + x^2 - 3x - 4$
2. Replace x with 1.	$P(1) = 2(1)^3 + (1)^2 - 3(1) - 4$
3. Simplify and write the answer.	$= 2 + 1 - 3 - 4$ $= -4$

b. **1.** Write the expression.

2. Replace x with -2.

3. Simplify and write the answer.

b. $P(x) = 2x^3 + x^2 - 3x - 4$

$P(-2) = 2(-2)^3 + (-2)^2 - 3(-2) - 4$

$= 2(-8) + (4) + 6 - 4$
$= -16 + 4 + 6 - 4$
$= -10$

c. **1.** Write the expression.

2. Replace x with a.

3. No further simplification is possible.

c. $P(x) = 2x^3 + x^2 - 3x - 4$

$P(a) = 2a^3 + a^2 - 3a - 4$

d. **1.** Write the expression.

2. Replace x with $2b$.

3. Simplify and write the answer.

d. $P(x) = 2x^3 + x^2 - 3x - 4$

$P(2b) = 2(2b)^3 + (2b)^2 - 3(2b) - 4$

$= 2(8b^3) + 4b^2 - 6b - 4$
$= 16b^3 + 4b^2 - 6b - 4$

e. **1.** Write the expression.

2. Replace x with $(x + 1)$.

3. Expand the right-hand side and collect like terms.

4. Write the answer.

e. $P(x) = 2x^3 + x^2 - 3x - 4$

$P(x + 1) = 2(x + 1)^3 + (x + 1)^2 - 3(x + 1) - 4$

$= 2(x + 1)(x + 1)(x + 1) + (x + 1)(x + 1) - 3(x + 1) - 4$
$= 2(x + 1)(x^2 + 2x + 1) + x^2 + 2x + 1 - 3x - 3 - 4$
$= 2(x^3 + 2x^2 + x + x^2 + 2x + 1) + x^2 - x - 6$
$= 2(x^3 + 3x^2 + 3x + 1) + x^2 - x - 6$
$= 2x^3 + 6x^2 + 6x + 2 + x^2 - x - 6$
$= 2x^3 + 7x^2 + 5x - 4$

Exercise 15.5 Polynomial values

| 15.5 Quick quiz **on** | 15.5 Exercise |

Individual pathways

■ PRACTISE	■ CONSOLIDATE	■ MASTER
1, 3, 7, 11, 12, 15	2, 5, 8, 9, 13, 16	4, 6, 10, 14, 17

Fluency

 For questions **1** to **6**, $P(x) = 2x^3 - 3x^2 + 2x + 10$. Calculate the following.

1. a. $P(0)$ **b.** $P(1)$

2. a. $P(2)$ **b.** $P(3)$

3. a. $P(-1)$ **b.** $P(-2)$

4. a. $P(-3)$ **b.** $P(a)$

5. a. $P(2b)$ **b.** $P(x+2)$

6. a. $P(x-3)$ **b.** $P(-4y)$

Understanding

7. For the polynomial $P(x) = x^3 + x^2 + x + 1$, calculate the following showing your full working.
    a. $P(1)$
    b. $P(2)$
    c. $P(-1)$
    d. The remainder when $P(x)$ is divided by $(x-1)$.
    e. The remainder when $P(x)$ is divided by $(x-2)$.
    f. The remainder when $P(x)$ is divided by $(x+1)$.

8. For the polynomial $P(x) = x^3 + 2x^2 + 5x + 2$, calculate the following showing your full working.
    a. $P(1)$
    b. $P(2)$
    c. $P(-2)$
    d. The remainder when $P(x)$ is divided by $(x-1)$.
    e. The remainder when $P(x)$ is divided by $(x-2)$.
    f. The remainder when $P(x)$ is divided by $(x+2)$.

9. For the polynomial $P(x) = x^3 - x^2 + 4x - 1$, calculate the following showing your full working.
    a. $P(1)$
    b. $P(2)$
    c. $P(-2)$
    d. The remainder when $P(x)$ is divided by $(x-1)$.
    e. The remainder when $P(x)$ is divided by $(x-2)$.
    f. The remainder when $P(x)$ is divided by $(x+2)$.

10. For the polynomial $P(x) = x^3 - 4x^2 - 7x + 3$, calculate the following showing your full working.
    a. $P(1)$
    b. $P(-1)$
    c. $P(-2)$
    d. The remainder when $P(x)$ is divided by $(x-1)$.
    e. The remainder when $P(x)$ is divided by $(x+1)$.
    f. The remainder when $P(x)$ is divided by $(x+2)$.

Communicating, reasoning and problem solving

11. Copy and complete:
    a. A quick way of determining the remainder when $P(x)$ is divided by $(x+8)$ is to calculate _____.
    b. A quick way of determining the remainder when $P(x)$ is divided by $(x-7)$ is to calculate _____.
    c. A quick way of determining the remainder when $P(x)$ is divided by $(x-a)$ is to calculate _____.

12. If $P(x) = 2(x-3)^5 + 1$, determine:
    a. $P(2)$ **b.** $P(-2)$ **c.** $P(a)$

13. If $P(x) = -2x^3 - 3x^2 + x + 3$, evaluate:
    a. $P(a) + 1$ **b.** $P(a+1)$.

14. When $x^2 + bx + 2$ is divided by $(x-1)$, the remainder is $b^2 - 4b + 7$. Determine the possible values of b.

15. If $P(x) = 2x^3 - 3x^2 + 4x + c$, determine the value of c if $P(2) = 20$.

16. If $P(x) = 3x^3 - 2x^2 - x + c$ and $P(2) = 8P(1)$, calculate the value of c.

17. If $P(x) = 5x^2 + bx + c$ and $P(-1) = 12$ while $P(2) = 21$, determine the values of b and c.

LESSON
15.6 The remainder and factor theorems

LEARNING INTENTION

At the end of this lesson you should be able to:
- identify factors of polynomials using the factor theorem
- apply the remainder theorem to identify factors of polynomials
- determine the remainder when a polynomial is divided by a linear expression.

15.6.1 The remainder theorem

eles-4979

- In the previous exercise, you may have noticed that:
  the remainder when $P(x)$ is divided by $(x-a)$ is equal to $P(a)$.
- This fact is summarised in the **remainder theorem**.

> **The remainder theorem**
>
> When $P(x)$ is divided by $(x-a)$, the remainder $R(x) = P(a)$.

- If $P(x) = x^3 + x^2 + x + 1$ is divided by $(x-2)$, the quotient is $x^2 + 3x + 7$ and the remainder is $P(2)$, which equals 15. That is:

$$\left(x^3 + x^2 + x + 1\right) \div (x-2) = x^2 + 3x + 7 + \frac{15}{x-2}$$

$$\text{and } \left(x^3 + x^2 + x + 1\right) = \left(x^2 + 3x + 7\right)(x-2) + 15$$

> **Dividing a polynomial by a linear factor**
>
> If $P(x)$ is divided by $(x-a)$, the quotient is $Q(x)$ and the remainder is $R(x) = P(a)$, we can write:
>
> $$P(x) \div (x-a) = Q(x) + \frac{R(x)}{(x-a)}$$
>
> $$\Rightarrow P(x) = (x-a)\,Q(x) + R(x)$$

General form of a polynomial

$$P(x) = D(x)\,Q(x) + R(x)$$

where: $P(x)$ is the polynomial

$D(x)$ is the divisor

$Q(x)$ is the quotient

$R(x)$ is the remainder.

WORKED EXAMPLE 8 Calculating remainders using the remainder theorem

Without actually dividing, determine the remainder when $x^3 - 7x^2 - 2x + 4$ is divided by:

a. $(x - 3)$

b. $(x + 6)$.

THINK	WRITE
a. 1. Name the polynomial.	a. Let $P(x) = x^3 - 7x^2 - 2x + 4$.
2. The remainder when $P(x)$ is divided by $(x - 3)$ is equal to $P(3)$.	$\begin{aligned} R(x) &= P(3) \\ &= 3^3 - 7(3)^2 - 2(3) + 4 \\ &= 27 - 7(9) - 6 + 4 \\ &= 27 - 63 - 6 + 4 \end{aligned}$
3. Write the remainder.	The remainder is -38.
b. 1. The remainder when $P(x)$ is divided by $(x + 6)$ is equal to $P(-6)$.	b. $\begin{aligned} R(x) &= P(-6) \\ &= (-6)^3 - 7(-6)^2 - 2(-6) + 4 \\ &= -216 - 7(36) + 12 + 4 \\ &= 216 - 252 + 12 + 4 \end{aligned}$
2. Write the remainder.	The remainder is -452.

▶ 15.6.2 The factor theorem

eles-4980

- The remainder when 12 is divided by 4 is zero, since 4 is a factor of 12.
- Similarly, if the remainder ($R(x)$) when $P(x)$ is divided by $(x - a)$ is zero, then $(x - a)$ is a factor of $P(x)$.
- Since $R(x) = P(a)$, determine the value of a that makes $P(a) = 0$, then $(x - a)$ is a factor.

This is summarised in the **factor theorem**.

The factor theorem

If $P(a) = 0$, then $(x - a)$ is a factor of $P(x)$.

- $P(x)$ could be factorised as follows:

$$P(x) = (x - a)Q(x), \text{ where } Q(x) \text{ is 'the other' factor of } P(x).$$

- $Q(x)$ is the quotient when $P(x)$ is divided by the linear factor $(x - a)$.

WORKED EXAMPLE 9 Applying the factor theorem to determine constants

$(x-2)$ is a factor of $x^3 + kx^2 + x - 2$. Determine the value of k.

THINK	WRITE
1. Name the polynomial.	Let $P(x) = x^3 + kx^2 + x - 2$.
2. The remainder when $P(x)$ is divided by $(x - 2)$ is equal to $P(2) = 0$.	$0 = P(2)$ $\quad = 2^3 + k(2)^2 + 2 - 2$ $0 = 8 + 4k$
3. Solve for k and write its value.	$4k = -8$ $\quad k = -2$

WORKED EXAMPLE 10 Using the remainder theorem to determine unknown coefficients of a cubic polynomial

The polynomial $P(x) = x^3 + 2x^2 + ax + b$ has a remainder 15 when divided by $(x - 3)$ and a remainder 23 when divided by $(x + 1)$.
Determine the values of a and b.

THINK	WRITE
1. Use the remainder theorem for each divisor given.	For divisor $(x - 3)$, $P(3) = 15$. For divisor $(x + 1)$, $P(-1) = 23$.
2. Express $P(3)$ and $P(-1)$ in terms of the unknowns a and b.	$P(3) = 3^3 + 2(3)^2 + a(3) + b$ $P(3) = 45 + 3a + b$ $P(-1) = (-1)^3 + 2(-1)^2 + a(-1) + b$ $P(-1) = 1 - a + b$
3. Write the pair of simultaneous equations to solve.	$15 = 45 + 3a + b$ $23 = 1 - a + b$
4. Solve the pair of simultaneous equations.	$a = -13$ and $b = 9$
5. Verify your answer.	$P(x) = x^3 + 2x^2 - 13x + 9$ $P(3) = 3^3 + 2(3)^2 - 13(3) + 9 = 15 \checkmark$ $P(-1) = (-1)^3 + 2(-1)^2 - 13(-1) + 9$ $\quad\quad = 23 \checkmark$
6. Write your answer.	$a = -13$ and $b = 9$

Exercise 15.6 The remainder and factor theorems

15.6 Quick quiz on	15.6 Exercise

Individual pathways

■ PRACTISE	■ CONSOLIDATE	■ MASTER
1, 3, 5, 8, 9, 16, 18, 21	2, 4, 6, 10, 11, 14, 17, 19, 22	7, 12, 13, 15, 20, 23, 24

Fluency

WE8 For questions **1** and **2**, without actually dividing, determine the remainder when $x^3 + 3x^2 - 10x - 24$ is divided by:

1. a. $x - 1$ b. $x + 2$ c. $x - 3$ d. $x + 5$

2. a. $x - 0$ b. $x - k$ c. $x + n$ d. $x + 3c$

For questions **3** to **7**, determine the remainder when the first polynomial is divided by the second without performing long division.

3. a. $x^3 + 2x^2 + 3x + 4$, $(x - 3)$ b. $x^3 - 4x^2 + 2x - 1$, $(x + 1)$

4. a. $x^3 + 3x^2 - 3x + 1$, $(x + 2)$ b. $x^3 - x^2 - 4x - 5$, $(x - 1)$

5. a. $x^3 + x^2 + 8$, $(x - 5)$ b. $-3x^3 - 2x^2 + x + 6$, $(x + 1)$

6. a. $-x^3 + 8$, $(x + 3)$ b. $x^3 - 3x^2 - 2$, $(x - 2)$

7. a. $2x^3 + 3x^2 + 6x + 3$, $(x + 5)$ b. $x^3 + 2x^2$, $(x - 7)$

Understanding

8. **WE9** The remainder when $x^3 + kx + 1$ is divided by $(x + 2)$ is -19. Calculate the value of k.

9. The remainder when $x^3 + 2x^2 + mx + 5$ is divided by $(x - 2)$ is 27. Determine the value of m.

10. The remainder when $x^3 - 3x^2 + 2x + n$ is divided by $(x - 1)$ is 1. Calculate the value of n.

11. The remainder when $ax^3 + 4x^2 - 2x + 1$ is divided by $(x - 3)$ is -23. Determine the value of a.

12. The remainder when $x^3 - bx^2 - 2x + 1$ is divided by $(x + 1)$ is 0. Calculate the value of b.

13. The remainder when $-4x^2 + 2x + 7$ is divided by $(x - c)$ is -5. Determine a possible whole number value of c.

14. The remainder when $x^2 - 3x + 1$ is divided by $(x + d)$ is 11. Calculate the possible values of d.

15. The remainder when $x^3 + ax^2 + bx + 1$ is divided by $(x - 5)$ is -14. When the cubic polynomial is divided by $(x + 1)$, the remainder is -2. Determine the values of a and b.

16. **MC** Answer the following. *Note:* There may be more than one correct answer.
    a. When $x^3 + 2x^2 - 5x - 5$ is divided by $(x + 2)$, the remainder is:
       A. -5 **B.** -2 **C.** 2   **D.** 5
    b. Choose a factor of $2x^3 + 15x^2 + 22x - 15$ from the following.
       A. $(x - 1)$ **B.** $(x - 2)$ **C.** $(x + 3)$ **D.** $(x + 5)$
    c. When $x^3 - 13x^2 + 48x - 36$ is divided by $(x - 1)$, the remainder is:
       A. -3 **B.** -2 **C.** -1 **D.** 0
    d. Select a factor of $x^3 - 5x^2 - 22x + 56$ from the following.
       A. $(x - 2)$ **B.** $(x + 2)$ **C.** $(x - 7)$ **D.** $(x + 4)$

17. Determine one factor of each of the following cubic polynomials.

   a. $x^3 - 3x^2 + 3x - 1$

   b. $x^3 - 7x^2 + 16x - 12$

   c. $x^3 + x^2 - 8x - 12$

   d. $x^3 + 3x^2 - 34x - 120$

Communicating, reasoning and problem solving

For questions 18 and 19, without actually dividing, show that the first polynomial is exactly divisible by the second (that is, the second polynomial is a factor of the first).

18. a. $x^3 + 5x^2 + 2x - 8$, $(x - 1)$

   b. $x^3 - 7x^2 - x + 7$, $(x - 7)$

   c. $x^3 - 7x^2 + 4x + 12$, $(x - 2)$

   d. $x^3 + 2x^2 - 9x - 18$, $(x + 2)$

19. a. $x^3 + 3x^2 - 9x - 27$, $(x + 3)$

   b. $-x^3 + x^2 + 9x - 9$, $(x - 1)$

   c. $-2x^3 + 9x^2 - x - 12$, $(x - 4)$

   d. $3x^3 + 22x^2 + 37x + 10$, $(x + 5)$

20. Prove that each of the following is a linear factor of $x^3 + 4x^2 - 11x - 30$ by substituting values into the cubic function: $(x + 2)$, $(x - 3)$, $(x + 5)$.

21. When $(x^3 + ax^2 - 4x + 1)$ and $(x^3 - ax^2 + 8x - 7)$ are each divided by $(x - 2)$, the remainders are equal. Determine the value of a.

22. **WE10** When $x^4 + ax^3 - 4x^2 + b$ and $x^3 - ax^2 - 7x + b$ are each divided by $(x - 2)$, the remainders are 26 and 8 respectively. Calculate the values of a and b.

23. Both $(x - 1)$ and $(x - 2)$ are factors of $P(x) = x^4 + ax^3 - 7x^2 + bx - 30$. Determine the values of a and b and the remaining two linear factors.

24. The remainder when $2x - 1$ is divided into $6x^3 - x^2 + 3x + k$ is the same as when it is divided into $4x^3 - 8x^2 - 5x + 2$. Calculate the value of k.

LESSON
15.7 Factorising polynomials

LEARNING INTENTION

At the end of this lesson you should be able to:
- factorise polynomials using long division
- factorise polynomials using short division or inspection.

15.7.1 Using long division

eles-4981

- Once one factor of a polynomial has been found (using the factor theorem as in the previous section), long division may be used to find other factors.
- In the case of a cubic polynomial, one — possibly two — other factors may be found.

COMMUNICATING — COLLABORATIVE TASK: A handy way to determine possible factors of a monic cubic polynomial

Equipment: pen, paper, calculator

1. In pairs, observe the following factorised polynomials.

$$x^3 - 4x^2 - 11x + 30 = (x+3)(x-2)(x-5)$$

$$x^3 + 6x^2 + 3x - 10 = (x+5)(x+2)(x-1)$$

$$x^3 + 8x^2 + x - 42 = (x+7)(x+3)(x-2)$$

$$x^3 - 12x^2 + x + 132 = (x+3)(x-4)(x-11)$$

$$x^3 - 9x^2 - 2x + 18 = (x^2 - 2)(x-9)$$

$$x^3 + 3x^2 - 16x + 12 = (x+6)(x-1)(x-2)$$

Can you recognise a pattern between the constant term in pink on the left-hand side and the values in blue on the right-hand side?

Hint: For each polynomial, write the factors of the constant term.

2. As a class, discuss any pattern observed.
3. Consider the polynomial $P(x) = x^3 - 9x^2 - 73x + 273$. List the factors of 273.
4. Which values of a would you try first so that $P(a) = 0$? Use your calculator to check your suggestions.
5. Discuss your findings as a class. Could this information be useful when trying to determine a factor of a monic cubic polynomial?

WORKED EXAMPLE 11 Factorising polynomials using long division

Use long division to factorise the following.

a. $x^3 - 5x^2 - 2x + 24$ b. $x^3 - 19x + 30$ c. $-2x^3 - 8x^2 + 6x + 4$

THINK	WRITE
a. 1. Name the polynomial.	a. $P(x) = x^3 - 5x^2 - 2x + 24$
2. Look for a value of x such that $P(x) = 0$. For cubics containing a single x^3, try the factors of the constant term (24 in this case). The factors of 24 are $1, 2, 3, 4, 6, 8, 12$ and 24. Thus, you can try $\pm 1, \pm 2, \pm 3, \pm 4, \pm 6, \pm 8, \pm 12$ and ± 24. Try $P(1)$. $P(1) \neq 0$, so $(x-1)$ is not a factor. Try $P(2)$. $P(2) \neq 0$, so $(x-2)$ is not a factor. Try $P(-2)$.	$P(1) = 1^3 - 5 \times 1^2 - 2 \times 1 + 24$ $= 1 - 5 - 2 + 24$ $= 18$ $\neq 0$ $P(2) = 2^3 - 5 \times 2^2 - 2 \times 2 + 24$ $= 8 - 20 - 4 + 24$ $\neq 0$ $P(-2) = (-2)^3 - 5 \times (-2)^2 - 2 \times (-2) + 24$ $= -8 - 20 + 4 + 24$ $= -28 + 28$ $= 0$
$P(-2)$ does equal 0, so $(x+2)$ is a factor.	So, $(x+2)$ is a factor.

3. Divide $(x+2)$ into $P(x)$ using long division to determine a quadratic factor.

$$\begin{array}{r} x^2 - 7x + 12 \\ x+2\overline{)x^3 - 5x^2 - 2x + 24} \\ \underline{x^3 + 2x^2} \\ -7x^2 - 2x \\ \underline{-7x^2 - 14x} \\ 12x + 24 \\ \underline{12x + 24} \\ 0 \end{array}$$

4. Write $P(x)$ as a product of the two factors found so far.

$$P(x) = (x+2)(x^2 - 7x + 12)$$

5. Factorise the quadratic factor if possible.

$$P(x) = (x+2)(x-3)(x-4)$$

b. 1. Name the polynomial.
Note: There is no x^2 term, so include $0x^2$.

b. $P(x) = x^3 - 19x + 30$

$$P(x) = x^3 + 0x^2 - 19x + 30$$

2. Look at the last term in $P(x)$, which is 30. This suggests it is worth trying $P(5)$ or $P(-5)$. Try $P(-5)$. $P(-5) = 0$ so $(x+5)$ is a factor.

$$\begin{aligned} P(-5) &= (-5)^3 - 19 \times (-5) + 30 \\ &= -125 + 95 + 30 \\ &= 0 \end{aligned}$$
So, $(x+5)$ is a factor.

3. Divide $(x+5)$ into $P(x)$ using long division to find a quadratic factor.

$$\begin{array}{r} x^2 - 5x + 6 \\ x+5\overline{)x^3 + 0x^2 - 19x + 30} \\ \underline{x^3 + 5x^2} \\ -5x^2 - 19x \\ \underline{-5x^2 - 25x} \\ 6x + 30 \\ \underline{6x + 30} \\ 0 \end{array}$$

4. Write $P(x)$ as a product of the two factors found so far.

$$P(x) = (x+5)(x^2 - 5x + 6)$$

5. Factorise the quadratic factor if possible.

$$P(x) = (x+5)(x-2)(x-3)$$

c. 1. Write the given polynomial.

c. Let $P(x) = -2x^3 - 8x^2 + 6x + 4$
$$= -2(x^3 + 4x^2 - 3x - 2)$$

2. Take out a common factor of -2. (We could take out $+2$ as the common factor, but taking out -2 results in a positive leading term in the part still to be factorised.)

3. Let $Q(x) = x^3 + 4x^2 - 3x - 2$. (We have already used P earlier.)

Let $Q(x) = x^3 + 4x^2 - 3x - 2$.

4. Evaluate $Q(1)$.
$Q(1) = 0$, so $(x-1)$ is a factor.

$$\begin{aligned} Q(1) &= 1 + 4 - 3 - 2 \\ &= 0 \end{aligned}$$
So, $(x-1)$ is a factor.

5. Divide $(x - 1)$ into $Q(x)$ using long division to determine a quadratic factor.

$$
\begin{array}{r}
x^2 + 5x + 2 \\
x - 1 \overline{) x^3 + 4x^2 - 3x - 2} \\
\underline{x^3 - x^2} \\
5x^2 - 3x \\
\underline{5x^2 - 5x} \\
2x - 2 \\
\underline{2x - 2} \\
0
\end{array}
$$

6. Write the original polynomial $P(x)$ as a product of the factors found so far.

In this case, it is not possible to further factorise $P(x)$.

$$P(x) = -2(x - 1)(x^2 + 5x + 2)$$

- *Note:* In some of these examples, $P(x)$ may have been factorised without long division by finding all three values of x that make $P(x) = 0$ (and hence the three factors).

▶ 15.7.2 Using short division, or by inspection

- Short division, or factorising by inspection, is a quicker method than long division.

- Consider $P(x) = x^3 + 2x^2 - 13x + 10$
  Using the factor theorem:
  $P(1) = 0$ so $(x - 1)$ is a factor.

  $\therefore P(x) = (x - 1) Q(x)$, where $Q(x)$ is the quadratic quotient.

  So, $P(x) = (x - 1)\left(ax^2 + bx + c\right)$.

Expanding and equating:

Term in x^3: $x \times ax^2$	$(x - 1)\left(ax^2 + bx + c\right)$	Equating with $P(x)$: $ax^3 = x^3$ $\therefore a = 1$
Terms in x^2: $-1 \times ax^2 + bx^2$	$(x - 1)\left(ax^2 + bx + c\right)$	Equating with $P(x)$: $-1 \times 1x^2 + bx^2 = 2x^2$ $-1 + b = 2$ $\therefore b = 3$
Constant term: $-1 \times c$	$(x - 1)\left(ax^2 + bx + c\right)$	Equating with $P(x)$: $-c = 10$ $\therefore c = -10$

$$\therefore P(x) = (x - 1)\left(x^2 + 3x - 10\right)$$

Factorising the quadratic gives:

$$P(x) = (x - 1)(x + 5)(x - 2)$$

- *Note:* In this example, the values of a and c can be seen simply by inspecting $P(x)$.
  Hence, $P(x) = (x - 1)\left(x^2 + bx - 10\right)$, leaving only the value of b unknown.

- The following worked example is a repeat of a previous one, but explains the use of short, rather than long, division.

Use short division to factorise $x^3 - 5x^2 - 2x + 24$.

THINK	WRITE
1. Name the polynomial.	Let $P(x) = x^3 - 5x^2 - 2x + 24$
2. Look for a value of x such that $P(x) = 0$. Try $P(-2)$.	$\begin{aligned} P(-2) &= (-2)^3 - 5 \times (-2)^2 - 2 \times (-2) + 24 \\ &= -8 - 20 + 4 + 24 \\ &= -28 + 28 \\ &= 0 \end{aligned}$
$P(-2)$ does equal 0, so $(x + 2)$ is a factor.	So, $(x + 2)$ is a factor.
3. Look again at the original and equate the factorised form to the expanded form. The values of a and c can be determined simply by inspection. Since the coefficient of the x^3 term is 1, $a = 1$. Since the constant term is 24, $c = 12$.	$\begin{aligned} x^3 - 5x^2 - 2x + 24 &= (x + 2)\left(ax^2 + bx + c\right) \\ &= (x + 2)\left(x^2 + bx + 12\right) \end{aligned}$
4. Expand the brackets and equate the coefficients of the x^2 terms. We can then solve for b.	$\begin{aligned} x^3 - 5x^2 - 2x + 24 &= x^3 + bx^2 + 12x + 2x^2 + 2bx + 24 \\ -5x^2 &= bx^2 + 2x^2 \\ -5x^2 &= (b + 2)x^2 \\ b + 2 &= -5 \\ b &= -7 \\ P(x) &= (x + 2)\left(x^2 - 7x + 12\right) \end{aligned}$
5. Factorise the expression in the second pair of brackets if possible.	$P(x) = (x + 2)(x - 3)(x - 4)$

Exercise 15.7 Factorising polynomials

learn **on**

15.7 Quick quiz   **on**   **15.7 Exercise**

Individual pathways

■ PRACTISE	■ CONSOLIDATE	■ MASTER
1, 2, 8, 11, 15, 18	3, 4, 5, 9, 12, 16, 19	6, 7, 10, 13, 14, 17, 20

Fluency

WE11 For questions **1** to **7** apply long division to factorise each dividend.

1. a. $x + 1 \overline{\smash{)}\, x^3 + 10x^2 + 27x + 18}$ b. $x + 2 \overline{\smash{)}\, x^3 + 8x^2 + 17x + 10}$

2. a. $x + 9 \overline{\smash{)}\, x^3 + 12x^2 + 29x + 18}$ b. $x + 1 \overline{\smash{)}\, x^3 + 8x^2 + 19x + 12}$

3. a. $x+3\overline{)x^3+14x^2+61x+84}$

   b. $x+7\overline{)x^3+12x^2+41x+42}$

4. a. $x+2\overline{)x^3+4x^2+5x+2}$

   b. $x+3\overline{)x^3+7x^2+16x+12}$

5. a. $x+5\overline{)x^3+14x^2+65x+100}$

   b. $x\overline{)x^3+13x^2+40x}$

6. a. $x\overline{)x^3+7x^2+12x}$

   b. $x+5\overline{)x^3+10x^2+25x}$

7. a. $x+1\overline{)x^3+6x^2+5x}$

   b. $x+6\overline{)x^3+6x^2}$

WE12 For questions **8** to **10**, factorise the following as fully as possible.

8. a. x^3+x^2-x-1 b. x^3-2x^2-x+2 c. $x^3+7x^2+11x+5$ d. $x^3+x^2-8x-12$

9. a. $x^3+9x^2+24x+16$ b. $x^3-5x^2-4x+20$ c. x^3+2x^2-x-2 d. x^3-7x-6

10. a. $x^3+8x^2+17x+10$ b. x^3+x^2-9x-9 c. $x^3-x^2-8x+12$ d. $x^3+9x^2-12x-160$

Understanding

For questions **11** to **14**, factorise as fully as possible.

11. a. $2x^3+5x^2-x-6$
    b. $3x^3+14x^2+7x-4$
    c. $3x^3+2x^2-12x-8$
    d. $4x^3+35x^2+84x+45$

12. a. x^3+x^2+x+1
    b. $4x^3+16x^2+21x+9$
    c. $6x^3-23x^2+26x-8$
    d. $7x^3+12x^2-60x+16$

13. a. $3x^3-x^2-10x$
    c. $3x^3-6x^2-24x$
    b. $4x^3+2x^2-2x$
    d. $-2x^3-12x^2-18x$

14. a. $-x^3-7x^2-12x$
    c. $-2x^3+10x^2-12x$
    b. $-x^3-3x^2+x+3$
    d. $-5x^3+24x^2-36x+16$

Communicating, reasoning and problem solving

15. Factorise $x^4-9x^2-4x+12$.

16. Factorise $-x^5+6x^4+11x^3-84x^2-28x+240$.

17. Two of the factors of x^3+px^2+qx+r are $(x+a)$ and $(x+b)$. Determine the third factor.

18. Factorise $x^5-5x^4+5x^3+5x^2-6x$.

19. $(x-1)$ and $(x-2)$ are known to be factors of $x^5+ax^4-2x^3+bx^2+x-2$. Determine the values of a and b and hence fully factorise this fifth-degree polynomial.

20. The polynomial $x^4-6x^3+13x^2-12x-32$ has three factors, one of which is x^2-3x+8. Evaluate the other two factors.

LESSON
15.8 Solving polynomial equations

LEARNING INTENTION

At the end of this lesson you should be able to:
- solve polynomial equations by applying the Null Factor Law
- determine the solutions or roots of the equation $P(x) = 0$.

▶ 15.8.1 Solving polynomial equations

eles-4983

- To solve the polynomial equation of the form $P(x) = 0$:

  Step 1: factorise $P(x)$
  Step 2: apply the Null Factor Law
  Step 3: state the solutions.

- The Null Factor Law applies to polynomial equations just as it does for quadratics.
- If $P(x)$ is of degree n, then $P(x) = 0$ has up to n solutions.
  Solving each of these equations produces the solutions (roots). These solutions or roots are also known as the zeroes of the polynomial, because for these values of x the polynomial is equal to zero.

> **Solving $P(x) = 0$**
>
> If $P(x) = (x - a)(x - b)(x - c) \ldots (x - p)(x - q)$,
>
> the solutions or roots of the equation $P(x) = 0$
>
> are $x = a, b, c, \ldots, p, q$.

- If $P(x) = k(lx - a)(mx - b)(nx - c) = 0$, then the solutions can be found as follows.
  Let each factor equal zero.

$$lx - a = 0 \qquad mx - b = 0 \qquad nx - c = 0$$

Solving each of these equations produces the solutions.
Note: The coefficient k used in this example does not produce a solution because $k \neq 0$.

$$x = \frac{a}{l} \qquad x = \frac{b}{m} \qquad x = \frac{c}{n}.$$

WORKED EXAMPLE 13 Solving polynomial equations

Solve:
a. $x^3 = 9x$
b. $-2x^3 + 4x^2 + 70x = 0$
c. $2x^3 - 11x^2 + 18x - 9 = 0.$

THINK	WRITE
a. 1. Write the equation.	a. $x^3 = 9x$
2. Rearrange so all terms are on the left.	$x^3 - 9x = 0$
3. Take out a common factor of x.	$x(x^2 - 9) = 0$
4. Factorise the quadratic expression using the difference of two squares.	$x(x + 3)(x - 3) = 0$
5. Use the Null Factor Law to solve.	$x = 0, x + 3 = 0$ or $x - 3 = 0$
6. Write the values of x.	$x = 0, x = -3$ or $x = 3$

b. 1. Write the equation.

    2. Take out a common factor of $-2x$.

    3. Factorise the quadratic expression.

    4. Use the Null Factor Law to solve.

    5. Write the values of x.

b. $-2x^3 + 4x^2 + 70x = 0$

$-2x(x^2 - 2x - 35) = 0$

$-2x(x - 7)(x + 5) = 0$

$-2x = 0,\ x - 7 = 0 \text{ or } x + 5 = 0$

$x = 0,\ x = 7 \text{ or } x = -5$

c. 1. Name the polynomial.

    2. Use the factor theorem to determine a factor (search for a value a such that $P(a) = 0$). Consider factors of the constant term (that is, factors of 9 such as 1 and 3). The simplest value to try is 1.

    3. Use long or short division to determine another factor of $P(x)$.

c. Let $P(x) = 2x^3 - 11x^2 + 18x - 9$.

$P(1) = 2 - 11 + 18 - 9$

$\quad\ = 0$

So $(x - 1)$ is a factor.

$$
\begin{array}{r}
2x^2 - 9x + 9 \\
x - 1\overline{)\,2x^3 - 11x^2 + 18x - 9} \\
\underline{2x^3 - 2x^2} \\
-9x^2 + 18x \\
\underline{-9x^2 + 9x} \\
9x - 9 \\
\underline{9x - 9} \\
0
\end{array}
$$

    4. Factorise the quadratic factor.

$P(x) = (x - 1)(2x^2 + 9x - 9)$

$P(x) = (x - 1)(2x - 3)(x - 3)$

$(x - 1)(2x - 3)(x - 3) = 0$

    5. Use the Null Factor Law to solve.

$x - 1 = 0,\ 2x - 3 = 0 \text{ or } x - 3 = 0$

    6. Write the values of x.

$x = 1,\ x = \dfrac{3}{2} \text{ or } x = 3$

Exercise 15.8 Solving polynomial equations

learnon

15.8 Quick quiz on

15.8 Exercise

Individual pathways

■ PRACTISE	■ CONSOLIDATE	■ MASTER
1, 5, 8, 10, 14, 17	2, 6, 9, 11, 12, 15, 18	3, 4, 7, 13, 16, 19

Fluency

 For questions **1** to **4**, solve the following.

1. **a.** $x^3 - 4x = 0$ **b.** $x^3 - 16x = 0$ **c.** $2x^3 - 50x = 0$

2. **a.** $-3x^3 + 81 = 0$ **b.** $x^3 + 5x^2 = 0$ **c.** $x^3 - 2x^2 = 0$

3. **a.** $-4x^3 + 8x = 0$ **b.** $12x^3 + 3x^2 = 0$ **c.** $4x^2 - 20x^3 = 0$

4. **a.** $x^3 - 5x^2 + 6x = 0$ **b.** $x^3 - 8x^2 + 16x = 0$ **c.** $x^3 + 6x^2 = 7x$

WE13c For questions **5** to **7**, apply the factor theorem to solve the following.

5. **a.** $x^3 - x^2 - 16x + 16 = 0$ **b.** $x^3 - 6x^2 - x + 30 = 0$
   c. $x^3 - x^2 - 25x + 25 = 0$ **d.** $x^3 + 4x^2 - 4x - 16 = 0$

6. **a.** $x^3 - 4x^2 + x + 6 = 0$ **b.** $x^3 - 4x^2 - 7x + 10 = 0$
   c. $x^3 + 6x^2 + 11x + 6 = 0$ **d.** $x^3 - 6x^2 - 15x + 100 = 0$

7. **a.** $x^3 - 3x^2 - 6x + 8 = 0$ **b.** $x^3 + 2x^2 - 29x + 42 = 0$
   c. $2x^3 + 15x^2 + 19x + 6 = 0$ **d.** $-4x^3 + 16x^2 - 9x - 9 = 0$

8. **MC** *Note:* There may be more than one correct answer.
   Select a solution to $x^3 - 7x^2 + 2x + 40 = 0$ from the following.

   A. $x = 5$ **B.** $x = -4$ **C.** $x = -2$ **D.** $x = 1$

9. **MC** A solution of $x^3 - 9x^2 + 15x + 25 = 0$ is $x = 5$. Select the number of other (distinct) solutions there are.

   A. 0   **B.** 1   **C.** 2   **D.** 3

Understanding

10. Solve $P(x) = 0$ for each of the following.
    a. $P(x) = x^3 + 4x^2 - 3x - 18$ **b.** $P(x) = 3x^3 - 13x^2 - 32x + 12$
    c. $P(x) = -x^3 + 12x - 16$ **d.** $P(x) = 8x^3 - 4x^2 - 32x - 20$

11. Solve $P(x) = 0$ for each of the following.
    a. $P(x) = x^4 + 2x^3 - 13x^2 - 14x + 24$ **b.** $P(x) = -72 - 42x + 19x^2 + 7x^3 - 2x^4$
    c. $P(x) = x^4 + 2x^3 - 7x^2 - 8x + 12$ **d.** $P(x) = 4x^4 + 12x^3 - 24x^2 - 32x$

12. Solve each of the following equations.
    a. $x^3 - 3x^2 - 6x + 8 = 0$ **b.** $x^3 + x^2 - 9x - 9 = 0$ **c.** $3x^4 + 3x^3 - 18x = 0$

13. Solve each of the following equations.
    a. $2x^4 + 10x^3 - 4x^2 - 48x = 0$ **b.** $2x^4 + x^3 - 14x^2 - 4x + 24 = 0$ **c.** $x^4 - 2x^3 + 1 = 0$

Communicating, reasoning and problem solving

14. Solve for a if $x = 2$ is a solution of $ax^3 - 6x^2 + 3x - 4 = 0$.

15. Solve for p if $x = \dfrac{p}{2}$ is a solution of $x^3 - 5x^2 + 2x + 8 = 0$.

16. Show that it is possible for a cuboid of side lengths x cm, $(x - 1)$ cm and $(x + 2)$ cm to have a volume that is 4 cm^3 less than twice the volume of a cube of side length x cm.
    Comment on the shape of such a cuboid.

17. Solve the following equation for x.
$$x^3 + 8 = x(5x - 2)$$

18. Solve the following equation for x.
$$2\left(x^3 + 5\right) = 13x(x - 1)$$

19. Solve the following equation for z.
$$z(z - 1)^3 = -2(z^3 - 5z^2 + z + 3)$$

LESSON
15.9 Review

15.9.1 Topic summary

Polynomials

- Polynomials are expressions with only non-negative integer powers.
- The degree of a polynomial is the highest power of the variable that it contains.
- The leading term is the term with highest power of the variable.
- Monic polynomials have a leading coefficient of 1.
- Polynomials are often denoted $P(x)$.
- The value of a polynomial can be determined by substituting the x-value into the expression.
  e.g.
  $P(x) = x^4 - 3x^2 + 8$ is a monic polynomial of degree 4.
  The coefficient of the x^2 term is -3 and the constant term is 8.

The remainder theorem

- When $P(x)$ is divided by $(x - a)$, the remainder, R, is given by:
  $$R = P(a)$$
  e.g.
  The remainder when $P(x) = 2x^3 + 3x^2 - 4x - 5$ is divided by $(x + 2)$ is $P(-2)$.
  $$\begin{aligned} P(-2) &= 2(-2)^3 + 3(-2)^2 - 4(-2) - 5 \\ &= 2(-8) + 3(4) + 8 - 5 \\ &= -16 + 12 + 8 - 5 \\ &= -1 \end{aligned}$$

POLYNOMIALS (PATH)

Operations on polynomials

- To add or subtract polynomials simply add or subtract like terms.
  e.g.
  $$\begin{aligned} &(2x^3 - 5x + 1) + (-6x^3 + 8x^2 + 3x - 11) \\ &= (2x^3 - 6x^3) + 8x^2 + (-5x + 3x) + (1 - 11) \\ &= -4x^3 + 8x^2 - 2x - 10 \end{aligned}$$
- To multiply polynomials use the same methods as with quadratic expressions. Use FOIL and then simplify.
- For polynomials of degree 3 and higher you may need to use FOIL multiple times.
  e.g.
  $$\begin{aligned} (x + 3)(x - 1)(2x + 4) &= (x + 3)(2x^2 + 4x - 2x - 4) \\ &= (x + 3)(2x^2 + 2x - 4) \\ &= 2x^3 + 2x^2 - 4x + 6x^2 + 6x - 12 \\ &= 2x^3 + 8x^2 + 2x - 12 \end{aligned}$$

Factorising polynomials

- The factor theorem states the following:
  If $P(a) = 0$, then $(x - a)$ is a factor of $P(x)$.
- This can be used to find a factor, and then other factors can be found using the methods used to factorise quadratics.
  e.g.
  $$\begin{aligned} P(x) &= x^3 - 2x^2 - 5x + 6 \\ P(1) &= (1)^3 - 2(1)^2 - 5(1) + 6 \\ &= 1 - 2 - 5 + 6 \\ &= 0 \end{aligned}$$
  Therefore, $(x - 1)$ is a factor of $P(x)$.
  $$\begin{aligned} P(x) &= (x - 1)(x^2 - x - 6) \\ &= (x - 1)(x - 3)(x + 2) \end{aligned}$$

Long division

- Polynomials can be divided using long division.
- The example below shows the division of $P(x) = x^3 + 2x^2 - 13x + 10$ by $x - 3$.

$$\require{enclose}\begin{array}{r} x^2 + 5x + 2 \quad \leftarrow \text{Quotient} \\ x - 3 \enclose{longdiv}{x^3 + 2x^2 - 13x + 10} \\ \underline{x^3 - 3x} \\ 5x^2 - 13x \\ \underline{5x^2 - 15x} \\ 2x + 10 \\ \underline{2x - 6} \\ 16 \quad \leftarrow \text{Remainder} \end{array}$$

- The result is:
  $P(x) = (x - 3)(x^2 + 5x + 2) + 16$

Solving polynomial equations

- To solve a polynomial equation:
  1. express in the form $P(x) = 0$
  2. factorise $P(x)$
  3. solve using the Null Factor Law.
  e.g.
  $$\begin{aligned} 2x^3 + 7x^2 &= 9 \\ 2x^3 + 7x^2 - 9 &= 0 \end{aligned}$$
  Let $P(x) = 2x^3 + 7x^2 - 9$.
  $$\begin{aligned} P(1) &= 2(1)^3 + 7(1)^2 - 9 \\ &= 2 + 7 - 9 \\ &= 0 \end{aligned}$$
  Therefore, $(x - 1)$ is a factor of $P(x)$.
  $$\begin{aligned} P(x) &= (x - 1)(2x^2 + 9x + 9) \\ &= (x - 1)(2x - 3)(x + 3) \end{aligned}$$
  Using the Null Factor Law:
  $$x = 1, x = -\frac{3}{2} \text{ or } x = -3$$

15.9.2 Project

Investigating polynomials

A polynomial is a function involving the sum of integer powers of a variable (for example, $y = -4x^3 + 3x^2 - 4$). The highest power of the variable determines the degree of the polynomial. In the case of the given example, the degree is 3.

A polynomial of the first degree is a linear function (for example, $y = 3x - 8$), and a second-degree function is a quadratic (for example, $y = 5x^2 - 6x + 7$). Let us investigate how the degree of a polynomial affects the shape of its graph.

In order to simplify the graphing of these functions, the polynomials will be expressed in factor form. A graphics calculator or some other digital technology will make the graphing process less tedious.

It will be necessary to adjust the window of the calculator from time to time in order to capture the relevant features of the graph.

1. Consider the following polynomials.
   a. $y_1 = (x + 1)$
   b. $y_2 = (x + 1)(x - 2)$
   c. $y_3 = (x + 1)(x - 2)(x + 3)$
   d. $y_4 = (x + 1)(x - 2)(x + 3)(x - 4)$
   e. $y_5 = (x + 1)(x - 2)(x + 3)(x - 4)(x + 5)$
   f. $y_6 = (x + 1)(x - 2)(x + 3)(x - 4)(x + 5)(x - 6)$

   For each of the functions:
   i. give the degree of the polynomial
   ii. sketch the graph, marking in the x-intercepts
   iii. describe how the degree of the polynomial affects the shape of the graph.
   Complete question 1 on a separate sheet of paper.

2. Let us now look at the effect that the exponent of each factor has on the shape of the graph of the polynomial. Consider the following functions.
   a. $y_1 = (x + 1)(x - 2)(x + 3)$
   b. $y_2 = (x + 1)^2(x - 2)(x + 3)$
   c. $y_3 = (x + 1)^2(x - 2)^2(x + 3)$
   d. $y_4 = (x + 1)^2(x - 2)(x + 3)^3$
   e. $y_5 = (x + 1)^3(x - 2)(x + 3)^4$
   f. $y_6 = (x + 1)^5(x - 2)^3(x + 3)^2$
   i. On a separate sheet of paper, draw a sketch of each of the polynomials, marking in the x-intercepts.
   ii. Explain how the power of the factor affects the behaviour of the graph at the x-intercept.

3. Create and draw a sketch of polynomials with the following given characteristics. Complete your graphs on a separate sheet of paper.
   a. A first-degree polynomial that:
      i. crosses the x-axis
      ii. does not cross the x-axis.
   b. A second-degree polynomial that:
      i. crosses the x-axis twice
      ii. touches the x-axis at one and only one point.
   c. A third-degree polynomial that crosses the x-axis:
      i. three times
      ii. twice
      iii. once.
   d. A fourth-degree polynomial that crosses the x-axis:
      i. four times
      ii. three times
      iii. twice
      iv. once.

4. Considering the powers of factors of polynomials, write a general statement outlining the conditions under which the graph of a polynomial will pass through the x-axis or just touch the x-axis.

 Resources

Interactivities Crossword (int-2875)
              Sudoku puzzle (int-3892)

Exercise 15.9 Review questions

learn on

Fluency

1. **MC** Select which of the following is *not* a polynomial.
   A. $x^3 - \dfrac{x^2}{3} + 7x - 1$
   B. $a^4 + 4a^3 + 2a + 2$
   C. $\sqrt{x^2 + 3x + 2}$
   D. 5

2. Consider the polynomial $y = -\dfrac{1}{7}x^4 + x^5 + 3$.
   a. State the degree of y.
   b. State the coefficient of x^4.
   c. State the constant term.
   d. Determine the leading term.

3. **MC** The expansion of $(x + 5)(x + 1)(x - 6)$ is:
   A. $x^3 - 30$
   B. $x^3 + 12x^2 - 31x + 30$
   C. $x^3 - 31x - 30$
   D. $x^3 + 5x^2 - 36x - 30$

4. **MC** $x^3 + 5x^2 + 3x - 9$ is the expansion of:
   A. $(x + 3)^3$
   B. $x(x + 3)(x - 3)$
   C. $(x - 1)(x + 3)^2$
   D. $(x - 1)(x + 1)(x + 3)$

5. Expand each of the following.
   a. $(x - 2)^2(x + 10)$
   b. $(x + 6)(x - 1)(x + 5)$
   c. $(x - 7)^3$
   d. $(5 - 2x)(1 + x)(x + 2)$

6. **MC** Consider the following long division.

$$
\begin{array}{r}
x^2 + x + 2 \\
x - 4 \overline{\smash{\big)}\, x^3 + 5x^2 + 6x - 1} \\
\underline{x^3 + 4x^2} \\
x^2 + 6x \\
\underline{x^2 + 4x} \\
2x - 1 \\
\underline{2x + 8} \\
-9
\end{array}
$$

a. The quotient is:

    A. -9 **B.** 9 **C.** $x + 4$ **D.** $x^2 + x + 2$

b. The remainder is:

    A. -9 **B.** 2 **C.** 4 **D.** $2x - 1$

Understanding

7. Determine the quotient and remainder when the first polynomial is divided by the second in each case.

    a. $x^3 + 2x^2 - 16x - 3,\ (x + 2)$

    b. $x^3 + 3x^2 - 13x - 7,\ (x - 3)$

    c. $-x^3 + x^2 + 4x - 7,\ (x + 1)$

8. **MC** If $P(x) = x^3 - 3x^2 + 7x + 1$, then $P(-2)$ equals:

    A. -34 **B.** -33 **C.** -9 **D.** 9

9. If $P(x) = -3x^3 + 2x^2 + x - 4$, calculate:

    a. $P(1)$

    b. $P(-4)$

    c. $P(2a)$

10. Without dividing, determine the remainder when $x^3 + 3x^2 - 16x + 5$ is divided by $(x - 1)$.

11. Show that $(x + 3)$ is a factor of $x^3 - 2x^2 - 29x - 42$.

12. Factorise $x^3 + 4x^2 - 100x - 400$.

13. Solve:

    a. $(2x + 1)(x - 3)^2 = 0$

    b. $x^3 - 9x^2 + 26x - 24 = 0$

    c. $x^4 - 4x^3 - x^2 + 16x - 12 = 0$

Communicating, reasoning and problem solving

14. Let $P(x) = a_n x^n + a_{n-1} x^{n-1} + \ldots + a_1 x + a_0$ be a polynomial where the coefficients are integers. Also let $P(w) = 0$ where w is an integer. Show that w is a factor of a_0.

15. Evaluate the area of a square whose sides are $(2x - 3)$ cm. Expand and simplify your answer. If the area is $16\,\text{cm}^2$, determine the value of x.

16. A window is in the shape of a semicircle above a rectangle. The height of the window is $(6x+1)$ cm and its width is $(2x+2)$ cm.

    a. Evaluate the total area of the window.
    b. Expand and simplify your answer.
    c. Determine the perimeter of the window.

17. Answer the following questions.
    a. Determine the volume of a cube of side $(x+4)$ cm.
    b. Evaluate the surface area of the cube.
    c. Determine the value of x for which the volume and surface are numerically equal.
    d. Calculate the value of x if the numerical value of the volume is 5 less than the numerical value of the surface area.

18. Determine the quotient and remainder when $mx^2 + nx + q$ is divided by $(x - p)$.

19. When $P(x)$ is divided by $(x - n)$, the quotient is $x^2 - 2x + n$ and the remainder is $(n + 1)$. Evaluate the value of $P(x)$.

20. Given that the polynomial $P(x) = x^3 + 3x^2 + 5x + 15$ has one integer root, solve $P(x) \geq 0$.

on To test your understanding and knowledge of this topic, go to your learnON title at www.jacplus.com.au and complete the **post-test**.

Answers

Topic 15 Polynomials (Path)

15.1 Pre-test

1. a. 2   b. 3   c. 1   d. 0
2. A
3. $a = 2, b = -1$
4. D
5. A
6. a. 3   b. -3 c. -1 d. 2
7. A
8. $x^2 + x$
9. 9
10. 18
11. D
12. $b = 2$
13. B
14. C
15. $x = -1, -\dfrac{1}{2}$ or -6

15.2 Polynomials

1. a. i. 3   ii. x
   b. i. 7   ii. x
   c. i. 2   ii. x
2. a. i. 6   ii. x
   b. i. 8   ii. y
   c. i. 5   ii. u
3. a. i. 5   ii. e
   b. i. 1   ii. g
   c. i. 6   ii. f
4. a. Polynomial 3b   b. Polynomial 1c
   c. Polynomial 1a   d. Polynomials 1a, 2a and 2b
5. a. N   b. P   c. P
6. a. N   b. N   c. P
7. a. P   b. N   c. N
8. a. 3   b. x c. 4
   d. 5   e. $3x$ f. $-2x^3$
9. a. 7   b. w c. 7
   d. 0   e. -9 f. $6w^7$
10. a. 4   b. 1   c. x^4 d. 1
11. a. $4x^3 + 2x^2 - 10x + 18$ b. $3x^4 - 3x^3 - x^2 + 7x - 7$
    c. $5x^3 - 4x^2 - 13x - 6$
12. a. 6
    b. t
    c. 6
    d. Sample responses can be found in the worked solutions in the online resources.
    e. Sample responses can be found in the worked solutions in the online resources.

13. a. 5 units to the right of the origin
    b. 4 units to the right of the origin
    c. The body moves towards the origin, then away.
14. $a = 1, b = -4$
15. $a = 4, b = -6$
16. $a = \pm 3, b = \pm 2$

15.3 Adding, subtracting and multiplying polynomials

1. a. $x^4 + 2x^3 - x^2 - 10$
   b. $x^6 + 2x^4 - 3x^3 + 9x^2 + 5$
   c. $5x^3 - 5x^2 + 7x - 13$
   d. $2x^4 + 3x^3 + 12x^2 - 4x + 14$
   e. $x^5 + 13x^4 - 10$
2. a. $x^4 + 2x^2 + 2x + 4$ b. $x^6 - x^5 + x^3 + x^2 + 2$
   c. $5x^7 - 4x^3 + 5x$ d. $10x^4 - 7x^2 + 20x + 5$
   e. $2x^3 + 6x^2 - 10x + 15$
3. a. $x^3 + 7x^2 + 6x$ b. $x^3 - 7x^2 - 18x$
   c. $x^3 + 8x^2 - 33x$ d. $2x^3 + 10x^2 + 12x$
   e. $48x - 3x^3$
4. a. $5x^3 + 50x^2 + 80x$ b. $x^3 + 4x^2$
   c. $2x^3 - 14x^2$ d. $-30x^3 - 270x^2$
   e. $-7x^3 - 56x^2 - 112x$
5. a. $x^3 + 12x^2 + 41x + 42$ b. $x^3 - 3x^2 - 18x + 40$
   c. $x^3 + 3x^2 - 36x + 32$ d. $x^3 - 6x^2 + 11x - 6$
   e. $x^3 + 6x^2 - x - 6$
6. a. $x^3 + 5x^2 - 49x - 245$ b. $x^3 + 4x^2 - 137x - 660$
   c. $x^3 + 3x^2 - 9x + 5$ d. $x^3 - 12x^2 + 21x + 98$
   e. $x^3 + x^2 - x - 1$
7. a. $x^3 + 13x^2 + 26x - 112$
   b. $3x^3 + 26x^2 + 51x - 20$
   c. $4x^4 + 3x^3 - 37x^2 - 27x + 9$
   d. $10x^3 - 49x^2 + 27x + 36$
   e. $-6x^3 - 71x^2 - 198x + 35$
8. a. $21x^4 - 54x^3 - 144x^2 + 96x$
   b. $54x^3 + 117x^2 - 72x$
   c. $24x^3 - 148x^2 + 154x + 245$
   d. $20x^4 - 39x^3 - 50x^2 + 123x - 54$
   e. $4x^3 + 42x^2 + 146x + 168$
9. a. $x^3 + 6x^2 + 12x + 8$ b. $x^3 + 15x^2 + 75x + 125$
   c. $x^3 - 3x^2 + 3x - 1$
10. a. $x^4 - 12x^3 + 54x^2 - 108x + 81$
    b. $8x^3 - 72x^2 + 216x - 216$
    c. $81x^4 + 432x^3 + 864x^2 + 768x + 256$
11. $(2a + 5b)x + (2b - 5c)$
12. $x^4 + (a - 4b)x^3 + (2a - 4ab + 3b^2)x^2 +$
    $(2a^2 - 2ab + 3ab^2)x - 2a^2b$
13. $a = 1, b = -12, c = 54, d = -108, e = 81$
14. $8x^3 - 45x^2 + 78x - 43$
15. $\dfrac{1}{8}(11x^3 - 105x^2 + 73x - 27)$
16. $a = 2, b = 3$

17. $a = 1$, $b = -2$ and $c = 1$

18. $a = 1$, $b = 4$ and $c = -1$

15.4 Long division of polynomials

1. a. $x^2 + 2x$, 9 **b.** $x^2 + x + 3$, -2
   c. $x^2 + 3x - 6$, 19 **d.** $x^2 - x + 5$, -17

2. a. $x^2 + 4x - 6$, 14 **b.** $x^2 + 1$, 2
   c. $x^2 + 5$, -36 **d.** $x^2 - x + 6$, -11

3. a. $x^2 + 4x + 3$, -3 **b.** $x^2 + 4x + 13$, 48
   c. $x^2 + 3x - 3$, -11 **d.** $x^2 - 3x + 7$, 5

4. a. $x^2 - 2x - 3$, -17 **b.** $x^2 - 6x + 3$, -4
   c. $x^2 + 14x + 72$, 359 **d.** $x^2 + 8x + 27$, 104

5. a. $3x^2 - 7x + 20$, -35 **b.** $4x^2 - 8x + 18$, -22
   c. $2x^2 - 3x + 3$, 7

6. a. $2x^2 - 9$, 35 **b.** $4x^2 + 2x - 3$, -1
   c. $3x^2 + x - 1$, -2

7. a. $3x^2 - 2x + 1$, 5 **b.** $2x^2 + 5x - 6$, -7
   c. $4x^2 - 7x - 2$, -3

8. a. $x^2 - 4x + 3$, 8 **b.** $x^2 + x - 6$, -11
   c. $3x^2 + 2x + 1$, 13

9. a. $-x^2 - 5x - 2$, -14 **b.** $-3x^2 - 2x + 4$, -3
   c. $-x^2 + 5x + 6$, 9 **d.** $-2x^2 + 7x - 1$, 1

10. a. $x^2 - x - 2$, 3 **b.** x^2, -7
   c. $x^2 - x - 2$, -8 **d.** $-x^2 - x - 8$, 0

11. a. $5x - 2$, 7 **b.** $2x^2 - 2x + 10$, -54
   c. $-2x^2 - 4x - 9$, -16 **d.** $-2x^2 + 4x - 1$, 1

12. a. $x^3 + 2x^2 + 5x - 2$, -2
   b. $x^3 + 2x^2 - 9x - 18$, 0
   c. $x^4 - 3x^3 + 6x^2 - 18x + 58$, -171

13. a. $2x^5 - 4x^4 + 7x^3 - 13x^2 + 32x - 69$, 138
   b. $6x^3 + 17x^2 + 53x + 155$, 465
   c. $x^3 - \dfrac{7}{3}x^2 + \dfrac{7}{9}x + 3\dfrac{20}{27}$, $-3\dfrac{20}{27}$

14. Quotient: $x^3 - 4x^2 + 8x - 12$
   Remainder: $(a + 72)$

15. Quotient $= ax + (b + ad)$
   Remainder $= Rc + d(b + ad)$

16. $q = \dfrac{2p}{c}$

17. $a = -3$

18. $a = 6$

19. $a = 3$, $b = -5$

15.5 Polynomial values

1. a. 10 **b.** 11

2. a. 18 **b.** 43

3. a. 3 **b.** -22

4. a. -77 **b.** $2a^3 - 3a^2 + 2a + 10$

5. a. $16b^3 - 12b^2 + 4b + 10$
   b. $2x^3 + 9x^2 + 14x + 18$

6. a. $2x^3 - 21x^2 + 74x - 77$
   b. $-128y^3 - 48y^2 - 8y + 10$

7. a. 4 **b.** 15 **c.** 0
   d. 4 **e.** 15 **f.** 0

8. a. 10 **b.** 28 **c.** -8
   d. 10 **e.** 28 **f.** -8

9. a. 3 **b.** 11 **c.** -21
   d. 3 **e.** 11 **f.** -21

10. a. -7 **b.** 5 **c.** -7
   d. -7 **e.** 5 **f.** -7

11. a. $P(-8)$ **b.** $P(7)$ **c.** $P(a)$

12. a. -1 **b.** -6249 **c.** $2(a - 3)^5 + 1$

13. a. $-2a^3 - 3a^2 + a + 4$ **b.** $-2a^3 - 9a^2 - 11a - 1$

14. $b = 1, 4$

15. $c = 8$

16. $c = 2$

17. $b = -2$, $c = 5$

15.6 The remainder and factor theorems

1. a. -30 **b.** 0 **c.** 0 **d.** -24

2. a. -24 **b.** $k^3 + 3k^2 - 10k - 24$
   c. $-n^3 + 3n^2 + 10n - 24$ **d.** $-27c^3 + 27c^2 + 30c - 24$

3. a. 58 **b.** -8

4. a. 11 **b.** -9

5. a. 158 **b.** 6

6. a. 35 **b.** -6

7. a. -202 **b.** 441

8. 6

9. 3

10. 1

11. -2

12. 2

13. 2

14. $-5, 2$

15. $a = -5$, $b = -3$

16. a. D **b.** C, D **c.** D **d.** A, C, D

17. a. $(x - 1)$ **b.** $(x - 3)$ or $(x - 2)$
   c. $(x - 3)$ or $(x + 2)$ **d.** $(x - 6)$ or $(x + 4)$ or $(x + 5)$

18. a–d. Sample responses can be found in the worked solutions in the online resources.

19. a–d. Sample responses can be found in the worked solutions in the online resources.

20. Sample responses can be found in the worked solutions in the online resources.

21. $a = 2$

22. $a = 3$, $b = 2$

23. $a = -5$, $b = 41$, $(x + 3)$ and $(x - 5)$

24. $k = -4$

15.7 Factorising polynomials

1. a. $(x + 1)(x + 3)(x + 6)$ **b.** $(x + 1)(x + 2)(x + 5)$

2. a. $(x + 1)(x + 2)(x + 9)$ **b.** $(x + 1)(x + 3)(x + 4)$

3. a. $(x + 3)(x + 4)(x + 7)$ **b.** $(x + 2)(x + 3)(x + 7)$

4. a. $(x+1)^2(x+2)$ b. $(x+2)^2(x+3)$

5. a. $(x+4)(x+5)^2$ b. $x(x+5)(x+8)$

6. a. $x(x+3)(x+4)$ b. $x(x+5)^2$

7. a. $x(x+1)(x+5)$ b. $x^2(x+6)$

8. a. $(x-1)(x+1)^2$ b. $(x-2)(x-1)(x+1)$
   c. $(x+1)^2(x+5)$ d. $(x-3)(x+2)^2$

9. a. $(x+1)(x+4)^2$ b. $(x-5)(x-2)(x+2)$
   c. $(x-1)(x+1)(x+2)$ d. $(x-3)(x+1)(x+2)$

10. a. $(x+1)(x+2)(x+5)$ b. $(x-3)(x+1)(x+3)$
    c. $(x-2)^2(x+3)$ d. $(x-4)(x+5)(x+8)$

11. a. $(2x+3)(x-1)(x+2)$ b. $(3x-1)(x+1)(x+4)$
    c. $(3x+2)(x-2)(x+2)$ d. $(4x+3)(x+3)(x+5)$

12. a. $(x+1)(x^2+1)$ b. $(x+1)(2x+3)^2$
    c. $(x-2)(2x-1)(3x-4)$ d. $(7x-2)(x-2)(x+4)$

13. a. $x(x-2)(3x+5)$ b. $2x(x+1)(2x-1)$
    c. $3x(x-4)(x+2)$ d. $-2x(x+3)^2$

14. a. $-x(x+4)(x+3)$ b. $-(x-1)(x+1)(x+3)$
    c. $-2x(x-3)(x-2)$ d. $-(x-2)^2(5x-4)$

15. $(x-1)(x+2)(x+2)(x-3)$

16. $-(x-2)(x+2)(x+3)(x-4)(x-5)$

17. $(x-p+(a+b))$

18. $x(x-1)(x+1)(x-2)(x-3)$

19. $a=-2, b=4, (x-1)^2(x+1)^2(x-2)$

20. The other two factors are $(x-4)$ and $(x+1)$.

15.8 Solving polynomial equations

1. a. $-2, 0, 2$ b. $-4, 0, 4$ c. $-5, 0, 5$

2. a. 3 b. $-5, 0$ c. $0, 2$

3. a. $-\sqrt{2}, 0, \sqrt{2}$ b. $-\dfrac{1}{4}, 0$ c. $0, \dfrac{1}{5}$

4. a. $0, 2, 3$ b. $0, 4$ c. $-7, 0, 1$

5. a. $-4, 1, 4$ b. $-2, 3, 5$
   c. $-5, 1, 5$ d. $-4, -2, 2$

6. a. $-1, 2, 3$ b. $-2, 1, 5$
   c. $-3, -2, -1$ d. $-4, 5$

7. a. $-2, 1, 4$ b. $-7, 2, 3$
   c. $-6, -\dfrac{1}{2}, -1$ d. $-\dfrac{1}{2}, \dfrac{3}{2}, 3$

8. A, C

9. B

10. a. $-3, 2$ b. $-2, \dfrac{1}{3}, 6$

    c. $-4, 2$ d. $-1, \dfrac{5}{2}$

11. a. $-4, -2, 1, 3$ b. $-2, -\dfrac{3}{2}, 3, 4$

    c. $-3, -2, 1, 2$ d. $-4, -1, 0, 2$

12. a. $-2, 1, 4$ b. $-3, -1, 3$ c. $-3, 0, 2$

13. a. $-4, -3, 0, 2$ b. $-2, \dfrac{3}{2}, 2$ c. $-1, 1$

14. $a = 2.75$

15. $p = -2, 4, 8$

16. $x = 1.48$ (to 2 decimal places)

17. $x = -1, 4$ and 2

18. $x = \dfrac{-1}{2}, 2, 5$

19. $z = -1, 1, -2$ and 3

Project

1. a. i. 1
   ii.
   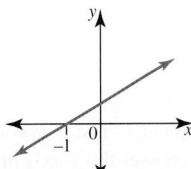
   iii. The graph is linear and crosses the x-axis once (at $x = -1$).

   b. i. 2
   ii.
   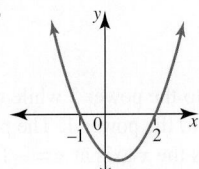
   iii. The graph is quadratic and crosses the x-axis twice (at $x = -1$ and $x = 2$).

   c. i. 3
   ii.
   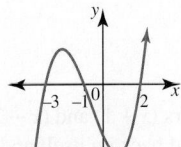
   iii. The graph is a curve and crosses the x-axis 3 times (at $x = -1$, $x = 2$ and $x = -3$).

   d. i. 4
   ii.
   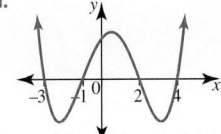
   iii. The graph is a curve and crosses the x-axis 4 times (at $x = -1$, $x = 2$, $x = -3$ and $x = 4$).

   e. i. 5
   ii.
   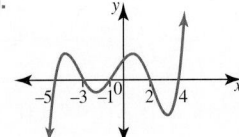
   iii. The graph is a curve and crosses the x-axis 5 times (at $x = -1$, $x = 2$, $x = -3$, $x = 4$ and $x = -5$).

f. i. 6

ii.

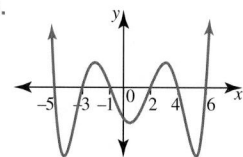

iii. The graph is a curve and crosses the x-axis 6 times (at $x = -1$, $x = 2$, $x = -3$, $x = 4$, $x = -5$ and $x = 6$).

2. a. i.

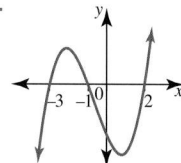

ii. Each factor is raised to the power 1. The polynomial is of degree 3 and the graph crosses the x-axis in 3 places(-3, -1 and 2).

b. i.

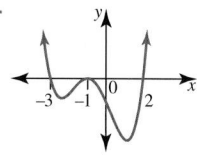

ii. The factor $(x + 1)$ is raised to the power 2 while the other two factors are raised to the power 1. The power 2 causes the curve not to cross the x-axis at $x = -1$ but to be curved back on itself.

c. i.

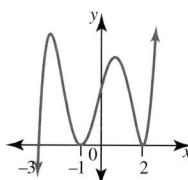

ii. The power 2 on the two factors $(x + 1)$ and $(x - 2)$ causes the curve to be directed back on itself and not to cross the x-axis at those two points ($x = -1$ and $x = 2$).

d. i.

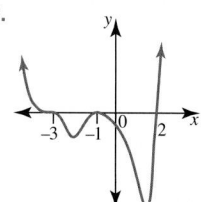

ii. The power 3 on the factor $(x + 3)$ causes the curve to run along the axis at that point then to cross the axis (at $x = -3$).

e. i.

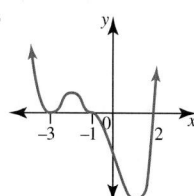

ii. The power 3 on the factor $(x + 1)$ causes the curve to run along the axis at $x = -1$, then cross the axis. The power 4 on the factor $(x + 3)$ causes the curve to be directed back on itself without crossing the axis at $x = -3$.

f. i.

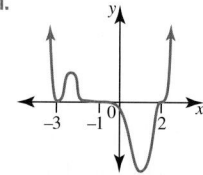

ii. The power 5 on the factor $(x + 1)$ causes the curve to run along the axis at $x = -1$, then cross the axis.

3. Answers will vary. Possible answers could be as follows.
   a. i. $y = 3x + 2$
      ii. $y = 4$
   b. i. $y = (x + 1)(x + 2)$
      ii. $y = (x + 1)^2$
   c. i. $y = (x + 1)(x + 2)(x + 3)$
      ii. Not possible
      iii. $y = (x + 1)^2(x + 2)$
   d. i. $y = (x + 1)(x + 2)(x + 3)(x + 4)$
      ii. Not possible
      iii. $y = (x + 1)^2(x + 2)(x + 3)$, $y = (x + 1)^3(x + 2)$
      iv. Not possible

4. If the power of the factor of a polynomial is an odd integer, the curve will pass through the x-axis. If the power is 1, the curve passes straight through. If the power is 3, 5..., the curve will run along the x-axis before passing through it. On the other hand, an even power of a factor causes the curve to just touch the x-axis then move back on the same side of the x-axis.

15.9 Review questions

1. C
2. a. 5 **b.** $-\dfrac{1}{7}$ **c.** 3 **d.** x^5
3. C
4. C
5. a. $x^3 + 6x^2 - 36x + 40$
   b. $x^3 + 10x^2 + 19x - 30$
   c. $x^3 - 21x^2 + 147x - 343$
   d. $-2x^3 - x^2 + 11x + 10$
6. a. D   **b.** A
7. a. $x^2 - 16, 29$ **b.** $x^2 + 6x + 5, 8$ **c.** $-x^2 + 2x + 2, -9$
8. B
9. a. -4
   b. 216
   c. $-24a^3 + 8a^2 + 2a - 4$
10. -7
11. Sample responses can be found in the worked solutions in the online resources.
12. $(x - 10)(x + 4)(x + 10)$
13. a. $-\dfrac{1}{2}, 3$ **b.** 2, 3, 4 **c.** $-2, 1, 2, 3$
14. For example, given $P(x) = x^3 - x^2 - 34x - 56$ and $P(7) = 0 \Rightarrow (x - 7)$ is a factor and 7 is a factor of 56.
15. $4x^2 - 12x + 9; x = -\dfrac{1}{2}, \dfrac{7}{2}$

16. a. Area $= \left(\dfrac{1}{2}\pi + 10\right)x^2 + (\pi + 10)x + \dfrac{\pi}{2}$

b. Area $= \left(\dfrac{1}{2}\pi + 10\right)x^2 + (\pi + 10)x + \dfrac{\pi}{2}$

c. Perimeter $= (12 + \pi)x + (2 + \pi)$

17. a. $(x + 4)^3$

b. $6(x + 4)^2$

c. $x = 2$

d. $-3, \dfrac{-3 + 3\sqrt{5}}{2}$

18. $mx + (n + mp); q + p(n + mp)$

19. $x^3 - (2 + n)x^2 + 3nx - (n^2 - n - 1)$

20. $x \geq -3$

16 Logarithms (Path)

LESSON SEQUENCE

LESSON
16.1 Overview

Why learn this?

In 1949 Willard Libby and his team at the University of Chicago made a highly significant discovery. By measuring the amount of radiation emitted by decaying carbon particles, Libby devised a method for determining the age of organic materials. The method, known as carbon dating, is based on the mathematics of exponential functions, because the radioactive isotope carbon-14 undergoes exponential decay.

Carbon is present in the tissues of living organisms and is replenished through natural means throughout the life of the organism. However, this replenishment ceases at death and the carbon starts to decay.

A mass spectrometer can measure the radiation emitted by the decaying carbon remaining in the dead organism. Once this data is substituted into the exponential model, the date of death can be estimated for samples up to 60 000 years old. As the half-life of carbon-14 is 5730 years, carbon dating is less reliable over very long time periods.

The ability to date carbon was revolutionary and resulted in changing several historical dates and theories. For example, the discovery of fossil remains in 1969 and 1974 at Lake Mungo, a World Heritage site in New South Wales, radically altered the estimates of the length of time Indigenous Australians have inhabited this country. Logarithms are not only used in carbon dating, they also appear when considering the exponential growth of bacteria, or to measure pH levels, or noise levels, or to model asset returns in finance.

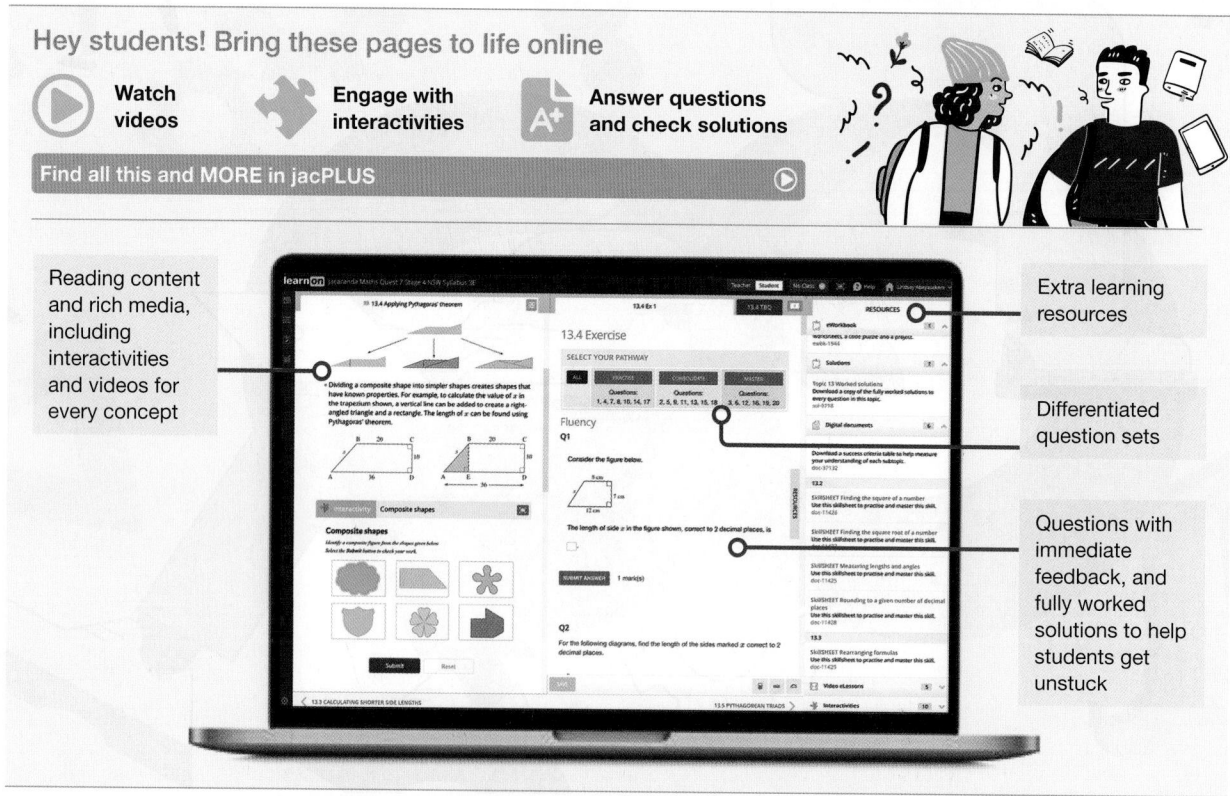

Exercise 16.1 Pre-test

1. Write the equation $5^2 = 25$ in logarithmic form.

2. Write the equation $\sqrt{9} = 3$ in logarithmic form.

3. Write the equation $\log_2(16) = 4$ in index form.

4. True or false? The statement $p = q^w$ is equivalent to $w = \log_q(p)$.

5. Evaluate: $\log_2(128)$

6. Evaluate: $\log_{36}(6)$

7. Determine the value of x if $\log_x(64) = 3$

8. Determine the value of x if $\log_6(x) = -2$

9. Solve the following equation for x: $x = \log_{\frac{1}{4}}(16)$

10. Simplify the following expression: $\log_6(2) + \log_6(3)$.

11. Simplify the following expression: $\log_6(648) - \log_6(3)$

12. Simplify the following expression: $\log_2\left(\dfrac{1}{4}\right) + \log_2(32) - \log_2(8)$

13. True or false? The logarithmic curve is asymptotic to the x-axis.

14. **MC** Choose the correct value for x in the equation: $3 + \log_2(3) = \log_2(x)$
    A. $x = 0$
    B. $x = 3$
    C. $x = 9$
    D. $x = 24$

15. **MC** The equation $y = 5^{3x}$ is equivalent to:
    A. $x = \dfrac{1}{3}\log_5(y)$
    B. $x = \log_5\left(y^{\frac{1}{3}}\right)$
    C. $x = \log_5(y) - 3$
    D. $x = 5^{(y-3)}$

LESSON
16.2 Logarithms

LEARNING INTENTION

At the end of this lesson you should be able to:
- represent expressions given in index form as logarithms and vice versa
- evaluate logarithms and use logarithms in scale measurement.

▶ 16.2.1 Logarithms

eles-4676

- The index, power or exponent in the statement $y = a^x$ is also known as a **logarithm** (or log for short).

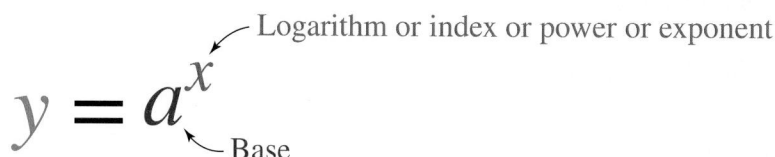

$$y = a^x \qquad \text{Logarithm or index or power or exponent}$$
$$\text{Base}$$

- This statement $y = a^x$, where $a > 0$ and $a \neq 1$, can be written in an alternative form as $\log_a y = x$, which is read as 'the logarithm of y to the base a is equal to x'. These two statements are equivalent.

Index form		Logarithmic form
$a^x = y$ $(a > 0,\ a \neq 1)$	$\Leftrightarrow$	$\log_a (y) = x$ $y > 0$

For example, $3^2 = 9$ can be written as $\log_3(9) = 2$. The log form would be read as 'the logarithm of 9, to the base of 3, is 2'. In both forms, the base is 3 and the logarithm is 2.
- The logarithm of a number to any positive base a is the index to which a is raised to give this number.
- Logarithms take in large numbers and output small numbers (powers).

WORKED EXAMPLE 1 Converting to logarithmic form

Write the following in logarithmic form.
a. $10^4 = 10\,000$ **b.** $6^x = 216$

THINK	WRITE
a. 1. Write the given statement.	**a.** $10^4 = 10\,000$
2. Identify the base (10) and the logarithm (4) and write the equivalent statement in logarithmic form. (Use $a^x = y \Leftrightarrow \log_a y = x$, where the base is a and the log is x.)	$\log_{10}(10\,000) = 4$
b. 1. Write the given statement.	**b.** $6^x = 216$
2. Identify the base (6) and the logarithm (x) and write the equivalent statement in logarithmic form.	$\log_6(216) = x$

WORKED EXAMPLE 2 Converting to index form

Write the following in index form.

a. $\log_2(8) = 3$

b. $\log_{25}(5) = \dfrac{1}{2}$

THINK	WRITE
a. 1. Write the statement.	a. $\log_2(8) = 3$
2. Identify the base (2) and the log (3), and write the equivalent statement in index form. Remember that the log is the same as the index.	$2^3 = 8$
b. 1. Write the statement.	b. $\log_{25}(5) = \dfrac{1}{2}$
2. Identify the base (25) and the log $\left(\dfrac{1}{2}\right)$, and write the equivalent statement in index form.	$25^{\frac{1}{2}} = 5$

- In the previous examples, we found that:

$$\log_2(8) = 3 \Leftrightarrow 2^3 = 8 \quad \text{and} \quad \log_{10}(10\,000) = 4 \Leftrightarrow 10^4 = 10\,000.$$

We could also write $\log_2(8) = 3$ as $\log_2(2^3) = 3$ and $\log_{10}(10\,000) = 4$ as $\log_2\left(10^4\right) = 4$.

- Can this pattern be used to work out the value of $\log_3(81)$? We need to find the power when the base of 3 is raised to that power to give 81.
- A negative number cannot be expressed as a power of a positive base, thus the logarithm of a negative number is undefined.

WORKED EXAMPLE 3 Evaluating a logarithm

Evaluate $\log_3(81)$.

THINK	WRITE
1. Write the log expression.	$\log_3(81)$
2. Express 81 in index form with a base of 3.	$= \log_3(3^4)$
3. Write the value of the logarithm.	$= 4$

16.2.2 Using logarithmic scales in measurement

eles-4677

- Logarithms can also be used to display data sets that cover a range of values which vary greatly in size.

  For example, when measuring the amplitude of earthquake waves, some earthquakes will have amplitudes of around 10 000, whereas other earthquakes may have amplitudes of around 10 000 000 (1000 times greater).
- Rather than trying to display this data on a linear scale, we can take the logarithm of the amplitude, which gives us the magnitude of each earthquake.

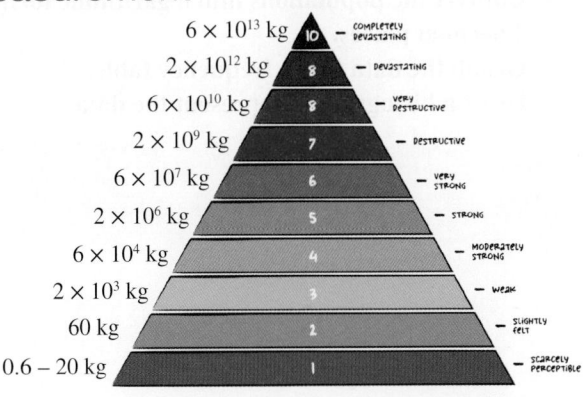

- The Richter scale uses the magnitudes of earthquakes to display the difference in their power. The illustration shown displays the equivalent energy released, in kg of TNT, by Earthquakes of different magnitudes.
- The logarithm that is used in these scales is the logarithm with base 10, which means that an increase by 1 on the scale, is an increase of 10 in the actual value.
- The logarithm with base 10 is often written simply as $\log(x)$, with the base omitted.

WORKED EXAMPLE 4 Real world application of logarithms

Convert the following amplitudes of earthquakes into values on the Richter scale, correct to 1 decimal place.
a. 1989 Newcastle earthquake: amplitude 398 000
b. 2010 Canterbury earthquake: amplitude 12 600 000
c. 2010 Chile earthquake: amplitude 631 000 000

THINK	WRITE
a. Use a calculator to calculate the logarithmic value of the amplitude. Round the answer to 1 decimal place. Write the answer in words.	a. $\log(398\ 000) = 5.599\ldots$ $\qquad\qquad\qquad = 5.6$ The 1989 Newcastle earthquake rated 5.6 on the Richter scale.
b. Use a calculator to calculate the logarithmic value of the amplitude. Round the answer to 1 decimal place. Write the answer in words.	b. $\log(12\ 600\ 000) = 7.100\ldots$ $\qquad\qquad\qquad\ = 7.1$ The 2010 Canterbury earthquake rated 7.1 on the Richter scale.
c. Use a calculator to calculate the logarithmic value of the amplitude. Round the answer to 1 decimal place. Write the answer in words.	c. $\log(631\ 000\ 000) = 8.800\ldots$ $\qquad\qquad\qquad\ = 8.8$ The 2010 Chile earthquake rated 8.8 on the Richter scale.

Displaying logarithmic data in histograms

- If we are given a data set in which the data vary greatly in size, we can use logarithms to transform the data into more manageable figures, and then group the data into intervals to provide an indication of the spread of the data.

WORKED EXAMPLE 5 Creating a histogram with a log scale

The following table displays the population of 10 different towns and cities in Victoria (using data from the 2011 census).
a. Convert the populations into logarithmic form, correct to 2 decimal places.
b. Group the data into a frequency table.
c. Draw a histogram to represent the data.

Town or city	Population
Benalla	9328
Bendigo	76 051
Castlemaine	9124
Echuca	12 613
Geelong	143 921
Kilmore	6 142
Melbourne	3 707 530
Stawell	5734
Wangaratta	17 377
Warrnambool	29 284

THINK

a. Use a calculator to calculate the logarithmic values of all of the populations. Round the answers to 2 decimal places.

WRITE

a.

Town or city	log(population)
Benalla	3.97
Bendigo	4.88
Castlemaine	3.96
Echuca	4.10
Geelong	5.16
Kilmore	3.79
Melbourne	6.57
Stawell	3.76
Wangaratta	4.24
Warrnambool	4.67

b. Group the logarithmic values into class intervals and create a frequency table.

b.

log(population)	Frequency
3 – < 4	4
4 – < 5	4
5 – < 6	1
6 – < 7	1

c. Construct a histogram of the data set.

c.

DISCUSSION

Observe the following four circles. In both rows, the one on the right contains 10 more dots than the one on the left. Is this fact reflected on how you perceive the difference between the two circles on each row? Is it more difficult to perceive the relative difference between the left-hand side circle and the right-hand side circle in the bottom row?

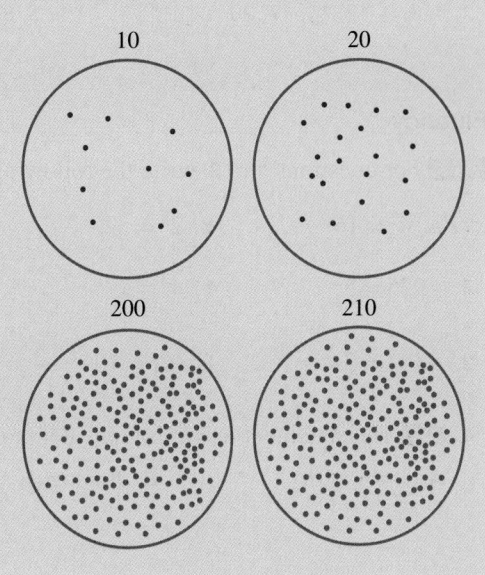

- Logarithmic scales are useful in human perception, where the relationship between a stimulus and our perception of it is logarithmic: a large increase of the impulse leads to a small perceived increase.
- For instance, the decibel scale is used to express the relative difference of power intensity between acoustic signals. 0 decibel (0 dB) corresponds to the power intensity of the lower sound perceived by the human ear, a sound 100 dB has 10^{10} times that power intensity.
- The brightness of stars, the acidity levels (pH scale) are other examples of logarithmic scales. A solution with a pH of 2.3 for instance (as in your stomach) is 100 000 times more acidic than a solution of pH 7.3 (as in your blood).

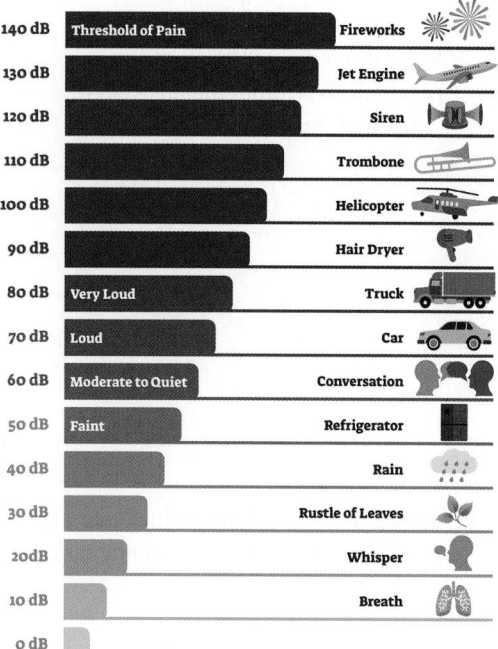

DECIBEL SCALE

140 dB	Threshold of Pain	Fireworks
130 dB		Jet Engine
120 dB		Siren
110 dB		Trombone
100 dB		Helicopter
90 dB		Hair Dryer
80 dB	Very Loud	Truck
70 dB	Loud	Car
60 dB	Moderate to Quiet	Conversation
50 dB	Faint	Refrigerator
40 dB		Rain
30 dB		Rustle of Leaves
20dB		Whisper
10 dB		Breath
0 dB		

 Resources

Interactivity Logarithms (int-6194)

Exercise 16.2 Logarithms

learn**on**

16.2 Quick quiz **on**	16.2 Exercise

Individual pathways

■ PRACTISE	■ CONSOLIDATE	■ MASTER
1, 4, 5, 9, 12, 14, 17, 20	2, 6, 8, 10, 13, 15, 18, 21	3, 7, 11, 16, 19, 22

Fluency

WE1 For questions **1** to **3**, write the following in logarithmic form.

1. **a.** $4^2 = 16$ **b.** $2^5 = 32$ **c.** $3^4 = 81$ **d.** $6^2 = 36$ **e.** $1000 = 10^3$

2. **a.** $25 = 5^2$ **b.** $4^3 = x$ **c.** $5^x = 125$ **d.** $7^x = 49$ **e.** $p^4 = 16$

3. **a.** $9^{\frac{1}{2}} = 3$ **b.** $0.1 = 10^{-1}$ **c.** $2 = 8^{\frac{1}{3}}$ **d.** $2^{-1} = \dfrac{1}{2}$ **e.** $4^{\frac{3}{2}} = 8$

4. **MC** The statement $w = h^t$ is equivalent to:

    A. $w = \log_t(h)$ **B.** $h = \log_t(w)$ **C.** $t = \log_w(h)$ **D.** $t = \log_h(w)$

For questions **5** to **7**, write the following in index form.

5. a. $\log_2(16) = 4$ b. $\log_3(27) = 3$ c. $\log_{10}(1\,000\,000) = 6$ d. $\log_5(125) = 3$

6. a. $\log_{16}(4) = \dfrac{1}{2}$ b. $\log_4(64) = x$ c. $\dfrac{1}{2} = \log_{49}(7)$ d. $\log_3(x) = 5$

7. a. $\log_{81}(9) = \dfrac{1}{2}$ b. $\log_{10}(0.01) = -2$ c. $\log_8(8) = 1$ d. $\log_{64}(4) = \dfrac{1}{3}$

8. **MC** The statement $q = \log_r(p)$ is equivalent to:

    A. $q = r^p$ **B.** $p = r^q$ **C.** $r = p^q$ **D.** $r = q^p$

WE3 For questions **9** to **11**, evaluate the following logarithms.

9. a. $\log_2(16)$ b. $\log_4(16)$ c. $\log_{11}(121)$ d. $\log_{10}(100\,000)$

10. a. $\log_3(243)$ b. $\log_2(128)$ c. $\log_5(1)$ d. $\log_9(3)$

11. a. $\log_3\left(\dfrac{1}{3}\right)$ b. $\log_6(6)$ c. $\log_{10}\left(\dfrac{1}{100}\right)$ d. $\log_{125}(5)$

12. Write the value of each of the following.
    a. $\log_{10}(1)$ b. $\log_{10}(10)$ c. $\log_{10}(100)$

13. Write the value of each of the following.
    a. $\log_{10}(1000)$ b. $\log_{10}(10\,000)$ c. $\log_{10}(100\,000)$

Understanding

14. Use your results to question **12** and **13** to answer the following.
    a. Between which two whole numbers would $\log_{10}(7)$ lie?
    b. Between which two whole numbers would $\log_{10}(4600)$ lie?
    c. Between which two whole numbers would $\log_{10}(85)$ lie?

15. a. Between which two whole numbers would $\log_{10}(12\,750)$ lie?
    b. Between which two whole numbers would $\log_{10}(110)$ lie?
    c. Between which two whole numbers would $\log_{10}(81\,000)$ lie?

16. **WE4** Convert the following amplitudes of earthquakes into values on the Richter scale, correct to 1 decimal place.

    a. 2016 Northern Territory earthquake: amplitude $1\,260\,000$.
    b. 2011 Christchurch earthquake: amplitude $2\,000\,000$.
    c. 1979 Tumaco earthquake: amplitude $158\,000\,000$.

Communicating, reasoning and problem solving

17. a. If $\log_{10}(g) = k$, determine the value of $\log_{10}(g^2)$. Justify your answer.
    b. If $\log_x(y) = 2$, determine the value of $\log_y(x)$. Justify your answer.
    c. By referring to the equivalent index statement, explain why x must be a positive number given $\log_4(x) = y$, for all values of y.

18. Calculate each of the following logarithms.

    a. $\log_2(64)$ b. $\log_3\left(\dfrac{1}{81}\right)$ c. $\log_{10}(0.00001)$

19. Calculate each of the following logarithms.

    a. $\log_3(243)$ **b.** $\log_4\left(\dfrac{1}{64}\right)$ **c.** $\log_5\left(\sqrt{125}\right)$

20. For each of the following, determine the value of x.

    a. $\log_x\left(\dfrac{1}{243}\right) = -5$ **b.** $\log_x(343) = 3$ **c.** $\log_{64}(x) = -\dfrac{1}{2}$

21. Simplify $10^{\log_{10}(x)}$.

22. Simplify the expression $3^{2-\log_3(x)}$.

LESSON
16.3 Logarithm laws

LEARNING INTENTION

At the end of this lesson you should be able to:
- apply the laws of logarithms to evaluate and simplify expressions.

▶ 16.3.1 Logarithm laws

eles-6268

- Recall the index laws:

First index law: $a^m \times a^n = a^{m+n}$	Second index law: $\dfrac{a^m}{a^n} = a^{m-n}$
Third index law: $a^0 = 1$	Fourth index law: $(a^m)^n = a^{mn}$
Fifth index law: $(ab)^m = a^m b^m$	Sixth index law: $\left(\dfrac{a}{b}\right)^m = \dfrac{a^m}{b^m}$
Seventh index law: $a^{-m} = \dfrac{1}{a^m}$ where $a \neq 0$	Eighth index law: $a^{\frac{m}{n}} = \sqrt[n]{a^m}$

- The index laws can be used to produce a set of equivalent logarithm laws and logarithmic results.

Laws of logarithms

- If $x = a^m$ and $y = a^n$, then $\log_a(x) = m$ and $\log_a(y) = n$ (equivalent log form).

    Now $xy = a^m \times a^n$

    or $xy = a^{m+n}$ (First Index Law).

    So $\log_a(xy) = m + n$ (equivalent log form)

    or $\log_a(xy) = \log_a(x) + \log_a(y)$ (substituting for m and n).

Logarithm Law 1

$$\log_a(x) + \log_a(y) = \log_a(xy)$$

- This means that the sum of two logarithms with the same base is equal to the logarithm of the product of the numbers.

WORKED EXAMPLE 6 Adding logarithms

Evaluate $\log_{10}(20) + \log_{10}(5)$.

THINK	WRITE
1. Since the same base of 10 used in each log term, use $\log_a(x) + \log_a(y) = \log_a(xy)$ and simplify.	$\log_{10}(20) + \log_{10}(5) = \log_{10}(20 \times 5)$ $= \log_{10}(100)$
2. Evaluate. (Remember that $100 = 10^2$.)	$= 2$

- If $x = a^m$ and $y = a^n$, then $\log_a(x) = m$ and $\log_a(y) = n$ (equivalent log form).

Now $$\frac{x}{y} = \frac{a^m}{a^n}$$

or $$\frac{x}{y} = a^{m-n}$$ (Second Index Law).

So $$\log_a\left(\frac{x}{y}\right) = m - n$$ (equivalent log form)

or $$\log_a\left(\frac{x}{y}\right) = \log_a(x) - \log_a(y)$$ (substituting for m and n).

Logarithm Law 2

$$\log_a(x) - \log_a(y) = \log_a\left(\frac{x}{y}\right)$$

- This means that the difference of two logarithms with the same base is equal to the logarithm of the quotient of the numbers.

WORKED EXAMPLE 7 Subtracting logarithms

Evaluate $\log_4(20) - \log_4(5)$.

THINK	WRITE
1. Since the same base of 4 is used in each log term, use $\log_a(x) - \log_a(y) = \log_a\left(\frac{x}{y}\right)$ and simplify.	$\log_4(20) - \log_4(5) = \log_4\left(\frac{20}{5}\right)$ $= \log_4(4)$
2. Evaluate. (Remember that $4 = 4^1$.)	$= 1$

WORKED EXAMPLE 8 Simplifying multiple logarithm terms

Evaluate $\log_5(35) + \log_5(15) - \log_5(21)$.

THINK	WRITE
1. Since the first two log terms are being added, use $\log_a(x) + \log_a(y) = \log_a(xy)$ and simplify.	$\log_5(35) + \log_5(15) - \log_5(21)$ $= \log_5(35 \times 15) - \log_5(21)$ $= \log_5(525) - \log_5(21)$
2. To calculate the difference between the two remaining log terms, use $\log_a(x) - \log_a(y) = \log_a\left(\dfrac{x}{y}\right)$ and simplify.	$= \log_5\left(\dfrac{525}{21}\right)$ $= \log_5(25)$
3. Evaluate. (Remember that $25 = 5^2$.)	$= 2$

- Once you have gained confidence in using the first two laws, you can reduce the number of steps of working by combining the application of the laws. In Worked example 8, we could write:

$$\log_5(35) + \log_5(15) - \log_5(21) = \log_5\left(\frac{35 \times 15}{21}\right)$$
$$= \log_5(25)$$
$$= 2$$

- If $x = a^m$, then $\log_a(x) = m$ (equivalent log form).

Now	$x^n = (a^m)^n$	
or	$x^n = a^{mn}$	(Fourth Index Law)
So	$\log_a(x^n) = mn$	(equivalent log form)
or	$\log_a(x^n) = \left(\log_a(x)\right) \times n$	(substituting for m)
or	$\log_a(x^n) = n\log_a(x)$	

Logarithm Law 3

$$\log_a(x^n) = n\log_a(x)$$

- This means that the logarithm of a number raised to a power is equal to the product of the power and the logarithm of the number.

WORKED EXAMPLE 9 Simplifying a logarithm of a number raised to a power

Evaluate $2\log_6(3) + \log_6(4)$.

THINK	WRITE
1. The first log term is not in the required form to use the log law relating to sums. Use $\log_a(x^n) = n\log_a(x)$ to rewrite the first term in preparation for applying the first log law.	$2\log_6(3) + \log_6(4) = \log_6\left(3^2\right) + \log_6(4)$ $= \log_6(9) + \log_6(4)$
2. Use $\log_a(x) + \log_a(y) = \log_a(xy)$ to simplify the two log terms to one.	$= \log_6(9 \times 4)$ $= \log_6(36)$
3. Evaluate. (Remember that $36 = 6^2$.)	$= 2$

⏵ 16.3.2 Logarithmic results

eles-6269

- As

$$a^0 = 1 \qquad \text{(Third Index Law)}$$
$$\log_a(1) = 0 \qquad \text{(equivalent log form)}$$

Logarithmic result 1

$$\log_a(1) = 0$$

- This means that the logarithm of 1 with any base is equal to 0.

- As

$$a^1 = a$$
$$\log_a(a) = 1 \qquad \text{(equivalent log form)}$$

Logarithmic result 2

$$\log_a(a) = 1$$

- This means that the logarithm of any number a with base a is equal to 1.

- Now $\qquad \log_a\left(\dfrac{1}{x}\right) = \log_a\left(x^{-1}\right) \qquad \text{(Seventh Index law)}$

 or $\qquad \log_a\left(\dfrac{1}{x}\right) = -1 \times \log_a(x) \qquad \text{(using the fourth log law)}$

 or $\qquad \log_a\left(\dfrac{1}{x}\right) = -\log_a(x).$

Logarithmic result 3

$$\log_a\left(\dfrac{1}{x}\right) = -\log_a(x)$$

- Now $\qquad \log_a(a^x) = x\log_a(a) \qquad \text{(using the third log law)}$

 or $\qquad \log_a(a^x) = x \times 1 \qquad \text{(using the fifth log law)}$

 or $\qquad \log_a(a^x) = x.$

Logarithmic result 4

$$\log_a(a^x) = x$$

- Now

$$a^x = b \iff x = \log_a(b)$$ (definition of log)

or $$\log_c(a^x) = \log_c(b)$$ (taking logs of both sides)

or $$x\log_c(a) = \log_c(b)$$ (using the third log law)

$$x = \frac{\log_c(b)}{\log_c(a)}$$

$$\log_a(b) = \frac{\log_c(b)}{\log_c(a)}$$

Logarithmic result 5 (change of base)

$$\log_a(b) = \frac{\log_c(b)}{\log_c(a)}$$

WORKED EXAMPLE 10 Applying the laws of logarithms to evaluate and simplify expressions

Evaluate $\log_a(18)$ given $\log_a(2) = 0.431$ and $\log_a(3) = 0.683$.

THINK	WRITE
1. Express 18 as the product or power of 2 and 3.	$18 = 2 \times 3^2$
2. Use the logarithmic law 1.	$\log_a(xy) = \log_a(x) + \log_a(y)$ $\log_a(18) = \log_a(2 \times 3^2)$ $\log_a(18) = \log_a(2) + \log_a(3^2)$
3. Use the logarithmic law 3.	$\log_a(x^n) = n\log_a(x)$ $\log_a(3^2) = 2\log_a(3)$
4. Substitute the values given for $\log_a(2)$ and $\log_a(3)$.	$\log_a(18) = \log_a(2) + 2\log_a(3)$ $\log_a(18) = 0.431 + 2 \times 0.683$
5. Write the answer.	$\log_a(18) = 1.797$

 Resources

Interactivities The first law of logarithms (int-6195)

The second law of logarithms (int-6196)

The third law of logarithms (int-6197)

The fourth law of logarithms (int-6198)

The fifth law of logarithms (int-6199)

The sixth law of logarithms (int-6200)

The seventh law of logarithms (int-6201)

16.3 Quick quiz on	16.3 Exercise

Individual pathways

■ PRACTISE	■ CONSOLIDATE	■ MASTER
1, 2, 3, 6, 9, 12, 13, 15, 18, 19, 23, 27, 30	4, 7, 10, 14, 16, 20, 22, 24, 28, 31	5, 8, 11, 17, 21, 25, 29, 32

Fluency

1. Use a calculator to evaluate the following, correct to 5 decimal places.

   a. $\log_{10}(50)$
   b. $\log_{10}(25)$
   c. $\log_{10}(5)$
   d. $\log_{10}(2)$

2. Use your answers to question 1 to show that each of the following statements is true.

   a. $\log_{10}(25) + \log_{10}(2) = \log_{10}(50)$
   b. $\log_{10}(50) - \log_{10}(2) = \log_{10}(25)$
   c. $\log_{10}(25) = 2\log_{10}(5)$
   d. $\log_{10}(50) - \log_{10}(25) - \log_{10}(2) = \log_{10}(1)$

WE6 For questions 3 to 5, evaluate the following.

3. a. $\log_6(3) + \log_6(2)$
   b. $\log_4(8) + \log_4(8)$

4. a. $\log_{10}(25) + \log_{10}(4)$
   b. $\log_8(32) + \log_8(16)$

5. a. $\log_6(108) + \log_6(12)$
   b. $\log_{14}(2) + \log_{14}(7)$

WE7 For questions 6 to 8, evaluate the following.

6. a. $\log_2(20) - \log_2(5)$
   b. $\log_3(54) - \log_3(2)$

7. a. $\log_4(24) - \log_4(6)$
   b. $\log_{10}(30\,000) - \log_{10}(3)$

8. a. $\log_6(648) - \log_6(3)$
   b. $\log_2(224) - \log_2(7)$

WE8 For questions 9 to 11, evaluate the following.

9. a. $\log_3(27) + \log_3(2) - \log_3(6)$
   b. $\log_4(24) - \log_4(2) - \log_4(6)$

10. a. $\log_6(78) - \log_6(13) + \log_6(1)$
    b. $\log_2(120) - \log_2(3) - \log_2(5)$

11. a. $\log_7(15) + \log_7(3) - \log_7(315)$
    b. $\log_9(80) - \log_9(8) - \log_9(30)$

12. Evaluate $2\log_4(8)$.

WE9 For questions 13 to 17, evaluate the following.

13. a. $2\log_{10}(5) + \log_{10}(4)$
    b. $\log_3(648) - 3\log_3(2)$

14. a. $4\log_5(10) - \log_5(80)$
    b. $\log_2(50) + \dfrac{1}{2}\log_2(16) - 2\log_2(5)$

15. a. $\log_8(8)$
    b. $\log_5(1)$
    c. $\log_2\left(\dfrac{1}{2}\right)$
    d. $\log_4\left(4^5\right)$

16. **a.** $\log_6\left(6^{-2}\right)$ **b.** $\log_{20}(20)$

    c. $\log_2(1)$ **d.** $\log_3\left(\dfrac{1}{9}\right)$

17. **a.** $\log_4\left(\dfrac{1}{2}\right)$ **b.** $\log_5\left(\sqrt{5}\right)$

    c. $\log_3\left(\dfrac{1}{\sqrt{3}}\right)$ **d.** $\log_2\left(8\sqrt{2}\right)$

18. **WE10** Evaluate $\log_a(200)$ given $\log_a(2) = 0.356$ and $\log_a(5) = 0.827$.

Understanding

For questions **19** to **21**, use the logarithm laws to simplify each of the following.

19. **a.** $\log_a(5) + \log_a(8)$ **b.** $\log_a(12) + \log_a(3) - \log_a(2)$

    c. $4\log_x(2) + \log_x(3)$ **d.** $\log_x(100) - 2\log_x(5)$

20. **a.** $3\log_a(x) - \log_a\left(x^2\right)$ **b.** $5\log_a(a) - \log_a\left(a^4\right)$

    c. $\log_x(6) - \log_x(6x)$ **d.** $\log_a\left(a^7\right) + \log_a(1)$

21. **a.** $\log_p\left(\sqrt{p}\right)$ **b.** $\log_k\left(k\sqrt{k}\right)$

    c. $6\log_a\left(\dfrac{1}{a}\right)$ **d.** $\log_a\left(\dfrac{1}{\sqrt[3]{a}}\right)$

22. **MC** *Note:* There may be more than one correct answer.

    a. The equation $y = 10^x$ is equivalent to:

        A. $x = 10^y$ **B.** $x = \log_{10}(y)$

        C. $x = \log_x(10)$ **D.** $x = \log_y(10)$

    b. The equation $y = 10^{4x}$ is equivalent to:

        A. $x = \log_{10}\left(\sqrt{4y}\right)$ **B.** $x = \log_{10}\left(\sqrt[4]{y}\right)$

        C. $x = 10^{\frac{1}{4}y}$ **D.** $x = \dfrac{1}{4}\log_{10}(y)$

    c. The equation $y = 10^{3x}$ is equivalent to:

        A. $x = \dfrac{1}{3}\log_{10}(y)$ **B.** $x = \log_{10}\left(y^{\frac{1}{3}}\right)$

        C. $x = \log_{10}(y) - 3$ **D.** $x = 10^{y-3}$

    d. The equation $y = ma^{nx}$ is equivalent to:

        A. $x = \dfrac{1}{n}a^{my}$ **B.** $x = \log_a\left(\dfrac{m}{y}\right)^n$

        C. $x = \dfrac{1}{n}\left(\log_a(y) - \log_a(m)\right)$ **D.** $x = \dfrac{1}{n}\log_a\left(\dfrac{y}{m}\right)$

For questions **23** to **25**, simplify, and evaluate where possible, each of the following without a calculator.

23. a. $\log_2(8) + \log_2(10)$ **b.** $\log_3(7) + \log_3(15)$

   c. $\log_{10}(20) + \log_{10}(5)$ **d.** $\log_6(8) + \log_6(7)$

24. a. $\log_2(20) - \log_2(5)$ **b.** $\log_3(36) - \log_3(12)$

   c. $\log_5(100) - \log_5(8)$ **d.** $\log_2\left(\dfrac{1}{3}\right) + \log_2(9)$

25. a. $\log_4(25) + \log_4\left(\dfrac{1}{5}\right)$ **b.** $\log_{10}(5) - \log_{10}(20)$ **c.** $\log_3\left(\dfrac{4}{5}\right) - \log_3\left(\dfrac{1}{5}\right)$

   d. $\log_2(9) + \log_2(4) - \log_2(12)$ **e.** $\log_3(8) - \log_3(2) + \log_3(5)$ **f.** $\log_4(24) - \log_4(2) - \log_4(6)$

26. **MC** **a.** The expression $\log_{10}(xy)$ is equal to:

   A. $\log_{10}(x) \times \log_{10}(y)$ **B.** $\log_{10}(x) - \log_{10}(y)$
   C. $\log_{10}(x) + \log_{10}(y)$ **D.** $y \log_{10}(x)$

   b. The expression $\log_{10}(x^y)$ is equal to:

   A. $x \log_{10}(y)$ **B.** $y \log_{10}(x)$
   C. $10 \log_x(y)$ **D.** $\log_{10}(x) + \log_{10}(y)$

   c. The expression $\dfrac{1}{3} \log_2(64) + \log_2(10)$ is equal to:

   A. $\log_2(40)$ **B.** $\log_2(80)$
   C. $\log_2\left(\dfrac{64}{10}\right)$ **D.** 1

Communicating, reasoning and problem solving

27. For each of the following, write the possible strategy you intend to use.

   a. Evaluate $\left(\log_3(81)\right)\left(\log_3(27)\right)$.

   b. Evaluate $\dfrac{\log_a(81)}{\log_a(3)}$.

   c. Evaluate $5^{\log_5(7)}$.

   In each case, explain how you obtained your final answer.

28. Simplify $\log_5(10) + 2\log_5(2) - 3\log_5(10)$.

29. Simplify $\log_2\left(\dfrac{8}{125}\right) - 3\log_2\left(\dfrac{3}{5}\right) - 4\log_2\left(\dfrac{1}{2}\right)$.

30. Simplify $\log_a\left(a^5 + a^3\right) - \log_a\left(a^4 + a^2\right)$.

31. If $2\log_a(x) = 1 + \log_a(8x - 15a)$, determine the value of x in terms of a where a is a positive constant and x is positive.

32. Solve the following for x:

$$\log_3(x + 2) + \log_3(x - 4) = 3$$

LESSON
16.4 Solving equations

LEARNING INTENTION

At the end of this lesson you should be able to:
- simplify and solve equations involving logarithms using the logarithm laws and index laws.

▶ 16.4.1 Solving equations with logarithms

eles-4679
- The equation $\log_a(y) = x$ is an example of a general **logarithmic equation**.
- Laws of logarithms and indices are used to solve these equations.

WORKED EXAMPLE 11 Solving by converting to index form

Solve for x in the following equations.

a. $\log_2(x) = 3$ **b.** $\log_6(x) = -2$ **c.** $\log_3(x^4) = -16$ **d.** $\log_5(x-1) = 2$

THINK	WRITE
a. 1. Write the equation.	**a.** $\log_2(x) = 3$
2. Rewrite using $a^x = y \Leftrightarrow \log_a(y) = x$.	$2^3 = x$
3. Rearrange and simplify.	$x = 8$
b. 1. Write the equation.	**b.** $\log_6(x) = -2$
2. Rewrite using $a^x = y \Leftrightarrow \log_a(y) = x$.	$6^{-2} = x$
3. Rearrange and simplify.	$x = \dfrac{1}{6^2}$ $= \dfrac{1}{36}$
c. 1. Write the equation.	**c.** $\log_3\left(x^4\right) = -16$
2. Rewrite using $\log_a(x^n) = n\log_a(x)$.	$4\log_3(x) = -16$
3. Divide both sides by 4.	$\log_3(x) = -4$
4. Rewrite using $a^x = y \Leftrightarrow \log_a(y) = x$.	$3^{-4} = x$
5. Rearrange and simplify.	$x = \dfrac{1}{3^4}$ $= \dfrac{1}{81}$

---

d. 1. Write the equation.

$$\textbf{d. } \log_5(x-1) = 2$$

2. Rewrite using $a^x = y \Leftrightarrow \log_a(y) = x$.

$$5^2 = x - 1$$

3. Solve for x.

$$x - 1 = 25$$
$$x = 26$$

WORKED EXAMPLE 12 Solving for the base of a logarithm

Solve for x in $\log_x(25) = 2$, given that $x > 0$.

THINK	WRITE
1. Write the equation.	$\log_x(25) = 2$
2. Rewrite using $a^x = y \Leftrightarrow \log_a(y) = x$.	$x^2 = 25$
3. Solve for x. *Note:* $x = -5$ is rejected as a solution because $x > 0$.	$x = 5$ (because $x > 0$)

WORKED EXAMPLE 13 Evaluating logarithms

Solve for x in the following.

a. $\log_2(16) = x$ **b.** $\log_3\left(\dfrac{1}{3}\right) = x$ **c.** $\log_9(3) = x$

THINK	WRITE
a. 1. Write the equation.	**a.** $\log_2(16) = x$
2. Rewrite using $a^x = y \Leftrightarrow \log_a(y) = x$.	$2^x = 16$
3. Write 16 with base 2.	$= 2^4$
4. Equate the indices.	$x = 4$
b. 1. Write the equation.	**b.** $\log_3\left(\dfrac{1}{3}\right) = x$
2. Rewrite using $a^x = y \Leftrightarrow \log_a(y) = x$.	$3^x = \dfrac{1}{3}$ $= \dfrac{1}{3^1}$
3. Write $\dfrac{1}{3}$ with base 3.	$3^x = 3^{-1}$
4. Equate the indices.	$x = -1$
c. 1. Write the equation.	**c.** $\log_9(3) = x$
2. Rewrite using $a^x = y \Leftrightarrow \log_a(y) = x$.	$9^x = 3$
3. Write 9 with base 3.	$\left(3^2\right)^x = 3$

4. Remove the grouping symbols.	$3^{2x} = 3^1$
5. Equate the indices.	$2x = 1$
6. Solve for x.	$x = \dfrac{1}{2}$

WORKED EXAMPLE 14 Solving equations with multiple logarithm terms

Solve for x in the equation $\log_2(4) + \log_2(x) - \log_2(8) = 3$.

THINK	WRITE
1. Write the equation.	$\log_2(4) + \log_2(x) - \log_2(8) = 3$
2. Simplify the left-hand side. Use $\log_a(x) + \log_a(y) = \log_a(xy)$ and $\log_a(x) - \log_a(y) = \log_a\left(\dfrac{x}{y}\right)$.	$\log_2\left(\dfrac{4 \times x}{8}\right) = 3$
3. Simplify.	$\log_2\left(\dfrac{x}{2}\right) = 3$
4. Rewrite using $a^x = y \Leftrightarrow \log_a(y) = x$.	$2^3 = \dfrac{x}{2}$
5. Solve for x.	$x = 2 \times 2^3$ $= 2 \times 8$ $= 16$

16.4.2 Solving exponential equations using logarithms

eles-6270

- When solving an equation such as $\log_2(8) = x$, it could be rewritten in index form as $2^x = 8$. This can be written with the same base of 2 to produce $2^x = 2^3$ with the solutions $x = 3$.
- Can we do this to solve the equation $\log_2(7) = x$ or the equivalent equation $2^x = 7$?
- Logarithm Law 8, or the Change of base law, allows us to calculate approximations to the solution.
- Your calculator has base 10, so this is the most commonly used base for this solution technique.
- Worked example 14 illustrates using the Change of base law.

WORKED EXAMPLE 15 Solving equations giving approximate solutions

Solve for x, correct to 3 decimal places, $2^x = 7$.

THINK	WRITE
1. Write the equation	$2^x = 7$
2. Convert to the equivalent log equation for the exact solution	$x = \log_2(7)$
3. Change the base to 10	$x = \dfrac{\log_{10}(7)}{\log_{10}(2)}$
4. Evaluate, using a calculator	$x = 2.80735...$
5. Write the answer correct to 3 decimal places	$x \approx 2.807$

Without using a calculator, solve for t, in exact form, $4^{t+1} = \dfrac{1}{8\sqrt{2}}$

THINK	WRITE
1. Write the equation	$4^{t+1} = \dfrac{1}{8\sqrt{2}}$
2. Convert to the equivalent log equation for the exact solution	$t+1 = \log_4\left(\dfrac{1}{8\sqrt{2}}\right)$
3. Express $8\sqrt{2}$ as a power of 2	$8\sqrt{2} = 2^3 \times 2^{\frac{1}{2}} = 2^{\frac{7}{2}}$
4. Use the laws of logarithms to simplify $\log_4\left(\dfrac{1}{8\sqrt{2}}\right)$	$\log_4\left(\dfrac{1}{8\sqrt{2}}\right) = -\log_4\left(8\sqrt{2}\right)$ $= -\log_4\left(2^{\frac{7}{2}}\right)$ $= -\dfrac{7}{2}\log_4(2)$ Thus $t+1 = -\dfrac{7}{2}\log_4(2)$ $\log_4(2) = \log_4\left(4^{\frac{1}{2}}\right)$
5. Use the logarithmic result $\log_a a^x = x$ to determine $\log_4(2)$	$\log_4(2) = \dfrac{1}{2}$
6. Substitute the exact value of $\log_4(2)$ in the equation	$t+1 = -\dfrac{7}{2}\log_4(2) = -\dfrac{7}{2} \times \dfrac{1}{2}$ $t+1 = -\dfrac{7}{4}$
7. Solve for t	$t = -\dfrac{7}{4} - 1 = -\dfrac{11}{4}$
8. Write the answer in exact form	$t = -\dfrac{11}{4}$

Digital technology

Graphing applications, such as Desmos or GeoGebra, can be used to solve equations involving exponentials or logarithms, or to check the solution(s).

For instance, Desmos is used here to check the solution to Worked example 16.

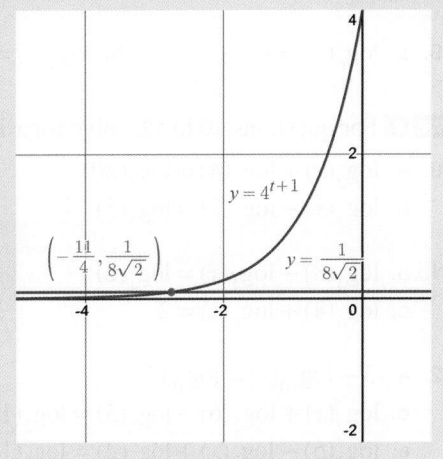

Exercise 16.4 Solving equations

learn on

16.4 Quick quiz on

16.4 Exercise

Individual pathways

■ PRACTISE	■ CONSOLIDATE	■ MASTER
1, 4, 7, 10, 14, 17, 21, 24	2, 5, 8, 11, 15, 18, 22, 25	3, 6, 9, 12, 13, 16, 19, 20, 23, 26

Fluency

WE11 For questions **1** to **3**, solve for x in the following equations.

1. **a.** $\log_5(x) = 2$ **b.** $\log_3(x) = 4$ **c.** $\log_2(x) = -3$ **d.** $\log_4(x) = -2$ **e.** $\log_{10}(x^2) = 4$

2. **a.** $\log_2(x^3) = 12$ **b.** $\log_3(x+1) = 3$ **c.** $\log_5(x-2) = 3$ **d.** $\log_4(2x-3) = 0$ **e.** $\log_{10}(2x+1) = 0$

3. **a.** $\log_2(-x) = -5$ **b.** $\log_3(-x) = -2$ **c.** $\log_5(1-x) = 4$ **d.** $\log_{10}(5-2x) = 1$

WE12 For questions **4** to **6**, solve for x in the following equations, given that $x > 0$.

4. **a.** $\log_x(9) = 2$ **b.** $\log_x(16) = 4$ **c.** $\log_x(25) = \dfrac{2}{3}$

5. **a.** $\log_x(125) = \dfrac{3}{4}$ **b.** $\log_x\left(\dfrac{1}{8}\right) = -3$ **c.** $\log_x\left(\dfrac{1}{64}\right) = -2$

6. **a.** $\log_x(6^2) = 2$ **b.** $\log_x(4^3) = 3$

WE13 For questions **7** to **9**, solve for x in the following equations.

7. **a.** $\log_2(8) = x$ **b.** $\log_3(9) = x$ **c.** $\log_5\left(\dfrac{1}{5}\right) = x$

8. **a.** $\log_4\left(\dfrac{1}{16}\right) = x$ **b.** $\log_4(2) = x$ **c.** $\log_8(2) = x$

9. **a.** $\log_6(1) = x$ **b.** $\log_8(1) = x$ **c.** $\log_{\frac{1}{2}}(2) = x$ **d.** $\log_{\frac{1}{3}}(9) = x$

WE14 For questions **10** to **12**, solve for x in the following.

10. **a.** $\log_2(x) + \log_2(4) = \log_2(20)$ **b.** $\log_5(3) + \log_5(x) = \log_5(18)$
    c. $\log_3(x) - \log_3(2) = \log_3(5)$ **d.** $\log_{10}(x) - \log_{10}(4) = \log_{10}(2)$

11. **a.** $\log_4(8) - \log_4(x) = \log_4(2)$ **b.** $\log_3(10) - \log_3(x) = \log_3(5)$
    c. $\log_6(4) + \log_6(x) = 2$ **d.** $\log_2(x) + \log_2(5) = 1$

12. **a.** $3 - \log_{10}(x) = \log_{10}(2)$ **b.** $5 - \log_4(8) = \log_4(x)$
    c. $\log_2(x) + \log_2(6) - \log_2(3) = \log_2(10)$ **d.** $\log_2(x) + \log_2(5) - \log_2(10) = \log_2(3)$
    e. $\log_3(5) - \log_3(x) + \log_3(2) = \log_3(10)$ **f.** $\log_5(4) - \log_5(x) + \log_5(3) = \log_5(6)$

13. **MC** **a.** The solution to the equation $\log_7(343) = x$ is:

    A. $x = 2$ **B.** $x = 3$ **C.** $x = 1$ **D.** $x = 0$

  b. If $\log_8(x) = 4$, then x is equal to:

    A. 4096   **B.** 512   **C.** 64   **D.** 2

  c. Given that $\log_x(3) = \dfrac{1}{2}$, x must be equal to:

    A. 3   **B.** 6   **C.** 81   **D.** 9

  d. If $\log_a(x) = 0.7$, then $\log_a\left(x^2\right)$ is equal to:

    A. 0.49   **B.** 1.4   **C.** 0.35   **D.** 0.837

Understanding

For questions **14** to **16**, solve for x in the following equations.

14. **WE15**

  a. $2^x = 128$ **b.** $3^x = 9$ **c.** $7^x = \dfrac{1}{49}$ **d.** $9^x = 1$ **e.** $5^x = 625$

15. **a.** $64^x = 8$ **b.** $6^x = \sqrt{6}$ **c.** $2^x = 2\sqrt{2}$ **d.** $3^x = \dfrac{1}{\sqrt{3}}$ **e.** $4^x = 8$

16. **a.** $9^x = 3\sqrt{3}$ **b.** $2^x = \dfrac{1}{4\sqrt{2}}$ **c.** $3^{x+1} = 27\sqrt{3}$ **d.** $2^{x-1} = \dfrac{1}{32\sqrt{2}}$ **e.** $4^{x+1} = \dfrac{1}{8\sqrt{2}}$

WE14 For questions **17–19**, solve for x in the following equations, giving your answers correct to 3 decimal places.

17. **a.** $2^x = 11$ **b.** $2^x = 0.6$ **c.** $3^x = 20$ **d.** $3^x = 1.7$

18. **a.** $5^x = 8$ **b.** $0.7^x = 3$ **c.** $0.4^x = 5$ **d.** $3^{x+2} = 12$

19. **a.** $7^{-x} = 0.2$ **b.** $8^{-x} = 0.3$ **c.** $10^{-2x} = 7$ **d.** $8^{2-x} = 0.75$

20. **WE16** Without using a calculator, solve for t, in exact form, $8^{t-1} = \dfrac{\sqrt{2}}{2}$

Communicating, reasoning and problem solving

21. The apparent brightness of stars is measured on a logarithmic scale called magnitude, in which lower numbers mean brighter stars. The relationship between the ratio of apparent brightness of two objects and the difference in their magnitudes is given by the formula:

$$m_2 - m_1 = -2.5 \log_{10}\left(\frac{b_2}{b_1}\right)$$

where m is the magnitude and b is the apparent brightness. Determine how many times brighter a magnitude 2.0 star is than a magnitude 3.0 star.

22. The decibel (dB) scale for measuring loudness, d, is given by the formula $d = 10\log_{10}(I \times 10^{12})$, where I is the intensity of sound in watts per square metre.

   a. Determine the number of decibels of sound if the intensity is 1.
   b. Evaluate the number of decibels of sound produced by a jet engine at a distance of 50 metres if the intensity is 10 watts per square metre.
   c. Determine the intensity of sound if the sound level of a pneumatic drill 10 metres away is 90 decibels.
   d. Determine how the value of d changes if the intensity is doubled. Give your answer correct to the nearest decibel.
   e. Evaluate how the value of d changes if the intensity is 10 times as great.
   f. Determine by what factor does the intensity of sound have to be multiplied in order to add 20 decibels to the sound level.

23. The Richter scale is used to describe the energy of earthquakes. A formula for the Richter scale is:
$R = \dfrac{2}{3}\log_{10}(K) - 0.9$, where R is the Richter scale value for an earthquake that releases K kilojoules (kJ) of energy.

   a. Determine the Richter scale value for an earthquake that releases the following amounts of energy:
      i. 1000 kJ   ii. 2000 kJ   iii. 3000 kJ   iv. 10 000 kJ   v. 100 000 kJ   vi. 1 000 000 kJ
   b. Does doubling the energy released double the Richter scale value? Justify your answer.
   c. Determine the energy released by an earthquake of:
      i. magnitude 4 on the Richter scale
      ii. magnitude 5 on the Richter scale
      iii. magnitude 6 on the Richter scale.
   d. Explain the effect (on the amount of energy released) of increasing the Richter scale value by 1.
   e. Explain why an earthquake measuring 8 on the Richter scale so much more devastating than one that measures 5.

24. Solve for x.
   a. $3^{x+1} = 7$
   b. $3^{x+1} = 7^x$

25. Solve the following for x.
   $(27 \times 3^x)^3 = 81^x \times 3^2$

26. Solve $\left\{x: (3^x)^2 = 30 \times 3^x - 81\right\}$.

LESSON
16.5 Graphing logarithmic and exponential functions

LEARNING INTENTION

At the end of this lesson you should be able to:
- graph the functions $y = a^x$ and $y = \log_a(x)$
- compare and contrast these two functions
- recognise if the function is increasing and decreasing.

16.5.1 Graphs of $y = a^x$ and $y = \log_a(x)$, for $a > 1$

eles-6306

The graph of $y = a^x$ where $a > 1$

- This graph was introduced in Topic 8 Non-linear relationships
- When sketching such a graph, consider a table of values for the function with the equation $y = 2^x$ and plot the points.

x	-3	-2	-1	0	1	2	3
$y = 2^x$	$\dfrac{1}{8}$	$\dfrac{1}{4}$	$\dfrac{1}{2}$	1	2	4	8

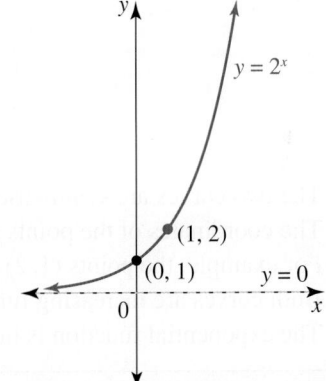

- The graph has many significant features. These include:
  - the y-intercept is $(0, 1)$
  - the value of y is always positive, $y > 0$
  - as x increases, y increases which can be written as when $x \to \infty, y \to \infty$
  - as x decreases, y approaches but never reaches the x-axis which can be written as when $x \to -\infty, y \to 0$
  - it has a horizontal asymptote, $y = 0$, the x-axis
  - it is an increasing function, at an increasing rate.

The graph of $y = \log_a(x)$ where $a > 1$

- This graph is related to the graph of $y = a^x$
- Before sketching this graph, consider a table of values for the function with equation $y = \log_2(x), x > 0$

x	$\dfrac{1}{8}$	$\dfrac{1}{4}$	$\dfrac{1}{2}$	1	2	4	8
$y = \log_2(x)$	-3	-2	-1	0	1	2	3

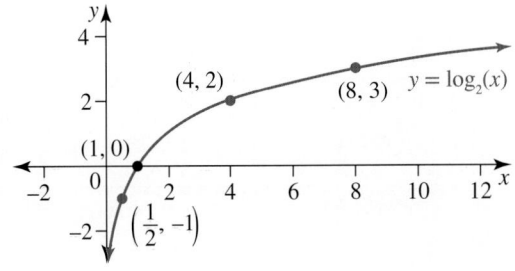

- The graph has many significant features. These include:
  - the x-intercept is $(1, 0)$
  - the value of x is always positive, $x > 0$
  - as x increases, y increases which can be written as when $x \to \infty$, $y \to \infty$
  - as x approaches zero, y approaches negative infinity which can be written as when $x \to 0$, $y \to -\infty$
  - it has a vertical asymptote, $x = 0$, the y-axis
  - it is an increasing function, at a decreasing rate.

Graphing on the same Cartesian plane

- When sketching $y = a^x$ and $y = \log_a(x)$ on the same Cartesian plane, ensure the axes have the same scale so the related features can be easily seen.
- Consider again $y = 2^x$ and $y = \log_2(x)$, $x > 0$
  - Plot the points from their table of values on the same Cartesian plane.

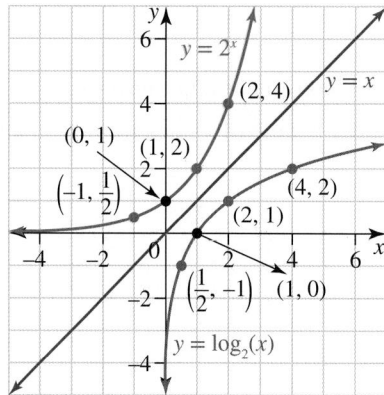

- The two curves are symmetrical about the line $y = x$.
- The coordinates of the points are reflected in the axis of symmetry, $y = x$.
  For example, the points $(1, 2)$ and $(2, 1)$ are reflections.
- Both curves are increasing functions, since as $x \to \infty$, $y \to \infty$
- The exponential function is increasing at a faster rate than the logarithmic function.

Digital technology

Graphing applications, such as Desmos or GeoGebra, can be used to graph logarithmic and exponential functions.

For instance, GeoGebra is used here to graph the functions in Worked example 17.

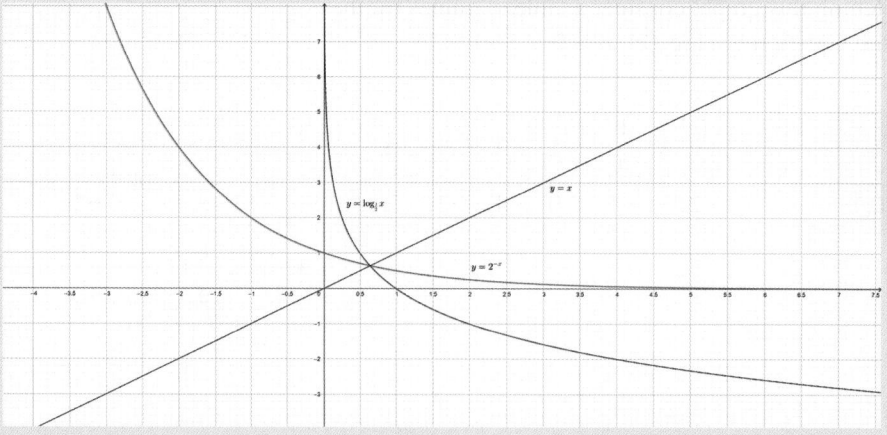

▶ 16.5.2 Graphs of $y = a^x$ and $y = \log_a(x)$, for $0 < a < 1$

eles-6271

The graph of $y = a^x$ where $0 < a < 1$

- This graph was also introduced in Topic 8
- It can be written as $y = a^{-x}$, $a > 1$
- The basic shape of this exponential function for different values of a are shown below.

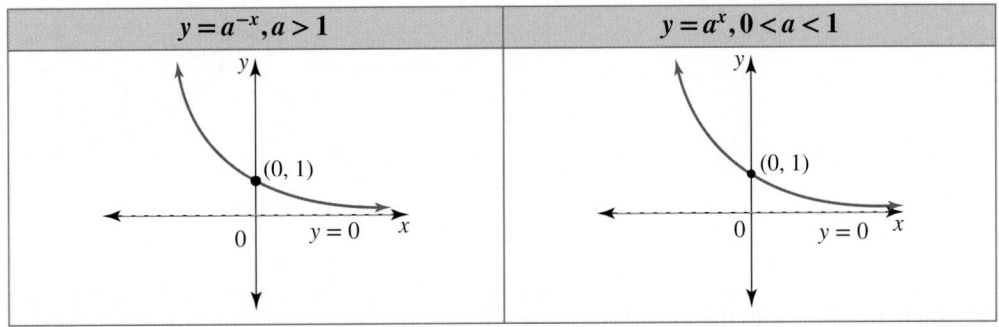

$y = a^{-x}, a > 1$	$y = a^x, 0 < a < 1$

- The graph has many significant features. These include;
  - the y-intercept is $(0, 1)$
  - the value of y is always positive, $y > 0$
  - it has a horizontal asymptote, $y = 0$, or the x-axis
  - it is a decreasing function, as $x \to \infty$, $y \to 0$

The graph of $y = \log_a(x)$ where $0 < a < 1$

- The logarithmic and exponential curves have $y = x$ as a line of symmetry, or line of reflection, where the x and y values are interchanged.
- This gives significant features of the logarithmic curve, which include:
  - the x-intercept is $(1, 0)$
  - the value of x is always positive, $x > 0$
  - it has a vertical asymptote, $x = 0$, or the y-axis
  - it is a decreasing function, as $x \to \infty$, $y \to -\infty$
- The graph of $y = \log_a(x)$ where $0 < a < 1$ is shown at right

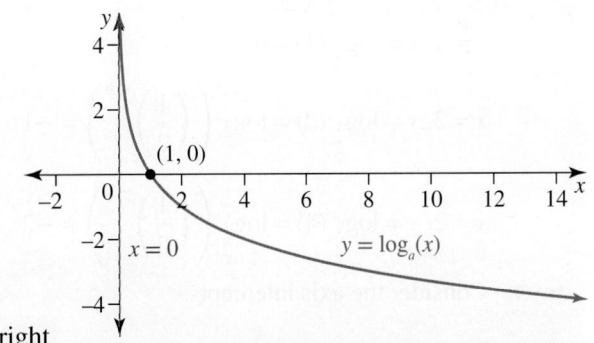

WORKED EXAMPLE 17 Graphing logarithmic curve given the exponential curve

a. Given the graph of $y = 2^{-x}$ or its equivalent $y = \left(\dfrac{1}{2}\right)^x$ sketch on the same axes the graph of $y = \log_{\frac{1}{2}}(x)$.

b. Compare these two graphs.

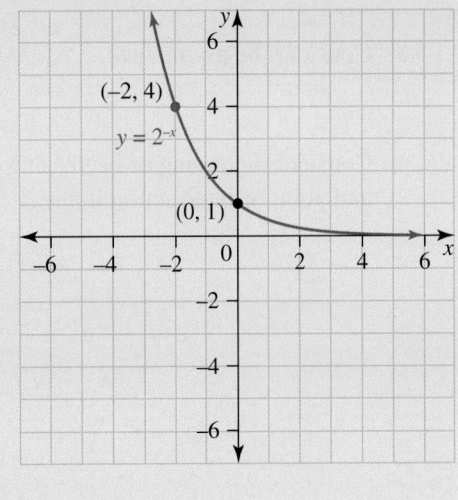

a. 1. Using the line of symmetry, (or reflection), reflect the points on the exponential curve over the line $y = x$ by interchanging the x and y values of the points to give points on the logarithmic curve:

$(-2, 4) \rightarrow (4, -2)$

$(-1, 2) \rightarrow (2, -1)$

$(0, 1) \rightarrow (1, 0)$

a.

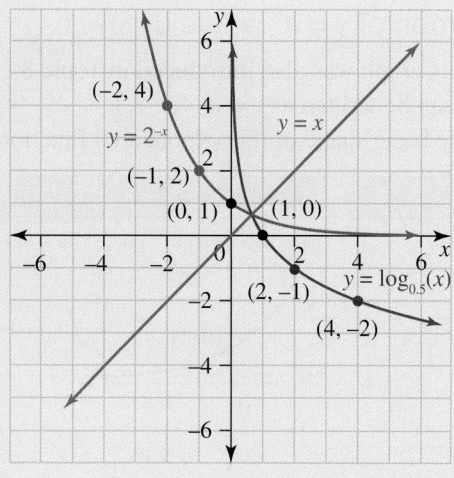

2. Sketch the graph, together with the line of symmetry, $y = x$.

3. Or: substitute points into the rule giving a table of values.

$x = \dfrac{1}{4}, y = \log_{\frac{1}{2}}\left(\dfrac{1}{4}\right) = 2$

$x = \dfrac{1}{2}, y = \log_{\frac{1}{2}}\left(\dfrac{1}{2}\right) = 1$

$x = 1, y = \log_{\frac{1}{2}}(1) = 0$

$x = 2, y = \log_{\frac{1}{2}}(2) = \log_{\frac{1}{2}}\left(\left(\dfrac{1}{2}\right)^{-1}\right) = -1$

$x = 2, y = \log_{\frac{1}{2}}(4) = \log_{\frac{1}{2}}\left(\left(\dfrac{1}{2}\right)^{-2}\right) = -2$

b. 1. Consider the axis intercepts.

b. The exponential curve has a y-intercept of $(0, 1)$ and the logarithmic curve has an x-intercept of $(1, 0)$.

2. Consider the restrictions on the x and y values.

The exponential curve is defined for $y > 0$ whereas the logarithmic curve is defined for $x > 0$.

3. Consider the asymptotes.

The exponential curve is asymptotic to the x-axis, which the logarithmic curve is asymptotic to the y-axis.

4. Consider increasing or decreasing functions and point of intersection if any.

Both curves are decreasing functions and intersect on the line $y = x$.

COMMUNICATING — COLLABORATIVE TASK: restrictions on $\log(x^2) = 2\log(x)$?

Equipment: pen, paper, graphing application, calculator

Mai and Ryan are working together to solve the following equation: $\log_3(x^2) = 1$

Ryan writes the following:

$$\log_3(x^2) = 1$$
$$2\log_3(x) = 1$$
$$\log_3(x) = \frac{1}{2}$$
$$x = 3^{\frac{1}{2}}$$
$$x = \sqrt{3}$$

Ryan tells Mai that the answer to: $\log_3(x^2) = 1$ is $x = \sqrt{3}$.

Mai uses Desmos to graph $y = \log_3(x^2)$ and $y = 1$, and determines the point(s) of intersection and tells Ryan that the answer to: $\log_3(x^2) = 1$ is $x = \sqrt{3}$ or $x = -\sqrt{3}$.

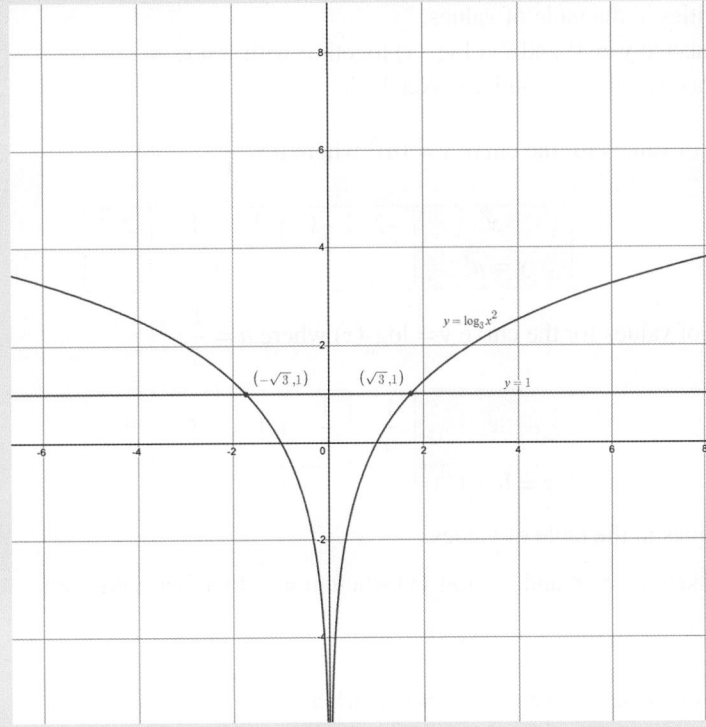

1. In pairs, discuss whether you think Mai or Ryan is correct.
2. Use a calculator to check whether $\log_3\left(\left(-\sqrt{3}\right)^2\right) = 1$ and $\log_3\left(\left(\sqrt{3}\right)^2\right) = 1$.

   Can you do the same with $2\log_3\left(\sqrt{3}\right) = 1$ and $2\log_3\left(-\sqrt{3}\right) = 1$?
3. Ryan remembers that the logarithm of a negative number is undefined. As a class, discuss whether this means that the restriction that $x > 0$ should be added to the logarithmic law $\log_a(x^n) = n\log_a(x)$.

| **16.5 Quick quiz** on | **16.5 Exercise** |

Individual pathways

■ PRACTISE	■ CONSOLIDATE	■ MASTER
1, 2, 4, 7	3, 5, 9, 10	6, 8, 11, 12

Fluency

1. a. Complete the table of values for the curve $y = 3^x$.

x	−2	−1	0	1	2
$y = 3^x$					

b. Complete the table of values for the curve $y = \log_3(x)$.

x	$\dfrac{1}{9}$	$\dfrac{1}{3}$	1	3	9
$y = \log_3(x)$					

c. Discuss the similarities in the table of values.
d. On the same axes, sketch $y = 3^x$ and $y = \log_3(x)$ together with $y = x$.
e. Compare and contrast the graphs you have sketched.

2. a. Complete the table of values for the curve $y = (a)^x$ where $a = \dfrac{1}{3}$.

x	−2	−1	0	1	2
$y = a^x$					

b. Complete the table of values for the curve $y = \log_a(x)$ where $a = \dfrac{1}{3}$.

x	$\dfrac{1}{9}$	$\dfrac{1}{3}$	1	3	9
$y = \log_a(x)$					

c. Discuss the similarities in the table of values.
d. On the same axes, sketch $y = a^x$ and $y = \log_a(x)$ where $a = \dfrac{1}{3}$ together with $y = x$.
e. Compare and contrast the graphs you have sketched.

3. a. Complete the table of values for the curve $y = (a)^x$ where $a = \dfrac{1}{2}$.

x	−2	−1	0	1	2
$y = a^x$					

b. Complete the table of values for the curve $y = \log_a(x)$ where $a = \dfrac{1}{2}$.

x	$\dfrac{1}{4}$	$\dfrac{1}{2}$	1	2	4
$y = \log_a(x)$					

c. Discuss the similarities in the table of values.

d. On the same axes, sketch $y = a^x$ and $y = \log_a(x)$ where $a = \dfrac{1}{2}$.

e. Compare and contrast the graphs you have sketched.

Understanding

4. **WE17** Use a graphing application of your choice to answer the following.

   a. Given the graph of $y = 5^{-x}$ or its equivalent $y = \left(\dfrac{1}{5}\right)^x$, sketch on the same axes the graph of $y = \log_{\frac{1}{5}}(x)$.

   b. Compare these two graphs.

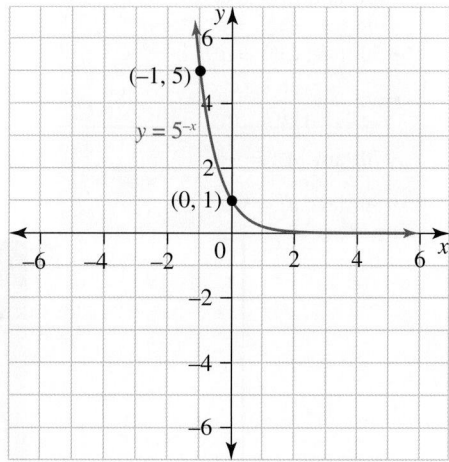

5. a. Complete a suitable table of values for $y = 10^x$ and $y = \log_{10}(x)$.
   b. Using a graphing application of your choice, sketch the two graphs on the same Cartesian plane.
   c. Compare and contrast the graphs you have sketched.

6. a. Complete a suitable table of values for $y = a^x$ and $y = \log_a(x)$, where $a = \dfrac{1}{10}$.
   b. Using a graphing application of your choice, sketch the two graphs on the same Cartesian plane.
   c. Compare and contrast the graphs you have sketched.

Communicating, reasoning and problem solving

7. a. On the same axes, using a graphing application of your choice, sketch $y = \log_2(x)$ and $y = \log_4(x)$.
   b. Compare the two logarithmic graphs you have sketched.

8. a. On the same axes, using a graphing application of your choice, sketch $y = \log_{\frac{1}{2}}(x)$ and $y = \log_{\frac{1}{4}}(x)$.
   b. Compare the two logarithmic graphs you have sketched.

9. a. On the same axes, using a graphing application of your choice, sketch $y = \log_5(x)$ and $y = \log_{\frac{1}{5}}(x)$.
   b. Discuss the similarities and differences between these two graphs.

10. Discuss how you would recognise if logarithmic curves were increasing or decreasing functions.

11. a. By plotting points, using a graphing application of your choice, sketch, on the same axes, $y = \log_2(x)$, $y = 2\log_2(x)$ and $y = 3\log_2(x)$.
    b. Discuss the similarities and differences between these two graphs.

12. a. By plotting points, using a graphing application of your choice, sketch, on the same axes, $y = -\log_2(x)$, $y = -2\log_2(x)$ and $y = -3\log_2(x)$.
    b. Discuss the similarities and differences between these two graphs.

LESSON
16.6 Review

16.6.1 Topic summary

Index laws

- 1st law: $a^m \times a^n = a^{m+n}$
- 2nd law: $a^m \div a^n = a^{m-n}$
- 3rd law: $a^0 = 1, a \neq 0$
- 4th law: $(a^m)^n = a^{m \times n} = a^{mn}$
- 5th law: $(ab)^n = a^n b^n$
- 6th law: $\left(\dfrac{a}{b}\right)^n = \dfrac{a^n}{b^n}$
- 7th law: $a^{-n} = \dfrac{1}{a^n}$
- 8th law: $a^{\frac{1}{n}} = \sqrt[n]{a}$

Logarithms

- Index form: $y = a^x$
- Logarithmic form: $\log_a(y) = x$
- Log laws:
  - $\log_a(x) + \log_a(y) = \log_a(xy)$
  - $\log_a(x) - \log_a(y) = \log_a\left(\dfrac{x}{y}\right)$
  - $\log_a(x)^n = n\log_a(x)$
  - $\log_a(1) = 0$
  - $\log_a(a) = 1$
  - $\log_a\left(\dfrac{1}{x}\right) = -\log_a(x)$
  - $\log_a(a^x) = x$
  - $\log_a(b) = \dfrac{\log_c(b)}{\log_c(a)}$
- Each log law is equivalent to one of the index laws.

LOGARITHMS (PATH)

Graphs of logarithms

- The graph of $y = \log_a(x)$ where $a > 1$ is related to the graph of $y = a^x$.

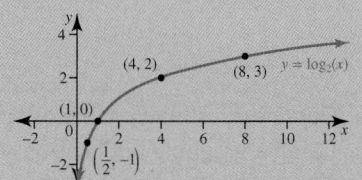

- The graph of $y = \log_a(x)$ where $0 < a < 1$ is related to the graph of $y = a^{-x}$, $a > 1$

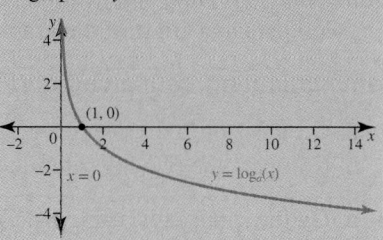

Solving logarithmic equations

- Simplify both sides of the equation so there is at most a single logarithm on each side.
- Switch to index form or log form as required.
  e.g.
  $\log_3(x) = 4 \implies x = 3^4$
  $5^x = 7 \implies \log_5(7) = x$

16.6.2 Project

Other number systems

Throughout history, different systems have been used to aid with counting. Ancient tribes are known to have used stones, bones and knots in rope to help keep count. The counting system that is used around the world today is called the Hindu-Arabic system. This system had its origin in India around 300–200BC . The Arabs brought this method of counting to Europe in the Middle Ages.

The Hindu–Arabic method is known as the decimal or base 10 system, as it is based on counting in lots of ten. This system uses the digits 0, 1, 2, 3, 4, 5, 6, 7, 8 and 9. Notice that the largest digit is one less than the base number, that is, the largest digit in base 10 is 9. To make larger numbers, digits are grouped together. The position of the digit tells us about its value. We call this *place value*.

For example, in the number 325, the 3 has a value of 'three lots of a hundred', the 2 has a value of 'two lots of ten' and the 5 has a value of 'five lots of units'. Another way to write this is:

$$3 \times 100 + 2 \times 10 + 5 \times 1 \text{ or } 3 \times 10^2 + 2 \times 10^1 + 5 \times 10^0$$

In a decimal system, every place value is based on the number 10 raised to a power. The smallest place value (units) is described by 10^0, the tens place value by 10^1, the hundreds place value by 10^2, the thousands by 10^3 and so on.

Computers do not use a decimal system. The system for computer languages is based on the number 2 and is known as the binary system. The only digits needed in the binary system are the digits 0 and 1.

Can you see why?

Decimal number	0	1	2	3	4	5	6	7	8	9	10	11	12	13
Binary number	0	1	10	11	100	101	110	111	1000	1001	1010	1011	1100	1101

Consider the decimal number 7. From the table above, you can see that its binary equivalent is 111. How can you be sure this is correct?

$$111 = 1 \times 2^2 + 1 \times 2^1 + 1 \times 2^0 = 4 + 2 + 1 = 7$$

Notice that this time each place value is based on the number 2 raised to a power. You can use this technique to change any binary number into a decimal number.

(The same pattern applies to other bases, for example, in base 6 the place values are based on the number 6 raised to a power.)

Binary operations

When adding in the decimal system, each time the addition is greater than 9, we need to 'carry over' into the next place value. In the example below, the units column adds to more than 9, so we need to carry over into the next place value.

$$\begin{array}{r} {}^{1}17 \\ +\ 13 \\ \hline 30 \end{array}$$

The same is true when adding in binary, except we need to 'carry over' every time the addition is greater than 1.

$$\begin{array}{r} {}^{1}01 \\ +\ 01 \\ \hline 10 \end{array}$$

1. Perform the following binary additions.

   a. $\quad 11_2$
   $\quad +\,01_2$
   ____

   b. $\quad 111_2$
   $\quad +\,110_2$
   ____

   c. $\quad 1011_2$
   $\quad +\ \ 101_2$
   ____

2. Perform the following binary subtractions. Remember that if you need to borrow a number from a column on the left-hand side, you will actually be borrowing a 2 (not a 10).

   a. $\quad 11_2$
   $\quad -\,01_2$
   ____

   b. $\quad 111_2$
   $\quad -\,110_2$
   ____

   c. $\quad 1011_2$
   $\quad -\ \ 101_2$
   ____

3. Try some multiplication. Remllember to carry over lots of 2.

   a. $\quad 11_2$
   $\quad \times 01_2$
   ____

   b. $\quad 111_2$
   $\quad \times 110_2$
   ____

   c. $\quad 1011_2$
   $\quad \times\ \ 101_2$
   ____

4. What if our number system had an 8 as its basis (that is, we counted in lots of 8)? The only digits available for use would be 0, 1, 2, 3, 4, 5, 6 and 7. (Remember the maximum digit is 1 less than the base value.)
   Give examples to show how numbers would be added, subtracted and multiplied using this base system. Remember that you would 'carry over' or 'borrow' lots of 8.

5. The hexadecimal system has 16 as its basis. Investigate this system. Explain how it would be possible to have 15, for example, in a single place position.
   Give examples to show how the system would add, subtract and multiply.

 Resources

Interactivities Crossword (int-2872)
Sudoku puzzle (int-3891)

Fluency

1. Evaluate the following.
   a. $\log_{12}(18) + \log_{12}(8)$
   b. $\log_{4}(60) - \log_{4}(15)$
   c. $\log_{9}\left(9^{8}\right)$
   d. $2\log_{3}(6) - \log_{3}(4)$

2. Use the logarithm laws to simplify each of the following.
   a. $\log_{a}(16) + \log_{a}(3) - \log_{a}(2)$
   b. $\log_{x}\left(x\sqrt{x}\right)$
   c. $4\log_{a}(x) - \log_{a}\left(x^{2}\right)$
   d. $5\log_{x}\left(\dfrac{1}{x}\right)$

3. Solve for x in the following, given that $x > 0$.
   a. $\log_{2}(x) = 9$
   b. $\log_{5}(x) = -2$
   c. $\log_{x}(25) = 2$
   d. $\log_{x}\left(2^{6}\right) = 6$
   e. $\log_{3}(729) = x$
   f. $\log_{7}(1) = x$

4. Solve for x in the following.
   a. $\log_{5}(4) + \log_{5}(x) = \log_{5}(24)$
   b. $\log_{3}(x) - \log_{3}(5) = \log_{3}(7)$

5. Solve for x in the following equations.
   a. $6^{x} = \dfrac{1}{36}$
   b. $7^{x} = \dfrac{1}{\sqrt{7}}$
   c. $2^{x+1} = 8\sqrt{2}$

6. Solve for x in the following equations, correct to 3 decimal places.
   a. $2^{x} = 25$
   b. $0.6^{x} = 7$
   c. $9^{-x} = 0.84$

7. a. Express $5^{4} = 625$ as a logarithm statement.
   b. Express $\log_{36}(6) = \dfrac{1}{2}$ as an index statement.

Understanding

8. a. Solve the equation $10^{x} = 8.52$, expressing the exponent x to 2 significant figures.
   b. Solve the equation $\log_{3}(x) = -1$ for x.

9. Use the logarithm laws to evaluate the following.
   a. $\log_{5}(5 \div 5)$
   b. $\log_{10}(5) + \log_{10}(2)$
   c. $\log_{3}\left(\dfrac{1}{3}\right)$
   d. $\log_{2}(32)$
   e. $\log_{4}\left(\dfrac{7}{32}\right) - \log_{4}(14)$
   f. $\log_{6}(9) + \log_{6}(8) - \log_{6}(2)$

10. Use the logarithm laws to simplify the following.

a. $\log_3 \left(x^3\right) - \log_3 \left(x^2\right)$

b. $\log_a \left(2x^5\right) + \log_a \left(\dfrac{x}{2}\right)$

c. $-\dfrac{1}{2}\log_{10}(a) + \log_{10}\left(\sqrt{a}\right)$

d. $\dfrac{1}{2}\log_b \left(16a^4\right) - \dfrac{1}{3}\log_b \left(8a^3\right)$

e. $3\log_{10}(x) - 2\log_{10}\left(x^3\right) + \dfrac{1}{2}\log_{10}\left(x^5\right)$

f. $\dfrac{\log_3 \left(x^6\right)}{\log_3 \left(x^2\right)}$

11. Use the logarithm laws to evaluate the following.

a. $\log_9(3) + \log_9(27)$

b. $\log_9(3) - \log_9(27)$

c. $2\log_2(4) + \log_2(6) - \log_2(12)$

d. $\log_5 \left(\log_3(3)\right)$

Communicating, reasoning and problem solving

12. Express the following as single logarithms.

a. $\log_{10}(2) + \log_{10}(7)$

b. $\log_5(4) + \log_5(11)$

c. $\log_3(20) - \log_3(2)$

d. $\log_{10}(32) - \log_{10}(4)$

e. $2\log_2(5)$

f. $-3\log_5(2)$

13. Simplify the following.

a. $\log_7(7)$

b. $\log_6(3 \div 3)$

c. $\log_2(8)$

d. $\log_{10}\left(1 + 3^2\right)$

e. $\log_9(3)$

f. $2\log_2\left(\dfrac{1}{4}\right)$

14. Solve the following for x.

a. $\log_5(x - 1) = 2$

b. $\log_2(2x + 1) = -1$

c. $\log_x\left(\dfrac{1}{49}\right) = -2$

d. $\log_x(36) - \log_x(4) = 2$

15. Solve the following for x.

a. $\log_{10}(x + 5) = \log_{10}(2) + 3\log_{10}(3)$

b. $\log_4(2x) + \log_4(5) = 3$

c. $\log_{10}(x + 1) = \log_{10}(x) + 1$

d. $2\log_6(3x) + 3\log_6(4) - 2\log_6(12) = 2$

16. If $\log_2(3) - \log_2(2) = \log_2(x) + \log_2(5)$, solve for x, using logarithm laws.

17. Solve the equation $\log_3(x) + \log_3(2x + 1) = 1$ for x.

18. Solve the equation $\log_6(x) - \log_6(x - 1) = 2$ for x.

19. a. Sketch, on the same Cartesian plane, $y = 4^x$ and $y = \log_4(x)$.
    b. Compare and contrast the graphs sketched in part **a**.

20. a. Sketch, on the same Cartesian plane, the graphs of $y = \log_a(x)$ when $a = 4$ and $a = \dfrac{1}{4}$.
    b. Compare and contrast the graphs sketched in part **a**.

on To test your understanding and knowledge of this topic, go to your learnON title at www.jacplus.com.au and complete the **post-test**.

Answers

Topic 16 Logarithms (Path)

16.1 Pre-test

1. $\log_5(25) = 2$

2. $\log_9(3) = \dfrac{1}{2}$

3. $2^4 = 16$

4. True

5. 7

6. $\dfrac{1}{2}$

7. $x = 3$

8. $\dfrac{1}{36}$

9. $x = -2$

10. 1

11. 3

12. 0

13. False

14. D

15. A

16.2 Logarithms

1. a. $\log_4(16) = 2$ b. $\log_2(32) = 5$
   c. $\log_3(81) = 4$ d. $\log_6(36) = 2$
   e. $\log_{10}(1000) = 3$

2. a. $\log_5(25) = 2$ b. $\log_4(x) = 3$
   c. $\log_5(125) = x$ d. $\log_7(49) = x$
   e. $\log_p(16) = 4$

3. a. $\log_9(3) = \dfrac{1}{2}$ b. $\log_{10}(0.1) = -1$
   c. $\log_8(2) = \dfrac{1}{3}$ d. $\log_2\left(\dfrac{1}{2}\right) = -1$
   e. $\log_4(8) = \dfrac{3}{2}$

4. D

5. a. $2^4 = 16$ b. $3^3 = 27$
   c. $10^6 = 1\,000\,000$ d. $5^3 = 125$

6. a. $16^{\frac{1}{2}} = 4$ b. $4^x = 64$
   c. $49^{\frac{1}{2}} = 7$ d. $3^5 = x$

7. a. $81^{\frac{1}{2}} = 9$ b. $10^{-2} = 0.01$
   c. $8^1 = 8$ d. $64^{\frac{1}{3}} = 4$

8. B

9. a. 4   b. 2
   c. 2   d. 5

10. a. 5   b. 7
    c. 0   d. $\dfrac{1}{2}$

11. a. -1 b. 1
    c. -2 d. $\dfrac{1}{3}$

12. a. 0   b. 1   c. 2

13. a. 3   b. 4   c. 5

14. a. 0 and 1   b. 3 and 4   c. 1 and 2

15. a. 4 and 5   b. 2 and 3   c. 4 and 5

16. a. 6.1   b. 6.3   c. 8.2

17. a. $\log_{10}(g) = k$ implies that $g = k$ so $g^2 = \left(10^k\right)^2$. That is, $g^2 = 10^{2k}$, therefore, $\log_{10}\left(g^2\right) = 2k$.

    b. $\log_x(y) = 2$ implies that $y = x^2$, so $x = y^{\frac{1}{2}}$ and therefore $\log_y(x) = \dfrac{1}{2}$.

    c. The equivalent exponential statement is $x = 4^y$, and we know that 4^y is greater than zero for all values of y. Therefore, x is a positive number.

18. a. 6   b. -4 c. -5

19. a. 5   b. -3 c. $\dfrac{3}{2}$

20. a. 3   b. 7   c. $\dfrac{1}{8}$

21. x

22. $\dfrac{9}{x}$

16.3 Logarithm laws

1. a. 1.698 97   b. 1.397 94
   c. 0.698 97   d. 0.301 03

2. Sample responses can be found in the worked solutions in the online resources.

3. a. 1   b. 3

4. a. 2   b. 3

5. a. 4   b. 1

6. a. 2   b. 3

7. a. 1   b. 4

8. a. 3   b. 5

9. a. 2   b. $\dfrac{1}{2}$

10. a. 1   b. 3

11. a. -1 b. $-\dfrac{1}{2}$

12. 3

13. a. 2   b. 4

14. a. 3   b. 3

15. a. 1   b. 0
    c. -1 d. 5

16. a. -2 b. 1
    c. 0   d. -2

17. a. $-\dfrac{1}{2}$ **b.** $\dfrac{1}{2}$

c. $-\dfrac{1}{2}$ **d.** $\dfrac{7}{2}$

18. 2.722

19. a. $\log_a(40)$ **b.** $\log_a(18)$
 c. $\log_x(48)$ **d.** $\log_x(4)$

20. a. $\log_a(x)$ **b.** 1
 c. -1 **d.** 7

21. a. $\dfrac{1}{2}$ **b.** $\dfrac{3}{2}$

c. -6 **d.** $-\dfrac{1}{3}$

22. a. B **b.** B, D
 c. A, B **d.** C, D

23. a. $\log_2(80)$ **b.** $\log_3(105)$
 c. $\log_{10}(100) = 2$ **d.** $\log_6(56)$

24. a. $\log_2(4) = 2$ **b.** $\log_3(3) = 1$
 c. $\log_5(12.5)$ **d.** $\log_2(3)$

25. a. $\log_4(5)$ **b.** $\log_{10}\left(\dfrac{1}{4}\right)$

c. $\log_3(4)$ **d.** $\log_2(3)$

e. $\log_3(20)$ **f.** $\log_4(2) = \dfrac{1}{2}$

26. a. C **b.** B **c.** A

27. a. 12 (Evaluate each logarithm separately and then calculate the product.)
 b. 4 (First simplify the numerator by expressing 81 as a power of 3.)
 c. 7 (Let $y = 5^{\log_5(7)}$ and write an equivalent statement in logarithmic form.)

28. -2

29. $7 - 3\log_2(3)$

30. 1

31. $x = 3a, \ 5a$

32. 7

16.4 Solving equations

1. a. 25 **b.** 81 **c.** $\dfrac{1}{8}$

d. $\dfrac{1}{16}$ **e.** $100, -100$

2. a. 16 **b.** 26 **c.** 127
 d. 2 **e.** 0

3. a. $-\dfrac{1}{32}$ **b.** $-\dfrac{1}{9}$ **c.** -624
 d. -2.5

4. a. 3 **b.** 2 **c.** 125

5. a. 625 **b.** 2 **c.** 8

6. a. 6 **b.** 4

7. a. 3 **b.** 2 **c.** -1

8. a. -2 **b.** $\dfrac{1}{2}$ **c.** $\dfrac{1}{3}$

9. a. 0 **b.** 0 **c.** -1
 d. -2

10. a. 5 **b.** 6 **c.** 10
 d. 8

11. a. 4 **b.** 2 **c.** 9
 d. $\dfrac{2}{5}$

12. a. 500 **b.** 128 **c.** 5
 d. 6 **e.** 1 **f.** 2

13. a. B **b.** A **c.** D
 d. B

14. a. 7 **b.** 2 **c.** -2
 d. 0 **e.** 4

15. a. $\dfrac{1}{2}$ **b.** $\dfrac{1}{2}$ **c.** $\dfrac{3}{2}$

d. $-\dfrac{1}{2}$ **e.** $\dfrac{3}{2}$

16. a. $\dfrac{3}{4}$ **b.** $-\dfrac{5}{2}$ **c.** $\dfrac{5}{2}$

d. $-\dfrac{9}{2}$ **e.** $-\dfrac{11}{4}$

17. a. 3.459 **b.** -0.737
 c. 2.727 **d.** 0.483

18. a. 1.292 **b.** -3.080
 c. -1.756 **d.** 0.262

19. a. 0.827 **b.** 0.579
 c. -0.423 **d.** 2.138

20. $t = \dfrac{5}{6}$

21. Approximately 2.5 times brighter.

22. a. 120 **b.** 130
 c. 0.001 **d.** 3 dB are added.
 e. 10 dB are added. **f.** 100

23. a. i. 1.1 **ii.** 1.3 **iii.** 1.418
 iv. 1.77 **v.** 2.43 **vi.** 3.1

b. No; see answers to **23a i** and **ii** above.

c. i. 22 387 211 KJ
 ii. 707 945 784 KJ
 iii. 22 387 211 386 KJ.

d. The energy is increased by a factor of 31.62.

e. It releases 31.62^3 times more energy.

24. a. $x = 0.7712$ **b.** $x = 1.2966$

25. $x = 7$

26. $x = 1, 3$

16.5 Graphing logarithmic and exponential functions

1. a.

x	-2	-1	0	1	2
$y = 3^x$	$\dfrac{1}{9}$	$\dfrac{1}{3}$	1	3	9

b.

x	$\dfrac{1}{9}$	$\dfrac{1}{3}$	1	3	9
$y = \log_3(x)$	-2	-1	0	1	2

c. The similarities are the interchange between the x and y values of points.
For example, the point $(1, 3)$ on the exponential curve becomes the point $(3, 1)$ on the logarithmic curve.

d.

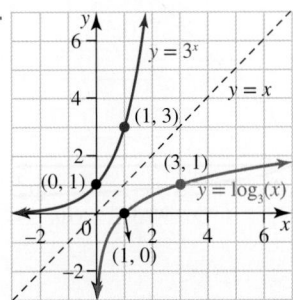

e. The y-intercept of the exponential curve is $(0, 1)$ and the x-intercept of the logarithmic curve is $(1, 0)$
The exponential curve is asymptotic to the x-axis, the logarithmic curve is asymptotic to the y-axis.
Both curves are **increasing** as $x \to \infty$.
The line $y = x$ is a line of symmetry between the two curves.

2. a.

x	-2	-1	0	1	2
$y = a^x$	9	3	1	$\dfrac{1}{3}$	$\dfrac{1}{9}$

where $a = \dfrac{1}{3}$

b.

x	$\dfrac{1}{9}$	$\dfrac{1}{3}$	1	3	9
$y = \log_a(x)$	2	1	0	-1	-2

where $a = \dfrac{1}{3}$

c. The similarities are the interchange between the x and y values of points.
For example, the point $(-2, 9)$ on the exponential curve becomes the point $(9, -2)$ on the logarithmic curve.

d.

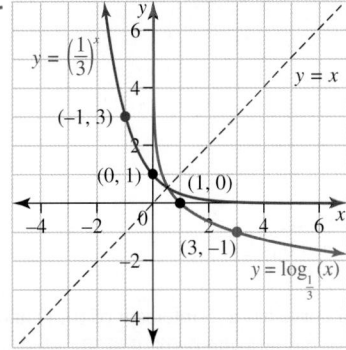

e. The y-intercept of the exponential curve is $(0, 1)$ and the x-intercept of the logarithmic curve is $(1, 0)$
The exponential curve is asymptotic to the x-axis, the logarithmic curve is asymptotic to the y-axis.
Both curves are **decreasing** as $x \to \infty$.
They intersect on the line $y = x$.

3. a.

x	-2	-1	0	1	2
$y = a^x$	4	2	1	$\dfrac{1}{2}$	$\dfrac{1}{4}$

where $a = \dfrac{1}{2}$

b.

x	$\dfrac{1}{4}$	$\dfrac{1}{2}$	1	2	4
$y = \log_a(x)$	2	1	0	-1	-2

where $a = \dfrac{1}{2}$

c. The similarities are the interchange between the x and y values of points.
For example, the point $(-2, 4)$ on the exponential curve becomes the point $(4, -2)$ on the logarithmic curve.

d.

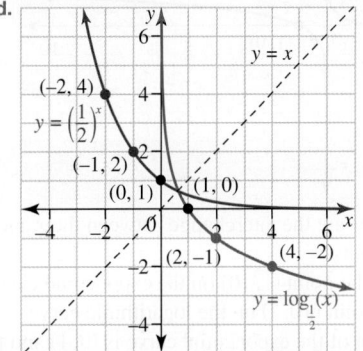

e. The y-intercept of the exponential curve is $(0, 1)$ and the x-intercept of the logarithmic curve is $(1, 0)$
The exponential curve is asymptotic to the x-axis, the logarithmic curve is asymptotic to the y-axis.
Both curves are **decreasing** as $x \to \infty$.
The curves intersect on the line of symmetry, $y = x$

4. a.

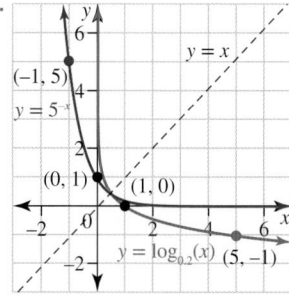

b. The exponential curve has a y-intercept of $(0, 1)$ and the logarithmic curve has an x-intercept of $(1, 0)$.

The exponential curve is defined for $y > 0$ whereas the logarithmic curve is defined for $x > 0$.

The exponential curve is asymptotic to the x-axis, which the logarithmic curve is asymptotic to the y-axis.

Both curves are decreasing functions and intersect on the line $y = x$.

5. a.

x	-1	0	1
$y = 10^x$	0.1	1	10

x	$\dfrac{1}{10}$	1	10
$y = \log_{10}(x)$	-1	0	1

b.

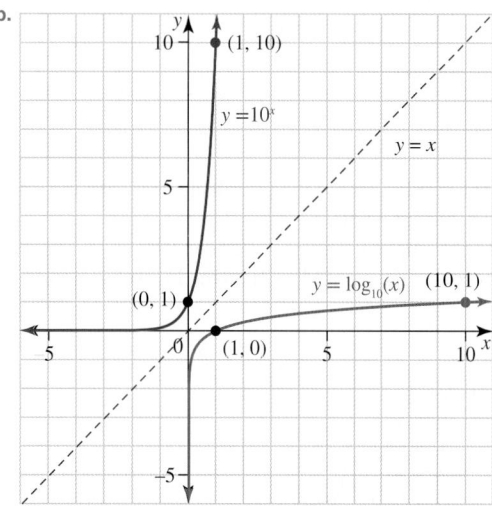

c. The similarities are the interchange between the x and y values of points.

For example, the point $(1, 10)$ on the exponential curve becomes the point $(10, 1)$ on the logarithmic curve.

The y-intercept of the exponential curve is $(0, 1)$ and the x-intercept of the logarithmic curve is $(1, 0)$

The exponential curve is asymptotic to the x-axis, the logarithmic curve is asymptotic to the y-axis.

Both curves are **increasing** as $x \to \infty$.

The line $y = x$ is a line of symmetry between the two curves.

6. a.

x	-1	0	1
$y = a^x$	10	1	$\dfrac{1}{10}$

where $a = \dfrac{1}{10}$

x	$\dfrac{1}{10}$	1	10
$y = \log_a(x)$	1	0	-1

where $a = \dfrac{1}{10}$

b.

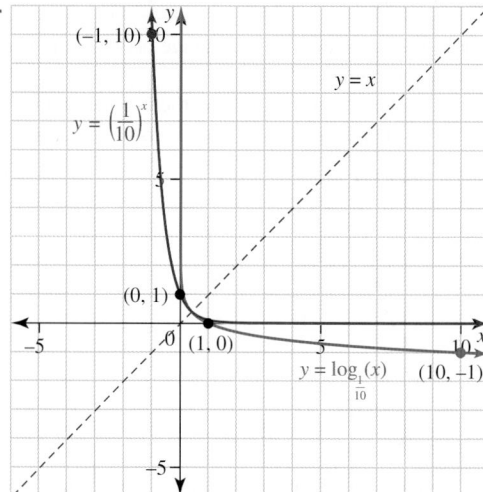

c. The similarities are the interchange between the x and y values of points.

For example, the point $(-1, 10)$ on the exponential curve becomes the point $(10, -1)$ on the logarithmic curve when reflected over the line $y = x$.

The y-intercept of the exponential curve is $(0, 1)$ and the x-intercept of the logarithmic curve is $(1, 0)$

The exponential curve is asymptotic to the x-axis, the logarithmic curve is asymptotic to the y-axis.

Both curves are **decreasing** as $x \to \infty$.

The curves intersect on the line of symmetry, $y = x$

7. a.

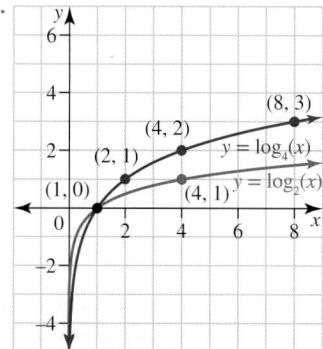

b. Both logarithmic curves are increasing, defined for $x > 0$, have an x-intercept of $(1, 0)$ and a vertical asymptote at $y = 0$. For $x > 1$, the $y = \log_4(x)$ increases at a slower rate than the curve $y = \log_2(x)$.

8. a.

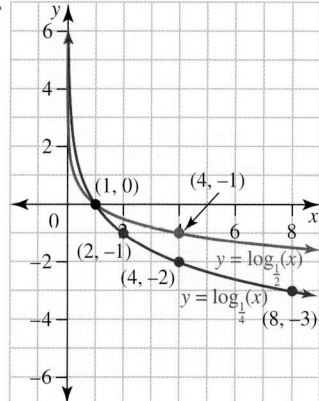

b. Both logarithmic curves are decreasing, defined for $x > 0$, have an x-intercept of $(1, 0)$ and a vertical asymptote at $y = 0$. For $x > 1$, the $y = \log_{\left(\frac{1}{4}\right)}(x)$ increases at a slower rate than the curve $y = \log_{\left(\frac{1}{2}\right)}(x)$.

9. a.

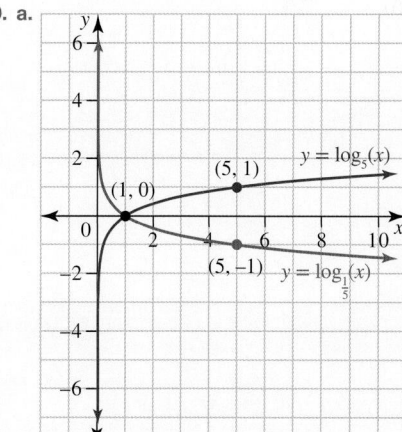

b. Both logarithmic curves defined for $x > 0$, have an x-intercept of $(1, 0)$ and a vertical asymptote at $y = 0$. They are the same shape but are a reflection over the x-axis.

10. Increasing logarithmic curves have the rule $y = \log_a(x)$ where $a > 1$.
Decreasing logarithmic curves have the rule $y = \log_a(x)$ where $0 < a < 1$.

11. a.

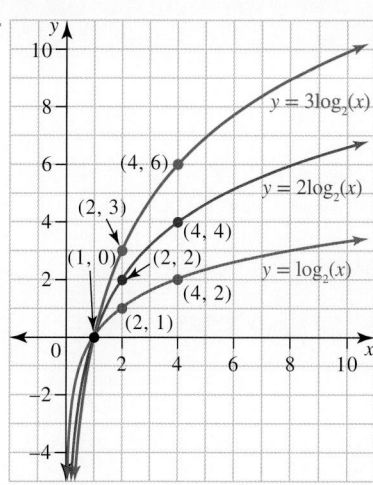

b. All curves have an x-intercept of $(1, 0)$, are defined for $x > 0$, with a vertical asymptote of $y = 0$. All the curves are increasing functions.
The differences are in the dilation factors, the greater the dilation from the x-axis, the steeper the curve. The dilation factor is the constant multiple of the function.

12. a.

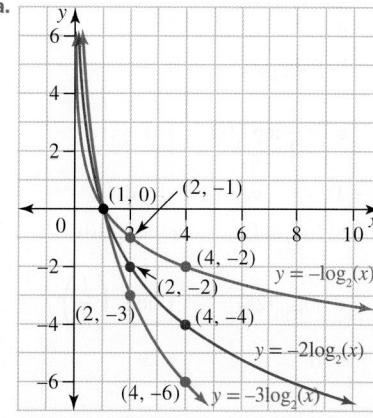

b. All curves have an x-intercept of $(1, 0)$, are defined for $x > 0$, with a vertical asymptote of $y = 0$. All the curves are decreasing functions.
The differences are in the dilation factors, the greater the dilation from the x-axis, the steeper the curve. Also, the reflection in the x-axis as the rules have the negative sign. The dilation factor is the constant multiple of the function.

Project

1. **a.** 100_2 **b.** 1101_2 **c.** 10000_2

2. **a.** 10_2 **b.** 1_2 **c.** 110_2

3. **a.** 11_2 **b.** 101010_2 **c.** 110111_2

4. Sample responses can be found in the worked solutions in the online resources. The digits in octal math are 0, 1, 2, 3, 4, 5, 6, and 7. The value "eight" is written as "1 eight and 0 ones", or 10_8.

5. Sample responses can be found in the worked solutions in the online resources. The numbers 10, 11, 12, 13, 14 and 15 are allocated the letters A, B,C, D, E and F respectively.

16.6 Review questions

1. **a.** 2   **b.** 1
   c. 8   **d.** 2

2. **a.** $\log_a(24)$ **b.** $\dfrac{3}{2}$

   c. $\log_a\left(x^2\right)$ or $2\log_a(x)$ **d.** -5

3. **a.** 512   **b.** $\dfrac{1}{25}$ **c.** 5

   d. 2   **e.** 6   **f.** 0

4. **a.** 6   **b.** 35

5. **a.** -2 **b.** $-\dfrac{1}{2}$ **c.** $\dfrac{5}{2}$

6. **a.** 4.644   **b.** -3.809 **c.** 0.079

7. a. $4 = \log_5(625)$ **b.** $6 = 36^{\frac{1}{2}}$

8. a. $x \approx 0.93$ **b.** $x = \dfrac{1}{3}$

9. a. 0 **b.** 1 **c.** −1

  d. 5 **e.** −3 **f.** 2

10. a. $\log_3(x)$ **b.** $6\log_a(x)$ **c.** 0

  d. $\log_b(2a)$ **e.** $-\dfrac{1}{2}\log_{10}(x)$ **f.** 3

11. a. 2 **b.** −1 **c.** 3 **d.** 0

12. a. $\log_{10}(14)$ **b.** $\log_5(44)$ **c.** $\log_3(10)$

  d. $\log_{10}(8)$ **e.** $\log_{10}(25)$ **f.** $\log_5\left(\dfrac{1}{8}\right)$

13. a. 1 **b.** 0 **c.** 3

  d. 1 **e.** $\dfrac{1}{2}$ **f.** −4

14. a. $x = 26$ **b.** $x = -\dfrac{1}{4}$

  c. $x = 7$ **d.** $x = 3$

15. a. $x = 49$ **b.** $x = 6.4$

  c. $x = \dfrac{1}{9}$ **d.** $x = 3$

16. $x = \dfrac{3}{10}$

17. $x = 1$

18. $x = \dfrac{36}{35}$

19. a.

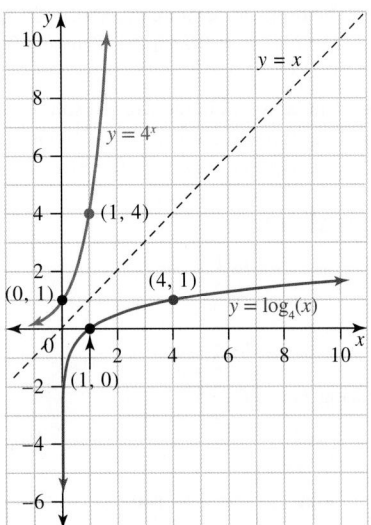

  b. The two graphs are symmetrical about the line $y = x$.
  Both are increasing functions.
  The exponential curve is asymptotic to the x-axis and
  cuts the y-axis at $(0, 1)$.
  The logarithmic curve is asymptotic to the y-axis and cuts
  the x-axis at $(1, 0)$.

20. a.

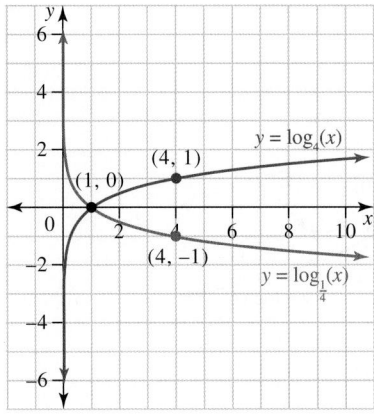

  b. The two graphs are a reflection over the x-axis.
  One is an increasing function, when $a = 4$
  The other is a decreasing function, when $a = \dfrac{1}{4}$
  Both curves are asymptotic to the y-axis and cut the
  x-axis at $(1, 0)$.

17 Functions and other graphs (Path)

LESSON SEQUENCE

LESSON
17.1 Overview

Why learn this?

Functions and relations are broad and interesting topics of study. They are topics with many real-world applications and are essential notions to understand as you head towards higher studies in mathematics. You will have already seen some functions and relations in your maths classes; linear equations, quadratics and polynomials are all examples of functions, and circles are examples of relations.

In your previous study of quadratics you learned about graphs with an x^2 term, but have you wondered what a graph would look like if it had an x^3 term or an x^4 term? You will be learning about these and other graphs in this topic.

An understanding of how to apply and use functions and relations is relevant to many professionals. Medical teams working to map the spread of diseases, engineers designing complicated structures such as the Sydney Opera House, graphic designers creating a new logo, video game designers developing a new map for their game — all require the use and understanding of functions and relations.

This topic builds on what you already know and extends it into new areas of mathematics. By the end of this topic, you will know all about different types of functions and relations, and how to graph them, interpret them and transform them.

Hey students! Bring these pages to life online

▶ Watch videos

🧩 Engage with interactivities

A+ Answer questions and check solutions

Find all this and MORE in jacPLUS ▶

Reading content and rich media, including interactivities and videos for every concept

Extra learning resources

Differentiated question sets

Questions with immediate feedback, and fully worked solutions to help students get unstuck

1. **MC** Choose the type of relation that the graph represents.
   A. One-to-one relation
   B. One-to-many relation
   C. Many-to-one relation
   D. Many-to-many relation

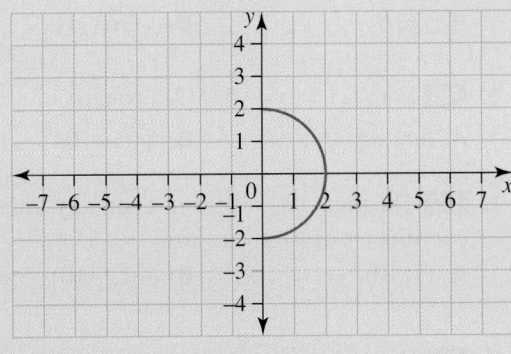

2. **MC** A function is a relation that is one-to-one or:

   A. many-to-one   B. many-to-many
   C. one-to-many   D. one-to-two

3. The graph below is a function. State whether this is true or false.

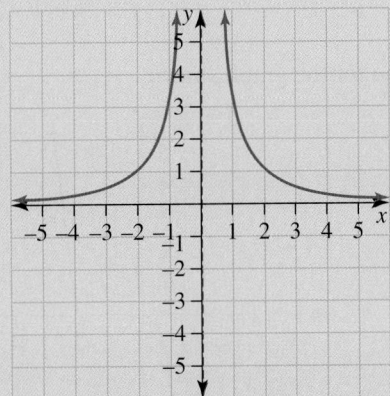

4. **MC** Select the correct domain of the relation shown in the graph.
   A. $x \in R$
   B. $x \in [-1, 2]$
   C. $x \in [2, 8]$
   D. $x \in [0, 8]$

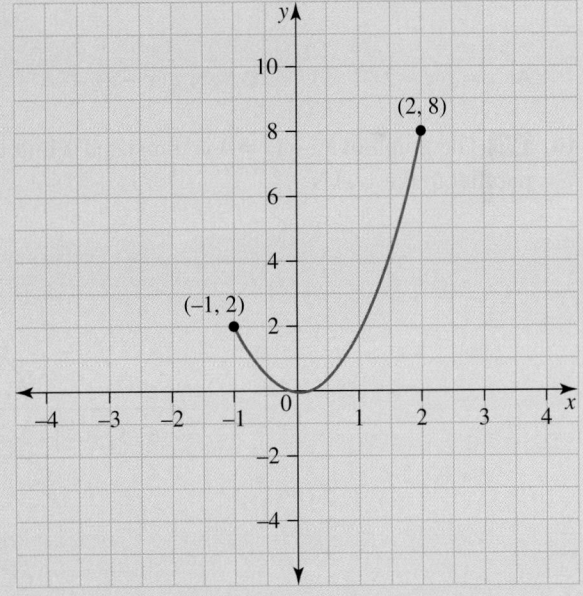

5. **MC** Select the correct range of the function $f(x) = \dfrac{1}{x-2} + 1$.

A. $y \in R$ B. $y \in R \backslash \{1\}$ C. $y \in R \backslash \{-1\}$ D. $y \in R \backslash \{2\}$

6. **MC** For $f(x) = 3x^2 - 5$, $f(2a)$ equals:

A. $6a^2 - 5$ B. $6a^2 - 10a$ C. $12a^2 - 5$ D. $12a^2 - 10a$

7. **MC** The point $(3, -10)$ is horizontally translated 4 units to the left and then reflected in the y-axis. The coordinates of its image are:

A. $(7, 10)$ B. $(-1, -10)$ C. $(-1, 10)$ D. $(1, -10)$

8. **MC** For $f(x) = \dfrac{x^2 - 2}{x}$, $f(a-1)$ equals:

A. $\dfrac{a^2 - 1}{a - 1}$ B. $\dfrac{(a-1)^2 - 1}{a - 1}$ C. $\dfrac{(a-1)^2 - 2}{a - 1}$ D. $\dfrac{(a-1)^2 - 2}{a - 1} - 1$

9. **MC** Select the correct equation for the graph shown.

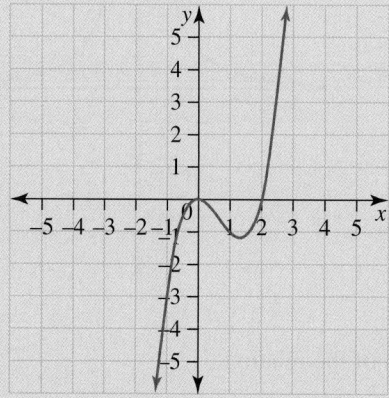

A. $y = x(x-2)^2$ B. $y = x(x-2)$ C. $y = x(x+2)^2$ D. $y = x^2(x-2)$

10. **MC** The graph of $x^2 + y^2 = 4$ is translated 1 unit to the left parallel to the x-axis and 2 units upwards, parallel to the y-axis.

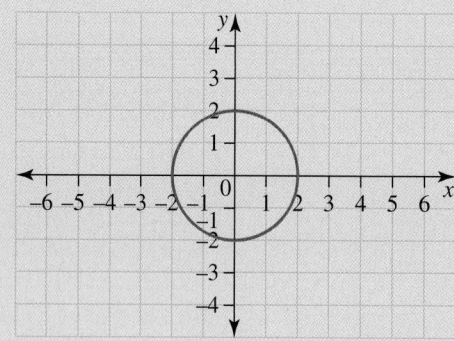

Select the new transformed equation.

A. $(x+1)^2 + (y+2)^2 = 7$ B. $(x+1)^2 + (y-2)^2 = 7$
C. $(x+1)^2 + (y+2)^2 = 4$ D. $(x+1)^2 + (y-2)^2 = 4$

11. [MC] Consider the function $f(x) = (x^2 - 9)(x - 2)(1 - x)$.
The graph of $f(x)$ is best represented by:

A.
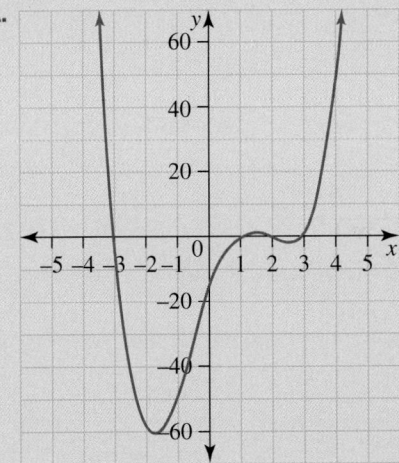

B.
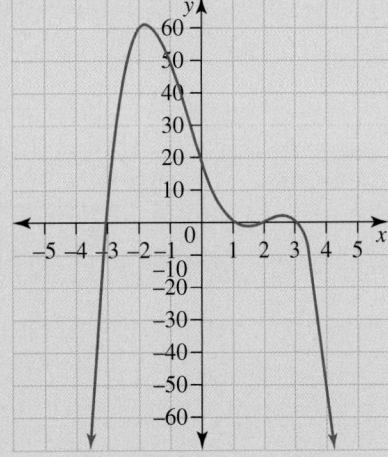

C.
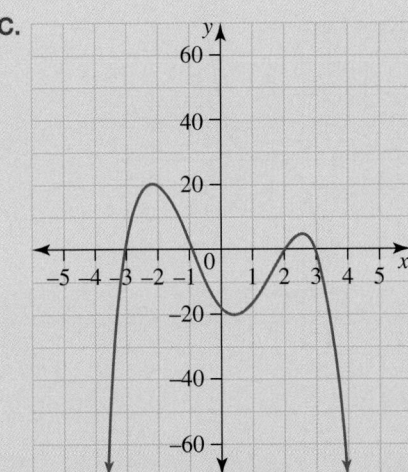

D.
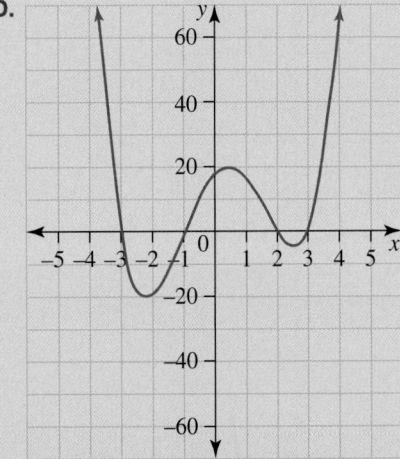

12. [MC] The quartic function has two x-intercepts at -1 and 4 and passes through the point $(0, -8)$.
Select the equation that best represents the function.

A. $f(x) = -\dfrac{1}{2}(x + 1)^2(x - 4)^2$

B. $f(x) = \dfrac{1}{2}(x + 1)^2(x - 4)^2$

C. $f(x) = -\dfrac{1}{2}(x - 1)^2(x + 4)^2$

D. $f(x) = -2(x + 1)^2(x - 4)^2$

13. **MC** If the graph shown is represented by the equation $y = f(x)$, select the correct graph for the equation $y = f(x - 1)$.

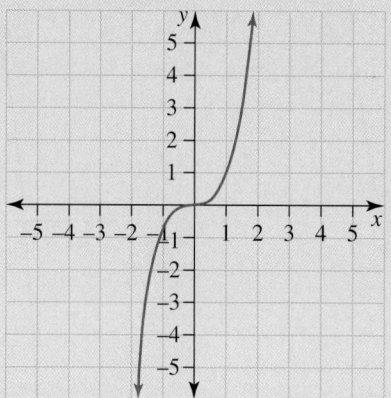

A.

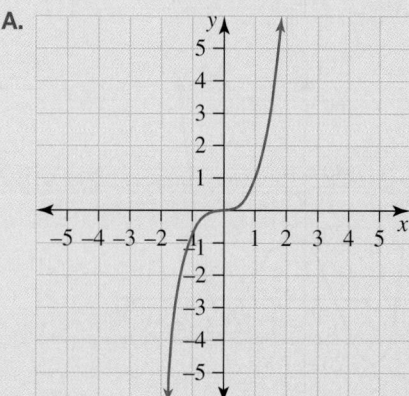

B.

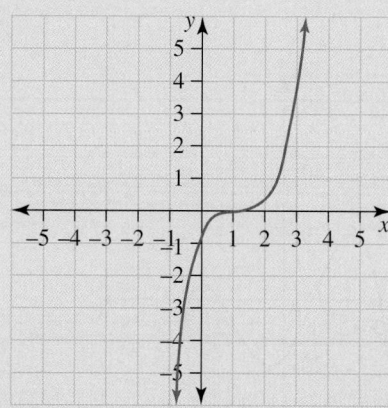

C.

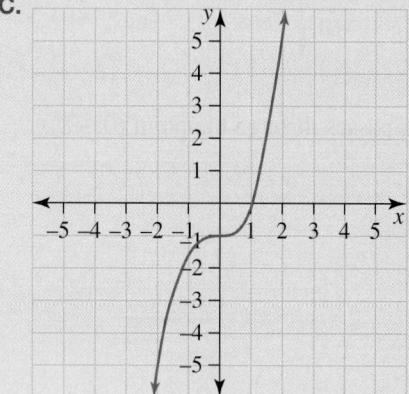

D.

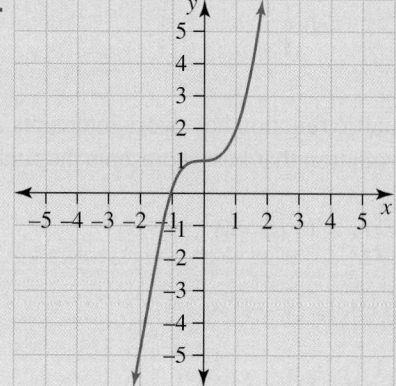

14. **MC** If the graph shown is represented by the equation $y = f(x)$, select the correct graph for the equation $y = -f(x) + 2$.

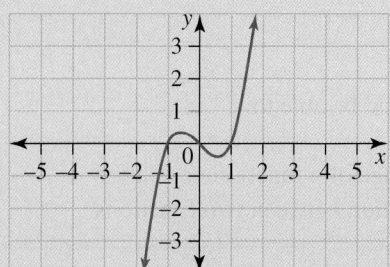

A.
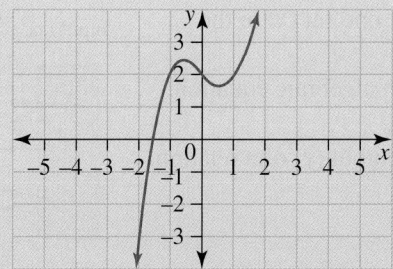

B.
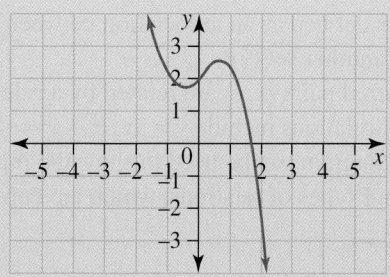

C.
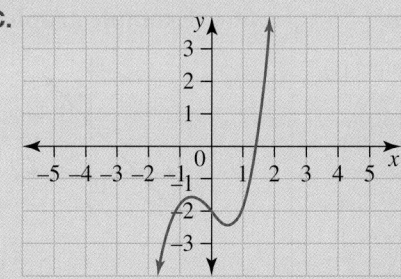

D.
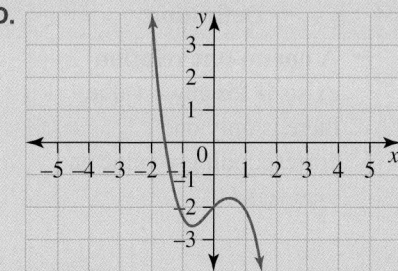

15. **MC** Consider the sketch of $y = f(x)$ and the graph of a transformation of $y = f(x)$.

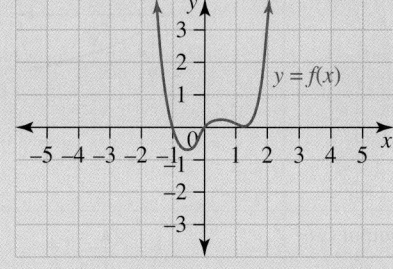
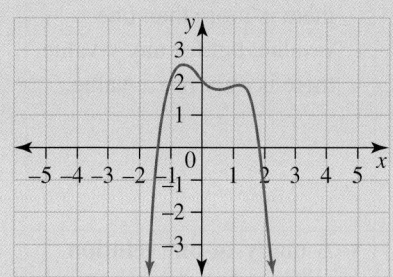

Select the possible equation in terms of $f(x)$ for the transformation of $y = f(x)$.

A. $y = f(x) + 2$

B. $y = f(x + 2)$

C. $y = -f(x) - 2$

D. $y = -f(x) + 2$

LESSON
17.2 Functions and relations

LEARNING INTENTION

At the end of this lesson you should be able to:
- identify the type of a relation
- determine the domain and range of a function or relation
- identify the points of intersection between two functions.

▶ 17.2.1 Types of relations

eles-4984

- A **relation** is defined as an association between the elements of one set (x) to the elements of another set (y).
- A set of ordered pairs (x, y) are related by a rule expressed as an algebraic equation. Examples of relations include $y = 3x$, $x^2 + y^2 = 4$ and $y = 2^x$.
- A mapping diagram can be used to show the input (x) and the output (y) of a relation.
- There are four types of relations, which are defined as follows.

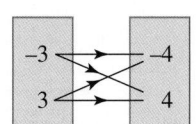

Types of relations	Definition	Example mapping diagram	Example graph
One-to-one relations	• A **one-to-one relation** exists if for any x-value there is only one corresponding y-value and vice versa.	Input (x) Output (y) −2 → 1 −1 → 2 0 → 3 1 → 4 2 → 5	
One-to-many relations	• A **one-to-many relation** exists if for any x-value there is more than one y-value, but for any y-value there is only one x-value.	Input (x) Output (y) 0 → 0 1 → −1 → 1 4 → −2 → 2 9 → −3 → 3	
Many-to-one relations	• A **many-to-one relation** exists if there is more than one x-value for any y-value but for any x-value there is only one y-value.	Input (x) Output (y) 0 → 0 −1 → 1 1 → −2 → 4 2 → −3 → 9 3 →	

| Many-to-many relations | • A **many-to-many relation** exists if there is more than one x-value for any y-value and vice versa. | 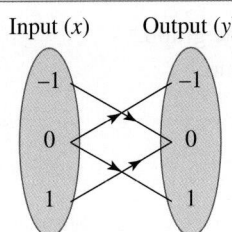 | 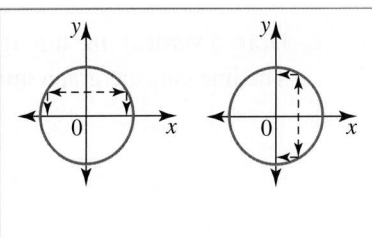 |

Determining the type of a relation

To determine the type of a relation:
  • Draw a horizontal line through the graph so that it cuts the graph the maximum number of times. Determine whether the number of cuts is one or many.
  • Draw a vertical line through the graph so that it cuts the graph the maximum number of times. Determine whether the number of cuts is one or many.

WORKED EXAMPLE 1 Identifying the types of relations

State the type of relation that each graph represents.

a.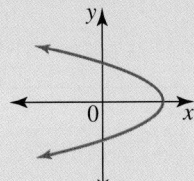

b.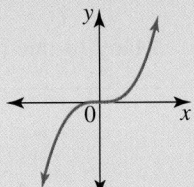

c.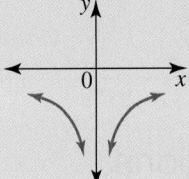

THINK

a. 1. Draw a horizontal line through the graph. The line cuts the graph **one** time.

2. Draw a vertical line through the graph. The line cuts the graph **many** times.

b. 1. Draw a horizontal line through the graph. The line cuts the graph **one** time.

WRITE

a.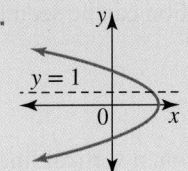

One-to- _____ relation

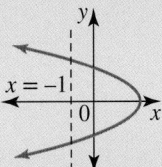

One-to-many relation

b.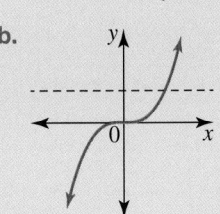

One-to-____ relation

2. Draw a vertical line through the graph. The line cuts the graph **one** time.

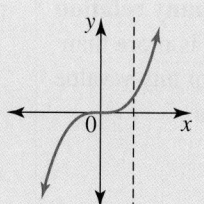

One-to-one relation

c. 1. Draw a horizontal line through the graph. The line cuts the graph **many** times.

c.

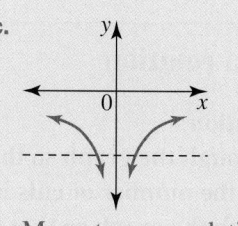

Many-to-_____ relation

2. Draw a vertical line through the graph. The line cuts the graph **one** time.

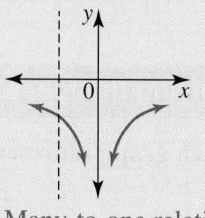

Many-to-one relation

▶ 17.2.2 Functions

eles-4985

- Relations that are one-to-one or many-to-one are called **functions**. That is, a function is a relation for which any x-value there is at most one y-value.
- A one-to-one or many-to-one relation can be seen from a mapping diagram or looking at the function's graph.

Vertical line test

- To determine if a graph is a function, a vertical line is drawn anywhere on the graph. If it does not intersect with the curve more than once, then the graph is a function.

  For example, in each of the two graphs below, each vertical line intersects the graph only once.

1.

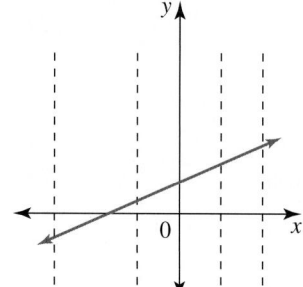

2.

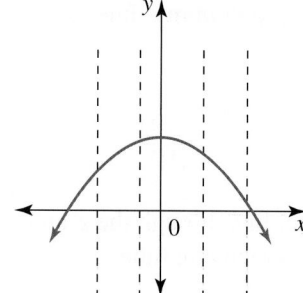

WORKED EXAMPLE 2 Identifying whether a relation is a function

State whether or not each of the following relations are functions.

a.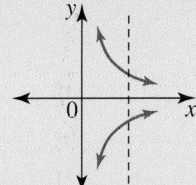

b.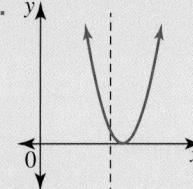

THINK

a. It is possible for a vertical line to intersect with the curve more than once.

b. It is not possible for any vertical line to intersect with the curve more than once.

WRITE

a. Not a function

b. Function

Function notation

- Consider the relation $y = 2x$, which is a function.
  The y-values are determined from the x-values, so we say 'y is a function of x', which is abbreviated to $y = f(x)$.
  So, the rule $y = 2x$ can also be written as $f(x) = 2x$.
- For a given function $y = f(x)$, the value of y when $x = 1$ is written as $f(1)$, the value of y when $x = 5$ is written as $f(5)$, the value of y when $x = a$ as $f(a)$, etc.
- For the function $f(x) = 2x$:
  when $x = 1$, $y = f(1)$
  $ = 2 \times 1$
  $ = 2.$
  when $x = 2$, $y = f(2)$
  $ = 2 \times 2$
  $ = 4$, and so on.
- A mapping diagram can be used to show the input (x) in a particular function $y = (2x)$ and what out (y) it gives.

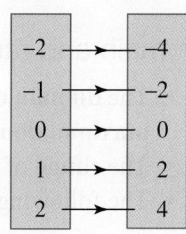

WORKED EXAMPLE 3 Evaluating a function using function notation

If $f(x) = x^2 - 3$, calculate:

a. $f(1)$
b. $f(a)$
c. $3f(2a)$
d. $f(a) + f(b)$
e. $f(a + b)$.

THINK

a. 1. Write the rule.

2. Substitute $x = 1$ into the rule.

3. Simplify and write the answer.

b. 1. Write the rule.

2. Substitute $x = a$ into the rule.

WRITE

a. $f(x) = x^2 - 3$

$f(1) = 1^2 - 3$

$ = 1 - 3$
$ = -2$

b. $f(x) = x^2 - 3$

$f(a) = a^2 - 3$

c. 1. Write the rule.

     c. $f(x) = x^2 - 3$

2. Substitute $x = 2a$ into the rule and simplify.

$$f(2a) = (2a)^2 - 3$$
$$= 2^2 a^2 - 3$$
$$= 4a^2 - 3$$

3. Multiply the answer by 3 and simplify.

$$3f(2a) = 3(4a^2 - 3)$$

4. Write the answer.

$$= 12a^2 - 9$$

d. 1. Write the rule.

     d. $f(x) = x^2 - 3$

2. Evaluate $f(a)$.

$$f(a) = a^2 - 3$$

3. Evaluate $f(b)$.

$$f(b) = b^2 - 3$$

4. Evaluate $f(a) + f(b)$.

$$f(a) + f(b) = a^2 - 3 + b^2 - 3$$

5. Write the answer.

$$= a^2 + b^2 - 6$$

e. 1. Write the rule

     e. $f(x) = x^2 - 3$

2. Evaluate $f(a + b)$.

$$f(a + b) = (a + b)^2 - 3$$
$$= (a + b)(a + b) - 3$$

3. Write the answer.

$$= a^2 + 2ab + b^2 - 3$$

Domain and range

- The **domain** of a function is the set of all allowable values of x. It is sometimes referred to as the **maximal domain**.
- The **range** of a function is the set of y-values produced by the function.
- The following examples show how to determine the domain and range of some graphs.

Graph	Domain	Range
	The domain is all x values except 0. Domain: $x \in R \setminus \{0\}$	The range is all y values except 0. Range: $y \in R \setminus \{0\}$

Graph	Domain	Range
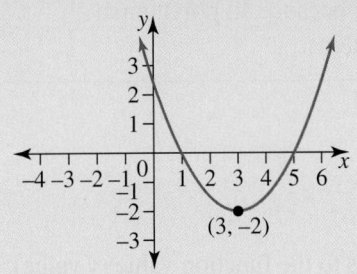	The domain is all x values. Domain: $x \in R$	The range is all y values that are greater than or equal to -3. Range: $y \geq -3$

WORKED EXAMPLE 4 Stating domain and range, and whether a relation is a function

For each of the following, state the domain and range, and whether the relation is a function or not.

a.

b.
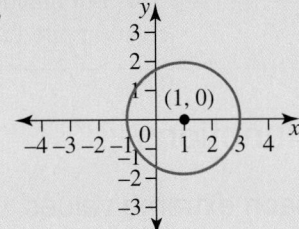

c. $y = 4 - x^3$

THINK

a. 1. State the domain.

2. State the range.

3. Use the vertical line test.

b. 1. State the domain and range.

2. Use the vertical line test.

WRITE

a. Reading from left to right horizontally in the direction of the x-axis, the graph uses every possible x-value. The domain is $(-\infty, \infty)$ or R.

Reading from bottom to top vertically in the direction of the y-axis, the graph's y-values start at -2 and increase from there. The range is $[-2, \infty)$ or $\{y : y \geq -2\}$.

This is a function, since any vertical line cuts the graph exactly once.

b. The domain is $[-1, 3]$; the range is $[-2, 2]$.

This is not a function, as a vertical line can cut the graph more than once.

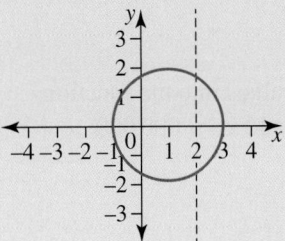

c. 1. State the domain.

c. $y = 4 - x^3$
This is the equation of a polynomial, so its domain is R.

2. State the range.

It is the equation of a cubic polynomial with a negative coefficient of its leading term, so as $x \to \pm\infty$, $y \to \mp\infty$. The range is R.

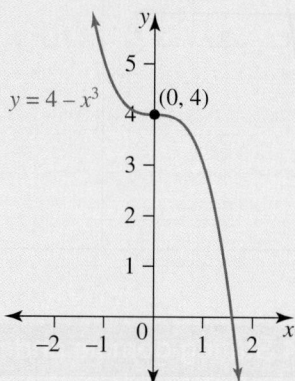

3. Is the relation a function?

This is a function, because all polynomial relations are functions.

▶ 17.2.3 Identifying features of functions

eles-4986

Behaviour of functions as they approach extreme values

- We can identify features of certain functions by observing what happens to the function value (y value) when x approaches a very small value such as 0 ($x \to 0$) or a very large value such as ∞ ($x \to \infty$).

WORKED EXAMPLE 5 Identifying end behaviour of a function

Describe what happens to these functions as the value of x increases, that is, as $x \to \infty$.

a. $y = x^2$ **b.** $f(x) = 2^{-x}$ **c.** $f(x) = \dfrac{1}{x} + 1$

THINK

a. 1. Write the function.

2. Substitute large x values into the function, such as $x = 10\,000$ and $x = 1\,000\,000$.

3. Write a conclusion.

b. 1. Write the function.

2. Substitute large x values into the function, such as $x = 10\,000$ and $x = 1\,000\,000$.

3. Write a conclusion.

c. 1. Write the function.

WRITE

a. $y = x^2$

$f(10\,000) = 100\,000\,000$
$f(1\,000\,000) = 1 \times 10^{12}$

As $x \to \infty$, $f(x)$ also increases; that is, $f(x) \to \infty$.

b. $f(x) = 2^{-x}$

$f(10\,000) \approx 0$
$f(1\,000\,000) \approx 0$

As $x \to \infty$, $f(x) \to 0$.

c. $f(x) = \dfrac{1}{x} + 1$

2. Substitute large x values into the function, such as $x = 10\,000$ and $x = 1\,000\,000$.	$f(10\,000) = 1.0001$ $f(1\,000\,000) = 1.000\,001$
3. Write a conclusion.	As $x \to \infty, f(x) \to 1$.

Points of intersection

- A **point of intersection** between two functions is a point at which the two graphs cross paths.
- To determine points of intersection, equate the two graphs and solve to calculate the coordinates of the points of intersection. For example, there are three points of intersection between the linear relationship and the graph of cubic function shown.

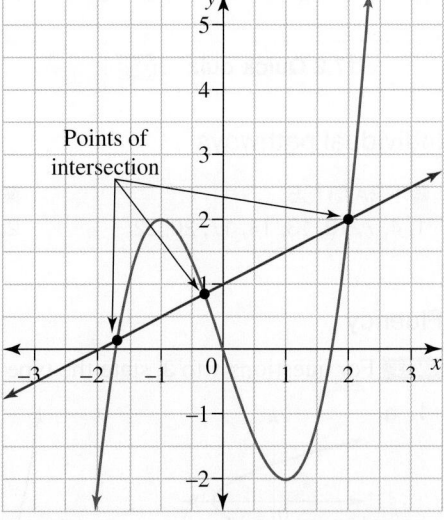

Points of intersection

WORKED EXAMPLE 6 Determining points of intersection

Determine any points of intersection between $f(x) = 2x + 1$ and $g(x) = \dfrac{1}{x}$.

THINK	WRITE
1. Write the two equations.	$f(x) = 2x + 1$ $g(x) = \dfrac{1}{x}$
2. Points of intersection are common values between the two curves. To solve the equations simultaneously, equate both functions.	For points of intersection: $2x + 1 = \dfrac{1}{x}$
3. Rearrange the resulting equation and solve for x.	$2x^2 + x = 1$ $2x^2 + x - 1 = 0$ $(2x - 1)(x + 1) = 0$ $x = \dfrac{1}{2} \text{ or } -1$
4. Substitute the x values into either function to calculate the y values.	$f\left(\dfrac{1}{2}\right) = 2 \times \dfrac{1}{2} + 1 = 2$ $f(-1) = 2 \times (-1) + 1 = -1$
5. Write the coordinates of the two points of intersection.	The points of intersection are $\left(\dfrac{1}{2}, 2\right)$ and $(-1, -1)$.

 Resources

 Interactivities Relations (int-6208)
Evaluating functions (int-6209)

Exercise 17.2 Functions and relations

learn

17.2 Quick quiz on	**17.2 Exercise**

Individual pathways

■ PRACTISE	■ CONSOLIDATE	■ MASTER
1, 4, 7, 10, 13, 14, 17, 20, 22	2, 5, 8, 11, 12, 15, 18, 23	3, 6, 9, 16, 19, 21, 24, 25

Fluency

WE1 For questions **1** to **3**, state the type of relation that each graph represents.

1. a. 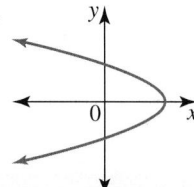   b. 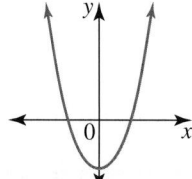   c. 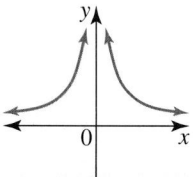   d.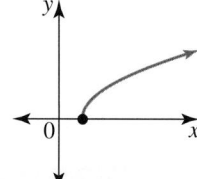

2. a. 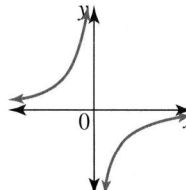   b. 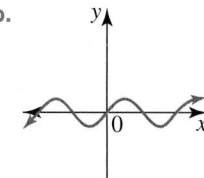   c. 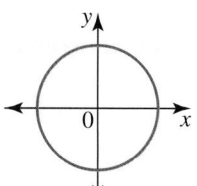   d.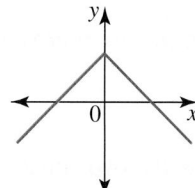

3. a. 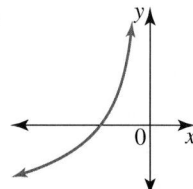   b. 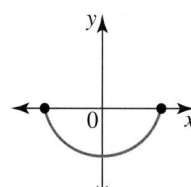   c. 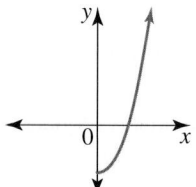   d.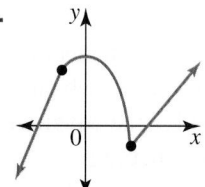

4. **WE2** Use the vertical line test to determine which of the relations in question **1** are functions.

5. Use the vertical line test to determine which of the relations in question **2** are functions.

6. Use the vertical line test to determine which of the relations in question **3** are functions.

7. **WE3** If $f(x) = 3x + 1$, calculate:

   a. $f(0)$ b. $f(2)$ c. $f(-2)$ d. $f(5)$

8. If $g(x) = \sqrt{x + 4}$, calculate:

   a. $g(0)$ b. $g(-3)$ c. $g(5)$ d. $g(-4)$

9. If $g(x) = 4 - \dfrac{1}{x}$, calculate:

   a. $g(1)$ **b.** $g\left(\dfrac{1}{2}\right)$ **c.** $g\left(-\dfrac{1}{2}\right)$ **d.** $g\left(-\dfrac{1}{5}\right)$

10. If $f(x) = (x+3)^2$, calculate:

   a. $f(0)$ **b.** $f(-2)$ **c.** $f(1)$ **d.** $f(a)$

11. If $h(x) = \dfrac{24}{x}$, calculate:

   a. $h(2)$ **b.** $h(4)$ **c.** $h(-6)$ **d.** $h(12)$

12. For each of the following, state the domain and range.

   a.

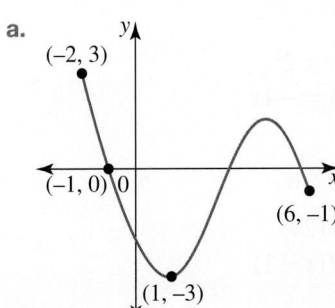

   b.

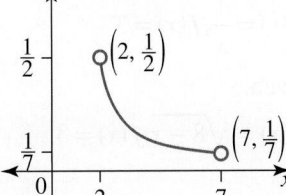

13. **WE4** For each of the following, state the domain and range, and whether the relation is a function or not.

   a.

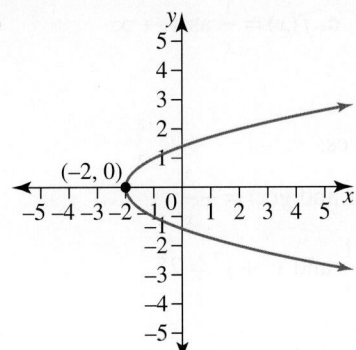

   b.

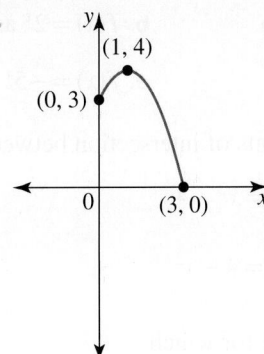

14. Sketch the graph of $y = (x+2)^2 - 3$, and state its domain and range.

Understanding

15. Consider the relation $y = 2x - 1, -2 \le x \le 6$.

   a. Sketch the graph of the relation.
   b. State the domain and range.
   c. Explain, with a reason, whether or not this relation is a function.

16. Consider the relation $y = (x-2)^3 + 4, -2 \le x \le 4$.

    a. Sketch the graph of the relation.

    b. State the domain and range.

    c. Explain, with a reason, whether or not this relation is a function.

17. State which of the following relations are functions.

    a. 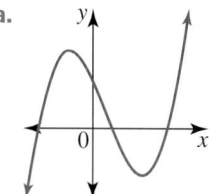   **b.** $x^2 + y^2 = 9$ **c.** $y = 8x - 3$ **d.**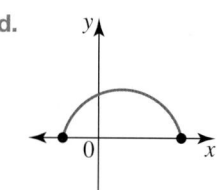

18. State which of the following relations are functions.

    a. $y = 2x + 1$ **b.** $y = x^2 + 2$ **c.** $y = 2^x$

    d. $x^2 + y^2 = 25$ **e.** $x^2 + 4x + y^2 + 6y = 14$ **f.** $y = -4x$

19. Given that $f(x) = \dfrac{10}{x} - x$, determine:

    a. $f(2)$ **b.** $f(-5)$ **c.** $f(2x)$

    d. $f(x^2)$ **e.** $f(x+3)$ **f.** $f(x-1)$

20. Calculate the value (or values) of x for which each function has the value given.

    a. $f(x) = 3x - 4, f(x) = 5$ **b.** $g(x) = x^2 - 2, g(x) = 7$ **c.** $f(x) = \dfrac{1}{x}, f(x) = 3$

21. Calculate the value (or values) of x for which each function has the value given.

    a. $h(x) = x^2 - 5x + 6, h(x) = 0$ **b.** $g(x) = x^2 + 3x, g(x) = 4$ **c.** $f(x) = \sqrt{8-x}, f(x) = 3$

Communicating, reasoning and problem solving

22. **WE5** Describe what happens to:

    a. $f(x) = x^2 + 3$ as $x \to \infty$ **b.** $f(x) = 2^x$ as $x \to -\infty$ **c.** $f(x) = \dfrac{1}{x}$ as $x \to \infty$

    d. $f(x) = x^3$ as $x \to -\infty$ **e.** $f(x) = -5^x$ as $x \to -\infty$

23. **WE6** Determine any points of intersection between the following curves.

    a. $f(x) = 2x - 4$ and $g(x) = x^2 - 4$ **b.** $f(x) = -3x + 1$ and $g(x) = -\dfrac{2}{x}$

    c. $f(x) = x^2 - 4$ and $g(x) = 4 - x^2$ **d.** $f(x) = \dfrac{3}{4}x - 6\dfrac{1}{4}$ and $x^2 + y^2 = 25$

24. Determine the value(s) of for which:

    a. $f(x) = x^2 + 7$ and $f(x) = 16$ **b.** $g(x) = \dfrac{1}{x-2}$ and $g(x) = 3$ **c.** $h(x) = \sqrt{8+x}$ and $h(x) = 6$.

25. Consider the function defined by the rule $f(x) = (x-1)^2 + 2$.

    a. State the range of the function.

    b. Determine the type of mapping for the function.

    c. Sketch the graph of the function stating where it cuts the y-axis and its turning point.

    d. Select the largest domain where x is positive such that f is a one-to-one function.

LESSON
17.3 Graphing cubic functions

LEARNING INTENTION

At the end of this lesson you should be able to:
- plot the graph of a cubic function using a table of values
- sketch the graph of a cubic function by calculating its intercepts
- determine the equation of a cubic function by inspection.

▶ 17.3.1 Cubic functions

eles-4989

- **Cubic functions** are **polynomials** where the highest power of the variable is three or the product of pronumeral makes up three.
- Some examples of cubic functions are $y = x^3$, $y = (x+1)(x-2)(x+3)$ and $y = 2x^2(4x-1)$.
- The following worked examples show how the graphs of cubic functions can be created by plotting points.

WORKED EXAMPLE 7 Plotting a cubic function using a table of values

Plot the graph of $y = x^3 - 1$ by completing a table of values.

THINK	WRITE/DRAW
1. Prepare a table of values, taking x-values from -3 to 3. Fill in the table by substituting each x-value into the given equation to determine the corresponding y-value.	

x	-3	-2	-1	0	1	2	3
y	-28	-9	-2	-1	0	7	26

2. Draw a set of axes and plot the points from the table. Join them with a smooth curve.

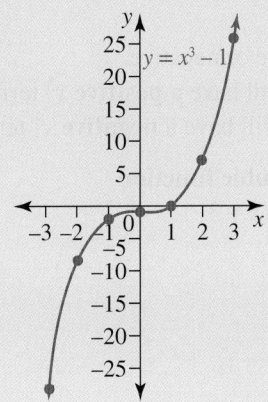

WORKED EXAMPLE 8 Plotting a cubic function using a table of values

Plot the curve of $y = x(x-2)(x+2)$ by completing a table of values.

THINK	WRITE/DRAW
1. Prepare a table of values, taking x-values from -3 to 3. Fill in the table by substituting each x-value into the given equation.	

x	-3	-2	-1	0	1	2	3
y	-15	0	3	0	-3	0	15

▶

2. Draw a set of axes and plot the points from the table. Join them with a smooth curve.

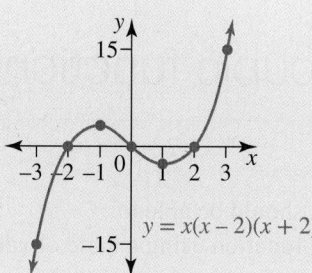

$y = x(x - 2)(x + 2)$

17.3.2 Sketching cubic functions

eles-4990

- Graphs of cubic functions have either *two turning points* or *one point of inflection*.
  - These two types of graphs are shown. Note that a turning point is a point where the gradient of the graph changes from decreasing to increasing or vice versa.
  - For the purposes of this topic, we will only consider the points of inflection where the graph momentarily flattens out.

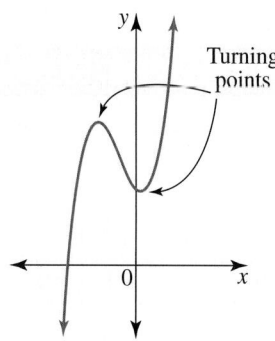

Turning points

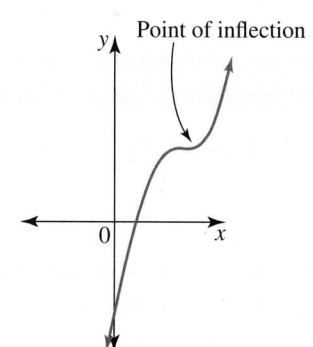

Point of inflection

- Cubic functions can be positive or negative.
  - A positive cubic function will have a **positive** x^3 term and will have a general **upward** slope.
  - A negative cubic function will have a **negative** x^3 term and will have a general **downward** slope.

Example of a positive cubic function
$y = (x + 2)(x - 1)(x + 3)$

Example of a negative cubic function
$y = (x + 2)(1 - x)(x + 3)$

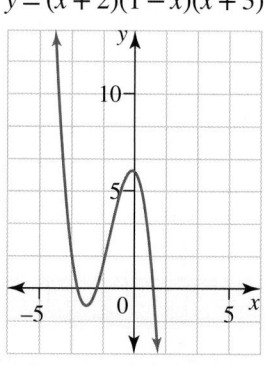

- To sketch the graph of a cubic function:
  1. determine the y-intercept by setting $x = 0$ and solving for y
  2. determine the x-intercepts by setting $y = 0$ and solving for x (you will need to use the Null Factor Law)
  3. draw the intercepts on the graph, then use those points to sketch the cubic graph.

- There are three special cases when sketching a cubic graph in factorised form. These are shown below.

$y = k(x-a)(x-b)(x-c)$	$y = k(x-a)(x-b)^2$	$y = k(x-a)^3$
Cubic functions of this form will - cut the x-axis at $x = a, b, c$. 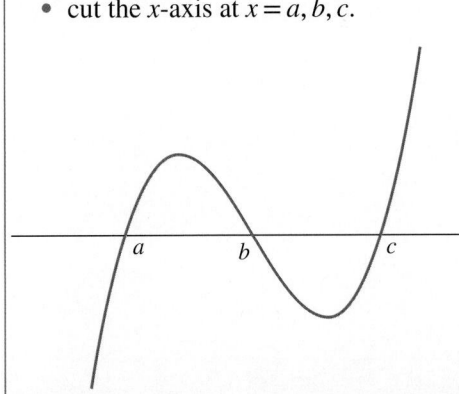	Cubic functions of this form will - cut the x-axis at $x = a$ - have a turning point on the x-axis at $x = b$. 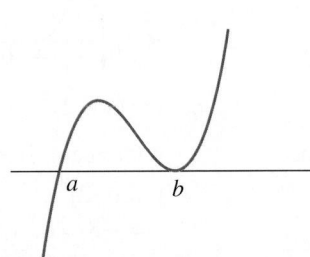	Cubic functions of this form will - have a point of inflection on the x-axis at $x = a$. 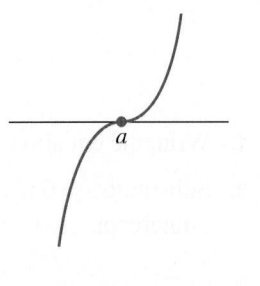

WORKED EXAMPLE 9 Sketching a cubic graph by determining intercepts

Sketch the following, showing all intercepts.
a. $y = (x-2)(x-3)(x+5)$ b. $y = (x-6)^2(4-x)$ c. $y = (x-2)^3$

THINK	WRITE/DRAW
a. 1. Write the equation.	a. $y = (x-2)(x-3)(x+5)$
2. The y-intercept occurs where $x = 0$. Substitute $x = 0$ into the equation.	y-intercept: if $x = 0$, $y = (-2)(-3)(5)$ $= 30$ Point: $(0, 30)$
3. Solve $y = 0$ to calculate the x-intercepts.	x-intercepts: if $y = 0$, $x - 2 = 0, x - 3 = 0$ or $x + 5 = 0$ $x = 2, x = 3$ or $x = -5$ Points: $(2, 0), (3, 0), (-5, 0)$
4. Combine the above steps to sketch.	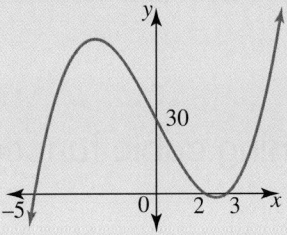
b. 1. Write the equation.	b. $y = (x-6)^2(4-x)$
2. Substitute $x = 0$ to calculate the y-intercept.	y-intercept: if $x = 0$, $y = (-6)^2(4)$ $= 144$ Point: $(0, 144)$
3. Solve $y = 0$ to calculate the x-intercepts.	x-intercept: if $y = 0$, $x - 6 = 0$ or $4 - x = 0$ $x = 6$ or $x = 4$ Point: $(6, 0), (4, 0)$

4. Combine all information and sketch the graph.
   Note: The curve just touches the x-axis at $x = 6$. This occurs with a double factor such as $(x - 6)^2$.

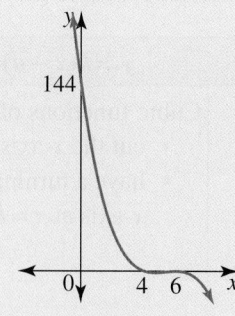

c. 1. Write the equation.

   2. Substitute $x = 0$ to calculate the y-intercept.

   3. Solve $y = 0$ to calculate the x-intercepts.

   4. Combine all information and sketch the graph.
      Note: The point of inflection is at $x = 2$. This occurs with a triple factor such as $(x - 2)^3$.

c. $y = (x - 2)^3$

y-intercept: if $x = 0$,
$y = (-2)^3$
$= -8$

x-intercept: if $x = 0$,

$x - 2 = 0$
$x = 2$

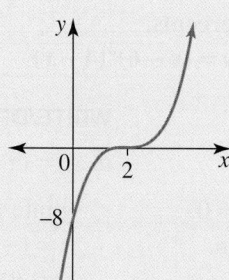

 Resources ─────────────────────────────────────

 Interactivity Cubic polynomials (int-2566)

Exercise 17.3 Graphing cubic functions

learn on

17.3 Quick quiz on	17.3 Exercise

Individual pathways

■ PRACTISE	■ CONSOLIDATE	■ MASTER
1, 4, 9, 10, 13, 16	2, 5, 7, 11, 14, 17	3, 6, 8, 12, 15, 18

Fluency

WE7,8&9 For questions **1** to **3**, sketch the following, showing all intercepts.

1. a. $y = (x - 1)(x - 2)(x - 3)$
   c. $y = (x + 6)(x + 1)(x - 7)$

   b. $y = (x - 3)(x - 5)(x + 2)$
   d. $y = (x + 4)(x + 9)(x + 3)$

2. **a.** $y = (x+8)(x-11)(x+1)$
   c. $y = (2x-5)(x+4)(x-3)$
   b. $y = (2x-6)(x-2)(x+1)$
   d. $y = (3x+7)(x-5)(x+6)$

3. **a.** $y = (4x-3)(2x+1)(x-4)$
   c. $y = (x-3)^2(x-6)$
   b. $y = (2x+1)(2x-1)(x+2)$
   d. $y = (x+2)(x+5)^2$

For questions **4** to **6**, sketch the following (a mixture of positive and negative cubics).

4. **a.** $y = (2-x)(x+5)(x+3)$
   c. $y = (x+8)(x-8)(2x+3)$
   b. $y = (1-x)(x+7)(x-2)$
   d. $y = (x-2)(2-x)(x+6)$

5. **a.** $y = x(x+1)(x-2)$
   c. $y = 3(x+1)(x+10)(x+5)$
   b. $y = -2(x+3)(x-1)(x+2)$
   d. $y = -3x(x-4)^2$

6. **a.** $y = 4x^2(x+8)$
   c. $y = (6x-1)^2(x+7)$
   b. $y = (5-3x)(x-1)(2x+9)$
   d. $y = -2x^2(7x+3)$

7. **MC** Select a reasonable sketch of $y = (x+2)(x-3)(2x+1)$ from the following.

A.
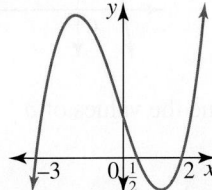

B.
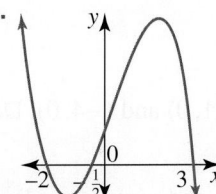

C.
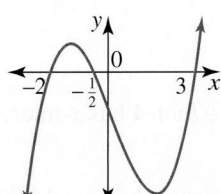

D.
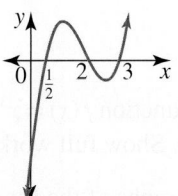

8. **MC** The graph shown could be that of:

   A. $y = x^2(x+2)$
   B. $y = (x+2)^3$
   C. $y = (x-2)(x+2)^2$
   D. $y = (x-2)^2(x+2)$

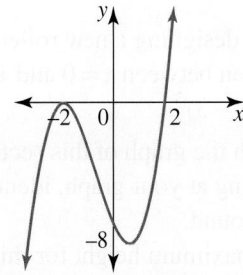

9. **MC** The graph shown has the equation:

   A. $y = (x+1)(x+2)(x+3)$
   B. $y = (x+1)(x-2)(x+3)$
   C. $y = (x-1)(x+2)(x+3)$
   D. $y = (x-1)(x+2)(x-3)$

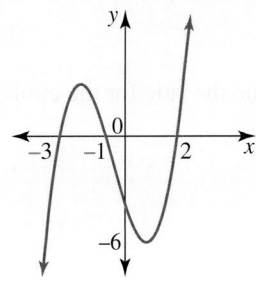

Understanding

10. **MC** If a, b and c are positive numbers, the equation of the graph shown could be:

    A. $y = (x-a)(x-b)(x-c)$
    B. $y = (x+a)(x-b)(x+c)$
    C. $y = (x+a)(x+b)(x-c)$
    D. $y = (x-a)(x+b)(x-c)$

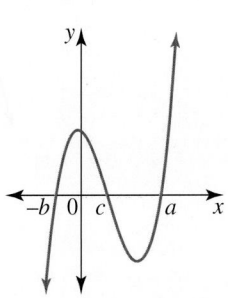

11. Sketch the graph of each of the following.
    a. $y = x(x-1)^2$
    b. $y = -(x+1)^2(x-1)$

12. Sketch the graph of each of the following.
    a. $y = (2-x)(x^2-9)$
    b. $y = -x(1-x^2)$

Communicating, reasoning and problem solving

13. For the graph shown, explain whether:
    a. the gradient is positive, negative or zero to the left of the point of inflection.
    b. the gradient is positive, negative or zero to the right of the point of inflection.
    c. the gradient is positive, negative or zero at the point of inflection.
    d. this is a positive or negative cubic graph.

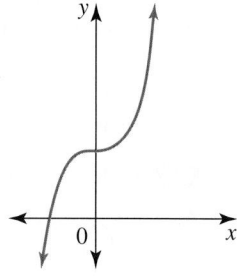

14. The function $f(x) = x^3 + ax^2 + bx + 4$ has x-intercepts at $(1, 0)$ and $(-4, 0)$. Determine the values of a and b. Show full working.

15. The graphs of the functions $f(x) = x^3 + (a+b)x^2 + 3x - 4$ and $g(x) = (x-3)^3 + 1$ touch. Express a in terms of b.

16. Susan is designing a new rollercoaster ride using maths. For the section between $x = 0$ and $x = 3$, the equation of the ride is $y = x(x-3)^2$.

    a. Sketch the graph of this section of the ride.
    b. Looking at your graph, identify where the ride touches the ground.
    c. The maximum height for this section is reached when $x = 1$. Use algebra to calculate the maximum height.

17. Determine the rule for the cubic function shown.

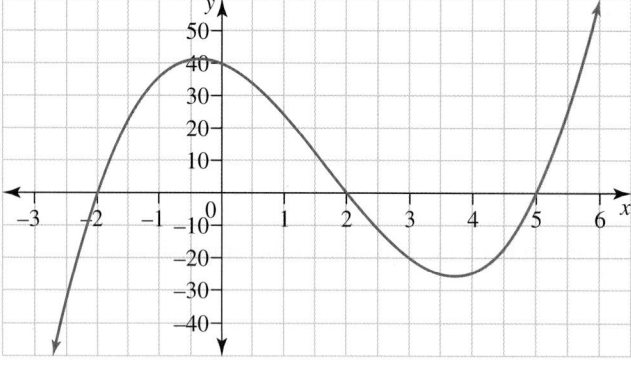

18. A girl uses 140 cm of wire to make a frame of a cuboid with a square base as shown.

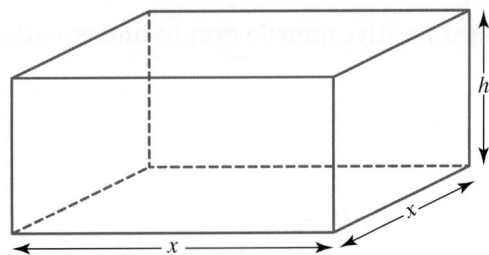

The base length of the cuboid is x cm and the height is h cm.

a. Explain why the volume cm^3 is given by $V = 35x^2 - 2x^3$.
b. Determine possible values that x can assume.
c. Evaluate the volume of the cuboid when the base area is 81 cm^2.
d. Sketch the graph of V versus x.
e. Use technology to determine the coordinates of the maximum turning point. Explain what these coordinates mean.

LESSON
17.4 Graphing quartic functions

LEARNING INTENTION

At the end of this lesson you should be able to:
- factorise the equation of a quartic function
- sketch the graph of a quartic function from a factorised equation
- determine the equation of a quartic function by inspection.

17.4.1 Quartic functions

eles-4991

- **Quartic function** are polynomials where the highest power of the variable is 4 or the product of pronumeral makes up four.
- Some examples of quartic functions are $y = x^4$ and $y = (x + 1)(x - 2)(x + 3)(x - 4)$.
- There are three types of quartic functions: those with *one turning point*, those with *three turning points* and those with *one turning point and one point of inflection*.
- The table below includes the standard types of positive quartic equations.

Type of quartic function	Standard positive quartic graphs and equations with $k > 0$
One turning point	$y = kx^4$ $y = x^2(kx^2 + c), c \geq 0$

(continued)

(continued)

Type of quartic function	Standard positive quartic graphs and equations with $k > 0$		
Three turning points	$y = kx^2(x-b)(x-c)$	$y = k(x-b)^2(x-c)^2$	$y = k(x-b)(x-c)(x-d)(x-e)$
One turning point and one point of inflection	$y = k(x-b)(x-c)^3$		

- There are also negative equivalents to all of the above graphs when $k < 0$. The negative graphs have the same shape, but are reflected across the x-axis. For example, the graph of $y = kx^4$ for $k < 0$ is shown.

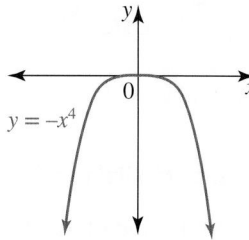

$y = -x^4$

▶ 17.4.2 Sketching quartic functions

eles-4992

- To sketch a quartic function:
  1. factorise the equation so that the equation is in the form matching one of the standard quartics. To factorise, use the **factor theorem** and long division. (If the equation is already factorised, skip this step.)
  2. calculate all x- and y-intercepts
  3. draw the intercepts on the graph, then use those points to sketch the cubic graph.

WORKED EXAMPLE 10 Factorising then sketching the graph of a quartic function

Sketch the graph of $y = x^4 - 2x^3 - 7x^2 + 8x + 12$, showing all intercepts.

THINK

1. Calculate the y-intercept.

2. Let $P(x) = y$.

3. Determine two linear factors of the quartic expressions, if possible, using the factor theorem.

WRITE/DRAW

When $x = 0$, $y = 12$.
The y-intercept is 12.

Let $P(x) = x^4 - 2x^3 - 7x^2 + 8x + 12$

$P(1) = (1)^4 - 2(1)^3 - 7(1)^2 + 8(1) + 12$
$= 12$
$\neq 0$
$P(-1) = (-1)^4 - 2(-1)^3 - 7(-1)^2 + 8(-1) + 12$
$= 0$
$(x + 1)$ is a factor.

$$P(2) = (2)^4 - 2(2)^3 - 7(2)^2 + 8(2) + 12$$
$$= 0$$
$(x - 2)$ is a factor.

4. Calculate the product of the two linear factors
$(x + 1)(x - 2) = x^2 - x - 2$

5. Use long division to divide the quartic by the quadratic factor $x^2 - x - 2$.

$$
\begin{array}{r}
x^2 - x - 6 \\
x^2 - x - 2 \overline{) x^4 - 2x^3 - 7x^2 + 8x + 12} \\
\underline{x^4 - x^3 - 2x^2} \\
-x^3 - 5x^2 + 8x \\
\underline{-x^3 + x^2 + 2x} \\
-6x^2 + 6x + 12 \\
\underline{-6x^2 + 6x + 12} \\
0
\end{array}
$$

6. Express the quartic in factorised form.
$$y = (x + 1)(x - 2)(x^2 - x - 6)$$
$$= (x + 1)(x - 2)(x - 3)(x + 2)$$

7. To calculate the x- intercepts, solve $y = 0$.
If $0 = (x + 1)(x - 2)(x - 3)(x + 2)$
$x = -1, 2, 3, -2$.

8. State the x-intercepts.
The x-intercepts are $-2, -1, 2, 3$.

9. Sketch the graph of the quartic.

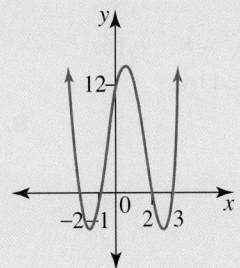

on Resources

★ **Interactivity** Quartic functions (int-6213)

Exercise 17.4 Graphing quartic functions

learn on

17.4 Quick quiz on	17.4 Exercise

Individual pathways

■ PRACTISE	■ CONSOLIDATE	■ MASTER
1, 4, 7, 10, 11, 16	2, 5, 9, 12, 13, 17	3, 6, 8, 14, 15, 18

Fluency

WE10 For questions **1** to **3**, sketch the graph of each of the following showing all intercepts. You may like to verify the shape of the graph using digital technology.

1. **a.** $y = (x - 2)(x + 3)(x - 4)(x + 1)$

   b. $y = (x^2 - 1)(x + 2)(x - 5)$

2. **a.** $y = 2x^4 + 6x^3 - 16x^2 - 24x + 32$ **b.** $y = x^4 + 4x^3 - 11x^2 - 30x$

3. **a.** $y = x^4 - 4x^2 + 4$ **b.** $y = 30x - 37x^2 + 15x^3 - 2x^4$

For questions **4** to **6**, sketch each of the following.

4. **a.** $y = x^2(x-1)^2$ **b.** $y = -(x+1)^2(x-4)^2$

5. **a.** $y = -x(x-3)^3$ **b.** $y = (2-x)(x-1)(x+1)(x-4)$

6. $y = (x-a)(b-x)(x+c)(x+d)$, where $a, b, c, d > 0$

Understanding

7. **MC** A quartic touches the x-axis at $x = -3$ and $x = 2$. It crosses the y-axis at $y = -9$. A possible equation is:

   A. $y = \dfrac{1}{4}(x+3)^2(x-2)^2$ **B.** $y = -\dfrac{1}{6}(x+3)^3(x-2)$

   C. $y = -\dfrac{3}{8}(x+3)(x-2)^3$ **D.** $y = -\dfrac{1}{4}(x+3)^2(x-2)^2$

8. **MC** Consider the function $f(x) = x^4 - 8x^2 + 16$. When factorised, $f(x)$ is equal to:

   A. $(x+2)(x-2)(x-1)(x+4)$ **B.** $(x+3)(x-2)(x-1)(x+1)$

   C. $(x-2)^3(x+2)$ **D.** $(x-2)^2(x+2)^2$

9. **MC** Consider the function $f(x) = x^4 - 8x^2 + 16$.
   The graph of $f(x)$ is best represented by:

   A. 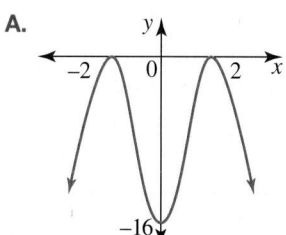   **B.**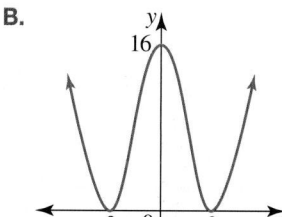

   C. 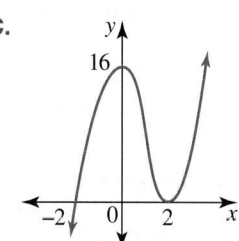   **D.**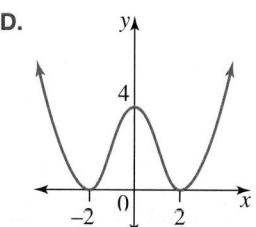

Communicating, reasoning and problem solving

For questions **10** to **12**, sketch the graph of each of the following functions. Verify your answers using digital technology.

10. **a.** $y = x(x-1)^3$ **b.** $y = (2-x)(x^2-4)(x+3)$

11. **a.** $y = (x+2)^3(x-3)$ **b.** $y = 4x^2 - x^4$

12. **a.** $y = -(x-2)^2(x+1)^2$ **b.** $y = x^4 - 6x^2 - 27$

13. The functions $y = (a - 2b)x^4 - 3x - 2$ and $y = x^4 - x^3 + (a + 5b)x^2 - 5x + 7$ both have an x-intercept of 1. Determine the value of a and b. Show your working.

14. Sketch the graph of each of the following functions. Verify your answers using digital technology.
    a. $y = x^4 - x^2$
    b. $y = 9x^4 - 30x^3 + 13x^2 + 20x + 4$

15. Patterns emerge when we graph polynomials with repeated factors, that is, polynomials of the form $P(x) = (x - a)^n, n > 1$. Discuss what happens if:
    a. n is even
    b. n is odd.

16. The function $f(x) = x^4 + ax^3 - 4x^2 + bx + 6$ has x-intercepts $(2, 0)$ and $(-3, 0)$. Determine the values of a and b.

17. A carnival ride has a piece of the track modelled by the rule

$$h = -\frac{1}{300}x(x - 12)^2(x - 20) + 15, \ 0 \le x \le 20$$

where x metres is the horizontal displacement from the origin and h metres is the vertical displacement of the track above the horizontal ground.

    a. Determine how high above the ground level the track is at the origin.
    b. Use technology to sketch the function. Give the coordinates of any stationary points (that is, turning points or points of inflection).
    c. Evaluate how high above ground level the track is when $x = 3$.

18. Determine the rule for the quartic function shown.

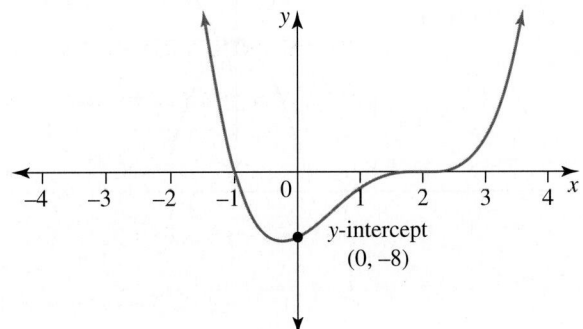

LESSON
17.5 Transformations

LEARNING INTENTION

At the end of this lesson you should be able to:
- sketch the graph of a function that has undergone some transformations
- describe a transformation using the words translation, dilation and reflection.

▶ 17.5.1 General transformations

eles-4993

- When the graph of a function has been moved, stretched and/or flipped, this is called a **transformation**.
- There are three types of transformations:
  - **dilations** are stretches of graphs to make them thinner or wider
  - **reflections** are when a graph is flipped in the x- or y-axis
  - **translations** are movements of graphs left, right, up or down.
- The following table summarises transformations of the general function $f(x)$ with examples given for transformations of the basic quadratic function $y = x^2$ shown here:

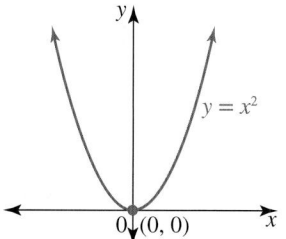

Transformation	Equation and explanation	Example(s)
Dilation	$y = kf(x)$ This is a dilation by a factor of k in the x direction. If $k > 1$ the graph becomes thinner, if $0 < k < 1$ the graph becomes wider.	$y = 2x^2$ $y = x^2$ (0, 0) $y = \frac{1}{4}x^2$ $y = x^2$ (0, 0)
Reflection	$y = -f(x)$ This is a reflection in the x-axis. $y = f(-x)$ This is a reflection in the y-axis.	$y = x^2$ (0, 0) $y = -x^2$

Vertical translation	$y = f(x) + c$ This is a vertical translation of c units. If c is positive the translation is up, if c is negative the translation is down.	── original function ── translated function $y = x^2 + 2$, $y = x^2$, $(0, 2)$ $y = x^2$, $y = x^2 - 3$, $(0, -3)$
Horizontal translation	$y = f(x - b)$ This is a horizontal translation of b units. If b is positive the translation is right, if b is negative the translation is left.	$y = x^2$, $(0, 4)$, $y = (x - 2)^2$, $(2, 0)$ $y = (x + 1)^2$, $y = x^2$, $(0, 1)$, $(-1, 0)$

- Note that the graph $y = x^2$ does not change when reflected across the y-axis because it is symmetrical about the y-axis.
- With knowledge of the transformations discussed in this section, it is possible to generate many other graphs without knowing the equation of the original function.

WORKED EXAMPLE 11 Sketching transformations

Use the sketch of $y = f(x)$ shown to sketch:
a. $y = f(x) + 1$ **b.** $y = f(x) - 1$ **c.** $y = -f(x)$ **d.** $y = f(-x)$.

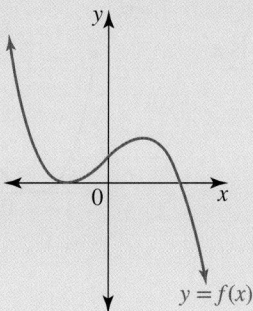

$y = f(x)$

THINK

a. 1. Sketch the original $y = f(x)$.

2. Look at the equation $y = f(x) + 1$. This is a translation of one unit in the vertical direction: that is one unit up.

WRITE/DRAW

a.

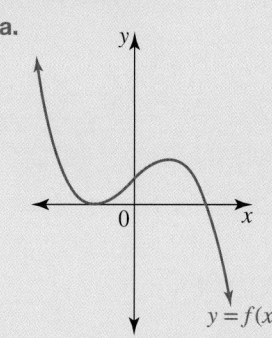

$y = f(x)$

3. Sketch the graph of $y = f(x) + 1$ using a similar scale to the original.

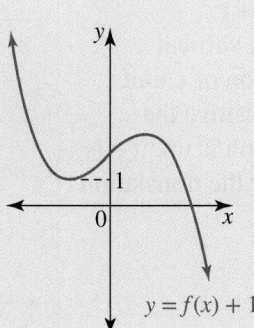

b. 1. Sketch the original $y = f(x)$.
   2. Look at the equation $y = f(x) - 1$. This is a translation of negative one unit in the vertical direction; that is, one unit down.

b.

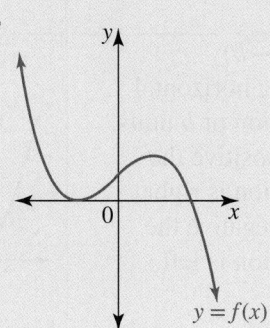

3. Sketch the graph of $y = f(x) - 1$ using a similar scale to the original.

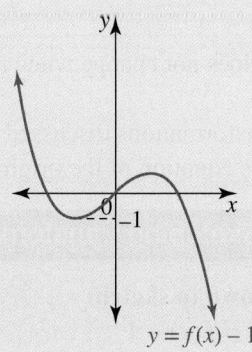

c. 1. Sketch the original $y = f(x)$.
   2. Look at the equation $y = -f(x)$. This is a reflection across the x-axis.

c.

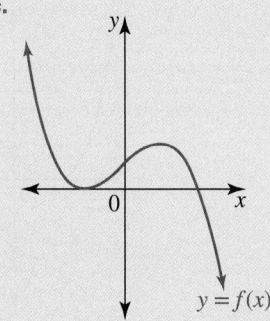

3. Sketch the graph of $y = -f(x)$ using a similar scale to the original.

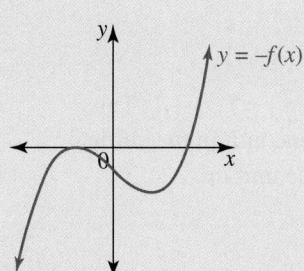

d. 1. Sketch the function $y = f(x)$

2. Look at the equation $y = f(-x)$. This is a reflection in the y-axis.

d.

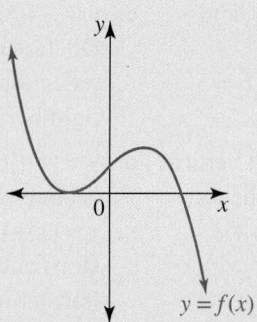

3. Sketch the graph of $y = f(-x)$ using a similar scale to the original.

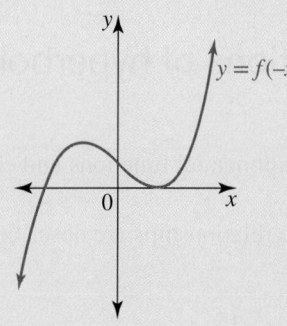

- When applying transformations to a graph, the order can be important; generally, they should be applied in the following order:
  - dilation
  - reflection
  - translation

Transformations of $y = f(x)$

To transform the graph $y = f(x)$ to the graph of $y = kf(x - b) + c$
- **k gives the dilation factor from the x-axis**
  - if $k < 0$, there is a reflection in the x-axis
- **b gives the horizontal translation**
  - if $b > 0$, horizontal translation to the right
  - if $b < 0$, horizontal translation to the left
- **c gives the vertical translation**
  - if $c > 0$, vertical translation upwards
  - if $c < 0$, vertical translation downwards

WORKED EXAMPLE 12 Describing transformations

Describe the transformations applied to the graph of $y = x^3$ to obtain the graph:

a. $y = 2x^3$ **b.** $y = 3x^3 - 1$ **c.** $y = -4(x - 5)^3$ **d.** $y = (x + 6)^3 + 7$

THINK

a. Identify the dilation factor, $k = 2$

b. Identify the dilation factor, $k = 2$ and vertical translation $c = -1$

WRITE

a. $y = 2x^3$
Dilation of factor 2 from the x-axis

b. $y = 3x^3 - 1$
Dilation of factor 3 from the x-axis followed by a vertical translation downwards by 1 unit.

c. Identify the dilation factor, and reflection in the x-axis, and horizontal translation $b = 5$

c. $y = -4(x - 5)^3$

Dilation of factor 4 from the x-axis followed by a reflection in the x-axis, then a horizontal translation to the right by 5 units.

d. Identify the horizontal and vertical translations, rewriting in the correct form $y = kf(x - b) + c$

d. $y = (x + 6)^3 + 7$

is equivalent to

$y = (x - (-6))^3 + 7$

Horizontal translation to the left by 6 units and a vertical translation upwards by 7 units.

⏵ 17.5.2 Transformations of hyperbolas, exponential functions and circles

eles-4994

- Graphs of hyperbolas, exponential functions and circles were introduced in topic 8, non-linear relationships.
- Transformations of these relationships are described below.

Hyperbolas

- The standard hyperbola $y = \dfrac{1}{x}$ is transformed to $y = \dfrac{k}{x - b} + c$.

The hyperbola $y = \dfrac{k}{x - b} + c$

This is the standard hyperbola $y = \dfrac{1}{x}$ under the transformations:

- **dilation by a factor of k**
- **reflection in the x-axis if $k < 0$**
- **horizontal translation of b units**
  - **to the right if $b > 0$**
  - **to the left if $b < 0$**
- **vertical translation of c units**
  - **upwards if $c > 0$**
  - **downwards if $c < 0$**

- Below are the graphs of the standard hyperbola $y = \dfrac{1}{x}$ and the transformed hyperbola $y = \dfrac{2}{x - 3} + 1$

$y = \dfrac{1}{x}$	$y = \dfrac{2}{x - 3} + 1$

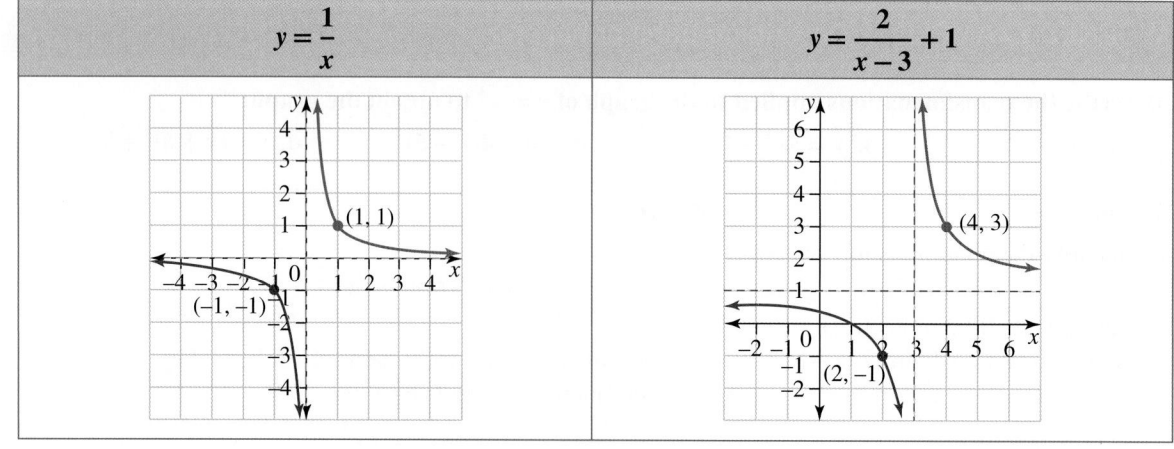

The transformations are:
- dilation by a factor of 2
- horizontal translation of 3 to the right
- vertical translation of 1 upwards.

Note: Hyperbolas may be written in various forms. It is generally easier to see transformations if the x is positive.

For example: $\dfrac{2}{(4-x)} = \dfrac{2}{(-x+4)} = \dfrac{-2}{(x-4)}$ is a dilation, reflection and horizontal translation.

Exponential functions

- The standard exponential function $y = a^x, a > 1$ is transformed to $y = ka^{(x-b)} + c$.

The exponential function $y = ka^{(x-b)} + c$

This is the standard exponential function $y = a^x, a \neq 1$ under the transformations:
- **dilation by a factor of k**
- **reflection in the x-axis if $k < 0$**
- **horizontal translation of b units**
    - **to the right if $b > 0$**
    - **to the left if $b < 0$**
- **vertical translation of c units**
    - **upwards if $c > 0$**
    - **downwards if $c < 0$**

- The exponential function $y = a^{-x}, a > 1$ is the reflection of $y = a^x$ in the y-axis.
- It transforms to $y = ka^{-(x-b)} + c$
- Below are graphs of $y = 2^x$ and the transformed exponential graph $y = -2^{(x-3)} + 1$

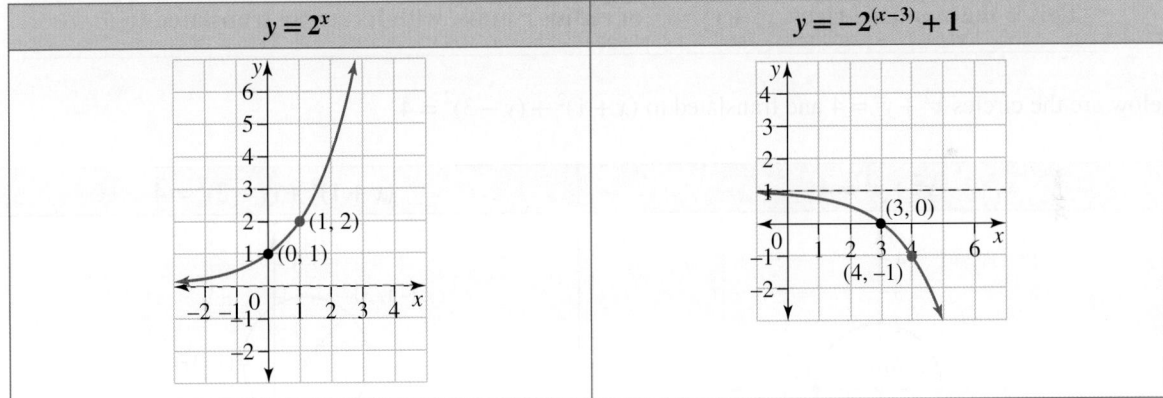

$y = 2^x$	$y = -2^{(x-3)} + 1$

The transformations are:
- reflection in the x-axis
- horizontal translation of 3 to the right
- vertical translation of 1 upwards.

- Below are graphs of $y = 2^{-x}$ and the transformed exponential graph $y = 3 \times 2^{-(x+4)} + 1$

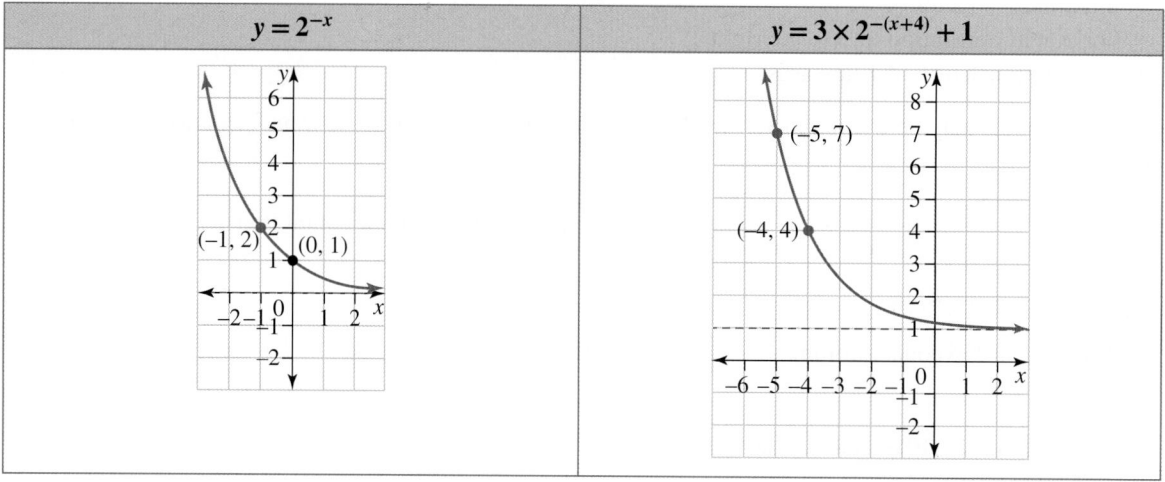

$y = 2^{-x}$	$y = 3 \times 2^{-(x+4)} + 1$

The transformations are:
- reflection of $y = 2^x$ in the y-axis to transform to $y = 2^{-x}$
- dilation by a factor of 3
- horizontal translation of 4 to the left
- vertical translation of 1 upwards.

Circles

- The circle $x^2 + y^2 = r^2$ has a centre $(0, 0)$ and radius r units.
- The translated circle $(x - a)^2 + (y - b)^2 = r^2$ has centre (a, b) and radius r units.

> **The circle $(x - a)^2 + (y - b)^2 = r^2$**
>
> **This is the standard circle $x^2 + y^2 = r^2$ of radius r units, with its centre translated to (a, b)**

Below are the circles $x^2 + y^2 = 4$ and translated to $(x + 1)^2 + (y - 3)^2 = 4$

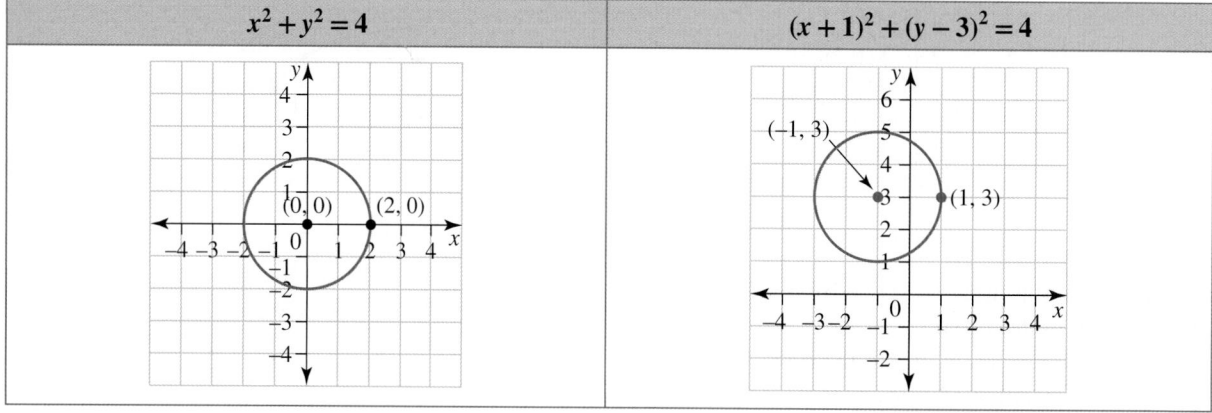

$x^2 + y^2 = 4$	$(x + 1)^2 + (y - 3)^2 = 4$

The transformations are:
- horizontal translation of 1 to the left
- vertical translation of 3 upwards.
- centre $(-1, 3)$, radius 2.

Describe the transformations for the following graphs.

a. **A transformation of the graph $x^2 + y^2 = 4$ to the graph $(x-3)^2 + (y+7)^2 = 4$.**

b. **A transformation of the graph $y = \dfrac{1}{x}$ to the graph $y = \dfrac{2}{-x-4} + 5$.**

THINK	WRITE
a. 1. Using the standard translated formula $(x-a)^2 + (y-b)^2 = r^2$ identify the value of a for the horizontal translation and the value of b for the vertical translation.	a. $a = 3, b = -7$. The graph is translated 3 units right in the positive direction and 7 units down in the negative direction. The transformed circle has a centre of $(3, -7)$, radius 2.
b. 1. Rewrite in the translated form $y = \dfrac{k}{x-b} + c$	b. $y = \dfrac{2}{-x-4} + 5$ is equivalent to:
2. Identify the value of k for the dilation factor and reflection.	$y = \dfrac{-2}{(x-(-4))} + 5$ $k = -2$ The graph is dilated by a factor of 2 from the x-axis and reflected in the x-axis.
3. Identify the values of b and c for the translations.	$b = -4, c = 5$ The graph is translated 4 units horizontally to the left and 5 units vertically upwards.

COMMUNICATING — COLLABORATIVE TASK: Graphing and comparing polynomial curves

Equipment: pen, paper, graphing application

1. As a pair, use a graphing application to graph $y = x^n$ for values of $n = 2$, 3, 4, 5 and 6.
2. Compare the shape of these curves and describe the connection between the value of n in $y = x^n$ and the shape of the curve. (Hint: Consider odd and even values of n.)
3. In your pair, use a graphing application to graph curves of the form $y = kx^n + c$ and $y = k(x-b)^n$ for values of $k = -3, -2, -1, -\dfrac{1}{2}, \dfrac{1}{2}, 2, 3$, $b = -3, -2, -1, 1, 2, 3$ and $n = 2$, 3, 4, 5 and 6.
4. Describe the transformations in part 3 from $y = kx^n$.

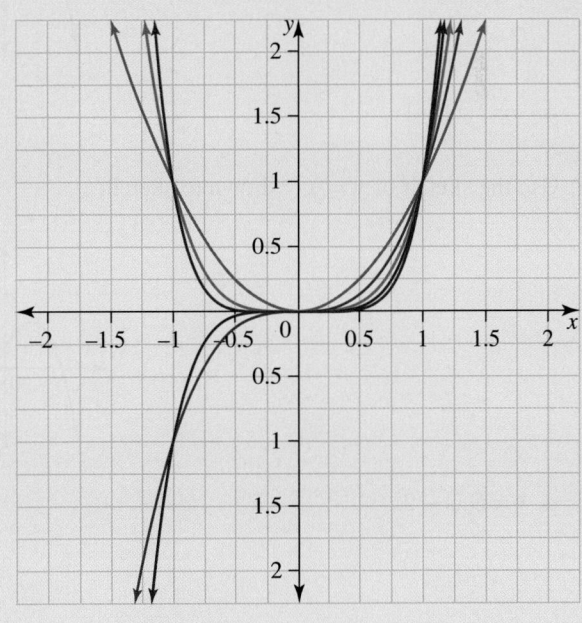

 **Resources**

Interactivities Horizontal translations of parabolas (int-6054)
Vertical translations of parabolas (int-6055)
Dilation of parabolas (int-6096)
Exponential functions (int-5959)
Reflection of parabolas (int-6151)
Hyperbolas (int-6155)
Translations of circles (int-6214)
Transformations of exponentials (int-6216)
Transformations of cubics (int-6217)
The rectangular hyperbola (int-2573)

Exercise 17.5 Transformations

learn

| 17.5 Quick quiz on | 17.5 Exercise |

Individual pathways

■ PRACTISE	■ CONSOLIDATE	■ MASTER
1, 4, 7, 10, 13, 16	2, 5, 8, 11, 14, 17, 18	3, 6, 9, 12, 15, 19, 20

Fluency

1. **WE11** Use the sketch of $y = f(x)$ shown to sketch:

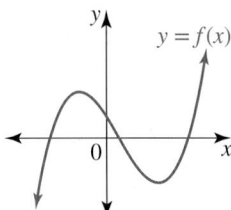

   a. $y = f(x) + 1$ **b.** $y = -f(x)$

2. Use the sketch of $y = f(x)$ shown to sketch:

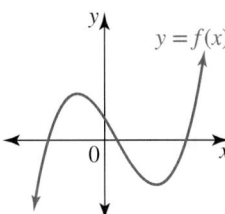

   a. $y = f(x) - 2$ **b.** $y = 2f(x)$.

3. Consider the sketch of $y = f(x)$ shown. Sketch:

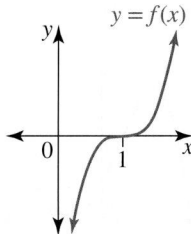

a. $y = f(x) + 1$ b. $y = -f(x)$ c. $y = f(x + 2)$.

4. **WE12** Describe the transformations applied to the graph of $y = x^3$ to obtain the graph:

a. $y = 5x^3$ b. $y = (x - 5)^3$ c. $y = x^3 - 5$

5. Describe the transformations applied to the graph of $y = x^3$ to obtain the graph:

a. $y = 4x^3 + 1$ b. $y = (x + 4)^3 + 1$ c. $y = -x^3 - 4$

6. Describe the transformations applied to the graph of $y = x^3$ to obtain the graph:

a. $y = -2(x + 6)^3$ b. $y = 7 - (x - 6)^3$ c. $y = 5(6 - x)^3 + 8$

7. **WE13** Describe the transformations for the following graphs.

a. A transformation of the graph $x^2 + y^2 = 9$ to the graph $(x - 2)^2 + (y - 1)^2 = 9$.

b. A transformation of the graph $y = \dfrac{1}{x}$ to the graph $y = \dfrac{3}{x - 2} + 7$.

8. Describe the transformations for the following graphs.

a. A transformation of the graph $y = \dfrac{1}{x}$ to the graph $y = \dfrac{5}{x - 1} + 2$.

b. A transformation of the graph $y = 3^x$ to the graph $y = 3^{(-x+2)} + 4$.

9. Describe the transformations for the following graphs.

a. A transformation of the graph $y = \dfrac{1}{x}$ to the graph $y = \dfrac{4}{-x + 3} + 10$.

b. A transformation of the graph $y = 5^x$ to the graph $y = -\left(5^{(-x+7)} - 6\right)$.

Understanding

10. Draw any polynomial $y = f(x)$. Discuss the similarities and differences between the graphs of $y = f(x)$ and $y = -f(x)$.

11. Draw any polynomial $y = f(x)$. Discuss the similarities and differences between the graphs of $y = f(x)$ and $y = 2f(x)$.

12. Draw any polynomial $y = f(x)$. Discuss the similarities and differences between the graphs of $y = f(x)$ and $y = f(x) - 2$.

Communicating, reasoning and problem solving

13. Consider the sketch of $y = f(x)$ shown below.

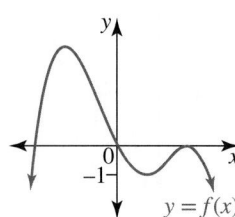

Give a possible equation for each of the following in terms of $f(x)$.

a.

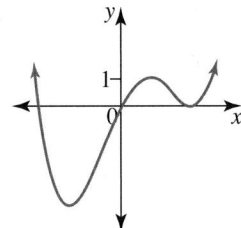

b.

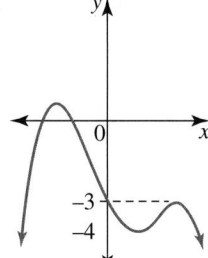

c.

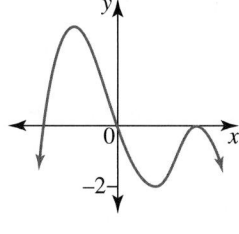

14. $y = x(x-2)(x-3)$ and $y = -2x(x-2)(x-3)$ are graphed on the same set of axes. Describe the relationship between the two graphs using the language of transformations.

15. If $y = -hr^{-(x+f)} - r$, explain what translations take place from the original graph, $y = r^x$.

Problem solving

16. a. Sketch the graph of $y = 3 \times 2^x$.
    b. If the graph of $y = 3 \times 2^x$ is transformed into the graph $y = 3 \times 2^x + 4$, describe the transformation.
    c. Sketch the graph of $y = 3 \times 2^x + 4$.
    d. Determine the coordinates of the y-intercept of $y = 3 \times 2^x + 4$.

17. The graph of $y = \dfrac{1}{x}$ is reflected in the y-axis, dilated by a factor of 2 parallel to the x-axis, translated 2 units to the left and up 1 unit. Determine the equation of the resultant curve. Give the equations of any asymptotes.

18. a. Sketch the graph of $y = \dfrac{1}{x}$.

    b. If the graph of $y = \dfrac{1}{x}$ is transformed into the graph $y = \dfrac{2}{(x+4)}$, describe the transformations.

    c. Sketch the graph of $y = \dfrac{2}{(x+4)}$.

    d. For the graph of $y = \dfrac{2}{(x+4)}$, state the equations of the asymptotes and the coordinates of any axis intercepts.

19. a. Sketch the graph of $y = 3^x$.
    b. If the graph of $y = 3^x$ is transformed into the graph $y = 2 \times 3^x + 1$, describe the transformations.
    c. Sketch the graph of $y = 2 \times 3^x + 1$.
    d. Using transformations, sketch the graph of $y = 2 \times 3^{(x-4)}$, stating the equations of any asymptotes and the coordinates of any axis intercepts.

20. The graph of an exponential function is shown.

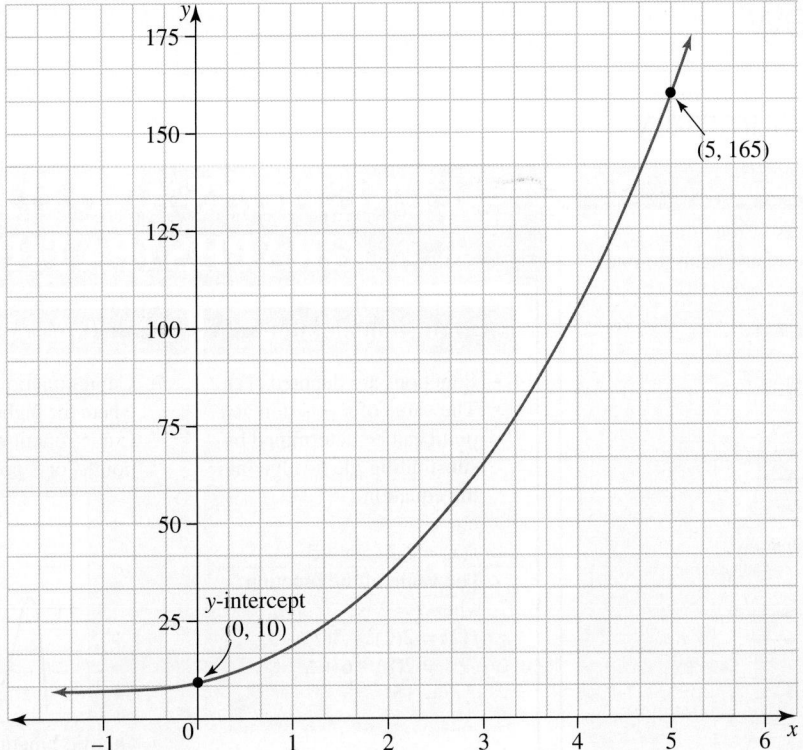

Its general rule is given by $y = k(2^x) + c$.

a. Determine the values of k and c.

b. Describe any transformations that had to be applied to the graph of $y = 2^x$ to achieve this graph.

LESSON
17.6 Review

17.6.1 Topic summary

Relations

- There are four types of relations: one-to-one, one-to-many, many-to-one and many-to-many.

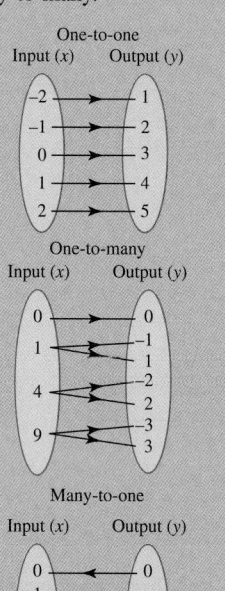

- Relations that are one-to-one or many-to-one are called functions.

Domain and range of functions

- The domain of a function is the set of allowable values of x.
- The range of a function is the set of y-values it produces.
- Consider the function $y = x^2 - 1$.
  The domain of the function is $x \in R$.
  The range of the function is $y \geq -1$.

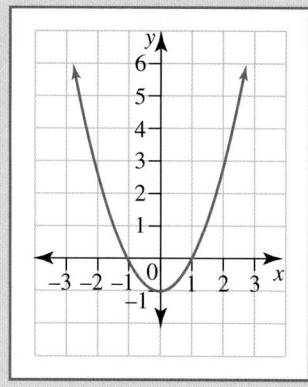

FUNCTIONS AND OTHER GRAPHS (PATH)

Function notation

- Functions are denoted $f(x)$.
- The value of a function at a point can be determined by substituting the x-value into the equation.
  e.g.
  $f(x) = 2x^3 - 3x + 5$
  The value of the function when $x = 2$ is:
  $f(2) = 2(2)^3 - 3(2) + 5$
  $= 2(8) - 6 + 5$
  $= 15$

Transformations

- There are 3 types of transformations that can be applied to functions and relations:
  - Reflections
  - Dilations
  - Translations
- The equation of a hyperbola $\left(f(x) = \dfrac{1}{x}\right)$ when transformed is
  $f(x) = \dfrac{k}{x-b} + c$.
- The equation of an exponential $(f(x) = a^x)$ when transformed is
  $f(x) = ka^{(x-b)} + c$.
- The equation of a circle $(x^2 + y^2 = r^2)$ when transformed is
  $(x - a)^2 + (y - b)^2 = r^2$.

Types of functions

- Cubic functions are functions where the highest power of x is 3. Cubic functions have 2 turning points, or 1 point of inflection.

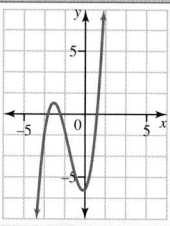

- Quartic functions are functions where the highest power of x is 4. Quartic functions have 1 turning point, 3 turning points or 1 turning point and 1 point of inflection.

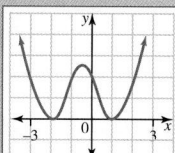

- Hyperbolas are of the form $y = \dfrac{k}{x}$

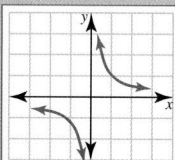

- Exponential functions are of the form $f(x) = k \times a^x$. The y-intercept is $y = k$.

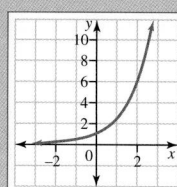

17.6.2 Project

Shaping up!

Many beautiful patterns are created by starting with a single function or relation and transforming and repeating it over and over.

In this task, you will apply what you have learned about functions, relations and transformations (dilations, reflections and translations) to explore mathematical patterns.

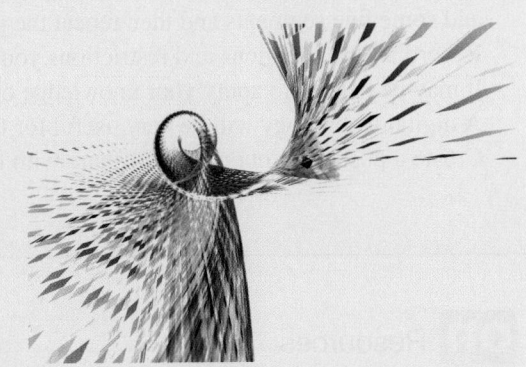

Exploring patterns using transformations

1. **a.** On the same set of axes, draw the graphs of:
   i. $y = x^2 - 4x + 1$
   ii. $y = x^2 - 3x + 1$
   iii. $y = x^2 - 2x + 1$
   iv. $y = x^2 + 2x + 1$
   v. $y = x^2 + 3x + 1$
   vi. $y = x^2 + 4x + 1$

   b. Describe the pattern formed by your graphs. Use mathematical terms such as intercepts, turning points, shape and transformations.

What you have drawn is referred to as a family of curves — curves in which the shape of the curve changes if the values of a, b and c in the general equation $y = ax^2 + bx + c$ change.

   c. Explore the family of parabolas formed by changing the values of a and c. Comment on your findings.
   d. Explore exponential functions belonging to the family of curves with equation $y = ka^x$, families of cubic functions with equations $y = ax^3$ or $y = ax^3 + bx^2 + cx + d$, and families of quartic functions with equations $y = ax^4$ or $y = ax^4 + bx^3 + cx^2 + dx + e$. Comment on your findings.
   e. Choose one of the designs shown below and recreate it (or a simplified version of it). Record the mathematical equations used to complete the design.

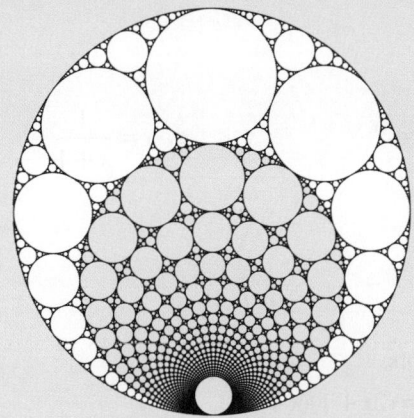

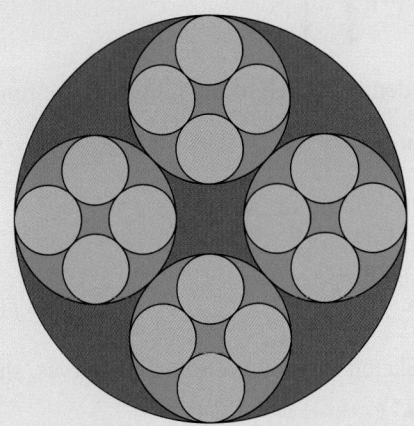

Coming up with your design

2. Use what you know about transformations to functions and relations to create your own design from a basic graph. You could begin with a circle, add some line segments and then repeat the pattern with some changes. Record all the equations and restrictions you use.
It may be helpful to apply your knowledge of inverse functions too.
A digital technology will be very useful for this task.
Create a poster of your design to share with the class.

 Resources

 Interactivities Crossword (int-2878)
Sudoku puzzle (int-3893)

Exercise 17.6 Review questions

learnon

Fluency

1. State which of the following are functions.

   a.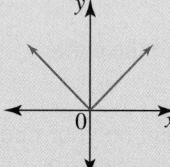

   b.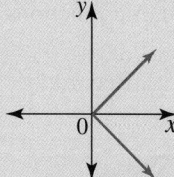

2. Identify which of the following are functions.

   a. $y = 2x - 7$ b. $x^2 + y^2 = \sqrt{30}$ c. $y = 2^x$ d. $y = \dfrac{1}{x+1}$

3. If $f(x) = \sqrt{4 - x^2}$:
   calculate:
   a. $f(0)$ b. $f(1)$ c. $f(2)$

4. Sketch each of the following curves, showing all intercepts.
   a. $y = (x - 1)(x + 2)(x - 3)$ b. $y = (2x + 1)(x + 5)^2$

5. Give an example of the equation of a cubic that would just touch the x-axis and cross it at another point.

Understanding

6. Match each equation with its type of curve.

   a. $y = x^2 + 2$

   b. $x^2 + y^2 = 9$

   c. $f(x) = \dfrac{2}{x+2}$

   d. $g(x) = 6^{-x}$

   e. $h(x) = (x+1)(x-3)(x+5)$

   A. circle

   B. cubic

   C. exponential

   D. parabola

   E. hyperbola

7. **MC** The equation for the graph shown could be:

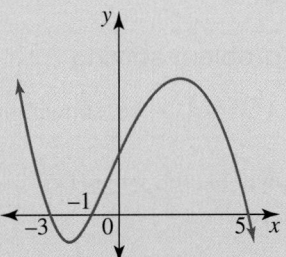

   A. $y = (x-5)(x+1)(x+3)$ **B.** $y = (x-3)(x-1)(x+5)$

   C. $y = (x-3)(x+1)(x+5)$ **D.** $y = (5-x)(1+x)(3+x)$

8. **MC** Select which of the following shows the graph of $y = -2(x+5)^3 - 12$.

   A. 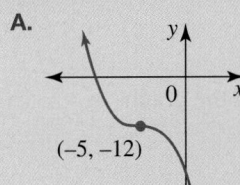   **B.** 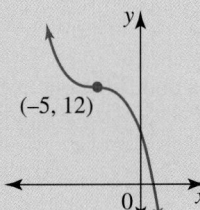   **C.** 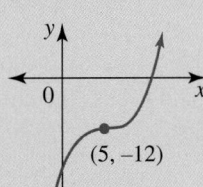   **D.**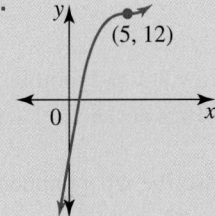

9. Sketch the following functions:

   a. $y = x(x-2)(x+11)$ b. $y = x^3 + 6x^2 - 15x + 8$ c. $y = -2x^3 + x^2$

10. **MC** The rule for the graph shown could be:

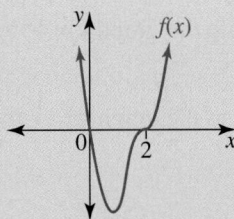

   A. $f(x) = x(x+2)^3$ **B.** $f(x) = -x(x-2)^2$

   C. $f(x) = x^2(x-2)^2$ **D.** $f(x) = x(x-2)^3$

11. **MC** The graph of $y = (x+3)^2(x-1)(x-3)$ is best represented by:

A.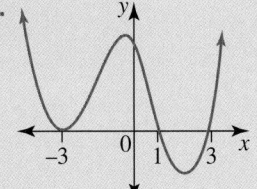

B.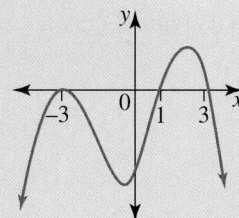

C.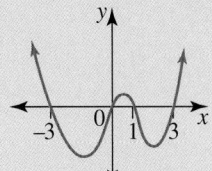

D.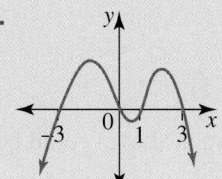

Communicating, reasoning and problem solving

12. Sketch the graph of $y = x^4 - 7x^3 + 12x^2 + 4x - 16$, showing all intercepts.

13. Consider the sketch of $y = f(x)$ shown. Sketch $y = -f(x)$.

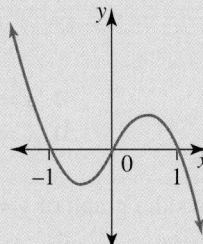

14. Draw any polynomial $y = f(x)$. Discuss the similarities and differences between the graphs of $y = f(x)$ and $y = f(x) + 3$.

15. Describe what happens to $f(x) = -2^x$ as $x \to \infty$ and $x \to -\infty$.

16. Determine any points of intersection between $f(x) = x^2 - 4$ and $g(x) = x^3 + x^2 - 12$.

17. Describe the transformations applied to the graph $y = x^4$ to obtain the graph $y = 5 - x^4$.

18. Describe the transformations applied to the graph $y = x^4$ to obtain the graph $y = (x-2)^4 + 1$.

19. Describe the transformations applied to the graph $y = 4^x$ to obtain the graph $y = 1 - 3 \times 4^x$.

20. Describe the transformations applied to the graph $y = \dfrac{1}{x}$ to obtain the graph $y = \dfrac{-2}{(x-1)} + 3$.

on To test your understanding and knowledge of this topic, go to your learnON title at www.jacplus.com.au and complete the **post-test**.

Answers

Topic 17 Functions and other graphs (Path)

17.1 Pre-test

1. B
2. A
3. True
4. B
5. B
6. C
7. D
8. C
9. D
10. D
11. B
12. A
13. B
14. B
15. D

17.2 Functions and relations

1. a. One-to-many   b. Many-to-one
   c. Many-to-one   d. One-to-one

2. a. One-to-one   b. Many-to-one
   c. Many-to-many   d. Many-to-one

3. a. One-to-one   b. Many-to-one
   c. One-to-one   d. Many-to-one

4. b, c, d

5. a, b, d

6. a, b, c, d

7. a. 1   b. 7   c. -5 d. 16

8. a. 2   b. 1   c. 3   d. 0

9. a. 3   b. 2   c. 6   d. 9

10. a. 9   b. 1
    c. 16   d. $a^2 + 6a + 9$

11. a. 12   b. 6   c. -4 d. 2

12. a. Domain: $-2 \leq x \leq 6$, range: $-3 \leq y \leq 3$
    b. Domain: $2 < x < 7$, range: $\dfrac{1}{7} < y < \dfrac{1}{2}$

13. a. Domain: $x \geq -2$, range: $y \in R$, not a function
    b. Domain: $0 \leq x \leq 3$, range: $0 \leq y \leq 4$, function

14.
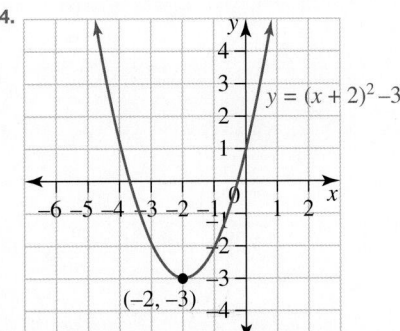

Domain: $x \geq -2$
Range: $y \geq -3$

15. a.
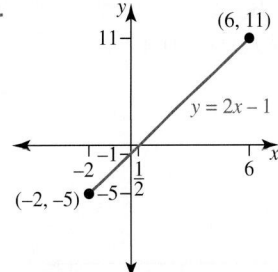

b. Domain: $-2 \leq x \leq 6$
   Range: $-5 \leq y \leq 11$

c. This is a function since the vertical line test cuts the graph exactly once.

16. a.
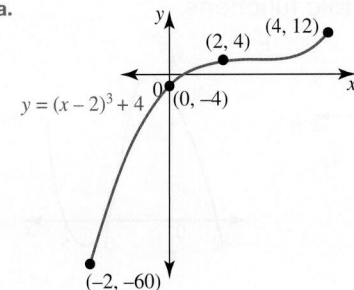

b. Domain: $-2 \leq x \leq 4$
   Range: $-60 \leq y \leq 12$

c. This is a function since the vertical line test cuts the graph exactly once.

17. a, c, d

18. a, b, c, f

19. a. 3   b. 3   c. $\dfrac{5}{x} - 2x$
    d. $\dfrac{10}{x^2} - x^2$ e. $\dfrac{10}{x+3} - x - 3$ f. $\dfrac{10}{x-1} - x + 1$

20. a. 3   b. -3 or 3   c. $\dfrac{1}{3}$

21. a. 2 or 3   b. -4 or 1   c. -1

22. a. $f(x) \to \infty$ b. $f(x) \to 0$ c. $f(x) \to 0$
    d. $f(x) \to -\infty$ e. $f(x) \to 0$

23. a. $(0, -4), (2, 0)$
    b. $(1, -2), \left(-\dfrac{2}{3}, 3\right)$
    c. $(2, 0), (-2, 0)$
    d. $(3, -4)$

24. a. $x = \pm 3$
    b. $x = 2\dfrac{1}{3}$
    c. $x = 28$

25. a. Ran $= [2, \infty)$
    b. Many-to-one

c.

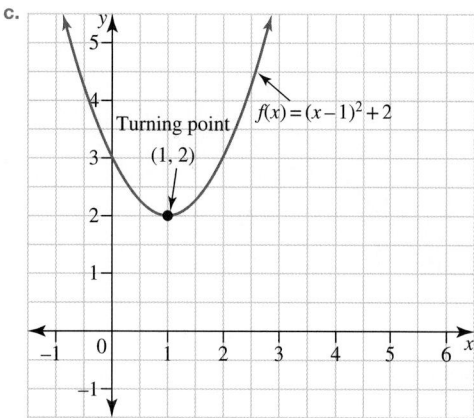

Turning point (1, 2)

$f(x) = (x-1)^2 + 2$

d. Dom $= [1, \infty)$

17.3 Graphing cubic functions

1. a.

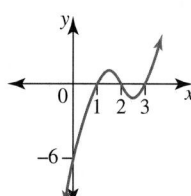

b.

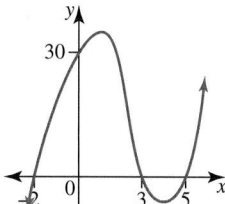

c.

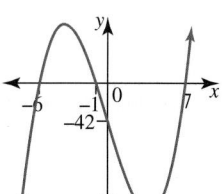

d.

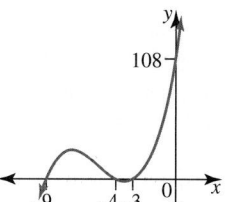

2. a.

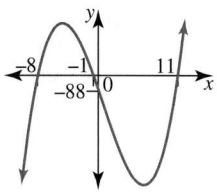

b.

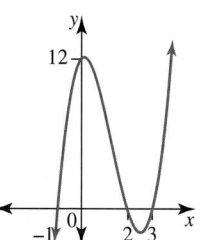

c.

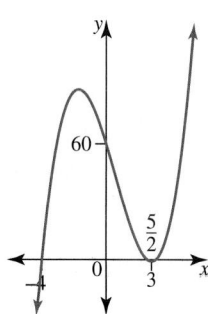

d.

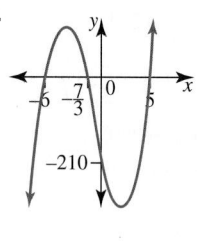

3. a.

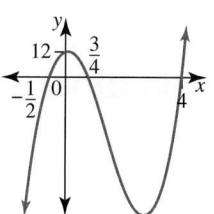

b.

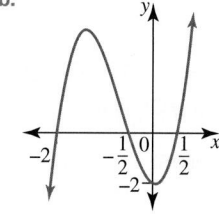

c.

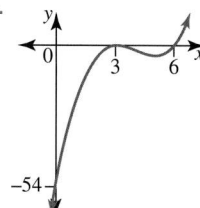

d.

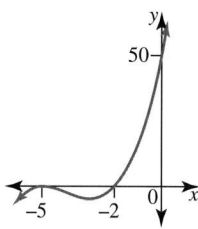

4. a.

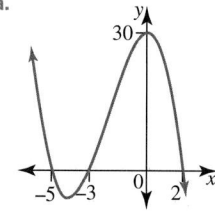

b.

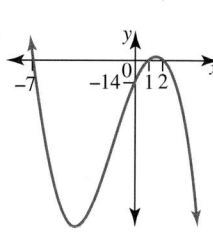

c.

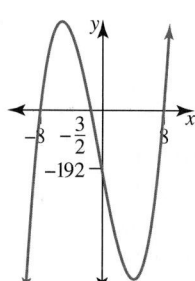

d.

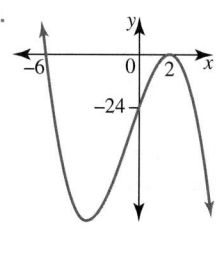

5. a.

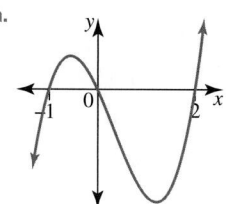

b.

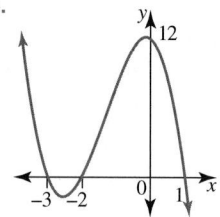

c.

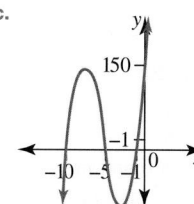

d.

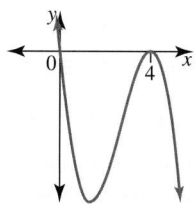

6. a.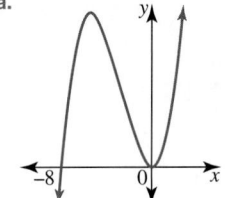

b.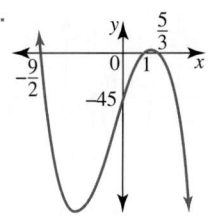

c.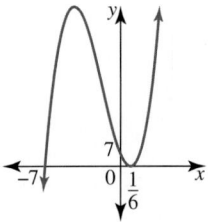

d.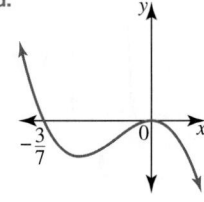

7. C

8. C

9. B

10. D

11. a.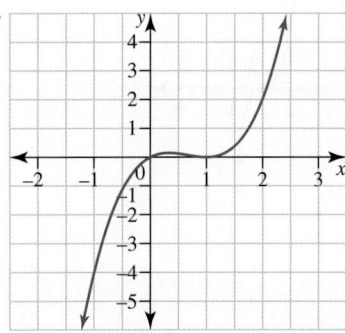

b.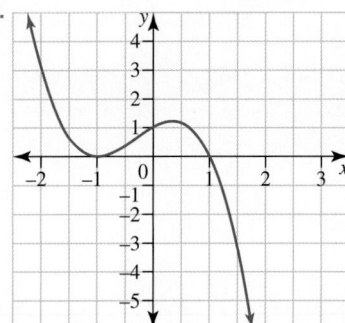

12. a.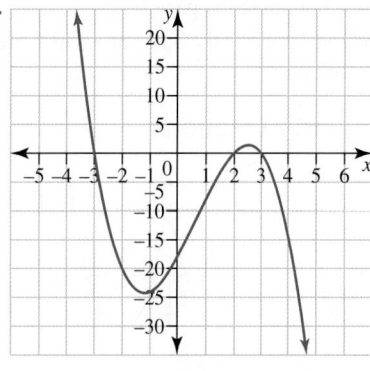

b.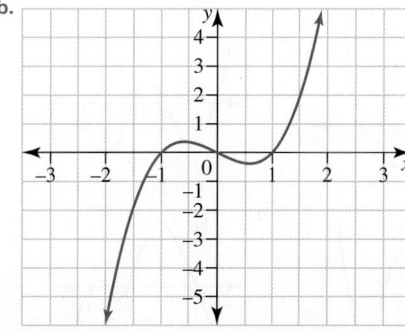

13. a. Positive   **b.** Positive
c. Zero   **d.** Positive

14. $a = 2, b = -7$

15. $a = \dfrac{-(27 + 11b)}{11}$

16. a.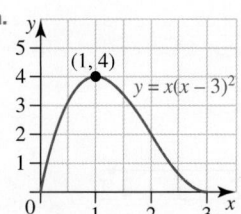

b. $x = 0$ and $x = 3$

c. 4 units

17. $y = 2(x + 2)(x - 2)(x - 5)$

18. a. Sample responses can be found in the worked solutions in the online resources.

     b. $0 < x < 17.5$

     c. 1377 cm^3

d. See figure at the bottom of the page.*

e. (11.6667, 1587.963); this is the value of x which creates the maximum volume.

17.4 Graphing quartic functions

1. a. 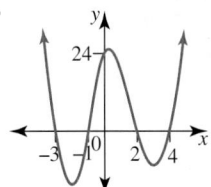 **b.**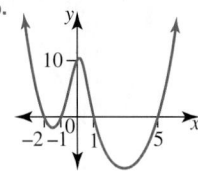

b.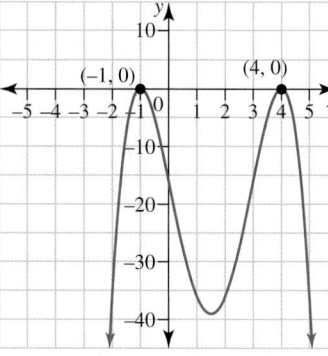

2. a. 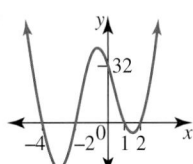 **b.**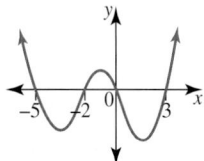

5. a.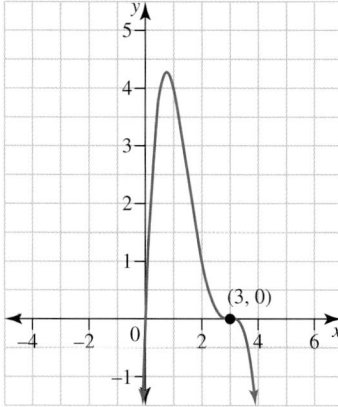

3. a. 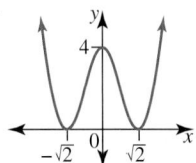 **b.**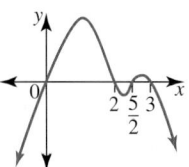

4. a.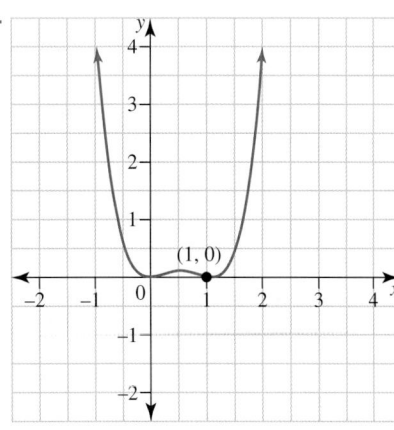

---

*18. d.

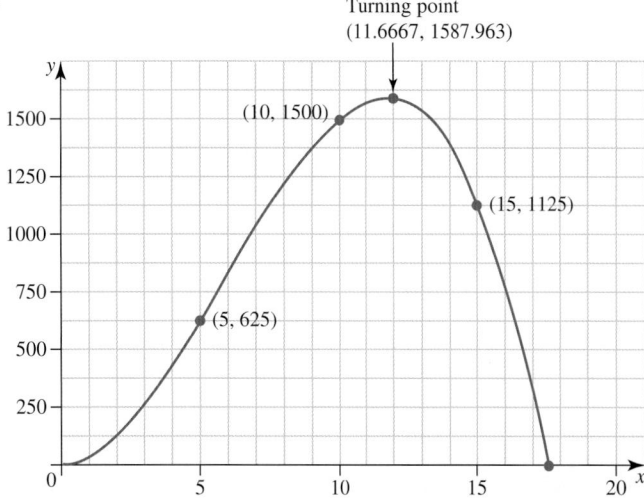

b.

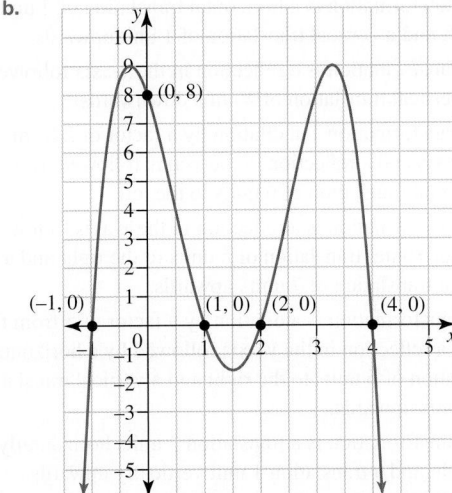

6.

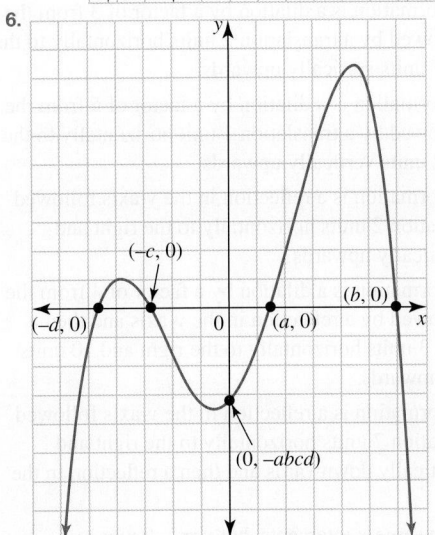

7. D

8. D

9. B

10. a.

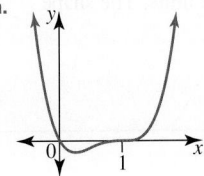

b.

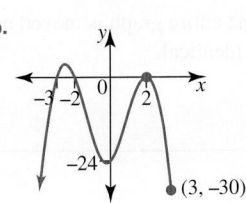

11. a.

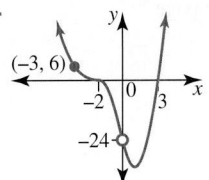

b.

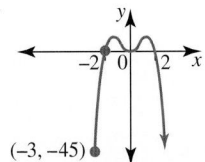

12. a.

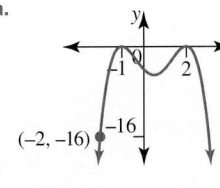

b.

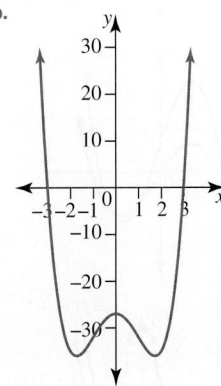

13. $a = 3, b = -1$

14. a.

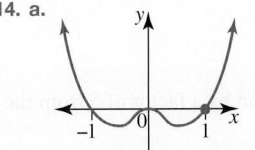

b.

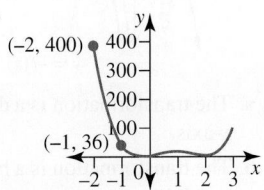

15. a. If n is even, the graph touches the x-axis.

b. If n is odd, the graph cuts the x-axis.

16. $a = 4, b = -19$

17. a. 15 m

b. See figure at the bottom of the page.*

c. 28.77 m

18. $y = (x + 1)(x - 2)^3$

17.5 Transformations

1.

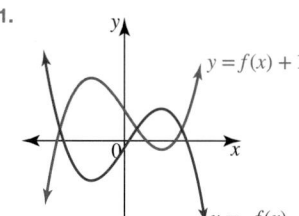

2.

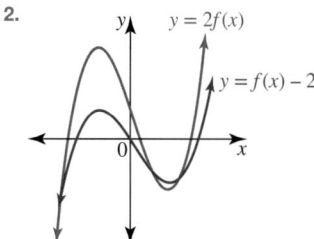

3.

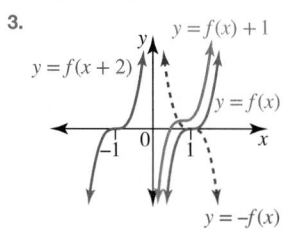

4. a. The transformation is a dilation by a factor of 5 from the x-axis.

b. The transformation is a horizontal translation of 5 units to the right.

c. The transformation is a vertical translation of 5 units downwards.

5. a. The transformation is a dilation by a factor of 4 from the x-axis followed by a vertical translation of 1 unit upwards.

b. The transformation is a horizontal translation of 4 units to the left and a vertical translation of 1 unit upwards.

c. The transformation is a reflection in the x-axis followed by a vertical translation of 4 units downwards.

6. a. The transformation is a dilation by a factor of 2 from the x-axis and a reflection in the x-axis followed by a horizontal translation of 6 units to the left.

b. The transformation is a reflection in the x-axis followed by a horizontal translation of 6 units to the right and a vertical translation of 7 units upwards.

c. The transformation is a dilation by a factor of 5 from the x-axis a reflection in the y-axis followed by a horizontal translation of 6 units to the right and a vertical translation of 8 units upwards.

7. a. The transformation is a translation 2 units horizontally to the right and a translation 1 unit vertically upwards.

b. The transformation is a dilation by a factor of 3 from the x-axis followed by a translation 2 units horizontally to the right and 7 units vertically upwards.

8. a. The transformation is a dilation by a factor of 5 from the x-axis followed by a translation 1 unit horizontally to the right and 2 units vertically upwards.

b. The transformation is a reflection in the y-axis followed by a translation 2 units horizontally to the right and 4 units vertically upwards.

9. a. The transformation is a dilation by a factor of 4 from the x-axis followed by a reflection in the y-axis and then a translation 3 units horizontally to the right and 10 units vertically upwards.

b. The transformation is a reflection in the y-axis followed by a translation 7 units horizontally to the right and 6 units vertically downwards and then a reflection in the x-axis.

10. They have the same x-intercepts, but $y = -f(x)$ is a reflection of $y = f(x)$ in the x-axis.

11. They have the same x-intercepts, but the y-values in $y = 2f(x)$ are all twice as large.

12. The entire graph is moved down 2 units. The shape is identical.

***17. b.**

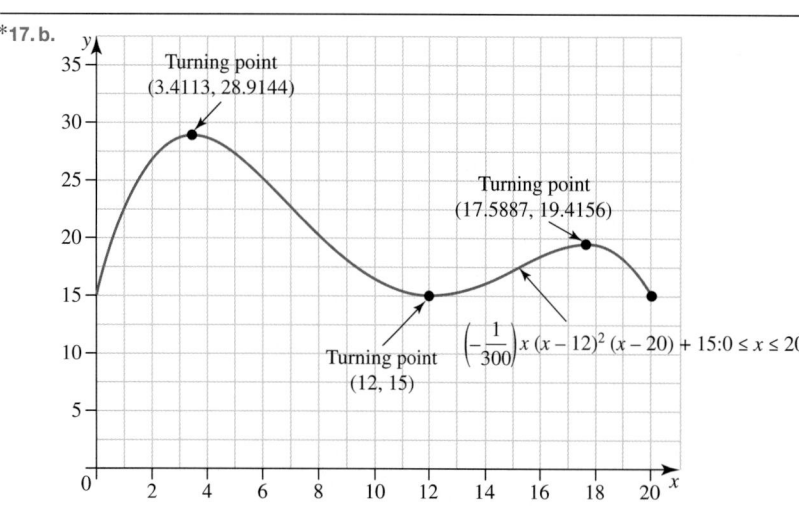

13. a. $y = -f(x)$ **b.** $y = f(x) - 3$ **c.** $y = 2f(x)$

14. The original graph has been reflected in the x-axis and dilated by a factor of 2. The location of the intercepts remains unchanged.

15. Dilation by a factor of h from the x-axis, reflection in the x-axis, reflection in the y-axis, translation of f units left, translation of r units down.

16. a.

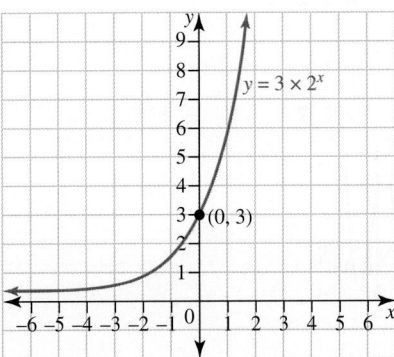

b. Translation 4 units vertically.

c.

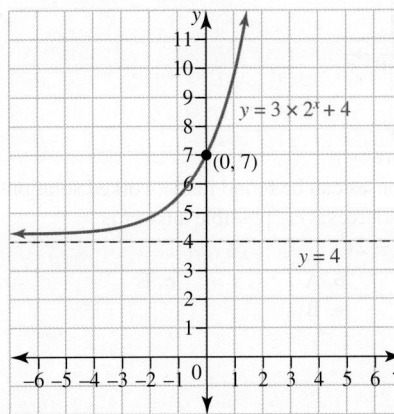

d. $(0, 7)$

17. $y = -\dfrac{2}{(x+2)} + 1, x = -2, y = 1$

18. a. $y = \dfrac{1}{x}$

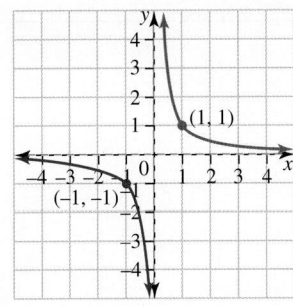

b. The transformations are a dilation by a factor of 2 from the x-axis and a horizontal translation of 4 units to the left.

c. $y = \dfrac{2}{(x+4)}$

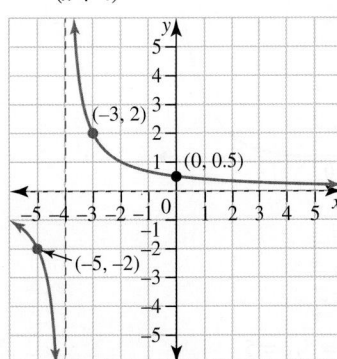

d. Asymptotes are: $x = -4$, $y = 0$
  y-intercept: $(0, 0.5)$

19. a. $y = 3^x$

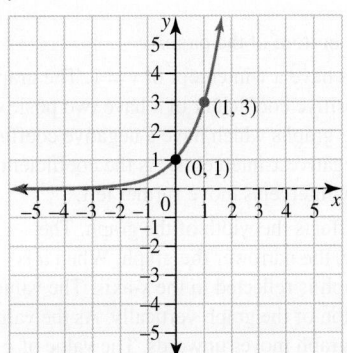

b. The translations are a dilation by a factor of 2 from the x-axis and vertically translated upwards by 1 unit.

c. $y = 2 \times 3^x + 1$

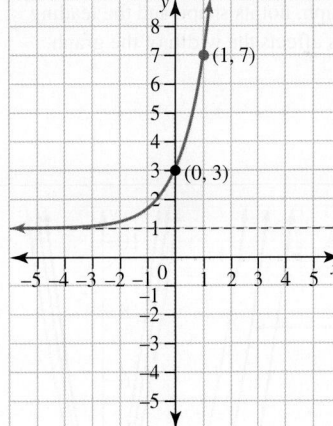

d. Asymptote: $y = 0$

y-intercept: $\left(0, \dfrac{2}{81}\right)$

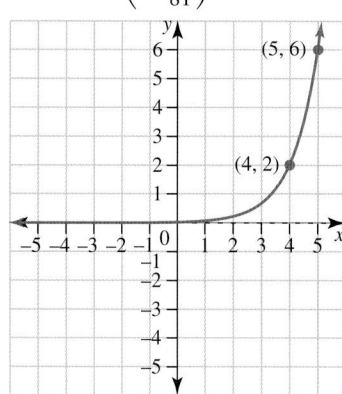

20. a. $y = 5(2^x) + 5, k = 5, c = 5$

b. Dilation by a factor of 5 from the x-axis and translation of 5 units up.

Project

1. a. See figure at the bottom of the page.*

b. All of the graphs have a y-intercept of $y = 1$. The graphs which have a positive coefficient of x have two positive x-intercepts. The graphs which have a negative coefficient of x have two negative x-intercepts. As the coefficient of x decreases the x-intercepts move further left.

c. The value of a affects the width of the graph. The greater the value, the narrower the graph. When a is negative, the graph is reflected in the x-axis. The value of c affects the position of the graph vertically. As the value of c increases, the graph moves upwards. The value of c does not affect the shape of the graph.

d. Students may find some similarities in the effect of the coefficients on the different families of graphs. For example, the constant term always affects the vertical position of the graph, not its shape and the leading coefficient always affects the width of the graph.

e. Student responses will vary. Students should aim to recreate the pattern as best as possible and describe the process in mathematical terms. Sample responses can be found in the worked solutions in the online resources.

2. Student responses will vary. Students should try to use their imagination and create a pattern that they find visually beautiful.

17.6 Review questions

1. a

2. a, c, d

3. a. 2   **b.** $\sqrt{3}$ **c.** 0

4. a.

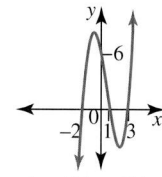

$y = (x - 1)(x + 2)(x - 3)$

b.

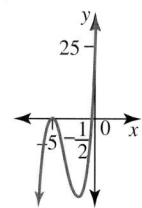

$y = (2x + 1)(x + 5)^2$

5. Sample responses can be found in the worked solutions in the online resources. One possible answer is $y = (x - 1)(x - 2)^2$.

6. a. D   **b.** A   **c.** E
    d. C   **e.** B

7. D

8. A

---

*1. a.

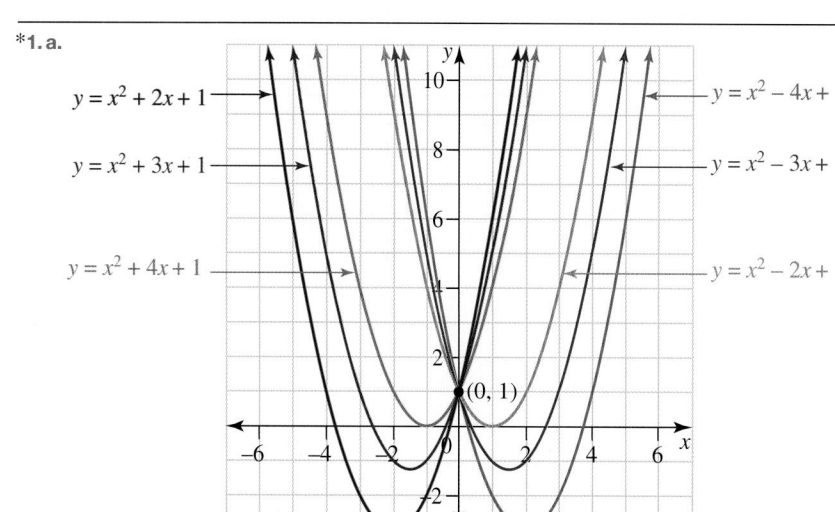

9. a.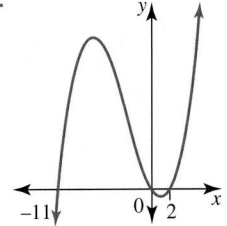

b.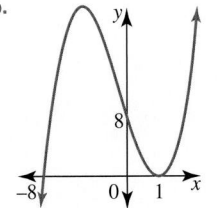

c.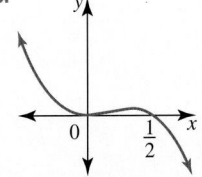

10. D

11. A

12.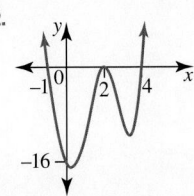

13.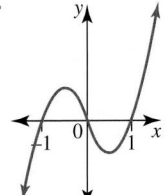

14. The entire graph is moved up 3 units. The shape is identical.

15. As $x \to \infty, f(x) \to -\infty$
As $x \to -\infty, f(x) \to 0$

16. $(2, 0)$

17. The transformations are a reflection in the x-axis and a vertical translation upwards by 5 units.

18. The transformations are a horizontal translation to the right by 2 units and a vertical translation upwards by 1 unit.

19. The transformations are dilation by a factor of 3 from the x-axis, reflected in the x-axis, with a vertical translation upwards by 1 unit.

20. The transformations are a dilation by a factor of 2 from the x-axis, reflected in the x-axis, with a horizontal translation of 1 unit to the right followed by a vertical translation upwards by 3 units.

18 Circle geometry (Path)

LESSON
18.1 Overview

Why learn this?

For thousands of years, humans have been fascinated by circles. Since they first looked upwards towards the sun and moon, which, from a distance at least, looked circular, artists have created circular monuments to nature. The most famous circular invention, one that has been credited as the most important invention of all, is the wheel. The potter's wheel can be traced back to around 3500 BC, approximately 300 years before wheels were used on chariots for transportation. Our whole transportation system revolves around wheels — bicycles, cars, trucks, trains and planes. Scholars as early as Socrates and Plato, Greek philosophers of the fourth century BCE, have been fascinated with the sheer beauty of the properties of circles. Many scholars made a life's work out of studying them, most famously Euclid, a Greek mathematician. It is in circle geometry that the concepts of congruence and similarity, studied earlier, have a powerful context. Today, we see circles in many different areas. Some buildings are now constructed based on circular designs. Engineers, designers and architects understand the various properties of circles. Road systems often have circular interchanges, and amusement parks usually include ferris wheels. As with the simple rectangle, circles are now part of our everyday life. Knowing the various properties of circles helps with our understanding and appreciation of this simple shape.

Hey students! Bring these pages to life online

- ▶ **Watch videos**
- 🧩 **Engage with interactivities**
- A+ **Answer questions and check solutions**

Find all this and MORE in jacPLUS ▶

Reading content and rich media, including interactivities and videos for every concept

Extra learning resources

Differentiated question sets

Questions with immediate feedback, and fully worked solutions to help students get unstuck

1. State the name of the area of the circle between a chord and the circumference.

2. **MC** Select the name of a line that touches the circumference of a circle at one point only.
   A. Secant   **B.** Radius   **C.** Chord   **D.** Tangent

3. Calculate the angle of x within the circle shown.

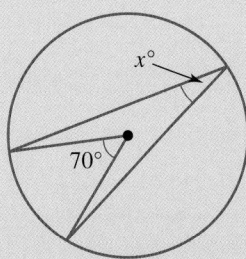

4. **MC** Select the correct angles for y and z in the circle shown.

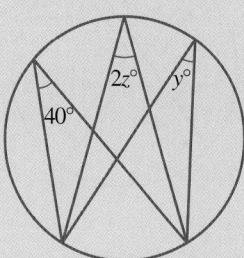

     A. $y = 40°$ and $z = 80°$ **B.** $y = 20°$ and $z = 40°$
     C. $y = 40°$ and $z = 80°$ **D.** $y = 40°$ and $z = 20°$

5. Determine the value of x in the circle shown.

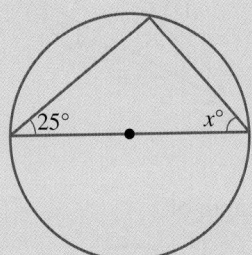

6. Determine the value of y in the shape shown.

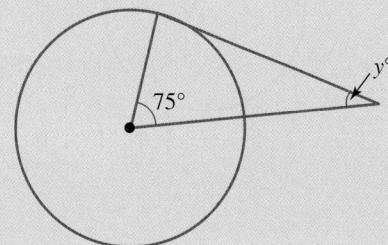

7. MC Select the correct values of x, y and z from the following list.

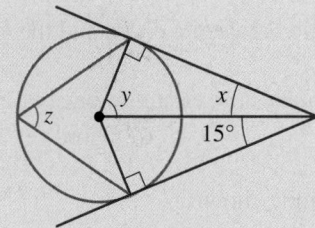

A. $x = 15°$, $y = 75°$ and $z = 150°$
B. $x = 15°$, $y = 30°$ and $z = 150°$
C. $x = 15°$, $y = 15°$ and $z = 37.5°$
D. $x = 15°$, $y = 75°$ and $z = 75°$

8. Calculate the length of p in the circle.

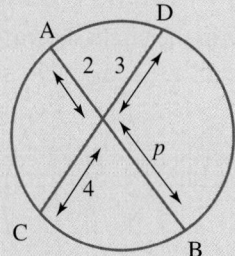

9. Calculate the length of m, correct to 1 decimal place.

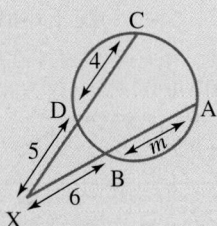

10. MC Choose the correct value for the length of x.

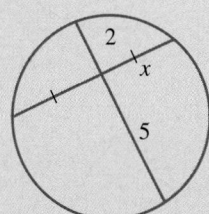

A. $\sqrt{10}$
B. $\sqrt{5}$
C. 2
D. 10

11. Determine the values of m and p.

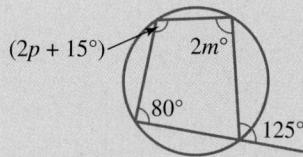

12. MC Select the correct values for x and y.

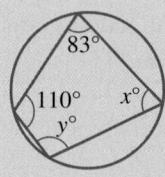

A. $x = 110°$ and $y = 83°$ **B.** $x = 83°$ and $y = 110°$

C. $x = 70°$ and $y = 97°$ **D.** $x = 97°$ and $y = 70°$

13. MC Choose which of the following correctly states the relationships between x, y and z in the diagram.

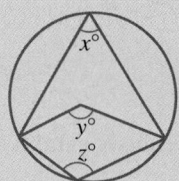

A. $x = 2y$ and $z + y = 180°$ **B.** $y = \dfrac{x}{2}$ and $z + y = 180°$

C. $x = 2y$ and $z + x = 180°$ **D.** $y = 2x$ and $z + x = 180°$

14. MC Choose the correct value for x.

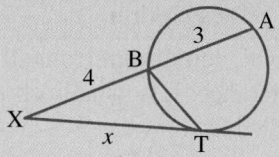

A. $2\sqrt{5}$ **B.** 28   **C.** 7   **D.** $2\sqrt{7}$

15. MC $AB = 9$ and P divides AB in the ratio $2 : 3$.
If $PO = 3$ where O is the centre of the circle, select the exact length of the diameter of the circle.

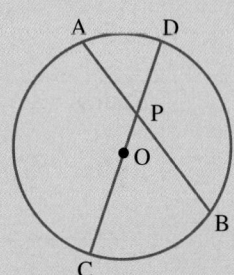

A. $\dfrac{6\sqrt{79}}{5}$ **B.** $\dfrac{4\sqrt{79}}{2}$ **C.** $\dfrac{3\sqrt{79}}{5}$ **D.** $\dfrac{6\sqrt{81}}{5}$

LESSON
18.2 Angles in a circle

LEARNING INTENTION

At the end of this lesson you should be able to:
- determine relationships between the angles at the centre and the circumference of a circle
- understand and use the angle in a semicircle being a right angle
- apply the relationships between tangents and radii of circles
- identify and prove angle properties of circles.

▶ 18.2.1 Circles

eles-4995

- A **circle** is a connected set of points that lie a fixed distance (the **radius**) from a fixed point (the **centre**).
- In circle geometry, there are many theorems that can be used to solve problems. It is important that we are also able to prove these theorems.

Steps to prove a theorem

Step 1. State the aim of the proof.

Step 2. Use given information and previously established theorems to establish the result.

Step 3. Give a reason for each step of the proof.

Step 4. State a clear conclusion.

Parts of a circle

Part (name)	Description	Diagram
Centre	The middle point, equidistant from all points on the circumference. It is usually shown by a dot and labelled O.	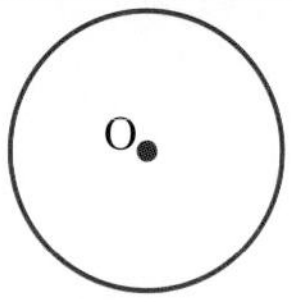 O•
Circumference	The outside length or the boundary forming the circle. It is the circle's perimeter.	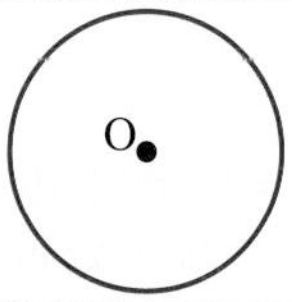 O•
Radius Radii	A straight line from the centre to any point on the circumference. Plural of radius.	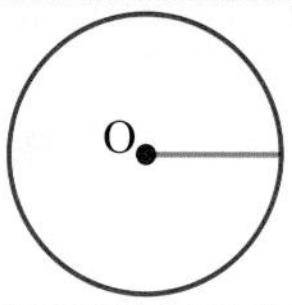 O•—

Part (name)	Description	Diagram
Diameter	A straight line from one point on the circumference to another, passing through the centre.	
Chord	A straight line from one point on the circumference to another.	
Segment	The area of the circle between a chord and the circumference. The smaller segment is called the minor segment and the larger segment is the major segment.	
Sector	An area of a circle enclosed by 2 radii and the circumference.	
Arc	A portion of the circumference.	
Tangent	A straight line that touches the circumference at one point only.	
Secant	A chord extended beyond the circumference on one side.	

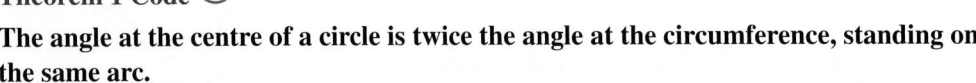

▶ 18.2.2 Angles in a circle

eles-6272

- In the diagram at right, chords AC and BC form the angle ACB. Arc AB has **subtended** angle ACB.
- **Theorem 1 Code**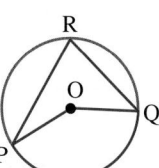

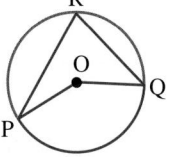

The angle at the centre of a circle is twice the angle at the circumference, standing on the same arc.

Proof:

Let $\angle PRO = x$ and $\angle QRO = y$

$\qquad RO = PO = QO$ (radii of the same circle are equal)

$\quad \angle RPO = x$

and $\angle RQO = y$

$\quad \angle POM = 2x$ (exterior angle of triangle)

and $\angle QOM = 2y$ (exterior angle of triangle)

$\quad \angle POQ = 2x + 2y$

$\qquad\quad = 2(x + y)$

which is twice the size of $\angle PRQ = x + y$.

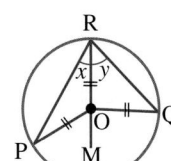

The angle subtended at the centre of a circle is twice the angle subtended at the circumference, standing on the same arc.

- **Theorem 2 Code**

All angles that have their vertex on the circumference and are subtended by the same arc are equal.

Proof:

Join P and Q to O, the centre of the circle.

Let $\angle PSQ = x$

$\quad \angle POQ = 2x$ (angle at the centre is twice the angle at the circumference)

$\quad \angle PRQ = x$ (angle at the circumference is half the angle of the centre)

$\quad \angle PSQ = \angle PRQ$.

Angles at the circumference subtended by the same arc are equal.

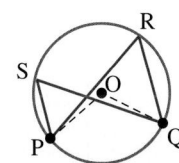

WORKED EXAMPLE 1 Determining angles in a circle

Determine the values of the pronumerals x and y in the diagram, giving reasons for your answers.

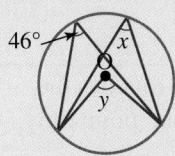

THINK	WRITE
1. Angles x and 46° are angles subtended by the same arc and both have their vertex on the circumference.	$x = 46°$ 
2. Angles y and 46° stand on the same arc. The 46° angle has its vertex on the circumference and y has its vertex at the centre. The angle at the centre is twice the angle at the circumference.	$y = 2 \times 46°$ $\quad = 92°$

- **Theorem 3 Code**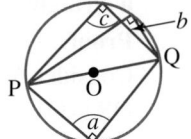

  Angles subtended by the diameter, that is, angles in a semicircle, are right angles.

  In the diagram, PQ is the diameter. Angles a, b and c are right angles. This theorem is in fact a special case of Theorem 1.

  Proof:

  $\angle POQ = 180°$ (straight line)

  Let S refer to the angle at the circumference subtended by the diameter. In the figure, S could be at the points where a, b and c are represented on the diagram.

  $\angle PSQ = 90°$ (angle at the circumference is half the angle at the centre)

  Angles subtended by a diameter are right angles.

eles-4996

▶ 18.2.3 Tangents to a circle

- In the diagram, the tangent touches the circumference of the circle at the point of contact.

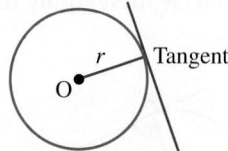

- **Theorem 4 Code**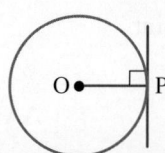

  A tangent to a circle is perpendicular to the radius of the circle at the point of contact on the circumference.

  In the diagram, the radius is drawn to a point, P, on the circumference. The tangent to the circle is also drawn at P. The radius and the tangent meet at right angles, that is, the angle at P equals 90°.

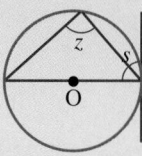

WORKED EXAMPLE 2 Determining angles when tangents are drawn to a circle

Determine the values of the pronumerals in the diagram, giving a reason for your answer.

THINK	WRITE
1. Angle z is subtended by the diameter. Use an appropriate theorem to state the value of z.	$z = 90°$ 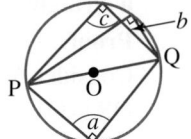
2. Angle s is formed by a tangent and a radius, drawn to the point of contact. Apply the corresponding theorem to find the value of s.	$s = 90°$ 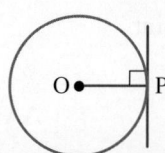

- **Theorem 5 Code**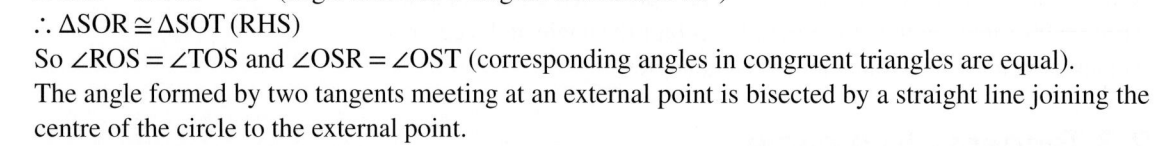

  The angle formed by two tangents meeting at an external point is bisected by a straight line joining the centre of the circle to that external point.

  Proof:

  Consider $\triangle SOR$ and $\triangle SOT$.

  $OR = OT$ (radii of the same circle are equal)

  OS is common.

  $\angle ORS = \angle OTS = 90°$ (angle between a tangent and radii is 90°)

  $\therefore \triangle SOR \cong \triangle SOT$ (RHS)

  So $\angle ROS = \angle TOS$ and $\angle OSR = \angle OST$ (corresponding angles in congruent triangles are equal).

  The angle formed by two tangents meeting at an external point is bisected by a straight line joining the centre of the circle to the external point.

WORKED EXAMPLE 3 Determining angles using properties of tangents

Given that BA and BC are tangents to the circle, determine the values of the pronumerals in the diagram. Give reasons for your answers.

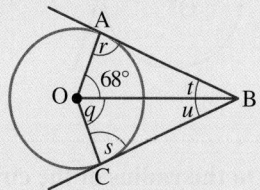

THINK

1. Angles r and s are angles formed by the tangent and the radius, drawn to the same point on the circle. State their size.

2. In the triangle ABO, two angles are already known and so angle t can be found using our knowledge of the sum of the angles in a triangle.

3. $\angle ABC$ is formed by the two tangents, so the line BO, joining the vertex B with the centre of the circle, bisects this angle. This means that angles t and u are equal.

4. All angles in a triangle have a sum of 180°. $\triangle AOB$ and $\triangle COB$ are congruent triangles using RHS, they are both right angled triangles, the hypotenuse is common and $OA = OB$, radii.

WRITE

$s = r = 90°$

$\triangle ABO: t + 90° + 68° = 180°$
$$t + 158° = 180°$$
$$t = 22°$$

$\angle ABO = \angle CBO$
$\angle ABO = t = 22°, \angle CBO = u$
$$u = 22°$$

In $\triangle AOB$ and $\triangle COB$

$r + t + 68° = 180°$

$s + u + q = 180°$

$r = s = 90°$ (proved previously)

$t = u = 22°$ (proved previously

$\therefore q = 68°$

DISCUSSION

What are the common steps in proving a theorem?

 Resources

Interactivities Circle theorem 1 (int-6218)
Circle theorem 2 (int-6219)
Circle theorem 3 (int-6220)
Circle theorem 4 (int-6221)
Circle theorem 5 (int-6222)

Exercise 18.2 Angles in a circle

learn on

18.2 Quick quiz on	18.2 Exercise

Individual pathways

■ PRACTISE	■ CONSOLIDATE	■ MASTER
1, 4, 6, 8, 13, 17	2, 5, 9, 11, 14, 18	3, 7, 10, 12, 15, 16, 19

Fluency

WE1 For questions **1** to **3**, calculate the values of the pronumerals in each of the following, giving reasons for your answers.

1. a.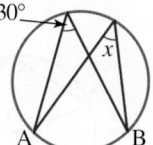

   b.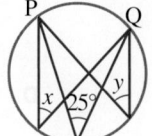

   c.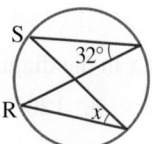

2. a.

   b.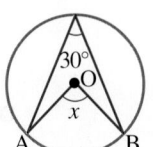

   c.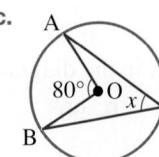

3. a.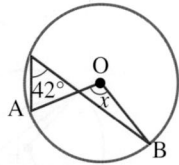

   b.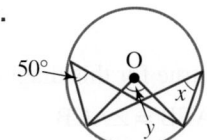

   c.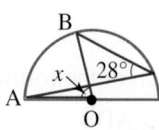

4. **WE2** Determine the values of the pronumerals in each of the following figures, giving reasons for your answers.

   a.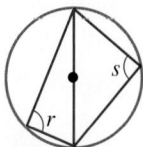

   b.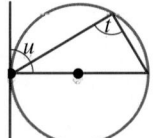

   c.

5. Calculate the values of the pronumerals in each of the following figures, giving reasons for your answers.

a.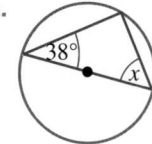

b.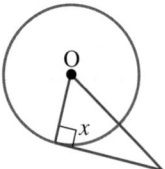

c.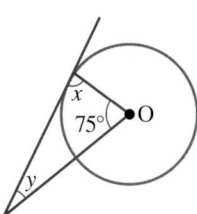

Understanding

6. **WE3** Given that AB and DB are tangents, determine the values of the pronumerals in each of the following, giving reasons for your answers.

a.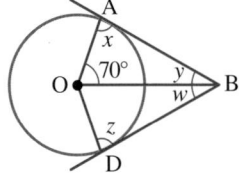

b.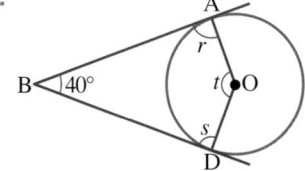

c.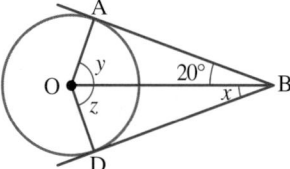

7. Given that AB and DB are tangents, determine the values of the pronumerals in each of the following, giving reasons for your answers.

a.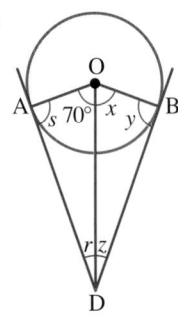

b.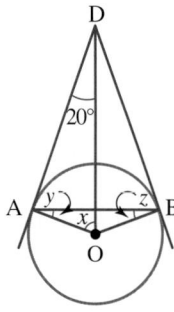

c.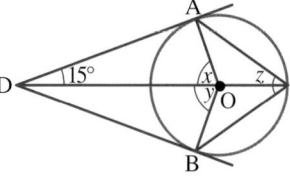

8. **MC** The value of x in the diagram is:

   A. 240°   B. 120°   C. 90°   D. 60°

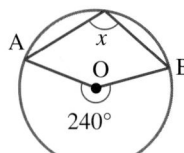

9. **MC** The value of x in the diagram is:

   A. 50°   B. 90°   C. 100°   D. 80°

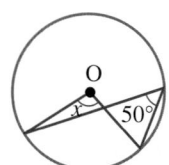

10. **MC** Choose which of the following statements is true for this diagram.

   A. ∠ACB = 2 × ∠ADB   B. ∠AEB = ∠ACB
   C. ∠ACB = ∠ADB   D. ∠AEB = ∠ADB

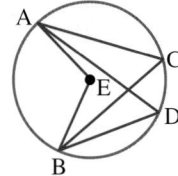

11. **MC** In the diagram shown, determine which angle is subtended by the same arc as ∠APB.
    Note: There may be more than one correct answer.

    A. ∠APC   **B.** ∠BPC   **C.** ∠ABP   **D.** ∠ADB

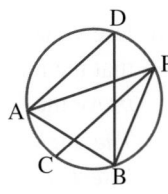

12. **MC** For the diagram shown, determine which of the statements is true.
    Note: There may be more than one correct answer.

    A. 2∠AOD = ∠ABD   **B.** ∠AOD = 2∠ACD
    C. ∠ABF = ∠ABD   **D.** ∠ABD = ∠ACD

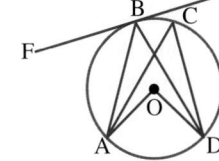

Communicating, reasoning and problem solving

13. Values are suggested for the pronumerals in the diagram shown. AB is a
    tangent to a circle and O is the centre.
    In each case give reasons to justify suggested values.

    a. $s = t = 45°$ b. $r = 45°$
    c. $u = 65°$ d. $m = 25°$
    e. $n = 45°$

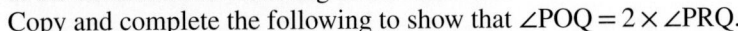

14. Set out below is the proof of this result: The angle at the centre of a circle is twice the angle
    at the circumference standing on the same arc.
    Copy and complete the following to show that ∠POQ = 2 × ∠PRQ.
    Construct a diameter through R. Let the opposite end of the diameter be S.
    Let ∠ORP = x and ∠ORQ = y.
    OR = OP (_____)
    ∠OPR = x (_____)
    ∠SOP = $2x$ (exterior angle equals _____)
    OR = OQ (_____)
    ∠OQR = _____ (_____)
    ∠SOQ = _____ (_____)
    Now ∠PRQ = _____ and ∠POQ = _____.
    Therefore ∠POQ = 2 × ∠PRQ.

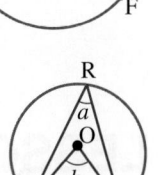
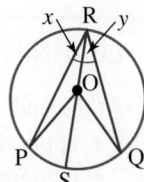

15. Prove that the segments formed by drawing tangents from an external point to a circle are equal in length.

16. Use the figure shown to prove that angles subtended by the same arc are equal.

17. Determine the value of x in each of the following diagrams.

    a.
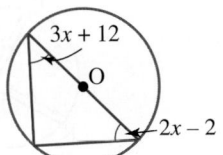

    b.
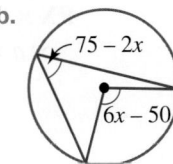

18. Use your knowledge of types of triangles, angles in triangles and the fact that the radius of a circle meets the tangent to the circle at right angles *to prove* the following theorem:

The angle formed between two tangents meeting at an external point is bisected by a line from the centre of the circle to the external point.

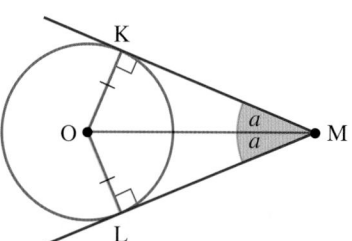

19. WX is the diameter of a circle with centre at O. Y is a point on the circle and WY is extended to Z so that OY = YZ. Prove that angle ZOX is three times angle YOZ.

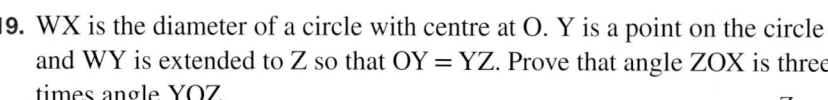

LESSON
18.3 Intersecting chords, secants and tangents

LEARNING INTENTION

At the end of this lesson you should be able to:
- apply the intersecting chords, secants and tangents theorems
- recognise that a radius will bisect a chord at right angles
- understand the concept of the circumcentre of a triangle
- identify and prove chord, secant and tangent properties of circles.

▶ 18.3.1 Intersecting chords

eles-4997

- In the diagram at right, chords PQ and RS intersect at X.

- **Theorem 6 Code** ⊗

  If the two chords intersect inside a circle, then the point of intersection divides each chord into two segments so that the product of the lengths of the segments for both chords is the same.

$$PX \times QX = RX \times SX$$
$$\text{or } a \times b = c \times d$$

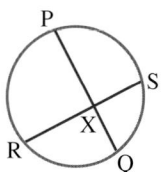

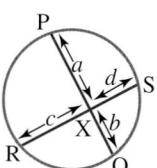

Proof:
Join PR and SQ.
Consider $\triangle PRX$ and $\triangle SQX$.
$\angle PXR = \angle SXQ$ (vertically opposite angles are equal)
$\angle RSQ = \angle RPQ$ (angles at the circumference standing on the same arc are equal)
$\angle PRS = \angle PQS$ (angles at the circumference standing on the same arc are equal)
$\triangle PRX \sim \triangle SQX$ (**equiangular**)
$$\frac{PX}{SX} = \frac{RX}{QX} \text{ (ratio of sides in similar triangles is equal)}$$
or, $PX \times QX = RX \times SX$

WORKED EXAMPLE 4 Determining values using intersecting chords

Determine the value of the pronumeral m.

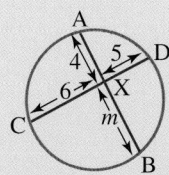

THINK

1. Chords AB and CD intersect at X. Point X divides each chord into two parts so that the products of the lengths of these parts are equal. Write this as a mathematical statement.

2. Identify the lengths of the line segments.

3. Substitute the given lengths into the formula and solve for m.

WRITE

$AX \times BX = CX \times DX$

$AX = 4$, $BX = m$, $CX = 6$, $DX = 5$

$4m = 6 \times 5$
$m = \dfrac{30}{4}$
$= 7.5$

18.3.2 Intersecting secants

eles-4998

- In the diagram at right, chords CD and AB are extended to form secants CX and AX respectively. They intersect at X.

- **Theorem 7 Code**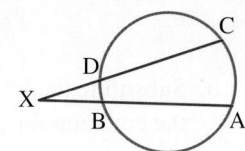

  If two secants intersect outside the circle as shown, then the following relationship is always true:

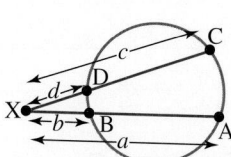

$$\mathbf{AX \times XB = XC \times DX}$$
$$\text{or } a \times b = c \times d.$$

Proof:
Join D and A to O, the centre of the circle.
Let $\angle DCA = x$
$\quad \angle DOA = 2x$ (angle at the centre is twice the angle at the circumference
$\qquad\qquad\qquad$ standing on the same arc)
Reflex $\angle DOA = 360° - 2x$ (angles in a revolution add to 360°)
$\quad\quad \angle DBA = 180° - x$ (angle at the centre is twice the angle at the circumference
$\qquad\qquad\qquad\qquad$ standing on the same arc)
$\quad\quad \angle DBX = x$ (angle sum of a straight line is 180°)
$\quad\quad \angle DCA = \angle DBX$

Consider $\triangle BXD$ and $\triangle CXA$.
$\angle BXD$ is common.
$\quad\quad \angle DCA = \angle DBX$ (shown previously)
$\quad\quad \angle XAC = \angle XDB$ (angle sum of a triangle is 180°)
$\quad\quad \angle AXC \sim \triangle DXB$ (equiangular)
$$\frac{AX}{DX} = \frac{XC}{XB}$$
or, $AX \times XB = XC \times DX$

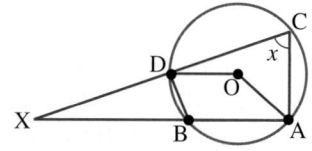

WORKED EXAMPLE 5 Determining pronumerals using intersecting secants.

Determine the value of the pronumeral y.

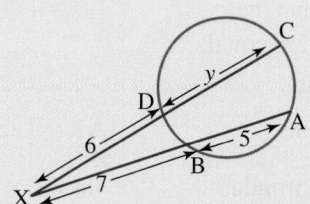

THINK	WRITE
1. Secants XC and AX intersect outside the circle at X. Write the rule connecting the lengths of XC, DX, AX and XB.	$XC \times DX = AX \times XB$
2. State the length of the required line segments.	$XC = y + 6 \quad DX = 6$ $AX = 7 + 5 \quad XB = 7$ $\quad\quad = 12$
3. Substitute the length of the line segments and solve the equation for y.	$(y + 6) \times 6 = 12 \times 7$ $6y + 36 = 84$ $6y = 48$ $y = 8$

18.3.3 Intersecting tangents

eles-4999

- In the diagram, the tangents AC and BC intersect at C.

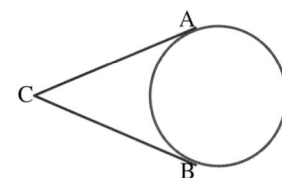

- **Theorem 8 Code**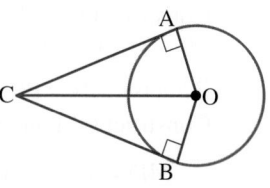
  If two tangents meet outside a circle, then the lengths from the external point to where they meet the circle are equal.
  Proof:
  Join A and B to O, the centre of the circle.
  Consider $\triangle OCA$ and $\triangle OCB$.
  OC is common.
  $\qquad$ OA = OB (radii of the same circle are equal)
  $\angle OAC = \angle OBC$ (radius is perpendicular to tangent through the point of contact)
  $\triangle OCA \cong \triangle OCB$ (RHS)
  $\qquad$ AC = BC (corresponding sides of congruent triangles are equal)

  If two tangents meet outside a circle, the lengths from the external point to the point of contact are equal.

WORKED EXAMPLE 6 Determining the pronumeral using lengths of tangents.

Determine the value of the pronumeral m.

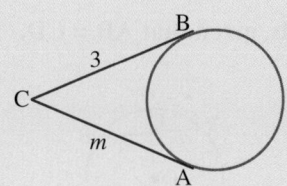

THINK	WRITE
1. BC and AC are tangents intersecting at C. State the rule that connects the lengths BC and AC.	AC = BC 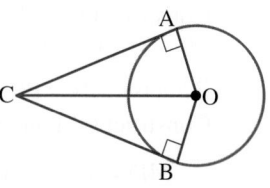
2. State the lengths of BC and AC.	AC = m, BC = 3
3. Substitute the required lengths into the equation to find the value of m.	$m = 3$

18.3.4 Chords and radii

eles-5000

- In the diagram at right, the chord AB and the radius OC intersect at X at 90°; that is, $\angle OXB = 90°$. OC bisects the chord AB; that is, AX = XB.
- **Theorem 9 Code**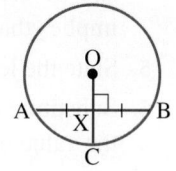
  If a radius and a chord intersect at right angles, then the radius bisects the chord.
  Proof:
  Join OA and OB.
  Consider $\triangle OAX$ and $\triangle OBX$.
  OA = OB (radii of the same circle are equal)
  $\angle OXB = \angle OXA$ (given)
  OX is common.
  $\triangle OAX \cong \triangle OBX$ (RHS)
  AX = BX (corresponding sides in congruent triangles are equal)
  If a radius and a chord intersect at right angles, then the radius bisects the chord.

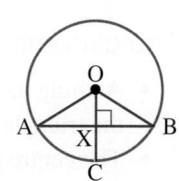

- The converse is also true:
  If a radius bisects a chord, the radius and the chord meet at right angles.

- **Theorem 10 Code**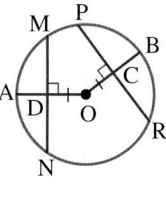

  Chords equal in length are equidistant from the centre.

  This theorem states that if the chords MN and PR are of equal length, then OD = OC.

  Proof:

  Construct OA⊥MN and OB⊥PR.

  Then OA bisects MN and OB bisects PR (Theorem 9)

  Because MN = PR, MD = DN = PC = CR.

  Construct OM and OP, and consider ΔODM and ΔOCP.

  MD = PC (shown above)
  OM = OP (radii of the same circle are equal)
  ∠ODM = ∠OCP = 90° (by construction)
  ΔODM ≅ ΔOCP (RHS)
  So OD = OC (corresponding sides in congruent triangles are equal)

  Chords equal in length are equidistant from the centre.

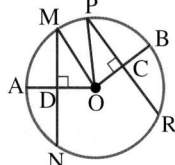

WORKED EXAMPLE 7 Determining pronumerals using theorems on chords

Determine the values of the pronumerals, given that AB = CD.

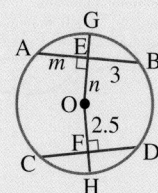

THINK	WRITE
1. Since the radius OG is perpendicular to the chord AB, the radius bisects the chord.	AE = EB
2. State the lengths of AE and EB.	AE = m, EB = 3
3. Substitute the lengths into the equation to find the value of m.	$m = 3$
4. AB and CD are chords of equal length and OE and OF are perpendicular to these chords. This implies that OE and OF are equal in length.	OE = OF
5. State the lengths of OE and OF.	OE = n, OF = 2.5
6. Substitute the lengths into the equation to find the value of n.	$n = 2.5$

The circumcentre of a triangle

- A circle passing through the three vertices of a triangle is called the **circumcircle** of the triangle.
- The centre of this circle is called the **circumcentre** of the triangle.

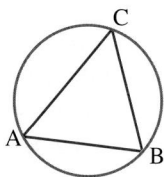

- The circumcentre is located by:
  Step 1: drawing any triangle ABC and label the vertices

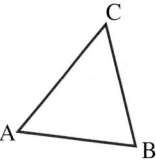

  Step 2: constructing the perpendicular bisectors of the three sides

  Step 3: let the bisectors intersect at O.

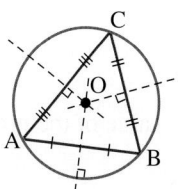

- This means OA = OB = OC, by congruent triangles.
- A circle, centre O, can be drawn through the vertices A, B and C.
- The point O is the circumcentre of the triangle.

DISCUSSION

What techniques will you use to prove circle theorem?

 Resources

 Interactivities Circle theorem 6 (int-6223)
Circle theorem 7 (int-6224)
Circle theorem 8 (int-6225)
Circle theorem 9 (int-6226)
Circle theorem 10 (int-6227)

Exercise 18.3 Intersecting chords, secants and tangents   learn on

18.3 Quick quiz on	18.3 Exercise

Individual pathways

■ PRACTISE	■ CONSOLIDATE	■ MASTER
1, 4, 9, 14	2, 5, 7, 10, 12	3, 6, 8, 11, 13, 15

Fluency

1. **WE4** Determine the value of the pronumeral in each of the following.

a.
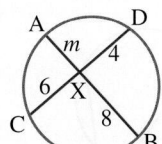

b.
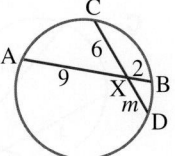

c.
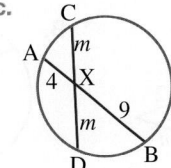

2. **WE5** Determine the value of the pronumeral in each of the following.

a.
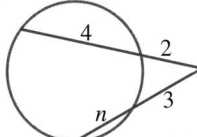

b.
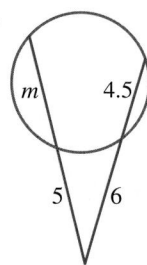

3. Determine the value of the pronumeral in each of the following.

a.
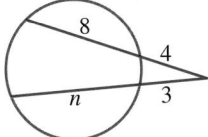

b.
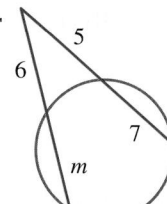

4. **WE6** Determine the values of the pronumerals in each of the following.

a.
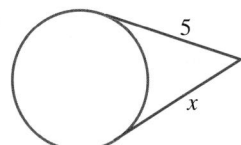

b.
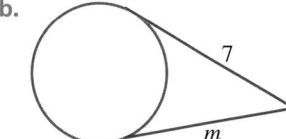

c.
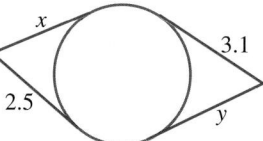

5. **WE7** Determine the value of the pronumeral in each of the following.

a.
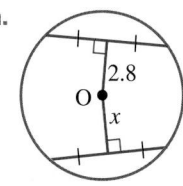

b.
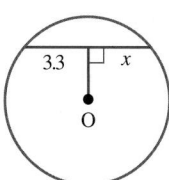

6. Determine the value of the pronumeral in each of the following.

a.
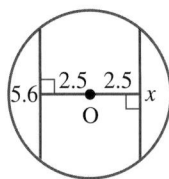

b.
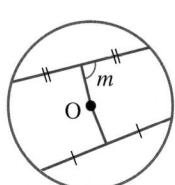

Understanding

7. **MC** Select which of the following figures allows the value of m to be determined by solving a linear equation.

Note: There may be more than one correct answer.

A.
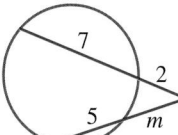

B.
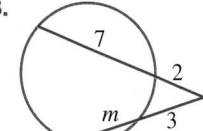

C.
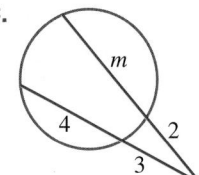

D.
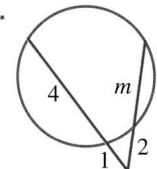

8. Calculate the length ST in the diagram.

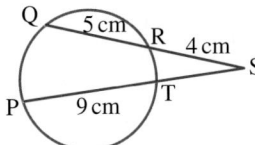

Communicating, reasoning and problem solving

9. Prove the result: If a radius bisects a chord, then the radius meets the chord at right angles. Remember to provide reasons for your statements.

10. Prove the result: Chords that are an equal distance from the centre are equal in length. Provide reasons for your statements.

11. Prove that the line joining the centres of two intersecting circles bisects their common chord at right angles. Provide reasons for your statements.

12. Prove the result: Equal chord of a circle subtend equal angles at the centre.

13. Prove that when two circles touch, their centres and the point of contact are collinear.

14. Determine the values of the pronumerals in each of the following diagrams.

a.

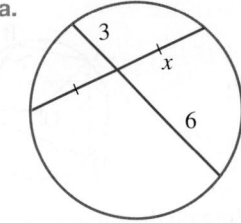

b.

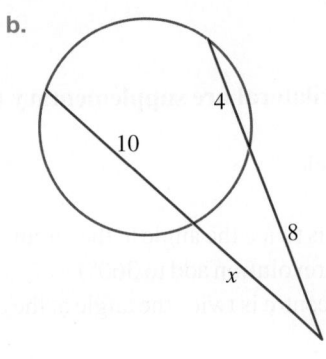

c.

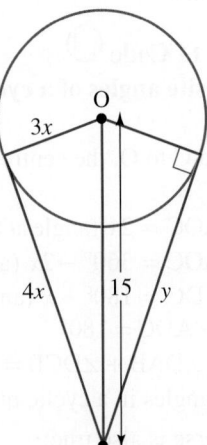

15. AOB is the diameter of the circle. CD is a chord perpendicular to AB and meeting AB at M.

a. Explain why M is the midpoint of CD.

b. If $CM = c$, $AM = a$ and $MB = b$, prove that $c^2 = ab$.

c. Explain why the radius of the circle is equal to $\dfrac{a+b}{2}$.

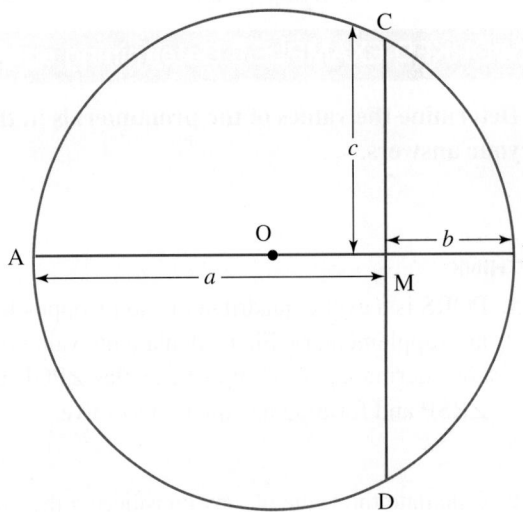

LESSON
18.4 Cyclic quadrilaterals

LEARNING INTENTION

At the end of this lesson you should be able to:
- recognize a cyclic quadrilateral
- identify and prove properties of cyclic quadrilaterals
- apply properties of cyclic quadrilaterals to determine unknown angles.

▶ 18.4.1 Quadrilaterals in circles

eles-5001

- A **cyclic quadrilateral** has all four vertices on the circumference of a circle; that is, the quadrilateral is inscribed in the circle.
- In the diagram at right, points A, B, C and D lie on the circumference; hence, ABCD is a cyclic quadrilateral.
- It can also be said that points A, B, C and D are **concyclic**; that is, the circle passes through all the points.

- **Theorem 11 Code**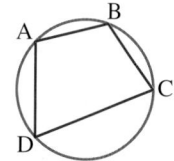
  The opposite angles of a cyclic quadrilateral are supplementary (add to 180°).
  Proof:
  Join A and C to O, the centre of the circle.
  Let $\angle ABC = x$.
  Reflex $\angle AOC = 2x$ (angle at the centre is twice the angle at the circumference standing on the same arc)
  Reflex $\angle AOC = 360° - 2x$ (angles in a revolution add to 360°)
  $\angle ADC = 180° - x$ (angle at the centre is twice the angle at the circumference standing on the same arc)
  $\angle ABC + \angle ADC = 180°$
  Similarly, $\angle DAB + \angle DCB = 180°$.
  Opposite angles in a cyclic quadrilateral are supplementary.
- The converse is also true:
  If opposite angles of a quadrilateral are supplementary, then the quadrilateral is cyclic.

WORKED EXAMPLE 8 Determining angles in a cyclic quadrilateral

Determine the values of the pronumerals in the diagram below. Give reasons for your answers.

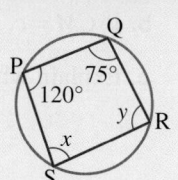

THINK	WRITE
1. PQRS is a cyclic quadrilateral, so its opposite angles are supplementary. First calculate the value of x by considering a pair of opposite angles $\angle PQR$ and $\angle RSP$ and forming an equation to solve.	$\angle PQR + \angle RSP = 180°$ (The opposite angles of a cyclic quadrilateral are supplementary.) $\angle PQR = 75°, \angle RSP = x$ $x + 75° = 180°$ $x = 105°$
2. Calculate the value of y by considering the other pair of opposite angles ($\angle SPQ$ and $\angle QRS$).	$\angle SPQ + \angle QRS = 180°$ 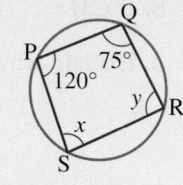 $\angle SPQ = 120°, \angle QRS = y$ $y + 120° = 180°$ $y = 60°$

- **Theorem 12 Code**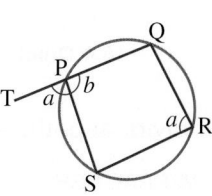
  The exterior angle of a cyclic quadrilateral is equal to the interior opposite angle.
  Proof:
  ∠QPS + ∠QRS = 180° (opposite angles of a cyclic quadrilateral)
  ∠QPS + ∠SPT = 180° (adjacent angles on a straight line)
  Therefore ∠SPT = ∠QRS.
  The exterior angle of a cyclic quadrilateral is equal to the interior opposite angle.

WORKED EXAMPLE 9 Determining pronumerals in a cyclic quadrilateral

Determine the value of the pronumerals x and y in the diagram below.

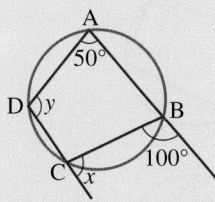

THINK	WRITE
1. ABCD is a cyclic quadrilateral. The exterior angle, x, is equal to its interior opposite angle, ∠DAB.	$x = ∠DAB$, $∠DAB = 50°$ So $x = 50°$.
2. The exterior angle, 100°, is equal to its interior opposite angle, ∠ADC.	$∠ADC = 100°$, $∠ADC = y$ So $y = 100°$.

DISCUSSION

Observe the constructions with isosceles trapezoids. Do you think isosceles trapezoids are always cyclic quadrilaterals? What about non-isosceles trapezoids?

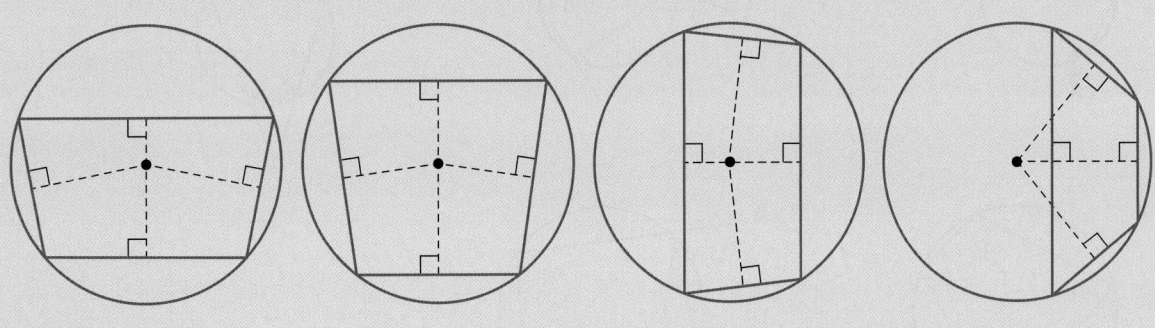

---

on Resources

Interactivities Circle theorem 11 (int-6228)
Circle theorem 12 (int-6229)

Exercise 18.4 Cyclic quadrilaterals

learnon

| 18.4 Quick quiz on | 18.4 Exercise |

Individual pathways

■ PRACTISE	■ CONSOLIDATE	■ MASTER
1, 4, 5, 10, 13, 16	2, 3, 6, 11, 14, 17	7, 8, 9, 12, 15, 18

Fluency

WE8 In questions **1** and **2**, calculate the values of the pronumerals in each of the following.

1. a.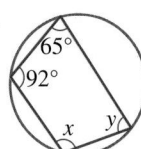

   b.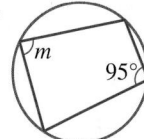

   c.

2. a.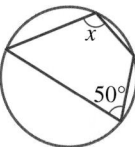

   b.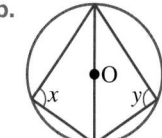

   c.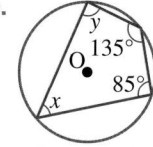

WE9 In questions **3** to **6**, calculate the values of the pronumerals in each of the following.

3. a.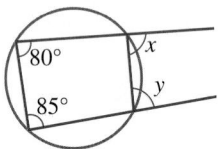

   b.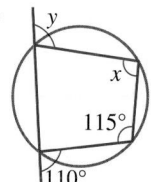

   c.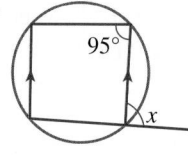

4. a.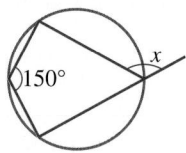

   b.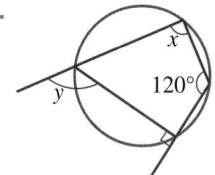

   c.

5. a.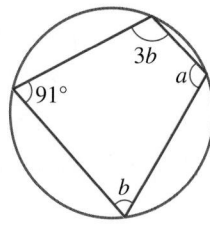

   b.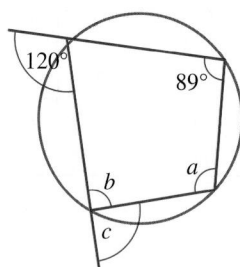

6. a.

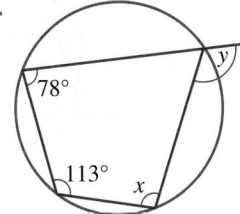

b.

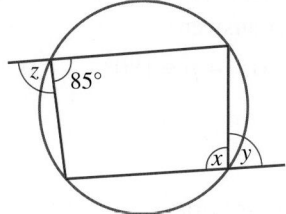

c.

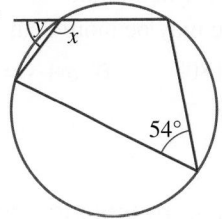

d.

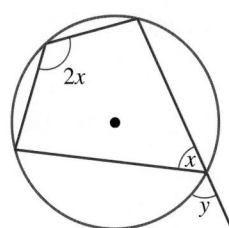

e.

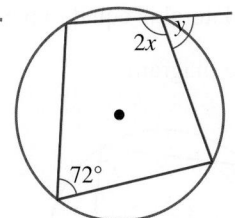

f.

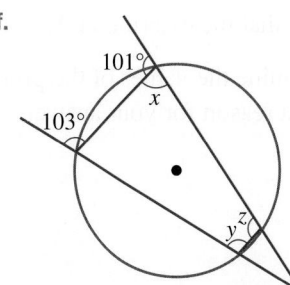

Understanding

7. **MC** Which of the following statements is true for this diagram?

   A. $r = q$
   B. $r + n = 180°$
   C. $m + n = 180°$
   D. $r + m = 180°$

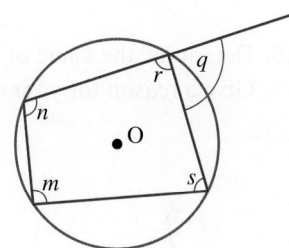

8. **MC** Which of the following statements is *not* true for this diagram?

   A. $b + f = 180°$
   B. $a = f$
   C. $e = d$
   D. $c + f = 180°$

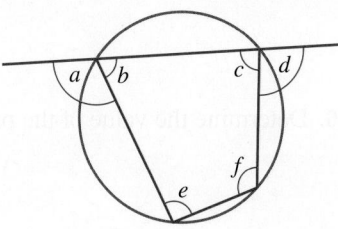

9. **MC** Choose which of the following correctly states the relationship between x, y and z in the diagram.

   A. $x = y$ and $x = 2z$
   B. $x = 2y$ and $y + z = 180°$
   C. $z = 2x$ and $y = 2z$
   D. $x + y = 180°$ and $z = 2x$

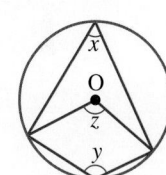

10. Follow the steps below to set out the proof that the opposite angles of a cyclic quadrilateral are equal.

    a. Calculate the size of ∠DOB.
    b. Calculate the size of the reflex angle DOB.
    c. Calculate the size of ∠BCD.
    d. Calculate ∠DAB + ∠BCD.

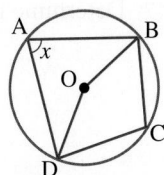

11. **MC** Choose which of the following statements is always true for the diagram shown.
    Note: There may be more than one correct answer.

    A. $r = t$
    B. $r = p$
    C. $r = q$
    D. $r = s$

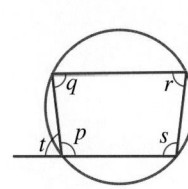

12. [MC] Choose which of the following statements is correct for the diagram shown.
Note: There may be more than one correct answer.

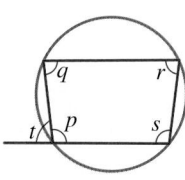

A. $r + p = 180°$ **B.** $q + s = 180°$ **C.** $t + p = 180°$ **D.** $t = r$

Communicating, reasoning and problem solving

13. Prove that the exterior angle of a cyclic quadrilateral is equal to the interior opposite angle.

14. Determine the values of the pronumerals in the diagram.
Give a reason for your answer.

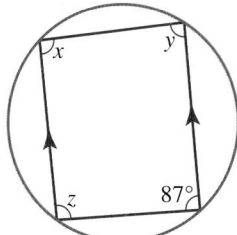

15. Determine the value of the pronumeral x in the diagram.
Give a reason for your answer.

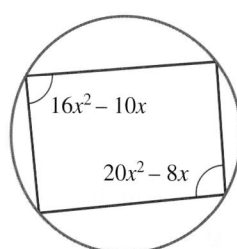

16. Determine the value of the pronumeral x in the diagram shown.

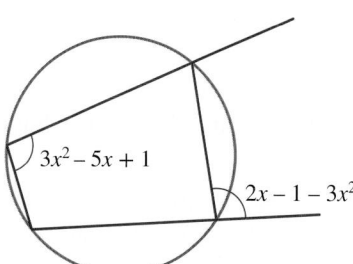

17. Determine the value of each pronumeral in the diagram shown.

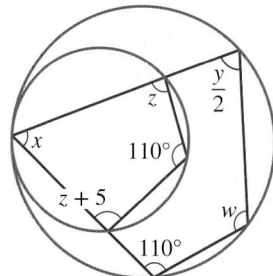

18. $\angle FAB = 70°$, $\angle BEF = a°$, $\angle BED = b°$ and $\angle BCD = c°$.

   a. Calculate the values of a, b and c.

   b. Prove that CD is parallel to AF.

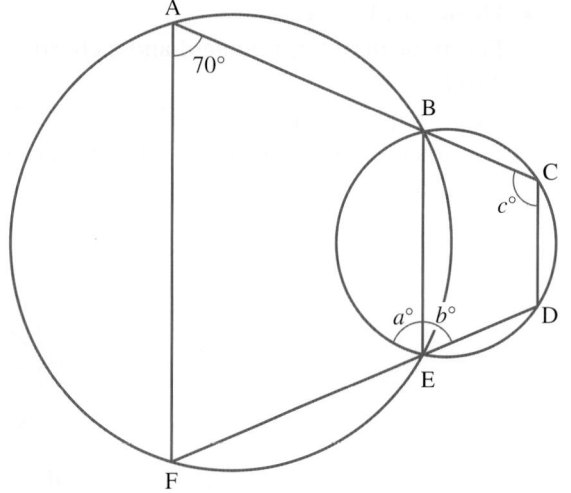

LESSON
18.5 Tangents, secants and chords

LEARNING INTENTION

At the end of this lesson you should be able to:
- apply the alternate segment theorem to evaluate pronumerals
- determine the lengths of tangents and secants when they intersect
- identify and prove tangent and secant properties of circles.

▶ 18.5.1 The alternate segment theorem

- Consider the figure shown. Line BC is a tangent to the circle at the point A.
- A line is drawn from A to anywhere on the circumference, point D.
  The angle $\angle BAD$ defines a segment (the shaded area).
  The unshaded part of the circle is called the **alternate segment** to $\angle BAD$.

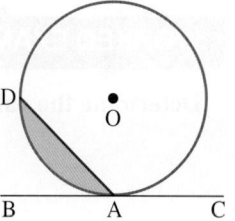

- Now consider angles subtended by the chord AD in the alternate segment, such as the angles marked in pink and blue.
- The alternate segment theorem states that these are equal to the angle that made the segment, namely:

$$\angle BAD = \angle AED \text{ and } \angle BAD = \angle AFD$$

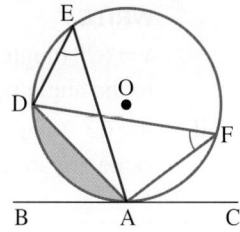

TOPIC 18 Circle geometry (Path) **965**

- **Theorem 13 Code**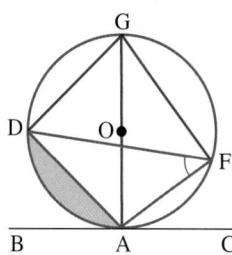

The angle between a tangent and a chord is equal to the angle in the alternate segment.
Proof:
We are required to prove that $\angle BAD = \angle AFD$.
Construct the diameter from A through O, meeting the circle at G.
Join G to the points D and F.

$\angle BAG = \angle CAG = 90°$ (radii $\perp$ tangent at point of contact)
$\angle GFA = 90°$ (angle in a semicircle is 90°)
$\angle GDA = 90°$ (angle in a semicircle is 90°)
Consider $\triangle GDA$. We know that $\angle GDA = 90°$.
$\angle GDA + \angle DAG + \angle AGD = 180°$
$\qquad 90° + \angle DAG + \angle AGD = 180°$
$\qquad\qquad \angle DAG + \angle AGD = 90°$
$\angle BAG$ is also a right angle.
$\angle BAG = \angle BAD + \angle DAG = 90°$
Equate the two results.
$\angle DAG + \angle AGD = \angle BAD + \angle DAG$
Cancel the equal angles ($\angle DAG$) on both sides.
$\angle AGD = \angle BAD$
Now consider the fact that both triangles DAG and DAF are subtended from the same chord (DA).
$\angle AGD = \angle AFD$ (Angles in the same segment standing on the same arc are equal).
Equate the two equations.
$\angle AFD = \angle BAD$

WORKED EXAMPLE 10 Determining pronumerals using the alternate segment theorem

Determine the value of the pronumerals x and y, giving reasons.

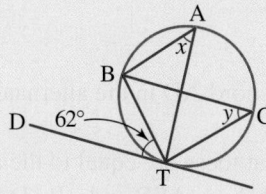

THINK

1. Use the alternate segment theorem to calculate x.

2. The value of y is the same as x because x and y are subtended by the same chord BT.

WRITE

$x = 62°$ (angle between a tangent and a chord is equal to the angle in the alternate segment.)

$y = 62°$ (angles in the same segment standing on the same arc are equal.)

18.5.2 Tangents and secants

eles-5003

- **Theorem 14 Code**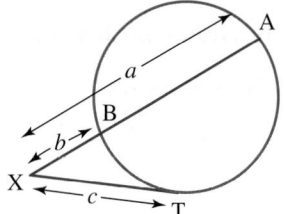
  If a tangent and a secant intersect as shown, the following relationship is always true:
  $$XA \times XB = (XT)^2 \text{ or } a \times b = c^2.$$
  Proof:
  Join BT and AT.
  Consider $\triangle TXB$ and $\triangle AXT$.
  $\angle TXB$ is common.
  $\angle XTB = \angle XAT$ (angle between a tangent and a chord is equal to the angle
  in the alternate segment)
  $\angle XBT = \angle XTA$ (angle sum of a triangle is $180°$)
  $\triangle TXB \sim \triangle AXT$ (equiangular)
  So, $\dfrac{XB}{XT} = \dfrac{XT}{XA}$
  or, $XA \times XB = (XT)^2$.

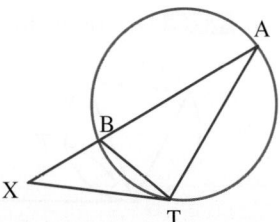

WORKED EXAMPLE 11 Determining pronumerals with intersecting tangents and secants

Determine the value of the pronumeral m.

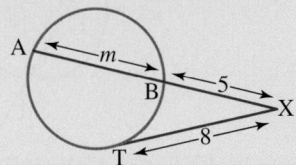

THINK	WRITE
1. Secant XA and tangent XT intersect at X. Write the rule connecting the lengths of XA, XB and XT.	$XA \times XB = (XT)^2$ 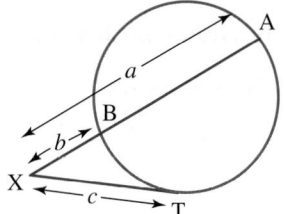
2. State the values of XA, XB and XT.	$XA = m + 5, \ XB = 5, \ XT = 8$
3. Substitute the values of XA, XB and XT into the equation and solve for m.	$(m + 5) \times 5 = 8^2$ $5m + 25 = 64$ $5m = 39$ $m = 7.8$

DISCUSSION

Describe the alternate segment of a circle.

 Resources

 Interactivities Circle theorem 13 (int-6230)
Circle theorem 14 (int-6231)

18.5 Quick quiz on	18.5 Exercise

Individual pathways

■ PRACTISE	■ CONSOLIDATE	■ MASTER
1, 2, 3, 9, 13, 14, 18, 22	4, 5, 6, 10, 15, 16, 17, 19, 23	7, 8, 11, 12, 20, 21, 24

Fluency

1. **WE10** Determine the values of the pronumerals in the following diagrams.

   a.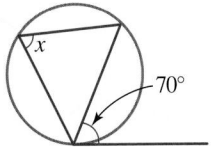

   b.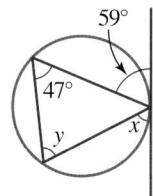

2. **WE11** Calculate the values of the pronumerals in the following diagrams.

   a.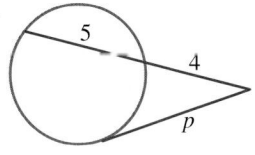

   b.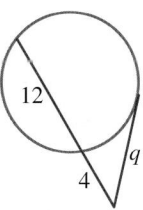

3. Line AB is a tangent to the circle as shown in the figure. Calculate the values of the angles labelled x and y.

   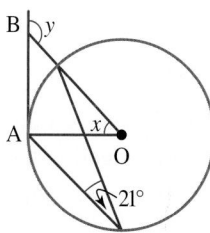

Questions **4** to **6** refer to the figure shown. The line MN is a tangent to the circle, and EA is a straight line. The circles have the same radius.

4. **MC** If $\angle DAC = 20°$, then $\angle CFD$ and $\angle FDG$ are respectively:

   A. 70° and 50°   B. 70° and 40°

   C. 40° and 70°   D. 70° and 70°

5. **MC** A triangle similar to FDA is:

   A. FDG   B. FGB

   C. EDA   D. GDE

6. State six different right angles.

7. Calculate the values of the angles x and y in the figure shown.

   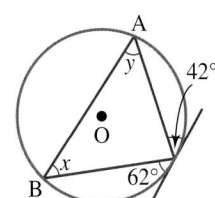

Understanding

8. Show that if the sum of the two given angles in question **7** is 90°, then the line AB must be a diameter.

9. Calculate the value of x in the figure shown, given that the line underneath the circle is a tangent.

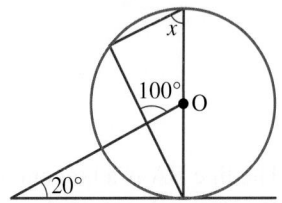

10. In the figure shown, express x in terms of a and b. This is the same diagram as in question **9**.

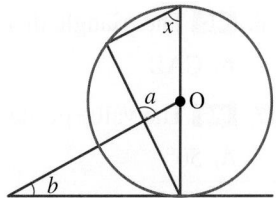

11. Two tangent lines to a circle meet at an angle y, as shown in the figure. Determine the values of the angles x, y and z.

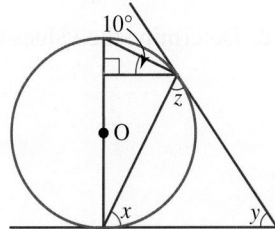

12. Solve question **11** in the general case (see the figure) and show that $y = 2a$. This result is important for space navigation (imagine the circle to be the Earth) in that an object at y can be seen by people at x and z at the same time.

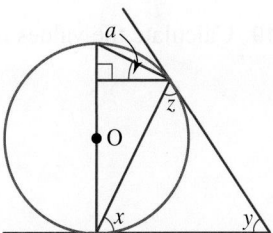

13. In the figure shown, determine the values of the angles x, y and z.

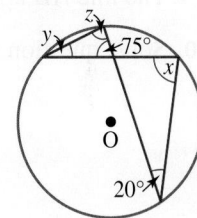

14. **MC** Examine the figure shown. The angles x and y (in degrees) are respectively:

    A. 51 and 99
    B. 51 and 129
    C. 39 and 122
    D. 51 and 122

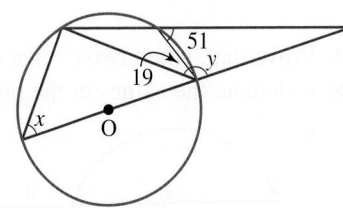

Questions **15** to **17** refer to the figure shown.

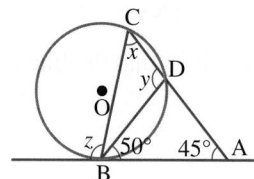

The line BA is a tangent to the circle at point B. Line AC is a chord that meets the tangent at A.

15. Determine the values of the angles x and y.

16. **MC** The triangle that is similar to triangle BAD is:

    A. CAB   **B.** BCD   **C.** BDC   **D.** AOB

17. **MC** The value of the angle z is:

    A. 50°   **B.** 85°   **C.** 95°   **D.** 100°

Communicating, reasoning and problem solving

18. Determine the values of the angles x, y and z in the figure shown. The line AB is tangent to the circle at B.

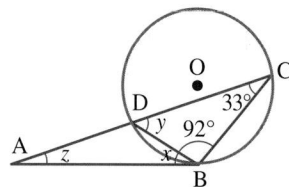

19. Calculate the values of the angles x, y and z in the figure shown.

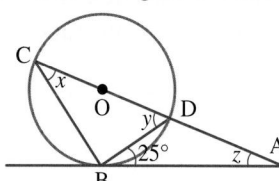

    The line AB is tangent to the circle at B. The line CD is a diameter.

20. Solve question **19** in the general case; that is, express angles x, y and z in terms of a (see the figure).

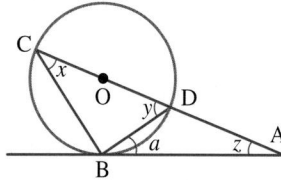

21. Prove that, when two circles touch, their centres and the point of contact are collinear.

22. Calculate the values of the pronumerals in the following.

    a.   **b.**   **c.**

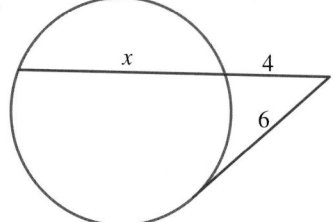 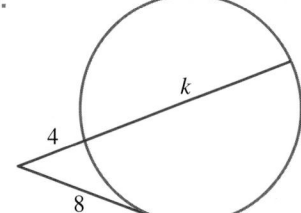

23. Determine the value of the pronumerals in the following.

a.

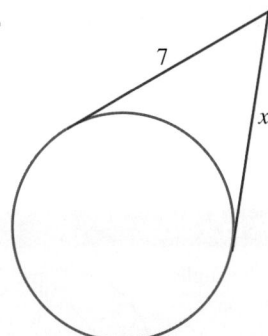

b.

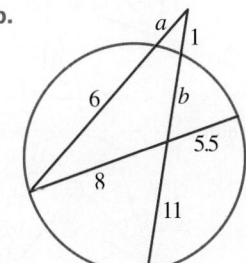

c.

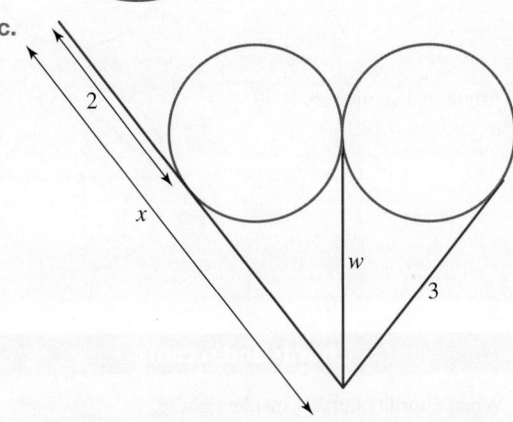

24. Calculate the values of a, b and c in each case.

a. $\angle BCE = 50°$ and $\angle ACE = c$

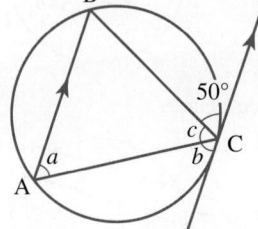

b.

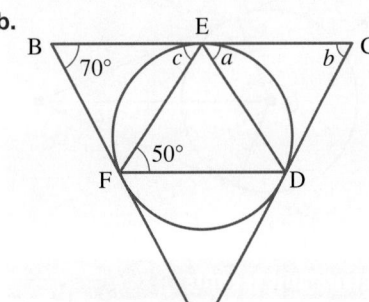

LESSON
18.6 Review

18.6.1 Topic summary

Cyclic quadrilaterals

- **Opposite angles** of cyclic quadrilateral are supplementary.
  e.g. $a + b = 180°$
- **Exterior angle** of a cyclic quadrilateral equals the interior opposite angle.

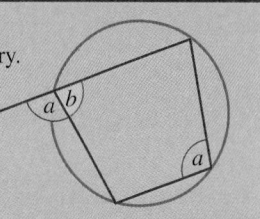

CIRCLE GEOMETRY (PATH)

Tangents to a circle

- Tangent is perpendicular to the radius at the point of contact.

- Lengths of tangents from an external point are equal.
  $a = b$

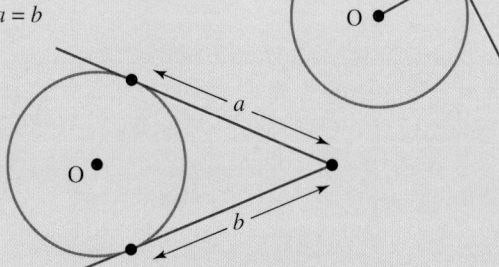

- Line from the centre bisects the external angle between two tangents.

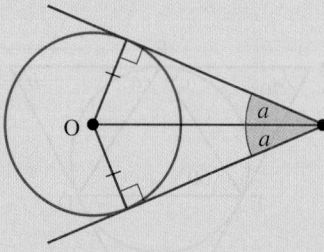

Tangents and secants

- Angle between a tangent and a chord equals the angle in the alternate segment.
- If tangent and segment intersect then: $a \times b = c^2$

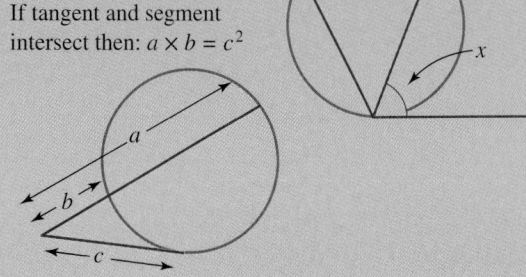

Angles in a circle

- Angle at centre is twice the angles at circumference standing on the same arc.
  e.g. $b = 2a$

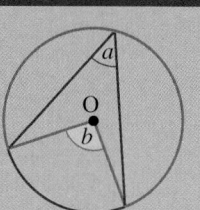

- Angles at circumference standing on the same arc are equal.
  $a = b = c$

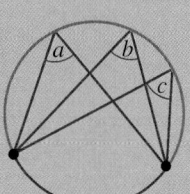

- Angle in a semicircle is 90°.
  $a = 90°, b = 180°$

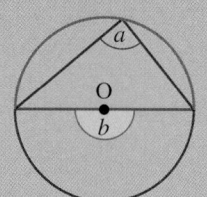

Chords and secants

- When chords intersect inside a circle, the product of the lengths of the segments are equal.
  $a \times b = c \times d$

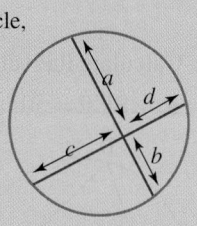

- When chords intersect outside a circle, then the following is true:
  $a \times b = c \times d$

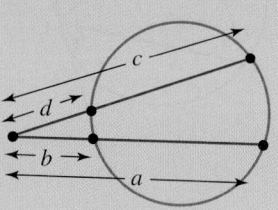

- Perpendicular bisector of a chord passes through the centre of the circle.

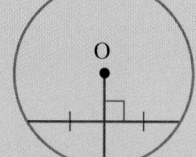

18.6.2 Project

Variation of distance

The Earth approximates the shape of a sphere. Lines of longitude travel between the North and South poles, while lines of latitude travel east–west, parallel to the equator. While the lines of longitude are all approximately the same length, this is not the case with lines of latitude. The line of latitude at the equator is the maximum length and these lines decrease in length on approaching both the North and South poles.

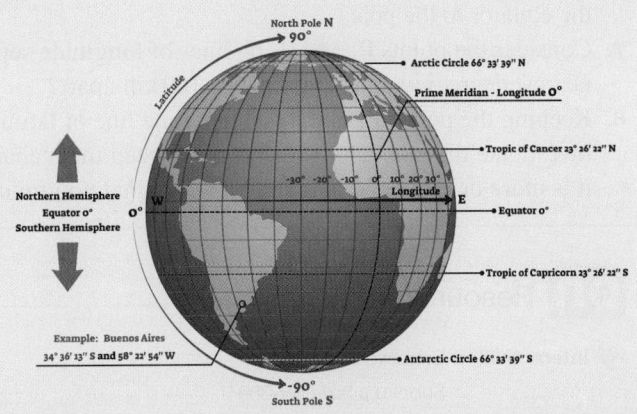

This investigation looks at how the distance between points on two given lines of longitude and the same line of latitude changes as we move from the equator to the pole.

Consider two lines of longitude, 0° and 100°E. Take two points, P_1 and P_2, lying on the equator on lines of longitude 0° and 100°E respectively.

The distance (in km) between two points on the same line of latitude is given by the formula:

$$\text{Distance} = \text{angle sector between the two points} \times 111 \times \cos(\text{degree of latitude})$$

1. The size of the angle sector between P_1 and P_2 is 100° and these two points lie on 0° latitude. The distance between the points would be calculated as $100 \times 111 \times \cos(0°)$. Determine this distance.

2. Move the two points to the 10° line of latitude. Calculate the distance between P_1 and P_2 in this position. Round your answer to the nearest kilometre.

3. Complete the following table showing the distance (rounded to the nearest kilometre) between the points P_1 and P_2 as they move from the equator towards the pole.

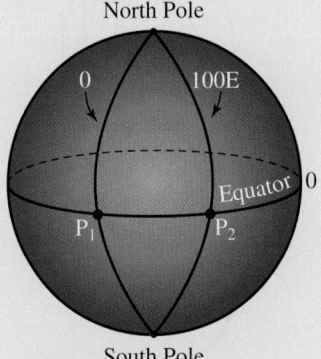

Latitude	Distance between P_1 and P_2 (km)
0°	
10°	
20°	
30°	
40°	
50°	
60°	
70°	
80°	
90°	

4. Describe what happens to the distance between P_1 and P_2 as we move from the equator to the pole. Is there a constant change? Explain your answer.

5. You would perhaps assume that, at a latitude of 45°, the distance between P_1 and P_2 is half the distance between the points at the equator. This is not the case. At what latitude does this occur?

6. Using grid paper, sketch a graph displaying the change in distance between the points in moving from the equator to the pole.

7. Consider the points P_1 and P_2 on lines of longitude separated by 1°. On what line of latitude (to the nearest degree) would the points be 100 km apart?

8. Keeping the points P_1 and P_2 on the same line of latitude, and varying their lines of longitude, investigate the rate that the distance between them changes from the equator to the pole. Explain whether it is more or less rapid in comparison to what you found earlier.

 Resources

 Interactivities Crossword (int-2881)

Sudoku puzzle (int-3894)

Exercise 18.6 Review questions

learn on

Fluency

For questions **1** to **3**, determine the values of the pronumerals in each of the diagrams.

1. a.
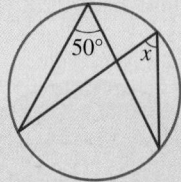

b.
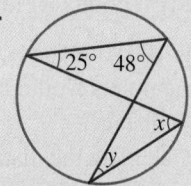

c.
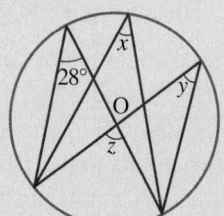

d.
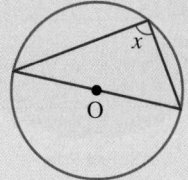

e.
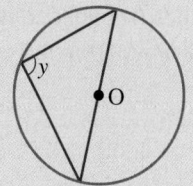

f.
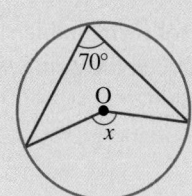

2. a.
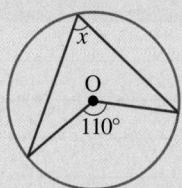

b.
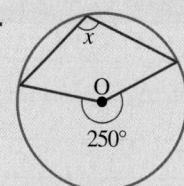

c.
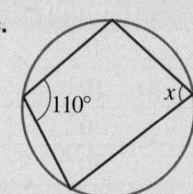

d.
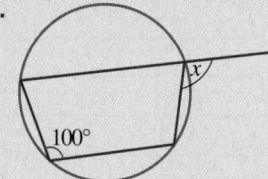

e.
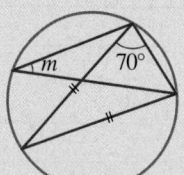

f.
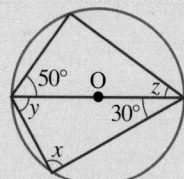

3. a.

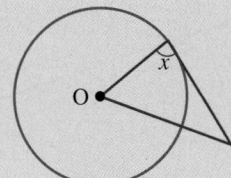

b.

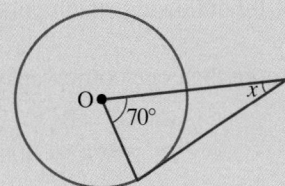

c.

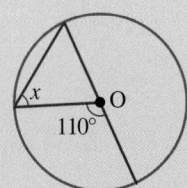

d.

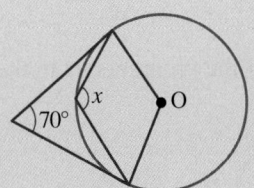

4. Determine the value of m in each of the following.

a.

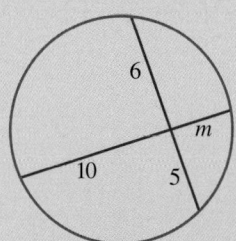

b.

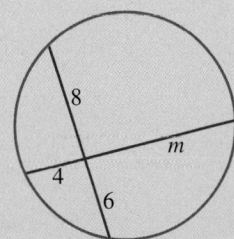

c.

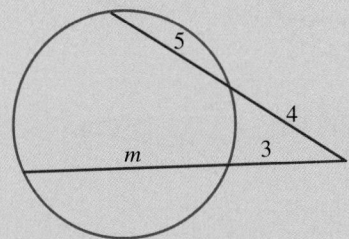

d.

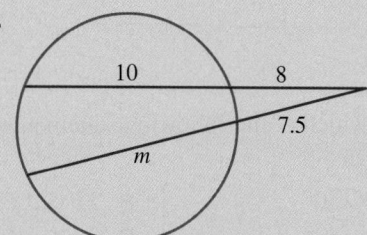

Understanding

5. **MC** Choose for which of the following figures it is possible to get a reasonable value for the pronumeral. *Note*: There may be more than one correct answer.

A.

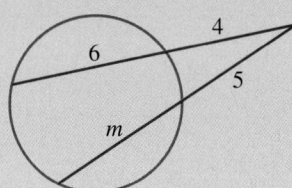

B.

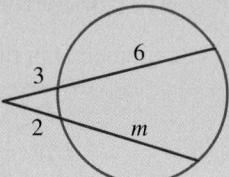

C.

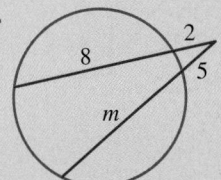

D.

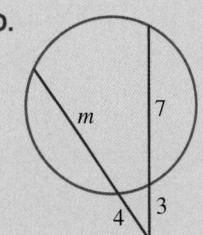

6. **MC** Choose which of the following statements is true for the diagram shown.

   Note: There may be more than one correct answer.

   A. AO = BO   **B.** AC = BC

   C. ∠OAC = ∠OBC   **D.** ∠AOC = 90°

   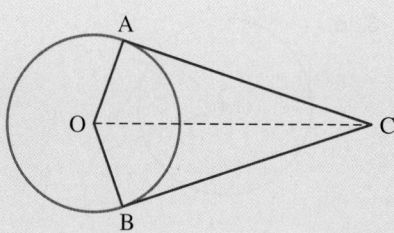

7. Determine the values of the pronumerals in the following figures.

   a.
   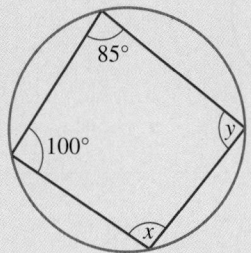

   b.
   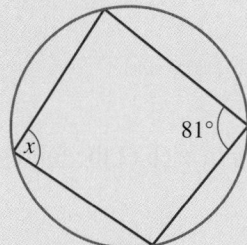

   c.
   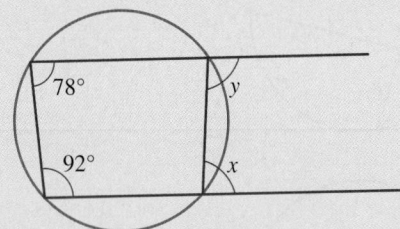

   d.
   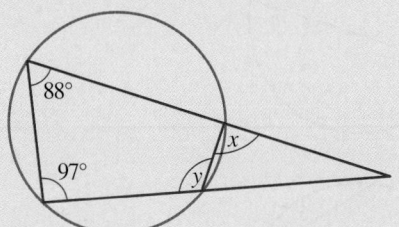

8. **MC** Choose which of the following statements is *not* always true for the diagram shown.

   A. ∠a + ∠c = 180°   **B.** ∠b + ∠d = 180°

   C. ∠e + ∠c = 180°   **D.** ∠a + ∠e = 180°

   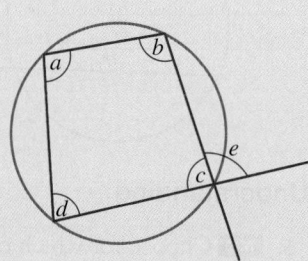

Communicating, reasoning and problem solving

9. Determine the values of the pronumerals in the following figures.

   a.
   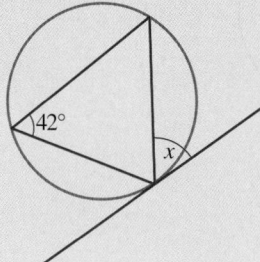

   b.
   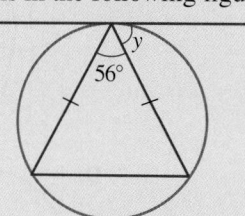

   c.
   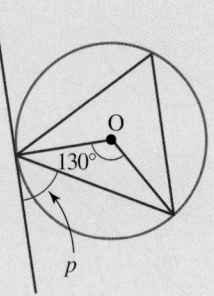

10. Two chords, AB and CD, intersect at E as shown. If AE = CE, prove that EB = ED.

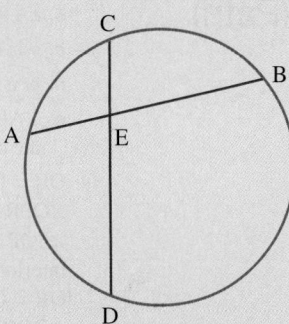

11. Two circles intersect at X and Y. Two lines, AXB and CXD, intersect one circle at A and C, and the other at B and D, as shown. Prove that ∠AYC = ∠BYD.

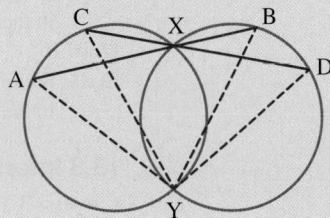

12. Name at least five pairs of equal angles in the following diagram.

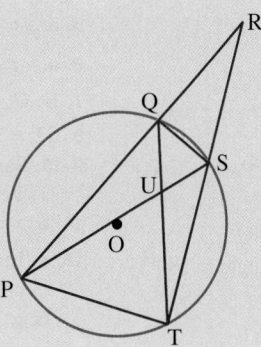

on To test your understanding and knowledge of this topic, go to your learnON title at www.jacplus.com.au and complete the **post-test**.

Answers

Topic 18 Circle geometry (Path)

18.1 Pre-test

1. Segment
2. D
3. $35°$
4. D
5. $65°$
6. $15°$
7. D
8. 6
9. 1.5
10. A
11. $m = 55°$, $p = 50°$
12. C
13. D
14. D
15. A

18.2 Angles in a circle

1. a. $x = 30°$ (theorem 2)
   b. $x = 25°$, $y = 25°$ (theorem 2 for both angles)
   c. $x = 32°$ (theorem 2)
2. a. $x = 40°$, $y = 40°$ (theorem 2 for both angles)
   b. $x = 60°$ (theorem 1)
   c. $x = 40°$ (theorem 1)
3. a. $x = 84°$ (theorem 1)
   b. $x = 50°$ (theorem 2); $y = 100°$ (theorem 1)
   c. $x = 56°$ (theorem 1)
4. a. $s = 90°$, $r = 90°$ (theorem 3 for both angles)
   b. $u = 90°$ (theorem 4); $t = 90°$ (theorem 3)
   c. $m = 90°$, $n = 90°$ (theorem 3 for both angles)
5. a. $x = 52°$ (theorem 3 and angle sum in a triangle $= 180°$)
   b. $x = 90°$ (theorem 4)
   c. $x = 90°$ (theorem 4); $y = 15°$ (angle sum in a triangle $= 180°$)
6. a. $x = z = 90°$ (theorem 4); $y = w = 20°$ (theorem 5 and angle sum in a triangle $= 180°$)
   b. $s = r = 90°$ (theorem 4); $t = 140°$ (angle sum in a quadrilateral $= 360°$)
   c. $x = 20°$ (theorem 5); $y = z = 70°$ (theorem 4 and angle sum in a triangle $= 180°$)
7. a. $s = y = 90°$ (theorem 4); $x = 70°$ (theorem 5); $r = z = 20°$ (angle sum in a triangle $= 180°$)
   b. $x = 70°$ (theorem 4 and angle sum in a triangle $= 180°$); $y = z = 20°$ (angle sum in a triangle $= 180°$)
   c. $x = y = 75°$ (theorem 4 and angle sum in a triangle $= 180°$); $z = 75°$ (theorem 1)
8. B
9. C
10. C
11. D
12. B, D
13. a. Base angles of a right-angled isosceles triangle.
    b. $r + s = 90°$, $s = 45° \Rightarrow r = 45°$
    c. u is the third angle in $\triangle$ ABD, which is right-angled.
    d. m is the third angle in $\triangle$ OCD, which is right-angled.
    e. $\angle$AOC and $\angle$AFC stand on the same arc with $\angle$AOC at the centre and $\angle$AFC at the circumference.
14. OR = OP (radii of the circle)
    $\angle$OPR $= x$ (equal angles lie opposite equal sides)
    $\angle$SOP $= 2x$ (exterior angle equals the sum of the two interior opposite angles)
    OR = OQ (radii of the circle)
    $\angle$OQR $= y$ (equal angles lie opposite equal sides)
    $\angle$SOQ $= 2y$ (exterior angle equals the sum of the two interior opposite angles)
    Now $\angle$PRQ $= x + y$ and $\angle$POQ $= 2x + 2y = 2(x + y)$.
    Therefore $\angle$POQ $= 2 \times \angle$PRQ.
15, 16. Sample responses can be found in the worked solutions in the online resources.
17. a. $16°$ b. $20°$
18, 19. Sample responses can be found in the worked solutions in the online resources.

18.3 Intersecting chords, secants and tangents

1. a. $m = 3$ b. $m = 3$ c. $m = 6$
2. a. $n = 1$ b. $m = 7.6$
3. a. $n = 13$ b. $m = 4$
4. a. $x = 5$ b. $m = 7$ c. $x = 2.5$, $y = 3.1$
5. a. $x = 2.8$ b. $x = 3.3$
6. a. $x = 5.6$ b. $m = 90°$
7. B, C, D
8. ST $= 3$ cm
9–13. Sample responses can be found in the worked solutions in the online resources.
14. a. $x = 3\sqrt{2}$ b. $x = 6$ c. $x = 3$, $y = 12$
15. a. Line from centre perpendicular to the chord bisects the chord, giving M as the midpoint.
    b, c. Sample responses can be found in the worked solutions in the online resources.

18.4 Cyclic quadrilaterals

1. a. $x = 115°$, $y = 88°$ b. $m = 85°$
   c. $n = 25°$
2. a. $x = 130°$ b. $x = y = 90°$ c. $x = 45°$, $y = 95°$
3. a. $x = 85°$, $y = 80°$
   b. $x = 110°$, $y = 115°$
   c. $x = 85°$
4. a. $x = 150°$ b. $x = 90°$, $y = 120°$
   c. $m = 120°$, $n = 130°$
5. a. $a = 89°$, $b = 45°$
   b. $a = 120°$, $b = 91°$, $c = 89°$
6. a. $x = 102°$, $y = 113°$
   b. $x = 95°$, $y = 85°$, $z = 95°$
   c. $x = 126°$, $y = 54°$
   d. $x = 60°$, $y = 120°$

e. $x = 54°$, $y = 72°$

f. $x = 79°$, $y = 101°$, $z = 103°$

7. D

8. D

9. D

10. a. $2x$ **b.** $360° - 2x$

     c. $180° - x$ **d.** $180°$

11. A

12. A, B, C, D

13. Sample responses can be found in the worked solutions in the online resources.

14. $x = 93°$, $y = 87°$, $z = 93°$

15. $x = -2$ or $\dfrac{5}{2}$

16. $x = \dfrac{2}{3}$ or $\dfrac{1}{2}$

17. $w = 110°$, $x = 70°$, $y = 140°$, $z = 87.5°$

18. a. $a = 110°$, $b = 70°$ and $c = 110°$

     b. Sample responses can be found in the worked solutions in the online resources.

18.5 Tangents, secants and chords

1. a. $x = 70°$ **b.** $x = 47°$, $y = 59°$

2. a. $p = 6$ **b.** $q = 8$

3. $x = 42°$, $y = 132°$

4. B

5. D

6. MAC, NAC, FDA, FBA, EDG, EBG.

7. $x = 42°$, $y = 62°$

8. Sample responses can be found in the worked solutions in the online resources.

9. $60°$

10. $x = 180° - a - b$

11. $x = 80°$, $y = 20°$, $z = 80°$

12. Sample responses can be found in the worked solutions in the online resources.

13. $x = 85°$, $y = 20°$, $z = 85°$

14. D

15. $x = 50°$, $y = 95°$

16. A

17. C

18. $x = 33°$, $y = 55°$, $z = 22°$

19. $x = 25°$, $y = 65°$, $z = 40°$

20. $x = a$, $y = 90° - a$, $z = 90° - 2a$

21. Sample responses can be found in the worked solutions in the online resources.

22. a. $x = 5$ **b.** $k = 12$ **c.** $m = 6$, $n = 6$

23. a. $x = 7$ **b.** $b = 4$, $a = 2$ **c.** $w = 3$, $x = 5$

24. a. $a = 50°$, $b = 50°$ and $c = 80°$

     b. $a = 50°$, $b = 70°$ and $c = 70°$

Project

1. 11 100 km

2. 10 931 km

3.

Latitude	Distance between P_1 and P_2 (km)
0°	11 100
10°	10 931
20°	10 431
30°	9613
40°	8503
50°	7135
60°	5550
70°	3796
80°	1927
90°	0

4. The distance between P_1 and P_2 decreases from 11 100 km at the equator to 0 km at the pole. The change is not constant. The distance between the points decreases more rapidly on moving towards the pole.

5. Latitude $60°$

6.

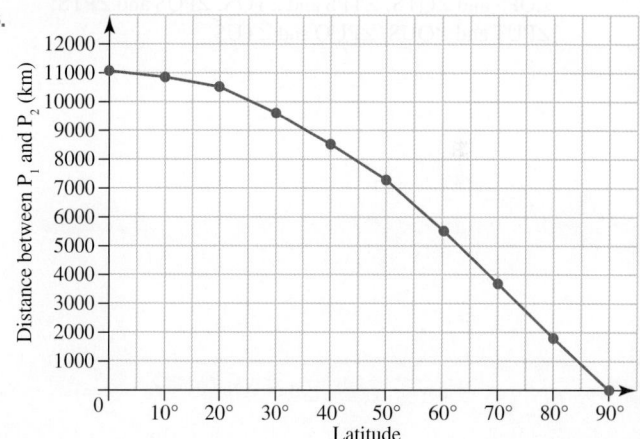

7. Latitude $26°$

8. Sample responses can be found in the worked solutions in the online resources. Students need to investigate the rate that the distance between them changes from the equator to the pole and also comparison with the earlier values.

18.6 Review questions

1. a. $x = 50°$ **b.** $x = 48°$, $y = 25°$

     c. $x = y = 28°$, $z = 56°$ **d.** $x = 90°$

     e. $y = 90°$ **f.** $y = 140°$

2. a. $x = 55°$ **b.** $x = 125°$

     c. $x = 70°$ **d.** $x = 100°$

     e. $m = 40°$ **f.** $x = 90°$, $y = 60°$, $z = 40°$

3. a. $x = 90°$ **b.** $x = 20°$

     c. $x = 55°$ **d.** $x = 125°$

4. a. $m = 3$ **b.** $m = 12$

     c. $m = 9$ **d.** $m = 11.7$

5. A, B, D

6. A, B, C

7. a. $x = 95°$, $y = 80°$ b. $x = 99°$
   c. $x = 78°$, $y = 92°$ d. $x = 97°$, $y = 92°$

8. D

9. a. $x = 42°$ b. $y = 62°$ c. $p = 65°$

10. $CE \times ED = AE \times EB$

AE = CE (given)
∴ ED = EB

11. $\angle AYC = \angle AXC$

$\angle BXD = \angle BYD$

But $\angle AXC = \angle BXD$

⋈

$\Rightarrow \angle AYC = \angle BYD$

12. $\angle PQT$ and $\angle PST$, $\angle PTS$ and $\angle RQS$, $\angle TPQ$ and $\angle QSR$,
    $\angle QPS$ and $\angle QTS$, $\angle TPS$ and $\angle TQS$, $\angle PQS$ and $\angle PTS$,
    $\angle PUT$ and $\angle QUS$, $\angle PUQ$ and $\angle TUS$

19 Introduction to networks (Path)

LESSON SEQUENCE

LESSON
19.1 Overview

Why learn this?

Networks are used to show how things are connected. The study of networks and decision mathematics is a branch of graph theory.

The mathematician Leonhard Euler (1707–83) is usually credited with being the founder of graph theory. He famously used it to solve a problem known as the 'Seven Bridges of Königsberg'.

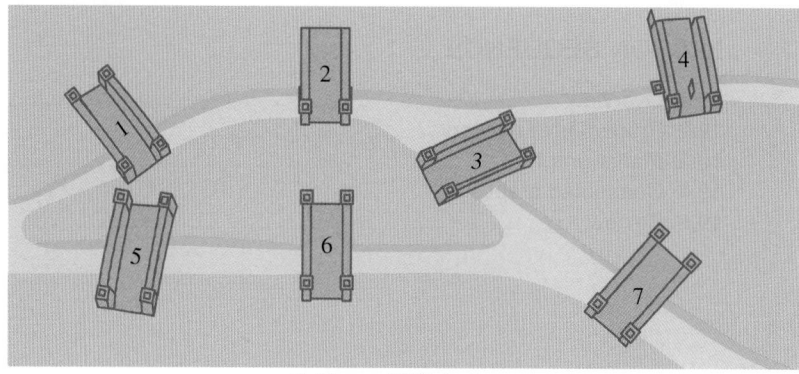

Bridges of Königsberg

For a long time it had been pondered whether it was possible to travel around the European city of Königsberg (now called Kaliningrad) in such a way that the seven bridges would only have to be crossed once each.

In the branch of mathematics known as graph theory, diagrams involving points and lines are used as a planning and analysis tool for systems and connections.

Networks have been used to deliver mail, land people on the moon, organise train timetables and improve the flow of traffic. The social connections between a group of friends, the network of airports in Australian or the kinship systems of First Nations Australians can be represented using network diagrams. Graph theory and networks have also been applied to a wide range of disciplines from social networks where they are used to examine the structure of relationships and social identities, to biological networks, which analyse molecular networks.

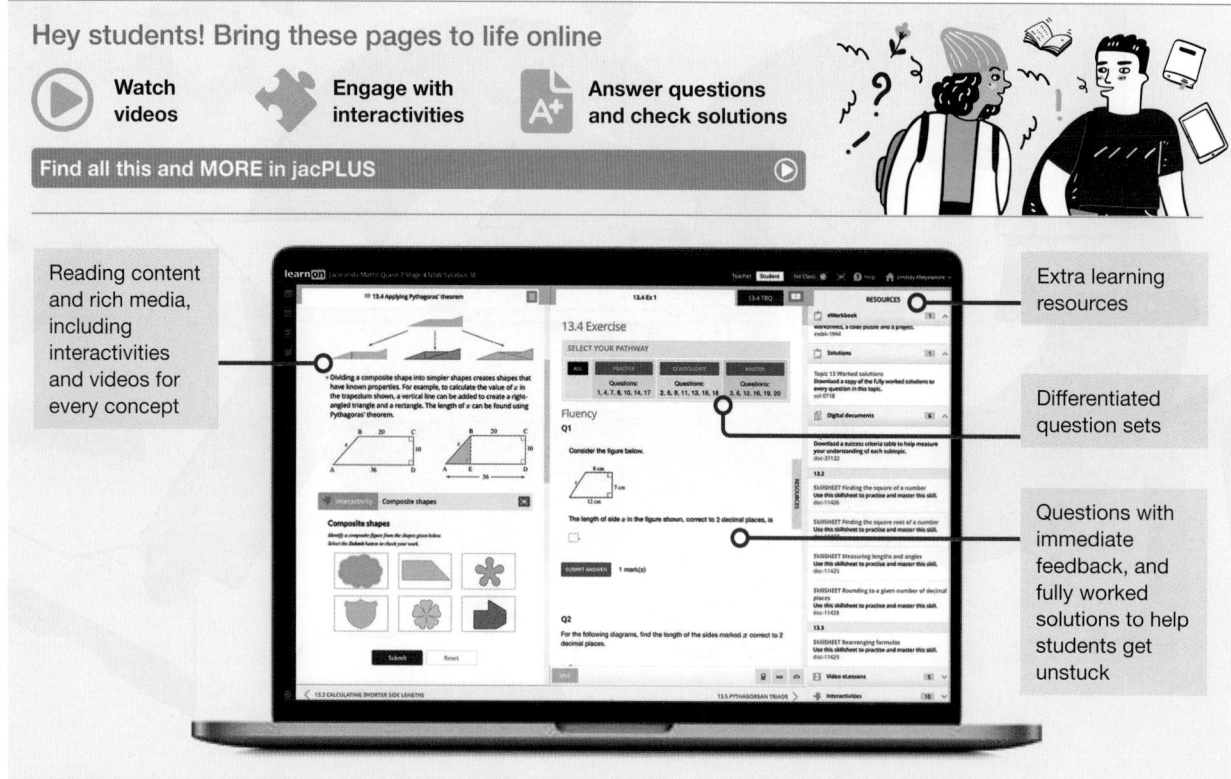

1. Consider the following diagram.

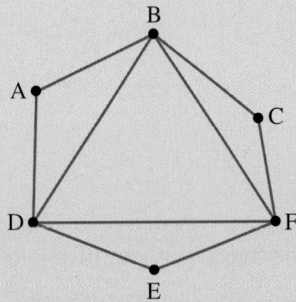

State how many lines join directly to point B.

2. Consider the following diagram.

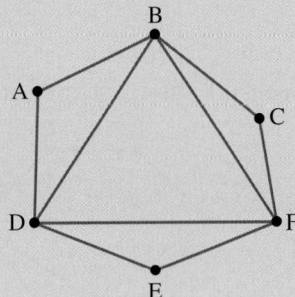

State the number of vertices.

3. Consider the following diagram.

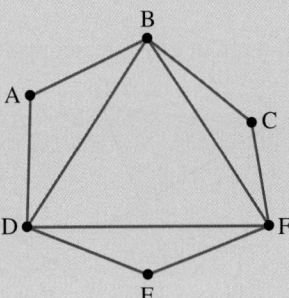

State the number of edges.

4. Consider the following diagram.

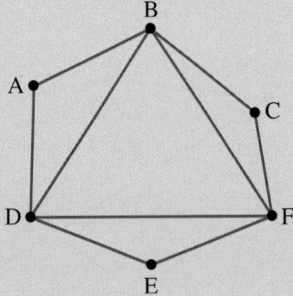

Imagine that the diagram represents walking paths in a park. Each letter represents a viewing platform. Beginning at A, trace a path that you can walk along so that you visit each viewing platform only once.

5. Consider the following diagram.
   Starting from A, is it possible to walk through the park so that you travel on each path exactly once? If so, list a path you could follow.

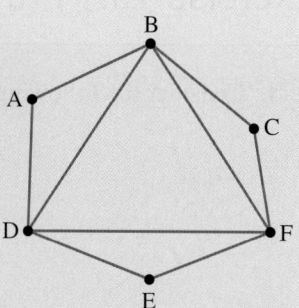

6. Consider the following diagram.
   Explain whether the diagram can be traced without lifting the pen off the paper.

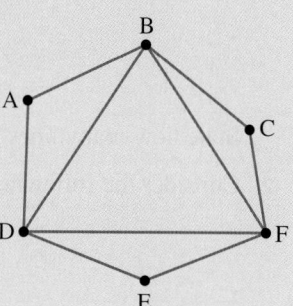

7. Determine an Euler circuit for the network shown, starting at A.

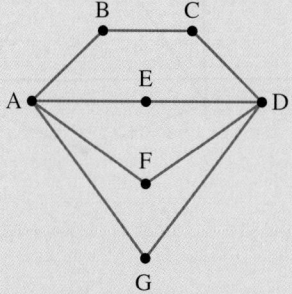

8. Redraw the following planar graph without any intersecting edges

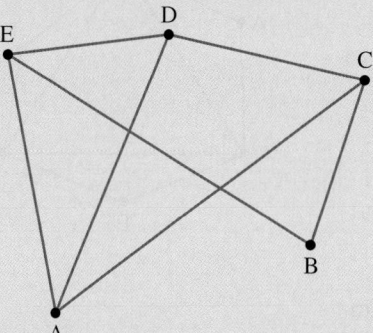

9. Consider the following diagram.

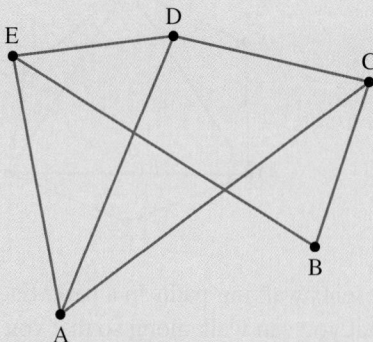

Determine the number of faces of the planar graph.

10. Consider the following diagram.
State which vertex in the network shown has the highest degree.

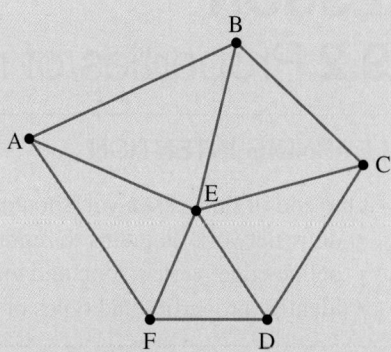

11. In a planar graph, the number of vertices = 7 and the number of edges = 13. Therefore, calculate the number of faces.

12. Consider the following diagram.

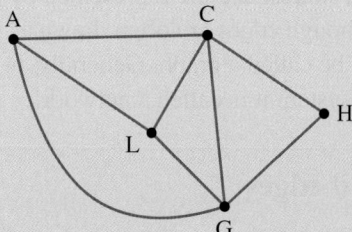

Determine the degree of each vertex.

13. Consider the following diagram.

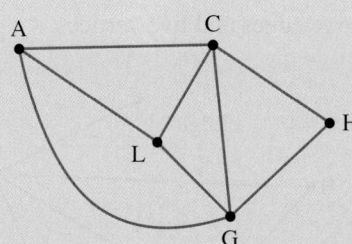

Identify a circuit starting at vertex A.

14. Determine if the following graph is planar or not planar.

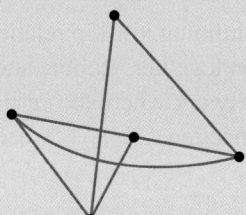

15. Using the network shown, determine what two edges should be added to the network so that it has an Euler circuit.

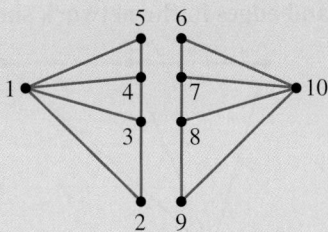

LESSON
19.2 Properties of networks

LEARNING INTENTION

At the end of this lesson you should be able to:
- draw network diagrams to represent connections
- define edge, vertex, loop and the degree of a vertex
- identify properties and types of graphs, including connected and isomorphic graphs.

▶ 19.2.1 Networks

eles-6275

- A network is a collection of objects interconnected by lines that can represent systems in the real world.
- The objects are called **vertices** or **nodes**, and are represented by points. The connections are called **edges** and are represented by lines. Although edges are often drawn as straight lines, this is not necessary.
- A diagram of a network can also be called a graph. Generally, in a mathematical setting it is called a graph but when related to a real-world system it is called a network.

> **Vertices and edges**
>
> **In a network, the points are called vertices (or nodes) and the lines ae called edges, with each edge joining a pair of vertices.**

- The network diagram shown has five edges and five vertices.

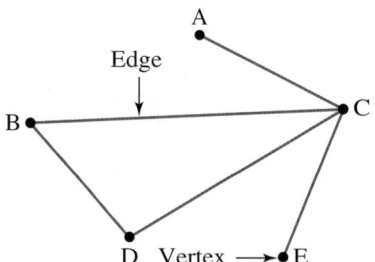

- Consider a simple network diagram outlining kinship relationships; Savannah and Binda have two children together; Robbie and Isabella. This family can be represented as a network where the vertices represent people, while the edges indicate a family connection.

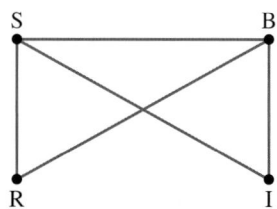

WORKED EXAMPLE 1 Identifying the number of vertices and edges

Determine the number of vertices and edges in the network shown.

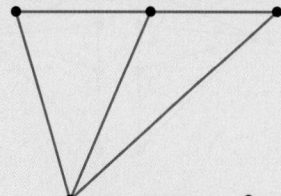

THINK	WRITE/DRAW
1. Vertices are objects that are represented by points in the network. Count the vertices (black dots).	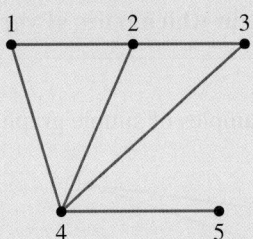
2. State the answer.	There are 5 vertices.
3. Edges are the lines joining the vertices (points). Count the edges (blue lines) and write the answer.	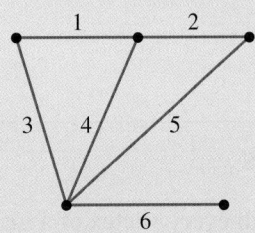
4. State the answer.	There are 6 edges.

WORKED EXAMPLE 2 Using network diagrams to represent connections

Draw the network that represents the family tree showing Aaron his two parents Brendan and Cara and his grandparents, David and Evelyn (paternal) and Feza and Gamila (maternal).

THINK	WRITE/DRAW
1. List the objects (people) in the network.	A for Aaron, B for Brendan, C for Cara, D for David, E for Evelyn, F for Feza, G for Gamila.
2. Draw them in rows to indicate the generations.	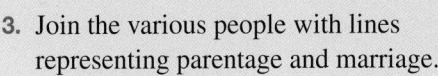
3. Join the various people with lines representing parentage and marriage. D is married to E. F is married to G. D and E are the parents of B. F and G are the parents of C. B is married to C. B and C are the parents of A.	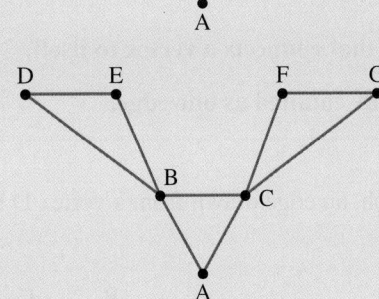

▶ ## 19.2.2 Types and properties of graphs

eles-6276

- A graph is a series of points and lines that can be used to represent the connections that exist in various settings.
- The term graph is generally used when referring to the diagrams but the term network is used to relate graphs to real-world systems. However, the terms graph and network are interchangeable.

Simple graphs

A simple graph is one in which pairs of vertices are connected by at most one edge.

- The graphs below are examples of simple graphs.

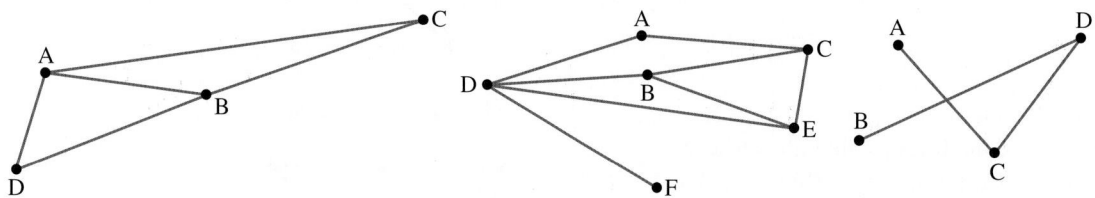

Connected graphs

If it is possible to reach every vertex of a graph by moving along the edges, it is called a connected graph; otherwise, it is a disconnected graph.

- The graph below left is an example of a connected graph, whereas the graph below right is not connected (it is disconnected).

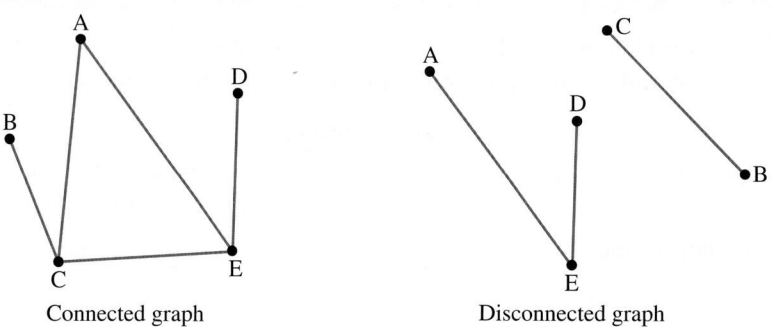

Connected graph   Disconnected graph

A loop

A loop is an edge that connects a vertex to itself.

Note: **A loop is only counted as one edge.**

- In the following graph, an edge drawn from a vertex D back to itself is known a loop.

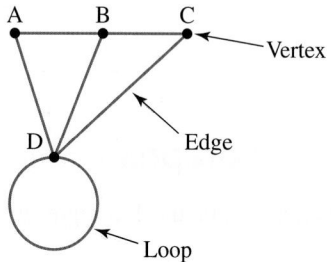

The diagram represents a system of paths and gates in a large park. Draw a graph to represent the possible ways of travelling to each gate in the park.

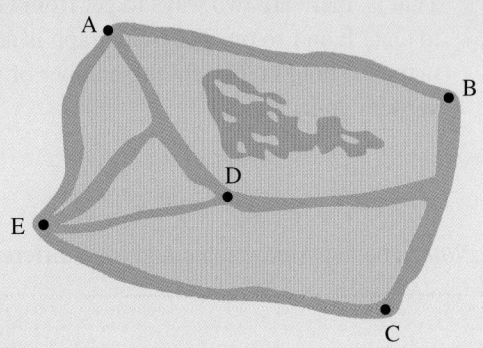

THINK

1. Identify, draw and label all possible vertices.

2. Draw edges to represent all the direct connections between the identified vertices.

3. Identify all the other unique ways of connecting vertices.

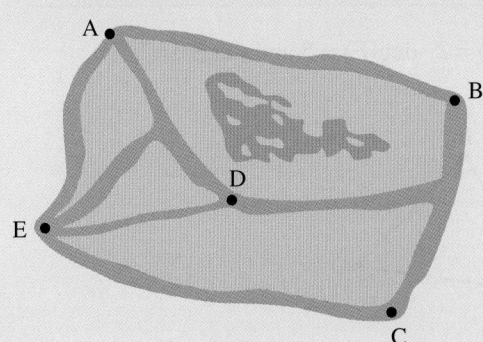

WRITE/DRAW

Represent each gate as a vertex.

A•

•B

D
•

E•

•C

Direct pathways exist for
A–B, A–D, A–E, B–C, C–E and D–E.

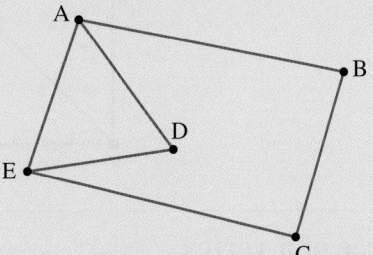

Other unique pathways exist for
A–E, D–E, B–D and C–D.

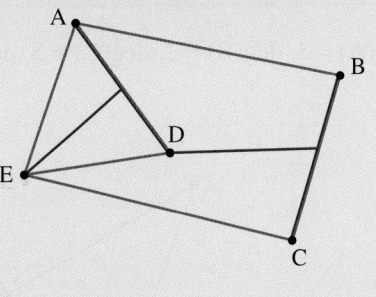

4. Draw the final graph.

That is, there are two ways to get from Gate A to Gate E and so on. The final graph is not a map but should represent the number of ways vertices connect.

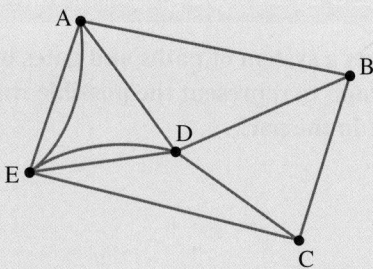

Note: The network graph could look different, but the connections should all be represented.

▶ 19.2.3 The degree of a vertex

eles-6277

- The total number of edges that are directly connected to a particular vertex is known as the **degree** of the vertex.
- The notation deg(V) is used, where V represents the vertex.
- If there is a loop at a vertex it counts twice to the degree of that vertex. Remember, a loop counts as one edge but counts twice to the degree of the vertex.

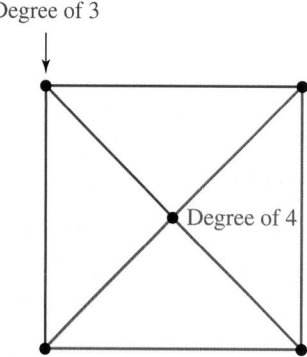

The degree of a vertex

The degree of a vertex in a graph is the number of edges directly connected to that vertex, *except* that a loop at a vertex contributes twice to the degree of that vertex.

- In the graph, $\deg(A) = 2$, $\deg(B) = 2$, $\deg(C) = 5$, $\deg(D) = 2$, $\deg(E) = 5$ and $\deg(F) = 2$.

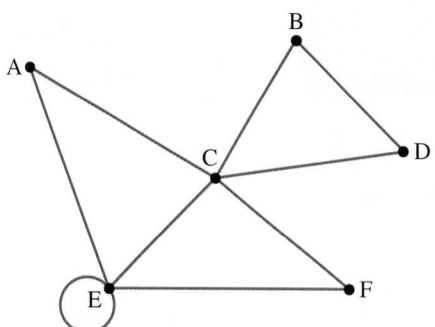

State the degree of each vertex in the network shown

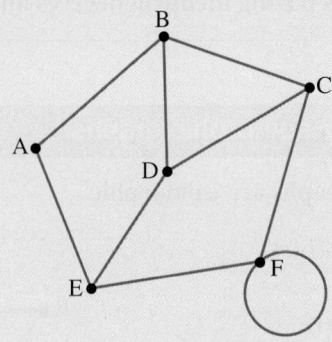

THINK	WRITE
1. Starting with vertex A, count its number of edges. The number of edges is equal to the degree of that vertex.	$\deg(A) = 2$
2. Repeat step 1 for all other vertices, being careful to include the loop at F as contributing twice to the degree.	$\deg(B) = 3$ $\deg(C) = 3$ $\deg(D) = 3$ $\deg(E) = 3$ $\deg(F) = 4$

19.2.4 Isomorphic graphs

eles-6278

- Consider the following graphs.

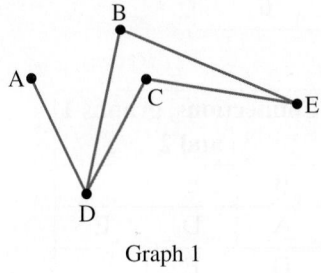

Graph 1

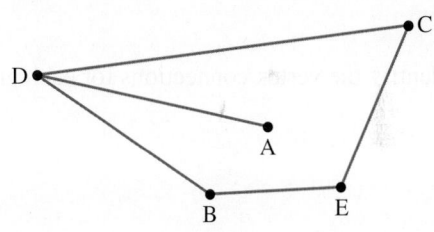

Graph 2

- For the two graphs, the connections for each vertex can be summarised as shown in the table.

Graph 1

Vertex	Connections		
A	D		
B	D	E	
C	D	E	
D	A	B	C
E	B	C	

Graph 2

Vertex	Connections		
A	D		
B	D	E	
C	D	E	
D	A	B	C
E	B	C	

- Although the graphs don't look exactly the same, they could represent exactly the same information. Both graphs have the same amount of edges as they do vertices, such graphs are known as isomorphic graphs

Isomorphic graphs

Isomorphic graphs have the same number of vertices and edges, with corresponding vertices having identical degrees and connections.

WORKED EXAMPLE 5 Identifying isomorphic graphs

Confirm whether the following two graphs are isomorphic.

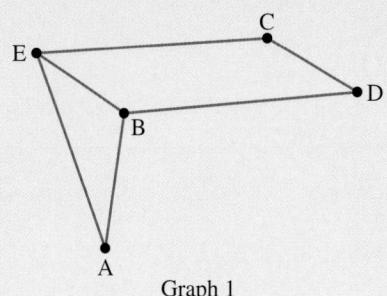

Graph 1

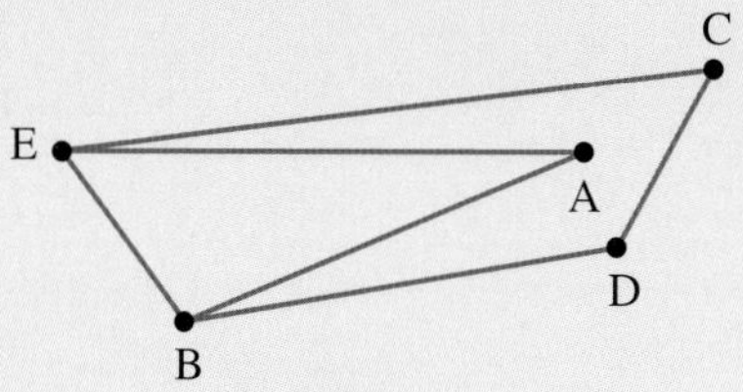

Graph 2

THINK

1. Identify the degree of the vertices for each graph.

2. Identify the number of edges for each graph.

3. Identify the vertex connections for each graph.

4. Comment on the two graphs.

WRITE

Graph	A	B	C	D	E
Graph 1	2	3	2	2	3
Graph 2	2	3	2	2	3

Graph	Edges
Graph 1	6
Graph 2	6

Vertex	Connections, graphs 1 and 2		
A	B	E	
B	A	D	E
C	D	E	
D	B	C	
E	A	B	C

The two graphs are isomorphic as they have the same number of vertices and edges, with corresponding vertices having identical degrees and connections.

COMMUNICATING — COLLABORATIVE TASK: Networks in real life contexts.

Networks can be used to represent real life situations.
1. In pairs, create or research a network such as a social network, a supply chain network or a communication infrastructure network.
2. Determine whether the network is a connected graph or not.
3. Determine the number of:
   a. degrees of each vertex
   b. vertices
   c. edges.
4. Does the network contain any loops?

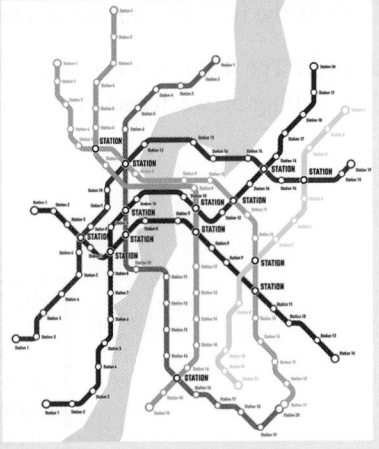

Exercise 19.2 Properties of networks

learnon

19.2 Quick quiz on	19.2 Exercise

Individual pathways

■ PRACTISE	■ CONSOLIDATE	■ MASTER
1, 3, 7, 9, 11, 14, 17	2, 5, 10, 12, 15, 18	4, 6, 8, 13, 16, 19

Fluency

1. **WE1** Determine the number of vertices and edges in the following networks.

   a. 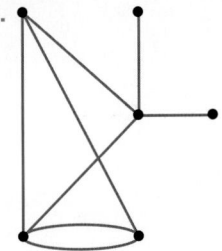   b.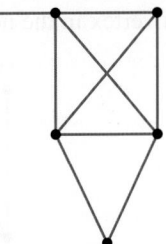

2. Determine the number of vertices and edges in the following networks.

   a. 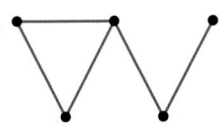   b. 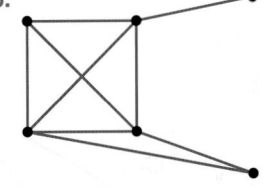   c. 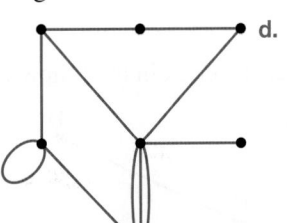   d.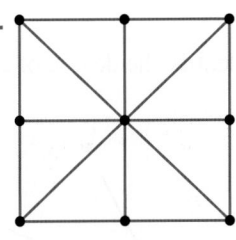

3. **WE2** Draw the network that represents the following family tree.
   Henry and Ida marry and have one child, Jane.
   Jane marries Kenneth and they have one child, Louise.
   Louise marries Mark and they have two children, Neil and Otis.

4. Four towns, Joplin, Amarillo, Flagstaff and Bairstow, are connected to each other as follows: Joplin to Amarillo; Joplin to Flagstaff; Amarillo to Bairstow; Amarillo to Flagstaff; Flagstaff to Bairstow. Draw the network represented by these connections.

5. **WE3** The diagram shows the plan of a floor of a house. Draw a graph to represent the possible ways of travelling between each room on the floor.

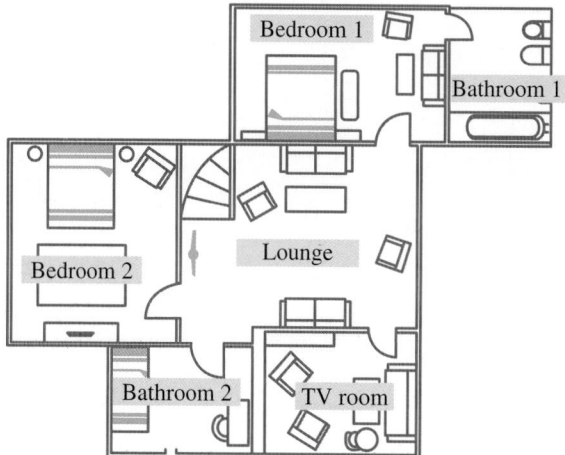

6. Draw a graph to represent the following tourist map.

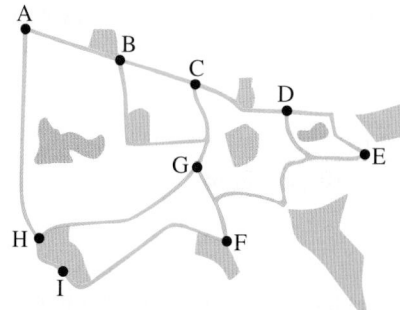

7. **WE4** State the degree of each vertex in the network shown.

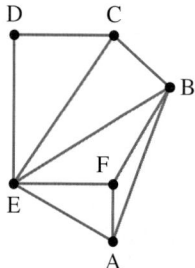

8. Identify the degree of each vertex in the following graphs.

a.

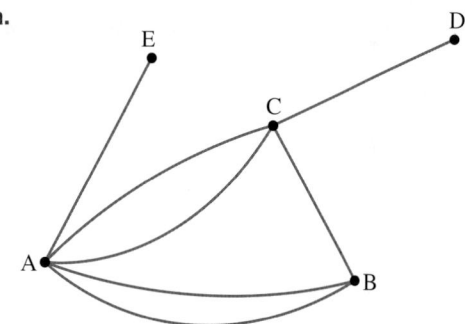

b.

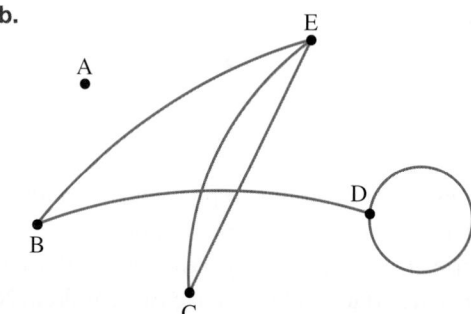

c.

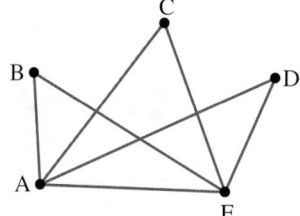

d.

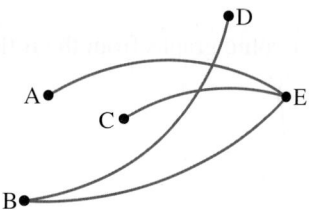

9. **WE5** Confirm whether the following pairs of graphs are isomorphic.

a.

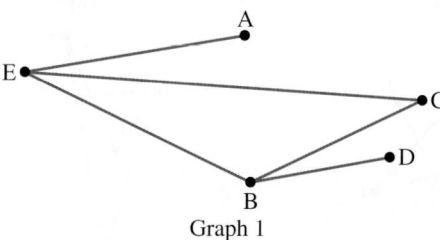

Graph 1   Graph 2

b.

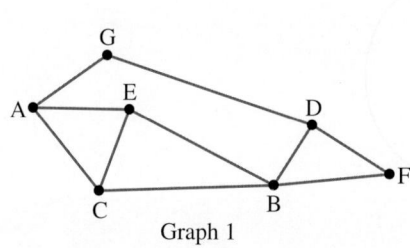

Graph 1

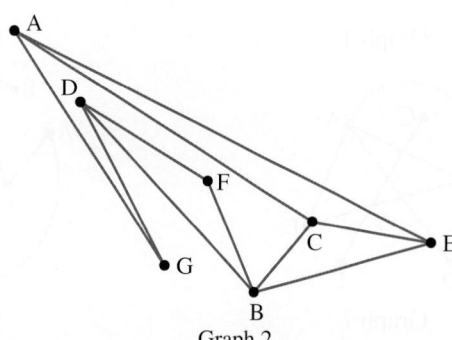

Graph 2

10. Explain why the following pairs of graphs are not isomorphic.

a.

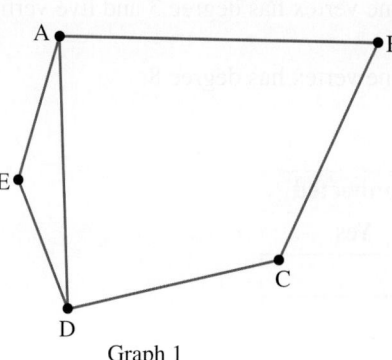

Graph 1

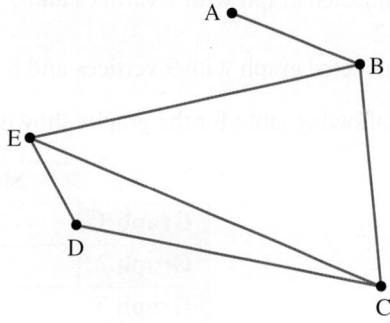

Graph 2

b.

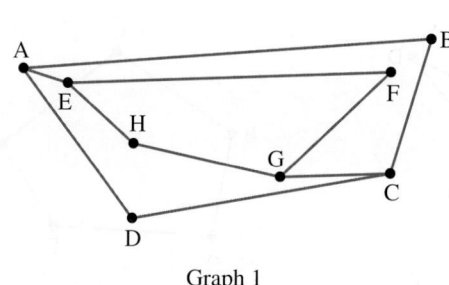

Graph 1

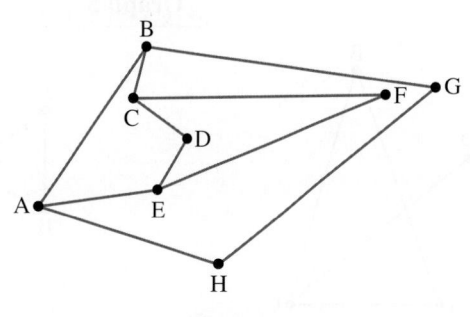

Graph 2

Understanding

11. Identify pairs of isomorphic graphs from the following.

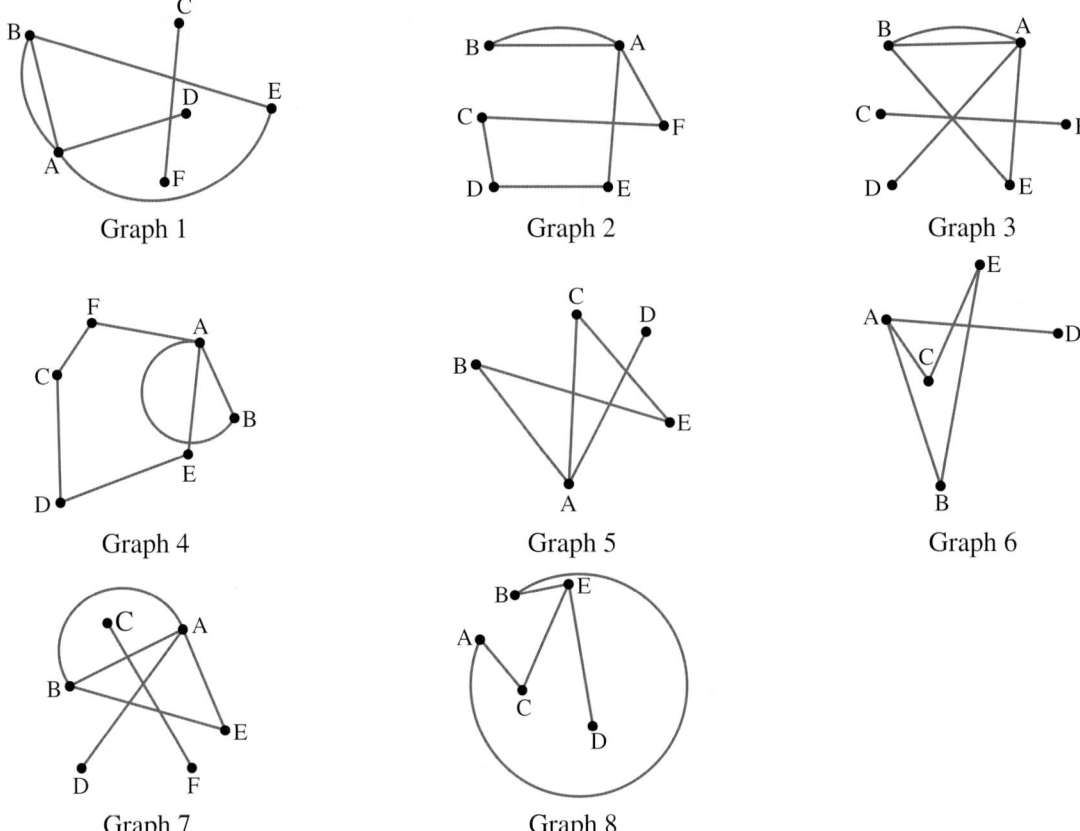

Graph 1   Graph 2   Graph 3

Graph 4   Graph 5   Graph 6

Graph 7   Graph 8

12. Draw a graph of:

   a. a simple, connected graph with 6 vertices and 7 edges
   b. a simple, connected graph with 7 vertices and 7 edges, where one vertex has degree 3 and five vertices have degree 2
   c. a simple, connected graph with 9 vertices and 8 edges, where one vertex has degree 8.

13. Complete the following table for the graphs shown.

	Simple	Connected
Graph 1	Yes	Yes
Graph 2		
Graph 3		
Graph 4		
Graph 5		

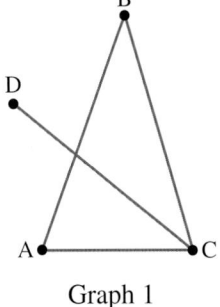

Graph 1

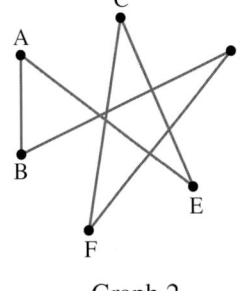

Graph 2

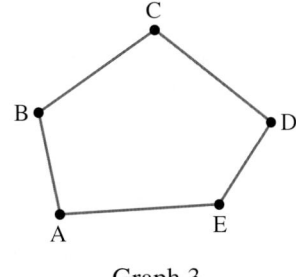

Graph 3

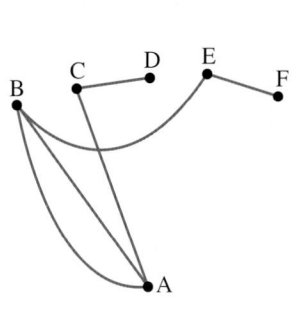

Graph 4   Graph 5

Communicating, reasoning and problem solving

14. Four streets are connected to each other as follows.
    Draw a network represented by these connections.
    Princess Street — Bird Avenue
    Princess Street — Chatlie Street
    Charlie Street — Dundas Street
    Princess Street — Dundas Street

15. A round robin tournament occurs when each team plays all other teams once only.
    At the beginning of the school year, five schools play a round robin competition in table tennis.

    a. Draw a graph to represent the games played.
    b. State what the total number of edges in the graph indicates.

16. Consider a network of 4 vertices, where each vertex is connected to each of the other 3 vertices with a single edge (no loops, isolated vertices or parallel edges).

    a. List the vertices and edges.
    b. Construct a diagram of the network.
    c. List the degree of each vertex.

17. The diagram shows a map of some of the main suburbs of Beijing.

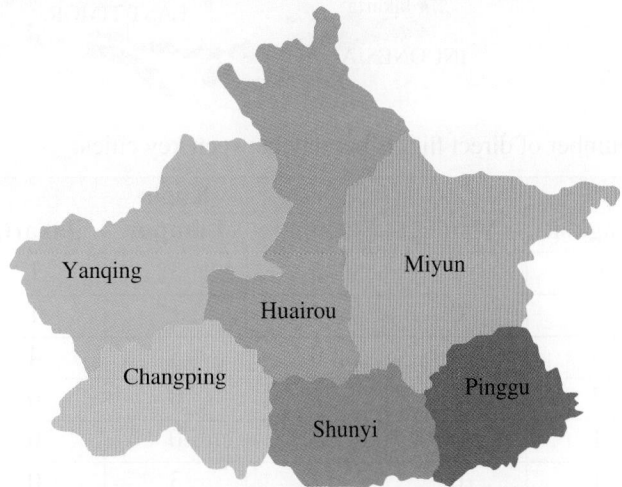

    a. Draw a graph to represent the shared boundaries between the suburbs.
    b. State which suburb has the highest degree.
    c. State the type of graph.

18. By indicating the passages with edges and the intersections and passage endings with vertices, draw a graph to represent the maze shown.

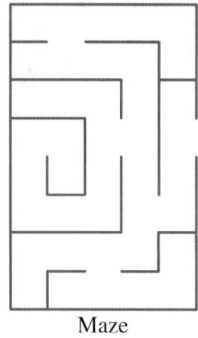

Maze

19. Jetways Airlines operates flights in South-East Asia.

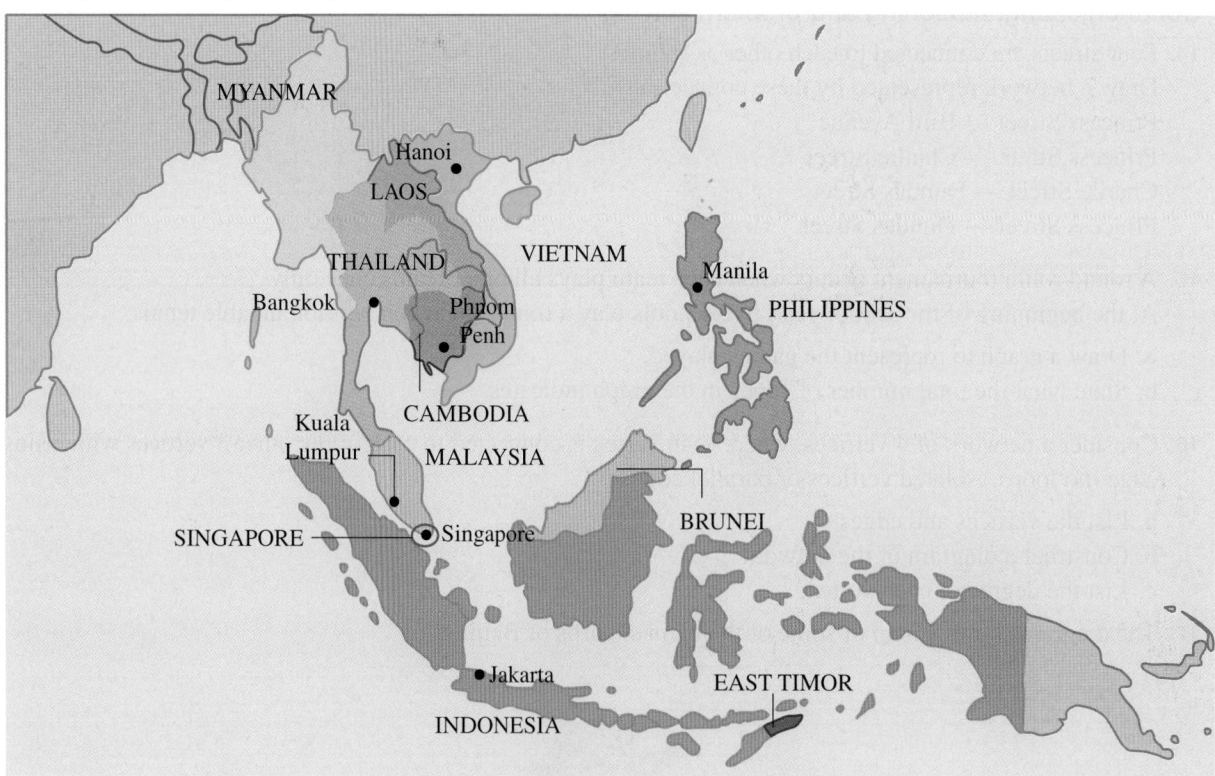

The table indicates the number of direct flights per day between key cities.

From: To:	Bangkok	Manila	Singapore	Kuala Lumpur	Jakarta	Hanoi	Phnom Penh
Bangkok	0	2	5	3	1	1	1
Manila	2	0	4	1	1	0	0
Singapore	5	4	0	3	4	2	3
Kuala Lumpur	3	1	3	0	0	3	3
Jakarta	1	1	4	0	0	0	0
Hanoi	1	0	2	3	0	0	0
Phnom Penh	1	0	3	3	0	0	0

a. Draw a graph to represent the number of direct flights.
b. State whether this graph would be considered directed or undirected. Explain why.
c. State the number of ways you can travel from Hanoi to Bangkok.

LESSON
19.3 Planar graphs

LEARNING INTENTION

At the end of this lesson you should be able to:
- identify and draw planar graphs
- apply Euler's formula for planar graphs.

▶ 19.3.1 Planar and non-planar graphs

eles-6279

- A **planar** graph is a graph that can be drawn in the plane so no edges cross each other.
- In some networks it is possible to redraw graphs so that they have no intersecting edges.
  For example, in the graph below it is possible to redraw one of the intersecting edges so that it still represents the same information.

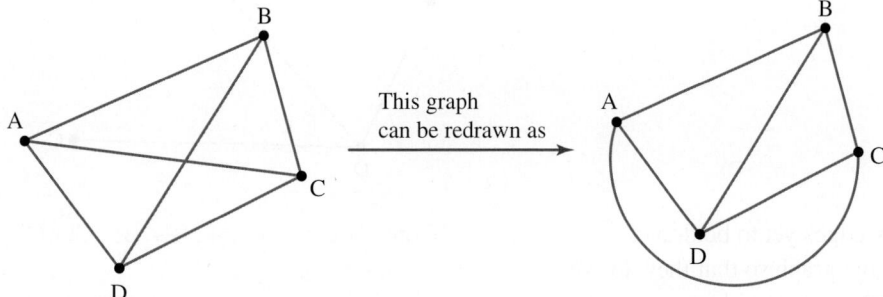

- **Non-planar** graphs are graphs that can never be drawn in the plane without some edges crossing.
  For example, the graph shown cannot be redrawn with any edges crossing.

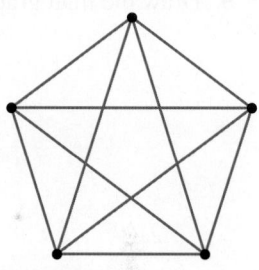

WORKED EXAMPLE 6 Redrawing a graph to make it planar

Redraw the following graph so that it has no intersecting edges.

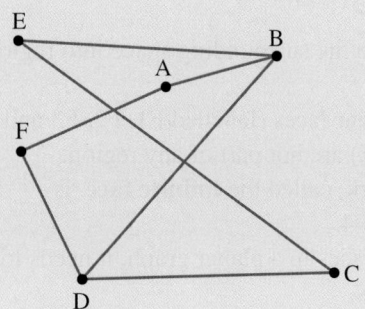

THINK	WRITE/DRAW
1. List all connections in the original graph.	Connections: AB; AF; BD; BE; CD; CE; DF
2. Draw all vertices and any section(s) of the graph that have no intersecting edges.	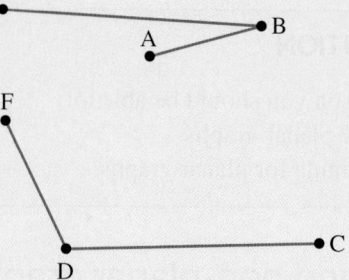
3. Draw any further edges that don't create intersections. Start with edges that have the fewest intersections in the original drawing.	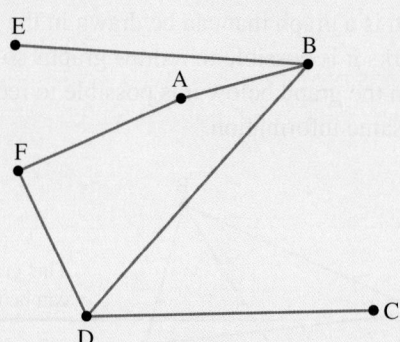
4. Identify any edges yet to be drawn and redraw the graph so that they do not intersect with the other edges.	Connections: ~~AB~~; ~~AF~~; ~~BD~~; BE; ~~CD~~; CE; ~~DF~~
5. Draw the final graph.	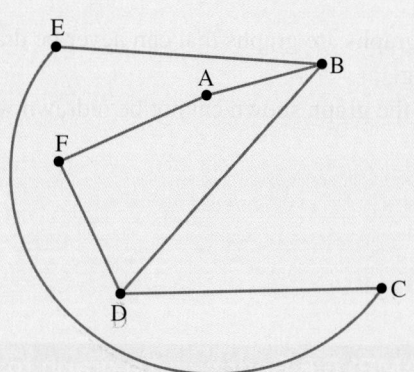

▶ 19.3.2 Faces in a planar graph

eles-6280

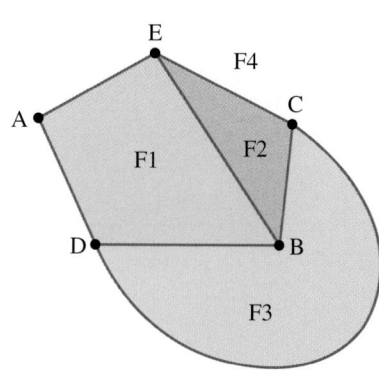

- For a planar graph, the edges divide the surrounding space into regions, also known as **faces**.
- For the network shown, there are four faces (labelled F1, F2, F3 and F4). The vertices (A, B, C, D and E) are not part of any region.
- The space outside the entire network, called the **infinite face**, is counted as a face; in this case it is F4.
- In order to identify the number of faces in a planar graph, it needs to be re-drawn with no intersecting edges.

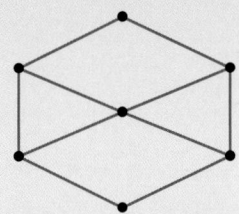

WORKED EXAMPLE 7 Determining the number of faces in a planar graph

Identify the number of vertices, edges and faces in the planar graph shown.

THINK

1. Label the vertices and count them.

WRITE/DRAW

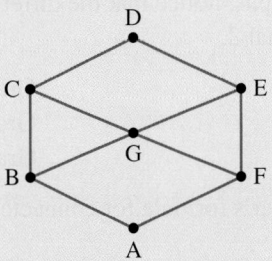

Number of vertices = 7
Number of edges = 10

2. Count the edges. The easiest way to do this is to cross off every edge that has been counted with a small mark. (This will guarantee that no edge is missed and no edge is counted twice.)

3. Colour in each face with a different colour. Count the faces. Do not forget the face outside the network.

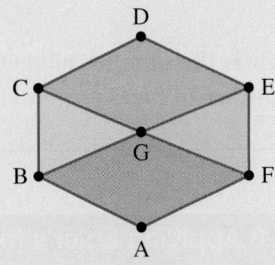

Number of faces = 5

4. Write the answer.

The network has 7 vertices, 10 edges and 5 faces.

19.3.3 Euler's formula

eles-6281

- The famous mathematician Leonhard Euler (pronounced 'oil-er') discovered the relationship between the number of faces, edges and vertices for all connected planar graphs.
- Consider the following group of connected planar graphs.

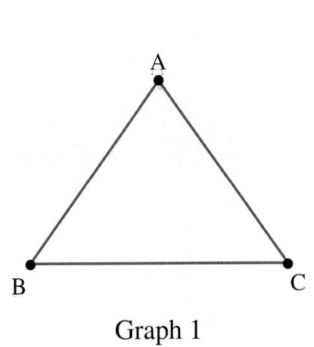

Graph 1

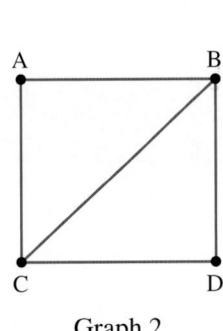

Graph 2

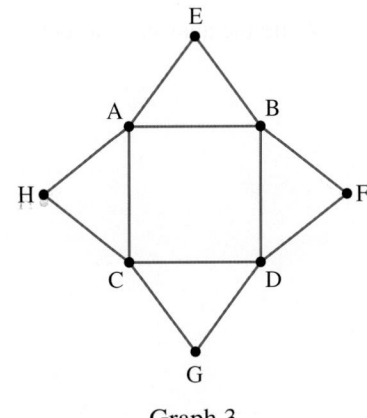

Graph 3

- The number of vertices, edges and faces for each graph is summarised in the following table.

	Vertices	Edges	Faces
Graph 1	3	3	2
Graph 2	4	5	3
Graph 3	8	12	6

Note: Do not forget to include the face outside the graph.

- For each of these graphs, notice that the difference between the vertices and edges added to the number of faces will always equal 2.

$$\text{Graph 1: } 3 - 3 + 2 = 2$$
$$\text{Graph 2: } 4 - 5 + 3 = 2$$
$$\text{Graph 3: } 8 - 12 + 6 = 2$$

This is known as Euler's formula for connected planar graphs.

Euler's formula

For any connected planar graph:

$$v - e + f = 2$$

where v is the number of vertices, e is the number of edges and f is the number of faces (regions).

WORKED EXAMPLE 8 Applying Euler's formula

Determine the number of faces in a connected planar graph of 7 vertices and 10 edges.

THINK	WRITE
1. Substitute the given values into Euler's formula.	$v - e + f = 2$ $7 - 10 + f = 2$
2. Solve the equation for the unknown value.	$7 - 10 + f = 2$ $f = 2 - 7 + 10$ $f = 5$
3. Write the answer as a sentence.	There will be 5 faces in a connected planar graph with 7 vertices and 10 edges.

19.3 Quick quiz on	19.3 Exercise

Individual pathways

■ PRACTISE	■ CONSOLIDATE	■ MASTER
1, 4, 8, 12, 13	2, 5, 6, 9, 14, 16	3, 7, 10, 11, 15, 17

Fluency

1. **WE6** Redraw the following graph so that it has no intersecting edges.

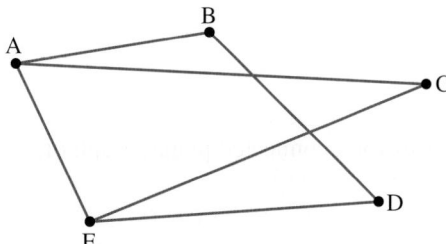

2. Convert the following graph to a planar graph.

3. Redraw the following network diagram so that it is a planar graph.

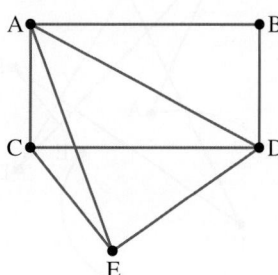

4. **WE7** Identify the number of vertices, edges and faces in the network shown.

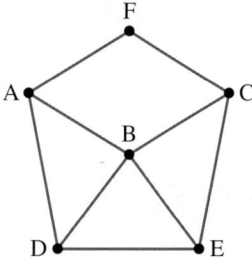

5. Identify the number of vertices, edges and faces in the networks below.

a.
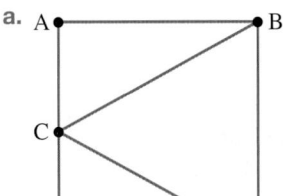

b.
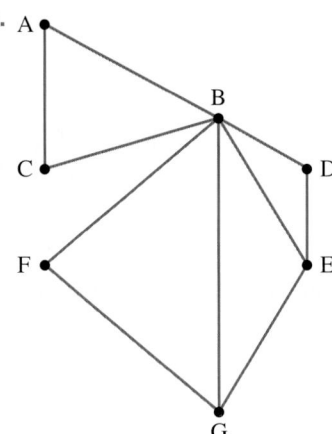

6. **WE8** Determine the number of faces for a connected planar graph of:
   a. 8 vertices and 10 edges
   b. 11 vertices and 14 edges.

7. a. For a connected planar graph of 5 vertices and 3 faces, state the number of edges.
   b. For a connected planar graph of 8 edges and 5 faces, state the number of vertices.

Understanding

8. Redraw the following graph to show that it is planar.

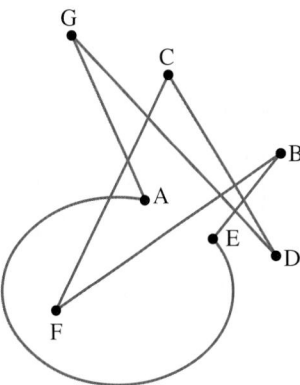

9. Identify which of the following graphs are *not* planar.

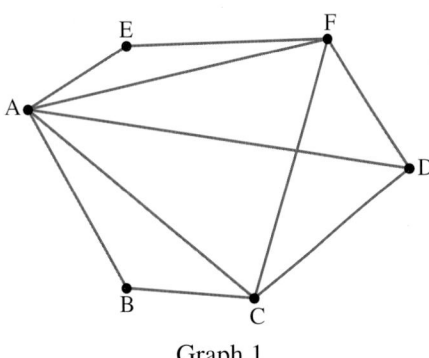

Graph 1

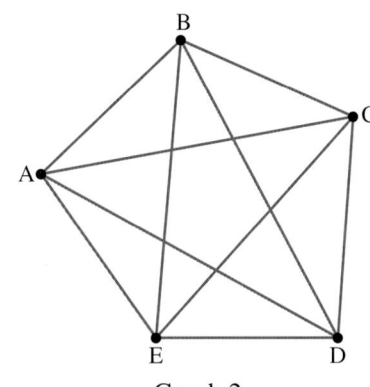

Graph 2

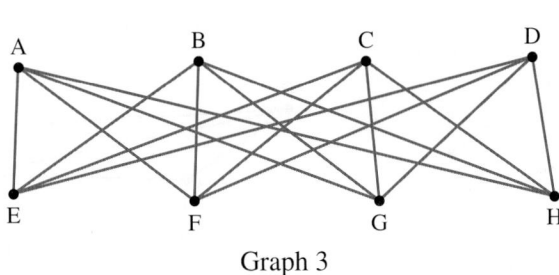

Graph 3

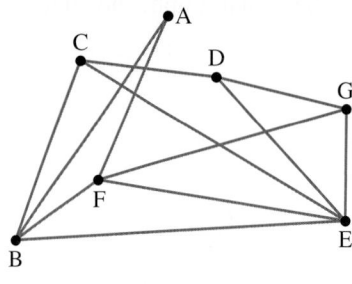

Graph 4

10. Consider the following network.

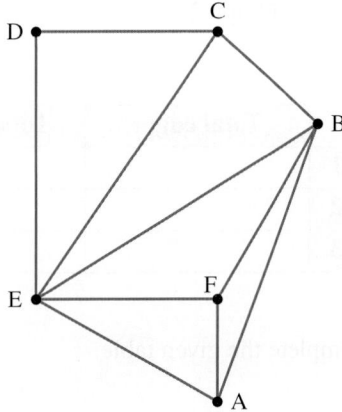

a. Identify the number of vertices, edges and faces in the shown network.

b. Confirm Euler's formula for the shown network.

11. For each of the following planar graphs, identify the number of faces.

a.

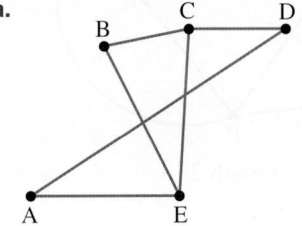

b.

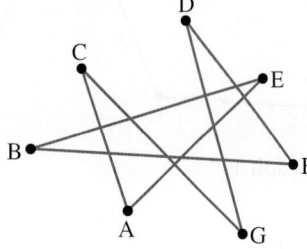

Communicating, reasoning and problem solving

12. Construct a connected planar graph with:

    a. 6 vertices and 5 faces

    b. 11 edges and 9 faces.

13. Represent the following 3-dimensional shape as a planar graph.

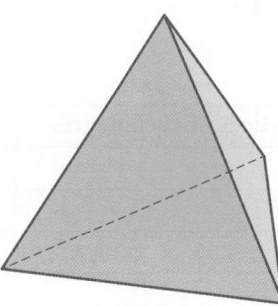

Tetrahedron

14. Use the planar graphs shown to complete the given table.

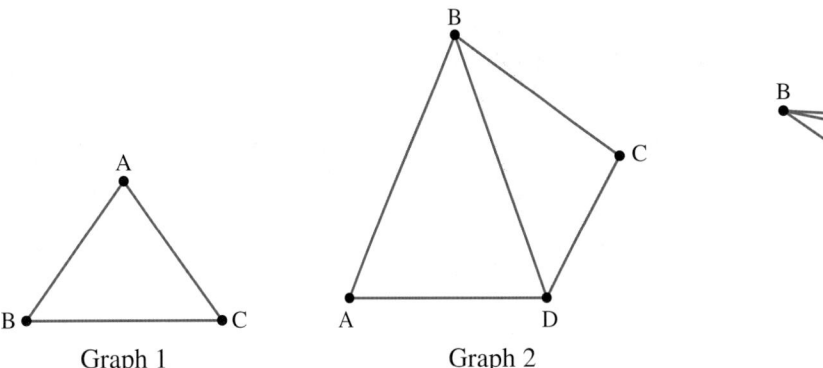

Graph 1   Graph 2   Graph 3

	Total edges	Total degrees
Graph 1		
Graph 2		
Graph 3		

15. a. Use the planar graphs shown to complete the given table.

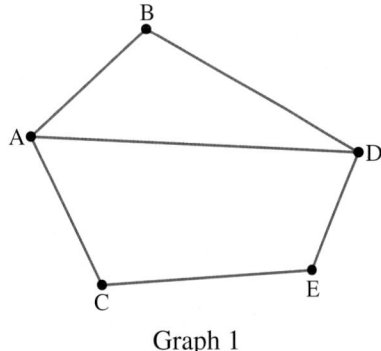

Graph 1

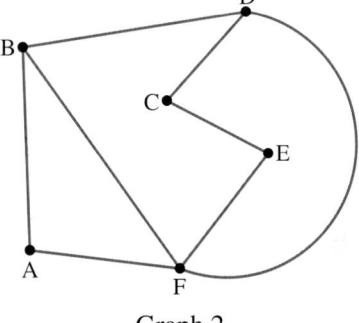

Graph 2

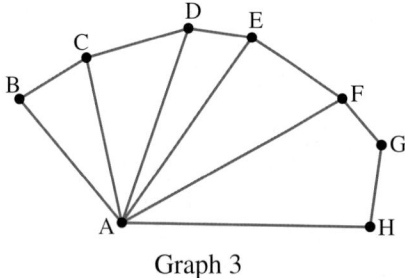

Graph 3

	Total vertices of even degree	Total vertices of odd degree
Graph 1		
Graph 2		
Graph 3		

b. Comment on any pattern evident from the table.

16. Represent the following 3-dimensional shape as a planar graph.

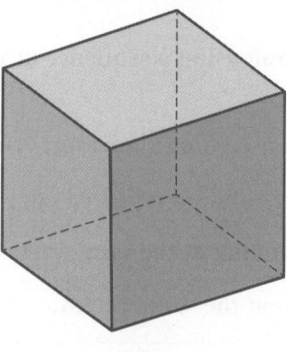

Cube

17. The table displays the most common methods of communication for a group of people.

	WhatsApp	Snapchat
Adam	Ethan	Ethan
Michelle		Sophie, Ethan
Liam	Michelle	Ethan, Sophie

a. Display the information for the entire table in a graph.
b. State who would be the best person to introduce Adam.
c. Display the Snapchat information in a separate graph.

LESSON
19.4 Connected graphs

LEARNING INTENTION

At the end of this lesson you should be able to:
- define walks, trails, paths, circuits and cycles in the context of traversing a graph.

▶ 19.4.1 Traversable graphs

eles-6282

- Networks can represent many applications, such as social networks, electrical wiring, internet and communications.
- Sometimes it may be important to follow a sequence that goes through all vertices only once, for example a salesperson who wishes to visit each town once.
- Sometimes it may be important to follow a sequence that use all edges only once, such as a road repair gang repairing all the roads in a suburb.
- Movement through a simple connected graph is described in terms of starting and finishing at specified vertices by travelling along the edges.
- The definitions of the main terms used when describing movement across a network are as follows.

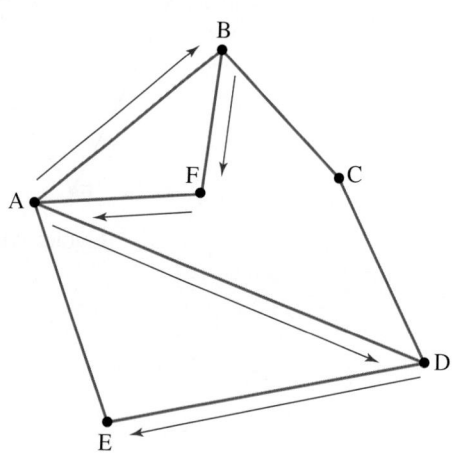

Route: ABFADE

Moving around a graph

Walk: Any route taken through a graph using a sequence of vertices through edges. A walk can include repeated edges.

Trail: A walk where any edge used is used once; however, vertices may repeat.

Path: A walk with no repeated vertices and no repeated edges.

Cycle: A closed path beginning and ending at the same vertex.

Circuit: A trail beginning and ending at the same vertex.

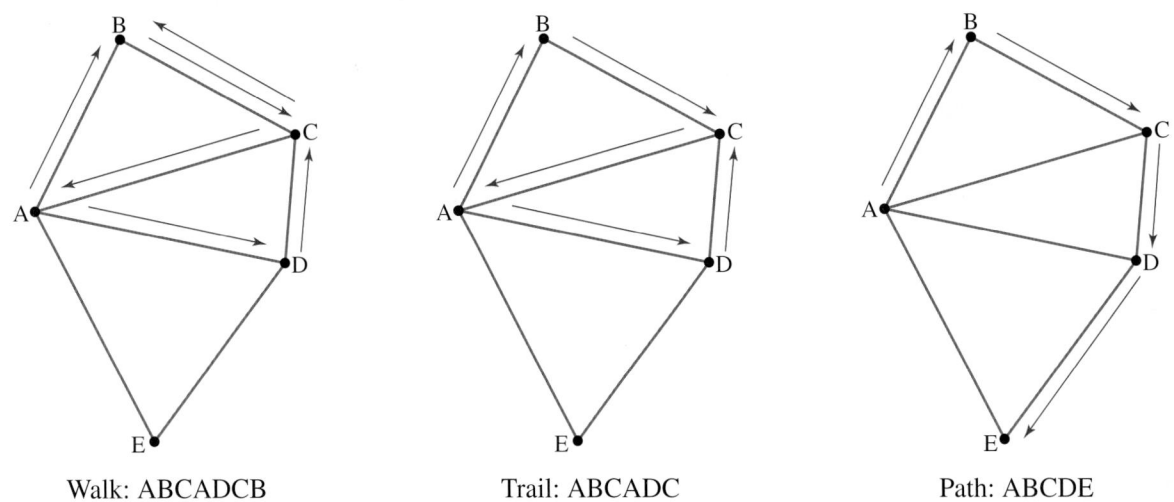

Walk: ABCADCB   Trail: ABCADC   Path: ABCDE

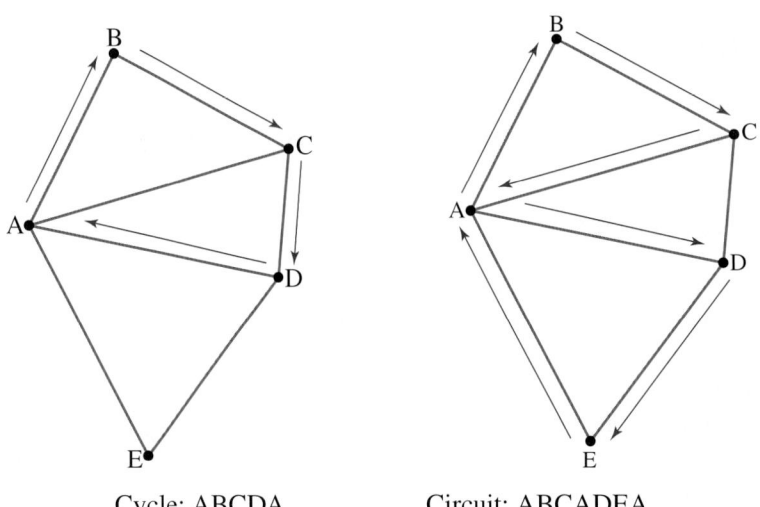

Cycle: ABCDA   Circuit: ABCADEA

In the network shown, identify two different routes: one cycle and one circuit.

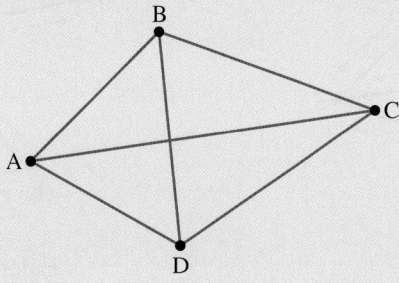

THINK	WRITE
1. For a cycle, identify a route that doesn't repeat a vertex and doesn't repeat an edge, apart from the start/finish.	Cycle: ABDCA

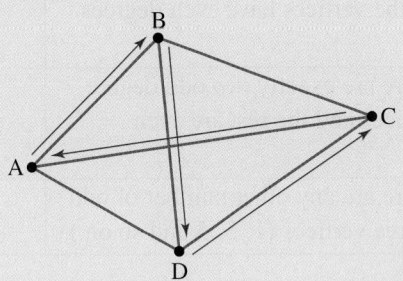

| 2. For a circuit, identify a route that doesn't repeat an edge and ends at the starting vertex. | Circuit: ADBCA |

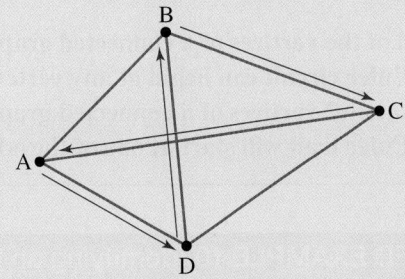

19.4.2 Eulerian trails and circuits

eles-6283

- In some practical situations it is most efficient if a route travels along each edge only once, for example parcel deliveries and garbage collections.
- If it is possible to move around a network using each edge only once, the route is known as a **Euler trail** or **Eulerian circuit**.

Euler trails and circuits

An Euler trail is a trail in which every edge is used once.

An Euler circuit is an Euler trail that starts and ends at the same vertex.

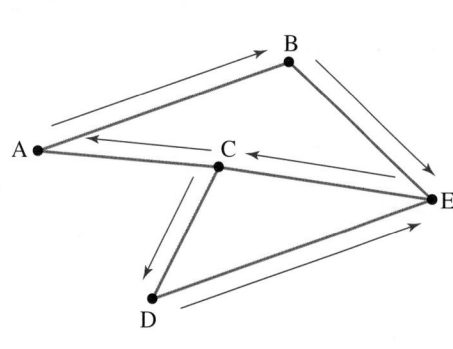

Euler trail: CDECABE

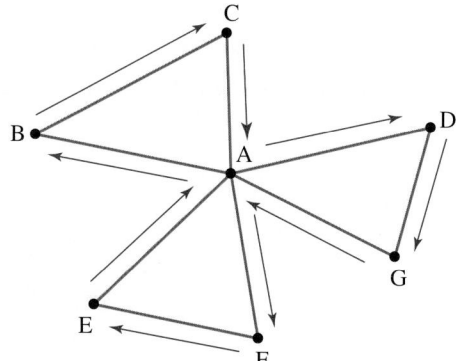

Euler circuit: ABCADGAFEA

- By determining the degree of each vertex and whether each degree is even or odd, it can be established if the network is an Euler trail or Eulerian circuit.

Degree of vertices	Euler trail	How?	Euler circuit
All the vertices have even degrees.	Yes	Start and finish at any vertex.	Yes
There are exactly *two* odd degree vertices and the rest are even.	Yes	Start and finish at the two odd degree vertices.	No
There are any other number of odd degree vertices (1, 3, 5 and so on.)	No		No

Existence conditions for Euler trails and Euler circuits

- **If all of the vertices of a connected graph have an even degree, then an Euler circuit exists. An Euler circuit can begin at any vertex.**
- **If exactly 2 vertices of a connected graph have an odd degree, then an Euler trail exists. An Euler trail will start at one of the odd degree vertices and finish at the other.**

WORKED EXAMPLE 10 Identifying an Euler trail

Determine whether there is an Euler trail through the network shown and, if so, give an example.

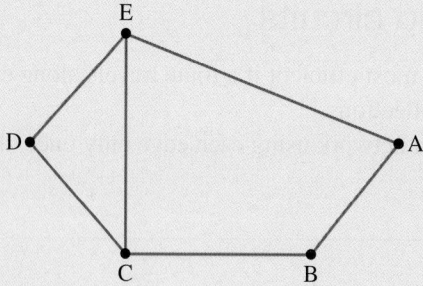

THINK	**WRITE/DRAW**
1. Determine the degree of each vertex by counting the number of edges connected to it.	Vertex A has degree = 2. Vertex B has degree = 2. Vertex C has degree = 3. Vertex D has degree = 2. Vertex E has degree = 3.
2. Count the number of odd degree vertices and hence state whether there is an Euler trail through the network.	Number of odd vertices = 2 Therefore, an Euler trail exists.
3. Since there are exactly 2 vertices with odd degrees (C and E), an Euler trail has to start and finish with these; say, begin at C and end at E. Attempt to find a trail that uses each edge. *Note:* Although each edge must be used exactly once, vertices may be used more than once. To ensure that each edge has been used, label them as you go.	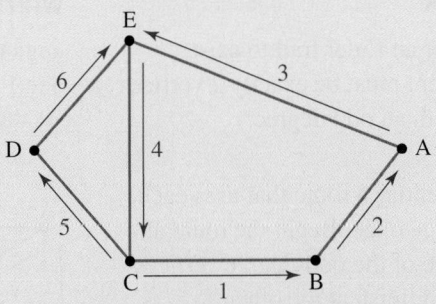
4. List the sequence of vertices along the trail.	An Euler trail is C–B–A–E–C–D–E.

DISCUSSION

Examine Euler's Seven Bridges of Königsberg network problem shown.

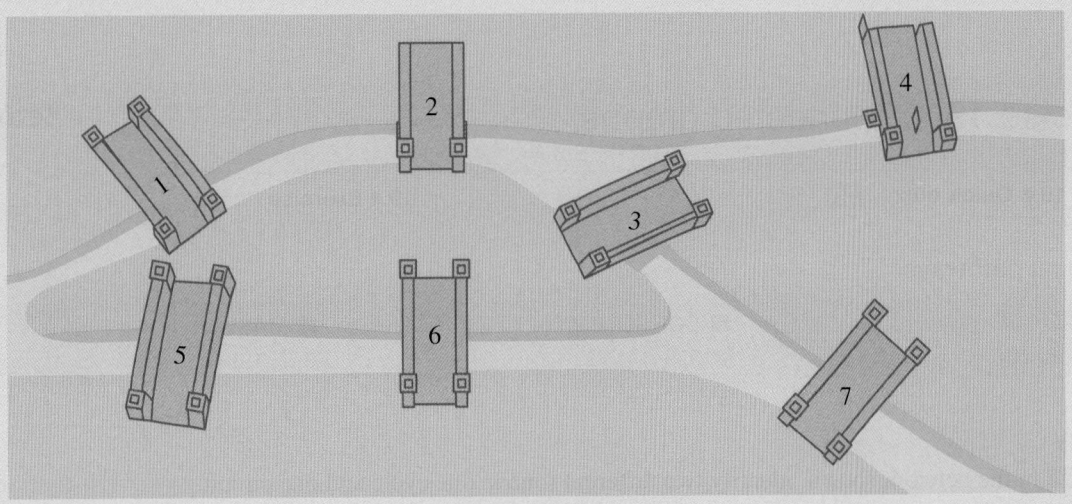

Bridges of Königsberg

Choosing any starting point, can all bridges only be crossed once?

How does Euler's Seven Bridges of Königsberg relate to the definition of an Eulerian trail or circuit?

WORKED EXAMPLE 11 Describing an Euler trail

Identify an Euler trail in the network shown.

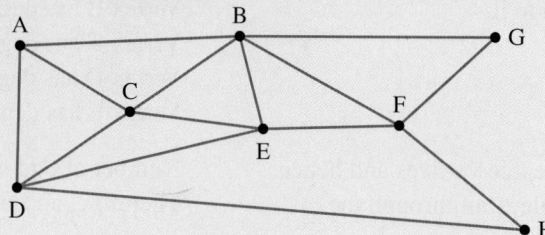

THINK	WRITE/DRAW
1. For an Euler trail to exist, there must be exactly 2 vertices with an odd degree.	$\deg(A) = 3, \deg(B) = 5, \deg(C) = 4, \deg(D) = 4, \deg(E) = 4,$ $\deg(F) = 4, \deg(G) = 2, \deg(H) = 2$ As there are only two odd-degree vertices, an Euler trail must exist.
2. Identify a route that uses each edge once. Begin the route at one of the odd-degree vertices and finish at the other.	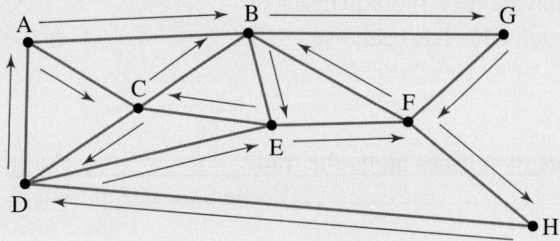 Euler trail: ABGFHDEFBECDACB
3. Write the answer.	Euler trail: ABGFHDEFBECDACB (other answers are possible)

Exercise 19.4 Connected graphs

learnon

19.4 Quick quiz on	19.4 Exercise

Individual pathways

■ PRACTISE	■ CONSOLIDATE	■ MASTER
1, 4, 6, 8, 10	2, 5, 9, 11, 13	3, 7, 12, 14

Fluency

1. **WE9** In the network shown, identify two different routes: one cycle and one circuit.

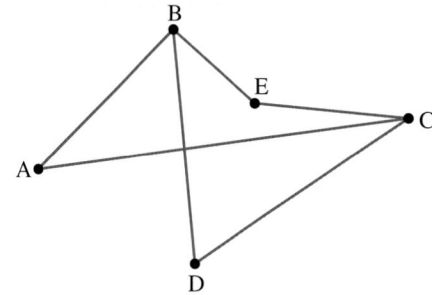

2. In the network shown, identify three different routes: one path, one cycle and one circuit.

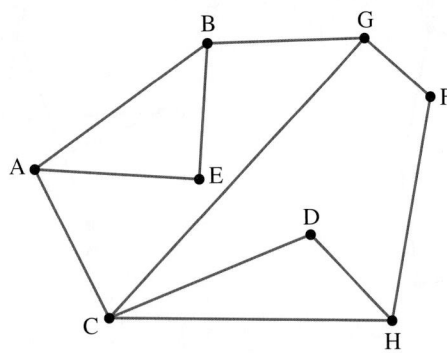

3. State which of the terms *walk*, *trail*, *path*, *cycle* and *circuit* could be used to describe the following routes on the graph shown.

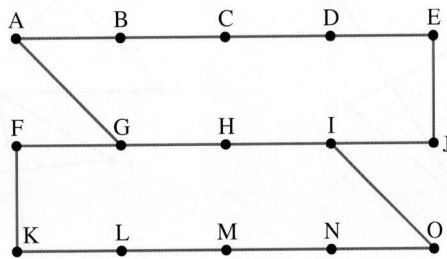

    a. AGHIONMLKFGA   **b.** IHGFKLMNO

    c. HIJEDCBAGH   **d.** FGHIJEDCBAG

4. `WE10` Determine whether there is an Euler trail through the network shown and, if so, give an example.

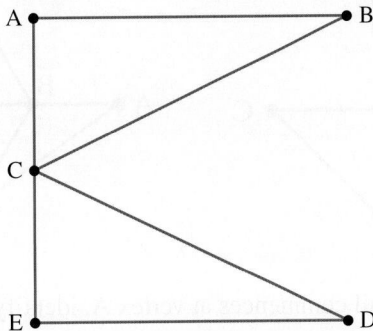

5. Starting at vertex R, determine an Euler trail for the planar graph shown.
    (*Hint:* What vertex should the trail end at?)

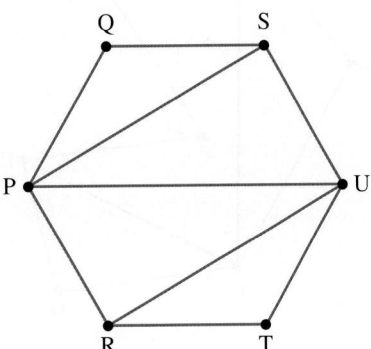

6. **WE11** Identify an Euler trail in each of the following graphs.

a.

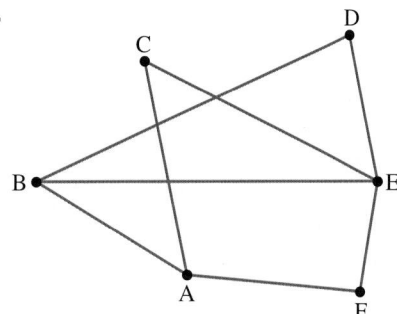

b.

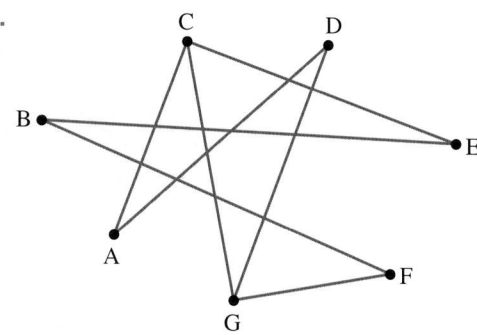

7. Identify an Euler circuit in each of the following graphs, if one exists.

a.

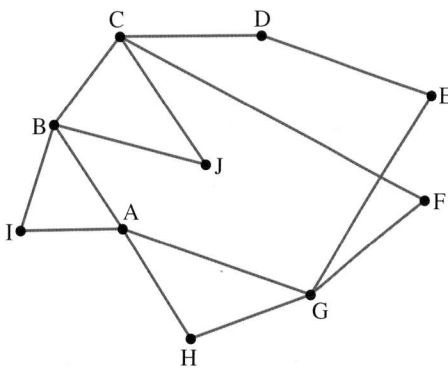

b.

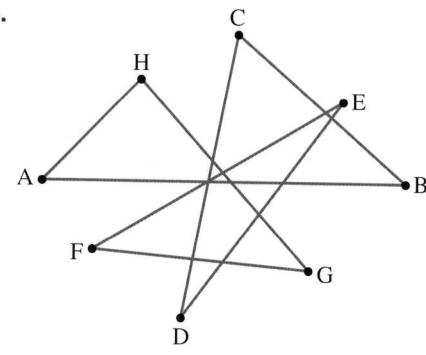

Understanding

8. Identify an Euler circuit for each of the following networks shown.

a.

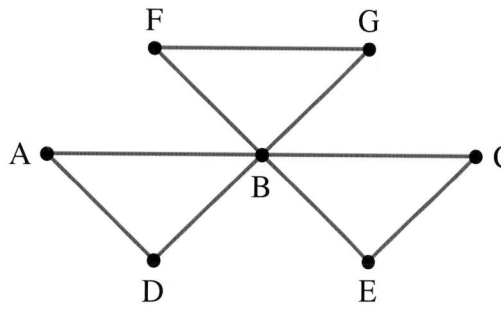

b.

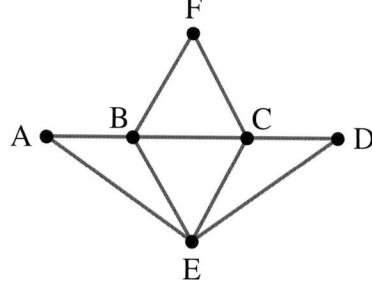

9. In the following graph, if an Euler trail commences at vertex A, identify the vertices at which it could finish.

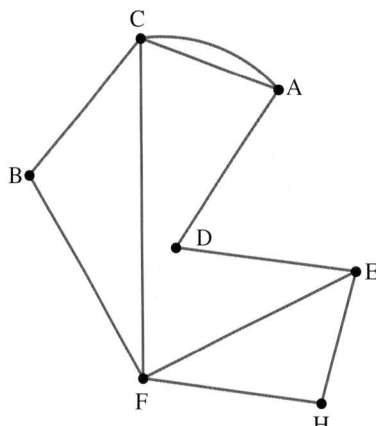

Communicating, reasoning and problem solving

10. A road inspection crew must travel along each road shown on the map exactly once.

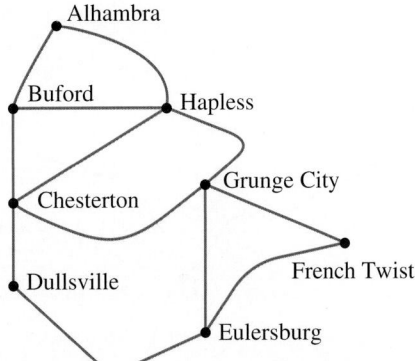

a. State from either of which two cities the crew must begin its tour.
b. Determine a path using each road once.
c. State which cities are visited most often and why.

11. In a computer network, the file server (F) is connected to all the other computers as shown.

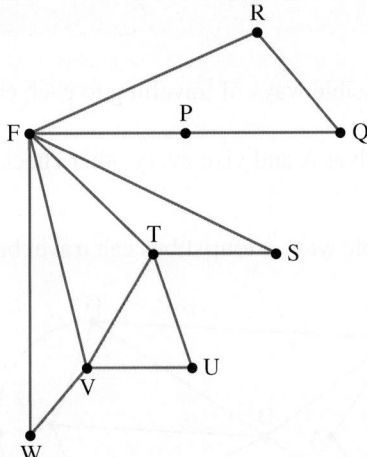

Craig, the technician, wishes to test that each connecting cable is functioning properly. He wishes to route a signal, starting at F, so that it travels down each cable exactly once and then returns to the file server. Determine such a circuit for the network configuration shown.

12. On the map shown, a school bus route is indicated in yellow. The bus route starts and ends at the school indicated.

a. Draw a graph to represent the bus route.
b. Students can catch the bus at stops that are located at the intersections of the roads marked in yellow.
   Determine whether it is possible for the bus to collect students by driving down each section of the route only once.
   Explain your answer.

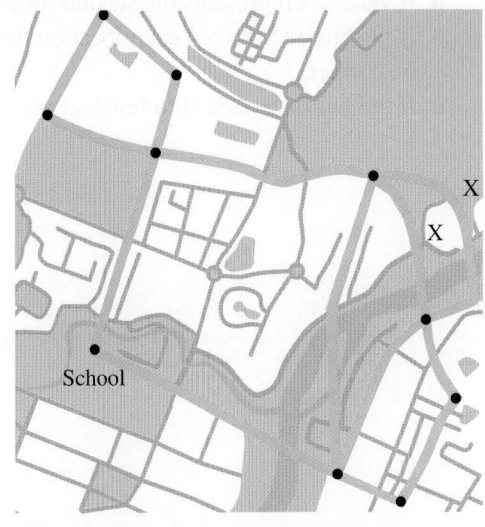

13. The map of an orienteering course is shown. Participants must travel to each of the nine checkpoints along any of the marked paths.

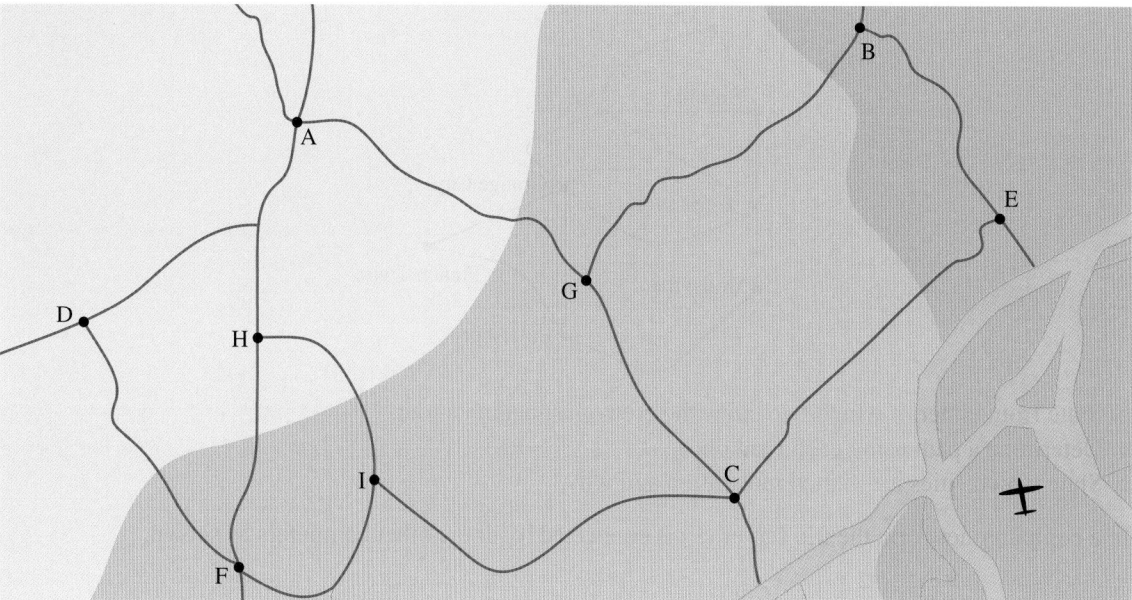

   a. Draw a graph to represent the possible ways of travelling to each checkpoint.
   b. State the degree of checkpoint H.
   c. If participants must start and finish at A and visit every other checkpoint only once, identify two possible routes they could take.

14. The graph shown outlines the possible ways a tourist bus can travel between eight locations.

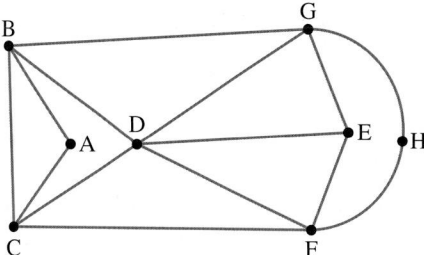

   a. If vertex A represents the second location visited, list the possible starting points.
   b. If the bus also visited each location only once, state which of the starting points listed in part **a** could not be correct.
   c. If the bus also needed to finish at vertex D, list the possible paths that could be taken.

LESSON
19.5 Review

19.5.1 Topic summary

Properties of networks

- In a graph, the lines are called edges and the points are called vertices (or nodes), with each edge joining a pair of vertices.
- A simple graph is one in which pairs of vertices are connected by at most one edge.
- If it is possible to reach every vertex of a graph by moving along the edges, the graph is a *connected graph*.
- If it is only possible to move along the edges of a graph in one direction, the graph is a *directed graph* (or digraph) and the edges are represented by arrows. Otherwise, it is an *undirected graph*.
- The degree of a vertex = the number of edges directly connected to that vertex, except that a loop at a vertex contributes twice to the degree of that vertex.

INTRODUCTION TO NETWORKS (PATH)

Planar graphs

- If a graph can be drawn with no intersecting edges, then it is a *planar graph*.

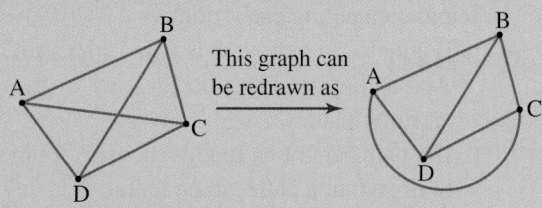
This graph can be redrawn as

- Euler's formula identifies the relationship between the vertices (v), edges (e) and faces (f) in a planar graph: $v - e + f = 2$.

Connected graphs

- *Walk:* Any route taken through a network, including routes that repeat edges and vertices, e.g. ALCAG
- *Trail:* A walk with no repeated edges, e.g. ALCGHC
- *Path:* A walk with no repeated vertices and no repeated edges, e.g. ALCGH
- *Cycle:* A path beginning and ending at the same vertex, e.g. ALCGA
- *Circuit:* A trail beginning and ending at the same vertex, e.g. ALCGHCA
- An Euler trail is a trail in which every edge is used once.
- An Euler circuit is a circuit in which every edge is used once.
- If all of the vertices of a connected graph are even, then an Euler circuit exists.
- If exactly two vertices of a connected graph are odd, then an Euler trail exists.

19.5.2 Project

Networks at school

Work in a small group to analyse the pathways in your school.

1. On a map of your school, identify all the buildings that students may use at school.
2. Represent this information as a network diagram. Use edges to represent the pathways in your school.
3. As a group, determine whether a path can be found connecting each room.
4. a. If a path can be found, is it an Euler path? Add additional edges if necessary to create an Euler path.
   b. If a path cannot be found, choose a room to omit so that a path can be found. Add any paths necessary to create an Euler path.

5. If possible, identify an Eulerian circuit within the building you are in now (or the building you use for your Maths class).
6. Make recommendations about the pathways currently found in your school. Do you believe that any more should be created?

 Resources

 Interactivities Crossword (int-9204)
Sudoku puzzle (int-9205)

Exercise 19.5 Review questions

learnon

Fluency

1. **MC** The minimum number of edges in a connected graph with eight vertices is:

   A. 5   **B.** 6   **C.** 7   **D.** 8

2. For the network shown:
   a. determine the number of vertices (v)
   b. determine the number of edges (e)
   c. determine the number of faces (f)
   d. verify Euler's formula.

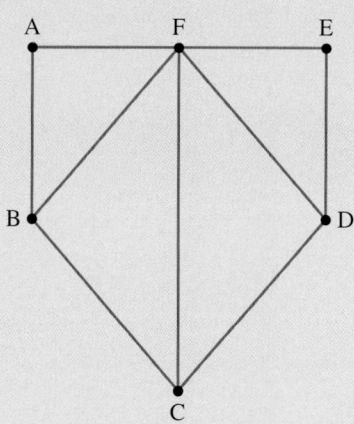

3. **MC** In the graph shown, identify the number of edges.

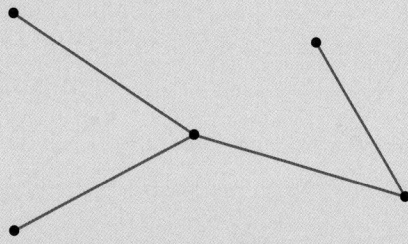

    A. 2   **B.** 3   **C.** 4   **D.** 5

4. **MC** A connected graph with 9 vertices has 10 faces. The number of edges in the graph is:

    A. 15   **B.** 16   **C.** 17   **D.** 18

Understanding

5. a. For the network shown, determine an Euler path.

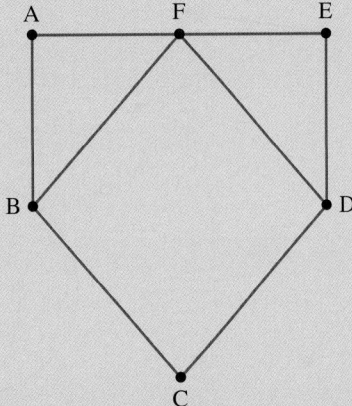

    b. Explain why is there no Euler circuit.

6. **MC** The number of faces in the planar graph shown is:

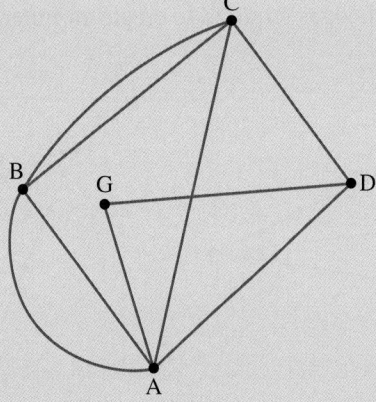

    A. 6   **B.** 7   **C.** 8   **D.** 9

7. a. Identify whether the following graphs are planar or not planar.

i.

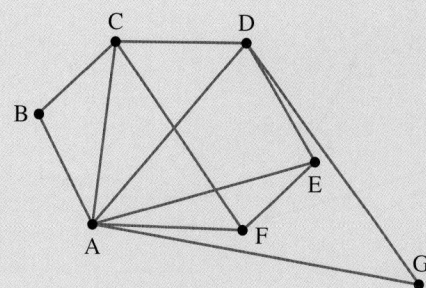

ii.

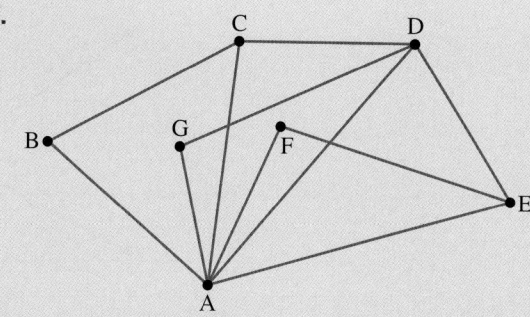

b. Redraw the graphs that are planar without any intersecting edges.

8. Identify which of the following graphs are isomorphic.

a.

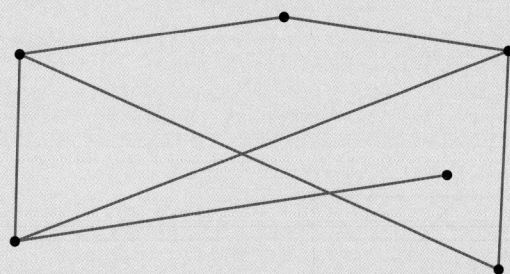

b.

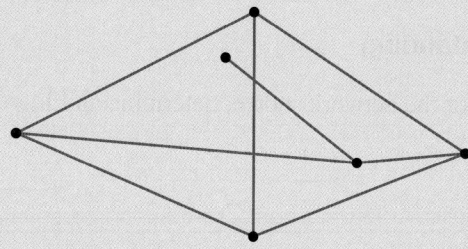

c.

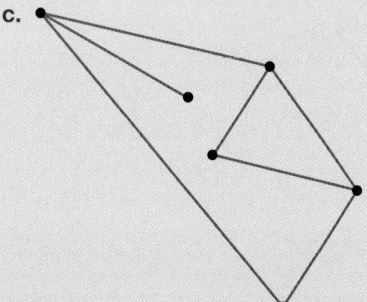

d.

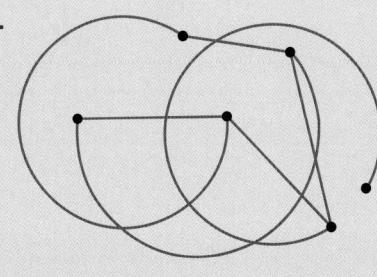

9. For each of the following graphs:
  i. add the minimum number of edges required to create an Euler trail
  ii. state the Euler trail created.

a.

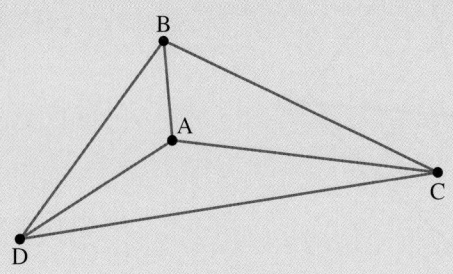

b.

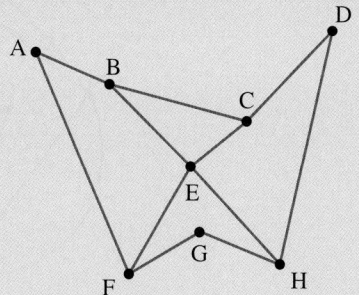

Communicating, reasoning and problem solving

10. The graph shown represents seven towns labelled A–G in a particular district. The lines (edges) represent roads linking the towns (vertices), with distances given in km.

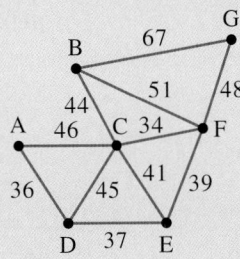

a. Does the graph represent a connected planar graph?

b. Determine the number of:
   i. vertices
   ii. edges
   iii. faces (regions) for the given graph.

c. Using your answers from part b, determine whether Euler's formula holds for the given graph.

d. i. Write the degree of each vertex.
   ii. Calculate the sum of the degrees of all the vertices and compare with the number of edges.
   iii. How many odd-degree vertices are there?

11. The city of Kaliningrad in Russia is situated on the Pregel River. It has two large islands that are connected to each other and to the mainland by a series of bridges. In the early 1700s, when the city was called Königsberg, there were seven bridges. The challenge was to walk a route that crossed each bridge exactly once. In 1735, Swiss mathematician Leonard Euler (pronounced 'oiler') used networks to prove whether this was possible.

a. Draw a network with four vertices (to represent the areas of land) and seven edges to represent the seven connecting bridges of Königsberg in 1735.

b. Use your knowledge of networks to explain whether it is possible to walk a route that crosses each bridge only once.

c. The diagrams below show the bridges in Kaliningrad in 1735 and the bridges as they are now. Use your knowledge of networks to explain whether it is possible now to walk a route that crosses each bridge only once.

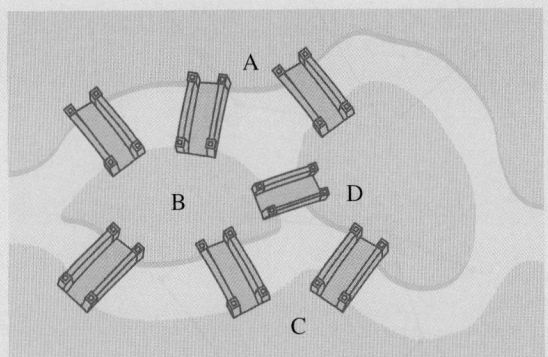

Bridges as they were in 1735

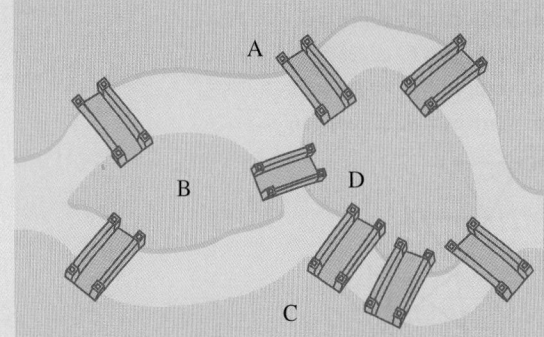

Bridges as they are now

on To test your understanding and knowledge of this topic, go to your learnON title at www.jacplus.com.au and complete the **post-test**.

Answers

Topic 19 Introduction to networks (Path)

19.1 Pre-test

1. 4
2. 6
3. 9
4. One possible path is from A is A–B–C–F–E–D
5. Yes it is possible. One possible path from A is:
   A–D–E–F–D–B–F–C–B–A
6. The diagram is able to be traced from any vertex.
7. A–B–C–D–E–A–F–D–G–A
8.

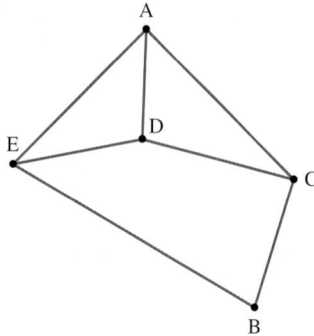

9. 4
10. E
11. 8
12. deg (A) = 3, deg (C) = 4, deg (L) = 3, deg (G) = 4, deg (H) = 2
13. Sample responses can be found in the worked solutions in the online resources.
14. Planar when redrawn.
15. Join 4 to 7 and 3 to 8

19.2 Properties of networks

1. a. Vertices = 6, edges = 8
   b. Vertices = 7, edges = 9
2. a. Vertices = 5, edges = 5
   b. Vertices = 6, edges = 9
   c. Vertices = 7, edges = 11
   d. Vertices = 9, edges = 16
3.

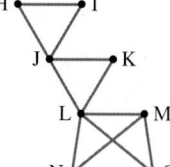

4.

5.

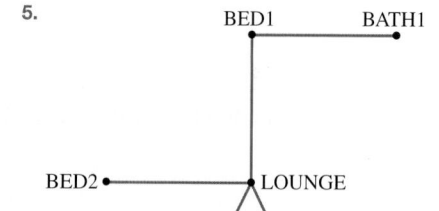

6.

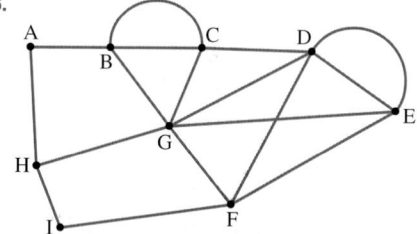

7. Degree of A = 3, B = 4, C = 3, D = 2, E = 5, F = 3
8. a. deg (A) = 5; deg (B) = 3; deg (C) = 4; deg (D) = 1; deg (E) = 1
   b. deg (A) = 0; deg (B) = 2; deg (C) = 2; deg (D) = 3; deg (E) = 3
   c. deg (A) = 4; deg (B) = 2; deg (C) = 2; deg (D) = 2; deg (E) = 4
   d. deg (A) = 1; deg (B) = 2; deg (C) = 1; deg (D) = 1; deg (E) = 3
9. a. Yes   b. Yes
10. a. Different degrees and connections
    b. Different degrees and connections
11. Graphs 2 and 4; Graphs 5 and 6; Graph 1 and 7
12. Answer will vary. Possible answers are shown.
    a.

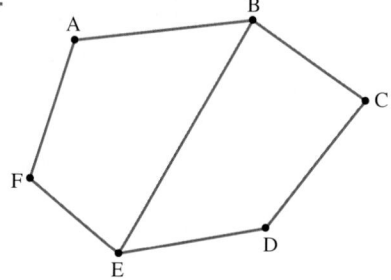

    b.

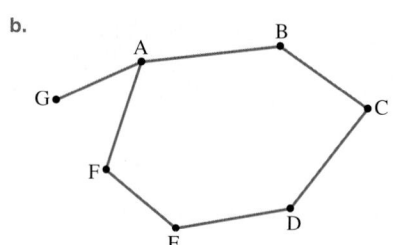

c.

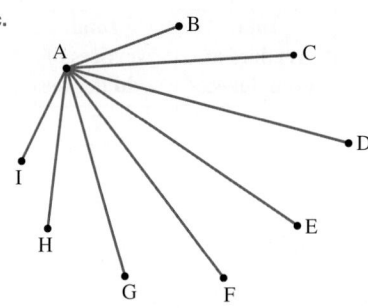

13.

Graph	Simple	Connected
Graph 1	Yes	Yes
Graph 2	Yes	Yes
Graph 3	Yes	Yes
Graph 4	No	Yes
Graph 5	Yes	Yes

14.

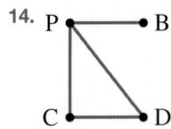

15. a.

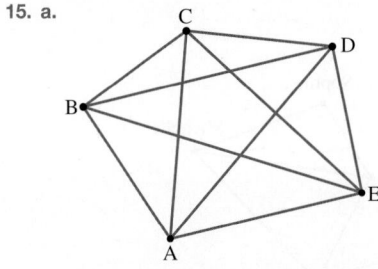

b. The total number of edges represents the total games played, as each edge represents one team playing another.

16. a. $v = \{1, 2, 3, 4\}$
$e = \{(1, 2), (1, 3), (1, 4), (2, 3), (2, 4), (3, 4)\}$

b.

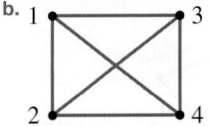

c. Degree $(1) = 3$
Degree $(2) = 3$
Degree $(3) = 3$
Degree $(4) = 3$

17. a.

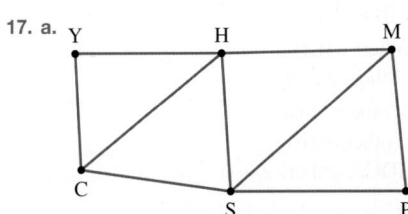

b. deg (Huairou) = deg (Shunyi) = 4

c. This is a simple, connected graph as there are no loops or multiple edges, and all vertices are reachable.

18.

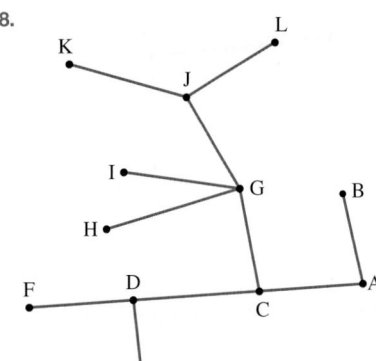

19. a.

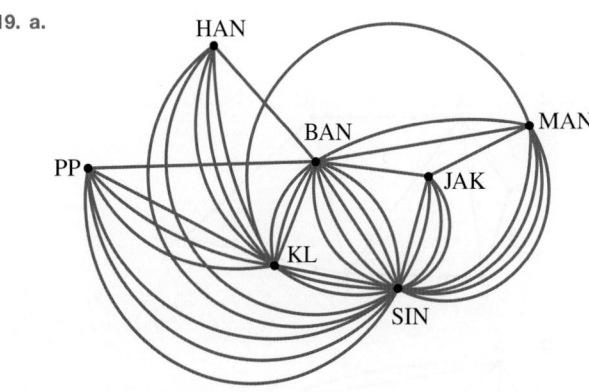

b. Directed as it would be important to know the direction of the flight.

c. 7

19.3 Planar graphs

1.

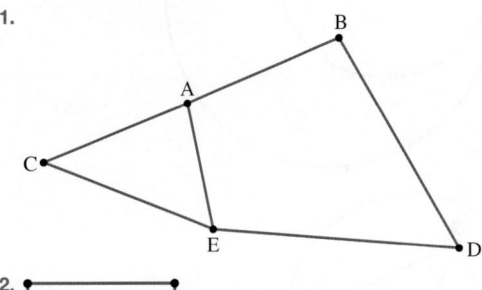

2.

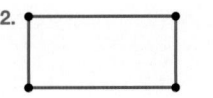

3.

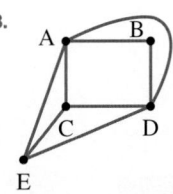

4. Vertices $= 6$, edges $= 9$, faces $= 5$

5. a. Vertices $= 5$, edges $= 7$, faces $= 4$

  b. Vertices $= 7$, edges $= 10$, faces $= 5$

6. a. 4   **b.** 5

7. a. 6   **b.** 5

8.

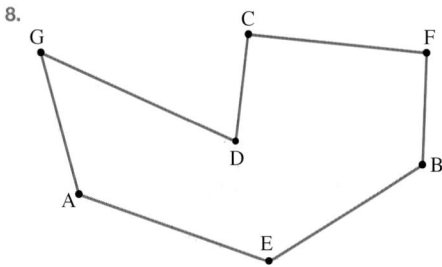

9. Graph 3 is not planar as it cannot be redrawn without intersecting edges.

10. a. Vertices = 6, edges = 10, faces = 6

b. Please see the worked solutions.

11. a. 3

b. 2

12. a.

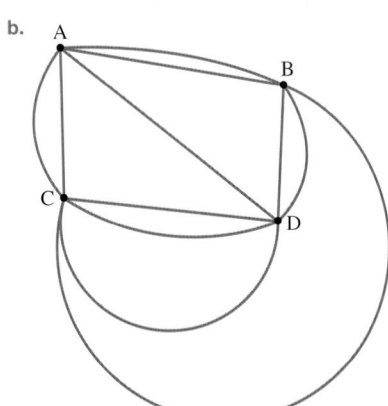

b.

13.

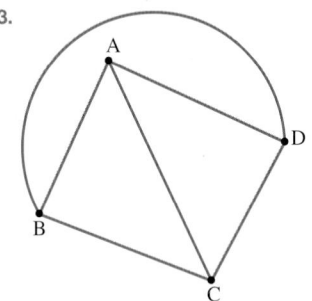

Tetrahedron

14.

Graph	Total edges	Total degrees
Graph 1	3	6
Graph 2	5	10
Graph 3	8	16

15. a.

Graph	Total vertices of even degree	Total vertices of odd degree
Graph 1	3	2
Graph 2	4	2
Graph 3	4	4

b. No clear pattern evident

16.

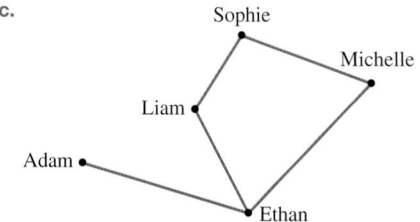

Cube

17. a.

b. Ethan and Michelle, as they are in contact via more methods.

c.

19.4 Connected graphs

1. Cycle: ABECA (others exist)
Circuit: BECDB (others exist)

2. Path: ABGFHDC (others exist)
Cycle: DCGFHD (others exist)
Circuit: AEBGFHDCA (others exist)

3. a. Walk

b. Walk, trail and path

c. Walk, trail, path, cycle and circuit

d. Walk and trail

4. A–B–C–D–E–C–A (also a circuit)
5. R–T–U–R–P–U–S–P–Q–S
6. a. Euler: AFEDBECAB
   b. Euler: GFBECGDAC
7. a. Euler: AIBAHGFCJBCDEGA
   b. Euler: ABCDEFGHA (other exist)
8. Other circuits are possible.
   a. B–A–D–B–E–C–B–F–G–B
   b. A–E–D–C–E–B–C–F–B–A
9. E
10. a. Buford, Eulersburg
    b. B–A–H–B–C–H–G–C–D–E–G–F–E
    c. Chesterton, Hapless, Grunge City, because they all have the highest degree
11. Other circuits are possible.
    F–P–Q–R–F–S–T–U–V–T–F–V–W–F
12. a.

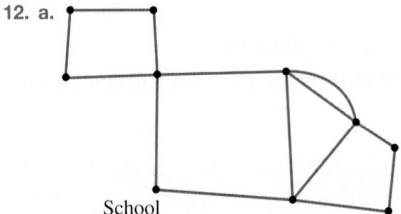

School

   b. Yes, because the degree of each intersection or corner point is an even number.

13. a.

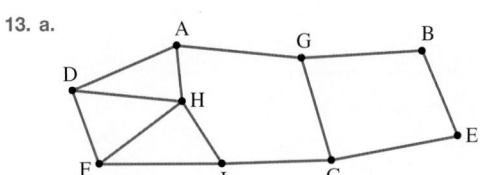

   b. deg (H) = 4
   c. ADHFICEBGA or AHDFICEBGA
14. a. B: BCA
    C: CBA
    D: DBA or DCA
    F: FCA
    G: GBA
    Therefore, B, C, D, F or G
   b. B or C, as once A is reached, the next option is returning to the start.
   c. None possible.

Project

The steps to analyse the pathways in your school:
1. On a map of your school, identify all the buildings that students may use at school.
   • Make a list of all the buildings in your school that are accessible to students.
2. Represent this information as a network diagram. Use edges to represent the pathways in your school.
   • Draw a node for each building and connect them with edges to represent the pathways between them.

3. As a group, determine whether a path can be found connecting each room.
   • Try to find a path that connects all the rooms in your school.
4. a. If a path can be found, is it an Euler path? Add additional edges if necessary to create an Euler path.
   • An Euler path is a path that visits every edge exactly once. If the path you found satisfies this condition, then it is an Euler path. Otherwise, you can add additional edges to create an Euler path.
   b. If a path cannot be found, choose a room to omit so that a path can be found. Add any paths necessary to create an Euler path.
   • If you cannot find a path that connects all the rooms, choose a room to omit and try again. You can add additional paths to create an Euler path.
5. If possible, identify an Euler circuit within the building you are in now (or the building you use for your Maths class).
   • To identify an Euler circuit within a building, draw a network diagram of the rooms in the building and the pathways between them. If the building has an Euler circuit, then it is possible to walk from any room in the building and return to that room by following a sequence of pathways that visits each room exactly once.
6. Make recommendations about the pathways currently found in your school. Do you believe that any more should be created?
   • Based on your analysis of the pathways in your school, you can make recommendations on whether additional pathways should be created.
   For example, if you found that there are rooms that are not connected to the rest of the buildings, you could recommend adding new pathways to connect them. Alternatively, if you found that some pathways are rarely used or lead to congestion, you could recommend creating new pathways to alleviate traffic. Your recommendations should take into account the safety, efficiency, and convenience of the pathways for students and staff.

19.5 Review questions
1. C
2. a. $v = 6$
   b. $e = 9$
   c. $f = 5$
   d. $6 - 9 + 5 = 2$
3. C
4. C
5. a. B–A–F–E–D–F–B–C–D
   b. Because there are 2 vertices of odd degree.
6. A

7. a. i. Planar   **ii.** Planar

b. i.

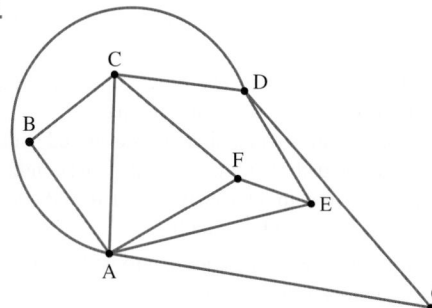

ii.

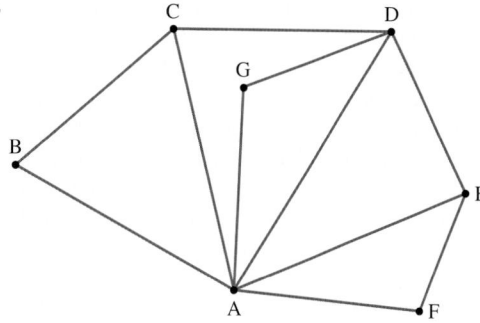

8. a and d

9. a. i.

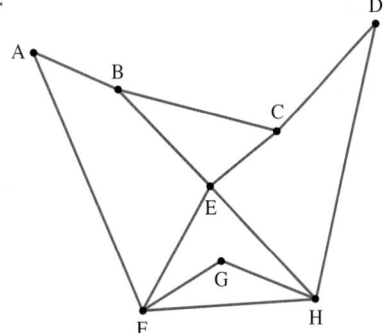

ii. ABDBCADC

b. i.

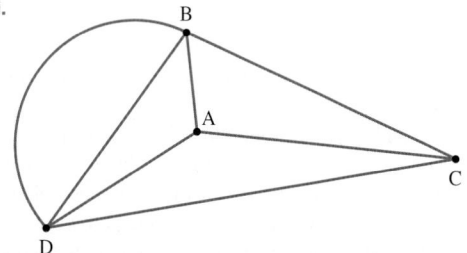

ii. BAFEHGFHDCEBC

10. a. Yes, because the edges do not cross each other.

   b. i. 7

      ii. 11

      iii. 6

   c. Yes, $v - e + f = 2$.

   d. i. Degree A = 2, degree B = 3, degree C = 5, degree D = 3, degree E = 3, degree F = 4, degree G = 2

      ii. Sum = 22

      iii. 4

11. a.

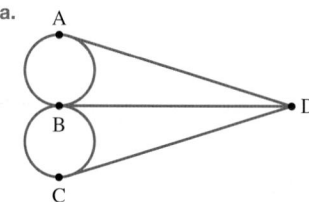

   b. No. All vertices have an odd degree.

   c. Yes. It is possible as there are 2 odd vertices and 2 even vertices.

Semester review 2

The learnON platform is a powerful tool that enables students to complete revision independently and allows teachers to set mixed and spaced practice with ease.

Student self-study

Review the **Course Content** to determine which topics and lessons you studied throughout the year. Notice the green bubbles showing which elements were covered.

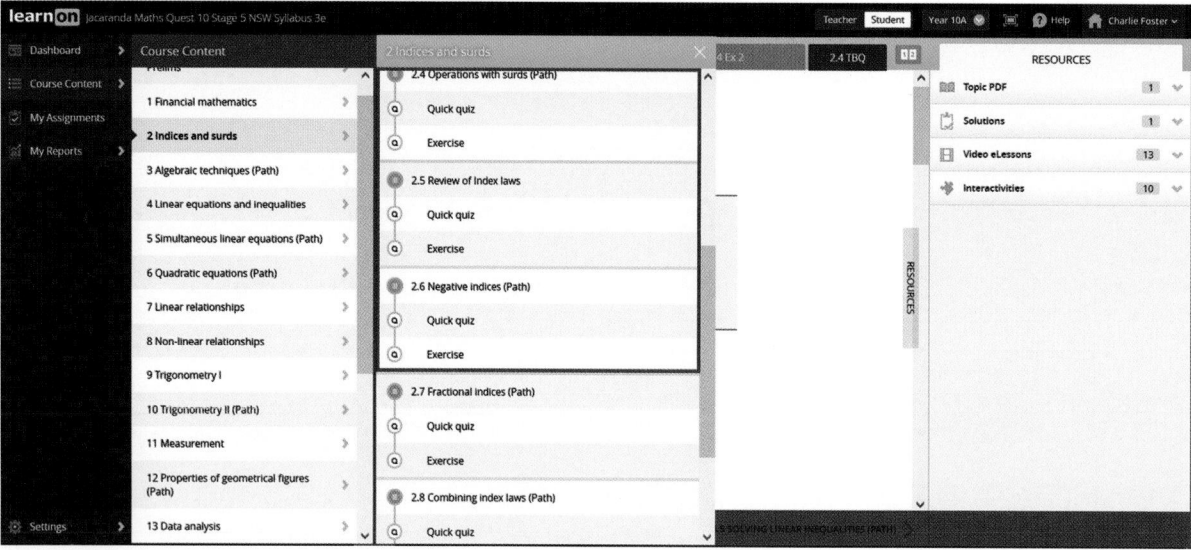

Review your results in **My Reports** and highlight the areas where you may need additional practice.

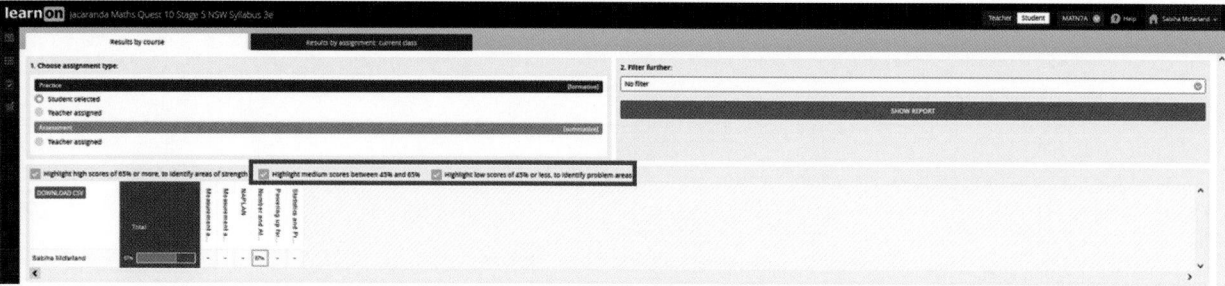

Use these and other tools to help identify areas of strengths and weakness and target those areas for improvement.

Teachers

It is possible to set questions that span multiple topics. These assignments can be given to individual students, to groups or to the whole class in a few easy steps.

Go to **Menu** and select **Assignments** and then **Create Assignment**. You can select questions from one or many topics simply by ticking the boxes as shown below.

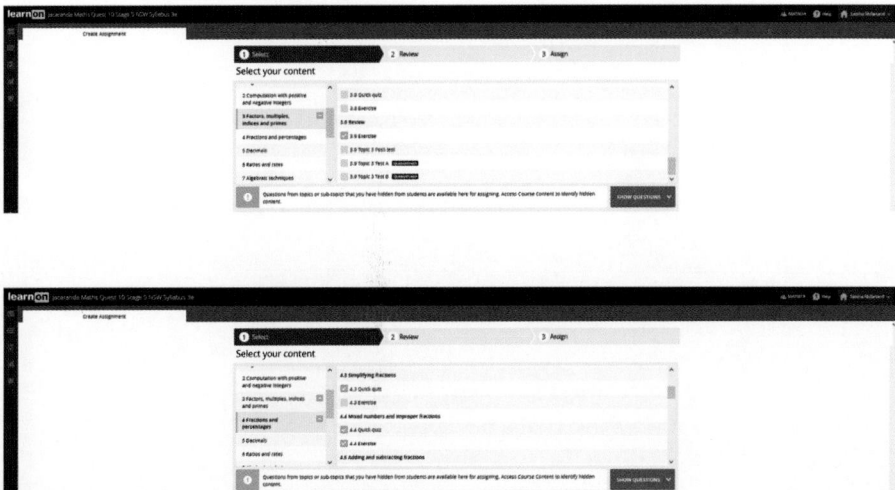

Once your selections are made, you can assign to your whole class or subsets of your class, with individualised start and finish times. You can also share with other teachers.

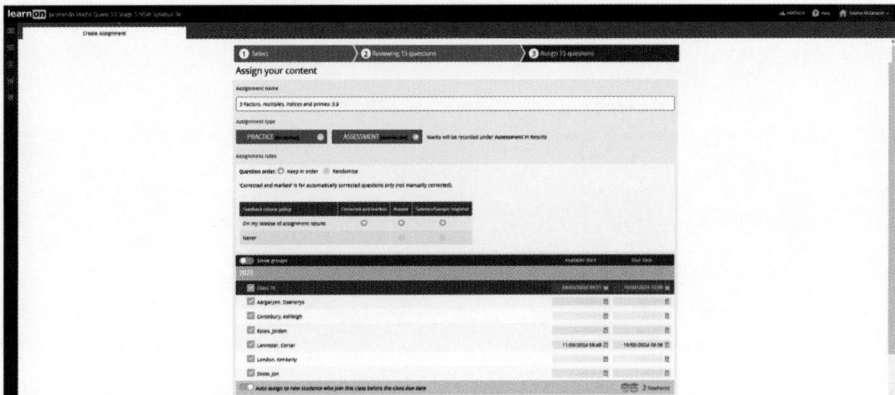

More instructions and helpful hints are available at www.jacplus.com.au.

GLOSSARY

3-dimensional a shape that occupies space (a solid); that is, one that has dimensions in three directions — length, width and height.

Addition Law of probability for the events A and B, the formula for the probability of $A \cup B$ is known as the Addition Law of probability and is given by the formula: $\Pr(A \cup B) = \Pr(A) + \Pr(B) - \Pr(A \cap B)$.

algebraic fractions fractions that contain pronumerals (letters).

algorithm a step-by-step set of tasks to solve a particular problem. A program is an implementation of an algorithm.

alternate segment in the diagram shown, the angle $\angle$BAD defines a segment (the shaded area). The unshaded part of the circle is called the alternate segment to $\angle$BAD.

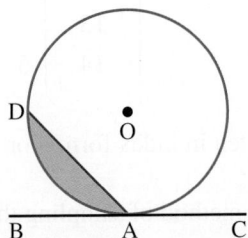

amplitude half the distance between the maximum and minimum values of a trigonometric function.

angle of depression the angle measured down from the horizontal line (through the observation point) to the line of sight.

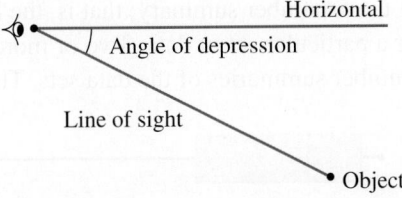

angle of elevation the angle measured up from the horizontal line (through the observation point) to the line of sight.

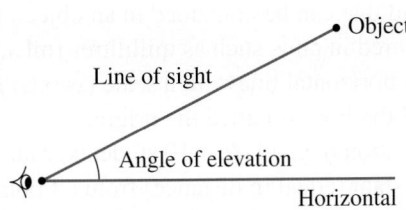

area the amount of flat surface enclosed by the shape. It is measured in square units, such as square metres, m², or square kilometres, km².

Associative Law a method of combining two numbers or algebraic expressions is associative if the result of the combination of these objects does not depend on the way in which the objects are grouped. Addition and multiplication obey the Associative Law, but subtraction and division are not associative.

asymptote a line that a graph approaches but never meets.

axis of symmetry a line through a shape so that each side is a mirror image.

back-to-back stem-and-leaf plot a method for comparing two data distributions by attaching two sets of 'leaves' to the same 'stem' in a stem-and-leaf plot; for example, comparing the pulse rate before and after exercise. A key should always be included.

Key: 8 | 6 = 86

Leaf	Stem	Leaf
Before exercise		After exercise
9 8 8 8	6	
8 6 6 4 1 1 0	7	
8 8 6 2	8	6 7 8 8
6 0	9	0 2 2 4 5 8 9 9
4	10	0 4 4
0	11	8
	12	4 4
	13	
	14	6

base the digit at the bottom of a number written in index form. For example, in 6^4, the base is 6. This tells us that 6 is multiplied by itself four times.

bias designing a questionnaire or choosing a method of sampling that would not be representative of the population as a whole.

bivariate data sets of data where each piece is represented by two variables.

Boolean a JavaScript data type with two possible values: true or false. JavaScript Booleans are used to make logical decisions.

boundary line indicates whether the points on a line satisfy the inequality.

box plot a graphical representation of the 5-number summary; that is, the lowest score, lower quartile, median, upper quartile and highest score, for a particular set of data. Two or more box plots can be drawn on the same scale to visually compare the five-number summaries of the data sets. These are called parallel box plots.

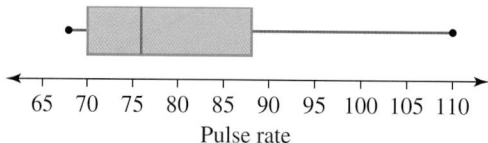

canvas a defined area on a web page where graphics can be drawn with JavaScript.

capacity the maximum amount of fluid that can be contained in an object. It is usually applied to the measurement of liquids and is measured in units such as millilitres (mL), litres (L) and kilolitres (kL).

Cartesian plane the area formed by a horizontal line with a scale (x-axis) joined to a vertical line with a scale (y-axis). The point of intersection of the lines is called the origin.

census collection of data from a population (e.g. all Year 10 students) rather than a sample.

centre middle point of a circle, equidistant (equal in distance) from all points on its circumference.

centre of enlargement the point from which the enlargement of an image is measured.

character in programming, a string of length 1. A JavaScript character is used to represent a letter, digit or symbol.

circle the general equation of a circle with centre (0, 0) and radius r is $x^2 + y^2 = r^2$.

circumcentre the centre of a circle drawn so that it touches all three vertices of a triangle.

circumcircle a circle drawn so that it touches all three vertices of a triangle.

Closure Law when an operation is performed on an element (or elements) of a set, the result produced must also be an element of that set.

coincident lines that lie on top of each other.

collinear points points that all lie on the same straight line.

Commutative Law a method of combining two numbers or algebraic expressions is commutative if the result of the combination does not depend on the order in which the objects are given. For example, the addition of 2 and 3 is commutative, since $2 + 3 = 3 + 2$. However, subtraction is not commutative, since $2 - 3 \neq 3 - 2$.

compass bearing directions measured in degrees from the north–south line in either a clockwise or anticlockwise direction. To write the compass bearing we need to state whether the angle is measured from the north or south, the size of the angle and whether the angle is measured in the direction of east or west; for example, N27°W, S32°E.

complement the complement of a set, A, written A', is the set of elements that are in ξ but not in A.

complementary angles two angles that add to 90°; for example, 24° and 66° are complementary angles.

complementary events events that have no common elements and together make up the sample space. If A and A' are complementary events, then $\Pr(A) + \Pr(A') = 1$.

completing the square the process of writing a general quadratic expression in turning point form.

composite figure a figure made up of more than one basic shape.

compound interest the interest earned by investing a sum of money (the principal) when each successive interest payment is added to the principal for the purpose of calculating the next interest payment. The formula used for compound interest is $FV = PV(1 + r)^n$, where FV is the future value of the investment, PV is the present value or initial amount invested, r is the interest rate per compounding period (as a decimal) and n is the number of compounding periods. The compound interest is calculated by subtracting the principal from the amount: $CI = FV - PV$.

compounded value the value of the investment with accrued interest included.

compounding period the period of time over which interest is calculated.

concave polygon a polygon with at least one reflex interior angle.

concyclic points that lie on the circumference of a circle.

conditional probability where the probability of an event is conditional (depends) on another event occurring first. For two events A and B, the conditional probability of event B, given that event A occurs, is denoted by $P(B|A)$ and can be calculated using the formula:

$$P(B|A) = \frac{P(A \cap B)}{P(A)}, P(A) \neq 0.$$

conjugate surds surds that, when multiplied together, result in a rational number. For example, $\left(\sqrt{a} + \sqrt{b} \right)$ and $\left(\sqrt{a} - \sqrt{b} \right)$ are conjugate surds, because $\left(\sqrt{a} + \sqrt{b} \right) \times \left(\sqrt{a} - \sqrt{b} \right) = a - b$.

congruent triangles there are five standard congruence tests for triangles: SSS (side, side, side), SAS (side, included angle, side), ASA (two angles and one included side), AAS (two angles and one non-included side) and RHS (right angle, hypotenuse, side).

console a special region in a web browser for monitoring the running of JavaScript programs.

constant of proportionality used to prove that a proportionality relationship (direct or inverse) exists between two or more variables (or quantities).

convex polygon a polygon with no interior reflex angles.

coordinates a pair of values (typically x and y) that represent a point on the screen.

correlation a measure of the relationship between two variables. Correlation can be classified as linear, non-linear, positive, negative, weak, moderate or strong.

cosine ratio the ratio of the adjacent side to the hypotenuse in a right-angled triangle; $\cos(\theta) = \dfrac{\text{adjacent}}{\text{hypotenuse}}$.

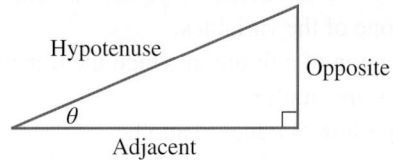

cosine rule in any triangle ABC, $a^2 = b^2 + c^2 - 2bc \cos(A)$.

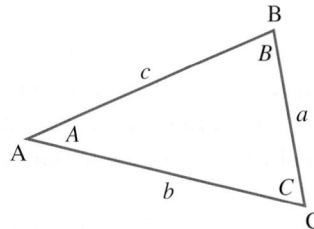

cubic function the basic form of a cubic function is $y = ax^3$. These functions can have 1, 2 or 3 roots.

cumulative frequency curve a line graph that is formed when the cumulative frequencies of a set of data are plotted against the end points of their respective class intervals and then joined up by straight-line segments. It is also called an ogive.

cumulative frequency the total of all frequencies up to and including the frequency for a particular score in a frequency distribution.

cyclic quadrilateral a quadrilateral that has all four vertices on the circumference of a circle. That is, the quadrilateral is inscribed in the circle.

data various forms of information.

degree the degree of a polynomial in x is the highest power of x in the expression.

degrees a unit used to measure the size of an angle.

denominator the lower number of a fraction that represents the number of equal fractional parts a whole has been divided into.

dependent events successive events in which one event affects the occurrence of the next.

dependent variable this is the variable that is impacted by the other variable. This is also known as the response variable.

Depreciation the reduction in the value of an item as it ages over a period of time. The formula used is $S = V_0(1 - r)^n$, where S is the salvage value of the asset, V_0 is its initial value, r is the percentage the item depreciates each year (expressed as a decimal) and n is the number of years the item has depreciated.

deviation the difference between a data value and the mean.

difference of two squares type of expansion of a pair of brackets $(a + b)(a - b) = a^2 - b^2$.

dilation occurs when a graph is made thinner or wider.

discriminant referring to the quadratic equation $ax^2 + bx + c = 0$, the discriminant is given by $\Delta = b^2 - 4ac$. It is the expression under the square-root sign in the quadratic formula and can be used to determine the number and type of solutions of a quadratic equation.

domain the set of all allowable values of x.

dot plot this graphical representation uses one dot to represent a single observation. Dots are placed in columns or rows, so that each column or row corresponds to a single category or observation.

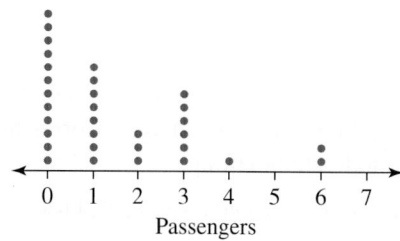

elimination method a method used to solve simultaneous equations. This method combines the two equations into a third equation involving only one of the variables.

enlargement a scaled-up (or down) version of a figure in which the transformed figure is in proportion to the original figure; that is, the two figures are similar.

equate the process of writing one expression as equal to another.

equation a statement that asserts that two expressions are equal in value. An equation must have an equal sign. For example, $x + 4 = 12$.

equiangular when two or more shapes have all corresponding angles equal.

equilateral triangle a triangle with all sides equal in length, and all angles equal to $60°$.

event space a list of all the possible outcomes obtained from a probability experiment. It is written as ξ or S, and the list is enclosed in a pair of curled brackets $\{\}$. It is also called the sample space.

experimental probability the probability of an event based on the outcomes of experiments, simulations or surveys.

exponential decay a quantity that decreases by a constant percentage in each fixed period of time. This growth can be modelled by exponential functions of the type $y = ka^x$, where $0 < a < 1$.

exponential function relationships of the form $y = ka^x$, where $a \neq 1$, are called exponential functions with base a.

exponential growth a quantity that grows by a constant percentage in each fixed period of time. This growth can be modelled by exponential functions of the type $y = ka^x$, where $a > 1$.

explanatory variable this is the variable that is not impacted by the other variable. This is also known as the independent variable.

extrapolation the process of predicting a value of a variable outside the range of the data.

factor theorem if $P(x)$ is a polynomial, and $P(a) = 0$ for some number a, then $P(x)$ is divisible by $(x - a)$.

FOIL a diagrammatic method of expanding a pair of brackets. The letters in FOIL represent the order of the expansion: First, Outer, Inner and Last.

frequency the number of times a particular score appears.

function a process that takes a set of x-values and produces a related set of y-values. For each distinct x-value, there is only one related y-value. They are usually defined by a formula for $f(x)$ in terms of x; for example, $f(x) = x^2$.

future value the future value of a loan or investment.

half plane a region that represents all the points that satisfy an inequality.

Heron's formula this formula is used to calculate the area of a triangle when all three sides are known. The formula is $A = \sqrt{s(s-a)(s-b)(s-c)}$, where a, b and c are the lengths of the sides and s is the semi-perimeter or $s = \dfrac{a+b+c}{2}$.

histogram a graph that displays continuous numerical variables and does not retain all original data.

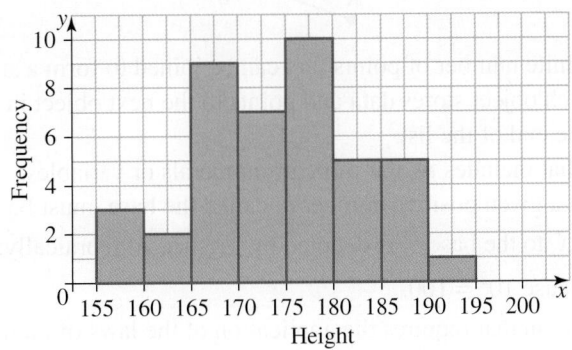

hyperbola the graph of $y = \dfrac{1}{x}$ is a rectangular hyperbola with asymptotes on the x- and y-axes.

hypotenuse the longest side of a right-angled triangle. It is the side opposite the right angle.

Identity Law when 0 is added to an expression or the expression is multiplied by 1, the value of the expression does not change. For example, $x + 0 = x$ and $x \times 1 = x$.

image the enlarged (or reduced) figure produced.

independent events successive events that have no effect on each other.

independent variable this is the variable that is not impacted by the other variable. This is also known as the explanatory variable.

index (index notation) the number that indicates how many times the base is being multiplied by itself when an expression is written in index form, also known as an *exponent* or *power*; (algorithms) an integer that points to a particular item in an array.

inequality when one algebraic expression or one number is greater than or less than another.

inequation similar to equations, but contain an inequality sign instead of an equal sign. For example, $x = 3$ is an equation, but $x < 3$ is an inequation.

integers these include the positive and negative whole numbers, as well as zero; that is, $\ldots, -3, -2, -1, 0, 1, 2, \ldots$

interpolation the process of predicting a value of a variable from within the range of the data.

interquartile range (IQR) the difference between the upper (or third) quartile, Q_{upper} (or Q_3), and the lower (or first) quartile, Q_{lower} (or Q_1); that is, $IQR = Q_{upper} - Q_{lower} = Q_3 - Q_1$. It is the range of approximately the middle half of the data.

inverse function when a function is reflected across the line $y = x$.

Inverse Law when the additive inverse of a number or pronumeral is added to itself, the sum is equal to 0. When the multiplicative inverse of a number or pronumeral is multiplied by itself, the product is equal to 1. So, $x + (-x) = 0$ and $x \times \dfrac{1}{x} = 1$.

inverse proportion describes a particular relationship between two variables (or quantities); that is, as one variable increases, the other decreases. The rule used to relate the two variables is $y = \dfrac{k}{x}$.

irrational number numbers that cannot be written as fractions. Examples of irrational numbers include surds, π and non-terminating, non-recurring decimals.

isosceles triangle a triangle with exactly two sides equal in length.

lay-by a method used to purchase an item whereby the purchaser makes regular payments to the retailer, who retains the item until the complete price is paid.

leading coefficient the coefficient of the leading term in a polynomial.

leading term the term in a polynomial that contains the highest power of x.

line of best fit a straight line that best fits the data points of a scatterplot that appear to follow a linear trend. It is positioned on the scatterplot so that there is approximately an equal number of data points on either side of the line, and so that all the points are as close to the line as possible.

line segment a line segment or interval is a part of a line with end points.

linear graph consist of an infinite number of points that can be joined to form a straight line.

linked list a list of objects. Each object stores data and points to the next object in the list. The last object points to a terminator to indicate the end of the list.

literal equation an equation that includes two or more pronumerals or variables.

logarithm the power to which a given positive number b, called the base, must be raised in order to produce the number x. The logarithm of x, to the base b, is denoted by $\log_b(x)$. Algebraically: $\log_b(x) = y \leftrightarrow b^y = x$; for example, $\log_{10}(100) = 2$ because $10^2 = 100$.

logarithmic equation an equation that requires the application of the laws of indices and logarithms to solve.

loop in JavaScript, a process that executes the same code many times with different data each time.

many-to-many relation a relation in which one range value may yield more than one domain value and vice versa.

many-to-one relation a function or mapping that takes the same value for at least two different elements of its domain.

matrix a rectangular array of numbers arranged in rows and columns.

maximal domain the limit of the x-values that a function can have.

mean one measure of the centre of a set of data. It is given by $mean = \dfrac{\text{sum of all scores}}{\text{number of scores}}$ or $\bar{x} = \dfrac{\sum x}{n}$. When data are presented in a frequency distribution table, $\bar{x} = \dfrac{\sum (f \times x)}{n}$.

measures of central tendency mean, median and mode.

measures of spread range, interquartile range, standard deviation.

median one measure of the centre of a set of data. It is the middle score for an odd number of scores arranged in numerical order. If there is an even number of scores, the median is the mean of the two middle scores when they are ordered. Its location is determined by the rule $\dfrac{n+1}{2}$. For example, the median value of the set 1 3 3 4 5 6 8 9 9 is 5, while the median value for the set 1 3 3 4 5 6 8 9 9 10 is the mean of 5 and 6 (5.5).

midpoint the midpoint of a line segment is the point that divides the segment into two equal parts. The coordinates of the midpoint M between the two points $P(x_1, y_1)$ and $Q(x_2, y_2)$ is given by the formula $M = \left(\dfrac{x_1 + x_2}{2}, \dfrac{y_1 + y_2}{2} \right).$

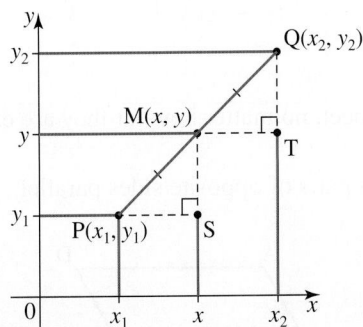

mode one measure of the centre of a set of data. It is the score that occurs most often. There may be no mode, one mode or more than one mode (two or more scores occur equally frequently).

monic any polynomial with a leading coefficient of 1.

Multiplication Law of probability if events A and B are independent, then: $P(A \text{ and } B) = P(A) \times P(B)$ or $P(A \cap B) = P(A) \times P(B).$

mutually exclusive events that cannot occur together. On a Venn diagram, two mutually exclusive events will appear as disjoint sets. If events A and B are mutually exclusive, then $P(A \cap B) = 0.$

natural numbers the set of positive integers, or counting numbers; that is, the set 1, 2, 3, …

negatively skewed showing larger amounts of data as the values of the data increase.

nested loop a loop within a loop. The outer loop contains an inner loop. The first iteration of the outer loop triggers a full cycle of the inner loop until the inner loop completes. This triggers the second iteration of the outer loop, which triggers a full cycle of the inner loop again. This process continues until the outer loop finishes a full cycle.

number a JavaScript data type that represents a numerical value.

object a general JavaScript data type that can have many properties. Each property is a name–value pair so that the property has a name to reference a value.

ogive a graph formed by joining the top right-hand corners of the columns of a cumulative frequency histogram.

one-dimensional array a simple array of values in which the values can be of any type except for another array.

one-to-many relation a relation in which there may be more than one range value for one domain value but only one domain value for each range value.

one-to-one relation refers to the relationship between two sets such that every element of the first set corresponds to one and only one element of the second set.

outlier a piece of data that is considerably different from the rest of the values in a set of data; for example, 24 is the outlier in the set of ages {12, 12, 13, 13, 13, 13, 13, 14, 14, 24}. Outliers sit $1.5 \times IQR$ or greater away from Q_1 or Q_3.

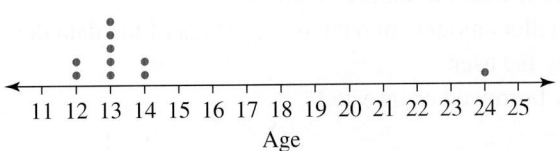

parabola the shape of the graph of a quadratic function. For example, the typical shape is that of the graph of $y = x^2$.

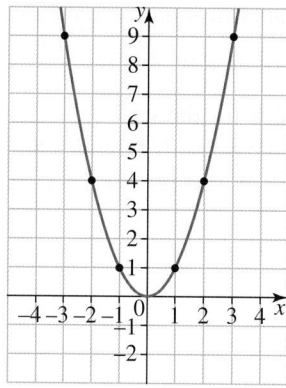

parallel parallel lines in a plane never meet, no matter how far they are extended. Parallel lines have the same gradient.

parallelogram a quadrilateral with both pairs of opposite sides parallel.

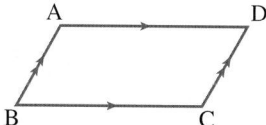

percentile the value below which a given percentage of all scores lie. For example, the 20th percentile is the value below which 20% of the scores in the set of data lie.

perfect square an algebraic expression or a whole number that is multiplied by itself.

period the distance between repeating peaks or troughs of periodic functions.

periodic function a function that has a graph that repeats continuously in cycles, for example, graphs of $y = \sin(x)$ and $y = \cos(x)$.

perpendicular perpendicular lines are at right angles to each other. The product of the gradients of two perpendicular lines is -1.

pi (π) the Greek letter π represents the ratio of the circumference of any circle to its diameter. The number π is irrational, with an approximate value of $\frac{22}{7}$, and a decimal value of $\pi = 3.141\,59\ldots$.

point of intersection a point where two or more lines intersect. The solution to a pair of simultaneous equations can be found by graphing the two equations and identifying the coordinates of the point of intersection.

pointer in JavaScript, a variable that points to a JavaScript object or array. Multiple pointers can point to the same object or array.

polygon a plane figure bounded by line segments.

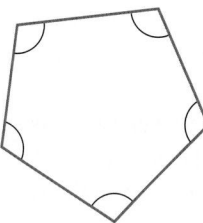

polynomial an expression containing only non-negative integer powers of a variable.

population the whole group from which a sample is drawn.

positively skewed showing smaller amounts of data as the values of the data decrease.

primary data data collected by the user.

principal an amount of money borrowed or invested.

probability the likelihood or chance of a particular event (result) occurring.

$$\text{Pr(event)} = \frac{\text{number of favourable outcomes}}{\text{number of possible outcomes}}$$

The probability of an event occurring ranges from 0 (impossible — will not occur) to 1 (certainty — will definitely occur).

proof an argument that shows why a statement is true.

property references a value on an object. A complex object may have many properties. Each property has a unique name on the object.

quadratic equation an equation in the form $ax^2 + bx + c = 0$, where a, b and c are real numbers.

quadratic formula gives the roots of the quadratic equation $ax^2 + bx + c = 0$. It is expressed as

$$x = \frac{-b \pm \sqrt{b^2 - 4ac}}{2a}.$$

quantile percentiles expressed as decimals. For example, the 95th percentile is the same as the 0.95 quantile.

quartic function the basic form of a quartic function is $y = ax^4$. If the value of a is positive, the curve is upright, whereas a negative value of a results in an inverted graph. A maximum of 4 roots can result.

quartile values that divide an ordered set into four (approximately) equal parts. There are three quartiles: the first (or lower) quartile, Q_1; the second quartile (or median), Q_2; and the third (or upper) quartile, Q_3.

radius the straight line from a circle's centre to any point on its circumference.

range (functions and relations) the set of y-values produced by the function; (statistics) the difference between the highest and lowest scores in a set of data; that is, range = highest score − lowest score.

rational number numbers that can be written as fractions, where the denominator is not zero.

real numbers rational and irrational numbers combine to form the set of real numbers.

reciprocal when a number is multiplied by its reciprocal, the result is 1.

recurring decimal a decimal which has one or more digits repeated continuously; for example, 0.999 .... They can be expressed exactly by placing a dot or horizontal line over the repeating digits; for example,

$8.343\,434 = 8.\overset{..}{3}\overset{}{4}$ or $8.\overline{34}$.

reflection when a graph is flipped in the x- or y-axis.

regression line a line of best fit that is created using technology.

regular polygon a polygon with sides of the same length and interior angles of the same size.

relation a set of ordered pairs.

relative frequency represents the frequency of a particular score divided by the total sum of the frequencies. It is given by the rule:

$$\text{relative frequency of a score} = \frac{\text{frequency of the score}}{\text{total sum of frequencies}}.$$

remainder theorem if a polynomial $P(x)$ is divided by $(x - a)$, where a is any real number, the remainder is $P(a)$.

required region the region that contains the points that satisfy an inequality.

response variable this is the variable that is impacted by the other variable. This is also known as the dependent variable.

rhombus a parallelogram with all sides equal.

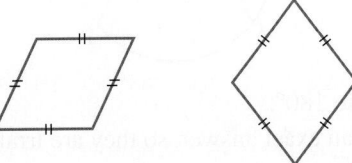

sample part of a population chosen so as to give information about the population as a whole.

sample space a list of all the possible outcomes obtained from a probability experiment. It is written as ξ or S, and the list is enclosed in a pair of curled brackets { }. It is also called the event space.

scale factor the ratio of the corresponding sides in similar figures, where the enlarged (or reduced) figure is referred to as the image and the original figure is called the object.

$$\text{scale factor} = \frac{\text{image length}}{\text{object length}}$$

scatterplot a graphical representation of bivariate data that displays the degree of correlation between two variables. Each piece of data on a scatterplot is shown by a point. The x-coordinate of this point is the value of the independent variable and the y-coordinate is the corresponding value of the dependent variable.

secondary data data collected by others.

similar triangles triangles that have similar shape but different size. There are four standard tests to determine whether two triangles are similar: AAA (angle, angle, angle), SAS (side, angle, side), SSS (side, side, side) and RHS (right angle, hypotenuse side).

simple interest formula the interest accumulated when the interest payment in each period is a fixed fraction of the principal. The formula used is $I = Prn$, where I is the interest earned or paid (in \$) when a principal of \$$P$ is invested or borrowed at an interest rate of $r\%$ p.a. for a period of n years.

simultaneous occurring at the same time.

simultaneous equations the equations of two (or more) linear graphs that have the same solution.

sine ratio the ratio of the opposite side to the hypotenuse in a right-angled triangle; $\sin(\theta) = \dfrac{\text{opposite}}{\text{hypotenuse}}$.

sine rule in any triangle ABC, $\dfrac{a}{\sin(A)} = \dfrac{b}{\sin(B)} = \dfrac{c}{\sin(C)}$.

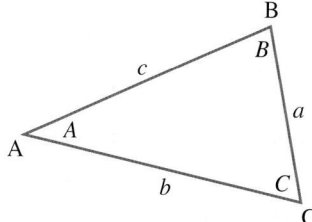

standard deviation a measure of the variability of spread of a data set. It gives an indication of the degree to which the individual data values are spread around the mean.

string a JavaScript data type that represents text.

subject of an equation the variable that is expressed in terms of the other variables. In the equation $y = 3x + 4$, the variable y is the subject.

substitution method a method used to solve simultaneous equations. It is useful when one (or both) of the equations has one of the variables as the subject.

subtended In the diagram shown, chords AC and BC form the angle ACB. Arc AB has subtended angle ACB.

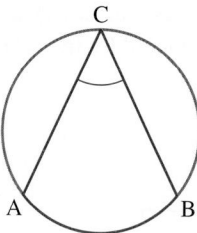

supplementary angles angles that add to 180°.

surd roots of numbers that do not have an exact answer, so they are irrational numbers. Surds themselves are exact numbers; for example, $\sqrt{6}$ or $\sqrt[3]{5}$.

symmetrical the identical size, shape and arrangement of parts of an object on opposite sides of a line or plane.

system of equations a set of two or more equations with the same variables.

tangent ratio the ratio of the opposite side to the adjacent side in a right-angled triangle; $\tan(\theta) = \dfrac{\text{opposite}}{\text{adjacent}}$.

terminating decimal a decimal which has a fixed number of places; for example, 0.6 and 2.54.

theorem rules or laws.

theoretical probability given by the rule $P(\text{event}) = \dfrac{\text{number of favourable outcomes}}{\text{number of possible outcomes}}$ or $P(E) = \dfrac{n(E)}{n(S)}$, where $n(E) =$ number of times or ways an event, E, can occur and $n(S) =$ number of elements in the sample space or number of ways all outcomes can occur, given all the outcomes are equally likely.

time series a sequence of measurements taken at regular intervals (daily, weekly, monthly and so on) over a certain period of time. They are used for analysing general trends and making predictions for the future.

total surface area (TSA) the area of the outside surface(s) of a 3-dimensional figure.

transformation changes that occur to the basic parabola $y = x^2$ in order to obtain another graph. Examples of transformations are translations, reflections or dilations. Transformations can also be applied to non-quadratic functions.

translation movements of graphs left, right, up or down.

transversal a line that meets two or more other lines in a plane.

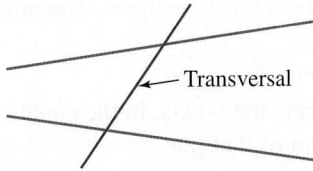

tree diagram branching diagram that lists all the possible outcomes of a probability experiment. This diagram shows the outcomes when a coin is tossed twice.

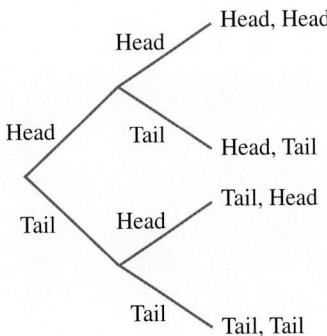

trial the number of times a probability experiment is conducted.

trigonometric ratios three different ratios of one side of a triangle to another. The three ratios are the sine, cosine and tangent.

true bearing a direction which is written as the number of degrees (3 digits) from north in a clockwise direction, followed by the word *true* or T; for example, due east would be 090° true or 090°T.

turning point a point at which a graph changes direction (either up or down).

two-dimensional array an array of one-dimensional arrays.

two-step chance experiment a probability experiment that involves two trials.

two-way table tables that list all the possible outcomes of a probability experiment in a logical manner.

Hair colour	Hair type		Total
Red	1	1	2
Brown	8	4	12
Blonde	1	3	4
Black	7	2	9
Total	17	10	27

unit circle a circle with its centre at the origin and having a radius of 1 unit.

univariate data data relating to a single variable.

variable a named container or memory location that holds a value.

vertex the point at which the graph of a quadratic function (parabola) changes direction (either up or down).

vertically opposite angles when two lines intersect, four angles are formed at the point of intersection, and two pairs of vertically opposite angles result. Vertically opposite angles are equal.

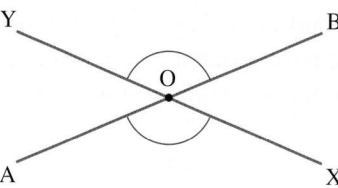

volume the amount of space a 3-dimensional object occupies. The units used are cubic units, such as cubic centimetres (cm^3) and cubic metres (m^3).

***x*-intercept** the point where a graph intersects the *x*-axis.

***y*-intercept** the point where a graph intersects the *y*-axis. In the equation of a straight line, $y = mx + c$, the constant term, c, represents the *y*-intercept of that line.

INDEX

A

abstraction 152
addition
  algebraic fractions 119–24
  logarithms 851
  polynomials 806–8
  surds 48–9
Addition Law of probability 765
algebra 152
  difference of two squares 100–1
  dividing algebraic fractions
    127–31
  further expansions 105
  highest common factor 108
  multiplying algebraic fractions
    126
  perfect squares 101–2
algebraic expressions
  converting worded problems into
    131–3
algebraic fractions
  adding and subtracting 119–24
  definition 119
  dividing 127
  equations with 160–2
  multiplying 126
  multi-step equations 160
  simplifying 124–6
algebraic techniques 97–150
  factorising
    by grouping in pairs 108–11
    non-monic quadratics 116–19
    special products 111–16
    special binomial products 100–5
alternate angle rule 474
alternate segment theorem 965–7
ambiguous case 509–13
amplitude 536
angle of depression 473–8
  definition 473
  to solve problems 474
angle of elevation 473–8
  definition 473
  to solve problems 474
angles
  in degrees and minutes 469
  proofs and theorems 617–19
  sums of 617
  using inverse trigonometric ratios
    468
angles at a point 617
angles in circle 944–52

B

anticlockwise 527
arbelos 549–50
arc 945
area 568–76
  composite figures 570–2
  definition 568
  formula 568–70
  of plane figures, calculation 569
areas of triangles 520–7
  three sides are known 522–4
asymptotes 379, 536
  exponential equations 379
  hyperbola 388
  tangent 536
axis of symmetry 347

base 57
bearings 478–85
  compass bearings 478
  problems with 2 stages 480
  solving trigonometric problems
    479
  true bearings 479
binary operations 874
binomial common factor 108
binomial products 51–105
  applications of 131–7
  expanding 100–7
  with surds 51
bisected 948
bivariate data 674–84
  correlation 676–7
  definition 674
  drawing conclusions from
    correlation 677–80
  lines of best fit by eye 684–95
    meaning and making predictions
      688
    predictions using lines of best fit
      686–91
  scatterplots 674–6
boundary line 312
buying on terms 8–12

C

capacity 594–5
Cartesian plane 312–14
  definition 313
  inequalities on 316
  required region on 314–16

cash 5
census 713
centre 944
centre of enlargement 629
chord
  circle geometry 955–6
  description 945
circle 393–8
  angles in 944–52
  centre and radius 393
  equation 393
  graph of 394
  non-linear relationships 393–8
circle geometry 939–80
  alternate segment theorem 965–7
  angles in circles 944–52
  chords and radii 955–6
  cyclic quadrilaterals 960–5
  intersecting chords 952–3
  intersecting secants 953–4
  intersecting tangents 954–5
  tangents and secants 967
  tangents to a circle 947–9
circuit 1008
circumcentre
  definition 956
  triangle 956–7
circumcircle 956
circumference 944
coincident 189–92
collinear points 306–12
combining index laws 74–80
  simplifying complex expressions
    involving multiple steps 76
  simplifying expressions in multiple
    steps 75
  simplifying expressions with
    multiple fractions 77
common binomial factor 108
compass bearings 478–9
complement 743, 764
complementary angles 530, 617
complementary events 743, 764
completing the square
  quadratic equations by 228–9
composite figure 570
composite solids, volume of 593–4
compounded value 13
compounding period 14–17
compound interest 13–20
  amount of 15
  compounding period 16–17

INDEX   1068